Important Physical Constants

Constant	Symbol	Value
Velocity of light (*vacuo*)	c	2.9979×10^{10} cm sec^{-1}
Planck constant	h	6.6256×10^{-27} erg sec = 4.1356×10^{-15} e Vsec
Avogadro number	N	6.0226×10^{23} particles mol^{-1}
Faraday constant	F	96487 C mol^{-1}
Gas constant	R	8.3143×10^{7} erg deg^{-1} mol^{-1}
		1.9872 cal deg^{-1} mol^{-1}
		8.2054×10^{-2} liter atm deg^{-1} mol^{-1}
Boltzmann constant	k	1.3805×10^{-16} erg deg^{-1}
Rest mass of the electron	m_c	9.1090×10^{-28} g
Electronic charge	e	-4.8033×10^{-10} esu
		-1.6021×10^{-19} C

Energy Conversion Factors

	Ergs	Joules	Calories	Liter Atmospheres	Electron Volts
1 erg =	1	10^{-7}	2.3901×10^{-8}	9.8687×10^{-10}	6.2418×10^{11}
1 joule =	10^{7}	1	2.3901×10^{-1}	9.8687×10^{-3}	6.2418×10^{18}
1 calorie =	4.1840×10^{7}	4.1840	1	4.1291×10^{-2}	2.6116×10^{19}
1 liter atmosphere =	1.0133×10^{9}	1.0133×10^{2}	24.218	1	6.3248×10^{20}
1 electron volt =	1.6021×10^{-12}	1.6021×10^{-19}	3.8291×10^{-20}	1.5811×10^{-21}	1

Principles of

Instrumental Analysis

Third Edition

Douglas A. Skoog   Stanford University

Jim Luo
Jan. 5. 1989

Saunders Golden Sunburst Series

 Saunders College Publishing

Philadelphia   New York   Chicago   San Francisco
Montreal   Toronto   London   Sydney   Tokyo

Text Typeface: 10/12 Baskerville
Compositor: Bi-Comp, Inc.
Aquisitions Editor: John J. Vondeling
Project Editors: Maureen Iannuzzi and Patrice Smith
Copyeditor: Will Eaton
Art and Design Director: Carol C. Bleistine
Design Assistant: Virginia A. Bollard
Text Design: Arlene Putterman
Cover Design: Lawrence R. Didona
Text Artwork: Anco/Boston
Managing Editor and Production Manager: Tim Frelick
Assistant Production Manager: Maureen Iannuzzi

Library of Congress Cataloging in Publication Data

Skoog, Douglas Arvid, 1918–
     Principles of instrumental analysis.

     Includes bibliographical references and index.

     1. Instrumental analysis.   I. Title.
QD79.I5S58   1984   543'.08   84-10704
ISBN 0-03-001229-5

Principles of Instrumental Analysis, Third Edition   ISBN 0-03-01229-5

Printed in the United States of America

 89   016   987

Saunders College Publishing
Holt, Rinehart and Winston
The Dryden Press

Preface

Today, physical and biological scientists have available an impressive array of powerful and elegant tools for obtaining qualitative and quantitative information about the composition of matter. Students of chemistry, biochemistry, geology, health related sciences, and environmental sciences must develop an appreciation for these tools and how they are used to solve analytical problems. It is to such students that this book is addressed.

It is the author's belief that the choice and efficient use of analytical instruments requires an understanding of the fundamental principles upon which modern measuring devices are based. Only then can intelligent choices be made among the many possible ways of solving an analytical problem; only then can an appreciation be developed for the pitfalls that accompany most physical measurements; and only then can a feel be gained for the potential limitations of measurements in terms of sensitivity and accuracy. As with the earlier editions of this text, it is the goal of the third edition to provide the student with an introduction to the principles of spectroscopic, electrometric, and chromatographic methods of analysis, as well as to engender an appreciation of the kinds of instruments that are currently available and their strengths and limitations.

The users of the second edition of this text will recognize that the new edition is generally organized in much the same way as its predecessor. Thus, two early chapters deal with electronics, microprocessors, microcomputers, and signal-to-noise enhancement. Ten chapters follow that treat the general principles and the practical aspects of the various methods of optical spectroscopy. Next are found chapters on nuclear magnetic resonance, X-ray spectrometry, electron spectroscopy (a new chapter), radiochemical methods, and mass spectrometry. The discussion then turns to the field of electrochemistry, which is covered in five chapters. Next is a brief chapter on thermal methods followed by four chapters dealing with the various aspects of chromatography. The final chapter, a new one, is devoted to automation and automatic methods. As in the second edition, introductory chapters on optical spectroscopy, electroanalytical chemistry, and chromatography precede the chapters dealing with individual methods.

While the second and third editions of this text share a common organization, numerous changes, additions, and deletions will be encountered, which reflect the rapid and exciting innovations that have appeared in the last five years. A brief outline of the important differences between the two editions follows.

The introductory chapter has been expanded to include material, formerly found in the appendix, on statistical treatment of data; a section on regression analysis; and one on sensitivity and detection limits. Most of the material in the second edition on electricity and electrical circuits has been deleted or moved to appendixes in the new edition in order to make room for treatment of integrated circuitry, digital electronics, and software methods of signal-to-noise enhancement. A new chapter (Chapter 3) is devoted entirely to an elementary presentation of microprocessors and microcomputers for the control of instruments and for data manipulation.

New material in the chapters on optical spectroscopy include multichannel detectors and multiplex instruments; fiber optics; photoacoustic instruments and techniques; chemiluminescence and phosphorimetric methods; computer spectral search systems; fluorescence and Fourier transform infrared detectors for liquid chromatography; plasma sources for atomic fluorescence measurements; and resonance and coherent anti-Stokes Raman spectroscopy.

The many new and recent innovations in mass spectroscopic ion sources and the much expanded use of mass spectrometers as chromatographic detectors have necessitated major additions to the chapter devoted to these topics. The new material deals with field ionization, fast atom bombardment, secondary ion, chemical ionization, and plasma sources; Fourier transform mass spectrometers; computerized spectral searches; tandem mass spectrometers; elemental analysis by inductively coupled plasma/mass spectrometry; surface analysis by mass spectroscopy; and the ion trap detector for gas chromatography.

Several parts of the chapter on electrochemistry have been deleted or condensed. Deletions include the sections devoted to differential potentiometric titrations, electrogravimetry, applications of classical polarography, voltammetry at stationary solid electrodes, and oscillometry. Sections that have been condensed include those on applications of potentiometric and amperometric titrations.

Major additions to the chapter on liquid chromatography have been made in response to the remarkable growth of this field. New materials include bonded-phase and reversed-phase packings; microbore columns, electrochemical and fluorescent detectors; systematic methods for choice of mobile and stationary phases; and supercritical fluid chromatography. In order to accommodate these additions, it was found desirable to separate the material on planar chromatography as a new chapter.

Finally, a new chapter on automatic and automated methods, including flow injection analysis, has been added.

It is worth noting that certain parts of this text also appear in two other books co-authored by the writer.[1] The principal overlap occurs in the chapters on electrochemistry. Duplication to a lesser extent is found in the chapters on ultraviolet and visible absorption spectroscopy and atomic emission spectroscopy.

I wish to acknowledge with great thanks the significant contributions of the following who read all or large parts of the manuscript in detail and offered numerous useful suggestions: Professor M. F. Byrant, University of Georgia; Professor F. J. Holler, University of Kentucky; Professor J. J. Leary, James Madison University; and Professor H. Veening, Bucknell University.

Obviously, the co-author of the first two editions of this book, Professor D. M. West of San Jose State University, deserves much credit for his contribution to the overall organization and presentation found in those editions and this edition as well. Unfortunately, he chose not to participate actively in the preparation of this edition because of other professional pressures. I want, however, to thank him for his many earlier contributions and for his encouragement in the task of producing this edition.

Douglas A. Skoog
Stanford University

[1] D. A. Skoog and D. M. West, *Fundamentals of Analytical Chemistry,* 4th ed., Saunders: Philadelphia, 1982; and *Analytical Chemistry: An Introduction,* 3d ed., Holt, Rinehart and Winston, 1979.

Contents Overview

1   Introduction   1
2   Elementary Electronics   30
3   Microcomputers and Microprocessors in Chemical Instrumentation   80
4   Electromagnetic Radiation and its Interaction with Matter   90
5   Instruments for Optical Spectroscopy   112
6   An Introduction to Absorption Spectroscopy   160
7   Molecular Ultraviolet and Visible Absorption Spectroscopy   182
8   Molecular Fluorescence, Phosphorescence, and Chemiluminescence Spectroscopy   225
9   Atomic Spectroscopy Based upon Flame and Electrothermal Atomization   250
10   Emission Spectroscopy Based upon Plasma, Arc, and Spark Atomization   292
11   Infrared Absorption Spectroscopy   315
12   Raman Spectroscopy   362
13   Miscellaneous Optical Methods   379
14   Nuclear Magnetic Resonance Spectroscopy   407
15   X-Ray Spectroscopy   456
16   Electron Spectroscopy   487
17   Radiochemical Methods   502
18   Mass Spectrometry   523
19   An Introduction to Electroanalytical Chemistry   567
20   Potentiometric Methods   600
21   Coulometric Methods   644
22   Voltammetry and Polarography   664
23   Conductometric Methods   704
24   Thermal Methods   713
25   An Introduction to Chromatographic Separations   727
26   Gas Chromatography   757
27   High-Performance Liquid Chromatography   784
28   Planar Chromatography   837
29   Automated Methods of Analysis   851
Appendix 1   Simple Electrical Measurements   880
Appendix 2   Alternating Currents and Electrical Reactance   887
Appendix 3   Activity Coefficients   900
Appendix 4   Some Standard and Formal Electrode Potentials   904
Appendix 5   Compounds Recommended for the Preparation of Standard Solutions of Some Common Elements   907
Answers to Problems   909
Index   928

Contents

1 Introduction   1

1A Types of Analytical Methods, 2
1B Instruments for Analysis, 3
1C Uncertainties in Instrumental Measurements, 5
1D Sensitivity and Detection Limit for Instruments, 22

2 Elementary Electronics   30

2A Semiconductors, 31
2B Power Supplies and Regulators, 37
2C Operational Amplifiers, 39
2D Digital Electronics, 54
2E Readout Devices, 62
2F Noise, 65
2G Signal-to-Noise Enhancement, 68

3 Microcomputers and Microprocessors in Chemical Instrumentation   80

3A Introduction, 80
3B Components of a Computer, 83
3C Computer Programming, 87
3D Applications of Computers, 88

4 Electromagnetic Radiation and its Interaction with Matter   90

4A Electromagnetic Radiation as Waves, 91
4B Quantum-Mechanical Properties of Electromagnetic Radiation, 104

5 Instruments for Optical Spectroscopy   112

5A Types of Methods, 112
5B Radiation Sources, 115
5C Wavelength Selectors; Filters, 120

5D   Wavelength Selectors; Monochromators, 123
5E   Sample Containers, 135
5F   Radiation Detectors, 135
5G   Signal Processors and Readouts, 143
5H   Fiber Optics, 144
5I   Instrument Designs, 145

6   An Introduction to Absorption Spectroscopy   160
6A   Terms Employed in Absorption Spectroscopy, 160
6B   Quantitative Aspects of Absorption Measurements, 162
6C   Photometer and Spectrophotometer Designs, 175

7   Molecular Ultraviolet and Visible Absorption Spectroscopy   182
7A   The Magnitude of Molar Absorptivities, 182
7B   Absorbing Species, 183
7C   Components of Instruments for Absorption Measurements in the Ultraviolet, Visible, and Near Infrared Regions, 193
7D   Some Typical Instruments, 195
7E   Application of Absorption Measurement to Qualitative Analysis, 207
7F   Quantitative Analysis by Absorption Measurements, 208
7G   Photometric Titrations, 215
7H   Photoacoustic Spectroscopy, 217

8   Molecular Fluorescence, Phosphorescence, and Chemiluminescence Spectroscopy   225
8A   Theory of Fluorescence and Phosphorescence, 226
8B   Instruments for Measuring Fluorescence and Phosphorescence, 236
8C   Applications of Photoluminescence Methods, 241
8D   Chemiluminescence, 244

9   Atomic Spectroscopy Based upon Flame and Electrothermal Atomization   250
9A   Theory of Atomic Spectroscopy, 251
9B   Flame Atomization, 257
9C   Atomizers for Atomic Spectroscopy, 261
9D   Atomic Absorption Spectroscopy, 266
9E   Flame Emission Spectroscopy, 279
9F   Atomic Fluorescence Spectroscopy, 286

10   Emission Spectroscopy Based upon Plasma, Arc, and Spark Atomization   292
10A   Spectra From Higher Energy Sources, 293
10B   Emission Spectroscopy Based on Plasma Sources, 294
10C   Atomic Fluorescence Methods Based upon Plasma Atomization, 304
10D   Emission Spectroscopy Based on Arc and Spark Sources, 306

11 Infrared Absorption Spectroscopy   315

11A Theory of Infrared Absorption, 316
11B Infrared Instrument Components, 325
11C Some Typical Instruments, 329
11D Sample Handling Techniques, 340
11E Qualitative Applications of Infrared Absorption, 343
11F Quantitative Applications, 351
11G Far-Infrared Spectroscopy, 355
11H Infrared Emission Spectroscopy, 355

12 Raman Spectroscopy   362

12A Theory of Raman Spectroscopy, 363
12B Instrumentation, 371
12C Applications of Raman Spectroscopy, 373
12D Application of Other Types of Raman Spectroscopy, 376

13 Miscellaneous Optical Methods   379

13A Nephelometry and Turbidimetry, 379
13B Refractometry, 382
13C Polarimetry, 386
13D Optical Rotatory Dispersion and Circular Dichroism, 399

14 Nuclear Magnetic Resonance Spectroscopy   407

14A Theory of Nuclear Magnetic Resonance, 407
14B Environmental Effects on NMR Spectra, 417
14C Experimental Methods of NMR Spectroscopy, 431
14D Applications of Proton NMR, 440
14E Application of NMR to Isotopes Other Than the Proton, 444
14F Electron Spin Resonance Spectroscopy, 447

15 X-Ray Spectroscopy   456

15A Fundamental Principles, 456
15B Instrument Components, 464
15C X-Ray Fluorescence Methods, 474
15D X-Ray Absorption Methods, 481
15E X-Ray Diffraction Methods, 482
15F The Electron Microprobe, 484

16 Electron Spectroscopy   487

16A Principles of Electron Spectroscopy, 488
16B Instrumentation, 492
16C Applications of Electron Spectroscopy, 495

17 Radiochemical Methods   502

17A Radioactive Isotopes, 502
17B Instrumentation, 510

17C Neutron Activation Methods, 511
17D Isotopic Dilution Methods, 517
17E Radiometric Methods, 520

18 Mass Spectrometry 523
18A The Mass Spectrometer, 524
18B Mass Spectra, 542
18C Identification of Pure Compounds by Mass Spectrometry, 547
18D Quantitative Applications of Mass Spectroscopy, 559
18E Molecular Secondary Ion Mass Spectroscopy, 565

19 An Introduction to Electroanalytical Chemistry 567
19A Electrochemical Cells, 567
19B Cell Potentials, 573
19C Electrode Potentials, 575
19D Calculation of Cell Potentials from Electrode Potentials, 587
19E Effect of Current on Cell Potentials, 590

20 Potentiometric Methods 600
20A Reference Electrodes, 600
20B Metallic Indicator Electrodes, 603
20C Membrane Indicator Electrodes, 605
20D Instruments for Measuring Cell Potentials, 624
20E Direct Potentiometric Measurements, 627
20F Potentiometric Titrations, 631

21 Coulometric Methods 644
21A Current-Voltage Relationships During an Electrolysis, 644
21B An Introduction to Coulometric Methods of Analysis, 650
21C Potentiostatic Coulometry, 652
21D Coulometric Titrations (Amperostatic Coulometry), 655

22 Voltammetry and Polarography 664
22A Theory of Classical Polarography, 665
22B Instrumentation, 677
22C Modified Voltammetric Methods, 681
22D Application of Polarography, 693
22E Voltammetry at Solid Electrodes, 696
22F Amperometric Titrations, 697

23 Conductometric Methods 704
23A Electrical Conductance in Solutions of Electrolytes, 704
23B The Measurement of Conductance, 706
23C Conductometric Titrations, 708
23D Applications of Direct Conductance Measurements, 711

24 *Thermal Methods*   713
24A *Thermogravimetric Methods, 713*
24B *Differential Thermal Analysis and Differential Scanning Calorimetry, 716*
24C *Enthalpimetric Methods, 721*

25 *An Introduction to Chromatographic Separations*   727
25A *A General Description of Chromatography, 727*
25B *The Rate Theory of Chromatography, 733*
25C *Summary of Important Relationships for Chromatography, 747*
25D *Qualitative and Quantitative Analysis by Chromatography, 747*
25E *Computerized Chromatography, 751*

26 *Gas Chromatography*   757
26A *Principles of Gas-Liquid Chromatography, 758*
26B *Instruments for Gas-Liquid Chromatography, 759*
26C *The Stationary Phase, 769*
26D *Applications of Gas-Liquid Chromatography, 774*
26E *Gas-Solid Chromatography, 779*
26F *Examples of Applications of Gas Chromatography, 781*

27 *High-Performance Liquid Chromatography*   784
27A *Scope of HPLC, 785*
27B *Column Efficiency in Liquid Chromatography, 786*
27C *Instruments for Liquid Chromatography, 788*
27D *Chromatographic Mobile Phases, 799*
27E *Partition Chromatography, 801*
27F *Adsorption Chromatography, 815*
27G *Ion-Exchange Chromatography, 817*
27H *Size-Exclusion Chromatography, 824*
27I *Comparison of High-Performance Liquid Chromatography with Gas-Liquid Chromatography, 830*
27J *Supercritical-Fluid Chromatography, 830*

28 *Planar Chromatography*   837
28A *Scope of Thin-Layer Chromatography, 837*
28B *Principles of Thin-Layer Chromatography, 838*
28C *The Practice of Thin-Layer Chromatography, 840*
28D *Applications of Thin-Layer Chromatography, 846*
28E *Paper Chromatography, 847*
28F *Electrophoresis and Electrochromatography, 847*

29 *Automated Methods of Analysis*   851
29A *An Overview of Automatic Instruments and Automation, 851*
29B *Automation of Sampling and Preliminary Sample Treatment, 853*
29C *Continuous Flow Methods, 858*
29D *Discrete Methods, 871*
29E *Automatic Analysis Based on Multilayer Films, 875*

Appendix 1   Simple Electrical Measurements   880
Appendix 2   Alternating Currents and Electrical Reactance   887
Appendix 3   Activity Coefficients   900
Appendix 4   Some Standard and Formal Electrode Potentials   904
Appendix 5   Compounds Recommended for the Preparation of Standard Solutions of Some Common Elements   907
Answers to Problems   909
Index   928

1

Introduction

A chemical analysis provides information about the composition of a sample of matter. The results of some analyses are qualitative and yield useful clues from which the molecular or atomic species, the structural features, or the functional groups in the sample can be deduced. Other analyses are quantitative; here, the results take the form of numerical data in units such as percent, parts per million, or moles per liter. In both types of analyses, the required information is obtained by measuring a physical property that is characteristically related to the component of interest (the *analyte*).

It is convenient to describe properties that are useful for determining chemical composition as *analytical signals*; examples of such signals include intensity of emitted or absorbed light, electrical or thermal conductance, weight, volume, and refractive index. None of these signals is unique to a given species. For example, all metallic elements in a sample will ordinarily emit ultraviolet and visible radiation when heated to a suffi-

ciently high temperature in an electric arc; all charged species conduct electricity; and all of the components in a mixture contribute to its refractive index, weight, and volume. Therefore, all analyses require a separation. In some instances, the separation step involves physical isolation of the individual chemical components in the sample prior to generation of the analytical signal; in others, the signal is generated or observed for the entire sample, following which the signal from the analyte is isolated from those for the other components of the sample. Some signals are susceptible to the latter treatment, while others are not. For example, when a substance is heated in an electric arc, the wavelength distribution for the radiation of each metallic species is unique to that species; separation of the wavelengths in a suitable device (a monochromator) thus makes possible the identification of each component without physical separation. On the other hand, no general method exists for distinguishing the con-

1

ductance of sodium ions from that of potassium ions. Here, a physical separation is required if conductance is to serve as the signal for the analysis of one of these species in a sample that also contains the other.

1A Types of Analytical Methods

Table 1–1 lists many of the common signals that are useful for analytical purposes. Note that the first six signals involve either the emission of radiation or the interaction of radiation with matter. The next three signals are electrical. Finally, five miscellaneous signals are grouped together. Also listed in the table are the names of analytical methods that are based on the various signals.

It is of interest to note that until perhaps 1920, most analyses were based on the last two signals listed in Table 1–1—namely, mass and volume. As a consequence, gravimetric and volumetric procedures have come to be known as *classical* methods of

analysis, in contrast to the other procedures, which are known as *instrumental methods*.

Beyond the chronology of their development, few features clearly distinguish instrumental methods from classical ones. Some instrumental techniques are more sensitive than classical techniques, but many are not. With certain combinations of elements or compounds, an instrumental method may be more specific; with others, a gravimetric or a volumetric approach may be less subject to interference. Generalizations on the basis of accuracy, convenience, or expenditure of time are equally difficult to draw. Nor is it necessarily true that instrumental procedures employ more sophisticated or more costly apparatus; indeed, use of a modern automatic balance in a gravimetric analysis involves more complex and refined instrumentation than is required for several of the methods found in Table 1–1.

In addition to the methods listed in the second column of Table 1–1, there exists

Table 1–1
Some Analytical Signals

Signal	Analytical Methods Based on Measurement of Signal
Emission of radiation	Emission spectroscopy (X-ray, UV, visible, electron, Auger); fluorescence and phosphorescence spectroscopy (X-ray, UV, visible); radiochemistry
Absorption of radiation	Spectrophotometry (X-ray, UV, visible, IR); photoacoustic spectroscopy; nuclear magnetic resonance and electron spin resonance spectroscopy
Scattering of radiation	Turbidimetry; nephelometry; Raman spectroscopy
Refraction of radiation	Refractometry; interferometry
Diffraction of radiation	X-Ray and electron diffraction methods
Rotation of radiation	Polarimetry; optical rotatory dispersion; circular dichroism
Electrical potential	Potentiometry; chronopotentiometry
Electrical current	Polarography; amperometry; coulometry
Electrical resistance	Conductometry
Mass-to-charge ratio	Mass spectrometry
Rate of reaction	Kinetic methods
Thermal properties	Thermal conductivity and enthalpy methods
Mass	Gravimetric analysis
Volume	Volumetric analysis

another group of analytical procedures that are employed for separation and resolution of closely related compounds. Common separation methods include chromatography, distillation, extraction, ion exchange, fractional crystallization, and selective precipitation. One of the signals listed in Table 1–1 is ordinarily used to complete the analysis following the separation step. Thus, for example, thermal conductivity, volume, refractive index, and electrical conductance have all been employed in conjunction with various chromatographic methods.

This text deals with most of the instrumental methods listed in Table 1–1, as well as with many of the most widely employed separation procedures. Little space is devoted to the classical methods, since their use is normally treated in introductory analytical chemistry courses.

Table 1–1 suggests that the chemist faced with an analytical problem may have a bewildering array of methods from which to choose. The amount of time spent on the analytical work and the quality of the results are critically dependent on this choice. In arriving at a decision as to which method to choose, the chemist must take into account the complexity of the materials to be analyzed, the concentration of the species of in-

terest, the number of samples to be analyzed, and the accuracy required. His choice will then depend upon a knowledge of the basic principles underlying the various methods available to him, and thus their strengths and limitations. The development of this type of knowledge represents a major goal of this book.

1B Instruments for Analysis

In the broadest sense, an instrument for chemical analysis converts a signal that is usually not directly detectable and understandable by man to a form that is. Thus, an instrument can be viewed as a communication device between the system under study and the scientist.

1B-1 COMPONENTS OF INSTRUMENTS

Regardless of its complexity, an instrument generally contains no more than four fundamental components. As shown schematically in Figure 1–1, these components are a signal generator, an input transducer or detector, a signal processor, and an output transducer or readout. A general description of these components follows.

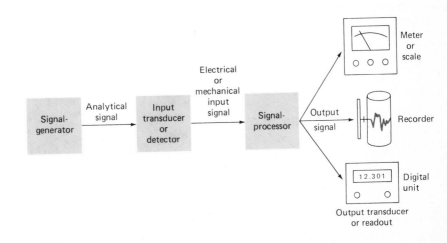

FIGURE 1-1 Components of a typical instrument.

Signal Generators.   Signal generators produce analytical signals from the components of the sample. The generator may simply be the sample itself. For example, the signal for an analytical balance is the mass of a component of the sample; for a pH meter, the signal is the activity of hydrogen ions in a solution. In many other instruments, however, the signal generator is more elaborate. For example, the signal generator of an infrared spectrophotometer includes, in addition to the sample, a source of infrared radiation, a monochromator, a beam chopper and splitter, a sample holder, and a radiation attenuator.

The second column of Table 1–2 lists some examples of signal generators.

Input Transducers or Detectors.   A *transducer* is a device that converts one type of signal to another. Examples include a thermocouple, which converts a radiant heat signal into an electrical voltage, and the bellows in an aneroid barometer, which as it contracts or expands transduces a pressure signal into the motion of a pointer on a scale. Most of the transducers that we will encounter convert analytical signals into an electrical voltage,

current, or resistance because of the ease with which electrical signals can be amplified and modified to drive readout devices. Note, however, that the transduced signals for the last two instruments in Table 1–2 are nonelectrical.

Signal Processors.   The signal processor modifies the transduced signal in such a way as to make it more convenient for operation of the readout device. Perhaps the most common modification is that of amplification—that is, multiplication of the signal by a constant greater than unity. In a two-pan analytical balance, the motion of the beam is amplified by the pointer, whose displacement is significantly greater than that of the beam itself. Amplification by a photographic film is enormous; here, a single photon may produce as many as 10^{12} silver atoms. Electrical signals can, of course, be readily amplified by a factor of 10^6 or more.

A variety of other modifications are commonly carried out on electrical signals. In addition to being amplified, signals are often multiplied by a constant smaller than unity (*attenuated*), integrated, differentiated, added or subtracted, or increased exponen-

Table 1–2
Some Examples of Instrument Components

Instrument	Signal Generator	Analytical Signal	Input Transducer	Transduced Signal	Signal Processor	Readout
Photometer	Tungsten lamp, glass filter, sample	Attenuated light beam	Photocell	Electrical current	None	Current meter
Atomic emission spectrometer	Flame, monochromator, chopper, sample	UV or visible radiation	Photomultiplier tube	Electrical potential	Amplifier, demodulator	Chart recorder
Coulometer	Dc source, sample	Cell current	Electrodes	Electrical current	Amplifier	Chart recorder
pH meter	Sample	Hydrogen ion activity	Glass-calomel electrodes	Electrical potential	Amplifier, digitizes	Digital unit
X-Ray powder diffractometer	X-Ray tube, sample	Diffracted radiation	Photographic film	Latent image	Chemical developer	Black images on film
Color comparator	Sunlight, sample	Color	Eye	Optic nerve signal	Brain	Visual color response

tially. Other operations may involve conversion into an alternating current, rectification to give a direct current, comparison of the transduced signal with one from a standard, and transformation from a current to a voltage or the converse.

1B-2 ELECTRONICS, MICROPROCESSORS, AND COMPUTERS IN INSTRUMENTATION

Most modern instruments for analysis are based upon sophisticated electronic devices such as operational amplifiers, integrated circuits, analog-to-digital and digital-to-analog converters, counters, microprocessors, and computers. In order to develop an appreciation of both the power and the limitations of such instruments, it is necessary for the scientist to have at least a qualitative understanding of how these devices function. For this reason, Chapters 2 and 3 of this text are devoted to elementary descriptions of analog and digital electronic circuits, microprocessors, and computers. Subsequent chapters then treat the various types of analytical instruments and their applications.

1C   Uncertainties in Instrumental Measurements

The data from physical measurements are always plagued by uncertainties or errors. An estimate of the magnitude of these uncertainties is of prime importance if experimental results are to have meaning. Unfortunately, no simple, generally applicable methods exist that will provide a measure of the reliability of experimental data with absolute certainty; indeed, the work required in evaluating the quality of data is frequently comparable to the effort that goes into obtaining them. Ultimately, the scientist can only make a *judgment* as to the probable accuracy of a measurement; with experience, judgments of this kind tend to become harsher and less optimistic.

1C-1   PRECISION AND ACCURACY

Two terms are widely used in discussion of reliability of data, *precision* and *accuracy*.

Precision.   Precision describes the reproducibility of results; that is, the agreement between the numerical values of two or more measurements that have been made *in exactly the same way*. Generally, the precision of experimental measurements is readily obtained.

Several methods exist for expressing precision of a set of data, one being the deviation from the mean (or arithmetic average) value for the set. The mean $\bar{x}$ for a set of n replicate measurements is the sum of the data in the set divided by n. The absolute deviation from the mean for a particular datum x_i is then $(x_i - \bar{x})$. The relative deviation from the mean is the ratio between the absolute deviation and the mean; that is, $(x_i - \bar{x})/\bar{x}$. Often it is expressed as a percent or as parts per thousand (ppt).

The *spread* or *range* (w) of a set of data is the numerical difference between the highest and lowest result and also constitutes a measure of precision. The most significant and useful measures of precision are the *standard deviation* and the *variance* of data. These terms will be defined subsequently.

Accuracy.   Accuracy refers to the closeness of a measurement to its true or accepted value, x_t. The *absolute error* in a measured quantity x_i is given by $(x_i - x_t)$. Note that for both deviation from the mean and error, a positive value indicates that the measured quantity is greater than the reference value. The relative error of a result is given by $(x_i - x_t)/x_t$.

The accuracy of a measurement is, of course, the only true criterion of its reliability. Unfortunately, accuracy can never be determined unambiguously because such a determination requires a sure knowledge of that which is being sought, namely, the true value. Consequently, at best a scientist can only estimate the accuracy of data; this estimate is generally based on past experience, analysis of standard samples, the scientific literature, and common sense. It is important to appreciate that the precision of a measurement often is not a reliable measure of its accuracy because of the existence of more than one type of error.

1C-2 TYPES OF ERROR

Measurement uncertainties fall into two broad categories: *determinate* or *systematic errors* and *indeterminate* or *random errors*. In some instances it is difficult or impossible to determine into which category a given error falls; nevertheless, the concept is useful.

Determinate Errors. Determinate errors are those that have a definite value and an assignable cause; in principle (but not always in practice) the experimentalist can measure, account, and correct for these errors. One common source of determinate uncertainties can be traced to *instrumental errors*. Examples include decreases in the voltage of battery-operated power supplies with use, increases in circuit resistances due to oxidation of electrical contacts, temperature effects on detectors, vibrations on the positions of pointers, and currents induced in electronic equipment from 110-V power lines. Often these errors can be identified and corrected by calibration.

Method errors are often introduced from nonideal chemical and physical behavior of reagents and reactions upon which an analysis is based. Possible sources include slowness of chemical reactions, instability of reagents, contaminants in reagents, and chemical interferences. Method errors are frequently more difficult to detect and rectify than are instrumental errors.

Personal errors are introduced into a measurement by the experimentalist. Examples include misreading the position of a meter needle, transposing numbers when entering data, inadvertent spillages, and bias. The

last is a source of error that must be actively fought by most scientists.

Indeterminate Errors in Instrumental Measurements. As implied by the name, indeterminate errors arise from uncertainties in a measurement from sources that are unknown and uncontrollable by the experimenter. Indeterminate errors often manifest themselves in the form of small, random fluctuations (called *noise*[1]) in the readout device of instruments. These small but detectable variations represent the accumulation of a very large number of uncertainties arising in various parts of the instrument and system under study. These uncertainties are not detectable individually, and it is only their accumulated effect that is recognizable as noise.

One of the characteristics of the effect of indeterminate errors is randomness. In some instances, most of the individual uncertainties may, by accident, occur in a positive direction; the net effect is a relatively large positive fluctuation of the readout device. At other times, the individual signals may by chance be largely negative, leading to a smaller than average net signal. The most probable event, however, is for the number and size of the negative and positive noise signals to be about the same, thus giving a reading that approaches an average or mean value.

The random behavior of indeterminate errors makes it possible to treat their effects by the methods of statistics. The remainder of this section is devoted to an introduction to statistical techniques.[2]

---

[1] This terminology is derived from radio engineering, where the presence of unwanted signal fluctuation was recognized by the ear as static or noise.

[2] Two excellent monographs on statistics are *Statistical Methods in Research and Production*, 4th ed., O. L. Davies and P. L. Goldsmith, Eds., Longman Group Limited: New York, 1972; W. J. Dixon and F. J. Massey, Jr., *Introduction to Statistical Analysis*, 3d ed., McGraw-Hill: New York, 1969.

1C-3 DISTRIBUTION OF INDETERMINATE UNCERTAINTIES

For a large number of replicate observations, the accumulated uncertainties leading to random noise usually cause the results to be distributed in the symmetrical way shown in Figure 1–2. This distribution curve, which relates frequency of occurrence to deviations from the mean, is termed *Gaussian* or *normal*. The bell-shaped Gaussian curve can be described by the equation

$$y = \frac{e^{-(x_i - \mu)^2/2\sigma^2}}{\sigma\sqrt{2\pi}} = \frac{e^{-z^2/2}}{\sigma\sqrt{2\pi}} \qquad (1–1)$$

In this equation, x_i represents values of individual measurements, and μ is the arithmetic mean for an infinite number of such measurements. The quantity $(x_i - \mu)$ is thus the deviation from the mean; y is the frequency of occurrence for each value of $(x_i - \mu)$. The symbol π has its usual meaning, and e is the base for natural, or Napierian, logarithms. The parameter σ is called the *standard deviation* and is a constant that has a unique value for any set of data comprising a large number of measurements. The *breadth* of the normal error curve, which increases as the precision of a measurement decreases, is directly related to σ. Thus, σ is widely employed to define the precision of measurements.

The exponential term in Equation 1–1 can be simplified by introducing the variable

$$z = \frac{x_i - \mu}{\sigma} \qquad (1–2)$$

which then gives the deviation from the mean in units of standard deviations.

The Standard Deviation.   Equation 1–1 indicates that a unique distribution curve exists for each value of the standard deviation. Regardless of the size of σ, however, it can be shown that 68.3% of the area beneath the curve lies within one standard deviation $(\pm 1\sigma)$ of the mean, μ. Thus, 68.3% of the values lie within these boundaries. Approximately 95.5% of all values will be within $\pm 2\sigma$; 99.7% will be within $\pm 3\sigma$. Values of $(x_i - \mu)$ corresponding to $\pm 1\sigma$, $\pm 2\sigma$, and $\pm 3\sigma$ are indicated by broken vertical lines in Figure 1–2.

These properties of the normal error curve are useful because they permit statements to be made about the probable magnitude of the indeterminate error in a given measurement *provided the standard deviation of the method of measurement is known*. Thus, if σ were available, one could say that the chances are 68.3 out of 100 that the indeterminate error associated with any given single measurement is smaller than $\pm 1\sigma$, that the chances are 95.5 out of 100 that the error is less than $\pm 2\sigma$, and so forth. Clearly, the standard deviation for a method of measurement is a useful quantity for estimating and reporting the probable size of indeterminate errors.

For an infinite set of data, the standard

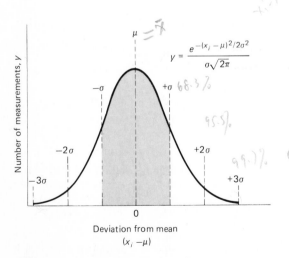

FIGURE 1-2   A normal or Gaussian distribution curve.

deviation is given by

$$\sigma = \sqrt{\frac{\sum_{i=1}^{N}(x_i - \mu)^2}{N}}$$

$$= \sqrt{\frac{\sum_{i=1}^{N} x_i^2 - (\sum_{i=1}^{N} x_i)^2/N}{N}} \qquad (1\text{--}3)$$

Here, the sum of the squares of the individual deviations from the mean $(x_i - \mu)$ is divided by the total number of measurements in the set, N. Extraction of the square root of this quotient gives σ. The identity on the right is less cumbersome to use when calculating σ with a hand calculator.

Another precision term widely employed by statisticians is the *variance* which is equal to σ^2. Most experimental scientists prefer to employ σ rather than σ^2 because the units of the standard deviation are the same as those of the quantity measured. The variance, on the other hand, has the advantage of additivity. That is, if several independent causes of variation exist in a system, the total variance σ_T^2 is the sum of the individual variances.

The Coefficient of Variation, CV.   The coefficient of variation is the standard deviation expressed as the percent of the arithmetic mean. Thus,

$$CV = (\sigma/\mu) \times 100 \qquad (1\text{--}4)$$

The coefficient of variation is useful for comparing the variability of groups of measurements having widely different mean values. For example, consider the following data from the determination of manganese in two different samples.

% Mn = 0.906 ($\pm$0.014)

% Mn = 9.06 ($\pm$0.14)

Here, the figures in parentheses are standard deviations, which differ widely. The coefficients of variation for the two determinations are, however, identical (1.5%).

The Standard Error of a Mean.   As was noted earlier, the standard deviation permits statements to be made about the probability of any single observation lying within some interval about the true mean. Generally, the mean of several observations provides a more reliable estimate of the true mean than a single observation; this implies that the standard deviation for a mean is smaller than that for a single observation. The standard deviation of a mean is known as its *standard error.*

It can be shown that the standard error σ_m of n observations is

$$\sigma_m = \sigma/\sqrt{n} \qquad (1\text{--}5)$$

where σ is defined by Equation 1–3.

The Standard Deviation For Small Sets of Data.   It has been found that direct application of classical statistics, based on Equations 1–1 and 1–3, to a small number of replicate measurements often leads to false conclusions regarding the probable magnitude of the indeterminate error. Fortunately, modifications of the relationships have been developed to permit valid statements about the random error associated with as few as two or three values.

Equations 1–1 and 1–3 are not directly applicable to a small number of replicate measurements because μ, the mean value of an infinitely large number of measurements (and the true value *in the absence of determinate error*), is never known. In its stead, we are forced to employ $\bar{x}$, the mean of a relatively small number of data. More often than not, $\bar{x}$ will differ somewhat from μ. This difference is, of course, the result of the indeterminate error whose probable magnitude we are trying to assess. It is important to note that *any error in $\bar{x}$ causes a corresponding error in σ* (Equation 1–3). Thus, with a small set of data, not only is the mean $\bar{x}$ likely to differ from μ but, *equally important, the estimate of the standard deviation also may be misleading*. In short, we have *two* uncertainties to cope with, the one residing in the mean and the other in the standard deviation.

It has been found experimentally that application of Equation 1–3 to small sets of data ($N < 20$) leads to standard deviations that are on the average smaller than they should be. This negative bias in σ is a general phenomenon and is attributable to the fact that both a mean and a standard deviation must be extracted from the same small set. It can be shown that this bias can be largely eliminated by substituting the *number of degrees of freedom* $(N - 1)$ for N in Equation 1–3. That is, we define the standard deviation for a small number of measurements as

$$s = \sqrt{\frac{\sum_{i=1}^{N}(x_i - \bar{x})^2}{N - 1}}$$

$$= \sqrt{\frac{\sum_{i=1}^{N} x_i^2 - (\sum_{i=1}^{N} x_i)^2/N}{N - 1}} \quad (1\text{-}6)$$

Here again, the expression on the right is easier to use when calculating s with a hand calculator.[3] Note that Equation 1–6 differs from Equation 1–3 in three regards. First, the denominator is now $(N - 1)$. Second, $\bar{x}$, the measured mean for the small set, replaces the true but unknown mean μ. Finally, to emphasize that the resulting standard deviation is but an approximation of the true value, it is given the symbol s rather than σ. In order to distinguish between σ and s in this text we shall refer to the former as the *population standard deviation* because the hypothetical set of all possible observations of the type being referred to is known as the *population* or *universe* of values. The quantity s will then be called the sample

standard deviation since it is only part of the population. As the sample size increases, its properties will resemble more and more closely those of the population.

In an analogous way, μ will be called the population mean and $\bar{x}$ the sample mean.

The rationale for the use of $(N - 1)$ in Equation 1–6 is as follows. When μ is unknown, we calculate two quantities, $\bar{x}$ and s, from our set of replicate data. The need to establish the mean $\bar{x}$ from the data removes one degree of freedom. That is, if their signs are retained, the individual deviations from $\bar{x}$ must total zero; once $(N - 1)$ deviations have been established, the final one is necessarily known. Thus, only $(N - 1)$ deviations provide independent measures of the precision for the set.

EXAMPLE 1-1.   The following data were obtained for the determination of iron in a soil sample by a colorimetric method: 1.67, 1.63, and 1.70 ppm. Calculate the standard deviation for the measurement.

x_i	$(x_i)^2$
1.67	2.7889
1.63	2.6569
1.70	2.8900
$\Sigma x_i = 5.00$	$\Sigma (x_i)^2 = 8.3358$

$$s = \sqrt{\frac{8.3358 - (5.00)^2/3}{3 - 1}}$$

$$= 0.0351 = 0.04 \text{ ppm}$$

Generally, to avoid rounding errors, it is wise to avoid rounding intermediate results and round only at the end.

1C-4 CONFIDENCE INTERVALS

The true mean value (μ) of a measurement is a constant that must always remain unknown. With the aid of statistical theory, however, limits may be set about the experi-

---

[3] It is important to know that application of the expression on the right can lead to serious errors when a handheld calculator or small computer is used to derive s and the data contain five or more significant figures. See: P. M. Wanek, *et al.*, *Anal. Chem.*, **1982**, 54, 1877 and H. E. Solberg, *Anal. Chem.*, **1983**, 55, 1611.

mentally determined mean ($\bar{x}$) within which we may expect to find the true mean with a given degree of probability; the limits obtained in this manner are called *confidence limits*. The interval defined by these limits is known as the *confidence interval*.

Some of the properties of the confidence interval are worthy of mention. For a given set of data, the size of the interval depends in part upon the odds for correctness desired. Clearly, for a prediction to be absolutely correct, we would have to choose an interval about the mean large enough to include all conceivable values that x_i might take. Such an interval, of course, has no predictive value. On the other hand, the interval does not need to be this large if we are willing to accept the probability of being correct 99 times in 100; it can be even smaller if 95% probability is acceptable. In short, as the probability for making a correct prediction is made less favorable, the interval included by the confidence limits becomes smaller.

The confidence interval, which is derived from the standard deviation s for the method of measurement, also depends in magnitude upon the certainty with which s is known. Often, the chemist will have reason to believe that the experimental value for s is an excellent approximation of σ. In other situations, however, a considerable uncertainty in s may exist. Under these circumstances, the confidence interval will necessarily be broader.

Methods for Obtaining a Good Approximation of σ.   Fluctuations in the calculated value for s decrease as the number of measurements N in Equation 1–6 increases; in fact, it is proper to assume that s and σ are, for all practical purposes, identical when N is greater than about 20. It thus becomes feasible for the chemist to obtain a good approximation of σ when the method of measurement is not too time-consuming and when an adequate amount of sample is available. For example, if it were necessary to measure the pH of numerous solutions in the course of an in-

vestigation, it might prove worthwhile to evaluate s in a series of preliminary experiments. This particular measurement is simple, requiring only that a pair of rinsed and dried electrodes be immersed in the test solution; the potential across the electrodes serves as a measure of the pH. To determine s, 20 to 30 portions of a buffer solution of fixed pH could be measured, following exactly all steps of the procedure. Normally, it would be safe to assume that the indeterminate error in this test would be the same as that in subsequent measurements and that the value of s calculated by means of Equation 1–6 would be a valid and accurate measure of the theoretical σ.

For analyses that are time-consuming, the foregoing procedure is not ordinarily practical. Here, however, precision data from a series of samples accumulated in the course of time can often be pooled to provide an estimate of s which is superior to the value for any individual subset. Again, one must assume the same sources of indeterminate error among the samples. This assumption is usually valid provided the samples have similar compositions and each has been analyzed identically. To obtain a pooled estimate of s, deviations from the mean for each subset are squared; the squares for all of the subsets are then summed and divided by an appropriate number of degrees of freedom. The pooled s is obtained by extracting the square root of the quotient. One degree of freedom is lost for each subset. Thus, the number of degrees of freedom for the pooled s is equal to the total number of measurements minus the number of subsets. An example of this calculation follows.

EXAMPLE 1–2.   The mercury in samples of seven fish taken from Lake Erie was determined by a method based upon absorption of radiation by elemental mercury. The results are given in the accompanying table. Calculate a standard deviation for the method based upon the pooled precision data.

Sample Number	Number of Replications	Results, Hg Content, ppm	Mean, ppm Hg	Sum of Squares of Deviations from Mean
1	3	1.80, 1.58, 1.64	1.67	0.0258
2	4	0.96, 0.98, 1.02, 1.10	1.02	0.0116
3	2	3.13, 3.35	3.24	0.0242
4	6	2.06, 1.93, 2.12, 2.16, 1.89, 1.95	2.02	0.0611
5	4	0.57, 0.58, 0.64, 0.49	0.57	0.0114
6	5	2.35, 2.44, 2.70, 2.48, 2.44	2.48	0.0685
7	4	1.11, 1.15, 1.22, 1.04	1.13	0.0170
Number of measurements	28			Sum of squares = 0.2196

The values in columns 4 and 5 for sample 1 were calculated as follows:

(x_i)	$|(x_i - \bar{x}_1)|$	$(x_i - \bar{x}_1)^2$
1.80	0.127	0.0161
1.58	0.093	0.0086
1.64	0.033	0.0011
3|5.02		$\Sigma = 0.0258$

$\bar{x}_1 = 1.673 = 1.67$

The other data in column 5 were obtained similarly. Then

$$s = \sqrt{\frac{0.2196}{28 - 7}}$$

$$= 0.10 \text{ ppm Hg}$$

Note that one degree of freedom has been lost for each of the 7 samples. Because the remaining degrees of freedom number greater than 20, however, this estimate of s can be considered to be a good approximation of σ.

Confidence Interval When s is a Good Approximation of σ. As indicated earlier (p. 7), the breadth of the normal error curve is determined by σ. For any given value of σ the area under a part of the normal error curve *relative* to the total area can be related to the parameter z by means of Equation 1–2. This ratio of areas (usually expressed as a percent) is called the *confidence level,* and it measures the probability for the absolute deviation $(x - \mu)$ being equal to or less than $z\sigma$. Thus, the area under the curve encompassed by $z = \pm 1.96\sigma$ corresponds to 95% of the total area. Here, the confidence level is 95%, and we may state that 95 times out of 100 the calculated value of $(x - \mu)$ for a large number of measurements will be equal to or less than $\pm 1.96\sigma$. Table 1–3 lists confidence intervals for various values of z.

The confidence limit (*C.L.*) for a single measurement can be obtained by rearranging Equation 1–2 and remembering that z can be either plus or minus. Thus,

$$C.L. \text{ for } \mu = x \pm z\sigma \qquad (1\text{–}7)$$

Table. 1-3
Confidence Levels for Various Values of z

Confidence Level, %	z
50	0.67
68	1.00
80	1.29
90	1.64
95	1.96
99	2.58
99.7	3.00
99.9	3.29

EXAMPLE 1-3. Calculate the 50% and the 95% confidence limits for the first entry (1.80 ppm Hg) in Example 1–2.

Here, we calculated that $s = 0.10$ ppm Hg and had sufficient data to assume $s \rightarrow \sigma$. From Table 1–3 we see that $z = 0.67$ and 1.96 for the two confidence levels. Thus, from Equation 1–7

$$50\% \ C.L. \text{ for } \mu = 1.80 \pm 0.67 \times 0.10$$
$$= 1.80 \pm 0.07$$

$$95\% \ C.L. \text{ for } \mu = 1.80 \pm 1.96 \times 0.10$$
$$= 1.80 \pm 0.20$$

The chances are 50 in 100 that μ, the true mean (and *in the absence of determinate error* the true value), will be in the interval between 1.73 and 1.87 ppm Hg; there is a 95% chance that it will be in the interval between 1.60 and 2.00 ppm Hg.

Equation 1–2 applies to the result of a single measurement. Application of Equation 1–5 shows the confidence interval is decreased by $\sqrt{n}$ for the average of n replicate measurements. Thus, a more general form of Equation 1–7 is

$$C.L. \text{ for } \mu = \bar{x} \pm \frac{z\sigma}{\sqrt{n}} \tag{1–8}$$

EXAMPLE 1-4. Calculate the 50% and the 95% confidence limits for the mean value (1.67 ppm Hg) for sample 1 in Example 1–2. Again, $s \rightarrow \sigma = 0.10$.

For the three measurements,

$$50\% \ C.L. = 1.67 \pm \frac{0.67 \times 0.10}{\sqrt{3}}$$

$$= 1.67 \pm 0.04$$

$$95\% \ C.L. = 1.67 \pm \frac{1.96 \times 0.10}{\sqrt{3}}$$

$$= 1.67 \pm 0.11$$

Thus, the chances are 50 in 100 that the true mean will lie in the interval of 1.63 to 1.71 ppm Hg and 95 in 100 that it will be between 1.56 and 1.78 ppm.

A consideration of Equation 1–8 indicates that the confidence interval for an analysis can be halved by employing the mean of four measurements. Sixteen measurements would be required to narrow the limit by another factor of two. It is apparent that a point of diminishing return is rapidly reached in acquiring additional data. Thus, the chemist ordinarily takes advantage of the relatively large gain afforded by averaging two to four measurements but can seldom afford the time required for further increases in confidence.

In data analysis, it is essential to keep in mind always that confidence intervals based on Equation 1–8 apply only *in the absence of determinate errors.*

Confidence Limits When σ is Unknown. Frequently, a chemist must make use of an unfamiliar method where limitations in time or amount of available sample preclude an accurate estimation of σ. Here, a single set of replicate measurements must provide not only a mean value but also a precision estimate. As indicated earlier, s calculated from a small set of data may be subject to considerable uncertainty; thus, the confidence limits must be broadened under these circumstances.

To account for the potential variability of s, use is made of the parameter t, which is defined as

$$t = \frac{x - \mu}{s} \tag{1–9}$$

In contrast to z in Equation 1–2, t is dependent not only on the desired confidence level but also upon the number of degrees of freedom available in the calculation of s. Table 1–4 provides values for t for a few degrees of freedom; much more extensive tables are found in various mathematical handbooks. Note that the values for t be-

Table 1-4
Values of t for Various Levels of Probability

Degrees of Freedom	Factor for Confidence Interval, %				
	80	90	95	99	99.9
1	3.08	6.31	12.7	63.7	637
2	1.89	2.92	4.30	9.92	31.6
3	1.64	2.35	3.18	5.84	12.9
4	1.53	2.13	2.78	4.60	8.60
5	1.48	2.02	2.57	4.03	6.86
6	1.44	1.94	2.45	3.71	5.96
7	1.42	1.90	2.36	3.50	5.40
8	1.40	1.86	2.31	3.36	5.04
9	1.38	1.83	2.26	3.25	4.78
10	1.37	1.81	2.23	3.17	4.59
11	1.36	1.80	2.20	3.11	4.44
12	1.36	1.78	2.18	3.06	4.32
13	1.35	1.77	2.16	3.01	4.22
14	1.34	1.76	2.14	2.98	4.14
∞	1.29	1.64	1.96	2.58	3.29

come equal to those for z (Table 1–3) as the number of degrees of freedom becomes infinite.

The confidence limit for the mean $\bar{x}$ of n replicate measurements can be derived from t by an equation analogous to Equation 1–8; that is,

$$\text{C.L. for } \mu = \bar{x} \pm \frac{ts}{\sqrt{n}} \qquad (1\text{--}10)$$

EXAMPLE 1-5. A chemist obtained the following data for the alcohol content of a sample of blood; percent ethanol = 0.084, 0.089, and 0.079. Calculate the 95% confidence limit for the mean assuming (a) no additional knowledge about the precision of the method and (b) that on the basis of previous experiences $s \rightarrow \sigma = 0.006\%$ ethanol.

(a) $\Sigma x_i = 0.084 + 0.089 + 0.079 = 0.252$

$\Sigma x_i^2 = 0.007056 + 0.007921$
$\qquad\qquad + 0.006241 = 0.021218$

$$s = \sqrt{\frac{0.021218 - (0.252)^2/3}{3 - 1}} = 0.005$$

Here, $\bar{x} = 0.252/3 = 0.084$. Table 1–4 indicates that $t = \pm 4.30$ for two degrees of freedom and 95% confidence. Thus,

$$95\% \text{ C.L.} = \bar{x} \pm \frac{ts}{\sqrt{n}}$$

$$= 0.084 \pm \frac{4.3 \times 0.005}{\sqrt{3}}$$

$$= 0.084 \pm 0.012$$

(b) Because a good value of σ is available,

$$95\% \text{ C.L.} = \bar{x} \pm \frac{z\sigma}{\sqrt{n}}$$

$$= 0.084 \pm \frac{1.96 \times 0.006}{\sqrt{3}}$$

$$= 0.084 \pm 0.007$$

Note that a sure knowledge of σ decreased the confidence interval by almost half.

propagation

1C-5 PROPAGATION OF MEASUREMENT UNCERTAINTIES

A typical instrumental method of analysis involves several experimental measurements each of which is subject to an indeterminate uncertainty and each of which contributes to the indeterminate error of the final result. For the purpose of showing how such indeterminate uncertainties affect the outcome of an analysis, let us assume that a result x is dependent upon the experimental variables $p, q, r, \ldots$, each of which fluctuates in a random and independent way. That is, x is a function f of $p, q, r, \ldots$ so that we may write

$$x = f(p,q,r, \ldots) \qquad (1-11)$$

The uncertainty dx_i (that is, the deviation from the mean) in the ith measurement of x will depend upon the size and sign of the corresponding uncertainties dp_i, dq_i, dr_i, $\ldots$. That is,

$$dx_i = f(dp_i, dq_i, dr_i, \ldots) = f(x_i - \mu)$$

The variation in dx as a function of the uncertainties in $p, q, r, \ldots$ can be derived by taking the total differential of Equation 1–11. That is,

$$dx = \left(\frac{\delta x}{\delta p}\right)_{q,r,\ldots} dp + \left(\frac{\delta x}{\delta q}\right)_{p,r,\ldots} dq$$
$$+ \left(\frac{\delta x}{\delta r}\right)_{p,q,\ldots} dr + \cdots \quad (1-12)$$

In order to relate the various terms in Equation 1–12 to the standard deviation of $x, p, q,$ and r as defined by Equation 1–3, it is necessary to square the foregoing equation. Thus,

$$(dx)^2 = \left[\left(\frac{\delta x}{\delta p}\right)_{q,r,\ldots} dp + \left(\frac{\delta x}{\delta q}\right)_{p,r,\ldots} dq\right.$$
$$\left. + \left(\frac{\delta x}{\delta r}\right)_{p,q,\ldots} dr + \cdots\right]^2 \quad (1-13)$$

This equation must then be summed between the limits of $i = 1$ to $i = N$, where N again is the total number of replicate measurements (Equation 1–3).

In squaring Equation 1–12, two types of terms from the right-hand side of the equation emerge. The first are type 1 terms, such as

$$\left[\left(\frac{\delta x}{\delta p}\right) dp\right]^2, \left[\left(\frac{\delta x}{\delta q}\right) dq\right]^2, \left[\left(\frac{\delta x}{\delta r}\right) dr\right]^2$$

Because they are squares, type 1 terms *will always be positive and can never cancel*. In contrast, type 2 terms (called *cross terms*) may be either positive or negative in sign. Examples are

$$\left(\frac{\delta x}{\delta p}\right)\left(\frac{\delta x}{\delta q}\right) dpdq, \left(\frac{\delta x}{\delta p}\right)\left(\frac{\delta x}{\delta r}\right) dpdr$$

If dp, dq, and dr represent independent and random uncertainties, some of the cross terms will be negative and others positive. Thus, the *summation of all such terms should approach zero*, particularly when N is large.

As a consequence of the canceling tendency of type 2 terms, the square of Equation 1–12 can be assumed to be made up exclusively of type 1 terms. Summing these terms gives

$$\Sigma_{i=1}^{N}(dx_i)^2 = \left(\frac{\delta x}{\delta p}\right)^2 \Sigma_{i=1}^{N}(dp_i)^2$$
$$+ \left(\frac{\delta x}{\delta q}\right)^2 \Sigma_{i=1}^{N}(dq_i)^2 + \left(\frac{\delta x}{\delta r}\right)^2 \Sigma_{i=1}^{N}(dr_i)^2 + \cdots$$

Dividing through by N gives

$$\frac{\Sigma(dx_i)^2}{N} = \left(\frac{\delta x}{\delta p}\right)^2 \frac{\Sigma(dp_i)^2}{N} + \left(\frac{\delta x}{\delta q}\right)^2 \frac{\Sigma(dq_i)^2}{N}$$
$$+ \left(\frac{\delta x}{\delta r}\right)^2 \frac{\Sigma(dr_i)^2}{N} + \cdots \quad (1-14)$$

From Equation 1–3, however, we see that

$$\frac{(dx_i)^2}{N} = \frac{(x_i - \mu)^2}{N} = \sigma_x^2$$

where σ_x^2 is the variance of x. Similarly,

$$\frac{(dp_i)^2}{N} = \sigma_p^2$$

and so forth. Thus, Equation 1–14 can be written in terms of the variances of the quantities; that is,

$$\sigma_x^2 = \left(\frac{\delta x}{\delta p}\right)^2 \sigma_p^2 + \left(\frac{\delta x}{\delta q}\right)^2 \sigma_q^2$$
$$+ \left(\frac{\delta x}{\delta r}\right)^2 \sigma_r^2 + \cdots \quad (1\text{–}15)$$

It is of interest to apply Equation 1–15 to the derivation of the standard deviations for the results of various kinds of arithmetic manipulations of experimental data.

Addition and Subtraction.   Consider the following relationship

$$x = p + q - r$$

where we wish to derive the standard deviation σ (or s) in x arising from the standard deviations in the three quantities on the right. Applying Equation 1–15, we write

$$\sigma_x^2 = \left(\frac{\delta x}{\delta p}\right)^2 \sigma_p^2 + \left(\frac{\delta x}{\delta q}\right)^2 \sigma_q^2 + \left(\frac{\delta x}{\delta r}\right)^2 \sigma_r^2$$

but

$$\frac{\delta x}{\delta p} = \frac{\delta x}{\delta q} = 1 \quad \text{and} \quad \frac{\delta x}{\delta r} = -1$$

Therefore,

$$\sigma_x^2 = \sigma_p^2 + \sigma_q^2 + \sigma_r^2$$

As a generality, then, it may be stated that the *absolute* variance of a sum or difference is simply the sum of the *absolute* variances of the numbers making up the sum or difference; the corresponding absolute standard deviation is

$$\sigma_x = \sqrt{\sigma_p^2 + \sigma_q^2 + \sigma_r^2} \quad (1\text{–}16)$$

Note that when the standard deviations in Equation 1–16 have been obtained from a limited number of data, the expression can better be written in the form

$$s_x = \sqrt{s_p^2 + s_q^2 + s_r^2}$$

This same substitution applies to Equations 1–17 through 1–20 as well.

EXAMPLE 1-6.   Calculate the standard deviation for the result of the following calculation (the numbers in parentheses are standard deviations).

$$x = 0.50(\pm 0.02) + 4.10(\pm 0.03)$$
$$- 1.97(\pm 0.05) = 2.63$$

Here,

$$\sigma_x = \sqrt{(\pm 0.02)^2 + (\pm 0.03)^2 + (\pm 0.05)^2}$$
$$= \pm 0.062$$

Hence,

$$x = 2.63(\pm 0.06)$$

Note that in Example 1–6, the uncertainty could be as large as ± 0.10 if the individual uncertainties all happened to bear the same sign. On the other hand, the uncertainty could also be as small as zero under fortuitous circumstances ($+0.02 + 0.03 - 0.05 = 0.00$). The calculated value of σ_x provides a more probable measure of the uncertainty than either the maximum or minimum.

Multiplication and Division.   Consider the relationship

$$x = p \cdot q/r$$

Application of Equation 1–15 here gives

$$\sigma_x^2 = (q/r)^2 \sigma_p^2 + (p/r)^2 \sigma_q^2 + (-pq/r^2)^2 \sigma_r^2$$

Dividing this equation by the square of the original relationship yields

$$(\sigma_x)_r^2 = \left(\frac{\sigma_x}{x}\right)^2$$

$$= \left(\frac{\sigma_p}{p}\right)^2 + \left(\frac{\sigma_q}{q}\right)^2 + \left(\frac{\sigma_r}{r}\right)^2 \qquad (1-17)$$

Thus for multiplication and division, the sum of *relative* variances yields the *relative* variance of the result $(\sigma_x)_r^2$. The *absolute* standard deviation is then

$$\sigma_x = x\sqrt{\left(\frac{\sigma_p}{p}\right)^2 + \left(\frac{\sigma_q}{q}\right)^2 + \left(\frac{\sigma_r}{r}\right)^2} \qquad (1-18)$$

EXAMPLE 1-7.   Calculate (a) the coefficient of variation and (b) the absolute standard deviation for x, where

$$x = \frac{4.10(\pm0.02) \times 0.0050(\pm0.0001)}{1.97(\pm0.04)}$$

$$= 0.01041$$

(a) To apply Equation 1–17 we write

$$\left(\frac{\sigma_p}{p}\right)^2 = \left(\frac{\pm0.02}{4.10}\right)^2 \quad = 0.2 \times 10^{-4}$$

$$\left(\frac{\sigma_q}{q}\right)^2 = \left(\frac{\pm0.0001}{0.0050}\right)^2 \quad = 4.0 \times 10^{-4}$$

$$\left(\frac{\sigma_r}{r}\right)^2 = \left(\frac{\pm0.04}{1.97}\right)^2 \quad = 4.1 \times 10^{-4}$$

$$\text{Sum of variances} = 8.3 \times 10^{-4}$$

and

$$\sigma_x/x = (\sigma_x)_r = \sqrt{8.3 \times 10^{-4}}$$

$$= 0.029$$

$$CV = 0.029 \times 100 = 2.9\%$$

(b) To obtain the absolute standard deviation, we multiply x by the relative

standard deviation. That is,

$$\sigma_x = 0.01041 \times 0.029 = 0.0003$$

and the result can be written as

$$x = 0.0104 \ (\pm0.0003)$$

Exponential Calculations.   Consider the relationship

$$x = p^y$$

Application of Equation 1–15 gives

$$\sigma_x^2 = (\delta p^y/\delta p)^2\sigma_p^2 = (yp^{y-1})^2\sigma_p^2$$

Dividing by the square of the original relationship gives

$$\left(\frac{\sigma_x}{x}\right)^2 = (\sigma_x)_r^2 = \left(\frac{yp^{y-1}}{p^y}\right)^2\sigma_p^2 = \left(\frac{y}{p}\right)^2\sigma_p^2$$

and

$$\frac{\sigma_x}{x} = y\frac{\sigma_p}{p} \quad \text{or} \quad (\sigma_x)_r = y(\sigma_p)_r \qquad (1-19)$$

Note that the relative standard deviation of the result is simply the relative standard deviation of the original number multiplied by the exponent.

EXAMPLE 1-8.   The standard deviation in measuring the diameter d of a sphere is ±0.02 cm. What is the standard deviation in its volume V if $d = 2.15$ cm?

$$V = \frac{4}{3}\pi\left(\frac{d}{2}\right)^3 = \frac{4}{3}\pi\left(\frac{2.15}{2}\right)^3 = 5.20 \text{ cm}^3$$

Applying Equation 1–19 gives

$$\sigma_V/V = (\sigma_V)_r = 3 \times 0.02/2.15$$

$$= 0.028 \quad \text{or} \quad 2.8\%$$

$$\sigma_V = 5.20 \times 0.028 = 0.15 \text{ cm}$$

Thus,

$$V = 5.2(\pm 0.2) \text{ cm}^3$$

It is important to note that the propagation of error in taking a number to a power is different from the error propagation in multiplication. For example, consider the uncertainty in the square of $4.0(\pm 0.2)$. Here, the relative error in the result 16.0 is given by

$$\frac{\sigma_x}{x} = 2 \times \frac{0.2}{4.0} = 0.10 \quad \text{or} \quad 10\%$$

Consider now the case when x is the product of two *independently measured* numbers, which by chance happen to have values of $x_1 = 4.0(\pm 0.2)$ and $x_2 = 4.0(\pm 0.2)$. In this case, the relative error of the product $x_1 x_2 = 16.0$ is given by Equation 1–18. That is,

$$\frac{\sigma}{x} = \sqrt{(0.2/4)^2 + (0.2/4)^2} = 0.071 \quad \text{or} \quad 7\%$$

The reason for this apparent anomaly lies in the fact that in the second case, the signs associated with each number can be the same or different. If both happened to be the same, then the uncertainty would be identical to that encountered in the first case where the signs *must* be the same. On the other hand, the possibility exists that the signs could be opposite in which case the relative uncertainties would tend to cancel. Thus, the probable uncertainty lies between the maximum (10%) and zero.

Logarithms and Antilogarithms. Consider the expression

$$x = \log p = 0.434 \ln p$$

where ln symbolizes the natural logarithm. Applying Equation 1–15 yields

$$\sigma_x^2 = \left[\frac{\delta(0.434 \ln p)}{\delta p}\right]^2 \sigma_p^2 = \left(\frac{0.434}{p}\right)^2 \sigma_p^2$$

or

$$\sigma_x = 0.434 \frac{\sigma_p}{p} = 0.434(\sigma_p)_r \qquad (1\text{–}20)$$

Here, the *absolute* standard deviation x is determined by the *relative* standard deviation of p.

EXAMPLE 1-9. Calculate the absolute standard deviation of the results of the following computations. The numbers in parentheses are absolute standard deviations.

$$x = \log[3.00 \times 10^2(\pm 0.02 \times 10^2)]$$
$$= 2.477$$

$$p = \text{antilog}[45.7(\pm 0.2)]$$
$$= 5.012 \times 10^{45}$$

(a) Application of Equation 1-20 gives

$$\sigma_x = \pm 0.434 \frac{0.02 \times 10^2}{3.00 \times 10^2} = \pm 0.003$$

and

$$x = 2.477(\pm 0.003)$$

(b) Rearranging Equation 1–20 gives

$$\frac{\sigma_p}{p} = \frac{\sigma_x}{0.434} = 2.304\sigma_x$$

$$\sigma_p = p \times 2.304\sigma_x$$
$$= 5.012 \times 10^{45} \times 2.304 \times 0.2$$
$$= 2.3 \times 10^{45}$$

and

$$p = 5.0 \times 10^{45}(\pm 2.3 \times 10^{45})$$

or better

$$p = 5 \times 10^{45}(\pm 2 \times 10^{45})$$

Note in part (b) of Example 1–9 that a large absolute error is associated with the antilogarithm of a number with few digits beyond the decimal point. The large uncertainty in such cases arises from the fact that the numbers to the left of the decimal point (the characteristic 45) in the original number serve only to locate the decimal point. All of the information about the digits making up the antilogarithm are contained in the mantissa (0.7), which in the example was known to only one significant figure. For this same reason, the logarithm of a number will appear to be made up of more significant figures than the original number. Thus, in part (a) of the example, it was found that log $3.00 \times 10^2 = 2.477$. Here, however, the initial 2 only indicates the location of the decimal point in the original number and hence is not significant; information about 3.00 is contained in the three digits 477. Thus, agreement exists between the number of significant figures in the original number and in the answer.

The following example demonstrates the application of some of the relationships just derived to the estimation of the standard deviation for the result of an instrumental analysis.

EXAMPLE 1-10. The chloride in a 0.1200 (±0.0002) g sample was determined by titration with electrolytically generated silver ions. An end point was reached after 167.4(±0.3) sec with a current of 20.00(±0.04) mA. A blank required 13.2(±0.3) sec with the same current. The numbers in parentheses are the absolute standard deviations for each measurement. The % Cl in the sample is given by

$$x = \% \text{ Cl} = \frac{(T - T_0)I}{W} \times k$$

$$= \frac{(167.4 - 13.2) \times 20.00}{0.1200} k$$

$$= 2.570 \times 10^4 \times 3.6744 \times 10^{-5}$$

$$= 0.94432$$

The value for k (3.6744×10^{-5}) has a negligible uncertainty associated with it.

Calculate (a) the absolute standard deviation for % Cl and (b) its coefficient of variation.

(a) We first derive the uncertainty associated with the difference $(T - T_0)$. Here,

$$T - T_0 = 167.4 - 13.2 = 154.2$$

From Equation 1–16,

$$s_{(T-T_0)} = \sqrt{(0.3)^2 + (0.3)^2} = \pm 0.42$$

The relative standard deviation for % Cl is then obtained from the relative variances of $T - T_0, I$, and W. That is,

$$(s_{T-T_0})_r^2 = \left(\frac{0.42}{154.2}\right)^2 = 7.4 \times 10^{-6}$$

$$(s_I)_r = \left(\frac{0.04}{20.00}\right)^2 = 4.0 \times 10^{-6}$$

$$(s_W)_r = \left(\frac{0.0002}{0.1200}\right)^2 = 2.8 \times 10^{-6}$$

$$\text{Sum of variances} = 14.2 \times 10^{-6}$$

The relative standard deviation for % Cl is then

$$(s_x)_r = \sqrt{14.2 \times 10^{-6}} = 3.8 \times 10^{-3}$$

To obtain the absolute uncertainty in the result, we write

$$s_x = (s_x)_r \times x$$
$$= 3.8 \times 10^{-3} \times 0.944$$
$$= 0.0036\% \text{ Cl}$$

and the result could be reported as

$$\% \text{ Cl} = 0.994 \pm 0.004$$

(b) The coefficient of variation is given by

$$CV = (s_x)_r \times 100 = 3.8 \times 10^{-3} \times 100$$
$$= 0.38 \quad \text{or} \quad 0.4\%$$

1C-6 UNCERTAINTIES IN CALIBRATION CURVES

Most analytical methods are based upon calibration data obtained by measurements performed on a series of standards containing known concentrations of the analyte. The data are frequently plotted to give a calibration curve such as that shown in Figure 1–3. Typically such plots approximate a straight line; it is seldom, however, that all of the data will fall exactly on that line because of the indeterminate errors in the measuring process. Thus, the investigator must try to derive a "best" straight line through the points. Statistics provides a mechanism for objectively obtaining such a line and also for specifying the uncertainties associated with its use for subsequent analyses. The technique involved is called a *regression analysis* by statisticians. Here, we will be treating only the most simple regression procedure, *linear regression*, by means of the *method of least squares*.

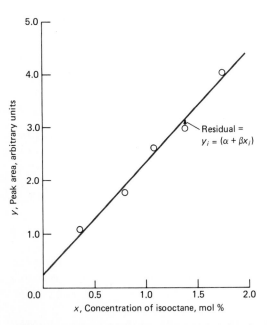

FIGURE 1–3 Calibration curve for determining isooctane in hydrocarbon mixtures.

Assumptions. In applying the method of least squares to the derivation of a calibration curve, two assumptions will be made. The first is that a linear relationship does indeed exist between analyte concentration and the measured variable. Second, it will be assumed that no significant error exists in the composition of the standards—that is, the concentrations of the standards are known exactly. Thus, the deviations of points from the straight line shown in Figure 1–3 are entirely a consequence of the indeterminate error in y, the area of chromatographic peaks. Both of these assumptions are appropriate for most analytical calibrations.

Derivation of a Least-Squares Line. The assumed linear relationship between the measured or dependent variable y and the independent variable, the analyte concentration x, is described by

$$y = \alpha + \beta x$$

where α is the intercept or value of y when x is zero and β is the slope of the line. The method of least squares permits evaluation of α and β as well as standard deviations for these quantities.

The method of least squares is based upon the derivation of that straight line for which the squares of the deviations for the individual points from that line *(on the y axis)* Q_i, are minimized. Here, Q_i is defined as

$$Q_i = [y_i - (\alpha + \beta x_i)]^2$$

where the subscripts i refer to the various individual pairs of points. Note in Figure 1–3 that the deviations or *residuals* represent the *vertical* displacements of experimental data from the least-squares line.

Although the mathematical equations necessary for a least-squares analysis are readily derived, we shall, in the interest of saving space, simply present them and focus instead upon their use.[4]

For convenience, we shall define three quantities S_{xx}, S_{yy}, and S_{xy} as follows:

$$S_{xx} = \Sigma(x_i - \bar{x})^2 = \Sigma x_i^2 - (\Sigma x_i)^2/n \qquad (1\text{--}21)$$

$$S_{yy} = \Sigma(y_i - \bar{y})^2 = \Sigma y_i^2 - (\Sigma y_i)^2/n \qquad (1\text{--}22)^5$$

$$S_{xy} = \Sigma(x_i - \bar{x})(y_i - \bar{y})$$
$$= \Sigma x_i y_i - \Sigma x_i \Sigma y_i/n \qquad (1\text{--}23)$$

Here, x_i and y_i are individual pairs of data for x and y that are used to define the least-squares line. The quantity n is the number of pairs of data used in preparation of the calibration curve and $\bar{x}$ and $\bar{y}$ are the average values for the variables. That is,

$$\bar{x} = \Sigma x_i/n$$

$$\bar{y} = \Sigma y_i/n$$

Note that S_{xx} and S_{yy} are simply the sum of the squares of the deviations from the means for the individual values for x and y. In order to ease the computational process, the equivalent expressions shown on the right are ordinarily used.

Five useful quantities can be derived from S_{xx}, S_{yy}, and S_{xy}: the intercept α; the slope β; the standard deviation for β, s_β; the *standard deviation about the regression, s_r*; and the standard deviation, s_c, for the results of an analysis based upon this calibration curve. These

---

[4] For detailed discussions of regression analysis and the least-squares method, see: *Statistical Methods in Research and Production*, 4th ed., O. L. Davies and P. L. Goldsmith, Eds., Chapter 7, Longman Group Limited: New York, 1972; W. J. Dixon and F. J. Massey, Jr., *Introduction to Statistical Analysis*, 3d ed., Chapter 11, McGraw-Hill: New York, 1969.

[5] The student should be careful to distinguish between Σx_i^2 and $\Sigma(x_i)^2$ or Σy_i^2 and $\Sigma(y_i)^2$. The first is obtained by first squaring the value of x_i or y_i and then summing. For the second, the values of x_i and y_i are first summed; the sums are then squared.

quantities are given by

$$\beta = S_{xy}/S_{xx} \qquad (1\text{--}24)$$

$$\alpha = \bar{y} - \beta\bar{x} \qquad (1\text{--}25)$$

$$s_r = \sqrt{\frac{S_{yy} - \beta^2 S_{xx}}{n - 2}} \qquad (1\text{--}26)$$

$$s_\beta = \sqrt{s_r^2/S_{xx}} \qquad (1\text{--}27)$$

$$s_c = \frac{s_r}{\beta}\sqrt{\frac{1}{m} + \frac{1}{n} + \frac{(y_c - \bar{y})^2}{\beta^2 S_{xx}}} \qquad (1\text{--}28)$$

The standard deviation about regression, s_r, is the standard deviation for y when the deviations are measured not from the mean of y (as is usually the case) but instead from the derived straight line. That is,

$$s_r = \sqrt{\frac{[y_i - (\alpha + \beta x_i)]^2}{N - 2}}$$

Here, the number of degrees of freedom is $(N - 2)$ since one degree is lost in the calculation of β and one in determining α.

Equation 1–28 permits the calculation of the standard deviation from the mean $\bar{y}_c$ of a set of m replicate analyses when a calibration curve that contains n points is used; recall that $\bar{y}$ is the mean value of y for the n calibration data.

EXAMPLE 1-11. The first two columns of Table 1–5 contain the experimental data that are plotted in Figure 1–3. Carry out a least-squares analysis of the data to obtain the regression line.

Columns 3, 4, and 5 of the table contain computed values for x_i^2, y_i^2, and $x_i y_i$; their sums appear as the last entry of each column. Note that the number of figures carried in the computed values should be the *maximum allowed by the calculator; that is, rounding should not be performed until the end*.

We now use Equations 1–21, 1–22, and

1–23 to give

$$S_{xx} = \Sigma x_i^2 - (\Sigma x_i)^2/n$$
$$= 6.90201 - (5.365)^2/5$$
$$= 1.145365$$

$$S_{yy} = \Sigma y_i^2 - (\Sigma y_i)^2/n$$
$$= 36.3775 - (12.51)^2/5$$
$$= 5.07748$$

$$S_{xy} = \Sigma x_i y_i - \Sigma x_i \Sigma y_i/n$$
$$= 15.81992 - 5.365 \times 12.51/5$$
$$= 2.39669$$

Substitution of these quantities into Equations 1–24 through 1–27 yields

$$\beta = 2.39669/1.145365 = 2.0925 = 2.09$$

$$\alpha = \frac{12.51}{5} - 2.0925 \times \frac{5.365}{5}$$

$$= 0.2567 = 0.26$$

Thus, the equation for the least-squares line is

$$y = 0.26 + 2.09x$$

The standard deviation about the regression is

$$s_r = \sqrt{\frac{S_{yy} - \beta^2 S_{xx}}{n - 2}}$$

$$= \sqrt{\frac{5.07748 - (2.0925)^2 \times 1.145365}{5 - 2}}$$

$$= 0.144 = 0.14$$

and the standard deviation of the slope is

$$s_\beta = \sqrt{s_r^2/S_{xx}} = \sqrt{(0.144)^2/1.145365} = 0.13$$

The confidence limit for the slope can be derived using t from Table 1–4. Here, the number of degrees of freedom is two less than the number of points because one degree of freedom was lost in calculating α and one for β. The 90% confidence limit (*C.L.*) in this example is

$$90\% \; C.L. = 2.09 \pm ts_\beta$$
$$= 2.09 \pm 2.35 \times 0.13$$
$$= 2.09 \pm 0.31$$

EXAMPLE 1–12.   The calibration curve derived in Example 1–11 was used to calculate the concentration of isooctane in a hydrocarbon mixture when a peak area of 2.65 was observed. Compute the mole percent isooctane and the standard deviation for the result assuming the area was (a) the result of a single measurement and (b) the mean of four measurements.

Table 1–5
Calibration Data for a Chromatographic Method for the Determination of Isooctane in a Hydrocarbon Mixture

Mole Percent Isooctane, x_i	Peak Area, y_i	x_i^2	y_i^2	$x_i y_i$
0.352	1.09	0.12390	1.1881	0.38368
0.803	1.78	0.64481	3.1684	1.42934
1.08	2.60	1.16640	6.7600	2.80800
1.38	3.03	1.90140	9.1809	4.18140
1.75	4.01	3.06250	16.0801	7.01750
5.365	12.51	6.90201	36.3775	15.81992

In either case

$$x = \frac{y - 0.26}{2.09} = \frac{2.65 - 0.26}{2.09} = 1.14 \text{ mol \%}$$

(a) Substituting into Equation 1–28, we obtain

$$s_c = \frac{0.144}{2.09} \sqrt{\frac{1}{1} + \frac{1}{5} + \frac{(2.65 - 12.51/5)^2}{(2.09)^2 \times 1.145}}$$

$$= 0.076 \quad \text{or} \quad 0.08 \text{ mol \%}$$

(b) For the mean of four measurements

$$s_c = \frac{0.144}{2.09} \sqrt{\frac{1}{4} + \frac{1}{5} + \frac{(2.65 - 12.51/5)^2}{(2.09)^2 \times 1.145}}$$

$$= 0.046 \quad \text{or} \quad 0.05 \text{ mol \%}$$

1D Sensitivity and Detection Limit for Instruments

Two terms that appear widely in the literature as figures of merit for instruments and instrumental methods are sensitivity and detection limit. Although there appears to be a consensus among scientists as to the qualitative definition of these terms, the same can not be said when it comes to defining them mathematically.[6]

1D-1 SENSITIVITY

Most chemists agree that the sensitivity of an instrument or a method measures its ability to discriminate between small differences in analyte concentration. Two factors limit sensitivity, the slope of the calibration curve and the reproducibility or precision of the measuring device. For two methods having equal precision, the one having the steeper calibration curve will be the more sensitive. A corollary to this statement is that if two methods have calibration curves with equal slopes, the one exhibiting the better precision will be the more sensitive.

The simplest quantitative definition of sensitivity, and the one accepted by the International Union of Pure and Applied Chemists (IUPAC), is *calibration sensitivity, m*, which is the slope of the calibration curve *at the concentration of interest*. For linear calibration curves, the calibration sensitivity is independent of the concentration C and can be derived from the relationship

$$S = mC + S_{bl} \tag{1–29}$$

where S and S_{bl} are the instrumental signals for the analyte and a blank respectively. The calibration sensitivity as a figure of merit suffers from the fact that it fails to take into account one of the two factors that determines sensitivity, namely precision.

Mandel and Stiehler[7] recognized the need to include precision in a meaningful mathematical statement of sensitivity and proposed the following definition for *analytical sensitivity γ*:

$$\gamma = m/s_S \tag{1–30}$$

Here, m is again the slope and s_S is the standard deviation of the signals.

The analytical sensitivity offers the advantage that it is relatively insensitive to amplification factors. For example, increasing the gain of an instrument by a factor of five will produce a fivefold increase in m. Ordinarily, however, this increase will be accompanied by a corresponding increase in s_S, thus leaving the analytical sensitivity more or less constant. A second advantage of analytical sensitivity is that it is independent of the measurement units for S.

A disadvantage of analytical sensitivity is

---

[6] For example, see: J. D. Ingle, Jr., *J. Chem. Educ.*, **1974**, *51*, 100.

[7] J. Mandel and R. D. Stiehler, *J. Res. Natl. Bur. Std.*, **1964**, *A53*, 155.

that it may be concentration dependent since s_S often varies with concentration.

1D-2 DETECTION LIMIT

The most generally accepted qualitative definition of detection limit is that it is the minimum concentration or weight of analyte that can be detected at a known confidence level. This limit depends upon the ratio of the magnitude of the analytical signal to the size of the statistical fluctuations in the blank signal. That is, unless the analytical signal is larger by some multiple k than the variations in the blank, certain detection of the analytical signal is not possible. Thus, as the limit of detection is approached, the analytical signal and its standard deviation approach the blank signal S_{bl} and its standard deviation s_b. The minimum distinguishable analytical signal S_m is then taken as a multiple k of the standard deviation of the blank. That is,

$$S_m = \bar{S}_{bl} + ks_{bl} \qquad (1-31)$$

Experimentally S_m can be determined by performing 20 to 30 blank determinations, preferably over an extended period of time. The resulting data are then treated statistically to obtain $\bar{S}_{bl}$ and s_{bl}. Finally, the slope from Equation 1–29 is used to convert S_m to C_m, which is defined as the detection limit. That is, the detection limit is given by

$$C_m = \frac{Sm - S_{bl}}{m} \qquad (1-32)$$

As pointed out by Ingle[8], numerous alternatives, based correctly or incorrectly on t and z statistics (p. 11 and 12), have been used to determine a value for k in Equation 1–31. Kaiser[9] argues that a reasonable value

for the constant is $k = 3$. He points out that it is wrong to assume a strictly normal distribution of results from blank measurements and that when $k = 3$, the confidence level of detection will be 95% in most cases. He further argues that little is to be gained by using a larger value of k and thus a greater confidence level. Long and Winefordner[10], in a recent discussion of detection limits, also recommend the use of $k = 3$.

EXAMPLE 1-13. A least-squares analysis of calibration data for the determination of lead based upon its flame emission spectrum yielded the equation

$$I = 0.0312 + 1.12\ C_{Pb}$$

where C_{Pb} is the lead concentration in parts per million and I is a measure of the relative intensity of the lead emission. The following replicate data were then obtained.

Concn Pb, ppm	No. of Replications	Mean Value of I	s
10.0	10	11.62	0.15
1.00	10	1.14	0.025
0.000	24	0.0296	0.0082

Calculate (a) the calibration sensitivity, (b) the analytical sensitivity at 1 and 10 ppm of Pb, and (c) the detection limit.

(a) By definition, the calibration sensitivity m is the slope of the straight line. Thus,

$$m = 1.12$$

(b) At 10 ppm Pb,

$$\gamma = m/s_S = 11.62/0.15 = 77$$

---

[8] See footnote 6, page 22.

[9] H. Kaiser, *Anal. Chem.*, **1970,** *42*, 53A-58A.

[10] G. L. Long and J. D. Winefordner, *Anal. Chem.*, **1983,** 55, 712A.

At 1 ppm Pb,

$$\gamma = 1.14/0.025 = 46$$

(c) Applying Equation 1–31

$$I_m = 0.0296 + 3 \times 0.0082 = 0.054$$

Substituting into Equation 1–32 gives

$$C_m = \frac{0.054 - 0.0296}{1.12} = 0.022 \text{ ppm Pb}$$

PROBLEMS

1–1. Consider the following sets of data:

A	B	C	D	E
61.45	3.27	12.06	2.7	9.961
61.53	3.26	12.14	2.4	10.004
61.32	3.24		2.6	10.002
	3.24		2.9	9.973
	3.28			9.986
	3.23			

Calculate: (a) the mean; (b) the absolute and relative standard deviation (ppt); (c) the number of degrees of freedom; (d) the coefficient of variation.

1–2. The accepted value for the quantity that provided each of the sets of data in Problem 1–1 is: A 61.71, B 3.28, C 12.23, D 2.75, E 11.241. Calculate: (a) the absolute error for the mean of each set; (b) the percent relative error for each mean.

1–3. A particular method for the analysis of copper yields results that are low by 0.5 mg. What will be the percent relative error due to this source if the weight of copper in a sample is
(a) 25 mg?   (b) 100 mg?   (c) 250 mg?   (d) 500 mg?

1–4. The method described in Problem 1–3 is to be used to analyze an ore that contains about 4.8% copper. What minimum sample weight should be taken if the relative error due to a 0.5-mg loss is to be smaller than
(a) 0.1%?   (b) 0.5%?   (c) 0.8%?   (d) 1.2%?

1–5. A constant solubility loss of 1.8 mg Se is associated with a gravimetric determination of the element. A sample containing approximately 18% Se was analyzed by this method. Calculate the relative error (in parts per thousand) in the result if the sample taken for the analysis weighed 0.400 g.

1–6. Following are data from a continuing study of calcium ion in the blood plasma of several individuals:

Subject	Mean Calcium Content, mg/100 mL	No. of Observations	Deviation of Individual Results from Mean Values
1	3.16	5	0.14, 0.09, 0.06, 0.00, 0.11
2	4.08	4	0.07, 0.12, 0.10, 0.01
3	3.75	5	0.13, 0.05, 0.08, 0.14, 0.07
4	3.49	3	0.10, 0.13, 0.07
5	3.32	6	0.07, 0.10, 0.11, 0.03, 0.14, 0.05

(a) Calculate s for each set of values.

(b) Pool the data and calculate s for the analytical method.

1–7. A method for determining the particulate lead content of air samples is based upon drawing a measured quantity of air through a filter and performing the analysis on circles cut from the filter. Calculate the individual values for s as well as a pooled value for the accompanying data.

Sample	μg Pb/m^3 Air
1	1.5, 1.2, 1.3
2	2.0, 2.3, 2.3, 2.2
3	1.8, 1.7, 1.4, 1.6
4	1.6, 1.3, 1.2, 1.5, 1.6

1–8. Based on extensive past experience, it is known that the standard deviation for an analytical method for gold in sea water is 0.025 ppb. Calculate the 99% confidence limit for an analysis using this method, based on

(a) a single measurement.

(b) three measurements.

(c) five measurements.

1–9. An established method of analysis for chlorinated hydrocarbons in air samples has a standard deviation of 0.030 ppm.

(a) Calculate the 95% confidence limit for the mean of four measurements obtained by this method.

(b) How many measurements should be made if the 95% confidence limit is to be ±0.017?

1–10. The standard deviation in a method for the analysis of carbon monoxide in automotive exhaust gases has been found, on the basis of extensive past experience, to be 0.80 ppm.

(a) Estimate the 90% confidence limit for a triplicate analysis.

(b) How many measurements would be needed for the 90% confidence limit for the set to be 0.50 ppm?

1–11. The method described in Problem 1–10 was significantly modified and found to have a standard deviation of 0.70, based upon four measurements. Establish the 90% confidence limit for the modified method.

1–12. A method for the analysis of potassium in blood serum yielded the accompanying data:

Trial	Concentration, mg K$^+$/100 mL
1	15.3
2	15.6
3	15.4
4	16.3

Calculate the 90% confidence interval for the set, assuming that these are the only data available.

1–13. Estimate the absolute and relative standard deviations in the results of the accompanying calculations (the numbers in parentheses are standard deviations associated with the individual data). Round the result to the appropriate number of significant figures.
(a) $16.9286(\pm 0.0001) + 16.8797(\pm 0.0001) = 33.8083$
(b) $16.9286(\pm 0.0001) - 16.8797(\pm 0.0001) = 0.0489$
(c) $[0.0354(\pm 0.003) + 7.147(\pm 0.002) - 2.8610(\pm 0.0003)]^{1/5}$
    $= 1.34006$
(d) $[3.69(\pm 0.02) \times 10^{-3} + 7.87(\pm 0.08) \times 10^{-4}]^2 = 2.004 \times 10^{-5}$

1–14. Estimate the absolute and relative standard deviations in the results of the accompanying calculations (the numbers in parentheses are standard deviations associated with the individual data). Round the result to the appropriate number of significant figures.
(a) $64.4(\pm 0.2) \times 0.381(\pm 0.007) = 24.5364$
(b) $\sqrt{18.18(\pm 0.03)} \times 4.764(\pm 0.009) = 9.3064$
(c) $26.94(\pm 0.08) \div 0.0496(\pm 0.0004) = 543.1452$
(d) $[0.9194(\pm 0.0008) \div 46.18(\pm 0.03)]^4 = 1.57109 \times 10^{-7}$

1–15. Estimate the absolute and relative standard deviations in the results of the accompanying calculations (the numbers in parentheses are standard deviations associated with the individual data). Round the result to the appropriate number of significant figures.

(a) $\dfrac{29.67(\pm 0.02) - 8.51(\pm 0.01)}{6.36(\pm 0.02) + 4.83(\pm 0.02)} = 1.89097$

(b) $\dfrac{7.614(\pm 0.008) - 6.923(\pm 0.005)}{14.2468(\pm 0.0002) - 13.6719(\pm 0.0001)} = 1.20195$

(c) $\left(\dfrac{3.44(\pm 0.01) \times 10^{-5}}{0.100(\pm 0.004) + 0.250(\pm 0.001)}\right)^{1/3} = 0.0461491$

(d) $\dfrac{[44.41(\pm 0.02) - 3.12(\pm 0.01)] \times 0.2048(\pm 0.0006)}{12.6349(\pm 0.0001) - 12.2775(\pm 0.0001)} = 23.6603$

(e) $\left(\dfrac{765(\pm 1) \times 3.564(\pm 0.004)}{192.5(\pm 0.2)}\right)^5 = 5.69957 \times 10^5$

1–16. Estimate the absolute standard deviation in the result derived from the following operations (the numbers in parentheses are absolute standard deviations for the numbers they follow). Report the result to the appropriate number of significant figures.
 (a) $x = \log 878(\pm 4) = 2.94349$
 (b) $x = \log 0.4957(\pm 0.0004) = -0.30478$
 (c) $p = $ antilogarithm $3.64(\pm 0.01) = 4365.16$
 (d) $p = $ antilogarithm $-7.191(\pm 0.002) = 6.44169 \times 10^{-8}$

1–17. Sulfate in natural water can be determined by measuring the turbidity that results when an excess of $BaCl_2$ is introduced into a measured quantity of the sample. A turbidimeter, the instrument used for this analysis, was calibrated with a series of standard Na_2SO_4 solutions whose concentrations were known exactly. The following data were obtained.

mg SO_4^{2-}/L, C_x	Turbidimeter Reading, R
0.00	0.06
5.00	1.48
10.00	2.28
15.0	3.98
20.0	4.61

Assume that a linear relationship exists between the instrument reading and concentration.
 (a) Plot the data and draw a straight line through them by eye.
 (b) Derive a least-squares equation for the relationship between the variables.
 (c) Compare the straight line from the relationship derived in (b) with that in (a).
 (d) Calculate the standard deviation for the slope and about the regression of the least-squares line.
 (e) Calculate the concentration of sulfate in a sample yielding a turbidimeter reading of 3.67. Calculate the absolute standard deviation of the result and the coefficient of variation.
 (f) Repeat the calculations in (e) assuming that the 3.67 was a mean of 6 turbidimeter readings.
 (g) Twenty blank determinations yielded an average reading of 0.12 with a standard deviation of 0.27. Calculate the detection limit of the procedure.

(h) Twelve analyses of the standard containing 10.0 mg SO_4^{2-}/L yielded a mean turbidimeter reading of 2.31 with a standard deviation of 0.10. Calculate the calibration and analytical sensitivities of the procedure.

1–18. The following data were obtained in calibrating a calcium ion electrode for the determination of pCa. A linear relationship between the potential E and pCa is known to exist.

pCa	E, mV
5.00	−53.8
4.00	−27.7
3.00	+ 2.7
2.00	+31.9
1.00	+65.1

(a) Plot the data and draw a line through the points by eye.
(b) Derive a least-squares expression for the best straight line through the points. Plot this line.
(c) Calculate the standard deviation for the slope of the least-squares line.
(d) Calculate the standard deviation about the regression from the least-squares line.
(e) Calculate the pCa of a serum solution in which the electrode potential was found to be 20.3 mV. Calculate the absolute and relative standard deviations for pCa if the result was from a single voltage measurement.
(f) Calculate the absolute and relative standard deviations for pCa if the millivolt reading in (e) was the mean of two replicate measurements. Repeat the calculation based upon the mean of eight measurements.
(g) Calculate the molar calcium ion concentration for the sample described in (e).
(h) Calculate the absolute and relative standard deviations in the calcium ion concentration if the measurement was performed as described in (f).

1–19. The following are polarographic diffusion currents for standard solutions of methyl vinyl ketone (MVK).

Concn MVK, mmol/L	Current, μA
0.500	3.76
1.50	9.16
2.50	15.03
3.50	20.42
4.50	25.33
5.50	31.97

(a) Derive a least-squares expression assuming the variables bear a linear relationship to one another.

(b) Plot the least-squares line as well as the experimental points.

(c) Calculate the uncertainty of the points around the line as well as the uncertainty of the slope.

(d) Two samples containing MVK yielded currents of 6.3 and 27.5 μA. Calculate the concentration of MVK in each solution.

(e) Assume that the results in (d) represent a single measurement as well as the mean of four measurements. Calculate the respective absolute and relative standard deviations.

(f) Sixteen standard solutions containing 1.50 mmol/L of the analyte gave a mean current of 9.35 μA with a standard deviation of 0.22 μA. Similarly, the standard containing 5.50 mmol/L gave a mean current of 32.16 μA with a standard deviation of 0.47 μA. Calculate the calibration and analytical sensitivities at the two concentrations.

(g) Nineteen blank determinations gave a mean current of 0.99 μA with a standard deviation of 0.036 μA. Calculate the detection limit for the method.

Elementary Electronics

Electronic circuits, by definition, contain one or more nonlinear devices such as transistors, semiconductor diodes, and vacuum or gas-filled tubes.[1] In contrast to circuit components such as resistors, capacitors, and inductors, the input and output voltages or currents of nonlinear devices are not proportional to one another. As a consequence, nonlinear components can be made to change an electrical signal from ac to dc (*rectification*) or the reverse, to amplify or to attenuate a voltage or current (*amplitude modulation*, AM), or to alter the frequency of an ac signal (*frequency modulation*, FM).

Historically, the vacuum tube was the predominant nonlinear device used in electronic circuitry. In the fifties, however, tubes were suddenly and essentially completely displaced by semiconductor based diodes and transistors, which have the advantages of low cost, low power consumption, small heat generation, long life, and compactness. The era of the individual or discrete transistor was remarkably short, however, and electronics is now based largely upon integrated circuits, which contain as many as hundreds of thousands of transistors, resistors, capacitors, and conductors formed on a single tiny semiconductor chip. Integrated circuits permit the scientist or engineer to design and construct relatively sophisticated instruments without having a detailed knowledge of electronic circuitry.

The emphasis of this chapter is on the properties and uses of integrated circuits in

---

[1] For further information about modern electronics, see: H. V. Malmstadt, C. G. Enke, and S. R. Crouch, *Electronics and Instrumentation for Scientists,* Benjamin/Cummings: Menlo Park, CA, 1981; A. J. Diefenderfer, *Principles of Electronic Instrumentation,* 2d ed., Saunders: Philadelphia, 1979; R. J. Smith, *Electronics: Circuits and Devices,* 2d ed., Wiley: New York, 1980; B. H. Vassos and G. W. Ewing, *Analog and Digital Electronics for Scientists,* 2d ed., Wiley-Interscience: New York, 1980; J. J. Brophy, *Basic Electronics for Scientists,* 3d ed., McGraw-Hill: New York, 1977.

instrument design. At the outset, however, a brief description of semiconductors and semiconductor-based devices is presented. As preparation for this material, some readers may find it helpful to review the material in Appendix 2 on alternating currents and electrical reactance.

2A Semiconductors

A semiconductor is a crystalline material having a conductivity between that of a conductor and an insulator. Many types of semiconducting materials exist, including elementary silicon and germanium, intermetallic compounds (such as silicon carbide), and a variety of organic compounds. Two semiconducting materials, which have found widest application for electronic devices, are crystalline silicon and germanium; we shall limit our discussions to these substances.

2A-1 PROPERTIES OF SILICON AND GERMANIUM SEMICONDUCTORS

Silicon and germanium are Group IV elements and thus have four valence electrons available for bond formation. In a silicon crystal, each of these electrons is immobilized by combination with an electron from another silicon atom to form a covalent bond. Thus, in principle, no free electrons exist in crystalline silicon, and the material would be expected to be an insulator. In fact, however, sufficient thermal agitation occurs at room temperature to liberate an occasional electron from its bonded state, leaving it free to move through the crystal lattice and thus to conduct electricity. This thermal *excitation* of an electron leaves a positively charged region, termed a *hole,* associated with the silicon atom. The hole, however, like the electron is mobile and thus also contributes to the electrical conductance of the crystal. The mechanism of hole

movement is stepwise; a bound electron from a neighboring silicon atom jumps to the electron-deficient region and thereby leaves a positive hole in its wake. Thus, conduction by a semiconductor involves motion of thermal electrons in one direction and holes in the other.

The conductivity of a silicon or germanium crystal can be greatly enhanced by *doping,* a process whereby a tiny, controlled amount of an impurity is introduced by diffusion into the heated germanium or silicon crystal. Typically, a silicon or germanium semiconductor is doped with a Group V element such as arsenic or antimony or a Group III element such as indium or gallium. When an atom of a Group V element replaces a silicon atom in the lattice, one unbound electron is introduced into the structure; only a small thermal energy is then needed to free this electron for conduction. Note that the resulting positive Group V ion does not provide a *mobile* hole inasmuch as there is little tendency for electrons to move from a covalent silicon bond to this nonbonding position. A semiconductor that has been doped so that it contains nonbonding electrons is termed an *n*-type (negative type) because electrons are the *majority carriers* of current. Positive holes still exist as in the undoped crystal (associated with silicon atoms), but their number is small with respect to the number of electrons; thus, holes represent *minority carriers* in an *n*-type semiconductor.

A *p*-type (positive type) semiconductor is formed when silicon or germanium is doped with a Group III element, which contains only three valence electrons. Here, positive holes are introduced when electrons from adjoining silicon atoms jump to the vacant orbital associated with the impurity atom. Note that this process imparts a negative charge to the Group III atoms. Movement of the holes from silicon atom to silicon atom, as described earlier, constitutes a current in which the majority carrier is positive. Positive holes are less mobile than free electrons; thus, the conductivity of a *p*-type semiconductor is inherently less than that of an *n*-type.

2A-2 SEMICONDUCTOR DIODES

A *diode* is a nonlinear device that has greater conductance in one direction than in another. Useful diodes are manufactured by forming adjacent *n*- and *p*-type regions within a single germanium or silicon crystal; the interface between these regions is termed a *pn* junction.

Properties of a *pn* Junction.   Figure 2–1**a** is a cross section of one type of *pn* junction, which is formed by diffusing an excess of a *p*-type impurity such as indium into a minute silicon chip that has been doped with an *n*-type impurity such as antimony. A junction of this kind permits ready flow of positive charge from the *p* region through the *n* region (or flow of negative charge in the reverse direction); it offers a high resistance to the flow of positive charge in the other direction and is thus a *current rectifier*.

Figure 2–1**b** illustrates the symbol employed in circuit diagrams to denote the presence of a diode. The arrow points in the direction of low resistance to positive currents.

Figure 2–1**c** shows the mechanism of conduction of electricity when the *p* region is made positive with respect to the *n* region by

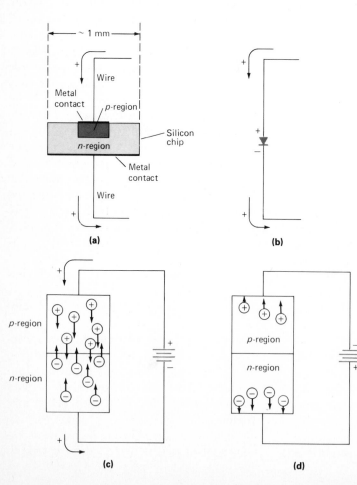

FIGURE 2-1 A *pn* junction diode. (**a**) Physical appearance of one type formed by diffusion of a *p*-type impurity into an *n*-type semiconductor; (**b**) symbol for; (**c**) current under forward bias; (**d**) resistance to current under reverse bias.

application of a potential; this process is called *forward biasing*. Here, the positive holes in the *p* region and the excess electrons in the *n* region (the majority carriers in both regions) move under the influence of the electric field toward the junction, where they can combine with and thus annihilate each other. The negative terminal of the battery injects new electrons into the *n* region, which can then continue the conduction process; the positive terminal, on the other hand, extracts electrons from the *p* region thus creating new holes which are free to migrate toward the *pn* junction.

When the diode is *reverse-biased*, as in Figure 2–1**d**, the majority carriers in each region drift away from the junction to leave a *depletion layer,* which contains few charges. Only the small concentration of minority carriers present in each region drift toward the junction and thus create a current. Consequently, conductance under reverse bias is typically 10^{-6} to 10^{-8} that of conductance under forward bias.

Current-Voltage Curves for Semiconductor Diodes. Figure 2–2 shows the behavior of a typical semiconductor diode under forward and reverse bias. With forward bias, the current increases nearly exponentially with voltage; often currents of several amperes are the result. Under reverse bias, a current on the order of microamperes is observed over a considerable voltage range; in this region, conduction is by the minority carriers. Ordinarily, this reverse current is of no consequence. As the reverse potential is increased, however, a *breakdown voltage* is ultimately reached where the reverse current increases abruptly to very high values. Here, holes and electrons, formed by the rupture of covalent bonds of the semiconductor, are accelerated by the field to produce additional electrons and holes by collision. In addition, quantum mechanical tunneling of electrons through the junction layer contributes to the enhanced conductance. This conduction, if sufficiently large, may result in heating and damaging of the diode. The voltage at which the sharp increase in current occurs under reverse bias is called the *Zener breakdown voltage*. By controlling the thickness and type of the junction layer, Zener voltages ranging from a few volts to several hundred volts can be realized. As we shall see, this phenomenon has practical application in electronics.

2A-3 TRANSISTORS

The transistor is the basic semiconductor amplifying device and performs the same function as a vacuum amplifier tube—that is, it provides an output signal that is usually significantly greater than the input. Several types of transistors are available; two of the most widely used of these, the *bipolar* and the *field-effect transistor,* will be described here.

Bipolar Transistors. Bipolar transistors consist of two back-to-back semiconductor diodes. The *pnp* transistor consists of an *n*-type region sandwiched between two *p*-type regions; the *npn* type has the reverse struc-

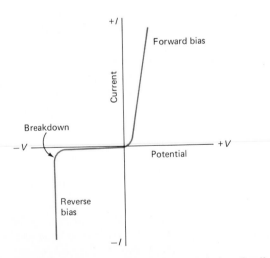

FIGURE 2-2 Current-voltage characteristics of a silicon semiconductor diode. Note that, for the sake of clarity, the small current under reverse bias before breakdown *has been greatly exaggerated.*

ture. Bipolar transistors are constructed in a variety of ways, two of which are illustrated in Figure 2–3. The symbols for the *pnp* and the *npn* type of transistor are shown on the right in Figure 2–3 (sometimes the circle is omitted). The arrow on the emitter lead indicates the direction of flow of positive charge. Thus, in the *pnp* type, positive charge flows from the emitter to the base; the reverse is true for the *npn* type.

Electrical Characteristics of a Bipolar Transistor.   The discussion that follows will focus upon the behavior of a *pnp*-type bipolar transistor. It should be appreciated that the *npn* type acts analogously except for the direction of the flow of electricity, which is opposite.

When a transistor is to be used in an electronic device, one of its terminals is connected to the input and the second serves as the output; the third terminal is connected to both and is the *common* terminal. Three configurations are thus possible: a common-emitter, a common-collector, and a common-base. The common-emitter configuration has the widest application in amplification and is the one we shall consider in detail.

Figure 2–4 illustrates the current amplification that occurs when a *pnp* transistor is employed in the common-emitter mode. Here, a small dc input current I_B, which is to be amplified, is introduced in the emitter-base circuit; this current is labeled as the base current in the figure. As we shall show later, an ac current can also be amplified by introducing it in series with I_B. After amplification, the dc component can then be removed by a filter.

The emitter-collector circuit is powered by a dc power supply, such as that described in Section 2B. Typically, the power supply will provide a potential between 9 and 30 V.

Note that, as shown by the breadth of the arrows, the collector or output current I_C is significantly larger than the base input current I_B. Furthermore, the magnitude of the collector current is directly proportional to the input current. That is,

$$I_C = \beta I_B \tag{2-1}$$

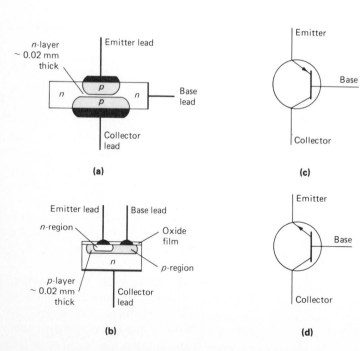

(a)

(b)

(c)

(d)

FIGURE 2-3 Two types of bipolar transistors. Construction details are shown in (**a**) for a *pnp* alloy junction transistor and in (**b**) for an *npn* planar transistor. Symbols for a *pnp* and an *npn* bipolar transistor are shown in (**c**) and (**d**), respectively. Note that alloy junction transistors may also be fabricated as *npn* types and planar transistors as *pnp*.

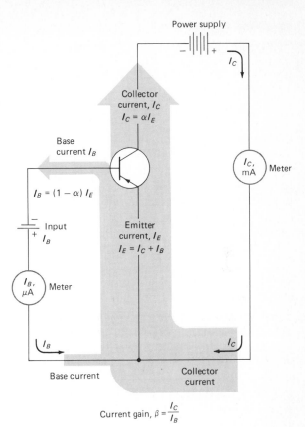

Power supply

I_C

Collector
current, I_C
$I_C = \alpha I_E$

Base
current I_B

I_{C},
mA   Meter

$I_B = (1 - \alpha) I_E$

$-$
Input
$+$ I_B

Emitter
current, I_E
$I_E = I_C + I_B$

I_B,
μA Meter

I_B

I_C

Base current

Collector
current

FIGURE 2–4 Currents in a common-emitter circuit with a *pnp* transistor. Ordinarily, $\alpha = 0.95$ to 0.995 and $\beta = 20$ to 200.

Current gain, $\beta = \dfrac{I_C}{I_B}$

where the proportionality constant β is the *current gain,* which measures the current amplification that has occurred. Values for β for typical transistors range from 20 to 200.

Mechanism of Amplification with a Bipolar Transistor. It should be noted that the emitter-base interface of the transistor shown in Figure 2–4 constitutes a forward-biased *pn* junction similar in behavior to that shown in Figure 2–1c, while the base-collector region is a reverse-biased *np* junction similar to the circuit shown in Figure 2–1d. Under forward bias, a significant current I_B develops when an input signal of a few tenths of a volt is applied (see Figure 2–2). In contrast, passage of electricity across the reverse-biased collector-base junction is inhibited by the migration of majority carriers away from the junction, as shown in Figure 2–1d.

In the manufacture of a *pnp* transistor, the *p* region is purposely much more heavily doped than the *n* region. As a consequence, the concentration of holes in the *p* region is a hundredfold or more greater than the concentration of mobile electrons in the *n* layer. Thus, the fraction of the current that takes the form of a movement of positive holes is perhaps one hundred times greater than the fraction in the form of electrons.

Turning again to Figure 2–4, it is apparent that holes are formed at the *p*-type emitter junction through removal of electrons by the two dc sources, namely, the input and the power supply. These holes can then move into the very thin *n*-type base region where some will combine with the electrons from the input source; the base current I_B is the result. The majority of the holes will, however, drift through the narrow base layer and be attracted to the negatively charged collector junction, where they can combine with electrons from the power supply; the collector current I_C is the result.

It is important to appreciate that the magnitude of the collector current is determined by the number of current-carrying holes available in the emitter. This number, however, is a fixed multiple of the number of electrons supplied by the input base current. Thus, when the base current doubles, so also does the collector current. This relationship leads to the current amplification exhibited by a bipolar transistor.

Field-Effect Transistors (FET). Several types of field-effect transistors have been developed and are widely used in integrated circuits. One of these, the insulated-gate field-effect transistor, was the outgrowth of the need to increase the input resistance of amplifiers. Typical insulated-gate field-effect transistors have input impedances that range from 10^9 to 10^{14} Ω. This type of transistor is most commonly referred to as a MOSFET, which is the acronym for metal oxide semiconductor field-effect transistor.

Figure 2–5a shows the structural features of an *n-channel* MOSFET. Here, two isolated *n* regions are formed in a *p*-type substrate. Covering both regions is a thin layer of highly insulating silicon dioxide which may be further covered with a protective layer of silicon nitride. Holes are etched through these layers so that electrical contact can be made to the two *n* regions. Two additional contacts are formed, one to the substrate and the other to the surface of the insulating layer. The latter is termed the gate because the potential of this electrode determines the magnitude of the positive current between the drain and the source. Note that the insulating layer of silicon dioxide between the gate lead and the substrate accounts for the high impedance of a MOSFET.

In the absence of a gate potential, essentially no current develops between drain and source because one of the two *pn* junctions is always reverse-biased regardless of the sign of the potential V_{DS}. MOFSET devices are designed to operate in either an *enhancement* or a *depletion mode*. The former type is shown in Figure 2–5a where current enhancement is brought about by application of a positive potential to the gate. As shown, this positive potential induces a negative substrate *channel* immediately below the layer of silicon dioxide that covers the gate electrode. The number of negative charges here, and thus the current, in-

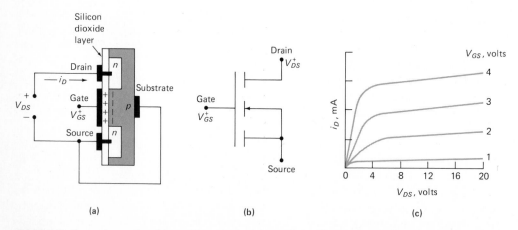

(a)   (b)   (c)

FIGURE 2–5 An *n*-channel enhancement mode MOSFET. (**a**) Structure; (**b**) symbol; (**c**) performance characteristics.

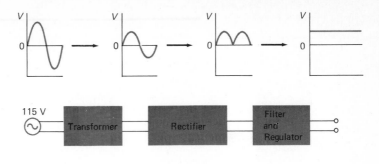

FIGURE 2-6 Diagram showing the components of a power supply and their effect on a signal.

creases as the gate voltage V_{GS} increases. The magnitude of this effect is shown in Figure 2–5c. Also available are *p*-channel enhancement mode MOSFET devices in which the *p* and *n* regions are reversed from that shown in Figure 2–5**a**.

Depletion mode MOSFET devices are designed to conduct in the *absence* of a gate voltage and to become nonconducting as potential is applied to the gate. An *n*-channel MOSFET of this type is similar in construction to the transistor shown in Figure 2–5**a** except that the two *n* regions are now connected by a narrow channel of *n*-type semiconductor. Application of a negative voltage at V_{DS} repels electrons out of the channel and thus decreases the conduction through the channel.

2B Power Supplies and Regulators

Generally, laboratory instruments require dc power to operate amplifiers and other reactive components. The most convenient source of electrical power, however, is 115-V ac furnished by public utility companies. As shown in Figure 2–6, laboratory power supply units increase or decrease the potential from the house supply, rectify the current so that it has a single polarity, and finally, smooth the output to give a signal that approximates dc. Most power supplies also contain a voltage regulator which maintains the output voltage at a constant desired level.

2B-1 TRANSFORMERS

Alternating current is readily increased or decreased in voltage by means of a power transformer such as that shown schematically in Figure 2–7. The varying magnetic field formed around the *primary* coil in this device from the 115-V alternating current induces alternating currents in the *secondary* coils; the potential V_x across each is given by

$$V_x = 115 \times N_2/N_1$$

where N_2 and N_1 are the number of turns in the secondary and primary coils, respectively. Power supplies with multiple taps, as in Figure 2–7, are available commercially; many voltage combinations can be had.

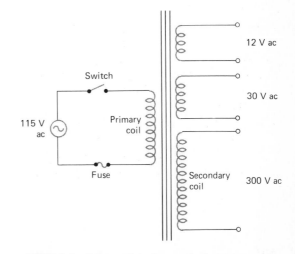

FIGURE 2-7 Schematic of a typical power transformer with multiple secondary windings.

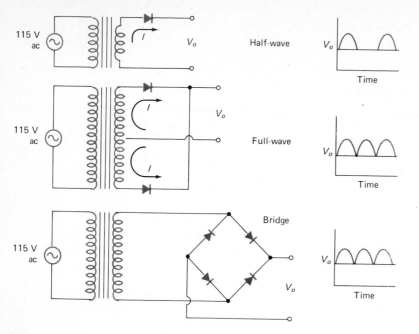

FIGURE 2–8   Three types of rectifiers.

Thus, a single transformer can serve as a power supply for several components of an instrument.

2B-2   RECTIFIERS

Figure 2–8 shows three types of rectifiers and their output-signal forms. Each uses transistor diodes (see Section 2A–2) to block current in one direction while permitting it in the other.

2B-3   FILTERS

In order to minimize the current fluctuations shown in Figure 2–8, the output of a rectifier is usually filtered by placing a large

capacitance in parallel with the load R_L as shown in Figure 2–9. The charge and discharge of the capacitor has the effect of decreasing the variations to a relatively small *ripple*. In some applications, an inductor in series and a capacitor in parallel with the load serve as a filter; this type of filter is known as an *L section*. By suitable choice of capacitance and inductance, the peak-to-peak ripple can be reduced to the millivolt range or lower.

2B-4   VOLTAGE REGULATORS

Often, instrument components require dc voltages that are constant and independent of the current. Voltage regulators serve this

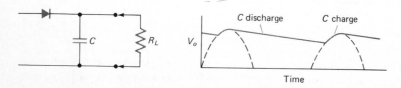

FIGURE 2-9   Filtering the current from a rectifier.

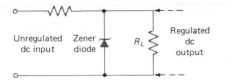

FIGURE 2-10 A voltage regulator.

purpose. Figure 2–10 illustrates a simple voltage regulator that employs a *Zener diode*, a *pn* junction which has been designed to operate under breakdown conditions; note the special symbol for this type of diode. In Figure 2–2 (p. 33), it is seen that at a certain reverse bias, a transistor diode undergoes an abrupt breakdown whereupon the current changes precipitously. For example, under breakdown conditions, a current change of 20 to 30 mA may result from a potential change of 0.1 V or less. Zener diodes with a variety of specified breakdown voltages are available commercially.

For voltage regulators, a Zener diode is chosen such that it always operates under breakdown conditions; that is, the input voltage to be regulated is greater than the breakdown voltage. For the regulator shown in Figure 2–10, an increase in voltage results in an increase in current through the diode. Because of the steepness of the current-voltage curve in the breakdown region (Figure 2–2), however, the voltage drop across the diode, and thus the load, is virtually constant.

2C *Operational Amplifiers*

The electrical signal from a transducer is ordinarily so small that it must be amplified, often millions of times, before it is sufficient to operate meters, recorders, or other devices. Seldom is a single stage of amplification, such as that shown in Figure 2–4, adequate to meet these needs, and a satisfactory output can only be achieved by cascading several amplifier stages.

Operational amplifiers are a class of amplifiers having the following properties: (1) large open-loop gains (10^4 to 10^6); (2) high input impedances (1 to 10^6 MΩ); (3) low output impedances (10 to 100 Ω); and (4) essentially zero output for zero input (ideally <0.1 mV output). Most operational amplifiers do exhibit a small output voltage with zero input due to circuit characteristics or component instabilities. The *offset voltage* for an operational amplifier is the input voltage required to produce a zero output potential. Often, operational amplifiers are provided with a "balance" adjustment to reduce the offset to some minimal figure.

Operational amplifiers derive their name from their original applications in analog computers, where they were employed to perform such mathematical operations as summing, multiplying, differentiating, and integrating. These operations also play an important part in modern instrumentation. In addition to their mathematic role, however, operational amplifiers find general application in the precise measurement of voltage, current, and resistance—the typical signals from the transducers employed for chemical measurements.[2]

2C-1 GENERAL CHARACTERISTICS OF OPERATIONAL AMPLIFIERS

Operational amplifiers ordinarily contain several amplifying stages each of which may be made up of several resistors, capacitors, and transistors. Negative feedback, which is discussed in the next section, is used extensively in their design. Early operational amplifiers employed vacuum tubes and were

---

[2] For a more complete discussion of applications of operational amplifiers, see: R. Kalvoda, *Operational Amplifiers in Chemical Instrumentation*, Halsted Press: New York, 1975.

consequently bulky and expensive. In contrast, modern commercial operational amplifiers are tiny; typically, they are manufactured on a single thin silicon chip having surface dimensions that are a few millimeters on a side. Through integrated-circuit technology, resistors, capacitors, and transistors are formed on the surface of the chip by photolithographic techniques. After leads have been soldered in place, the entire amplifier is encased in a plastic housing that has dimensions of a centimeter or less; the package, of course, does not include the power supply.

Contemporary operational amplifiers, in addition to being compact, are remarkably reliable and inexpensive. Their cost ranges typically from less than a dollar to several dollars. Manufacturers offer a large number of operational amplifiers, each differing in such properties as gain, input and output impedance, operating voltages, speed, and maximum power.

Figure 2–11 is the equivalent circuit of a widely used, commercially available opera-

tional amplifier made up of some 20 bipolar transistors. It has a gain of 2×10^5, an input resistance of 2 MΩ, an output resistance of 75 Ω, and a voltage offset of 1 to 5 mV.

Symbols for Operational Amplifiers. In circuit diagrams, the operational amplifier is ordinarily depicted as a triangle having, as a minimum, an input and an output terminal (Figure 2–12**d**). Often, additional terminals are indicated as in (**a**), (**b**), and (**c**) of the figure. A symbol as complete as that in Figure 2–12**a** is seldom encountered. Here, the power supply leads (V_c^+ and V_c^-) are shown; ordinarily the power supply connections are deleted, their presence being assumed.

Figure 2-12**b** depicts a common operational amplifier configuration in which the noninverting terminal (labeled +) is grounded. The input and output potentials are then with respect to ground. Figure 2–12**c** shows an operational amplifier being used in the differential mode. Here, the difference between two input potentials is amplified and appears as an output with respect to ground.

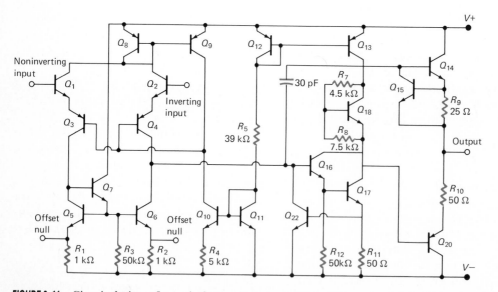

FIGURE 2-11 Circuit design of a typical operational amplifier. (Courtesy of Fairchild Camera & Instrument Corporation)

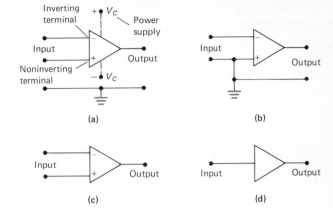

FIGURE 2-12 Symbols for operational amplifiers. More detail than is usually provided is shown in (**a**). The circuit in (**b**) is a common design in which the input, the inverting terminal, and the output are all grounded. The circuit in (**c**) is a differential amplifier in which neither input is grounded. A shorthand representation of an operational amplifier circuit is shown in (**d**).

The symbol shown in Figure 2-12**d** is occasionally encountered. Unfortunately, its use may lead to confusion because it is not evident whether the input is to the inverting or the noninverting terminal; thus, we shall avoid this symbol.

Inverting and Noninverting Terminals. It is important to realize that the *positive and negative signs show the inverting and noninverting terminals* of the amplifier and *do not* imply that they are necessarily to be connected to positive and negative inputs. Thus, if the negative terminal of a rectifier is connected to the minus or inverting terminal, the output of the amplifier is positive with respect to it; if, on the other hand, the positive terminal of the source is connected to the minus input of the amplifier, a negative output results. An ac signal input into the inverting terminal yields an output *that is 180 deg out of phase*. The positive input terminal of an amplifier, on the other hand, yields an in-phase signal or a dc signal of the same polarity as the input.

Ground Connections. Electronic circuits normally have a low-resistance wire or foil that interconnects one terminal of each signal source and sometimes the power supply; other components of the circuit may also be connected to this common line, called the *ground* of the system. The ground provides a common return for all currents to their sources. As a consequence, all voltages in the circuit are with reference to the common ground, which is indicated by the three parallel lines at the bottom of Figure 2–12**b**. Note that one side of the input and output signals, as well as the noninverting terminal are similarly connected and are thus at ground potential.

2C-2 *CIRCUITS EMPLOYING OPERATIONAL AMPLIFIERS*

Operational amplifiers are employed in circuit networks that contain various combinations of capacitors, resistors, and other electrical components. Under ideal conditions, the output of the amplifier is determined entirely by the nature of the network and its components and is *independent of the operational amplifier itself*. Thus, it is important to examine some of the various networks that employ operational amplifiers.

Feedback Circuits. Often it is desirable to return a fraction of the output signal from an amplifier to the input terminal. The fractional signal is called *feedback*. Figure 2–13 shows an operational amplifier with a feedback loop consisting of the resistor R_f. Note that in this figure only a single terminal is shown for the input and output voltages v_i and v_o, the second terminal being assumed

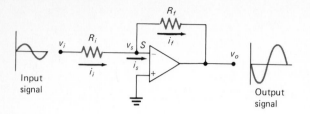

FIGURE 2–13   An operational amplifier circuit with negative feedback.

to be at ground potential.[3] Note also that the output signal is 180 deg out of phase with the input and thus tends to reduce the net voltage v_s to the inverting terminal. For this reason, the circuit is described as having *negative feedback*.

From Ohm's law, the input current i_i in the circuit shown in Figure 2–13 is given by

$$i_i = \frac{v_i - v_s}{R_i}$$

Similarly, the feedback current i_f is given by

$$i_f = \frac{v_s - v_o}{R_f}$$

As mentioned earlier, one of the characteristics of an operational amplifier is its very high resistance. Thus, the current into the operational amplifier i_s is negligible with respect to i_i and i_f; therefore, from Kirchhoff's current law, the latter two currents must be essentially equal. That is $i_i \cong i_f$, and

$$\frac{v_i - v_s}{R_i} = \frac{v_s - v_o}{R_f} \qquad (2-2)$$

The output potential for the operational amplifier is equal to the potential at point S (v_s) multiplied by the gain of the amplifier β. That is,

$$v_o = -\beta v_s$$

---

[3] Here, we follow the common practice of indicating ac voltages and currents in lower case italics, that is, v and i.

The negative sign arises because we are dealing with the inverting terminal. Substitution of this relationship into the previous equation gives upon rearrangement

$$\frac{v_o}{v_i} = -\frac{\beta R_f}{\beta R_i + R_i + R_f}$$

Because β is very large (10^4 to 10^6) for operational amplifiers, βR_i is generally much larger than either R_i and R_f, and the foregoing equation simplifies to

$$\frac{v_o}{v_i} \cong -\frac{R_f}{R_i} \qquad (2-3)$$

Thus, the gain v_o/v_i of a typical operational amplifier with negative feedback depends only upon R_f and R_i and is *independent* of fluctuations in the performance characteristics and gain of the amplifier itself.

Point S in the circuit shown in Figure 2–13 is called the *summing point* (for the currents i_f, i_i, and i_s). It is important to appreciate that when Equation 2–3 applies, the voltage at point S with respect to ground is necessarily negligible compared with either v_i or v_o. That this statement is correct can be seen by noting that the only condition under which Equation 2–3 follows from 2–2 is when $v_s \ll v_o$ and v_i. Thus, the potential at point S must approach ground potential; as a consequence, point S is said to be at *virtual ground* and we may write

$$V_- \cong -V_+ = v_s$$

where V_- and V_+ are the potentials at the

inverting and noninverting terminals respectively. Thus, the operational amplifier acts in such a way that these equalities are maintained.

Frequency Response of a Negative Feedback Circuit.   The gain of a typical operational amplifier decreases rapidly in response to high-frequency input signals. This frequency dependence arises from small capacitances that develop at *pn* junctions. The effect for a typical amplifier is shown in Figure 2–14, where the curve labeled open-loop gain represents the behavior of the amplifier *in the absence of the feedback resistor R_f in Figure 2–13.* Note that both the ordinate and abscissa are log scales and that the open-loop gain for this particular amplifier decreases rapidly at frequencies greater than about 100 Hz.

In contrast, an operational amplifier employing external negative feedback, as in Figure 2–13, has a constant gain or *bandwidth* that extends from dc to over 10^5 Hz. In this region, the gain depends only upon R_f/R_i, as shown by Equation 2–3. For many purposes, the frequency independence of the negative feedback inverting circuit is of

great importance and more than offsets the loss in amplification.

The Voltage-Follower Circuit.   Figure 2–15 depicts a *voltage follower,* a circuit in which the input is fed into the *noninverting* terminal, and a feedback loop is provided which involves the inverting terminal. The result is an amplifier with a gain of approximately unity. Because the input is not inverted, the output potential is the same as the input. The input impedance is the open-loop impedance for the amplifier, which can be very large (10^6 MΩ or more) when field-effect transistors are employed. The output impedance, on the other hand, is low (<1 Ω). As will be shown later, this impedance transformation is valuable in the measurement of high-impedance sources with low-impedance measuring devices.

2C–3   *AMPLIFICATION AND MEASUREMENT OF TRANSDUCER SIGNALS*

Operational amplifiers find general application to the amplification and measurement of the electrical signals from transducers. These signals, which are often concentration-dependent, include current, potential, and resistance (or conductance). This section includes simple applications of operational amplifiers to the measurement of each type of signal.

Current Measurement.   The accurate measurement of small currents is important to such analytical methods as voltammetry, coulometry, photometry, and gas chromatography.

As pointed out in Appendix 1, an important concern that arises in all physical measurements, including that of current, is

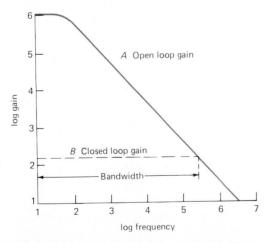

FIGURE 2-14   Frequency response of a typical operational amplifier: *A* without negative feedback and *B* with negative feedback.

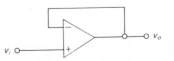

FIGURE 2-15   Voltage-follower circuit.

whether the measuring process will, in itself, alter significantly the signal being measured, thus leading to an error. It is inevitable that any measuring process will perturb a system under study in such a way that the quantity actually measured differs from its original value before the measurement. All that can be hoped is that the perturbation can be kept sufficiently small. For a current measurement, this consideration requires that the internal resistance of the measuring device be minimized.

A low-resistance current measuring device is readily obtained by deleting the resistor R_i in Figure 2–13 and using the current to be measured as the signal input. An arrangement of this kind is shown in Figure 2–16, where a small direct current I_x is generated by a phototube, a transducer that converts radiant energy such as light into an electrical current. When the cathode of the phototube is maintained at a potential of about −90 V, absorption of radiation by its surface results in the ejection of electrons, which are then accelerated to the wire anode; a current results that is directly propor-

tional to the power of the radiant beam.

If the conclusions reached in the discussion of feedback are applied to this circuit, we may write

$$I_x = I_f + I_s \cong I_f$$

In addition, the point S is at virtual ground so that the potential V_o corresponds to the potential drop across the resistor R_f. Therefore, from Ohm's law,

$$V_o = -I_f R_f = -I_x R_f$$

and

$$I_x = -V_o/R_f = kV_o$$

Thus, the potential measurement V_o gives the current, provided R_f is known. By making R_f reasonably large, the accurate measurement of small currents is feasible. For example, if R_f is 100 kΩ, a 1−µA current results in a potential of 0.1 V, a quantity that is readily measured with a high degree of accuracy.

As shown by the following example, an important property of the circuit shown in Figure 2–16 is its low resistance with respect to the current from the transducer. Thus, the meter is driven not by the transducer but by the amplified current from the external power supply of the operational amplifier. The result is a minimal measuring error.

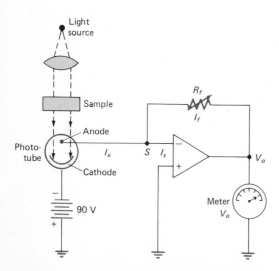

FIGURE 2-16 Application of an operational amplifier to the measurement of a small photocurrent, I_x.

EXAMPLE 2-1.   Assume that R_f in Figure 2–16 is 200 kΩ, the internal resistance of the phototube is 5.0×10^4 Ω, and the amplifier gain is 1.0×10^5. Calculate the relative error in the current measurement that results from the presence of the measuring circuit.

Here, the resistance of the measuring circuit R_m is the resistance between the summing point S and ground. This resistance is given by Ohm's law. That is,

$$R_m = \frac{V_s}{I_x}$$

We may also write (p. 42)

$$I_f = \frac{V_s - V_o}{R_f} = I_x$$

Combining these two equations yields upon rearrangement

$$R_m = \frac{V_s R_f}{V_s - V_o} = \frac{R_f}{1 - V_o/V_s}$$

where $-V_o/V_s$ is the amplifier gain β (p. 34). Substituting numerical values gives

$$R_m = 200 \times 10^3/(1 + 1.0 \times 10^5) = 2.0 \; \Omega$$

Equation A1–1 in Appendix 1 shows that the relative error in a current measurement is given by

$$\text{rel error} = \frac{R_m}{R_T + R_m}$$

where R_T is the resistance of the circuit in the absence of the resistance of the measuring device R_m. Thus,

$$\text{rel error} = \frac{2.0}{5.0 \times 10^4 + 2.0}$$
$$= -4.0 \times 10^{-5} \quad \text{or} \quad -0.004\%$$

The instrument shown in Figure 2–16 is called a photometer; it measures the attenuation of a light beam by absorption brought about by a colored analyte in a solution; this parameter is related to the concentration of the species responsible for the absorption. Photometers are described in detail in Section 7D–1.

Potential Measurements. Potential measurements are used extensively for the determination of temperature and the concentration of ions in solution. In the former application, the transducer is a thermocouple; in the latter, it is an ion-sensitive electrode.

In Example A1-3 in Appendix 1, it is demonstrated that accurate potential measurements require that the resistance of the measuring device be large with respect to the internal resistance of the voltage source to be measured. The need for a high-resistive measuring device becomes particularly acute in the determination of pH with a glass electrode. For this reason, the basic feedback circuit shown in Figure 2–13, which typically has an internal resistance of perhaps $10^5 \; \Omega$, is not satisfactory for voltage measurements. On the other hand, a feedback circuit can be combined with the voltage-follower circuit shown in Figure 2–15 to give a very high-impedance voltage-measuring device. An example of such a circuit is shown in Figure 2–17. The first stage involves a voltage follower, which provides an impedance of as much as $10^{12} \; \Omega$. An inverting amplifier circuit follows, which amplifies the input by R_f/R_i or 20. An amplifier, such as this, with a resistance of 100 MΩ or more is often called an *electrometer*.

Resistance or Conductance Measurements. Electrolytic cells and temperature-responsive devices, such as thermistors and bolometers, are common examples of transducers whose

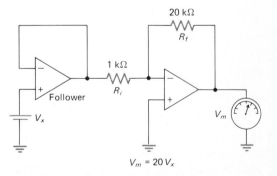

$$V_m = 20 V_x$$

FIGURE 2-17 A high-impedance circuit for voltage amplification.

electrical resistance or conductance varies in response to an analytical signal. These devices are employed for conductometric and thermometric titrations, for infrared absorption and emission measurements, and for temperature control in a variety of analytical applications.

The circuit shown in Figure 2–13 provides a convenient means for measurement of the resistance or conductance of a transducer. Here, a constant potential ac source is employed for v_i and the transducer is substituted for either R_i or R_f in the circuit. The amplified output potential v_o, after rectification, is then measured with a suitable meter, potentiometer, or recorder. Thus, if the transducer is substituted for R_f in Figure 2–13, the output, as can be seen from rearrangement of Equation 2–3, is

$$R_x = -\frac{v_o R_i}{v_i} = k v_o \qquad (2\text{--}4)$$

where R_x is the resistance to be measured and k is a constant which can be calculated if R_i and v_i are known; alternatively, k can be determined by a calibration wherein R_x is replaced by a standard resistor.

If conductance rather than resistance is of interest, the transducer conveniently replaces R_i in the circuit. Here, from Equation 2–3, it is found that

$$\frac{1}{R_x} = G_x = -\frac{v_o}{v_i R_f} = k' v_o \qquad (2\text{--}5)$$

where G_x is the desired conductance. Note that in either type of measurement, the value of k, and thus the range of the measured values, can be readily varied by employing a variable resistor R_i or R_f.

Figure 2–18 illustrates two simple applications of operational amplifiers for the measurement of conductance or resistance. In (a), the conductance of a cell for a conductometric titration is of interest. Here, an alternating-current input signal v_i of perhaps 5 to 10 V is provided from the secondary of a transformer. The output signal is then rectified and measured as a potential. The variable resistance R_f provides a means for varying the range of conductances that can be recorded or read. Calibration is provided by switching the standard resistor R_s into the circuit in place of the conductivity cell.

Figure 2–18**b** illustrates how the circuit in Figure 2–13 can be applied to the measurement of a ratio of resistances or conductances. Here, the absorption of radiant energy by a sample is being compared with that for a reference solution. The two photodiode transducers (Section 5F–2) replace R_f and R_i in Figure 2–13. A dc power supply serves as the source of power and the output potential M, as seen from Equation 2–3, is

$$V_o = M = -V_i \frac{R_o}{R} = -V_i \frac{P_o}{P} \qquad (2\text{--}6)$$

Typically, the resistance of a photoconductive cell is directly proportional to the radiant power, P, of the radiation striking it. Thus, the meter reading M is proportional to the ratio of the power of the two beams (P_o/P).

Comparison of Transducer Outputs.   It is frequently desirable to compare a signal generated by an analyte to a reference signal, as in Figure 2–18**b**. A difference amplifier, such as that shown in Figure 2–19, can also be applied for this purpose. Here, the amplifier is being employed for a temperature measurement. Note that the two input resistors (R_i) have equal resistances; similarly, the feedback resistor and the resistor between the noninverting terminal and ground (both labeled R_k) are also alike.

Applying Ohm's law to the circuit shown in Figure 2–19 gives

$$I_1 = \frac{V_1 - V_-}{R_i}$$

and

$$I_f = \frac{V_- - V_o}{R_k}$$

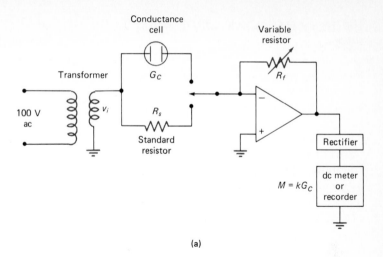

(a)

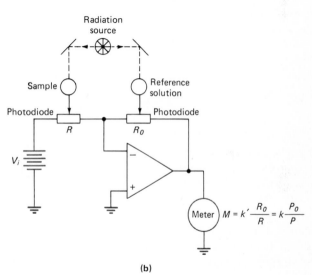

FIGURE 2-18   Two simple circuits for transducers with conductance or resistance outputs.

(b)

But as we have shown earlier I_1 and I_f are approximately equal. Thus,

$$\frac{V_1 - V_-}{R_i} = \frac{V_- - V_o}{R_k}$$

The argument demonstrating that point S in Figure 2–13 is at virtual ground (p. 42) can be applied to Figure 2–19, where it becomes apparent that

$$V_- \cong -V_+$$

Substitution of this relationship into the

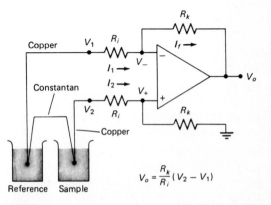

$$V_o = \frac{R_k}{R_i}(V_2 - V_1)$$

FIGURE 2-19   A circuit for the amplification of differences.

previous equation yields, upon rearrangement,

$$-V_+ = \frac{V_1 R_k - V_o R_i}{R_i + R_k}$$

The potential V_+ can also be written in terms of V_2 by means of the voltage-divider equation. Thus,

$$-V_+ = \frac{V_2 R_k}{R_i + R_k}$$

Combining the last two equations gives, upon rearrangement,

$$V_o = \frac{R_k}{R_i}(V_2 - V_1) \qquad (2\text{--}7)$$

Thus, it is the difference between the two signals that is amplified.

Any extraneous potential *common to the two input terminals* shown in Figure 2–19 will be subtracted and not appear in the output. Thus, any slow drift in the output of the transducers or any 60-cycle currents induced from the laboratory power lines will be eliminated from V_o. This useful property accounts for the widespread use of a differential circuit in the first amplifier stage of many instruments.

The transducers shown in Figure 2–19 are a pair of *thermocouple junctions*, one of which is immersed in the sample and the second in a reference solution (often an ice bath) held at constant temperature. A temperature-dependent contact potential develops at each of the two junctions formed from wires made of copper and an alloy called constantan (other metal pairs are also employed). The potential developed is roughly 5 mV per 100°C temperature difference.

2C–4 APPLICATION OF OPERATIONAL AMPLIFIERS TO VOLTAGE AND CURRENT CONTROL

Operational amplifiers are readily employed to generate constant-potential or constant-current signals.

Constant-Voltage Source. Several instrumental methods require a power source whose potential is precisely known and from which reasonable currents can be obtained without alteration of this potential. A circuit that meets these qualifications is termed a *potentiostat.*

Several reference sources are available to provide an accurately known voltage. One of these is the *Weston standard cell*, which can be represented as follows:

$$Cd(Hg)_x | CdSO_4 \\ \cdot (8/3)H_2O(sat'd), Hg_2SO_4(sat'd) | Hg$$

One of the electrodes is a cadmium amalgam represented as $Cd(Hg)_x$; the other is mercury. The electrolyte is a solution saturated with cadmium and mercury sulfates. The half reactions as they occur at the two electrodes are

$$Cd(Hg)_x \rightarrow Cd^{2+} + xHg + 2e$$
$$Hg_2SO_4 + 2e \rightarrow 2Hg + SO_4^{2-}$$

The potential of the Weston cell at 25°C is 1.0183 V. Also on the market are inexpensive Zener stabilized integrated circuits that provide voltages that are constant to a few hundredths of a percent. None of these sources will, however, maintain its potential when a large current is required.

Figure 2–20 illustrates how a reference cell can be employed to provide a standard voltage source from which relatively large currents can be drawn. Note that in both circuits, the standard source appears in the feedback loop of an operational amplifier.

Recall that point S in Figure 2–20a is at virtual ground. For this to be so, it is necessary that $V_o = V_{std}$. That is, the current

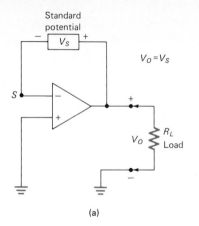

$V_O = V_S$

(a)

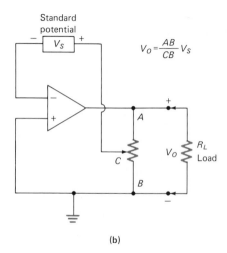

$$V_O = \frac{AB}{CB} V_S$$

(b)

FIGURE 2–20 Constant-potential sources.

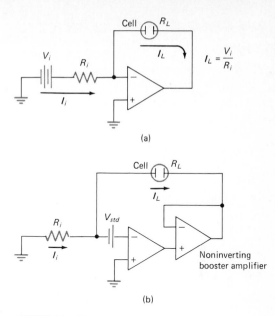

FIGURE 2-21 Constant-current sources.

through the load resistance R_L must be such that $IR_L = V_{std}$. It is important to appreciate, however, that this current arises from the power source of the operational amplifier and *not the standard cell*. Thus, the standard cell controls V_o but provides essentially none of the current through the load.

Figure 2–20**b** illustrates a modification of the circuit in (**a**) which permits the output voltage of the potentiostat to be fixed at a level that is a known multiple of the output voltage of the standard cell.

Constant-Current Sources. Constant-current

sources, called *amperostats,* find application in several analytical instruments. These devices are usually employed to maintain a constant current through an electrochemical cell. An amperostat reacts to a change in input power or a change in internal resistance of the cell by altering its output potential in such a way as to maintain the current at a predetermined level.

Figure 2–21 shows two amperostats. The first requires a voltage input V_i whose potential is constant in the presence of a current. Recall from our earlier discussion that

$$I_L = I_i = \frac{V_i}{R_i}$$

Thus, the current will be constant and independent of the resistance of the cell, provided that V_i and R_i remain constant.

Figure 2–21**b** is an amperostat which employs a standard cell (V_{std}) to maintain a constant current; no significant current is, however, present in the standard. The noninverting booster amplifier develops relatively large currents in the cell.

2C-5 APPLICATION OF OPERATIONAL AMPLIFIERS TO MATHEMATICAL OPERATIONS

As shown in Figure 2–22, substitution of various combinations of resistors, capacitors, and transistors for R_i and R_f in the circuit shown in Figure 2–13 permit various mathematical operations to be performed on electrical signals as they are generated by an analytical instrument. For example, the output from an instrument for chromatography usually takes the form of a Gaussian peak when the electrical signal from a detector is plotted as a function of time. Integration of this peak to find its area is necessary in order to find the analyte concentration. The operational amplifier shown in Figure 2–22**c** is capable of performing this integration automatically thus giving a signal that is directly proportional to analyte concentration.

Multiplication and Division by a Constant. Figure 2–22**a** shows how an input signal V_i can be multiplied by a constant whose magnitude is $-R_f/R_i$. The equivalent of division by a constant occurs when this ratio is less than unity.

Addition or Subtraction. Figure 2–22**b** illustrates how an operational amplifier can produce an output signal that is the sum of several input signals. Because the impedance of the amplifier is large and because the output must furnish a sufficient current I_f to keep the summing point S at virtual ground, we may write

$$I_f \cong I_1 + I_2 + I_3 + I_4 \qquad (2\text{–}8)$$

But $I_f = -V_o/R_f$, and we may thus write

$$V_o = -R_f \left(\frac{V_1}{R_1} + \frac{V_2}{R_2} + \frac{V_3}{R_3} + \frac{V_4}{R_4} \right) \qquad (2\text{–}9)$$

If $R_f = R_1 = R_2 = R_3 = R_4$, then the output signal is the sum of the four inputs (but opposite in sign).

$$-V_o = (V_1 + V_2 + V_3 + V_4)$$

To obtain an average of the four signals, let $R_1 = R_2 = R_3 = R_4 = R = 4R_f$. Substituting into Equation 2–9 gives

$$V_o = -\frac{R}{4} \left(\frac{V_1}{R} + \frac{V_2}{R} + \frac{V_3}{R} + \frac{V_4}{R} \right)$$

and V_o becomes the average of the four inputs. Thus,

$$V_o = -(V_1 + V_2 + V_3 + V_4)/4 \qquad (2\text{–}10)$$

Clearly, a weighted average can be obtained by varying the ratios of the resistances of the input resistors.

Subtraction can be performed by the circuit in Figure 2–22**b** by introducing an inverter in parallel with one or more of the resistors, thus changing the sign of one or more of the inputs.

Integration. Figure 2–22**c** illustrates a circuit for integrating a variable input signal v_i with respect to time. When the reset switch is open and the hold switch is closed,

$$i_i = i_f \qquad (2\text{–}11)$$

Substituting Ohm's law and Equation A2–8 (Appendix 2) into this equation yields upon rearranging

$$dv_o = -\frac{v_i}{R_iC} dt \qquad (2\text{–}12)$$

Integrating between times t_1 and t_2 gives

$$\int_{v_{o1}}^{v_{o2}} dv_o = -\frac{1}{R_iC} \int_{t_1}^{t_2} v_i dt$$

or

$$v_{o2} - v_{o1} = -\frac{1}{R_iC} \int_{t_1}^{t_2} v_i dt \qquad (2\text{–}13)$$

The integral is ordinarily obtained by first opening the hold switch and closing the reset switch to discharge the capacitor thus

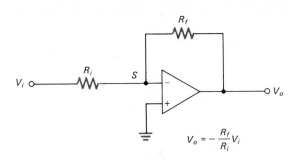

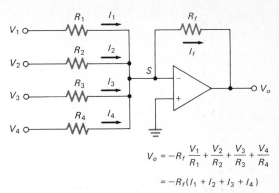

$$V_o = -\frac{R_f}{R_i}V_i$$

$$V_o = -R_f\left(\frac{V_1}{R_1} + \frac{V_2}{R_2} + \frac{V_3}{R_3} + \frac{V_4}{R_4}\right)$$

$$= -R_f(I_1 + I_2 + I_3 + I_4)$$

(a) Multiplication or Division

(b) Addition or Subtraction

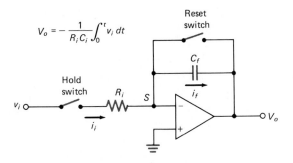

$$V_o = -\frac{1}{R_i C_i}\int_0^t v_i\,dt$$

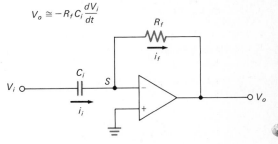

$$V_o \cong -R_f C_i \frac{dV_i}{dt}$$

(c) Integration

(d) Differentiation

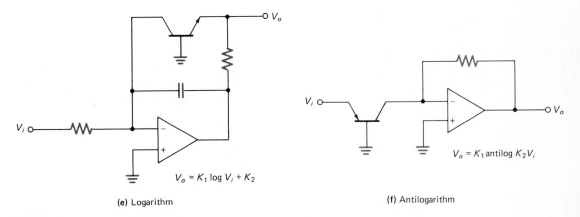

$$V_o = K_1 \log V_i + K_2$$

$$V_o = K_1\,\text{antilog}\,K_2 V_i$$

(e) Logarithm

(f) Antilogarithm

FIGURE 2–22 Mathematical operations with operational amplifiers.

making $v_{o1} = 0$ when $t_1 = 0$. Equation 2–13 then simplifies to

$$V_o = -\frac{1}{R_iC} \int_0^t v_i dt \qquad (2\text{–}14)$$

To begin the integration, the reset switch is opened and the hold switch closed. The integration is stopped at time t by opening the hold switch. The integral over the period of 0 to t is then given by $-v_o$.

Differentiation. Figure 2–22**d** is a simple circuit for differentiation. Note that it differs from the integration circuit only in the respect that the positions of C and R have been reversed. Proceeding as in the previous derivation, we may write

$$C_i \frac{dV_i}{dt} = -\frac{V_o}{R_f}$$

or

$$V_o = -R_fC_i \frac{dV_i}{dt} \qquad (2\text{–}15)$$

The circuit shown in Figure 2–22**d** is not, in fact, practical for most chemical applications, where the rate of change in the transducer signal is low. For example, differentiation is a useful way to treat the data from a potentiometric titration; here, the potential change of interest occurs over a period of a second or more ($f \leq 1$ Hz). The input signal will, however, contain extraneous 60-, 120-, and 240-Hz potentials (see Figure 2–36), which are induced by the ac power supply. In addition, signal fluctuations resulting from incomplete mixing of the reagent and analyte solutions are often encountered. Unfortunately, the output of the circuit in Figure 2–22**d** has a strong frequency dependence; as a consequence, the output voltage from the extraneous signals often becomes as great or greater than that from the low frequency transducer signal, even

though the magnitude of the former voltage relative to the latter is small.

This problem is readily overcome by introducing a small parallel capacitance C_f in the feedback circuit and a small series resistor R_i in the input circuit to filter the high-frequency voltages. These added elements are kept small enough so that significant attenuation of the analytical signal does not occur.

Generation of Logarithms and Antilogarithms. A number of transducers produce an electrical response which is exponentially related to concentration. Thus, it is the logarithm of the signal that is directly proportional to concentration. A simple way of obtaining an output voltage that is the logarithm of the input signal is based upon the logarithmetic relationship that exists between current and voltage in a forward biased pn junction. That is,

$$I = K_1 e^{K_2 V}$$

Figure 2–22**e** shows a circuit that takes advantage of this relationship. It is readily shown that the output in this instance is given by

$$V_o = K_1 \log V_i + K_2 \qquad (2\text{–}16)$$

Figure 2–22**f** illustrates the analogous circuit for obtaining the antilogarithm of an input voltage. Both circuits require that the input be of fixed polarity. Furthermore, the accuracy of the output is heavily dependent upon the thermal characteristics of the pn junctions.

2C-6 APPLICATION OF OPERATIONAL AMPLIFIERS TO SWITCHING

Another important and widespread application of transistors and operational amplifiers is to switching.

Figures 2–23**a** and 2–23**b** show the be-

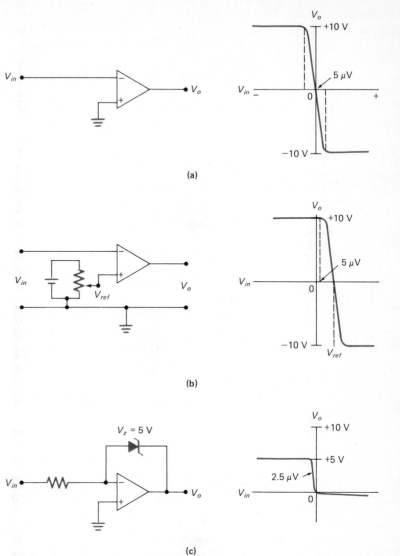

FIGURE 2-23 Switching circuits.

havior of two comparator circuits. In the first, the input potential is compared with ground while in the second the comparison is with a reference potential E_{ref}. Note that in the absence of feedback, the region of linear behavior between the input and output potentials is restricted to a range of only $\pm 5 \ \mu\text{V}$ in input potential. In the first case, this range brackets zero or ground potential and in the second, the reference potential.

Outside of this small range, an amplifier is *saturated,* and the output potential is independent of the input potential. The potential at which saturation occurs depends upon the characteristics of the amplifier and the potential of the power supply; typically, the saturation potential is about ± 10 V as shown.

Figure 2–23c shows a comparator circuit in which the potential excursions are limited by the insertion in the feedback circuit of a Zener diode with a breakdown voltage of 5 V. In this instance a change in input potential of less than 5 μV causes the circuit to pass from a conducting to an essentially nonconducting state. That is, the circuit is a kind of electronic switch that has just two states: one in which the output is +5 V and one in which the output is at most a few tenths of a volt.

Electronic switches, such as those shown in Figure 2–23, respond at much greater rates than do mechanical switches; it is this property that makes them so important for digital electronic and computer applications.

2D Digital Electronics

During the past 10 to 15 years, many of the analog devices described in the previous section have been supplemented or replaced by digital designs in which the circuits have but two stable conditions represented by transistors that are fully conducting or completely cut off.[4] Digital circuits offer several advantages over their analog counterparts. One of the most important is their lower susceptibility to environmental noise such as 60-Hz induced currents from 110-V power lines. Furthermore, digital circuits generally are more stable and suffer less from drift or flicker noise. Another advantage of digital signals is that they are more readily transmitted over long distances and stored for later processing; for example, the satellite photographs of Saturn were transmitted to

---

[4] For a detailed treatment of this subject, see: H. V. Malmstadt, C. G. Enke, and S. R. Crouch, *Electronics and Instrumentation for Scientists*, Benjamin/Cummings: Menlo Park, CA, 1981; A. J. Diefenderfer, *Principles of Electronic Instrumentation*, Saunders: Philadelphia, 1979; and B. H. Vassos and G. W. Ewing, *Analog and Digital Electronics for Scientists*, 2d ed., Wiley: New York, 1980.

Earth in digital form, stored, and subsequently processed to yield prints, a form of analog information. Finally, a most important property of digital information and devices is that they are compatible with computers and can take advantage of these powerful devices.

2D-1 DIGITAL AND ANALOG SIGNALS

Chemical signals are of two types, *continuous* and *discrete*. By far the more common are continuous or *analog signals*, which vary continuously and can take any value within a wide range. An example is the pH of an aqueous solution, which can assume any conceivable value between roughly −1 and 15. The potential of a glass electrode immersed in an aqueous solution responds continuously to pH variations and is thus an example of an analog electric signal.

An obvious example of a discrete chemical signal is the radiation produced by the decay of a radioactive species. Here, the signal consists of a series of pulses of energy, which can be converted to electrical pulses and counted. The resulting information, which can be expressed as integers or whole numbers, is usually called *digital* information.

It is important to appreciate that whether information is analog or digital may depend upon how it is observed. For example, the yellow radiation produced by heating sodium ions in a flame is often measured with a photodetector that converts the radiant energy into an analog electric current, which can vary continuously over a considerable range. It is known, however, that this radiation is quantized and is thus emitted as energy pulses called photons each of which results from an individual atomic event. The response time of the typical detector is so slow, however, that the individual photons are not detected. Instead, only the average number in a given time interval is measured. At low radiation intensity, however, a properly designed detector can respond to the individual photons; here, a digital signal is produced, which consists of a series of electric pulses, which can be counted.

Digital devices offer sufficient advantages to make it worthwhile in many cases to convert an analog or continuous signal to a digital one made up of a series of equal-size voltage pulses, the number of which is directly proportional to the magnitude of the original signal. These pulses are then counted electronically and displayed in digital form or, alternatively, are converted back to an analog form for driving meters or analog recorders. In either case, an important part of the digital devices is the pulse counter.

2D-2 *ELECTRONIC COUNTERS AND COUNTING*

In a typical digital measurement, a high-speed *electronic counter* is used to count the number of signals that occur within a specified set of boundary conditions. Examples of signals and boundary conditions include number of photons or α particles emitted by the analyte per second, number of drops of titrant per mole of analyte, or the number of steps of a stepper motor per milliliter of reagent delivered from a syringe.

Counting such signals electronically requires that they first be transduced to provide a series of pulses of more or less equal voltage. Ultimately these pulses are converted by the counter to a decimal number for display. It turns out, however, that the decimal system for representing numbers is not a convenient one to use in the counting operation itself because this system requires 10 different electrical signals to represent the digits of 0 through 9. For this reason, electronic counting is performed by binary numbers; here, only two digits, 0 and 1, are required to represent any number. In electronic counters, the 0 is generally represented by a voltage signal of 0 ± 0.5 V and

the 1 by a voltage of typically 5 ± 1 V or the reverse.

The Binary Number System. Each digit in the decimal numbering system represents the coefficient of some power of 10. Thus, the number 3076 can be written as

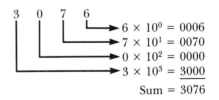

$$6 \times 10^0 = 0006$$
$$7 \times 10^1 = 0070$$
$$0 \times 10^2 = 0000$$
$$3 \times 10^3 = \underline{3000}$$
$$\text{Sum} = 3076$$

Similarly, each digit in the binary system of numbers corresponds to a coefficient of a power of 2.

Interconversion of Binary and Decimal Numbers. Table 2–1 illustrates the relationship between a few decimal and binary numbers. The examples that follow illustrate methods for interconversions between the systems.

Table 2–1 Relationship Between Some Decimal and Binary Numbers	
Decimal Number	**Binary Representation**
0	0
1	1
2	10
3	11
4	100
5	101
6	110
7	111
8	1000
9	1001
10	1010
12	1100
15	1111
16	10000
32	100000
64	1000000

EXAMPLE 2-2. Convert 101011 in the binary system to a decimal number.

Binary numbers can be expressed in terms of base 2. Thus,

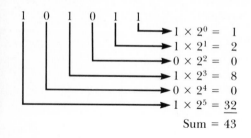

$$1 \times 2^0 = 1$$
$$1 \times 2^1 = 2$$
$$0 \times 2^2 = 0$$
$$1 \times 2^3 = 8$$
$$0 \times 2^4 = 0$$
$$1 \times 2^5 = \underline{32}$$
$$\text{Sum} = 43$$

EXAMPLE 2-3. Convert 710 to a binary number.

As a first step, we determine the largest power of 2 that is less than 710. Thus, since $2^{10} = 1024$,

$$2^9 = 512 \quad \text{and} \quad 710 - 512 = 198$$

The process is repeated for 198

$$2^7 = 128 \quad \text{and} \quad 198 - 128 = 70$$

Continuing, we find

$$2^6 = 64 \quad \text{and} \quad 70 - 64 = 6$$
$$2^2 = 4 \quad \text{and} \quad 6 - 4 = 2$$
$$2^1 = 2 \quad \text{and} \quad 2 - 2 = 0$$

The binary number is then derived as follows:

$$1 \ 0 1 \ 1 \ 0 \ 0 \ 0 \ 1 \ 1 \ 0$$
$$2^9 \ \_ \ 2^7 \ 2^6 \ \_ \ \_ \ \_ \ 2^2 \ 2^1 \ \_$$

It is worthwhile noting that in the binary numbering system, the digit lying farthest to the right in a number is termed the *least significant digit;* the one on the far left is the *most significant digit.* It is also important to note that each digit, be it 1 or 0, in a binary

number is termed a *bit,* which is a contraction for binary digit.

Binary Arithmetic. Arithmetic with binary numbers is similar to, but simpler than, decimal arithmetic. For addition, only four combinations are possible.

0	0	1	1
+0	+1	+0	+ 1
0	1	1	10

Note that in the last sum, a 1 is carried over to the next higher power of 2. Similarly, for multiplication,

0	0	1	1
×0	×1	×0	×1
0	0	0	1

The following example illustrates the use of these operations.

EXAMPLE 2-4. Perform the following calculations with binary arithmetic: (a) 7 + 3, (b) 19 + 6, (c) 7 × 3, and (d) 22 × 5.

(a)
```
    7      111
  + 3    + 11
   10     1010
```
(b)
```
   19    10011
  + 6    + 110
   25    11001
```
(c)
```
    7      111
  × 3    × 11
   21      111
          111
         10101
```
(d)
```
   22    10110
  × 5    × 101
  110    10110
         00000
         10110
        1101110
```

Note that a carry operation, similar to that in the decimal system, is used. Thus, in (a) the sum of the two ones in the right column is equal to 0 plus 1 to carry to the next column. Here, the sum of the three ones is 1 plus 1 to carry to the next column. Finally, this carry combines with the one in the next column to give 0 plus 1 as the most significant digit.

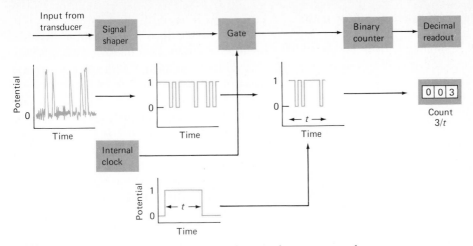

FIGURE 2-24 A counter for determining voltage pulses per second.

2D-3 COUNTERS

Figure 2–24 is a block diagram of an instrument for counting the number of electrical pulses that are received from a transducer per unit of time. The voltage signal from the transducer first passes into a shaper that removes the small background signals and converts the large pulses to square waves having the same frequency as the input signal; often the signal is inverted in this process. The resulting signal is then fed into a gate wherein the output from an internal clock provides an exact time interval t during which counting occurs. Finally, the binary output of the counter is converted to a decimal number for readout.

Signal Shapers. Figure 2–25a shows a circuit for a typical signal shaper. It makes use of a voltage comparator such as that shown in Figure 2–23b to convert the input signal to the square wave form shown in Figure 2–25c. In Figure 2–23c, it is seen that the output of a comparator is at one of two voltage levels, high or low; these two levels, often termed *logic states*, have been designated as 1 and 0 respectively in Figures 2–24 and 2–25. Typically, the difference in potential between these states is 5 V. When the comparator input voltage V_{in} is less than the

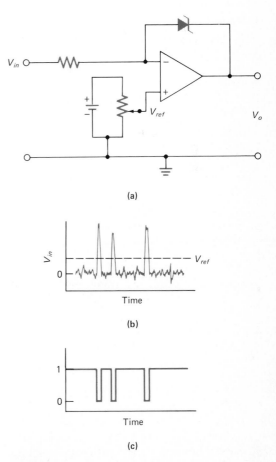

FIGURE 2-25 A wave shaper: (**a**) circuit; (**b**) input signal; (**c**) output signal.

reference voltage V_{ref}, the output is in the logic state 1. On the other hand, when V_{in} is greater than V_{ref}, the output is in logic state 0. Note that the comparator responds only to signals greater than V_{ref} and ignores the fluctuation in the background signal.

Binary Counters. Electronic counters employ a series of binary circuits to count electrical pulses. These circuits are basically electronic switches, such as those shown in Figure 2–23, that have two logic states, on and off or 1 and 0. Each binary can then be used to represent one digit of a binary number (or the coefficient of a power of 2). Two binary circuits, either in series or in parallel, can have four possible outputs, 0/0, 0/1, 1/0, and 1/1. It is readily shown that three binaries have 8 different combinations while four have 16. Thus, n binary circuits have 2^n easily distinguishable output combinations. By using sufficient binaries, the number of significant figures in a count can be made as large as desired. Thus, seven binaries have 128

states, which would provide a count that is accurate to 1 part in 128 or something better than 1% relative.

A convenient form of binary circuits for counting is the so-called *JK flip-flop*. This circuit changes its output level whenever the input signal changes from logic state 1 to 0; *no change in output* is associated with an input change of 0 to 1. Flip-flop circuits are switches that are made up of a suitable combination of diodes and/or transistors. Details of their construction can be found in the references in footnote 4, page 54.

Figure 2–26 shows how four flip-flops can be arranged to give a counting device that can count from 1 through 15. Additional flip-flops will extend the range to larger numbers. Thus, five flip-flops will provide a range of 1 to 31 and six a range from 1 to 63. Figure 2–27 shows the wave forms of the signals as they appear at flip-flops *A*, *B*, *C*, and *D* in the counter shown in Figure 2–26. As shown by the top wave form in Fig-

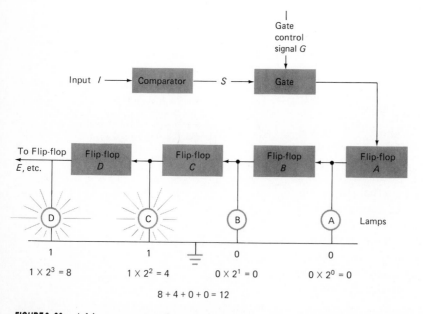

FIGURE 2–26   A binary counter for numbers 0 through 15. The count shown is binary 1011 or decimal 12.

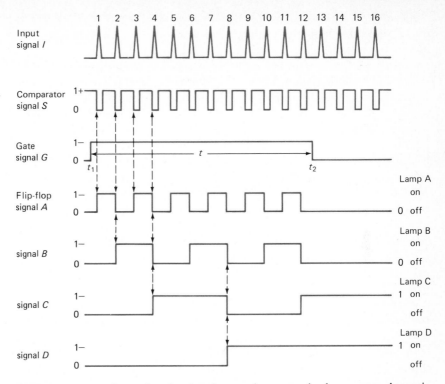

FIGURE 2–27 Wave forms for the signals at various spots in the counter shown in Figure 2–26. Here the count during period t is binary 1011 or decimal 12.

ure 2–27, the input signal I is a series of voltage pulses of equal magnitude and frequency, which are first converted in the comparator to the square wave form shown as signal S; here, the 1 state corresponds to +5 V and the 0 to 0 V. The gate signal G starts the counting as pulse one as the output of the comparator goes from 1 to 0. Ultimately, the gate signal terminates the counting after a preset time t, which, as is shown in the figure, corresponds to 12 counts.

Before counting is initiated, all of the flip-flops are brought to their 0 state by grounding the output terminals. At this point all of the lamps in Figure 2–26 are off. At the start of counting, flip-flop A passes from 0 to 1 as a consequence of the comparator signal changing from 1 to 0 (see first vertical dot-

ted line, Figure 2–27). At this point lamp A comes on and remains on until signal S again shifts from 1 to 0 (see second vertical line). This switching on and off of lamp A continues until time t_2; note that lamp A is off at this point.

As can be seen in Figure 2–27, the output from flip-flop A is also a square wave having a frequency that is exactly one-half that of the input signal S. It is apparent from the figure that the response of flip-flop B to the signal from flip-flop A is exactly analogous to the response of A to the signal S. Thus, B is a square wave that has a frequency just one-half that of A and one-fourth that of S. Similarly, the output from binary C has a frequency that is one-eighth that of S while the frequency of signal D is one-sixteenth that of S.

After time t_2, it is seen that flip-flops C and D end up in logic state 1, while A and B are in the 0 condition. These states correspond to the binary number 1100, which is 12 in the decimal system. Figure 2–26 also demonstrates how the binary count could be read out directly by means of lamps.

Decimal Counting. For most counters, the display is not in the binary form shown by the lamps in Figure 2–26 but in the more convenient and easily comprehended decimal format. Several systems have been developed for binary to decimal conversion. The most common system is the so-called 8421 system or the *binary coded decimal system.* Here, each digit in a decimal number is represented by a set of four binaries such as that shown in Figure 2–26. Only nine of the possible 16 states are then employed. The system is so arranged that after a count of nine, the output of all of the binaries are returned to 0 with the 1 to 0 transition from the D binary being fed into the A binary of the next set of four binaries. Since a set of four binaries is used for one decade of the decimal system, the set is often called a *decade counting unit* (DCU). Figure 2–28 illustrates how four decade counting units can yield a decimal number with four significant figures (6395).

2D-4 SOME OTHER DIGITAL ELECTRONIC COMPONENTS

In addition to counters, digital equipment requires certain other components.

Scalers. From Figure 2–27, and the accompanying discussion, it is apparent that one output pulse is produced from flip-flop D for every 16 input pulses. Thus, by cascading flip-flops it is possible to reduce the number of pulses by a known fraction. Decimal counting units can also be cascaded. Here, each unit reduces the number of counts by a factor of 10. The process of reducing a count by a known fraction is called scaling and becomes important when the frequency of a signal is greater than a counting device can accommodate. In this situation, a *scaler* is introduced between the signal source and the counter. Scalers have been widely used in conjunction with electromechanical counters that can respond to only perhaps 100 pulses per second.

Clocks. Many digital applications require a highly reproducible and accurately known frequency source to be used in conjunction with the measurement of time. Generally, electronic frequency sources are based upon quartz crystals that exhibit the *piezoelectric effect.* Piezoelectric crystals are deformed me-

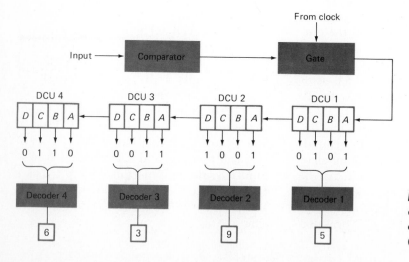

FIGURE 2–28 A binary coded decimal counter using four decade counting units (DCU).

chanically when subjected to an electrical field. The inverse also occurs; that is, when the crystal is deformed by a mechanical force, a potential develops across the crystal. A thin quartz plate, sandwiched between conducting electrodes, vibrates when the electrodes are connected to an ac source. These vibrations, however, produce an electrical signal that interacts with the current from the source. The vibrations and signals reach a maximum at the natural resonant frequency of the crystal. The resonant frequency depends upon the mass and dimensions of the crystal. By varying these parameters electrical output frequencies that range from 10 kHz to 10 MHz or greater can be obtained. Typically, these frequencies are constant to 100 ppm. With special precautions crystal oscillators can be constructed for time standards that are accurate to 1 part in 10 million.

The use of a series of decade scalers with a quartz oscillator provides a clock whose frequency can be varied in steps from perhaps 0.1 Hz to 1 MHz.

Digital to Analog Converters (DAC). Digital signals are often converted to their analog counterparts for the control of instruments or for display by readout devices such as meters or analog recorders. Figure 2–29 illustrates the principle of one of the common ways of accomplishing this conversion, which is based upon a *weighted-resistor ladder* network. Note that the circuit is similar to the sum-

ming circuit shown in Figure 2–22**b** with four resistors weighted in the ratio $8:4:2:1$. From the discussion of summing circuits it is apparent that the output V_a is given by

$$V_a = -V\left(\frac{D}{1} + \frac{C}{2} + \frac{B}{4} + \frac{A}{8}\right) \qquad (2\text{--}17)$$

where V is the voltage associated with logic state 1 and D, C, B, and A designate the logic states (0 or 1) for a 4-bit binary number in which A is the least significant digit (p. 56) and D is the most significant one. Table 2–2 shows the analog output from the weighted-resistor ladder shown in Figure 2–29 when V is $+4$ V.

The resolution of a digital to analog converter depends upon the number of input bits the device will accommodate. Thus, a 10-bit DAC has 2^{10} or 1024 output voltages and, therefore, a resolution of 1 part in 1024.

Analog to Digital Converters (ADC). The output from most transducers used in analytical instruments is an analog signal. To realize the advantages of digital electronics, it is necessary to convert the analog signal to a digital form. Numerous methods exist for this kind of conversion. One will be described here,

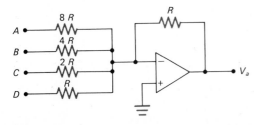

FIGURE 2-29 A 4-bit digital-to-analog converter. Here, A, B, C, and D are $+4$ V for logic state 1 and 0 V for logic state 0.

Table 2–2
Analog Output from the Digital-to-Analog Converter in Figure 2–29

Binary Number BCDA	Decimal Equivalent	$V_a{}^a$
0000	0	0 V
0001	1	−0.5 V
0010	2	−1.0 V
0011	3	−1.5 V
0100	4	−2.0 V
0101	5	−2.5 V

[a] Here, logic state 1 corresponds to +4 V.

namely a converter suitable for voltage measurements.

Figure 2–30 is a simplified schematic diagram of a device for converting an unknown analog voltage V_{in} into a digitized voltage V_o. Here, an n-bit binary counter, controlled by the signals from a quartz clock, is used to drive an n-bit digital to analog converter similar to that described in the previous section. The output of the latter is the *staircase* voltage output V_{DAC} shown in the lower part of the figure. Each step of this signal corresponds to some voltage increment, say 1 mV. The output of the DAC is compared with the unknown analog input voltage V_{in} by means of a comparator circuit. When the two voltages become identical, the comparator shifts logic from 1 to 0 or the reverse, which in turn stops the counter. The count then corresponds to the input voltage in units of millivolts. Closing the reset switch sets the counter back to zero in preparation for the count of a new voltage, which is begun by opening the reset switch.

2E Readout Devices

In this section, three common readout devices are described, namely, the cathode-ray tube (CRT), the laboratory recorder, and the alphanumeric display unit.

2E-1 THE CATHODE-RAY TUBE

The oscilloscope is a most useful and versatile laboratory instrument that employs a cathode-ray tube as a readout device. Figure 2–31 is a schematic drawing showing the main components of the tube. Here, the display is formed by the interaction of electrons in a focused beam with a phosphorescent coating on the interior of the large curved surface of the tube. The electron beam is formed at a heated cathode, which is maintained at ground potential; an anode with a potential of the order of kilovolts accelerates the electrons through a control grid and a second anode that serves to focus the beam on the screen. In the absence of

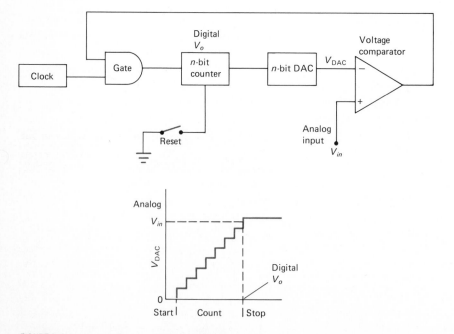

FIGURE 2–30   An analog-to-digital converter.

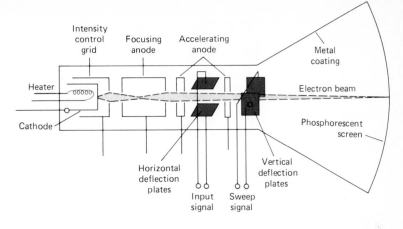

FIGURE 2-31   Schematic of a cathode-ray tube.

input signals, the beam appears as a small bright dot in the center of the screen.

Input signals are applied to two sets of plates, one of which deflects the beam horizontally and the other vertically. Thus, *x-y* plotting of two related signals becomes possible. Because the screen is phosphorescent, the movement of the dot appears as a lighted continuous trace that fades after a brief period.

The most common way of operating the cathode-ray tube is to cause the dot to sweep periodically across the central horizontal axis of the tube by applying a sawtooth sweep signal to the vertical deflection plates. The signal to be measured is then applied across the vertical plates. If the signal is dc, it will simply displace the horizontal line to a position above or below the central horizontal axis.

In order to have a repetitive signal, such as a sine wave, displayed on the screen, it is essential that each sweep begins at an identical place on the wave; that is, for example, at a maximum, a minimum, or a zero crossing. Synchronization is usually realized by mixing a portion of the test signal with the sweep signal in such a way as to produce a voltage spike for, say, each maximum or some multiple thereof. This spike then serves to trigger the sweep. Thus, the wave

form can be observed as a continuous image on the screen.

2E-2 RECORDERS[5]

The typical laboratory recorder is an example of a *servosystem*, a null device that compares two signals and then makes a mechanical adjustment that reduces their difference to zero; that is, a servosystem continuously seeks the null condition.

In the laboratory recorder, shown schematically in Figure 2–32, the signal to be recorded, V_x, is continuously compared with the output from a potentiometer powered by a reference signal, V_{ref}. In most modern recorders, the reference signal is generated by a temperature-compensated Zener diode rectifier circuit (p. 39), which provides a constant potential indefinitely. Any difference in potential between the potentiometer output and V_x is converted to a 60-cycle ac current by a mechanical or electronic chopper; the resulting signal is then amplified sufficiently to activate a small phase-sensitive electrical motor that is mechanically geared or linked (by a pulley arrangement

---

[5] For a discussion of laboratory recorders, see: G. W. Ewing, *J. Chem. Educ.*, **1976**, *53*, A361, A407.

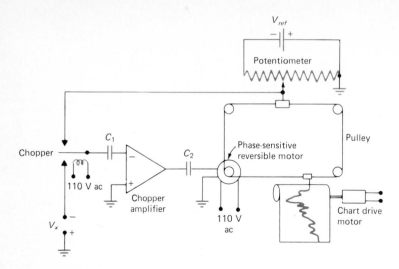

FIGURE 2-32 Schematic diagram of a self-balancing recording potentiometer.

in Figure 2–32) to both a recorder pen and the sliding contact of the potentiometer. The direction of rotation of the motor is such that the potential difference between the potentiometer and V_x is decreased to zero, whereupon the motor stops.

To understand the directional control of the motor, it is important to note that a reversible ac motor has two sets of coils, one of which is fixed (the stator) and the other of which rotates (the rotor). One of these, say the rotor, is powered from the 110-V house line and thus has a continuously fluctuating magnetic field associated with it. The output from the ac amplifier, on the other hand, is fed to the coils of the stator. The magnetic field induced here interacts with the rotor field and causes the rotor to turn. The direction of motion depends upon the *phase* of the stator current with respect to that of the rotor; the phase of the stator current, however, differs by 180 deg, depending upon whether V_x is greater or smaller than the signal from V_{ref}. Thus, the amplified difference signal can be caused to drive the servomechanism to the null state from either direction.

In most laboratory recorders, the paper is moved at a fixed speed. Thus, a plot of signal intensity as a function of time is obtained. In X-Y recorders, the paper is fixed

as an individual sheet mounted on a flat bed. The paper is traversed by an arm that moves along the X axis. The pen travels along the arm in the Y direction. The arm drive and the pen drive are connected to the X and Y inputs, respectively, thus permitting both to vary continuously. Often recorders of this type are equipped with two pens, thus allowing the simultaneous plotting of two functions on the Y axis. An example of an application of this kind is to chromatography where it is desirable to have a plot of the detector output as a function of time as well as the time integral of this output.

A modern laboratory recorder has several chart speeds ranging typically from 0.1 to 20 cm/min. Most provide a choice of several voltage ranges from 1 mV full scale to several volts. Generally, the precision of these instruments is on the order of a few tenths of a percent of full scale.

Digital recorders are becoming more and more widely used. Here, the pen is driven by a stepper motor, which responds to digitized voltage signals by turning some precise fraction of a rotation for each voltage pulse.

2E-3 ALPHANUMERIC DISPLAYS

The output from digital equipment is most conveniently displayed in terms of decimal numbers and letters, that is, in *alphanumeric* form. The seven-segment readout device is based on the principle that any alphanumeric character can be represented by lighting an appropriate combination of seven segments arranged as shown in Figure 2–33. Here, for example, a five is formed when segments *a*, *f*, *g*, *c*, and *d* are lighted; the letter C is observed when segments *a*, *d*, *e*, and *f* are displayed. Perhaps the most common method of lighting a seven-segment display is to fashion each segment as a light-emitting diode (LED). A typical LED consists of a *pn* junction shaped as one of the segments and prepared from gallium arsenide, which is doped with phosphorus. Under forward bias, the junction emits red radiation as a consequence of recombinations of minority carriers in the junction region. Each of the seven segments is connected to a decoder logic circuit so that it is activated at the proper time.

Seven-segment liquid crystal display units (LCD) are also widely encountered. Here, a small amount of a liquid crystal is contained in a thin, flat optical cell, the walls of which are coated with a conducting film. Application of an electrical field to a certain region of the cell causes a change in alignment of the molecules in the liquid crystal and a consequent change in its optical appearance.[6]

2F Noise[7]

The ultimate accuracy and detection limit of an analytical method is determined by the inevitable presence of unwanted *noise*,

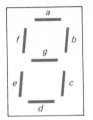

FIGURE 2-33 A seven-segment display.

which is superimposed on the analyte signal causing it to fluctuate in a random way. Figure 2–34**a**, which is a strip chart recording of a tiny dc signal (10^{-15} A), illustrates the typical appearance of noise on a physical measurement. Figure 2–34**b** is a theoretical plot showing the same current in the absence of noise. Unfortunately, at low signal strengths, a plot of the latter kind is never experimentally realizable because some types of noise arise from fundamental thermodynamic and quantum effects that can never be totally eliminated.

Ordinarily, amplification of a small noisy signal, such as that shown in Figure 2–34**a**, provides no improvement in the detection limit or precision of a measurement because both the noise and the information bearing signal are amplified to the same extent; as a matter of fact, the situation is often worsened by amplification as a consequence of added noise introduced by the amplifying device.

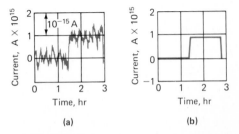

FIGURE 2-34 Effect of noise on a current measurement. (**a**) Experimental strip chart recording of a 10^{-15} A direct current. (**b**) Mean of the fluctuations. (Adapted from: T. Coor, *J. Chem. Educ.*, **1968**, *45*, A594. With permission.)

[6] For discussions of the properties and applications of liquid crystals, see: G. H. Brown and P. P. Crooker, *Chem. Eng. News*, **1983**, *Jan. 31*, 24; G. H. Brown, *J. Chem. Educ.*, **1983**, *60*, 900.

[7] For a more detailed discussion of noise, see: T. Coor, *J. Chem. Educ.*, **1968**, *45*, A533, A583; G. M. Hieftje, *Anal. Chem.*, **1972**, *44* (6), 81A; **1972**, *44* (7), 69A; and H. V. Malmstadt, C. G. Enke, and S. R. Crouch, *Electronics and Instrumentation for Scientists*, Chapter 14, Benjamin/Cummings: Menlo Park, CA, 1981.

In an instrumental method of analysis both instrumental and chemical noise are encountered. Examples of sources of the latter include variability in the completeness of a reaction, side reactions, interferences by components of the sample matrix, and uncontrolled temperature effects on the rate of chemical reactions.

This section deals exclusively with noise that is generated in various components of instruments. Chemical noise, which is associated with the chemical behavior of analytes, will be dealt with in later chapters devoted to specific analytical methods.

2F–1 SIGNAL-TO-NOISE RATIO

It is apparent from Figure 2–34 and the foregoing discussion that noise becomes increasingly important as its magnitude approaches that of the analyte signal. Thus, the *signal-to-noise ratio (S/N)* is a much more useful figure of merit for describing the quality of an instrument or an instrumental method than noise itself.

For a dc signal, the ultimate noise typically takes the form of the time variation of the signal about the mean as shown in Figure 2–34**a**. Here, the noise N is conveniently defined as the standard deviation of the signal while the signal S is given by the mean. The signal-to-noise ratio is then the reciprocal of the relative standard deviation of the measured signal. That is,

$$\frac{S}{N} = \frac{\text{mean}}{\text{standard deviation}}$$

$$= \frac{1}{\text{relative standard deviation}} \quad (2\text{–}18)$$

For an ac signal, the relationship between measurement precision and S/N is less straightforward. This relationship will not be dealt with here since most ac signals are converted to dc before display; Equation 2–18 then applies.

As a rule, visual observation of a signal becomes impossible when S/N is smaller than perhaps 2 or 3. Figure 2–35 illustrates this rule. The upper plot is a recorded nuclear magnetic resonance spectrum for progesterone with a signal-to-noise ratio of approximately 4.3; in the lower plot, the ratio is 43. At the lower ratio, the presence of some but not all of the peaks is apparent.

2F–2 SOURCES OF INSTRUMENTAL NOISE

Noise is associated with each component of an instrument—that is, with the source, the transducer, the signal processor, and the readout. Furthermore, the noise from each of these components may be of several types and arise from several sources. Thus, the noise that is finally observed is a complex composite, which usually cannot be fully characterized. Certain kinds of noise are recognizable, however, and a consideration of their properties is useful.

Instrumental noise can be divided into four general categories. Two, *thermal* or

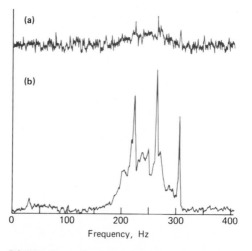

FIGURE 2-35 Effect of signal-to-noise ratio on the NMR spectrum of progesterone: (**a**) S/N = 4.3; (**b**) S/N = 43. (Adapted from: R. R. Ernst and W. A. Anderson, *Rev. Sci. Inst.*, **1966,** 37, 101. With permission.)

Johnson noise and *shot noise*, are well understood, and quantitative statements can be made about the magnitude of each. It is clear that neither Johnson nor shot noise can ever be totally eliminated from an instrumental measurement. Two other types of noise are also recognizable, *environmental noise* or *interference* and *flicker* or *1/f* noise. Their sources are not always well defined nor understood. In principle, however, they can be eliminated (and in practice, minimized) by appropriate instrument design.

Johnson Noise. Johnson or thermal noise owes its source to the thermal agitation of electrons or other charge carriers in resistors, capacitors, radiation detectors, electrochemical cells, and other resistive elements in an instrument. This agitation, or motion of charge particles, is random and periodically creates charge inhomogeneities within the conducting element. These inhomogeneities in turn create voltage fluctuations, which then appear in the readout as noise. It is important to note that Johnson noise is present even in the absence of current in a resistive element.

The magnitude of Johnson noise is readily derived from thermodynamic considerations[8] and is given by

$$v_{\text{rms}} = \sqrt{4kTR\Delta f} \qquad (2\text{–}19)$$

where v_{rms} is the root-mean-square noise voltage lying in a frequency bandwidth of Δf Hz, k is the Boltzmann constant (1.38×10^{-23} J/deg), T is the absolute temperature, and R is the resistance in ohms of the resistive element.

It is important to note that Johnson noise, while dependent upon the frequency bandwidth, is *independent* of frequency itself; thus, it is sometimes termed *white noise* by analogy to white light, which contains all visible frequencies. It is also noteworthy that Johnson noise is independent of the physical size of the resistor.

Two methods are available to reduce Johnson noise in a system—narrow the bandwidth and lower the temperature. The former is often applied to amplifiers and other electronic components, where filters can be used to yield narrower bandwidths. Thermal noise in photomultipliers and other detectors can be attenuated by cooling. For example, lowering the temperature of a detector from room to liquid-nitrogen temperature will halve the Johnson noise.

Shot Noise. Shot noise is encountered wherever a current involves the movement of electrons or other charged particles across a junction. In the typical electronic circuit, these junctions are found at p and n interfaces; in photocells and vacuum tubes, the junction consists of the evacuated space between the anode and cathode. The currents in such devices involve a series of quantized events, namely, the transfer of individual electrons across the junction. These events are random, however, and the rate at which they occur are thus subject to statistical fluctuations, which are described by the equation

$$i_{\text{rms}} = \sqrt{2Ie\Delta f} \qquad (2\text{–}20)$$

where i_{rms} is the root-mean-square current fluctuation associated with the average direct current I, e is the charge on the electron (-1.6×10^{-19} C), and Δf is again the bandwidth of frequencies being considered. Like Johnson noise, shot noise has a "white" spectrum.

Flicker Noise. Flicker noise is characterized as having a magnitude that is inversely proportional to the frequency f of the signal being observed; it is sometimes termed *1/f* (*one-over-f*) noise as a consequence. The causes of flicker noise are not well understood; its ubiquitous presence, however, is recognizable by its frequency dependence.

---

[8] For example, see: T. Coor, *J. Chem. Educ.*, **1968**, *45*, A534.

Flicker noise becomes significant at frequencies lower than about 100 Hz. The long-term *drift* observed in dc amplifiers, meters, and galvanometers is a manifestation of flicker noise.

Environmental Noise.   Environmental noise is a composite of noises arising from the surroundings. Figure 2–36 suggests typical sources of environmental noise in a university laboratory.

Much environmental noise occurs because each conductor in an instrument is potentially an antenna capable of picking up electromagnetic radiation and converting it to an electrical signal. Numerous sources of electromagnetic radiation exist in the environment, including ac power lines, radio and TV stations, ignition systems in gasoline engines, arcing switches, brushes in electrical motors, lightning, and ionospheric disturbances. Note that some of these sources, such as power lines and radio stations, cause noise with limited-frequency bandwidths.

It is also noteworthy that the noise spectrum shown in Figure 2–36 contains a large, continuous noise region at low frequencies. This noise has the properties of flicker noise; its sources are unknown. Superimposed upon the flicker noise are noise peaks associated with yearly and daily temperature fluctuations and other periodic phenomena associated with the use of a laboratory building.

Finally, two quiet-frequency regions in which environmental noises are low are indicated in Figure 2–36. Often, signals are converted to these frequencies to reduce noise during signal processing.

2G Signal-to-Noise Enhancement

Many laboratory measurements require only minimal effort to maintain the signal-to-noise ratio at an acceptable level, because the signals are relatively strong and the requirements for precision and accuracy are low. Examples include the weight determinations made in the course of a chemical synthesis or the color comparison made in determining the chlorine content of the water in a swimming pool. For both examples, the signal is large relative to the noise and the requirements for accuracy are minimal. When the need for sensitivity and accuracy increases, however, the S/N ratio often be-

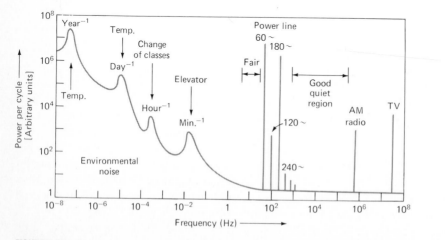

FIGURE 2-36   Some sources of environmental noises in a university laboratory. Note the frequency dependence. (From: T. Coor, *J. Chem. Educ.*, **1968**, *45*, A540. With permission.)

comes the limiting factor to the precision of a measurement.

Two general methods are available for improving the signal-to-noise ratio of an instrumental method, namely *hardware* and *software*. The first involves noise reductions by incorporation into the instrument design components such as filters, choppers, shields, modulators, and synchronous detectors, which remove or attenuate the noise without affecting the analytical signal significantly. Software methods are based upon various digital computer algorithms that permit extraction of signals from noisy environments. As a minimum, software methods require sufficient hardware to condition the output signal from the instrument and convert it from analog to digital form; obviously, a computer and readout system are also necessary.

2G-1 SOME HARDWARE DEVICES FOR NOISE REDUCTION

This section contains a brief discussion of several hardware devices and techniques used for signal-to-noise enhancement.

Grounding and Shielding. Noise arising from environmentally generated electromagnetic radiation can often be substantially reduced by shielding, grounding, and minimizing the lengths of conductors. Shielding consists of surrounding a circuit, or some of the wires in a circuit, with a conducting material that is attached to earth ground. Electromagnetic radiation is then absorbed by the shield rather than by the enclosed conductors; noise generation in the instrument circuit is thus avoided in principle.

Shielding becomes particularly important when the output of a high-impedance transducer, such as the glass electrode, is being amplified. Here, even minuscule induced currents give rise to relatively large voltage drops and thus to large voltage fluctuations.

Difference Amplifiers. Any noise generated in the transducer circuit is particularly critical because it appears in an amplified form in the instrument readout. To attenuate this type of noise, most instruments employ a difference amplifier, such as that shown in Figure 2–19, for the first stage of amplification. An ac signal induced in the transducer circuit generally appears in phase at both the inverting and noninverting terminals; cancellation then occurs at the output.

Analog Filtering. One of the most common methods of improving the signal-to-noise ratio in analytical instruments is by use of low-pass analog filters such as that shown in Figure A2–7b in Appendix 2. The reason for this widespread application is that the majority of analyte signals are dc with bandwidths extending over a range of only a few herz. Thus, a low-pass filter with an output such as that shown in Figure A2–8b will effectively remove any high-frequency component of the signal including those arising from Johnson or shot noise. Figure 2–37 illustrates the use of a low-pass *RC* filter for reducing environmental and Johnson noise from a slowly varying dc signal.

High-pass analog filters such as that shown in Figure A2–7a also find considerable application in analytical instruments

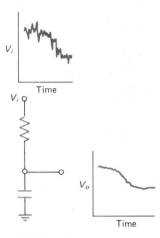

FIGURE 2-37 Use of a low-pass filter with a large time constant to remove noise from a slowly changing dc voltage.

where the analyte signal is ac. Here, the filter reduces the effect of drift and other low-frequency flicker noise.

Narrow-band electronic filters are also available to attenuate noise. We have pointed out that the magnitude of fundamental noise is directly proportional to the square root of the frequency bandwidth of a signal. Thus, significant noise reduction can be achieved by restricting the input signal to a narrow band of frequencies and employing an amplifier that is tuned to this band. It should be appreciated, however, that if the signal generated from the analyte varies with time, the band passed by the filter must be sufficiently wide to carry all of the information provided by the signal; obviously, the frequency ranges of the filter and analyte signal must also be the same.

Modulation. Direct amplification of a low-frequency or dc signal is particularly troublesome because of amplifier drift and flicker noise. Often, this $1/f$ noise is several times larger than the types of noise that predominate at higher frequencies. For this reason, low-frequency or dc signals from transducers are often converted to a higher frequency, where $1/f$ noise is less troublesome. This process is called *modulation*. After amplification, the modulated signal can be freed from amplifier $1/f$ noise by filtering with a high-pass filter; demodulation and filtering with a low-pass filter then produces an amplified dc signal suitable for driving a readout device.

Figure 2–38 is a schematic diagram showing the flow of a signal through such a system. Here, the original dc current is modulated to give a narrow-band 400 Hz signal, which is then amplified by a factor of 10^5. As shown, in the center of the figure, amplification introduces $1/f$ and power-line noise; much of this noise can, however, be removed with the aid of a suitable high-pass filter (see Figure A2–7**a**, Appendix 2). Demodulation of this filtered signal results in the amplified dc signal shown in the right of the figure.

Signal Chopping; the Chopper Amplifier. The *chopper amplifier* provides one means for accomplishing the signal flow shown in Figure 2–38. In this device, the input signal is converted to a square-wave form by an electronic or mechanical *chopper*. Chopping can be performed either on the source itself or on the electrical signal from the transducer.

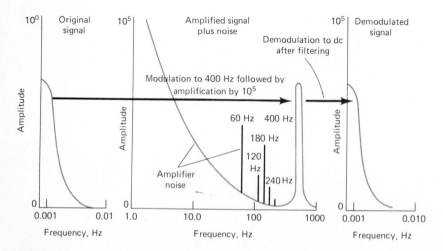

FIGURE 2–38   Amplification of a dc signal with a chopper amplifier. (Adapted from: T. Coor, *J. Chem. Educ.*, **1968**, *45*, A586. With permission.)

In general, it is desirable to chop the signal as close to its source as possible because only the noise arising *after chopping* is removed by the process.

Infrared spectroscopy provides an example of the use of a mechanical chopper for signal modulation. Noise is a major concern in detecting and measuring infrared radiation because both source intensity and detector sensitivity are low. As a consequence, the electrical signal from an infrared transducer is generally small and requires large amplification. Furthermore, infrared transducers, which are heat detectors, respond to thermal radiation from their surroundings; that is, they suffer from serious environmental noise effects.

In order to minimize these noise problems, the beams from infrared sources are generally chopped by imposition of a slotted rotating disk in the beam path. The rotation of this chopper produces a radiant signal that fluctuates periodically between zero and some maximum intensity. After interaction with the sample, the signal is converted by the transducer to a square-wave ac electrical signal whose frequency depends upon the size of the slots and the rate at which the disk rotates. Environmental noise associated with infrared measurement is generally dc; thus, it can be significantly reduced by use of a high-pass filter prior to amplification of the electrical signal.

Another example of the use of a chopper is shown in Figure 2–39. This device is a *chopper amplifier,* which employs an ac-driven electromagnet to operate a switch that, in its closed position, shorts the input or the output signal to ground. The appearance of the signal at various stages is shown above the circuit diagram. The transducer input is assumed to be a 6-mV dc signal (*A*). The vibrating switch converts the input to an approximately square-wave signal with an amplitude of 6 mV (*B*). Amplification produces an ac signal with an amplitude of 6 V (*C*), which, however, is shorted to ground periodically; as shown in (*D*), shorting also reduces the amplitude of the signal to 3 V. Finally, the *RC* filter serves to smooth the

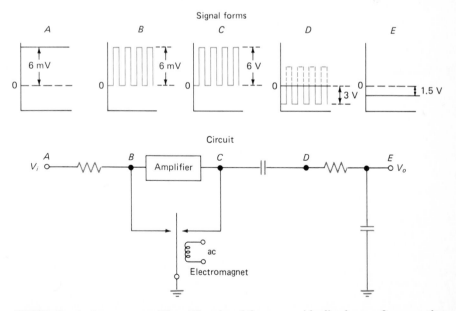

FIGURE 2–39 A chopper amplifier. The signal forms are idealized wave forms at the various indicated points in the circuit.

signal and produce a 1.5-V dc output. The synchronous demodulation process has the effect of rejecting the noise generated within the amplifier.

Lock-in Amplifiers.[9]   Lock-in amplifiers permit the recovery of small signals even when S/N is unity or less. Generally, a lock-in amplifier requires a reference signal that is coherent with the signal to be amplified. That is, the

reference signal must be of the same frequency as the analytical signal and, equally important, must bear a fixed phase relationship to the latter. Figure 2–40**a** shows a system which employs an optical chopper to provide coherent analytical and reference signals. The reference signal is provided by a lamp and can be quite intense, thus freeing it from potential environmental interferences. The reference and signal beams are chopped synchronously by the rotating slotted wheel, thus providing signals that are identical in frequency and have a fixed phase angle to one another.

---

[9] See: T. C. O'Haver, *J. Chem. Educ.*, **1972**, *49*, A131, A211.

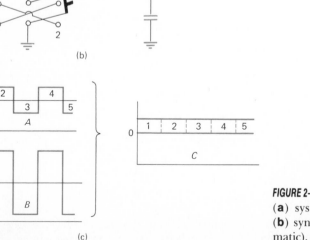

(a)

(b)

(c)

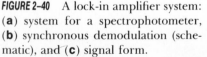

FIGURE 2–40   A lock-in amplifier system: (**a**) system for a spectrophotometer, (**b**) synchronous demodulation (schematic), and (**c**) signal form.

The synchronous demodulator acts in a manner analogous to the double-pole–double-throw switch shown in Figure 2–40**b**. Here, the reference signal controls the switching so that the polarity of the analytical signal is reversed periodically to provide a rectified dc signal, as shown to the right in Figure 2–40**c**. The ac noise is then removed by a low-pass filtering system.

A lock-in amplifier is generally relatively free of noise because only those signals that are "locked-in" to the reference signal are amplified; that is, those that are coherent. All other frequencies are rejected by the system.

2G-2 SOFTWARE METHODS

With the widespread availability of microprocessors and microcomputers, many of the signal-to-noise enhancement devices described in the previous section are being replaced or supplemented by digital computer software algorithms. Among these are programs for various types of averaging, digital filtering, Fourier transformation, and correlation techniques. Generally, these procedures are applicable to nonperiodic or irregular wave forms, such as an absorption spectrum, or to signals having no synchronizing or reference wave.

Some of these common software procedures are discussed briefly here.

Ensemble Averaging.[10] In ensemble averaging, successive sets of data (*arrays*), each of which adequately describes the analyte wave form, are collected and summed point by point as an array in the memory of a computer (or in a series of capacitors for hardware averaging). After the collection and summation is complete, the data are averaged by dividing the sum for each point by the number of scans performed. Figure 2–41 illustrates ensemble averaging of a simple absorption spectrum.

Ensemble averaging owes its effectiveness to the fact that individual noise signals N_n, insofar as they are random, tend to cancel one another. Consequently, their average N is given by a relationship analogous to Equation 1–5 (p. 8). That is,

$$N = N_n/\sqrt{n}$$

where n is the number of arrays averaged. The signal-to-noise ratio for the averaged arrays is then given by

$$\frac{S}{N} = \frac{S_n/n}{N_n/\sqrt{n}} = \frac{S_n}{N_n}\sqrt{n} \qquad (2-21)$$

where S_n/n is the average signal. It should be noted that this same signal-to-noise enhancement is realized in the boxcar averag-

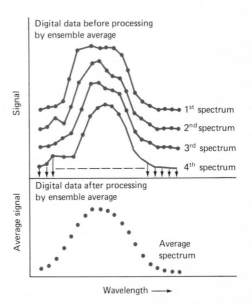

FIGURE 2-41 Ensemble averaging of a spectrum. (From: D. Binkley and R. Dessy, *J. Chem. Educ.,* **1979,** *56,* 150. With permission.)

[10] For a more extended description of the various types of signal averaging, see: D. Binkley and R. Dessy, *J. Chem. Educ.,* **1979,** *56,* 148; R. L. Rowell, *J. Chem. Educ.,* **1978,** *56,* 148; G. Dulaney, *Anal. Chem.,* **1975,** *47,* 24A.

ing and digital filtering, which are described in subsequent sections.

To realize the advantage of ensemble averaging and still extract all of the information available in an analyte wave form, it is necessary to measure points at a frequency that is at least twice as great as the highest frequency component of the wave form. Much greater sampling frequencies, however, provide no additional information but include more noise. Furthermore, it is highly important to sample the wave-form reproducibility (that is, at the same point each time). For example, if the wave form is a visible absorption spectrum, each scan of the spectrum must start at *exactly* the same wavelength and the rate of wavelength change must be identical for each sweep. Generally, the former is realized by means of a synchronizing pulse, which is derived from the wave form itself. This pulse then initiates the recording of the wave form.

Ensemble averaging can produce dramatic improvements in signal-to-noise ratios as demonstrated by the three NMR spectra in Figure 2–42. Here, only a few of the absorption peaks are discernible in the single scan because their magnitudes are roughly the same as the recorder excursions due to random noise. The improvement with added scans is obvious.

Ensemble averaging can be performed by either hardware or software techniques. The latter is now the more common with the signals at various points being accumulated in a computer memory for subsequent processing and display.

Boxcar Averaging.   Boxcar averaging is a digital procedure for *smoothing* irregularities in a wave form, the assumption being made that these irregularities are the consequence of noise. That is, it is assumed that the analog analytical signal varies only slowly with time and that the average of a small number of adjacent points is a better measure of the signal than any of the individual points. Fig-

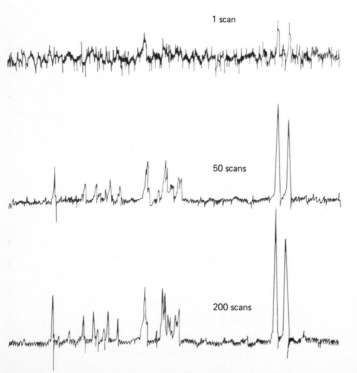

1 scan

50 scans

200 scans

FIGURE 2–42   Effect of signal averaging. Note that the vertical scale is smaller as the number of scans increases. That is, the noise grows in absolute value with increased number of scans; its value relative to the analytical signal decreases, however.

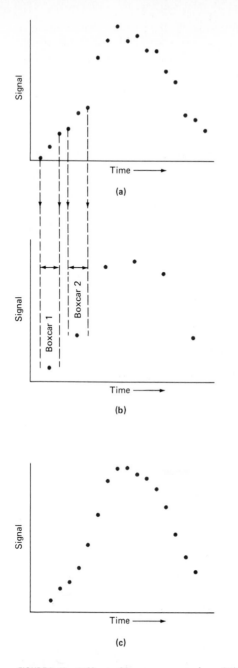

ure 2–43**b** illustrates the effect of the technique on the data plotted in Figure 2–43**a**. The first point on the boxcar plot is the mean of points 1, 2, and 3 on the original curve; point 2 is the average of points 4, 5, and 6, and so forth. In practice 2 to 50 points are averaged to generate a final point. Most often this averaging is performed by a computer in real time, that is, as the data is being collected (in contrast to ensemble averaging, which requires storage of the data for subsequent processing). Clearly, detail is lost by boxcar averaging, and its utility is limited for complex signals which change rapidly as a function of time. It is of considerable importance, however, for square-wave or repetitive pulsed outputs where only the average amplitude is important.

Figure 2–43**c** shows a moving-window boxcar average of the data in Figure 2–43**a**. Here, the first point is the average of original points 1, 2, and 3; the second boxcar point is an average of points 2, 3, and 4, and so forth. Here, only the first and last points are lost. The size of the boxcar again can vary over a wide range.

Digital Filtering.   The moving-window, boxcar method just described is a kind of *linear* filtering wherein it is assumed that an approximately linear relationship exists among the points being sampled in each boxcar. More complex polynomial relationships can, however, be assumed to derive a center point for each window.

Digital filtering can also be carried by a Fourier transform procedure. Here the original signal, which varies as a function of time (a *time-domain signal*), is converted to a *frequency-domain signal* in which the independent variable is now frequency rather than time. This transformation, which is discussed in Section 5I–5, is accomplished mathematically on a digital computer by a *Fourier transform* procedure. The frequency signal is then multiplied by the frequency response of a digital filter, which has the effect of removing a certain frequency re-

FIGURE 2–43   Effect of boxcar averaging. (**a**) Original data. (**b**) Data after boxcar averaging. (**c**) Data after moving-window boxcar averaging. Reprinted with permission from G. Dulaney, *Anal. Chem*, **1975**, *47*, 28A. (Copyright 1975 American Chemical Society.)

gion of the transformed signal. The filtered time-domain signal is then recovered by an inverse Fourier transform.

Correlation. Correlation methods are beginning to find application for the processing of data from analytical instruments. These procedures provide powerful tools for performing such tasks as extracting signals that appear to be hopelessly lost in noise, smoothing noisy data, comparing a spectrum of an analyte with stored spectra of pure compounds, and resolving overlapping or unresolved peaks in spectroscopy and chromatography.[11] Correlation meth-ods are based upon complex mathematical data manipulations that can only be carried out conveniently by means of a digital computer.

Correlation methods will not be discussed in this text. The interested reader should consult the references given in footnote 11.

[11] For a more detailed discussion of correlation methods, see: G. Horlick and G. M. Hieftje, in *Contemporary Topics in Analytical and Clinical Chemistry*, D. M. Hercules, *et al.*, Eds., Vol. 3, p. 153–216, Plenum Press: New York, 1978. For a briefer discussion, see: G. M. Hieftje and G. Horlick, *American Laboratory*, **1981**, *13* (3), 76.

Problems

2–1. Assume the following values for the component of the circuit in Figure 2–13: $R_i = 1.00$ kΩ, $R_f = 30.0$ kΩ, $\beta = 200$, and $V_i = 0.910$ mV dc. Calculate (a) V_o; (b) I_i; (c) I_f.

2–2. Determine the relative error that results from using Equation 2–3 to calculate v_o rather than the more exact expressions if $R_i = 2.00$ MΩ, $R_f = 40.0$ MΩ, and $\beta = 5 \times 10^4$.

2–3. Design a circuit having an output given by

$$-V_o = 3V_1 + 5V_2 - 6V_3$$

2–4. Design a circuit for calculating the average value of three input voltages multiplied by 1000.

2–5. Design a circuit to perform the calculation

$$-y = \frac{1}{10} (5x_1 + 3x_2)$$

2–6. Design a circuit to perform the calculation

$$-V_o = 4V_i + 1.00 \times 10^3 I_i$$

2–7. For the following circuit:

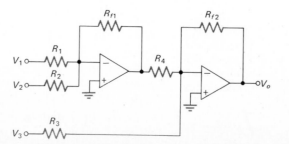

(a) write an expression giving the output voltage in terms of the three input voltages and the various resistances.

(b) indicate the mathematical operation performed by the circuit when $R_1 = R_{f1} = 200$ kΩ; $R_4 = R_{f2} = 400$ kΩ; $R_2 = 50$ kΩ; $R_3 = 10$ kΩ.

2–8. Show the algebraic relationship between the voltage input and output for the following circuit.

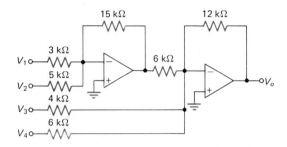

2–9. Derive an expression for the output potential of the following circuit.

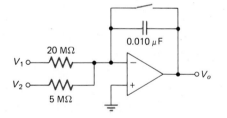

2–10. Show that when the four resistances are equal, the following circuit becomes a subtracting circuit.

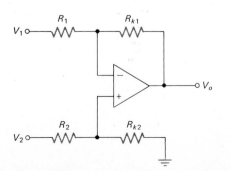

2–11. Derive a relationship between V_o and V_i for the following circuit.

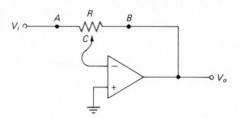

2–12. The linear slide wire AB has a length of 100 cm. Where along its length should contact C be placed in order to provide a potential of exactly 3.00 V?

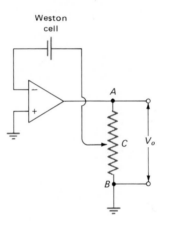

2–13. Design a circuit that will produce the following output:

$$v_o = 4.0 \int_0^t v_1 \, dt + 5.0 \int_0^t v_2 \, dt$$

2–14. Design a circuit that will produce the following output:

$$v_o = 2.0 \int_0^t v_1 \, dt - 6.0(v_2 + v_3)$$

2–15. Plot the output voltage of an integrator after 1, 3, 5, and 7 sec after the start of integration if the input resistor is 2.0 MΩ, the feedback capacitor is 0.25 μF, and the input voltage is 4.0 mV.

2–16. The following data were obtained for repetitive weighings of a 2.001 g standard weight by means of a top-loading balance:

2.000	2.000	1.997
1.998	2.002	1.999
2.004	2.003	2.001

(a) Calculate the signal-to-noise ratio for the balance assuming the noise is random.

(b) How many measurements would have to be averaged to increase S/N to 1000?

2–17. The following data were obtained for a voltage measurement, in mV, on a noisy system: 1.37, 1.84, 1.35, 1.47, 1.10, 1.73, 1.54, 1.08.

(a) What is the signal-to-noise ratio assuming the noise is random?

(b) How many measurements would have to be averaged to increase S/N to 10?

2–18. The resistance of a dc circuit is 1500 Ω. (a) What is the maximum resistance that a current-measuring device can have if the current in this circuit is to be measured with a relative error of less than 2%? The expected current range is 0 to 20 μA. (b) Devise an operational amplifier circuit employing a 0- to 10-mV meter that will meet the requirements specified in (a).

2–19. The resistance of a dc circuit is 200 Ω. (a) What is the maximum resistance that a current-measuring device can have if the current in this circuit is to be measured with a relative error of less than 5%? The expected current range is 0 to 50 μA. (b) Devise an operational amplifier circuit employing a 0- to 10-mV meter that will meet the requirements specified in (a).

3

Microcomputers

and Microprocessors

in Chemical Instrumentation

Microcomputers and microprocessors have become an integral part of many modern laboratory instruments where they serve to control operating conditions, to process data, and to communicate analytical results to the scientist. The interfacing of these devices to instruments is too large and complex a subject to be treated in detail in this text. Thus, the discussion in this chapter is limited to a general summary of computer terminology, the architecture and properties of microprocessors and microcomputers, and the advantage gained by employing these remarkable devices.[1]

---

[1] For further information, see: H. V. Malmstadt, C. G. Enke, and S. R. Crouch, *Electronics and Instrumentation for Scientists*, Chapters 12–14, Benjamin/Cummings: Menlo Park, CA, 1981; A. Osborne, *An Introduction to Microcomputers*, 2d ed., vol. 1, Osborne/McGraw-Hill: Berkeley, CA, 1980; S. P. Perone and D. O. Jones, *Digital Computers in Scientific Instrumentation*, McGraw-Hill: New York, 1973; C. A. Odgin, *Microcomputer Design*, Prentice Hall: Edgewood Cliffs, N.J., 1978; *Scientific American*, **1977,** *237* (3), 62–230.

3A Introduction

A microprocessor is a large-scale integrated circuit made up of tens and even hundreds of thousands of transistors, resistors, switches, and other circuit elements miniaturized to fit on a single silicon chip a few millimeters square. A microprocessor often serves as an arithmetic and logic component, called the *central processing unit* (CPU), of a digital microcomputer. Microprocessors also find widespread use for the control of the operation of such diverse items as analytical instruments, automobile ignition systems, microwave ovens, cash registers, and electronic games.

The development of inexpensive and tiny microprocessor chips may well be the greatest technical development of the last half-century and one of the greatest in all of man's history. For example, the first electronic digital computer, developed in 1946, was a 30-ton monster that occupied 1500 square feet, cost millions of dollars, employed 18,500 vacuum tubes (one of which burned out, on the average, every seven minutes), and consumed 130 kW of power. Yet it had no more computing power and was considerably slower than a modern microcomputer consisting of a few microprocessor chips that cost less than $20 and are roughly twice the size of one of the capital letters on this page.

Microcomputers consist of one or more microprocessors combined with other circuit components that provide memory storage, timing, input, and output functions. Microcomputers (and microprocessors as well) are finding ever-increasing use for the control of analytical instruments and for processing, storing, and displaying the data derived therefrom. At least two reasons exist for connecting a computer to an analytical instrument. The first is that partial or complete automation of measurements becomes possible. Ordinarily, automation leads to more rapid data acquisition, which shortens the time required for an analysis or increases its precision by providing time for additional replicate measurements to be made. Automation, moreover, frequently provides better control over experimental variables than a human operator can achieve; more precise and accurate data are the result.

A second reason for interfacing computers with instruments is to take advantage of their tremendous computational and data-handling capabilities. These capabilities make possible the routine use of techniques that would ordinarily be impractical because of their excessive computational time requirements. Notable among such applications is the use of Fourier transform calculations, signal averaging, and correlation techniques in spectroscopy to extract small analytical signals from noisy environments.

3A-1 COMPUTER TERMINOLOGY

One of the problems facing the newcomer to the field of computers and computer applications is the bewildering array of new terms and acronyms (such as CPU, ALU, PROM, and SSIC), which are encountered and which, unfortunately, are often not defined even in elementary presentations. Some of the most important of these terms and abbreviations are defined here; others will appear later in the chapter.

Analog Versus Digital Computers. In Section 2C–5 it was shown how operational amplifier circuits can perform various mathematical operations including arithmetic calculations, integration, differentiation, and exponentiation. These circuits can be readily combined in such a way as to provide versatile *analog* computers, which are based upon analog or continuously variable electrical signals. In contrast, *digital* computers respond to the same type of discrete signals (that is, 1 and 0) that were described in the section dealing with digital electronics (Section 2D). Digital computers are much more powerful than their analog counterparts with respect to accuracy, precision, freedom from noise, and data storage capacity. As a consequence, the term computer has come to mean digital computer in most contexts.

A digital computer owes its power to the fact that it can perform a huge number of elementary operations in a brief period—that is, several hundred thousand per second. These operations are performed in response to commands called *instructions*. A list of instructions written in order of execution in some programming language is termed a *program*.

Bits, Bytes, and Words. The binary digits 0 and 1, which are used in digital electronics and computers, are called *bits*. As in digital electronics, bits are represented in a computer by two electrical states differing from one another by 5 to 10 V. Generally, a combination of eight bits is called a *byte*. A series of bytes arranged in sequence to represent a piece of data or an instruction is called a *word*. The number of bits (or bytes) per word depends upon the size of the computer; some common sizes include 8, 16, 32, and 64 bits or the equivalent 1, 2, 4, and 8 bytes. Microcomputers and microprocessors often use 8- or 16-bit words, although 32-bit units are now coming into use and 64-bit devices do not seem far away.

To illustrate these definitions, the binary 10011011 is an 8-bit or 1-byte word that can be used to represent the decimal number 155.

Registers. The basic component from which digital computers are constructed is the *register,* a physical device that can store a complete byte or a word. A binary coded decimal counter made up of four decade-counting units, such as that shown in Figure 2–28, can serve as a register that is capable of holding a 16-bit word or 2 bytes.

A datum contained in a register can be manipulated in a number of ways. For example a register can be *cleared,* a process by which the register is reset to all zeros; the *ones complement* of a register can be taken (every 1 changed to 0 and conversely); or the content of one register can be transferred to another. Also, the content of one register can be added, subtracted, multiplied, or divided by that of another; a register in which these processes are performed is called an *accumulator.* It has been proven that the proper sequence of register operations can solve any computational or informational processing problem no matter how complex provided only that an *algorithm* exists for the solution. An algorithm is a detailed statement of the individual steps required to arrive at a solution. One or more algorithms then makes up a *computer program.*

Hardware and Software. Computer *hardware* consists of the physical devices making up a computer. Examples include disk drives, printers, clocks, memory units, and registers for performing arithmetic or logic operations. The collection of programs and instructions to the computer (including the tapes or disks for their storage) are the *software.* Hardware and software are equally

important to the successful application of computers, and the initial cost of the latter may be as great as the former.

3A-2 OPERATIONAL MODES OF COMPUTERIZED INSTRUMENTS

Figure 3–1 suggests three ways in which computers can be used in conjunction with analytical measurements. In the *off-line* method shown in 3–1a, the data are collected by a human operator and subsequently transferred to the computer for data processing by means of magnetic tape, punched cards, or a keyboard. The *on-line* method shown in Figure 3–1b differs from the off-line procedure in the respect that direct communication between the instrument and the computer is made possible by

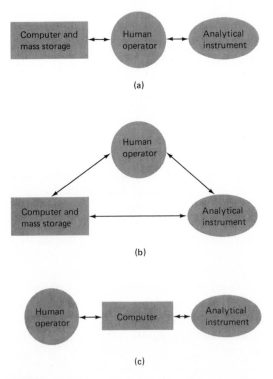

FIGURE 3-1 Three methods of using computers for analytical measurements: (**a**) off-line; (**b**) on-line; (**c**) in-line.

means of an electronic *interface* wherein the signal from the instrument is shaped, digitized, and stored by the computer. Here, the computer remains a distinct entity with provision for mass storage of data and instructions for processing these data; off-line operation is also possible with this arrangement.

Many modern instruments are configured as shown in Figure 3–1c. In this *in-line* arrangement, a microcomputer or microprocessor is inbedded in the instrument. Here, the operator communicates with and directs the instrument operation via the computer. He does not, however, necessarily program the computer (often he has the option, however), this having been done by the manufacturer.

In in-line and on-line operations, the data are often transferred to the computer in *real time,* that is, as they are generated by the instrument. Often, the rate of data produced by an instrument is low enough so that only a small fraction of the computer's time is occupied in data acquisition; under this circumstance the periods between data collection can be used for processing the information in various ways. For example, data processing may involve calculation of a concentration, curve smoothing, combining the datum with previously collected and stored data for subsequent averaging, and plotting or printing out the result. *Real-time processing* involves data treatment performed simultaneously with data acquisition. Real-time processing has two major advantages. First, it may reduce significantly the amount of data storage space required, thus making possible the use of a less sophisticated and less expensive computer. Second, if sufficient time exists between the data acquisition points, the processed signal may be used to adjust instrument parameters to improve the quality of future output signals. For example, potentiometric titration curves can be obtained automatically with microprocessor controlled instru-

ments, which employ a motor-driven syringe to control the rate of reagent flow. Often in such instruments, the data are processed in real time to give the first derivative of the potential with respect to volume. This parameter, which reaches a maximum at the equivalence point, is then used in turn to control the flow rate. For example, in the early parts of the titration curve, where the rate of potential change is low and the derivative small, the reagent is added rapidly. As the equivalence point is approached, however, an increase in the derivative function occurs, and this increase signals the microprocessor to operate the motor at a slower speed to reduce the flow rate. The reverse process occurs beyond the equivalence. Thus, the instrument performs the titration in much the same way as a human operator, with rapid additions of reagent being made before and after the equivalence point region and slow additions in the immediate vicinity of equivalence.

3B Components of a Computer

Figure 3–2 is a block diagram showing the major components (the hardware) of a computer and its peripheral devices.

3B-1 CENTRAL PROCESSING UNIT (CPU)

The heart of a computer is the central processing unit, which in the case of a microcomputer is a microprocessor chip. A microprocessor is made up of a control unit and an arithmetic logic unit. The former controls the overall sequence of operations by means of instructions from a program stored in the memory unit. The control unit receives information from the input device, fetches instructions and data from the memory, transmits instructions to the arithmetic unit for *execution,* and transmits the results of computations in the arithmetic unit to the output and often to memory as well.

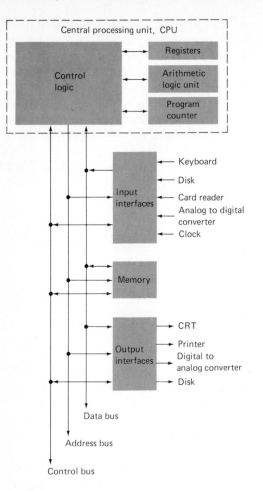

FIGURE 3-2 Basic components of a digital computer including peripheral devices.

The arithmetic unit (ALU) of a CPU is made up of a series of equal-size registers or accumulators wherein the intermediate results of binary arithmetic are accumulated.

3B-2 BUSES

The various parts of a computer and its memory and peripheral devices are joined by *buses*, each of which is made up of a number of transmission lines. For rapid communication among the various parts of a computer, all of the digital signals making

up a word are usually transmitted simultaneously by the parallel lines of the bus. The number of lines in the internal buses of the CPU is then equal to the size of the word processed by the computer. For example, the internal bus for an 8-bit CPU will require eight parallel transmission lines each of which transmits one of the eight bits.

Data are carried into and out of the CPU by a data bus (see Figure 3–2). Both the origin and destination of the signals in the data bus line are specified by the address bus. An address bus with 16 lines can directly address 2^{16} or 65,536 registers or other locations within the computer or its memory. The control bus carries control and status information to and from the CPU. These transfers are sequenced by timing signals carried in the control bus.

External serial buses, which require only two lines per peripheral, are also used to connect communication terminals and remote instruments to the computer. They suffer from the slow rate at which information can be transmitted.

3B-3 MEMORY

In a microcomputer, the memory is a storage area that is directly accessible by the CPU. Because the memory contains both data and program information, the memory must be accessed by the CPU at least once for each program step. The time required to retrieve a piece of information from the memory is called its *access time;* access times vary from tens of nanoseconds (1 nsec = 10^{-9} sec) to hundreds.

Memory Chips.   The basic unit of a memory chip is a cell, which is capable of existing in one of two states and thus capable of storing one bit of information. Typically, thousands of these cells are contained on a single silicon memory chip. Figure 3–3 illustrates the functions associated with an individual memory cell. With a READ command from the CPU, the logic state (1 or 0) appears as

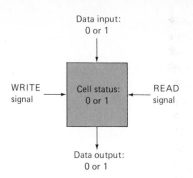

FIGURE 3-3 An individual computer memory cell for storage of one bit.

one of two possible voltages at the output. A WRITE command allows the 1 or 0 voltage from the input terminal to displace the voltage already present in the cell and to store the new voltage in its place.

A common way of organizing the individual cells is a square array on a silicon chip. Each cell then consists of a tiny transistorized flip-flop circuit, which has two states, or a transistor-capacitor circuit, which stores information as a charge (or no charge). A common type of memory chip contains 4096 of these cells arranged in 64 columns and 64 rows. A chip with this number of storage spaces is called a 4 K chip, where K is a unit corresponding to 1024 individual cells. The maximum number of cells on a single silicon chip has increased continuously since such devices were invented. Currently, 64 K chips are commonly available, 256 K chips are beginning to be shipped to customers in small quantities, and 1 megabyte chips have been announced.[2]

At this point, it is desirable to warn the reader that the definition of K in the previous paragraph applies to *memory chips only.* For computer memories, K is a unit that describes memory capacity in terms of number of bytes (or sometimes words). Thus, a 4 K-byte memory would be capable of storing

4096 bytes, each of which is made up of 8 bits. Such a memory could be constructed from eight 4 K-bit chips connected by a bus consisting of 8 parallel transmission lines. The total number of bits would then be $4 \times 1024 \times 8 = 32{,}768$.

Memory Cell Addresses. In order to be useful, each cell in a memory chip must have an *address* or location that can be uniquely specified by the CPU when it gives a READ or WRITE command. These addresses are transmitted to the memory by the address bus shown in Figure 3–2. Figure 3–4 illustrates how one member of a square array of cells can be uniquely addressed. Here, 4 address lines are required for the 16 cells ($16 = 2^4$). For the 4096 array described earlier, the address bus would have to carry 12 ($4096 = 2^{12}$) parallel lines.

Word Storage. Thus far, we have considered how individual bits of information can be addressed, stored, and retrieved. A computer memory, however, is designed so that the CPU reads or writes in terms of bytes (or words). Figure 3–5 illustrates one way of organizing eight of the 4 K-bit chips we have just described, so that the 8 bits making up a 1-byte word can be stored or retrieved si-

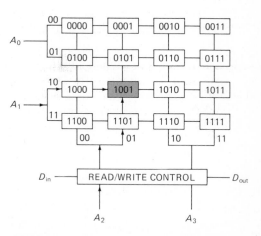

FIGURE 3-4 A 16-bit memory module with 4 address lines A_1, A_2, A_3, and A_4. The shaded cell is addressed when $A_1 = 0$ and $A_2 = 1$.

---

[2] See: *Science,* **1984,** *224,* 590.

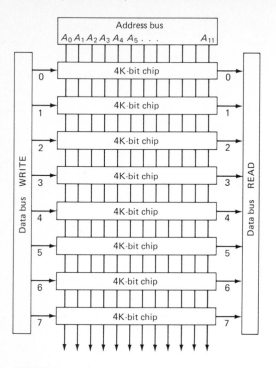

FIGURE 3–5 Organization of a 4 K-byte memory, which will store 4096 8-bit or 1-byte words.

multaneously. Again a 12-line bus is used for the address signal.

The sequence of steps for a write operation with a memory unit such as that in Figure 3–5 is as follows. A digital address is first placed on the address bus followed by the datum on the data bus. The order WRITE is then sent to the memory through the control bus. The memory then transfers the datum on the data bus to the specified memory location. The reverse process occurs for the READ command.

Types of Memories. Two types of memories are contained in most microprocessors and computers, namely *random access memory* (RAM) and *read only memory* (ROM). The term random access is somewhat misleading, however, because read only memories are also capable of random access, a term which indicates that all locations in the memory are equally accessible and can be

reached at about the same speed. Thus, *read/write* memory is a better and more descriptive term for a RAM.

RAM. Several types of RAM devices are available. In the simplest, each storage cell consists of a single tiny capacitor and a bipolar transistor. The latter serves as a binary switch that determines whether the capacitor is charged or discharged by an input signal to give either a 1 or 0 state. The capacitance of each capacitor is remarkably small, being on the order of 5×10^{-14} F. The single transistor device suffers from the disadvantage that its memory is *volatile* (not permanent). That is, the charge on the capacitor is lost whenever the memory unit is read. In addition, the typical capacitor has a significant leakage current, which leads to its discharge in as little as a few milliseconds. Because of these properties, additional circuits must be provided that refresh the memory unit by recharging it each time it is read; a circuit to offset the leakage current is also required. Even with these provisions, memory is lost whenever the power is turned off.

Storage cells with two, three, four, or six transistors offer additional advantages. Some, which employ field-effect rather than bipolar transistors, do not require continuous refreshing and draw so little current that small dry-cells will power them for weeks. Memories of this sort are called static memories and are found in pocket calculators and computers that retain data and instructions even when the device is turned off.

ROM. Read only memories contain permanent instructions and data, which have been placed there at the time of their manufacture. These memories are truly static in the sense that they retain their original states for the life of the computer or calculator. The contents of a ROM cannot be altered by reprogramming. A variant of ROM is the *erasable program read only* memory (EPROM or erasable PROM) in which the program contents can be erased by exposure to ultraviolet radiation. After this treatment, the memory can be reprogrammed by means of special equipment. A recent development are ROMs that can be reprogrammed more easily by electrical signals. They are designated as EAROMs (electrically alterable ROMs).

A ROM cell is simpler and more compact than its RAM counterpart. ROM devices can thus be made with large capacities—typically 8 to 64 K bits. Furthermore, several such devices can be arranged on a single chip in a configuration analogous to that for the several chips shown in Figure 3–5.

In microcomputers and handheld calculator systems, ROM devices are used to store programs needed for performing various mathematical operations, such as obtaining logs, exponentials, and trigonometric functions; performing statistical calculations (means, standard deviations, and least squares); and providing various methods of data presentation (fixed point, scientific or engineering).

Bulk Storage Devices. In addition to semiconductor memories, computers are often equipped with bulk storage devices. The simplest of these is a magnetic tape, which is similar to an audio cassette. One tape can store 10^6 to 10^7 bits with access times of 10 to 100 sec. More sophisticated bulk storage is obtained by various types of magnetic disks. Some of these can store more than 10^8 bits with access times in the 0.1 to 0.5 sec range.

3B–4 INPUT/OUTPUT SYSTEMS

Input/output devices provide the means by which the user (or his instrument) communicates with the computer. Familiar input devices include the electric printer, magnetic tapes or disks, punch cards, and the transduced signals from analytical instruments. Output devices include recorders, printers, cathode-ray tubes (CRT), and meters. It is important to appreciate that many

of these devices provide or use an *analog* signal while, as we have pointed out, the computer can respond only to digital signals. Thus, an important part of the input/output system is an *analog-to-digital converter* (ADC) for providing data in a form that the computer can use, and a *digital-to-analog converter* (DAC) for converting the output from the computer to a usable analog signal (examples of these devices were described in Section 2D–4).

3C Computer Programming

Communication with a computer entails setting an enormous aggregation of switches to appropriate off or on (0 and 1) positions. A program consists of a set of instructions indicating how these switches are to be set for each step in a program. These instructions must be written in a form to which the computer can respond—that is, a binary *machine code*. Machine coding is tedious and time consuming and is often prone to errors. For this reason, *assembly languages* have been developed in which the switch-setting steps are assembled into groups, which can be designated by code words. For example, the abbreviation for subtract might be SUB and might correspond to 101 in machine language. Clearly, SUB is a good deal easier for the programmer to remember than 101.

Assembly programming, while simpler than machine programming, is still difficult and tedious. As a consequence, a number of high-level languages, such as FORTRAN, BASIC, COBOL, APL, PL-1, and PASCAL, have been developed. These languages, which are easily learned, have been designed to make communication with the computer relatively straightforward. Here, instructions in the high-level language are translated by a computer program (called a *compiler*) into machine language, which can then control the computer for computations. Unfortunately, loss of efficiency ac-

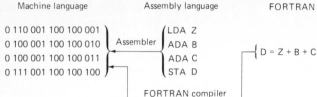

Machine language   Assembly language   FORTRAN

0 110 001 100 100 001	LDA Z
0 100 001 100 100 010	ADA B
0 100 001 100 100 011	ADA C
0 111 001 100 100 100	STA D

D = Z + B + C

FORTRAN compiler

FIGURE 3-6 Relationships among machine, assembly, and a high-level language. (From S. P. Perone, *J. Chromatog. Sci.,* **1969,** *7,* 715. With permission.)

companies the use of higher-level languages.

Figure 3–6 illustrates the application of the three kinds of languages for obtaining a sum.

3D   Applications of Computers

Computer interactions with analytical instruments are of two types, *passive* and *active.* In passive applications, the computer does not participate in the control of the experiment but is used only for data handling, processing, storing, file searching, or display. In an active interaction, the output from the computer controls the sequence of steps required for operation of the instrument. For example, in a spectroscopic determination, the computer may choose the proper source, cause this source to be activated and its intensity adjusted to an appropriate level, cause the radiation to pass through the sample and then a blank, control the monochromator so that a proper wavelength is chosen, adjust the detector response, and record the intensity level. In addition, the computer may be programmed to use the data as it is being collected to vary experimental conditions in such a way as to improve the quality of subsequent data. Instruments with computer control are said to be *automated.*

3D-1   PASSIVE APPLICATIONS

Data processing by a computer may involve relatively simple mathematical operations such as calculation of concentrations, data averaging, least-square analysis, statistical analysis, and integration to obtain peak areas. More complex calculations may involve solution of several simultaneous equations, curve fitting, averaging, and Fourier transformations.

Data storage is another important passive function of computers. For example, a powerful tool for the analysis of complex mixtures results when gas/liquid chromatography (GLC) is linked with mass spectrometry (MS). The former separates mixtures on the basis of the time required for the individual components to appear at the end of suitably packed columns. Mass spectrometry permits identification of each component according to the mass of the fragments formed when the compound is bombarded with a beam of electrons. Equipment for GLC/MS may produce data for as many as 100 spectra in less than an hour, with each spectrum being made up of tens to hundreds of peaks. Conversion of these data to an interpretable form (a graph) in real time is often impossible. Thus, the data are often stored in digital form for subsequent processing and presentation in graphical form.

Identification of a species from its mass spectrum involves a search of files of spectra for pure compounds until a match is found; manually this process is time consuming but can be accomplished quickly by using a computer. Here, the spectra of pure compounds, stored in bulk storage, are searched until spectra are found that are similar to the analyte. Several thousands of spectra can be scanned in a minute or less. Such a search usually produces several possible compounds. Further comparison of spectra by the scientist often makes identification possible. This technique is discussed in more detail in later chapters.

Another important passive application utilizes the high-speed data fetching and correlating capabilities of the computer. Thus, for example, the computer can be called upon to display on a cathode-ray screen the mass spectrum of any one of the components after it has exited from a gas-chromatographic column.

3D-2 ACTIVE APPLICATIONS

In active applications only part of the computer's time is devoted to data collection, the rest being employed for data processing and control. Thus, active applications are real-time operations. Most modern instruments contain one or more microprocessors that perform control functions. Examples would include adjustment of the slit width and wavelength settings of a monochromator, the temperature of a chromatographic column, the potential applied to an electrode, the rate of addition of a reagent, and the time at which the integration of a peak is to begin. Referring again to the GLC/MS instrument considered in the last section, a computer is often used to initiate collection of mass spectral data each time a compound is sensed at the end of the chromatographic column.

Computer control can be relatively simple, as in the examples just cited, or more complex. For example, the determination of the concentration of elements by atomic emission involves the measurement of the heights of emission peaks, which are found at wavelengths that are characteristic for each element. Here, the computer can cause a monochromator to rapidly sweep a range of wavelengths until a peak is detected. The rate of sweep is then slowed to better determine the exact wavelength at which the maximum output signal is obtained. Intensity measurements are made at this point until an average is obtained that gives a suitable signal-to-noise ratio. The computer then causes the instrument to repeat this operation for each peak of interest in the spectrum. Finally, the computer calculates and prints out the concentrations of the elements present.

Because of its great speed, a computer can often control variables more efficiently than can a human operator. Furthermore, with some experiments, a computer can be programmed to alter the way in which the measurement is being made, according to the nature of the initial data. Here, a feedback loop is employed in which the signal output is fed back through the computer and serves to control and optimize the way in which later measurements are performed.

Electromagnetic Radiation and

Its Interaction with Matter

This chapter provides a brief review of the fundamental properties of electromagnetic radiation and the mechanisms by which it interacts with matter. The material serves as an introduction to Chapters 5 through 16.[1]

Electromagnetic radiation is a type of energy that is transmitted through space at enormous velocities. It takes numerous forms, the most easily recognizable being light and radiant heat. Less obvious manifestations include X-ray, ultraviolet, microwave, and radio radiations.

Many of the properties of electromagnetic radiation are conveniently described by means of a classical wave model, which employs such parameters as wavelength, frequency, velocity, and amplitude. In contrast to other wave phenomena, such as sound, electromagnetic radiation requires no supporting medium for its transmission; thus, it readily passes through a vacuum.

The wave model fails to account for phenomena associated with the absorption and emission of radiant energy; for these processes, it is necessary to view electromagnetic radiation as a stream of discrete particles or wave packets of energy called *photons*. The energy of a photon is proportional to the frequency of the radiation. These dual views of radiation as particles and as waves are not mutually exclusive but, rather, complementary. Indeed, the duality is found to apply to the behavior of streams of electrons or other elementary particles such as protons as well and is completely rationalized by wave mechanics.

---

[1] For a further review of these topics, see: E. J. Bair, *Introduction to Chemical Instrumentation,* Chapters 1 and 2, McGraw-Hill: New York, 1962; E. J. Bowen, *The Chemical Aspects of Light,* 2d ed., The Clarendon Press: Oxford, 1946; E. J. Meehan, in *Treatise on Analytical Chemistry,* 2d ed., P. J. Elving, E. J. Meehan, and I. M. Kolthoff, Eds., Part I, vol. 7, Chapter 1, Wiley: New York, 1981; and F. Grum, in *Physical Methods of Chemistry,* A. Weissberger and B. W. Rossiter, Eds., Vol. I, Part IIIB, pp. 214–305, Wiley-Interscience: New York, 1972.

4A Electromagnetic Radiation as Waves

This section deals with the wave description of electromagnetic radiation and the properties that are most easily understood by treating radiation as waves, namely interference, diffraction, transmission, refraction, reflection, scattering, and polarization.

4A-1 WAVE PROPERTIES

For many purposes, electromagnetic radiation is conveniently treated as an electrical field force that oscillates in a direction that is 90 deg to the direction of wave propagation. Associated with the electrical field, and at 90 deg to it and the direction of propagation, is a magnetic field.

Figure 4–1 depicts the simplest form of electromagnetic radiation. Here, the radiation is represented as a series of vectors whose lengths are proportional to the electrical field that is associated with the radiation at any instant. These vectors are shown as being in the plane of the paper and at 90 deg to the direction of propagation of the radiation. The graph then represents the variation in the field strength as a function of time as the radiation passes a fixed point in space. Alternatively, the vector can be plotted as a function of distance, with time held constant.

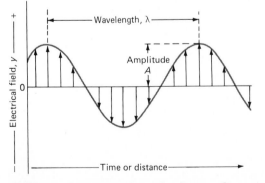

FIGURE 4–1 Representation of a beam of monochromatic, plane-polarized radiation. The arrows represent the electrical vectors.

In Figure 4–1, the electrical vector serves as the ordinate; a plot of the magnetic vector would be identical in every regard except that the ordinate would be rotated 90 deg around the abscissa so that the oscillations are perpendicular to the page. Normally, only the electrical vector is plotted because it is the electrical field that is responsible for such phenomena as transmission, reflection, refraction, and absorption of radiation. The plot in Figure 4–1 is for *monochromatic plane-polarized* radiation. Here, the first term indicates that the radiation has a single frequency; the second implies that it oscillates in but a single plane.

Radiation that is monochromatic and plane polarized can only be approximated in the laboratory.

Wave Parameters. Referring to Figure 4–1, the time required for the passage of successive maxima through a fixed point in space is called the *period p* of the radiation. The *frequency ν* is the number of oscillations of the field that occurs per second[2] and is equal to $1/p$. It is important to realize that *the frequency is determined by the source and remains invariant* regardless of the media traversed by the radiation. In contrast, the *velocity* of propagation v_i, the rate at which the wave front moves through a medium, is *dependent* upon both the composition of the medium and the frequency; the subscript i is sometimes employed to indicate this frequency dependence. Another parameter of interest is the *wavelength* λ_i, which is the linear distance between successive maxima or minima of a wave.[3] Multiplication of the frequency

---

[2] The common unit of frequency is the *hertz* Hz, which is equal to one cycle per second.

[3] The units commonly used for describing wavelength differ considerably in the various spectral regions. For example, the ångström unit, Å (10^{-10} m), is convenient for X-ray and short ultraviolet radiation; the nanometer, nm (10^{-9} m), is employed with visible and ultraviolet radiation; the micrometer, μm (10^{-6} m), is useful for the infrared region.

in cycles per second by the wavelength in centimeters per cycle gives the velocity of propagation in centimeters per second; that is,

$$v_i = \nu\lambda_i \qquad (4-1)$$

In a vacuum, the velocity of radiation becomes independent of frequency and is at its maximum. This velocity, given the symbol c, has been accurately determined to be 2.99792×10^{10} cm/sec. Thus, for a vacuum,

$$c = \nu\lambda \cong 3.00 \times 10^{10} \text{ cm/sec} \qquad (4-2)$$

In any medium containing matter, the rate of propagation is less because of interactions between the electromagnetic field of the radiation and the bound electrons in the atoms or molecules present. Since the radiant frequency is invariant and fixed by the source, the *wavelength must decrease* as radiation passes from a vacuum to some other medium (Equation 4–1). This effect is illustrated in Figure 4–2 for a beam of visible radiation. Note that the wavelength shortens nearly 200 nm as it passes into glass; a reverse change occurs as the radiation again enters air.

It is also noteworthy that the velocity of radiation in air differs only slightly from c (about 0.03% less); thus, except for the most precise calculations, Equation 4–2 is applicable to air as well as to a vacuum.

The *wavenumber* σ is defined as the number of waves per centimeter and is yet another way of describing electromagnetic radiation. When the wavelength *in vacuo* is expressed in centimeters, the wavenumber is equal to $(1/\lambda)$ cm^{-1}.

Radiant Power or Intensity.   The *power P* of radiation is the energy of the beam reaching a given area per second; the *intensity I* is the power per unit solid angle. These quantities are related to the square of the amplitude A (see Figure 4–1). Although it is not strictly correct to do so, power and intensity are often used synonymously.

Mathematical Description of a Wave.   With time as a variable, the wave in Figure 4–1 can be described by an equation analogous to Equation A2–4 (Appendix 2). That is,

$$y = A\sin(\omega t + \phi) \qquad (4-3)$$

where y is the electric field (Figure 4–1), A is the *amplitude* or maximum value for y, t is time, and ϕ is the phase angle, a term which is defined in Section A2–1 in Appendix 2. The angular velocity of the vector, ω, is related to the frequency of the radiation ν by the equation

$$\omega = 2\pi\nu$$

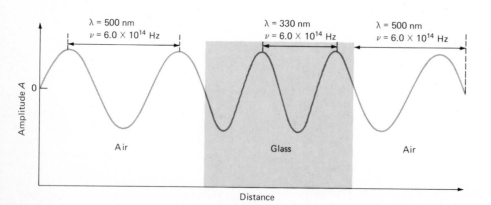

FIGURE 4-2   Effect of change of medium on a monochromatic beam of radiation.

Substitution of this relationship into Equation 4–3 yields

$$y = A \sin(2\pi\nu t + \phi) \qquad (4-4)$$

Superposition of Waves. *The principle of superposition* states that when two or more waves traverse the same space, a displacement occurs which is the sum of the displacements caused by the individual waves. This principle applies to electromagnetic waves, where the displacements involve an electrical force field, as well as to several other types of waves, where atoms or molecules are displaced. When n electromagnetic waves of the same frequency but differing amplitudes and phase angles pass some point in space simultaneously, the principle of superposition and Equation 4–4 permits us to write

$$y = A_1 \sin(2\pi\nu t + \phi_1) + A_2 \sin(2\pi\nu t + \phi_2) \\ + \cdots + A_n \sin(2\pi\nu t + \phi_n) \quad (4-5)$$

where y is the resultant force field.

The solid line in Figure 4–3a shows the application of Equation 4–5 to two waves of identical frequency but somewhat different amplitude and phase angle. Note that the resultant is a sine wave with the same frequency but larger amplitude than either of the component waves. Figure 4–3b differs from 4–3a in that the difference in phase angle is greater; here, the resultant ampli-

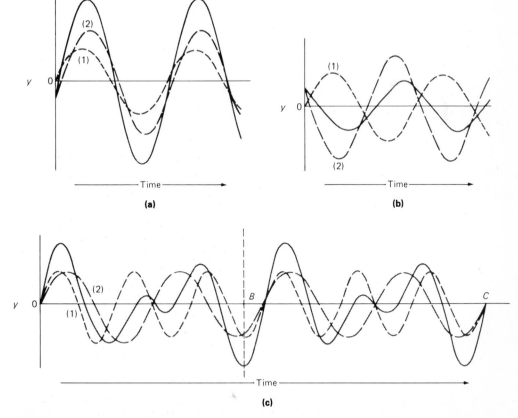

FIGURE 4-3 Superposition of sinusoidal waves: (**a**) $A_1 < A_2$, $(\phi_1 - \phi_2) = -20°$, $v_1 = v_2$; (**b**) $A_1 < A_2$, $(\phi_1 - \phi_2) = -200°$, $v_1 = v_2$; (**c**) $A_1 = A_2$, $\phi_1 = \phi_2$, $v_1 = 1.5\, v_2$. In each instance, the solid curve is the resultant from combination of the two dashed curves.

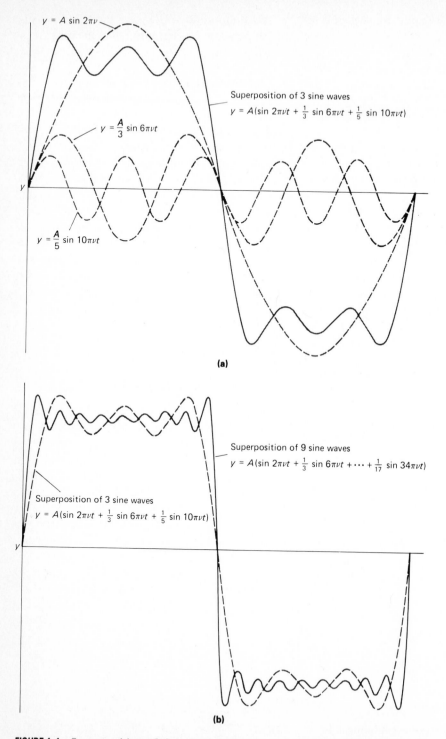

FIGURE 4–4   Superposition of sine waves to form a square wave: (**a**) combination of three sine waves, and (**b**) combination of three, as in (**a**), and nine sine waves.

tude is smaller than those of the component waves. Clearly, a maximum amplitude for the resultant will occur when the two waves are completely in phase; this situation prevails whenever the phase difference between waves ($\phi_1 - \phi_2$) is 0 deg, 360 deg, or an integer multiple of 360 deg. Under these circumstances, maximum *constructive interference* is said to occur. A maximum *destructive interference* occurs when ($\phi_1 - \phi_2$) is equal to 180 deg or 180 deg plus an integer multiple of 360 deg. The property of interference plays an important role in many instrumental methods based on electromagnetic radiation.

Figure 4–3c depicts the superposition of two waves with the same amplitude but different frequency. The resulting wave is no longer sinusoidal; it does, however, exhibit a periodicity. Thus, the wave from 0 to B in the figure is identical to that from B to C.

An important aspect of superposition is that the calculation can be reversed by a mathematical operation called a *Fourier transformation*. Jean Fourier, an early French mathematician (1768–1830), demonstrated that any wave motion, regardless of complexity, can be described by a sum of simple sine or cosine terms such as those shown in Equation 4–5. For example, the square wave form widely encountered in electronics can be represented by an equation having the form

$$y = A(\sin 2\pi v t + \frac{1}{3} \sin 6\pi v t + \frac{1}{5} \sin 10\pi v t$$

$$+ \cdots + \frac{1}{n} \sin 2n\pi v t) \quad (4\text{--}6)$$

where the n takes values of 7, 9, 11, 13, and so forth. A graphical representation of the summation process is shown in Figure 4–4. The solid line in Figure 4–4a is the sum of three sine waves differing in amplitude in the ratio of 5 : 3 : 1 and in frequency in the ratio of 1 : 3 : 5. Note that the resultant is already beginning to approximate the shape of a square wave. As shown by the solid line in Figure 4–4b, the resultant more closely approaches a square wave when nine waves are incorporated.

Mathematically, resolution of a complex wave form into its sine or cosine components is tedious and time consuming; modern computers, however, have made it practical to exploit the power of the Fourier transformation on a routine basis. The application of this technique will be considered in the discussion of several types of spectroscopy.

4A-2 THE ELECTROMAGNETIC SPECTRUM

As shown in Figure 4–5, the electromagnetic spectrum encompasses an enormous range of wavelengths and frequencies. Note that the ranges are so great that logarithmic scales are required. The figure also depicts qualitatively the major spectral regions. The divisions are based upon the methods required to generate and detect the various kinds of radiation. Several overlaps are evident. Note that the visible portion of the spectrum to which the human eye is sensitive is tiny when compared with other spectral regions.

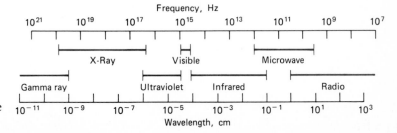

FIGURE 4-5 Regions of the electromagnetic spectrum.

Spectroscopy is a branch of science that involves the study of electromagnetic radiation and its applications. Table 4–1 lists the wavelength and frequency ranges for the regions of the spectrum that are important for analytical purposes and also gives the names of the various spectroscopic methods associated with each. The last column of the table lists the types of nuclear, atomic, or molecular quantum transitions of the analyte that serve as the bases for the procedure.

4A-3 DIFFRACTION OF RADIATION

All types of electromagnetic radiation exhibit *diffraction*, a process in which a parallel beam of radiation is bent as it passes by a sharp barrier or through a narrow opening. Figure 4–6 illustrates the process. Diffraction is a wave property, which can be observed not only for electromagnetic radia-

tion but also for mechanical or acoustical waves. For example, diffraction is readily demonstrated in the laboratory by mechanically generating waves of constant frequency in a tank of water and observing the wave crests before and after they pass through a rectangular opening or slit. When the slit is wide relative to the wavelength of the motion (Figure 4–6**a**), diffraction is slight and difficult to detect. On the other hand, when the wavelength and the slit opening are of the same order of magnitude, as in Figure 4–6**b**, diffraction becomes pronounced. Here, the slit or opening behaves as a new source from which waves radiate in a series of nearly 180-deg arcs. Thus, the direction of the wave front appears to bend as a consequence of passing the two edges of the slit.

Diffraction is a consequence of interference. This relationship is most easily understood by considering an experiment, per-

Table 4-1
Common Spectroscopic Methods

Type Spectroscopy	Usual Wavelength Range[a]	Usual Frequency Range, Hz	Type Quantum Transition
Gamma ray emission	<0.1 Å	$>3 \times 10^{19}$	Nuclear
X-Ray absorption emission, fluorescence, and diffraction	$0.1 - 100$ Å	$3 \times 10^{19} - 3 \times 10^{16}$	Inner electron
Vacuum ultraviolet absorption	$5 - 180$ nm	$6 \times 10^{16} - 2 \times 10^{15}$	Bonding electrons
Ultraviolet-visible absorption, emission, and fluorescence	$180 - 780$ nm	$2 \times 10^{15} - 4 \times 10^{14}$	Bonding electrons
Infrared absorption and Raman scattering	$0.78 - 300$ μm	$4 \times 10^{14} - 1 \times 10^{12}$	Rotation/vibration of molecules
Microwave absorption	$0.75 - 3.75$ mm	$4 \times 10^{11} - 8 \times 10^{10}$	Rotation of molecules
Electron spin resonance	3 cm	1×10^{10}	Spin of electrons in a magnetic field
Nuclear magnetic resonance	$0.6 - 10$ m	$5 \times 10^{8} - 3 \times 10^{7}$	Spin of nuclei in a magnetic field

[a] 1 Å $= 10^{-10}$ m $= 10^{-8}$ cm
  1 nm $= 10^{-9}$ m $= 10^{-7}$ cm
  1 μm $= 10^{-6}$ m $= 10^{-4}$ cm

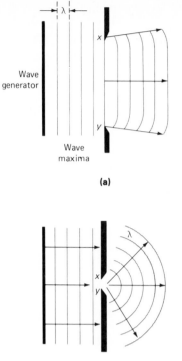

FIGURE 4-6 Propagation of waves through a slit: (**a**) $xy \gg \lambda$; (**b**) $xy \cong \lambda$.

formed first by Thomas Young in 1800, by which the wave nature of light was unambiguously demonstrated. As shown in Figure 4–7, a parallel beam of radiation is allowed to pass through a narrow slit A (or in Young's experiment, a pin-hole) whereupon it is diffracted and illuminates more or less equally two closely spaced slits or pinholes B and C; the radiation emerging from these slits is then observed on the screen lying in a plane XY. If the radiation is monochromatic, a series of dark and light images perpendicular to the plane of the page is observed.

Figure 4–7**b** shows the intensities of the various bands reaching the screen. If, as in this diagram, the slit widths approach the wavelength of radiation, the band intensities

decrease only gradually with increasing distances from the central band. With wider slits, the decrease is much more pronounced.

The existence of the central band E, which lies in the shadow of the opaque material separating the two slits, is readily explained by noting that the paths from B to E and C to E are identical. Thus, constructive interference of the diffracted rays from the two slits occurs, and an intense band is observed. With the aid of Figure 4–7**c**, the conditions for maximum constructive interference, which result in the other light bands, are readily derived. The angle of diffraction θ is the angle from the normal, formed by the dotted line extending from a point O, halfway between the slits, to the point of maximum intensity D. The solid lines BD and CD represent the light paths from the slits B and C to this point. Ordinarily, the distance $\overline{OE}$ is enormous compared to the distance between the slits $\overline{BC}$; as a consequence, the lines BD, OD, and CD are, for all practical purposes, parallel. Line BF is perpendicular to CD and forms the triangle BCF, which is, to a close approximation, similar to DOE; consequently, the angle CBF is equal to the angle of diffraction θ. We may then write

$$\overline{CF} = \overline{BC} \sin \theta$$

Because $\overline{BC}$ is so very small compared to $\overline{OE}$, $\overline{FD}$ closely approximates $\overline{BD}$, and the distance $\overline{CF}$ is a good measure of the difference in path lengths of beams $\overline{BD}$ and $\overline{CD}$. For the two beams to be in phase at D, it is necessary that $\overline{CF}$ correspond to the wavelength of the radiation; that is,

$$\lambda = \overline{CF} = \overline{BC} \sin \theta$$

Reinforcement would also occur when the additional path length corresponds to 2λ, 3λ, and so forth. Thus, a more general ex-

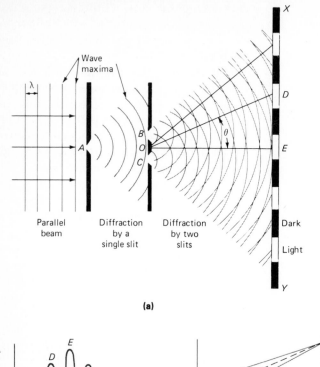

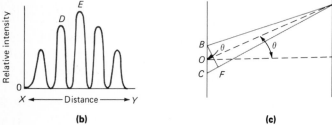

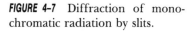

FIGURE 4-7 Diffraction of mono-chromatic radiation by slits.

pression for the light bands surrounding the central band is

$$n\lambda = \overline{BC} \sin \theta \tag{4-7}$$

where **n** is an integer called the *order* of interference.

The linear displacement $\overline{DE}$ of the diffracted beam along the plane of the screen is a function of the distance $\overline{OE}$ between screen and the plane of the slits, as well as the spacing between the slits $\overline{BC}$; that is,

$$\overline{DE} = \overline{OD} \sin \theta$$

Substitution into Equation 4–7 gives

$$n\lambda = \frac{\overline{BC}\,\overline{DE}}{\overline{OD}} \cong \frac{\overline{BC}\,\overline{DE}}{\overline{OE}} \tag{4-8}$$

Equation 4–8 permits the calculation of the wavelength from the three measurable quantities.

EXAMPLE 4-1. Suppose that the screen in Figure 4–7 is 2.00 m from the plane of the slits and that the slit spacing is 0.300 mm. What is the wavelength of radiation if the fourth band is located 15.4 mm from the central band?

Substituting into Equation 4–8 gives

$$4\lambda = \frac{0.300 \times 15.4}{2 \times 1000} \text{ mm}$$

$$\lambda = 5.78 \times 10^{-4} \text{ mm} \quad \text{or} \quad 578 \text{ nm}$$

4A–4 COHERENT RADIATION

In order to produce a diffraction pattern such as that shown in Figure 4–7**a**, it is necessary that the electromagnetic waves traveling from slits B and C to any given point on the screen (such as D or E) have sharply defined phase differences that remain entirely constant with time; that is, the radiation from slits B and C must be *coherent*. The conditions for coherence are: (1) the two sources of radiation must have identical frequency and wavelength (or sets of frequencies and wavelengths); and (2) the phase relationships between the two beams must remain constant with time. The necessity for these requirements can be demonstrated by illuminating the two slits in Figure 4–7**a** with individual filament lamps. Under these circumstances, the well-defined light and dark patterns disappear and are replaced by a more or less uniform illumination of the screen. This behavior is a consequence of the *incoherent* character of filament sources (many other sources of electromagnetic radiation are incoherent as well).

In incoherent sources, light is emitted by individual atoms or molecules, and the resulting beam is the summation of countless individual events, each of which lasts on the order of 10^{-8} sec. Thus, a beam of radiation from this type of source is not continuous, as is the case in microwave or laser radiation; instead, it is composed of a series of *wave trains* that are a few meters in length at most. Because the processes that produce a train are random, the phase differences among the trains must also be variable. A wave train from slit B may arrive at a point on the screen in phase with a wave train from C so that constructive interference occurs; an instant later, the trains may be totally out of phase at the same point, and destructive interference occurs. Thus, the radiation at all points on the screen is governed by the random phase variations among the wave trains; uniform illumination, which represents an average for the trains, is the result.

Sources do exist which produce electromagnetic radiation in the form of trains with essentially infinite length and constant frequency. Examples include radio-frequency oscillators, microwave sources, and optical lasers. Various mechanical sources, such as a two-pronged vibrating tapper in a water-containing riffle tank, produce a mechanical analog of coherent radiation. When two coherent sources are substituted for slit A in the experiment shown in Figure 4–7, a regular diffraction pattern is observed.

Diffraction patterns can be obtained from random sources, such as a tungsten filament, provided that an arrangement similar to that shown in Figure 4–7**a** is employed. Here, the very narrow slit A assures that the radiation reaching B and C emanates from the same small region of the source. Under these circumstances, the various wave trains exiting from slits B and C have a constant set of frequencies and phase relationships to one another and are thus coherent. If the slit at A is widened so that a larger part of the source is sampled, the diffraction pattern becomes less pronounced because the two beams are only partially coherent. If slit A is made sufficiently wide, the incoherence may become great enough to produce only a constant illumination across the screen.

4A–5 TRANSMISSION AND REFRACTION OF RADIATION

It is observed experimentally that the rate at which radiation is propagated through a transparent substance is less than its velocity in a vacuum; furthermore, the rate depends upon the kinds and concentrations of atoms, ions, or molecules in the medium. It follows from these observations that the radiation must interact in some way with the matter. Because a frequency change is not observed, however, the interaction *cannot* involve a permanent energy transfer.

The *refractive index* of a medium is a measure of its interaction with radiation and is defined by

$$n_i = \frac{c}{v_i} \qquad (4-9)$$

where n_i is the refractive index at a specified frequency i, v_i is the velocity of the radiation in the medium, and c is its velocity *in vacuo*.

The interaction involved in the transmission process can be ascribed to the alternating electrical field of the radiation, which causes the bound electrons of the particles contained in the medium to oscillate with respect to their heavy (and essentially fixed) nuclei; periodic polarization of the particles thus results. Provided the radiation is not absorbed, the energy required for polarization is only momentarily retained (10^{-14} to 10^{-15} sec) by the species and is reemitted without alteration as the substance returns to its original state. Since there is no net energy change in this process, the frequency of the emitted radiation is unchanged, but the rate of its propagation has been slowed by the time required for retention and reemission to occur. Thus, transmission through a medium can be viewed as a stepwise process involving polarized atoms, ions, or molecules as intermediates.

One would expect the radiation from each polarized particle in a medium to be emitted in all directions. If the particles are small, however, it can be shown that destructive interference prevents the propagation of significant amounts in any direction other than that of the original light path. On the other hand, if the medium contains large particles (such as polymer molecules or colloidal particles), this destructive effect

is incomplete and a portion of the beam is *scattered* as a consequence of the interaction step. Scattering is considered in a later section of this chapter.

Dispersion. As we have noted, the velocity of radiation in matter is frequency-dependent; since c in Equation 4–9 is independent of this parameter, the refractive index of a substance must also change with frequency. The variation in refractive index of a substance with frequency or wavelength is called its *dispersion*. The dispersion of a typical substance is shown in Figure 4–8. Clearly, the relationship is complex; generally, however, dispersion plots exhibit two types of regions. In the *normal dispersion* region, there is a gradual increase in refractive index with increasing frequency (or decreasing wavelength). *Anomalous dispersion* regions are those frequency ranges in which a sharp change in refractive index is observed. Anomalous dispersion always occurs at frequencies that correspond to the natural harmonic frequency associated with some part of the molecule, atom, or ion of the substance. At such a frequency, permanent energy transfer from the radiation to

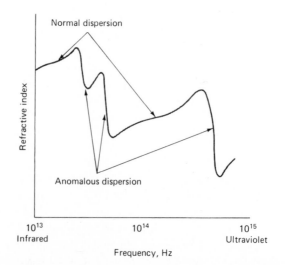

FIGURE 4–8 Typical dispersion curve.

the substance occurs and *absorption* of the beam is observed. Absorption is discussed in a later section.

Dispersion curves are important in the choice of materials for the optical components of instruments. A substance that exhibits normal dispersion over the wavelength region of interest is most suitable for the manufacture of lenses, for which a high and relatively constant refractive index is desirable. Chromatic aberration is minimized through the choice of such a material. In contrast, a substance with a refractive index that is not only large but also highly frequency-dependent is selected for the fabrication of prisms. The applicable wavelength region for the prism thus approaches the anomalous dispersion region for the material from which it is fabricated.

Refraction of Radiation. When radiation passes from one medium to another of differing physical density, an abrupt change in direction of the beam is observed as a consequence of differences in the velocity of the radiation in the two media. This *refraction* of a beam is illustrated in Figure 4–9. The extent of refraction is given by the relationship

$$\frac{\sin \theta_1}{\sin \theta_2} = \frac{n_2}{n_1} = \frac{v_1}{v_2} \qquad (4-10)$$

If the medium M_1 is vacuum, v_1 becomes c and n_1 is unity (Equation 4–9); thus, the refractive index n_2 of M_2 is simply the ratio of the sines of the two angles.

4A-6 REFLECTION AND SCATTERING OF RADIATION

Reflection. When radiation crosses an interface between media with differing refractive indexes, reflection occurs. The fraction reflected becomes larger with increasing differences in refractive index; for a beam traveling normally to the interface, the fraction reflected is given by

$$\frac{I_r}{I_0} = \frac{(n_2 - n_1)^2}{(n_2 + n_1)^2} \qquad (4-11)$$

where I_0 is the intensity of the incident beam and I_r is the reflected intensity; n_1 and n_2 are the refractive indexes of the two media.

EXAMPLE 4-2. Calculate the percent loss of intensity due to reflection of a perpendicular beam of yellow light as it passes through a glass cell containing water. Assume that for yellow radiation the refractive index of glass is 1.50, of water is 1.33, and of air is 1.00.

The total reflective loss will be the sum of the losses occurring at each of the interfaces. For the first interface (air to glass), we can write

$$\frac{I_{r1}}{I_0} = \frac{(1.50 - 1.00)^2}{(1.50 + 1.00)^2} = 0.040$$

$$I_{r1} = 0.040 \, I_0$$

The beam intensity is reduced to $(I_0 - 0.040 \, I_0) = 0.960 \, I_0$. Reflection loss at the

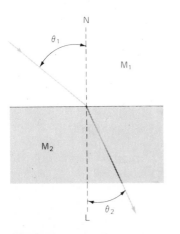

FIGURE 4-9 Refraction of light in passing from a less dense medium M_1 into a more dense medium M_2, where its velocity is lower.

glass-to-water interface is then given by

$$\frac{I_{r2}}{0.960\,I_0} = \frac{(1.50 - 1.33)^2}{(1.50 + 1.33)^2} = 0.0036$$

$$I_{r2} = 0.0035\,I_0$$

The beam intensity is further reduced to $(0.960\,I_0 - 0.0035\,I_0) = 0.957\,I_0$. At the water-glass interface

$$\frac{I_{r3}}{0.957\,I_0} = \frac{(1.50 - 1.33)^2}{(1.50 + 1.33)^2} = 0.0036$$

$$I_{r3} = 0.0034\,I_0$$

and the beam intensity becomes $0.954\,I_0$. Finally, the reflection at the second glass-to-air interface will be

$$\frac{I_{r4}}{0.954\,I_0} = \frac{(1.50 - 1.00)^2}{(1.50 + 1.00)^2} = 0.040$$

$$I_{r4} = 0.038\,I_0$$

The total reflection loss I_{rt} is

$$I_{rt} = 0.040\,I_0$$
$$+\ 0.0035\,I_0 + 0.0034\,I_0 + 0.038\,I_0$$

and

$$\frac{I_{rt}}{I_0} = 0.849 \quad \text{or} \quad 8.5\%$$

It will become evident in later chapters that losses such as those shown in this example are of considerable significance in various optical instruments.

Reflective losses at a polished glass or quartz surface increase only slightly as the angle of the incident beam increases up to about 60 deg. Beyond this figure, however, the percentage of radiation that is reflected increases rapidly and approaches 100% at 90 deg or grazing incidence.

Scattering. As noted earlier, the transmission of radiation in matter can be pictured as a momentary retention of the radiant energy, which causes a brief polarization of the ions, atoms, or molecules present. Polarization is followed by reemission of radiation in all directions as the particles return to their original state. When the particles are small with respect to the wavelength of the radiation, destructive interference removes nearly all of the reemitted radiation except that which travels in the original direction of the beam; the path of the beam appears to be unaltered as a consequence of the interaction. Careful observation, however, reveals that a very small fraction of the radiation is transmitted at all angles from the original path and that the intensity of this *scattered radiation* increases with particle size. With particles of colloidal dimensions, scattering becomes sufficiently intense to be seen by the naked eye (the Tyndall effect).

Scattering by molecules or aggregates of molecules with dimensions significantly smaller than the wavelength of radiation is called *Rayleigh scattering;* its intensity is readily related to wavelength (an inverse fourth-power effect), the dimensions of the scattering particles, and their polarizability. An everyday manifestation of scattering is the blueness of the sky, which results from the greater scattering of the shorter wavelengths of the visible spectrum.

Scattering by larger particles, as in a colloidal suspension, is much more difficult to treat theoretically; the intensity varies roughly as the inverse square of wavelength.

Measurements of scattered radiation can be used to determine the size and shape of polymer molecules and colloidal particles. The phenomenon is also utilized in *nephelometry,* an analytical method considered in Chapter 13.

Raman Scattering. The Raman effect differs from ordinary scattering in that part of the scattered radiation suffers quantized frequency changes. These changes are the result of vibrational energy level transitions occurring in the molecule as a consequence of the polarization process. Raman spectroscopy is discussed in Chapter 12.

4A-7 POLARIZATION

Ordinary radiation consists of a bundle of electromagnetic waves in which the amplitude of vibrations is equally distributed among a series of planes centered along the path of the beam. Viewed end on, a beam of monochromatic light can be visualized as an infinite set of electrical vectors that fluctuate in length from zero to some maximum amplitude A. Figure 4–10**b** depicts these vectors as they can be envisaged at various time intervals during the passage of one wavelength of radiation.

Figure 4–11**a** shows a few of the vectors at the instant that the wave is at its maximum. The vector in any one plane, say *XY,* can be resolved into two mutually perpendicular components *AB* and *CD* as shown in Figure 4–11**b.** If the two components for each plane are combined, the resultant has the appearance shown in Figure 4–11**c.** Note that Figure 4–11**c** has a different scale from Figure 4–11**a** or 4–11**b** to keep its size within reason.

Removal of one of the two resultant

planes of vibration in Figure 4–11**c** produces a beam that is *plane polarized.* The vibration of the electrical vector of a plane-polarized beam, then, occupies a single plane in space. Figure 4–10**c** shows an end-on view of a beam of plane-polarized radiation after various time intervals.

Plane-polarized electromagnetic radiation is produced by certain radiant energy sources. For example, the radio waves emanating from an antenna commonly have this characteristic. Presumably, the radiation from a single atom or molecule is also polarized; however, since common light sources contain large numbers of these particles in all orientations, the resultant is a beam that vibrates equally in all directions around the axis of travel.

The absorption of radiation by certain types of matter is dependent upon the plane of polarization of the radiation. For example, when properly oriented to a beam, anisotropic crystals selectively absorb radiations vibrating in one plane but not in the other. Thus, a layer of anisotropic crystals (see Section 13C–1 for a description of anisotropic substances) absorbs all of the com-

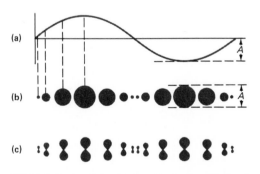

FIGURE 4-10   Unpolarized and plane-polarized radiation: (**a**) cross-sectional view of a beam of monochromatic radiation, (**b**) successive end-on view of the radiation in *A* if it is unpolarized, and (**c**) successive end-on views of the radiation of *A* if it is plane-polarized on the vertical axis.

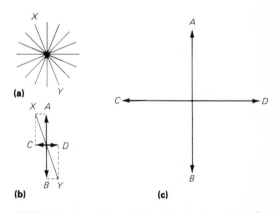

FIGURE 4-11   (**a**) A few of the electrical vectors of a beam traveling perpendicular to the page. (**b**) The resolution of a vector in plane *XY* into two mutually perpendicular components. (**c**) The resultant when all vectors are resolved (not to scale).

ponents of, say, *CD* in Figure 4–11c and transmits nearly completely the components of *AB*. A polarizing sheet, regardless of orientation, removes approximately half of the radiation from an unpolarized beam, and transmits the other half as a plane-polarized beam. The plane of polarization of the transmitted beam is dependent upon the orientation of the sheet with respect to the incident beam. When two polarizing sheets, oriented at 90 deg to one another, are placed perpendicular to the beam path, essentially no radiation is transmitted. Rotation of one results in a continuous increase in transmission until a maximum is reached when molecules of the two sheets have the same alignment.

The way in which radiation is reflected, scattered, transmitted, and refracted by certain substances is also dependent upon direction of polarization. As a consequence, a group of analytical methods have been developed whose selectivity is based upon the interaction of these substances with light of a particular polarization. These methods are considered in detail in Section 13C.

4B   Quantum-Mechanical Properties of Electromagnetic Radiation

When electromagnetic radiation is absorbed or emitted, a permanent transfer of energy to the absorbing medium or from the emitting object occurs. In order to describe these phenomena, it is necessary to treat electromagnetic radiation not as bundles of waves but instead as a stream of discrete particles called *photons* or *quanta*. The energy of a photon depends upon the frequency of the radiation as given by the well-known equation

$$E = h\nu \qquad\qquad (4\text{--}12)$$

where h is Planck's constant (6.63×10^{-27} erg sec). In terms of wavelength and

wavenumber

$$E = hc/\lambda = hc\sigma$$

Note that both frequency and wavenumber are directly proportional to energy. Thus, for quantum calculations, these parameters are more easily used than is wavelength. On the other hand, most instruments for studying electromagnetic radiation are based upon their wave properties; consequently, wavelength finds wider use in the instrumental laboratory.

The tables inside of the end pages of this text provide conversion factors among the various units for wavelength, frequency, and energy.

EXAMPLE 4-3.   Calculate the energy of (a) a 5.3 Å X-ray beam and (b) a 530 nm beam of visible radiation.

(a) The energy of radiation in the X-ray region is commonly expressed in electron volts, the energy acquired by an electron being accelerated through a potential of one volt. In the conversion table inside the front cover of the book, we see that Planck's constant can be converted from the unit erg sec to the unit eV sec by multiplication by 6.2×10^{11}. Thus,

$$\begin{aligned} h &= 6.6 \times 10^{-27} \times 6.2 \times 10^{11} \\ &= 4.1 \times 10^{-15} \text{ eV sec} \end{aligned}$$

The frequency of the X-radiation is given by

$$\begin{aligned} \nu &= \frac{c}{\lambda} \\ &= \frac{3.0 \times 10^{10} \text{ cm sec}^{-1}}{5.3 \text{ Å} \times 10^{-8} \text{ cm/Å}} \\ &= 5.7 \times 10^{17} \text{ sec}^{-1} \end{aligned}$$

Then,

$$E = h\nu$$

$$= 4.1 \times 10^{-15} \text{ eV sec} \times \frac{5.7 \times 10^{17}}{\text{sec}}$$

$$= 2.3 \times 10^3 \text{ eV}$$

(b) Here,

$$\nu = \frac{3.0 \times 10^{10} \text{ cm sec}^{-1}}{530 \text{ nm} \times 10^{-7} \text{ nm/cm}}$$

$$= \frac{5.7 \times 10^{14}}{\text{sec}}$$

$$E = 4.1 \times 10^{-15} \text{ eV sec} \times \frac{5.7 \times 10^{14}}{\text{sec}}$$

$$= 2.3 \text{ eV}$$

4B-1 THE PHOTOELECTRIC EFFECT

The need for a quantum model to describe the behavior of electromagnetic radiation can be seen by consideration of the *photoelectric effect*. When sufficiently energetic radiation impinges on a metallic surface, electrons are emitted. The energy of the emitted electrons is found to be related to the frequency of the incident radiation by the equation

$$E = h\nu - w \qquad (4\text{--}13)$$

where w, the *work function*, is the work required to remove the electron from the metal to a vacuum. While E is directly *dependent upon the frequency*, it is found to be totally *independent of the intensity* of the beam; an increase in intensity merely causes an increase in the *number of electrons emitted* with energy E.

Calculations indicate that no single electron could acquire sufficient energy for ejection if the radiation striking the metal were uniformly distributed over the surface; nor could any electron accumulate enough energy for its removal in a reasonable length of time. Thus, it is necessary to assume that the energy is not uniformly distributed over the beam front, but rather is concentrated at certain points or in packets of energy.

The work w required to cause emission of electrons is characteristic of the metal. The alkali metals possess low work functions and emit electrons when exposed to radiation in the visible region. Metals to the right of the alkali metals in the periodic chart have larger work functions and require the more energetic ultraviolet radiation to exhibit the photoelectric effect. As we shall note in later chapters, the photoelectric effect has great practical importance in the detection of radiation using phototubes.

4B-2 ABSORPTION OF RADIATION

When radiation passes through a layer of solid, liquid, or gas, certain frequencies may be selectively removed by the process of *absorption*. Here, electromagnetic energy is transferred to the atoms or molecules constituting the sample; as a result, these particles are promoted from a lower-energy state to higher-energy states, or *excited states*. At room temperature, most substances are in their lowest energy or *ground state*. Absorption then ordinarily involves a transition from the ground state to higher-energy excited states.

Atoms, molecules, or ions have only a limited number of discrete, quantized energy levels; for absorption of radiation to occur, the energy of the exciting photon must exactly match the energy difference between the ground state and one of the excited states of the absorbing species. Since these energy differences are unique for each species, a study of the frequencies of absorbed radiation provides a means of characterizing the constituents of a sample of matter. For this purpose, a plot of absorbance as a function of wavelength or frequency is experimentally derived (absorbance, a measure of the decrease in radiant power, is defined in Section 6A–2). Typical *absorption spectra* are shown in Figure 4–12.

The general appearance of an absorption spectrum will depend upon the complexity, the physical state, and the environment of the absorbing species. It is convenient to recognize two types of spectra, namely, those associated with atomic absorption and those resulting from molecular absorption.

Atomic Absorption.   The passage of polychromatic ultraviolet or visible radiation through a medium consisting of monatomic particles, such as gaseous mercury or sodium, results in the absorption of but a few well-defined frequencies (see Figure 4–12a). The relative simplicity of such spectra is due to the small number of possible energy states for the particles. Excitation can occur only

by an electronic process in which one or more of the electrons of the atom is raised to a higher-energy level. Thus, with sodium, excitation of the 3s electron to the 3p state requires energy corresponding to a wavenumber of 1.697×10^4 cm^{-1}. As a result, sodium vapor exhibits a sharp absorption peak at 589.3 nm (yellow light). Several other narrow absorption lines, corresponding to other permitted electronic transitions, are also observed. For example, the peak at about 285 nm in Figure 4–12**a** results from the excitation of the 3s electron in sodium to the excited 5p state, a process requiring significantly greater energy than excitation to the 3p state.

Ultraviolet and visible wavelengths have sufficient energy to cause transitions of the outermost or bonding electrons only. X-Ray frequencies, on the other hand, are several orders of magnitude more energetic and are capable of interacting with electrons that are closest to the nuclei of atoms. Absorption peaks corresponding to electronic transitions of these innermost electrons are thus observed in the X-ray region.

Regardless of the wavelength region involved, atomic absorption spectra typically consist of a limited number of narrow peaks. Spectra of this type are discussed in connection with X-ray absorption (Chapter 15) and atomic absorption spectroscopy (Chapter 9).

Molecular Absorption.   Absorption by polyatomic molecules, particularly in the condensed state, is a considerably more complex process because the number of energy states is greatly enhanced. Here, the total energy of a molecule is given by

$$E = E_{\text{electronic}} + E_{\text{vibrational}} + E_{\text{rotational}} \qquad (4\text{–}14)$$

where $E_{\text{electronic}}$ describes the electronic energy of the molecule, and $E_{\text{vibrational}}$ refers to the energy of the molecule resulting from various atomic vibrations. The third term in Equation 4–14 accounts for the energy asso-

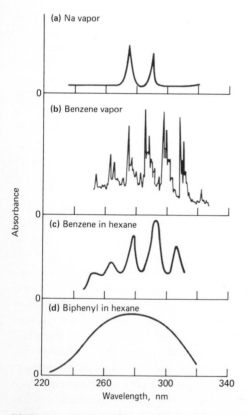

FIGURE 4-12   Some typical ultraviolet absorption spectra.

ciated with the rotation of the molecule about its center of gravity. For each electronic energy state of the molecule, there normally exist several possible vibrational states, and for each of these, in turn, numerous rotational states. As a consequence, the number of possible energy levels for a molecule is much greater than for an atomic particle.

Figure 4–13 is a graphical representation of the energies associated with a few of the electronic and vibrational states of a molecule. The heavy line labeled E_0 represents the electronic energy of the molecule in its ground state (its state of lowest electronic

energy); the lines labeled E_1 and E_2 represent the energies of two excited electronic states. Several vibrational energy levels (e_0, e_1, . . . , e_n) are shown for each of these electronic states.

As can be seen in Figure 4–13, the energy difference between the ground state and an electronically excited state is large relative to the energy differences between vibrational levels in a given electronic state (typically, the two differ by a factor of 10 to 100).

Figure 4–13**a** depicts with arrows some of the transitions which result from absorption of radiation. Visible radiation causes excitation of an electron from E_0 to any of the

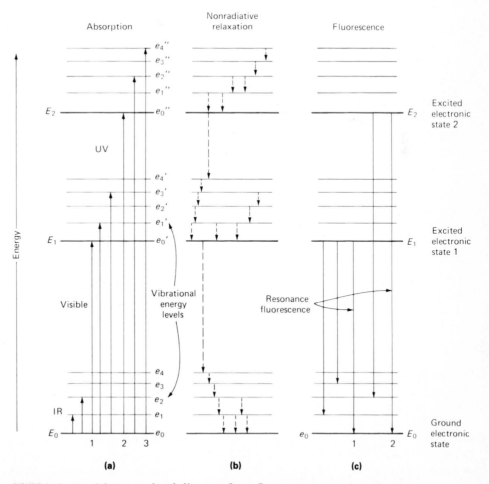

FIGURE 4-13 Partial energy level diagram for a fluorescent organic molecule.

vibrational levels associated with E_1. The frequencies absorbed would be given by

$$\nu_n = \frac{1}{h}(E_1 + e'_n - E_0) \qquad (4-15)$$

Similarly, the frequencies of absorbed ultraviolet radiation would be given by

$$\nu_n = \frac{1}{h}(E_2 + e''_n - E_0) \qquad (4-16)$$

Finally, as shown in Figure 4–13**a**, the less energetic near- and mid-infrared radiation can only bring about transition among the vibrational levels of the ground state. Here,

$$\nu_n = \frac{1}{h}(e_n - e_0) \qquad (4-17)$$

Although they are not shown, several rotational energy levels are associated with each vibrational level. The energy difference between these is small relative to the energy difference between vibrational levels; transitions between a ground and an excited rotational state is brought about by radiation in the 0.01 to 1 cm range, which includes microwave and longer infrared radiations.

In contrast to atomic absorption spectra, which consist of a series of sharp, well-defined lines, molecular spectra in the ultraviolet and visible regions are ordinarily characterized by absorption bands that often encompass a substantial wavelength range (see Figure 4–12**b,c**). Molecular absorption also involves electronic transitions. As shown by Equations 4–15 and 4–16, however, several closely spaced absorption lines will be associated with each electronic transition, owing to the existence of numerous vibrational states. Furthermore, as we have mentioned, many rotational energy levels are associated with each vibrational state. As a consequence, the spectrum for a molecule ordinarily consists of a series of closely spaced absorption peaks that constitute an *absorption band*, such as those shown for benzene vapor in Figure 4–12**b**. Unless a high-resolution instrument is employed, the individual peaks may not be detected, and the spectra will appear as smooth bands such as those shown in Figure 4–12**c**. Finally, in the condensed state, and in the presence of solvent molecules, the individual bands tend to broaden to give *continuous spectra* such as that shown in Figure 4–12**d**. Solvent effects are considered in later chapters.

Pure vibrational absorption can be observed in the infrared region, where the energy of radiation is insufficient to cause electronic transitions. Here, spectra exhibit narrow, closely spaced absorption peaks resulting from transitions among the various vibrational quantum levels (see the transition labeled IR at the bottom of Figure 4–13**a**). Variations in rotational levels may give rise to a series of peaks for each vibrational state. However, rotation is often hindered or prevented in liquid or solid samples; thus, the effects of these small energy differences are not ordinarily detected in such samples.

Pure rotational spectra for gases can be observed in the microwave region.

Absorption Induced by a Magnetic Field. When electrons or the nuclei of certain elements are subjected to a strong magnetic field, additional quantized energy levels are generated as a consequence of magnetic properties of these elementary particles. The differences in energy between the induced states are small, and transitions between the states are brought about only by absorption of long-wavelength (or low-frequency) radiation. With nuclei, radio waves ranging from 30 to 500 MHz are generally involved, while for electrons, microwaves with a frequency of about 9500 MHz are absorbed.

Absorption by nuclei or by electrons in magnetic fields is studied by *nuclear magnetic resonance* (NMR) and *electron spin resonance*

(ESR) techniques, respectively; these methods are considered in Chapter 14.

Relaxation Processes. Ordinarily, the lifetime of an atom or molecule excited by absorption of radiation is brief because several *relaxation processes* exist which permit its return to the ground state. As shown in Figure 4–11**b**, nonradiative relaxation involves the loss of energy in a series of small steps, the excitation energy being converted to kinetic energy by collision with other molecules. A minute increase in the temperature of the system results.

As shown in Figure 4–11**c**, relaxation can also occur by emission of fluorescent radiation. Still other relaxation processes are discussed in Chapters 8 and 14.

4B-3 EMISSION OF RADIATION

Electromagnetic radiation is produced when excited particles (ions, atoms, or molecules) return to lower-energy levels or to their ground states. Excitation can be brought about by a variety of means, including bombardment with electrons or other elementary particles, exposure to a high-potential alternating current spark, heat treatment in an arc or a flame, or absorption of electromagnetic radiation.

Radiating particles that are well separated from one another, as in the gaseous state, behave as independent bodies and often produce radiation containing only a relatively few specific wavelengths. The resulting spectrum is then *discontinuous* and is termed a *line spectrum*. A *continuous spectrum*, on the other hand, is one in which all wavelengths are represented over an appreciable range, or one in which the individual wavelengths are so closely spaced that resolution is not feasible by ordinary means. Continuous spectra result from excitation of: (1) solids or liquids, in which the atoms are so closely packed as to be incapable of independent behavior; or (2) complicated molecules possessing many closely related energy states. Continuous spectra also arise when the energy changes involve particles with unquantized kinetic energies.

Both continuous spectra and line spectra are of importance in analytical chemistry. The former are frequently employed in methods based on the interaction of radiation with matter, such as spectrophotometry. Line spectra, on the other hand, are important because they permit the identification and determination of the emitting species.

Thermal Radiation. When solids are heated to incandescence, the continuous radiation that is emitted is more characteristic of the temperature of the emitting surface than of the material of which that surface is composed. Radiation of this kind (called *black-body radiation*) is produced by the innumerable atomic and molecular oscillations excited in the condensed solid by the thermal energy. Theoretical treatment of black-body radiation leads to the following conclusions: (1) the radiation exhibits a maximum emission at a wavelength that varies inversely with the absolute temperature; (2) the total energy emitted by a black body (per unit of time and area) varies as the fourth power of temperature; and (3) the emissive power at a given temperature varies inversely as the fifth power of wavelength. These relationships are reflected in the behavior of several experimental radiation sources shown in Figure 4–14; the emission from these sources approaches that of the ideal black body. Note that the energy peaks in Figure 4–14 shift to shorter wavelengths with increasing temperature. It is clear that very high temperatures are needed to cause a thermally excited source to emit a substantial fraction of its energy as ultraviolet radiation.

Heated solids are used to produce infrared, visible, and longer-wavelength ultraviolet radiation for analytical instruments.

Emission of Gases. Atoms, ions, or molecules in the gaseous state can often be excited, by

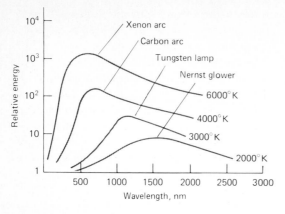

FIGURE 4-14 Black-body radiation curves.

electrical discharge or by heat, to produce radiation in the ultraviolet and visible regions. The process involves promotion of the outermost electrons of a species to an excited electronic state; radiation is then emitted as the excited species returns to the ground state. Atomic emission spectra consist of a series of discrete lines whose energies correspond to the energy differences between the various electronic states. For molecules, emission spectra are usually more complicated because there can exist several vibrational and rotational states for each possible electronic energy level; instead of a single line for each electronic transition, numerous closely spaced lines form an emission band.

A true continuous spectrum is sometimes produced from the excitation of gaseous molecules. For example, when deuterium or hydrogen at low pressure is subjected to an electric discharge, a continuous spectrum from about 400 to 200 nm is produced that is a useful source for absorption spectrophotometry (see Section 7C–1).

Emission of X-Ray Radiation. Radiation in the X-ray region is normally generated by the bombardment of a metal target with a stream of high-speed electrons. The electron beam causes the innermost electrons in the atoms of the target material to be raised to higher energy levels or to be ejected entirely. The excited atoms or ions then return to the ground state by various stepwise electronic transitions that are accompanied by the emission of photons, each having an energy $h\nu$. The consequence is the production of an X-ray spectrum consisting of a series of lines characteristic of the target material. This discrete spectrum is superimposed on a continuum of nonquantized radiation given off when some of the high-speed electrons are partially decelerated as they pass through the target material.

Fluorescence and Phosphorescence. Fluorescence and phosphorescence are analytically important emission processes in which atoms or molecules are excited by absorption of a beam of electromagnetic radiation; radiant emission then occurs as the excited species return to the ground state.

Fluorescence occurs much more rapidly than phosphorescence and is generally complete after about 10^{-5} sec (or less) from the time of excitation; it is most easily observed at a 90-deg angle to the excitation beam. Phosphorescence emission takes place over periods longer than 10^{-5} sec, and may indeed continue for minutes or even hours after irradiation has ceased.

Resonance fluorescence describes a process in which the emitted radiation is identical in frequency to the radiation employed for excitation. The lines labeled 1 and 2 in Figures 4–13a and 4–13c illustrate this type of fluorescence. Here, the species is excited to the energy states E_1 or E_2 by radiation having an energy of $(E_1 - E_0)$ or $(E_2 - E_0)$. After a brief period, emission of radiation of identical energy occurs, as depicted in Figure 4–13c. Resonance fluorescence is most commonly produced by *atoms* in the gaseous state which do not have vibrational energy states superimposed on electronic energy levels.

Nonresonance fluorescence is brought about by irradiation of *molecules* in solution or in the gaseous state. As shown in Figure

4–13**a**, absorption of radiation promotes the molecules into any of the several vibrational levels associated with the two excited electronic levels. The lifetimes of these excited vibrational states are, however, only of the order of 10^{-15} sec, which is much smaller than the lifetimes of the excited electronic states (10^{-8} sec). Therefore, on the average, vibrational relaxation occurs before electronic relaxation. As a consequence, the energy of the emitted radiation is smaller than that of the absorbed by an amount equal to the vibrational excitation energy. For example, for the absorption labeled 3 in Figure 4–13**a**, the absorbed energy would be equal to $(E_2 - E_0 + e''_4 - e''_0)$; the energy of the fluorescent radiation, on the other hand,

would again be given by $(E_2 - E_0)$; the emitted radiation would necessarily have a lower frequency or greater wavelength than the radiation that excited the fluorescence. (This shift in wavelength to lower frequencies is sometimes called the *Stokes shift*.) Clearly, both resonance and nonresonance radiation can accompany fluorescence of molecules; the latter tends to predominate, however, because of the much larger number of vibrationally excited states.

Phosphorescence occurs when an excited molecule relaxes to a metastable excited electronic state (called the *triplet state*), which has an average lifetime of greater than about 10^{-5} sec. The nature of this type of excited state is discussed in Chapter 8.

PROBLEMS

4–1. Calculate the frequency in hertz and the wavenumber for
 (a) an X-ray beam with a wavelength 2.7 Å.
 (b) an emission line for Cu at 211.0 nm.
 (c) the line at 694.3 nm produced by a ruby laser.
 (d) the output of a CO_2 gas laser at 10.6 μm.
 (e) an infrared absorption peak at 14.7 μm.
 (f) a microwave beam of wavelength 1.86 cm.

4–2. Calculate the wavelength in centimeters for
 (a) the radio frequency of an airport tower at 119.3 MHz.
 (b) a radio navigation beam having a frequency of 327.4 kHz.
 (c) an NMR signal at 101 MHz.
 (d) an EPR signal at 9500 MHz.

4–3. Calculate the energy of each of the radiations in Problem 4–1 in (a) erg/photon, (b) kcal/mol, and (c) eV.

4–4. Calculate the energy of each of the radiations in Problem 4–2 in (a) erg/photon, (b) kcal/mol, and (c) eV.

4–5. Calculate the velocity, wavelength, and frequency of the sodium D line ($\lambda = 5890$ Å in a vacuum) in
 (a) chloroform vapor ($n_D = 1.0014$).
 (b) chloroform ($n_D = 1.4459$).
 (c) a dense flint glass ($n_D = 1.890$).
 (d) 3 M KCl ($n_D = 1.360$).

4–6. Calculate the reflective loss when a 589-nm beam of radiation
 (a) passes through a piece of dense flint glass ($n_D = 1.59$).
 (b) passes through a cell having glass windows ($n_D = 1.43$) and containing an alcohol solution with a refractive index of 1.32.

Instruments for Optical

Spectroscopy

The term *spectroscopy* was originally used to describe a branch of science that was based upon the resolution of visible radiation into its component wavelengths. With the passage of time, however, the meaning of the term has been broadened to encompass studies involving the entire electromagnetic spectrum and even some techniques that do not involve electromagnetic radiation. Examples of the last are mass, electron, and acoustic spectroscopy.

The first spectroscopic instruments were developed for use in the visible region and were thus called *optical instruments*. This term has now been extended to include instruments designed for the ultraviolet and infrared regions as well; while not strictly correct, the terminology is nevertheless useful in that it emphasizes the many features that are common to the instruments used for these three important spectral regions.

The purpose of this chapter is to describe those components of optical instruments that are common to all of the various optical spectroscopic techniques and which are dealt with in Chapters 6 through 13. The instruments for spectroscopic studies in regions more energetic than the ultraviolet and less energetic than the infrared have characteristics that differ substantially from optical instruments and are considered separately in Chapters 14 and 15.

5A Types of Methods

Optical spectroscopic methods are based upon the phenomena of *emission, absorption, fluorescence, phosphorescence,* and *scattering.* While the instruments for each differ somewhat in configuration, many of their basic components are remarkably similar. Furthermore, the required properties of these components are the same regardless of whether they are applied to the ultraviolet, visible, or infrared portion of the spectrum.[1]

Typical spectroscopic instruments contain five components, including: (1) a stable source of radiant energy; (2) a transparent container for holding the sample; (3) a device that isolates a restricted region of the spectrum for measurement; (4) a radiation detector or transducer, which converts radiant energy to a usable signal (usually electrical); and (5) a signal processor and *readout*,

which displays the transduced signal on a meter scale, an oscilloscope face, or a recorder chart. Figure 5–1 shows the arrangement of these components for the five types of spectroscopic measurements mentioned earlier. As can be seen in the figure, the configuration of components (3), (4), and (5) is the same for each type of instrument.

Emission spectroscopy differs from the other types in the respect that no external radiation source is required; the sample itself is the emitter. Here, the sample container is an arc, a spark, or a flame, which both holds the sample and causes it to emit characteristic radiation.

Absorption as well as fluorescence, phosphorescence, and scattering spectroscopy require an external source of radiant energy. In the former, the beam from the source passes through the sample into the wavelength selector (in some instruments,

---

[1] For a more complete discussion of the components of optical instruments, see: R. P. Bauman, *Absorption Spectroscopy*, Chapters 2 and 3, Wiley: New York, 1962; E. J. Meehan, in *Treatise on Analytical Chemistry*, 2d ed., P. J. Elving, E. J. Meehan, and I. M. Kolthoff, Eds., Part I, vol. 7, Chapter 3, Wiley: New York, 1981; and *Physical Methods of Chemistry*, A. Weissberger and B. W. Rossiter, Eds., Part III B, vol. 1, Chapters 1–5, Wiley-Interscience: New York, 1972.

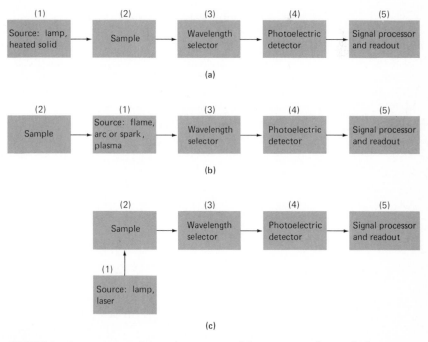

FIGURE 5-1 Components for various types of instruments for optical spectroscopy: (**a**) absorption spectroscopy, (**b**) emission spectroscopy, and (**c**) fluorescence, phosphorescence, and scattering spectroscopy.

the position of sample and selector is reversed). In the latter three, the source induces the sample, held in a container, to emit characteristic fluorescent, phosphorescent, or scattered radiation, which is measured at an angle (usually 90 deg) with respect to the source.

Figure 5–2 summarizes the characteristics of the first four components shown in Figure 5–1. It is clear that instrument compo-

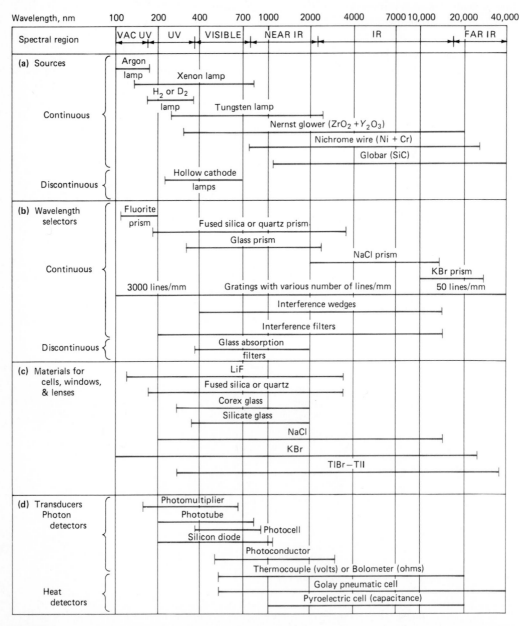

FIGURE 5–2 Components and materials for spectroscopic instruments. (Adapted from a figure by Professor A. R. Armstrong, College of William and Mary. With permission.)

nents differ in detail, depending upon the wavelength region within which they are to be used. Their design also depends on the primary use of the instrument; that is, whether it is to be employed for qualitative or quantitative analysis and whether it is to be applied to atomic or molecular spectroscopy. Nevertheless, the general function and performance requirements of each type of component are similar, regardless of wavelength region and application.

5B Radiation Sources

In order to be suitable for spectroscopic studies, a source must generate a beam of radiation with sufficient power for ready detection and measurement. In addition, its output should be stable. Typically, the radiant power of a source varies exponentially with the potential of the electrical supply. Thus, a regulated power source is often needed to provide the required stability. The problem of source stability is circumvented in some instruments by splitting the output radiation into a reference beam and a sample beam. The first passes directly to a transducer while the second interacts with the sample and is then focused on a matched transducer. (Alternatively, some instruments have only a single transducer which is irradiated alternately by the sample and reference beams.) The ratio of the outputs of the two transducers serves as the analytical parameter. The effect of fluctuations in the source output is largely canceled by this means.

Both continuous and line sources are used in optical spectroscopy.[2] The former finds application in molecular absorption methods. The latter is employed in fluorescence and atomic absorption spectroscopy.

Figure 5–2**a** lists the most widely used spectroscopic sources.

5B-1 CONTINUOUS SOURCES

Continuous sources find widespread use in absorption and fluorescence spectroscopy. For the ultraviolet region, the most common source is the deuterium lamp. High-pressure, gas-filled arc lamps containing argon, xenon, or mercury serve when a particularly intense source is required. The common infrared sources are inert solids heated to 1500 to 2000°K, a temperature at which the maximum radiant output occurs at 1.5 to 1.9 μm (see Figure 4–14).

Details on the construction and behavior of these various continuous sources will be found in the chapter dealing with specific types of spectroscopic methods.

5B-2 LINE SOURCES

Sources that emit a few discrete lines find use in atomic absorption spectroscopy, Raman spectroscopy, refractometry, and polarimetry.

Metal Vapor Lamps. Two of the most common line sources are the familiar mercury and sodium vapor lamps, which provide a relatively few sharp lines in the ultraviolet and visible regions. Hollow cathode lamps provide line spectra for a large number of elements. Their use has, however, been confined to atomic absorption and atomic fluorescence spectroscopy. Discussion of line sources is thus deferred to Chapter 9.

5B-3 LASERS

The first laser was constructed in 1960.[3] Since that time, chemists have found nu-

---

[2] For a review of sources for optical spectroscopy, see: A. Hell, *Anal. Chem.,* **1971,** *43* (5), 79A; and S. Z. Lewin, *J. Chem. Educ.,* **1965,** *42,* 9165.

[3] For a more complete discussion of lasers, see: G. M. Hieftje, J. C. Travis, and F. E. Lytle, *Lasers in Chemical Analysis,* Humana Press: Clifton, N.J., 1981; and *Analytical Laser Spectroscopy,* N. Omenetto, Ed., Wiley, New York, 1979.

merous useful applications for these sources in high-resolution spectroscopy, kinetic studies of processes with lifetimes in the range of 10^{-9} to 10^{-12} sec, the detection and determination of extremely small concentrations of species in the atmosphere, and the induction of isotopically selective reactions.[4] In addition, laser sources have become important in several routine analytical methods, including Raman spectroscopy, molecular absorption spectroscopy, emission spectroscopy, and as part of instruments for Fourier transform infrared spectroscopy.

The term laser is an acronym for *light amplification* by *stimulated emission* of *radiation*. As a consequence of their light-amplifying properties, lasers produce spatially narrow, extremely intense beams of radiation. The process of stimulated emission produces a beam of highly monochromatic (bandwidths of 0.01 nm or less) and remarkably coherent (Section 4A–4) radiation. Because of these unique properties, lasers have become important sources for use in the ultraviolet, visible, and infrared regions of the spectrum. A limitation of early lasers was that the radiation from a given source was restricted to a relatively few discrete wavelengths or lines. Recently, however, dye lasers have become available; *tuning* of these sources provides a narrow band of radiation

---

[4] For reviews of some of these applications, see: S. R. Leone, *J. Chem. Educ.*, **1976**, *53*, 13; J. E. Wright and M. J. Wirth, *Anal. Chem.*, **1980**, *52*, 988A, 1087A; and A. Schawlow, *Science*, **1982**, *217*, 9.

at any chosen wavelength within the range of the source.

Figure 5–3 is a schematic representation showing the components of a typical laser source. The heart of the device is a *lasing medium*. It may be a solid crystal such as ruby, a semiconductor such as gallium arsenide, a solution of an organic dye, or a gas. The lasing material is often activated or *pumped* by radiation from an external source so that a few photons of proper energy will trigger the formation of a cascade of photons of the same energy. Pumping can also be carried out by an electrical current or by an electrical discharge. Thus, *gas lasers* usually do not have the external radiation source shown in Figure 5–3; instead, the power supply is connected to a pair of electrodes contained in a cell filled with the gas.

A laser normally functions as an oscillator in the sense that the radiation produced from the lasing action is caused to pass back and forth through the medium numerous times by means of a pair of mirrors as shown in Figure 5–3. Additional photons are generated with each passage, thus leading to enormous amplification. The repeated passage also produces a beam that is highly parallel because nonparallel radiation escapes from the sides of the medium after being reflected a few times (see Figure 5–3).

In order to obtain a usable laser beam, one of the mirrors is coated with a sufficiently thin layer of reflecting material so that a fraction of the beam is transmitted rather than reflected (see Figure 5–3).

Laser action can be understood by consid-

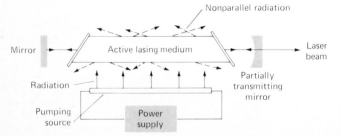

FIGURE 5-3 Schematic representation of a typical laser source.

ering the four processes depicted in Figure 5–4, namely, (**a**) pumping, (**b**) spontaneous emission (fluorescence), (**c**) stimulated emission, and (**d**) absorption. For purposes of illustration, we will focus on just two of several electronic energy levels that will exist in the atoms, ions, or molecules making up the laser material; as shown in the figure, the two electronic levels have energies E_y and E_x. Note that the higher electronic state is shown as having several slightly different vibrational energy levels, E_y, E_y', E_y'', and so

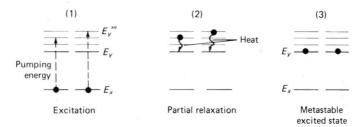

(**a**) Pumping (excitation by electrical, radiant, or chemical energy)

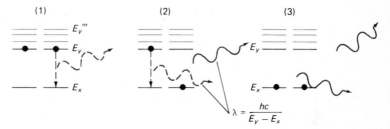

(**b**) Spontaneous Emission

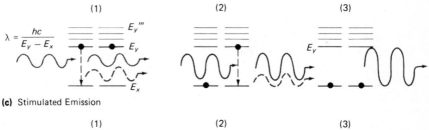

(**c**) Stimulated Emission

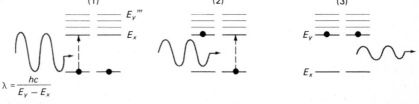

(**d**) Absorption

FIGURE 5–4 Four processes important in laser action: (**a**) pumping (excitation by electrical, radiant, or chemical energy), (**b**) spontaneous emission, (**c**) stimulated emission, and (**d**) absorption.

forth. We have not shown additional levels for the lower electronic state, although such often exist.

Pumping. Pumping, which is necessary for laser action, is a process by which the active species of a laser is excited by means of an electrical discharge, passage of an electrical current, exposure to an intense radiant source, or interaction with a chemical species. During pumping, several of the higher electronic and vibrational energy levels of the active species will be populated. In diagram **a**−1 of Figure 5−4, one atom or molecule is shown as being promoted to an energy state E_y''; the second is excited to the slightly higher vibrational level $E_y''{}'$. The lifetime of excited *vibrational* states is brief, however; after 10^{-13} to 10^{-15} sec, relaxation to the lowest excited vibrational level (E_y in diagram **a**−3) occurs with the production of an undetectable quantity of heat. Some excited electronic states of laser materials have lifetimes considerably longer (often 1 msec or more) than their excited vibrational counterparts; long-lived states are sometimes termed *metastable* as a consequence.

Spontaneous Emission. As was pointed out in the discussion of fluorescence in Section 4B−3, a species in an excited electronic state may lose all or part of its excess energy by spontaneous emission of radiation. This process is depicted in the diagrams shown in Figure 5−4**b**. Note that the wavelength of the fluorescent radiation is exactly equal to the energy difference between the two electronic states, $E_y - E_x$. It is also important to note that the instant at which emission occurs and the path of the resulting photon varies from excited electron to excited electron. That is, spontaneous emission is a random process; thus, as shown in Figure 5−4, the fluorescent radiation produced by one of the particles in diagram **b**−1 differs in direction and phase from that produced by the second particle (diagram **b**−2). Spontaneous emission, therefore, yields *incoherent* monochromatic radiation.

Stimulated Emission. Stimulated emission, which is the basis of laser behavior, is depicted in Figure 5−4**c**. Here, the excited laser particles are struck by externally produced photons having precisely the same energies ($E_y - E_x$) as the photons produced by spontaneous emission. Collisions of this type cause the excited species to relax *immediately* to the lower energy state and to emit simultaneously a photon of exactly the same energy as the photon that stimulated the process. More important, *the emitted photon is precisely in phase with the photon that triggered the event. That is, the stimulated emission is totally coherent with the incoming radiation.*

Absorption. The absorption process, which competes with stimulated emission, is depicted in Figure 5−4**d**. Here, two photons with energies exactly equal to $E_y - E_x$ are absorbed to produce the metastable excited state shown in diagram **d**−3; note that the state shown in diagram **d**−3 is identical to that attained by pumping (diagram **a**−3).

Population Inversion and Light Amplification. In order to have light amplification in a laser, it is necessary that the number of photons produced by stimulated emission exceed the number lost by absorption. This condition will prevail only when the number of particles in the higher energy state exceeds the number in the lower; that is, a *population inversion* from the normal distribution of energy states must exist. Population inversions are brought about by pumping. Figure 5−5 contrasts the effect of incoming radiation on a noninverted population with an inverted one.

Three- and Four-Level Laser Systems. Figure 5−6 shows simplified energy diagrams for the two common types of laser systems. In the three-level system, the transition responsible for laser radiation is between an excited state E_y and the ground state E_0; in a four-level system, on the other hand, radiation is generated by a transition from E_y to a state E_x that has a greater energy than the ground state. Furthermore, it is necessary that tran-

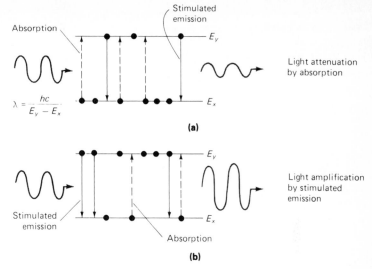

FIGURE 5-5 Passage of radiation through (**a**) a noninverted population and (**b**) an inverted population.

sitions between E_x and the ground state be rapid.

The advantage of the four-level system is that the population inversions necessary for laser action are more readily achieved. To understand this fact, note that at room temperature a large majority of the laser particles will be in the ground-state energy level E_0 in both systems. Sufficient energy must thus be provided to convert more than 50% of the lasing species to the E_y level of a three-level system. In contrast, it is only necessary to pump sufficiently to make the number of particles in the E_y energy level exceed the *number in* E_x of a four-level system. The lifetime of a particle in the E_x state is brief, however, because the transition to E_0 is fast; thus, the number in the E_x state

will generally be negligible with respect to the number having energy E_0 and also (with a modest input of pumping energy) with respect to the number in the E_y state. That is, the four-level laser usually achieves a population inversion with a small expenditure of pumping energy.

Some Examples of Useful Lasers. The first successful laser, and one that still finds widespread use, was a three-level device in which a ruby crystal was the active medium. Ruby is primarily Al_2O_3 but contains approximately 0.05% chromium(III) distributed among the aluminum(III) lattice sites, which accounts for the red coloration. The chromium(III) ions are the active lasing material. In early lasers, the ruby was machined into a rod about 4 cm in length and 0.5 cm

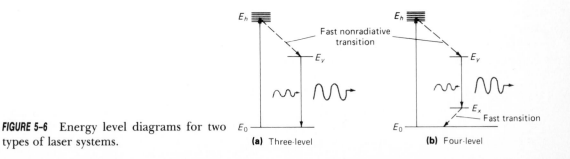

FIGURE 5-6 Energy level diagrams for two types of laser systems.

in diameter. A flash tube (often a low-pressure xenon lamp) was coiled around the cylinder to produce intense flashes of light ($\lambda = 694.3$ nm). Because the pumping was discontinuous, a pulsed beam was produced. Continuous wave ruby sources are now available.

A variety of gas lasers are sold commercially. An important example is the argon ion laser, which produces intense lines in the green (514.5 nm) and blue (488.0 nm) regions. This laser is a four-level device in which argon ions are formed by an electrical or radio-frequency discharge. The input energy is sufficient to excite the ions from their ground state, with a principal quantum number of 3, to various $4p$ states. Laser activity then involves transitions to the $4s$ state. Another important gas laser is the He/Ne source, which emits red radiation at 632.8 nm.

Dye lasers[5] have become important radiation sources in chemistry because they are tunable over a range of 20 to 50 nm; that is, dye lasers provide bands of radiation of a chosen wavelength with widths of a few hundredths of a nanometer or less.

The active materials in dye lasers are solutions of organic compounds capable of fluorescing in the ultraviolet, visible, or infrared regions. In contrast to ruby or gas lasers, however, the lower energy level for laser action (E_x in Figure 5–6) is not a single energy but a band of energies arising from the superposition of a large number of small vibrational and rotational energy states upon the base electronic energy state. Electrons in E_y may then undergo transitions to any of these states, thus producing photons of slightly different energies.

Tuning of dye lasers can be readily accomplished by replacing the nontransmit-ting mirror shown in Figure 5–3 with a monochromator equipped with a reflection grating or a Littrow-type prism (Section 5D–3) which will reflect only a narrow bandwidth of radiation into the laser medium; the peak wavelength can be varied by rotation of the grating or prism. Emission is then stimulated for only part of the fluorescent spectrum, namely, the wavelength reflected from the monochromator.

5C   Wavelength Selectors; Filters

For most spectroscopic analyses, radiation consisting of a limited, narrow, continuous group of wavelengths called a band is required.[6] A narrow bandwidth tends to enhance the sensitivity of absorbance measurements, may provide selectivity to both emission and absorption methods, and is frequently a requirement from the standpoint of obtaining a linear relationship between the optical signal and concentration (Section 6B–3). Two types of wavelength selectors are encountered, *filters* and *monochromators*. The latter, which are discussed in the next section, permit continuous variation in the wavelength whereas the former provide but a limited, selected wavelength region.

Ideally, the output from a wavelength selector, monochromator or filter, would be radiation of a single wavelength or frequency. No existing wavelength selector even approaches this ideal; instead, a Gaussian-shaped distribution of wavelengths such as those shown in Figure 5–7 is obtained. Here, the percent of incident radiation of a given wavelength that is transmitted by the selector is plotted as a function of wavelength. The *effective bandwidth*, which is defined in Figure 5–7, is an inverse measure

---

[5] For further information, see: R. B. Green, *J. Chem. Educ.*, **1977**, *54* (9), A365, (10), A407; and G. M. Hieftje, *Amer. Lab.*, **1983**, *15* (5), 66.

[6] Note that the term *band* in this context has a somewhat different meaning from that used in describing types of spectra in Chapter 4.

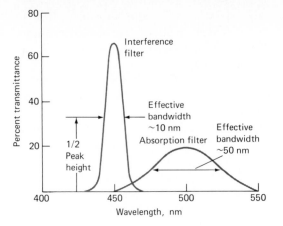

FIGURE 5-7 Effective bandwidths for two types of filters.

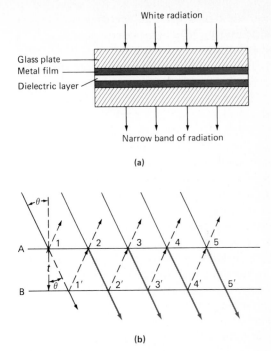

FIGURE 5-8 (**a**) Schematic cross section of an interference filter. Note that the drawing is not to scale and the three central bands are much narrower than shown. (**b**) Schematic to show the conditions for constructive interference.

of the quality of the device, a narrower bandwidth representing better performance. Figure 5–7 shows the characteristics of typical filters; the output from a monochromator has the same shape but ordinarily a considerably narrower bandwidth.

Absorption and interference filters are employed for wavelength selection (the latter are sometimes called Fabry-Perot filters). The former are restricted to the visible region of the spectrum. Interference filters, on the other hand, are available for ultraviolet, visible, and well into the infrared region.

5C-1 INTERFERENCE FILTERS

As the name implies, interference filters rely on optical interference to provide relatively narrow bands of radiation (see Figure 5–8). An interference filter consists of a transparent dielectric[7] (frequently calcium fluoride

or magnesium fluoride) that occupies the space between two semitransparent metallic films. This array is sandwiched between two plates of glass or other transparent materials (see Figure 5–8a). The thickness of the dielectric layer is carefully controlled and determines the wavelength of the transmitted radiation. When a perpendicular beam of collimated radiation strikes this array, a fraction passes through the first metallic layer while the remainder is reflected. The portion that is passed undergoes a similar partition upon striking the second metallic film. If the reflected portion from this second interaction is of the proper wavelength, it is partially reflected from the inner side of the first layer in phase with incom-

---

[7] Dielectrics are a class of substances that are nonconductors because they contain no (or only a few) free electrons. Generally, dielectrics are optically transparent in contrast to electrically conducting solids, which either absorb radiation or reflect it strongly.

ing light of the same wavelength. The result is that this particular wavelength is reinforced, while most other wavelengths, being out of phase, suffer destructive interference.

The relationship between the thickness of the dielectric layer t and the transmitted wavelength λ can be found with the aid of Figure 5–8b. For purposes of clarity, the incident beam is shown as arriving at an angle θ from the perpendicular. At point 1, the radiation is partially reflected and partially transmitted to point 1′ where partial reflection and transmission again takes place. The same process occurs at 2, 2′, and so forth. For reinforcement to occur at point 2, the distance traveled by the beam reflected at 1′ must be some multiple of its wavelength in the medium λ'. Since the path length between surfaces can be expressed as $t/\cos\theta$, the condition for reinforcement is that $n\lambda' = 2t/\cos\theta$ where n is a small whole number.

In ordinary use, θ approaches zero and $\cos\theta$ approaches unity so that the equation derived from Figure 5–8 simplifies to

$$n\lambda' = 2t \tag{5–1}$$

where λ' is the wavelength of radiation *in the dielectric* and t is the thickness of the dielectric. The corresponding wavelength in air is given by

$$\lambda = \lambda'n$$

where n is the refractive index of the dielectric medium. Thus, the wavelengths of radiation transmitted by the filter are

$$\lambda = \frac{2tn}{n} \tag{5–2}$$

The integer n is the *order* of interference.

The glass layers of the filter are often selected to absorb all but one of the reinforced bands; transmission is thus restricted to a single order.

Figure 5–9 illustrates the performance characteristics of typical interference filters. Ordinarily, filters are characterized, as shown, by the wavelength of their transmittance peaks, the percentage of incident radiation transmitted at the peak (their *percent transmittance*, Section 6A–1), and their effective bandwidths.

Interference filters are available throughout the ultraviolet and visible regions and up to about 14 μm in the infrared. Typically, effective bandwidths are about 1.5% of the wavelength at peak transmittance, although this figure is reduced to 0.15% in some narrow-band filters; these have maximum transmittances of about 10%.

5C-2 INTERFERENCE WEDGES

An interference wedge consists of a pair of mirrored, partially transparent plates sepa-

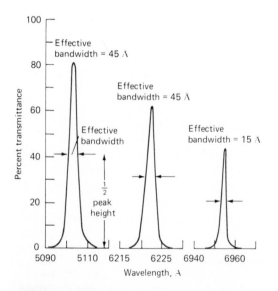

FIGURE 5-9 Transmission characteristics of typical interference filters.

rated by a wedge-shaped layer of a dielectric material. The length of the plates ranges from about 50 to 200 mm. The radiation transmitted varies continuously from one end to the other as the thickness of the wedge varies. By choosing the proper linear position along the wedge, a bandwidth of about 20 nm can be isolated.

Interference wedges are available for the visible region (400 to 700 nm), the near-infrared region (1000 to 2000 nm), and for several parts of the infrared region (2.5 to 14.5 μm). They can serve in place of prisms or gratings in monochromators.

5C–3 ABSORPTION FILTERS

Absorption filters, which are generally less expensive than interference filters, have been widely used for band selection in the visible region. These filters function by absorbing certain portions of the spectrum. The most common type consists of colored glass or of a dye suspended in gelatin and sandwiched between glass plates. The former have the advantage of greater thermal stability.

Absorption filters have effective bandwidths that range from perhaps 30 to 250 nm (see Figures 5–7 and 5–10). Filters that provide the narrowest bandwidths also absorb a significant fraction of the desired radiation, and may have a transmittance of 0.1 or less at their band peaks. Glass filters with transmittance maxima throughout the entire visible region are available commercially.

Cut-off filters have transmittances of nearly 100% over a portion of the visible spectrum, but then rapidly decrease to zero transmittance over the remainder. A narrow spectral band can be isolated by coupling a cut-off filter with a second filter (see Figure 5–10).

It is apparent from Figure 5–7 that the performance characteristics of absorption filters are significantly inferior to those of

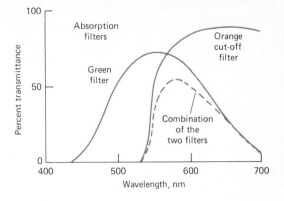

FIGURE 5-10 Comparison of various types of filters for visible radiation.

interference-type filters. Not only are their bandwidths greater, but for narrow bandwidths, the fraction of light transmitted is also less. Nevertheless, absorption filters are totally adequate for many applications.

5D Wavelength Selectors; Monochromators

For many spectroscopic methods, it is necessary or desirable to be able to vary the wavelength of radiation continuously over a considerable range. *Monochromators* fulfill this need.

Monochromators for ultraviolet, visible, and infrared radiation are all similar in mechanical construction in the sense that they employ slits, lenses, mirrors, windows, and prisms or gratings. To be sure, the materials from which these components are fabricated will depend upon the wavelength region of intended use (see Figure 5–2**b** and **c**).

5D–1 COMPONENTS OF A MONOCHROMATOR

All monochromators contain an entrance slit, a collimating mirror (or less commonly, a lens) to produce a parallel beam of radiation, a prism or grating as a dispersing element[8], and a focusing mirror or lens which projects a series of rectangular images of the entrance slit upon a plane surface (the *focal plane*). In addition, most monochromators have entrance and exit *windows*, which are designed to protect the components from dust and corrosive laboratory fumes.

Figure 5–11 shows the optical design of two monochromators, one employing a grating for dispersal of radiation and the other a prism. A source of radiation containing but two wavelengths, λ_1 and λ_2, is assumed for purposes of illustration. This radiation enters the monochromators via a narrow rectangular opening or slit, is collimated, and then strikes the surface of the dispersing element at an angle. For the grating monochromator, angular dispersion results from diffraction, which occurs at the reflective surface; for the prism, refraction at the two faces results in angular dispersal of the radiation, as shown. In both designs, the dispersed radiation is focused on the focal plane *AB* where it appears as two images of the entrance slit (one for λ_1 and one for λ_2).

5D–2 GRATING MONOCHROMATORS

Dispersion of ultraviolet, visible, and infrared radiation can be brought about by directing a polychromatic beam through a *transmission grating* or onto the surface of a *reflection grating;* the latter is by far the more common practice.

---

[8] Much less generally employed are *interference wedges* (Section 5C–2).

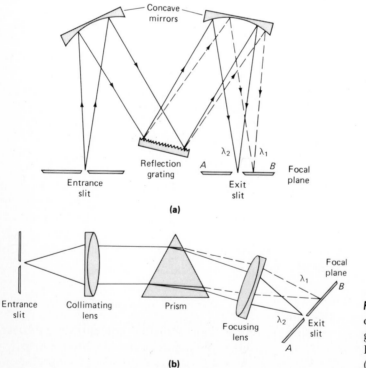

(a)

(b)

FIGURE 5–11 Two types of monochromators: (**a**) Czerney-Turner grating monochromator; (**b**) Bunsen prim monochromator. (In both instances, $\lambda_1 > \lambda_2$.)

The first step in the manufacture of reflection gratings involves the preparation of a *master grating* from which numerous *replica gratings* can be formed. The master grating consists of a large number of parallel and closely spaced grooves ruled on a hard, polished surface with a suitably shaped diamond tool. A magnified cross-sectional view of a few typical grooves is shown in Figure 5–12. For the ultraviolet and visible region, a grating will contain from 300 to 2000 grooves/mm with 1200 to 1400 being most common. For the infrared region, 10 to 200 grooves/mm are encountered; for spectrophotometers designed for the most widely used infrared range of 5 to 15 μm, a grating with about 100 grooves/mm is suitable. The construction of a good master grating is tedious, time-consuming, and expensive, because the grooves must be identical in size, exactly parallel, and equally spaced over the length of the grating (3 to 10 cm).

Replica gratings are formed from a master grating by evaporating a film of aluminum onto the latter after it has been coated with a parting agent, which permits ready separation of the aluminum from the master. A glass plate is cemented to the aluminum; the plate and film can then be lifted from the master mold giving the finished replica grating.

The Echellette Grating. Figure 5–12 is a schematic representation of an *echellette-type* grating, which is grooved or *blazed* such that it has relatively broad faces from which reflection occurs and narrow unused faces. This geometry provides highly efficient diffraction of radiation. Each of the broad faces can be considered to be a point source of radiation; thus interference among the reflected beams 1, 2, and 3 can occur. In order for the interference to be constructive, it is necessary that the path lengths differ by an integral multiple **n** of the wavelength λ of the incident beam.

In Figure 5–12, parallel beams of monochromatic radiation 1 and 2 are shown striking the grating at an incident angle i to the *grating normal*. Maximum constructive interference is shown as occurring at the reflected angle r. It is evident that beam 2 travels a greater distance than beam 1 and that this difference is equal to $(\overline{CD} - \overline{AB})$. For constructive interference to occur, this

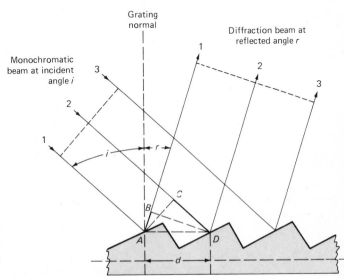

FIGURE 5–12 Schematic diagram illustrating the mechanism of diffraction from an echellette-type grating.

difference must equal **n**λ. That is,

$$n\lambda = (\overline{CD} - \overline{AB})$$

where **n**, a small whole number, is called the diffraction *order*. Note, however, that angle *CAD* is equal to angle *i* and that angle *BDA* is identical with angle *r*. Therefore, from simple trigonometry, we may write

$$\overline{CD} = d \sin i$$

where *d* is the spacing between the reflecting surfaces. It is also seen that

$$\overline{AB} = -d \sin r$$

The minus sign, by convention, indicates that the angle of reflection *r* lies on the opposite side of the grating normal from the incident angle *i* (as in Figure 5–12); angle *r* is positive when it is on the same side. Substitution of the last two expressions into the first gives the condition for constructive interference. Thus,

$$n\lambda = d(\sin i + \sin r) \qquad (5-3)$$

Equation 5–3 suggests that several values of λ exist for a given diffraction angle *r*. Thus, if a first-order line (**n** = 1) of 800 nm is found at *r*, second-order (400 nm) and third-order (267 nm) lines also appear at this angle. Ordinarily, the first-order line is the most intense; indeed, it is possible to design gratings that concentrate as much as 90% of the incident intensity in this order. The higher-order lines can generally be removed by filters. For example, glass, which absorbs radiation below 350 nm, eliminates the higher-order spectra associated with first-order radiation in most of the visible region. The example which follows illustrates these points.

EXAMPLE 5–1. An echellette grating containing 1450 blazes per millimeter was irradiated with a polychromatic beam at an incident angle 48 deg to the grating normal. Calculate the wavelengths of radiation that would appear at an angle of reflection of +20, +10, 0, and −10 deg (angle *r*, Figure 5–12).

To obtain *d* in Equation 5–3, we write

$$d = \frac{1 \text{ mm}}{1450 \text{ blazes}} \times 10^6 \frac{\text{nm}}{\text{mm}} = 689.7 \frac{\text{nm}}{\text{blaze}}$$

When *r* in Figure 5–12 equals +20 deg

$$\lambda = \frac{689.7}{n}(\sin 48 + \sin 20) = \frac{748.4}{n} \text{ nm}$$

and the wavelengths for the first-, second-, and third-order reflections are 748, 374, and 249 nm, respectively.

Similarly, when angle *r* is −10 deg

$$\lambda = \frac{689.7}{n}[\sin 48 + \sin (-10)] = 392.8 \text{ nm}$$

Further calculations of a similar kind yield the following data:

r, deg	**n** = 1	**n** = 2	**n** = 3
	Wavelength (nm) for		
20	748	374	249
10	632	316	211
0	513	256	171
−10	393	196	131

Concave Gratings. Gratings can be formed on a concave surface in much the same way as on a plane surface. A concave grating permits the design of a monochromator without auxiliary collimating and focusing mirrors or lenses. That is, the concave grating both disperses and focuses the radiation on the exit slit. Such an arrangement is advantageous in terms of cost; in addition, the reduction in number of optical surfaces increases the energy throughput of a monochromator containing a concave grating.

Holographic Gratings.[9] One of the products from the emergence of laser technology is an optical (rather than mechanical) technique for forming gratings on plane or concave glass surfaces. *Holographic gratings* produced in this way are appearing in ever increasing numbers in modern optical instruments, even some of the less expensive ones. Holographic gratings, because of their greater perfection with respect to line shape and dimensions provide spectra that are freer from stray radiation and ghosts (double images).

In the preparation of holographic gratings, the beams from a pair of identical lasers are brought to bear at suitable angles upon a glass surface coated with photoresist. The resulting interference fringes from the two beams sensitize the photoresist so that it can be dissolved away leaving a grooved structure that can be coated with aluminum or other reflecting substance to produce a reflection grating. The spacing of the grooves can be changed by changing the angle of the two laser beams with respect to one another. Nearly perfect, large (~50 cm) gratings having as many as 6000 lines/mm can be manufactured in this way at a relatively low cost.

The Echelle Grating. The echelle grating is similar in appearance to the echellette grating shown in Figure 5–12. In contrast, however, the groove density is much lower (~80/mm). In addition, the shorter faces of the groove are used thus increasing the

blaze angle and the angle of reflection significantly. These changes have a profound effect on the dispersion properties of the grating and make it particularly useful for multielement emission analysis. The discussion of this grating will, therefore, be postponed to the chapters on atomic emission methods (Section 10B–3).

5D-3 PRISM MONOCHROMATORS

Prisms can be used to disperse ultraviolet, visible, and infrared radiation. The material used for their construction will differ, however, depending upon the wavelength region.

Construction Materials. Table 5–1 lists the properties of the common substances for fabricating monochromator components. For the ultraviolet, visible, and near-infrared regions (up to about 2700 nm), a quartz prism is often employed; as will be demonstrated later, however, glass provides better resolution for the same size prism for wavelengths between 350 and 2000 nm. As shown in column 5 of Table 5–1, several prisms are required to cover the entire infrared region.

Types of Prisms and Prism Monochromators. Figure 5–13 shows the two most common types of prism designs. The first is a 60-deg prism, which is ordinarily fabricated from a single block of material. When crystalline (but not fused) quartz is the construction material, however, the prism is usually formed by cementing two 30-deg prisms together, as shown in Figure 5–13**a**; one is fabricated from right-handed quartz and the second from left-handed quartz. In this way, the optically active quartz causes no net polar-

[9] See: J. Flamand, A. Grillo, and G. Hayat, *Amer. Lab.,* **1975,** 7 (5), 47; and J. M. Lerner, *et al., Proc. Photo-Opt. Instrum. Eng.,* **1980,** *240,* pp. 72, 82.

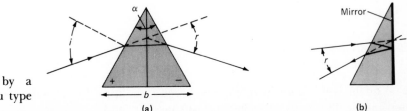

FIGURE 5-13 Dispersion by a prism: (**a**) quartz Cornu type and (**b**) Littrow type.

Table 5–1
Construction Material for Spectrophotometer Optics

Material	Transmittance Range, μm	Refractive Index[a]	Angular Dispersion[b]	Useful Range for Prisms, μm	Hardness and Chemical Resistance
Fused silica	0.18–3.3	1.46	0.52×10^{-4}	0.18–2.7	Excellent
Quartz	0.20–3.3	1.54	0.63×10^{-4}	0.20–2.7	Excellent
Flint glass	0.35–2.2	1.66	1.70×10^{-4}	0.35–2	Excellent
Calcium fluoride (fluorite)	0.12–12	1.43	0.33×10^{-4}	5–9.4	Good
Lithium fluoride	0.12–6	1.39	0.29×10^{-4}	2.7–5.5	Poor
Sodium chloride	0.3–17	1.54	0.94×10^{-4}	8–16	Poor
Potassium bromide	0.3–29	1.56	1.45×10^{-4}	15–28	Poor
Cesium iodide	0.3–70	1.79	—	15–55	Poor
KRS-5 (TlBr-TlI)	1–40	2.63	—	24–40	Good

[a] At 0.589 μm.
[b] Radians/μm at 0.589 μm

ization of the emitted radiation; this type of prism is called a *Cornu prism.* Figure 5–11**b** shows a *Bunsen monochromator,* which employs a 60-deg prism, likewise often made of quartz.

A *Littrow prism,* which permits more compact monochromator designs, is a 30-deg prism with a mirrored back (see Figure 5–13**b**). It is seen that refraction occurs twice at the same interface; the performance characteristics of the Littrow prism are thus similar to those of the 60-deg prism in a Bunsen mount. With quartz prisms, polarization is canceled by the reversal of the radiation path. A typical Littrow-type monochromator is shown in Figure 7–11.

5D–4 PERFORMANCE CHARACTERISTICS OF MONOCHROMATORS

The quality of a monochromator is dependent upon the purity of its radiant output, its ability to resolve adjacent wavelengths, its light gathering power, and its spectral bandwidth. The last property is discussed in Section 5D–5.

Spectral Purity. The exit beam of a monochromator is usually contaminated with small amounts of *scattered* or *stray* radiation having wavelengths far different from that of the instrument setting. This unwanted radiation can be traced to several sources. Among these are reflections of the beam from various optical parts and the monochromator housing; the former arise from mechanical imperfections, particularly in gratings, introduced during manufacture. Scattering by dust particles in the atmosphere or on the surfaces of optical parts also causes stray radiation to reach the exit slit. Generally, the effects of spurious radiation are minimized by introducing baffles in appropriate spots in the monochromator and by coating interior surfaces with flat

Table 5–2
Design Characteristics of Instruments for Optical Spectroscopy

Major Design Types	Number of Channels	Sub Class	Common Examples
Temporal	1	Nondispersive →	→ Interchangeable filter → Tunable laser
		Dispersive →	→ Sequential linear scan → Sequential slew scan
Spatial	Many	Nondispersive →	→ Multiple filter and detector systems
		Dispersive →	→ Photographic plate → Multiple detector system → Silicon diode arrays → Vidicon tubes
Multiplex	1	Nondispersive →	→ Fourier transform system → Correlation methods
		Dispersive →	→ Hadamard transform systems

slit. Spectra could then be obtained by rotating the dispersing element manually or mechanically while monitoring the output of the detector.

A *sequential linear scan instrument* often contains a motor-driven prism or grating system that sweeps the spectral region of interest at a constant rate. The paper drive of a recorder can be synchronized with the motion of the dispersing element, thus providing a wavelength scale based upon time. Often, much time is wasted in scanning spectral regions that contain little or no spectral information.

A *sequential slew scan instrument* may be similar to the one just described except that it is programmed to recognize and remain at significant spectral features such as peaks until a suitable signal-to-noise ratio is attained. Regions in which the radiant power is not changing rapidly (that is, where the derivative of power with respect to time approaches zero) are scanned at high speed until the next peak is approached. Slew scan instruments often require very precise methods for locating peak maxima and logic circuits to control the scan rate. Note, however, that the simple filter photometer described in the section on nondispersive instruments provides a very simple means of slew scanning.

Sequential slew scan instruments, while more complex and expensive than linear scan counterparts, can provide data more rapidly and efficiently.

5I–4 SPATIAL DESIGNS

Spatial instruments are based upon multiple detectors or channels to obtain information about different parts or elements of the spectrum *simultaneously*.

Nondispersive Systems. As an example of a nondispersive spatial device, a photometer for the simultaneous determination of sodium, potassium, and lithium has been described in which the radiation from a flame containing the sample is allowed to illuminate three slits arranged at different angles from the source. A photomultiplier tube (and associated electronics and readout) is placed at each slit, as well as interference filters that selectively transmit the peak radiation for one of the elements; simultaneous monitoring of the concentration of the three elements is thus possible.

Dispersive Systems. The classic dispersive spatial instrument is the spectrograph in which a photographic plate or film, located at the focal plane of a monochromator, stores all of the elements of a spectrum simultaneously. Unfortunately, to retrieve this information requires a good deal of time for film processing as well as determining the degree of blackening of the photosensitive surface.

The direct-reading spectrometer, which has been widely applied in the metals industry and by geologists and geochemists for the simultaneous quantitative determination of a dozen or more elements based on the intensity of emission lines, consists of a monochromator with a series of exit slits and photomultiplier tubes fixed at suitable locations on the focal plane. The output of each phototube is collected in a capacitor for readout when the excitation process is complete. The size of the slit and bulk of the photomultiplier tube severely limits the number of channels that can be observed. In addition, adjusting such instruments from one set of elements to another is difficult or impossible. Instruments of this kind are generally quite costly.

A promising new type of spatial dispersive instruments is based on the silicon diodes or vidicons discussed in Section 5F–3. These detectors, when placed at the focal plane of a monochromator, provide as many as 1000 individual detectors whose outputs can be amplified, processed, and read out nearly simultaneously.

Multichannel dispersive instruments are generally more complex and more costly than single channel temporal instruments.

Most are microprocessor controlled and provide data in a variety of forms. They do offer distinct advantages in the speed in which spectra can be obtained without degradation in signal-to-noise ratio. This speed, of course, results from the fact that all elements of the spectra are being measured simultaneously rather than one element at a time as in the single channel case. Multichannel instruments can also provide enhanced sensitivity and precision because the speed of measurement makes signal averaging more practical. As was pointed out in Section 2G–2, signal averaging often permits the extraction of a minute signal from a noisy environment.

In many applications, multichannel instruments also offer the advantage of significantly smaller sample consumption. Again, these instruments are generally more expensive than most single channel devices.

5I–5 MULTIPLEX DESIGNS

The term multiplex comes from communication theory, where it is used to describe systems in which many sets of information are transported simultaneously through a single channel. As the name implies, multiplex analytical instruments are single channel devices in which all elements of the signal are observed *simultaneously*. In order to determine the magnitude of each of these elements it is necessary to modulate the analyte signal in a way that permits subsequent decoding of the signal to give its component parts or elements.[20]

---

[20] Not all writers reserve the term multiplexing for systems in which all elements of a signal are observed simultaneously. For example, some use the term to indicate any system in which a number of independent signals can be encoded and transmitted over a single transmitting medium. By this definition, most of the instruments in Table 5–2 are multiplex devices. (See: K. W. Busch and L. D. Benton, *Anal. Chem.*, **1983**, *55*, 445A.) In this text, the first, more restrictive, definition will be used.

Most multiplex instruments depend upon the *Fourier transform* for signal decoding and are consequently often called Fourier transform instruments. Such instruments are by no means confined to optical spectroscopy. Indeed, Fourier transform devices have been described for nuclear magnetic resonance, mass, and microwave spectroscopy as well as for certain types of electroanalytical measurements. Several of these instruments will be described in some detail in subsequent chapters.

In optical spectroscopy, decoding has also been performed by the so-called *Hadamard transform*. This procedure has not found widespread application to date; thus, it will be given only brief mention in Section 11C-2.

Fourier transform spectroscopy was first developed by astronomers in the early 1950s in order to study the infrared spectra of distant stars; only by the Fourier technique could the very weak signals from these sources be isolated from environmental noise. The first chemical applications of Fourier transform spectroscopy, which were reported approximately a decade later, were to the energy-starved far-infrared region; by the late 1960s, instruments for chemical studies in both the far-infrared (10 to 400 cm^{-1}) and the ordinary-infrared regions were available commercially.[21] Descriptions of Fourier transform instruments for the ultraviolet and visible spectral regions can also be found in the literature, but their adoption has not been widespread to date. For this type of radiation, the advantages of the method are not great enough to justify the expense and complexity of the equipment.

An Inherent Advantage of Fourier Transform Spectroscopy. For purposes of this discussion, it is convenient to think of an experimentally derived spectrum as being made up of *m* individual transmittance measurements at equally spaced frequency or wavelength intervals called *resolution elements*. The quality of the spectrum—that is, the amount of spectral detail—increases as the number of resolution elements becomes larger or as the frequency intervals between measurements become smaller.[22] Thus, in order to increase spectral quality, *m* must be made larger; clearly, increasing the number of resolution elements must also increase the time required for obtaining a spectrum with a scanning instrument.

Consider, for example, the derivation of an infrared spectra from 500 to 5000 cm^{-1}. If resolution elements of 3 cm^{-1} were chosen, *m* would be 1500; if 0.5 second were required for recording the transmittance of each resolution element, 750 sec or 12.5 min would be needed to obtain the spectrum. Reducing the width of the resolution element to 1.5 cm^{-1} would be expected to provide significantly greater spectral detail; it would also double the number of resolution elements as well as the time required for their measurement.

For most optical instruments, particularly those designed for the infrared region, decreasing the width of the resolution element has the unfortunate effect of decreasing the signal-to-noise ratio because narrower slits must be used, which lead to weaker signals

---

[21] For a more complete discussion of Fourier transform infrared spectroscopy, consult the following references: P. R. Griffiths, *Chemical Fourier Transform Spectroscopy*, Wiley: New York, 1975; *Transform Techniques in Chemistry*, P. R. Griffiths, Ed., Plenum Press: New York, 1978; *Fourier Transform Infrared Spectroscopy: Applications to Chemical Systems*, J. R. Ferraro and L. J. Basile, Eds., Vol. I, 1978 and Vol. II, 1979, Academic Press: New York; A. G. Marshall, *Fourier, Hadamard and Hilbert Transforms in Chemistry*, Plenum Press: New York, 1982. For a brief review of recent developments, see: P. R. Griffiths, *Science*, **1983**, *222*, 297.

---

[22] With a recording spectrophotometer, of course, individual point-by-point measurements are not made; nevertheless, the idea of a resolution element is useful, and the ideas generated from it apply to recording instruments as well.

reaching the transducer. For infrared detectors, the reduction in signal is not accompanied by a corresponding decrease in noise. Therefore, a degradation in signal-to-noise ratio results.

In Section 2G–2, it was pointed out that marked improvements in signal-to-noise ratios accompany signal averaging. Here, it was shown (Equation 2–21) that the signal-to-noise ratio S/N for the average of n measurements is given by

$$\frac{S}{N} = \frac{S_n}{N_n} \times \sqrt{n}$$

where S_n and N_n are the averaged signal and noise. The application of signal averaging to conventional spectroscopy is, unfortunately, costly in terms of time. Thus, in the example just considered, 750 sec were required to obtain a spectrum of 1500 resolution elements. To improve the signal-to-noise ratio by a factor of 2 would require averaging 4 spectra, which would then require 4×750 sec or 50 min.

Fourier transform spectroscopy differs from conventional spectroscopy in that all of the resolution elements for a spectrum are measured *simultaneously*, thus reducing enormously the time required to derive a spectrum at any chosen signal-to-noise ratio. An entire spectrum of 1500 resolution elements can then be recorded in about the time required to observe just one element by conventional spectroscopy (0.5 sec in our earlier example). This large decrease in observation time is often used to markedly enhance the signal-to-noise ratio of Fourier transform measurements. For example, in the 750 sec required to derive the spectrum in the earlier sample, 1500 Fourier transform spectra could be recorded and averaged. According to Equation 2–21, the improvement in signal-to-noise ratio would in principle be $\sqrt{1500}$ or about 39. This inherent advantage of Fourier transform spec-

troscopy was first recognized by P. Fellgett in 1958 and has become known as the *Fellgett advantage*. It is worth noting here that for several reasons, the theoretical $\sqrt{n}$ improvement in S/N is seldom realized. Nonetheless, major gains in signal-to-noise ratios are generally observed with the Fourier transform technique.

Time Domain Spectroscopy. Conventional spectroscopy can be termed *frequency domain* spectroscopy in that radiant power data are recorded as a function of frequency (or the inversely related wavelength). In contrast, *time domain* spectroscopy, which can be achieved by the Fourier transform, is concerned with changes in radiant power with time. Figure 5–28 illustrates the difference.

The plots in (**c**) and (**d**) are conventional spectra of two monochromatic sources with frequencies ν_1 and ν_2 Hz. The curve in (**e**) is the spectrum of a source containing both frequencies. In each case, some measure of the radiant power, $P(\nu)$ is plotted with respect to the frequency in hertz. The symbol in parentheses is added to emphasize the frequency dependence of the power; time domain power will be indicated by $P(t)$.

The curve in Figure 5–28**a** shows the time domain spectra for each of the monochromatic sources. The two have been plotted together in order to make the small frequency difference between them more obvious. Here, the instantaneous power $P(t)$ is plotted as a function of time. The curve in Figure 5–28**b** is the time domain spectrum of the source containing the two frequencies. As is shown by the horizontal arrow, the plot exhibits a periodicity or *beat* as the two waves go in and out of phase.

Examination of Figure 5–29 reveals that the time domain spectrum for a source containing several wavelengths is considerably more complex than those shown in Figure 5–28. Because a large number of wavelengths are involved, a full cycle is not realized in the time shown. To be sure, a pattern of beats can be observed as certain wave-

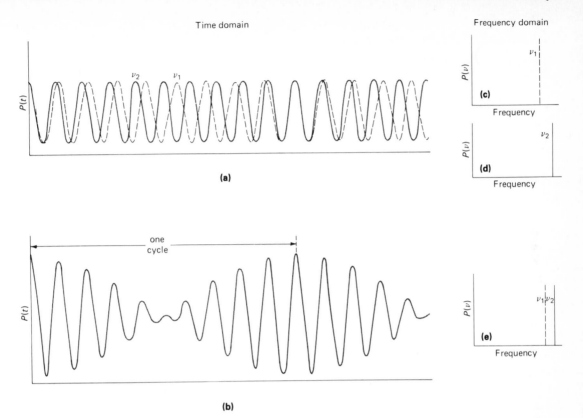

Time domain

Frequency domain

$P(t)$

ν_2 ν_1

(a)

$P(\nu)$

ν_1

(c)

Frequency

$P(\nu)$

ν_2

(d)

Frequency

$P(t)$

one
cycle

(b)

$P(\nu)$

$\nu_1\nu_2$

(e)

Frequency

FIGURE 5–28   Illustrations of time domain plots (**a**) and (**b**); frequency domain plots (**c**), (**d**), and (**e**).

lengths pass in and out of phase. In general, the signal power decreases with time as a consequence of the various closely spaced wavelengths becoming more and more out of phase.

It is important to appreciate that a time domain spectrum contains the same information as does a spectrum in the frequency domain, and in fact, one can be converted to the other by mathematical manipulations. Thus, Figure 5–28**b** was derived from Figure 5–28**e** by means of the equation

$$P(t) = k(\cos 2\pi\nu_1 t + \cos 2\pi\nu_2 t) \qquad (5\text{–}18)$$

where k is a constant and t is the time. The difference in frequency between the two lines was approximately 10% of ν_2.

The interconversion of time and fre-

quency domain spectra becomes exceedingly complex and mathematically tedious when more than a few lines are involved; the operation is only practical with a high-speed computer.

Methods of Obtaining Time Domain Spectra. Time domain spectra, such as those shown in Figures 5–28 and 5–29, cannot be acquired experimentally with radiation of the frequency that is useful for spectroscopy (10^{15} Hz for ultraviolet to 10^7 Hz for nuclear magnetic resonance) because transducers that will respond to power variations at these enormous frequencies do not exist. Thus, a typical transducer yields a signal that corresponds to the average power of a high-frequency signal and not to its periodic variation. To obtain time domain spectra requires, therefore, a method of converting

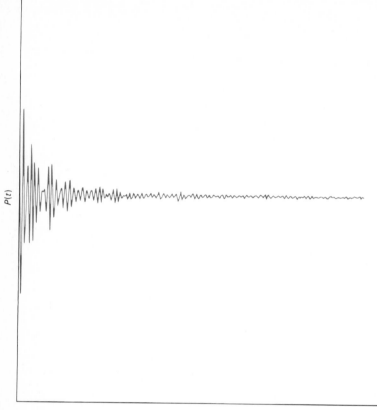

FIGURE 5-29 Time domain spectrum of a source made up of several wavelengths.

(or *modulating*) a high-frequency signal to one of measurable frequency without distorting the time relationships carried in the signal; that is, the frequencies in the modulated signal must be directly proportional to those in the original. Different signal-modulation procedures are employed for the various wavelength regions of the spectrum. The Michelson interferometer is used extensively for measurements in the optical region.

The Michelson Interferometer. The device used for modulating optical radiation is an interferometer similar in design to one first described by Michelson in 1891. The Michelson interferometer is a device that splits a beam of radiation into two beams of nearly equal power and then recombines them in such a way that intensity variations of the combined beam can be measured as a function of differences in the lengths of the paths of the two halves. Figure 5–30 is a schematic diagram of such a device as it is used for optical Fourier transform spectroscopy.

As shown in the figure, a beam of radiation from a source is collimated and impinges on a beam splitter, which transmits

EXAMPLE 5-2. Compare the size of (1) a fused silica prism; (2) a glass prism; and (3) a grating with 1200 lines/mm that would be required to resolve two lithium emission lines at 460.20 and 460.30 nm. Average values for the dispersion ($dn/d\lambda$) of fused silica and glass in the region of interest are 1.3×10^{-4} and 3.6×10^{-4} nm^{-1}, respectively.

For the two prisms, we employ Equation 5–13

$$R = \frac{\lambda}{\Delta\lambda} = b\frac{dn}{d\lambda}$$

or

$$b = \frac{\lambda}{\Delta\lambda} \times \frac{1}{dn/d\lambda}$$

$$= \frac{460.25}{460.30 - 460.20} \times \frac{1}{dn/d\lambda}$$

For fused silica,

$$b = \frac{460.25 \text{ nm}}{0.10 \text{ nm}} \times \frac{10^{-7} \text{ cm/nm}}{1.3 \times 10^{-4} \text{ nm}^{-1}}$$

$$= 3.5 \text{ cm}$$

and for glass,

$$b = \frac{460.25}{0.10} \times \frac{1}{3.6 \times 10^{-4}} \times 10^{-7} = 1.3 \text{ cm}$$

For the grating, we employ Equation 5–12

$$R = \frac{\lambda}{\Delta\lambda} = \mathbf{n}N$$

For the first-order spectrum ($\mathbf{n} = 1$)

$$N = \frac{460.25}{0.10 \times 1} = 4.60 \times 10^3 \text{ lines}$$

$$\text{grating length} = \frac{4.6 \times 10^3 \text{ lines}}{1200 \text{ lines/mm}} \times \frac{0.1 \text{ cm}}{\text{mm}}$$

$$= 0.38 \text{ cm}$$

Thus, in theory, the separation can be accomplished with a silica prism with a base of 3.5 cm, a glass prism with a base of 1.3 cm, or 0.38 cm of grating.

Light Gathering Power of Monochromators. In order to increase the signal-to-noise ratio it is necessary that the radiant energy reaching the detector be as large as possible. The *f/number* or *speed* provides a measure of the ability of a monochromator to collect the radiation emerging from the entrance slit. The *f/number* is defined by the equation

$$f = F/d \qquad (5-14)$$

where F is the focal length of the collimating

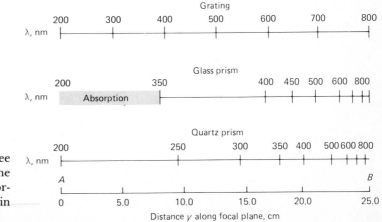

FIGURE 5–15 Dispersion for three types of monochromators. The points *A* and *B* on the scale correspond to the points shown in Figure 5–11.

mirror (or lens) and d is its diameter. The light gathering power of an optical device increases as the inverse square of the f/number. Thus, an $f/2$ lens gathers four times more light than an $f/4$ lens. The f/numbers for many monochromators lie in the 1 to 10 range.

5D-5 MONOCHROMATOR SLITS

The slits of a monochromator play an important role in determining its performance characteristics and quality. Slit jaws are formed by carefully machining two pieces of metal to give sharp edges, as shown in Figure 5–16. Care is taken to assure that the edges of the slit are exactly parallel to one another and that they lie on the same plane.

In some monochromators, the openings of the two slits are fixed; more commonly, the spacing can be adjusted with a micrometer mechanism.

The entrance slit (see Figure 5–11) of a monochromator serves as a radiation source; its image is ultimately focused on the surface containing the exit slit. If the radiation source consists of a few discrete wavelengths, a series of rectangular images appears on this surface as bright lines, each corresponding to a given wavelength. A particular line can be brought to focus on the exit slit by rotating the dispersing element. If the entrance and exit slits are of the same size (as is usually the case), the image of the entrance slit will in theory just fill the exit-slit opening when the setting of the

monochromator corresponds to the wavelength of the radiation. Movement of the monochromator mount in one direction or the other results in a continuous decrease in emitted intensity, zero being reached when the entrance-slit image has been displaced by its full width.

Effect of Slit Width on Resolution. Figure 5–17 illustrates the situation in which monochromatic radiation of wavelength λ_2 strikes the exit slit. Here, the monochromator is set for λ_2 and the two slits are identical in width. The image of the entrance slit just fills the exit slit. Movement of the monochromator to a setting of λ_1 or λ_3 results in the image being moved completely out of the slit. The lower half of the figure shows a plot of the radiant power emitted as a function of monochromator setting. Note that the *bandwidth* is de-

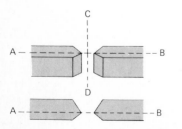

FIGURE 5–16 Construction of slits.

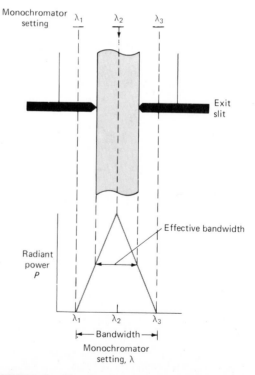

FIGURE 5-17 Illumination of an exit slit by monochromatic radiation λ_2 at various monochromator settings. Exit and entrance slits are identical.

fined as the span of monochromator settings (in units of wavelength) needed to move the image of the entrance slit across the exit slit. If polychromatic radiation were employed, it would also represent the span of wavelengths from the exit slit for a given monochromator setting.

The *effective bandwidth*, which is one-half the bandwidth when the two slit widths are identical, is seen to be the range of wavelengths exiting the monochromator at a given wavelength setting. The effective bandwidth can be related to the reciprocal linear dispersion by writing Equation 5–5 in the form

$$D^{-1} = \frac{\Delta \lambda}{\Delta y}$$

where $\Delta\lambda$ and Δy are now finite intervals of wavelength and linear distance along the focal plane respectively. As shown by Figure 5–17, when Δy is equal to the slit width w, $\Delta\lambda$ is the effective bandwidth. That is,

$$\Delta\lambda_{\text{eff}} = wD^{-1} \qquad (5\text{–}15)$$

Figure 5–18 illustrates the relationship between the effective bandwidth of an instrument and its ability to resolve spectral peaks. Here, the exit slit of a grating monochromator is illuminated with a beam composed of just three, equally spaced wavelengths, λ_1, λ_2, and λ_3; each wavelength is assumed to be of the same intensity. In the top figure, the effective bandwidth of the instrument is exactly equal to the difference

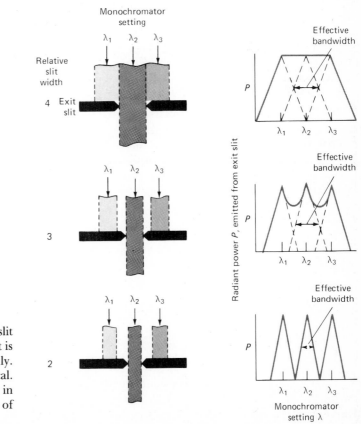

FIGURE 5–18 The effect of the slit width on spectra. The entrance slit is illuminated with λ_1, λ_2, and λ_3 only. Entrance and exit slits are identical. Plots on the right show changes in emitted power as the setting of monochromator is varied.

in wavelength between λ_1 and λ_2 or λ_2 and λ_3. When the monochromator is set at λ_2, radiation of this wavelength just fills the slit. Movement of the monochromator in either direction diminishes the transmitted intensity of λ_2, but increases the intensity of one of the other lines by an equivalent amount. As shown by the solid line in the plot to the right, no spectral resolution of the three wavelengths is achieved.

In the middle drawing of Figure 5–18, the effective bandwidth of the instrument has been reduced by narrowing the openings of the exit and entrance slits to three-quarters that of their original dimensions. The solid line in the plot on the right shows that partial resolution of the three lines results. When the effective bandwidth is decreased to one-half the difference in wavelengths of the three beams, complete resolution is obtained, as shown in the bottom drawing. Thus, complete resolution of two lines is feasible only if the slit width is adjusted so that the effective bandwidth of the monochromator is equal to one-half their wavelength difference.

> *EXAMPLE 5–3.* A grating monochromator with a reciprocal linear dispersion of 1.2 nm/mm is to be used to separate the sodium lines at 589.0 and 589.6. In theory, what slit width would be required?
>
> Complete resolution of the two lines requires that
>
> $$\Delta\lambda_{\text{eff}} = \frac{1}{2}(589.6 - 589.0) = 0.3 \text{ nm}$$
>
> Substitution into Equation 5–15 after rearrangement gives
>
> $$w = \frac{\Delta\lambda_{\text{eff}}}{D^{-1}} = \frac{0.3 \text{ nm}}{1.2 \text{ nm/mm}} = 0.25 \text{ mm}$$

It is important to note that slit widths calculated as in Example 5–3 are theoretical. Imperfections, which are present in most monochromators, are such that slit widths narrower than theoretical are often required to achieve a desired resolution.

Choice of Slit Widths. The effective bandwidth of a monochromator depends upon the dispersion of the grating or prism as well as the width of the entrance and exit slits. Most monochromators are equipped with variable slits so that the effective bandwidth can be changed. The use of minimal slit widths is desirable where the resolution of narrow absorption or emission bands is needed. On the other hand, a marked decrease in the available radiant power accompanies a narrowing of slits, and accurate measurement of this power becomes more difficult. Thus, wider slit widths may be used for quantitative analysis than for qualitative work, where spectral detail is important.

5D-6 A COMPARISON OF GRATING AND PRISM MONOCHROMATORS

Gratings offer several advantages over prisms as dispersing elements. Perhaps the most important is the wavelength independence of dispersion, which makes the design of a monochromator considerably more simple. Furthermore, the fixed dispersion makes it easy to scan an entire spectrum at constant bandwidth since the slits need only be adjusted at the outset. With a prism instrument, in contrast, much narrower slits are required at short wavelengths than at long if a constant bandwidth is to be realized. Figure 5–19 illustrates the slit width changes that are required in order to maintain a constant effective bandwidth of 1 nm with a monochromator equipped with a Littrow prism. Thus at 300 nm, a slit width of about 0.3 mm suffices; at 700 nm, the width must be reduced by a factor of 10 to 0.03 mm.

Another advantage of gratings can be seen from Example 5–2, namely, that better dispersion can be expected for the same size of dispersing element. In addition, reflec-

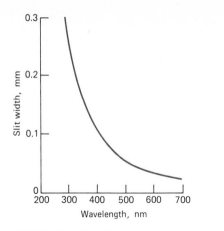

FIGURE 5-19 The slit width needed to maintain a constant effective bandwidth of 1 nm from a quartz Littrow prism monochromator.

tion gratings provide a means of dispersing radiation in the far ultraviolet and far infrared regions where absorption prevents the use of prisms.

Gratings have the disadvantage of producing somewhat greater amounts of stray radiation as well as higher-order spectra. These disadvantages are ordinarily not serious, particularly with holographic gratings, because the extraneous radiation can be minimized through the use of filters and by proper instrument design.

For a number of years, high-quality gratings were more expensive than good prisms. This disadvantage has now largely disappeared. Thus, the majority of modern optical instruments are based upon gratings rather than prisms.

5E Sample Containers

Sample containers are required for all spectroscopic studies except emission spectroscopy. In common with the optical elements of monochromators, the *cells* or *cuvettes* that hold the samples must be made of material that passes radiation in the spectral region of interest. Thus, as shown in Figure 5-2,

quartz or fused silica is required for work in the ultraviolet region (below 350 nm); both of these substances are transparent in the visible region and up to about 3 μm in the infrared region as well. Silicate glasses can be employed in the region between 350 and 2000 nm. Plastic containers have also found application in the visible region. Crystalline sodium chloride is the most common substance employed for cell windows in the infrared region; the other infrared transparent materials listed in Table 5-1 (p. 128) may also be used for this purpose.

5F Radiation Detectors

5F-1 INTRODUCTION

The detectors for early spectroscopic instruments were the human eye or a photographic plate or film. These detection devices have been largely supplanted by transducers that convert radiant energy into an electrical signal; our discussion will be confined to those more modern detectors. Brief consideration of photographic detection will be found in Section 10D-5.

Properties of the Ideal Detector. The ideal detector would have a high sensitivity, a high signal-to-noise ratio, and a constant response over a considerable range of wavelengths. In addition, it would exhibit a fast response time and a minimal output signal in the absence of illumination. Finally, the electrical signal produced by the transducer would be directly proportional to the radiant power P. That is,

$$S = kP \qquad (5-16)$$

where S is the electrical response in terms of current, voltage, or resistance and k is the calibration sensitivity (Section 1D-1).

Many detectors exhibit a small, constant response, known as a *dark current*, in the absence of radiation. In those cases, the response is described by the relationship

$$S = kP + k_d \qquad (5-17)$$

where k_d represents the dark current, which is ordinarily constant for short periods. Instruments with transducers that produce a dark current are usually equipped with a compensating circuit that reduces k_d to zero; Equation 5–16 then applies.

Types of Radiation Transducers. As indicated in Figure 5–2**d**, two general types of radiation transducers are encountered; one responds to photons, the other to heat. All photon or *photoelectric* detectors have an active surface, which is capable of absorbing radiation. In some types, the absorbed energy causes emission of electrons and the development of a photocurrent. In others, the radiation promotes electrons into conduction bands; detection here is based on the resulting en-

hanced conductivity (*photoconduction*). Only ultraviolet, visible, and near–infrared radiation have sufficient energy to cause these processes to occur. Photoelectric detectors also differ from heat detectors in that their electrical signal results from a series of individual events (absorption of one photon), the probability of which can be described by the use of statistics. In contrast, heat transducers, which are required for the detection of infrared radiation, are nonquantized sensors.

As will be shown in Section 6B–4, the distinction between photon and heat detectors is important because shot noise (p. 67) often limits the behavior of the former while Johnson noise may limit the latter. As a consequence, the indeterminate errors associated with the two types of detectors are fundamentally different.

Figure 5–20 shows the relative spectral response of the various kinds of detectors

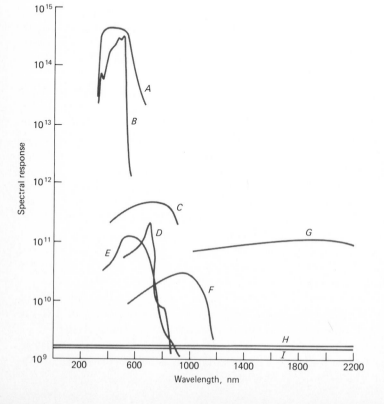

FIGURE 5-20   Relative response of various types of photoelectric transducers (*A–G*) and heat transducers (*H–I*): *A* photomultiplier tube; *B* CdS photoconductivity cell; *C* GaAs photovoltaic cell; *D* CdSe photoconductivity cell; *E* Se/SeO photovoltaic cell; *F* silicon photodiode; *G* PbS photoconductivity cell; *H* thermocouple; *I* Golay cell. (Adapted from P. W. Kruse, L. N. McGlauchlin and R. B. Quistan, *Elements of Infrared Technology,* pp. 424–425, Wiley: New York, 1962. Reprinted by permission of John Wiley & Sons, Inc.)

that are useful for ultraviolet, visible, and infrared spectroscopy. The ordinate function is inversely related to the noise of the detector and directly related to the square root of its surface area. Note that the relative sensitivity of the two heat transducers (curves *H* and *I*) is independent of wavelength but significantly lower than the sensitivity of photoelectric transducers. On the other hand, photon detectors are often far from ideal with respect to constant response versus wavelength.

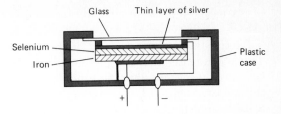

FIGURE 5–21 Schematic of a typical barrier-layer cell.

5F-2 PHOTON DETECTORS

Several types of photon detectors are available, including (1) *photovoltaic cells,* in which the radiant energy generates a current at the interface of a semiconductor layer and a metal; (2) *phototubes,* in which radiation causes emission of electrons from a photosensitive solid surface; (3) *photomultiplier tubes,* which contain a photoemissive surface as well as several additional surfaces that emit a cascade of electrons when struck by electrons from the photosensitive area; (4) *photoconductivity detectors,* in which absorption of radiation by a semiconductor produces electrons and holes, thus leading to enhanced conductivity; and (5) *silicon photodiodes,* in which photons increase the conductance across a reverse-biased *pn* junction.

Photovoltaic or Barrier-Layer Cells. The photovoltaic cell is used primarily to detect and measure radiation in the visible region. The typical cell has a maximum sensitivity at about 550 nm; the response falls off to perhaps 10% of the maximum at 350 and 750 nm (see Figure 5–20*E*). Its range approximates that of the human eye.

The photovoltaic cell consists of a flat copper or iron electrode upon which is deposited a layer of semiconducting material, such as selenium or copper(I) oxide (see Figure 5–21). The outer surface of the semiconductor is coated with a thin transparent metallic film of gold, silver, or lead, which serves as the second or collector electrode; the entire array is protected by a transparent envelope. When radiation of sufficient energy reaches the semiconductor, covalent bonds are broken, with the result that conduction electrons and holes are formed. The electrons then migrate toward the metallic film and the holes toward the base upon which the semiconductor is deposited. The liberated electrons are free to migrate through the external circuit to interact with these holes. The result is an electrical current of a magnitude that is proportional to the number of photons striking the semiconductor surface. Ordinarily, the currents produced by a photovoltaic cell are large enough to be measured with a galvanometer or microammeter; if the resistance of the external circuit is kept small (<400 Ω), the photocurrent is directly proportional to the power of the radiation striking the cell. Currents on the order of 10 to 100 μA are typical.

The barrier-layer cell constitutes a rugged, low-cost means for measuring radiant power. No external source of electrical energy is required. On the other hand, the low internal resistance of the cell makes the amplification of its output less convenient. Consequently, although the barrier-layer cell provides a readily measured response at high levels of illumination, it suffers from lack of sensitivity at low levels. Another disadvantage of the barrier-type cell is that it exhibits *fatigue;* that is, its current output decreases gradually during continued illumination; proper circuit design and choice of experimental conditions minimize this effect.

Barrier-type cells find use in simple, portable instruments where ruggedness and low cost are important. For routine analyses, these instruments often provide perfectly reliable analytical data.

Vacuum Phototubes.[12]   A second type of photo-electric device is the vacuum phototube, which consists of a semicylindrical cathode and a wire anode sealed inside an evacuated transparent envelope (see Figure 5–22). The concave surface of the electrode supports a layer of photoemissive material (Section 4B–1) that tends to emit electrons upon being irradiated. When a potential is applied across the electrodes, the emitted electrons flow to the wire anode generating a photocurrent, which is generally about one-tenth as great as that associated with a photovoltaic cell for a given radiant intensity. In contrast, however, amplification is easily accomplished (see Figure 2–16, p. 44) since the phototube has a high electrical resistance.

---

[12] For a discussion of vacuum phototubes and photo-multiplier tubes, see: F. E. Lytle, *Anal. Chem.*, **1974**, *46*, 545A.

The number of electrons ejected from a photoemissive surface is directly proportional to the radiant power of the beam striking that surface. As the potential applied across the two electrodes of the tube is increased, the fraction of the emitted electrons reaching the anode rapidly increases; when the saturation potential is achieved, essentially all of the electrons are collected at the anode. The current then becomes independent of potential and directly proportional to the radiant power. Phototubes are usually operated at a potential of about 90 V, which is well within the saturation region.

A variety of photoemissive surfaces are used in commercial phototubes. Typical examples are shown in Figure 5–23. From the user's standpoint, photoemissive surfaces fall into four categories: highly sensitive, red sensitive, ultraviolet sensitive, and flat response. The most sensitive cathodes are bialkali types such as number 117 in Figure 5–23; it is made up of potassium, cesium, and antimony. Red-sensitive materials are multialkali types (Na/K/Cs/Sb, for example) or Ag/O/Cs formulations. The behavior of the latter is shown as S-11 in the figure. Compositions of Ga/In/As extend the red region

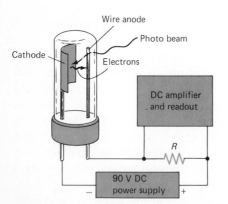

FIGURE 5-22   A phototube and accessory circuit. The photocurrent induced by the radiation causes a potential drop across *R*, which is then amplified to drive a meter or recorder.

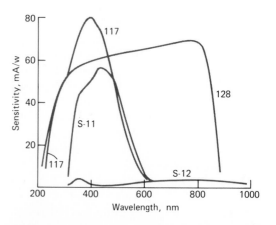

FIGURE 5-23   Spectral response of some typical photoemissive surfaces. (From: F. E. Lytle, *Anal. Chem.*, **1974**, *46*, 546A. Copyright 1974 American Chemical Society.)

up to about 1.1 μm. Most formulations are ultraviolet sensitive provided the tube is equipped with transparent windows. Flat responses are obtained with Ga/As compositions such as that labeled 128 in Figure 5–23.

Phototubes frequently produce a small dark current (see Equation 5–17), which results from thermally induced electron emission and natural radioactivity from ^{40}K in the glass housing of the tube.

Photomultiplier Tubes. For the measurement of low radiant power, the *photomultiplier* tube offers advantages over the ordinary photo-

tube.[13] Figure 5–24 is a schematic diagram of such a device. The cathode surface is similar in composition of those of the phototubes described in Figure 5–23, electrons being emitted upon exposure to radiation. The tube also contains additional electrodes (nine in Figure 5–24) called *dynodes*. Dynode 1 is maintained at a potential 90 V more positive than the cathode, and electrons are

---

[13] For a detailed discussion of the theory and applications of photomultipliers, see: R. W. Engstrom, *Photomultiplier Handbook*, RCA Corporation: Lancaster PA, 1980.

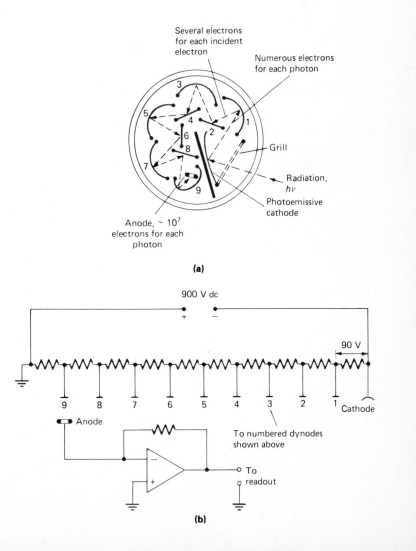

FIGURE 5-24 Photomultiplier tube: (**a**) cross section of the tube; (**b**) electrical circuit.

accelerated toward it as a consequence. Upon striking the dynode, each photoelectron causes emission of several additional electrons; these, in turn, are accelerated toward dynode 2, which is 90 V more positive than dynode 1. Again, several electrons are emitted for each electron striking the surface. By the time this process has been repeated nine times, 10^6 to 10^7 electrons have been formed for each photon; this cascade is finally collected at the anode. The resulting current is then electronically amplified and measured.

As shown by Figure 5–20A, photomultipliers are highly sensitive to ultraviolet and visible radiation; in addition, they have extremely fast time responses. Often, the sensitivity of an instrument with a photomultiplier detector is limited by its dark current emission. Because thermal emission is the major source of dark current electrons, the performance of a photomultiplier can be enhanced by cooling. In fact, thermal dark currents can be virtually eliminated by cooling the detector to $-30°C$. Detector housings, which can be cooled by circulation of an appropriate coolant, are available commercially.

Photoconductivity Detectors. The most sensitive detectors for monitoring radiation in the near-infrared region of about 0.75 to 3 μm are semiconductors whose resistances decrease when radiation within this range is absorbed. Crystalline semiconductors are formed from the sulfides, selenides, and stibnides of such metals as lead, cadmium, gallium, and indium. Absorption of radiation by these materials promotes some of their bound electrons into an energy state in which they are free to conduct electricity. The resulting change in conductivity can then be measured with a Wheatstone bridge arrangement (Figure A1–6, Appendix 1) or, alternatively, with a circuit such as that shown in Figure 2–18**b** (p. 47).

The most widely used photoconductive material is lead sulfide, which is sensitive in the region between 0.8 and 2 μm (12,500 to 5000 cm^{-1}). A thin layer of this compound is deposited on glass or quartz plates to form the cell. The entire assembly is then sealed in an evacuated container to protect the semiconductor from reaction with the atmosphere. The sensitivity of a lead sulfide detector is shown in Figure 5–20G. Figure 5–20B depicts the responses of a CdS photoconductor.

Silicon Diode Detectors. A silicon diode detector consists of a reverse-biased pn junction formed on a silicon chip. As shown in Figure 2–1**d** (Section 2A–2), the reverse bias creates a depletion layer which reduces the conductance of the junction to nearly zero. If radiation is allowed to impinge on the depletion layer, however, holes and electrons are formed, which provide a current that is proportional to radiant power.

A silicon diode detector is more sensitive than a simple vacuum phototube but less sensitive than a photomultiplier tube (see Figure 5–20F). Most photodiodes have spectral ranges from about 400 to 1100 nm although some respond to radiation as short as 200 nm.

5F–3   MULTICHANNEL PHOTON DETECTORS

Multichannel photon detectors consist of an array of tiny photosensitive devices that are capable of converting an optical image into a charge pattern that can be read out as a video electrical signal. Image sensors of this kind, which were initially developed for television, have recently found use as detectors for spectroscopic instruments. In this application, the image sensor is located at the focal plane of a monochromator. When the dispersed radiation strikes the multichannel detector, a charge pattern develops which is related to the intensity of radiation along the focal plane. This charge pattern is detected and stored for ultimate conversion to a spectrum. In effect then, all of the elements of the spectrum are observed *simulta-*

neously rather than serially. Thus, a multichannel detector functions in the same way as a photographic plate, providing a series of images of the entrance slit each of which corresponds to a different band of wavelengths. Recording all of the elements of a spectrum simultaneously offers several advantages, which will be described subsequently.

Several types of multichannel detectors are available from commercial sources.[14] Two of these, the silicon *vidicon* and the silicon *diode array* appear to have the most promise for spectroscopic applications; both are based upon the silicon diode detector.

---

[14] For a discussion of multichannel detectors, see: Y. Talmi, *Appl. Spectrosc.*, **1982**, *36*, 1; Y. Talmi, *Multichannel Image Detectors*, Vol. 1, 1979 and Vol. 2, 1982, American Chemical Society: Washington, D.C.; and G. Horlick and E. G. Codding, *Contemp. Top. Anal. Clin. Chem.* **1977**, *1*, 195.

Vidicons. Vidicon is a generic name for image sensing vacuum tubes, which find widespread use in television imaging. Figure 5–25**a** is a schematic diagram of a vidicon tube. The target of the tube is a silicon diode array with a diameter of 16 mm and containing over 15,000 photodiodes per square millimeter. Figure 5–25**b** shows a magnified cross section and end view of a part of the target array of photodiodes. Each photodiode consists of a cylindrical section of *p*-type silicon surrounded by an insulating layer of silicon dioxide. Each diode is thus electrically independent of neighboring diodes; all are connected to the common *n*-type layer. It should be emphasized that the schematic has been greatly magnified; indeed, the distance between centers of the *p* regions is only about 8 μm.

When the surface of the target is swept by the electron beam, each of the *p*-type cylin-

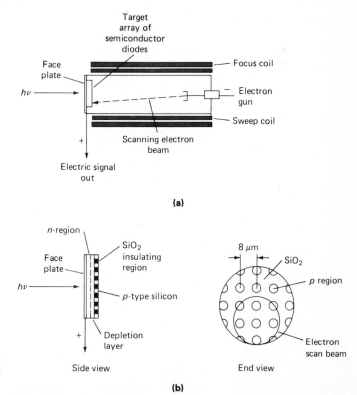

FIGURE 5–25 A vidicon tube detector: (**a**) vidicon tube and (**b**) target array of silicon diodes.

ders becomes successively charged to the potential of the beam and acts as a small charged capacitor, which remains charged (except for leakage) until a photon strikes the *n*-type surface opposite to it. Then, holes are created which diffuse toward the other surface and discharge several of the capacitors that lie along the extension of the path of the incident beam. The electron beam recharges these capacitors when it again reaches them in the course of its sweeps (see the end view, Figure 5–25b). This charging current is then amplified and serves as the signal.

The width of the electron beams is such ($\sim$20 μm) that the surface of the target is effectively divided into 512 lines or channels. The signal from each of these channels can be stored separately in a computer memory. When the vidicon tube is placed at the focal plane of a monochromator, the signal from each channel corresponds to a band of radiation of a different wavelength. It, therefore, becomes possible to obtain a spectrum without movement of the monochromator. That is, scanning is accomplished electronically rather than optically, and channel numbers rather than mono-

chromator settings serve as a measure of wavelength for spectra.

Silicon Diode Arrays. Silicon diode arrays are made up of numerous pairs of photosensitive silicon diodes, which were described earlier, and companion storage capacitors all formed on a silicon chip. The number of diode-capacitor pairs on a single chip varies from manufacturer to manufacturer; some common figures are 211, 256, 512, 1024, 2048, and 4096. Typically, the diodes have widths ranging from 15 to 50 μm and heights of about 500 μm. The lengths of the chip range from about 1 to 6 cm.

In addition to the paired photodiodes and capacitors, the chip also contains an integrated circuit for scanning each pair to give an output signal for computer processing. Figure 5–26 is a simplified schematic diagram of a photodiode array integrated circuit. Note that in parallel with each photodiode is a companion 10 pF storage capacitor. Each pair is sequentially connected to a common output line via a transistor switch. The shift register sequentially closes each of these switches momentarily causing the capacitor to be charged to −5 V. Radiation impinging on any of the diode

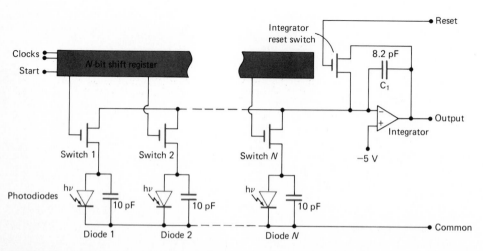

FIGURE 5–26   Block diagram of a photodiode array detector chip.

surfaces causes partial discharge of the corresponding capacitor. This lost charge is replaced during the next cycle. The resulting charging current is integrated by the preamplifier circuit, which produces a voltage that is proportional to the radiant intensity. After amplification, the analog signal from the preamplifier passes into an analog-to-digital converter and to a microprocessor that controls the readout.

An instrument consisting of a monochromator, a multichannel detector, and a computer is called an *optical multichannel analyzer*. It is of considerable importance because it allows a portion or all of a spectrum to be recorded essentially simultaneously. A disadvantage is that the physical size of the state of the art detectors places a limit on either the wavelength range or the resolution of such an analyzer. Additionally, the silicon photodetectors are not as sensitive as a photomultiplier detector. Manufacturers of photodiode arrays are now, however, offering image intensifiers, which appear to provide sensitivities that approach that of a photomultiplier.

Instruments based upon arrays and vidicon detectors are described in Sections 7D–3 and 9E–2.

5F–4 HEAT DETECTORS

The measurement of infrared radiation is difficult as a result of the low intensity of available sources and the low energy of the infrared photon. As a consequence of these properties, the electrical signal from an infrared detector is small, and its measurement requires large amplification. It is usually the detector system that limits the sensitivity and the precision of an infrared instrument.[15]

The convenient phototubes discussed earlier are generally not applicable in the infrared because the photons in this region lack the energy to cause photoemission of electrons. Thus, thermal detectors and detection based upon photoconduction (p. 140) are required (the latter is limited to infrared radiation of wavelengths less than about 2.5 to 3 μm). Neither of these is as satisfactory as the photomultiplier tube.

A discussion of thermal detectors will be deferred to Chapter 11, which deals with infrared methods.

5G Signal Processors and Readouts

The signal processor is ordinarily an electronic device that amplifies the electrical signal from the detector; in addition, it may alter the signal from dc to ac (or the reverse), change the phase of the signal, and filter it to remove unwanted components. Furthermore, the signal processor may be called upon to perform such mathematical operations on the signal as differentiation, integration, or conversion to a logarithm.

Several types of readout devices, described in Chapters 2 and Appendix 1, are found in modern instruments. Some of these include the d'Arsonval meter, digital meters, the scales of potentiometers, recorders, and cathode-ray tubes.

5G–1 PHOTON COUNTING

Frequently, the output from the photoelectric detectors described in the previous section is processed and displayed by analog techniques. That is, the average current, potential, or conductance associated with the detector is amplified and recorded or fed into a suitable meter. As was indicated in Chapter 2, analog signals such as these vary continuously; in spectroscopy, they are generally proportional to the average radiant power of the incident beam.

---

[15] For further information on infrared detectors, see: G. W. Ewing, *J. Chem. Educ.*, **1971**, *48* (9), A521; and H. Levinstein, *Anal. Chem.*, **1969**, *41* (14), 81A.

In some instances, it is possible and advantageous to employ direct digital techniques wherein electrical pulses produced by individual photons are counted. Here, radiant power is proportional to the number of pulses rather than to an average current or potential.

Counting techniques have been used for many years for measuring the power of X-ray beams and of radiation produced by the decay of radioactive species (these techniques are considered in detail in Chapter 15). Only relatively recently, however, has photon counting been applied to ultraviolet and visible radiation.[16] Here, the output of a photomultiplier tube is employed. In the previous section, it was indicated that a single photon striking the cathode of a photomultiplier ultimately leads to a cascade of 10^6 to 10^7 electrons, which comprises a pulse of current that can be amplified and counted.

Generally, the equipment for photon counting is similar to that shown in Figure 2–24 in which a comparator rejects pulses unless they exceed some predetermined minimum voltage (see Figure 2–25). Such a device is useful because dark current and instrument noise are often significantly smaller than the signal pulse and are thus not counted; an improved signal-to-noise ratio results.

Photon counting has a number of advantages over analog signal processing, including improved signal-to-noise ratio, sensitivity to low radiation levels, improved precision for a given measurement time, and lowered sensitivity to voltage and temperature changes. The required equipment is, however, more complex and expensive; the technique has not yet been widely applied for routine measurements in the ultraviolet and visible regions.

5H   Fiber Optics

In the late nineteen-sixties, analytical instruments began to appear on the market that contained fiber optics for transmitting radiation and images from one component of the instrument to another. These useful devices, sometimes called light pipes, have added a new dimension to optical instrument designs.[17]

Optical fibers are fine strands of glass or plastic that are capable of transmitting radiation for considerable distances (several hundred feet or more). The diameter of optical fibers range from 2 μm to as large as 0.6 cm. Where images are to be transmitted, bundles of fibers, fused on the ends, are employed. A major application of these fiber bundles has been to medical diagnoses, where their flexibility permits transmission of images of organs through tortuous pathways to the physician. Light pipes are used not only for observation, but also for illumination of objects; here, the ability to illuminate without heating is often of considerable importance.

Light transmission in an optical fiber takes place by total internal reflection as shown in Figure 5–27. In order for total internal reflections to occur, it is necessary that the transmitting fiber be coated with a material that has a somewhat smaller refractive index than the material from which the fiber is constructed. Thus, a typical glass fiber will consist of a core having a refractive index of about 1.6 and a glass sheath with a refractive index of approximately 1.5. Typical plastic fibers consist of a polymethylmethacrylate core ($n_1 = 1.5$) and a polymer coating of refractive index of 1.4.

A fiber, such as that shown in Figure 5–27, will transmit radiation contained in a

[16] For a review of photon counting, see: H. V. Malmstadt, M. L. Franklin, and G. Horlick, *Anal. Chem.*, **1972**, *44* (8), 63A.

[17] For a review of applications of fiber optics, see: I. Chabay, *Anal. Chem.*, **1982**, *54*, 1071A; J. K. Crum, *Anal. Chem.*, **1969**, *41*, 26A; J. I. Peterson and G. G. Vurek, *Science*, **1984**, *224*, 123.

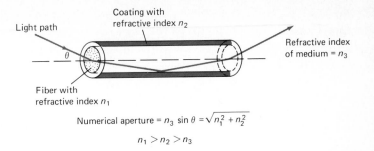

$$\text{Numerical aperture} = n_3 \sin \theta = \sqrt{n_1^2 + n_2^2}$$

$$n_1 > n_2 > n_3$$

FIGURE 5-27 Schematic drawing showing the light path through an optical fiber.

limited incident cone having a half-angle shown as θ in the figure. Incident radiation at greater angles is not reflected from but is transmitted by the sheath. The numerical aperture of the fiber provides a measure of the magnitude of the cone.

By suitable choice of construction materials, fibers that will transmit ultraviolet, visible, or infrared radiation can be manufactured. Several examples of their application to conventional analytical instruments will be found in the chapters that follow. Less conventional applications of optical fibers have been for the construction of optical sensors, sometimes called *optrodes*.[18] These devices consist of a reagent phase immobilized on the end of a fiber optic. Interaction of the analyte with the reagent creates a change in absorbance, reflectance, or luminescence, which is then transmitted to a detector via the optical fiber. At the time of writing, this technique is in its infancy and thus will not be discussed further.

5I Instrument Designs

Several hundred models of optical spectroscopic instruments are found on the market. The choice among these for a given application can be confusing even to the experienced spectroscopist. Some of this confu-

sion can be relieved by recognizing that optical instruments regardless of whether they are to be used for absorption, emission, or fluorescence measurements, can be classified into a relatively limited number of general design types each having its characteristic advantages and disadvantages. The purpose of this section is to describe the basic types of optical instruments and the performance characteristic of each. Discussions of specific types of instruments will be found in subsequent chapters dealing with various spectroscopic methods.

5I-1 THE NAMES OF VARIOUS OPTICAL INSTRUMENTS

A *spectroscope* is an optical instrument used for the visual identification of emission lines. It consists of a monochromator, such as that shown in Figure 5–11a, in which the exit slit is replaced by an eyepiece that can be moved along the focal plane. The wavelength of an emission line can then be determined from the angle between the incident and dispersed beam when the line is centered on the eyepiece.

A *colorimeter* is an instrument for absorption measurements in which the human eye serves as the detector. One or more color-comparison standards are required.

A *photometer* consists of a source, a filter, and a photoelectric detector plus a signal processor and readout. Filter photometers are commercially available for absorption measurements in the ultraviolet, visible, and

---

[18] See: W. Rudolf Seitz, *Anal. Chem.*, **1984**, *56*, 16A.

infrared regions, as well as emission and fluorescence in the first two wavelength regions. Photometers designed for fluorescence measurements are also called *fluorometers*.

A *spectrograph* is similar in construction to the two monochromators shown in Figure 5–11 except that the focal plane is made up of a holder for a photographic film or plate. Here, all of the elements of the dispersed radiation are recorded *simultaneously*. Recently, as mentioned in Section 5F–3, instruments have become available in which a multichannel electronic detector, such as a vidicon or a diode array, replaces the film holder in a spectrograph. Some writers have chosen to include these devices in the term spectrograph.

Monochromators with a fixed slit in the focal plane (such as those in Figure 5–11) are called *spectrometers*. A spectrometer equipped with a photoelectric detector is called a *spectrophotometer*. In contrast to spectrographs, spectrometers or spectrophotometers are single channel devices in which each element of the spectrum is viewed serially, not simultaneously. A spectrophotometer for fluorescence analysis is sometimes called a *spectrofluorometer*. Spectrophotometers are employed for absorbance measurements in the ultraviolet, visible, and infrared regions, and for fluorescence measurements in the former two.

5I–2 TYPES OF DESIGN OF OPTICAL INSTRUMENTS

Spectroscopic instruments are designed to provide bands of radiation of known wavelength and in most instances, to give information about the intensity or power of these bands. Three basic instrument designs for accomplishing these purposes are recognizable: temporal designs, spatial designs, and multiplex designs. Within each of these major categories are two subclasses, nondispersive and dispersive. The characteristics of these design categories, which are illustrated in Table 5–2, are described in the sections that follow.[19]

5I–3 TEMPORAL DESIGNS

Temporal instruments operate with a single detector and are often termed *single channel* devices as a consequence. In such systems, successive radiation bands are examined sequentially in time.

Nondispersive Instruments.   An example of a nondispersive temporal instrument would be a photometer equipped with a series of narrow band filters of appropriate wavelength. Such an instrument could, for example, be used for the quantitative determination of each of the alkali metals by injection of a solution of the sample into a flame. By interchanging filters, the intensity of a line for each of the elements could be determined sequentially and used to calculate its concentration.

Tunable lasers (Section 5B–3) also permit the construction of a nondispersive temporal instrument for determining a portion of the absorption or emission spectrum of a compound. Here, a photomultiplier tube could provide a series of light intensity data as the laser is tuned serially from one wavelength to the next.

Nondispersive instruments generally offer the advantage of simplicity, low cost, higher energy throughput (often leading to better signal-to-noise ratios) and lower levels of stray radiation. On the other hand, they often do not provide important spectral detail over a wide range of wavelengths, which is particularly important for qualitative and structural studies.

Dispersive Instruments.   The two monochromators depicted in Figure 5–11 could be operated as temporal dispersive instruments by locating a photoelectric detector at the exit

---

[19] This classification scheme is taken from: J. D. Winefordner, J. J. Fitzgerald, and N. Omenetto, *Appl. Spectrosc.,* **1975,** *29,* 369.

Table 5-2
Design Characteristics of Instruments for Optical Spectroscopy

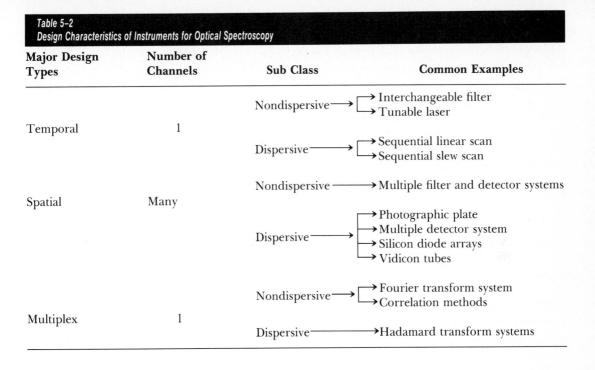

Major Design Types	Number of Channels	Sub Class	Common Examples
Temporal	1	Nondispersive →	Interchangeable filter / Tunable laser
		Dispersive →	Sequential linear scan / Sequential slew scan
Spatial	Many	Nondispersive →	Multiple filter and detector systems
		Dispersive →	Photographic plate / Multiple detector system / Silicon diode arrays / Vidicon tubes
Multiplex	1	Nondispersive →	Fourier transform system / Correlation methods
		Dispersive →	Hadamard transform systems

slit. Spectra could then be obtained by rotating the dispersing element manually or mechanically while monitoring the output of the detector.

A *sequential linear scan instrument* often contains a motor-driven prism or grating system that sweeps the spectral region of interest at a constant rate. The paper drive of a recorder can be synchronized with the motion of the dispersing element, thus providing a wavelength scale based upon time. Often, much time is wasted in scanning spectral regions that contain little or no spectral information.

A *sequential slew scan instrument* may be similar to the one just described except that it is programmed to recognize and remain at significant spectral features such as peaks until a suitable signal-to-noise ratio is attained. Regions in which the radiant power is not changing rapidly (that is, where the derivative of power with respect to time approaches zero) are scanned at high speed until the next peak is approached. Slew scan instruments often require very precise methods for locating peak maxima and logic circuits to control the scan rate. Note, however, that the simple filter photometer described in the section on nondispersive instruments provides a very simple means of slew scanning.

Sequential slew scan instruments, while more complex and expensive than linear scan counterparts, can provide data more rapidly and efficiently.

5I-4 SPATIAL DESIGNS

Spatial instruments are based upon multiple detectors or channels to obtain information about different parts or elements of the spectrum *simultaneously*.

Nondispersive Systems. As an example of a nondispersive spatial device, a photometer for the simultaneous determination of sodium, potassium, and lithium has been described in which the radiation from a flame containing the sample is allowed to illuminate three slits arranged at different angles from the source. A photomultiplier tube (and associated electronics and readout) is placed at each slit, as well as interference filters that selectively transmit the peak radiation for one of the elements; simultaneous monitoring of the concentration of the three elements is thus possible.

Dispersive Systems.   The classic dispersive spatial instrument is the spectrograph in which a photographic plate or film, located at the focal plane of a monochromator, stores all of the elements of a spectrum simultaneously. Unfortunately, to retrieve this information requires a good deal of time for film processing as well as determining the degree of blackening of the photosensitive surface.

The direct-reading spectrometer, which has been widely applied in the metals industry and by geologists and geochemists for the simultaneous quantitative determination of a dozen or more elements based on the intensity of emission lines, consists of a monochromator with a series of exit slits and photomultiplier tubes fixed at suitable locations on the focal plane. The output of each phototube is collected in a capacitor for readout when the excitation process is complete. The size of the slit and bulk of the photomultiplier tube severely limits the number of channels that can be observed. In addition, adjusting such instruments from one set of elements to another is difficult or impossible. Instruments of this kind are generally quite costly.

A promising new type of spatial dispersive instruments is based on the silicon diodes or vidicons discussed in Section 5F–3. These detectors, when placed at the focal plane of a monochromator, provide as many as 1000 individual detectors whose outputs can be amplified, processed, and read out nearly simultaneously.

Multichannel dispersive instruments are generally more complex and more costly than single channel temporal instruments.

Most are microprocessor controlled and provide data in a variety of forms. They do offer distinct advantages in the speed in which spectra can be obtained without degradation in signal-to-noise ratio. This speed, of course, results from the fact that all elements of the spectra are being measured simultaneously rather than one element at a time as in the single channel case. Multichannel instruments can also provide enhanced sensitivity and precision because the speed of measurement makes signal averaging more practical. As was pointed out in Section 2G–2, signal averaging often permits the extraction of a minute signal from a noisy environment.

In many applications, multichannel instruments also offer the advantage of significantly smaller sample consumption. Again, these instruments are generally more expensive than most single channel devices.

5I–5 MULTIPLEX DESIGNS

The term multiplex comes from communication theory, where it is used to describe systems in which many sets of information are transported simultaneously through a single channel. As the name implies, multiplex analytical instruments are single channel devices in which all elements of the signal are observed *simultaneously*. In order to determine the magnitude of each of these elements it is necessary to modulate the analyte signal in a way that permits subsequent decoding of the signal to give its component parts or elements.[20]

---

[20] Not all writers reserve the term multiplexing for systems in which all elements of a signal are observed simultaneously. For example, some use the term to indicate any system in which a number of independent signals can be encoded and transmitted over a single transmitting medium. By this definition, most of the instruments in Table 5–2 are multiplex devices. (See: K. W. Busch and L. D. Benton, *Anal. Chem.*, **1983**, *55*, 445A.) In this text, the first, more restrictive, definition will be used.

Most multiplex instruments depend upon the *Fourier transform* for signal decoding and are consequently often called Fourier transform instruments. Such instruments are by no means confined to optical spectroscopy. Indeed, Fourier transform devices have been described for nuclear magnetic resonance, mass, and microwave spectroscopy as well as for certain types of electroanalytical measurements. Several of these instruments will be described in some detail in subsequent chapters.

In optical spectroscopy, decoding has also been performed by the so-called *Hadamard transform*. This procedure has not found widespread application to date; thus, it will be given only brief mention in Section 11C–2.

Fourier transform spectroscopy was first developed by astronomers in the early 1950s in order to study the infrared spectra of distant stars; only by the Fourier technique could the very weak signals from these sources be isolated from environmental noise. The first chemical applications of Fourier transform spectroscopy, which were reported approximately a decade later, were to the energy-starved far-infrared region; by the late 1960s, instruments for chemical studies in both the far-infrared (10 to 400 cm^{-1}) and the ordinary-infrared regions were available commercially.[21] Descriptions of Fourier transform instruments for the ultraviolet and visible spectral regions can also be found in the literature, but

their adoption has not been widespread to date. For this type of radiation, the advantages of the method are not great enough to justify the expense and complexity of the equipment.

An Inherent Advantage of Fourier Transform Spectroscopy. For purposes of this discussion, it is convenient to think of an experimentally derived spectrum as being made up of m individual transmittance measurements at equally spaced frequency or wavelength intervals called *resolution elements*. The quality of the spectrum—that is, the amount of spectral detail—increases as the number of resolution elements becomes larger or as the frequency intervals between measurements become smaller.[22] Thus, in order to increase spectral quality, m must be made larger; clearly, increasing the number of resolution elements must also increase the time required for obtaining a spectrum with a scanning instrument.

Consider, for example, the derivation of an infrared spectra from 500 to 5000 cm^{-1}. If resolution elements of 3 cm^{-1} were chosen, m would be 1500; if 0.5 second were required for recording the transmittance of each resolution element, 750 sec or 12.5 min would be needed to obtain the spectrum. Reducing the width of the resolution element to 1.5 cm^{-1} would be expected to provide significantly greater spectral detail; it would also double the number of resolution elements as well as the time required for their measurement.

For most optical instruments, particularly those designed for the infrared region, decreasing the width of the resolution element has the unfortunate effect of decreasing the signal-to-noise ratio because narrower slits must be used, which lead to weaker signals

---

[21] For a more complete discussion of Fourier transform infrared spectroscopy, consult the following references: P. R. Griffiths, *Chemical Fourier Transform Spectroscopy*, Wiley: New York, 1975; *Transform Techniques in Chemistry*, P. R. Griffiths, Ed., Plenum Press: New York, 1978; *Fourier Transform Infrared Spectroscopy: Applications to Chemical Systems*, J. R. Ferraro and L. J. Basile, Eds., Vol. I, 1978 and Vol. II, 1979, Academic Press: New York; A. G. Marshall, *Fourier, Hadamard and Hilbert Transforms in Chemistry*, Plenum Press: New York, 1982. For a brief review of recent developments, see: P. R. Griffiths, *Science*, **1983**, *222*, 297.

[22] With a recording spectrophotometer, of course, individual point-by-point measurements are not made; nevertheless, the idea of a resolution element is useful, and the ideas generated from it apply to recording instruments as well.

reaching the transducer. For infrared detectors, the reduction in signal is not accompanied by a corresponding decrease in noise. Therefore, a degradation in signal-to-noise ratio results.

In Section 2G–2, it was pointed out that marked improvements in signal-to-noise ratios accompany signal averaging. Here, it was shown (Equation 2–21) that the signal-to-noise ratio *S/N* for the average of *n* measurements is given by

$$\frac{S}{N} = \frac{S_n}{N_n} \times \sqrt{n}$$

where S_n and N_n are the averaged signal and noise. The application of signal averaging to conventional spectroscopy is, unfortunately, costly in terms of time. Thus, in the example just considered, 750 sec were required to obtain a spectrum of 1500 resolution elements. To improve the signal-to-noise ratio by a factor of 2 would require averaging 4 spectra, which would then require 4×750 sec or 50 min.

Fourier transform spectroscopy differs from conventional spectroscopy in that all of the resolution elements for a spectrum are measured *simultaneously*, thus reducing enormously the time required to derive a spectrum at any chosen signal-to-noise ratio. An entire spectrum of 1500 resolution elements can then be recorded in about the time required to observe just one element by conventional spectroscopy (0.5 sec in our earlier example). This large decrease in observation time is often used to markedly enhance the signal-to-noise ratio of Fourier transform measurements. For example, in the 750 sec required to derive the spectrum in the earlier sample, 1500 Fourier transform spectra could be recorded and averaged. According to Equation 2–21, the improvement in signal-to-noise ratio would in principle be $\sqrt{1500}$ or about 39. This inherent advantage of Fourier transform spec-

troscopy was first recognized by P. Fellgett in 1958 and has become known as the *Fellgett advantage*. It is worth noting here that for several reasons, the theoretical $\sqrt{n}$ improvement in *S/N* is seldom realized. Nonetheless, major gains in signal-to-noise ratios are generally observed with the Fourier transform technique.

Time Domain Spectroscopy. Conventional spectroscopy can be termed *frequency domain* spectroscopy in that radiant power data are recorded as a function of frequency (or the inversely related wavelength). In contrast, *time domain* spectroscopy, which can be achieved by the Fourier transform, is concerned with changes in radiant power with time. Figure 5–28 illustrates the difference.

The plots in (**c**) and (**d**) are conventional spectra of two monochromatic sources with frequencies ν_1 and ν_2 Hz. The curve in (**e**) is the spectrum of a source containing both frequencies. In each case, some measure of the radiant power, $P(\nu)$ is plotted with respect to the frequency in hertz. The symbol in parentheses is added to emphasize the frequency dependence of the power; time domain power will be indicated by $P(t)$.

The curve in Figure 5–28**a** shows the time domain spectra for each of the monochromatic sources. The two have been plotted together in order to make the small frequency difference between them more obvious. Here, the instantaneous power $P(t)$ is plotted as a function of time. The curve in Figure 5–28**b** is the time domain spectrum of the source containing the two frequencies. As is shown by the horizontal arrow, the plot exhibits a periodicity or *beat* as the two waves go in and out of phase.

Examination of Figure 5–29 reveals that the time domain spectrum for a source containing several wavelengths is considerably more complex than those shown in Figure 5–28. Because a large number of wavelengths are involved, a full cycle is not realized in the time shown. To be sure, a pattern of beats can be observed as certain wave-

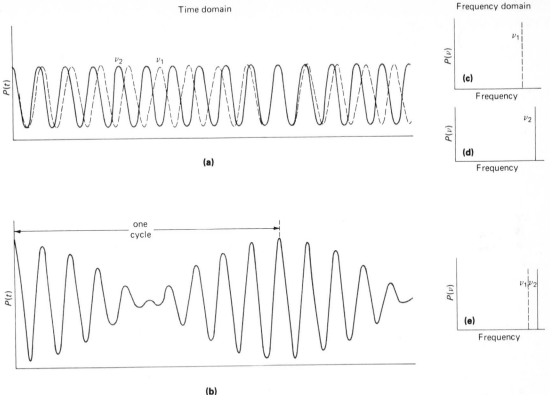

FIGURE 5-28 Illustrations of time domain plots (**a**) and (**b**); frequency domain plots (**c**), (**d**), and (**e**).

lengths pass in and out of phase. In general, the signal power decreases with time as a consequence of the various closely spaced wavelengths becoming more and more out of phase.

It is important to appreciate that a time domain spectrum contains the same information as does a spectrum in the frequency domain, and in fact, one can be converted to the other by mathematical manipulations. Thus, Figure 5–28**b** was derived from Figure 5–28**e** by means of the equation

$$P(t) = k(\cos 2\pi\nu_1 t + \cos 2\pi\nu_2 t) \qquad (5\text{–}18)$$

where k is a constant and t is the time. The difference in frequency between the two lines was approximately 10% of ν_2.

The interconversion of time and fre-quency domain spectra becomes exceedingly complex and mathematically tedious when more than a few lines are involved; the operation is only practical with a high-speed computer.

Methods of Obtaining Time Domain Spectra. Time domain spectra, such as those shown in Figures 5–28 and 5–29, cannot be acquired experimentally with radiation of the frequency that is useful for spectroscopy (10^{15} Hz for ultraviolet to 10^7 Hz for nuclear magnetic resonance) because transducers that will respond to power variations at these enormous frequencies do not exist. Thus, a typical transducer yields a signal that corresponds to the average power of a high-frequency signal and not to its periodic variation. To obtain time domain spectra requires, therefore, a method of converting

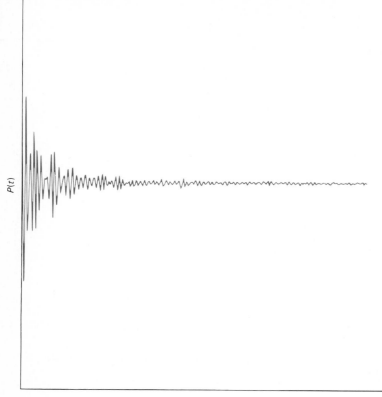

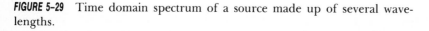

Time

FIGURE 5–29 Time domain spectrum of a source made up of several wavelengths.

(or *modulating*) a high-frequency signal to one of measurable frequency without distorting the time relationships carried in the signal; that is, the frequencies in the modulated signal must be directly proportional to those in the original. Different signal-modulation procedures are employed for the various wavelength regions of the spectrum. The Michelson interferometer is used extensively for measurements in the optical region.

The Michelson Interferometer. The device used for modulating optical radiation is an interferometer similar in design to one first described by Michelson in 1891. The Michelson interferometer is a device that splits a beam of radiation into two beams of nearly equal power and then recombines them in such a way that intensity variations of the combined beam can be measured as a function of differences in the lengths of the paths of the two halves. Figure 5–30 is a schematic diagram of such a device as it is used for optical Fourier transform spectroscopy.

As shown in the figure, a beam of radiation from a source is collimated and impinges on a beam splitter, which transmits

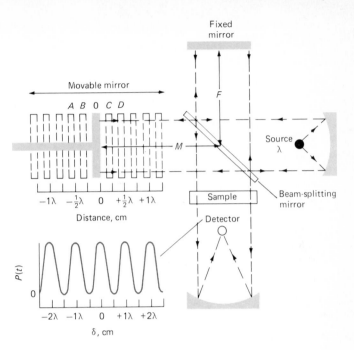

FIGURE 5-30 Schematic diagram of a Michelson interferometer illuminated by a monochromatic source.

approximately half of the radiation and reflects the other half. The resulting twin beams are then reflected from mirrors, one of which is fixed and the other of which is movable. The beams then meet again at the beam splitter, with half of each beam being directed toward the sample and detector and the other two halves being directed back toward the source. Only the two halves passing through the sample to the detector are employed for analytical purposes, although the other halves contain the same information about the source.

Horizontal motion of the movable mirror will cause the power of the radiation reaching the detector to fluctuate in a predictable manner. When the two mirrors are equidistant from the splitter (position 0 in Figure 5–30), the two parts of the recombined beam will be totally in phase and the power will be a maximum. For a monochromatic source, motion of the movable mirror in either direction by a distance equal to exactly one-quarter wavelength (position B or C in

the figure) will change the path length of the corresponding reflected beam by one-half wavelength (one-quarter wavelength for each direction). Under this circumstance, destructive interference will reduce the radiant power of the recombined beams to zero. Further motion to A and D will bring the two halves back in phase so that constructive interference can again occur.

The difference in path lengths for the two beams, $2(M - F)$ in the figure is termed the *retardation* δ. A plot of the output power from the detector versus δ is called an *interferogram;* for monochromatic radiation, the interferogram takes the form of a cosine curve such as that shown in the lower left of Figure 5–30 (cosine rather than sine because the power is always a maximum when δ is zero and the two paths are identical).

In the typical scanning interferometer, the mirror is motor-driven at a constant velocity of v_M cm/sec. Since peaks occur at higher multiples of the wavelength of the incident radiation, the frequency f of the *interferogram* is given by

$$f = \frac{2v_M \text{ cm/sec}}{\lambda \text{ cm}} \qquad (5\text{--}19)$$

Here, λ is the wavelength of the incident beam (in cm) and the factor 2 is required because the reflection process doubles the pathway of the radiation.

In some instances, it is convenient to express Equation 5–19 in terms of wavenumber of the incident radiation σ rather than in wavelength. Thus,

$$f = 2v_M\sigma \qquad (5\text{--}20)$$

The relationship between the *optical frequency* of the radiation and the frequency of the interferogram is readily obtained by substitution of $\lambda = c/v$ into Equation 5–19. Thus,

$$f = \frac{2v_M}{c}v \qquad (5\text{--}21)$$

where v is the frequency of the radiation and c is the velocity of light (3×10^{10} cm/sec). When v_M is constant, it is evident that the *interferogram frequency f is directly proportional to the optical frequency v*. Furthermore, the proportionality constant will generally be a very small number. For example, if the mirror is driven at a rate of 1.5 cm/sec,

$$\frac{2v_M}{c} = \frac{2 \times 1.5 \text{ cm/sec}}{3 \times 10^{10} \text{ cm/sec}} = 10^{-10}$$

and

$$f = 10^{-10}\,v$$

As shown by the following example, the frequency of visible and infrared radiation is readily modulated into the audio range by a Michelson interferometer.

EXAMPLE 5–4.   Calculate the frequency range of a modulated signal from a Michelson interferometer with a mirror velocity of 0.20 cm/sec, for visible radiation of 700 nm and infrared radiation of 16 μm (4.3×10^{14} to 1.9×10^{13} Hz).

Employing Equation 5–19, we find

$$f_1 = \frac{2 \times 0.20 \text{ cm/sec}}{700 \text{ nm} \times 10^{-7} \text{ cm/nm}} = 5700 \text{ Hz}$$

$$f_2 = \frac{2 \times 0.20}{16 \times 10^{-4}} = 250 \text{ Hz}$$

Certain types of visible and infrared transducers are capable of following fluctuations in signal power that fall into the audio-frequency range. Thus, it becomes possible to record a modulated time domain spectrum which reflects exactly the appearance of the very high-frequency time domain spectra of a visible or infrared source. Figure 5–31 shows three examples of such time domain interferograms on the left and their frequency domain counterparts on the right.

Fourier Transformation of Interferograms.   The cosine wave of the interferogram shown in Figure 5–31a (and also in Figure 5–30) can be described in theory by the equation

$$P(\delta) = \frac{1}{2} P(\sigma) \cos 2\pi f t \qquad (5\text{--}22)$$

where $P(\sigma)$ is the radiant power of the infrared beam incident upon the interferometer and $P(\delta)$ is the amplitude or power of the interferogram signal. The parenthetical symbols emphasize that one power (σ) is in the frequency domain and the other (δ) in the time domain. In practice, the foregoing equation is modified to take into account the fact that the interferometer ordinarily will not split the source exactly in half and that the detector response and the amplifier behavior are frequency-dependent. Thus, it is

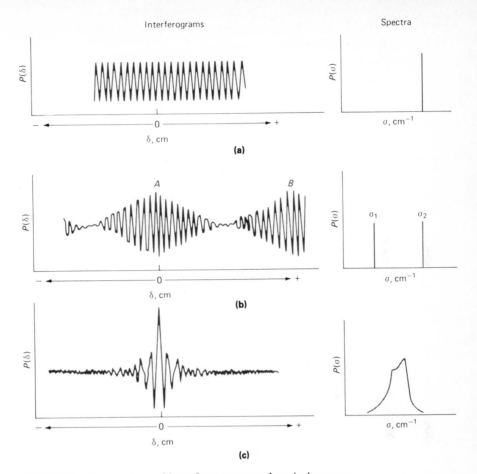

Interferograms   Spectra

(a)

(b)

(c)

FIGURE 5-31 Comparison of interferograms and optical spectra.

useful to introduce a new variable $B(\sigma)$ which depends upon $P(\sigma)$ but also takes these other factors into account. Therefore, we rewrite the equation in the form

$$P(\delta) = B(\sigma) \cos 2\pi ft \qquad (5\text{--}23)$$

Substitution of Equation 5–20 into Equation 5–23 leads to

$$P(\delta) = B(\sigma) \cos 2\pi 2 v_M \sigma t \qquad (5\text{--}24)$$

But the mirror velocity can be expressed in terms of retardation or

$$v_M = \frac{\delta}{2t}$$

Substitution of this relationship into Equation 5–24 gives

$$P(\delta) = B(\sigma) \cos 2\pi \delta \sigma$$

which expresses the magnitude of the interferogram signal as a function of the retardation factor and the wavenumber of the optical input signal.

The interferograms shown in part (**b**) of

Figure 5–31 can be described by two terms, one for each wavenumber. Thus,

$$P(\delta) = B_1(\sigma) \cos 2\pi\delta\sigma_1$$
$$+ B_2(\sigma) \cos 2\pi\delta\sigma_2 \quad (5\text{–}25)$$

For a continuous source, such as in part (c) of Figure 5–31, the interferogram can be represented as a sum of an infinite number of cosine terms. That is,

$$P(\delta) = \int_{n=-\infty}^{+\infty} B(\sigma) \cos 2\pi\sigma\delta_n \cdot d\sigma \quad (5\text{–}26)$$

The Fourier transform of this integral is

$$P(\sigma) = \int_{n=-\infty}^{+\infty} B(\delta) \cos 2\pi\sigma\delta_n d\sigma \quad (5\text{–}27)$$

Infrared Fourier transform spectroscopy consists of recording $P(\delta)$ as a function of δ (Equation 5–26) and then mathematically transforming this relation to one that gives $P(\sigma)$ as a function of σ (the frequency spectrum) as shown by Equation 5–27.

Equations 5–26 and 5–27 cannot be employed as written because they assume that the beam contains radiation from zero to infinite wavenumbers and a mirror drive of infinite length. Furthermore, Fourier transformations with a computer require that the detector output be digitized; that is, the output must be sampled periodically and stored in digital form. Equation 5–26 however, demands that the sampling intervals $d\delta$ be infinitely small; that is, $d\delta \rightarrow 0$.

From a practical standpoint, only a finite-sized sampling interval can be summed over a finite retardation range (a few centimeters). These limitations have the effect of limiting the resolution of a Fourier transform instrument and restricting its frequency range.

Resolution of a Fourier Transform Spectrometer. The resolution of a Fourier transform spectrometer can be described in terms of the difference in wavenumber between two lines that can be just separated by the instrument. That is,

$$\Delta\sigma = \sigma_1 - \sigma_2 \quad (5\text{–}28)$$

where σ_1 and σ_2 are wavenumbers for a pair of barely resolvable infrared lines.

It is possible to show that in order to resolve two lines, it is necessary to scan the time domain spectrum long enough so that one complete cycle or beat for the two lines is completed; only then will all of the information contained in the spectra have been recorded. For example, resolution of the two lines σ_1 and σ_2 in Figure 5–31b would require recording the interferogram from the maximum A at zero retardation to the maximum B where the two waves are again in phase. The maximum at B occurs, however, when $\delta\sigma_2$ is larger than $\delta\sigma_1$ by 1 in Equation 5–25. That is, when

$$\delta\sigma_2 - \delta\sigma_1 = 1$$

or

$$\sigma_2 - \sigma_1 = \frac{1}{\delta}$$

Substitution into Equation 5–28 reveals that the resolution is given by

$$\Delta\sigma = \sigma_2 - \sigma_1 = \frac{1}{\delta} \quad (5\text{–}29)$$

EXAMPLE 5-5. What length of mirror drive will provide a resolution of 0.1 cm^{-1}?
   Substituting into Equation 5–29 gives

$$0.1 = \frac{1}{\delta}$$

$$\delta = 10 \text{ cm}$$

The mirror motion required is one-half the retardation, or 5 cm.

Fourier Transform Instruments. Details about modern Fourier transform optical spectrometers are found in Chapter 11. An integral part of these instruments is a sophisticated computer for controlling data acquisition, storing data, signal averaging, and performing the Fourier transformations. Four advantages accrue to the use of these instruments. One is the Fellgett advantage, which was described earlier. A second known as the *Jaquinot advantage,* is the large energy throughput of interferometric instruments, which have few optical elements and no slits to attenuate radiation. The third advantage is high wavelength precision, which makes signal averaging feasible. A final advantage is the ease and convenience with which the Fourier transformed data can be manipulated within the computer.

A disadvantage of Fourier transform instruments is their cost both initially and for maintenance.

PROBLEMS

5–1. Why must the slit width of a prism monochromator be varied to provide constant effective bandwidths whereas a nearly constant slit width may be used with a grating monochromator?

5–2. Why do quantitative and qualitative analyses often require different monochromator slit widths?

5–3. The Wien displacement law states the wavelength maximum in micrometers for black-body radiation is given by the following relationship

$$\lambda_{max}T = 2.90 \times 10^3$$

where T is the temperature in °K. Calculate the wavelength maximum for a black body that has been heated to (a) 4000°K, (b) 2000°K, and (c) 1000°K.

5–4. Stefan's law states that the total energy E_T emitted by a black body per unit time and per unit area is given by

$$E_T = \alpha T^4$$

where α has a value of 5.69×10^{-8} W m^{-2} K^{-4}.

Calculate the total energy output in W/m^2 for each of the black bodies described in Problem 5–3.

5–5. Relationships described in Problems 5–3 and 5–4 may be of help in solving the following.
   (a) Calculate the wavelength of maximum emission of a tungsten filament bulb operated at the usual temperature 2870°K and at a temperature of 3000°K.
   (b) Calculate the total energy output of the bulb in terms of W/cm^2.

5–6. Contrast spontaneous and stimulated emission.

5–7. Describe the advantage of a four-level laser system over a three-level type.

5–8. Define the term *effective bandwidth* of a filter.

5–9. An interference filter is to be constructed for isolation of the CS_2 absorption band at 4.54 μm.
   (a) If it is to be based upon first-order interference, what should be the thickness of the dielectric layer (refractive index 1.34)?
   (b) What other wavelengths would be transmitted?

5–10. A 10.0-cm interference wedge is to be built that has a linear dispersion from 400 to 700 nm. Describe details of its construction. Assume that a dielectric with a refractive index of 1.32 is to be employed.

5–11. Why is glass better than fused silica as a prism construction material for a monochromator to be used in the region of 400 to 800 nm?

5–12. For a grating, how many lines per millimeter would be required in order for the first-order diffraction line for λ = 500 nm to be observed at a reflection angle of −40 deg when the angle of incidence is 60 deg?

5–13. Consider an infrared grating with 72.0 lines per millimeter and 10.0 nm of illuminated area. Calculate the first-order resolution ($\lambda/\Delta\lambda$) of this grating. How far apart (in cm^{-1}) must two lines centered at 1000 cm^{-1} be if they are to be resolved?

5–14. For the grating described in Problem 5–13, calculate the wavelengths on the first- and second-order diffraction spectra at reflective angles of (a) −20 deg, (b) 0 deg, and (c) +20 deg. Assume the incident angle is 50 deg.

5–15. With the aid of Figure 5–2 suggest instrument components and materials for constructing an instrument that would be well suited for each of the following purposes:
   (a) The investigation of the fine structure of absorption bands in the region of 450 to 750 nm.
   (b) For obtaining absorption spectra in the far infrared (20 to 50 μm).
   (c) A portable device for determining the iron content of natural water based upon the absorption of radiation by the red $Fe(SCN)^{2+}$ complex.
   (d) The routine determination of nitrobenzene in air samples based upon its absorption peak at 11.8 μm.
   (e) For determining the wavelengths of flame emission lines for metallic elements in the region from 200 to 780 nm.
   (f) For spectroscopic studies in the vacuum ultraviolet region.
   (g) For spectroscopic studies in the near infrared.

5–16. What is the speed of a lens having a diameter of 4.2 cm and a focal length of 8.1 cm?

5–17. Compare the light gathering power of the lens described in Problem 5–16 with one having a diameter of 2.6 cm and a focal length of 8.1 cm.

5–18. A monochromator had a focal length of 1.6 m and a collimating mirror with a diameter of 2.0 cm. The dispersing device was a grating with 1250 lines/mm. For first-order diffraction,
(a) what was the resolving power of the monochromator if a collimated beam illuminated 2.0 cm of the grating?
(b) what is the first- and second-order reciprocal linear dispersion of the monochromator described above?

5–19. A monochromator having a focal length of 0.65 m was equipped with an echellette grating having 2000 blazes per millimeter.
(a) Calculate the reciprocal linear dispersion of the instrument for first-order spectra.
(b) If 3.0 cm of the grating were illuminated, what is the first-order resolving power of the monochromator?
(c) At approximately 560 nm, what minimum wavelength difference could in theory be completely resolved by the instrument?

5–20. Describe the basis for radiation detection with a silicon diode detector.

5–21. Distinguish among (a) a spectroscope, (b) a spectrograph, and (c) a spectrophotometer.

5–22. A Michelson interferometer had a mirror velocity of 1.25 cm/sec. What would be the frequency of the interferogram for (a) UV radiation of 300 nm, (b) visible radiation of 700 nm, (c) infrared radiation of 7.5 μm, and (d) infrared radiation of 20 μm?

5–23. What length of mirror drive in a Michelson interferometer is required to produce a resolution sufficient to separate
(a) infrared peaks at 20.34 and 20.35 μm?
(b) infrared peaks at 2.500 and 2.501 μm?

An Introduction to Absorption

Spectroscopy

This chapter provides introductory material that is applicable to ultraviolet-visible, infrared, and atomic absorption spectroscopy.[1] Detailed discussions of these methods appear in later chapters.

6A Terms Employed in Absorption Spectroscopy

Table 6–1 lists the common terms and symbols employed in absorption spectroscopy.

In recent years, a considerable effort has been made by the American Society for Testing Materials (ASTM) and others to develop a standard nomenclature; the terms and symbols listed in the first two columns of Table 6–1 are based on recommendations by ASTM and *Analytical Chemistry*. Column 3 contains alternative symbols that will be found in the older literature. A standard nomenclature seems most worthwhile in order to avoid ambiguities; the reader is, therefore, urged to learn and use the recommended terms and symbols.

6A-1 TRANSMITTANCE

Figure 6–1 depicts a beam of parallel radiation before and after it has passed through a layer of solution having a thickness of b cm and a concentration c of an absorbing species. As a consequence of interactions between the photons and absorbing particles, the power of the beam is attenuated from P_0 to P. The *transmittance* T of the solution is

---

[1] For more detailed treatment of absorption spectroscopy, see: E. J. Meehan, in *Treatise on Analytical Chemistry*, 2d ed., P. J. Elving, E. J. Meehan, and I. M. Kolthoff, Eds., Part I, vol. 7, Chapter 2, Wiley: New York, 1981; J. R. Edisbury, *Practical Hints on Absorption Spectrometry*, Plenum Press: New York, 1968; F. Grum, in *Physical Methods of Chemistry*, A. Weissberger and B. W. Rossiter, Eds., Vol. I, Part III B, Chapter 3, Wiley-Interscience: New York, 1972; G. F. Lothian, *Absorption Spectrophotometry*, 3d ed., Adam Hilger Ltd.: London, 1969; and J. E. Crooks, *The Spectrum in Chemistry*, Academic Press: London, 1978.

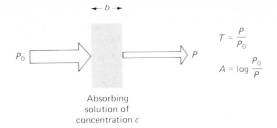

FIGURE 6-1 Attenuation of a beam of radiation by an absorbing solution.

then the fraction of incident radiation transmitted by the solution. That is,

$$T = P/P_0 \qquad (6\text{-}1)$$

Transmittance is often expressed as a percentage.

6A-2 ABSORBANCE

The absorbance of a solution is defined by the equation

$$A = -\log_{10} T = \log \frac{P_0}{P} \qquad (6\text{-}2)$$

Note that, in contrast to transmittance, the absorbance of a solution increases as attenuation of the beam becomes larger.

6A-3 ABSORPTIVITY AND MOLAR ABSORPTIVITY

As will be shown presently, absorbance is directly proportional to the path length through the solution and the concentration of the absorbing species. That is,

$$A = abc \qquad (6\text{-}3)$$

where a is a proportionality constant called the *absorptivity*. The magnitude of a will clearly depend upon the units used for b and c. When the concentration is expressed in moles per liter and the cell length is in centimeters, the absorptivity is called the *molar absorptivity* and given the special symbol

Table 6-1
Important Terms and Symbols Employed in Absorption Measurement

Term and Symbol[a]	Definition	Alternative Name and Symbol
Radiant power, P, P_0	Energy of radiation (in ergs) impinging on a 1-cm² area of a detector per second	Radiation intensity, I, I_0
Absorbance, A	$\log \dfrac{P_0}{P}$	Optical density, D; extinction, E
Transmittance, T	$\dfrac{P}{P_0}$	Transmission, T
Path length of radiation, in cm b	—	l, d
Absorptivity,[b] a	$\dfrac{A}{bc}$	Extinction coefficient, k
Molar absorptivity,[c] ε	$\dfrac{A}{bc}$	Molar extinction coefficient

[a] Terminology recommended by *Analytical Chemistry* and ASTM. See: *Anal. Chem.*, **1984**, 56, 125–126.
[b] c is preferably expressed in g/L, but other specified concentration units may be used; b is preferably expressed in cm, but other units of length may be used.
[c] c is expressed in units of mol/L.

ε. Thus, when b is in centimeters and c is in moles per liter,

$$A = \varepsilon bc \qquad (6-4)$$

where ε has the units L cm^{-1} mol^{-1}.

6A-4 MEASUREMENT OF TRANSMITTANCE AND ABSORBANCE

Many manual photometers and spectrophotometers are equipped with a display that has a linear scale extending from 0 to 100% T. In order to make such an instrument direct reading in percent transmittance, two preliminary adjustments are carried out, namely the *0% T or dark current adjustment* and the *100% T adjustment*.

The 0% T adjustment is performed with the detector screened from the source by a mechanical shutter. As noted earlier, many detectors exhibit a small dark current in the absence of radiation; the 0% T adjustment then involves application of a counter signal of such a magnitude as to give a zero reading on the readout.

The 100% T adjustment is made with the shutter open and the solvent in the light path. Normally the solvent is contained in a cell that is, as nearly as possible, identical to the one containing the samples. The 100% T adjustment may involve increasing or decreasing the radiation output of the source electrically; alternatively, the power of the beam may be varied with an adjustable diaphragm or by appropriate positioning of a comb or optical wedge, which attenuates the beam to a varying degree depending upon its position with respect to the beam. This adjustment is carried out in such a way as to give a scale reading of exactly 100% T. Effectively, this step sets P_0 as 100. When the solvent cell is replaced by the cell containing the sample, the scale then gives the percent transmittance directly, that is,

$$\% \, T = \frac{P}{P_0} \times 100 = \frac{P}{100} \times 100 = P$$

Obviously, an absorbance scale can also be scribed on the readout device. Such a scale will be nonlinear unless the output is converted to a logarithmic function electronically.

The use of the solvent in a matched cell for the 100% T adjustment tends to compensate for diminution of the radiant power in the sample cell due to reflection losses at the air/cell and cell/solution interfaces (see Example 4–2, Section 4A–6); this procedure also tends to correct for any attenuation due to absorption by the cell walls or scattering by the solvent. The success of this technique is highly dependent upon the closeness of the match between the two cells. When similar cells are not available, a single cell must be used. Here, the solvent is replaced by the analyte solution after the 100% T adjustment. Clearly, the instrument response must remain constant during the time required for emptying the cell of solvent and rinsing and filling with the analyte solution.

6B Quantitative Aspects of Absorption Measurements

This section is devoted to an examination of Equation 6–4 ($A = \varepsilon bc$), with particular attention to causes of deviations from this relationship. In addition, consideration is given to the effects that uncertainties in the measurement of P and P_0 have on absorbance (and thus concentration).

6B-1 BEER'S LAW

Equations 6–3 and 6–4 are statements of *Beer's law*. These relationships can be rationalized as follows.[2] Consider the block of absorbing matter (solid, liquid, or gas) shown

---

[2] The discussion that follows is based on a paper by F. C. Strong, *Anal. Chem.*, **1952**, *24*, 338. For a rigorous derivation of the law, see: D. J. Swinehart, *J. Chem. Educ.*, **1972**, *39*, 333.

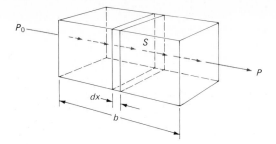

FIGURE 6-2 Attenuation of radiation with initial power P_0 by a solution containing c moles per liter of absorbing solute and with a path length of b cm. $P < P_0$.

in Figure 6–2. A beam of parallel monochromatic radiation with power P_0 strikes the block perpendicular to a surface; after passing through a length b of the material, which contains n absorbing particles (atoms, ions, or molecules), its power is decreased to P as a result of absorption. Consider now a cross section of the block having an area S and an infinitesimal thickness dx. Within this section there are dn absorbing particles; associated with each particle, we can imagine a surface at which photon capture will occur. That is, if a photon reaches one of these areas by chance, absorption will follow immediately. The total projected area of these capture surfaces within the section is designated as dS; the ratio of the capture area to the total area, then, is dS/S. On a statistical average, this ratio represents the probability for the capture of photons within the section.

The power of the beam entering the section, P_x, is proportional to the number of photons per square centimeter per second, and dP_x represents the quantity removed per second within the section; the fraction absorbed is then $-dP_x/P_x$, and this ratio also equals the average probability for capture. The term is given a minus sign to indicate that P undergoes a decrease. Thus,

$$-\frac{dP_x}{P_x} = \frac{dS}{S} \tag{6-5}$$

Recall, now, that dS is the sum of the capture areas for particles within the section; it must therefore be proportional to the number of particles, or

$$dS = adn \tag{6-6}$$

where dn is the number of particles and a is a proportionality constant, which can be called the *capture cross section*. Combining Equations 6–5 and 6–6 and summing over the interval between zero and n, we obtain

$$-\int_{P_0}^{P} \frac{dP_x}{P_x} = \int_{0}^{n} \frac{adn}{S}$$

which, upon integration, gives

$$-\ln \frac{P}{P_0} = \frac{an}{S}$$

Upon converting to base 10 logarithms and inverting the fraction to change the sign, we obtain

$$\log \frac{P_0}{P} = \frac{an}{2.303S} \tag{6-7}$$

where n is the total number of particles within the block shown in Figure 6–2. The cross-sectional area S can be expressed in terms of the volume of the block V in cm³ and its length b. Thus,

$$S = \frac{V}{b} \text{ cm}^2$$

Substitution of this quantity into Equation 6–7 yields

$$\log \frac{P_0}{P} = \frac{anb}{2.303V} \tag{6-8}$$

Note that n/V has the units of concentration (that is, number of particles per cubic centimeter); we can readily convert n/V to moles per liter. Thus, the number of moles is given by

$$mol = \frac{n \text{ particles}}{6.02 \times 10^{23} \text{ particles/mol}}$$

and c in mol/L is given by

$$c = \frac{n}{6.02 \times 10^{23}} mol \times \frac{1000 \text{ cm}^3/\text{L}}{V \text{ cm}^3}$$

$$= \frac{1000n}{6.02 \times 10^{23} V} \text{ mol/L}$$

Combining this relationship with Equation 6–8 yields

$$\log \frac{P_0}{P} = \frac{6.02 \times 10^{23} \, abc}{2.303 \times 1000}$$

Finally, the constants in this equation can be collected into a single term ε to give

$$\log \frac{P_0}{P} = \varepsilon bc = A \qquad (6\text{–}9)$$

which is, of course, a statement of Beer's law.

6B-2 APPLICATION OF BEER'S LAW TO MIXTURES

Beer's law also applies to a solution containing more than one kind of absorbing substance. Provided there is no interaction among the various species, the total absorbance for a multicomponent system is given by

$$A_{\text{total}} = A_1 + A_2 + \cdots + A_n$$
$$= \varepsilon_1 b c_1 + \varepsilon_2 b c_2 + \cdots + \varepsilon_n b c_n \quad (6\text{–}10)$$

where the subscripts refer to absorbing components $1, 2, \ldots, n$.

6B-3 LIMITATIONS TO THE APPLICABILITY OF BEER'S LAW

Few exceptions are found to the generalization that absorbance is linearly related to path length. On the other hand, deviations from the direct proportionality between the measured absorbance and concentration when b is constant are frequently encountered. Some of these deviations are fundamental and represent real limitations of the law. Others occur as a consequence of the manner in which the absorbance measurements are made or as a result of chemical changes associated with concentration changes; the latter two are sometimes known, respectively, as *instrumental deviations* and *chemical deviations*.

Real Limitations to Beer's Law. Beer's law is successful in describing the absorption behavior of dilute solution only; in this sense, it is a limiting law. At high concentrations (usually $>0.01 \, M$), the average distance between the species responsible for absorption is diminished to the point where each affects the charge distribution of its neighbors. This interaction, in turn, can alter their ability to absorb a given wavelength of radiation. Because the extent of interaction depends upon concentration, the occurrence of this phenomenon causes deviations from the linear relationship between absorbance and concentration. A similar effect is sometimes encountered in solutions containing low absorber concentrations but high concentrations of other species, particularly electrolytes. The close proximity of ions to the absorber alters the molar absorptivity of the latter by electrostatic interactions; the effect is lessened by dilution.

While the effect of molecular interactions is ordinarily not significant at concentrations below $0.01 \, M$, some exceptions are encountered among certain large organic ions or molecules. For example, the molar absorptivity at 436 nm for the cation of methylene blue is reported to increase by 88% as the dye concentration is increased from 10^{-5} to $10^{-2} \, M$; even below $10^{-6} \, M$, strict adherence to Beer's law is not observed.

Deviations from Beer's law also arise because ε is dependent upon the refractive index of the solution.[3] Thus, if concentration changes cause significant alterations in the refractive index n of a solution, departures from Beer's law are observed. A correction for this effect can be made by substitution of the quantity $\varepsilon n/(n^2 + 2)^2$ for ε in Equation 6–9. In general, this correction is never very large and is rarely significant at concentrations less than 0.01 M.

Apparent Chemical Deviations. Apparent deviations from Beer's law arise when an analyte dissociates, associates, or reacts with a solvent to produce a product having a different absorption spectrum from the analyte. A common example of this behavior is found with acid/base indicators. For example, the color change associated with a typical indicator HIn arises from shifts in the equilibrium

$$HIn \rightleftharpoons H^+ + In^-$$

color 1   color 2

Example 6–1 demonstrates how the shift in this equilibrium with dilution results in deviation from Beer's law.

EXAMPLE 6–1. The molar absorptivities of the weak acid HIn ($K_a = 1.42 \times 10^{-5}$) and its conjugate base In⁻ at 430 and 570 nm were determined by measurements of strongly acidic and strongly basic solutions of the indicator (where essentially all of the indicator was in the HIn and In⁻ form respectively). The results were

	ε'_{430}	ε''_{570}
HIn	6.30×10^2	7.12×10^3
In⁻	2.06×10^4	9.61×10^2

Derive absorbance data for unbuffered solutions having total indicator concentra-

---

[3] G. Kortum and M. Seiler, *Angew. Chem.,* **1939**, *52,* 687.

tions ranging from 2×10^{-5} to 16×10^{-5} M.

Let us calculate the molar concentrations of [HIn] and [In⁻] of the two species in a solution in which the total concentration of indicator is 2.00×10^{-5} M. Here,

$$HIn \rightleftharpoons H^+ + In^-$$

and

$$K_a = 1.42 \times 10^{-5} = \frac{[H^+][In^-]}{[HIn]}$$

From the equation for the dissociation process, we may write

$$[H^+] = [In^-]$$

Furthermore, the sum of the concentrations of the two indicator species must equal the total molar concentration of the indicator. Thus,

$$[In^-] + [HIn] = 2.00 \times 10^{-5}$$

Substitution of these relationships into the expression for K_a gives

$$\frac{[In^-]^2}{2.00 \times 10^{-5} - [In^-]} = 1.42 \times 10^{-5}$$

Rearrangement yields the quadratic expression

$$[In^-]^2 + 1.42 \times 10^{-5}[In^-] - 2.84 \times 10^{-10} = 0$$

The positive solution to this equation is

$$[In^-] = 1.12 \times 10^{-5}$$

$$[HIn] = 2.00 \times 10^{-5} - 1.12 \times 10^{-5} = 0.88 \times 10^{-5}$$

We are now able to calculate the absorbance at the two wavelengths. Thus, substituting into Equation 6–10 gives

$$A_{430} = \varepsilon'_{In}b[In^-] + \varepsilon'_{HIn}b[HIn]$$

$$\begin{aligned} A_{430} &= 2.06 \times 10^4 \times 1.00 \times 1.12 \times 10^{-5} \\ &+ 6.30 \times 10^2 \times 1.00 \times 0.88 \times 10^{-5} \\ &= 0.236 \end{aligned}$$

Similarly at 570 nm,

$$\begin{aligned} A_{570} &= 9.61 \times 10^2 \times 1.00 \times 1.12 \times 10^{-5} \\ &+ 7.12 \times 10^3 \times 1.00 \times 0.88 \times 10^{-5} \\ &= 0.073 \end{aligned}$$

Additional data, obtained in the same way, are shown in Table 6–2.

Figure 6–3 is a plot of the data shown in Table 6–2, which illustrates the kinds of departures from Beer's law that arise when the absorbing system is capable of undergoing dissociation or association. Note that the direction of curvature is opposite at the two wavelengths.

Apparent Instrumental Deviations with Polychromatic Radiation. Strict adherence to Beer's law is observed only with truly monochromatic radiation; this observation is yet another manifestation of the limiting character of the law. Unfortunately, the use of radiation that is restricted to a single wavelength is seldom practical because devices that isolate portions of the output from a continuous source produce a more or less symmetric band of wavelengths around the desired one (see Figures 5–7 and 5–9, for example).

The following derivation shows the effect of polychromatic radiation on Beer's law.

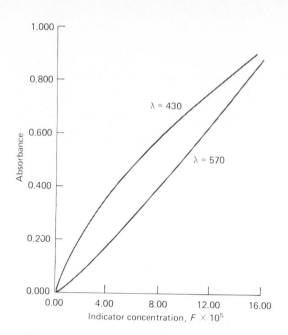

FIGURE 6-3 Chemical deviations from Beer's law for unbuffered solutions of the indicator HIn. For data, see Table 6–2.

Consider a beam consisting of just two wavelengths λ' and λ''. Assuming that Beer's law applies strictly for each of these individually, we may write for radiation λ'

$$A' = \log \frac{P'_0}{P'} = \varepsilon'bc$$

or

$$P'_0/P' = 10^{\varepsilon'bc}$$

Table 6-2
Concentration and Absorbance Data Derived by the Technique Shown in Example 6–1

M_{HIn}	[HIn]	[In$^-$]	A_{430}	A_{570}
2.00×10^{-5}	0.88×10^{-5}	1.12×10^{-5}	0.236	0.073
4.00×10^{-5}	2.22×10^{-5}	1.78×10^{-5}	0.381	0.175
8.00×10^{-5}	5.27×10^{-5}	2.73×10^{-5}	0.596	0.401
12.00×10^{-5}	8.52×10^{-5}	3.48×10^{-5}	0.771	0.640
16.00×10^{-5}	11.9×10^{-5}	4.11×10^{-5}	0.922	0.887

and

$$P' = P_0' \, 10^{-\varepsilon'bc}$$

Similarly, for λ''

$$P'' = P_0'' \, 10^{-\varepsilon''bc}$$

When an absorbance measurement is made with radiation composed of both wavelengths, the power of the beam emerging from the solution is given by $(P' + P'')$ and that of the beam from the solvent by $(P_0' + P_0'')$. Therefore, the measured absorbance is

$$A_M = \log \frac{(P_0' + P_0'')}{(P' + P'')}$$

Substituting for P' and P'' yields

$$A_M = \log \frac{(P_0' + P_0'')}{(P_0' \, 10^{-\varepsilon'bc} + P_0'' \, 10^{-\varepsilon''bc})}$$

or

$$A_M = \log(P_0' + P_0'') \\ - \log(P_0' \, 10^{-\varepsilon'bc} + P_0'' \, 10^{-\varepsilon''bc})$$

Now, when $\varepsilon' = \varepsilon''$, this equation simplifies to

$$A_M = \varepsilon'bc$$

and Beer's law is followed. As shown in Figure 6–4, however, the relationship between A_M and concentration is no longer linear when the molar absorptivities differ; moreover, greater departures from linearity can be expected with increasing differences between ε' and ε''. This derivation can be expanded to include additional wavelengths; the effect remains the same.

It is an experimental fact that deviations from Beer's law resulting from the use of a polychromatic beam are not appreciable, provided the radiation used does not encompass a spectral region in which the ab-

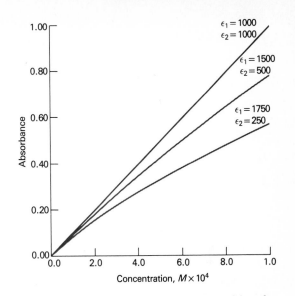

FIGURE 6-4 Derivations from Beer's law with polychromatic light. Here, two wavelengths or radiation λ_1 and λ_2 have been assumed for which the absorber has the indicated molar absorptivities.

sorber exhibits large changes in absorption as a function of wavelength. This observation is illustrated in Figure 6–5.

Instrumental Deviations in the Presence of Stray Radiation.   We have noted earlier (Chapter 5) that the radiation exiting from a monochromator is ordinarily contaminated with small amounts of scattered or stray radiation, which reaches the exit slit owing to scattering and reflections from various internal surfaces. Stray radiation often differs greatly in wavelength from that of the principal radiation and, in addition, may not have passed through the sample.

When measurements are made in the presence of stray radiation, the observed absorbance is given by

$$A' = \log \frac{P_0 + P_s}{P + P_s}$$

where P_s is the power of nonabsorbed stray radiation. Figure 6–6 shows a plot of A' versus concentration for various ratios between

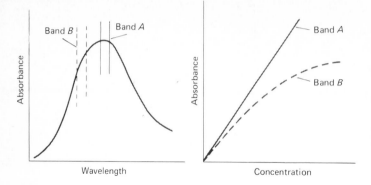

FIGURE 6-5 The effect of polychromatic radiation upon the Beer's law relationship. Band *A* shows little deviation since ε does not change greatly throughout the band. Band *B* shows marked deviations since ε undergoes significant changes in this region.

P_s and P_0. It is noteworthy that at high concentrations and at longer path lengths, stray radiation can also cause deviations from the linear relationship between absorbance and path length.[4]

Note also that the instrumental deviations illustrated in Figures 6–5 and 6–6 result in absorbances that are smaller than theoretical. It can be shown that instrumental deviations always lead to negative absorbance errors.[5]

6B-4   THE EFFECT OF INSTRUMENTAL NOISE ON THE PRECISION OF SPECTROPHOTOMETRIC ANALYSES

The accuracy and precision of spectrophotometric analyses are often limited by the uncertainties or noise associated with the instrument.[6] A general discussion of instrumental noise and signal-to-noise optimization is found in Section 2F; the reader may find it helpful to review this material before undertaking a detailed study of this section.

As was pointed out earlier, a spectrophotometric measurement entails three steps: a 0% *T* adjustment, a 100% *T* adjustment, and a measurement of % *T* with the sample in the radiation path. The noise associated with each of these steps combines to give a net uncertainty for the final value obtained for *T*. The relationship between the noise

---

[4] For a discussion of the effects of stray radiation, see: M. R. Sharpe, *Anal. Chem.*, **1984**, *56*, 339A.

[5] E. J. Meehan, in *Treatise on Analytical Chemistry*, 2d ed., P. J. Elving, E. J. Meehan and I. M. Kolthoff, Eds., Part I, vol. 7, p. 73, Wiley: New York, 1981.

[6] See: L. D. Rothman, S. R. Crouch, and J. D. Ingle, Jr., *Anal. Chem.*, **1975**, *47*, 1226; J. D. Ingle, Jr. and S. R. Crouch, *Anal. Chem.*, **1972**, *44*, 1375; H. L. Pardue, T. E. Hewitt, and M. J. Milano, *Clin. Chem.*, **1974**, *20*, 1028; J. O. Erickson and T. Surles, *Amer. Lab.*, **1976**, *8* (6), 41; *Optimum Parameters for Spectrophotometry*, Varian Instruments Division: Palo Alto, CA, 1977.

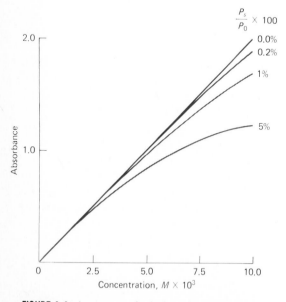

FIGURE 6-6 Apparent deviation from Beer's law brought about by various amounts of stray radiation.

encountered in the measurement of T and the uncertainty in concentration can be derived by writing Beer's law in the form

$$c = -\frac{1}{\varepsilon b}\log T = \frac{-0.434}{\varepsilon b}\ln T \qquad (6\text{--}11)$$

In order to relate the standard deviation in $c(\sigma_c)$ to the standard deviation in $T(\sigma_T)$, we proceed as in Section 1C–5 by taking the partial derivative of this equation with respect to T holding b and c constant. That is,

$$\frac{\delta c}{\delta T} = \frac{-0.434}{\varepsilon b T}$$

Application of Equation 1–15 gives

$$\sigma_c^2 = \left(\frac{\delta c}{\delta T}\right)^2 \sigma_T^2 = \left(\frac{-0.434}{\varepsilon b T}\right)^2 \sigma_T^2 \qquad (6\text{--}12)$$

Dividing Equation 6–12 by the square of Equation 6–11 gives

$$\left(\frac{\sigma_c}{c}\right)^2 = \left(\frac{\sigma_T}{T \ln T}\right)^2$$

$$\frac{\sigma_c}{c} = \frac{\sigma_T}{T \ln T} = \frac{0.434\,\sigma_T}{T \log T} \qquad (6\text{--}13)$$

For a limited number of measurements, we replace the population standard deviations σ_c and σ_T with the sample standard deviations s_c and s_T and obtain

$$\frac{s_c}{c} = \frac{0.434\,s_T}{T \log T} \qquad (6\text{--}14)$$

This equation relates the *relative* standard deviation of c to the absolute standard deviation of the transmittance measurement s_T. Experimentally, s_T can be evaluated by making, say, 20 replicate transmittance measurements ($N = 20$) of the transmittance of a solution in exactly the same way and substituting the data into Equation 1–6, page 9.

It is clear from an examination of Equation 6–14 that the uncertainty in a photometric concentration measurement varies in a complex way with the magnitude of the transmittance. The situation is even more complicated than is suggested by Equation 6–14, however, because the uncertainty s_T is, under many circumstances, also *dependent upon T.*

In a detailed theoretical and experimental study, Rothman, Crouch, and Ingle have described several sources of instrumental uncertainties and shown their net effect on the precision of absorbance or transmittance measurements.[7] These uncertainties fall into three categories depending upon how they are affected by the magnitude of the photocurrent and thus T. For *Case I uncertainties,* the precision is independent of T; that is, s_T is equal to a constant k_1. For *Case II uncertainties,* the precision is directly proportional to $\sqrt{T^2 + T}$. Finally, *Case III uncertainties* are directly proportional to T. Table 6–3 summarizes information about the sources of these three types of uncertainty and the kinds of instruments where each is likely to be encountered.

Case I: $s_T = k_1$. Case I uncertainties are often encountered with less expensive ultraviolet and visible spectrophotometers or photometers that are equipped with meters or digital readouts with limited resolution. For example, a typical instrument may be equipped with a meter having a 5- or 7-in. scale which is readable to 0.2 to 0.5% of full scale. Here, the absolute uncertainty in T is the same from one end of the scale to the other. A similar limitation in readout resolution is found in some digital instruments.

Infrared and near-infrared spectrophotometers also exhibit Case I behavior. With these, the limiting random error usually arises from Johnson noise in the thermal detector. Recall (p. 67) that this type of noise is

---

[7] L. D. Rothman, S. R. Crouch, and J. D. Ingle, Jr., *Anal. Chem.,* **1975**, *47*, 1226.

independent of the magnitude of the photo-current; indeed, fluctuations are observed even in the absence of radiation and, there-fore, net current.

Dark current and amplifier noise are usu-ally small compared with other sources of noise in photometric and spectrophotomet-ric instruments and become important only under conditions of low photocurrents where the lamp intensity or the photodetec-tor sensitivity is low. For example, such con-ditions are often encountered near the wavelength extremes for an instrument.

The precision of concentration data ob-tained with an instrument that is limited by Case I noise can be obtained directly by substituting an experimentally determined value for $s_T = k_1$, into Equation 6–14. Clearly, the precision of a particular *concen-tration* determination depends upon the magnitude of T even though the instrumen-tal precision is independent of T. The third column of Table 6–4 shows data obtained with Equation 6–14 when an absolute stand-ard deviation s_T of ± 0.003 or $\pm 0.3\% \, T$ was

assumed. A plot of the data is shown by curve A in Figure 6–7.

An indeterminate uncertainty of 0.3% T is typical of many moderately priced spec-trophotometers or photometers. Clearly, concentration errors of 1 to 2% relative are to be expected with such instruments. It is also evident that precision at this level can only be realized if the absorbance of the sample lies between about 0.1 and 1.

Case II: $s_T = k_2\sqrt{T^2 + T}$. This type of uncer-tainty often limits the precision of the high-est quality instruments. It has its origin in shot noise (p. 67), which must be expected whenever the passage of electricity involves transfer of charge across a junction, such as the movement of electrons from the cath-ode to the anode of a photomultiplier tube. Here, an electric current results from a se-ries of discrete events (emission of electrons from a cathode) the number of which per unit time is distributed in a random way about a mean value. The magnitude of the current fluctuations is proportional to the square root of current (see Equation 2–20,

Table 6–3
Types and Sources of Uncertainties in Transmittance Measurements

Category	Characterized by[a]	Typical Sources	Likely to be Important in
Case I	$s_T = k_1$	Limited readout resolution	Inexpensive photometers and spectrophotometers having small transmittance meter scales
		Heat detector Johnson noise	IR and near IR spectrophotometers and photometers
		Dark current and amplifier noise	Regions where source intensity and detector sensitivity are low
Case II	$s_T = k_2\sqrt{T^2 + T}$	Photon detector shot noise	High-quality UV/visible spectrophotometers
Case III	$s_T = k_3 T$	Cell positioning uncertainties	High-quality UV/visible and IR spectrophotometers
		Source flicker	Inexpensive photometers and spectrophotometers

[a] k_1, k_2, and k_3 are constants for a given system.

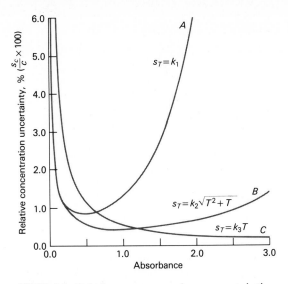

FIGURE 6-7 Relative concentration uncertainties arising from various categories of instrumental noise: A Case I; B Case II; C Case III. The data are taken from Table 6–4.

p. 67). The effect of shot noise on s_c is readily derived, and is given by the relationship

$$\frac{s_c}{c} = \frac{0.434\, k_2}{\log T} \sqrt{\frac{1}{T} + 1} \qquad (6\text{--}15)$$

The data in column 4 of Table 6–4 were obtained with the aid of Equation 6–15. Figure 6–7, curve B is a plot of such data. Note the much larger range of absorbances that can be encompassed without serious loss of accuracy when shot noise, rather than Johnson, limits the precision. This increased range represents a major advantage of photon-type detectors over thermal types, which are represented by curve A in the figure. As with Johnson-noise-limited instruments, shot-limited instruments do not give very reliable concentration data at transmittances greater than 95% (or A < 0.02).

Case III: $s_T = k_3 T$. One source of noise of this type is the slow drift in the radiant output of the source; this type of noise can be

Table 6–4
Relative Precision of Concentration Measurements as a Function of Transmittance and Absorbance for Three Categories of Instrument Noise

Transmittance, T	Absorbance, A	Relative Standard Deviation in Concentration[a]		
		Case I Noise[b]	Case II Noise[c]	Case III Noise[d]
0.95	0.022	± 6.2	±8.4	±25.3
0.90	0.046	± 3.2	±4.1	±12.3
0.80	0.097	± 1.7	±2.0	± 5.8
0.60	0.222	± 0.98	±0.96	± 2.5
0.40	0.398	± 0.82	±0.61	± 1.4
0.20	0.699	± 0.93	±0.46	± 0.81
0.10	1.00	± 1.3	±0.43	± 0.56
0.032	1.50	± 2.7	±0.50	± 0.38
0.010	2.00	± 6.5	±0.65	± 0.28
0.0032	2.50	±16.3	±0.92	± 0.23
0.0010	3.00	±43.4	±1.4	± 0.19

[a] $s_c \times 100/c$
[b] From Equation 6–14 with $s_T = k_1 = \pm0.0030$
[c] From Equation 6–15 with $k_2 = \pm0.0030$
[d] From Equation 6–16 with $k_3 = \pm0.013$

called *source flicker noise* (p. 67). The effects of fluctuations in the intensity of a source can be minimized by the use of a constant voltage power supply or a split beam arrangement (p. 176). With many instruments, source flicker noise does not limit performance.

An important and widely encountered noise source, which is proportional to transmittance, results from failure to position sample and reference cells reproducibly with respect to the beam during replicate transmittance measurements. All cells have minor imperfections. As a consequence, reflection and scattering losses vary as different sections of the cell window are exposed to the beam; small variations in transmittance result. Rothman, Crouch, and Ingle have shown that this uncertainty often is the most common limitation to the accuracy of high-quality ultraviolet-visible spectrophotometers. It is also a serious source of uncertainty in infrared instruments.

One method of reducing the effect of cell positioning with a double-beam instrument is to leave the cells in place during calibration and analysis; new standards and samples are introduced after washing and rinsing the cell in place with a syringe. Care must be taken to avoid touching or jarring the cells during this process.

The effect of uncertainties that are proportional to transmittance on analytical results can be obtained by substitution of $s_T = k_3 T$ into Equation 6–14, which gives

$$\frac{s_c}{c} = -\frac{0.434\, k_3}{\log T} \tag{6–16}$$

Column 5 of Table 6–4 contains data obtained from Equation 6–16 when k_3 is assumed to have a value of 0.013, which approximates the value observed in the Rothman, Crouch, and Ingle study. The data are plotted as curve C in Figure 6–7.

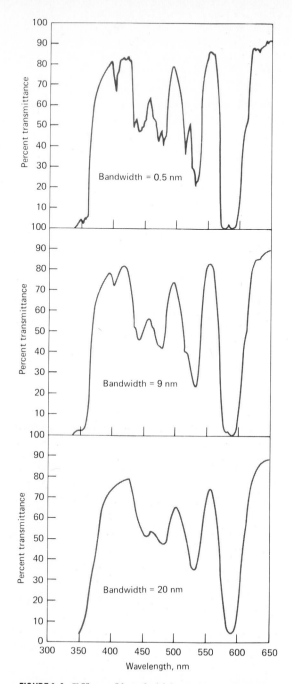

FIGURE 6–8 Effect of bandwidth on spectral detail. The sample was a didymium glass. (Spectra provided through the courtesy of Amsco Instrument Company [formerly G. K. Turner Associates] Carpinteria, CA.)

6B-5 EFFECT OF SLIT WIDTH ON ABSORPTION MEASUREMENTS

As shown in Section 5D–5, narrow slit widths are required to resolve complex spectra.[8] For example, Figure 6–8 illustrates the loss of detail that accompanies the use of wider slits. In this example, the transmittance spectrum of a didymium glass was obtained at slit settings that provided effective bandwidths of 0.5, 9, and 20 nm. The progressive loss of spectral detail is clear. For qualitative studies, such losses often loom important.

Figure 6–9 illustrates a second effect of slit width on spectra made up of narrow peaks. Here, the spectrum of a praseodymium chloride solution was obtained at slit settings of 1.0, 0.5, and 0.1 mm. Note that the peak absorbance values increase significantly (by as much as 70% in one instance) as the slit width decreases. At slit settings less than about 0.14 mm, absorbances were found to become independent of slit width. Careful inspection of Figure 6–8 reveals the same type of effect. In both sets of spectra, the areas under the individual peaks are the same, but wide slit widths result in broader lower peaks.

It is evident from both of these illustrations that quantitative measurement of narrow absorption bands demands the use of narrow slit widths or, alternatively, very reproducible slit-width settings.

Unfortunately, a decrease in slit width is accompanied by a second-order power reduction in the radiant energy; at very narrow settings spectral detail may be lost owing to an increase in the signal-to-noise

---

[8] For a discussion of the effects of slit width on spectra, see: *Optimum Parameters for Spectrophotometry*, Varian Instruments Division, Palo Alto, CA, 1977; and F. C. Strong III, *Anal. Chem.*, **1976,** *48,* 2155.

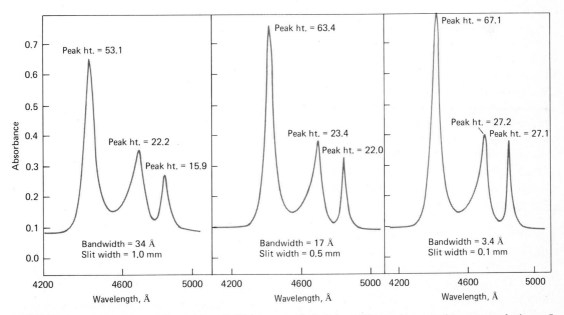

FIGURE 6-9 Effect of slit width and bandwidth on peak heights. Here, the sample was a solution of praseodymium chloride. (From *Optimum Spectrophotometer Parameters, Application Report AR* **14–2,** Cary Instruments: Monrovia, CA. With permission.)

ratio. The situation becomes particularly serious in spectral regions where the output of the source or the sensitivity of the detector is low. Under such circumstances, noise in either of these components or their associated electronic circuits may result in partial or total loss of spectral fine structure.

In general, it is good practice to narrow slits no more than is necessary for resolution of the spectrum at hand. With a variable slit spectrophotometer, proper slit adjustment can be determined by obtaining spectra at progressively narrower slits until peak heights become constant. Generally, constant peak heights are observed when the effective bandwidth of the instrument is 0.1 or less of the effective bandwidth of the absorption peak.

6B-6 EFFECT OF SCATTERED RADIATION AT WAVELENGTH EXTREMES OF A SPECTROPHOTOMETER

We have already noted that scattered radiation may cause instrumental deviations from Beer's law. When measurements are attempted at the wavelength extremes of an instrument, the effects of stray radiation may be even more serious, and on occasion lead to the appearance of false absorption peaks. For example, consider a spectrophotometer for the visible region equipped with glass optics, a tungsten source, and a photovoltaic cell detector. At wavelengths below about 380 nm, the windows, cells, and prism begin to absorb radiation, thus reducing the energy reaching the transducer. The output of the source falls off rapidly in this region as well; so also does the sensitivity of the photoelectric device. Thus, the total signal for the 100% T adjustment may be as low as 1 to 2% of that in the region between 500 and 650 nm.

The scattered radiation, however, is often made up of wavelengths to which the instrument is highly sensitive. Thus, its effects can be enormously magnified. Indeed, in some

instances the output signal produced by the stray radiation may exceed that from the monochromator beam; under these circumstances, the measured absorbance is as much that for the stray radiation as for the radiation to which the instrument is set.

An example of a false peak appearing at the wavelength extremes of a visible-region spectrophotometer is shown in Figure 6–10. The spectrum of a solution of cerium(IV) obtained with an ultraviolet-visible spectrophotometer, sensitive in the region of 200 to 750 nm, is shown by curve B. Curve A is a spectrum of the same solution obtained with a simple visible spectrophotometer. The apparent maximum shown in curve A arises from the instrument responding to stray wavelengths longer than 400 nm, which (as can be seen from the spectra) are not absorbed by the cerium(IV) ions.

This same effect is sometimes observed

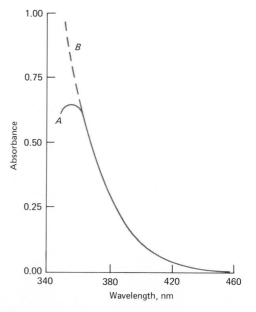

FIGURE 6-10 Spectra of cerium(IV) obtained with a spectrophotometer having glass optics (A) and quartz optics (B). The false peak in A arises from transmission of stray radiation of longer wavelengths.

with ultraviolet-visible instruments when attempts are made to measure absorbances at wavelengths lower than about 200 nm.

6C Photometer and Spectrophotometer Designs

A host of instruments are available commercially for absorbance measurements in the ultraviolet, visible, and infrared spectral regions. Several typical instruments will be described in the chapters that follow. In this section, some of the general design characteristics of all of these instruments are presented.

6C-1 DESIGN VARIABLES

Two important instrument design variables are: (1) the arrangement in space of the sample and reference cells with respect to the wavelength selector and (2) the timing of the measurement of P_0 and P as well as the dark current. The second consideration determines whether the instrument will be a single- or double-beam device.

Location of Sample and Reference Cells. The sample and reference cells can be placed either between the source and the wavelength selector or between the wavelength selector and the detector. Each arrangement has advantages and disadvantages. In the former configuration, all of the radiation passes through the sample. Scattering of radiation by components of the sample does little harm in this arrangement because the scattered radiation will be subsequently rejected by the selector and will thus not reach the detector. On the other hand, any fluorescent species in the sample will be induced to radiate by the shorter wavelengths from the source; the fluorescence will produce an emission spectrum that will interfere with absorption measurements. Furthermore, the shorter and more energetic wavelengths often cause photodecomposition of the analyte thus interfering with the measurement.

When the sample and reference solution are placed between the wavelength selector and detector, the possibility of fluorescence and photodecomposition are minimized because the shorter wavelengths have been removed before the radiation reaches the sample. Here, however, errors due to scattering may arise.

For the ultraviolet and visible regions, fluorescence and photosensitivity are likely to be more serious problems than scattering. Thus, quite generally, instruments for these regions are designed so that the sample and reference cells are located after the wavelength selector. On the other hand, infrared radiation is generally not sufficiently energetic to cause serious fluorescence or decomposition problems. Scattering, however, is more likely. Thus, in infrared photometers and spectrophotometers, the sample cell is located immediately after the source.

6C-2 SINGLE-BEAM INSTRUMENTS

As the name implies, single-beam instruments have but one beam of radiation, which passes from the source through the wavelength selector to the detector (see Figure 6–11a). The determination of transmittance then involves three successive steps that are separated in time: (1) the 0% T setting with a shutter in place; (2) the measurement of P_0 (or the instrumental adjustment to 100% T); and (3) the measurement of P. The reliability of transmittance data depends upon the constancy of the instrumental characteristics during the time required to complete the three steps. Therefore, these measurements are always completed as expeditiously as possible.

6C-3 DOUBLE-BEAM INSTRUMENTS

Double-beam or dual channel absorption instruments have two radiation beams, one of which passes through the reference solvent

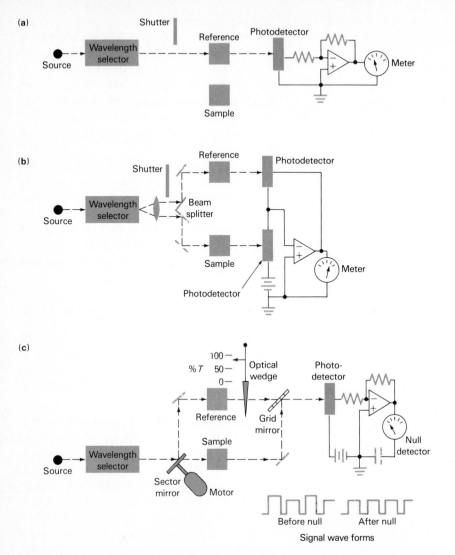

FIGURE 6-11 Instrument designs for photometers and spectrophotometers: (**a**) single-beam design; (**b**) dual channel design with beams separated in space but simultaneous in time; (**c**) double-beam design in which beams alternate between two channels.

and the other through the sample. In one type of instrument, the two paths are separated in space but are simultaneous in time; in the other they alternate in time but more closely coincide in space.

The first type of dual channel instruments is illustrated in Figure 6–11b. Here,

the two channels in space are formed by a V-shaped mirror called a beam splitter. One beam is reflected through the reference solution to a detector while simultaneously the second passes through the sample to a second matched detector. The ratio of the two beam powers may then be determined by an electronic circuit similar to that shown in Figure 2–18**b** (p. 47). In this instance, the measurement is a two-step operation involving first the zero adjustment with a shutter in place between the selector and the beam splitter and then the transmittance determination. Here, reliability is dependent upon the constancy of the dark current during the two steps and the similarity in the response characteristics of the two detectors.

Figure 6–11**c** illustrates the second type of dual channel arrangement. Note that in contrast to the other two designs, no shutter as such is used. Here, a rotating sector mirror serves to direct all of the radiation from the selector to the reference cell and the sample cell alternately. Thus, the dual channels are, in a sense, based on time. The pulses of radiation are recombined by means of a grid mirror which transmits one and reflects the other to the detector. The motor-driven sector mirror, which is circular in cross section, is made up of several pie-shaped segments, half of which are mirrored and half of which are open to permit unobstructed passage of radiation. The mirrored sections are held in place by blackened metal frames which serve to periodically interrupt the beam and prevent its reaching the detector. The detector circuit is programmed to use these periods to perform the dark current adjustment by means of a feedback loop.

Instruments such as that shown in Figure 6–11**c** are usually of the null type in which the beam passing through the reference cell is attenuated mechanically until its intensity just equals that of the sample beam; alternatively, the nulling can be done electronically by a feedback loop that acts on the electrical pulse corresponding to the reference cycle. The instrument in Figure 6–11**c** is equipped with an optical wedge, the transmission of which increases linearly along its length. Thus, the null point is reached by moving this wedge vertically in the beam until the two electrical pulses are identical.

Wave forms for the electrical signal are shown before and after balance. The output of the detector is time shared so that the minima in the waves are used for dark current adjustments. During the remaining time the output is a square-wave ac signal superimposed on a dc signal until null is reached; then the output appears as a dc signal only. Thus, a circuit such as that shown in Figure 6–11**c**, which responds only to an ac signal, can be employed to detect the null point.

An instrument such as that just described provides a signal that is free from drift in the source and detector without requiring carefully matched detectors and detector circuits.

6C–4 DOUBLE-BEAM VERSUS SINGLE-BEAM INSTRUMENTS

Because the measurement of P_0 and P is made simultaneously, or nearly so, a double-beam instrument compensates for the drift (flicker noise) in the radiant power of the source as well as all but the most short-term electrical noise from the detector and its associated circuitry. As a consequence, good performance can be obtained with lower quality electronic components. In addition, the double-beam design lends itself well to the continuous recording of transmittances or absorbances as a spectrum is swept. Offsetting these advantages, however, is the greater number and complexity of components required for double-beam operations and the electrical and mechanical noise introduced by the beam switching apparatus.

Most sophisticated, modern instruments, which are designed for recording absorbance or transmittance over a considerable spectral range, are double beam (in time). As will be shown in subsequent chapters, however, simple instruments embodying this principle are also encountered.

The single-beam design offers distinct advantages in terms of simplicity and concomitant lower cost. Furthermore, the signal-to-noise ratios of these simpler instruments are inherently greater for several reasons. First, the radiant energy reaching the detector of single-beam instruments is generally significantly larger than for their double-beam counterparts, thus leading to enhanced S/N ratios. In all spectrophotometers, light losses occur where the radiation interfaces with instrument components such as mirrors, dispersing elements, and windows. For example, a beam may be attenuated by 10 to 30% upon being reflected from a single mirror. Double-beam instruments often have as many as 10 to 15 of these elements, which leads to large net power losses. In contrast, a typical single-beam instrument will have perhaps 3 to 6 surfaces at which beam attenuation can occur. A double-beam instrument also suffers loss of throughput because as much as half the photons are lost in measuring the dark current; the remaining photons are then split in two in order to measure P and P_0 simultaneously or alternately.

In addition to having a weaker signal, double-beam instruments also tend to suffer from the effects of greater noise. For the twin detector type, the detector noise is doubled. Single detector instruments, on the other hand, require a motor-driven mechanical device which introduces vibrational noise as well as electrical and magnetic field noise. As a consequence, it has been shown that if the drift in the source and in the detector dark current can be minimized, better S/N ratios can be realized with a single-beam design.[9]

Single-beam instruments are particularly well suited for quantitative analyses wherein measurements at a single wavelength are employed. Here, the recording advantage of the double-beam instrument is often not important; the simplicity and high signal-to-noise ratio of the single-beam design constitutes a real advantage.

As will be described later, at least two manufacturers have developed single-beam instruments that permit the rapid recording of entire spectra thus combining the advantage of the two instrument designs.

---

[9] See: W. Kaye, D. Barber, and R. Marasco, *Anal. Chem.*, **1980**, *52*, 437A.

PROBLEMS

6–1. Convert the following absorbance data into percent transmittance: (a) 0.212, (b) 1.025, (c) 0.016, (d) 0.549, (e) 0.867, (f) 1.732.

6–2. Convert the following percent transmittance data into absorbance: (a) 32.6, (b) 21.7, (c) 90.0, (d) 2.76, (e) 61.7, (f) 30.0.

6–3. Calculate the percent transmittance of solutions with half the absorbance of those in Problem 6–1.

6–4. Calculate the absorbance of solutions with half the percent transmittance of those in Problem 6–2.

6–5. Calculate the percent transmittance after the solutions in Problem 6–2 have been diluted to twice their original volume.

6–6. Use the data provided to evaluate the missing quantities. Wherever necessary, assume that the molecular weight of the absorbing species is 280.

	A	$\% T$	ε	b, cm	c, M	c	a
(a)	0.350			1.80	5.25×10^{-4}	mg/L	
(b)		20.7		2.50	6.91×10^{-3}	ppm	
(c)	1.212			0.996		g/L	0.250
(d)		25.3	5.42×10^3		3.00×10^{-4}	mg/100 mL	
(e)			2.75×10^4	4.44		3.33 ppm	
(f)			7.65×10^3	1.46	7.84×10^{-5}	mg/L	
(g)		55.7		0.100		6.72 ppm	
(h)	0.974		5.56×10^3	1.00		g/L	
(i)		84.3		1.25		mg/L	0.631
(j)		7.72	9.16×10^2		9.76×10^{-3}	mg/L	

6–7. A solution that was $6.72 \times 10^{-3} M$ in X had a transmittance of 0.112 when measured in a 2.00-cm cell. What concentration of X would be required for the transmittance to be increased by a factor of 3 when a 1.00-cm cell was used?

6–8. A compound had a molar absorptivity of 1.41×10^3 L cm^{-1} mol^{-1}. What concentration of the compound would be required to produce a solution having a transmittance of 8.42% in a 2.50-cm cell?

6–9. At 580 nm, the wavelength of its maximum absorption, the complex FeSCN^{2+} has a molar absorptivity of 7.00×10^3 L cm^{-1} mol^{-1}. Calculate
(a) the absorbance of a $1.54 \times 10^{-4} M$ solution of the complex at 580 nm when measured in a 1.25-cm cell.
(b) the transmittance of a $2.68 \times 10^{-4} M$ solution employing the same cell as in (a).
(c) the absorbance of a solution that has half the transmittance of that described in (a).

6–10. A solution containing the thiourea complex of bismuth(III) has a molar absorptivity of 9.3×10^3 L cm^{-1} mol^{-1} at 470 nm.
(a) What will be the absorbance of a $2.00 \times 10^{-5} M$ solution of the complex when measured in a 1.10-cm cell?
(b) What will be the percent transmittance of the solution described in (a)?
(c) What concentrations of bismuth could be determined by means of this complex if 1.00-cm cells are to be used and the absorbance is to be kept between 0.100 and 1.500?

6–11. In ethanol, acetone has a molar absorptivity of 2.75×10^3 L cm^{-1} mol^{-1} at 366 nm. What range of acetone concentrations could be determined if the percent transmittances of the solutions are to be limited to a range of 10 to 90% and a 1.0-cm cell is to be used?

6–12. In neutral aqueous solution, it is found that log ε for phenol at 211 nm is 4.12. What range of phenol concentrations could be determined spectrophotometrically if absorbance values in a 2.00-cm cell are to be limited to a range of 0.100 to 1.50?

6–13. The equilibrium constant for the reaction

$$2CrO_4^{2-} + 2H^+ \rightleftharpoons Cr_2O_7^{2-} + H_2O$$

has a value of 4.2×10^{14}. The molar absorptivities for the two principal species in a solution of $K_2Cr_2O_7$ are

λ	$\varepsilon_1(CrO_4^{2-})$	$\varepsilon_2(Cr_2O_7^{2-})$
345	1.84×10^3	10.7×10^2
370	4.81×10^3	7.28×10^2
400	1.88×10^3	1.89×10^2

Four solutions were prepared by dissolving the following number of moles of $K_2Cr_2O_7$ in water and diluting to 1.00 L with a pH 5.60 buffer: 4.00×10^{-4}, 3.00×10^{-4}, 2.00×10^{-4}, and 1.00×10^{-4}. Derive theoretical absorbance values (1.00-cm cells) for each and plot the data for: (a) 345 nm, (b) 370 nm, (c) 400 nm.

6–14. A species Y has a molar absorptivity of 3000. Derive absorbance data (1.00-cm cells) for solutions of Y that are 4.00×10^{-4}, 3.00×10^{-4}, 2.00×10^{-4}, and 1.00×10^{-4} M in Y, assuming that the radiation employed was contaminated with the following percent of nonabsorbed radiation: (a) 0.000, (b) 0.300, (c) 2.00, (d) 6.00. Plot the data.

6–15. The complex that is formed between gallium (III) and 8-hydroxy-quinoline has an absorption maximum at 393 nm. A 1.29×10^{-4} M solution of the complex has a transmittance of 14.6% when measured in a 1.00-cm cell at this wavelength. Calculate the molar absorptivity of the complex.

6–16. A 50.0-mL aliquot of well water is treated with an excess of KSCN and diluted to 100.0 mL. Calculate the ppm of iron(III) in the sample if the diluted solution has an absorbance of 0.394 at 580 nm when measured in a 2.50-cm cell; see Problem 6–9.

6–17. A 2.83×10^{-4} M solution of potassium permanganate has a molar absorbance of 0.510 when measured in a 0.982-cm cell at 520 nm. Calculate
(a) the molar absorptivity for $KMnO_4$ at this wavelength.
(b) the absorptivity when the concentration is expressed in ppm.
(c) the molar concentration of permanganate in a solution that has an absorbance of 0.697 when measured in a 1.50-cm cell at 520 nm.
(d) the transmittance of the solution in (c).
(e) the absorbance of a solution that has twice the transmittance of the solution in (c).

6–18. Chromium(III) forms a complex with diphenylcarbazide whose molar absorptivity is 4.17×10^4 at 540 mm. Calculate

(a) the absorbance of a $7.68 \times 10^{-6}\ M$ solution of the complex at 540 nm when measured in a 1.00-cm cell.

(b) the transmittance of the solution described in (a).

(c) the path length needed to cause a $2.56 \times 10^{-6}\ M$ solution of the complex to have the same absorbance as the solution described in (a).

(d) the molar concentration of the complex in a solution that has an absorbance of 0.649 at 540 nm when measured in a 1.00-cm cell.

(e) the concentration of the complex in a solution that has a transmittance of 0.649 at 540 nm when measured in a 1.00-cm cell.

6–19. A portable photometer with a linear response to radiation registered 73.6 μA with a blank solution in the light path. Replacement of the blank with an absorbing solution yielded a response of 24.9 μA. Calculate

(a) the percent transmittance of the sample solution.

(b) the absorbance of the sample solution.

(c) the transmittance to be expected for a solution in which the concentration of the absorber is one-third that of the original sample solution.

(d) the transmittance to be expected for a solution that has twice the concentration of the sample solution.

6–20. Titanium forms a yellow complex with hydrogen peroxide which can be used for colorimetric determination of that element. The color of an unknown solution containing hydrogen peroxide was visually compared with that of a standard containing the same amount of peroxide and 25.0 ppm Ti. A visual match was observed in flat-bottom tubes when the length of the liquid in the standard tube was 23.7 cm and that in the analyte was 28.0 cm. Calculate the concentration of titanium in the unknown.

6–21. The meter of an inexpensive spectrophotometer had a 5-inch scale scribed in linear units from 0 to 100% T. The scale, which limited the precision of the instrument, could be read to about $\pm 0.5\%\ T$. Calculate the relative precision of concentration determinations for an absorbance of: (a) 0.020, (b) 0.050, (c) 0.100, (d) 0.400, (e) 0.800, (f) 1.200, (g) 2.000.

Molecular

Ultraviolet and Visible

Absorption Spectroscopy

Absorption measurements based upon ultraviolet or visible radiation find widespread application for the qualitative and quantitative determination of molecular species. Several of these applications are considered in this chapter.[1]

7A The Magnitude of Molar Absorptivities

Empirically, molar absorptivities that range from zero up to a maximum of the order of 10^5 are observed in ultraviolet and visible

absorption spectroscopy. For any particular peak, the magnitude of ε depends upon the capture cross section (Equation 6–6) of the species and the probability for an energy absorbing transition to occur. The relationship between ε and these parameters has been shown to be[2]

$$\varepsilon = 8.7 \times 10^{19} PA$$

where P is the transition probability and A is the cross section target area in square centimeters. The area for typical organic molecules has been estimated from electron diffraction and X-ray studies to be about 10^{-15} cm^2 while transition probabilities vary from zero to one. For quantum mechanically allowed transitions, values of P range from 0.1 to 1 which leads to strong absorption bands ($\varepsilon_{max} = 10^4$ to 10^5). Peaks having molar absorptivities less than about 10^3 are classified as being of low intensity. They result

[1] Some useful references on absorption methods include: E. J. Meehan, in *Treatise on Analytical Chemistry*, 2d ed., P. J. Elving, E. J. Meehan, and I. M. Kolthoff, Eds., Part I, vol. 7, Chapters 1–3, Wiley: New York, 1981; R. P. Bauman, *Absorption Spectroscopy*, Wiley: New York, 1962; F. Grum, in *Physical Methods of Chemistry*, A. Weissberger and B. W. Rossiter, Eds., Volume I, Part III B, Chapter 3, Wiley-Interscience: New York, 1972; H. H. Jaffé and M. Orchin, *Theory and Applications of Ultraviolet Spectroscopy*, Wiley: New York, 1962; G. F. Lothian, *Absorption Spectrophotometry*, 3d ed., Adam Hilger Ltd.: London, 1969.

[2] E. A. Braude, *J. Chem. Soc.*, **1950**, 379.

from so-called forbidden transitions, which have probabilities of occurrence that are less than 0.01.

7B   Absorbing Species

The absorption of ultraviolet or visible radiation by an atomic or molecular species M can be considered to be a two-step process, the first of which involves excitation as shown by the equation

$$M + h\nu \rightarrow M^*$$

The product of the reaction between M and the photon $h\nu$ is an electronically excited particle symbolized by M^*. The lifetime of the excited particle is brief (10^{-8} to 10^{-9} sec), its existence being terminated by any of several *relaxation* processes. The most common type of relaxation involves conversion of the excitation energy to heat; that is,

$$M^* \rightarrow M + heat$$

Relaxation may also occur by decomposition of M^* to form new species; such a process is called a *photochemical reaction*. Alternatively, relaxation may involve fluorescent or phosphorescent reemission of radiation. It is important to note that the lifetime of M^* is usually so very short that its concentration at any instant is ordinarily negligible. Furthermore, the amount of thermal energy evolved by relaxation is usually not detectable. Thus, absorption measurements create a minimal disturbance of the system under study (except when photochemical decomposition occurs).

The absorption of ultraviolet or visible radiation generally results from excitation of bonding electrons; as a consequence, the wavelengths of absorption peaks can be correlated with the types of bonds that exist in the species under study. Molecular absorption spectroscopy is, therefore, valuable for identifying functional groups in a molecule. More important, however, are the applications of ultraviolet and visible absorption spectroscopy to the quantitative determination of compounds containing absorbing groups.

For purposes of this discussion, it is useful to recognize three types of electronic transitions and to categorize absorbing species on this basis. The three include transitions involving: (1) π, σ, and n electrons, (2) d and f electrons, and (3) charge-transfer electrons.

7B-1   ABSORBING SPECIES CONTAINING π, σ, AND n ELECTRONS

Absorbing species containing π, σ, and n electrons include organic molecules and ions as well as a number of inorganic anions. Our discussion will deal largely with the former, although brief mention will be made of absorption by certain inorganic systems as well.

All organic compounds are capable of absorbing electromagnetic radiation because all contain valence electrons that can be excited to higher energy levels. The excitation energies associated with electrons forming most single bonds are sufficiently high that absorption by them is restricted to the so-called vacuum ultraviolet region ($\lambda < 185$ nm), where components of the atmosphere also absorb strongly. The experimental difficulties associated with the vacuum ultraviolet are formidable; as a result, most spectrophotometric investigations of organic compounds have involved the wavelength region greater than 185 nm. Absorption of longer-wavelength ultraviolet and visible radiation is restricted to a limited number of functional groups (called *chromophores*) that contain valence electrons with relatively low excitation energies.

The electronic spectra of organic molecules containing chromophores are usually complex, because the superposition of vibrational transitions on the electronic transitions leads to an intricate combination of overlapping lines; the result is a broad band of absorption that often appears to be continuous. The complex nature of the spectra makes detailed theoretical analysis difficult or impossible. Nevertheless, qualitative or semiquantitative statements concerning the types of electronic transitions responsible for a given absorption spectrum can be deduced from molecular orbital considerations.

Types of Absorbing Electrons.[3] The electrons that contribute to absorption by an organic molecule are: (1) those that participate directly in bond formation between atoms and are thus associated with more than one atom; (2) nonbonding or unshared outer electrons that are largely localized about such atoms as oxygen, the halogens, sulfur, and nitrogen.

Covalent bonding occurs because the electrons forming the bond move in the field about two atomic centers in such a manner as to minimize the repulsive coulombic forces between these centers. The nonlocalized fields between atoms that are occupied by bonding electrons are called *molecular orbitals* and can be considered to result from the overlap of atomic orbitals. When two atomic orbitals combine, either a low-energy *bonding molecular orbital* or a high-energy *antibonding molecular orbital* results. The electrons of a molecule occupy the former in the ground state.

The molecular orbitals associated with single bonds in organic molecules are designated as *sigma* (σ) *orbitals,* and the corresponding electrons are σ electrons. As shown in Figure 7–1**a**, the distribution of charge density of a sigma orbital is rotation-

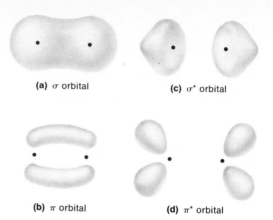

(a) σ orbital   **(c)** σ^* orbital

(b) π orbital   **(d)** π^* orbital

FIGURE 7-1 Electron distribution in sigma and pi molecular orbitals.

ally symmetric around the axis of the bond. Here, the average negative charge density arising from the motion of the two electrons around the two positive nuclei is indicated by the degree of shading.

The double bond in an organic molecule contains two types of molecular orbitals: a *sigma* (σ) orbital corresponding to one pair of the bonding electrons and a *pi* (π) *molecular orbital* associated with the other pair. Pi orbitals are formed by the parallel overlap of atomic *p* orbitals. Their charge distribution is characterized by a *nodal plane* (a region of low-charge density) along the axis of the bond and a maximum density in regions above and below the plane (see Figure 7–1**b**).

Also shown, in Figures 7–1**c** and 7–1**d**, are the charge-density distributions for antibonding sigma and pi orbitals; these orbitals are designated by σ^* and π^*.

In addition to σ and π electrons, many organic compounds contain nonbonding electrons. These unshared electrons are designated by the symbol *n*. An example showing the three types of electrons in a simple organic molecule is shown in Figure 7–2.

As shown in Figure 7–3, the energies for the various types of molecular orbitals differ

[3] For further details, see: H. H. Jaffé and M. Orchin, *Theory and Applications of Ultraviolet Spectroscopy,* Wiley: New York, 1962; C. N. R. Rao, *Ultra-Violet and Visible Spectroscopy: Chemical Applications,* 3d ed., Butterworths: London, 1975; R. P. Bauman, *Absorption Spectroscopy,* Chapters 6 and 8, Wiley: New York, 1962.

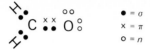

FIGURE 7-2 Types of molecular orbitals in formaldehyde.

125 nm. Ethane has an absorption peak at 135 nm, which must also arise from the same type of transition, but here, electrons of the C—C bond appear to be involved. Because the strength of the C—C bond is less than that of the C—H bond, less energy is required for excitation; thus, the absorption peak occurs at a longer wavelength.

Absorption maxima due to $\sigma \rightarrow \sigma^*$ transitions are never observed in the ordinary, accessible ultraviolet region; for this reason, no further discussion will be devoted to this type of absorption.

$n \rightarrow \sigma^*$ Transitions. Saturated compounds containing atoms with unshared electron pairs (nonbonding electrons) are capable of $n \rightarrow \sigma^*$ transitions. In general, these transitions require less energy than the $\sigma \rightarrow \sigma^*$ type and can be brought about by radiation in the region of between 150 and 250 nm, with most absorption peaks appearing below 200 nm. Table 7–1 shows absorption data for some typical $n \rightarrow \sigma^*$ transitions. It will be seen that the energy requirements for such transitions depend primarily upon the kind of atomic bond and to a lesser extent upon the structure of the molecule. The molar absorptivities (ε) associated with this type of absorption are low to intermedi-

significantly. Quite generally, the energy level of a nonbonding electron lies between those of the bonding and the antibonding π and σ orbitals.

Electronic transitions among certain of the energy levels can be brought about by the absorption of radiation. As shown in Figure 7–3, four types of transitions are possible: $\sigma \rightarrow \sigma^*$, $n \rightarrow \sigma^*$, $n \rightarrow \pi^*$, and $\pi \rightarrow \pi^*$.

$\sigma \rightarrow \sigma^*$ Transitions. Here, an electron in a bonding σ orbital of a molecule is excited to the corresponding antibonding orbital by the absorption of radiation. The molecule is then described as being in the σ,σ^* excited state. Relative to other possible transitions, the energy required to induce a $\sigma \rightarrow \sigma^*$ transition is large (see Figure 7–3), corresponding to radiant frequencies in the vacuum ultraviolet region. Methane, for example, which contains only single C—H bonds and can thus undergo only $\sigma \rightarrow \sigma^*$ transitions, exhibits an absorption maximum at

Table 7–1 Some Examples of Absorption Due To $n \rightarrow \sigma^*$ Transitions[a]		
Compound	$\lambda_{max}(nm)$	ε_{max}
H_2O	167	1480
CH_3OH	184	150
CH_3Cl	173	200
CH_3I	258	365
$(CH_3)_2S$[b]	229	140
$(CH_3)_2O$	184	2520
CH_3NH_2	215	600
$(CH_3)_3N$	227	900

[a] Samples in vapor state.
[b] In ethanol solvent.

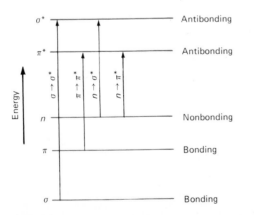

FIGURE 7-3 Electronic molecular energy levels.

ate in magnitude and usually range between 100 and 3000 L cm^{-1} mol^{-1}.

Absorption maxima for formation of the n,σ^* state tend to shift to shorter wavelengths in the presence of polar solvents such as water or ethanol. The number of organic functional groups with $n \rightarrow \sigma^*$ peaks in the readily accessible ultraviolet region is relatively small.

$n \rightarrow \pi^*$ and $\pi \rightarrow \pi^*$ Transitions. Most applications of absorption spectroscopy to organic compounds are based upon transitions for n or π electrons to the π^* excited state because the energies required for these processes bring the absorption peaks into an experimentally convenient spectral region (200 to 700 nm). Both transitions require the presence of an unsaturated functional group to provide the π orbitals. Strictly speaking, it is to these unsaturated absorbing centers that the term chromophore applies.

The molar absorptivities for peaks associated with excitation to the n,π^* state are generally low and ordinarily range from 10 to 100 L cm^{-1} mol^{-1}; values for $\pi \rightarrow \pi^*$ transitions, on the other hand, normally fall in the range between 1000 and 10,000. Another characteristic difference between the two types of absorption is the effect exerted by the solvent on the wavelength of the peaks. Peaks associated with $n \rightarrow \pi^*$ transitions are generally shifted to shorter wavelengths (a *hypsochromic* or *blue shift*) with increasing polarity of the solvent. Usually, but not always, the reverse trend (a *bathochromic* or *red shift*) is observed for $\pi \rightarrow \pi^*$ transitions. The hypsochromic effect apparently arises from the increased solvation of the unbonded electron pair, which lowers the energy of the n orbital. The most dramatic effects of this kind (blue shifts of 30 nm or more) are seen with polar hydrolytic solvents, such as water or alcohols, in which hydrogen-bond formation between the solvent protons and the nonbonded electron pair is extensive. Here, the energy of the n

orbital is lowered by an amount approximately equal to the energy of the hydrogen bond. When an $n \rightarrow \pi^*$ transition occurs, however, the remaining single n electron cannot maintain the hydrogen bond; thus, the energy of the n,π^* *excited* state is not affected by this type of solvent interaction. A blue shift, also roughly corresponding to the energy of the hydrogen bond, is therefore observed.

A second solvent effect that undoubtedly influences both $\pi \rightarrow \pi^*$ and $n \rightarrow \pi^*$ transitions leads to a bathochromic shift with increased solvent polarity. This effect is small (usually less than 5 nm) and as a result, is completely overshadowed in $n \rightarrow \pi^*$ transitions by the hypsochromic effect just discussed. With the bathochromic shift, attractive polarization forces between the solvent and the absorber tend to lower the energy levels of both the unexcited and the excited states. The effect on the excited state is greater, however, and the energy differences thus become smaller with increased solvent polarity; small bathochromic shifts result.

Organic Chromophores. Table 7–2 lists common organic chromophores and the approximate location of their absorption maxima. These data can serve only as rough guides for the identification of functional groups, since the positions of maxima are also affected by solvent and structural details of the molecule containing the chromophores. Furthermore, the peaks are ordinarily broad because of vibrational effects; the precise determination of the position of a maximum is thus difficult.

Effect of Conjugation of Chromophores. In the molecular-orbital treatment, π electrons are considered to be further delocalized by the conjugation process; the orbitals thus involve four (or more) atomic centers. The effect of this delocalization is to lower the energy level of the π^* orbital and give it less antibonding character. Absorption maxima are shifted to longer wavelengths as a consequence.

As seen from the data in Table 7–3, the absorptions of multichromophores in a single organic molecule are approximately additive, provided the chromophores are separated from one another by more than one single bond. Conjugation of chromophores, however, has a profound effect on spectral properties. For example, it is seen in Table 7–3 that 1,3–butadiene, CH_2=CHCH=CH_2, has a strong absorption band that is displaced to a longer wavelength by 20 nm as compared with the corresponding peak for an unconjugated diene. When three double bonds are conjugated, the bathochromic effect is even larger.

Conjugation between the doubly bonded oxygen of aldehydes, ketones, and carboxylic acids and an olefinic double bond gives rise to similar behavior (see Table 7–3). Analogous effects are also observed when two carbonyl or carboxylate groups are conjugated with one another. For α–β unsaturated aldehydes and ketones, the weak absorption peak due to $n \rightarrow \pi^*$ transitions is shifted to longer wavelengths by 40 nm or more. In addition, a strong absorption peak corresponding to a $\pi \rightarrow \pi^*$ transition appears. This latter peak occurs only in the vacuum ultraviolet if the carbonyl group is not conjugated.

The wavelengths of absorption peaks for conjugated systems are sensitive to the types of groups attached to the doubly bonded

Table 7-2
Absorption Characteristics of Some Common Chromophores

Chromophore	Example	Solvent	λ_{max}(nm)	ε_{max}	Type of Transition
Alkene	$C_6H_{13}CH$=CH_2	n-Heptane	177	13,000	$\pi \rightarrow \pi^*$
Alkyne	$C_5H_{11}C$≡C—CH_3	n-Heptane	178	10,000	$\pi \rightarrow \pi^*$
			196	2000	—
			225	160	—
Carbonyl	$CH_3\overset{O}{\overset{\|}{C}}CH_3$	n-Hexane	186	1000	$n \rightarrow \sigma^*$
			280	16	$n \rightarrow \pi^*$
	$CH_3\overset{O}{\overset{\|}{C}}H$	n-Hexane	180	large	$n \rightarrow \sigma^*$
			293	12	$n \rightarrow \pi^*$
Carboxyl	$CH_3\overset{O}{\overset{\|}{C}}OH$	Ethanol	204	41	$n \rightarrow \pi^*$
Amido	$CH_3\overset{O}{\overset{\|}{C}}NH_2$	Water	214	60	$n \rightarrow \pi^*$
Azo	CH_3N=NCH_3	Ethanol	339	5	$n \rightarrow \pi^*$
Nitro	CH_3NO_2	Isooctane	280	22	$n \rightarrow \pi^*$
Nitroso	C_4H_9NO	Ethyl ether	300	100	—
			665	20	$n \rightarrow \pi^*$
Nitrate	$C_2H_5ONO_2$	Dioxane	270	12	$n \rightarrow \pi^*$

atoms. Various empirical rules have been developed for predicting the effect of such substitutions upon absorption maxima and have proved useful for structural determinations.[4]

Absorption by Aromatic Systems. The ultraviolet spectra of aromatic hydrocarbons are characterized by three sets of bands that originate from $\pi \rightarrow \pi^*$ transitions. For example, benzene has a strong absorption peak at 184 nm ($\varepsilon_{max} \sim 60,000$); a weaker band, called the E_2 band, at 204 nm ($\varepsilon_{max} = 7900$); and a still weaker peak, termed the B band, at 256 ($\varepsilon_{max} = 200$). The long-wavelength band of benzene, and many other aromatics, contains a series of sharp peaks (see Figure 4–12**b**, p. 106) due to the superposition of vibrational transitions upon the basic electronic transitions. Polar solvents tend to reduce or eliminate this fine structure as do certain types of substitution.

---

[4] For a summary of these rules, see: R. M. Silverstein, G. C. Bassler, and T. C. Morrill, *Spectrometric Identification of Organic Compounds*, 4th ed., pp. 310–329, Wiley: New York, 1981.

All three of the characteristic bands for benzene are strongly affected by ring substitution; the effects on the two longer-wavelength bands are of particular interest because they can be readily studied with ordinary spectrophotometric equipment. Table 7–4 illustrates the effects of some common ring substituents.

By definition, an *auxochrome* is a functional group that does not itself absorb in the ultraviolet region but has the effect of shifting chromophore peaks to longer wavelengths as well as increasing their intensities. It is seen in Table 7–4 that —OH and —NH$_2$ have an auxochromic effect on the benzene chromophore, particularly with respect to the B band. Auxochromic substituents have at least one pair of n electrons capable of interacting with the π electrons of the ring. This interaction apparently has the effect of stabilizing the π^* state, thereby lowering its energy; a bathochromic shift results. Note that the auxochromic effect is more pronounced for the phenolate anion than for phenol itself, probably because the anion has an extra pair of unshared elec-

Table 7–3
Effect of Multichromophores on Absorption

Compound	Type	λ_{max}(nm)	ε_{max}
$CH_3CH_2CH_2CH{=}CH_2$	Olefin	184	$\sim 10,000$
$CH_2{=}CHCH_2CH_2CH{=}CH_2$	Diolefin (unconjugated)	185	$\sim 20,000$
$H_2C{=}CHCH{=}CH_2$	Diolefin (conjugated)	217	21,000
$H_2C{=}CHCH{=}CHCH{=}CH_2$	Triolefin (conjugated)	250	—
$CH_3CH_2CH_2CH_2\overset{\overset{\text{O}}{\|}}{C}CH_3$	Ketone	282	27
$CH_2{=}CHCH_2CH_2\overset{\overset{\text{O}}{\|}}{C}CH_3$	Unsaturated ketone (unconjugated)	278	30
$CH_2{=}CH\overset{\overset{\text{O}}{\|}}{C}CH_3$	Unsaturated ketone (conjugated)	324	24
		219	3600

trons to contribute to the interaction. With aniline, on the other hand, the nonbonding electrons are lost by formation of the anilinium cation, and the auxochromic effect disappears as a consequence.

Absorption by Inorganic Anions. A number of inorganic anions exhibit ultraviolet absorption peaks that are a consequence of $n \rightarrow \pi^*$ transitions. Examples include nitrate (313 nm), carbonate (217 nm), nitrite (360 and 280 nm), azido (230 nm), and trithiocarbonate (500 nm) ions.

7B-2 ABSORPTION INVOLVING d AND f ELECTRONS

Most transition-metal ions absorb in the ultraviolet or visible region of the spectrum. For the lanthanide and actinide series, the absorption process results from electronic transitions of $4f$ and $5f$ electrons; for elements of the first and second transition-metal series, the $3d$ and $4d$ electrons are responsible for absorption.

Absorption by Lanthanide and Actinide Ions. The ions of most lanthanide and actinide elements absorb in the ultraviolet and visible regions. In distinct contrast to the behavior of most inorganic and organic absorbers, their spectra consist of narrow, well-defined, and

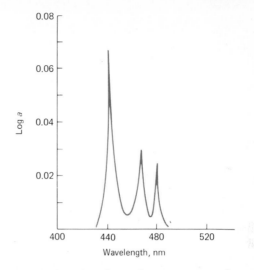

FIGURE 7–4 The absorption spectrum of a praseodymium chloride solution; a = absorptivity in $L \ cm^{-1} \ g^{-1}$.

characteristic absorption peaks, which are little affected by the type of ligand associated with the metal ion. Portions of a typical spectrum are shown in Figure 7–4.

The transitions responsible for absorption by elements of the lanthanide series appear to involve the various energy levels of $4f$ electrons, while it is the $5f$ electrons of

Table 7–4					
Absorption Characteristics of Aromatic Compounds					
Compound		**E$_2$ Band**		**B Band**	
		λ_{max}(nm)	ε_{max}	λ_{max}(nm)	ε_{max}
Benzene	C_6H_6	204	7900	256	200
Toluene	$C_6H_5CH_3$	207	7000	261	300
m-Xylene	$C_6H_4(CH_3)_2$	—	—	263	300
Chlorobenzene	C_6H_5Cl	210	7600	265	240
Phenol	C_6H_5OH	211	6200	270	1450
Phenolate ion	$C_6H_5O^-$	235	9400	287	2600
Aniline	$C_6H_5NH_2$	230	8600	280	1430
Anilinium ion	$C_6H_5NH_3^+$	203	7500	254	160
Thiophenol	C_6H_5SH	236	10,000	269	700
Naphthalene	$C_{10}H_8$	286	9300	312	289
Styrene	$C_6H_5CH=CH_2$	244	12,000	282	450

the actinide series that interact with radiation. These inner orbitals are largely screened from external influences by electrons occupying orbitals with higher principal quantum numbers. As a consequence, the bands are narrow and relatively unaffected by the nature of the solvent or the species bonded by the outer electrons.

Absorption by Elements of the First and Second Transition-Metal Series. The ions and complexes of the 18 elements in the first two transition series tend to absorb visible radiation in one if not all of their oxidation states. In contrast to the lanthanide and actinide elements, however, the absorption bands are often broad (Figure 7–5) and are strongly influenced by chemical environmental factors. An example of the environmental effect is found in the pale blue color of the aquo copper(II) ion and the much darker blue of the copper complex with ammonia.

Metals of the transition series are characterized by having five partially occupied d orbitals ($3d$ in the first series and $4d$ in the second), each capable of accommodating a pair of electrons. The electrons in these or-

bitals do not generally participate in bond formation; nevertheless, it is clear that the spectral characteristics of transition metals involve electronic transitions among the various energy levels of these d orbitals.

Two theories have been advanced to rationalize the colors of transition-metal ions and the profound influence of chemical environment on these colors. The *crystal-field theory*, which will be discussed briefly, is the simpler of the two and is adequate for a qualitative understanding. The more complex molecular-orbital treatment, however, provides a better quantitative treatment of the phenomenon.[5]

Both theories are based upon the premise that the energies of d orbitals of the transition-metal ions in solution are not identical and that absorption involves the transition of electrons from a d orbital of lower energy to one of higher energy. In the absence of an external electric or magnetic field (as in the dilute gaseous state), the energies of the five d orbitals are identical, and absorption of radiation is not required for an electron to move from one orbital to another. On the other hand, when complex formation occurs in solution between the metal ion and water or some other ligand, splitting of the d-orbital energies results. This effect results from the differential forces of electrostatic repulsion between the electron pair of the donor and the electrons in the various d orbitals of the central metal ion. In order to understand this splitting of energies, the spatial distribution of electrons in the various d orbitals must be considered.

The electron-density distribution of the five d orbitals around the nucleus is shown in Figure 7–6. Three of the orbitals, termed d_{xy}, d_{xz}, and d_{yz}, are similar in every regard except for their spatial orientation. Note that these orbitals occupy spaces *between* the

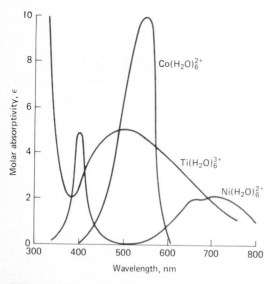

FIGURE 7–5 Absorption spectra of some transition-metal ions.

[5] For a nonmathematical discussion of these theories, see: L. E. Orgel, *An Introduction to Transition Metal Chemistry, Ligand-Field Theory*, Wiley: New York, 1960.

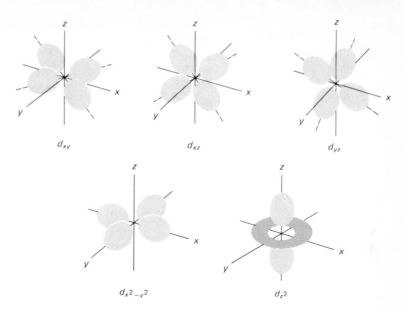

FIGURE 7-6 Electron density distribution in various *d* orbitals.

three axes; consequently, they have minimum electron densities along the axes and maximum densities on the diagonals between axes. In contrast, the electron densities of the $d_{x^2-y^2}$ and the d_{z^2} orbitals are directed along the axes.

Let us now consider a transition-metal ion that is coordinated to six molecules of water (or some other ligand). These ligand molecules or ions can be imagined as being symmetrically distributed around the central atom, one ligand being located at each end of the three axes shown in Figure 7–6; the resulting octahedral structure is the most common orientation for transition-metal complexes. The negative ends of the water dipoles are pointed toward the metal ion, and the electrical fields from these dipoles tend to exert a repulsive effect on all of the *d* orbitals, *thus increasing their energy;* the orbitals are then said to have become *destabilized*. The maximum charge density of the d_{z^2} orbital lies along the bonding axis. The negative field of a bonding ligand therefore has a greater effect on this orbital than upon the d_{xy}, d_{xz}, and d_{yz} orbitals, whose charge densities do not coincide with the bonding axes.

These latter orbitals will be destabilized equally, inasmuch as they differ from one another only in the matter of orientation. The effect of the electrical field on the $d_{x^2-y^2}$ orbital is less obvious, but quantum calculations have shown that it is destabilized to the same extent as the d_{z^2} orbital. Thus, the energy-level diagram for the octahedral configuration (Figure 7–7) shows that the energies of all of the *d* orbitals rise in the presence of a ligand field but, in addition, that the *d* orbitals are split into levels differing in energy by Δ. Also shown are energy diagrams for complexes involving four coordinated bonds. Two configurations are encountered; the *tetrahedral*, in which the four groups are symmetrically distributed around the metal ion; and the *square planar*, in which the four ligands and the metal ion lie in a single plane. Unique *d*-orbital splitting patterns for each configuration can be deduced by arguments similar to those used for the octahedral structure.

The magnitude of Δ (Figure 7–7) depends upon a number of factors, including the valence state of the metal ion and the position of the parent element in the peri-

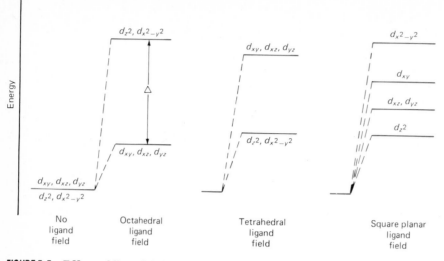

FIGURE 7-7   Effect of ligand field on d-orbital energies.

odic table. An important variable attributable to the ligand is the *ligand field strength,* which is a measure of the extent to which a complexing group will split the energies of the d electrons; that is, a complexing agent with a high ligand field strength will cause Δ to be large.

It is possible to arrange the common ligands in the order of increasing ligand field strengths: $I^- < Br^- < Cl^- < F^- < OH^- < C_2O_4^{2-} \sim H_2O < SCN^- < NH_3 <$ ethylene-

diamine $< o$-phenanthroline $< NO_2^- < CN^-$. With only minor exceptions, this order applies to all transition-metal ions and permits qualitative predictions about the relative positions of absorption peaks for the various complexes of a given transition-metal ion. Since Δ increases with increasing field strength, the wavelength of the absorption maxima decreases. This effect is demonstrated by the data in Table 7–5.

Table 7-5
Effect of Ligands on Absorption Maxima Associated with d–d Transitions

Central Ion	λ_{max}(nm) for the indicated ligands				
	Increasing Ligand Field Strength $\rightarrow$				
	$6Cl^-$	$6H_2O$	$6NH_3$	$3en^a$	$6CN^-$
Cr(III)	736	573	462	456	380
Co(III)	—	538	435	428	294
Co(II)	—	1345	980	909	—
Ni(II)	1370	1279	925	863	—
Cu(II)	—	794	663	610	—

[a] en = ethylenediamine, a bidentate ligand

7B-3 CHARGE-TRANSFER ABSORPTION

For analytical purposes, species that exhibit *charge-transfer absorption*[6] are of particular importance because molar absorptivities are very large ($\varepsilon_{max} > 10,000$). Thus, these complexes provide a highly sensitive means for detecting and determining absorbing species. Many inorganic complexes exhibit charge-transfer absorption and are therefore called *charge-transfer complexes*. Common examples of such complexes include the thiocyanate and phenolic complexes of iron(III), the *o*-phenanthroline complex of iron(II), the iodide complex of molecular iodine, and the ferro-ferricyanide complex responsible for the color of Prussian blue.

In order for a complex to exhibit a charge-transfer spectrum, it is necessary for one of its components to have electron-donor characteristics and for the other component to have electron-acceptor properties. Absorption of radiation then involves transfer of an electron from the donor to an orbital that is largely associated with the acceptor. As a consequence, the excited state is the product of a kind of internal oxidation-reduction process. This behavior differs from that of an organic chromophore, where the electron in the excited state is in the *molecular* orbital formed by two or more atoms.

A well-known example of charge-transfer absorption is observed in the iron(III)/thiocyanate complex. Absorption of a photon results in the transfer of an electron from the thiocyanate ion to an orbital associated with the iron(III) ion. The product is thus an excited species involving predominantly iron(II) and the neutral thiocyanate radical SCN. As with other types of electronic excitation, the electron, under ordi-

nary circumstances, returns to its original state after a brief period. Occasionally, however, dissociation of the excited complex may occur producing photochemical oxidation-reduction products.

As the tendency for electron transfer increases, less radiant energy is required for the charge-transfer process, and the resulting complexes absorb at longer wavelengths. For example, thiocyanate ion is a better electron donor (reducing agent) than is chloride ion; thus, the absorption of the iron(III)/thiocyanate complex occurs in the visible region, whereas the absorption maximum for the corresponding yellow chloride complex is in the ultraviolet region. Presumably the iodide complex of iron(III) would absorb at still longer wavelengths. Such an absorption peak has not been observed, however, because the electron-transfer process is complete, giving iron(II) and iodine as products.

In most charge-transfer complexes involving a metal ion, the metal serves as the electron acceptor. An exception is the *o*-phenanthroline complex of iron(II) or copper(I), where the ligand is the acceptor and the metal ion is the donor. Other examples of this type are known.

Organic compounds form many interesting charge-transfer complexes. An example is quinhydrone (a 1 : 1 complex of quinone and hydroquinone), which exhibits strong absorption in the visible region. Other examples include iodine complexes with amines, aromatics, and sulfides, among others.

7C Components of Instruments for Absorption Measurements in the Ultraviolet, Visible, and Near Infrared Regions

Instruments for measuring the absorption of ultraviolet, visible, and near infrared radiation are made up of one or more (1)

---

[6] For a brief discussion of this type of absorption, see: C. N. R. Rao, *Ultra-Violet and Visible Spectroscopy: Chemical Applications*, 3d ed., Chapter 11, Butterworths: London, 1975.

sources, (2) wavelength selectors, (3) sample containers, (4) radiation detectors, and (5) signal processors and readout devices. Details regarding the design and performance of components (2), (4), and (5) have been described in considerable detail in Chapter 5 and thus do not warrant further discussion here. It is desirable, however, to consider briefly the characteristics of sources and sample containers for the region of 190 to 2500 nm.

7C-1 SOURCES

For the purposes of molecular absorption measurements, a continuous source is required whose power does not change sharply over a considerable range of wavelengths.

Deuterium (or Hydrogen) Lamps.   A truly continuous spectrum in the ultraviolet region is produced by electrical excitation of deuterium or hydrogen at low pressure. The mechanism by which a continuous spectrum is produced involves initial formation of an excited molecular species followed by dissociation of the excited molecule to give two atomic species plus an ultraviolet photon. That is, for hydrogen

$$H_2 + E_e \rightarrow H_2^* \rightarrow H' + H'' + h\nu$$

where E_e is the electrical energy absorbed by the molecule. The energetics for the overall process can be represented by the equation

$$E_e = E_{H_2^*} = E_{H'} + E_{H''} + h\nu$$

Here, $E_{H_2^*}$ is the *fixed quantized energy* of H_2^* while $E_{H'}$ and $E_{H''}$ are the *kinetic energies* of the two hydrogen atoms. The sum of the latter two can vary continuously from zero to $E_{H_2^*}$. Thus the energy and the frequency of the photon can also vary continuously. That is, when the two kinetic energies are by chance small, $h\nu$ will be large, and conversely. The consequence is a true continuous spectrum from about 160 nm to the beginning of the visible region.

Most modern lamps of this type contain deuterium and are of a low-voltage type in which an arc is formed between a heated, oxide-coated filament and a metal electrode. The heated filament provides electrons to maintain a direct current when about 40 V is applied; a regulated power supply is required for constant intensities.

An important feature of deuterium discharge lamps is the shape of the aperture between the two electrodes, which constricts the discharge to a narrow path. As a consequence, an intense ball of radiation about 1 to 1.5 mm in diameter is produced. Deuterium gives a somewhat larger and brighter ball than hydrogen, which accounts for the widespread use of the former.

Both deuterium and hydrogen lamps produce a useful continuous spectrum in the region of 160 to 375 nm. At longer wavelengths (>360 nm), the lamps produce emission lines, which are superimposed on the continuous spectrum. For many applications, these lines represent a nuisance; they can be useful, however, for wavelength calibration of absorption instruments.

Quartz windows must be employed in deuterium and hydrogen lamps since glass absorbs strongly at wavelengths less than about 350 nm.

Tungsten Filament Lamps.   The most common source of visible and near infrared radiation is the tungsten filament lamp. The energy distribution of this source approximates that of a black body and is thus temperature dependent. Figure 4–14 (p. 110) illustrates the behavior of the tungsten filament lamp at 3000°K. In most absorption instruments, the operating filament temperature is 2870°K; the bulk of the energy is thus emitted in the infrared region. A tungsten filament lamp is useful for the wavelength region between 350 and 2500 nm. The lower limit is imposed by the glass envelope that houses the filament.

In the visible region, the energy output of a tungsten lamp varies approximately as the fourth power of the operating voltage. As a consequence, close voltage control is required for a stable radiation source. Constant voltage transformers or electronic voltage regulators are usually employed to obtain the required stability. As an alternative, the lamp can be operated from a 6-V storage battery, which provides a remarkably stable voltage source if it is maintained in good condition.

Tungsten/halogen lamps contain a small quantity of iodine within a quartz envelope that houses the tungsten filament. Quartz is required because of the high operating temperature (~3500°K) of the lamp. The lifetime of a tungsten/halogen lamp is more than double that of the ordinary lamp. This added life results from the reaction of the iodine with gaseous tungsten that forms by sublimation and ordinarily limits the life of the filament; the product is the volatile WI_2. When molecules of this compound strike the filament, decomposition occurs, which redeposits tungsten. Tungsten/halogen lamps are significantly more efficient and extend the output range well into the ultraviolet. For these reasons, they are found in many modern spectroscopic instruments.

Xenon Arc Lamps. This lamp produces intense radiation by the passage of current through an atmosphere of xenon. The spectrum is continuous over the range between about 250 and 600 nm, with the peak intensity occurring at about 500 nm (see Figure 4–14, p. 110). In some instruments, the lamp is operated intermittently by regular discharges from a capacitor; high intensities are obtained.

7C-2 SAMPLE CONTAINERS

In common with the other optical elements of an absorption instrument, the cells or cuvettes that hold the sample and solvent must be constructed of a material that passes radiation in the spectral region of interest. Thus, as shown in Figure 5–2 (p. 114), quartz or fused silica is required for work in the ultraviolet region (below 350 nm); both of these substances are transparent in the visible region and to about 3 μm in the infrared region as well. Silicate glasses can be employed in the region between 350 and 2000 nm. Plastic containers have also found application in the visible region.

The best cells have windows that are perfectly normal to the direction of the beam in order to minimize reflection losses. The most common cell length for studies in the ultraviolet and visible regions is 1 cm; matched, calibrated cells of this size are available from several commercial sources. Other path lengths, from 0.1 cm (and shorter) to 10 cm, can also be purchased. Transparent spacers for shortening the path length of 1-cm cells to 0.1 cm are also available.

For reasons of economy, cylindrical cells are sometimes employed in the ultraviolet and visible regions. Special care must be taken to duplicate the position of the cell with respect to the beam; otherwise, variations in path length and reflection losses at the curved surfaces can cause significant errors.

The quality of absorbance data is critically dependent upon the way the matched cells are used and maintained. Fingerprints, grease, or other deposits on the walls alter the transmission characteristics of a cell markedly. Thus, thorough cleaning before and after use is imperative; the surface of the windows must not be touched during the handling. Matched cells should never be dried by heating in an oven or over a flame—such treatment may cause physical damage or a change in path length. The cells should be calibrated against each other regularly with an absorbing solution.

7D Some Typical Instruments

The scientist interested in absorption measurements in the ultraviolet, visible, and near infrared regions has a hundred or more instrument makes and models to choose from. Some are simple and inexpensive (a few hundred dollars); others are complex, computerized devices costing $30,000 or more. For many applications, the simpler instruments provide information that is as satisfactory as that obtained by their more sophisticated counterparts, and as quickly. On the other hand, the more complicated instruments have been developed to perform tasks that are difficult, time-consuming, or impossible with the simpler ones.

In this section, some typical photometers and spectrophotometers are described. The choice made here among the scores of models available has not been based upon quality of performance or cost but rather to illustrate the wide variety of design variables that are encountered.

7D-1 PHOTOMETERS

The photometer provides a simple, relatively inexpensive tool for performing absorption analyses. Convenience, ease of maintenance, and ruggedness are properties of a filter photometer that may not be found in the more sophisticated spectrophotometers. Furthermore, photometers characteristically have high radiant energy throughputs and thus good signal-to-noise ratios even with relatively simple and inexpensive detectors and circuitry. Where high spectral purity is not important to a method (and often it is not), quantitative analyses can be performed as accurately with a photometer as with more complex instrumentation.

Visible Photometers.   Figure 7–8 presents schematic diagrams for two photometers. The first (Figure 7–8**a**) is a single-beam, direct-reading instrument consisting of a tungsten-filament lamp, a lens to provide a parallel beam of light, a filter, and a photovoltaic cell. The current produced is indicated with a microammeter, the face of which is ordinarily scribed with a linear scale from 0 to 100. The 0% T measurement involves mechanical or electrical adjustment of the meter needle while the shutter interrupts the incident beam. In some instruments, adjustment to obtain a full-scale (100% T) response with the solvent in the light path requires changing the voltage applied to the lamp. In others, the aperture size of a diaphragm located in the light path is altered. Since the signal from the photovoltaic cell is linear with respect to the power of the radiation it receives, the scale reading with the sample in the light path will be the percent transmittance (that is, the percent of full scale). Clearly, a logarithmic scale could be substituted to give the absorbance of the solution directly.

Figure 7–8**b** is a schematic representation of a double-beam, null-type photometer. Here, the light beam is split by a half-silvered mirror, which transmits about 50% of the radiation striking it and reflects the other 50%; note that this design is similar to that shown in Figure 6–11**b** which involves two beam paths in space. Half of the beam passes through the sample and thence to a photovoltaic cell; the other half passes through the solvent to a similar detector. The electric outputs from the two photovoltaic cells are passed through variable resistances; one of these is calibrated as a transmittance scale in linear units from 0 to 100. A sensitive galvanometer, which serves as a null detector, is connected across the two resistances. When the potential drop across AB is equal to that across CD, no electricity passes through the detector; under all other circumstances, a current is indicated. At the outset, the contact on the left is set to 0% T, the shutter is closed, and the pointer of the null detector is centered mechanically; the center mark then corresponds to zero current. Next, the solvent is placed in both cells, and contact A is set at 100 (the 100% T setting); with the shutter open, contact C is then adjusted until zero current is indicated. Replacement of the solvent with the sample

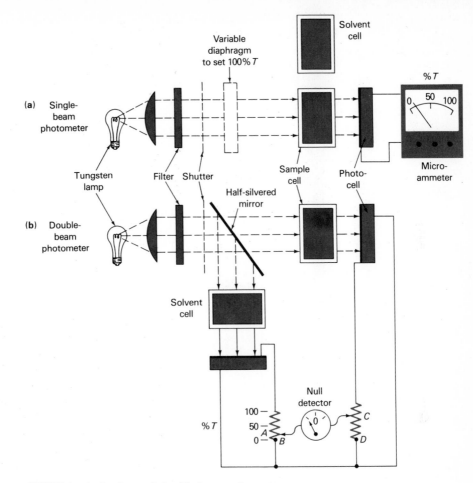

FIGURE 7-8   A single- and double-beam photometer.

in one cell results in a decrease in radiant power reaching the working phototube and a corresponding decrease in output potential; a current is then indicated by the galvanometer. The lack of balance is compensated for by moving A to a lower value. At balance, the percent transmittance is read directly from the scale.

Commercial photometers usually cost a few hundred dollars. The majority employ the double-beam principle to compensate for fluctuations in the source intensity due to voltage variations in the power supply.

Probe-Type Photometers.   Figure 7–9 is a schematic diagram of an interesting, commercially available dipping type photometer, which employs an optical fiber to transmit light from a source to a layer of solution lying between the glass seal at the end of the fiber and a mirror. The reflected radiation from the latter passes to a photodiode detector via a second glass fiber. The photometer uses an amplifier with an electronic chopper that is synchronized with the light source; as a result, the photometer does not respond to extraneous radiation. Six interference filters are provided, which can be interchanged by means of a knob located on the instrument panel. Custom filters are also available. Probe tips are manufactured from

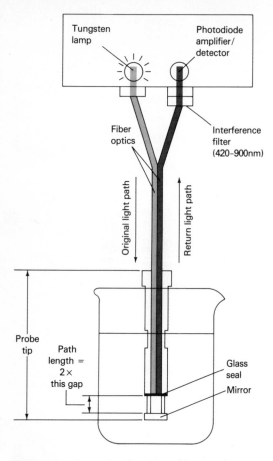

Tungsten lamp

Photodiode amplifier/ detector

Fiber optics

Interference filter (420-900nm)

Original light path

Return light path

Probe tip

Path length = 2× this gap

Glass seal

Mirror

FIGURE 7-9 Schematic diagram of a probe-type photometer. (Courtesy of Brinkman Instrument Company, Division of Sybron Corporation, Westbury, NY 11590.)

stainless steel, Swagelok® Stainless Steel, Pyrex®, and Acid-resistant Lexan® Plastic. Light path lengths that vary from 1 mm to 10 cm are available.

Absorbance is measured by first dipping the probe into the solvent and then into the solution to be measured. The device is particularly useful for photometric titrations (Section 7G).

Filter Selection for Absorption Analysis. Photometers are generally supplied with several filters, each of which transmits a different portion of the visible spectrum. Selection of the proper filter for a given application is important inasmuch as the sensitivity of the measurement is directly dependent upon this choice. The color of the light absorbed is the complement of the color of the solution itself. A liquid appears red, for example, because it transmits the red portion of the spectrum but absorbs the green. It is the intensity of green radiation that varies with concentration; a green filter should thus be employed. In general, then, the most suitable filter for a photometric analysis will be the color complement of the solution being analyzed. If several filters possessing the same general hue are available, the one that causes the sample to exhibit the greatest absorbance (or least transmittance) should be used.

Ultraviolet Absorption Photometers. Ultraviolet photometers often serve as detectors in high-performance liquid chromatography. In this application, a mercury-vapor lamp usually serves as a source, and the emission line at 254 nm is isolated by filters. This type of detector is described briefly in Section 27C–6.

Ultraviolet photometers are also available for continuously monitoring the concentration of one or more constituents of gas or liquid streams in industrial plants. The instruments are dual channel in space and often employ one of the emission lines of mercury, which has been isolated by a filter system. Typical applications include the determination of low concentration of phenol in waste water, monitoring the concentration of chlorine, mercury, or aromatics in gases, and the determination of the ratio of hydrogen sulfide to sulfur dioxide in the atmosphere.[7]

---

[7] For other examples, see: Bulletin 5A, DuPont 400 Photometric Analyzer, DuPont Instrument Systems, Wilmington, DE 19898.

7D-2 *SPECTROPHOTOMETERS*

Numerous spectrophotometers are available from commercial sources. Some have been designed for the visible region only; others are applicable in the ultraviolet and visible regions. A few have measuring capabilities from the ultraviolet through the near infrared (185 to 3000 nm).

Instruments for the Visible Region. Several spectrophotometers designed to operate within the wavelength range of about 380 to 800 nm are available from commercial sources. These instruments are frequently simple, single-beam grating instruments that are relatively inexpensive (less than $1000), rugged, and readily portable. At least one is battery operated and light and small enough to be hand held. The most common application of these instruments is for quantitative analysis, although several produce surprisingly good absorption spectra as well. Figure 7–10 depicts two typical examples.

The spectrophotometer shown in Figure 7–10a employs a tungsten filament light source, which is operated by a stabilized power supply that provides radiation of constant intensity. After diffraction by a simple reflection grating the radiation passes through the sample or reference cuvettes to a phototube. The amplified electrical signal from the detector then powers a meter with a 5½ in. scale calibrated in transmittance and absorbance.

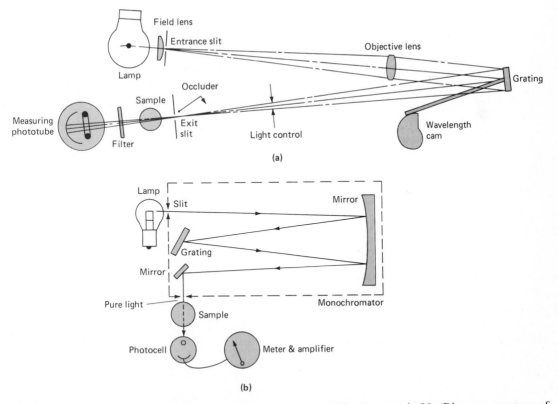

FIGURE 7-10 Two examples of simple spectrophotometers. (**a**) The Spectronic 20. (Diagram courtesy of Bausch and Lomb, Rochester, NY.) (**b**) The Turner 350. (Diagram courtesy of Amsco Instrument Company [formerly Turner Associates] Carpinteria, CA.)

The instrument is equipped with an occluder, which is a vane that automatically falls between the beam and the detector whenever the cuvette is removed from its holder; the 0% T adjustment can then be made. The light control device shown in Figure 7–10**a** consists of a V-shaped slot that can be moved in or out of the beam in order to set the meter to 100% T.

The range of the instrument shown in Figure 7–10**a** is from 340 to 625 nm; an accessory phototube extends this range to 950 nm. Other specifications for the instrument include a bandwidth of 20 nm and a wavelength accuracy of ±2.5 nm.

The instrument shown schematically in Figure 7–10**b** makes use of a tungsten-filament bulb as a source, a plane reflection grating in an *Ebert mounting* for dispersion, and a phototube detector that is sensitive in the range between 210 and 710 nm (the low wavelength limit of the instrument is about 350 nm unless a deuterium source is purchased). The readout device is a meter calibrated both in transmittance and absorbance; instruments with 4- or 7-in. scales are offered. The 0% T adjustment is accomplished by variation of the amplifier output while the 100% T adjustment is carried out by varying the output of the stabilized lamp power supply. Instrument specifications include a bandwidth of 8 nm, a wavelength accuracy of ±2 nm, and a photometric accuracy of 0.5% A.

Several accessories are offered with the instrument shown in Figure 7–10**b**. One, which includes a deuterium lamp, extends the range of the instrument to 210 nm; another provides an additional phototube that permits measurements to 1000 nm.

Single-Beam Instruments for the Ultraviolet-Visible Region. Several instrument manufacturers offer nonrecording single-beam instruments which can be used for both ultraviolet and visible measurements. The lower wavelength extremes for these instruments vary from 190 to 210 nm and the upper from 800 to 1000 nm. All are equipped with interchangeable tungsten and hydrogen or deuterium lamps. Most employ photomultiplier tubes as detectors and gratings for dispersion. Some are equipped with digital readout devices; others employ large meters. The prices for these instruments range from $2000 to $8000.

As might be expected, performance specifications vary considerably among instruments and are related, at least to some degree, to instrument price. Typically, bandwidths vary from 2 to 8 nm; wavelength accuracies of ±0.5 to ±2 nm are reported.

The optical designs for the various grating instruments do not differ greatly from those of the two instruments shown in Figure 7–10. One manufacturer, however, employs a concave rather than a plane grating; a simpler and more compact design results. Instruments equipped with holographic gratings (Section 5D–2) are also beginning to appear on the market.

Figure 7–11 is a schematic diagram of a high-quality, single-beam spectrophotometer for the ultraviolet-visible region. This instrument, the Beckman DU spectrophotometer, is of historic interest because it was the first ultraviolet-visible spectrometer to become commercially available to chemists. Before its appearance in 1941, obtaining an ultraviolet absorption spectrum was a major undertaking, requiring several hours of tedious labor. Its appearance heralded the beginning of widespread use of absorption spectroscopy by the average bench chemist.

The Beckman DU instrument is atypical of modern commercial instruments in the respect that its design is based upon a quartz Littrow prism. Radiation from a tungsten filament or a deuterium source is reflected through an adjustable slit to a collimating mirror and thence to this prism. Rotation of the prism permits focusing of the desired wavelengths on the exit slit and the sample or reference cell.

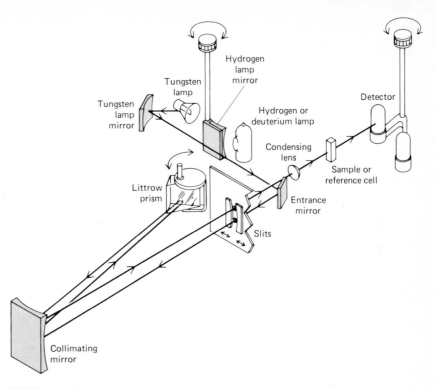

FIGURE 7-11   Schematic diagram of the Beckman DU®-2 Spectrophotometer. (By permission, Beckman Instruments, Inc., Fullerton, CA.)

The original model of the instrument shown in Figure 7–11 employed interchangeable detectors—one for the range of 190 to 625 nm and the other for 625 to 800 nm. A photometer modernization kit is now offered by one manufacturer[8] in which the phototubes are replaced by a single photomultiplier with the same range as the original two. The modified design achieves photometric accuracies as great as ±0.5% A by employing high-quality electronic components that are operated well below their rated capacities. Narrow effective bandwidths (less than 0.5 nm) can be obtained throughout the spectrum region by suitable variation of the slit adjustment (note that grating instruments discussed earlier are generally operated at fixed slit widths be-

cause of the linear characteristics of their dispersing elements).

Single-Beam Computerized Spectrophotometers.   One manufacturer is now offering a line of computerized, recording, single-beam spectrophotometers, which operate in the range of 190 to 800 nm (to 900 nm with an accessory).[9] With these instruments, a wavelength scan is first performed with the reference solution in the beam path. The resulting detector output is digitized in real time and stored in the memory of the computer. Samples are then scanned and absorbances calculated with the aid of the stored data. The complete spectrum is displayed on a cathode-ray tube within 2 sec of

---

[8] Gilford Instrument, Oberlin, Ohio 44074.

[9] W. Kaye, D. Barber, and R. Marasco, *Anal. Chem.,* **1980,** *52,* 437A; and V. A. Kohler and N. Brenner, *Amer. Lab.,* **1981,** *13* (9), 109.

data acquisition. Scan speeds as great as 1200 nm/min are feasible. The computer associated with the instrument provides several options with regard to data processing and presentation such as log absorbance, transmittance, derivatives, overlaid spectra, repetitious scans, concentration calculations, peak location and height determinations, and kinetic measurements.

As noted earlier (Section 6C–4), single-beam instruments have the inherent advantages of greater energy throughput, superior signal-to-noise ratios, and less cluttered sample compartments. On the other hand, the process of recording detector outputs for reference and sample solutions successively for subsequent absorbance or transmittance calculations has heretofore not been very satisfactory because of the drift or flicker noise in sources and detectors. The manufacturer of these new single-beam instruments claims to have eliminated these instabilities by means of a fundamentally new source design and a new electronic design that eliminates hysteresis or memory effects in the photodetector.

The photometric accuracies of the new instruments are reported to be $\pm 0.005\ A$ or $\pm 0.3\%\ T$ with a drift of less than $0.002\ A/hr$.

Bandwidths of 0.5, 1, and 2 nm are available by manual interchange of fixed slits.

Double-Beam Instruments for the Ultraviolet-Visible Region. Numerous double-beam spectrophotometers for the ultraviolet-visible region of the spectrum are now available. Generally, these instruments are more expensive than their single-beam counterparts, with the nonrecording variety ranging in cost from about $4000 to $15,000.

Figure 7–12 shows construction details of a relatively inexpensive, (~$5000), manual double-beam ultraviolet-visible spectrophotometer. In this instrument, the radiation is dispersed by a concave grating, which also focuses the beam on a rotating sector mirror. The instrument design is similar to that shown in Figure 6–11c (Section 6C–2).

The instrument has a wavelength range of 195 to 850 nm, a bandwidth of 4 nm, a photometric accuracy of $0.5\%\ T$, and a reproducibility of $0.2\%\ T$; stray radiation is less than 0.1% of P_0 at 240 and 340 nm. This instrument is typical of several similar ones offered by various instrument companies. Such instruments are well suited for quantitative work where derivation of an entire spectrum is not often required.

Figure 7–13 shows the optics of a more

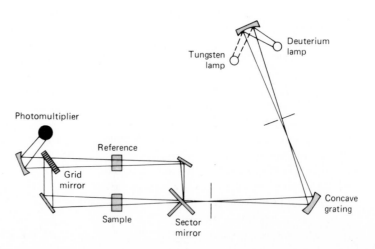

FIGURE 7-12 Schematic diagram of a typical manual double-beam spectrophotometer for the ultraviolet–visible region; the Hitachi Model 100-60. (Courtesy of Hitachi Scientific Instruments, Mountain View, CA.)

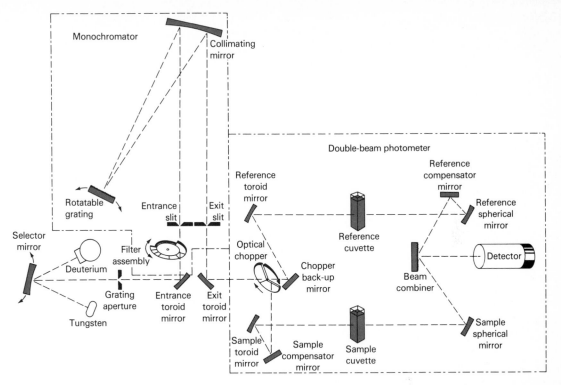

FIGURE 7–13 A double-beam recording spectrophotometer for the ultraviolet and visible regions; the Perkin-Elmer 57 Series. (Courtesy of Coleman Instruments Division, Oak Brook, IL 60521.)

sophisticated dual channel (in time), recording spectrophotometer, which employs a 45 × 45 mm plane grating having 1440 lines/mm. Its range is from 190 to 750 nm with the option of extending this range to 900 nm. Band widths of 0.2, 0.5, 1.0, and 3.00 mm can be chosen by exchange of slits. The instrument has a photometric accuracy of ±0.003 A; its stray radiation is less than 0.1% of P_0 at 220 and 340 nm. The performance of this instrument is significantly better than the double-beam instrument described previously; its price is correspondingly higher.

Double Dispersing Instruments. In order to enhance spectral resolution and achieve a marked reduction in scattered radiation, a number of instruments have been designed with two gratings or prisms serially ar-

ranged with an intervening slit; in effect then, these instruments consist of two monochromators in a series configuration.

The instrument shown in Figure 7–14 achieves the same performance characteristics with a single grating. Note that the radiation passing through entrance slit 1 is dispersed by the grating and travels through exit slit 1 to entrance slit 2. After a second dispersion by the grating, the beam emerges from exit slit 2 where it is split into a sample and a reference beam. The resolution of the instrument is reported to be 0.07 nm and the stray light 0.0008% from 220 to 800 nm. The wavelength range is 185 to 3125 nm. For wavelengths greater than 800 nm, a lead sulfide photoconducting detector is used with the tungsten source.

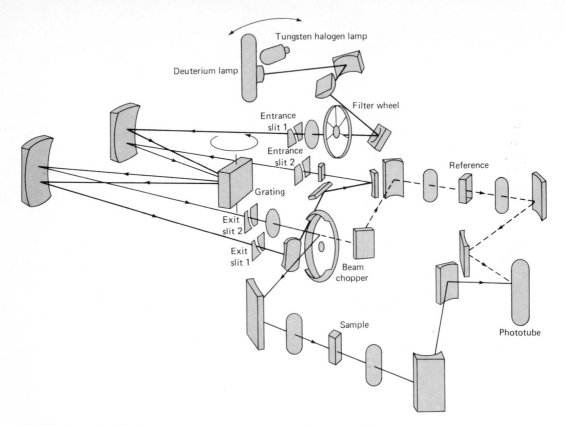

FIGURE 7-14 A double dispersing spectrophotometer. (Courtesy of Varian Instrument Division, Palo Alto, CA.)

7D-3 MULTICHANNEL ABSORPTION INSTRUMENTS

In the mid-nineteen-seventies, a number of scientific papers and review articles appeared in the literature describing applications of silicon diode arrays and vidicons (Section 5F–3) as detectors for spectrochemical measurements.[10] With these devices located at the focal plane of a monochromator, a spectrum can be obtained by electronic rather than mechanical scanning; all of the data points needed to define a spectrum can thus be gathered essentially simultaneously. The concept of multichannel instruments is attractive because of the potential speed at which spectra can be acquired as well as their applicability to simultaneous multicomponent determinations. Multichannel detectors as well as microprocessors for controlling and manipulating their outputs have been available from several commercial sources for a decade or more. Some of these sources also offer spectrometers, which can be conveniently coupled to these detector systems to give instruments for absorption, fluorescence, or

---

[10] Y. Talmi, *Appl. Spectrosc.*, **1982**, *36*, 1; Y. Talmi, *Anal. Chem.*, **1975**, *47*, 658A, 697A; Y. Talmi, *Multichannel Image Detectors*, Vol. 1, 1979, Vol. 2, 1983, American Chemical Society: Washington, D.C.; G. Horlick and E. G. Codding, *Contemp. Top. Anal. Clin. Chem.*, **1977**, *1*, 195.

emission spectral measurements. It was not until about 1980, however, that an electronic multichannel spectrometer designed specifically for absorption measurements in the ultraviolet-visible range became available to chemists from a commercial source.[11] This instrument, which has several unique design features, is described in the paragraphs that follow.

Figure 7–15 is a schematic diagram showing the optical design of a multichannel

(also called parallel access) spectrometer for absorption measurements in the region from 200 to 800 nm. Note that in contrast to other ultraviolet-visible absorption instruments, the sample cell is located between the source and the monochromator. This configuration is necessary, of course, if all of the elements of the spectrum are to be observed simultaneously.

One of the interesting features of the instrument in Figure 7–15 is the source system, which consists of a tungsten/halogen lamp and a see-through deuterium lamp configured in such a way as to produce a beam of radiation at the lamp housing slit that is continuous from 200 to 800 nm. Source switching has thus been eliminated.

---

[11] See: B. G. Willis, D. A. Fustier, and E. J. Bonelli, *Amer. Lab.*, **1981**, *13* (6), 62; and J. C. Millter, S. A. George, and B. G. Willis, *Science*, **1982**, *218*, 241. Also, the entire issue of the *Hewlett-Packard Journal*, **1980**, *31* (2) is devoted to this instrument.

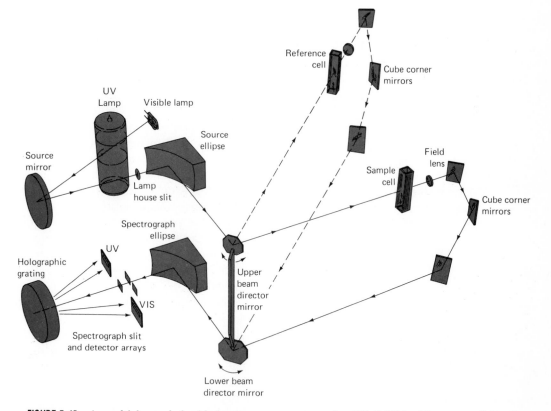

FIGURE 7-15 A multichannel double-beam spectrometer; the HP 8450A. (Courtesy of Hewlett-Packard Company, Palo Alto, CA 94305.)

Another feature worth noting is the double ellipsoidal optics, which includes a source ellipse and a spectrometer ellipse; this arrangement leads to a higher radiant energy throughput than could be realized with spherical optics. The lamp slit and sample compartment are located at the two foci of one ellipse while the sample compartment and the entrance slit to the monochromator are located at the foci of the second. This arrangement makes it possible to accommodate as many as three additional sample cells and cube corner mirrors in the area between the reference and sample cells shown in Figure 7–15. With all five beam paths in place, data for absorption spectra for four samples can be acquired successively in a matter of 6.4 sec.

A unique feature of this instrument is the servo-controlled beam director, which consists of a pair of flat mirrors mounted on a motor-driven vertical shaft. The upper mirror directs the beam alternately to the reference and then to the sample cell. The lower mirror reflects each beam in turn to the ellipsoidal entrance mirror of the spectrograph. The outer surfaces of both jaws of the entrance slit are mirrored in such a way that the edges of the beam are reflected to two large-area photodiodes (the beam width is approximately twice the slit width so that approximately half the radiation appears at the diode surfaces). The two diodes, which are not shown in the schematic, are used to control a servo mechanism that adjusts the position of the mirror shaft so that the beam is centered on the slit at all times (146 msec are required for the beam to be switched to within 3 sec of arc of its final position).

The monochromator consists of two holographic gratings formed on a single substrate. The gratings are tilted at 3.5 deg from one another so that the first order diffracted radiation from 200 to 400 nm is directed to one detector while that for 400 to 800 nm passes to the other.

The detectors are similar to those discussed in Section 5F–3. Each has 211 silicon diodes, storage capacitors, and transistor switches formed on a *pn* semiconductor chip that is about ½ in. in length; the dimensions of the active surface of each diode are 0.6×0.5 mm, the same as the dimensions of the slit image. Thus, the bandwidth of the instrument is 1 nm in the 200 to 400 nm region and 2 nm in the 400 to 800 nm range. The chips also contain the circuitry necessary for serially charging each capacitor. Integration of the charging current, amplification, and correction for noise is carried out by a preamplifier. The signal then passes through an analog-to-digital converter to a 16 bit microprocessor that converts the measurements to absorbance or transmittance in real time for display on a small cathode-ray tube. The data are also stored in a 32 K random access memory for subsequent plotting by a recorder. The microprocessor is programmed to carry out on command other useful functions such as computing concentrations and standard deviations, finding peak maxima, curve-fitting for calibrations with standards, and determining the concentrations of the individual component of mixtures of absorbing species. For the last application, absorption spectra for standard solutions of the individual components (up to 12) are acquired and stored in memory. Quantitative information about an unknown is then obtained by employing the stored data to determine the concentration of components necessary to synthesize a spectrum that best matches that of the unknown.

The actual measurement time required to obtain and process data for an entire spectrum is slightly more than 1 sec. The measurement sequence begins with a determination of dark current with the beam directed to a position where no radiation reaches the detector. The director is then commanded to switch the beam to the sample position. When the system verifies that the beam is in correct alignment, serial integration of the charging current to each of the diode capacitors is begun. When integration of all 422 is finished, the beam is switched to the reference cell and the process is repeated. Finally, the beam is returned to the dark position for an additional dark current measurement. If desired, the process can be repeated several times and the data averaged for improved signal-to-noise ratios.

It is noteworthy that samples are exposed to ultraviolet radiation for such short periods that even highly photosensitive analytes have shown little or no signs of photodecomposition. Furthermore, interference from fluorescence appears to be minimal; since fluorescence is emitted spherically, the fraction appearing in the beam path is less than the noise limit of the instrument. Finally, the reverse-optic design, with the sample and reference solutions located between the source and the monochromator, eliminates the need to house the sample and reference solutions in a light-tight compartment as is generally necessary with spectrophotometers.

The instrument shown in Figure 7–15 provides a powerful tool for such applications as the quantitative determination of the constituents of multicomponent mixtures, the study of transient intermediates in moderately fast reactions, kinetic studies, and the qualitative (and quantitative) determination of the eluted components from liquid chromatographic or other type columns. Disadvantages of the instrument are its somewhat limited resolution for spectrophotometric studies (1 nm in the ultraviolet and 2 nm in the visible regions) and its moderately high cost.

7E Application of Absorption Measurement to Qualitative Analysis

Ultraviolet and visible spectrophotometry have somewhat limited application for qualitative analysis because the number of absorption maxima and minima are relatively few. Thus, unambiguous identification is frequently impossible.

7E–1 QUALITATIVE TECHNIQUES

Solvents. In choosing a solvent, consideration must be given not only to its transparency, but also to its possible effects upon the absorbing system. Quite generally, polar solvents such as water, alcohols, esters, and ketones tend to obliterate spectral fine structure arising from vibrational effects; spectra that approach those of the gas phase (see Figure 7–16) are more likely to be observed in nonpolar solvents such as hydrocarbons. In addition, the positions of absorption maxima are influenced by the nature of the solvent. Clearly, the same solvent must be used when comparing absorption spectra for identification purposes.

Table 7–6 lists some common solvents and the approximate wavelength below which they cannot be used because of absorption. These minima depend strongly

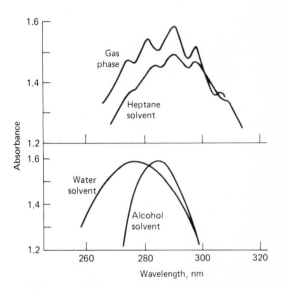

FIGURE 7-16 Effect of solvent on the absorption spectrum of acetaldehyde.

upon the purity of the solvent.[12] Common solvents for ultraviolet spectrophotometry include water, 95% ethanol, cyclohexane, and 1,4-dioxane. For the visible region, any colorless solvent is suitable.

Detection of Functional Groups. Even though it may not provide the unambiguous identification of an organic compound, an absorption spectrum in the visible and the ultraviolet regions is nevertheless useful for detecting the presence of certain functional groups that act as chromophores.[13] For example, a weak absorption band in the region of 280 to 290 nm, which is displaced toward shorter wavelengths with increased solvent polarity, strongly indicates the presence of the carbonyl group. A weak absorption band at about 260 nm with indications of vibrational fine structure constitutes evidence for the existence of an aromatic ring. Confirmation of the presence of an aromatic amine or a phenolic structure may be obtained by comparing the effects of pH on the spectra of solutions containing the sample with those shown in Table 7–4 for phenol and aniline.

7F   Quantitative Analysis by Absorption Measurements

Absorption spectroscopy is one of the most useful tools available to the chemist for quantitative analysis.[14] Important characteristics of spectrophotometric and photometric methods include: (1) wide applicability to both organic and inorganic systems; (2) typical sensitivities of 10^{-4} to 10^{-5} M (this range can often be extended to 10^{-6} to 10^{-7} M by certain modifications[15]); (3) moderate to high selectivity; (4) good accuracy (typically, relative uncertainties of 1 to 3% are encountered although with special precautions, errors can be reduced to a few tenths of a percent); (5) ease and convenience of data acquisition.

7F-1   SCOPE

The applications of quantitative, ultraviolet–visible absorption methods are not only numerous, but also touch upon every field in which quantitative chemical information is required. The reader can obtain a notion of the scope of spectrophotometry by con-

---

[12] Most major suppliers of reagent chemicals in the United States offer spectrochemical grades of solvents; these meet or exceed the requirements set forth in *Reagent Chemicals, American Chemical Society Specifications*, 6th ed., American Chemical Society: Washington, D.C., 1981.

[13] See: R. M. Silverstein, G. C. Bassler, and T. C. Morrill, *Spectrometric Identification of Organic Compounds*, 4th ed., Chapter 6, Wiley: New York, 1981.

---

[14] For a wealth of detailed, practical information on spectrophotometric practices, see: *Techniques in Visible and Ultraviolet Spectrometry*, Vol. I, *Standards in Absorption Spectroscopy*, C. Burgess and A. Knowles, Eds., Chapman and Hall: London, 1981; and J. R. Edisbury, *Practical Hints on Absorption Spectrometry*, Plenum Press: New York, 1968.

[15] See, for example: T. D. Harris, *Anal. Chem.*, **1982**, *54*, 741A.

**Table 7–6
Solvents for the Ultraviolet and the Visible Regions**

Solvent	Approximate[a] Transparency Minimum (nm)
Water	190
Ethanol	210
n-Hexane	220
Cyclohexane	210
Benzene	280
Diethyl ether	210
Acetone	330
1,4-Dioxane	220

[a] For 1-cm cells.

sulting a series of review articles published periodically in *Analytical Chemistry*[16] and from monographs on the subject.[17]

Applications to Absorbing Species. Tables 7–2, 7–3, and 7–4 list many common organic chromophoric groups. Spectrophotometric analysis for any organic compound containing one or more of these groups is potentially feasible; many examples of this type of analysis are found in the literature.

A number of inorganic species also absorb and are thus susceptible to direct determination; we have already mentioned the various transition metals. In addition, a number of other species also show characteristic absorption. Examples include nitrite, nitrate, and chromate ions; osmium and ruthenium tetroxides; molecular iodine; and ozone.

Applications to Nonabsorbing Species. Numerous reagents react selectively with nonabsorbing species to yield products that absorb strongly in the ultraviolet or visible regions.[18] The successful application of such reagents to quantitative analysis usually requires that the color-forming reaction be forced to near completion. It should be noted that color-forming reagents are frequently employed as well for the determination of absorbing species, such as transition-metal ions; the molar absorptivity of the product will frequently be orders of magnitude greater than that of the uncombined species.

A host of complexing agents find application for the determination of inorganic species. Typical inorganic reagents include thiocyanate ion for iron, cobalt, and molybdenum; the anion of hydrogen peroxide for titanium, vanadium, and chromium; and iodide ion for bismuth, palladium, and tellurium. Of even more importance are organic chelating agents which form stable, colored complexes with cations. Examples include *o*-phenanthroline for the determination of iron, dimethylglyoxime for nickel, diethyldithiocarbamate for copper, and diphenyldithiocarbazone for lead.

7F-2 PROCEDURAL DETAILS

The first steps in a photometric or spectrophotometric analysis involve the establishment of working conditions and the preparation of a calibration curve relating concentration to absorbance.

Selection of Wavelength. Spectrophotometric absorbance measurements are ordinarily made at a wavelength corresponding to an absorption peak because the change in absorbance per unit of concentration is greatest at this point; the maximum sensitivity is thus realized. In addition, the absorption curve is often flat in this region; under these circumstances, good adherence to Beer's law can be expected (Section 6B–3). Finally, the measurements are less sensitive to uncertainties arising from failure to reproduce precisely the wavelength setting of the instrument.

Variables that Influence Absorbance. Common variables that influence the absorption spectrum of a substance include the nature of the solvent, the pH of the solution, the temperature, high electrolyte concentrations, and the presence of interfering substances. The effects of these variables must be known; conditions for the analysis must be chosen

[16] See: L. G. Hargis and J. A. Howell, *Anal. Chem.*, **1984,** *56,* 225R; **1982,** *54,* 171R; **1980,** *52,* 306R; **1978,** *50,* 243R.

[17] See, for example: E. B. Sandell and H. Onishi, *Colorimetric Determination of Traces of Metals,* 4th ed., Interscience: New York, 1978; *Colorimetric Determination of Nonmetals,* 2d ed., D. F. Boltz and J. A. Howell, Eds., Wiley: New York, 1978; Z. Marczenko, *Spectrophotometric Determination of Elements,* Halsted Press: New York, 1975; M. Pisez and J. Bartos, *Colorimetric and Fluorometric Analysis of Organic Compounds and Drugs,* Marcel Dekker: New York, 1974; and F. D. Snell, *Photometric and Fluorometric Method of Analysis,* Parts 1 and 2, Metals; Part 3, Nonmetals, Wiley: New York, 1978–81.

[18] For example, see the first reference in footnote 17.

such that the absorbance will not be materially influenced by small, uncontrolled variations in their magnitudes.

Cleaning and Handling of Cells.   It is apparent that accurate spectrophotometric analysis requires the use of good-quality, matched cells. These should be regularly calibrated against one another to detect differences that can arise from scratches, etching, and wear. Equally important is the use of proper cell cleaning and drying techniques. Erickson and Surles[19] recommend the following cleaning sequence for the outside windows of cells. Prior to measurement, the cell surfaces are cleaned with a lens paper soaked in spectrograde methanol. The paper is held with a hemostat; after wiping, the methanol is allowed to evaporate, leaving the cell surfaces free of contaminants. The authors showed that this method was far superior to the usual procedure of wiping the cell surfaces with a dry lens paper, which apparently leaves lint and films on the surface.

Determination of the Relationship Between Absorbance and Concentration.   After deciding upon the conditions for the analysis, it is necessary to prepare a calibration curve from a series of standard solutions. These standards should approximate the overall composition of the actual samples and should cover a reasonable concentration range of the analyte. Seldom, if ever, is it safe to assume adherence to Beer's law and use only a single standard to determine the molar absorptivity. The results of an analysis should *never* be based on a literature value for the molar absorptivity.

Standard Addition Method.   Ideally, calibration standards should approximate the composition of the samples to be analyzed not only with respect to the analyte concentration but also with regard to the concentrations of the other species in the sample matrix in order to minimize the effects of various components of the sample on the measured absorbance. For example, the absorbance of many colored complexes of metal ions is decreased to a varying degree in the presence of sulfate and phosphate ions as a consequence of the tendency of these anions to form colorless complexes with metal ions. The color formation reaction is often less complete as a consequence, and lowered absorbances are the result. The matrix effect of sulfate and phosphate can often be counteracted by introducing into the standards amounts of the two species that approximate the amounts found in the samples. Unfortunately, when complex materials such as soils, minerals, and plant ash are being analyzed, however, preparation of standards that match the samples is often impossible or extremely difficult. When this is the case, the *standard addition method* is often helpful in counteracting matrix effects.

The standard addition method can take several forms.[20] The one most often chosen for photometric or spectrophotometric analyses, and the one that will be discussed here, involves adding one or more increments of a standard solution to sample aliquots of the same size. Each solution is then diluted to a fixed volume before measuring its absorbance. It should be noted that when the amount of sample is limited, standard additions can be carried out by successive introductions of increments of the standard to a single measured aliquot of the unknown. Measurements are made on the original and after each addition. This procedure is often more convenient for voltammetric and potentiometric measurements and will be discussed in later sections of the text.

Assume that several identical aliquots V_x of the unknown solution with a concentration C_x are transferred to volumetric flasks

[19] J. O. Erickson and T. Surles, *Amer. Lab.*, **1976**, *8* (6), 50.

[20] See: M. Bader, *J. Chem. Educ.*, **1980**, *57*, 703.

having a volume V_t. To each of these flasks is added a variable volume V_s mL of a standard solution of the analyte having a known concentration C_s. The color development reagents are then added, and each solution is diluted to volume. If Beer's law is followed, the absorbance of the solutions is described by

$$A_s = \frac{\varepsilon b V_x C_x}{V_t} + \frac{\varepsilon b V_s C_s}{V_t} \tag{7-1}$$

A plot of A_s as a function of V_s is a straight line of the form

$$A_s = \alpha + \beta V_s$$

where the slope β and the intercept α are given by

$$\beta = \frac{\varepsilon b C_s}{V_t}$$

and

$$\alpha = \frac{\varepsilon b V_x C_x}{V_t}$$

A least-square analysis (Section 1C–6) can be used to determine β and α; C_x can then be obtained from the ratio of these two quantities and the known values of C_s, V_x, and V_s. Thus,

$$\frac{\alpha}{\beta} = \frac{\varepsilon b V_x C_x / V_t}{\varepsilon b C_s / V_t} = \frac{V_x C_x}{C_s}$$

or

$$C_x = \frac{\alpha C_s}{\beta V_x} \tag{7-2}$$

An approximate value for the standard deviation in C_x can then be obtained by assuming that the uncertainties in C_s, V_s, and V_t are negligible with respect to those in α and β. Then, the relative variance of the result

$(s_c/C_x)^2$ is assumed to be the sum of the variances of α and β. That is,

$$\left(\frac{s_c}{C_x}\right)^2 = \left(\frac{s_r}{\alpha}\right)^2 + \left(\frac{s_\beta}{\beta}\right)^2$$

where s_r is the standard deviation about regression. Taking the square root of this equation gives

$$s_c = C_x \sqrt{\left(\frac{s_r}{\alpha}\right)^2 + \left(\frac{s_\beta}{\beta}\right)^2} \tag{7-3}$$

EXAMPLE 7–1. Ten-millimeter aliquots of a natural water sample were pipetted into 50.00-mL volumetric flasks. Exactly 0.00, 5.00, 10.00, 15.00, and 20.00 mL of a standard solution containing 11.1 ppm of Fe^{3+} were added to each followed by an excess of thiocyanate ion to give the red complex $Fe(SCN)^{2+}$. After dilution to volume, absorbances for the five solutions, measured with a photometer equipped with a green filter, were found to be 0.215, 0.424, 0.685, 0.826, and 0.967 respectively (0.982-cm cells). (a) What was the concentration of Fe^{3+} in the water sample? (b) Calculate a standard deviation for the slope, the standard deviation about regression, and the standard deviation for the concentration of Fe^{3+}.

(a) In this problem, $C_s = 11.1$ ppm, $V_x = 10.00$ mL, and $V_t = 50.00$ mL. A plot of the data, shown in Figure 7–17, demonstrates that Beer's law is obeyed.

To obtain the equation for the line in Figure 7–17 ($A = \alpha + \beta V_s$), the procedure illustrated in Example 1–11, Section 1C–6 is followed. The result is $\beta = 0.03812$ and $\alpha = 0.2422$, and thus

$$A = 0.2422 + 0.03812 \, V_s$$

Substituting into Equation 7–2 gives

$$C_x = \frac{0.2422 \times 11.1}{0.03812 \times 10.00} = 7.05 \text{ ppm } Fe^{3+}$$

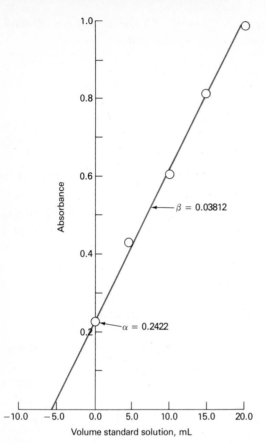

FIGURE 7-17 Data for standard addition method for the determination of Fe^{3+} as the SCN^- complex.

(b) Equations 1–26 and 1–27 give the standard deviation of the slope and about regression. That is, $s_r = 0.045$ and $s_c = 0.0029$.
Substituting into Equation 7–3 gives

$$s_c = 7.05 \sqrt{\left(\frac{0.045}{0.2422}\right)^2 + \left(\frac{0.0029}{0.0381}\right)^2}$$

$$= 0.2 \text{ ppm } Fe^{3+}$$

In the interest of saving time or sample, it is possible to perform a standard addition analysis using only two increments of sample. Here, a single addition of V_s mL of standard would be added to one of the two samples and we can write

$$A_1 = \frac{\varepsilon b V_x C_x}{V_t}$$

$$A_2 = \frac{\varepsilon b V_x C_x}{V_t} + \frac{\varepsilon b V_s C_s}{V_t}$$

where A_1 and A_2 are absorbances of the diluted sample and the diluted sample plus standard, respectively. Dividing the second equation by the first gives upon rearrangement

$$C_x = \frac{A_1 C_s V_s}{(A_2 - A_1)V_x}$$

Analysis of Mixtures of Absorbing Substances. The total absorbance of a solution at a given wavelength is equal to the sum of the absorbances of the individual components present (Section 6B–2). This relationship makes possible the quantitative determination of the individual constituents of a mixture, even if their spectra overlap. Consider, for example, the spectra of M and N, shown in Figure 7–18. Obviously, no wavelength

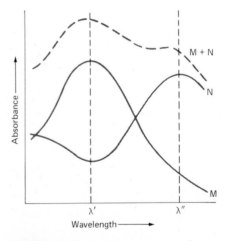

FIGURE 7-18 Absorption spectrum of a two-component mixture.

exists at which the absorbance of this mixture is due simply to one of the components; thus, an analysis for either M or N is impossible by a single measurement. However, the absorbances of the mixture at the two wavelengths λ' and λ'' may be expressed as follows:

$$A' = \varepsilon_M' b c_M + \varepsilon_N' b c_N \quad (\text{at } \lambda')$$

$$A'' = \varepsilon_M'' b c_M + \varepsilon_N'' b c_N \quad (\text{at } \lambda'')$$

The four molar absorptivities ε_M', ε_N', ε_M'', and ε_N'' can be evaluated from individual standard solutions of M and of N, or better, from the slopes of their Beer's law plots. The absorbances of the mixture, A' and A'', are experimentally determinable, as is b, the cell thickness. Thus, from these two equations, the concentrations of the individual constituents, c_M and c_N, can be readily calculated. These relationships are valid only if Beer's law is followed and if the two components behave independently of one another. The greatest accuracy in an analysis of this sort is attained by choosing wavelengths at which the differences in molar absorptivities are large.

Mixtures containing more than two absorbing species can be analyzed, in principle at least, if a further absorbance measurement is made for each added component. The uncertainties in the resulting data become greater, however, as the number of measurements increases. Some of the newer computerized spectrophotometers are capable of reducing these uncertainties by overdetermining the system. That is, these instruments use many more data points than unknowns and effectively match the entire spectrum of the unknown as closely as possible by deriving synthetic spectra assuming various concentrations of the components. The derived spectra are then compared with that of the analyte until a close match is found. The spectrum for standard solutions of each component is required, of course.

7F–3 DERIVATIVE AND DUAL-WAVELENGTH SPECTROPHOTOMETRY[21]

In derivative spectrophotometry, spectra are obtained by plotting the first or a higher order derivative of absorbance or transmittance with respect to wavelength as a function of wavelength. Often these plots reveal spectral detail that is lost in an ordinary spectrum. In addition, concentration measurements of an analyte in the presence of an interference can sometimes be made more easily or more accurately. Unfortunately, the advantages of derivative spectra are at least partially offset by a degradation in signal-to-noise ratio that accompanies obtaining derivatives. In many parts of the ultraviolet and visible regions, however, signal-to-noise ratio is not a serious limiting factor; it is here that derivative spectra are of most use. An additional disadvantage of derivative spectrophotometry is that the required equipment is generally more costly.

A variety of methods are used to obtain derivative spectra. For microprocessor controlled digital spectrophotometers, the differentiation can be performed numerically. With analog instruments, derivatives of spectral data can be obtained by a suitable operational amplifier circuit (see Section 2C–5). A third procedure is by wavelength modulation.

Wavelength Modulation Devices. Several procedures have been developed for wavelength modulation. In some, a wavelength interval of a few nanometers is swept rapidly and repetitively while the spectrum is being

---

[21] For additional information, see: T. C. O'Haver, *Anal. Chem.*, **1979**, *51*, 91A; J. E. Cahill, *Amer. Lab.*, **1979**, *11* (11), 79; J. E. Cahill and F. G. Padera, *Amer. Lab.*, **1980**, *12* (4), 101; and T. J. Porro, *Anal. Chem.*, **1972**, *44* (4), 93A.

scanned in the usual way. The amplitude of the resulting ac signal from the detector is a good approximation of the wavelength derivative. The repetitive, rapid sweep has been obtained by any of a number of mechanical means including oscillation or vibration of a mirror, slit, or dispersing element of a monochromator. A second scheme for wavelength modulation, and one that is offered by several instrument makers, involves the use of dual dispersing systems arranged in such a way that two beams of slightly different wavelengths (typically 1 or 2 nm) fall alternately onto a sample cell and its detector; no reference beam is used. The ordinate parameter is the difference between the alternate signals, which provides a good approximation of the derivative of absorbances as a function of wavelength ($\Delta A/\Delta\lambda$). Figure 7–19 is a schematic diagram of a dual wavelength instrument. Note that a reference cell is not employed and the sample cell is alternately exposed to radiation from each of the monochromators. Generally, dual wavelength instruments can also be operated in the single wavelength mode with the radiation passing alternately through a reference and standard cell.

Applications of Derivative Spectra.   Many of the most important applications of derivative spec-

troscopy in the ultraviolet and visible regions have been for qualitative identification of species in which the enhanced detail of a derivative spectrum makes it possible to distinguish among compounds having overlapping spectra. Figure 7–20 illustrates the way in which a derivative plot can reveal details of a spectrum consisting of three overlapping absorption peaks.

Dual-wavelength spectrophotometry has proven particularly useful for extracting ultraviolet-visible absorption spectra of analytes present in turbid solutions, where light scattering obliterates the details of an absorption spectrum. For example, three amino acids, tryptophan, tyrosine, and phenylalanine, contain aromatic side chains,

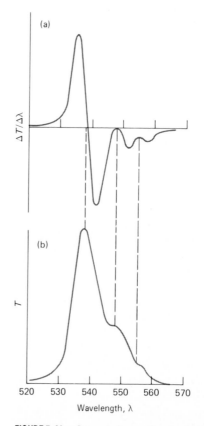

FIGURE 7–20 Comparison of a derivative curve (**a**) with a standard transmittance curve (**b**).

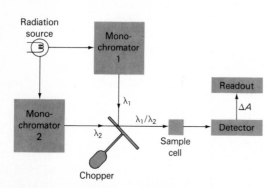

FIGURE 7–19 Schematic of a dual-wavelength spectrophotometer. The chopper causes λ_1 and λ_2 to pass through the sample alternately.

which exhibit sharp absorption peaks in the 240 to 300 nm range. These sharp peaks are not, however, apparent in spectra of typical protein preparations, such as bovine or egg albumin, because the large protein molecules scatter radiation severely yielding only a smooth absorption peak such as that shown in Figure 7–21**a**. As shown in curves (**b**) and (**c**), the aromatic fine structure is revealed in first and second derivative spectra.

Dual-wavelength spectrophotometry has also proved useful for analysis of an analyte in the presence of a spectral interference. Here, the instrument is operated in the non-scanning mode with absorbances being measured at two wavelengths at which the interference has identical molar absorptivities. In contrast, the analyte must absorb more strongly at one of these wavelengths than the other. The differential absorbance is

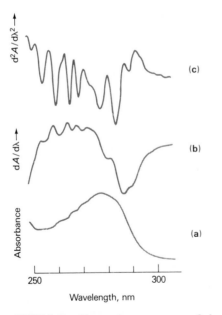

FIGURE 7–21 Absorption spectra of bovine albumin: (**a**) ordinary spectrum; (**b**) first derivative spectrum; (**c**) second derivative spectrum. (Reprinted with permission from: J. E. Cahill and F. G. Padera, *Amer. Lab.*, **1980,** *12* (4), 109. Copyright 1980 by International Scientific Communications, Inc.)

then directly proportional to the analyte concentration.

7G *Photometric Titrations*

Photometric or spectrophotometric measurements can be employed to advantage in locating the equivalence point of a titration provided the analyte, the reagent, or the titration product absorbs radiation.[22] Alternatively, an absorbing indicator can provide the absorbance change necessary for location of equivalence.

7G-1 *TITRATION CURVES*

A photometric titration curve is a plot of absorbance, corrected for volume changes, as a function of the volume of titrant. If conditions are chosen properly, the curve will consist of two straight-line regions with differing slopes, one occurring at the outset of the titration and the other located well beyond the equivalence-point region; the end point is taken as the intersection of extrapolated linear portions. Figure 7–22 shows some typical titration curves. Titration of a nonabsorbing species with a colored titrant that is decolorized by the reaction produces a horizontal line in the initial stages, followed by a rapid rise in absorbance beyond the equivalence point (Figure 7–22**a**). The formation of a colored product from colorless reactants, on the other hand, initially produces a linear rise in the absorbance, followed by a region in which the absorbance becomes independent of reagent volume (Figure 7–22**b**). Depending upon the absorption characteristics of the reactants and the products, the other curve

---

[22] For further information concerning this technique, see: J. B. Headridge, *Photometric Titrations*, Pergamon Press: New York, 1961; and M. A. Leonard, in *Comprehensive Analytical Chemistry*, G. Svehla, Ed., vol. 8, Chapter 3, Elsevier: New York, 1977.

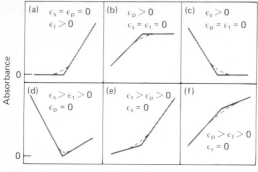

FIGURE 7–22 Typical photometric titration curves. Molar absorptivities of the substance titrated, the product, and the titrant are given by ε_s, ε_p, ε_t, respectively.

forms shown in Figure 7–22 are also possible.

In order to obtain a satisfactory photometric end point, it is necessary that the absorbing system(s) obey Beer's law; otherwise, the titration curve will lack the linear regions needed for extrapolation to the end point. Further, it is necessary to correct the absorbance for volume changes; here, the observed values are multiplied by $(V + v)/V$, where V is the original volume of the solution and v is the volume of added titrant.

7G–2 INSTRUMENTATION

Photometric titrations are ordinarily performed with a spectrophotometer or a photometer that has been modified to permit insertion of the titration vessel in the light path.[23] Alternatively, a probe-type cell, such as that shown in Figure 7–9 can be employed. After the zero adjustment of the meter scale has been made, radiation is allowed to pass through the solution of the analyte, and the instrument is adjusted by varying the source intensity or the detector sensitivity until a convenient absorbance reading is obtained. Ordinarily, no attempt is made to measure the true absorbance, since relative values are perfectly adequate for the purpose of end-point detection. Data for the titration are then collected without alteration of the instrument setting.

The power of the radiation source and the response of the detector must be reasonably constant during the period required for a photometric titration. Cylindrical containers are ordinarily used, and care must be taken to avoid any movement of the vessel that might alter the length of the radiation path.

Both filter photometers and spectrophotometers have been employed for photometric titrations. The latter are preferred, however, because their narrower bandwidths enhance the probability of adherence to Beer's law.

7G–3 APPLICATION OF PHOTOMETRIC TITRATIONS

Photometric titrations often provide more accurate results than a direct photometric analysis because the data from several measurements are pooled in determining the end point. Furthermore, the presence of other absorbing species may not interfere, since only a change in absorbance is being measured.

The photometric end point possesses the advantage over many other commonly used end points in that the experimental data are taken well away from the equivalence-point region. Thus, the titration reactions need not have such favorable equilibrium constants as those required for a titration that depends upon observations near the equivalence point (for example, potentiometric or indicator end points). For the same reason, more dilute solutions may be titrated.

---

[23] Titration flasks and cells for use in a Spectronic 20 spectrophotometer are available from the Kontes Manufacturing Corp., Vineland, N.J.

The photometric end point has been applied to all types of reactions.[24] Most of the reagents used in oxidation-reduction titrations have characteristic absorption spectra and thus produce photometrically detectable end points. Acid-base indicators have been employed for photometric neutralization titrations. The photometric end point has also been used to great advantage in titrations with EDTA and other complexing agents. Figure 7–23 illustrates the application of this end point to the successive titration of bismuth(III) and copper(II). At 745 nm, neither cation nor the reagent absorbs, nor does the more stable bismuth complex, which is formed in the first part of the titration; the copper complex, however, does absorb. Thus, the solution exhibits no absorbance until essentially all of the bismuth has been titrated. With the first formation of the copper complex, an increase in absorbance occurs. The increase continues until the copper equivalence point is reached. Fur-

ther reagent additions cause no further absorbance change. Clearly, two well-defined end points result.

The photometric end point has also been adapted to precipitation titrations. Here, the suspended solid product has the effect of diminishing the radiant power by scattering; titrations are carried to a condition of constant turbidity.

7H Photoacoustic Spectroscopy

Photoacoustic or optoacoustic spectroscopy, which was developed in the early 1970s, provides a means for obtaining ultraviolet and visible absorption spectra of solids, semisolids, or turbid liquids. Acquisition of spectra for these kinds of samples by ordinary methods is usually difficult at best and often impossible because of light scattering and reflection.

7H-1 THE PHOTOACOUSTIC EFFECT

Photoacoustic spectroscopy is based upon a light absorption effect that was first investigated in the 1880s by Alexander Graham Bell and others. This effect is observed when a gas in a closed cell is irradiated with a chopped beam of radiation of a wavelength that is absorbed by the gas. The absorbed radiation causes periodic heating of the gas, which in turn results in regular pressure fluctuations within the chamber. If the chopping rate lies in the acoustical frequency range, these pulses of pressure can be detected by a sensitive microphone. The photoacoustic effect has been used since the turn of the century for the analysis of absorbing gases and has recently taken on new importance for this purpose with the advent of tunable infrared lasers as sources. Of greater importance, however, has been the application of this phenomenon to the deri-

---

[24] See, for example, the review: A. L. Underwood, *Advances in Analytical Chemistry and Instrumentation*, C. N. Reilley, Ed., vol. 3, pp. 31–104, Interscience: New York, 1964.

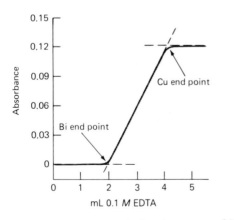

FIGURE 7-23 Photometric titration curve of 100 mL of a solution that was 2.0×10^{-3} *M* in Bi^{3+} and Cu^{2+}. Wavelength: 745 nm.

vation of absorption spectra of solids and turbid liquids.[25]

7H-2   PHOTOACOUSTIC SPECTRA

In photoacoustic studies of solids, the sample is placed in a closed cell containing air or some other *nonabsorbing* gas and a sensitive microphone. The solid is then irradiated with a chopped beam from a monochromator. The photoacoustic effect is observed *provided the radiation is absorbed by the solid;* the power of the resulting sound is directly related to the extent of absorption. Radiation reflected or scattered by the sample has

no effect on the microphone and thus does not interfere. This latter property is perhaps the most important characteristic of the method.

The source of the photoacoustic effect in solids appears to be similar to that in gases. That is, nonradiative relaxation of the absorbing solid causes a periodic heat flow from the solid to the surrounding gas; the resulting pressure fluctuations in the gas are then detected by the microphone.

7H-3   INSTRUMENTS

Figure 7–24 is a block diagram showing the components of a single-beam photoacoustic spectrometer. In this apparatus, the spectrum from the lamp is first recorded digitally followed by the spectrum for the sample. The stored lamp data are then used to correct the output from the sample for vari-

[25] For a review on applications, see: A. Rosencwaig, *Anal. Chem.,* **1975,** *47,* 592A; J. W. Lin and L. P. Dubek, *Anal. Chem.,* **1979,** *51,* 1627; J. F. McClelland, *Anal. Chem.,* **1983,** *55,* 89A; and A. Rosencwaig, *Photoacoustics and Photoacoustic Spectroscopy,* Wiley: New York, 1980.

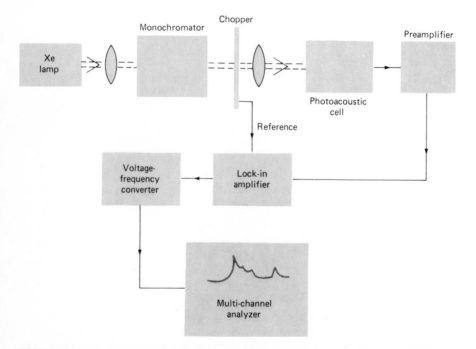

FIGURE 7-24 Block diagram of a single-beam photoacoustic spectrometer with digital data processing. (Reprinted with permission from: A. Rosencwaig, *Anal. Chem.,* **1975,** *47,* 593A. Copyright 1975 American Chemical Society.)

ations in the lamp output as a function of wavelength. With this technique, it is necessary to assume an absence of drift in the source and detector systems. Double-beam instruments, which largely avoid the drift problem, have also been described. One such instrument is equipped with a pair of matched cells (and detectors), one of which contains the sample and the other a reference material such as finely divided carbon. A commercially available instrument is also based upon the split-beam principle. In this instrument, however, about 8% of the output from the grating monochromator is directed to a pyroelectric detector, the rest passing into the sample cell. The output from the sample detector is compared with that from the pyroelectric detector to produce a spectrum that is corrected for variations in the lamp output as a function of wavelength as well as time.

7H–4 APPLICATIONS

Figure 7–25 illustrates one application of photoacoustic spectroscopy. Here, photoacoustic spectra of smears of whole blood, blood cells freed of plasma, and hemoglobin extracted from the cells are shown. Conventional spectroscopy, even with very dilute solutions of whole blood, does not yield satisfactory spectra because of the strong light-scattering properties of the blood cells, protein, and lipid molecules present. Photoacoustic spectroscopy clearly permits spectroscopic studies of blood without the necessity of a preliminary separation of these large molecules.

Figure 7–26 shows another application of photoacoustic spectroscopy. The five spectra on the left were for five organic compounds that had been separated on thin-layer chromatographic plates (Section 28B). These spectra were taken on the thin-layer plates themselves; for comparison, solution spectra are shown on the right. The similar-

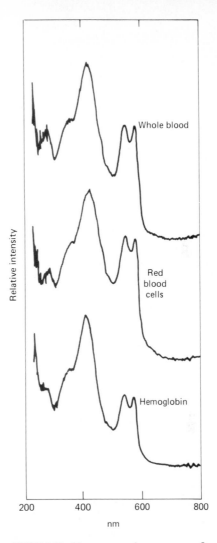

FIGURE 7-25 Photoacoustic spectra of smears of blood and blood components. (Reprinted with permission from: A. Rosencwaig, *Anal. Chem.,* **1975,** *47,* 596A. Copyright 1975 American Chemical Society.)

ity of the two makes rapid identification of the compound possible.

Other applications of the method have included the study of minerals; semiconductors; natural products, such as marine algae and animal tissue; coatings on surfaces; and catalytic surfaces.

Finally, photoacoustic measurements in the mid-infrared region have proved useful for qualitative identification of components in solids. Usually, Fourier transform techniques are necessary to obtain satisfactory *S/N* ratios. Photoacoustic cells are generally available as accessories for Fourier transform instruments.

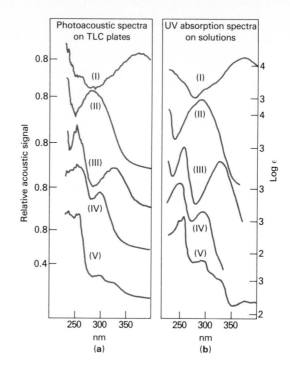

FIGURE 7-26 Spectra of spots on a thin-layer chromatogram (left) and of solution of the same compounds (right). The compounds are: (I) *p*-nitroaniline, (II) benzylidene acetone, (III) salicylaldehyde, (IV) 1-tetralone, and (V) fluorenone. (Reprinted with permission from: A. Rosencwaig, *Anal. Chem.,* **1975,** *47,* 600A. Copyright 1975 American Chemical Society.)

PROBLEMS

7–1. The logarithm of the molar absorptivity for acetone in ethanol is 2.75 L cm^{-1} mol^{-1} at 366 nm. Calculate the range of acetone concentrations that can be used if the percent transmittance is to be greater than 10% and less than 90% with a 1.50-cm cell.

7–2. The logarithm for the molar absorptivity of phenol in aqueous solution is 3.812 at 211 nm. Calculate the range of phenol concentrations that can be used if the absorbance is to be greater than 0.100 and less than 2.000 with a 1.25-cm cell.

7–3. A standard solution was put through appropriate dilutions to give the concentrations of iron shown below. The iron(II)-1,10-phenanthroline complex was then developed in 25.0-mL aliquots of these solutions, following which each was diluted to 50.0 mL. The following absorbances were recorded at 510 nm:

Concentration of Fe(II) in the Original Solutions, ppm	Absorbance, A (1.00-cm cells)
4.00	0.160
10.00	0.390
16.00	0.630
24.0	0.950
32.0	1.260
40.0	1.580

(a) Produce a calibration curve from these data.
(b) By the method of least squares (p. 19), derive an equation relating absorbance and concentration of iron(II).
(c) Calculate the standard deviation about regression.
(d) Calculate the standard deviation of the slope.

7–4. The method developed in Problem 7–3 was used for the routine determination of iron in 25.0-mL aliquots of ground water. Express the concentration (as ppm Fe) in samples that yielded the accompanying absorbance data (1.00-cm cell). Estimate standard deviations for the derived concentrations. Repeat the calculations assuming the absorbance data are means of three measurements.

(a) 0.143   (c) 0.068   (e) 1.512
(b) 0.675   (d) 1.009   (f) 0.546

7–5. Verify the results found in Example 7–1.

7–6. A 25.0-mL aliquot of an aqueous quinine solution was diluted to 50.0 mL and found to have an absorbance of 0.416 at 348 nm when measured in a 1.00-cm cell. A second 25.0-mL aliquot was mixed with 10.00 mL of a solution containing 23.4 ppm of quinine; after dilution to 50.0 mL, this solution had an absorbance of 0.610 (1.00-cm cell). Calculate the ppm of quinine in the sample.

7–7. A 6.81-g pesticide sample was decomposed by wet ashing and then diluted to 200.0 mL in a volumetric flask. The analysis was completed by treating aliquots of this solution as indicated.

| Volume of Sample Taken, mL | Reagent Volumes Used, mL | | | Absorbance, A, 545 nm (1.00-cm cells) |
	3.00 ppm Cu^{2+}	Ligand	H_2O	
50.0	0.00	20.0	30.0	0.376
50.0	4.00	20.0	26.0	0.697

Calculate the percentage of copper in the sample.

7–8. A simultaneous determination for cobalt and nickel can be based upon absorption by their respective 8-hydroxyquinolinol complexes. Molar absorptivities corresponding to their absorption maxima are

| | Molar Absorptivity, ε, at | |
	365 nm	700 nm
Co	3529	428.9
Ni	3228	10.2

Calculate the molar concentration of nickel and cobalt in each of the following solutions based upon the accompanying data:

	Absorbance, A, (1.00-cm cells)	
Solution	365 nm	700 nm
(a)	0.602	0.044
(b)	0.650	0.040
(c)	0.796	0.027
(d)	0.888	0.081
(e)	0.725	0.062

7–9. When measured with a 1.00-cm cell, a 8.50×10^{-5} M solution of species A exhibited absorbances of 0.129 and 0.764 at 475 and 700 nm respectively. A 4.65×10^{-5} M solution of species B gave absorbances of 0.567 and 0.083 under the same circumstances. Calculate the concentrations of A and B in solutions that yielded the accompanying absorbance data.

	Absorbance, A (1.25-cm cell)	
Solution	475 nm	700 nm
(a)	0.452	0.892
(b)	0.864	0.448
(c)	0.690	0.709
(d)	0.746	0.596

7–10. The acid-base indicator HIn undergoes the following reaction in dilute aqueous solution:

$$\underset{\text{color 1}}{\text{HIn}} \rightleftharpoons \text{H}^+ + \underset{\text{color 2}}{\text{In}^-}$$

The following absorbance data were obtained for a 5.00×10^{-4} M solution of HIn in 0.1 M NaOH and 0.1 M HCl. Measurements were made at a wavelength of 485 nm and 625 nm with 1.00-cm cells.

$$0.1 \; M \; \text{NaOH} \quad A_{485} = 0.052 \quad A_{625} = 0.823$$

$$0.1 \; M \; \text{HCl} \quad A_{485} = 0.454 \quad A_{625} = 0.176$$

In the NaOH solution, essentially all of the indicator is present as In^-; in the acidic solution, it is essentially all in the form of HIn.
(a) Calculate molar absorptivities for In^- and HIn at 485 and 625 nm.
(b) Calculate the acid dissociation constant for the indicator if a pH 5.00 buffer containing a small amount of the indicator exhibits

an absorbance of 0.472 at 485 nm and 0.351 at 625 nm (1.00-cm cells).

(c) What is the pH of a solution containing a small amount of the indicator that exhibits an absorbance of 0.530 at 485 nm and 0.216 at 635 nm (1.00-cm cells)?

(d) A 25.00-mL aliquot of a solution of purified weak organic acid HX required exactly 24.20 mL of a standard solution of a strong base to reach a phenolphthalein end point. When exactly 12.10 mL of the base were added to a second 25.00-mL aliquot of the acid, which contained a small amount of the indicator under consideration, the absorbance was found to be 0.306 at 485 nm and 0.555 at 625 nm (1.00-cm cells). Calculate the pH of the solution and K_a for the weak acid.

(e) What would be the absorbance of a solution at 485 and 625 nm (1.25-cm cells) that was 2.00×10^{-4} M in the indicator and was buffered to a pH of 6.000?

7–11. The absolute error in transmittance for a particular photometer is 0.005 and independent of the magnitude of T. Calculate the percent relative error in concentration that is caused by this source when

(a) $A = 0.642$.   (d) $T = 0.0612$.
(b) $T = 51.7\%$.   (e) $T = 99.35\%$.
(c) $A = 1.900$.   (f) $A = 0.0041$.

7–12. Maxima exist at 470 nm in the absorption spectrum for the bismuth(III)/thiourea complex and at 265 nm in the spectrum for the bismuth(III)/EDTA complex. Predict the shape of a curve for the photometric titration of

(a) bismuth(III) with thiourea (tu) at 470 nm.
(b) bismuth(III) with EDTA (H_2Y^{2-}) at 265 nm.
(c) the bismuth(III)/thiourea complex with EDTA at 470 nm. Reaction:

$$Bi(tu)_6^{3+} + H_2Y^{2-} \rightarrow BiY^- + 6tu + 2H^+$$

(d) the reaction in (c) at 265 nm.

7–13. Given the information that

$$Fe^{3+} + Y^{4-} \rightleftharpoons FeY^- \qquad K_f = 1.0 \times 10^{25}$$
$$Cu^{2+} + Y^{4-} \rightleftharpoons CuY^{2-} \qquad K_f = 6.3 \times 10^{18}$$

and the further information that, among the several reactants and products, only CuY^{2-} absorbs at 750 nm, describe how Cu(II) could be used as indicator for the photometric titration of Fe(III) with H_2Y^{2-}. Reaction:

$$Fe^{3+} + H_2Y^{2-} \rightarrow FeY^- + 2H^+$$

7–14. The chelate CuA_2^{2-} exhibits maximum absorption at 480 nm. When the chelating reagent is present in at least a 10-fold excess, the absorbance is dependent only upon the analytical concentration of Cu(II) and conforms to Beer's law over a wide range. A solution in which the analytical concentration of Cu^{2+} is $2.30 \times 10^{-4}\,M$ and that for A^{2-} is $8.60 \times 10^{-3}\,M$ has an absorbance of 0.690 when measured in a 1.00-cm cell at 480 nm. A solution in which the analytical concentrations of Cu^{2+} and A^{2-} are $2.30 \times 10^{-4}\,M$ and $5.00 \times 10^{-4}\,M$, respectively, has an absorbance of 0.540 when measured under the same conditions. Use this information to calculate a value for the formation constant K_f for the process

$$Cu^{2+} + 2A^{2-} \rightleftharpoons CuA_2^{2-}$$

7–15. Mixture of the chelating reagent B with Ni(II) gives rise to formation of the highly colored NiB_2^{2+}, solutions of which obey Beer's law over a wide range. Provided the analytical concentration of the chelating reagent exceeds that of Ni(II) by a factor of 5 (or more), the cation exists, within the limits of observation, entirely in the form of the complex. Use the accompanying data to evaluate the formation constant K_f for the process

$$Ni^{2+} + 2B \rightleftharpoons NiB_2^{2+}$$

Analytical concentration, M		Absorbance, A, 395 nm
Ni^{2+}	B	(1.00-cm cells)
2.50×10^{-4}	2.20×10^{-1}	0.765
2.50×10^{-4}	1.00×10^{-3}	0.360

8

Molecular Fluorescence,

Phosphorescence, and

Chemiluminescence Spectroscopy

Three related types of optical methods are considered in this chapter, namely *molecular fluorescence, phosphorescence,* and *chemiluminescence.* In each, molecules of the analyte are excited to give a species whose emission spectrum provides information for qualitative or quantitative analyses. The methods are known collectively as *luminescence* procedures.

Fluorescence and phosphorescence are alike in the respect that excitation is brought about by absorption of photons. As a consequence, the two phenomena are often referred to by the more general term *photoluminescence.* As will be shown later, fluorescence differs from phosphorescence in the respect that the electronic energy transitions responsible for the photoluminescence do not involve a change in electron spin. As a consequence, fluorescent radiation is short-lived with luminescence ceasing almost immediately ($<10^{-6}$ sec). In contrast, a change in electron spin accompanies phosphorescent emissions, which causes the radi-

ation to endure for an easily detectable time after termination of irradiation—often several seconds or longer. In most instances, photoluminescent radiation, be it fluorescence or phosphorescence, is longer in wavelength than the radiation used for its excitation.

The third type of luminescence, chemiluminescence, is based upon an excited species that is formed in the course of a chemical reaction. In some instances, the excited particles are the products of a reaction between the analyte and a suitable reagent (usually a strong oxidant such as ozone or hydrogen peroxide); the result is a spectrum characteristic of the *oxidation product* of the analyte rather than the analyte itself. In other instances, the analyte is not directly involved in the chemiluminescent reaction; instead, it is the catalytic or inhibiting effect of the analyte on the chemiluminescence that serves as the analytical parameter.

Measurement of the intensity of photoluminescent or chemiluminescent radiation

225

permits the quantitative determination of a variety of important inorganic and organic species in trace amounts. At the present time, the number of fluorometric methods is significantly larger than the number of applications of phosphorescent and chemiluminescent procedures.

One of the most attractive features of luminescent methods is their inherent sensitivity, with detection limits often being one to three orders of magnitude smaller than those encountered in absorption spectroscopy. Typical detection limits range between a few thousandths to perhaps one-tenth part per million. Another advantage of photoluminescent methods is their large linear concentration ranges, which are often significantly greater than those encountered in absorption methods. Finally, selectivity of luminescent procedures is at least as good and often is better than that of absorption methods. Luminescent methods, however, are much less widely applicable than absorption methods because of the relatively limited number of chemical systems that can be made to produce luminescent radiation.[1]

8A   Theory of Fluorescence and Phosphorescence

Fluorescent behavior occurs in simple as well as in complex gaseous, liquid, and solid chemical systems. The simplest kind of fluo-

rescence is that exhibited by dilute atomic vapors. For example, the $3s$ electrons of vaporized sodium atoms can be excited to the $3p$ state by absorption of radiation of wavelengths 5896 and 5890 Å. After approximately 10^{-8} sec, the electrons return to the ground state, and in so doing emit radiation of the same two wavelengths in all directions. This type of fluorescence, in which the absorbed radiation is reemitted without alteration, is known as *resonance radiation* or *resonance fluorescence.*

Molecular species also exhibit resonance fluorescence on occasion. Much more often, however, molecular fluorescence (or phosphorescence) occurs as bands of radiation that are centered at wavelengths that are longer than the resonance line. This shift towards longer wavelengths is termed the *Stokes shift.*

8A–1   FLUORESCENT AND PHOSPHORESCENT EXCITED STATES

The characteristics of fluorescent and phosphorescent spectra can be rationalized by means of the simple molecular orbital considerations described in Section 7B–1. However, an understanding of the difference between the two photoluminescent phenomena requires a review of *electron spin* and *singlet* and *triplet excited states.*

Electron Spin.   In order to explain the behavior of atoms and molecules in strong magnetic fields, it is necessary to assume that each electron in an atom or molecule has associated with it a magnetic field that arises from a spinning motion of the electron around its axis. Furthermore, it must be assumed that only two quantized spin states can exist, with the direction of rotation and thus the associated magnetic field being opposite for the two. In a molecular orbital containing two electrons, the spins of the two are necessarily opposed as a consequence of the Pauli exclusion principle. Under these circumstances, the spins are said to

---

[1] For further discussion of the theory and applications of fluorescence, phosphorescence, and luminescence, see: W. R. Seitz, in *Treatise on Analytical Chemistry,* 2d ed., P. J. Elving, E. J. Meehan, and I. M. Kolthoff, Eds., Part I, vol. 7, Chapter 4, Wiley: New York, 1981; J. R. Lakowicz, *Principles of Fluorescence Spectroscopy,* Plenum: New York, 1983; S. G. Schulman, *Fluorescence and Phosphorescence Spectroscopy: Physiochemical Principles and Practice,* Pergamon Press: New York, 1977; *Modern Fluorescence Spectroscopy,* E. L. Wehry, Ed., vol. 1–4, Plenum Press: New York, 1976–81; and J. D. Winefordner, S. G. Schulman, and T. C. O'Haver, *Luminescence Spectrometry in Analytical Chemistry,* Wiley-Interscience: New York, 1972.

be paired. Because of spin pairing, most molecules have no net magnetic field and are thus *diamagnetic*—that is, they are repelled by permanent magnetic fields. In contrast, free radicals, which contain an unpaired electron, have a magnetic moment and consequently are attracted into a magnetic field; free radicals are thus said to be *paramagnetic*.

Singlet/Triplet Excited States. A molecular *electronic* state in which all electron spins are paired is called a *singlet* state, and no splitting of energy level occurs when the molecule is exposed to a magnetic field (here, we neglect the effects of nuclear spin). The ground state for a free radical, on the other hand, is a *doublet* state because the odd electron can assume two orientations in a magnetic field, which imparts slightly different energies to the system.

When one of a pair of electrons of a molecule is excited to a higher energy level, a singlet or a *triplet* state is permitted. In the excited singlet state, the spin of the promoted electron is still paired with the ground-state electron; in the triplet state, however, the spins of the two electrons have become unpaired and are thus parallel. These states can be represented as follows where the arrows represent the direction of spin.

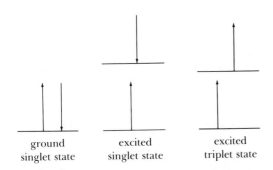

| ground | excited | excited |
| singlet state | singlet state | triplet state |

The nomenclature of singlet, doublet, and triplet derives from spectroscopic *multiplicity* considerations, which need not concern us here. Note that the excited triplet state is less energetic than the corresponding excited singlet state.

The properties of a molecule in the excited triplet state differ significantly from those of the excited singlet state. For example, a molecule is paramagnetic in the former and diamagnetic in the latter. More important, however, is the fact that a singlet-triplet transition (or the reverse), which also involves a change in electronic state, is a significantly less probable event than the corresponding singlet-singlet transition. As a consequence, the average lifetime of an excited triplet state may range from 10^{-4} to several seconds, as compared with an average lifetime of 10^{-6} to 10^{-8} sec for an excited singlet state. Furthermore, radiation-induced excitation of a ground-state molecule to an excited triplet state does not occur readily, and absorption peaks due to this process are several orders of magnitude less intense than the analogous singlet-singlet transition. We shall see, however, that an excited triplet state can be populated from an *excited* singlet state of certain molecules; the ultimate consequence of this process is often phosphorescent behavior.

Energy-Level Diagrams for Photoluminescent Molecules. Figure 8–1 is a partial energy-level diagram for a typical photoluminescent molecule. The lowest heavy horizontal line represents the ground-state energy of the molecule, which is normally a singlet state and is labeled S_0. At room temperature, this state represents the energies of essentially all of the molecules in a solution.

The upper heavy lines are energy levels for the ground vibrational states of three excited electronic states. The two lines on the left represent the first (S_1) and second (S_2) electronic *singlet* states. The one on the right (T_1) represents the energy of the first electronic *triplet* state. As is normally the case, the energy of the first excited triplet state is lower than the energy of the corresponding singlet state.

Numerous vibrational energy levels are

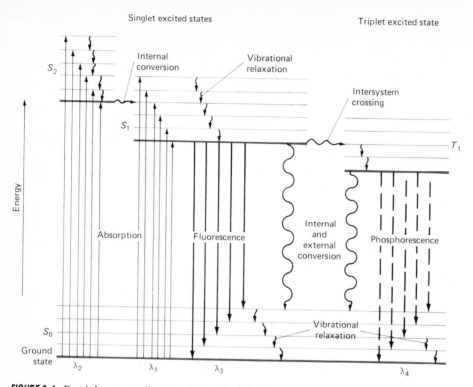

FIGURE 8-1 Partial energy diagram for a photoluminescent system.

associated with each of the four electronic states, as suggested by the lighter horizontal lines.

As shown in Figure 8–1, excitation of this molecule can be brought about by absorption of two bands of radiation, one centered about the wavelength $\lambda_1(S_0 \rightarrow S_1)$ and the second around the shorter wavelength $\lambda_2(S_0 \rightarrow S_2)$. Note that the excitation process results in conversion of the molecule to any of the several excited vibrational states. Note also that direct excitation to the triplet state is not shown because this transition does not occur to any significant extent since this process involves a change in multiplicity, an event which, as we have mentioned, has a low probability of occurrence (a transition of this type is sometimes called *forbidden*).

8A-2 DEACTIVATION PROCESSES

An excited molecule can return to its ground state by a combination of several mechanistic steps. As shown by the straight vertical arrows in Figure 8–1, two of these steps, fluorescence and phosphorescence, involve the release of a photon of radiation. The other deactivation steps, indicated by wavy arrows, are radiationless processes. The favored route to the ground state is the one that minimizes the lifetime of the excited state. Thus, if deactivation by fluorescence is rapid with respect to the radiationless processes, such emission is observed. On the other hand, if a radiationless path has a more favorable rate constant, fluorescence is either absent or less intense.

The photoluminescent phenomenon is

limited to a relatively small number of systems incorporating structural and environmental features that cause the rate of radiationless relaxation or deactivation processes to be slowed to a point where the emission reaction can compete kinetically. Information concerning emission processes is sufficiently complete to permit a quantitative accounting of their rates. Understanding of other deactivation routes, however, is rudimentary at best; for these processes, only qualitative statements or speculations about rates and mechanism can be put forth. Nevertheless, the interpretation of photoluminescence requires consideration of these other routes.

Rates of Absorption and Emission. The rate at which a photon of radiation is absorbed is enormous, the process requiring on the order of 10^{-14} to 10^{-15} sec. Fluorescent emission, on the other hand, occurs at a significantly slower rate. Here, an inverse relationship exists between the lifetime of the excited state and the molar absorptivity of the absorption peak corresponding to the excitation process. Thus, for molar absorptivities in the 10^3 to 10^5 range, lifetimes of excited states are 10^{-7} to 10^{-9} sec. For weakly absorbing systems, where the probability of the transition process is less, lifetimes may be as long as 10^{-6} to 10^{-5} sec. As we have noted, the average rate of a triplet to singlet transition is less than a corresponding singlet to singlet transition. Thus, phosphorescent emission requires times in the range of 10^{-4} to 10 sec or more.

Vibrational Relaxation. As shown in Figure 8–1, a molecule may be promoted to any of several vibrational levels during the electronic excitation process. In solution, however, the excess vibrational energy is immediately lost as a consequence of collisions between the molecules of the excited species and those of the solvent; the result is an energy transfer and a minuscule increase in temperature of the solvent. This relaxation process is so efficient that the average lifetime of a vibrationally excited molecule is 10^{-12} sec or less, a period significantly shorter than the average lifetime of an electronically excited state. As a consequence, fluorescence from solution, when it occurs, always involves a transition *from the lowest vibrational level of an excited state.* Several closely spaced peaks are produced, however, since the electron can return *to any one of the vibrational levels of the ground state* (Figure 8–1), whereupon it will rapidly fall to the lowest ground state by further vibrational relaxation.

A consequence of the efficiency of vibrational relaxation is that the fluorescent band for a given electronic transition is displaced toward lower frequencies or longer wavelengths from the absorption band (the Stokes shift); overlap occurs only for the resonance peak involving transitions between the lowest vibrational level of the ground state and the corresponding level of the excited state.

Internal Conversion. The term *internal conversion* is employed to describe intermolecular processes by which a molecule passes to a lower energy *electronic* state without emission of radiation. These processes are neither well defined nor well understood, but it is apparent that they are often highly efficient, since relatively few compounds exhibit fluorescence.

Internal conversion appears to be particularly efficient when two electronic energy levels are sufficiently close for the existence of an overlap in vibrational energy levels. This situation is depicted for the two excited singlet states in Figure 8–1. At the overlaps shown, the potential energies of the two excited states are identical; this equality apparently permits an efficient transition. Internal conversion through overlapping vibrational levels is usually more probable than the loss of energy by fluorescence from a higher excited state. Thus, referring again to Figure 8–1, excitation by radiation of λ_2

usually produces fluorescence of wavelength λ_3 to the exclusion of a band that would result from a transition between S_2 and S_0. Here, the excited molecule proceeds from the higher electronic state to the lowest vibrational state of the lower electronic excited state via a series of vibrational relaxations, an internal conversion, and then further relaxations. Under these circumstances, the fluorescence would be of λ_3 *only*, regardless of whether radiation of wavelength λ_1 or λ_2 was responsible for the excitation.

The mechanisms of the internal conversion process $S_1 \rightarrow S_0$ shown in Figure 8–1 are not well understood. The vibrational levels of the ground state may overlap those of the first excited electronic state; under such circumstances, deactivation will occur rapidly by the mechanism just described. This situation prevails with aliphatic compounds, for example, and accounts for the fact that these species seldom fluoresce; that is, deactivation by energy transfer through overlapping vibrational levels occurs so rapidly that fluorescence does not have time to occur.

Internal conversion may also result in the phenomenon of *predissociation*. Here, the electron moves from a higher electronic state to an upper vibrational level of a lower electronic state in which the vibrational energy is great enough to cause rupture of a bond. In a large molecule, there is an appreciable probability for the existence of bonds with strengths less than the electronic excitation energy of the chromophores. Rupture of these bonds can occur as a consequence of absorption by the chromophore followed by internal conversion of the electronic energy to vibrational energy associated with the weak bond.

Predissociation should be differentiated from *dissociation,* in which the absorbed radiation excites the electron of a chromophore directly to a sufficiently high vibrational level to cause rupture of the chromophoric bond; no internal conversion is involved. Dissociation processes also compete with the fluorescent process.

External Conversion.   Deactivation of an excited electronic state may involve interaction and energy transfer between the excited molecule and the solvent or other solutes. These processes are called *external conversions*. Evidence for external conversion includes the marked effect upon fluorescent intensity exerted by the solvent; furthermore, those conditions that tend to reduce the number of collisions between particles (low temperature and high viscosity) generally lead to enhanced fluorescence. The details of external conversion processes are not well understood.

Radiationless transitions to the ground state from the lowest excited singlet and triplet states (Figure 8–1) probably involve external conversions, as well as internal conversions.

Intersystem Crossing.   *Intersystem crossing* is a process in which the spin of an excited electron is reversed, and a change in multiplicity of the molecule results. As with internal conversion, the probability of this transition is enhanced if the vibrational levels of the two states overlap. The singlet-triplet transition shown in Figure 8–1 is an example; here, the lowest singlet vibrational state overlaps one of the upper triplet vibrational levels and a change in spin state is thus more probable.

Intersystem crossings are most common in molecules that contain heavy atoms, such as iodine or bromine (the *heavy-atom effect*). Apparently interaction between the spin and orbital motions becomes large in the presence of such atoms, and a change in spin is thus more favorable. The presence of paramagnetic species such as molecular oxygen in solution also enhances intersystem crossing and a consequent decrease in fluorescence.

Phosphorescence.   Deactivation may also involve phosphorescence. After intersystem crossing to an excited triplet state, further deactivation can occur either by internal or external conversion or by phosphorescence. A triplet-singlet transition is much less probable than a singlet-singlet conversion; thus, as has been noted, the average lifetime of the excited triplet state with respect to emission ranges from 10^{-4} to 10 sec or more. Thus, emission from such a transition may persist for some time after irradiation has been discontinued.

External and internal conversions compete so successfully with phosphorescence that this kind of emission is ordinarily observed only at low temperatures, in highly viscous media, or by molecules that are adsorbed on solid surfaces.

8A-3 *VARIABLES THAT AFFECT FLUORESCENCE AND PHOSPHORESCENCE*

Both molecular structure and chemical environment are influential in determining whether a substance will or will not fluoresce (or phosphoresce); these factors also determine the intensity of emission when photoluminescence does occur. The effects of some of these variables are considered briefly in this section.

Quantum Yield. The *quantum yield,* or *quantum efficiency,* for a fluorescent process is simply the ratio of the number of molecules that fluoresce to the total number of excited molecules (the quantum yield for phosphorescence can be defined in an analogous way). For a highly fluorescent molecule such as fluorescein, the quantum efficiency under some conditions approaches unity. Chemical species that do not fluoresce appreciably have efficiencies that approach zero.

From a consideration of Figure 8-1 and our discussion of deactivation processes, it is apparent that the fluorescent quantum yield ϕ for a compound must be determined by the relative rate constants k_x for the processes by which the lowest excited singlet state is deactivated—namely, fluorescence

(k_f), intersystem crossing (k_i), external conversion (k_{ec}), internal conversion (k_{ic}), predissociation (k_{pd}), and dissociation (k_d). We may express these relationships by the equation

$$\phi = \frac{k_f}{k_f + k_i + k_{ec} + k_{ic} + k_{pd} + k_d} \qquad (8\text{-}1)$$

where the k terms are the respective rate constants for the several processes enumerated above.

Equation 8-1 permits a qualitative interpretation of many of the structural and environmental factors that influence fluorescent intensity. Clearly, those variables that lead to high values for the fluorescence rate constant k_f and low values for the other k terms enhance fluorescence. The magnitude of k_f, the predissociation rate constant k_{pd}, and the dissociation rate constant k_d are mainly dependent upon chemical structure; the remaining constants are strongly influenced by environment and to a somewhat lesser extent by structure.

Transition Types in Fluorescence. It is important to note that fluorescence seldom results from absorption of ultraviolet radiation of wavelengths lower than 250 nm because such radiation is sufficiently energetic to cause deactivation of the excited states by predissociation or dissociation. For example, 200-nm radiation corresponds to about 140 kcal/mol; most molecules have at least some bonds that can be ruptured by energies of this magnitude. As a consequence, fluorescence due to $\sigma^* \rightarrow \sigma$ transitions is seldom observed; instead, such emission is confined to the less energetic $\pi^* \rightarrow \pi$ and $\pi^* \rightarrow n$ processes (see Figure 7-3, p. 185, for the relative energies associated with these transitions).

As we have noted, an electronically excited molecule ordinarily returns to its *lowest excited state* by a series of rapid vibrational relaxations and internal conversions that produce no emission of radiation. Thus, any fluorescence observed most commonly arises from a transition from the first excited electronic state to one of the ground vibrational levels. For the majority of fluorescent compounds then, radiation is produced by a transition involving either the n,π^* or the π,π^* excited state, depending upon which of these is the less energetic.

Quantum Efficiency and Transition Type. It is observed empirically that fluorescent behavior is more commonly found in compounds in which the lowest energy excited state is of a π,π^* type than in those with a lowest energy n,π^* state; that is, the quantum efficiency is greater for $\pi^* \to \pi$ transitions.

The greater quantum efficiency associated with the π,π^* state can be rationalized in two ways. First, the molar absorptivity of a $\pi \to \pi^*$ transition is ordinarily 100- to 1000-fold greater than for an $n \to \pi^*$ process, and this quantity represents a measure of transition probability in either direction. Thus, the inherent lifetime associated with a $\pi \to \pi^*$ transition is shorter (10^{-7} to 10^{-9} sec compared with 10^{-5} to 10^{-7} sec for an n,π^* state) and k_f in Equation 8–1 is larger.

It is also believed that the rate constant for intersystem crossing k_i is smaller for π,π^* excited states because the energy difference between the singlet-triplet states is larger; that is, more energy is required to unpair the electrons of the π,π^* excited state. As a consequence, overlap of triplet vibrational levels with those of the singlet state is less, and the probability of an intersystem crossing is smaller.

In summary, then, fluorescence is more commonly associated with π,π^* states than with n,π^* states because the former possess shorter average lifetimes (k_f is larger) and because the deactivation processes that compete with fluorescence are less likely to occur.

Fluorescence and Structure. The most intense and most useful fluorescent behavior is found in compounds containing aromatic functional groups with low-energy $\pi \to \pi^*$ transition levels. Compounds containing aliphatic and alicyclic carbonyl structures or highly conjugated double-bond structures may also exhibit fluorescence, but the number of these is small compared with the number in the aromatic systems.

Most unsubstituted aromatic hydrocarbons fluoresce in solution, the quantum efficiency usually increasing with the number of rings and their degree of condensation. The simple heterocyclics, such as pyridine, furan, thiophene, and pyrrole, do not exhibit fluorescent behavior; on the other hand, fused-ring structures ordinarily do. With nitrogen heterocyclics, the lowest-energy electronic transition is believed to involve an $n \to \pi^*$ system that rapidly converts to the triplet state and prevents fluorescence. Fusion of benzene rings to a heterocyclic nucleus, however, results in an increase in the molar absorptivity of the absorption peak. The lifetime of an excited state is shorter in such structures; fluorescence is thus observed for compounds such as quinoline, isoquinoline, and indole.

Substitution on the benzene ring causes shifts in the wavelength of absorption maxima and corresponding changes in the fluorescence peaks. In addition, substitution frequently affects the fluorescence efficiency; some of these effects are illustrated by the data for benzene derivatives in Table 8–1.

The influence of halogen substitution is striking; the decrease in fluorescence with increasing atomic number of the halogen is thought to be due in part to the heavy atom effect (p. 230), which increases the probability for intersystem crossing to the triplet state. Predissociation is thought to play an important role in iodobenzene and in nitro derivatives as well; these compounds have easily ruptured bonds that can absorb the excitation energy following internal conversion.

Substitution of a carboxylic acid or carbonyl group on an aromatic ring generally inhibits fluorescence. In these compounds, the energy of the n,π^* system is less than in the π,π^* system; as we have pointed out earlier, the fluorescent yield from the former type of system is ordinarily low.

Effect of Structural Rigidity. It is found empirically that fluorescence is particularly favored in molecules that possess rigid structures. For example, the quantum efficiencies for fluorene and biphenyl are nearly 1.0 and 0.2, respectively, under similar conditions of measurement. The difference in behavior

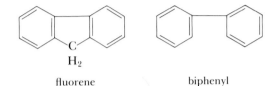

fluorene   biphenyl

appears to be largely a result of the increased rigidity furnished by the bridging methylene group in fluorene. Many similar

examples can be cited. In addition, enhanced emission frequently results when fluorescing dyes are adsorbed on a solid surface; here again, the added rigidity provided by the solid surface may account for the observed effect.

The influence of rigidity has also been invoked to account for the increase in fluorescence of certain organic chelating agents when they are complexed with a metal ion. For example, the fluorescent intensity of 8–hydroxyquinoline is much less than that of the zinc complex:

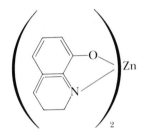

Table 8–1
Effect of Substitution on the Fluorescence of Benzene[a,b]

Compound	Formula	Wavelength of Fluorescence, nm	Relative Intensity of Fluorescence
Benzene	C_6H_6	270–310	10
Toluene	$C_6H_5CH_3$	270–320	17
Propylbenzene	$C_6H_5C_3H_7$	270–320	17
Fluorobenzene	C_6H_5F	270–320	10
Chlorobenzene	C_6H_5Cl	275–345	7
Bromobenzene	C_6H_5Br	290–380	5
Iodobenzene	C_6H_5I	—	0
Phenol	C_6H_5OH	285–365	18
Phenolate ion	$C_6H_5O^-$	310–400	10
Anisole	$C_6H_5OCH_3$	285–345	20
Aniline	$C_6H_5NH_2$	310–405	20
Anilinium ion	$C_6H_5NH_3^+$	—	0
Benzoic acid	C_6H_5COOH	310–390	3
Benzonitrile	C_6H_5CN	280–360	20
Nitrobenzene	$C_6H_5NO_2$	—	0

[a] In ethanol solution.
[b] Taken from W. West, *Chemical Applications of Spectroscopy* (*Techniques of Organic Chemistry*, Vol. IX, p. 730), Interscience Publishers, Inc.: New York, 1956. Reprinted by permission of John Wiley & Sons, Inc.

Lack of rigidity in a molecule probably causes an enhanced internal conversion rate (k_{ic} in Equation 8–1) and a consequent increase in the likelihood for radiationless deactivation. One part of a nonrigid molecule can undergo low-frequency vibrations with respect to its other parts; such motions undoubtedly account for some energy loss.

Temperature and Solvent Effects. The quantum efficiency of fluorescence by most molecules decreases with increasing temperature because the increased frequency of collisions at elevated temperatures improves the probability for deactivation by external conversion. A decrease in solvent viscosity also increases the likelihood of external conversion and leads to the same result.

The polarity of the solvent may also have an important influence. In Chapter 7, we pointed out that the energy for $n \rightarrow \pi^*$ transitions is often increased in polar solvents, while that for a $\pi \rightarrow \pi^*$ transition suffers the opposite effect. Such shifts may occasionally be great enough to lower the energy of the $\pi \rightarrow \pi^*$ process below that of the $n \rightarrow \pi^*$ transition; enhanced fluorescence results.

The fluorescence of a molecule is decreased by solvents containing heavy atoms or other solutes with such atoms in their structure; carbon tetrabromide and ethyl iodide are examples. The effect is similar to that which occurs when heavy atoms are substituted into fluorescent compounds; orbital spin interactions result in an increase in the rate of triplet formation and a corresponding decrease in fluorescence. Compounds containing heavy atoms are frequently incorporated into solvents when enhanced phosphorescence is desired.

Effect of pH on Fluorescence. The fluorescence of an aromatic compound with acidic or basic ring substituents is usually pH-dependent. Both the wavelength and the emission intensity are likely to be different for the ionized and nonionized forms of the compound. The data for phenol and aniline shown in Table 8–1 illustrate this effect.

The changes in emission of compounds of this type arise from the differing number of resonance species that are associated with the acidic and basic forms of the molecules. For example, aniline has several resonance forms while anilinum has but one. That is,

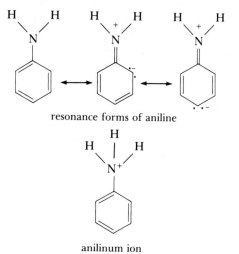

resonance forms of aniline

anilinum ion

The additional resonance forms lead to a more stable first excited state; fluorescence in the ultraviolet region is the consequence.

The fluorescent behavior of certain compounds as a function of pH has been used for the detection of end points in acid-base titrations. For example, fluorescence of the phenolic form of 1-naphthol-4-sulfonic acid is not detectable by the eye since it occurs in the ultraviolet region. When the compound is converted to the phenolate ion by the addition of base, however, the emission peak shifts to visible wavelengths where it can readily be seen. It is of interest that this change occurs at a different pH than would be predicted from the acid dissociation constant for the compound. The explanation of this discrepancy is that the acid dissociation constant for the *excited* molecule differs from that for the same species in its ground state. Changes in acid or base dissociation constants with excitation are common and are occasionally as large as four or five orders of magnitude.

It is clear from these observations that analytical procedures based on fluorescence frequently require close control of pH.

Effect of Dissolved Oxygen. The presence of dissolved oxygen often reduces the emission intensity of a fluorescent solution. This effect may be the result of a photochemically induced oxidation of the fluorescent species. More commonly, however, the quenching takes place as a consequence of the paramagnetic properties of molecular oxygen, which promotes intersystem crossing and conversion of excited molecules to the triplet state. Other paramagnetic species also tend to quench fluorescence.

Effect of Concentration on Fluorescent Intensity. The power of fluorescent radiation F is proportional to the radiant power of the excitation beam that is absorbed by the system. That is,

$$F = K'(P_0 - P) \tag{8--2}$$

where P_0 is the power of the beam incident upon the solution and P is its power after traversing a length b of the medium. The constant K' depends upon the quantum efficiency of the fluorescent process. In order to relate F to the concentration c of the fluorescing particle, we write Beer's law in the form

$$\frac{P}{P_0} = 10^{-\varepsilon bc} \tag{8--3}$$

where ε is the molar absorptivity of the fluorescent molecules and εbc is the absorbance A. By substitution of Equation 8–3 into Equation 8–2, we obtain

$$F = K'P_0(1 - 10^{-\varepsilon bc}) \tag{8--4}$$

The exponential term in Equation 8–4 can be expanded as a series to

$$F = K'P_0 \left[2.3\varepsilon bc - \frac{(2.3\varepsilon bc)^2}{2!} \right.$$
$$\left. + \frac{(2.3\varepsilon bc)^3}{3!} \cdots \right] \tag{8--5}$$

Provided $\varepsilon bc = A < 0.05$, all of the subsequent terms in the brackets become small with respect to the first; under these conditions, we may write

$$F = 2.3K'\varepsilon bcP_0 \tag{8--6}$$

or at constant P_0,

$$F = Kc \tag{8--7}$$

Thus, a plot of the fluorescent power of a solution versus concentration of the emitting species should be linear at low concentrations c. When c becomes great enough so that the absorbance is larger than about 0.05, the higher order terms in Equation 8–7 become important and linearity is lost; F then lies below an extrapolation of the straight-line plot.

Two other factors, also responsible for negative departures from linearity at high concentration, are *self-quenching* and *self-absorption*. The former is the result of collisions between excited molecules. Radiationless transfer of energy occurs, perhaps in a fashion analogous to the transfer to solvent molecules that occurs in an external conversion. Self-quenching can be expected to increase with concentration because of the greater probability of collisions occurring.

Self-absorption occurs when the wavelength of emission overlaps an absorption peak; fluorescence is then decreased as the emitted beam traverses the solution.

The effects of these phenomena are such that a plot relating fluorescent power to concentration may exhibit a maximum.

8A-4 EMISSION AND EXCITATION SPECTRA

Figure 8–2 shows three types of photoluminescence spectra for phenanthrene. An *excitation* spectrum is obtained by measuring luminescence intensity at a fixed wavelength while the excitation wavelengh is varied.

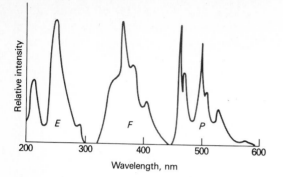

FIGURE 8–2 Spectra for phenanthrene: *E*, excitation; *F*, fluorescence; *P*, phosphorescence. (From: W. R. Seitz, in *Treatise on Analytical Chemistry*, 2d ed., P. J. Elving, E. J. Meehan, and I. M. Kolthoff, Eds., Part I, vol. 7, p. 169, Wiley: New York, 1981. Reprinted by permission of John Wiley & Sons, Inc.)

Fluorescence and phosphorescence spectra, on the other hand, involve excitation at a fixed wavelength while recording the emission intensity as a function of wavelength.

As has been pointed out earlier, photoluminescence usually occurs at wavelengths that are longer than the excitation wavelength. Furthermore, phosphorescence bands are generally found at higher wavelengths than fluorescence bands because the excited triplet state is, in most instances, lower in energy than the corresponding singlet state. In fact, the wavelength difference between the two provides a convenient measure of the energy difference between triplet and singlet states.

8B *Instruments for Measuring Fluorescence and Phosphorescence*

The various components of instruments for measuring photoluminescence are similar to those found in ultraviolet-visible photometers or spectrophotometers. Figure 8–3 shows a typical configuration for these componets in a *fluorometer* or a *spectrofluorometer*. Nearly all fluorescence instruments employ

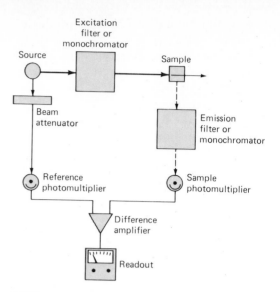

FIGURE 8–3 Components of a fluorometer or a spectrofluorometer.

double-beam optics as shown in order to compensate for fluctuations in the power of the source. The sample beam first passes through an excitation filter or a monochromator, which transmits radiation that will excite fluorescence but excludes or limits the emitted radiation. Fluorescent radiation is propagated from the sample in all directions but is most conveniently observed at right angles to the excitation beam; at other angles, increased scattering from the solution and the cell walls may cause large errors in the measurement of intensity. The emitted radiation reaches a photoelectric detector after passing through a second filter or monochromator that isolates a fluorescent peak for measurement.

The reference beam passes through an attenuator that reduces its power to approximately that of the fluorescent radiation (the power reduction is usually by a factor of 100 or more). The signals from the reference and sample phototubes are then fed into a difference amplifier whose output is displayed by a meter or recorder. Many fluorescence instruments are of the null type, this state being achieved by optical or electrical attenuators.

The sophistication, performance characteristics, and costs of fluorometers and spectrofluorometers differ as widely as do the corresponding instruments for absorption measurements. Fluorometers are analogous to absorption photometers in that filters are employed to restrict the wavelengths of the excitation and emission beams. Spectrofluorometers are of two types. The first employs a suitable filter to limit the excitation radiation and a grating or prism monochromator to isolate a peak of the fluorescent emission spectrum. Several commercial spectrophotometers can be purchased with adapters that permit their use as spectrofluorometers.

True spectrofluorometers are specialized instruments equipped with two monochromators, one of which permits variation in the wavelength of excitation and the other allows derivation of a fluorescence emission spectrum. Figure 8–4**a** shows an excitation spectrum for anthracene in which the fluorescence was measured at a fixed wavelength, while the excitation wavelength was varied. With suitable corrections for variations in source output intensity and detector response as a function of wavelength, an absolute excitation spectrum is obtained which closely resembles an absorption spectrum.

Figure 8–4**b** is the fluorescence spectrum for anthracene; here, the excitation wavelength was held constant while scanning the fluorescent emission. As is often the case, the two spectra bear an approximate mirror image relationship to one another because the vibrational energy differences for the ground and excited electronic states are roughly the same (see Figure 8–1).

The selectivity provided by spectrofluorometers is of prime importance to investigations concerned with the electronic and structural characteristics of molecules and is of value in both qualitative and quantitative analytical work as well. For concentration

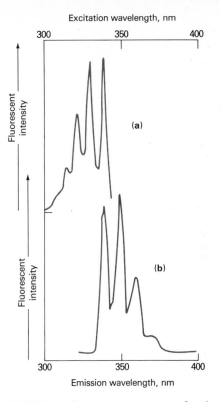

FIGURE 8–4 Fluorescence spectra for 1 ppm anthracene in alcohol: (**a**) excitation spectrum; (**b**) emission spectrum.

measurements, however, the information provided by simpler instruments is often entirely satisfactory. Indeed, relatively inexpensive fluorometers, which have been designed specifically to meet the measurement problems peculiar to fluorescence methods, are frequently as specific and selective as modified absorption spectrophotometers.

The discussion that follows is largely focussed on the simpler instruments for fluorescence analysis.

8B–1 COMPONENTS OF FLUOROMETERS AND SPECTROFLUOROMETERS

The components of fluorometers and spectrofluorometers differ only in detail from those of photometers and spectrophotometers; we need to consider only these differences.

Sources.   In most applications, a more intense source is needed than the tungsten or hydrogen lamp employed for the measurement of absorption. Indeed, as shown by Equation 8–6, the magnitude of the output signal, and thus the sensitivity, is directly proportional to the source power P_0. A mercury or xenon arc lamp is commonly employed.

The xenon arc lamp produces intense radiation by the passage of current through an atmosphere of xenon. The spectrum is continuous over the range between about 250 and 600 nm, with the peak intensity occurring at about 470 nm (see Figure 4–14, p. 110). In some instruments, regularly spaced flashes are obtained by discharging a capacitor through the lamp; higher intensities are realized in this way. In addition, the output of the phototubes is then ac, which can be readily amplified and processed.

Mercury arc lamps produce an intense line spectrum. High-pressure lamps (~8 atmospheres) give lines at 366, 405, 436, 546, 577, 691, and 773 nm. Low-pressure lamps, equipped with silica windows, additionally provide intense radiation at 254 nm. Inasmuch as fluorescent behavior can be induced in most fluorescing compounds by a variety of wavelengths, at least one of the mercury lines ordinarily proves suitable.

A recent important development has been the use of various types of lasers as excitation sources for photoluminescence measurement. Of particular interest is a tunable dye laser employing a pulsed nitrogen laser as the primary source. Monochromatic radiation between 360 and 650 nm is produced. Such a device eliminates the need for an excitation monochromator.

Filters and Monochromators.   Both interference and absorption filters have been employed in fluorometers. Most spectrofluorometers are equipped with grating monochromators.

Detectors.   The typical fluorescent signal is of low intensity; large amplification factors are thus required for its measurement. Photomultiplier tubes have come into widespread use as detectors in sensitive fluorescence instruments. Diode array detectors have also been proposed for spectrofluorometry.[2]

Cells and Cell Compartments.   Both cylindrical and rectangular cells fabricated of glass or silica are employed for fluorescence measurements. Care must be taken in the design of the cell compartment to reduce the amount of scattered radiation reaching the detector. Baffles are often introduced into the compartment for this purpose.

8B-2 INSTRUMENT DESIGNS

Fluorometers.   Figure 8–5 is a schematic diagram of a double-beam fluorometer that employs a mercury lamp and a single photomultiplier tube as a detector. Part of the radiation from the lamp passes through a filter to the sample. The fluorescent radiation then passes through a second filter to the detector. A reference beam is reflected off the mirrored surface of the light cam to a lucite light pipe, a fiber optical device that directs the beam to the photomultiplier tube. The rotating light interrupter causes this reference beam and the fluorescent beam to strike the detector surface alternately, thus producing an ac signal whenever the power of one differs from the other; the phase of the ac signal, however, will depend upon which is the stronger. A phase-sensitive device is employed to translate this difference and its sign into a deflection of a meter needle (not shown). The power of the reference beam can then be varied by rotation of the light cam, which mechanically increases or decreases the fraction of the reference beam that reaches the detector. The cam is equipped with a

---

[2] See: Y. Talmi, *Appl. Spectrosc.*, **1982**, *36*, 1.

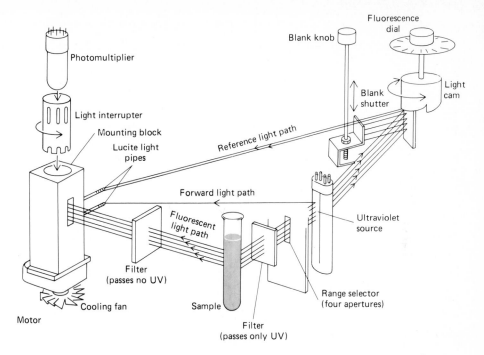

FIGURE 8-5 Optical design of the Turner Model 110 Fluorometer. (Courtesy of Amsco Instrument Company [formerly G. K. Turner Associates], Capinteria, CA.)

linear dial, each increment of which corresponds to an equal fraction of light.

Accurate adjustment of any null device requires that the condition of null be approachable from both directions. For a totally nonfluorescent sample, however, no light would reach the detector in one of the phases and the null point could then be approached from one direction only. To avoid the resultant error, a third beam of constant intensity (the forward light path) is directed to the detector *in phase* with the fluorescent beam so that, under all conditions, some radiation strikes the photomultiplier. Correction for the effect of the third beam on the measured fluorescence is accomplished by setting the fluorescence dial to zero with a solvent blank or a nonfluorescing dummy cuvette in the cell compartment; the intensity of the reference beam can then be varied by means of the blank shutter until an

optical null is indicated; this operation must be carried out frequently during a set of measurements.

The single-detector aspect of this double-beam instrument imparts high reproducibility, even with long-term drift in detector sensitivity and source output. Thus, only occasional checks of calibration curves with a standard are required.

The instrument just described is representative of the dozen or more fluorometers available commercially. Some of these are simpler single-beam instruments. The cost of such fluorometers ranges from a few hundred dollars to perhaps $5000.

Spectrofluorometers. Several instrument manufacturers offer spectrofluorometers capable of providing both excitation and emission spectra. The optical design of one of these, which employs two grating monochromators, is shown in Figure 8–6. Radia-

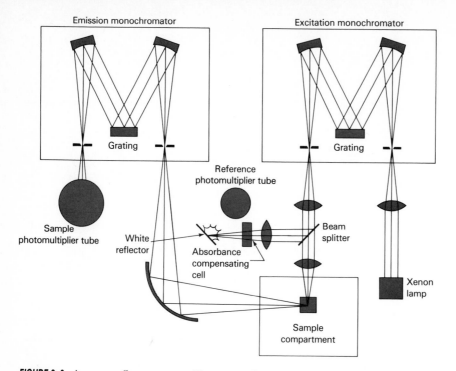

FIGURE 8–6 A spectrofluorometer. (Courtesy of American Instrument Company, Silver Springs, MD.)

tion from the first monochromator is split, part passing to a reference photomultiplier and part to the sample. The resulting fluorescent radiation, after dispersion by the second monochromator, is detected by a second photomultiplier.

An instrument such as that shown in Figure 8–6 provides perfectly satisfactory spectra for quantitative analysis. The emission spectra obtained will not, however, necessarily compare well with spectra from other instruments because the output depends not only upon the intensity of fluorescence but also upon the characteristics of the lamp, detector, and monochromators. All of these instrument characteristics vary with wavelength and differ from instrument to instrument.

A number of methods have been developed for obtaining a *corrected* spectrum, which is the true fluorescent spectrum freed

from instrumental effects; many of the newer and more sophisticated commercial instruments provide a means for obtaining corrected spectra directly.[3]

Phosphorimeters. Instruments that have been used for studying phosphorescence are similar in design to the fluorometers and spectrofluorometers just considered except that two additional components are required.[4] The first is a device that will alternately irradiate the sample and, after a suitable time delay, measure the phosphorescent intensity. Both mechanical and electronic devices are used, and many commercial fluores-

---

[3] For a summary of correction methods, see: N. Wotherspoon, G. K. Oster, and G. Oster, *Physical Methods of Chemistry*, A. Weissberger and B. R. Rossiter, Eds., vol. 1, Part III B, pp. 460–462 and pp. 473–478, Wiley-Interscience: New York, 1972.

[4] See: R. J. Hurtubise, *Anal. Chem.*, **1983**, *55*, 669A.

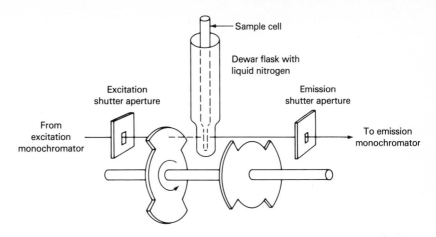

FIGURE 8-7 Schematic diagram of a device for alternately exciting and observing phosphorescence. (Reprinted with permission from: T. C. O'Haver and J. D. Winefordner, *Anal. Chem.,* **1966,** *38,* 603. Copyright 1966 American Chemical Society.)

cence instruments have accessories for phosphorescence measurements. An example of one type of mechanical device is shown in Figure 8–7.

Ordinarily, phosphorescence measurements are performed at liquid nitrogen temperature in order to prevent degradation of the output by collisional deactivation. Thus, as shown in Figure 8–7, a Dewar flask with quartz windows is ordinarily a part of a phosphorimeter. At the temperature used, the analyte exists as a solute in a solid, solvent glass (a common solvent is a mixture of diethylether, pentane, and ethanol).

8C Applications of Photoluminescence Methods

Fluorescence and phosphorescence methods are inherently applicable to lower concentration ranges than are spectrophotometric determinations and are thus among the most sensitive analytical techniques available to the scientist. The enhanced sensitivity arises from the fact that the concentration-related parameter for fluorometry and phosphorimetry F can be measured independently of the power of the source P_0. In contrast, a spectrophotometric measurement requires evaluation of both P_0 and P, because absorbance, which is proportional to concentration, is dependent upon the ratio between these two quantities. The sensitivity of a fluorometric method can be improved by increasing P_0 or by further amplifying the fluorescent signal. In spectrophotometry, in contrast, an increase in P_0 results in a proportionate change in P and therefore fails to affect A; thus, no improvement in sensitivity results. Similarly, amplification of the detector signal has the same effect on both P and P_0 and results in no net gain with respect to A. Thus, fluorometric methods generally have sensitivities that are one to three orders of magnitude better than the corresponding spectrophotometric procedures. On the other hand, the precision and accuracy of photoluminescence methods is usually poorer than spectrophotometric procedures by a factor of perhaps two to five. Generally, phosphorescence methods are less precise than their fluorescence counterparts.

8C–1 FLUOROMETRIC DETERMINATION OF INORGANIC SPECIES

Inorganic fluorometric methods are of two types. Direct methods involve the formation of a fluorescent chelate and the measurement of its emission. A second group is based upon the diminution of fluorescence resulting from the quenching action of the substance being determined. The latter technique has been most widely used for anion analysis.

Cations That Form Fluorescent Chelates. Two factors greatly limit the number of transition-metal ions that form fluorescent chelates. First, many of these ions are paramagnetic; this property increases the rate of intersystem crossing to the triplet state. Deactivation by fluorescence is thus unlikely, although phosphorescent behavior may be observed. A second reason is that transition-metal complexes are characterized by many closely spaced energy levels, which enhance the likelihood of deactivation by internal conversion. Nontransition-metal ions are less susceptible to the foregoing deactivation processes; it is for these elements that the principal inorganic applications of fluorometry are to be found. It is noteworthy that nontransition-metal cations are generally colorless and tend to form chelates which are also without color. Thus, fluorometry often complements spectrophotometry.

Fluorometric Reagents.[5] The most successful fluorometric reagents for cation analyses have aromatic structures with two or more donor functional groups that permit chelate formation with the metal ion. The structures of four common reagents follow:

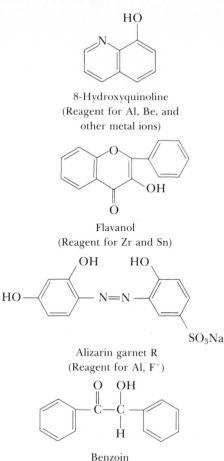

8-Hydroxyquinoline
(Reagent for Al, Be, and
other metal ions)

Flavanol
(Reagent for Zr and Sn)

Alizarin garnet R
(Reagent for Al, F⁻)

Benzoin
(Reagent for B, Zn, Ge, and Si)

Selected fluorometric reagents and their applications are presented in Table 8–2.

8C–2 FLUOROMETRIC DETERMINATION OF ORGANIC SPECIES

The number of applications of fluorometric analysis to organic problems is impressive. Weissler and White have summarized the most important of these in several tables.[6] Under a heading of *Organic and General Biochemical Substances* are found over

[5] For a more detailed discussion of fluorometric reagents, see: G. G. Guilbault, in *Comprehensive Analytical Chemistry*, G. Svehla, Ed., Vol. VIII, Chapter 2, pp. 167–178, Elsevier: New York, 1977; P. A. St. Johns, in *Trace Analysis*, J. D. Winefordner, Ed., pp. 263–271, Wiley: New York, 1976.

[6] A. Weissler and C. E. White, *Handbook of Analytical Chemistry*, L. Meites, Ed., pp. **6**–182 to **6**–196, McGraw-Hill: New York, 1963.

one hundred entries that include such diverse substances as adenine, anthranilic acid, aromatic polycyclic hydrocarbons, cysteine, guanidine, indole, naphthols, certain nerve gases, proteins, salicylic acid, skatole, tryptophan, uric acid, and warfarin. Some fifty medicinal agents, which can be determined fluorometrically, are listed. Included among these are adrenaline, alkyl-morphine, chloroquin, digitalis principles, lysergic acid diethylamide (LSD), penicillin, phenobarbital, procaine, and reserpine. Methods for the analysis of ten steroids and an equal number of enzymes and coenzymes are also found in these tables. Some of the plant products listed include chlorophyll, ergot alkaloids, rauwolfia serpentina alkaloids, flavonoids, and rotenone. Some eighteen listings for vitamins and vitamin products are also included; among these are ascorbic acid, folic acid, nicotinamide, pyridoxal, riboflavin, thiamin, vitamin A, and vitamin B_{12}.

Without question, the most important applications of fluorometry are in the analyses of food products, pharmaceuticals, clinical samples, and natural products. The sensitivity and selectivity of the method make it a particularly valuable tool in these fields.

8C-3 PHOSPHORIMETRIC METHODS

Phosphorescence and fluorescence methods tend to be complementary rather than competitive because strongly fluorescent compounds are weakly phosphorescent and vice versa.[7] For example, among condensed ring aromatic hydrocarbons, those containing heavier atoms such as halogens or sulfur are often strongly phosphorescent; on the other hand, the same compounds in the absence of the heavy atom are more fluorescent than phosphorescent.

Phosphorimetry has been used for determination of a variety of organic and bio-

---

[7] For a recent review of phosphorimetry, see: R. J. Hurtubise, *Anal. Chem.,* **1983,** *55,* 669A.

Table 8-2
Selected Fluorometric Methods for Inorganic Species[a]

Ion	Reagent	Wavelength, nm		Sensitivity $\mu g/mL$	Interference
		Absorption	Fluorescence		
Al^{3+}	Alizarin garnet R	470	500	0.007	Be, Co, Cr, Cu, F^-, NO_3^-, Ni, PO_4^{3-}, Th, Zr
F^-	Al complex of Alizarin garnet R (quenching)	470	500	0.001	Be, Co, Cr, Cu, Fe, Ni, PO_4^{3-}, Th, Zr
$B_4O_7^{2-}$	Benzoin	370	450	0.04	Be, Sb
Cd^{2+}	2-(*o*-Hydroxyphenyl)-benzoxazole	365	Blue	2	NH_3
Li^+	8-Hydroxyquinoline	370	580	0.2	Mg
Sn^{4+}	Flavanol	400	470	0.1	F^-, PO_4^{3-}, Zr
Zn^{2+}	Benzoin	—	Green	10	B, Be, Sb, Colored ions

[a] *Handbook of Analytical Chemistry,* L. Meites, Ed., **6**-178 to **6**-181, McGraw-Hill Book Company, Inc.: New York, 1963. With permission.

chemical species including such substances as nucleic acids, amino acids, pyrine and pyrimidine, enzymes, petroleum hydrocarbons, and pesticides. The method has not, however, found as widespread use as fluorometry perhaps because of the need for low temperatures and the generally poorer precision of phosphorescence measurements. On the other hand, the potentially greater selectivity of phosphorescence procedures is attractive.

During the past decade, considerable effort has been expended in the development of phosphorimetric methods that can be carried out at room temperature. These efforts have taken two directions. The first is based upon the enhanced phosphorescence that is observed for compounds adsorbed on solid surfaces, such as filter paper. In these applications, a solution of the analyte is dispersed on the solid following which the solvent is evaporated. The phosphorescence of the surface is then measured. Presumably the rigid matrix minimizes deactivation of the triplet state by external and internal conversions.

The second room temperature method involves solubilizing the analyte in detergent micelles in the presence of heavy metal ions. Apparently the micelles increase the proximity between the heavy metal ions and the phosphor thus enhancing phosphorescence.[8]

8C-4 APPLICATION OF FLUOROMETRY AND PHOSPHORIMETRY FOR DETECTION IN LIQUID CHROMATOGRAPHY

Photoluminescence measurements provide an important method for detecting and determining components of a sample as they appear at the end of a chromatographic column. This application will be discussed in Section 27C–6.

---

[8] See: L. J. Cline Love, M. Skrilec, and J. G. Habarta, *Anal. Chem.*, **1980**, *52*, 754; and M. Skrilec and L. J. Cline Love, *Anal. Chem.*, **1980**, *52*, 1559.

8D   Chemiluminescence

The application of chemiluminescence to analytical chemistry is a relatively recent development. The number of chemical reactions that produce chemiluminescence is small thus limiting the procedure to a relatively few species. Nevertheless, some of the compounds that do react to give chemiluminescence are important components of the environment. For these, the high selectivity, the simplicity, and the extreme sensitivity of the method account for its recent growth in usage.

8D-1   THE CHEMILUMINESCENCE PHENOMENON

Chemiluminescence is produced when a chemical reaction yields an electronically excited species, which emits light as it returns to its ground state. Chemiluminescent reactions are encountered in a number of biological systems, where the process is often termed *bioluminescence*. Examples of species that are bioluminescent include the firefly, the sea pansy, and certain jellyfish, bacteria, protozoa, and crustacea. The chemistry of the various natural bioluminescent processes is incompletely understood.

Over a century ago, it was discovered that several relatively simple nonbiological organic compounds also are capable of exhibiting chemiluminescence. Even with many of these, detailed mechanisms for the excitation reactions are largely unknown despite intensive study by many investigators.

The simplest type of reaction that produces chemiluminescence can be formulated as

$$A + B \rightarrow C^* + D$$

$$C^* \rightarrow C + h\nu$$

where C^* represents the excited state of the species C. Here, the luminescent spectrum is that of the reaction product C. Most chemiluminescence reactions are considerably more complicated than is suggested by the foregoing equations.

8D-2 MEASUREMENT OF CHEMILUMINESCENCE

The instrumentation for chemiluminescence measurements are remarkably simple and may consist only of a suitable reaction vessel and a photomultiplier tube. Generally, no wavelength restricting device is necessary since the only source of radiation is the chemical reaction between the analyte and reagent. Several instrument manufacturers offer chemiluminescence photometers.

The typical signal from a chemiluminescence experiment as a function of time rises rapidly to a maximum as mixing of reagent and analyte is complete; then a more or less exponential decay of signal follows. Usually, the signal is integrated for a fixed period of time and compared with standards treated in an identical way. Often a linear relationship between signal and concentration is observed over a concentration range of several orders of magnitude.

8D-3 ANALYTICAL APPLICATIONS OF CHEMILUMINESCENCE[9]

Chemiluminescence methods are generally highly sensitive because low light levels are readily monitored in the absence of noise. Furthermore, radiation attenuation by a filter or a monochromator is avoided. In fact, detection limits are usually determined not by detector sensitivity but rather by reagent purity. Typical detection limits lie in the parts per billion (or sometimes less) to parts per million range. Typical preci-

sions are difficult to judge from the present literature.

Analysis of Gases. Chemiluminescence methods for determining components of gases originated with the need for highly sensitive means for determining atmospheric pollutants such as ozone, oxides of nitrogen, and sulfur compounds. Undoubtedly, the most widely used of these methods is for the determination of nitric oxide; the reactions are

$$NO + O_3 \rightarrow NO_2^* + O_2$$

$$NO_2^* \rightarrow NO_2 + h\nu \ (\lambda = 600 \text{ to } 2800 \text{ nm})$$

Ozone from an electrogenerator and the atmospheric sample are drawn continuously into a reaction vessel, where the luminescent radiation is monitored by a photomultiplier tube. A linear response is reported for nitrous oxide concentrations of 1 ppb to 10,000 ppm. This procedure has become the predominant one for monitoring the concentration of this important atmospheric constituent from ground level to altitudes as high as 20 km.

The reaction of nitric oxide with ozone has also been applied to the determination of the higher oxides of nitrogen. For example, the nitrogen dioxide content of automobile exhaust gas has been determined by thermal decomposition of the gas at 700°C in a steel tube. The reaction is

$$NO_2 \rightleftharpoons NO + O$$

At least two manufacturers now offer an instrument for determination of nitrogen in solid or liquid materials containing 0.1 to 30% nitrogen. The samples are pyrolyzed in an oxygen atmosphere under conditions whereby the nitrogen is converted quantitatively to nitric oxide; the latter is then measured by the method just described.

Another important chemiluminescence method is used for monitoring atmospheric ozone. In this instance, the determination is based upon the luminescence produced

---

[9] For reviews of the applications of chemiluminescence to analytical chemistry, see: W. R. Seitz and M. P. Neary, *Anal. Chem.*, **1974**, *46*, 188A; D. B. Paul, *Talanta*, **1978**, *25*, 377; U. Isacsson and G. Wettermark, *Anal. Chim. Acta*, **1974**, *68*, 339; and *Bioluminescence and Chemiluminescence*, M. A. DeLuca and W. D. McElroy, Eds., Academic Press: New York, 1981.

when the analyte reacts with the dye rhodamine-B adsorbed on an activated silica gel surface. This procedure is sensitive to less than 1 ppb ozone; the response is linear up to 400 ppb ozone. Ozone can also be determined in the gas phase based on the chemiluminescence produced when the analyte reacts with ethylene. Both reagents are reported to be specific for ozone.

Problem 8–11 illustrates yet another application of chemiluminescence for the determination of sulfur-containing air pollutants such as sulfur dioxide or hydrogen sulfide.

Analysis for Inorganic Species in the Liquid Phase. Many of the analyses carried out in the liquid phase make use of organic chemiluminescent substances containing the group

$$\underset{\parallel}{\overset{O}{C}}\text{—NH—NHR}$$

These reagents react with oxygen, hydrogen peroxide, and many other strong oxidizing agents to produce a chemiluminescent oxidation product. Luminol provides the most common example of these compounds. Its reaction with oxygen is given

below. The emission produced matches the fluorescence spectrum of the product, 3-aminophthalate anion; the chemiluminescence appears blue in color.

Several metal ions exert a profound effect on the chemiluminescent intensity when luminol is mixed with hydrogen peroxide or oxygen in alkaline solution. In most cases, the effect is catalytic with enhanced peak intensities being observed. With a few cations, inhibition of luminescence occurs. Measurement of the increases or decreases in luminosity permit the determination of these ions at concentration levels that are generally below 1 ppm.

Analysis for Organic Species. A number of organic species have catalytic or inhibiting effects on the luminol reaction with hydrogen peroxide or oxygen, thus permitting their determination. Among these are amino acids, nerve gases, certain types of insecticides, hematins, napthols, and benzene derivatives containing —NO_2, —NH_2, and —OH groups. An important application of this effect is to the identification of blood stains; hemoglobin has a strong catalytic effect on luminol oxidation.

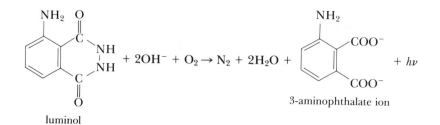

luminol   3-aminophthalate ion

PROBLEMS

8–1. Define the following terms: (a) fluorescence, (b) phosphorescence, (c) resonance fluorescence, (d) singlet state, (e) triplet state, (f) vibrational relaxation, (g) internal conversion, (h) external conversion, (i) intersystem crossing, (j) predissociation, (k) dissociation, (l) quantum yield, and (m) chemiluminescence.

8–2. Explain the difference between a fluorescence emission spectrum and a fluorescence excitation spectrum. Which more closely resembles an absorption spectrum?

8–3. Why is spectrofluorometry potentially more sensitive than spectro-photometry?

8–4. Which compound below is expected to have a greater fluorescent quantum yield? Explain.

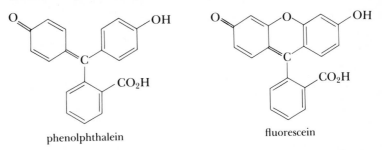

phenolphthalein   fluorescein

8–5. In which solvent would the fluorescence of naphthalene be expected to be greatest: 1-chloropropane, 1-bromopropane, or 1-iodopropane? Why?

8–6. Would the fluorescence of phenol be greater at pH 3 or pH 10? Explain.

8–7. The reduced form of nicotinamide adenine dinucleotide (NADH) is an important and highly fluorescent coenzyme. It has an absorption maximum at 340 nm and an emission maximum at 465 nm. Standard solutions of NADH gave the following fluorescence intensities:

Concn NADH, μmol/L	Relative Intensity
0.100	2.24
0.200	4.52
0.300	6.63
0.400	9.01
0.500	10.94
0.600	13.71
0.700	15.49
0.800	17.91

(a) Construct a calibration curve for NADH.
(b) Derive a least-squares equation for the plot in part (a).
(c) Calculate the standard deviation of the slope and about regression for the curve.
(d) An unknown exhibits a relative fluorescence of 12.16. Calculate the concentration of NADH.
(e) Calculate the relative standard deviation for the result in part (d).
(f) Calculate the relative standard deviation for the result in part (d) if the reading of 12.16 was the mean of three measurements.

8–8. The following volumes of a solution containing 1.10 ppm of Zn^{2+}

were pipetted into separatory funnels each containing 5.00 mL of an unknown zinc solution: 0.00, 1.00, 4.00, 7.00, and 11.00. Each was extracted with three 5-mL aliquots of CCl_4 containing an excess of 8-hydroxyquinoline. The extracts were then diluted to 25.0 mL and their fluorescence measured with a fluorometer. The results were

mL Std Zn^{2+}	Fluorometer Reading
0.000	6.12
4.00	11.16
8.00	15.68
12.00	20.64

(a) Plot the data.
(b) Derive by least squares an equation for the plot.
(c) Calculate the standard deviation of the slope and about regression.
(d) Calculate the concentration of zinc in the sample.
(e) Calculate a standard deviation for the result in part (d).

8–9. To four 10.0-mL aliquots of a water sample were added 0.00, 1.00, 2.00, 3.00 mL of a standard NaF solution containing 10.0 ppb F^-. Exactly 5.00 mL of a solution containing an excess of Al-acid Alizarin Garnet R complex, a strongly fluorescing complex, were added to each and the solutions were diluted to 50.0 mL. The fluorescent intensity of the four solutions plus a blank were

mL Sample	mL of Std F^-	Meter Reading
5.00	0.00	68.2
5.00	1.00	55.3
5.00	2.00	41.3
5.00	3.00	28.8

(a) Explain the chemistry of the analytical method.
(b) Plot the data.
(c) By least squares derive an equation relating the decrease in fluorescence to the volume of standard reagent.
(d) Calculate the standard deviation of the slope and about regression.
(e) Calculate the ppb F^- in the sample.
(f) Calculate the standard deviation of the datum in part (e).

8–10. Iron(II) ions catalyze the oxidation of luminol by H_2O_2. The intensity of the resulting chemiluminescence has been shown to increase linearly with iron(II) concentration from 10^{-10} to 10^{-8} M.

Exactly 1.00 mL of water was added to a 2.00-mL aliquot of an unknown Fe(II) solution, followed by 2.00 mL of a dilute H_2O_2

solution and 1.00 mL of an alkaline solution of luminol. The chemiluminescent signal from the mixture was integrated over a 10.0-sec
period and found to be 16.1.

To a second 2.00-mL aliquot of the sample was added 1.00 mL of
a 5.15×10^{-5} M Fe(II) solution followed by the same volume of
H_2O_2 and luminol. The integrated intensity was 29.6. Calculate the
Fe(II) molarity of the sample.

8–11. An important method for determining sulfur-bearing pollutants,
such as SO_2, H_2S, and CH_3SH, in the atmosphere involves heating
the gas sample in a hydrogen-rich flame and measuring the resulting chemiluminescence. The overall reaction for SO_2 is

$$4H_2 + 2SO_2 \rightleftharpoons S_2^* + 4H_2O$$

$$S_2^* \rightarrow S_2 + h\nu \quad (300 \text{ to } 425 \text{ nm})$$

Here, the radiation intensity is proportional to the concentration of
the excited sulfur dimer.

Derive an expression for the relationship between the concentration of SO_2 in the sample, the luminescent intensity, and the equilibrium constant for the first reaction.

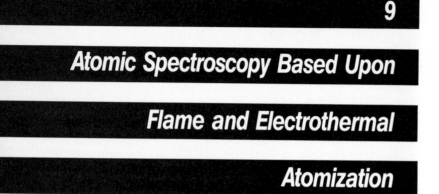

9

Atomic Spectroscopy Based Upon

Flame and Electrothermal

Atomization

Atomic spectroscopy is based upon absorption, fluorescence, or emission of electromagnetic radiation by atoms or ions. Two regions of the spectrum yield atomic information—the ultraviolet-visible and the X-ray. Applications of the latter are considered in Chapter 15.

Ultraviolet and visible atomic spectra are obtained by atomization, a process by which the molecular constituents of a sample are decomposed and converted to atomic particles. The emission, absorption, or fluorescence spectrum of the atomized species then serves as the basis for the analysis. To the extent that molecules and complex ions are absent, atomic spectra are simple relative to molecular spectra and consist of a number of narrow, discrete lines.

Table 9–1 categorizes the various atomic spectroscopic methods on the basis of the type of atomization procedure. Within the flame category are found absorption, emission, and fluorescence procedures. Electrothermal atomization has been used for absorption and fluorescence measurements but not extensively for emission work whereas the plasma sources have been applied primarily for emission and fluorescence techniques. The remaining atomization methods provide emission spectra only.

The procedures outlined in Table 9–1 frequently offer the advantages of high specificity, wide applicability, excellent sensitivity, speed, and convenience; they are among the most selective of all analytical methods. Perhaps 70 elements can be determined with sensitivities that fall in the parts-per-million to parts-per-billion range. An atomic spectral analysis can frequently be completed in a few minutes.

This chapter deals with the general theory of atomic spectroscopy as well as with spectral methods based upon flame and electrothermal atomization.[1] Note that these

---

[1] For a more detailed discussion of these topics, see: C. Th. J. Alkemade and R. Herrmann, *Fundamentals of Analytical Flame Spectroscopy,* Wiley: New York,

Table 9-1
Classification of Optical, Atomic Spectral Methods

Atomization Method	Typical Atomization Temperature, °C	Phenomenological Basis of Method	Common Name and Abbreviation for Method
Flame	1700–3150	Absorption	Atomic absorption spectroscopy, AAS
		Emission	Atomic emission spectroscopy, AES
		Fluorescence	Atomic fluorescence spectroscopy, AFS
Electrothermal	1200–3000	Absorption	Electrothermal atomic absorption spectroscopy
		Fluorescence	Electrothermal atomic fluorescence spectroscopy
Inductively coupled argon plasma	6000–8000	Emission	Inductively coupled plasma spectroscopy, ICP
		Fluorescence	Inductively coupled plasma fluorescence spectroscopy
Direct current argon plasma	6000–10000	Emission	DC argon plasma spectroscopy, DCP
Electric arc	4000–5000	Emission	Arc-source emission spectroscopy
Electric spark	40000 (?)	Emission	Spark-source emission spectroscopy

sources are the least energetic of all of those shown in Table 9–1, with temperatures in the 1200 to 3000°C range. The next chapter will be concerned with spectroscopy based upon the remaining more energetic atomization sources shown in the table.

9A *Theory of Atomic Spectroscopy*

When a sample is atomized by any of the procedures listed in Table 9–1, a substantial fraction of the metallic constituents are re-

1979; K. C. Thompson and R. J. Reynolds, *Atomic Absorption, Fluorescence, and Flame Emission Spectroscopy*, 2d ed., Wiley: New York, 1978; C. Th. J. Alkemade, *et al.*, *Metal Vapors in Flames*, Pergamon Press: Elmsford, N. Y., 1982; B. Magyar, *Guide-Lines to Planning Atomic Spectrometric Analysis*, Elsevier: New York, 1982.

duced to gaseous atoms; in addition, depending upon the temperature of the atomizer, a certain fraction of these atoms are ionized thus yielding a gaseous mixture of atoms and ions.

9A-1 SOURCES OF ATOMIC SPECTRA

The spectra of gaseous atomic particles (atoms or ions) consist of well-defined narrow lines arising from electronic transitions of the outermost electrons. For metals, the energies of these transitions are such as to yield ultraviolet, visible, and near-infrared radiation.

Energy Level Diagrams. The energy level diagram for the outer electrons of an element provides a convenient method for describing the processes upon which the various types of atomic spectroscopy are based. The diagram for sodium shown in Figure 9–1a is typical. Note that the energy scale is linear in units of electron volts (eV), with the $3s$ orbital being assigned a value of zero. The scale extends to about 5.2 eV, the energy necessary to remove the single $3s$ electron from the influence of the central atom, thus producing a sodium ion.

The energies of several atomic orbitals are indicated on the diagram by horizontal lines. Note that the p orbitals are split into two levels that differ but slightly in energy. This difference is rationalized by assuming that an electron spins about its own axis and that the direction of this motion may either be the same as or opposed to its orbital motion. Both the spin and the orbital motions create magnetic fields owing to the rotation

of the charge carried by the electron. The two fields interact in an attractive sense if these two motions are in the opposite direction; a repulsive force is generated when the motions are parallel. As a consequence, the energy of the electron whose spin opposes its orbital motion is slightly smaller than one in which the motions are alike. Similar differences exist in the d and f orbitals, but their magnitudes are ordinarily so slight as to be undetectable; thus, only a single energy level is indicated for d orbitals in Figure 9–1a.

The splitting of higher energy p, d, and f orbitals into two states is characteristic of all species containing a *single* external electron. Thus, the energy level diagram for the singly charged magnesium ion, shown in Fig-

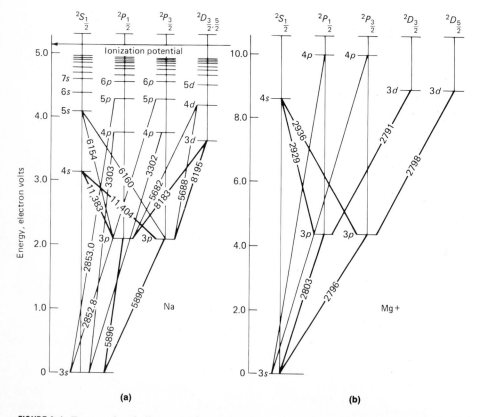

FIGURE 9–1 Energy level diagrams for (**a**) atomic sodium and (**b**) magnesium(I) ion. Note the similarity in pattern of lines (but not in actual wavelengths).

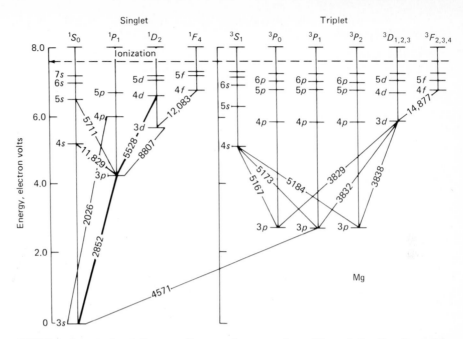

FIGURE 9-2 Energy level diagram for atomic magnesium. The relative line intensities are indicated very approximately by the width of the lines between states. Note that a singlet/triplet transition is considerably less probable than a singlet to singlet.

ure 9–1**b**, has much the same general appearance as that for the uncharged sodium atom. So also do the diagrams for the dipositive aluminum ion and the remainder of the alkali-metal atoms. It is important to note, however, that the energy difference between the $3p$ and $3s$ states is approximately twice as great for the magnesium ion as for the sodium atom because of the larger nuclear charge of the former.

A comparison of Figure 9–1**b** with Figure 9–2 shows that the energy levels, and thus the spectrum, of an ion are significantly different from that of its parent atom. For atomic magnesium, with two outer electrons, excited singlet and triplet states with different energies exist. In the excited singlet state the spins of the two electrons are opposed and said to be paired; in the triplet states the spins are unpaired or parallel (Section 8A–1). Using arrows to denote the direction of spin, the ground state and the two excited states can be represented as follows:

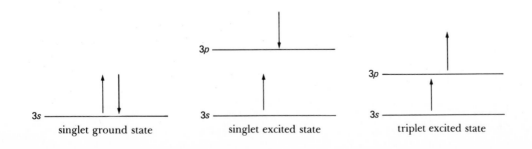

Note that, as is true of molecules, the triplet excited state is of lower energy than the corresponding singlet state.

The *p, d,* and *f* orbitals of the triplet state are split into three levels that differ slightly in energy. These splittings can be also rationalized by taking into account the interaction between the fields associated with the spins of the two outer electrons and the net field arising from the orbital motions of all the electrons. In the singlet state, the two spins are paired, and their respective magnetic effects cancel; thus, no energy splitting is observed. In the triplet state, however, the two spins are unpaired (that is, their spin moments lie in the same direction). The effect of the orbital magnetic moment on the magnetic field of the combined spins produces a splitting of the *p* level into a triplet. This behavior is characteristic of all of the alkaline-earth atoms, singly charged aluminum and beryllium ions, and so forth.

As the number of electrons outside the closed shell increases, the energy level diagrams become more and more complex. Thus, with three outer electrons, a splitting of energy levels into two and four states occurs; with four outer electrons, singlet, triplet, and quintet states exist.

Although correlation of atomic spectra with energy level diagrams for elements such as sodium and magnesium is relatively straightforward and amenable to theoretical interpretation, the same cannot be said for the heavier elements, and particularly the transition metals. These species have larger numbers of closely spaced energy levels; as a consequence, the number of absorption or emission lines can be enormous. For example, Harvey[2] has listed the number of lines observed in the arc and spark spectra of neutral and singly ionized atoms for a variety of elements. For the alkali metals, this number ranges from 30 for lithium to 645

for cesium; for the alkaline earths, magnesium has 173, calcium 662, and barium 472. Typical of the transition series, on the other hand, are chromium, iron, and cerium with 2277, 4757, and 5755 lines, respectively. Fewer lines are excited in lower-temperature atomizers, such as flames; still, the flame spectra of the transition metals are considerably more complex than the spectra of species with low atomic numbers.

Atomic Emission Spectra.   At room temperature, essentially all of the atoms of a sample of matter are in the ground state. For example, the single outer electron of metallic sodium occupies the $3s$ orbital under these circumstances. Excitation of this electron to higher orbitals can be brought about by the heat of a flame or an electric arc or spark. The lifetime of the excited atom is brief, however, and its return to the ground state is accompanied by the emission of a photon of radiation. The vertical lines in Figure 9–1**a** indicate some of the common electronic transitions that follow excitation of sodium atoms; the wavelength of the resulting radiation is also indicated. The two lines at 5890 and 5896 Å are the most intense under usual excitation conditions and are, therefore, the ones most commonly employed for analytical purposes. A slit width that is wide enough to permit simultaneous measurement of the two lines is frequently used.

Figure 9–3 shows a portion of a recorded emission spectrum for sodium. Excitation in this case resulted from spraying a solution of sodium chloride into an oxyhydrogen flame. Note the very large peak at the far right, which is off scale and corresponds to the $3p$ to $3s$ transitions at 5896 and 5890 Å shown in Figure 9–1**a**. The resolving power of the monochromator used was insufficient to separate the peaks. The much smaller peak at about 5700 Å is in fact two unresolved peaks that arise from the two $4d$ to $3p$ transitions also shown in the figure.

Atomic Absorption Spectra.   In a hot gaseous medium, sodium atoms are capable of *absorbing*

[2] C. E. Harvey, *Spectrochemical Procedures,* Chapter 4, Applied Research Laboratories: Glendale, CA, 1950.

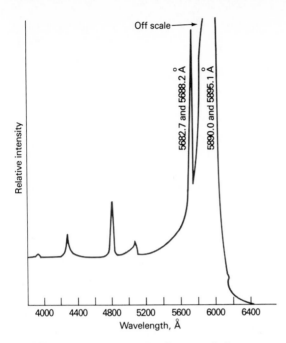

FIGURE 9-3 A portion of the flame emission spectrum for sodium.

radiation of wavelengths characteristic of electronic transitions from the $3s$ state to higher excited states. For example, sharp absorption peaks at 5890, 5896, 3302, and 3303 Å are observed experimentally; referring again to Figure 9–1a, it is apparent that each adjacent pair of these peaks corresponds to transitions from the $3s$ level to the $3p$ and the $4p$ levels, respectively. It should be mentioned that absorption due to the $3p$ to $5s$ transition is so weak as to go undetected because the number of sodium atoms in the $3p$ state is generally small at the temperature of a flame. Thus, typically, an atomic absorption spectrum consists predominately of *resonance lines,* which are the result of transitions from the *ground state* to upper levels.

Atomic Fluorescence Spectra. Atoms in a flame can be made to fluoresce by irradiation with an intense source containing wavelengths that are absorbed by the element. The fluorescence spectra is most conveniently mea-

sured at an angle of 90 deg to the light path. The observed radiation is most commonly the result of resonance fluorescence. For example, when magnesium atoms are exposed to an ultraviolet source, radiation of 2852 Å is absorbed as electrons are promoted from the $3s$ to the $3p$ level (see Figure 9–2); the resonance fluorescence emitted at this same wavelength may then be used for analysis. In contrast, when sodium atoms absorb radiation of wavelength 3303 Å, electrons are promoted to the $4p$ state (see Figure 9–1a). A radiationless transition to the two $3p$ states takes place more rapidly than resonance fluorescence. As a consequence, the observed fluorescence occurs at 5890 and 5896 Å. This behavior is analogous to that of molecules described in Section 8A–2. Figure 9–4 illustrates yet a third mechanism for atomic fluorescence. Here, some of the thallium atoms, excited in a flame, return to the ground state in two steps, including a fluorescent step producing a line at 5350 Å; radiationless deactivation to the ground state quickly follows. Resonance fluorescence at 3776 Å is also observed.

Line Widths. Atomic emission and absorption peaks are much narrower than the bands resulting from emission or absorption by molecules. The natural width of atomic lines can be shown to be about 10^{-4} Å. Two effects, however, combine to broaden the lines to a range from 0.02 to 0.05 Å.

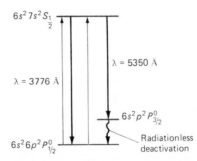

FIGURE 9-4 Energy level diagram for thallium showing the source of two fluorescence lines.

Doppler broadening arises from the rapid motion of the atomic particles in the plasma. Atoms that are moving toward the monochromator emit shorter wavelengths owing to the well-known Doppler shift; the effect is reversed for atoms moving away from the monochromator.

Doppler broadening is also observed for absorption lines. Those atoms traveling toward the source absorb radiation of slightly shorter wavelength than that absorbed by particles traveling perpendicular to the incident beam. The reverse is true of atoms moving away from the source.

Pressure broadening also occurs; here collisions among atoms cause small changes in the ground-state energy levels and a consequent broadening of peaks.

9A-2 MOLECULAR SPECTRA PRODUCED DURING ATOMIZATION

In flame atomization with hydrogen or hydrocarbon fuels, molecular absorption and emission bands are often encountered over certain wavelength ranges owing to the presence of such species as OH and CN radicals and C_2 molecules. In addition, some alkaline-earth and rare-earth metals form volatile oxides or hydroxides that also absorb and emit over broad spectral ranges. One example is shown in Figure 9–5 where the emission and absorption spectra of CaOH are shown. The dashed line in the figure shows the wavelength of the barium resonance line. The potential interference of calcium in the atomic absorption determination of barium can be avoided by employing a higher temperature which decomposes the CaOH molecule, thus causing the molecular absorption band shown in the figure to disappear.

Another example of a band spectrum can be seen in Figure 9–3. Here, the sodium lines at shorter wavelengths are superimposed on a continuum that arises from vibrational absorption due to molecular, organic decomposition products formed from

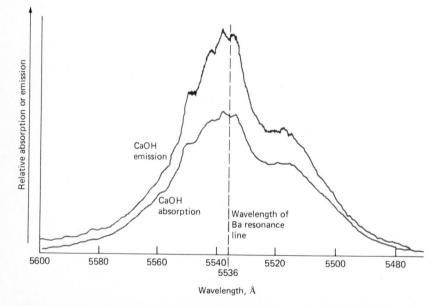

FIGURE 9-5 Molecular flame and flame absorption spectra for CaOH and Ba. (Adapted from L. Capacho-Delgado and S. Sprague, *Atomic Absorption Newsletter*, **1965**, *4*, 363. Courtesy of Perkin-Elmer Corporation, Norwalk, CT.)

the solvent, which was an alcohol/water mixture.

A band spectrum is useful for the determination of perhaps one-third of the elements that are amenable to emission analysis. For both emission and absorption spectroscopy, however, the presence of such bands represents a potential source of interference, which must be dealt with by proper choice of wavelength, by background correction, or by a change in atomization conditions.

9B Flame Atomization

In flame atomization, a solution of the sample is sprayed into a flame by means of a nebulizer. The high temperature causes formation of atoms, which can be observed by absorption or emission spectroscopy. The greatest source of uncertainty in flame spectroscopic methods arises from variations in the behavior of the flame. Thus, it is important to understand the characteristics of flames and the variables that affect these characteristics.

9B-1 FLAME TEMPERATURE

Table 9–2 lists the common fuels and oxidants employed in flame spectroscopy and the range of observed temperatures when the reactants are present in approximately stoichiometric proportions.

The temperatures provided by the burning of natural or manufactured gas in air are so low that only the alkali and alkaline-earth metals, with very low excitation energies, produce useful spectra. Acetylene/air mixtures give a somewhat higher temperature. Oxygen or nitrous oxide must be employed as the oxidant in order to excite the spectra of many metals; with the common fuels, temperatures of 2500 to 3100°C are obtained.

Effect of Temperature on Emission, Absorption, and Fluorescence. Emission intensities are strongly affected by flame temperature because this variable has a significant effect on the ratio between excited and unexcited atomic particles. The Boltzmann equation permits the calculation of this fraction. If N_j and N_0 are the number of atoms in an excited state and the ground state respectively, their ratio is given by

$$\frac{N_j}{N_0} = \frac{P_j}{P_0} \exp\left(-\frac{E_j}{kT}\right) \tag{9–1}$$

where k is the Boltzmann constant (1.38×10^{-16} erg/deg), T is the temperature in degrees Kelvin, and E_j is the energy difference in ergs between the excited state and the ground state. The quantities P_j and P_0 are statistical factors that are determined by the number of states having equal energy at each quantum level. As shown by the following example, the fraction of excited atoms

Table 9-2 Flame Temperatures		
Fuel	**Oxidant**	**Temperatures, °C**
Natural gas	Air	1700–1900
Natural gas	Oxygen	2700–2800
Hydrogen	Air	2000–2100
Hydrogen	Oxygen	2550–2700
Acetylene	Air	2100–2400
Acetylene	Oxygen	3050–3150
Acetylene	Nitrous oxide	2600–2800

in a typical gas flame ($T = 2500°K$) is very small.

EXAMPLE 9–1.   Calculate the ratio of sodium atoms in the $3p$ excited states to the number in the ground state at 2500 and 2510°K.

In order to calculate E_j in Equation 9–1, we employ an average wavelength of 5893 Å for the two sodium emission lines involving the $3p \rightarrow 3s$ transitions. To obtain the energy in ergs, we employ the conversion factors found inside the back cover.

$$\text{wavenumber} = \frac{1}{5893 \text{ Å} \times 10^{-8} \text{ cm/Å}}$$
$$= 1.697 \times 10^4 \text{ cm}^{-1}$$

$$E_j = \frac{1.697 \times 10^4}{\text{cm}} \times 1.986$$
$$\times 10^{-16} \text{ erg cm} = 3.37 \times 10^{-12} \text{ erg}$$

There are two quantum states in the $3s$ level and six in the $3p$. Thus,

$$\frac{P_j}{P_0} = \frac{6}{2} = 3$$

Substituting into Equation 9–1 yields

$$\frac{N_j}{N_0} = 3 \exp\left(-\frac{3.37 \times 10^{-12}}{1.38 \times 10^{-16} \times 2500}\right)$$

or

$$\frac{N_j}{N_0} = 3 \times 5.725 \times 10^{-5} = 1.72 \times 10^{-4}$$

Replacing 2500 with 2510 in the foregoing equations yields

$$\frac{N_j}{N_0} = 1.79 \times 10^{-4}$$

Example 9–1 demonstrates that a temperature fluctuation of only 10°K results in a 4% increase in the number of excited sodium atoms. A corresponding increase in

emitted power by the two lines would result. Thus, an analytical method based on the measurement of emission requires close control of flame temperature.

Absorption and fluorescence methods are theoretically less dependent upon temperature because both measurements are based upon initially *unexcited* atoms rather than thermally excited ones. In the example just considered, only about 0.017% of the sodium atoms were thermally excited at the temperature of a hydrogen/oxygen flame. An emission method is based upon this small fraction of the analyte. In contrast, absorption and fluorescence measurements use the 99.8% of the analyte present as unexcited sodium atoms to produce the analytical signals. Note also that while a 10°K temperature change causes a 4% increase in sodium ions, the corresponding *relative* change in percent of sodium atoms is inconsequential.

Temperature fluctuations actually do exert an indirect influence on flame absorption and fluorescence measurements in several ways. An increase in temperature usually increases the efficiency of the atomization process and hence the total number of atoms in the flame. In addition, line broadening and a consequent decrease in peak height occurs because the atomic particles travel at greater rates, which enhances the Doppler effect. Increased concentrations of gaseous species at higher temperatures also cause pressure broadening of the absorption lines. Finally, temperature variations influence the degree of ionization of the analyte and thus the concentration of unionized analyte upon which the analysis is usually based (see p. 275). Because of these several effects a reasonable control of the flame temperature is also required for quantitative absorption and fluorescence measurements.

The large ratio of unexcited to excited atoms in flames leads to another interesting comparison of the three atomic flame methods. Because atomic absorption and fluorescence methods are based upon a much larger population of particles, both procedures might be expected to be more sensitive than the emission procedure. This apparent advantage is offset in the absorption method, however, by the fact that an absorbance measurement involves evaluation of a difference ($A = \log P_0 - \log P$); when P and P_0 are nearly alike, larger relative errors must be expected in the difference. As a consequence, flame emission and flame absorption procedures tend to be complementary in sensitivity, the one being advantageous for one group of elements and the other for a different group. On the basis of active particle population, flame fluorescence methods should be the most sensitive of the three.

9B-2 FLAME STRUCTURES

Important regions of a flame include the *primary combustion zone,* the *interconal layer,* and the *outer cone* (see Figure 9–6). The appearance and relative size of these regions varies considerably with the fuel-to-oxidant ratio as well as the type of fuel and oxidant. The primary combustion zone is recognizable by its blue luminescence arising from band spectra of C_2, CH, and other radicals. Thermal equilibrium is ordinarily not reached in this region and it is, therefore, seldom used for flame spectroscopy.

The interconal area, which is relatively narrow in stoichiometric hydrocarbon flames, may reach several centimeters in height in fuel-rich acetylene/oxygen or acetylene/nitrous oxide sources. The zone is often rich in free atoms and is the most widely used part of the flame for spectroscopy. The outer cone is a secondary reaction zone where the products of the inner core are converted to stable molecular species.

A flame profile provides useful information about the processes that go on in different parts of a flame; it is a plot that reveals regions of the flame that have similar values for a parameter of interest. Some of these parameters include temperature, chemical composition, absorbance, and radiant or fluorescent intensity.

Temperature Profiles. Figure 9–7 is a temperature profile of a typical flame for atomic spectroscopy. The maximum temperature is

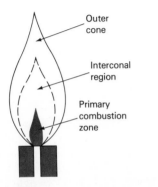

FIGURE 9-6 Schematic diagram of an acetylene/air flame.

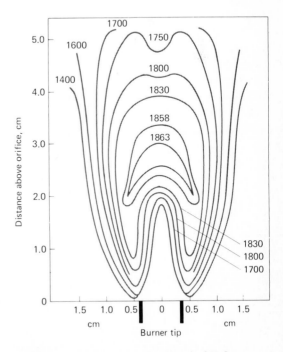

FIGURE 9-7 Temperature profiles (in °C) for a natural gas/air flame. (From: B. Lewis and G. van Elbe, *J. Chem. Phys.,* **1943,** *11,* 94. With permission.)

located somewhat above the primary combustion zone. Clearly, it is important—particularly for emission methods—to focus the same part of the flame on the entrance slit for all calibrations and analytical measurements.

Flame Absorbance Profiles. Figure 9–8 shows typical absorbance profiles for three elements. Magnesium exhibits a maximum in absorbance at about the middle of the flame because of two opposing effects. The initial increase in absorbance as the distance from the base becomes larger results from an increased number of magnesium atoms produced by the longer exposure to the heat of the flame. As the outer zone is approached, however, appreciable oxidation of the magnesium begins. This process leads to an eventual decrease in absorbance because the oxide particles formed are nonabsorbing at the wavelength used. To obtain maximum analytical sensitivity, then, the flame must be adjusted with respect to the beam until a maximum absorbance is obtained.

The behavior of silver, which is not readily oxidized, is quite different; here, a continuous increase in the number of atoms, and thus the absorbance, is observed from the base to the periphery of the flame. In contrast, chromium, which forms very stable oxides, shows a continuous decrease in absorbance beginning close to the burner tip; this observation suggests that oxide formation predominates from the start. Clearly, a different portion of the flame should be used for the analysis of each of these elements.

Emission Profiles. Figure 9–9 is a three-dimensional profile showing the emission intensity for a calcium line produced in a flame. Note that the emission maximum is found just above the primary combustion zone.

Figure 9–9 also demonstrates that the intensity of emission is critically dependent upon the rate at which the sample is introduced into the flame. Initially, the line intensity rises rapidly with increasing flow rate as a consequence of the increasing number of calcium atoms. A rather sharp maximum in intensity occurs, however, beyond which the increased flow of solution lowers the flame temperature and thus the absorbance.

Where molecular band spectra form the

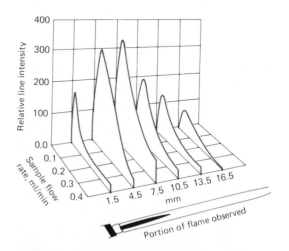

FIGURE 9-9 Flame profile for calcium line in a cyanogen/oxygen flame for different sample flow rates. (Reprinted with permission from: K. Fuwa, R. E. Thiers, B. L. Vallee, and M. R. Baker, *Anal. Chem.*, **1959**, *31*, 2041. Copyright 1959 American Chemical Society.)

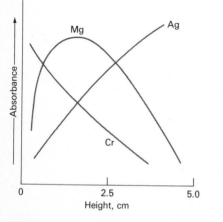

FIGURE 9-8 Flame absorbance profile for three elements.

basis for emission analysis, the maximum absorbance often appears in a lower part of the flame. For example, calcium produces a useful band in the region of 540 to 560 nm, probably due to the presence of CaOH in the flame (see Figure 9–5). The maximum intensity of this emission band occurs at the edge of the primary combustion zone and decreases rapidly in the interconal area as the molecules responsible for the emission dissociate at the higher temperatures of the latter region.

The more sophisticated instruments for flame emission spectroscopy are equipped with monochromators that sample the radiation from a relatively small part of the flame; adjustment of the position of the flame with respect to the entrance slit is thus critical. Filter photometers, on the other hand, employ a much larger portion of the flame; here, control of flame position is less important.

9C Atomizers for Atomic Spectroscopy

Two types of atomizers are encountered in atomic spectroscopy, *flame* and *electrothermal.* The former are employed for atomic emission, absorption, and fluorescence measurements. Nonflame or electrothermal atomizers have been largely confined to the latter two procedures.

9C-1 FLAME ATOMIZERS

The most common atomization device for atomic spectroscopy consists of a *nebulizer* and a *burner.*[3] The nebulizer converts a liquid sample into a fine spray or aerosol, which is then fed into the flame. Both *total consumption* (*turbulent flow*) and *premixed* (*laminar flow*) *burners* are encountered.

---

[3] For a detailed discussion of flame atomizers, see: R. D. Dresser, R. A. Mooney, E. M. Heithmar, and F. W. Pankey, *J. Chem. Educ.*, **1975,** *52,* A403.

Turbulent Flow Burners.   Figure 9–10 is a schematic diagram of a commercially available turbulent flow burner. Here, the nebulizer and burner are combined into a single unit. The sample is drawn up the capillary and nebulized by venturi action caused by the flow of gases around the capillary tip. Typical sample flow rates are 1 to 3 mL/min.

Turbulent flow burners offer the advantage of introducing a relatively large and representative sample into the flame. In contrast to laminar flow burners, no possibility of flashback and explosion exists. The disadvantages of turbulent flow burners include a relatively short flame path length and problems with clogging of the tip. In addition, these burners are noisy both from the electronic and auditory standpoint. Although sometimes used for emission and fluorescence analyses, turbulent flow burners find little use in present-day absorption work.

Laminar Flow Burners.   Figure 9–11 is a diagram of a typical commercial laminar flow burner. The sample is nebulized by the flow of oxidant past a capillary tip. The resulting aerosol is then mixed with fuel and flows past a series of baffles that remove all but the finest

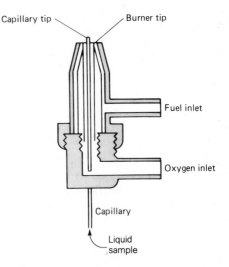

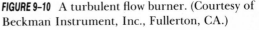

FIGURE 9-10 A turbulent flow burner. (Courtesy of Beckman Instrument, Inc., Fullerton, CA.)

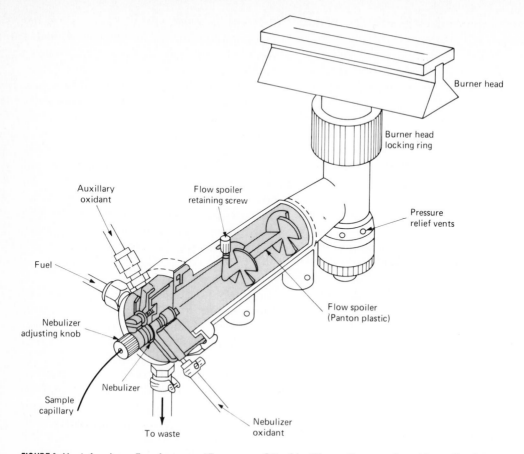

FIGURE 9–11 A laminar flow burner. (Courtesy of Perkin-Elmer Corporation, Norwalk, CT.)

droplets. As a result of the baffles, the majority of the sample collects in the bottom of the mixing chamber where it is drained to a waste container. The aerosol, oxidant, and fuel are then burned in a slotted burner that provides a flame which is usually 5 or 10 cm in length.

Laminar flow burners provide a relatively quiet flame and a significantly longer path length. These properties tend to enhance sensitivity and reproducibility. Furthermore, clogging is seldom a problem. Disadvantages include a lower rate of sample introduction (which may offset the longer path length advantage) and the possibility of selective evaporation of mixed solvents in the mixing chamber, which may lead to ana-

lytical uncertainties. Furthermore, the mixing chamber contains a potentially explosive mixture which can be ignited by a flashback. Note that the burner in Figure 9–11 is equipped with pressure relief vents for this reason. Furthermore, the burner head is sometimes held in place by stainless steel cables.

Fuel and Oxidant Regulators. An important variable that requires close control in flame spectroscopy is the flow rate of both oxidant and fuel. It is desirable to be able to vary each over a considerable range so that ideal atomization conditions can be found experimentally. Fuel and oxidant are ordinarily combined in approximately stoichiometric amounts. For the analysis of metals that form stable oxides, however, a flame that contains an excess of fuel may prove more desirable. Flow rates are ordinarily controlled by means of double-diaphragm pressure regulators followed by needle valves in the instrument housing. The most widely used device for measuring flow rates is the rotameter, which consists of a tapered, graduated, transparent tube that is mounted vertically with the smaller end down. A light-weight conical or spherical float is lifted by the gas flow; its vertical position is determined by the flow rate.

9C–2 ELECTROTHERMAL ATOMIZERS

In terms of reproducible behavior, flame atomization appears to be superior to all other methods that have been thus far developed with the possible exception of the inductively coupled plasma, which is described in Section 10B–1. In terms of sampling efficiency (and thus sensitivity), however, other atomization methods are markedly better. Two reasons for the lower sampling efficiency of the flame can be cited. First, a large portion of the sample flows down the drain (laminar burner) or is not completely atomized (turbulent burner). Second, the residence time of individual atoms in the optical path in the flame is brief ($\sim 10^{-4}$ sec).

Since 1970, electrothermal atomizers have appeared on the market.[4] These devices generally provide enhanced sensitivity because the entire sample is atomized in a short period, and the average residence time of the atoms in the optical path is a second or more. Fuller[5] has calculated that in principle, electrothermal devices should

enhance the sensitivity of atomic absorbance methods by a factor of about 10^6. Actual sensitivity enhancement with state-of-the-art, nonflame atomizers is found to range from about 100 to 4000. Fuller concludes from this large discrepancy that considerable scope for improvement exists in the design of electrothermal atomizers.

In a nonflame atomizer, a few microliters of sample are first evaporated at a low temperature and then ashed at a somewhat higher temperature on an electrically heated surface of carbon, tantalum, or other conducting material. The conductor can be formed as a hollow tube, a strip or rod, a boat, or a trough. After ashing, the current is rapidly increased to several hundred amperes, which causes the temperature to soar to perhaps 2000 to 3000°C; atomization of the sample occurs in a period of a few milliseconds to seconds. The absorption or fluorescence of the atomized particles can then be measured in the region immediately above the heated conductor.

Electrothermal Atomizer Designs. Figure 9–12**a** is a cross-sectional view of a commercial electrothermal atomizer. Atomization occurs in a cylindrical graphite tube which is open at both ends and has a central hole for introduction of sample by means of a micropipette. The tube is about 5 cm long and has an internal diameter of somewhat less than 1 cm. The interchangeable graphite tube fits snugly into a pair of cylindrical graphite electrical contacts located at the two ends of the tube. These contacts are held in a water-cooled metal housing. Two inert gas streams are provided. The external stream prevents the entrance of outside air and a consequent incineration of the tube. The internal stream flows into the two ends of the tube and out the central sample port. This stream not only excludes air but also serves to carry away vapors generated from the sample matrix during the first two heating stages.

Figure 9–12**b** illustrates the so-called

---

[4] See: R. D. Dresser, R. A. Mooney, E. M. Heithmar, and F. W. Plankey, *J. Chem. Educ.*, **1975**, *52*, A451, A503; R. E. Sturgeon, *Anal. Chem.*, **1977**, *49*, 1255A; S. R. Koirtyohann and M. L. Kaiser, *Anal. Chem.*, **1982**, *54*, 1515A; and C. W. Fuller, *Electrothermal Atomization for Atomic Absorption Spectroscopy*, The Chemical Society: London, 1978.

[5] See last reference, footnote 4.

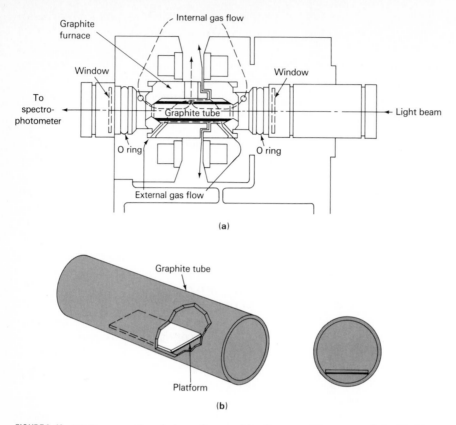

FIGURE 9-12 (**a**) Cross-sectional view of a graphite furnace. (Courtesy of the Perkin-Elmer Corp., Norwalk, CT.) (**b**) The L'vov platform and its position in the graphite furnace. (Reprinted with permission from: W. Slavin, *Anal. Chem.*, **1982**, *54*, 689A. Copyright 1982 American Chemical Society.)

L'vov platform, which is often used in graphite furnaces such as that shown in (**a**). The platform is also graphite and is located beneath the sample entrance port. The sample is evaporated and ashed on this platform in the usual way. When the tube temperature is raised rapidly, however, atomization is delayed since the sample is no longer directly on the furnace wall. As a consequence, atomization occurs in an environment in which the temperature is not changing so rapidly. More reproducible peaks are obtained as a consequence.

Figure 9–13 illustrates a second type of commercially available atomizer. Here, the sample is contained in a graphite cup (A) supported between two electrodes (B) of the same material. The cup is surrounded by a sheath of inert gas. The metal mounting (C) is water cooled.

It has been found empirically that some of the sample matrix effects and poor reproducibility associated with graphite furnace atomization can be alleviated by reducing the natural porosity of the graphite tube. During atomization, part of the analyte and matrix apparently diffuse into the tube, which slows the atomization process thus giving smaller analyte signals. To overcome this effect, most graphite surfaces are

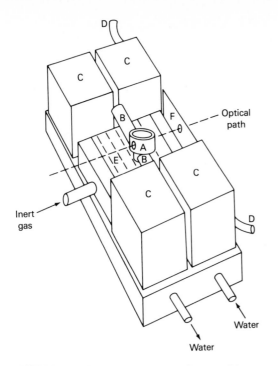

FIGURE 9–13 Schematic diagram of a graphite cup atomizer. (Courtesy of Varian Instrument Division, Palo Alto, CA.)

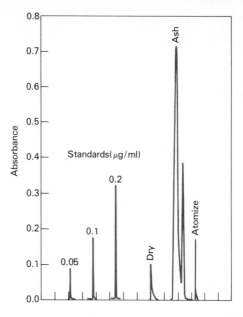

FIGURE 9-14 Typical output from a spectrophotometer equipped with an electrothermal atomizer. The times for drying and ashing are 20 and 60 sec respectively. (Courtesy of Varian Instrument Division, Palo Alto, CA.)

coated with a thin layer of pyrolitic carbon, which serves to seal the pores of the graphite tube. Pyrolitic graphite is a type of graphite that has been deposited layer by layer from a highly homogeneous environment. It is formed by passing a mixture of an inert gas and a hydrocarbon such as methane through the tube while it is held at an elevated temperature.

Output Signal.   At a wavelength at which absorbance (or fluorescence) occurs, the detector output rises to a maximum after a few seconds of ignition followed by a rapid decay back to zero as the atomization products escape into the surroundings. The change is rapid enough (often <1 sec) to require a high-speed recorder. Quantitative analyses are usually based on peak height although peak area has also been used.

Figure 9–14 shows typical output signals from an atomic absorption spectrophotometer equipped with a carbon rod atomizer. The series of peaks on the right show the absorbance at the wavelength of a lead peak as a function of time when a 2-μL sample of a canned orange juice was atomized. During both drying and ashing, peaks are produced probably due to particulate ignition products. The three peaks on the left are for lead standards employed for calibration. The sample peak on the far right indicates a lead concentration of 0.1 μg/mL of juice.

Electrothermal atomizers offer the advantage of unusually high sensitivity for small volumes of sample. Typically, sample volumes between 0.5 and 10 μL are employed; under these circumstances, absolute detection limits typically lie in the range of 10^{-10} to 10^{-13} g of analyte.

The relative precision of nonflame methods is generally in the range of 5 to 10%

compared with the 1 to 2% that can be expected for flame atomization. Furthermore, matrix interference problems, which are discussed in Section 9D–3, tend to be more severe with electrothermal than with flame atomizers.

9D   Atomic Absorption Spectroscopy[6]

Until recently, flame atomic absorption spectroscopy was the most widely used of all atomic spectral methods because of its simplicity, effectiveness, and relatively low cost. This position of preeminence is now being challenged, however, by inductively coupled plasma spectroscopy, an emission method described in Section 10B.

9D–1   RADIATION SOURCES FOR ATOMIC ABSORPTION METHODS

Analytical methods based on atomic absorption are potentially highly specific because atomic absorption lines are remarkably narrow (0.002 to 0.005 nm) and because electronic transition energies are unique for each element. On the other hand, the limited line widths create a problem not ordinarily encountered in molecular absorption spectroscopy. Recall that for Beer's law to be obeyed, it is necessary that the bandwidth of the source be narrow with respect to the width of an absorption peak (see Section 6B–3). Even good quality monochromators, however, have effective bandwidths that are significantly greater than the width of

atomic absorption lines. Consequently, nonlinear calibration curves are inevitable when atomic absorbance measurements are made with equipment designed for molecular absorption studies. Furthermore, the slopes of calibration curves obtained with such equipment are small because only a small fraction of the radiation from the monochromator slit is absorbed by the sample; poor sensitivities are the consequence.

The problem created by the limited width of atomic absorption peaks has been solved by the use of line sources having bandwidths even narrower than absorption peaks. For example, if the absorbance of the 589.6 nm peak of sodium is to serve as the basis for determining that element, a sodium emission peak at this same wavelength is isolated and used. In this instance, the line is produced by means of a sodium vapor lamp in which sodium atoms are excited by an electrical discharge. The other sodium lines emitted from the source are then removed with filters or with a relatively inexpensive monochromator. Operating conditions for the source are chosen such that Doppler broadening of the emitted lines is less than the broadening of the absorption peak that occurs in the flame. That is, the source temperature is kept below that of the flame. Figure 9–15 illustrates the principle of this procedure. Plot **a** is the *emission* spectrum of a typical atomic lamp source, which consists of four narrow lines. With a suitable filter or monochromator, all but one of these lines are removed. Figure 9–15**b** shows the absorption spectrum for the analyte between wavelengths λ_1 and λ_2. Note that the bandwidth is significantly greater than that of the emission peak. As shown in Figure 9–15**c**, passage of the line from the source through the flame reduces its intensity from P_0 to P; the absorbance is then given by log (P_0/P), which is linearly related to the concentration of the analyte in the sample.

A disadvantage of the procedure just described is that a separate lamp source is

---

[6] Reference books on atomic absorption spectroscopy include: W. Slavin, *Atomic Absorption Spectroscopy*, 2d ed., Interscience: New York, 1978; K. C. Thompson and R. J. Reynolds, *Atomic Absorption, Fluorescence, and Flame Emission Spectroscopy*, 2d ed., Wiley: New York, 1978; *Atomic Absorption Spectrometry*, J. E. Cantle, Ed., Elsevier: New York, 1982. For a short review describing the present and future state of atomic spectroscopy, see: W. Slavin, *Anal. Chem.*, **1982**, *54*, 685A.

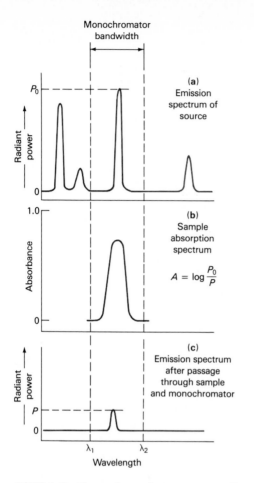

FIGURE 9–15 Absorption of a resonance line by atoms.

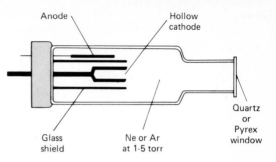

FIGURE 9–16 Schematic cross section of a hollow cathode lamp.

needed for each element (or sometimes group of elements).

Hollow Cathode Lamps. The most common source for atomic absorption measurements is the *hollow cathode lamp,* which consists of a tungsten anode and a cylindrical cathode sealed in a glass tube that is filled with neon or argon at a pressure of 1 to 5 torr (see Figure 9–16). The cathode is constructed of the metal whose spectrum is desired or serves to support a layer of that metal.

Ionization of the inert gas occurs when a potential on the order of 300 V is applied across the electrodes, and a current of about 5 to 10 mA is generated as ions and electrons migrate to the electrodes. If the potential is sufficiently large, the gaseous cations acquire enough kinetic energy to dislodge some of the metal atoms from the cathode surface and produce an atomic cloud; this process is called *sputtering*. A portion of the sputtered metal atoms are in excited states and thus emit their characteristic radiation as they return to the ground state. Eventually, the metal atoms diffuse back to the cathode surface or to the glass walls of the tube and are redeposited.

The cylindrical configuration of the cathode tends to concentrate the radiation in a limited region of the tube; this design also enhances the probability that redeposition will occur at the cathode rather than on the glass walls.

The efficiency of the hollow cathode lamp depends upon its geometry and the operating potential. High potentials, and thus high currents, lead to greater intensities. This advantage is offset somewhat by an increase in Doppler broadening of the emission lines. Furthermore, the greater currents result in an increase in the number of unexcited atoms in the cloud; the unexcited atoms, in turn, are capable of absorbing the radiation emitted by the excited ones. This *self-absorption* leads to lowered intensities, particularly at the center of the emission band.

A variety of hollow cathode tubes are

available commercially. The cathodes of some consist of a mixture of several metals; such lamps permit the analysis of more than a single element.

Electrodeless Discharge Lamps. Electrodeless discharge lamps are useful sources of atomic line spectra and provide radiant intensities that are usually one to two orders of magnitude greater than their hollow cathode counterparts. A typical lamp is constructed from a sealed quartz tube containing a few torr of an inert gas such as argon and a small quantity of the metal (or its salt) whose spectrum is of interest. The lamp contains no electrode but instead is energized by an intense field of radio-frequency or microwave radiation. Ionization of the argon occurs to give ions that are accelerated by the high-frequency component of the field until they gain sufficient energy to excite the atoms of the metal whose spectrum is sought.

Electrodeless discharge lamps are available commercially for several elements. Their performance does not appear to be as reliable as that of the hollow cathode lamp.

Source Modulation. In the typical atomic absorption instrument, it is necessary to eliminate interferences caused by emission of radiation by the flame. Much of this emitted radiation is, of course, removed by the monochromator. Nevertheless, emitted radiation corresponding in wavelength to the monochromator setting is inevitably present in the flame due to excitation and emission by analyte atoms. In order to eliminate the effects of flame emission, it is necessary to modulate the output of the source so that its intensity fluctuates at a constant frequency. The detector then receives two types of signal, an alternating one from the source and a continuous one from the flame. These signals are converted to the corresponding types of electrical response. A simple high-pass, *RC* filter (Section A2–4, Appendix 2) can then be employed to remove the unmodulated dc signal and pass the ac signal for amplification.

A simple and entirely satisfactory way of modulating the emission from the source is to interpose a circular metal disk or chopper in the beam between the source and the flame. Alternate quadrants of this disk are removed to permit passage of light. Rotation of the disk at constant speed provides a beam that is chopped to the desired frequency. As an alternative, the power supply for the source can be designed for intermittent or ac operation.

9D–2 INSTRUMENTS FOR ATOMIC ABSORPTION SPECTROSCOPY

Instruments for atomic absorption work are offered by numerous manufacturers; both single- and double-beam designs are available. The range of sophistication and cost (upward from a few thousand dollars) is substantial.

In general, the instrument must be capable of providing a sufficiently narrow bandwidth to isolate the line chosen for the measurement from other lines that may interfere with or diminish the sensitivity of the analysis. A glass filter suffices for some of the alkali metals, which have only a few widely-spaced resonance lines in the visible region. An instrument equipped with readily interchangeable interference filters is available commercially. A separate filter (and light source) is used for each element. Satisfactory results for the analysis of 22 metals are claimed. Most instruments, however, incorporate a good-quality ultraviolet-visible monochromator.

Generally, the detector and readout devices for atomic spectroscopy are similar to those already described for molecular spectroscopy in the ultraviolet-visible region. Photomultiplier detectors are found in most instruments. As pointed out earlier, electronic systems that are capable of discriminating between the modulated signal from the source and the continuous signal from the flame are required. Both null-point and direct-reading meters, calibrated in terms of absorbance or transmittance, are used. Many instruments are equipped with recorders to provide spectral plots. Some instruments give a digital readout of absorbance or, alternatively, analyte concentration.

Single-Beam Spectrophotometers. A typical single-beam instrument, such as that shown in Figure 9–17**a**, consists of several hollow cathode sources, a chopper or a pulsed power supply, an atomizer, and a simple grating spectrophotometer with a photomultiplier transducer. It is used in the same way as a single-beam instrument for molecular absorption work. Thus, the dark current is nulled with a shutter in front of the trans-

ducer. The 100% *T* adjustment is then made while a blank is aspirated into the flame (or ignited in a nonflame atomizer). Finally, the transmittance is obtained with the sample replacing the blank.

Double-Beam Spectrophotometers. Figure 9–17**b** is a schematic diagram of a typical double-beam (in time) instrument. The beam from the hollow cathode source is split by a mirrored chopper, one half passing through the flame and the other half around it. The two beams are then recombined by a half-silvered mirror and passed into a Czerney-Turner grating monochromator; a photomultiplier tube serves as the transducer. The output from the latter is fed to a lock-in

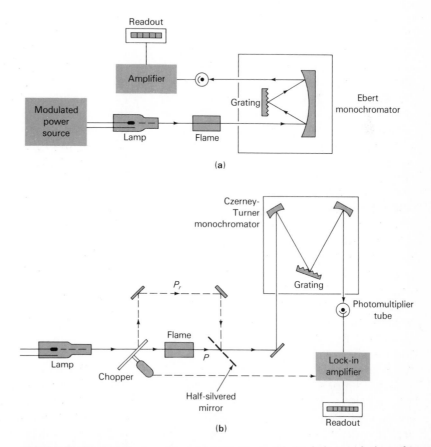

FIGURE 9–17 Typical flame spectrophotometers: (**a**) single-beam design; (**b**) double-beam design.

amplifier which is synchronized with the chopper drive. The ratio between the reference and sample signal is then amplified and fed to the readout which may be a meter, digital device, or recorder. Alternatively, the amplified signal from the reference beam may be attenuated by means of a potentiometer to match the sample signal; the transmittance or absorbance is then read from the position of the slide wire contact.

It should be noted that the reference beam in atomic double-beam instruments does not pass through the flame and thus does not correct for loss of radiant power due to absorption or scattering by the flame itself. Methods of correcting for these losses are discussed in the next section.

9D–3 SPECTRAL INTERFERENCES

Interferences of two types are encountered in atomic absorption methods using both flame and electrothermal atomization. *Spectral interferences* arise when the absorption or emission of an interfering species either overlaps or lies so close to the analyte absorption or emission that resolution by the monochromator becomes impossible. *Chemical interferences* result from various chemical processes occurring during atomization that alter the absorption characteristics of the analyte. A brief discussion of spectral interferences follows; sources of chemical interference are considered in the next section.

Because the emission lines of hollow cathode sources are so very narrow, interference due to overlapping lines is rare. For such an interference to occur, the separation between the two lines would have to be less than perhaps 0.1 Å. For example, a vanadium line at 3082.11 Å interferes in an analysis based upon the aluminum absorption line at 3082.15 Å. The interference is readily avoided, however, by employing the aluminum line at 3092.7 Å instead.

Spectral interferences also result from the presence of combustion products that ex-

hibit broad-band absorption or particulate products that scatter radiation. Both diminish the power of the transmitted beam and lead to positive analytical errors. Where the source of these products is the fuel and oxidant mixture alone, corrections are readily obtained from absorbance measurements while a blank is aspirated into the flame. Note that this correction must be employed with a double-beam as well as a single-beam instrument because the reference beam of the former does not pass through the flame (see Figure 9–17**b**).

A much more troublesome problem is encountered when the source of absorption or scattering originates in the sample matrix; here, the power of the transmitted beam, P, is reduced by the matrix components but the incident beam power, P_0, is not; a positive error in absorbance and thus concentration results. An example of a potential matrix interference due to absorption occurs in the determination of barium in alkaline-earth mixtures. As shown by the dashed line in Figure 9–5, the wavelength of the barium line used for atomic absorption analysis appears in the center of a broad absorption band for CaOH; clearly, interference of calcium in a barium analysis is to be expected. In this particular situation, the effect is readily eliminated by substituting nitrous oxide for air as the oxidant; the higher temperature decomposes the CaOH and eliminates the absorption band.

Spectral interference due to scattering by products of atomization is most often encountered when concentrated solutions containing elements such as Ti, Zr, and W—which form refractory oxides—are aspirated into the flame. Metal oxide particles with diameters greater than the wavelength of light appear to be formed; scattering of the incident beam results.

Fortunately, with flame atomization, spectral interferences by matrix products are not widely encountered and often can be avoided by variations in the analytical parameters, such as temperature and fuel-to-oxidant ratio. Alternatively, if the source of interference is known, an excess of the interfering substance can be added to both sample and standards; provided the excess is large with respect to the concentration from the sample matrix, the contribution of the latter will become insignificant. The added substance is sometimes called a *radiation buffer*.

The matrix interference problem is greatly exacerbated with electrothermal atomization and is one of the major causes of the poorer accuracy encountered with nonflame methods. Several sources of interference have been recognized. Among these is broad-band molecular absorption resulting from incomplete breakdown of the matrix in the brief time in which atomization takes place. In samples containing halogens, stable metal halides such as $CaCl_2$, $PbCl_2$, NaCl, and KCl are found to exist in atomized samples. These species exhibit broad absorption bands in the ultraviolet. Scattering by incompletely decomposed organic particles also occurs commonly. As a consequence, the need for background correction techniques is usually encountered with electrothermal atomization; it is occasionally desirable with flame atomization as well. A description of some of the most commonly used techniques follows.[7]

The Two-Line Correction Method. The two-line correction procedure requires the presence of a reference line from the source; this line should lie as close as possible to the analyte line but *must not be absorbed by the analyte*. If these conditions are met, it is assumed that any decrease in power of the reference line from that observed during calibration arises from absorption or scattering by the matrix products of the sample; this decrease is then used to correct the absorbance of the analyte line.

The reference line may be from an impurity in the lamp cathode, a neon or argon line from the gas contained in the lamp, or a nonresonant emission line of the element being determined.

Unfortunately, a suitable reference line is often not available.

The Continuous-Source Correction Method. Figure 9–18 illustrates a second method for background corrections, which is widely used. Here, a deuterium lamp provides a source of continuous radiation throughout the ultraviolet region. The configuration of the chopper is such that radiation from the continuous source and the hollow cathode lamp are passed alternately through the graphite-tube atomizer. The absorbance of the deuterium radiation is then subtracted from that of the analyte beam. The slit width is kept sufficiently wide so that the fraction of the continuous source that is absorbed by the atoms of the sample is negligible. Therefore, the attenuation of its power during passage through the atomized sample reflects only the broad-band absorption or scattering by the sample matrix compo-

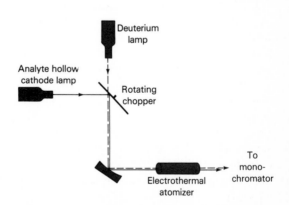

FIGURE 9–18 Schematic diagram of a continuum source background correction system. Note that the chopper can be dispensed with by alternately pulsing each lamp.

---

[7] For a critical discussion of the various methods for background correction, see; A. T. Zander, *Amer. Lab.,* **1976,** *8*(11), 11.

nents. A background correction is thus achieved.

Unfortunately, although most instrument manufacturers offer continuous source background correction systems, the performance of these devices is often imperfect leading to undercorrections in some systems and overcorrections in others. Several sources of uncertainty exist. One of these is the inevitable degradation of signal-to-noise that accompanies the addition of a lamp and chopper. Furthermore, the hot gaseous media are usually highly inhomogeneous both as to chemical composition and particulate distribution; thus if the two lamps are not in perfect alignment, an erroneous correction will result that can cause either positive or negative errors. Finally, the output of a deuterium lamp in the visible region is low enough to preclude the use of this correction procedure for wavelengths greater than about 350 nm.

Background Correction Based on the Zeeman Effect.[8] When an atomic vapor is exposed to a strong magnetic field (1 to 10 kG) a splitting of electronic energy levels of the atoms takes place, which leads to formation of several absorption lines for each electronic transition. These lines differ from one another by about 0.01 nm with the sum of the absorbances for the lines being exactly equal to that of the original line from which they were formed. This phenomenon, which is termed the *Zeeman effect,* is general for all atomic spectra. Several splitting patterns exist depending upon the type of electronic transition that is involved in the absorption process. The simplest splitting pattern, which is observed with singlet (p. 253) transitions, leads to a central or π *line* and two equally spaced satellite σ *lines*. The central line, which is at the original wavelength, has an absorbance that is twice that of each σ line. For more complex transitions, further splitting of the π and σ lines occurs.

Application of the Zeeman effect to atomic absorption instruments is based upon the differing response of the two types of absorption peaks to polarized radiation. The π peak absorbs only that radiation that is plane-polarized in a direction parallel to the external magnetic field; the σ peaks, in contrast, absorb only radiation polarized at 90 deg to the field.

Figure 9–19 shows details of an electrothermal atomic absorption instrument, which utilizes the Zeeman effect for background correction. Unpolarized radiation from an ordinary hollow cathode source (A) is passed through a rotating polarizer (B), which separates the beam into two components, plane-polarized at 90 deg to one another (C). These beams pass into a tube-type graphite furnace similar to the one shown in Figure 9–12. A permanent 11 kG magnet surrounds the furnace and splits the energy levels in such a way as to produce the three absorption peaks shown in (D). Note that the central peak absorbs only that radiation that is plane-polarized with the field. During that part of the cycle when the source radiation is polarized similarly, absorption of radiation by the analyte takes place. During the other half cycle, no analyte absorption can occur. Broad-band molecular absorption and scattering by the matrix products occur during both half cycles leading to the cyclical absorbance pattern shown in (F). The signal processor is programmed to subtract the absorbance during the perpendicular half-cycle from that for the parallel half-cycle, thus giving a background corrected value.

A second type of Zeeman effect instrument has been designed in which a magnet surrounds the hollow cathode source. Here, it is the *emission* spectra of the source that is split rather than the absorption spectrum of the sample. This instrument configuration provides an analogous correction. To date,

[8] For a detailed discussion of the application of the Zeeman effect to atomic absorption, see: S. D. Brown, *Anal. Chem.,* **1977,** *49* (14), 1269A; F. J. Fernandez, S. A. Myers, and W. Slavin, *Anal. Chem.,* **1980,** *52,* 741.

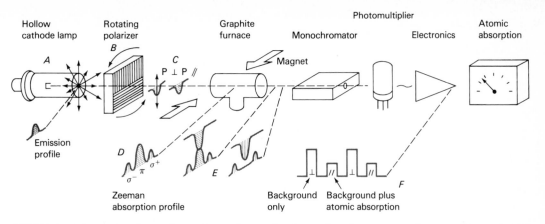

FIGURE 9-19 Schematic of an electrothermal atomic absorption instrument that provides a background correction based upon the Zeeman effect. (Courtesy of Hitachi Scientific Instruments, Mountain View, CA.)

most instruments are of the type illustrated in Figure 9–19.

Zeeman effect instruments appear to provide a more accurate correction for background than the methods described earlier. They are particularly useful for electrothermal atomizers and permit the direct determination of elements in samples such as urine and blood. The decomposition of organic material in these samples leads to large background corrections (background $A \geq 1$) and consequent susceptibility to significant error.

Background Correction Based on Source Self-Reversal. Recently, a remarkably simple means of background correction has been developed and marketed. It appears to offer all of the advantages of a Zeeman effect instrument at a cost that is less than even deuterium lamp instruments.[9] This method of correction is based upon the self-reversal or self-absorption behavior of radiation emitted from hollow cathode lamps when they are operated at high currents. As was mentioned earlier, high currents produce large concentrations of nonexcited atoms, which are capable of absorbing the radiation produced from the excited species. An additional effect of the

high currents is to significantly broaden the emission band of the excited species. The net effect is to produce a band that has a minimum in its center, which corresponds exactly in wavelength to that of the absorption peak (for an analogous example of self-reversal in flames, see Figure 9–24A).

In order to obtain corrected absorbances, the lamp is programmed to run at low current for several milliseconds and then at high current for about 300 μsec. The total absorbance is obtained during the low current operation and the absorbance due to background is provided by measurements during the second part of the cycle when radiation at the absorbance peak is at a minimum. An electronic package then subtracts the background absorbance from the total to give a corrected value. Recovery of the source when the current is reduced takes place in milliseconds. The measurement cycle can be repeated often enough to give satisfactory signal-to-noise ratios.

9D-4 CHEMICAL INTERFERENCES

Chemical interferences are more common than spectral ones. Their effects can frequently be minimized by a suitable choice of operating conditions.

---

[9] See: *Science*, **1983**, *220*, 183.

Both theoretical and experimental evidence suggest that many of the processes occurring in the mantle of a flame are in approximate equilibrium. As a consequence, it becomes possible to regard the burned gases as a solvent medium to which thermodynamic calculations can be applied. The equilibria of principal interest include formation of compounds of low volatility, dissociation reactions, and ionization.

Formation of Compounds of Low Volatility. Perhaps the most common type of interference is by anions which form compounds of low volatility with the analyte and thus reduce the rate at which it is atomized. Low results are the consequence. An example is the decrease in calcium absorbance that is observed with increasing concentrations of sulfate or phosphate. For example, for a fixed calcium concentration, the absorbance is found to fall off nearly linearly with increasing sulfate or phosphate concentrations until the anion to calcium ratio is about 0.5; the absorbance then levels off at about 30 to 50% of its original value and becomes independent of anion concentration.

Examples of cation interference have also been recognized. Thus, aluminum is found to cause low results in the determination of magnesium, apparently as a result of the formation of a heat-stable aluminum/magnesium compound (perhaps an oxide).

Interferences due to formation of species of low volatility can often be eliminated or moderated by use of higher temperatures. Alternatively, *releasing agents,* which are cations that react preferentially with the interference and prevent its interaction with the analyte, can be employed. For example, addition of an excess of strontium or lanthanum ion minimizes the interference of phosphate in the determination of calcium. The same two species have also been employed as releasing agents for the determination of magnesium in the presence of aluminum. In both instances, the strontium or lanthanum replaces the analyte in the compound formed with the interfering species.

Protective agents prevent interference by forming stable but volatile species with the analyte. Three common reagents for this purpose are EDTA, 8-hydroxyquinoline, and APDC (the ammonium salt of 1-pyrrolidinecarbodithioic acid). The presence of EDTA has been shown to eliminate the interference of aluminum, silicon, phosphate, and sulfate in the determination of calcium. Similarly, 8-hydroxyquinoline suppresses the interference of aluminum in the determination of calcium and magnesium.

Dissociation Equilibria. In the hot, gaseous environment of a flame or a furnace, numerous dissociation and association reactions lead to conversion of the metallic constituents to the elemental state. It seems probable that at least some of these reactions are reversible and can be treated by the laws of thermodynamics. Thus, in theory, it should be possible to formulate equilibria such as

$$MO \rightleftharpoons M + O$$

$$M(OH)_2 \rightleftharpoons M + 2OH$$

or, more generally,

$$MA \rightleftharpoons M + A$$

In practice, not enough is known about the nature of the chemical reactions in a flame to permit a quantitative treatment such as that for an aqueous solution. Instead, reliance must be placed on empirical observations.

Dissociation reactions involving metal oxides and hydroxides clearly play an important part in determining the nature of the emission or absorption spectra for an element. For example, the alkaline-earth oxides are relatively stable, with dissociation energies in excess of 5 eV. Molecular bands arising from the presence of metal oxides or hydroxides in the flame thus constitute a prominent feature of their spectra (see Figure 9–5). Except at very high temperatures, these bands are more intense than the lines for the atoms or ions. In contrast, the oxides and hydroxides of the alkali metals are much more readily dissociated so that line intensities for these elements are high, even at relatively low temperatures.

It seems probable that dissociation equilibria involving anions other than oxygen may also influence flame emission and absorption. For example, the line intensity for sodium is markedly decreased by the presence of HCl. A likely explanation is the mass-action effect on the equilibrium

$$NaCl \rightleftharpoons Na + Cl$$

Chlorine atoms formed from the added HCl decrease the atomic sodium concentration and thereby lower the line intensity.

Another example of this type of interference involves the enhancement of the absorption by vanadium when aluminum or titanium is present. The interference is significantly more pronounced in fuel-rich flames than in lean. These effects are readily explained by assuming that the three metals interact with such species as O and OH, which are always present in flames. If the oxygen-bearing species are given the general formula Ox, a series of equilibrium reactions can be postulated. Thus,

$$VOx \rightleftharpoons V + Ox$$
$$AlOx \rightleftharpoons Al + Ox$$
$$TiOx \rightleftharpoons Ti + Ox$$

In fuel-rich combustion mixtures, the concentration of Ox is sufficiently small that its concentration is lowered significantly when aluminum or titanium is present in the sample. This decrease causes the first equilibrium to shift to the right with an accompanying increase in metal concentration as well as absorbance. In lean mixtures, on the other hand, the concentration of Ox is apparently high relative to the total concentration of metal atoms. Thus, addition of aluminum or titanium scarcely changes the concentration of Ox. Therefore, the position of the first equilibrium is not disturbed significantly.

Ionization in Flames. Ionization of atoms and molecules is small in combustion mixtures that involve air as the oxidant, and generally can be neglected. In higher temperatures of flames where oxygen or nitrous oxide serves as the oxidant, however, ionization becomes important, and a significant concentration of free electrons exists as a consequence of the equilibrium

$$M \rightleftharpoons M^+ + e^- \tag{9–2}$$

where M represents a neutral atom or molecule and M^+ is its ion. We will focus upon equilibria in which M is a metal atom.

The equilibrium constant K for this reaction may take the form

$$K = \frac{[M^+][e^-]}{[M]} = \left(\frac{x^2}{1-x}\right) p \tag{9–3}$$

where the bracketed terms are activities, x is the fraction of M that is ionized, and p is the partial pressure of the metal in the gaseous solvent before ionization.

Table 9–3 shows the calculated degree of ionization for several common metals under conditions that approximate those used in flame emission spectroscopy. The temperatures correspond roughly to conditions that exist in air/fuel and oxygen/acetylene flames, respectively.

It is important to appreciate that treatment of the ionization process as an equilibrium—with free electrons as one of the products—immediately implies that the degree of ionization of a metal will be strongly influenced by the presence of other ionizable metals in the flame. Thus, if the medium contains not only species M, but species B as well, and if B ionizes according to the equation

$$B \rightleftharpoons B^+ + e^-$$

then the degree of ionization of M will be decreased by the mass-action effect of the electrons formed from B. Determination of the degree of ionization under these conditions requires a calculation involving the dissociation constant for B and the mass-balance expression

$$[e^-] = [B^+] + [M^+]$$

The presence of atom-ion equilibria in flames has a number of important consequences in flame spectroscopy. For example, the intensity of atomic emission or absorption lines for the alkali metals, particularly potassium, rubidium, and cesium, is affected by temperature in a complex way. Increased temperatures cause an increase in the population of excited atoms, according to the Boltzmann relationship (Section 9B–1); counteracting this effect,

however, is a decrease in concentration of atoms as a result of ionization. Thus, under some circumstances a decrease in emission or absorption may be observed in hotter flames. It is for this reason that lower excitation temperatures are usually specified for the analysis of alkali metals.

The effects of shifts in ionization equilibria can frequently be eliminated by addition of an *ionization suppressor,* which provides a relatively high concentration of electrons to the flame; suppression of ionization of the analyte results. The effect of a suppressor is demonstrated by the calibration curves for strontium shown in Figure 9–20. Note the marked steepening of these curves as strontium ionization is repressed by the increasing concentration of potassium ions and electrons. Note also the enhanced sensitivity that results from the use of nitrous oxide instead of air as the oxidant; the higher temperature achieved with nitrous oxide undoubtedly enhances the rate of decomposition and volatilization of the strontium compounds in the plasma.

Table 9–3
Degree of Ionization of Metals at Flame Temperatures[a]

| Element | Ionization Potential, eV | Fraction Ionized at the Indicated Pressure and Temperature | | | |
| | | $p = 10^{-4}$ atm | | $p = 10^{-6}$ atm | |
		2000°K	3500°K	2000°K	3500°K
Cs	3.893	0.01	0.86	0.11	>0.99
Rb	4.176	0.004	0.74	0.04	>0.99
K	4.339	0.003	0.66	0.03	0.99
Na	5.138	0.0003	0.26	0.003	0.90
Li	5.390	0.0001	0.18	0.001	0.82
Ba	5.210	0.0006	0.41	0.006	0.95
Sr	5.692	0.0001	0.21	0.001	0.87
Ca	6.111	3×10^{-5}	0.11	0.0003	0.67
Mg	7.644	4×10^{-7}	0.01	4×10^{-6}	0.09

[a] Data from B. L. Vallee and R. E. Thiers, in *Treatise on Analytical Chemistry,* I. M. Kolthoff and P. J. Elving, Eds., Part I, vol. 6, p. 3500, Interscience: New York, 1965. Reprinted with permission of John Wiley & Sons, Inc.

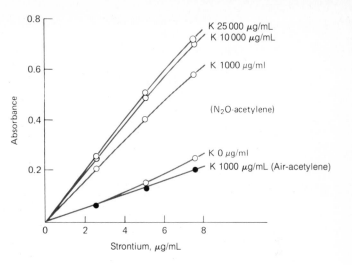

FIGURE 9-20 Effect of potassium concentration on the calibration curve for strontium. (Reprinted with permission from: J. A. Bowman and J. B. Willis, *Anal. Chem.*, **1967,** *39*, 1210. Copyright 1967 American Chemical Society.)

9D–5 ANALYTICAL TECHNIQUES FOR ATOMIC ABSORPTION SPECTROSCOPY

This section deals with some of the practical details that must be considered in carrying out an analysis based upon flame or electrothermal atomic absorption.

Preparation of Sample. A disadvantage of flame spectroscopic methods is the requirement that the sample be introduced into the excitation source in the form of a solution, most commonly an aqueous one. Unfortunately, many materials of interest, such as soils, animal tissue, plants, petroleum products, and minerals are not directly soluble in common solvents, and extensive preliminary treatment is often required to obtain a solution of the analyte in a form ready for atomization. Indeed, the decomposition and solution steps are often more time-consuming and introduce more errors than the spectroscopic measurement itself.

Decomposition of refractory materials such as those just cited usually require rigorous treatment of the sample at high temperatures with a concomitant potential for loss of the analyte by volatilization or as particulates in a smoke. Furthermore, the reagents used in decomposing a sample often introduce the kinds of chemical and spectral interferences that were discussed earlier. Additionally, the analyte element may be present in these reagents as an impurity. In fact, unless considerable care is taken, it is not uncommon in trace analyses to find that reagents are a larger source of the element of interest than the samples—a situation that can lead to serious error even with blank corrections.

Some of the common methods used for decomposing and dissolving samples for atomic absorption methods include treatment with hot mineral acids; oxidation with liquid reagents, such as sulfuric, nitric, or perchloric acids; combustion in an oxygen bomb or other closed container (to avoid loss of analyte); ashing at a high temperature; and high-temperature fusion with reagents such as boric oxide, sodium carbonate, sodium peroxide, or potassium pyrosulfate. Details of these procedures can be found in a monograph by Bock.[10]

One of the advantages of electrothermal atomization is that some materials can be atomized directly thus avoiding the solution step. For example, liquid samples such as blood, petroleum products, and organic sol-

---

[10] R. Bock, *A Handbook of Decomposition Methods in Analytical Chemistry,* Wiley: New York, 1979.

vents can be pipetted directly into the furnace for ashing and atomization. Solid samples, such as plant leaves, animal tissue, or some inorganic substances can be weighed directly into cup-type atomizers or into tantalum boats for introduction into tube-type furnaces. Calibration is, however, usually difficult and requires standards that approximate the sample in composition.

Calibration Standards.   Ideally, the standards for an atomic absorption analysis should not only contain the analyte element in exactly known concentrations but also closely approximate the sample as to matrix elements. Seldom can this ideal be realized and some of the procedures described earlier for minimizing matrix effects and chemical interferences must be resorted to.

Appendix 5 lists starting materials recommended for the preparation of standard solutions of many of the common elements.

Role of Organic Solvents in Flame Spectroscopy.   Early in the development of atomic absorption spectroscopy it was recognized that enhanced absorption peaks could be obtained from solutions containing low molecular weight alcohols, esters, and ketones. The effect was particularly pronounced with turbulent flow burners, with enhancements ranging from a factor of 2 to as great as 100. With premix burners the increases in sensitivity were considerably smaller and often not sufficient to offset the loss in sensitivity due to dilution from the added solvent.

The effect of organic solvents is largely attributable to an increased nebulizer efficiency; the lower surface tension of such solutions results in finer drop sizes and a consequent increase in the amount of sample that reaches the flame. In addition, more rapid solvent evaporation may also contribute to the effect. Leaner fuel-oxidant ratios must be employed with organic solvents in order to offset the presence of the added organic material. Unfortunately, however, the leaner mixture results in lower flame

temperatures and a consequent increase in the possibility for chemical interferences.

A most important analytical application of organic solvents to flame spectroscopy is the use of immiscible solvents such as methyl isobutyl ketone to extract chelates of metallic ions. The resulting extract is then nebulized directly into the flame. Here, the sensitivity is increased not only by the enhancement of absorption peaks due to the solvent but also by the fact that for many systems, only small volumes of the organic liquid are required to remove metal ions quantitatively from relatively large volumes of aqueous solution. This procedure has the added advantage that at least part of the matrix components are likely to remain in the aqueous solvent; a reduction in interference often results. Common chelating agents include ammonium pyrolidine-dithiocarbamate, diphenylthiocarbazone (dithizone), 8-hydroxyquinoline, and acetylacetone.

Hydride Generation Techniques.   Hydride generation techniques provide a method for introducing arsenic, antimony, tin, selenium, bismuth, and lead into an atomizer as a gas.[11] Such a procedure enhances the sensitivity by a factor of 10 to 100. Since several of these elements are highly toxic, their determination at low concentration levels is of considerable importance.

Rapid generation of volatile hydrides can generally be brought about by addition of an acidified aqueous solution of the sample to a small volume of 1% aqueous solution of sodium borohydride contained in a glass cell. After mixing for a brief period, the hydride is swept into the atomization chamber by an inert gas. The chamber is usually a silica tube heated to several hundred degrees in a tube furnace or in a flame. Radia-

---

[11] For a detailed discussion of these methods, see: W. B. Robbins and J. A. Caruso, *Anal. Chem.,* **1979,** *51,* 889A.

tion from the source passes through the tube to the monochromator and detector. The signal is a peak similar to that obtained with electrothermal atomization.

Calibration Curves. In theory, atomic absorption should follow Beer's law with absorbance being directly proportional to concentration. In fact, however, departures from linearity are often encountered, and it is foolhardy to perform an atomic absorption analysis without experimentally determining whether or not a linear relationship does, in fact, exist. Thus periodically, a calibration curve covering the range of concentrations found in the sample should be prepared. In addition, the number of uncontrolled variables in atomization and absorbance measurements are sufficiently large to warrant measurement of one standard solution each time an analysis is performed (even better is the use of two standards that bracket the analyte concentration). Any deviation of the standard from the original calibration curve can then be used to correct the analytical result.

Standard Addition Method. The standard addition method, which was described in Section 7F-2, is widely used in atomic absorption spectroscopy in order to partially or wholly counteract the chemical and spectral interferences introduced by the sample matrix.

9D-6 APPLICATIONS OF ATOMIC ABSORPTION SPECTROSCOPY

Atomic absorption spectroscopy is a sensitive means for the quantitative determination of more than 60 metals or metalloid elements. The resonance lines for the nonmetallic elements are generally located below 200 nm thus preventing their determination by convenient, nonvacuum spectrophotometers.

Detection Limits. Columns two and three of Table 9–4 provide information on detection limits for a number of common elements by flame and electrothermal atomic absorption. For comparison purposes, limits for some of the other atomic procedures are also included. Small differences among the quoted values are not significant. Thus, whereas an order of magnitude is probably meaningful, a factor of 2 or 3 certainly is not.

For many elements, detection limits for atomic absorption spectroscopy with flame atomization lie in the range of 1 to 20 ng/mL or 0.001 to 0.020 ppm; for electrothermal atomization, the corresponding figures are 0.002 to 0.01 ng/mL or 2×10^{-6} to 1×10^{-5} ppm. In a few cases, detection limits well outside these ranges are encountered.

Accuracy. Under usual conditions, the relative error associated with a flame absorption analysis is of the order of 1 to 2%. With special precautions, this figure can be lowered to a few tenths of one percent. Errors encountered with nonflame atomization usually exceed those for flame atomization by a factor of 5 to 10.

9E *Flame Emission Spectroscopy*

Atomic emission spectroscopy employing flames (also called flame emission spectroscopy or flame photometry) has found widespread application to elemental analysis.[12] Its most important uses have been in the determination of sodium, potassium, lithium, and calcium, particularly in biological fluids and tissues. For reasons of convenience, speed, and relative freedom from in-

---

[12] For a more complete discussion of the theory and applications of flame emission spectroscopy, see: *Flame Emission and Atomic Absorption Spectroscopy*, vol. 1: *Theory;* vol. 2: *Components and Techniques;* vol. 3: *Elements and Matrices*, J. A. Dean and T. C. Rains, Eds., Marcel Dekker: New York, 1969–75; A. Syty, in *Treatise on Analytical Chemistry*, 2d ed., P. J. Elving, E. J. Meehan, and I. M. Kolthoff, Eds., Part I, vol. 7, Chapter 7, Wiley: New York, 1981.

terferences, flame emission spectroscopy has become the method of choice for these otherwise difficult-to-determine elements. The method has also been applied, with varying degrees of success, to the determination of perhaps half the elements in the periodic table.

9E-1 INSTRUMENTATION

Instruments for flame emission work are similar in design to flame absorption instruments except that the flame now acts as the radiation source; the hollow cathode lamp and chopper are, therefore, unnecessary.

Many modern instruments are adaptable to either emission or absorption measurements.

Much of the early work in atomic emission analyses was accomplished with turbulent flow burners. Laminar flow burners, however, are becoming more and more widely used.

Spectrophotometers. For nonroutine analysis, a recording, ultraviolet/visible spectrophotometer with a resolution of perhaps 0.5 Å is desirable. The recording feature provides a simple means for making background corrections (see Figure 9–23).

Photometers. Simple filter photometers of-

Table 9–4					
Sensitivity (ng/mL)[a] for Selected Elements[b]					
Element	AAS,[c] Flame	AAS,[d] Electrothermal	AES,[c] Flame	AES,[c] ICP	AFS,[c] Flame
Al	30	0.005	5	2	5
As	100	0.02	0.0005	40	100
Ca	1	0.02	0.1	0.02	0.001
Cd	1	0.0001	800	2	0.01
Cr	3	0.01	4	0.3	4
Cu	2	0.002	10	0.1	1
Fe	5	0.005	30	0.3	8
Hg	500	0.1	0.0004	1	20
Mg	0.1	0.00002	5	0.05	1
Mn	2	0.0002	5	0.06	2
Mo	30	0.005	100	0.2	60
Na	2	0.0002	0.1	0.2	—
Ni	5	0.02	20	0.4	3
Pb	10	0.002	100	2	10
Sn	20	0.1	300	30	50
V	20	0.1	10	0.2	70
Zn	2	0.00005	0.0005	2	0.02

[a] Nanogram/milliliter = 10^{-3} μg/mL = 10^{-3} ppm.

[b] AAS = atomic absorption spectroscopy; AES = atomic emission spectroscopy; AFS = atomic fluorescence spectroscopy; ICP = inductively coupled plasma.

[c] Reprinted with permission from V. A. Fassel and R. N. Kniseley, *Anal. Chem.*, **1974**, *46*, 1111A. Copyright 1974 American Chemical Society.

[d] From C. W. Fuller, *Electrothermal Atomization for Atomic Absorption Spectroscopy*, pp. 65–83, The Chemical Society: London, 1977. With permission. The Royal Society of Chemistry, Burlington House: London.

ten suffice for routine determinations of the alkali and alkaline-earth metals. A low-temperature flame is employed to eliminate excitation of most other metals. As a consequence, the spectra are simple, and interference filters can be used to isolate the desired emission line.

Several instrument manufacturers supply flame photometers designed specifically for the analysis of sodium, potassium, and lithium in blood serum and other biological samples. In these instruments, the radiation from the flame is split into three beams of approximately equal power. Each then passes into a separate photometric system consisting of an interference filter (which transmits an emission line of one of the elements while absorbing those of the other two), a phototube, and an amplifier. The outputs can then be measured separately if desired. Ordinarily, however, lithium serves as an *internal standard* for the analysis. For this purpose, a fixed amount of lithium is introduced into each standard and sample. The ratios of outputs of the sodium and lithium transducer and the potassium and lithium transducer then serve as analytical parameters. This system provides improved accuracy because the intensities of the three lines are affected in the same way by most analytical variables, such as flame temperature, fuel flow rates, and background radiation. Clearly, lithium must be absent in the sample.

Automated Flame Photometers. Fully automated photometers now exist for the determination of sodium and potassium in clinical samples. In one of these, the samples are withdrawn sequentially from a sample turntable, dialyzed to remove protein and particulates, diluted with a lithium internal standard, and aspirated into a flame. Sample and reagent transport is accomplished with a roller-type pump. Air bubbles serve to separate samples. Results are printed out on a paper tape. Calibration is performed automatically after every nine samples.

9E-2 INSTRUMENTS FOR SIMULTANEOUS MULTIELEMENT ANALYSES

During the past decade, considerable effort has been made toward the development of instruments for rapid sequential or simultaneous flame determination of several elements in a single sample.[13] (We have already considered one example, the simple photometer for the simultaneous determination of sodium and potassium.) Some of these efforts have resulted in the development of computer-controlled monochromators that permit rapid sequential measurement of radiant power at several wavelengths corresponding to peaks for various elements. With such instruments, two to three seconds are required to move from one peak to the next. The detector photocurrent is then measured for one or two seconds. Thus, it is possible to determine the concentration of as many as ten elements per minute. Instruments of this type have been employed with all three types of flame methods, although the emission method has the distinct advantage of not requiring several sources.

Simultaneous, multielement, flame-emission analyses have been made possible by the use of the optical multichannel analyzers that were described earlier (p. 140). As an example,[14] a silicon diode vidicon tube was mounted on the optical plane originally occupied by the slit of an ordinary grating monochromator. The diameter of the tube surface was such that a 20-nm band of radiation was continuously monitored; by adjustment of the tube along the focal plane of the monochromator, various 20-nm bands of the spectrum could be observed. The resolution within the 20-nm band was such that lines 0.14 nm apart could be resolved.

[13] For a review of this topic, see: K. W. Busch and G. H. Morrison, *Anal. Chem.,* **1973,** *45* (8), 712A; and J. D. Winefordner, J. J. Fitzgerald, and N. Omenetto, *Appl. Spectrosc.,* **1975,** *29,* 369.

[14] K. W. Busch, N. G. Howell, and G. H. Morrison, *Anal. Chem.,* **1974,** *46,* 575.

Figure 9–21 shows a spectrum obtained simultaneously for eight elements that have emission peaks in the wavelength range of 388.6 to 408.6 nm. A nitrous oxide/acetylene flame was employed for excitation. Slightly more than half a minute was required to accumulate data for the eight analyses. A relative precision of better than 5% was obtained.

A limitation of instruments such as that just described is that in order to obtain sufficient optical resolution for emission work, only a relatively small portion of the spectrum (20 nm, for example, in the foregoing case) can be observed at any one time. This limitation has been largely overcome by the use of a two-dimensional dispersing system called an *Echelle grating* monochromator. This device is described in Section 10B–3, which deals with plasma sources for emission spectroscopy.

9E-3 INTERFERENCES

The interferences encountered in flame emission spectroscopy arise from the same sources as those in atomic absorption methods (see Sections 9D–3 and 9D–4); the severity of any given interference will often differ for the two procedures, however.

Spectral Line Interference.   Interference between two overlapping atomic *absorption* peaks occurs only in the occasional situation where the lines are within about 0.1 Å of one another. That is, the high degree of spectral specificity is more the result of the narrow line properties of the source than the high resolution of the monochromator. Atomic emission spectroscopy, in contrast, depends entirely upon the monochromator for selectivity; the probability of spectral interference due to line overlap is consequently greater. Figure 9–22 shows an emission

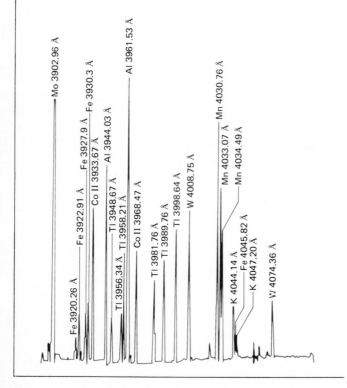

FIGURE 9-21 Multielement flame emission spectrum from 388.6 to 408.6 nm. (Reprinted with permission from: K. W. Busch, N. G. Howell, and G. H. Morrison, *Anal. Chem.*, **1974**, *46*, 578. Copyright 1974 American Chemical Society.)

spectrum for three transition elements, iron, nickel, and chromium. Note that several unresolved peaks exist and that care would have to be taken to avoid spectral interference in the analysis for any one of these elements.

Band Interference; Background Correction. Emission lines are often superimposed on bands emitted by oxides and other molecular species from the sample, the fuel, or the oxidant. An example appears in Figure 9–23. As shown in the figure, a background correction for band emission is readily made by scanning for a few nanometers on either side of the analyte peaks. For nonrecording instruments, a single measurement on either side of the peak suffices. The average of the two measurements is then subtracted from the total peak height.

Chemical Interferences. Chemical interferences in flame emission studies are essentially the same as those encountered in flame absorption methods. They are dealt with by judicious choice of flame temperature and the use of protective agents, releasing agents, and ionization suppressors.

Self-Absorption. The center of a flame is hotter than the exterior; thus, atoms that emit in the center are surrounded by a cooler region which contains a higher concentra-

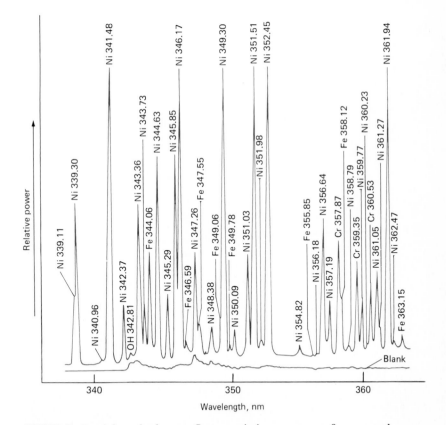

FIGURE 9-22 Partial oxyhydrogen flame emission spectrum for a sample containing 600 ppm Fe, 600 ppm Ni, and 200 ppm Cr. (Taken from: R. Herrmann and C. T. J. Alkemade, *Chemical Analysis by Flame Photometry*, 2d ed., p. 527, Interscience: New York, 1963. Reprinted by permission of John Wiley & Sons, Inc.)

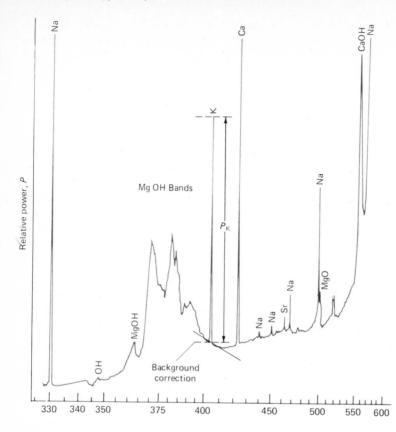

FIGURE 9-23 Flame emission spectrum for a natural brine showing the method used for correcting for background radiation. (Taken from: R. Herrmann and C. T. J. Alkemade, *Chemical Analysis by Flame Photometry*, 2d ed., p. 484, Interscience: New York, 1963. Reprinted by permission of John Wiley & Sons, Inc.)

tion of unexcited atoms; *self-absorption* of the resonance wavelengths by the atoms in the cooler layer will occur. Doppler broadening of the emission line is greater than the corresponding broadening of the resonance absorption line, however, because the particles are moving more rapidly in the hotter-emission zone. Thus, self-absorption tends to alter the center of a line more than its edges. In the extreme, the center may become less intense than the edges, or it may even disappear; the result is division of the emission maximum into what appears to be two peaks by *self-reversal*. Figure 9-24 shows an example of severe self-absorption and self-reversal.

Self-absorption often becomes troublesome when the analyte is present in high concentration. Under these circumstances, a nonresonance line, which cannot undergo self-absorption, may be preferable for an analysis.

Self-absorption and ionization sometimes result in S-shaped emission calibration curves with three distinct segments. At intermediate concentrations of potassium, for example, a linear relationship between intensity and concentration is observed (Figure 9-25). At low concentrations, curvature is due to the increased degree of ionization in the flame. Self-absorption, on the other hand, causes negative departures from a straight line at higher concentrations.

9E-4 ANALYTICAL TECHNIQUES

The analytical techniques for flame emission spectroscopy are similar to those described earlier for atomic absorption spectroscopy (p. 277). Both calibration curves and

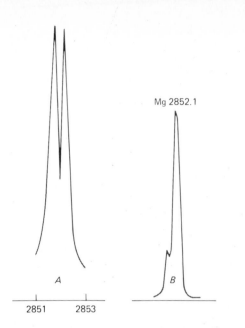

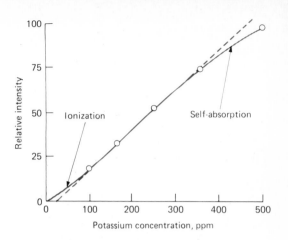

FIGURE 9–25 Effects of ionization and self-absorption on a calibration curve for potassium.

FIGURE 9–24 Curve *A* illustrates the self reversal that occurs with high concentration of Mg (2000 g/mL). Curve *B* shows the normal spectrum of 100 g/mL of Mg.

the standard addition method are employed. In addition, internal standards may be used to compensate for flame variables.

9E-5   COMPARISON OF ATOMIC EMISSION AND ATOMIC ABSORPTION METHODS

For purposes of comparison, the main advantages and disadvantages of the two widely used flame methods are listed in the paragraphs that follow.[15] The comparisons apply to versatile spectrophotometers that are readily adapted to the determination of numerous elements.

1. *Instruments.* A major advantage of the emission procedure is that the flame serves as the source. In contrast, absorption methods require an individual lamp for each element (or sometimes, for a limited group of elements). On the other hand, the quality of the monochromator for an absorption instrument does not have to be so great to achieve the same degree of selectivity because of the narrow lines emitted by the hollow cathode lamp.

2. *Operator Skill.* Emission methods generally require a higher degree of operator skill because of the critical nature of such adjustments as wavelength, flame zone sampled, and fuel-to-oxidant ratio.

3. *Background Correction.* Correction for band spectra arising from sample constituents is more easily, and often more exactly, carried out for emission methods.

4. *Precision and Accuracy.* In the hands of a skilled operator, uncertainties are about the same for the two procedures (± 0.5 to 1% relative). With less skilled personnel, atomic absorption methods have an advantage.

5. *Interferences.* The two methods suffer from similar chemical interferences. Atomic absorption procedures are less subject to spectral line interferences, although such interferences are usually easily recognized and avoided in emis-

---

[15] For an excellent comparison of the two methods, see: E. E. Pickett and S. R. Koirtyohann, *Anal. Chem.,* **1969**, *41* (14), 28A.

sion methods. Spectral band interferences were considered under background correction.

6. *Detection Limits.* The data in Table 9–5 provide a comparison of detection limits and emphasize the complementary nature of the two procedures.

9F   Atomic Fluorescence Spectroscopy

Beginning in about 1964, a significant research effort has been devoted to the development of analytical methods based upon atomic fluorescence.[16] This work has demonstrated clearly that this technique provides a useful and convenient means for quantitative determination of a reasonably large number of elements. To date, the procedure has not, however, found widespread use because of the overwhelming successes of atomic absorption and atomic emission (particularly the former) methods, which predated atomic fluorescence by more than a decade. As mentioned earlier, these suc-

cesses have led to the availability of absorption and emission instruments from numerous commercial sources. Until recently, only one commercial fluorescence instrument was on the market, and it for only a brief period.

The limited use of atomic fluorescence has not arisen so much from any inherent weakness of the procedure but rather because no distinct advantages have been demonstrated relative to the well established absorption and emission methods. Thus while fluorescence methods, particularly those based on electrothermal atomization, are somewhat more sensitive for perhaps five to ten elements, the procedure is also less sensitive and appears to have a smaller useful concentration range for several others. Furthermore, dispersive fluorescence instruments appear to be more complex and potentially more expensive to purchase and maintain.[17]

9F-1   INSTRUMENTATION

Two basic types of fluorescence instruments have been developed, dispersive and nondispersive.

---

[16] For further information on atomic fluorescence spectroscopy, see: C. Veillon, in *Trace Analysis*, J. D. Winefordner, Ed., Chapter VI, Wiley: New York, 1976; N. Omenetto and J. D. Winefordner, *Prog. Anal. At. Spectroscop.*, **1979**, *2*, 1; and J. C. Van Loon, *Anal. Chem.*, **1981**, *53*, 332A.

[17] See: W. B. Barnett and H. L. Kahn, *Anal. Chem.*, **1972**, *44*, 935.

Table 9–5
Comparison of Detection Limits for Various Elements By Flame Absorption and Flame Emission Methods[a]

Flame Emission More Sensitive	Sensitivity About the Same	Flame Absorption More Sensitive
Al, Ba, Ca, Eu, Ga, Ho, In, K, La, Li, Lu, Na, Nd, Pr, Rb, Re, Ru, Sm, Sr, Tb, Tl, Tm, W, Yb	Cr, Cu, Dy, Er, Gd, Ge, Mn, Mo, Nb, Pd, Rh, Sc, Ta, Ti, V, Y, Zr	Ag, As, Au, B, Be, Bi, Cd, Co, Fe, Hg, Ir, Mg, Ni, Pb, Pt, Sb, Se, Si, Sn, Te, Zn

[a] Adapted with permission from E. E. Pickett and S. R. Koirtyohann, *Anal. Chem.*, **1969**, *41* (14), 42A. Copyright 1969 American Chemical Society.

Dispersive Instruments. A dispersive system for atomic fluorescence measurements is made up of a modulated source, an atomizer (flame or nonflame), a monochromator or an interference filter system, a detector, and a signal processor and readout. With the exception of the source, most of these components are similar to those discussed in earlier parts of this chapter.

Sources. A continuous source would be desirable for atomic fluorescence measurements. Unfortunately, however, the output power of most continuous sources over a region as narrow as an atomic absorption line is so low as to restrict the sensitivity of the method severely.

In the early work on atomic fluorescence, conventional hollow cathode lamps often served as excitation sources. In order to enhance the output intensity without destroying the lamp, it was necessary to operate the lamp with short pulses of current that were greater than the lamp could tolerate for continuous operation. The detector, of course, was gated to observe the fluorescent signal only during pulses.

Perhaps the most widely used sources for atomic fluorescence have been the electrodeless discharge lamps (p. 268), which usually produce radiant intensities that exceed that of the hollow cathode lamp by an order of magnitude or two. Electrodeless lamps have been operated in both the continuous and pulsed modes. Unfortunately, this type of lamp is not available for some elements.

Lasers, with their high intensities and narrow bandwidths, are obviously the ideal source for atomic fluorescence measurements. Their high cost, however, has discouraged their widespread application to routine atomic fluorescence methods.

Nondispersive Instruments. In theory, no monochromator or filter should be necessary for atomic fluorescence measurements when an electrodeless discharge lamp or hollow cathode lamp serves as the excitation source since the emitted radiation is, in principle, that of a single element and will thus excite only atoms of that element. A nondispersive system then could be made up of only a source, an atomizer, and a detector. The advantages of such a system are several: (1) simplicity and low-cost instrumentation, (2) ready adaptability to multielement analysis, (3) high energy throughput and thus high sensitivity, and (4) simultaneous collection of energy from multiple lines, also enhancing sensitivity.

In order to realize these important advantages, it is necessary that the output of the source be free of contaminating lines from other elements; in addition, the atomizer should emit no significant background radiation. The latter consideration may be realized in some instances with electrothermal atomizers but certainly is not with typical flames. To overcome this latter problem, filters, located between the source and detector, have often been used to remove the bulk of the background radiation. Alternatively, solar-blind, photomultiplier detectors, which respond only to radiation below 320 nm in wavelength, have been applied. In this instance, the analyte must emit lines with wavelengths shorter than 320 nm.

Recently, a nondispersive instrument has been offered by a commercial source. Here, an inductively coupled plasma source, with a low background radiation, serves as the atomizer.[18] The instrument is described in Section 10C.

9F-2 INTERFERENCES

Interferences encountered in atomic fluorescence spectroscopy appear to be of the same type and of about the same magnitude as those found in atomic absorption spectroscopy.[19]

---

[18] See: D. R. Demers, D. A. Busch, and C. D. Allemand, *Amer. Lab.*, **1982**, *14* (3), 167.

[19] See: J. D. Winefordner and R. C. Elser, *Anal. Chem.*, **1971**, *43* (4), 24A.

9F–3   APPLICATIONS

Atomic fluorescence methods have been applied to the analysis of metals in such materials as lubricating oils, seawater, biological substances, graphite, and agricultural samples.[20] Detection limits for atomic fluorescence procedures are found in Table 9–4.

An important application of a nondispersive system has been for the determination of traces of mercury in environmental samples such as the atmosphere, river waters, and soils.[21] Here, mercury salts in an aqueous medium are reduced to the elemental form, swept out of solution with a stream of argon and passed over a trap containing silver, which retains the mercury as an amalgam. The analysis is completed by heating the amalgam and passing the released elemental mercury vapor through a flow cell. Here, fluorescence is excited by means of the 253.7 nm mercury line produced by a mercury pen lamp. The radiation is detected by a solar-blind cell. The procedure is significantly less susceptible to interference by organic compounds than a widely used method based upon atomic absorption. Furthermore, the equipment used is simple, inexpensive, and rugged.

---

[20] For a summary of applications, see: J. D. Winefordner, *J. Chem. Educ.,* **1978,** *55,* 72.

[21] For example, see: F. L. Cocoran, Jr., *Amer. Lab.,* **1974,** *6* (3), 69.

PROBLEMS

9–1. Describe the basic differences between atomic emission and atomic fluorescence spectroscopy.

9–2. Define the following terms: (a) atomization, (b) pressure broadening, (c) Doppler broadening, (d) turbulent flow nebulizer, (e) laminar flow nebulizer, (f) hollow cathode lamp, (g) sputtering, (h) self-absorption, (i) spectral interference, (j) chemical interference, (k) radiation buffer, (l) releasing agent, (m) protective agent, and (n) ionization suppressor.

9–3. Why is the CaOH spectrum in Figure 9–5 so much broader than the Ba resonance line?

9–4. Why is atomic emission more sensitive to flame instability than atomic absorption or fluorescence?

9–5. Describe the effects which are responsible for the three different absorbance profiles in Figure 9–8.

9–6. Why is a nonflame atomizer more sensitive than a flame atomizer?

9–7. Why is source modulation employed in atomic absorption spectroscopy?

9–8. Describe how a deuterium lamp can be employed to provide a background correction for an atomic absorption spectrum.

9–9. For the same concentration of nickel, the height of the absorption peak at 352.4 nm was found to be about 30% greater from a solution that contained 50% ethanol than from an aqueous solution. Explain.

9–10. The emission spectrum of a hollow cathode lamp for molybdenum was found to have a sharp peak at 313.3 nm as long as the lamp current was less than 50 mA. At higher currents, however, the peak developed a cup-like crater at its maximum. Explain.

9–11. For the flame shown in Figure 9–7 calculate the relative intensity of the 766.5 nm emission line for potassium at the following heights above the flame (assume no ionization):

    (a) 2.0 cm;   (b) 3.0 cm;   (c) 4.0 cm;   (d) 5.0 cm.

9–12. In a hydrogen/oxygen flame, an atomic absorption peak for iron was found to decrease in the presence of large concentrations of sulfate ion.

    (a) Suggest an explanation for this observation.

    (b) Suggest three possible methods for overcoming the potential interference of sulfate in a quantitative determination of iron.

9–13. The Doppler effect is one of the sources of line broadening in atomic absorption spectroscopy. Atoms moving toward the light source see a higher frequency than do atoms moving away from the source. The difference in wavelength, $\Delta\lambda$, experienced by an atom moving at speed v (compared to one at rest) is $\Delta\lambda/\lambda = v/c$, where c is the speed of light. Estimate the line width (in Å) of the sodium D line (5893 Å) when the absorbing atoms are at a temperature of 2000°K. The average speed of an atom is given by $v = \sqrt{8kT/\pi m}$, where k is Boltzmann's constant, T is temperature, and m is the mass.

9–14. For Na and Mg^+ atoms, compare the ratios of the number of particles in the $3p$ excited state to the number in the ground state in

    (a) a natural gas/air flame (1800°C).

    (b) a hydrogen/oxygen flame (2600°C).

    (c) an acetylene/oxygen flame (3100°C).

    (d) an inductively coupled plasma source (5700°C).

9–15. In higher-temperature sources, sodium atoms emit a doublet with an average wavelength of 1139 nm. The transition responsible is from the $4s$ to $3p$ state. Calculate the ratio of the number of excited atoms in the $4s$ to the number in the ground $3s$ state in

    (a) an acetylene/oxygen flame (3100°C).

    (b) the hottest part of an inductively coupled plasma source (~8000°C).

9–16. Assume that the peaks shown in Figure 9–14 were obtained for 2 μL aliquots of standards and sample. Calculate the parts per million of lead in the sample of canned orange juice.

9–17. Suggest sources of the two peaks in Figure 9–14 that appear during the drying and ashing processes.

9–18. In the concentration range of 500 to 2000 ppm of U, a linear relationship is found for absorbance at 351.5 nm and concentration. At lower concentrations the relationship becomes nonlinear unless about 2000 ppm of an alkali metal salt is introduced. Explain.

9–19. What is the purpose of an internal standard in flame emission methods?

9–20. A 5.00-mL sample of blood was treated with trichloroacetic acid to precipitate proteins. After centrifugation, the resulting solution was brought to a pH of 3 and extracted with two 5-mL portions of methyl isobutyl ketone containing the organic lead complexing agent APCD. The extract was aspirated directly into an air/acetylene flame yielding an absorbance at 283.3 nm of 0.502. Five-milliliter aliquots of standard solutions containing 0.400 and 0.600 ppm Pb were treated in the same way and yielded absorbances of 0.396 and 0.599. Calculate the ppm Pb in the sample assuming that Beer's law is followed.

9–21. The sodium in a series of cement samples was determined by flame emission spectroscopy. The flame photometer was calibrated with a series of standards containing 0, 20.0, 40.0, 60.0, and 80.0 μg Na_2O per mL. The instrument readings for these solutions were 3.1, 21.5, 40.9, 57.1, and 77.3.
   (a) Plot the data.
   (b) Derive a least-squares line for the data.
   (c) Calculate standard deviations for the slope and about regression for the line in (b).
   (d) The following data were obtained for replicate 1.000 g samples of cement that were dissolved in HCl, and diluted to 100.0 mL after neutralization.

	Blank	Sample A	Sample B	Sample C
		Emission Reading		
Replicate 1	5.1	28.6	40.7	73.1
Replicate 2	4.8	28.2	41.2	72.1
Replicate 3	4.9	28.9	40.2	spilled

Calculate the % Na_2O in each sample. What is the absolute and relative standard deviation for the average of each determination?

9–22. The chromium in an aqueous sample was determined by pipetting 10.0 mL of the unknown into each of five 50.0-mL volumetric flasks. Various volumes of a standard containing 12.2 ppm Cr were added to the flasks following which the solutions were diluted to volume.

Unknown, mL	Standard, mL	Absorbance
10.0	0.0	0.201
10.0	10.0	0.292
10.0	20.0	0.378
10.0	30.0	0.467
10.0	40.0	0.554

(a) Plot the data.
(b) Derive an equation for the relationship between absorbance and volume of standard.
(c) Calculate the standard deviation for the slope and about regression in (b).
(d) Calculate the ppm Cr in the sample.
(e) Calculate the standard deviation of the result in (d).

Emission Spectroscopy Based Upon

Plasma, Arc, and Spark Atomization

This chapter describes spectroscopic methods based upon the more energetic atomization sources listed in Table 9–1, namely the *inductively coupled plasma* (ICP), the *direct current plasma* (DCP), the *electric arc,* and the *electric spark.*[1] The latter two sources have been used for spectroscopic studies since the turn of the century and have found widespread application to elemental analysis beginning in the early nineteen-thirties. In contrast, plasma sources are relatively new, having been developed largely during the nineteen-seventies. A still more recent development has been the introduction by an instrument manufacturer of a nondispersive atomic fluorescence instrument based upon an inductively coupled plasma source.

Plasma, arc, and spark sources offer several benefits compared with the flame and electrothermal methods considered in the previous chapter. Among their advantages is lower interelement interference, which is a direct consequence of their higher temperatures. Second, good spectra can be obtained for most elements under a single set of excitation conditions; as a consequence, spectra for dozens of elements can be recorded *simultaneously.* This property is of particular importance for the multielement analysis of very small samples. Flame sources are less satisfactory in this regard because optimum excitation conditions vary widely from element to element; high temperatures are needed for excitation of some elements and low temperatures for others; reducing conditions work best in some analyses and oxidizing conditions in others; and finally, the region of the flame that gives rise to optimum line intensities varies from element to element. Another advantage of the more energetic sources is that they permit the determination of low concentrations of elements that tend to form refractory com-

---

[1] For more extensive treatment of emission spectroscopy, see: R. D. Sacks, in *Treatise on Analytical Chemistry*, 2d ed., P. J. Elving, E. J. Meehan, and I. M. Kolthoff, Eds., Part I, vol. 7, Chapter 6, Wiley: New York, 1981; T. Torok, J. Mika, and E. Gegus, *Emission Spectrochemical Analysis*, Crane Russak: New York, 1978.

pounds (that is, compounds that are highly resistant to decomposition by heat or other rigorous treatment) such as boron, phosphorus, tungsten, uranium, zirconium, and niobium. Finally, plasma sources often permit the determination of an element over several decades of concentration in contrast to one or two for most other spectroscopic procedures.

Despite these several advantages, it is unlikely that emission methods based upon high energy sources will ever displace flame and electrothermal atomic absorption. In fact, atomic emission and absorption methods appear to be complementary. Included among the advantages of atomic absorption procedures are simpler and less expensive equipment requirements, lower operating costs, somewhat greater precision (presently, at least), and procedures that require less operator skills to yield satisfactory results.

Plasma sources offer several advantages over the classical arc and spark emission method. Perhaps the most important is the much greater reproducibility of atomization conditions, which often leads to precisions that are better by a factor of ten or more. Furthermore, the equipment associated with these sources tends to be simpler, less expensive, and more versatile. The single major advantage of arc and spark sources is that they are more readily adapted to the direct atomization of such difficult samples as refractory minerals, ores, glasses, and alloys without extensive sample treatment. In contrast, flame and plasma procedures usually require decomposition of samples to give solutions (usually aqueous) for injection into the source.

10A   Spectra From Higher Energy Sources

An examination of the emission produced by an electrical arc or spark or by an argon plasma reveals three types of superimposed spectra: *continuous*, *band*, and *line*. For arc and spark sources, continuous radiation is emitted by the heated electrodes and perhaps also by hot particulate matter detached from the electrode surfaces. The frequency distribution of this radiation depends upon the temperature and approximates that of a black body (Section 4B–3). As will be pointed out later, the continuum found with plasma sources apparently arises from a recombination of thermally produced electrons with argon ions.

Band spectra, made up of a series of closely spaced lines, are also observed in certain wavelength regions, particularly with arc and spark sources. This type of emission is due to volatile molecular species, which produce bands as a consequence of the superposition of vibrational energy levels upon electronic levels. The cyanogen band, caused by the presence of CN radicals, is always observed when carbon electrodes are employed in an atmosphere containing nitrogen. Samples containing a high silicon content may yield an additional molecular band spectrum due to SiO. Another common source of bands is the OH radical. These bands may obscure line spectra of interest.

Emission spectroscopy is based upon the line spectra produced by excited atoms and ions. The source and nature of these spectra were discussed in Section 9A–1.

Arc, spark, and plasma sources are generally richer in lines than the sources described in Chapter 9 because of their greater energies. An electric arc is ordinarily less energetic than a spark; as a result, lines of neutral atoms tend to predominate in the former. On the other hand, spectra excited by a spark typically contain lines associated with excited *ions*. Similarly, many of the lines found in plasma spectra also arise from ions rather than atoms. As was pointed out in Section 9A–1, the spectrum of an ion is quite different from that of the atom from which it is formed.

10B Emission Spectroscopy Based on Plasma Sources

By definition, a plasma is a conducting gaseous mixture containing a significant concentration of cations and electrons. (The concentrations of the two are such that the net charge approaches zero.) In the argon plasma employed for emission analyses, argon ions and electrons are the principle conducting species although cations from the sample will also be present in lesser amounts. Argon ions, once formed in a plasma, are capable of absorbing sufficient power from an external source to maintain the temperature at a level at which further ionization sustains the plasma indefinitely; temperatures as great as 10,000°K are encountered. Three power sources have been employed in argon plasma spectroscopy.[2] One is a dc electrical source capable of maintaining a current of several amperes between electrodes immersed in a stream of argon. The second and third are powerful radio frequency and microwave frequency generators through which the argon flows. Of the three, the radio-frequency or *inductively coupled plasma* (ICP) source appears to offer the greatest advantage in terms of sensitivity and freedom from interference. On the other hand, the *dc plasma source* (DCP) has the virtue of simplicity and lower cost. Both will be described here.

10B-1 THE INDUCTIVELY COUPLED PLASMA SOURCE[3]

Figure 10–1 is a schematic drawing of an inductively coupled plasma source. It consists of three concentric quartz tubes through which streams of argon flow at a

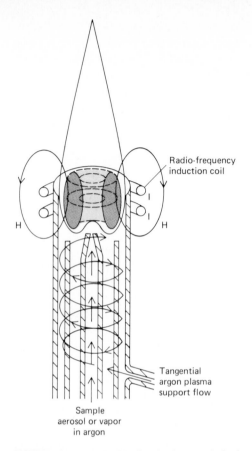

Radio-frequency induction coil

Tangential argon plasma support flow

Sample aerosol or vapor in argon

FIGURE 10–1   A typical inductively coupled plasma source. (From: V. A. Fassel, *Science,* **1978,** *202,* 185. With permission. Copyright 1978 by the American Association for the Advancement of Science.)

total rate between 11 and 17 L/min. The diameter of the largest tube is about 2.5 cm. Surrounding the top of this tube is a water-cooled induction coil that is powered by a radio-frequency generator, which is capable of producing 2 kW of energy at about 27 MHz. Ionization of the flowing argon is ini-

---

[2] For a detailed discussion of the various plasma sources, see: P. Tschöpel in *Comprehensive Analytical Chemistry,* G. Svehla, Ed., Vol. IX, Chapter 3, Elsevier: New York, 1979; R. D. Sacks, in *Treatise on Analytical Chemistry,* 2d ed., P. J. Elving, E. J. Meehan, and I. M. Kolthoff, Eds., Part I, vol. 7, pp. 516–526, Wiley: New York, 1981.

[3] For a more complete discussion, see: V. A. Fassel, *Science,* **1978,** *202,* 183; V. A. Fassel, *Anal. Chem.,* **1979,** *51,* 1290A; M. Thompson and J. N. Walsh, *Inductively Coupled Plasma Spectrometry,* Blackie: London, 1983; R. M. Barnes, *CRC Crit. Rev. Anal. Chem.,* **1978,** *7,* 203.

tiated by a spark from a Tesla coil. The resulting ions, and their associated electrons, then interact with the fluctuating magnetic field (labeled H in Figure 10–1) produced by the induction coil. This interaction causes the ions and electrons within the coil to flow in the closed annular paths depicted in the figure; ohmic heating is the consequence of their resistance to this movement.

The temperature of the plasma formed in this way is high enough to require thermal isolation from the outer quartz cylinder. This isolation is achieved by flowing argon tangentially around the walls of the tube as indicated by the arrows in Figure 10–1; the flow rate of this stream is 10 to 15 L/min. The tangential flow cools the inside walls of the center tube and centers the plasma radially.

Sample Injection. The sample is carried into the hot plasma at the head of the tubes by argon flowing at about 1 L/min through the central quartz tube. The sample may be an aerosol, a thermally generated vapor, or a fine powder.

The most widely used apparatus for sample injection is similar in construction to the nebulizer employed for flame methods. Figure 10–2 shows a typical arrangement. Here, the sample is nebulized by the flow of argon, and the resulting finely divided droplets are carried into the plasma. Aerosols have also been produced from liquids and solids by means of an ultrasonic nebulizer. Still another method of sample introduction involves deposition of the sample on a tantalum strip and vaporization by a large electrical current. The vapors are then swept into the plasma by the argon stream. No general, entirely practical method has yet been devised for the introduction of powder samples such as those encountered in minerals and ores.

Plasma Appearance and Spectra. The typical plasma has a very intense, brilliant white, nontransparent core topped by a flamelike tail. The core, which extends a few millime-

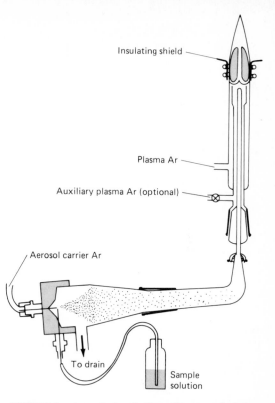

FIGURE 10–2 A typical nebulizer for sample injection into a plasma source. (From: V. A. Fassel, *Science*, **1978**, *202*, 186. With permission. Copyright 1978 by the American Association for the Advancement of Science.)

ters above the tube, is made up of a continuum upon which is superimposed the atomic spectrum for argon. The source of the continuum apparently arises from recombination of argon and other ions with electrons. In the region 10 to 30 mm above the core, the continuum fades, and the plasma is optically transparent. Spectral observations are generally made at a height of 15 to 20 mm above the induction coil. Here, the background radiation is remarkably free of argon lines and is well suited for analysis. Many of the most sensitive analyte lines in this region of the plasma are from ions such as Ca^+, Ca^{2+}, Cd^+, Cr^{2+}, and Mn^{2+}.

Analyte Atomization and Ionization. Figure 10–3 shows temperatures at various parts of the

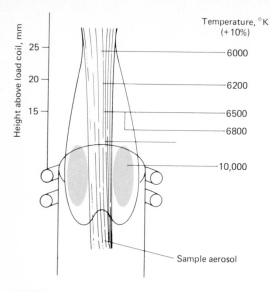

FIGURE 10-3   Temperatures in a typical inductively coupled plasma source. (From: V. A. Fassel, *Science*, **1978**, *202*, 187. With permission. Copyright 1978 by the American Association for the Advancement of Science.)

plasma. By the time the sample atoms have reached the observation point, they will have resided for about 2 msec at temperatures ranging from 6000 to 8000°K. These times and temperatures are roughly two to three times greater than those found in the hotter combustion flames (acetylene/nitrous oxide) employed in flame methods. As a consequence, atomization is more complete, and fewer chemical interference problems arise (p. 273). Surprisingly, ionization interference effects (p. 275) are small or nonexistent, probably because the electron concentration from ionization of the argon is large compared with that resulting from ionization of sample components.

Several other advantages are associated with the plasma source. First, atomization occurs in a chemically inert environment, which tends to enhance the lifetime of the analyte by preventing oxide formation. In addition, and in contrast to arc, spark, and flame, the temperature cross section of the plasma is relatively uniform; as a consequence, self-absorption and self-reversal effects (p. 283) are not encountered. Thus, linear calibration curves over several orders of magnitude of concentration are often observed.

10B-2   THE DIRECT CURRENT ARGON PLASMA SOURCE

Direct current plasma jets were first described in the 1920s and have been systematically investigated as sources for emission spectroscopy for more than two decades. It was not until recently, however, that a source based on this principle has been designed that can successfully compete with flame and inductively coupled plasma sources in terms of reproducible behavior.[4]

Figure 10-4 is a schematic diagram of a commercially available dc plasma source that is well-suited for excitation of emission spectra for a wide variety of elements. This plasma jet source consists of three electrodes arranged in an inverted Y configuration. A graphite anode is located in each arm of the Y and a tungsten cathode at the inverted base. Argon flows from the two anode blocks toward the cathode. The plasma jet is formed by bringing the cathode momentarily in contact with the anodes. Ionization of the argon occurs and a current develops (~14 A) that generates additional ions that sustain the current indefinitely. The temperature at the arc core is perhaps 10,000°K and at the viewing region 5000°K. The sample is aspirated into the area between the two arms of the Y, where it is atomized, excited, and viewed.

Spectra produced from the plasma jet tend to have fewer lines than those produced by the inductively coupled plasma, and the lines present are largely from atoms

[4] For additional details, see: G. W. Johnson, H. E. Taylor, and R. K. Skogerboe, *Anal. Chem.*, **1979**, *51*, 2403; *Spectrochim. Acta, Part B*, **1979**, *34*, 197; J. Reednick, *Amer. Lab.*, **1979**, *11* (3), 53.

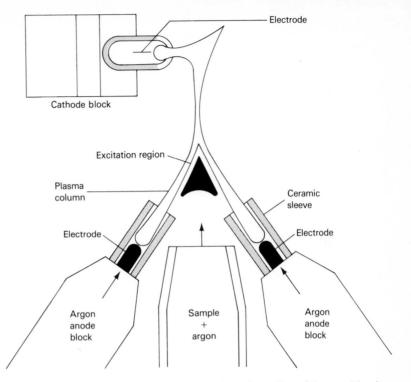

FIGURE 10-4   A three-electrode dc plasma jet. (Courtesy of SpectraMetrics, Inc., Haverhill, MA.)

rather than ions. Sensitivities achieved with the dc jet plasma appear to range from an order of magnitude lower to about the same as those found with the inductively coupled plasma. Reproducibilities of the two systems are similar. Significantly less argon is required for the dc plasma and the auxiliary power supply is simpler and less expensive. On the other hand, the graphite electrodes must be replaced every few hours while the inductively coupled plasma source requires little or no maintenance.

10B-3   INSTRUMENTS FOR PLASMA SPECTROSCOPY

Instruments for elemental emission analysis by plasma excitation are sold by several instrument makers. Their wavelength ranges vary considerably. Some encompass the entire ultraviolet-visible spectrum from 180 to 900 nm. Many do not operate above 500 to 600 nm inasmuch as useful lines for most elements occur at shorter wavelengths. A few instruments are equipped for vacuum operation, which extends the ultraviolet to 170 nm. This short wavelength region is important because it permits the determination of elements such as phosphorus, sulfur, and carbon.

Typical emission instruments have dispersions that range from 0.1 to 0.6 nm/mm and focal lengths of 0.5 to 3 m.

Types of Plasma Emission Instruments.   Instruments for emission spectroscopy are of two basic types, *sequential* and *simultaneous multichannel*.[5] The former, which are less complex

---

[5] For a discussion of types of atomic instruments, see: K. W. Busch and L. D. Benton, *Anal. Chem.*, **1983**, *55*, 445A.

and consequently often significantly less expensive, measure line intensities on a one-by-one basis. Here, the instrument is programmed to move (or is moved manually) from the line for one element to that of a second pausing long enough (a few seconds) at each to obtain a satisfactory signal-to-noise ratio. In contrast, multichannel instruments are designed to measure the intensities of emission lines for a large number of elements (sometimes as many as 50 or 60) *simultaneously*. Clearly, where several elements are to be determined, the excitation time with sequential instruments will be significantly greater. Thus, these instruments, while simpler, are costly in terms of sample consumption and time.

Sequential Instruments.   Figure 10–5 is an optical diagram of a versatile, sequential instrument, which can be used either for emission analyses with an inductively coupled plasma source or for atomic absorption analyses with a flame or graphite furnace atomizer.[6] The switch from emission to absorption

---

[6] For additional details, see: C. G. Fisher III, R. D. Ediger, and J. E. Delany, *Amer. Lab.*, **1981**, *13* (2), 115; T. J. Hanson and R. D. Ediger, *Amer. Lab.*, **1980**, *12* (3), 116.

mode involves movement of the movable mirror as shown. The holographic grating is driven by a stepping motor with each step corresponding to a change in wavelength of 0.007 nm. Two interchangeable gratings are available, one for the region of 175 to 460 nm and the second from 460 to 900 nm. In the emission mode, up to 20 elements can be determined at one time at a rate of three to four elements per minute.

Sequential instruments for emission methods based on the inductively coupled argon plasma source are offered by several other instrument manufacturers.

Multichannel Instruments.   Several manufacturers offer emission spectrometers in which numerous (as many as 60) photomultipliers are located behind fixed slits along the curved focal plane of a concave grating monochromator. The instrument shown schematically in Figure 10–6 is typical. Here the entrance slit, the exit slits, and the grating surface are located along the circumference of a *Rowland circle*, the curvature of which corresponds to the focal curve of the concave grating. Radiation from the several fixed slits are reflected by mirrors (two of which are shown) to photomultiplier tubes. The slits are factory-fixed to transmit lines for

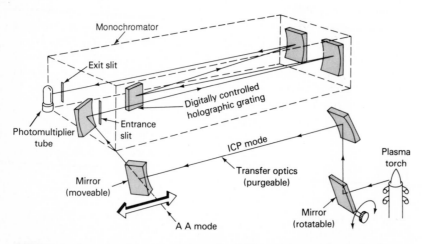

FIGURE 10–5   A sequential spectrometer for ICP emission and atomic absorption spectroscopy. (Courtesy of Perkin-Elmer Corp., Norwalk, CT.)

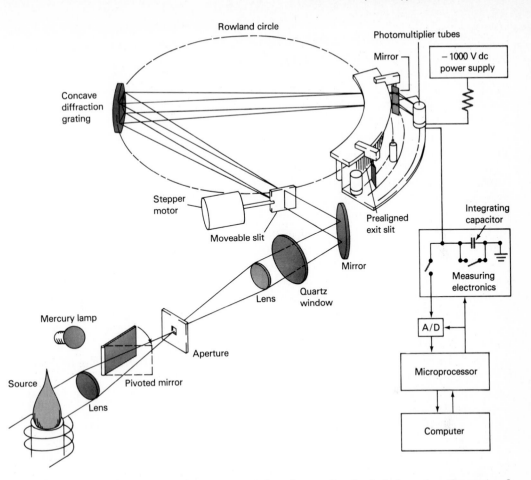

FIGURE 10–6 A plasma multichannel spectrometer based upon Rowland circle optics. (Courtesy of Baird Corp., Bedford, MA.)

elements chosen by the customer. In this instrument, the pattern of lines can, however, be changed relatively inexpensively to accommodate new elements or to delete others. The signals from the several photomultiplier tubes are fed into individual capacitor-resistor circuits for integration; the output voltages are then digitized, converted to concentrations, stored, and displayed. The entrance slit can be moved tangentially to the Rowland circle by means of a stepper motor. This device permits scanning through peaks and provides information for background corrections.

Spectrometers such as that shown in Figure 10–6 have been used both with plasma and with arc and spark sources. For rapid routine analyses, such instruments are often ideal. For example, in the production of alloys, quantitative determinations for 5 to 20 or more elements can be accomplished within five minutes of receipt of a sample; close control over the composition of a final product thus becomes possible.

In addition to speed, photoelectric multichannel spectrometers often offer the advantage of good analytical precision. Under ideal conditions reproducibilities of the order of 1% relative of the amount present have been demonstrated. Several of the newer instruments are provided with a second monochromator that permits spectral scanning thus adding a versatility that was absent in some earlier instruments.

Needless to say, multichannel instruments are more expensive (>$150,000) and generally not as versatile as the sequential instruments described in the previous section.

Multichannel Instruments Based Upon an Echelle Monochromator. The *echelle grating*, which was first described by G. R. Harrison in 1949, provides high dispersion and high resolution, which makes it particularly well suited for multichannel emission instruments.[7] Figure 10–7 shows a cross section of a typical echelle grating. It differs from the echellette grating shown in Figure 5–12 (p. 125) in several respects. First, in order to achieve a high angle of incidence, the blaze angle of an echelle grating is significantly greater than the conventional device, and the short side of the blaze is used rather than the long. Furthermore the grating is relatively coarse, having typically 300 or fewer grooves per millimeter for ultraviolet-visible radiation. Note that the angle of refraction r is much higher in the echelle grating than the echel-

---

[7] For a more detailed discussion of the echelle grating, see: P. N. Keliher and C. C. Wohlers, *Anal. Chem.*, **1976**, *48*, 333A; D. L. Anderson, A. R. Forster, and M. L. Parsons, *Anal. Chem.*, **1981**, *53*, 770; A. T. Zander and P. N. Keliher, *Appl. Spectrosc.*, **1979**, *33*, 499.

lette and approaches the angle of incidence i. That is,

$$r \cong i = \beta$$

Under these circumstances, Equation 5–3 (p. 126) for a grating becomes

$$\mathbf{n}\lambda = 2d \sin \beta \tag{10–1}$$

Equation 5–7 for reciprocal dispersion can then be written as

$$D^{-1} = \frac{2d \cos \beta}{\mathbf{n}F} \tag{10–2}$$

With a normal echellette grating, high dispersion or low reciprocal dispersion is obtained by making the groove width d small and the focal length F large. In contrast, the echelle grating achieves this same end by making both the angle β and the order of diffraction $\mathbf{n}$ large.

The advantages of the echelle grating are illustrated by the data in Table 10–1, which show the performance characteristics for two typical monochromators, one with a conventional echellette grating and the other with an echelle. Note that for the same focal length the linear dispersion and resolution are an order of magnitude greater; the light-gathering power of the echelle is also somewhat superior.

One of the problems encountered with the use of an echelle grating is that the linear dispersion at high orders of refraction is so great that to cover a reasonably broad spectral range it is necessary to use many successive orders. For example, one instrument designed to cover a range of 200 to 800 nm employs diffraction orders 28 to 118 (90 successive orders). Because these orders inevitably overlap, it is essential that a system of cross dispersion, such as that shown in Figure 10–8, be used to separate the orders. Here, the dispersed radiation from the grating is passed through a prism

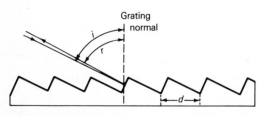

FIGURE 10–7 Echelle grating: i = angle of incidence; r = angle of reflection; d = groove spacing. In usual practice, $i \cong r = \beta = 63°26'$.

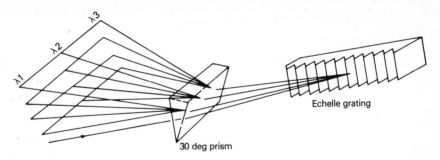

FIGURE 10-8   Two-dimensional dispersing element including an echelle grating and a 30-deg prism. (Courtesy of SpectraMetrics, Inc., Haverhill, MA.)

whose axis is at 90 deg to the grating. The effect of this arrangement is to produce at the focal plane a two-dimensional spectrum such as that shown schematically in Figure 10–9. In this figure, the location of wavelengths for 10 of a possible 70 orders are indicated by vertical lines. For any given order, the dispersion is approximately linear, but, as can be seen, the dispersion lessens at lower orders or at higher wavelengths. An actual two-dimensional spectrum from an echelle monochromator consists of a complex series of very short vertical lines lying along 50 to 100 horizontal axes, each axis corresponding to one diffraction order.

Figure 10–10 is a schematic diagram of a commercially available echelle spectrometer, which employs the dispersing elements shown in Figure 10–9. This instrument, which employs orders 28 to 118, has a focal length of 0.75 m, an average resolution of 0.003 nm, and an average reciprocal dispersion of 0.12 nm/mm. For qualitative work, spectra can be recorded with a Polaroid® camera located as shown. In this configuration, a plane mirror swings into place to direct the dispersed radiation to a 4 × 5 inch film. Clear plastic overlays, showing the location of prominent lines for the elements, permit identification of species responsible for lines on the developed film.

The monochromator shown in Figure

Table 10-1
Comparison of Performance Characteristics of a Conventional and Echelle Monochromator[a]

	Conventional	Echelle
Focal length	0.5 m	0.5 m
Groove density	1200/mm	79/mm
Diffraction angle, β	10°22′	63°26′
Order **n** (at 300 nm)	1	75
Resolution (at 300 nm), $\lambda/\Delta\lambda$	62,400	763,000
Reciprocal linear dispersion, D^{-1}	16 Å/mm	1.5 Å/mm
Light-gathering power, f	f/9.8	f/8.8

[a] With permission from P. E. Keliher and C. C. Wohlers, *Anal. Chem.*, **1976**, *48*, 334A. Copyright 1976 American Chemical Society.

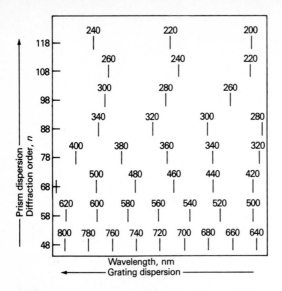

FIGURE 10-9   Schematic representation of the focal plane of an echelle monochromator showing the location of wavelengths for 10 of 70 orders.

first, a photomultiplier tube in a fixed position on the focal plane monitors intensities. Three-dimensional motion of the echelle permits radiation from 190 to 900 nm to be focused on the detector. Scanning and recording of data across the free spectral range of any given order is possible. A shorter scan provides a wavelength profile near a given line for background computation. In the multichannel instrument, 20 end-on photomultiplier tubes are arranged in a fixed hexagonal close packed configuration that occupies an approximately 80 cm² area at the focal plane. Interchangeable masks are then used to determine which set of elements are to be monitored simultaneously.

Instruments Based on Multichannel Photon Detectors. Several investigations leading to the development of multielement emission instruments based upon the multichannel photon detector described in Section 5F–3 are to be found in the literature. One such instrument, for flame emission spectroscopy, was noted in Section 9E–2. Others have been applied to plasma emission sources. The

10–10 serves as the dispersing element for a single channel sequential instrument or a multichannel instrument for simultaneous determination of up to 20 elements. In the

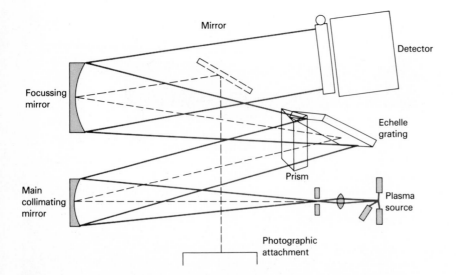

FIGURE 10-10   A spectrometer for plasma emission spectroscopy. (Courtesy of SpectraMetrics, Inc., Haverhill, MA.)

echelle monochromator with its enhanced resolution, would appear to offer considerable potential for the development of instruments of this kind.

10B-4 QUANTITATIVE APPLICATIONS OF PLASMA SOURCES[8]

Unquestionably, the inductively coupled and the direct current plasma sources yield significantly better quantitative analytical data than other emission sources. The excellence of these results stems from the high stability, low noise, low background, and freedom from interferences of the sources when operated under appropriate experimental conditions. The performance of the inductively coupled plasma source is somewhat better than that of the direct current plasma source in terms of detection limits. The latter, however, is less expensive to purchase and operate and is entirely adequate for many applications.

Analytical Techniques. The techniques for preparation of standards and samples and for preparation of calibration curves are similar to those described in Section 9D-5 for atomic absorption spectroscopy.

For the best results, the inductively coupled plasma source should be warmed-up for 15 to 30 min so that thermal equilibrium is reached. As in atomic absorption spectroscopy, one or more standards should be run periodically to correct for the effects of instrument drift. The improvement in precision that results from this procedure is illustrated by the data in Table 10-2. Note also the improved precision when higher concentrations of analyte are measured.[9]

Interferences. As was noted earlier, chemical interferences and matrix effects are significantly lower with plasma sources than with other atomizers. At low analyte concentrations, however, the background emission due to recombination of argon ions with electrons becomes large enough to require careful corrections. For single channel instruments, this correction is conveniently obtained from measurements on either side of the peak. Many multichannel instruments are equipped with a movable slit which permits similar corrections.

Figure 10-11 shows calibration curves for

[8] For useful discussions of the applications of plasma emission sources, see: *Applications of Inductively Coupled Plasmas to Emission Spectroscopy*, R. M. Barnes, Ed., The Franklin Institute Press: Philadelphia, 1978; and *Applications of Plasma Emission Spectrochemistry*, R. M. Barnes, Ed., Heyden: Philadelphia, 1979.

[9] For a recent discussion of precision in ICP spectrometry, see: R. L. Watters, Jr., *Amer. Lab.*, **1983,** *15* (3), 16.

Table 10-2
Effect of Standardization Frequency on Precision of ICP Data[a]

Frequency of Recalibration, hr	Relative Standard Deviation, %			
	Concentration Multiple Above Detection Limit			
	10^1 to 10^2	10^2 to 10^3	10^3 to 10^4	10^4 to 10^5
0.5	3–7	1–3	1–2	1.5–2
2	5–10	2–6	1.5–2.5	2–3
8	8–15	3–10	3–7	4–8

[a] Data from: R. M. Barnes, in *Applications of Inductively Coupled Plasmas to Emission Spectroscopy*, R. M. Barnes, Ed., p. 16, The Franklin Institute Press: Philadelphia, 1978. With permission.

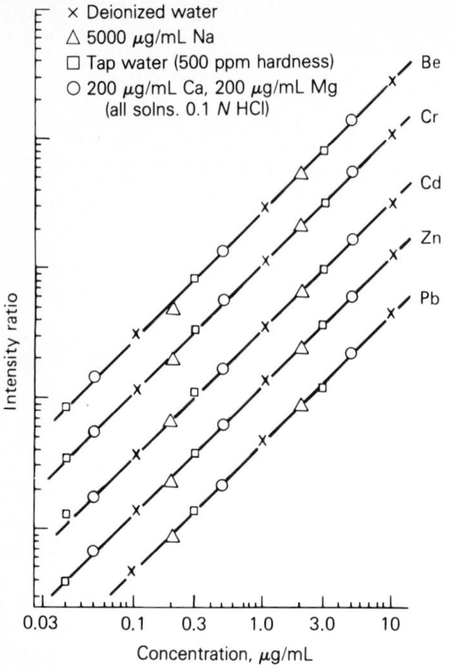

FIGURE 10–11 Calibration curves with an inductively coupled plasma source. Here, an yttrium line at 242.2 nm served as an internal standard. Notice the lack of interelement interference. (From: V. A. Fassel, *Science,* **1978,** *202,* 187. With permission. Copyright 1978 by the American Association for the Advancement of Science.)

several elements. Note that some of the data were obtained from solutions prepared by dissolving a compound of the element in pure water; others contained high concentrations of different species, indicating a freedom from interelement interferences. Note also that the curves cover a concentration range of nearly three orders of magnitude.

In general, the detection limits with the inductively coupled plasma source appear comparable to or better than other atomic spectral procedures. Table 10–3 compares the sensitivity of several of these methods. Note that more elements can be detected in the ten parts-per-billion (or less) range with plasma excitation than with other emission or absorption methods.

10C Atomic Fluorescence Methods Based Upon Plasma Atomization

In Section 9F, which was devoted to atomic fluorescence methods based upon flame and electrothermal sources, some of the potential advantages and disadvantages of fluorescence procedures were considered. Among the advantages of these techniques, as compared to atomic absorption methods,

Table 10–3
Comparison of Detection Limits for Several Atomic Spectral Methods[a]

	Number of Elements Detected at Concentrations of				
Method	**<1 ppb**	**1–10 ppb**	**11–100 ppb**	**101–500 ppb**	**>500 ppb**
Inductively coupled plasma emission	9	32	14	6	0
Atomic emission	4	12	19	6	19
Atomic fluorescence	4	14	16	4	6
Atomic absorption	1	14	25	3	14

[a] Detection limits correspond to a signal that is twice as great as the standard deviation for the background noise. Data abstracted with permission from: V. A. Fassel and R. N. Kniseley, *Anal. Chem.*, **1974**, *46* (13), 1111A. Copyright 1974 American Chemical Society.

are larger linear concentration ranges and greater sensitivities for several elements. In comparison with emission methods, atomic fluorescence appears to offer the potential for simpler and less costly instrumentation for multielement analysis and less spectral interferences.

A nondispersive instrument employing an inductively coupled plasma atomization has recently appeared on the market.[10] It is capable of determining 12 elements simultaneously based upon fluorescence that has been excited with hollow cathode lamps. The instrument design is considerably simpler than most simultaneous multielement instruments.

10C-1 INSTRUMENT DESIGN

Figure 10–12 is a schematic diagram of one of 12 source and detector modules that are arranged in a circle around a central inductively coupled plasma source. Each module consists of a hollow cathode excitation source, which emits radiation for one of the elements to be determined, an interference filter, and a photomultiplier detector. Here, in contrast to the usual design, fluorescence is observed at a vertical angle of about 45 deg. The lamp is pulsed at a frequency of about 500 Hz, and the detection system is synchronously gated to receive these pulses. The timing of the pulses for the 12 lamps is such that at any given instance, only one fluorescence signal is produced and detected.

Spectral specificity of the instrument results from the use of sources that produce only characteristic lines for each of the elements to be determined. The interference filter contributes little to this specificity but instead improves signal-to-noise ratios by limiting the background radiation from the plasma. The pulsed mode of the source also

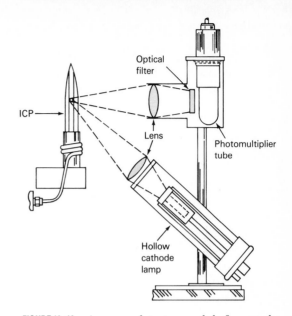

FIGURE 10–12 A source-detector module for a multielement fluorescence instrument based upon inductively coupled plasma atomization. (Courtesy of Baird Corporation, Bedford, MA.)

contributes to the freedom of the system from background effects.

The instrument is highly versatile because each module can be prealigned to observe fluorescence in a part of the plasma that is optimal for a given element. Simultaneous measurement of the fluorescence of 12 elements under the best conditions for each is thus possible. Furthermore, modules can be readily interchanged in a minute or less and are ready to operate after a ten-minute warm-up period.

10C-2 PERFORMANCE CHARACTERISTICS

The instrument just described has been applied for the determination of more than three dozen elements. From the standpoint of detection limits, the fluorescence procedure is comparable with flame emission and absorption methods for the alkali metals and outperforms plasma methods for these elements. On the other hand, the latter pro-

[10] D. R. Demers and C. D. Allemand, *Anal. Chem.,* **1981,** *53,* 1915; D. R. Demers, D. A. Busch, and C. D. Allemand, *Amer. Lab.,* **1982,** *14* (3), 167.

cedure is distinctly superior for the refractory metals such as molybdenum, tungsten, titanium, and vanadium. For the remaining elements, the fluorescence procedure appears roughly comparable in sensitivity to flame absorption and plasma emission methods.

As with plasma emission spectrometers, the fluorescence instrument just described provides linear responses over several decades of concentration for most elements (10^3 to 10^5) and is generally free of matrix effects. Relative precisions are comparable and lie in the 0.2 to 1% range. The major advantage of this new instrument is that it provides simultaneous multielement capabilities with equipment that is relatively inexpensive.

10D Emission Spectroscopy Based on Arc and Spark Sources

In arc and spark sources, sample excitation occurs in the gap between a pair of electrodes. Passage of electricity from the electrodes through the gap provides the necessary energy to atomize the sample and excite the resulting atoms to higher electronic states.[11]

10D-1 SAMPLE HANDLING

Samples for arc and spark source spectroscopy may be solids, liquids, or gases; for the former two, they must be distributed more or less regularly upon the surface of at least

---

[11] For additional details, see: T. Kantor, in *Comprehensive Analytical Chemistry*, G. Svehla, Ed., Vol. V, Chapter 1, Elsevier: New York, 1975; M. Pinta, *Modern Methods for Trace Element Analyses*, Chapters 2 and 3, Ann Arbor Science: Ann Arbor, Mich., 1978; P. W. J. M. Boumans, *Theory of Spectrochemical Excitation*, Plenum Press: New York, 1966; R. D. Sacks, in *Treatise on Analytical Chemistry*, 2d ed., P. J. Elving, E. J. Meehan, and I. M. Kolthoff, Eds., Part I, Vol. 7, Chapter 6, Wiley: New York, 1981.

one of the electrodes that serves as the source.

Metal Samples. If the sample is a metal or an alloy, one or both electrodes can be formed from the sample by milling, by turning, or by casting the molten metal in a mold. Ideally, the electrode will be shaped as a cylindrical rod that is one-eighth inch to one-quarter inch in diameter and tapered at one end. For some samples, it is more convenient to employ the polished, flat surface of a large piece of the metal as one electrode and a graphite or metal rod as the other. Regardless of the ultimate shape of the sample, care must be taken to avoid contamination of the surface while it is being formed.

Electrodes for Nonconducting Samples. For nonmetallic materials, the sample is often supported on an electrode whose emission spectrum will not interfere with the analysis. Carbon is an ideal electrode material for many applications. It can be obtained in a highly pure form, is a good conductor, has good heat resistance, and is readily shaped. Manufacturers offer carbon electrodes in many sizes, shapes, and forms. Frequently, one of the electrodes is a cylinder with a small crater drilled into one end; the sample is then packed into this cavity. The other electrode is commonly a tapered carbon rod with a slightly rounded tip. This configuration appears to produce the most stable and reproducible arc or spark. Figure 10–13 il-

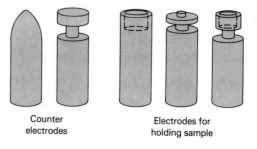

Counter electrodes   Electrodes for holding sample

FIGURE 10-13 Some typical graphite electrode shapes. Narrow necks are to reduce thermal conductivity.

lustrates some of the common electrode forms.

Silver or copper rods are also employed to hold samples when these elements are not of analytical interest. The surfaces of these electrodes must be cleaned and reshaped after each analysis.

Another common method for atomization of powdered samples is based upon *briquetting* or *pelleting* the sample. Here, the finely ground sample is mixed with a relatively large amount of powdered graphite, copper, or other conducting and compressible substance. The resulting mixture is then compressed at high pressure into the form of an electrode.

Excitation of the Constituents of Solutions. Several techniques are encountered for the excitation of the components of solutions or liquid samples. One common method is to evaporate a measured quantity of the solution in a small cup formed in the surface of a graphite or a metal electrode. Alternatively, a porous graphite electrode may be saturated by immersion in the solution; it is then dried before use.

10D-2 ARC SOURCES AND ARC SPECTRA

The usual arc source for a spectrochemical analysis is formed with a pair of graphite or metal electrodes spaced about 1 to 20 mm apart. The arc is initially ignited by a low current spark that causes momentary formation of ions for electrical conduction in the gap; once the arc is struck, thermal ionization maintains the current. Alternatively, the arc can be started by bringing the electrodes together to provide the heat for ionization; they are then separated to the desired distance.

In the typical arc, currents in the range of 1 to 30 A are used. A dc source usually has an open circuit voltage of about 200 V; ac arc source voltages range from 2200 to 4400 V.

Some Characteristics of Arc Sources. Electricity is carried in an arc by the motion of the electrons and ions formed by thermal ionization; the high temperature that develops is a result of the resistance to this motion by the cations in the arc gap. Thus, the arc temperature depends upon the composition of the plasma, which in turn depends upon the rate of formation of atomic particles from the sample and the electrodes. Little is known of the mechanisms by which a sample is dissociated into atoms and then volatilized in an arc. It can be shown experimentally, however, that the rates at which various species are volatilized differ widely. The spectra of some species appear early and then disappear; those for other species reach their maximum intensities at a later time. Thus, the composition of the plasma, and therefore the temperature, may undergo variation with time. Typically, the plasma temperature is 4000 to 5000°K.

The precision obtainable with an arc is generally poorer than that with a spark and much poorer than that with a plasma or flame. On the other hand, an arc source is more sensitive to traces of an element in a sample than is a spark source. In addition, because of its high temperature, chemical interferences, such as those found in flame spectroscopy, are much less common.

Controlled Atmosphere Arc Sources. When a carbon or graphite electrode is arced in air, intense bands due to CN molecules are emitted, which renders most of the region between 350 to 420 nm useless for elemental analysis. Unfortunately, several elements have their most sensitive lines in this region. To avoid or minimize this interference, arc excitation is often performed in an atmosphere of carbon dioxide, helium, or argon. The CN band emission is not completely eliminated under these conditions, however, unless the electrodes are heated under vacuum to drive off adsorbed nitrogen.

10D-3 SPARK SOURCES

Formation of a Spark. Figure 10–14 shows a typical electrical circuit for forming an alternating current in a spark gap. The transformer converts line power to 15,000 to 40,000 V, which then charges the capacitor. When the potential becomes large enough to break down the two air gaps shown on the right, a series of oscillating discharges follows. The voltage then drops until the capacitor is incapable of maintaining further current in the gaps, following which the cycle is repeated. The frequency and current of the discharge are determined by the magnitude of the capacitance, inductance, and resistance of the circuit; these parameters also affect the relative intensities of the neutral atom and ion lines as well as the background intensity.

Many spark sources employ a pair of spark gaps arranged in series (Figure 10–14). The analytical gap is formed by a pair of carbon or metal electrodes that contain the sample. The control gap may be formed from smoothly rounded and carefully spaced electrodes over which is blown a stream of air. As we have noted, conditions at the analytical gap are subject to continual change during operation. Accordingly, electrode spacing is arranged so that the voltage breakdown is determined by the more reproducible control gap. The employment of air-cooling tends to maintain constant operating conditions for the control gap; a higher degree of reproducibility is thus imparted to the analytical results than would otherwise be attained.

The *average* current with a high-voltage spark is usually significantly less than that of the typical arc, being on the order of a few tenths of an ampere. On the other hand, during the initial phase of the discharge the *instantaneous* current may exceed 1000 A; here, electricity is carried by a narrow streamer that involves but a minuscule fraction of the total space in the spark gap. The temperature within this streamer is estimated to be as great as 40,000°K. Thus, while the average electrode temperature of a spark source is much lower than that of an arc, the *energy* in the small volume of the streamer may be several times greater. As a consequence, ionic spectra are more pronounced in a high voltage spark than in an arc.

10D-4 LASER EXCITATION

A pulsed ruby laser provides sufficient energy to a small area on the surface of a sample to cause atomization and excitation of emission lines. This source is far from ideal, however, because of high background intensities, self-absorption, and weak line intensities. One technique that has shown promise is the *laser microprobe,* which is available commercially. Here, the laser is employed to vaporize the sample into a gap between two graphite electrodes, which serves as a spark excitation source. The device is applicable even to nonconducting materials and provides a surface analysis of a spot no greater than 50 μm in diameter.

10D-5 INSTRUMENTS FOR ARC AND
SPARK SOURCE SPECTROSCOPY

Because of their instability, arc and spark sources demand the use of a simultaneous multichannel instrument. In order to obtain reproducible data with these sources, it is necessary to integrate and average a signal

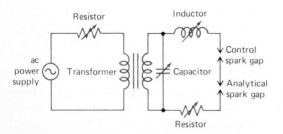

FIGURE 10–14 Power supply for a high-voltage spark source.

for at least 20 sec and often for a minute or more. When several elements must be determined, the time required and the sample consumed by a sequential procedure is usually unacceptable.

The classic multichannel emission instrument is the *spectrograph* in which detection of dispersed radiation is accomplished with a photographic film or plate located at the focal plane of a monochromator. Most spectrographic methods are based upon arc or spark excitation although this type of detection has also been applied more recently to inductively coupled dc plasma excitation.

Photographic Detection. A photographic emulsion can serve as a detector, an amplifier, and an integrating device for evaluating the intensity-time integral for a spectral line. When a photon of radiation is absorbed by a silver halide particle of the emulsion, the radiant energy is stored in the form of a latent image. Treatment of the emulsion with a reducing agent results in the formation of a multitude of silver atoms for each absorbed photon. This process is an example of chemical amplification. The number of black silver particles, and thus the darkness of the exposed area, is a function of the *exposure E*, which is defined by

$$E = I_\lambda t \qquad (10-3)$$

where I_λ is the intensity of radiation and t is the exposure time.[12] In obtaining spectra for quantitative purposes, t is kept constant for both sample and standards. Under these circumstances, exposure is directly proportional to the line intensity and thus to the concentration of the emitting species.

The blackness of a line on a photographic plate or film is expressed in terms of its *optical density D*, which is defined as

$$D = \log \frac{I_0}{I}$$

where I_0 is the intensity (power) of a beam of radiation after it has passed through an unexposed portion of the emulsion, and I is its intensity after being attenuated by the line. Note the similarity between optical density and absorbance; indeed, the former can be measured with an instrument that is similar to the photometer shown in Figure 7–8a. It is also noteworthy that in early spectrophotometry, what is now called absorbance was often termed optical density.

In order to convert the optical density of a line to its original intensity (or exposure), it is necessary to obtain an empirical *plate calibration curve*, which is a plot of optical density as a function of the logarithm of relative exposure. Such a plot is derived experimentally by exposing portions of a plate to radiation of constant intensity for various lengths of time; generally, the absolute intensity of the source will not be known, so that only relative exposures can be calculated from Equation 10–3.

Figure 10–15 shows typical calibration

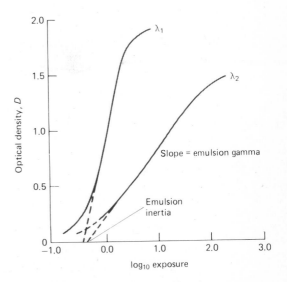

FIGURE 10-15 Typical plate calibration curves at two wavelengths.

---

[12] Here, we follow the convention of emission spectroscopists and employ the term intensity I, rather than the power P. The difference is small (p. 92), and for practical purposes, the terms can be considered synonomous.

curves obtained at two wavelengths. With the aid of such curves, the measured optical densities of various analytical lines can be related to their relative intensities. These relative intensities are the concentration-dependent parameter employed in quantitative spectrographic analyses. Several calibration curves are needed if a spectrum covers a large wavelength range because of the dependence of the slope on wavelength. (Figure 10–15).

Spectrographs. Figure 10–16 is a schematic drawing of an emission spectrograph, which employs a concave grating for dispersing the radiation from a source. Cornu and Littrow type quartz prisms have also found wide use in emission instruments. Generally provisions are made that allow vertical movement of the photographic plate or film holder (the camera) so that several spectra can be recorded successively (see Figure 10–17).

Photoelectric Detectors. Multichannel photoelectric instruments similar to that shown in Figure 10–6 are also used with arc and spark sources. These have found widespread use in the metals industry for production and quality control.

10D-6 QUALITATIVE APPLICATIONS OF ARC AND SPARK SOURCE SPECTROSCOPY

Arc and spark source emission spectroscopy with photographic detectors is one of the most widely used methods for identifying the elemental components of samples of matter. In the past, the procedure has also had extensive use for the routine determination of one or more elements in various types of samples. Now, however, quantitative determinations are more commonly performed by plasma or flame excitation and photoelectric detectors.

Photographic emission spectroscopy permits the detection of some 70 elements by brief arc (or sometimes spark) excitation of a few milligrams of sample. Detection limits vary from less than a part per billion for the alkaline and alkali-earth elements to several hundred times this figure for heavier elements such as tungsten, tantalum, and uranium, and the nonmetallic elements such as selenium, sulfur, phosphorus, and silicon. For the remaining elements detection limits range from 10 to 100 parts per billion.

A particularly important property of arc excitation is that it can be applied to most

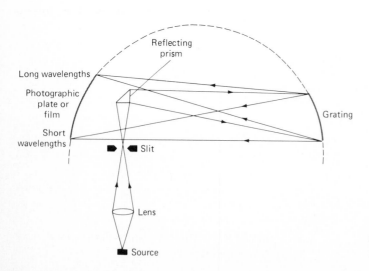

FIGURE 10-16 The Eagle mounting for a grating spectrograph.

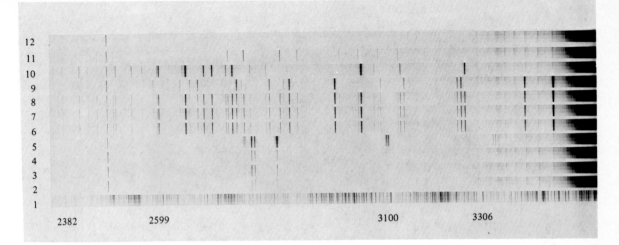

FIGURE 10-17 Typical spectra obtained with a 3.4-meter grating spectrograph. Numbers on horizontal axis are wavelengths in Å. Spectra: (1) iron standard; (2)–(5) casein samples; (6)–(8) Cd-Ge arsenide samples; (9)–(11) pure Cd, Ge, and As, respectively; (12) pure graphite electrode.

samples with little or no preliminary treatment. Furthermore, with photographic detection, the equipment is relatively simple and inexpensive particularly when compared to multichannel photoelectric instruments. Finally, it should be noted that with little extra trouble, semiquantitative information about the relative concentration of sample constituents can be obtained from the degree of blackening of the photographic emulsion.

Excitation Techniques. Ordinarily for qualitative and semiquantitative analyses, excitation times and arc currents are adjusted so that complete volatilization of the sample occurs; currents of 5 to 30 A for 20 to 200 sec are typical. Commonly, 2 to 50 mg of sample in the form of a powder, small chips, grindings, or filings, often mixed with a weighed amount of graphite, are packed into the cavity of graphite electrodes; solutions are evaporated onto the electrode surface. Usually the sample-containing electrode is made the anode with a second graphite counterelectrode acting as the cathode. Several spectra can be obtained on one plate by vertical movement of the camera after each excitation. Usually, one or more iron spectra are also obtained in order to align the plate or film with a master plate showing the location of characteristic lines.

Identification Methods. Figure 10–18 shows the appearance of a typical spectrum when projected adjacent to a master spectrum in a comparator-densitometer. The wavelength region under examination extends from approximately 3170 to 3310 Å. The upper three spectra are for a sample at three different exposures. The fourth spectrum is that of an iron electrode which was obtained immediately after the three earlier exposures. The bottom two spectra are iron spectra on the master plate. After aligning the master and sample plates by means of the many iron lines, identification of elements can be carried out. Usually, positive identification of an element requires matching of several lines. Note that the moderately intense lines at about 3174 and 3261 Å suggest that the sample contains tin. Copper, zinc, and perhaps vanadium also appear to be present in significant amounts. These

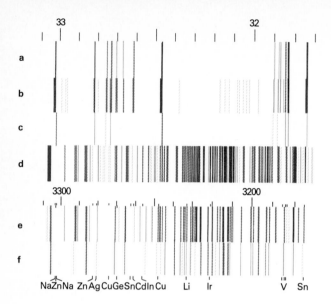

FIGURE 10-18 Projected spectra by a comparator-densitometer: (**a**), (**b**), and (**c**) Spectra of sample at three different exposures; (**d**) iron spectra on the sample plate; (**e**) and (**f**) iron spectra on the master plate.

conclusions would need to be confirmed by further observations at other wavelengths, however.

10D-7 QUANTITATIVE APPLICATIONS OF ARC AND SPARK SOURCE SPECTROSCOPY

Quantitative arc and spark analyses demand precise control of the many variables involved in sample preparation and excitation (and also in film processing with spectrographs). In addition, quantitative measurements require a set of carefully prepared standards for calibration; these standards should approximate as closely as possible the composition and physical properties of the samples to be analyzed.

Internal Standards and Data Treatment. As we have pointed out, the central problem of quantitative arc and spark methods is the very large number of variables that affect the blackness of the image of a spectral line on a photographic plate or the intensity of a line reaching the photoelectric detector. Most variables that are associated with the excitation and the photographic processes are difficult or impossible to control completely. In order to compensate for their effects, an *internal standard* is generally employed.

An internal standard is an element incorporated in a fixed concentration into each sample and each standard. The ratio of the relative intensity of an analyte line to that of a nearby internal standard line then serves as the analytical parameter. Often, a direct proportionality exists between this ratio and analyte concentration. Occasionally, however, the relationship is nonlinear; the resulting curve can still be employed for concentration determinations.

Criteria in the Choice of an Internal Standard. The ideal internal standard has the following properties.

1. Its concentration in samples and standards is always the same.
2. Its chemical and physical properties are as similar as possible to those of the element being determined; only under these circumstances will the internal standard provide adequate compensation for the variables associated with volatilization.
3. It should have an emission line that has about the same excitation energy as one

for the element being determined so that the two lines are similarly affected by temperature fluctuations in the source.

4. The ionization energies of the internal standard and the element of interest should be similar to assure that both have the same distribution ratio of atoms to ions in the source.

5. The lines of the standard and the analyte should be similar in intensity and should be in the same spectral region so as to provide adequate compensation for emulsion variables (or differences in detector response with photoelectric spectrometers).

It is seldom possible to find a line that will meet all of these criteria, and compromises must be made, particularly where the same internal standard is used for the determination of several elements.

If the samples to be analyzed are in solution, considerable leeway is available in the choice of internal standards, since a fixed amount can be introduced volumetrically. Here, an element must be chosen whose concentration in the sample is small relative to the amount to be added as the internal standard.

The introduction of a measured amount of an internal standard is seldom possible with metallic samples. Instead, the major element in the sample is chosen, and the assumption is made that its concentration is essentially invariant. For example, in the quantitative analysis of the minor constituents in a brass, either zinc or copper might be employed as the internal standard.

For powdered samples, the internal standard is sometimes introduced as a solid. Weighed quantities of the finely ground sample and the internal standard are thoroughly mixed prior to excitation.

The foregoing criteria provide theoretical guidelines for the selection of an internal standard; nevertheless, experimental verification of the effectiveness of a particular element and the line chosen is necessary. These experiments involve determining the effects of variation in excitation times, source temperatures, and development procedures on the relative intensities of the lines of the internal standard and the analytes.

Standard Samples. In addition to the choice of an internal standard, a most critical phase in the development of a quantitative emission method involves the preparation or acquisition of a set of standard samples from which calibration curves are prepared. For the ultimate in accuracy, the standards must closely approximate the samples both in chemical composition and physical form; their preparation often requires a large expenditure of time and effort—an expenditure that can be justified economically only if a large number of analyses is anticipated.

In some instances, standards can be synthesized from pure chemicals; solution samples are most readily prepared by this method. Standards for an alloy analysis might be prepared by melting together weighed amounts of the pure elements. Another common method involves chemical analysis of a series of typical samples encompassing the expected concentration range of the elements of interest. A set of standards is then chosen on the basis of these results.

The National Bureau of Standards has available a large number of carefully analyzed metals, alloys, and mineral materials; occasionally, suitable standards can be found among these. In addition, the United States Geological Survey and the Department of Agriculture also have available standard mineral and soil samples. Standard samples are also available from commercial sources.[13]

---

[13] For a listing of sources, see: R. E. Michaelis, *Report on Available Standard Samples and Related Materials for Spectrochemical Analysis, Am. Soc. Testing Materials Spec. Tech. Publ. No. 58–E,* American Society for Testing and Materials: Philadelphia, 1963.

The Method of Standard Additions.   When the number of samples is too small to justify extensive preliminary work, the method of standard additions described in Section 7F–2 may be employed.

Semiquantitative Methods.   Numerous semiquantitative spectrographic methods have been described which provide concentration data reliable to within 30 to 200% of the true amount of an element present in the sample.[14] These methods are useful where the preparation of good standards is not economical. Several such procedures are based upon the total vaporization of a measured quantity (1 to 10 mg) of sample in the arc.

The concentration estimate may then be based on a knowledge of the minimum amount of an element required to cause the appearance of each of a series of lines. In other methods, optical densities of elemental lines are measured and compared with the line of an added matrix material, or with the background. Concentration calculations are then based on the assumption that the line intensity is independent of the state in which the element occurs. The effects of other elements on the line intensity can be minimized by mixing a large amount of a suitable matrix material (a *spectroscopic buffer*) with the sample.

It is sometimes possible to estimate the concentration of an element that occurs in small amounts by comparing the blackness of several of its lines with a number of lines of a major constituent. Matching densities are then used to establish the concentration of the minor constituent.

---

[14] For a more complete description of some semiquantitative procedures, see: C. E. Harvey, *Spectrochemical Procedures,* Chapter 7, Applied Research Laboratories: Glendale, CA, 1950; T. Torok, J. Mika, and E. Gegus, *Emission Spectrochemical Analysis,* pp. 284–308, Crane Russak: New York, 1978.

PROBLEMS

10–1.   What is an internal standard and why is it used?

10–2.   Why are atomic emission methods with an inductively coupled plasma source better suited for multielement analysis than flame atomic absorption methods?

10–3.   Why do ion lines predominate in spark spectra and atom lines in arc and inductively coupled plasma spectra?

10–4.   What is the primary function of the optical interference filter shown in Figure 10–12?

10–5.   Calculate the theoretical reciprocal dispersion of an echelle grating having a focal length of 0.75 m, a groove density of 100 grooves/mm, a diffraction angle of 63°26′ when the diffraction order is (a) 30 and (b) 100.

10–6.   Why are arc sources often blanketed with a stream of an inert gas?

11

Infrared Absorption Spectroscopy

The infrared region of the spectrum encompasses radiation with wavenumbers ranging from about 12,800 to 10 cm^{-1} or wavelengths from 0.78 to 1000 μm.[1] From the standpoint of both application and instrumentation, it is convenient to subdivide the spectrum into *near-*, *middle-*, and *far*-infrared radiation; rough limits of each are shown in Table 11–1. The majority of analytical applications are confined to a portion of the middle region extending from 4000 to 400 cm^{-1} or 2.5 to 25 μm.

Infrared spectroscopy finds widespread application to qualitative and quantitative analyses.[2] Its single most important use has been for the identification of organic compounds whose spectra are generally complex and provide numerous maxima and minima that are useful for comparison purposes (see Figure 11–1). Indeed, in most instances, the infrared spectrum of an organic compound provides a unique fingerprint, which is readily distinguished from the absorption patterns of all other compounds; only optical isomers absorb in exactly the same way.

In addition to its application as a qualitative analytical tool, infrared measurements are finding increasing use for quantitative analysis as well. Here, the high selectivity of the method often makes possible the quantitative estimation of an analyte in a complex mixture with little or no prior separation steps. The most important analyses of this

---

[1] Until recently, the unit of wavelength that is 10^{-6} m was called the *micron*, μ; it is now more properly termed the micrometer, μm.

[2] For detailed discussion of infrared spectroscopy, see: N. B. Colthup, L. H. Daly, and S. E. Wiberley, *Introduction to Infrared and Raman Spectroscopy*, 2d ed., Academic Press: New York, 1975; K. Nakanishi and P. H. Solomon, *Infrared Absorption Spectroscopy*, 2d ed., Holden-Day: San Francisco, 1977; A. L. Smith, *Applied Infrared Spectroscopy*, Wiley: New York, 1979; and A. L. Smith, in *Treatise on Analytical Chemistry*, 2d ed., P. J. Elving, E. J. Meehan, and I. M. Kolthoff, Eds., Part I, vol. 7, Chapter 5, Wiley: New York, 1981.

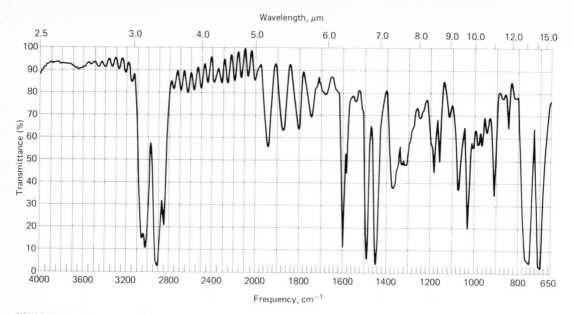

FIGURE 11-1   Infrared absorption spectrum of a thin polystyrene film recorded with a modern IR spectrophotometer. Note that the abscissa scale changes at 2000 cm^{-1}.

type have been of atmospheric pollutants from industrial processes.

11A   Theory of Infrared Absorption

11A-1   INTRODUCTION

A typical infrared spectrum, obtained with a double-beam recording spectrophotometer, is shown in Figure 11–1. In contrast to most ultraviolet and visible spectra, a bewildering array of maxima and minima are observed.

Spectral Plots. The plot shown in Figure 11–1 is a reproduction of the recorder output of a widely used commercial infrared spectrophotometer. As is ordinarily the case, the ordinate is linear in transmittance; it should be noted, however, that the chart paper of some manufacturers also contains a nonlinear absorbance scale for reference. Note also that the abscissa in this chart is linear in units of reciprocal centimeters; other instruments, particularly older ones, employ a scale that is linear in wavelength (an easy way to remember how to convert

Table 11-1			
Infrared Spectral Regions			
Region	Wavelength (λ) Range, μm	Wavenumber (σ) Range, cm^{-1}	Frequency (ν) Range, Hz
Near	0.78 to 2.5	12800 to 4000	3.8×10^{14} to 1.2×10^{14}
Middle	2.5 to 50	4000 to 200	1.2×10^{14} to 6.0×10^{12}
Far	50 to 1000	200 to 10	6.0×10^{12} to 3.0×10^{11}
Most used	2.5 to 15	4000 to 670	1.2×10^{14} to 2.0×10^{13}

from one scale to the other is the relationship $cm^{-1} \times \mu m = 10,000$). For comparison, a wavelength scale has been added on the upper axis of the original chart paper shown in Figure 11–1. Several brands of chart paper contain both wavelength and wavenumber scales; obviously, only one can be linear.

The preference for a linear wavenumber scale is based upon the direct proportionality between this quantity and both energy and frequency; the frequency of the absorbed radiation is, in turn, the molecular vibrational frequency actually responsible for the absorption process. Frequency, however, is seldom if ever employed as the abscissa, probably because of the inconvenient size of the unit (for example, the frequency scale of the plot in Figure 11–1 would extend from 1.2×10^{14} to 2.0×10^{13} Hz). A scale in terms of cm^{-1} is often referred to as a frequency scale; the student should bear in mind that this terminology is not strictly correct, the wavenumber being only proportional to frequency.

Finally, it should be noted that the horizontal scale of Figure 11–1 changes at 2000 cm^{-1}, with the units at higher wavenumbers being represented by half the linear distance of those at lower. This discontinuity is introduced for convenience, since much useful qualitative infrared detail appears at wavenumbers smaller than 2000 cm^{-1}.

Dipole Changes During Vibrations and Rotations. Electronic transitions require energies in the ultraviolet or visible regions; absorption of infrared radiation is thus confined largely to molecular species for which small energy differences exist between various vibrational and rotational states.

In order to absorb infrared radiation, a molecule must undergo a net change in dipole moment as a consequence of its vibrational or rotational motion. Only under these circumstances can the alternating electrical field of the radiation interact with the molecule and cause changes in the amplitude of one of its motions. For example, the charge distribution around a molecule such as hydrogen chloride is not symmetric, the chlorine having a higher electron density than the hydrogen. Thus, hydrogen chloride has a significant dipole moment and is said to be polar. The dipole moment is determined by the magnitude of the charge difference and the distance between the two centers of charge. As a hydrogen chloride molecule vibrates longitudinally, a regular fluctuation in dipole moment occurs, and a field is established which can interact with the electrical field associated with radiation. If the frequency of the radiation matches a natural vibrational frequency of the molecule, there occurs a net transfer of energy that results in a change in the *amplitude* of the molecular vibration; absorption of the radiation is the consequence. Similarly, the rotation of asymmetric molecules around their centers of mass results in a periodic dipole fluctuation; again, interaction with radiation is possible.

No net change in dipole moment occurs during the vibration or rotation of homonuclear species such as O_2, N_2, or Cl_2; consequently, such compounds cannot absorb in the infrared. With the exception of a few compounds of this type, all molecular species exhibit infrared absorption.

Rotational Transitions. The energy required to cause a change in rotational level is minute and corresponds to radiation of 100 μm and greater (<100 cm^{-1}). Because rotational levels are quantized, absorption by gases in this far-infrared region is characterized by discrete, well-defined lines. In liquids or solids, intramolecular collisions and interactions cause broadening of the lines into a continuum.

Vibrational-Rotational Transitions. Vibrational energy levels are also quantized, and the energy differences between quantum states correspond to the readily accessible regions of the infrared from about 13,000 to 675 cm^{-1} (0.78 to 15 μm). The infrared spectrum of a gas usually consists of a series of closely spaced lines, because there are several rotational energy states for each vibrational state. On the other hand, rotation is highly restricted in liquids and solids; in such samples, discrete vibrational-rotational lines disappear, leaving only somewhat broadened vibrational peaks. Our concern is primarily with the spectra of solutions, liquids, and solids, in which rotational effects are minimal.

Types of Molecular Vibrations. The relative positions of atoms in a molecule are not exactly fixed but instead fluctuate continuously as a consequence of a multitude of different types of vibrations. For a simple diatomic or triatomic molecule, it is easy to define the number and nature of such vibrations and relate these to energies of absorption. An analysis of this kind becomes difficult if not impossible for molecules made up of several atoms, not only because of the large number of vibrating centers, but also because interactions among several centers occur and must be taken into account.

Vibrations fall into the basic categories of *stretching* and *bending*. A stretching vibration involves a continuous change in the interatomic distance along the axis of the bond between two atoms. Bending vibrations are characterized by a change in the angle between two bonds and are of four types: *scissoring*, *rocking*, *wagging*, and *twisting*. The various types of vibrations are shown schematically in Figure 11–2.

All of the vibration types shown in Figure 11–2 may be possible in a molecule containing more than two atoms. In addition, interaction or *coupling* of vibrations can occur if the vibrations involve bonds to a single central atom. The result of coupling is a change in the characteristics of the vibrations involved.

In the treatment that follows, we shall first consider isolated vibrations employing a simple mechanical model called the *harmonic oscillator*. Modifications to the theory of the

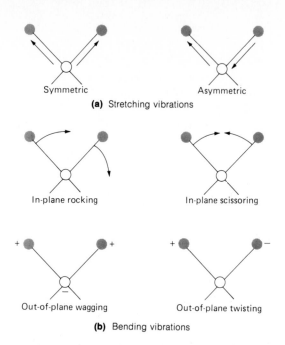

(a) Stretching vibrations

(b) Bending vibrations

FIGURE 11–2 Types of molecular vibrations. Note: + indicates motion from the page toward the reader; − indicates motion away from the reader.

harmonic oscillator, which are needed to describe a molecular system, will be taken up next. Finally, the effects of vibrational interactions in molecular systems will be discussed.

11A-2 MECHANICAL MODEL OF STRETCHING VIBRATIONS

The characteristics of an atomic stretching vibration can be approximated by a mechanical model consisting of two masses connected by a spring. A disturbance of one of these masses along the axis of the spring results in a vibration called a *simple harmonic motion*.

Let us first consider the vibration of a mass attached to a spring that is hung from an immovable object (see Figure 11–3a). If the mass is displaced a distance y from its equilibrium position by application of a

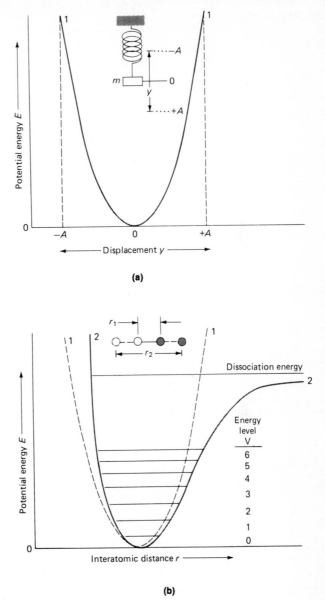

FIGURE 11-3 Potential energy diagrams. Curve 1, harmonic oscillator. Curve 2, anharmonic oscillator.

force along the axis of the spring, the restoring force is proportional to the displacement (Hooke's law). That is,

$$F = -ky \qquad (11\text{–}1)$$

where F is the restoring force and k is the *force constant*, which depends upon the stiff-

ness of the spring. The negative sign indicates that F is a restoring force.

Potential Energy of a Harmonic Oscillator. The potential energy E of the mass and spring can be considered to be zero when the mass is in its rest or equilibrium position. As the spring is compressed or stretched, however, the potential energy of this system increases by an amount equal to the work required to displace the mass. If, for example, the mass is moved from some position y to $(y + dy)$, the work and hence the change in potential energy dE is equal to the force F times the distance dy. Thus,

$$dE = -Fdy \qquad (11\text{–}2)$$

Combining Equations 11–2 and 11–1 yields

$$dE = kydy$$

Integrating between the equilibrium position ($y = 0$) and y gives

$$\int_0^E dE = k \int_0^y ydy$$

$$E = \frac{1}{2} ky^2 \qquad (11\text{–}3)$$

The potential-energy curve for a simple harmonic oscillation, derived from Equation 11–3, is plotted in Figure 11–3**a**. It is seen that the potential energy is a maximum when the spring is stretched or compressed to its maximum amplitude A, and decreases parabolically to zero at the equilibrium position.

Vibrational Frequency.   The *natural* vibrational frequency v_m of a mechanical oscillator, such as that shown in Figure 11–3**a**, is dependent upon the force constant of the spring and the mass of the attached body; it is, however, *independent* of the energy imparted to the system. That is, added energy merely increases the amplitude A of the vibration. The frequency of the vibration can be deduced as follows. Newton's law states that

$$F = ma$$

where m is the mass of the object and a is its acceleration. But acceleration is the second derivative of the displacement y with respect to time. That is,

$$F = m \frac{d^2y}{dt^2}$$

Substituting this relationship into Equation 11–1 gives upon rearranging

$$y = -\frac{m}{k} \frac{d^2y}{dt^2} \qquad (11\text{–}4)$$

A solution to this equation must be a periodic function such that its second derivative is equal to the original function times $-(k/m)$. A suitable cosine relationship meets this requirement. Thus, the instantaneous displacement of the mass at time t can be written as

$$y = A \cos 2\pi v_m t \qquad (11\text{–}5)$$

where v_m is the natural vibrational frequency and A is the maximum amplitude of the motion. The second derivative of Equation 11–5 is

$$\frac{d^2y}{dt^2} = -4\pi^2 v_m^2 A \cos 2\pi v_m t \qquad (11\text{–}6)$$

Substitution of Equations 11–5 and 11–6 into Equation 11–4 gives

$$A \cos 2\pi v_m t = \frac{4\pi^2 v_m^2 m}{k} A \cos 2\pi v_m t$$

The natural frequency of the oscillation is then

$$v_m = \frac{1}{2\pi} \sqrt{\frac{k}{m}} \qquad (11\text{–}7)$$

The equation just developed can be readily modified to describe the behavior of a system consisting of two masses m_1 and m_2 connected by a spring. Here, it is only necessary to substitute the *reduced mass* μ for the single mass m where

$$\mu = \frac{m_1 m_2}{m_1 + m_2} \qquad (11\text{--}8)$$

Thus, the vibrational frequency for such a system is given by

$$\upsilon_m = \frac{1}{2\pi} \sqrt{\frac{k}{\mu}} = \frac{1}{2\pi} \sqrt{\frac{k(m_1 + m_2)}{m_1 m_2}} \qquad (11\text{--}9)$$

Molecular Vibrations. The approximation is ordinarily made that the behavior of a molecular vibration is analogous to the mechanical model just described. Thus, the frequency of the molecular vibration is calculated from Equation 11–9 by substituting the masses of the two atoms for m_1 and m_2; the quantity k becomes the force constant for the chemical bond, which is a measure of its stiffness (but not necessarily its strength).

11A-3 QUANTUM TREATMENT OF VIBRATIONS

Harmonic Oscillators. The equations of ordinary mechanics, such as we have used thus far, do not completely describe the behavior of particles of atomic dimensions. For example, the quantized nature of molecular vibrational energies (and of course other atomic and molecular energies as well) does not appear in these equations. It is possible, however, to employ the concept of the simple harmonic oscillator for the development of the wave equations of quantum mechanics. Solutions of these equations for potential energies are found when

$$E = \left(\upsilon + \frac{1}{2}\right) \frac{h}{2\pi} \sqrt{\frac{k}{\mu}} \qquad (11\text{--}10)$$

where h is Planck's constant, and υ is *the vibrational quantum number*, which can take only positive integer values (including zero). Thus, in contrast to ordinary mechanics where vibrators can have any positive potential energy, quantum mechanics requires

that only certain discrete energies be assumed by a vibrator.

It is of interest to note that the term $(\sqrt{k/\mu})/2\pi$ appears in both the mechanical and the quantum equations; by substituting Equation 11–9 into 11–10, we find

$$E = \left(\upsilon + \frac{1}{2}\right) h\upsilon_m \qquad (11\text{--}11)$$

where υ_m is the vibrational frequency of the mechanical model.

We now assume that transitions in vibrational energy levels can be brought about by radiation, provided the energy of the radiation exactly matches the difference in energy levels ΔE between the vibrational quantum states (and provided also that the vibration causes a fluctuation in dipole). *This difference is identical between any pair of adjacent levels*, since υ in Equations 11–10 and 11–11 can assume only whole numbers; that is,

$$\Delta E = h\upsilon_m = \frac{h}{2\pi} \sqrt{\frac{k}{\mu}} \qquad (11\text{--}12)$$

At room temperature, the majority of molecules are in the ground state ($\upsilon = 0$); thus, from Equation 11–11,

$$E_0 = \frac{1}{2} h\upsilon_m$$

Promotion to the first excited state ($\upsilon = 1$) with energy

$$E_1 = \frac{3}{2} h\upsilon_m$$

requires radiation of energy

$$\left(\frac{3}{2} h\upsilon_m - \frac{1}{2} h\upsilon_m\right) = h\upsilon_m$$

The frequency of radiation υ that will bring

about this change is *identical to the classical vibration frequency of the bond* v_m. That is,

$$E_{radiation} = hv = \Delta E = hv_m = \frac{h}{2\pi}\sqrt{\frac{k}{\mu}}$$

or

$$v = v_m = \frac{1}{2\pi}\sqrt{\frac{k}{\mu}} \qquad (11-13)$$

If we wish to express the radiation in wavenumbers,

$$\sigma = \frac{1}{2\pi c}\sqrt{\frac{k}{\mu}} = 5.3 \times 10^{-12}\sqrt{\frac{k}{\mu}} \qquad (11-14)$$

where σ is the wavenumber of an absorption peak in cm^{-1}, k is the force constant for the bond in dynes/cm, c is the velocity of light in cm/sec, and μ is defined by Equation 11–8.

Equation 11–14 and infrared measurements permit the evaluation of the force constants for various types of chemical bonds. Generally, k has been found to lie in the range between 3×10^5 and 8×10^5 dynes/cm for most single bonds, with 5×10^5 serving as a reasonable average value. Double and triple bonds are found by this same means to have force constants of about two and three times this value, respectively. With these average experimental values, Equation 11–14 can be used to estimate the wavenumber of the fundamental absorption peak (the absorption peak due to the transition from the ground state to the first excited state) for a variety of bond types. The following example demonstrates such a calculation.

EXAMPLE 11-1. Calculate the approximate wavenumber and wavelength of the fundamental absorption peak due to the stretching vibration of a carbonyl group $C=O$.

The mass of the carbon atom is given by

$$m_1 = \frac{12 \text{ g/mol}}{6.0 \times 10^{23} \text{ atoms/mol}}$$

$$= 2.0 \times 10^{-23} \text{ g/atom}$$

Similarly for oxygen,

$$m_2 = 16/(6.0 \times 10^{23}) = 2.7 \times 10^{-23}$$

and the reduced mass μ is given by (Equation 11–8)

$$\mu = \frac{2.0 \times 10^{-23} \times 2.7 \times 10^{-23}}{(2.0 + 2.7) \times 10^{-23}}$$

$$= 1.1 \times 10^{-23}$$

As noted earlier, the force constant for the typical double bond is about 1×10^6 dynes/cm. Substituting this value and μ into Equation 11–14 gives

$$\sigma = 5.3 \times 10^{-12}\sqrt{\frac{1 \times 10^6}{1.1 \times 10^{-23}}}$$

$$= 1.6 \times 10^3 \text{ cm}^{-1}$$

The carbonyl stretching band is found experimentally to be in the region of 6.7 to 5.3 μm or 1500 to 1900 cm^{-1}.

Selection Rules. As given by Equations 11–11 and 11–12, the energy for a transition from energy level 1 to 2 or from level 2 to 3 should be identical to that for the 0 to 1 transition. Furthermore, quantum theory indicates that the only transitions that can take place are those in which the vibrational quantum number changes by unity; that is, the so-called selection rule states that $\Delta v = \pm 1$. Since the vibrational levels are equally spaced, only a single absorption peak should be observed for a given molecular vibration.

Anharmonic Oscillator. Thus far, we have considered the classical and quantum mechanical treatments of the harmonic oscillator. The potential energy of such a vibrator changes periodically as the distance between the masses fluctuates (Figure 11–3**a**). From qualitative considerations, however, it is apparent that this description of a molecular vibration is imperfect. For example, as the two atoms approach one another, coulombic repulsion between the two nuclei produces a force that acts in the same direction as the restoring force of the bond; thus, the potential energy can be expected to rise more rapidly than the harmonic approximation predicts. At the other extreme of oscillation, a decrease in the restoring force, and thus the potential energy, occurs as the interatomic distance approaches that at which dissociation of atoms takes place.

In theory, the wave equations of quantum mechanics permit the derivation of more nearly correct potential-energy curves for molecular vibrations. Unfortunately, however, the mathematical complexity of these equations precludes their quantitative application to all but the very simplest of systems. It is qualitatively apparent, however, that the curves must take the *anharmonic* form shown in Figure 11–3**b**. These curves depart from harmonic behavior by varying degrees, depending upon the nature of the bond and the atoms involved. Note, however, that the harmonic and anharmonic curves are nearly alike at low potential energies. This fact accounts for the success of the approximate methods described.

Anharmonicity leads to deviations of two kinds. At higher quantum numbers, ΔE becomes smaller (see Figure 11–3**b**), and the selection rule is not rigorously followed; as a result, transitions of $\Delta v = \pm 2$ or ± 3 are observed. Such transitions are responsible for the appearance of *overtone lines* at frequencies approximately two or three times that of the fundamental line; the intensity of overtone absorption is frequently low, and the peaks may not be observed.

Vibrational spectra are further complicated by the fact that two different vibrations in a molecule can interact to give absorption peaks with frequencies that are approximately the sums or differences of their fundamental frequencies. Again, the intensities of combination and difference peaks are generally low.

11A-4 VIBRATIONAL MODES

It is ordinarily possible to deduce the number and kinds of vibrations in simple diatomic and triatomic molecules and whether these vibrations will lead to absorption. Complex molecules may contain several types of atoms as well as bonds; for these, the multitude of possible vibrations gives rise to infrared spectra that are difficult, if not impossible, to analyze.

The number of possible vibrations in a polyatomic molecule can be calculated as follows. Three coordinates are needed to locate a point in space; to fix N points requires a set of three coordinates for each for a total of $3N$. Each coordinate corresponds to one degree of freedom for one of the atoms in a polyatomic molecule; for this reason, a molecule containing N atoms is said to have $3N$ *degrees of freedom*.

In defining the motion of a molecule, we need to consider: (1) the motion of the entire molecule through space (that is, the translational motion of its center of gravity); (2) the rotational motion of the entire molecule around its center of gravity; and (3) the motion of each of its atoms relative to the other atoms (in other words, its individual vibrations). Definition of translational motion requires three coordinates and uses up three degrees of freedom. Another three degrees of freedom are needed to describe the rotation of the molecule as a whole. The remaining $(3N - 6)$ degrees of freedom involve interatomic motion, and hence represent the number of possible vibrations within the molecule. A linear molecule is a special case since, by definition, all of the atoms lie on a single, straight line. Rotation about the bond axis is not possible, and two degrees of freedom suffice to describe rotational motion. Thus, the number of vibrations for a linear molecule is given by $(3N - 5)$. Each of the $(3N - 6)$ or $(3N - 5)$ vibrations is called a *normal mode*.

For each normal mode of vibration, there exists a potential-energy relationship such as that shown by the solid line in Figure 11–3**b**. The same selection rules discussed earlier apply for each of these. In addition, to the extent that a vibration approximates harmonic behavior, the differences between the energy levels of a given vibration are the same; that is, a single absorption peak should appear for each vibration in which there is a change in dipole.

In fact, however, the number of normal modes does not necessarily correspond exactly to the number of observed absorption peaks. The number of peaks is frequently less because: (1) the symmetry of the molecules is such that no change in dipole results from a particular vibration; (2) the energies of two or more vibrations are identical or nearly identical; (3) the absorption intensity is so low as to be undetectable by ordinary means; or (4) the vibrational energy is in a wavelength region beyond the range of the instrument. As we have pointed out, additional peaks arise from overtones as well as from combination or difference frequencies.

11A-5 VIBRATIONAL COUPLING

The energy of a vibration, and thus the wavelength of its absorption peak, may be influenced (or coupled) by other vibrators in the molecule. A number of factors that influence the extent of such coupling can be identified.

1. Strong coupling between stretching vibrations occurs only when there is an atom common to the two vibrations.
2. Interaction between bending vibrations requires a common bond between the vibrating groups.

3. Coupling between a stretching and a bending vibration can occur if the stretching bond forms one side of the angle that varies in the bending vibration.
4. Interaction is greatest when the coupled groups have individual energies that are approximately equal.
5. Little or no interaction is observed between groups separated by two or more bonds.
6. Coupling requires that the vibrations be of the same symmetry species.[3]

As an example of coupling effects, let us consider the infrared spectrum of carbon dioxide. If no coupling occurred between the two C=O bonds, an absorption peak would be expected at the same wavenumber as the peak for the C=O stretching vibration in an aliphatic ketone (about 1700 cm^{-1}, or 6 μm; see Example 11–1). Experimentally, carbon dioxide exhibits two absorption peaks, the one at 2330 cm^{-1} (4.3 μm) and the other at 667 cm^{-1} (15 μm).

Carbon dioxide is a linear molecule and thus has four normal modes ($3 \times 3 - 5$). Two stretching vibrations are possible; furthermore, interaction between the two can occur since the bonds involved are associated with a common carbon atom. As may be seen, one of the coupled vibrations is symmetric and the other is asymmetric.

Symmetric   Asymmetric

The symmetric vibration causes no change in dipole, since the two oxygen atoms simultaneously move away from or toward the central carbon atom. Thus, the symmetric vibration is infrared inactive. One oxygen approaches the carbon atom as the other

---

[3] For a discussion of symmetry operations and symmetry species, see: R. P. Bauman, *Absorption Spectroscopy*, Chapter 10, Wiley: New York, 1962; and F. A. Cotton, *Chemical Applications of Group Theory*, Wiley: New York, 1971.

moves away during asymmetric vibration. As a consequence, a net change in charge distribution occurs periodically; absorption at 2330 cm^{-1} results.

The remaining two vibrational modes of carbon dioxide involve scissoring, as shown below.

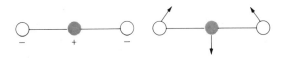

The two bending vibrations are the resolved components (at 90 deg to one another) of the bending motion in all possible planes around the bond axis. The two vibrations are identical in energy and thus produce but one peak at 667 cm^{-1}. (Quantum states that are identical, as these are, are said to be *degenerate*.)

It is of interest to compare the spectrum of carbon dioxide with that of a nonlinear, triatomic molecule such as water, sulfur dioxide, or nitric oxide. These molecules have $(3 \times 3 - 6)$, or 3, vibrational modes which take the following forms:

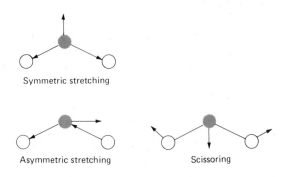

Symmetric stretching

Asymmetric stretching   Scissoring

Since the central atom is not in line with the other two, a symmetric stretching vibration will produce a change in dipole and will thus be responsible for infrared absorption. For example, stretching peaks at 3650 and 3760 cm^{-1} (2.74 and 2.66 μm) are observed for the symmetric and asymmetric vibrations of the water molecule. Only one component to the scissoring vibration exists for this non-linear molecule, since motion in the plane of the molecule constitutes a rotational degree of freedom. For water, the bending vibration causes absorption at 1595 cm^{-1} (6.27 μm).

The difference in behavior of linear and nonlinear triatomic molecules (two and three absorption peaks, respectively) illustrates how infrared-absorption spectroscopy can sometimes be used to deduce molecular shapes.

Coupling of vibrations is a common phenomenon; as a result, the position of an absorption peak corresponding to a given organic functional group cannot be specified exactly. For example, the C—O stretching frequency in methanol is 1034 cm^{-1} (9.67 μm); in ethanol it is 1053 cm^{-1} (9.50 μm) and in methylethylcarbinol it is 1105 cm^{-1} (9.05 μm). These variations result from a coupling of the C—O stretching with adjacent C—C or C—H vibrations.

While interaction effects may lead to uncertainties in the identification of functional groups contained in a compound, it is this very effect that provides the unique features of an infrared-absorption spectrum that are so important for the positive identification of a specific compound.

11B Infrared Instrument Components

Infrared photometers and spectrophotometers are similar in construction to the instruments already described for absorption studies in the ultraviolet and visible regions. The major components in these instruments are illustrated in Figure 5–1a, page 113.

11B-1 CONTINUOUS SOURCES OF INFRARED RADIATION

The common infrared source is an inert solid heated electrically to temperatures between 1500 and 2000°K. Continuous radiation approximating that of a black body results (see Figure 4–14, p. 110). The maximum radiant intensity at these temperatures occurs between 1.7 and 2 μm (5900 to 5000 cm^{-1}). At longer wavelengths, the intensity falls off continuously until it is about 1% of the maximum at 15 μm (667 cm^{-1}). On the short wavelength side, the decrease is much more rapid, and a similar reduction in intensity is observed at about 1 μm (10,000 cm^{-1}).

The Nernst Glower.   The Nernst glower is composed of rare earth oxides formed into a cylinder having a diameter of 1 to 2 mm and a length of perhaps 20 mm. Platinum leads are sealed to the ends of the cylinder to permit passage of electricity; temperatures between 1200 and 2000°K result. The Nernst glower has a large negative temperature coefficient of electrical resistance, and it must be heated externally to a dull red heat before the current is large enough to maintain the desired temperature. Because the resistance decreases with increasing temperature, the source circuit must be designed to limit the current; otherwise the glower rapidly becomes so hot that it is destroyed.

The Globar Source.   A Globar is a silicon carbide rod, usually about 50 mm in length and 5 mm in diameter. It also is electrically heated (1300 to 1500°K) and has the advantage of a positive coefficient of resistance. On the other hand, water cooling of the electrical contacts is required to prevent arcing. Spectral energies of the Globar and the Nernst glower are comparable except in the region below 5 μm, where the Globar provides a significantly greater output.

Incandescent Wire Source.   A source of somewhat lower intensity but longer life than the Globar or Nernst glower is a tightly wound spiral of nichrome wire heated to about 1100°K by an electrical current. A rhodium wire heater sealed in a ceramic cylinder has similar properties as a source.

The Mercury Arc.   For the far-infrared region of the spectrum ($\lambda > 50$ μm), none of the thermal sources just described provides sufficient energy for convenient detection.

Here, a high-pressure mercury arc is used. This device consists of a quartz-jacketed tube containing mercury vapor at a pressure greater than one atmosphere. Passage of electricity through the vapor forms an internal plasma source that provides continuous radiation in the far-infrared region.

The Tungsten Filament Lamp.   An ordinary tungsten filament lamp is a convenient source for the near-infrared region of 0.78 to 2.5 μm (12,800 to 4000 cm^{-1}).

The Carbon Dioxide Laser Source.   A tunable carbon dioxide laser has found application as an infrared source for monitoring the concentrations of certain atmospheric pollutants and for the determination of absorbing species in aqueous solutions.[4] A carbon dioxide laser produces a band of radiation in the 9 to 11 μm (1100 to 900 cm^{-1}) range, which consists of about 100 closely spaced discrete lines. As described on page 120, any one of these lines can be chosen by tuning the laser. Although the range of wavelength available is limited, the 9 to 11 μm region is one particularly rich in absorption bands (arising from interactive stretching modes). Thus, the source has proven useful for quantitative determination of a number of important species such as ammonia, butadiene, benzene, ethanol, nitrogen dioxide, and trichloroethylene. An important property of the laser source is the amount of energy available in each line, which is several orders of magnitude greater than that of black-body sources.

11B-2 *MONOCHROMATORS AND FILTERS*

Infrared wavelength selection can be accomplished by means of interference filters, prisms, or gratings.

Interference Filters.   Interference filters (Section 5C–1) for quantitative infrared analysis

---

[4] See: H. R. Jones and P. A. Wilks, Jr., *Amer. Lab.*, **1982**, *14* (3), 87; and L. B. Kreuzer, *Anal. Chem.*, **1974**, *46*, 239A.

of specific compounds are available commercially. For example, a filter with a transmission maximum of 9.0 μm is offered for the determination of acetaldehyde; one with a maximum at 13.4 μm is suggested for o-dichlorobenzene; and a filter exhibiting a transmission peak at 4.5 μm is employed for the determination of nitrous oxide. In general, these filters have effective bandwidths of about 1.5% of their peak wavelengths.

Filter Wedges. One instrument manufacturer[5] offers an instrument that provides narrow and continuously variable bands of infrared radiation by means of three filter wedges (Section 5C–2) formed as segments of a circle. The ranges of the three segments are 2.5 to 4.5, 4.5 to 8, and 8 to 14.5 μm. The wedges are mounted on a wheel which can be rotated to provide the desired wavelength at the slit (see Figure 11–13). The effective bandwidth of the device is about 1.5% of the wavelength that is transmitted to the slit.

Prisms. Several materials have been used for prism construction. Quartz is employed for the near-infrared region (0.8 to 3 μm), even though its dispersion characteristics for this region are far from ideal. It absorbs strongly beyond about 4 μm (2500 cm^{-1}). Crystalline sodium chloride is the most common prism material; its dispersion is high in the region between 5 and 15 μm (2000 and 670 cm^{-1}) and is adequate to 2.5 μm (4000 cm^{-1}). Beyond 20 μm (500 cm^{-1}), sodium chloride absorbs strongly and cannot be used. Crystalline potassium bromide and cesium bromide prisms are suitable for the far-infrared region (15 to 40 μm or 670 to 250 cm^{-1}), while lithium fluoride is useful in the near-infrared region (1 to 5 μm or 10,000 to 2000 cm^{-1}). Unfortunately, all common infrared-transmitting materials except quartz are easily scratched and are water-soluble. Desiccants or heaters are required to protect them from the condensation of moisture.

Reflection Gratings. At the present time all commercially available infrared spectrophotometers employ reflection gratings rather than prisms as dispersing elements. Generally, gratings provide better resolution because less loss of radiant energy accompanies their use; thus, narrower slit widths can be realized. Other advantages include more nearly linear dispersion and resistance to attack by water. An infrared grating is usually constructed from glass or plastic that has been coated with aluminum.

The disadvantage of a grating lies in the greater amounts of scattered radiation and the appearance of radiation of other spectral orders. In order to minimize these effects, gratings are blazed to concentrate the radiation into a single order. Filters (and occasionally prisms) are used in conjunction with the grating to minimize these problems.

11B-3 BEAM ATTENUATION

Infrared instruments are generally of a null type, in which the power of the reference beam is reduced, or *attenuated,* to match that of the beam passing through the sample. Attenuation is accomplished by imposing a device that removes a continuously variable fraction of the reference beam. The attenuator commonly takes the form of a fine-toothed comb, the teeth of which are tapered so that a linear relationship exists between the lateral movement of the comb and the decrease in power of the beam. Movement of the comb occurs when a difference in power of the two beams is sensed by the detector. This movement is synchronized with the recorder pen so that its position gives a measure of the relative power of the two beams and thus the transmittance of the sample. Many instruments also employ a beam attenuator in conjunction with the

---

[5] The Foxboro Company, Foxboro, MA 02035.

sample beam for the purpose of making the 100% T adjustment.

11B-4 INFRARED DETECTORS

Thermal detectors, whose responses depend upon the heating effect of radiation, are employed for detection of all but the shortest infrared wavelengths. With these devices, the radiation is absorbed by a small black body and the resultant temperature rise is measured. The radiant power level from a spectrophotometer beam is minute (10^{-7} to 10^{-9} W), so that the heat capacity of the absorbing element must be as small as possible if a detectable temperature change is to be produced. Every effort is made to minimize the size and thickness of the absorbing element and to concentrate the entire infrared beam on its surface. Under the best of circumstances, temperature changes are confined to a few thousandths of a degree Celsius.

The problem of measuring infrared radiation by thermal means is compounded by thermal effects or noise from the surroundings. For this reason, the detector is housed in a vacuum and is carefully shielded from thermal radiation emitted by other nearby objects. To further minimize the effects of extraneous heat sources, the beam from the source is always chopped. In this way, the analyte signal, after transduction, has the frequency of the chopper; with appropriate circuitry, this signal can then be separated from extraneous noise signals, which ordinarily vary only slowly with time.

Thermocouples.   In its simplest form, a thermocouple consists of the junctions formed when two pieces of a given metal are fused to either end of a dissimilar metal or a semiconducting metal alloy. A potential develops between the two thermocouple junctions which varies with the temperature *difference* between the two junctions (p. 47).

The detector junction for infrared radiation is formed from very fine wires of such metals as platinum and silver or antimony and bismuth. Alternatively, the detector junction may be constructed by evaporating the metals onto a nonconducting support. In either case, the junction is usually blackened (to improve its heat absorbing capacity) and sealed in an evacuated chamber with a window that is transparent to infrared radiation.

The reference junction, which is usually housed in the same chamber as the active junction, is designed to have a relatively large heat capacity; it is carefully shielded from the incident radiation. Because the analyte signal is chopped, only the difference in temperature between the two junctions is important; therefore, the reference junction does not need to be maintained at constant temperature.

A well-designed thermocouple detector is capable of responding to temperature differences of 10^{-6} °C. This figure corresponds to a potential difference of about 6 to 8 μV/ μW. The thermocouple of an infrared detector is a low-impedance device; it is frequently connected to a high-impedance preamplifier such as that shown in Figure 11–4.

Bolometers.   A bolometer is a type of resistance thermometer constructed of strips of metals such as platinum or nickel, or from a semiconductor; the latter devices are sometimes called *thermistors*. These materials exhibit a relatively large change in resistance as a function of temperature. The responsive element is kept small and blackened to absorb the radiant heat. Bolometers are not so extensively used as other infrared detectors.

Pyroelectric Detectors.   Certain crystals, such as lithium tantalate, barium titanate, and triglycine sulfate, possess temperature-sensitive dipole moments. When placed between metal plates, a temperature-sensitive capacitor is formed which can be employed

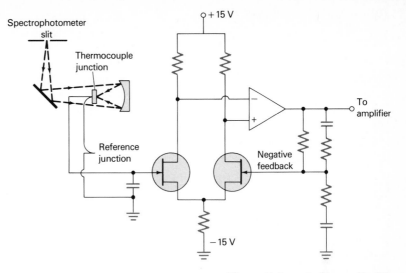

FIGURE 11-4 Thermocouple and preamplifier. (Adapted from: G. W. Ewing, *J Chem. Educ.*, **1971**, *48*, A521. With permission.)

for measuring the power of infrared radiation. Here, the transduced signal is capacitance.

The Golay Detector. The Golay detector is essentially a sensitive gas thermometer in which xenon is contained in a small cylindrical chamber, which contains a blackened membrane. One end of the cylinder is sealed with an infrared window; the other consists of a flexible diaphragm which is silvered on the outside. A light beam is reflected from this silvered surface to the cathode of a vacuum phototube. When infrared radiation enters the cell, the blackened membrane is warmed, which in turn heats the xenon by conduction. The resulting increase in pressure causes distortion of the silvered diaphragm. As a consequence, the fraction of the reflected light beam that strikes the active surface of the phototube is changed; a change in photocurrent results, which can be related to the power of the infrared beam.

The Golay cell is more expensive than other heat detectors and no more sensitive to near- and mid-infrared radiation; thus, it is seldom employed for these spectral regions. On the other hand, the Golay detector is significantly superior for radiation with wavelengths greater than 50 μm (<200 cm^{-1}); it finds general use in instruments designed for the far-infrared regions.

11C Some Typical Instruments

Several types of instruments for infrared absorption measurements are available from commercial sources. The most widely used are dispersive spectrophotometers based upon gratings, prisms, or interference wedges; these are used primarily for qualitative work. Interferometric multiplex instruments, employing the Fourier transform (Section 5I–5), are now finding more and more general applications for both qualitative and quantitative infrared measurements. Finally, a group of nondispersive photometers have been developed for quantitative determination of a variety of compounds.

11C-1 DISPERSIVE INSTRUMENTS

All dispersive infrared spectrophotometers from commercial sources are double-beam, recording instruments. As was pointed out in Section 6C–4, the double-beam design is less demanding with respect to the performance of sources and detectors—an important characteristic because of the relatively low stability of infrared sources and detectors and the need for large signal amplifications.

Generally, dispersive infrared spectrophotometers incorporate a low-frequency chopper (5 to 13 cycles per minute) to modulate the source output. This feature permits the detector to discriminate between the signal from the source and signals from extraneous radiation, such as infrared emission from various bodies surrounding the detector. Low chopping rates are demanded by the slow response times of most infrared detectors.

Several dozen infrared spectrophotometers, ranging in cost from $5000 to more than $30,000, are available from various instrument manufacturers. In general, the optical design of these instruments does not differ greatly from the ultraviolet-visible spectrophotometers discussed in the previous chapter except that the sample and reference compartment is always located between the source and the monochromator in infrared instruments. This arrangement is possible because infrared radiation, in contrast to ultraviolet-visible, is not sufficiently energetic to cause photochemical decomposition of the sample. Placing the sample and reference before the monochromator, however, has the advantage that most scattered radiation, generated within the cell compartment, is effectively removed by the monochromator and thus does not reach the detector.

Typical Instrument Design. Figure 11–5 shows schematically the arrangement of components in a typical infrared spectrophotometer. Note that three types of systems link the components: (1) a radiation linkage indicated by dashed lines; (2) a mechanical link-

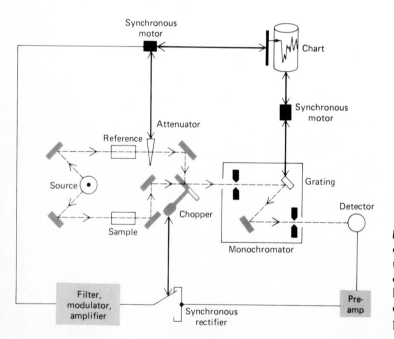

FIGURE 11-5 Schematic diagram of a double-beam spectrophotometer. Heavy dark line indicates mechanical linkage; light line indicates electrical linkage; dashed line indicates radiation path.

age shown by thick dark lines; and (3) an electrical linkage shown by narrow solid lines.

Radiation from the source is split into two beams, half passing into the sample-cell compartment and the other half into the reference area. The reference beam then passes through an attenuator and on to a chopper. The chopper consists of a motor-driven disk that alternately reflects the reference or transmits the sample beam into the monochromator. After dispersion by a prism or grating, the alternating beams fall on a detector and are converted to an electrical signal. The signal is amplified and passed to the synchronous rectifier, a device that is mechanically or electrically coupled to the chopper to cause the rectifier switch and the beam leaving the chopper to change simultaneously. If the two beams are identical in power, the signal from the rectifier is an unfluctuating direct current. If, on the other hand, the two beams differ in power, a fluctuating or ac current is produced, the phase of which is determined by which beam is the more intense. The current from the rectifier is filtered and further amplified

to drive a synchronous motor in one direction or the other, depending upon the phase of the input current. The synchronous motor is mechanically linked to both the attenuator and the pen drive of the recorder and causes both to move until a null is achieved. A second synchronous motor drives the chart and varies the wavelength simultaneously. There is frequently a mechanical linkage between the wavelength and slit drives so that the radiant power reaching the detector is kept approximately constant by variations in the slit width.

Figure 11–6 shows the optics of a simple inexpensive spectrophotometer manufactured by one of the well-known instrument companies. Radiation from the nichrome-wire source is split and passes through the sample compartment. Both beams have attenuators, one for the 100% T adjustment and the other for reducing the reference beam intensity to match that of the sample beam, thus providing a measure of the transmission of the sample. The beams are recombined by means of the segmented chopper, dispersed by the monochromator, and detected by a thermocouple. The ac

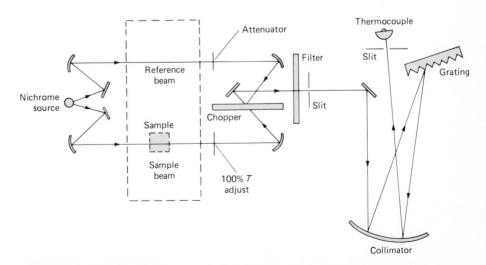

FIGURE 11-6 Schematic diagram of the Beckman AccuLab® Infrared Spectrophotometer. (Courtesy of Beckman Instruments, Inc., Fullerton, CA. With permission.)

output of the latter, after amplification, drives the reference attenuator to a null position; the pen drive is coupled mechanically to this movement.

Instrument Automation.   As with ultraviolet and visible instruments, newer infrared instruments usually contain microprocessor systems for partial automation of the measurement process. In the less-expensive instruments, automation includes slit programs, scan-time adjustments, calibration modes for checking the frequency scale, and gain adjustment for survey scans. More sophisticated instruments offer push-button control of the normal scanning parameters, a visual display of selected nonstandard parameters, automatic means for superimposing spectra, and computer storage of spectra for subsequent signal averaging.

11C-2 MULTIPLEX INSTRUMENTS; FOURIER TRANSFORM SPECTROMETERS

The theoretical basis and the inherent advantage of multiplex instruments was discussed in some detail in Section 5I-5, and the reader may find it worthwhile reviewing this section before proceeding further here. Two types of multiplex instruments have been described for the infrared region. In one, coding is accomplished by splitting the source into two beams whose path lengths can be varied periodically to give interference patterns; here, the Fourier transform is used for data processing. The second is the Hadamard transform spectrometer, which is a dispersive instrument that employs a moving mask at the focal plane of a monochromator for encoding the spectral data. Hadamard transform infrared instruments have not been widely adopted and will, therefore, not be discussed further in this text.[6]

---

[6] For a description of a Hadamard transform spectrometer, see: J. A. Decker, Jr., *Appl. Opt.*, **1971**, *16*, 510 and *Amer. Lab.*, **1972**, *4* (1), 29.

Fourier transform spectrometers are now offered by several instrument manufacturers. These instruments are particularly well-suited for analytical problems requiring high resolution, rapid scans, or where the analytical signal is weak. Fourier transform instruments are expensive, having prices that range from $20,000 to $150,000 or more.

Types of Fourier Transform Instruments.   Commercially available Fourier transform infrared instruments are of two types. The first, and more common, is based upon the Michelson interferometer, which is illustrated in Figure 5–30, page 153. A radically different design is offered by two companies, however. As shown in Figure 11–7, a periodic path difference in this case is generated by the motion of a scanning wedge rather than a moving mirror. This discussion will be concerned with the Michelson design only.

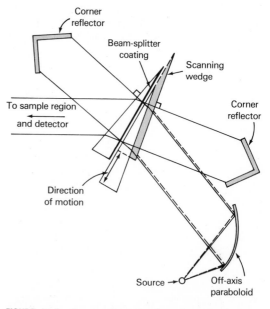

FIGURE 11-7 A moving wedge interferometer. (From: P. R. Griffiths, *Science*, **1983**, *222*, 299. With permission. Copyright 1983 by the American Association for the Advancement of Science.)

Drive Mechanism. A requirement for satisfactory interferograms (and thus satisfactory spectra) is that the speed of the moving mirror be relatively constant and its position exactly known at any instant. The planarity of the mirror must also remain constant during its entire sweep of 10 cm or more.

In the far-infrared region, where wavelengths range from 50 to 1000 μm, displacement of the mirror by a fraction of a wavelength, and accurate measurement of its position, can be accomplished by means of a motor-driven micrometer screw. A more precise and sophisticated mechanism is required for the mid- and near-infrared regions, however. Here, the mirror mount is generally floated on air cushions held within close-fitting stainless steel sleeves (see Figure 11–8). The mount is driven by an electromagnetic coil similar to the voice coil in a loudspeaker; an increasing current in the coil drives the mirror at constant velocity. After reaching its terminus, the mirror is returned rapidly to the starting point for the next sweep by a rapid reversal of the current. The length of travel varies from 2 to about 18 cm; the scan rates range from 0.05 to 4 cm/sec.

Two additional features of the mirror system are necessary for successful operation. The first is a means of sampling the interferogram at precisely spaced retardation intervals. The second is a method for determining exactly the zero-retardation point in order to permit signal averaging. If this point is not known precisely, the signals from repetitive sweeps would not be fully in phase; averaging would then tend to degrade rather than improve the signal.

The problem of precise signal sampling and signal averaging is accomplished in modern instruments by using three interferometers rather than one, with a single mirror mount holding the three movable mirrors. Figure 11–8 is a schematic diagram showing the arrangement employed by one

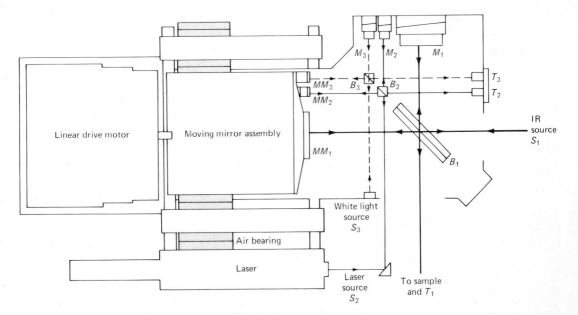

FIGURE 11-8 Interferometers in an infrared Fourier transform spectrometer. Subscripts 1 define the radiation path in the infrared interferometer; subscripts 2 and 3 refer to the laser and white-light interferometers, respectively. (Courtesy of Nicolet Analytical Instruments, Madison, WI.)

manufacturer. The components and radiation paths for each of the three interferometer systems are indicated by the subscripts 1, 2, and 3, respectively. System 1 is the infrared system that ultimately provides an interferogram similar to that shown as curve *A* in Figure 11–9. System 2 is a so-called *laser-fringe reference* system, which provides sampling-interval information. It consists of a helium neon laser S_2, an interferometric system including mirrors MM_2 and M_2, a beam splitter B_2, and a transducer T_2. The output from this system is sinusoidal, as shown in *C* of Figure 11–9. This signal is converted electronically to the square-wave form shown in *D*; sampling begins or terminates at each successive zero crossing. The laser-fringe reference system gives a highly reproducible and regularly spaced sampling interval. In most instruments, the laser signal is also employed to control the speed of the mirror-drive system at a constant level.

The third interferometer system, some-times called the *white-light* system, employs a tungsten source S_3 and transducer T_3 sensitive to visible radiation. Its mirror system is fixed to give a zero retardation that is displaced to the left from that for the analytical signal (see interferogram *B*, Figure 11–9). Because the source is polychromatic, its power at zero retardation is much larger than any signal before and after that point. Thus, this maximum can be employed to trigger the start of data sampling for each sweep at a highly reproducible point.

The triple mirror design of modern Fourier transform instruments leads to remarkable precision in determining frequencies, which significantly exceeds that realizable with conventional grating instruments. This high reproducibility is particularly important when many spectra are to be averaged.

Beam Splitters. Beam splitters are constructed of transparent materials with refractive indices such that approximately

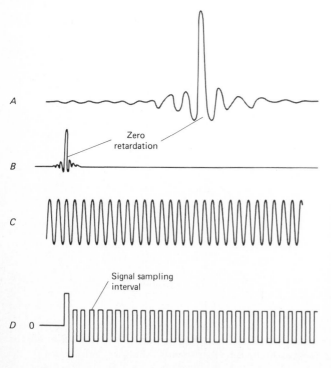

FIGURE 11-9 Time-domain signals for the three interferometers contained in a Fourier transform infrared instrument. Curve *A*: infrared signal; curve *B*: white-light signal; curve *C*: laser-fringe reference signal; curve *D*: square-wave electrical signal formed from the laser signal. (From: P. G. Griffiths, *Chemical Infrared Fourier Transform Spectroscopy*, p. 102, Wiley: New York, 1975. Reprinted by permission of John Wiley & Sons, Inc.)

50% of the radiation is reflected. A widely used material for the far-infrared region is a thin film of Mylar sandwiched between two plates of a low-refractive-index solid. Thin films of germanium or silicon deposited on cesium iodide or bromide, sodium chloride, or potassium bromide are satisfactory for the mid-infrared region. A film of iron(III) oxide is deposited on calcium fluoride for work in the near-infrared.

Sources and Detectors. The sources for Fourier transform infrared instruments are similar to those discussed earlier in this chapter. Generally, thermal detectors are not readily adapted to Fourier transform instruments because of their slow response times. In their stead, triglycine sulfate pyroelectric detectors were used in early instrument designs. By now, more sensitive liquid nitro-gen cooled mercury/cadmium telluride or indium antimonide photoconductive detectors are employed. Several other types of detectors are also beginning to make their appearance.

Double-Beam Design. Instruments for the far-infrared region are often of single-beam design. Most instruments for the higher-frequency mid-infrared range are double-beam; a typical design is shown in Figure 11–10. Here, movement of mirror B causes the beams to alternate between the reference and sample cell holders. Synchronous movement of mirror C then directs the beams to the triglycine sulfate pyroelectric detector. Movement of mirror A by 90 deg directs the beam externally where it can be employed as a gas-chromatographic detector.

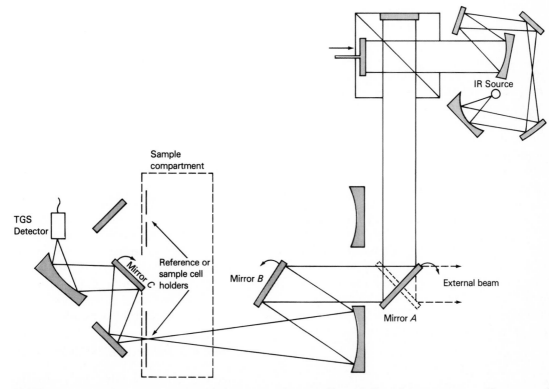

FIGURE 11-10 Optical design of a double-beam Fourier transform infrared spectrometer. (Courtesy of Digilab, Division of Bio-Rad Laboratories, Cambridge, MA.)

Performance Characteristics of Commercial Instruments. A number of instrument manufacturers offer several models of Fourier transform infrared instruments. The least expensive of these (~$25,000) has a range of 4800 to 400 cm^{-1} (2.1 to 25 μm) with a resolution of 4 cm^{-1}. This performance can be obtained with a scan time as brief as 1 sec. More expensive instruments (up to $150,000 or more) with interchangeable beam splitters, sources, and detectors offer expanded frequency ranges and higher resolutions. For example, one instrument is reported to produce spectra from the far-infrared (10 cm^{-1} or 1000 μm) through the visible region to 25,000 cm^{-1} or 400 nm. Resolutions for commercial instruments vary from 8 to 0.06 cm^{-1}. Several minutes are required to obtain a complete spectrum at the highest resolutions.

Advantages of Fourier Transform Spectrometers.[7] Over most of the mid-infrared spectral range, Fourier transform instruments appear to have signal-to-noise ratios that are better than those of a good-quality dispersive instrument by more than an order of magnitude. The enhanced signal-to-noise ratio can, of course, be traded for rapid scanning, with good spectra being attainable in a few tenths of a second in many cases. Interferometric instruments are also characterized by high resolutions (<0.1 cm^{-1}) and highly accurate and reproducible frequency determinations. The latter property is particularly helpful when spectra are to be subtracted for background correction.

A theoretical advantage of Fourier transform instruments is that their optics provide a much larger energy throughput (one or two orders of magnitude) than do dispersive instruments, which are limited in throughput by the necessity of narrow slit widths.

The potential gain here, however, may be partially offset by the lower sensitivity of the fast-response detector required for the interferometric measurements. Finally, it should be noted that the interferometer is free from the problem of stray radiation because each IR frequency is, in effect, chopped at a different frequency.

The Fourier transform method offers little or no advantage (other than shortened analysis time) over a good-quality grating spectrophotometer for routine qualitative applications in the region between 670 and 4000 cm^{-1} (15 and 2.5 μm). Furthermore, it suffers by comparison in terms of high initial cost and substantial maintenance problems. The latter arise because the quality of a Fourier transform spectrum degrades much more rapidly with instrument maladjustment than does a spectrum produced by grating instruments. It is noteworthy that the cost differential between the two types of instruments appears to be lessening with time. Furthermore, the reliability of Fourier instruments has improved markedly since their first appearance on the market.

The areas of chemistry where the extra performance of interferometric instruments appears to be particularly relevant include: (1) very high-resolution work which is encountered with gaseous mixtures having complex spectra resulting from the superposition of vibrational and rotational bands; (2) the study of samples with high absorbances; (3) the study of substances with weak absorption bands (for example, the study of compounds that are chemisorbed on catalyst surfaces); (4) investigations requiring fast scanning such as kinetic studies or detection of chromatographic effluents; (5) infrared emission studies.

11C-3 PHOTOACOUSTIC INFRARED INSTRUMENTS

In Section 7H, which deals briefly with the principles of photoacoustic measurements, it was noted that this technique has been

---

[7] For a comparison of performance characteristics of typical dispersive and interferometric instruments, see: D. H. Chenery and N. Sheppard, *Appl. Spectrosc.,* **1978,** *32,* 79. See also: *Anal. Chem.,* **1983,** *55,* 1054A.

used to obtain spectra for qualitative studies in the mid-infrared region.[8] As with ultraviolet and visible spectral studies, the technique is profitably applied to those solid and liquid samples that are difficult to handle by ordinary techniques because of their tendency to scatter radiation. In addition, the method has been used for detecting the components of mixtures separated by thin-layer and high-performance liquid chromatography. Most of this work has been carried out with Fourier transform instruments because of their better signal-to-noise characteristics. Most manufacturers offer photoacoustic cells as accessories for FTIR instruments.

Photoacoustic infrared spectroscopy has also been used for monitoring the concentrations of gaseous pollutants in the atmosphere.[9] Here, a tunable carbon dioxide laser source is used in conjunction with a photoacoustic cell. A system of this kind has been designed to analyze a mixture of 10 gases with a sensitivity of 1 ppb and a cycle time of 5 min.

11C-4 NONDISPERSIVE INSTRUMENTS

The typical dispersive, double-beam infrared spectrophotometer, which finds such wide use in chemical laboratories, is not very satisfactory for quantitative analyses for two reasons. First, the double-beam design, with its more complex electronic and switching devices, is inherently noisy. Second, the 0% T calibration and low-transmittance measurements with most double-beam instruments are subject to an error arising from their optical null design. The 0% T setting for such an instrument is obtained by blocking the sample beam from the detector. The

reference-beam attenuator is then driven in such a way as to decrease the power of the reference beam to zero also. Under these circumstances, essentially no energy reaches the detector and the exact null position cannot be located with precision. In practice, the intensity of the sample beam is slowly diminished by the gradual introduction of the shutter across its path; in this way, the tendency for the pen drive to overshoot zero is avoided, and a more accurate calibration is achieved. Figure 11–20 illustrates the tendency of a typical recording instrument to overshoot when measuring zero or low transmittances; here, the cause is the optical null design. Fortunately, this inevitable small error is not serious for qualitative work provided the transmittances being measured are greater than 5 to 10%.

Because of these limitations inherent in typical infrared dispersive spectrophotometers, a number of simple, rugged (and usually single-beam) instruments have been designed for quantitative work. Some are simple filter or nondispersive photometers; others are a type of spectrophotometer that employs filter wedges in lieu of a dispersing element.

Filter Photometers. Figure 11–11 is a schematic diagram of a portable (weight = 18 lb), infrared filter photometer designed for quantitative analysis of various organic substances in the atmosphere. The source is a nichrome-wire-wound ceramic rod; the transducer is a pyroelectric detector. A variety of interference filters, which transmit in the range between about 3000 to 750 cm^{-1} (3.3 to 13 μm), are available; each is designed for the analysis of a specific compound. The filters are readily interchangeable.

The gaseous sample is introduced into the cell by means of a battery-operated pump. The path length of the cell as shown is 0.5 m; a series of reflecting mirrors (not shown in Figure 11–11) permits increases in cell length to 20 m in increments of 1.5 m.

---

[8] For a recent review of applications of this technique, see: J. F. McClelland, *Anal. Chem.*, **1983**, *55*, 89A.

[9] See, for example: L. B. Kreuzer, *Anal. Chem.*, **1974**, *46*, 239A.

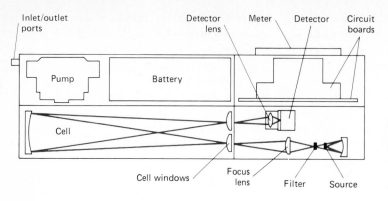

FIGURE 11-11 A portable, infrared photometer designed for gas analysis. (Courtesy of The Foxboro Company, Foxboro, MA.)

This feature greatly enhances the concentration range of the instrument.

The photometer is reported to be sensitive to a few tenths parts per million of such substances as acrylonitrile, chlorinated hydrocarbons, carbon monoxide, phosgene, and hydrogen cyanide.

Nonfilter Photometers. Photometers, which employ no wavelength-restricting device, are widely employed to monitor gas streams for a single component.[10] Figure 11–12 shows a typical nondispersive instrument designed to determine carbon monoxide in a gaseous mixture. The reference cell is a sealed container filled with a nonabsorbing gas; as shown in the figure, the sample flows through a second cell that is of similar length. The chopper blade is so arranged that the beams from identical sources are chopped simultaneously at the rate of about five times per second. Selectivity is obtained by filling both compartments of the *detector cell* with the gas being analyzed, here, carbon monoxide. The two chambers of the detector are separated by a thin, flexible, metal diaphragm that serves as one plate of a capacitor; the second plate is contained in the detector compartment on the left.

In the absence of carbon monoxide in the sample cell, the two detector chambers are heated equally by infrared radiation from the two sources. If the sample contains carbon monoxide, however, the right-hand beam is attenuated somewhat and the corresponding detector chamber becomes cooler with respect to its reference counterpart; the consequence is a movement of the dia-

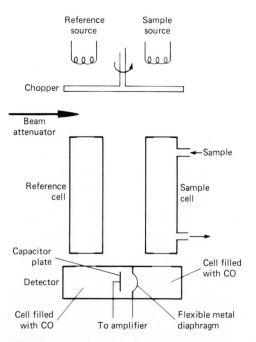

FIGURE 11-12 A nondispersive infrared photometer for monitoring carbon monoxide.

---

[10] For a description of process infrared measurements, see: M. S. Frant and G. LaButti, *Anal. Chem.*, **1980**, *52*, 1331A.

phragm to the right and a change in capacitance of the capacitor. This change in capacitance is sensed by the amplifier system, the output of which drives a servomotor that moves the beam attenuator into the reference beam until the two compartments are again at the same temperature. The instrument thus operates as a null device. The chopper serves to provide a dynamic, ac-type signal which is needed because the electrical system responds more reproducibly to an ac signal than to a slow dc drift.

The instrument is highly selective because heating of the detector gas occurs only from that narrow portion of the spectrum that is absorbed by the carbon monoxide in the sample. Clearly, the device can be adapted to the analysis of any infrared-absorbing gas.

11C-5 AUTOMATED INSTRUMENTS FOR QUANTITATIVE ANALYSIS

Figure 11–13 is a schematic diagram of a computer-controlled instrument designed specifically for quantitative infrared analyses. The wavelength selector, which consists of three filter wedges (p. 122) mounted in the form of a segmented circle, is shown in Figure 11–13**b**. The motor drive and potentiometric control permit rapid computer-controlled wavelength selection in the region between 4000 and 690 cm^{-1} (or 2.5 to 14.5 μm) with an accuracy of 0.4 cm^{-1}. The source and detector are similar to those described in the earlier section on filter photometers; note that a beam chopper is used here. The sample area can be readily adapted to solid, liquid, or gaseous samples.

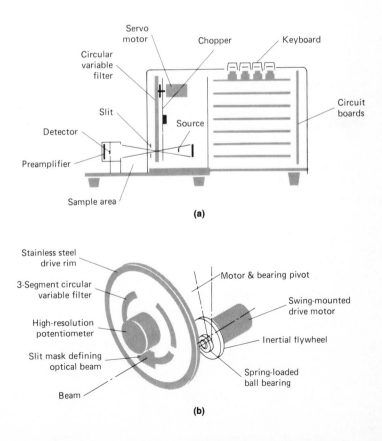

FIGURE 11-13 An infrared instrument for quantitative analysis. (**a**) Schematic of the instrument; (**b**) circular variable filter wheel. (Courtesy of The Foxboro Company, Foxboro, MA.)

The instrument can be programmed to determine the absorbance of a multicomponent sample at several wavelengths and then compute the concentration of each component.

11D   Sample Handling Techniques[11]

As we have seen, ultraviolet and visible spectra are most conveniently obtained from dilute solutions of the analyte. Absorbance measurements in the optimum range are obtained by suitably adjusting either the concentration or the cell length. Unfortunately, this approach is not generally applicable for infrared spectroscopy because no good solvents exist that are transparent throughout the region. As a consequence, techniques must often be employed for liquid and solid samples that make the accurate determination of molar absorptivities difficult if not impossible. Some of these techniques are discussed in the paragraphs that follow.

---

[11] For a more complete discussion, see: N. B. Colthup, L. H. Daly, and S. E. Wiberley, *Introduction to Infrared and Raman Spectroscopy*, 2d ed., p. 85, Academic Press: New York, 1975; R. P. Bauman, *Absorption Spectroscopy*, p. 184, Wiley: New York, 1962; K. Kiss-Eröss in *Comprehensive Analytical Chemistry*, G. Svehla, Ed., Vol. VI, Chapter 5, Elsevier: New York, 1976.

11D-1   GAS SAMPLES

The spectrum of a low-boiling liquid or gas can be obtained by permitting the sample to expand into an evacuated cell. For this purpose, a variety of cells are available with path lengths that range from a few centimeters to several meters. The longer path lengths are obtained in compact cells by providing reflecting internal surfaces, so that the beam makes numerous passes through the sample before exiting from the cell.

11D-2   SOLUTIONS

Solvents.   Figure 11–14 lists the more common solvents employed for infrared studies of organic compounds. It is apparent that no single solvent is transparent throughout the entire middle-infrared region.

Water and the alcohols are seldom employed, not only because they absorb strongly, but also because they attack alkali-metal halides, the most common materials used for cell windows. For these reasons also, care must be taken to dry the solvents shown in Figure 11–14 before use.

Cells.   Because of the tendency for solvents to absorb, infrared cells are ordinarily much narrower (0.01 to 1 mm) than those employed in the ultraviolet and visible regions. Light paths in the infrared range normally require sample concentrations from

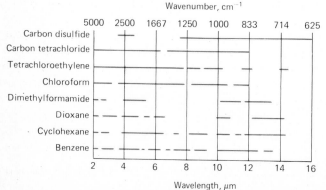

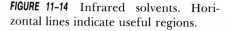

FIGURE 11-14   Infrared solvents. Horizontal lines indicate useful regions.

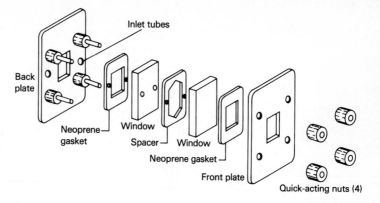

FIGURE 11-15 Expanded view of a demountable infrared cell for liquid samples. Teflon spacers ranging in thickness from 0.015 to 1 mm are available. (Courtesy of Perkin-Elmer, Norwalk, CT.)

0.1 to 10%. The cells are frequently demountable, with Teflon spacers to allow variation in path length (see Figure 11–15). Fixed-path-length cells can be filled or emptied with a hypodermic syringe.

Sodium chloride windows are most commonly employed; even with care, however, their surfaces eventually become fogged due to absorption of moisture. Polishing with a buffing powder returns them to their original condition.

The thickness of a narrow infrared cell can be determined by measuring the transmittance of the empty cell with air as the reference. Radiation reflected off the two walls of the cell interacts with the transmitted radiation to produce an interference pattern such as that shown in Figure 11–16. The thickness b of the cell in centimeters is obtained from the relationship

$$b = \frac{N}{2(\sigma_1 - \sigma_2)}$$

where N is the number of peaks between two wavenumbers σ_1 and σ_2.

It should be noted that interference fringes are ordinarily not seen when a cell is filled with liquid because the refractive index of most liquids approaches that of the window material; reflection is thus reduced (Equation 4–11, p. 101). On the other hand, interference can be observed between 2800

and 2000 cm^{-1} in Figure 11–1 (p. 316). Here, the sample is a sheet of polyethylene, which has a refractive index considerably different from air; consequently, significant reflection occurs at the two interfaces of the sheet.

11D-3 PURE LIQUIDS

When the amount of sample is small or when a suitable solvent is unavailable, it is common practice to obtain spectra on the pure (neat) liquid. Here, only a very thin film has a sufficiently short path length to produce satisfactory spectra. Commonly, a drop of the neat liquid is squeezed between two rock-salt plates to give a layer that has a

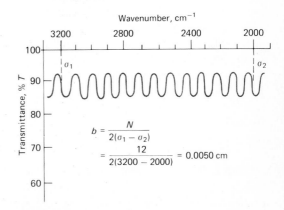

FIGURE 11-16 Determination of path length b from interference fringe produced by an empty cell.

thickness of 0.01 mm or less. The two plates, held together by capillarity, are then mounted in the beam path. Clearly, such a technique does not give particularly reproducible transmittance data, but the resulting spectra are usually satisfactory for qualitative investigations.

11D-4 SOLIDS

Spectra of solids for which no infrared-transparent solvent exists are derived by one of three techniques. The most recently developed of these methods is photoacoustic spectroscopy, which was described briefly in Section 11C–3. The other two procedures involve dispersing the solid in a liquid or solid matrix called a *mull*. For these techniques to be successful it is essential that the particle size of the suspended solid be smaller than the wavelength of the infrared beam; if this condition is not realized, a significant portion of the radiation is lost to scattering.

One method of forming a mull involves grinding 2 to 5 mg of the finely powdered sample (particle size <2 μm) in the presence of one or two drops of a heavy hydrocarbon oil (Nujol). If hydrocarbon bands are likely to interfere, Fluorolube, a halogenated polymer, can be used instead. In either case, the resulting mull is then examined as a film between flat salt plates.

In a second technique, a milligram or less of the finely ground sample is intimately mixed with about 100 mg of dried potassium bromide powder. Mixing can be carried out with a mortar and pestle, or better, in a small ball mill. The mixture is then pressed in a special die at 10,000 to 15,000 pounds per square inch to yield a transparent disk or mull. Best results are obtained if the disk is formed in a vacuum to eliminate occluded air. The disk is then held in the instrument beam for spectroscopic examination. The resulting spectra frequently exhibit bands at 2.9 and 6.1 μm (3450 to 1640 cm^{-1}) due to absorbed moisture.

11D-5 INTERNAL-REFLECTION SPECTROSCOPY[12]

When a beam of radiation passes from a more dense to a less dense medium, reflection occurs. The fraction of the incident beam that is reflected increases as the angle of incidence becomes larger; beyond a certain critical angle, reflection is complete. It has been shown both theoretically and experimentally that during the reflection process the beam acts as if, in fact, it penetrates a small distance into the less dense medium before reflection occurs.[13] The depth of penetration, which varies from a fraction of a wavelength up to several wavelengths, depends upon the wavelength, the index of refraction of the two materials, and the angle of the beam with respect to the interface. If the less dense medium absorbs the radiation, attenuation of the beam also occurs.

Figure 11–17 shows an apparatus that takes advantage of the internal-reflectance phenomenon for infrared-absorption measurements. As will be seen from the upper figure, the sample (here, a solid) is placed on opposite sides of a transparent crystalline material of high refractive index; a mixed crystal of thallium bromide/thallium iodide is frequently employed. By proper adjustment of the incident angle, the radiation undergoes multiple internal reflections before passing from the crystal to the detector. Absorption and attenuation takes place at each of these reflections.

Figure 11–17**b** is an optical diagram of a commercially available adapter that will fit

---

[12] For brief reviews of this technique, see: P. A. Wilks, Jr., *Amer. Lab.*, **1972,** *4* (11), 42 and *Amer. Lab.*, **1980,** *12* (6), 92. For a more comprehensive presentation, see: N. J. Harrick, *Internal Reflection Spectroscopy*, Interscience: New York, 1967.

[13] J. Fahrenfort, *Spectrochem. Acta*, **1961,** *17*, 698.

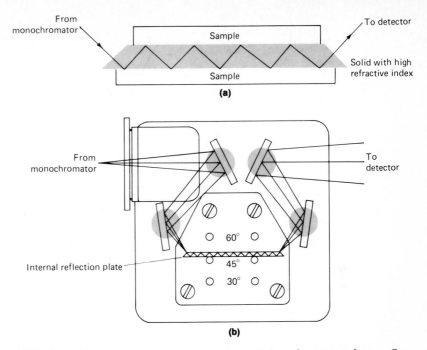

FIGURE 11-17 Internal reflectance apparatus. **(a)** Sample mounted on reflection plate; **(b)** internal reflection adapter. (Courtesy of The Foxboro Company, Foxboro, MA.)

into the cell area of most infrared spectrometers and will permit internal-reflectance measurements. Note that an incident angle of 30, 45, or 60 deg can be chosen. Cells for liquid samples are also available.

Internal-reflectance spectra are similar but not identical to ordinary absorption spectra. In general, while the same peaks are observed, their relative intensities differ. The absorbances, while dependent upon the angle of incidence, are independent of sample thickness, since the radiation penetrates only a few micrometers into the sample.

Internal-reflectance spectroscopy has been applied to many substances such as polymers, rubbers, and other solids. It is of interest to note that the resulting spectra are free from the interference fringes mentioned in the previous section.

11E Qualitative Applications of Infrared Absorption

We have noted that the approximate frequency (or wavenumber) at which an organic functional group, such as C=O, C=C, C—H, or C≡C, absorbs infrared radiation can be calculated from the masses of the atoms and the force constant of the bond between them (Equations 11–13 and 11–14). These frequencies, called *group frequencies,* are seldom totally invariant because of interactions with other vibrations associated with one or both of the atoms comprising the group. On the other hand, such interaction effects are frequently small; as a result, a range of frequencies can be assigned within which it is highly probable that the absorption peak for a given functional group will be found.

Group frequencies often make it possible to establish the probable presence or absence of a given functional group in a molecule.

11E-1 CORRELATION CHARTS

Over the years, a mass of empirical information has been accumulated concerning the frequency range within which various functional groups can be expected to absorb. *Correlation charts* provide a concise means for summarizing this information in a form that is useful for identification purposes. A number of correlation charts have been developed,[14] one of which is shown in Figure 11–18.

Correlation charts permit intelligent guesses to be made as to what functional groups are likely to be present or absent in a molecule. Ordinarily, it is impossible to identify unambiguously either the sources of all of the peaks in a given spectrum or the exact identity of the molecule. Instead, correlation charts serve as a starting point in the identification process.

11E-2 IMPORTANT SPECTRAL REGIONS IN THE INFRARED

The chemist interested in identifying an organic compound by the infrared technique usually examines certain regions of the spectrum in a systematic way in order to obtain clues as to the presence or absence of certain group frequencies. Some of the important regions are considered briefly.

Hydrogen Stretching Region 3700 to 2700 cm^{-1} (2.7 to 3.7 μm). The appearance of strong absorption peaks in this region usually results from a stretching vibration between hydrogen and some other atom. The motion is largely that of the hydrogen atom since it is so much lighter than the species with which it bonds; as a consequence, the absorption is not greatly affected by the rest of the molecule. Furthermore, the hydrogen stretching frequency is much higher than that for other chemical bonds, with the result that interaction of this vibration with others is usually small.

Absorption peaks in the region of 3700 to 3100 cm^{-1} (2.7 to 3.2 μm) are ordinarily due to various O—H and N—H stretching vibrations, with the former tending to appear at higher wavenumbers. The O—H bands are often broader than N—H bands and appear only in dilute, nonpolar solvents. Hydrogen bonding tends to broaden the peaks and move them toward lower wavenumbers.

Aliphatic C—H vibrations fall in the region between 3000 and 2850 cm^{-1} (3.3 to 3.5 μm). Most aliphatic compounds have a sufficient number of C—H bonds to make this a prominent peak. Any structural variation that affects the C—H bond strength will cause a shift in the maximum. For example, the band for Cl—C—H lies just above 3000 cm^{-1} (<3.3 μm), as do the bands for olefinic and aromatic hydrogen. The acetylenic C—H bond is strong and occurs at about 3300 cm^{-1} (3.0 μm). The hydrogen on the carbonyl group of an aldehyde usually produces a distinct peak in the region of 2745 to 2710 cm^{-1} (3.64 to 3.69 μm). Substitution of deuterium for hydrogen causes a shift to lower wavenumbers by the factor of approximately $1/\sqrt{2}$, as would be predicted from Equation 11–14; this effect has been employed to identify C—H stretching peaks.

---

[14] N. B. Colthup, *J. Opt. Soc. Am.*, **1950**, *40*, 397–400; A. D. Cross, *Introduction to Practical Infra-Red Spectroscopy*, 2d ed., pp. 56–62, Butterworths: Washington, 1964; R. N. Jones, *Infrared Spectra of Organic Compounds: Summary Charts of Principal Group Frequencies*, National Research Council of Canada: Ottawa, 1959; K. Nakanishi and P. H. Solomon, *Infrared Absorption Spectroscopy*, 2d ed., pp. 10–56, Holden-Day: San Francisco, 1977; and R. M. Silverstein, G. C. Bassler, and T. C. Morrill, *Spectrometric Identification of Organic Compounds*, 4th ed., pp. 166–172, Wiley: New York, 1981.

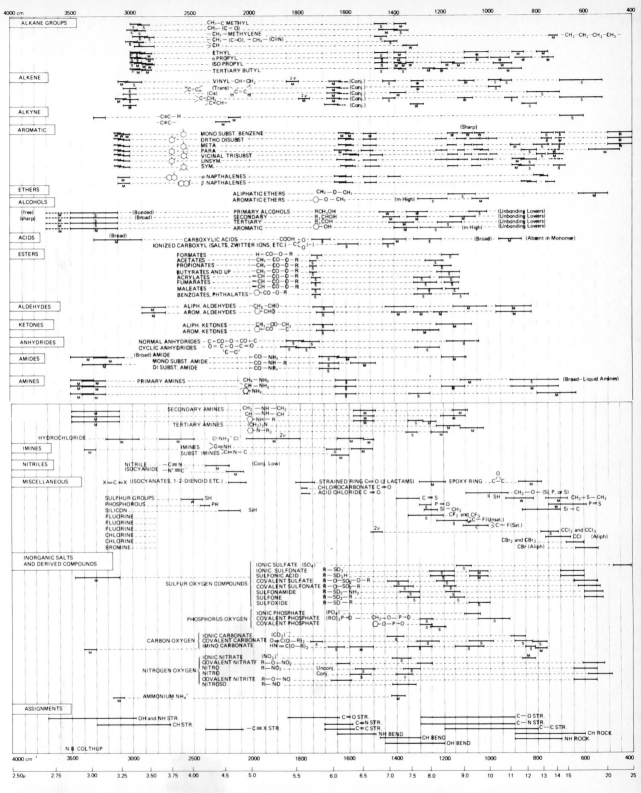

FIGURE 11-18 Correlation chart. (From: N. B. Colthup, *J. Optical Soc. Am.*, **1950**, *40*, 397. With permission.)

The Triple-Bond Region Between 2700 and 1850 cm^{-1} (3.7 to 5.4 μm). A limited number of groups absorb in this spectral region; their presence is thus readily apparent. Triple-bond stretching results in a peak at 2250 to 2225 cm^{-1} (4.44 to 4.49 μm) for —C≡N, at 2180 to 2120 cm^{-1} (4.59 to 4.72 μm) for —N$^+$≡C$^-$, and at 2260 to 2190 cm^{-1} (4.42 to 4.57 μm) for —C≡C—. Also present in this region are peaks for S—H at 2600 to 2550 cm^{-1} (3.85 to 3.92 μm), P—H at 2440 to 2350 cm^{-1} (4.10 to 4.26 μm), and Si—H at 2260 to 2090 cm^{-1} (4.42 to 4.78 μm).

The Double-Bond Region Between 1950 and 1450 cm^{-1} (5.1 to 6.9 μm). The carbonyl stretching vibration is characterized by absorption throughout this region. Ketones, aldehydes, acids, amides, and carbonates all have absorption peaks around 1700 cm^{-1} (5.9 μm). Esters, acid chlorides, and acid anhydrides tend to absorb at slightly higher wavenumbers; that is, 1770 to 1725 cm^{-1} (5.65 to 5.80 μm). Conjugation tends to lower the absorption peak by about 20 cm^{-1}. It is frequently impossible to determine the type of carbonyl that is present solely on the basis of absorption in this region; however, examination of additional spectral regions may provide the supporting evidence needed for clear-cut identification. For example, esters have a strong C—O—R stretching peak at about 1200 cm^{-1} (8.3 μm), while aldehydes have a distinctive hydrogen stretching peak just above 2700 cm^{-1} (3.7 μm), as noted previously.

Absorption peaks arising from C=C and C=N stretching vibrations are located in the 1690 to 1600 cm^{-1} (5.9 to 6.2 μm) range. Valuable information concerning the structure of olefins can be obtained from the exact position of such a peak.

The region between 1650 and 1450 cm^{-1} (6.1 to 6.9 μm) provides important information about aromatic rings. Aromatic compounds with a low degree of substitution exhibit four peaks near 1600, 1580, 1500, and 1460 cm^{-1} (6.25, 6.33, 6.67, and 6.85 μm). Variations of the spectra in this region with the number and arrangement of substituent groups are usually consistent but independent of the type of substituent; considerable structural information can thus be gleaned from careful study of aromatic absorption in the infrared region.

The "Fingerprint" Region Between 1500 and 700 cm^{-1} (6.7 to 14.3 μm). Small differences in the structure and constitution of a molecule result in significant changes in the distribution of absorption peaks in this region of the spectrum. As a consequence, a close match between two spectra in this fingerprint region (as well as others) constitutes strong evidence for the identity of the compounds yielding the spectra. Most single bonds give rise to absorption bands at these frequencies; because their energies are about the same, strong interaction occurs between neighboring bonds. The absorption bands are thus composites of these various interactions and depend upon the overall skeletal structure of the molecule. Exact interpretation of spectra in this region is seldom possible because of their complexity; on the other hand, it is this complexity that leads to uniqueness and the consequent usefulness of the region for final identification purposes.

A few important group frequencies are to be found in the fingerprint region. These include the C—O—C stretching vibration in ethers and esters at about 1200 cm^{-1} (8.3 μm) and the C—Cl stretching vibration at 700 to 800 cm^{-1} (14.3 to 12.5 μm). A number of inorganic groups such as sulfate, phosphate, nitrate, and carbonate also absorb at wavenumbers below 1200 cm^{-1} (>8.3 μm).

Limitations to the Use of Correlation Charts. The unambiguous establishment of the identity or the structure of a compound is seldom possible from correlation charts alone. Uncertainties frequently arise from overlapping group frequencies, spectral variations as a function of the physical state of the sample (that is, whether it is a solution, a mull, in a pelleted form, and so forth), and instrumental limitations.

In employing group frequencies, it is essential that the entire spectrum, rather than a small isolated portion, be considered and interrelated. Interpretation based on one part of the spectrum should be confirmed or rejected by study of other regions.

To summarize, then, correlation charts serve only as a guide for further and more careful study. Several excellent monographs describe the absorption characteristics of functional groups in detail.[15] A study of these characteristics, as well as the other physical properties of the sample, may permit unambiguous identification. Infrared spectroscopy, when used in conjunction with other methods such as mass spectroscopy, nuclear magnetic resonance, and elemental analysis, usually makes possible the positive identification of a species.

The examples that follow illustrate how infrared spectra are employed in the identification of pure compounds. In every case, confirmatory tests would be desirable; a comparison of the experimental spectrum with that of the pure compound would suffice for this purpose.

EXAMPLE 11-2. The spectrum in Figure 11–19 was obtained for a pure colorless liquid in a 0.01-mm cell; the liquid boiled at 190°C. Suggest a structure for the sample.

The four absorption peaks in the region of 1450 to 1600 cm^{-1} are characteristic of an aromatic system, and one quickly learns to recognize this grouping. The peak at 3100 cm^{-1} corresponds to a hydrogen-stretching vibration, which for an aromatic system is usually above 3000

cm^{-1}. For aliphatic hydrogens, this type of vibration causes absorption at or below 3000 cm^{-1} (see Figure 11–18). Thus, it would appear that the compound is predominately, if not exclusively, aromatic.

We then note the sharp peak at 2250 cm^{-1} and recall that only a few groups absorb in this region. From Figure 11–18, we see that these include $-C\equiv C-$, $-C\equiv CH$, $-C\equiv N$, and $Si-H$. The $-C\equiv CH$ group can be eliminated, however, since it has a hydrogen stretching frequency of about 3250 cm^{-1}.

The pair of strong peaks at 680 and 760 cm^{-1} suggests that the aromatic ring may be singly substituted, although the pattern of peaks in the region of 1000 to 1300 cm^{-1} is confusing and may lead to some doubt about this conclusion.

Two likely structures that fit the infrared data would appear to be $C_6H_5C\equiv N$ and $C_6H_5C\equiv C-C_6H_5$. The latter, however, is a solid at room temperature; in contrast, benzonitrile has a boiling point (191°C) similar to that of the sample. Thus, we tentatively conclude that the compound under investigation is $C_6H_5C\equiv N$.

EXAMPLE 11-3. A colorless liquid was found to have the empirical formula $C_6H_{12}O$ and a boiling point of 130°C. Its infrared spectrum (neat in a 0.025-mm cell) is given in Figure 11–20. Suggest a structure.

The presence of an aliphatic structure is suggested in Figure 11–20 by both the strong band at 3000 cm^{-1} and the empirical formula. The intense band at 1720 cm^{-1} and the single oxygen in the formula strongly indicate the likelihood of an aldehyde or a ketone. The band at about 2800 cm^{-1} could be attributed to the shift in stretching vibration associated with hydrogen that is bonded directly to a carbonyl carbon atom; if this interpretation is correct, the substance is an alde-

[15] N. B. Colthup, L. N. Daly, and S. E. Wilberley, *Introduction to Infrared and Raman Spectroscopy*, 2d ed., Academic Press: New York, 1975; and H. A. Szymanski, *Theory and Practice of Infrared Spectroscopy*, Plenum Press: New York, 1964.

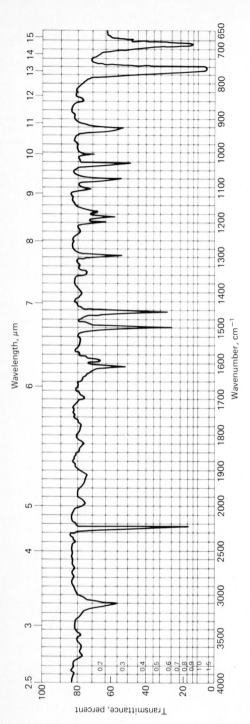

FIGURE 11-19 Spectrum of a colorless liquid. (From "Catalog of Selected Ultraviolet Spectral Data," Thermodynamics Research Center Data Project, Thermodynamics Research Center, Texas A&M University, College Station, Texas. Loose-leaf data sheets, extant 1975. With permission.)

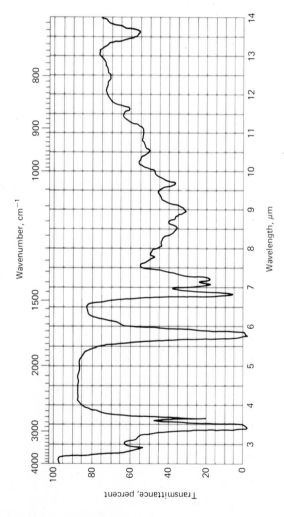

FIGURE 11-20 Infrared spectrum of an unknown. (From "Catalog of Selected Ultraviolet Spectral Data," Thermodynamics Research Center Data Project, Thermodynamics Research Center, Texas A&M University, College Station, Texas. Loose-leaf data sheets, extant 1975. With permission.)

hyde. The peak at 3450 cm⁻¹ is puzzling, for absorption in this region is usually the result of an N—H or an O—H stretching vibration; the possibility that the substance is an aliphatic alcohol must thus be considered. The empirical formula, however, is inconsistent with a structure that contains both an aldehyde and a hydroxyl group. Examination of the lower-frequency region fails to yield further information. This dilemma can be resolved by attributing absorption at 3450 cm⁻¹ to the strong O—H stretching band of water, which is often present as a contaminant. Thus, we conclude that the sample is probably an aliphatic aldehyde. We are unable to determine the extent of chain branching from the spectrum; to be sure, the peak at 730 cm⁻¹ suggests the presence of four or more methylene groups in a row. The boiling point of *n*-hexanal is 131°C; we are thus inclined to conclude that the sample is this compound. Confirmatory tests would be needed, however.

EXAMPLE 11-4. The spectrum shown in Figure 11–21 was obtained from a 0.5% solution of a liquid sample in a 0.5-mm cell. For the region of 2 to 8 μm, the solvent was CCl₄, and for the higher wavelengths, CS₂ was employed. The empirical formula of the compound was found to be $C_8H_{10}O$. Suggest a probable structure.

The empirical formula indicates that the compound is probably aromatic, and this supposition is borne out by the characteristic pattern of peaks in the 1450 to 1600 cm⁻¹ range. The C—H stretching bands, however, are somewhat lower in wavenumber than those for a purely aromatic system, which suggests the existence of some aliphatic groups as well. The broad band at 3300 cm⁻¹ appears important, and suggests an O—H or N—H stretching vibration. The formula permits us to conclude that the molecule must contain a phenolic or an alcoholic O—H group. The spectrum in the region of 1100 to 1400 cm⁻¹ is compatible with the

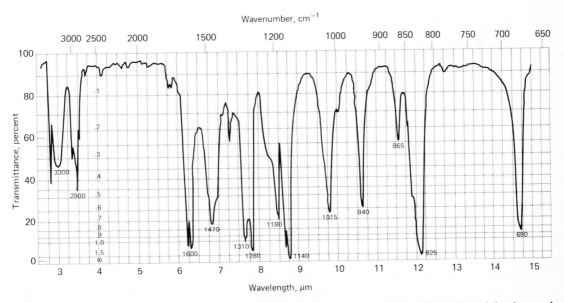

FIGURE 11-21 Spectrum of a sample of CCl₄ (2 to 8 μm). (From "Catalog of Selected Ultraviolet Spectral Data," Thermodynamics Research Data Project, Thermodynamics Research Center, Texas A&M University, College Station, Texas. Loose-leaf data sheets, extant 1975. With permission.)

postulated O—H group, but we are unable to decide from the spectral data whether the sample is a phenol or an alcohol.

From our observations thus far, and from the empirical formula, likely structures for the unknown appear to be:

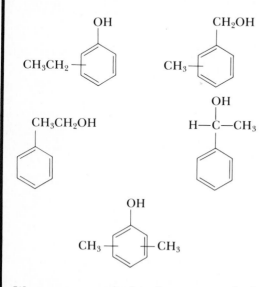

We now turn to the low-frequency end of the spectrum, where correlations of peaks with aromatic substitution patterns can often be found. The two strong peaks at 680 and 825 cm^{-1} and the weaker one at 865 cm^{-1} appear to best fit the pattern for a symmetrically trisubstituted benzene. Thus, 3,5-dimethylphenol seems the most logical choice. Confirmatory data would be needed, however, to establish the identity unambiguously.

11E-3 COLLECTIONS OF SPECTRA

As may be seen from the foregoing examples, correlation charts seldom suffice for the positive identification of an organic compound from its infrared spectrum. There are available, however, several catalogs of infrared spectra that assist in qualitative identification by providing comparison spectra for a large number of pure com-

pounds. These collections have become so extensive (as many as 100,000 spectra) as to require edge-punched cards, IBM cards, or magnetic tapes for efficient retrieval. Data presentation takes several forms including replica of spectra on notebook-size paper, cards, microfiches and microfilms, as well as magnetic tapes or disks. In some instances, digitized peak positions are employed in lieu of replicas of spectra. A list of sources of these collections is found in a paper by Gevantman.[16]

11E-4 COMPUTER SEARCH SYSTEMS

Instrument manufacturers are now offering computer search systems to assist the chemist in identifying compounds from stored infrared spectral data. The position and relative magnitudes of peaks in the spectrum of the analyte are determined and stored in memory to give a peak profile, which can then be compared with profiles of pure compounds stored on floppy disks. The computer then matches profiles and prints a list of compounds having spectra similar to that of the analyte. Usually the spectrum of the analyte and that of each potential match can then be displayed simultaneously on the CRT for comparison. The memory banks of these instruments are capable of storing profiles for several thousand pure compounds.

In 1980, the Sadtler Standard Infrared Collection and the Sadtler Commercial Infrared Collection became available as software packages. The first contains infrared data for approximately 65,000 compounds and the second for 35,000. Several manufacturers of Fourier transform instruments have now incorporated these packages into their instrument computers thus creating instantly available infrared libraries of nearly 100,000 compounds.

---

[16] L. H. Gevantman, *Anal. Chem.*, **1972**, *44* (7), 31A.

The Sadtler algorithm consists of a search system in which the spectrum of the unknown compound is first coded according to the location of its strongest absorption peak; then each additional strong band (% $T <$ 60%) in 10 regions 200 cm^{-1} wide from 4000 to 2100 cm^{-1} are coded by their location. Finally, the strong bands in 17 regions 100 cm^{-1} wide from 2100 to 400 cm^{-1} are coded in a similar way.[17]

The 100,000 compounds in the library are coded in this same way. The data are organized by the location of the strongest band with only those compounds having the same strongest band being considered in any sample identification. This procedure is rapid and produces a list of potential matches within a short period. For example, approximately 40 sec are required to search a base library of 25,000 compounds by this procedure. The output is in terms of a Sadtler grating number. The full spectrum can then be viewed by looking up the grating number in Sadtler's hard copy, microfilm, or microfiche spectral libraries.

Alternatively, the reference spectrum can be stored in microfiche format with a computer-compatible reader/printer system. This system allows the computer to seek and display on a viewing screen for comparison with the unknown, the full spectrum of any compound specified by the user.

11F Quantitative Applications

Quantitative infrared absorption methods differ somewhat from those discussed in the previous chapter because of the greater complexity of the spectra, the narrowness of the absorption bands, and the instrumental limitations of infrared instruments.

---

[17] For a fuller description of this system, see: R. H. Shaps and J. F. Sprouse, *Ind. Res. and Dev.*, **1981**, *23*, 168.

11F-1 DEVIATIONS FROM BEER'S LAW

With infrared radiation, instrumental deviations from Beer's law are more common than with ultraviolet and visible wavelengths because infrared absorption bands are relatively narrow. Furthermore, the low intensity of sources and low sensitivities of detectors in this region require the use of relatively wide monochromator slit widths; thus, the bandwidths employed are frequently of the same order of magnitude as the widths of absorption peaks. We have pointed out (Section 6B–3) that this combination of circumstances usually leads to a nonlinear relationship between absorbance and concentration. Calibration curves, determined empirically, are therefore often required for quantitative work.

11F-2 ABSORBANCE MEASUREMENTS

Matched absorption cells for solvent and solution are ordinarily employed in the ultraviolet and visible regions, and the measured absorbance is then found from the relation

$$A = \log \frac{P_{\text{solvent}}}{P_{\text{solution}}}$$

The use of the solvent in a matched cell as a reference absorber has the advantage of largely canceling out the effects of radiation losses due to reflection at the various interfaces, scattering and absorption by the solvent, and absorption by the container windows (p. 162). This technique is seldom practical for measurements in the infrared region because of the difficulty in obtaining cells whose transmission characteristics are identical. Most infrared cells have very short path lengths, which are difficult to duplicate exactly. In addition, the cell windows are readily attacked by contaminants in the atmosphere and the solvent; thus, their transmission characteristics change continually with use. For these reasons, a reference absorber is often dispensed with entirely in infrared work, and the intensity of the radiation passing through the sample is simply compared with that of the unobstructed beam; alternatively, a salt plate may be placed in the reference beam. In either case, the resulting transmittance is ordinarily less than 100%, even in regions of the spectrum where no absorption by the sample occurs; this effect is readily seen by examining the several spectra that have appeared earlier in this chapter.

For quantitative work, it is necessary to correct for the scattering and absorption by the solvent and the cell. Two methods are employed. In the so-called *cell-in/cell-out* procedure, spectra of the solvent and sample are obtained successively with respect to the unobstructed reference beam. The same cell is used for both measurements. The transmittance of each solution versus the reference beam is then determined at an absorption maximum of the analyte. These transmittances can be written as

$$T_0 = P_0/P_r$$

and

$$T_s = P/P_r$$

where P_r is the power of the reference beam and T_0 and T_s are the transmittances of the solvent and sample, respectively, against this reference. If P_r remains constant during the two measurements, then the transmittance of the sample with respect to the solvent can be obtained by division of the two equations. That is,

$$T = T_s/T_0 = P/P_0$$

An alternative way of obtaining P_0 and T is the *base-line* method, in which the solvent transmittance is assumed to be constant or at least to change linearly between the shoulders of the absorption peak. This technique is demonstrated in Figure 11–22.

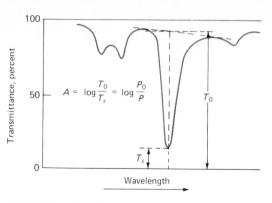

FIGURE 11–22 Base-line method for determination of absorbance.

11F–3 APPLICATIONS OF QUANTITATIVE INFRARED SPECTROSCOPY

With the exception of homonuclear molecules, all organic and inorganic molecular species absorb in the infrared region; thus, infrared spectrophotometry offers the potential for determining an unusually large number of substances. Moreover, the uniqueness of an infrared spectrum leads to a degree of specificity that is matched or exceeded by relatively few other analytical methods. This specificity has found particular application to analysis of mixtures of closely related organic compounds. Two examples that typify these applications follow.

Analysis of a Mixture of Aromatic Hydrocarbons. A typical application of quantitative infrared spectroscopy involves the resolution of C_8H_{10} isomers in a mixture which includes *o*-xylene, *m*-xylene, *p*-xylene, and ethylbenzene. The infrared absorption spectra of the individual components in the 12 to 15 μm range is shown in Figure 11–23; cyclohexane is the solvent. Useful absorption peaks for determination of the individual compounds occur at 13.47, 13.01, 12.58, and 14.36 μm, respectively. Unfortunately, however, the absorbance of a mixture at any one of these wavelengths is not entirely determined by the concentration of just one component because of overlapping absorption bands. Thus, molar absorptivities for each of the

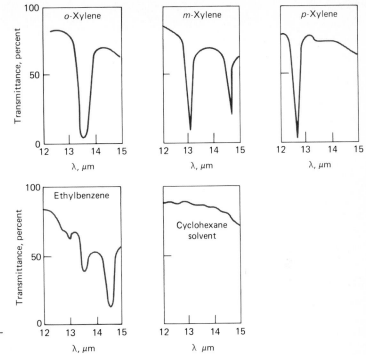

FIGURE 11-23 Spectra of C_8H_{10} isomers in cyclohexane.

four compounds must be determined at the four wavelengths. Then, four simultaneous equations can be written which permit the calculation of the concentration of each species from four absorbance measurements (see p. 212). Such calculations are most easily performed with a computer.

When the relationship between absorbance and concentration is nonlinear (as frequently occurs in the infrared region), the algebraic manipulations associated with an analysis of several components having overlapping absorption peaks are considerably more complex.[18]

Analysis of Air Contaminants. The recent proliferation of government regulations with respect to atmospheric contaminants has demanded the development of sensitive, rapid, and highly specific methods for a variety of chemical compounds. Infrared ab-

sorption procedures appear to meet this need better than any other single analytical tool.

Table 11-2 demonstrates the potential of infrared spectroscopy for the analysis of mixtures of gases. The standard sample of air containing five species in known concentration was analyzed with the computerized instrument shown in Figure 11-13; a 20-m gas cell was employed. The data were printed out within a minute or two after sample injection.

Table 11-3 shows potential applications of infrared filter photometers (such as that shown in Figure 11-11) for the quantitative determination of various chemicals in the atmosphere for the purpose of assuring compliance with OSHA regulations.

Of the more than 400 chemicals for which maximum tolerable limits have been set by the Occupational Safety and Health Administration, more than half appear to have absorption characteristics suitable for determination by means of infrared filter

---

[18] The treatment of infrared data for various types of mixtures is discussed in: R. P. Bauman, *Absorption Spectroscopy*, pp. 403–419, Wiley: New York, 1962.

Table 11-2
An Example of Infrared Analysis of Air Contaminants[a]

Contaminants	Concn, ppm	Found, ppm	Relative Error, %
Carbon monoxide	50	49.1	1.8
Methylethyl ketone	100	98.3	1.7
Methyl alcohol	100	99.0	1.0
Ethylene oxide	50	49.9	0.2
Chloroform	100	99.5	0.5

[a] Courtesy of The Foxboro Company, Foxboro, MA.

photometers or spectrophotometers. Obviously, among all of these absorbing compounds, peak overlaps are to be expected; yet the method should provide a moderately high degree of selectivity.

11F-4 DISADVANTAGES AND LIMITATIONS TO QUANTITATIVE INFRARED METHODS

Several disadvantages attend the application of infrared methods to quantitative analysis. Among these are the frequent nonadherence to Beer's law and the complexity of spectra; the latter enhances the probability of the overlap of absorption peaks. In addition, the narrowness of peaks and the effects of stray radiation make absorbance measurements critically dependent upon the slit width and the wavelength setting. Finally, the narrow cells required for many analyses are inconvenient to use and may lead to significant analytical uncertainties.

The analytical errors associated with a quantitative infrared analysis often cannot

Table 11-3
Some Examples of Infrared Vapor Analysis for OSHA Compliance[a]

Compound	Allowable Exposure, ppm[b]	λ, μm	Minimum Detectable Concentration, ppm[c]
Carbon disulfide	20	4.54	0.5
Chloroprene	25	11.4	4
Diborane	0.1	3.9	0.05
Ethylenediamine	10	13.0	0.4
Hydrogen cyanide	10	3.04	0.4
Methyl mercaptan	10	3.38	0.4
Nitrobenzene	1	11.8	0.2
Pyridine	5	14.2	0.2
Sulfur dioxide	5	8.6	0.5
Vinyl chloride	1	10.9	0.3

[a] Courtesy of The Foxboro Company, Foxboro, MA.
[b] 1977 OSHA exposure limits for 8-hour weighed average.
[c] For 20.25-m cell.

be reduced to the level associated with ultraviolet and visible methods, even with considerable care and effort.

11G Far-Infrared Spectroscopy

The far-infrared region is particularly useful for inorganic studies because absorption due to stretching and bending vibrations of bonds between metal atoms and both inorganic and organic ligands generally occur at frequencies lower than 650 cm^{-1} (>15 μm). For example, heavy-metal iodides generally absorb in the region below 100 cm^{-1}, while the bromides and chlorides have bands at higher frequencies. Absorption frequencies for metal-organic bonds are ordinarily dependent upon both the metal atom and the organic portion of the species.

Far-infrared studies of inorganic solids have also provided useful information about lattice energies of crystals and transition energies of semiconducting materials.

Molecules composed only of light atoms absorb in the far-infrared if they have skeletal bending modes that involve more than two atoms other than hydrogen. Important examples are substituted benzene derivatives, which generally show several absorption peaks. The spectra are frequently quite specific and useful for identifying a particular compound; to be sure, characteristic group frequencies also exist in the far-infrared region.

Pure rotational absorption by gases is observed in the far-infrared region, provided the molecules have permanent dipole moments. Examples include H_2O, O_3, HCl, and AsH_3. Absorption by water is troublesome; elimination of its interference requires evacuation or at least purging of the spectrometer.

Fourier transform spectrometers are particularly useful for far-infrared studies. The energy advantage of the interferometric system over a dispersive one generally results in a significant improvement in spectral quality. In addition, the application of gratings to this wavelength region is complicated by the overlapping of several orders of diffracted radiation.

11H Infrared Emission Spectroscopy

Upon being heated, molecules that absorb infrared radiation are also capable of emitting characteristic infrared wavelengths. The principal deterrent to the analytical application of this phenomenon has been the poor signal-to-noise characteristics of the infrared emission signal, particularly when the sample is at a temperature only slightly higher than its surroundings. With the interferometric method, interesting and useful applications are now appearing.

An early example of the application of infrared emission spectroscopy is found in a paper[19] which describes the use of a Fourier transform spectrometer for the identification of microgram quantities of pesticides. Samples were prepared by dissolving them in a suitable solvent and evaporating on a NaCl or KBr plate. The plate was then heated electrically near the spectrometer entrance. Pesticides such as DDT, malathion, and dieldrin were identified in amounts as low as 1 to 10 μg.

Equally interesting has been the use of the interferometric technique for the remote detection of components emitted from industrial stacks. In one of these applications,[20] an interferometer was mounted on an 8-inch reflecting telescope. With the telescope focused on the plume from an industrial plant, CO_2 and SO_2 were readily detected at a distance of several hundred feet.

[19] I. Coleman and M. J. D. Low, *Spectrochim. Acta*, **1966,** 22, 1293.

[20] M. J. D. Low and F. K. Clancy, *Env. Sci. Technol.,* **1967,** *1,* 73.

PROBLEMS

11–1. The infrared spectrum of CO shows a vibrational absorption peak at 2170 cm^{-1}.
   (a) What is the force constant for the CO bond?
   (b) At what wavenumber would the corresponding peak for ^{14}CO occur?

11–2. Gaseous HCl exhibits an infrared peak at 2890 cm^{-1} due to the hydrogen/chlorine stretching vibration.
   (a) Calculate the force constant for the bond.
   (b) Calculate the wavenumber of the absorption peak for DCl assuming the force constant is the same as that calculated in part (a).

11–3. Indicate whether the following vibrations will be active or inactive in the infrared spectrum.

Molecule	Motion
(a) CH_3—CH_3	C—C stretching
(b) CH_3—CCl_3	C—C stretching
(c) SO_2	Symmetric stretching
(d) CH_2=CH_2	C—H stretching:

(e) CH_2=CH_2 \qquad C—H stretching:

(f) CH_2=CH_2 \qquad CH_2 wag:

(g) CH_2=CH_2 \qquad CH_2 twist:

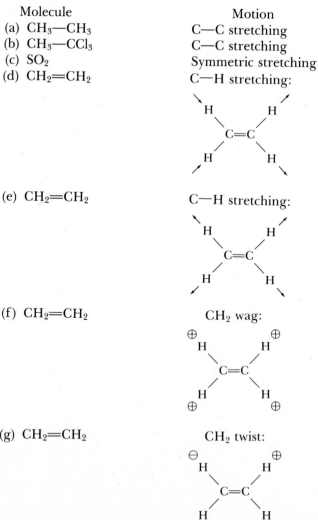

11–4. Calculate the absorption frequency corresponding to the —C—H stretching vibration treating the group as a simple diatomic C—H molecule. Compare the calculated value with the range found in correlation charts. Repeat the calculation for the deuterated bond.

11–5. The spectrum in Figure 11–24 was obtained for a liquid with an empirical formula of C_3H_6O. Identify the compound.

11–6. The spectrum in Figure 11–25 is that of a high-boiling liquid having an empirical formula $C_9H_{10}O$. Identify the compound as closely as possible.

11–7. The spectrum in Figure 11–26 is for an acrid-smelling liquid that boils at 52°C and has a molecular weight of about 56. What is the compound? What impurity is clearly present?

11–8. The spectrum in Figure 11–27 is that of a nitrogen-containing substance that boils at 97°C. What is the compound?

11–9. The spectrum in Figure 11–28 was obtained from CCl_4 and CS_2 solutions of a white crystalline compound having a melting point of 53°C and an empirical formula of $C_{12}H_{11}N$. Identify the compound.

11–10. The spectra in Figure 11–29 are for a pure liquid compound in cells of a different length; the compound contains only C, H, and O. What conclusion can be drawn as to the probable nature of the compound?

11–11. The spectrum in Figure 11–30 is for a pure liquid compound containing only C, H, and O. The boiling point of the liquid is about 177°C and its molecular weight is 106. Identify the compound.

11–12. The empirical formula for a liquid compound was found to be $C_6H_{12}O_2$. Its spectrum is shown in Figure 11–31. What conclusion can be drawn regarding the identity of the compound?

11–13. An empty cell showed 12 interference peaks in the wavelength range of 6.0 to 12.2 μm. Calculate the path length of the cell.

11–14. An empty cell exhibited 9.5 interference peaks in the region of 1250 to 1480 cm^{-1}. What was the path length of the cell?

11–15. What length of mirror drive in a Fourier transform spectrometer would be required to provide a resolution of (a) 0.020 cm^{-1} and (b) 2.0 cm^{-1}?

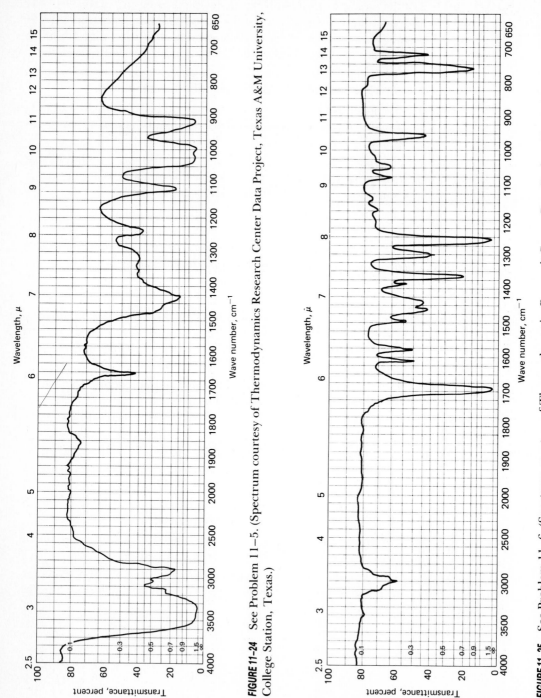

FIGURE 11-24 See Problem 11-5. (Spectrum courtesy of Thermodynamics Research Center Data Project, Texas A&M University, College Station, Texas.)

FIGURE 11-25 See Problem 11-6. (Spectrum courtesy of Thermodynamics Research Center Data Project, Texas A&M University, College Station, Texas.)

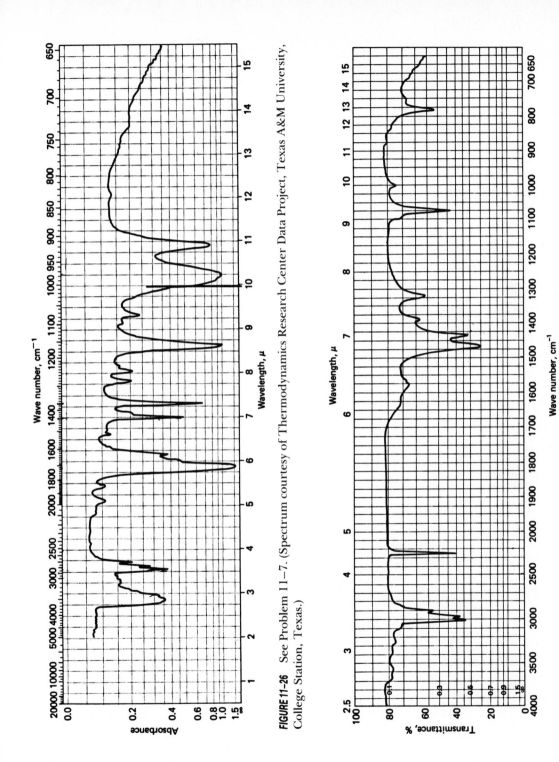

FIGURE 11-26 See Problem 11–7. (Spectrum courtesy of Thermodynamics Research Center Data Project, Texas A&M University, College Station, Texas.)

FIGURE 11-27 See Problem 11–8. (Spectrum courtesy of Thermodynamics Research Center Data Project, Texas A&M University, College Station, Texas.)

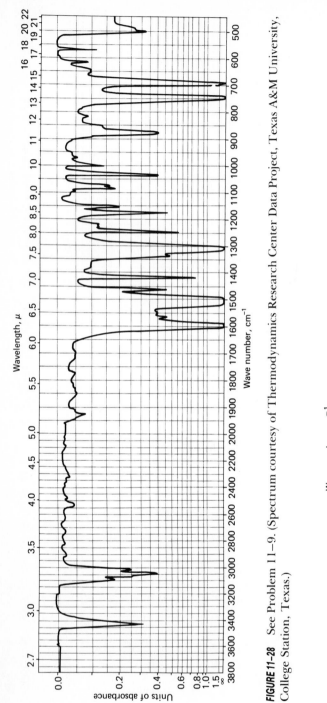

FIGURE 11-28 See Problem 11–9. (Spectrum courtesy of Thermodynamics Research Center Data Project, Texas A&M University, College Station, Texas.)

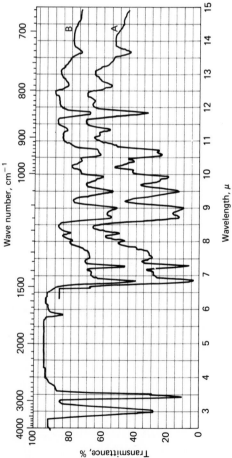

FIGURE 11-29 See Problem 11–10. (Spectrum courtesy of Thermodynamics Research Center Data Project, Texas A&M University, College Station, Texas.)

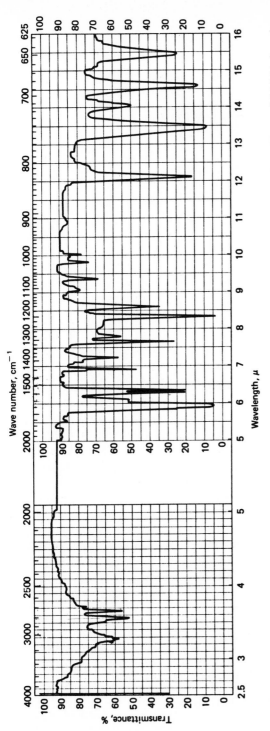

FIGURE 11-30 See Problem 11–11. (Spectrum courtesy of Thermodynamics Research Center Data Project, Texas A&M University, College Station, Texas.)

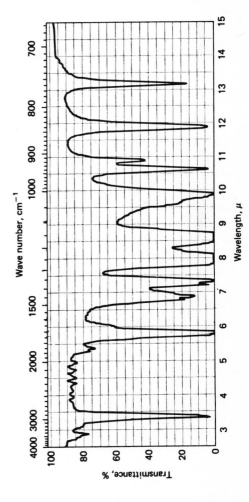

FIGURE 11-31 See Problem 11–12. (Spectrum courtesy of Thermodynamics Research Center Data Project, Texas A&M University, College Station, Texas.)

Raman Spectroscopy

When radiation passes through a transparent medium, the species present scatter a fraction of the beam in all directions (Section 4A–6). If the medium contains only particles of molecular dimensions, the scattered radiation is so weak as to be undetectable to the eye; this type of scattering is termed *Rayleigh scattering*. When the medium contains aggregates of particles with dimensions approximating that of the wavelength of the radiation, however, scattering becomes evident as the Tyndall effect or as a turbidity.

In 1928, the Indian physicist C. V. Raman discovered that the wavelength of a small fraction of the radiation scattered by certain molecules differs from that of the incident beam and furthermore that the shifts in wavelength depend upon the chemical structure of the molecules responsible for the scattering. He was awarded the 1931 Nobel prize in physics for this discovery and his systematic exploration of it.[1]

The theory of Raman scattering, which by now is well understood, shows that the phenomenon results from the same type of quantized vibrational changes that are associated with infrared absorption. Thus, the *difference* in wavelength between the incident and scattered radiation corresponds to wavelengths in the mid-infrared region. Indeed, the Raman scattering spectrum and infrared absorption spectrum for a given species often resemble one another quite closely. There are, however, enough differences between the kinds of groups that are infrared and Raman active to make the techniques complementary rather than

---

[1] For more complete discussions of the theory and practice of Raman spectroscopy, see: T. R. Gilson and P. J. Hendra, *Laser Raman Spectroscopy*, Wiley-Interscience: New York, 1970; *Raman Spectroscopy*, J. Lascombe and V. Huong, Eds., Wiley: New York, 1982; J. R. Durig and W. C. Harris, *Physical Methods of Chemistry*, Vol. 1, Part III B, Chapter 2, Wiley-Interscience: New York, 1972; and *Infrared and Raman Spectroscopy*, E. G. Brame and J. G. Grasselli, Eds., Dekker: New York, 1977.

competitive. For some problems, the infra-red method is the superior tool; for others, the Raman procedure offers more useful spectra.

An important advantage of Raman spectra over infrared lies in the fact that water does not cause interference; indeed, Raman spectra can be obtained from aqueous solutions. In addition, glass or quartz cells can be employed, thus avoiding the inconvenience of working with sodium chloride or other atmospherically unstable windows.

12A Theory of Raman Spectroscopy

Raman spectra are obtained by irradiating a sample with a powerful source of visible monochromatic radiation. A mercury arc was employed in early investigations; by now, this source has been superseded by high-intensity gas or solid lasers. During irradiation, the spectrum of the scattered radiation is measured at some angle (usually 90 deg) with a suitable visible-region spectrometer. At the very most, the intensities of Raman lines are 0.001% of the source; as a consequence, their detection and measurement are difficult. An exception to this statement is encountered with resonance Raman lines, which are considerably more intense. Resonance Raman spectroscopy is described near the end of this chapter.

12A-1 EXCITATION OF RAMAN SPECTRA

Figure 12–1 depicts a portion of a Raman spectrum, which was obtained by irradiating a sample of carbon tetrachloride with an intense beam of an argon ion laser having a wavelength of 488.0 nm (20492 cm^{-1}). The scattered radiation, which was observed at 90 deg to the incident beam, is of three types, namely *Stokes, anti-Stokes,* and *Rayleigh.* The last, whose wavelength is exactly that of the excitation source, is significantly

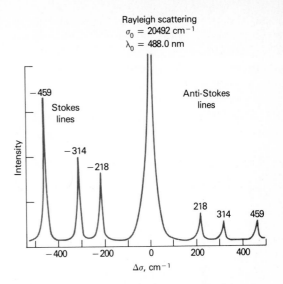

FIGURE 12-1 Raman spectrum for CCl_4 excited by laser radiation of $\lambda_0 = 488$ nm or $\sigma_0 = 20492$ cm^{-1}. The number above the peaks are the Raman shift, $\Delta\sigma = (\sigma_s - \sigma_0)$ cm^{-1}. (Reprinted with permission from: D. P. Strommen and K. Nakamato, *Amer. Lab.,* **1981,** *13* (10), 72. Copyright 1981 by International Scientific Communications, Inc.)

more intense than either of the other two types.

As is usually the case for Raman spectra, the abscissa of Figure 12–1 is the wavenumber shift $\Delta\sigma$, which is defined as the difference in wavenumbers (cm^{-1}) between the observed radiation and that of the source. Note that three Raman peaks are found on either side of the Rayleigh peak and that the pattern of shifts on the two sides are identical. That is, Stokes lines are found at wavenumbers that are 218, 314, and 459 cm^{-1} *smaller* than the Rayleigh peak while anti-Stokes peaks occur at 218, 314, and 459 cm^{-1} *greater* than the wavenumber of the source. It should also be noted that additional lines can be found at ±762 and 790 cm^{-1} as well. It is important to appreciate that the magnitude of Raman shifts are *independent of the wavelength of excitation.* Thus, shift patterns identical to those shown in Figure 12–1 would be observed for carbon tetrachloride regardless of whether excitation was carried out with a krypton ion laser (488.0 nm), a helium/neon laser (632.8 nm), or a ruby laser (694.3 nm).

Superficially, the appearance of Raman spectral lines at lower energies (longer wavelengths) is analogous to the Stokes shifts found in a fluorescence experiment (p. 226); for this reason, negative Raman shifts are called *Stokes shifts*. We shall see, however, that Raman and fluorescence spectra arise from fundamentally different processes; thus, the application of the same terminology to both fluorescent and Raman spectra is perhaps unfortunate.

Shifts toward higher energies are termed *anti-Stokes;* quite generally, anti-Stokes lines are appreciably less intense than the corresponding Stokes lines. For this reason only the Stokes part of a spectrum is generally used. Furthermore, the abscissa of the plot is often labeled simply frequency in cm^{-1} rather than wavenumber shift $\Delta\sigma$; the negative sign is also dispensed with. It is noteworthy that fluorescence may interfere seriously with the observation of Stokes shifts but not with anti-Stokes. With fluorescing samples, anti-Stokes signals may, therefore, be more useful despite their lower intensities.

12A–2   MECHANISM OF RAYLEIGH SCATTERING

As was pointed out in Section 4A–5, transmission of radiation through a transparent medium containing molecules can be viewed as a stepwise process in which the fluctuating electrical field of the radiation induces vibrational motion of electrons in the bonds between heavy and fixed atomic nuclei; the vibrational frequency corresponds to that of the radiation. Vibrating electrons, however, emit radiation in all directions (for example, radio waves are emitted from an antenna in which electrons are

in periodic motion); as a consequence, retention of a photon by a molecule is momentary (10^{-15} to 10^{-16} sec on the average). This retention time is long enough, however, to account for the slowing of the rate of propagation of radiation as it passes from a vacuum through a transparent medium.

The secondary waves from vibrating electrons, which are propagated in all directions, are coherent and thus interfere with one another and with the incident beam. The consequence is destructive interference of most of the waves except those with paths that parallel the incident beam. The small part of the radiation that does travel laterally from the incident beam (because of incomplete interference) is Rayleigh-type scattered radiation.

Although all transparent media exhibit some scattering, the amount varies widely and depends upon the wavelength of the radiation, the size of the scattering particles, and their polarizability. The wavelength dependence is large with the fraction scattered being inversely related to the fourth power of this parameter.

12A–3   MECHANISM OF NORMAL RAMAN SCATTERING

This section is devoted to a discussion of the theories of *normal Raman spectroscopy*. Here, excitation is carried out by radiation having a wavelength that is well away from any absorption peaks of the analyte. In *resonance Raman spectroscopy,* in contrast, radiation at or near the absorption peak is used. This technique is discussed in Section 12A–4.

Wave Theory.   The normal Raman effect can be partially understood with the aid of the classical wave theory, which was just used to rationalize Rayleigh scattering. Here, it is assumed that an electric moment m is induced in a molecular bond when it is subjected to the electric field E of a beam of electromagnetic radiation. The moment m is proportional to the field. That is,

$$m = \alpha E \qquad (12\text{--}1)$$

where the proportionality constant α is the *polarizability* of the bond, a measure of its deformability in an electric field.

For a light wave, the electric field varies as a cosine (or sine) function of time t (Section 4A–1). Thus, we may write

$$E = E_0 \cos 2\pi\nu t \qquad (12\text{--}2)$$

where E_0 is the maximum amplitude of the wave and ν is its frequency.

For illustrative purposes, it is convenient to consider the change in polarizability that takes place during stretching vibrations of a diatomic molecule. Here, the polarizability depends upon the interatomic distance r and becomes larger as the distance between atoms increases and smaller as the bond contracts. That is, the bond becomes more deformable as the separation between nuclei and electrons becomes greater. During molecular vibrations then, the polarizability of the bond fluctuates periodically at a rate that is given by the partial derivative of the polarizability as a function of distance $\partial\alpha/\partial r$. The polarizability at any instant can be approximated by

$$\alpha = \alpha_0 + (\partial\alpha/\partial r)r_0 \cos 2\pi\nu_m t$$

where r_0 is the equilibrium distance between atoms, α_0 is the polarizability of the bond at this distance, and ν_m is the frequency of the molecular vibration. Substitution of this equation and Equation 12–2 into Equation 12–1 yields

$$\begin{aligned} m &= [\alpha_0 + (\partial\alpha/\partial r)r_0 \cos 2\pi\nu_m t]E_0 \cos 2\pi\nu t \\ &= \alpha_0 E_0 \cos 2\pi\nu t \\ &\quad + (\partial\alpha/\partial r)r_0 E_0 \cos 2\pi\nu_m t \cos 2\pi\nu t \quad (12\text{--}3) \end{aligned}$$

From trigonometry, recall that

$$\cos x \cos y = \frac{1}{2}[\cos(x+y) + \cos(x-y)]$$

Applying this identity to Equation 12–3 gives

$$\begin{aligned} m &= \alpha_0 E_0 \cos 2\pi\nu t \\ &\quad + \frac{1}{2}(\partial\alpha/\partial r)r_0 E_0 \cos 2\pi t(\nu + \nu_m) \\ &\quad + \frac{1}{2}(\partial\alpha/\partial r)r_0 E_0 \cos 2\pi t(\nu - \nu_m) \quad (12\text{--}4) \end{aligned}$$

The component of the electric moment represented by the first term in Equation 12–4 produces Rayleigh scattering, which is seen to have the frequency ν of the incident beam. The second and third terms are electric moments for the scattered radiation where the incident frequency has been modulated by $\pm\nu_m$, the vibrational frequency of the bond. Thus, the incident radiation frequency modified by the molecular vibrational frequency, results in the observed Raman side bands. It is important to note that the existence of the Raman lines *requires* that the polarizability of the bond varies as a function of atomic distance. That is, $\partial\alpha/\partial r$ cannot be zero if a bond is to be Raman active.

The Particle Theory. The wave theory, while accounting for the existence of Stokes and anti-Stokes scattering, suggests erroneously that the intensities of the two types of radiation should be equal. As noted earlier, however, Stokes radiation is markedly more intense. Furthermore, the wave theory does not account for the fact that the intensity ratio of anti-Stokes to Stokes radiation increases significantly with temperature increases. To account for these discrepancies, it is necessary to invoke the particle theory of radiation and employ the techniques of quantum mechanics. The resulting expressions are complex and will not be discussed here. For our purposes a brief qualitative description based upon the particle properties of radiation will suffice.

In the quantum treatment, radiation is pictured as a stream of photons that are

scattered as a consequence of collisions between these radiant particles and the molecules of the medium. Most of these collisions
are *elastic* in the sense that no net transfer of
energy occurs, the only result of a collision
being a change in direction of the photon. A
few collisions are, however, *inelastic*. Here,
the vibrational energy of a bond is subtracted from or added to the energy of the
photon thus changing its frequency. These
processes are depicted in the energy level
diagram shown in Figure 12–2.

The heavy dark line on the left depicts the
energy change in a molecule upon being
struck by a photon having an energy of $h\nu$.
It is important to appreciate that the process
shown is *not quantized;* thus depending upon
the frequency of the radiation, the energy
of the molecule can assume any of an infinite number of values or states (called *virtual
states*) between the ground state and the first
excited electronic state shown in the upper
part of the diagram. The second and narrower arrow shows the type of change that
would occur if the molecule encountered by
the photon happened to be in the first vibrational level of the ground state. At room
temperature, the fraction of the molecules

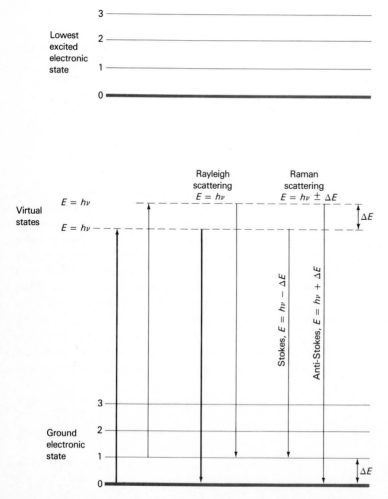

FIGURE 12-2 Origin of Rayleigh and Raman scattering.

in this state is small. Thus, as indicated by the width of the arrows, the probability of this process occurring is much smaller.

The middle set of arrows depicts the changes that produces Rayleigh scattering. Again the more probable change is shown by the wider arrow.

Finally, the energy changes that produce Stokes and anti-Stokes emission are depicted on the right. The two differ from the Rayleigh radiation by frequencies corresponding to $\pm\Delta E$, the energy of the first vibrational level of the ground state. Note that if the bond were infrared active, the energy of its absorption would also be ΔE. Thus, the Raman *frequency shift* and the infrared absorption *peak frequency* are identical.

Note also that the relative populations of the two upper energy states are such that Stokes emission is much favored over anti-Stokes. In addition, Rayleigh scattering has a considerably higher probability of occurring than Raman because the most probable event is the energy transfer to molecules in the ground state and reemission by the return of these molecules to the ground state. Finally, it should be noted that the ratio of anti-Stokes to Stokes intensities will increase with temperature because a larger fraction of the molecules will be in the first vibrationally excited state under these circumstances.

12A-4 RESONANCE RAMAN SCATTERING

Resonance Raman scattering refers to a phenomenon in which Raman line intensities are greatly enhanced by excitation with wavelengths that closely approach that of an *electronic* absorption peak of a molecule. Under this circumstance, the magnitudes of certain Raman peaks are enhanced by a factor of 10^2 to 10^6. As a consequence, resonance Raman spectra can be obtained routinely at analyte concentrations of 10^{-4} to $10^{-5}\ M$ (in contrast to normal Raman stud-

ies, which are ordinarily limited to concentrations greater than 0.1 *M*). In addition, since only Raman bands related to the electronic chromophore suffer intensity increases, resonance Raman spectra are markedly simpler than normal scattering spectra.

Figure 12–3**a** illustrates the energy changes responsible for resonance Raman scattering. This figure differs from the energy diagram for normal Raman scattering (Figure 12–2) in the respect that the electron is promoted into an excited vibronic state followed by an immediate relaxation to a vibrational level of the ground state. As shown in the figure, resonance Raman scattering differs from fluorescence in the respect that relaxation to the ground state is *not* preceded by prior relaxation to the lowest vibrational level of the excited state. The time scales for the two phenomena are also quite different with Raman relaxation occurring in less than 10^{-14} sec compared with the 10^{-6} to 10^{-8} sec for fluorescent emission.

The theory of resonance enhancement of Raman scattering is by now well developed but too complex for presentation here.[2]

12A-5 RAMAN ACTIVE VIBRATIONAL MODES

As we have just noted, for a given bond, the energy shifts observed in a Raman experiment should be identical to the *energies* of its infrared absorption bands, provided that the vibrational modes involved are active toward both infrared absorption and Raman scattering. Figure 12–4 illustrates the similarity of the two types of spectra; it is seen that several peaks with identical σ and $\Delta\sigma$ values exist for the two compounds. It is also noteworthy, however, that the relative size of the corresponding peaks is frequently

---

[2] For reviews of this theory, see: B. B. Johnson and W. L. Peticolas, *Ann. Rev. Phys. Chem.*, **1977**, *27*, 465; T. G. Spiro and P. Stein, *Ann. Rev. Phys. Chem.*, **1977**, *28*, 501.

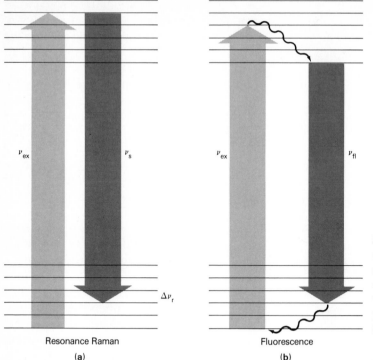

FIGURE 12-3 Energy diagram for (**a**) resonance Raman scattering and (**b**) fluorescent emission. Radiationless relaxation shown as wavy arrows. (Reprinted with permission from: M. D. Morris and D. J. Wallin, *Anal. Chem.,* **1979,** *51,* 185A. Copyright 1979 American Chemical Society.)

quite different; moreover, certain peaks that occur in one spectrum are absent in the other.

The differences between a Raman and an infrared spectrum are not surprising when it is considered that the basic mechanisms, although dependent upon the same vibrational modes, arise from processes that are mechanistically different. Infrared absorption requires that a vibrational mode of the molecule have a change in dipole or charge distribution associated with it. Only then can radiation of the same frequency interact with the molecule and promote it to an excited vibrational state. In contrast, scattering involves a momentary distortion of the electrons distributed around a bond in a molecule, followed by reemission of the radiation in all directions as the bond returns to its normal state. In its distorted form, the molecule is temporarily polarized; that is, it develops momentarily an induced dipole which disappears upon relaxation and reemission. The effectiveness of a bond toward scattering thus depends directly upon the ease with which the electrons of the bond can be distorted from their normal positions (that is, the *polarizability* of the bonds); polarizability decreases with increasing electron density, increasing bond strength, and decreasing bond length. The Raman shift in scattered radiation, then, requires that there be a *change in polarizability*—rather than a change in dipole—associated with the vibrational mode of the molecule; as a consequence, the Raman activity of a given mode may differ markedly from its infrared activity. For example, a homonuclear molecule such as nitrogen, chlorine, or hydrogen has no dipole moment either in its equilibrium position or when a stretching vibration causes a change in the distance between the two nuclei. Thus, absorption of radiation of the vibra-

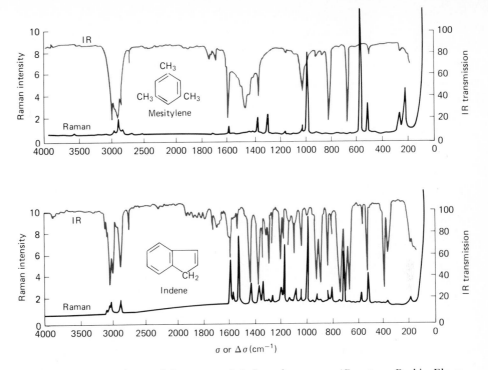

FIGURE 12-4 Comparison of Raman and infrared spectra. (Courtesy Perkin-Elmer Corp., Norwalk, CT.)

tion frequency cannot occur. On the other hand, the polarizability of the bond between the two atoms of such a molecule varies periodically in phase with the stretching vibrations, reaching a maximum at the greatest separation and a minimum at the closest approach. A Raman shift corresponding in frequency to that of the vibrational mode results.

It is of interest to compare the infrared and the Raman activities of coupled vibrational modes such as those described earlier (p. 324) for the planar carbon dioxide molecule. In the symmetric mode, no change in dipole occurs as the two oxygen atoms move away from or toward the central carbon atom; thus, this mode is infrared-inactive. The polarizability, however, fluctuates in phase with the vibration since distortion of bonds becomes easier as they lengthen and

more difficult as they shorten; Raman activity is associated with this mode.

In contrast, the dipole moment of carbon dioxide fluctuates in phase with the asymmetric vibrational mode; thus, an infrared absorption peak arises from this mode. On the other hand, as the polarizability of one of the bonds increases as it lengthens, the polarizability of the other decreases. Thus, the asymmetric stretching vibration is Raman inactive.

Often, as in the foregoing examples, parts of Raman and infrared spectra are complementary, each being associated with a different set of vibrational modes within a molecule. Other vibrational modes may be both Raman- and infrared-active. Here, the two spectra resemble one another with peaks involving the same energies. The relative intensities of corresponding peaks may differ, however, because the probability for the transition may be different for the two mechanisms.

12A-6 INTENSITY OF NORMAL RAMAN PEAKS

The intensity or power of a normal Raman peak depends in a complex way upon the polarizability of the molecule, the intensity of the source, and the concentration of the active group, as well as other factors. In the absence of absorption, the power of Raman emission increases with the fourth power of the frequency of the source; however, advantage can seldom be taken of this relationship because of the likelihood that ultraviolet irradiation will cause photodecomposition.

Raman intensities are usually directly proportional to the concentration of the active species. In this regard, Raman spectroscopy more closely resembles fluorescence than absorption, where the concentration-intensity relationship is logarithmic.

12A-7 RAMAN DEPOLARIZATION RATIOS

Raman measurements provide, in addition to intensity and frequency information, one additional parameter that is sometimes useful in determining the structure of molecules, namely the *depolarization ratio*. Here, it is important to carefully distinguish between the terms polarizability and polarization. The former term describes a *molecular* property having to do with the deformability of a bond. Polarization, in contrast, is a property of a beam of radiation and describes the plane in which the radiation vibrates.

When Raman spectra are excited by plane-polarized radiation (as they are when a laser source is used), the scattered radiation is found to be polarized to various degrees depending upon the type of vibration responsible for the scattering. The nature of this effect is illustrated in Figure 12–5 where radiation from a laser source is shown as being polarized in the yz plane. Part of the resulting scattered radiation is shown as being polarized parallel to the original beam, that is, in the xz plane; the intensity of this radiation is symbolized as $I_{\parallel}$. The remainder of the scattered beam is polarized in the xy plane, which is perpendicular to the polarization of the original beam; the intensity of this perpendicularly polarized radiation is shown as $I_{\perp}$. The depolar-

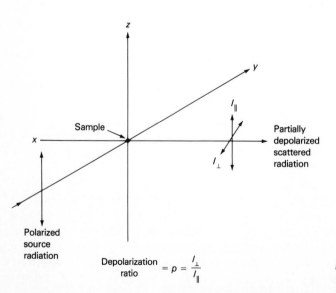

$$\text{Depolarization ratio} = p = \frac{I_{\perp}}{I_{\parallel}}$$

FIGURE 12-5 Depolarization resulting from Raman scattering.

ization ratio p is defined as

$$p = \frac{I_\perp}{I_\parallel} \tag{12-5}$$

Experimentally, the depolarization ratio is readily obtained by inserting a Nicol analyzer prism (Section 13C–1) between the sample and the monochromator. Spectra are then obtained with the prism oriented parallel with first the xz and then the xy plane shown in Figure 12–5.

The depolarization ratio is dependent upon the symmetry of the vibrations responsible for scattering. For example, the peak for carbon tetrachloride at 459 cm^{-1} (Figure 12–1) arises from a totally symmetric breathing vibration involving the simultaneous movement of the four tetrahedrally arranged chlorine atoms toward and away from the central carbon atom. The depolarization ratio is 0.005 indicating minimal depolarization (the 459 cm^{-1} line is thus said to be *polarized*). In contrast, the carbon tetrachloride peaks at 218 and 314 cm^{-1}, which arise from nonsymmetrical vibrations have depolarization ratios of about 0.75. From scattering theory it is possible to demonstrate that the depolarization for nonsymmetric vibrations is 3/4 while for symmetric vibrations, the ratio is always less than this number. The depolarization ratio is thus useful in correlating Raman lines with modes of vibration.

12B Instrumentation

Instrumentation for modern Raman spectroscopy consists of three components, namely, an intense source, a sample-illumination system, and a suitable spectrophotometer.

12B-1 SOURCES

The most widely used Raman source is probably a helium/neon laser, which oper-

ates in a continuous mode at a power of 50 mW. Laser radiation is produced at 632.8 nm; several other lower-intensity nonlasing lines accompany the principal line and must be removed by suitable narrow band filters. Alternatively, the effect of these lines can be eliminated by taking advantage of the fact that nonlasing lines diverge much more rapidly than the lasing line; thus, by making the distance between the source and the entrance slit large, the intensities of the former can be made to approach zero.

Argon-ion lasers, with lines at 488.0 and 514.5 nm, are also employed, particularly when higher sensitivity is required. Because the intensity of Raman scattering varies as the fourth power of the frequency of the exciting source, the argon line at 488 nm provides Raman lines that are nearly three times as intense as those excited by the helium/neon source, given the same input power.

A variety of other laser sources are available; undoubtedly new and improved sources will appear in the future. The need exists for several sources, inasmuch as one must be chosen that is not absorbed by the sample (except with resonance Raman measurements) or the solvent. Furthermore, care must be taken to avoid wavelengths that will cause the sample to fluoresce or to photodecompose.

12B-2 SAMPLE-ILLUMINATION SYSTEMS

Sample handling for Raman studies tends to be simpler than for infrared because the measured wavenumber differences are between two *visible* frequencies. Thus, glass can be employed for windows, lenses, and other optical components. In addition, the source is readily focused on a small area and the emitted radiation can be efficiently focused on a slit. Very small samples can be examined as a consequence. In fact, a common sample holder for liquid samples is an ordinary glass melting-point capillary.

Figure 12–6 shows two of many configu-

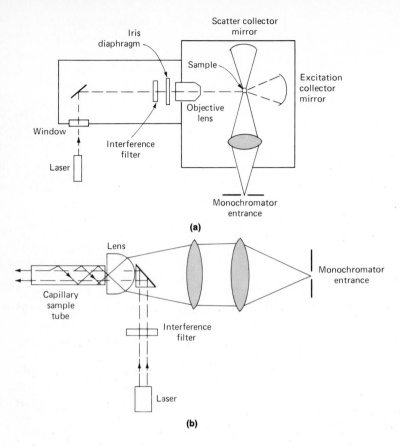

FIGURE 12-6 Two sample excitation systems.

rations for the handling of liquids. The size of the tube in (**b**) has been enlarged to show details of the reflection of the Raman radiation off the walls; in fact, the holder is a 1-mm o.d. glass capillary that is about 5 cm long.

For unusually weak signals such as might occur with a dilute gas sample, the cell is often placed between the mirrors of the laser source; enhanced excitation power results.

12B-3 RAMAN SPECTROMETERS

All early Raman studies were performed with prism spectrographs, with the spectra being recorded on photographic films or plates. Intensities were determined by measuring the blackness of the lines.

Several Raman spectrophotometers, which incorporate photomultipliers to record peak intensities, are now available. Their designs are not significantly different from the recording spectrophotometers discussed in Section 7D–2. Most employ double monochromators to minimize spurious radiation reaching the detector. In addition, a split-beam design is employed to compensate for the effects of fluctuations in the intensity of the source.

12B-4 COMPARISON OF RAMAN AND INFRARED SAMPLE-HANDLING TECHNIQUES

In terms of ease of sample preparation and handling, Raman spectroscopy possesses the great advantage of permitting use of glass cells instead of the more fragile and atmospherically less stable crystalline halides required for work in the infrared.

Substances of limited solubility can often be finely ground and tapped into an open-ended cavity for Raman examination; the more troublesome mull technique is required for infrared studies. Polymers can often be examined for Raman activity directly with no preliminary sample treatment; comparable infrared analyses may require that the polymer be compressed, molded, or cast into a thin film prior to examination.

Another important difference is that water is a weak scatterer of Raman radiation but a strong absorber of infrared. Thus, aqueous solutions of samples can be employed for Raman studies. This advantage is particularly pronounced for the study of biological systems, inorganic substances, and water pollution problems.

Liquid samples containing colloidal or suspended particles ordinarily scatter sufficient amounts of the laser beam to make observation of the Raman effect difficult or impossible. Such samples must be treated to remove solids before a Raman spectrum can be obtained.

12B-5 COMPARISON OF RAMAN AND INFRARED INSTRUMENTATION

H. J. Sloane has compared Raman and infrared spectroscopy on the basis of instrumentation, sample handling, and applicability.[3]

Optics. Raman spectrophotometers are somewhat less complicated from the standpoint that glass or quartz optics can be used throughout. Furthermore, Raman spectra in both the mid- and the far-infrared regions (4000 to 25 cm^{-1}) can be examined with a single optical system; in contrast, infrared studies require several gratings to cover the same range. The grating for a Ra-

man instrument, however, must be of higher quality because spectral artifacts arising from imperfections have more serious consequences.

Detectors. Raman instruments use photomultiplier detectors which offer distinct signal-to-noise advantages over the thermal detectors employed in infrared instruments (see p. 143).

Resolution. The resolution of the best infrared and Raman spectrometers is about the same ($\sim$0.2 cm^{-1}).

Cost. There are no low-cost Raman spectrophotometers equivalent to bench-top infrared spectrometers. The prices of Raman instruments have decreased significantly recently; it is to be hoped that instruments for routine applications will appear shortly at a cost of $5000 to $10,000.

12C Applications of Raman Spectroscopy

Raman spectroscopy has been applied to the qualitative and quantitative analysis of inorganic, organic, and biological systems.[4]

12C-1 RAMAN SPECTRA OF INORGANIC SPECIES[5]

The Raman technique is often superior to the infrared for investigating inorganic systems because aqueous solutions can be employed. In addition, the vibrational energies of metal-ligand bonds are generally in the range of 100 to 700 cm^{-1}, a region of the infrared that is experimentally difficult to study. These vibrations are frequently Raman-active, however, and peaks with $\Delta\sigma$ values in this range are readily observed. Ra-

---

[3] H. J. Sloane, *Appl. Spec.*, **1971**, *25*, 430.

[4] For detailed reviews of applications of Raman spectroscopy, see: D. L. Gerrard, *Anal. Chem.*, **1984**, *56*, 219R; D. J. Gardiner, *Anal. Chem.*, **1982**, *54*, 165R; **1980**, *52*, 96R; **1978**, *50*, 131R.

[5] For a review of applications, see: K. Nakamoto, *Infrared and Raman Spectra of Inorganic and Coordination Compounds*, 3d ed., Wiley: New York, 1978.

man studies are potentially useful sources of information concerning the composition, structure, and stability of coordination compounds. For example, numerous halogen and halogenoid complexes produce Raman spectra and thus are susceptible to investigation by this means. Metal-oxygen bonds are also active. Spectra for such species as VO_4^{3-}, $Al(OH)_4^-$, $Si(OH)_6^{2-}$, and $Sn(OH)_6^{2-}$ have been obtained; Raman studies have permitted conclusions regarding the probable nature of such species. For example, in perchloric acid solutions, vanadium(IV) appears to be present as $VO^{2+}(aq)$ rather than as $V(OH)_2^{2+}(aq)$; studies of boric acid solutions show that the anion formed by acid dissociation is the tetrahedral $B(OH)_4^-$ rather than $H_2BO_3^-$. Dissociation constants for strong acids such as H_2SO_4, HNO_3,

H_2SeO_4 and H_5IO_6 have been obtained by Raman measurements.

It seems probable that the future will see even wider use of Raman spectroscopy for theoretical studies of inorganic systems.

12C-2   RAMAN SPECTRA OF ORGANIC SPECIES

Raman spectra are similar to infrared spectra in that they have regions that are useful for functional group detection and fingerprint regions that permit the identification of specific compounds. Figure 12–7 shows a correlation chart that can be employed for functional group recognition. Dollish[6] has

[6] F. R. Dollish, W. G. Fateley, and F. F. Bentley, *Characteristic Raman Frequencies of Organic Compounds*, Wiley-Interscience: New York, 1971.

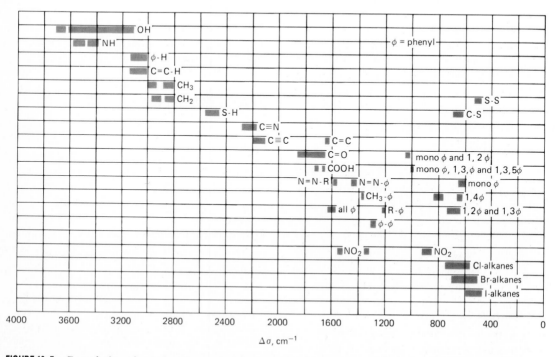

FIGURE 12-7   Correlation chart showing group frequencies for various organic functional groups. (Courtesy of Varian Instrument Division, Palo Alto, CA.)

published a comprehensive treatment of functional group frequencies. Catalogs of Raman spectra for organic compounds are also available.[7]

Raman spectra yield more information about certain types of organic compounds than do their infrared counterparts. For example, the double-bond stretching vibration for olefins results in weak and sometimes undetected infrared absorption. On the other hand, the Raman band (which, like the infrared band, occurs at about 1600 cm^{-1}) is intense, and its position is sensitive to the nature of substituents as well as to their geometry. Thus, Raman studies are likely to yield useful information about the olefinic functional group that may not be revealed by infrared spectra. This statement applies to cycloparaffin derivatives as well; these compounds have a characteristic Raman peak in the region of 700 to 1200 cm^{-1}. This peak has been attributed to a breathing vibration in which the nuclei move in and out symmetrically with respect to the center of the ring. The position of the peak decreases continuously from 1190 cm^{-1} for cyclopropane to 700 cm^{-1} for cyclooctane; Raman spectroscopy thus appears to be an excellent diagnostic tool for the estimation of ring size in paraffins. The infrared peak associated with this vibration is weak or nonexistent.

12C-3 BIOLOGICAL APPLICATIONS OF RAMAN SPECTROSCOPY

Raman spectroscopy has been applied widely for the study of biological systems.[8] The advantages of this technique include the small sample requirement, the minimal

sensitivity toward interference by water, the spectral detail, and the conformational and environmental sensitivity.

12C-4 QUANTITATIVE APPLICATIONS

Raman spectra tend to be less cluttered with peaks than infrared spectra. As a consequence, peak overlap in mixtures is less likely and quantitative measurements are simpler. In addition, Raman instrumentation is not subject to attack by moisture, and small amounts of water in a sample do not interfere. Despite these advantages, Raman spectroscopy has not yet been exploited widely for quantitative analysis.

An example of the potentialities of the method for the analysis of mixtures is provided in a paper by Nicholson,[9] in which a procedure for the determination of the constituents in an eight-component mixture is described. The components included benzene, isopropyl benzene, three diisopropyl benzenes, two triisopropyl derivatives, and 1,2,4,5-tetraisopropyl benzene. The power of the characteristic peaks, compared with the power of a reference peak (CCl_4), was assumed to vary linearly with volume percent of each component. Analysis of synthetic mixtures of all the components by the procedure produced results that agreed with the preparatory data to about 1% (absolute).

12D Application of Other Types of Raman Spectroscopy

With the development of tunable lasers, two new Raman spectroscopic methods were developed in the early 1970s. A brief discussion of the applications of each of these techniques follows.

---

[7] Samuel P. Sadtler and Sons, Inc., 2100 Arch Street, Philadelphia, PA.

[8] For reviews of biological applications, see: H. Fabian, et al., *Biophys.*, **1980**, *80*, 1; and B. P. Gaber, *Amer. Lab.*, **1977**, *9* (3), 15.

[9] D. E. Nicholson, *Anal. Chem.*, **1960**, *32*, 1634.

12D-1   RESONANCE RAMAN SPECTROSCOPY[10]

As noted in Section 12A–4, when Raman scattering is brought about by radiation with a frequency near or coincident with an electronic absorption peak of an analyte, the intensity of certain of the Raman lines is enhanced by as much as 10^6, thus making possible the study of solutions that are as dilute as 10^{-5} *M*. The enhancement is restricted to lines arising from electrons making up the chromophoric bonds. Consequently, resonance Raman spectra tend to contain many fewer lines than their normal Raman counterparts.

Resonance Raman spectroscopy became practical with the development of tunable dye lasers, which make possible the selection of an excitation wavelength from the ultraviolet through the near-infrared region. To minimize sample photodecomposition during excitation, it is common practice to rotate or flow the sample past the laser beam. Since the laser beam is focused on only a tiny fraction of the sample for a brief period, local heating and sample decomposition is largely avoided.

Perhaps the most important application of resonance Raman spectroscopy has been to the study of biological molecules under physiologically significant conditions; that is, in the presence of water and at low to moderate concentration levels. As an example, the technique has been used to determine the oxidation state and spin of iron atoms in hemoglobin and cytochrome-*c*. In these molecules, the resonance Raman bands are due solely to vibrational modes of the tetrapyrrole chromophore. None of the other bands associated with the protein is enhanced and at the concentrations normally used do not interfere as a consequence.

A major limitation to resonance Raman (as well as normal Raman) spectroscopy is interference by fluorescence either by the analyte itself or by other species present in the sample.

12D-2   COHERENT ANTI-STOKES RAMAN SPECTROSCOPY (CARS)[11]

This technique has been employed to overcome some of the drawbacks of conventional Raman spectroscopy, namely, its low efficiency, its limitation to the visible and near-ultraviolet regions, and its susceptibility to interference from fluorescence.

One way of carrying out a CARS experiment is to employ a laser to pump a tunable dye laser. The two lasers are arranged in such a way that part of the beam from the pumping laser of frequency ν_p is employed for pumping and part for sample excitation. The dye laser is also employed for sample excitation, and its frequency ν_d is varied until the difference between the two excitation frequencies is equal to the resonance frequency ν_r for one of the Raman lines. That is,

$$\nu_r = \nu_p - \nu_d$$

Under these circumstances, a beam of radiation is generated having a frequency ν_a that is given by

$$\nu_a = 2\nu_p - \nu_d$$

This radiation differs from normal Raman radiation in that it is *coherent* rather than scattered in all directions. Furthermore, it is

---

[10] For a brief review of this topic, see: M. D. Morris and D. J. Wallan, *Anal. Chem.*, **1979**, *51*, 182A; and D. P. Strommen and K. Nakamoto, *J. Chem. Educ.*, **1977**, *54*, 474.

[11] For brief reviews, see: R. F. Begley, A. B. Harvey, R. L. Beyer, and B. S. Hudson, *Amer. Lab.*, **1974**, *6* (11), 11; A. B. Harvey, *Anal. Chem.*, **1978**, *50* (9), 905A; and J. F. Verdieck, R. J. Hall, J. A. Shirley, and A. C. Eckbreth, *J. Chem. Educ.*, **1982**, *59*, 495. See also: *Chemical Applications of Nonlinear Raman Spectroscopy*, A. B. Harvey, Ed., Academic Press: New York, 1981

emitted at an angle such that it can be readily separated from the excitation beam without a monochromator. Finally, because it occurs in the anti-Stokes region (a similar beam occurs at $\nu_a = 2\nu_p + \nu_d$ in the Stokes region), interference from Stokes fluorescence by the sample is avoided.

The efficiency of coherent anti-Stokes emission is high, with as much as 1% of the excitation radiation being converted to the new frequency. Furthermore, because the beam is coherent, high detection efficiencies are realized. As a consequence of these two properties, sensitivity is enhanced.

PROBLEMS

12–1. At what wavelengths in nm would the Stokes and anti-Stokes Raman lines for carbon tetrachloride ($\Delta\sigma = 218, 314, 459, 762,$ and 790 cm^{-1}) appear if the source was
(a) a helium/neon laser (632.8 nm)?
(b) an argon ion laser (488.0 nm)?

12–2. Compare the relative intensities of one of the CCl_4 Raman lines when each of the two excitation sources described in Problem 12–1 is used.

12–3. Under what circumstances would a helium/neon laser be preferable to an argon ion laser as a Raman source?

12–4. For vibrational states, the Boltzmann equation can be written as

$$\frac{N_1}{N_0} = e^{-\Delta E/kT}$$

where N_0 and N_1 are the populations of the lower and higher energy states respectively, ΔE is the energy difference between the states, k is Boltzmann's constant, and T is the temperature in °K.

For temperatures of 20 and 40°C, calculate the ratios of the intensities of the anti-Stokes and Stokes lines for CCl_4 at
(a) 218 cm^{-1}. (b) 459 cm^{-1}. (c) 790 cm^{-1}.

12–5. The following Raman data were obtained for $CHCl_3$ with the polarizer of the spectrophotometer set (1) parallel to the plane of polarization of the laser and (2) at 90 deg to the plane of the source.

Calculate the depolarization ratio and indicate which Raman peaks are polarized.

| | $\Delta\sigma$, cm^{-1} | Relative Intensities | |
		(1) $I_{\parallel}$	(2) $I_{\perp}$
(a)	760	0.60	0.46
(b)	660	8.4	0.1
(c)	357	7.9	0.6
(d)	258	4.2	3.2

Miscellaneous Optical Methods

This chapter contains brief discussions of a few optical methods which, while important for some purposes, find less widespread use than those described earlier. Among these methods are *nephelometry* and *turbidimetry, refractometry, polarimetry, optical rotatory dispersion,* and *circular dichroism.*

13A Nephelometry and Turbidimetry

Nephelometry and turbidimetry are closely related analytical methods, which are based upon the scattering of radiation by a solution containing dispersed particulate matter.[1]

When light passes through a transparent medium in which solid particles are dispersed, part of the radiation is scattered in all directions, giving a turbid appearance to the mixture. The diminution of power of a collimated beam as a result of scattering by particles is the basis of *turbidimetric* methods. *Nephelometric* methods, on the other hand, are based upon the measurement of the scattered radiation, usually at a right angle to the incident beam. Nephelometry is generally more sensitive than turbidimetry for the same reasons that fluorometry is more sensitive than photometry (p. 241).

The choice between a nephelometric and a turbidimetric measurement depends upon the fraction of light scattered. When scattering is extensive, owing to the presence of many particles, a turbidimetric measurement is the more satisfactory. If scattering is minimal and the diminution in power of the incident beam is small, nephelometric measurements provide more satisfactory results.

13A-1 THEORY OF NEPHELOMETRY AND TURBIDIMETRY

It is important to appreciate that the scattering associated with nephelometry and turbidimetry is elastic (in contrast to Raman spec-

---

[1] For a more complete discussion, see: F. P. Hochgesang, in *Treatise on Analytical Chemistry,* I. M. Kolthoff and P. J. Elving, Eds., Part I, vol. 5, Chapter 63, Interscience: New York, 1964.

troscopy) and involves no net loss in radiant power; only the direction of propagation is affected. The intensity of radiation appearing at any angle depends upon the number of particles, their size and shape, the relative refractive indexes of the particles and the medium, and the wavelength of the radiation. The relationship among these variables is complex. A theoretical treatment is feasible but seldom applied to specific analytical problems because of its complexity. In fact, most nephelometric and turbidimetric procedures are highly empirical.

Effect of Concentration on Scattering. The attenuation of a parallel beam of radiation by scattering in a dilute suspension is given by the relationship

$$P = P_0 e^{-\tau b} \tag{13–1}$$

where P_0 and P are the power of the beam before and after passing through the length b of the turbid medium. The quantity τ is called the *turbidity coefficient*, or the turbidity; its value is often found to be linearly related to the concentration c of the scattering particles. As a consequence, a relationship analogous to Beer's law applies; that is,

$$\log_{10} \frac{P_0}{P} = kbc \tag{13–2}$$

where $k = 2.303 \, \tau/c$.

Equation 13–2 is employed in turbidimetric analysis in exactly the same way as Beer's law is used in photometric analysis. The relationship between $\log_{10} P_0/P$ and c is established with standard samples, the solvent being used as a reference to determine P_0. The resulting calibration curve is then used to determine the concentration of samples.

For nephelometric measurements, the power of the beam scattered at right angles to the incident beam is normally plotted against concentration; a linear relationship is frequently obtained. The procedure here is entirely analogous to a fluorometric method.

Effect of Particle Size on Scattering. The fraction of radiation scattered at any angle depends upon the size and shape of the particles responsible for the scattering; the effect is large. Since most analytical applications of scattering involve the generation of a colloidally dispersed phase in a solution, those variables that influence particle size during precipitation also affect both turbidimetric and nephelometric measurements. Thus, such factors as concentration of reagents, rate and order of mixing, length of standing, temperature, pH, and ionic strength are important experimental variables. Care must be exercised to reproduce all conditions likely to affect particle size during calibration and analysis.

Effect of Wavelength on Scattering. It has been shown experimentally that the turbidity coefficient varies with wavelength as given by the equation

$$\tau = s\lambda^{-t}$$

where s is a constant for a given system. The quantity t is dependent on particle size and has a value of 4 when the scattering particles are significantly smaller than the wavelength of radiation (Rayleigh scattering); for particles with dimensions similar to the wavelength (the usual situation in a turbidimetric analysis) t is found to be about 2.

For purposes of analysis, ordinary white light is employed. If the solution is colored, it is necessary to select a portion of the spectrum in which absorption by the medium is minimized.

13A-2 INSTRUMENTS

Instruments for nephelometric measurements are similar in design to the simple fluorometers described in Section 8B–2.[2] Turbidimetric measurements are usually performed with simple filter photometers

---

[2] See also: R. D. Vanous, *Amer. Lab.,* **1978,** *10* (7), 67.

such as those described in Section 7D–1. Rectangular cells are ordinarily employed. With the exception of those areas through which the radiation is transmitted, the cell walls are given a dull black coating to eliminate reflection of unwanted radiation to the detector.

13A-3 APPLICATIONS OF SCATTERING METHODS

Turbidimetric or nephelometric methods are widely used in the analysis of water, for the determination of clarity, and for the control of treatment processes. In addition, the concentration of a variety of ions can be determined by using suitable precipitating reagents. Conditions must be chosen so that the solid phase forms as a stable colloidal suspension. Surface-active agents (such as gelatin) are frequently added to prevent coagulation of the colloid. As noted earlier, reliable analytical data are obtained only when care is taken to control scrupulously all of the variables that affect particle size.

Table 13–1 lists some of the species that have been determined by turbidimetric or nephelometric methods. Perhaps most widely employed is the method for sulfate ion. Nephelometric methods permit the determination of concentrations as low as a few parts per million with a precision of 1 to 5%. Turbidimetric methods are reported to give the same degree of reproducibility with more concentrated solutions.

Turbidimetric measurements have also been employed for determining the end point in precipitation titrations. The apparatus can be very simple, consisting of a light source and a photocell located on opposite sides of the titration vessel. The photocurrent is then plotted as a function of volume of reagent. Ideally, the turbidity increases linearly with volume of reagent until the end point is reached, whereupon it remains essentially constant.[3]

---

[3] For an analysis of the factors affecting turbidity titration curves, see: E. J. Meehan and G. Chiu, *Anal. Chem.*, **1964,** *36*, 536.

Table 13-1
Some Turbidimetric and Nephelometric Methods[a]

Species	Method[b]	Suspensions	Reagent	Interferences
Ag	T, N	AgCl	NaCl	—
As	T	As	KH_2PO_2	Se, Te
Au	T	Au	$SnCl_2$	Ag, Hg, Pd, Pt, Ru, Se, Te
Ca	T	CaC_2O_4	$H_2C_2O_4$	Mg, Na, SO_4^{2-} (in high concentration)
Cl^-	T, N	AgCl	$AgNO_3$	Br^-, I^-
K	T	$K_2NaCo(NO_2)_6$	$Na_3Co(NO_2)_6$	SO_4^{2-}
Na	T, N	$NaZn(UO_2)_3(OAc)_9$	$Zn(OAc)_2$ and $UO_2(OAc)_2$	Li
SO_4^{2-}	T, N	$BaSO_4$	$BaCl_2$	Pb
Se	T	Se	$SnCl_2$	Te
Te	T	Te	NaH_2PO_2	Se, As

[a] Data taken from L. Meites, *Handbook of Analytical Chemistry*, p. **6**–175, McGraw-Hill Book Company, Inc.: New York, 1963. With permission.
[b] T = turbidimetric; N = nephelometric.

13B   Refractometry

When radiation passes through a transparent medium, interaction occurs between the electric field of the radiation and the bound electrons in the medium; as a consequence, the rate of propagation of the beam is less than in a vacuum (see Section 4A–5). The refractive index of a substance n_i at a wavelength i is given by the relationship

$$n_i = \frac{c}{v_i} \tag{13–3}$$

where v_i is the velocity of propagation of radiation in the medium and c is the velocity in vacuum (a constant under all conditions). The refractive index of most liquids lies between 1.3 and 1.8; it is 1.3 to 2.5 or higher for solids.[4]

13B–1   THE MEASUREMENT OF REFRACTIVE INDEX

The refractive index of a substance is ordinarily determined by measuring the change in direction (refraction) of collimated radiation as it passes from one medium to another. As was shown on page 101,

$$\frac{n_2}{n_1} = \frac{v_1}{v_2} = \frac{\sin \theta_1}{\sin \theta_2} \tag{13–4}$$

where v_1 is the velocity of propagation in the less dense medium M_1 and v_2 is the velocity in medium M_2; n_1 and n_2 are the corresponding refractive indexes and θ_1 and θ_2 are the angles of incidence and refraction, respectively (see Figure 4–9, p. 101). When M_1 is a vacuum, n_1 is unity because v_1 becomes equal to c in Equation 13–3. Thus,

$$n_2 = n_{vac} = \frac{c}{v_2} = \frac{\sin \theta_1}{\sin \theta_2} \tag{13–5}$$

---

[4] For a more complete discussion of refractometry, see: S. Z. Lewin and N. Bauer, in *Treatise on Analytical Chemistry*, I. M. Kolthoff and P. J. Elving, Eds., Part I, vol. 6, Chapter 70, Interscience: New York, 1965.

where n_{vac} is the *absolute refractive index* of M_2. Thus, n_{vac} can be obtained by measuring the two angles θ_1 and θ_2.

It is much more convenient to measure the refractive index with respect to some medium other than vacuum, and air commonly serves as a standard for this purpose. Most compilations of n for liquids and solids in the literature are with reference to air at laboratory temperatures and pressures. Fortunately, the change in the refractive index of air with respect to temperature and pressure is small enough so that a correction from ambient laboratory conditions to standard conditions is needed for only the most precise work. A refractive index n_D measured with respect to air with radiation from the D line of sodium can be converted to n_{vac} with the equation

$$n_{vac} = 1.00027 n_D$$

This conversion is seldom required.

It is usually necessary to measure a refractive index to an accuracy of at least 2×10^{-4}. Accuracies on the order of 6 to 7×10^{-5} may be required for the routine analysis of solutions. For the detection of impurities, a difference in refractive index between the sample and a pure standard is measured; here, the capability of detecting a difference on the order of 1×10^{-6} or better is needed.

13B–2   SPECIFIC AND MOLAR REFRACTION

The existence of a relationship between refractive index and density d has been recognized since the time of Newton, who observed that the ratio $(n^2 - 1)/d$ was approximately constant for a number of substances. It has further been recognized that the refractive index, although dependent upon density, must also be affected by the arrangement of electrons in a medium. Thus, if refractive indexes could be corrected for density differences, the resulting parameter, the *specific refraction*, should be a measure of electronic environment. Several empirical and theoretical formulas for specific refraction have been developed over the last century. The Lorentz and Lorenz relationship has a theoretical basis for certain classes of liquids and has been widely employed in structural studies. It defines the specific refraction r as

$$r = \frac{(n^2 - 1)}{(n^2 + 2)} \cdot \frac{1}{d} \qquad (13-6)$$

The *molar refraction R* is equal to rM, where M is the molecular weight of the substance.

Specific and molar refractions have proved useful for analytical purposes since they are found to vary in a systematic way within homologous series of compounds.

13B-3 VARIABLES THAT AFFECT REFRACTIVE INDEX MEASUREMENTS

Temperature, wavelength, and pressure are the most common experimentally controllable variables that affect a refractive index measurement.

Temperature. Temperature influences the refractive index of a medium primarily because of the accompanying change in density. For many liquids, the temperature coefficient lies in the range of -4 to -6×10^{-4} deg^{-1}. Water is an important exception, with a coefficient of about -1×10^{-4}; aqueous solutions behave similarly. Solids have temperature coefficients that are roughly an order of magnitude smaller than that of the typical liquid.

It is apparent from the foregoing that the temperature must be controlled closely for accurate refractive index measurements. For the average liquid, temperature fluctuations should be less than $\pm 0.2°C$ if fourth-place accuracy is required, and $\pm 0.02°C$ for measurements to the fifth place.

Wavelength of Radiation. As noted in Section 4A-5, the refractive index of a transparent medium gradually decreases with increasing wavelength; this effect is referred to as *normal dispersion*. In the vicinity of absorption bands, rapid changes in refractive index occur; here, the dispersion is referred to as *anomalous* (see Figure 4-8).

Dispersion phenomena make it essential that the wavelength employed be specified in quoting a refractive index. The D line from a sodium vapor lamp ($\lambda = 589$ nm) is most commonly used as a source in refractometry, and the corresponding refractive index is designated as n_D (often the temperature in °C is also indicated by a superscript; for example n_D^{20}). Other lines commonly employed for refractive index measurements include the C and F lines from a hydrogen source ($\lambda = 656$ nm and 486 nm, respectively) and the G line of mercury ($\lambda = 436$ nm).

Pressure. The refractive index of a substance increases with pressure because of the accompanying rise in density. The effect is most pronounced in gases, where the change in n amounts to about 3×10^{-4} per atmosphere; the figure is less by a factor of 10 for liquids; and it is yet smaller for solids. Thus, only for precise work with gases and for the most exacting work with liquids and solids is the variation in atmospheric pressure important.

13B-4 INSTRUMENTS FOR MEASURING REFRACTIVE INDEX

Two types of instruments for measuring refractive index are available from commercial sources. *Refractometers* are based upon measurement of the so-called *critical angle* or upon the determination of the displacement of an image. *Interferometers* utilize the interference phenomenon to obtain differential refractive indexes with very high precision. We shall consider only refractometers.

Critical Angle Refractometers. The most widely used instruments for the measurement of refractive index are of the critical angle type. The *critical angle* is defined as the angle of refraction in a medium when the angle of incident radiation approaches 90 deg (the *grazing angle*); that is, when θ_1 in Equation 13–4 approaches 90 deg, θ_2 becomes the critical angle θ_c. Thus,

$$\frac{n_2}{n_1} = \frac{\sin 90}{\sin \theta_c} = \frac{1}{\sin \theta_c} \qquad (13-7)$$

Figure 13–1**a** illustrates the critical angle that is formed when the critical ray approaches the surface of the medium M_2 at 90 deg to the normal and is then refracted at some point O on the surface. Note that if the medium could be viewed end-on, as in Figure 13–1**b**, the critical ray would appear as the boundary between a dark and a light field. It should be noted, however, that the illustration is unrealistic in that the rays are shown as entering the medium at but one point O; in fact, they would be expected to enter at all points along the surface and thus create an entire family of critical rays with the same angle θ_c. A condensing or focusing lens is needed to produce a single dark-light boundary such as shown in Figure 13–1**b**.

It is important to realize that the critical angle depends upon wavelength. Thus, if polychromatic radiation is employed, no single sharp boundary such as that in Figure 13–1**b** is observed. Instead, a diffuse chro-matic region between the light and dark areas develops, which makes precise establishment of the critical angle impossible. This difficulty is often overcome in refractometers by the use of monochromatic radiation. As a convenient alternative, many critical angle refractometers are equipped with a compensator that permits the use of radiation from a tungsten source, but compensates for the resulting dispersion in such a way as to give a refractive index in terms of the sodium D line. The compensator consists of one or two *Amici prisms,* as shown in Figure 13–2. The properties of this complex prism are such that the dispersed radiation is converged to give a beam of white light that travels in the path of the yellow sodium D line.

Abbé Refractometer. The Abbé instrument is undoubtedly the most convenient and widely used refractometer; Figure 13–3 shows a schematic diagram of its optical system. The sample is contained as a thin layer (~0.1 mm) between two prisms. The upper prism is firmly mounted on a bearing that permits its rotation by means of the side arm shown in dotted lines. The lower prism is hinged to the upper to allow separation for introduction of the sample. The lower prism face is rough-ground; when light is reflected into the prism, this surface effectively becomes the source for an infinite number of rays that pass through the sample at all angles. The radiation is refracted at the interface of the sample and the

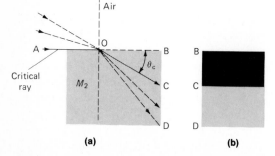

(a)   **(b)**

FIGURE 13-1   (**a**) Illustration of the critical angle θ_c and the critical ray AOC; (**b**) end-on view showing sharp boundary between the dark and light fields formed at the critical angle.

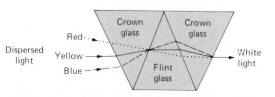

FIGURE 13-2   Amici prism for compensation of dispersion by sample. Note that the yellow radiation (sodium D line) suffers no net deviation from passage through the prism.

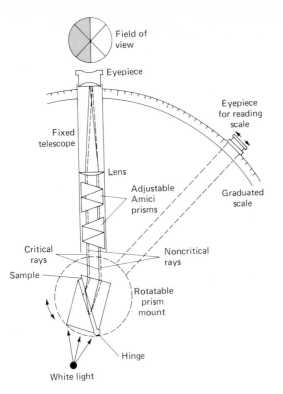

FIGURE 13-3 The Abbé refractometer.

smooth-ground face of the upper prism, whereupon it passes into the fixed telescope. Two Amici prisms, which can be rotated with respect to one another, serve to collect the divergent critical angle rays of different colors into a single white beam, which corresponds in path to that of the sodium D ray. The eyepiece of the telescope is equipped with cross hairs; in making a measurement, the prism angle is changed until the light-dark interface just coincides with the cross hairs. The position of the prism is then established from the fixed scale (which is normally graduated in units of n_D). Thermostating is accomplished by circulation of water through jackets surrounding the prisms.

The Abbé refractometer owes its popularity to its convenience, its wide range (n_D = 1.3 to 1.7), and to the minimal sample re-

quired. The accuracy of the instrument is about ±0.0002; its precision is half this figure. The most serious error in the Abbé instrument is caused by the fact that the nearly grazing rays are cut off by the arrangement of the two prisms; the boundary is thus less sharp than is desirable.

A *precision* Abbé refractometer, which diminishes the uncertainties of the ordinary instrument by a factor of about three, is also available; the improvement in accuracy is obtained by replacing the compensator with a monochromatic source and by using larger and more precise prism mounts. The former provides a much sharper critical boundary, and the latter permits a more accurate determination of the prism position.

Image Displacement Refractometer. The most straightforward method for determining refractive index involves measurement of the angles of incidence and refraction by means of a spectrometer arrangement somewhat like that shown in Figure 5–11**b** (p. 124). Liquid samples are contained in a prism-shaped container mounted at the center of a large circular metal table; solid samples are cut into the shape of a prism and are similarly mounted. A light source, a slit, and a collimator are employed to direct a parallel beam of radiation onto one surface of the prism. The refracted image of the slit is then viewed with a telescope mounted on the circle. Since the slit image can be made very sharp, the accuracy of the determination depends only upon the accuracy of the angular measurements and the control of temperature; uncertainties of 1×10^{-6} or smaller can be attained. Refractometers based on this principle are often used as detectors in liquid chromatography (see Figure 27–9, Section 27C–6).

13B-5 APPLICATIONS OF REFRACTOMETRY

In common with density, melting point, and boiling point, the refractive index is one of the classical physical constants that can be used to describe a chemical species. While it is a nonspecific property, few substances have identical refractive indexes at a given temperature and wavelength. Thus, this constant is useful for confirming the identity of a compound and measuring its purity.

A refractive index measurement is often the simplest, most convenient, and most rapid procedure for evaluating the composition of a binary liquid or a gaseous mixture. A linear relationship between refractive index and some concentration parameter frequently exists over at least a limited concentration range. A straight-line calibration curve is often obtained if the concentrations of aqueous solutions are expressed in grams of solute per 100 mL of solution. For mixtures of organic liquids, on the other hand, linearity is more often observed when volume percent is employed. Linearity, of course, is not a requirement for quantitative work since a suitable calibration curve can always be prepared.

Many quantitative applications of refractometry can be cited. For example, the method is widely used for the determination of the concentration of aqueous sugar solutions. Most of the common sugars have about the same effect on the refractive index of an aqueous solution; the procedure thus gives a measure of total carbohydrate concentration. Refractometric procedures have also proved useful in determining the concentration of sulfur in unvulcanized rubber and the bound styrene content of certain synthetic rubbers.

An important application of refractometry has been to the evaluation of apparatus and methods for separations such as distillation, extraction, adsorption chromatography, and diffusion. For example, the number of theoretical plates for a distillation column can be evaluated from refractive index measurements of fractions collected during the separation of a binary mixture such as heptane (b.p. = 98.4°C) and methyl-cyclohexane (b.p. = 100.9°C). The number of theoretical plates for the column can be calculated from the resulting analytical data.

Specific refraction and molar refraction data are also employed in quantitative analysis. For example, the specific refraction of certain silicate glasses is found to vary nearly linearly with the mole percent of SiO_2 over a range of 50 to 90%. Thus, accurate determination of the silicon content for such glasses is feasible from refractive index and density measurements. Similarly, specific refraction has been employed for the estimation of unsaturation in vegetable oils and for the degree of fluorination in paraffin oils. Molar refraction has been widely used in the petroleum industry for determining the percent carbon incorporated in aromatic structures in hydrocarbon mixtures.

For all applications, periodic calibration of refractometers is necessary. Standards for this purpose include purified liquids such as water (n_D^{20} = 1.33299), toluene (n_D^{20} = 1.4969), and methylcyclohexane (n_D^{20} = 1.4231). The latter two compounds can be obtained from the National Bureau of Standards as certified samples with indexes to five decimal places at 20, 25, and 30°C and for each of seven wavelengths. A glass test piece, supplied with most refractometers, can also be employed as a reference. The difference between the refractive index of the standard and the instrument scale reading is applied as an arithmetic correction to subsequent determinations. With the Abbé refractometer, the objective of the telescope can be adjusted mechanically so that the instrument indicates the proper refractive index for the standard.

13C   Polarimetry

Optical activity is a measure of the ability of certain substances to rotate plane-polarized light. This phenomenon, first reported for quartz in 1811, has been studied intensely since that time. By the mid-nineteenth century, many of the laws relating to optical activity had been formulated and later in the century played an important part in the development of the ideas of organic stereochemistry and structure. Some of the early concepts of optical activity have stood the test of time and remain essentially unaltered today. It is of interest, however, that despite this long history, the interactions of radiation with matter that cause the rotation of polarized light are less clearly understood than the processes responsible for absorption, emission, or refraction.

The term *polarimetry,* as it is used by most chemists, can be defined as the study of the rotation of polarized light by transparent substances. The direction and the extent of rotation (the *optical rotatory power*) is useful for both qualitative and quantitative analysis and for the elucidation of chemical structure as well.[5]

13C-1 TRANSMISSION AND REFRACTION OF RADIATION IN OPTICALLY ANISOTROPIC MEDIA

Optically *isotropic* substances transmit radiation at equal velocities in all directions regardless of the polarization of the radiation. Examples of isotropic material include homogeneous gases and liquids, solids that crystallize in the cubic form, and noncrystalline solids such as glasses and many polymers. Noncubic crystals, on the other hand, are *anisotropic* and may transmit polarized radiation at different velocities depending upon the angular relationship between the plane of polarization and a given axis of the crystal.

---

[5] For a more complete discussion of the various aspects of polarimetry, see: *Physical Methods of Chemistry,* A. Weissberger and B. W. Rossiter, Eds., vol. 1, Part IIIC, Chapters 1 and 2, Wiley: New York, 1972; and W. A. Struck and E. C. Olson, in *Treatise on Analytical Chemistry,* I. M. Kolthoff and P. J. Elving, Eds., Part I, vol. 6, Chapter 71, Interscience: New York, 1965.

Transmission of Polarized Radiation Through Anisotropic Crystal. Recall (Section 4A–5) that radiation is slowed as it passes through a medium containing atoms, ions, or molecules because its electrical vector interacts momentarily with the electrons of these particles, causing their temporary polarization. Reemission of the radiation occurs after 10^{-14} to 10^{-15} sec as the polarized particles return to their original state. A beam of radiation traversing an isotropic medium encounters a symmetrical distribution of electrons around its path of travel. Thus, the slowing of a beam of polarized radiation is the same regardless of the angle of the plane of polarization around the direction of travel. In contrast, the distribution of atomic or molecular particles around most paths through an anisotropic crystal is not symmetrical. Thus, radiation vibrating in one plane along this path encounters a different electron environment from that in another; a difference in rate of transmission results.

All anisotropic crystals have at least one axis, called the *optic axis,* around which there exists a symmetrical distribution of the particles making up the crystal. Polarized radiation travels along the optic axis at a constant rate regardless of its angle of polarization with respect to that axis.

Figure 13–4a depicts the electrical vectors at maximum amplitude for two beams of radiation that are in phase and polarized at 90 deg to one another. The arrows are the vectors for the beam which is vibrating in the plane of the paper. The dots represent the vector that is fluctuating in a plane perpendicular to the page. Figure 13–4a shows that the wavelengths of both beams are decreased equally and their velocities remain the same as they travel along the optic axis of an anisotropic crystal. Thus, the two beams remain in phase both within the crystal and as they emerge from it.

Figure 13–4b contrasts the behavior of the two beams when they strike the anisotropic crystal at an angle that is 90 deg to the

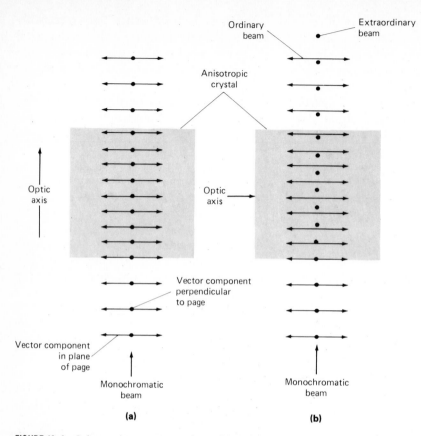

FIGURE 13-4 Schematic representation of passage of a monochromatic beam through the two axes of an anisotropic crystal of calcite. Arrows represent the electric vector component in the plane of the page. Dots represent the vector component in a plane perpendicular to the page.

optic axis. Here, the beam that is vibrating in the plane of the paper encounters an atomic or molecular environment similar to that depicted in Figure 13–4**a**; its wavelength and velocity behavior are thus similar to the two beams moving along the optic axis. The beam vibrating in the plane perpendicular to the page, however, encounters a less dense environment; thus, its velocity and wavelength are not attenuated as much, and the two beams become out of phase not only within the crystal but also after emerging from it.

The polarized beam that travels at the same velocity along the optic axis and perpendicular to it is called the *ordinary beam*; the faster moving beam is the *extraordinary beam*. (In some instances, the relative velocities of the two types of beams are reversed.) The velocity of the extraordinary beam varies continuously as the angle of its travel with respect to the optic axis is varied from 0 to 90 deg, and reaches a maximum (or minimum) at the latter angle.

Transmission of Unpolarized Radiation Through Anisotropic Crystals. When a beam of *unpolarized* monochromatic radiation passes through an anisotropic crystal at an angle to the optic axis, separation into an ordinary and extraordinary beam occurs. To understand this

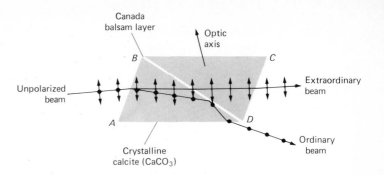

FIGURE 13-5 A Nicol prism for resolving an unpolarized beam into two beams plane-polarized at 90 deg to one another.

behavior, recall that the various electrical vectors of unpolarized monochromatic radiation can be resolved into two mutually perpendicular vectors, as shown in Figure 4–11 (p. 103); thus, a beam of ordinary radiation can be considered to be made up of two plane-polarized beams of equal amplitude whose planes are oriented at 90 deg to one another. Therefore, Figure 13–4 applies to an unpolarized beam as well as to two plane-polarized beams; here, it is only necessary to specify that the amplitudes of the latter beams are identical.

Double Refraction by Anisotropic Crystal. From the foregoing discussion, it is evident that the velocity, and thus the index of refraction of the extraordinary ray, in an anisotropic crystal is dependent upon direction, being identical with that for the ordinary ray along the optic axis and changing continuously to a maximum or a minimum along a perpendicular axis. Refractive indexes for the extraordinary ray are normally reported in terms of this perpendicular axis. Table 13–2 compares the refractive index for the

ordinary ray n_o with that for the extraordinary ray n_e in some common anisotropic crystals. In calcite, the extraordinary ray is clearly propagated at a greater rate than the ordinary ray; in quartz, the reverse is the case. Because anisotropic crystals have two characteristic refractive indexes, they can be made to refract the ordinary and the extraordinary rays at different angles; anisotropic crystals are thus *double-refracting*. This property provides a convenient means for separating an unpolarized beam into two beams that are plane-polarized at 90 deg to one another.

The Nicol Prism. Figure 13–5 depicts a Nicol prism, a device that exploits the double-refracting properties of crystalline calcite ($CaCO_3$) to produce plane-polarized radiation. The end faces of a natural crystal are trimmed slightly to give an angle of 68 deg, as shown; the crystal is then cut across its short diagonal. A layer of Canada balsam, a transparent substance with a refractive index intermediate between the two refractive indexes of calcite, is placed between the two crystal halves. This layer is totally reflecting for the ordinary ray with its greater refractive index but, as shown in the figure, transmits the extraordinary ray almost unchanged.

Pairs of Nicol prisms are employed in measurements involving the rotation of the plane of polarized light. One prism serves to produce a polarized extraordinary beam that is passed into the medium under study.

Table 13-2 Refractive Index Data for Selected Anisotropic Crystals		
Crystal	n_o	n_e
Calcite	1.6583	1.4864
Quartz	1.544	1.553
Ice	1.306	1.307

A second analyzer prism then determines the extent of rotation caused by the medium. If the two Nicol prisms are identically oriented with respect to the beam, and if the medium has no effect, the extraordinary ray, comprising nearly 50% of the original intensity, is emitted from the analyzer (see Figure 13–6**a**). If the polarizer is rotated (Figure 13–6**b**), only the vertical component *m'n'* of the beam *mn* emitted from the polarizer is transmitted through the analyzer, the horizontal component being now reflected from the Canada balsam layer. Here, less than 50% of the incident beam appears at the face of the analyzer. Rotation of the polarizer by 90 deg results in a beam from the analyzer that has no vertical component. Thus, no radiation is observed at the face of the analyzer. If a medium affecting the rotation of light is interposed between the two Nicol prisms, the relative orientations needed to achieve a maximum or a minimum in transmission are changed by an amount corresponding to the rotatory power of the medium.

13C-2 INTERFERENCE EFFECTS WITH POLARIZED RADIATION

To account for many of the experimental observations regarding the interactions of polarized radiation, it is necessary to assume that interference between polarized beams can occur, provided the beams are *coherent* (p. 99). The effect of interference can then be visualized by vector addition of the electromagnetic components of the individual beams.

Figure 13–7**a** illustrates the interference between two plane-polarized beams of equal amplitude that are in phase but oriented 90 deg to one another. Addition of the electrical vectors of the two beams is shown schematically in the end-on view to the right. Note that the numbered points along the axis MN have been projected onto the plane

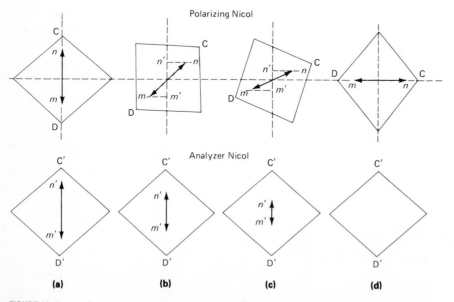

FIGURE 13-6 End view of a polarizer and analyzer Nicol. *mn* = electrical vector of beam transmitted by polarizer. *m'n'* = vertical component of beam transmitted by polarizer and analyzer.

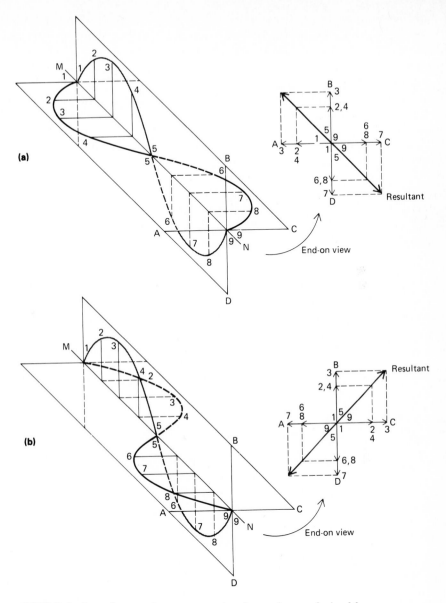

FIGURE 13-7 Interference between two in-phase plane-polarized beams.

ABCD which is perpendicular to MN, thus giving a two-dimensional representation of the resultant and its components. It will be seen that the resultant is a vector that oscillates in a plane oriented at 45 deg to the planes of the two component beams; *interference thus produces a single, plane-polarized beam.*

Figure 13–7b demonstrates interference when the phase relationship between the two plane-polarized beams differs by one-half wavelength (180 deg). Here, the waves are again in phase, and the resultant is a plane-polarized beam which, however, is perpendicular with respect to the resultant of Figure 13–7**a**.

Circularly and Elliptically Polarized Radiation. It is of interest now to examine the behavior of a plane-polarized beam of monochromatic radiation as it passes through an anisotropic crystal. In Figure 13–8**a**, the path of the incident beam is normal to the optic axis of the crystal, with the plane of the polarized radiation oriented at 45 deg to that axis (the angle of the plane of polarization is given by the arrow MP). As shown by Figure 13–7, however, the plane-polarized beam can be

considered to consist of two *coherent* components lying in perpendicular planes oriented along MB and MA. These components are also indicated in Figure 13–8. Upon entering the crystal, the component lying in the MA direction travels at the rate of an ordinary ray since this direction lies along the optic axis of the crystal; the orientation of component MB, on the other hand, corresponds to that for an extraordinary ray, and its rate of propagation is thus different. As a result of the velocity difference, *the two components are no longer coherent* and thus cannot interfere. That is, *within the crystal, the beam can be considered to consist of*

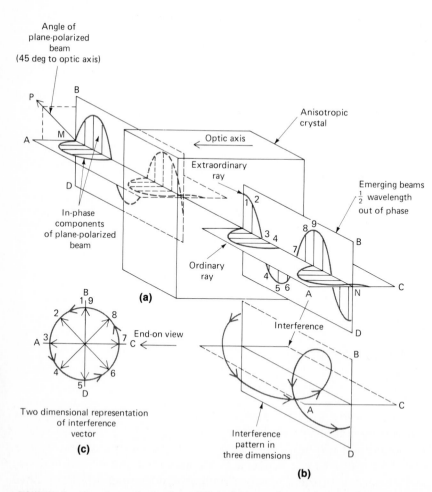

FIGURE 13-8   Circular polarization of light by an anisotropic crystal.

two components *having different velocities*; because of their incoherence, interference cannot occur.

When the two rays leave the crystal, their velocities again become equal in the isotropic air medium; thus, they can again interfere since they are once more coherent. The nature of the resultant will, however, depend upon the phase relationship between the two that exists at the instant they emerge from the crystal surface. This phase relationship is determined by the relative velocities of the two rays in the medium as well as by the length of traverse. If, for example, the path in the crystal is such that the two rays are completely in phase upon exiting, then constructive interference similar to that shown in Figure 13–7a occurs. That is, the resultant beam will be polarized at the same angle as the entering beam. If, on the other hand, the crystal thickness is such that the phase relationship between the two rays at the face is shifted exactly one-half wavelength, interference such as shown in Figure 13–7b results. Here, the plane of the exit beam is oriented 90 deg with respect to the entering beam.

Figure 13–8a shows the emerging waves as being one-quarter wavelength out of phase and indicates their relationship to one another, assuming that no interference takes place. In fact, however, interaction does occur as the two rays enter the air medium, and the path of the electrical vector for the resulting wave can be obtained by adding the two vectors. The resulting vector quantity is seen to travel in a helical pathway around the direction of travel (Figure 13–8b). If the vector sum is plotted in two-dimensional form (Figure 13–8c), a circle is obtained. This condition is in distinct contrast to the original, linearly polarized radiation in which the electrical vector lies in a single plane. A helical beam of this type is called *circularly polarized light*. Note that if the two waves had been out of phase one-quarter wavelength in the other sense, the

direction of travel of the vector would have been clockwise rather than counterclockwise.

Thus far, we have considered the nature of the exit beam from the anisotropic crystal when the phase difference created was 0, $\frac{1}{4}$, $\frac{1}{2}$, or some multiple of these fractions. If the light path in the crystal is such as to produce phase differences other than these, the path traced by the resultant electrical vector is an ellipse, and the radiation is called *elliptically* polarized light. Figure 13–9 summarizes the states that result when the components of a plane-polarized beam are emitted from an anisotropic crystal with various phase differences.

Anisotropic crystals of suitable length are employed experimentally to produce circularly polarized radiation. Such crystals are called *quarter-wave plates* and find use in circular dichroism studies.

Relationship Between Plane Polarized and Circularly Polarized Radiation. In the preceding section, we have seen that the behavior of plane-polarized radiation upon passage through an anisotropic crystal can be rationalized by consid-

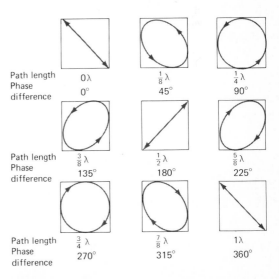

| Path length | 0λ | $\frac{1}{8}\lambda$ | $\frac{1}{4}\lambda$ |
| Phase difference | $0°$ | $45°$ | $90°$ |

| Path length | $\frac{3}{8}\lambda$ | $\frac{1}{2}\lambda$ | $\frac{5}{8}\lambda$ |
| Phase difference | $135°$ | $180°$ | $225°$ |

| Path length | $\frac{3}{4}\lambda$ | $\frac{7}{8}\lambda$ | 1λ |
| Phase difference | $270°$ | $315°$ | $360°$ |

FIGURE 13-9 Effect of an anisotropic crystal on plane-polarized radiation. Each diagram corresponds to a different path length in the crystal.

ering the beam to be the resultant of two plane-polarized rays that are in phase and oriented at 90 deg to one another. It is of equal importance to understand that plane-polarized radiation can also be treated as the interference product of *two coherent circular rays of equal amplitude that rotate in opposite directions.* Figure 13–10 shows how the vectors for the *d* and *l* circular components are added to produce the equivalent vectors of a plane-polarized beam. From the middle figure, it can be seen that each of the two rotating vectors describes a helical path around the axis of travel of the beam.

The rationalization of many phenomena to be considered in this chapter is based upon the idea that plane-polarized radiation consists of a *d* and an *l* circular component. Here, *d* (*dextrorotatory*) refers to the clockwise rotation as the beam approaches the observer; *l* (*levorotatory*) is the counterclockwise component.

13C-3 CIRCULAR DOUBLE REFRACTION

The rotation of plane-polarized light by an optically active species can be explained if one assumes that the rates of propagation of the *d* and the *l* circular components of a plane-polarized beam are different in the presence of such a species; that is, the refractive index of the substance with respect to *d* radiation (n_d) is different from that for *l* (n_l). Thus, optically active substances are anisotropic with respect to circularly polarized light and show *circular double refraction.* Note that the two circular components are no longer coherent in the anisotropic medium and cannot interfere until they again reach an isotropic medium.

The rotation of a beam of light by an optically active medium is shown schematically in Figure 13–11. Initially in Figure 13–11**a**, the beam is polarized in a vertical plane with the circular *d* and *l* components rotating at

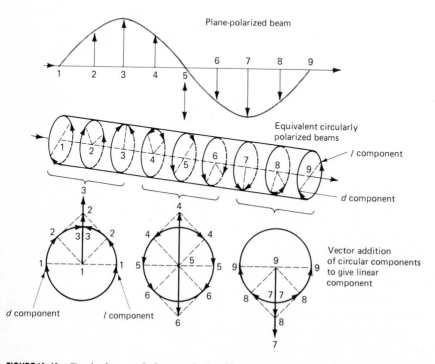

FIGURE 13-10   Equivalency of plane-polarized beam to two (*d*,*l*) circularly polarized beams.

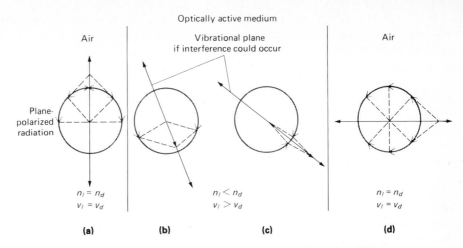

Optically active medium

FIGURE 13-11 Rotation of plane-polarized light in a medium in which $n_l < n_d$.

equal velocities. In this example, the beam of polarized radiation enters the anisotropic medium; here, the rate of propagation of the d component is slowed more than that of the l because $n_d > n_l$. Thus, at some location (Figure 13–11**b**) in the medium, the d vector will lag behind the l; if at this point the two rays could interfere, the resultant would still be a plane, but one that was rotated from the vertical. At a further point (Figure 13–11**c**), the d component would be still further retarded and a greater rotation would result. Figure 13–11 has been drawn to show that, upon emerging from the medium, retardation has been such that the resultant plane is now horizontal. As shown in Figure 13–11**c**, the d and l components are again propagated at identical rates in the isotropic air medium, and interference of the coherent rays occurs. Passage through the anisotropic medium has had the effect, however, of shifting the phase relationship such that the observed plane of polarization is now 90 deg with respect to the original.

Quantitative Relationships. It is readily shown that the rotation (α_λ) in degrees caused by an optically active substance is given by the relationship

$$\alpha_\lambda = \frac{180\, \mathbf{I'}}{\lambda} (n_l - n_d) \qquad (13\text{–}8)$$

where $\mathbf{I'}$ is the path length of the medium in centimeters and λ is the wavelength of the plane-polarized radiation (in vacuum), also in centimeters. The quantity $(n_l - n_d)$ is termed the *circular birefringence*. The following calculation demonstrates the magnitude of the circular birefringence required to bring about a typical rotation.

EXAMPLE 13-1. A solution contained in a 10-cm cell is found to rotate the plane-polarized radiation of the sodium D line by 100 deg. What is the difference in refractive index of the medium for the l and d circularly polarized components?

$\lambda = 589$ nm or 5.89×10^{-5} cm

Substituting into Equation 13–8 yields

$$100 \text{ deg} = \frac{180 \text{ deg} \times 10 \text{ cm}}{5.89 \times 10^{-5} \text{ cm}} (n_l - n_d)$$

$$(n_l - n_d) = 3.3 \times 10^{-6}$$

It is apparent from this calculation that a relatively small difference in refractive index has a large effect in terms of optical rotatory power. Note that if the refractive index for the sodium D line is about 1.5 (a typical value), then a 2.2 ppm difference between n_d and n_l is responsible for the 100-deg rotation shown in the example.

13C-4 OPTICALLY ACTIVE COMPOUNDS

Optical activity is associated with two types of species: (1) crystalline compounds, which lose their activity when the crystal is converted to a liquid, a gas, or a solution (quartz is the classic example of this manifestation of optical activity); (2) compounds in which the optical activity is inherent in the molecule itself and is observed regardless of the physical state of the compound. It is with this second type that we are concerned.

The structural requirements for an optically active molecule are well understood and treated in detail in organic chemistry textbooks. It is sufficient to note that the optically active forms of a molecule, called *enantiomers,* are mirror images which cannot be superimposed, regardless of the orientation of one with respect to the other. Enantiomers thus bear the same relationship to one another as the left to the right hand. The two isomers then rotate polarized light equally but in opposite directions; if one is in excess in a mixture or is isolated from the other, rotation is observed. The most common form of optical activity results from the presence of four different substituents on a tetrahedral carbon atom in an organic compound; this feature leads to two nonsuperimposable *chiral* molecules. Other types of asymmetric centers can also occur, both in organic and inorganic structures.[6]

---

[6] For a discussion of optical activity in inorganic systems, see: J. C. Bailar and D. H. Busch, *The Chemistry of the Coordination Compounds,* Reinhold: New York, 1956.

It is important to appreciate that the interactions which lead to rotation of polarized light are not peculiar to molecules with asymmetric centers but instead are characteristic of most molecules. However, rotation is not *observed* in noncrystalline samples that lack asymmetric character because the molecules are randomly oriented; rotation caused by a molecule in one orientation is canceled by the equal and opposite rotation of another oriented as its mirror image. For the same reason, samples containing *d* and *l* isomers in equal concentrations exhibit no net rotation because their individual effects cancel. It is only when one form of a chiral molecule is present in excess that a net rotation can be observed.

13C-5 VARIABLES THAT AFFECT OPTICAL ROTATION

The rotation of plane-polarized radiation by optically active compounds can range from several hundred to a few hundredths of a degree. Experimental variables that influence the observed rotation include the wavelength of radiation, the optical path length, the temperature, the density of the substance if undiluted, and its concentration if in solution. For solutions, the rotation may also vary with the kind of solvent.

The *specific rotation* or the *specific rotatory power* $[\alpha]_\lambda^T$ is widely employed to describe the rotatory characteristics of a liquid. It is defined as

$$[\alpha]_\lambda^T = \frac{\alpha}{\mathbf{I}c} \qquad (13\text{--}9)$$

where α is the observed rotation in degrees, $\mathbf{I}$ is the path length in decimeters, and c is the grams of solute in 100 cc of solution. The wavelength λ and temperature T are usually specified with a subscript and a superscript, as shown. Most specific rotations are measured at 20°C with sodium D line and are thus reported as $[\alpha]_D^{20}$. For a pure liquid, c is replaced by its density. By convention, counterclockwise, or *l* rotation as the observer faces the beam, is given the negative sign. Clockwise (*d*) rotation is positive.

The term *molecular rotation* [*M*] is also encountered; it is defined as

$$[M] = M[\alpha]/100 \qquad (13\text{--}10)$$

where *M* is the molecular weight.

It is frequently necessary to measure the optical rotatory power of a substance as a solute. Unfortunately, the specific rotation of a compound is nearly always found to vary with the nature of the solvent. Because of solubility considerations, no single standard solvent can be designated. In addition, the specific rotation in a given solvent may not be entirely independent of concentration, although variations in dilute solutions are usually small. Because of these effects, it is common practice to designate both the kind of solvent and the solute concentration in reporting a specific rotation.

The variation in specific rotation with temperature is approximately linear, but the temperature coefficient differs widely from substance to substance. For example, the specific rotation of tartaric acid solutions may vary as much as 10% per degree; on the other hand, the variation for sucrose is less than 0.1% per degree.[7]

The effect of wavelength on rotation is considered in a later section on optical rotatory dispersion.

13C-6 MECHANISM OF OPTICAL ROTATION

While the structural requirements for optical activity can be precisely defined, the mechanism by which a beam of circularly polarized radiation interacts with matter and is thus retarded is much less obvious. That is, we can predict with certainty that a compound such as 2-iodobutane is chiral because it possesses an asymmetric center. On the other hand, it is difficult to account for the specific rotations ($[\alpha]_D^{20} = \pm 32$ deg) for these isomers, compared with those of 2-butanol ($[\alpha]_D^{20} = \pm 13.5$ deg).

Several theories concerning the mechanism of optical rotation have been developed. Generally, these are couched in terms of quantum mechanics and are of sufficient complexity to evade all but the most mathematically oriented chemist. Furthermore, while in principle it may be possible to predict optical rotatory power for certain types of compounds with the aid of these theories, such calculations are not practical in terms of effort; nor has it been demonstrated that the values obtained from such calculations can be sufficiently precise to be useful.

13C-7 POLARIMETERS

The basic components of a polarimeter include a monochromatic light source, a polarizing prism, a sample tube, an analyzer prism with a circular scale, and a detector (see Figure 13–12). The eye serves as the detector for most polarimeters although photoelectric polarimeters are becoming more common.

Sources. Because optical rotation varies with wavelength, monochromatic radiation is employed. Historically, the sodium D line was obtained by introducing a sodium salt into a gas flame. Suitable filters then removed other lines and background radiation. Sodium vapor lamps with a filter to remove all but the D line are now employed. Mercury vapor lamps are also useful, the line at 546 nm being isolated by a suitable filter system.

Polarizer and Analyzer. Nicol prisms are most commonly employed to produce plane-polarized light and to determine the angle

---

[7] It is important to remember that most organic solvents have temperature coefficients of expansion of about 0.1% per deg C. Therefore, accurate measurements of $[\alpha]_\lambda^\theta$ for solutions require close control of temperature both during the measurement and *during the solution preparation.*

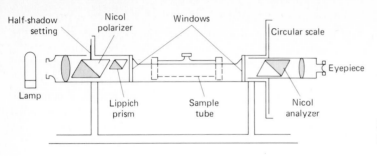

FIGURE 13-12   Typical visual polar-imeter.

through which the light has been rotated by the sample. In principle, the measurement could be made by first adjusting the two prisms to a crossed position that yields a minimum in light intensity in the absence of sample. With a sample in place, rotation of the beam would cause an increase in light intensity which could then be offset by rotation of the analyzer prism. The angular change required to minimize the intensity would correspond to the rotatory power of the sample. Unfortunately, however, the position of minimum intensity cannot be determined accurately with the eye (nor with a photoelectric detector, for that matter) because the rate of change in intensity per degree of rotation is at a minimum in this region. Therefore, polarimeters are equipped with *half-shadow* devices which permit the determination to be made by matching two halves of a field at a radiation intensity greater than the minimum.

Half-Shadow Devices.   Figure 13–12 shows a typical half-shadow device consisting of a small Nicol prism (called a *Lippich prism*) that intercepts about half the beam emerging from the polarizer. The position of the Lippich prism is adjusted to alter the plane of polarization by a few degrees; thus, in the absence of sample and with the analyzer Nicol prism at 90 deg with respect to the polarizer, a split, light-dark field is observed. The light portion, of course, corresponds to that half of the beam that has been rotated by the Lippich prism and the dark part of the field corresponds to the un-obstructed beam. The intensity of the two halves is then balanced by rotation of the analyzer. The analyzer scale is adjusted to read zero at this point. With the sample in place, the analyzer is rotated until the same balance is obtained. The rotation of the sample can then be read directly from the circular analyzer scale.

Other end point devices, which operate upon the same principle as the Lippich prism, permit the determination of optical rotatory power with a precision of 0.005 to 0.01 deg under ideal conditions. Photoelectric detectors yield a precision of about 0.001 deg.

Sample Tubes.   The sample for polarimetry is contained in a cylindrical tube, usually 5, 10, or 20 cm in length. The ends are plane-parallel glass disks that are either fused to the tube walls or held in place with screw-cap holders. For precise measurements, the tubes are surrounded by a jacket for temperature control. Tubes can be calibrated for length by measuring the rotation of a liquid of known rotatory power; nicotine/alcohol or sucrose/water mixtures are often used for this purpose.

13C-8   APPLICATIONS OF POLARIMETRY

Qualitative Analysis.   The optical rotation of a pure compound under a specified set of conditions provides a basic physical constant that is useful for identification purposes in the same way as its melting point, boiling point, or refractive index. Optical activity is characteristic of many naturally occurring substances such as amino acids, steroid, alkaloids, and carbohydrates; polarimetry represents a valuable tool for identifying such compounds.

Structural Determination. In this application, the change in optical rotation resulting from a chemical transformation is measured. Empirical correlations obtained from the study of known structures are then employed to deduce information about the unknown compound. Details of steroid structures, in particular, have been acquired from polarimetric measurements; similar information has been obtained for carbohydrates, amino acids, and other organic compounds.

Quantitative Analysis. Polarimetric measurements are readily adapted to the quantitative analysis of optically active compounds. Empirical calibration curves are used to relate optical rotation to concentration; these plots may be linear, parabolic, or hyperbolic.

The most extensive use of optical rotation for quantitative analysis is in the sugar industry. For example, if the only optically active constituent is sucrose, its concentration can be determined from a simple polarimetric measurement of an aqueous solution of the sample. The concentration is directly proportional to the measured rotation. If other optically active materials are present, a more complex procedure is required; here, the change in rotation resulting from the hydrolysis of the sucrose is determined. The basis for this analysis is shown by the equation

$$C_{12}H_{22}O_{11} + H_2O \xrightarrow{\text{acid}} C_6H_{12}O_6 + C_6H_{12}O_6$$

sucrose	glucose	fructose
$[\alpha]_D^{20} = +66.5°$	$+52.7°$	$-92.4°$

This reaction is termed an *inversion* because of the change in sign of the rotation that occurs. The concentration of sucrose is directly proportional to the difference in rotation before and after inversion.

13D Optical Rotatory Dispersion and Circular Dichroism

Optical rotatory dispersion and circular dichroism are two closely related physical methods that are based upon the interaction of circularly polarized radiation with an optically active species.[8] The former method measures the wavelength dependence of the molecular rotation of a compound. As we have already indicated, optical rotation at any wavelength depends upon the difference in refractive index of a substance toward d and l circularly polarized radiation—that is, upon the *circular birefringence* ($n_l - n_d$); this quantity is found to vary in a characteristic way as a function of wavelength. In contrast, circular dichroism depends upon the fact that the *molar absorptivity* of an optically active compound is different for the two types of circularly polarized radiation. Here, the wavelength dependence of ($\varepsilon_l - \varepsilon_d$) is studied, where ε_l and ε_d are the respective molar absorptivities.

The inequality of molar absorptivities was first reported by A. Cotton in 1895, and the whole complex relationship between absorptivity and refractive index differences is now termed the *Cotton effect*. The application of the Cotton effect to chemical investigations was delayed until convenient instruments for its study became available.

13D-1 GENERAL PRINCIPLES

Cotton employed solutions of potassium chromium tartrate, which absorb in the visible region, to demonstrate that right circu-

[8] For a more complete discussion of optical rotatory dispersion and circular dichroism, see: K. P. Wong, *J. Chem. Educ.*, **1974**, *51*, A573; **1975**, *52*, A9, A83; C. Djerassi, *Optical Rotatory Dispersion: Application to Organic Chemistry*, McGraw-Hill: New York, 1960; P. Crabbe, *Optical Rotatory Dispersion and Circular Dichroism*, Holden-Day: San Francisco, 1965; and A. Abu-Shumays and J. J. Duffield, *Anal. Chem.*, **1966**, *38* (7), 29A.

larly polarized radiation was not only re-fracted but also absorbed to a different extent than the left circularly polarized beam; that is, $\varepsilon_d \neq \varepsilon_l$. At the same time, Cotton observed that dramatic changes oc-curred in the circular birefringence ($n_l - n_d$), as well as in the difference in the molar absorptivities ($\varepsilon_l - \varepsilon_d$) of the beams in the region of an absorption maximum. These effects are shown in Figure 13–13. It is im-portant to appreciate that the ($n_l - n_d$) curve for a substance is similar in shape to curves showing the change in its refractive index as a function of the wavelength of unpolarized light (see Figure 4–8, p. 100); here, too, marked changes in the refractive index (anomalous dispersion) occur in the region of absorption.

13D-2 OPTICAL ROTATORY DISPERSION CURVES

An optical rotatory dispersion curve consists of a plot of specific or molecular rotation as a function of wavelength. Two types of cur-vature can be discerned. The first is the nor-mal dispersion range wherein [α] changes only gradually with wavelength. The sec-ond, the region of anomalous dispersion, occurs near an absorption peak. If one peak is isolated from others, the anomalous part of the dispersion curve will have the appear-ance of the curve labeled ($n_l - n_d$) in Figure 13–13. That is, the rotation undergoes rapid change to some maximum (or mini-mum) value, alters direction to a minimum (or maximum), and then finally reverts to values corresponding to normal dispersion. As indicated in Figure 13–13, a change in the sign of the rotation may accompany these changes.

If molecules have multiple absorption peaks, as is usually the case, overlapping re-gions of anomalous dispersion lead to opti-cal rotatory dispersion curves that are com-plex, as shown in Figure 13–14. Note that the ultraviolet absorption spectrum for the compound is also included for reference.

13D-3 CIRCULAR DICHROISM CURVES

In circular dichroism, one of the circular components of a plane-polarized beam is more strongly absorbed than the other. The effect of this differential absorption is to convert the plane-polarized radiation to an elliptically polarized beam. Figure 13–15 il-lustrates how two circular components of unequal amplitude, which result from the differential absorption by a medium, are combined to give a resultant that travels in an elliptical path. The l component of the original beam is shown as retarded more than the d component because $n_l > n_d$; on the other hand, the amplitude of the d com-ponent is less than that of the l component because we have assumed that its molar ab-sorptivity is greater; that is, $\varepsilon_d > \varepsilon_l$.

The angle of rotation α is taken as the angle between the major axis of the emer-gent elliptical beam and the plane of polar-

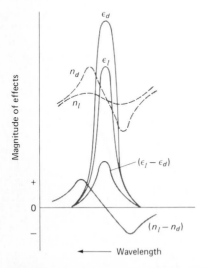

FIGURE 13-13 The Cotton effect. (Adapted from: W. Heller and H. G. Curme, in *Physical Methods of Chemistry*, A. Weissberger and B. W. Rossiter, Eds., Vol. 1, Part IIIC, p. 66, Wiley-Interscience: New York, 1972. Reprinted by permission of John Wiley & Sons, Inc.)

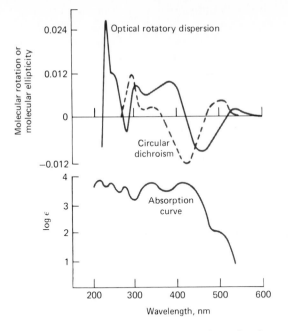

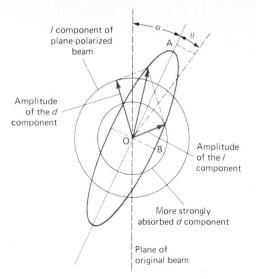

FIGURE 13-15   Elliptically polarized radiation after leaving a medium in which $\varepsilon_d > \varepsilon_l$ and $n_l > n_d$.

FIGURE 13-14   Optical rotatory dispersion, circular dichroism, and absorption curves for (+)-camphor trithione. (Adapted from: H. Wolf, E. Bunnenberg, C. Djerassi, A. Luttringhaus, and A. Stockhausen, *Justus Liebigs Ann. Chem.*, **1964**, *674*, 70. With permission.)

ization of the incident beam. The *ellipticity* is given by the angle θ; the tangent of θ is clearly equal to the ratio of the minor axis of the elliptical path to the major (that is, OB/OA).

It can be shown that the ellipticity is approximated by the relation

$$\theta = \frac{1}{4}(k_l - k_d) \qquad (13\text{--}11)$$

where k_l and k_d are the absorption coefficients[9] of the circularly polarized l and d radiation and θ is expressed in radians. The

---

[9] The absorption coefficient k is related to the more common molar absorptivity ε by the equation $k = 2.303\varepsilon c$, where c is the concentration in moles per liter.

quantity $(k_l - k_d)$ is termed the *circular dichroism*. The molecular ellipticity $[\theta]$ can be shown to be equal to

$$[\theta] = 3305(\varepsilon_l - \varepsilon_d) \qquad (13\text{--}12)$$

where $[\theta]$ has the units of degree-cm^2 per decimole and the ε's are the molar absorptivities of the respective circular components.

Circular dichroism curves consist of a plot of $[\theta]$ versus wavelength. Note that $[\theta]$ can be either negative or positive depending on the relative magnitudes of ε_l and ε_d. The dashed line in Figure 13–14 is a typical curve.

13D-4   INSTRUMENTATION

Optical Rotatory Dispersion.   A number of recording spectropolarimeters are now manufactured that provide optical rotatory dispersion curves in the ultraviolet and visible regions. Generally, these instruments are equipped with xenon arc lamp sources, which provide sufficient radiant power to offset the losses that occur in the polarizer and analyzer; a high-intensity source is also required because the wavelength region of greatest interest in optical rotatory dispersion studies is at or near absorption peaks, where maximum attenuation of the incident beam occurs.

Modern commercial spectropolarimeters ordinarily employ double monochromators of conventional design to minimize stray radiation. The output from the monochromator is passed through a polarizer, the sample, and an analyzer and is detected with a photomultiplier tube. The signal from the detector is amplified; it is then employed to adjust the analyzer position to compensate for rotation caused by the sample and to position a recorder pen as well. As with a visual polarimeter, the half-shade method is the most efficient way of determining the null position of the analyzer. In one instrument, the polarizer is mechanically rocked through a small angle at low frequency. The amplifier system of the detector responds to the resulting ac signal and adjusts the analyzer until the signal is symmetric around the null point. Another spectropolarimeter employs an ordinary double-beam spectrometer with two sets of polarizer-analyzer prisms. The two analyzers are offset from one another by a few degrees, and both beams are passed through the sample. The ratio of the power of the two beams is compared electronically and gives a measure of the optical rotation of the sample.

Circular Dichroism. A conventional spectrometer can be adapted to measure molecular ellipticity. From Beer's law, Equation 13–12 can be written in the form

$$[\theta] = \frac{3305}{bc}\left(\log\frac{P_{l0}}{P_l} - \log\frac{P_{d0}}{P_d}\right)$$

where P_{l0} and P_l represent the power of the circularly polarized *l* beam before and after it has passed through a solution of length *b*

and containing a molar concentration *c* of the sample. The terms P_d and P_{d0} have equivalent meanings for the *d* radiation. If now $P_{d0} = P_{l0}$, then

$$[\theta] = \frac{3305}{bc}\log\frac{P_d}{P_l} \tag{13–13}$$

Thus, the molecular ellipticity can be obtained directly by comparing the power of the two transmitted beams, provided the incident intensities of the incident *l* and *d* circularly polarized beams are identical.

In order to apply Equation 13–13 with an ordinary spectrophotometer, a device for producing *d* and *l* circularly polarized radiation must be provided. We have noted (p. 392) that circularly polarized radiation can be obtained by passing plane-polarized radiation through an anisotropic crystal which has a thickness such that the extraordinary and ordinary rays are one-quarter wavelength out of phase. Rotation of the optic axis of the quarter-wave plate by 90 deg yields either *d* or *l* circularly polarized radiation. For the measurement of circular dichroism with a single-beam spectrophotometer, a polarizer followed by a quarter-wave plate is inserted in the cell compartment of the instrument, provision being made to allow rotation of the plate by ±45 deg. The sample cell is then placed between the plate and the detector; with the plate set to produce *d* circular radiation, the instrument is set on 100% transmittance, or zero absorbance. The plate is then rotated 90 deg; the new absorbance reading then corresponds to log (P_d/P_l) for the sample. In order to cover a spectral range, a wedge-shaped quarter-wave plate is employed, which permits variations of its path length by variation in the position of the wedge with respect to the beam.

Other methods exist for producing circularly polarized radiation. One of these, a *Fresnel rhomb*, is incorporated as an adapter for a double-beam spectrophotometer; see

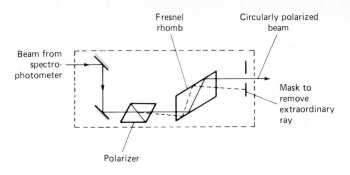

FIGURE 13-16 Spectrophotometer adapter for production of circularly polarized radiation.

Figure 13–16. When a polarized beam undergoes internal reflection in this device, one of the perpendicular components is retarded with respect to the other. The retardation depends upon the refractive index of the medium, the angle of incidence of the reflected beam, and the number of reflections. Quarter-wave retardation, and hence circular polarization, can be attained through proper adjustment of these variables.

The unit shown in Figure 13–16 is fitted into the sample compartment of a double-beam spectrophotometer, with a similar unit in the reference beam. The units are so adjusted that d radiation is emitted from one and l radiation from the other. The two beams then pass through identical cells containing the sample and their relative powers are compared photometrically.

Most modern instruments for circular dichroism studies use a *Pockels cell* for generating circularly polarized radiation. A Pockels cell consists of a uniaxial crystal of ammonium dihydrogen phosphate sandwiched between two transparent electrodes. The crystal is oriented so that radiation is transmitted along its main or z axis with the electrodes being perpendicular to this axis. Application of a potential induces a retardation that is directly proportional to the applied field. Thus, generation of either left-handed or right-handed circularly polarized radiation depends upon the field direction.

When a Pockels cell, powered by an ac signal, is inserted between the monochromator and the detector of a spectrophotometer, d and l circularly polarized beams of equal intensity fall on the detector and produce a dc current. If an optically active sample is then introduced into the light path, unequal absorption of the left- and right-handed components of the radiation occurs inducing an ac component to the detector output. The magnitude of this ac potential, which is readily amplified and measured, depends upon the difference in the d and l molar absorbances of the analyte. Thus, a signal proportional to the desired quantity $(\varepsilon_l - \varepsilon_d)$ is obtained directly.

13D-5 APPLICATIONS OF OPTICAL ROTATORY DISPERSION AND CIRCULAR DICHROISM

Optical rotatory dispersion and circular dichroism studies often provide spectral details for optically active compounds that are absent in their ultraviolet spectra. Thus, in the lower plot in Figure 13–14, the absorption spectrum is seen to consist of a group of overlapping peaks that would be difficult to interpret. On the other hand, the molecular rotation and ellipticity curves for the optically active groups are much more clearly defined and lend themselves to detailed analysis.

Optical rotatory dispersion data have been mainly applied to structural determinations in two major areas: (1) amino acids, polypeptides, and proteins; and (2) complex natural products such as steroids, terpenes, and antibiotics. Most of the structural conclusions from this work are empirical, being based upon spectral observations of known structures. The curves can provide information concerning the configuration of angular substituents at ring junctures, the location of ketone groups, conformational analysis of substituents exerting a vicinal action on an optically active chromophore, the degree of coiling of protein helices, and the type of substitution in amino acids.

The applications of circular dichroism are less developed than optical rotatory dispersion; it appears, however, that the technique will also provide much useful structural information regarding organic and biological systems as well as metal-ligand complexes.

PROBLEMS

13–1. For each of the following pure compounds calculate the specific and molar refraction.

Compound	n_D^{20}	Density at 20°C, g/cm^3
(a) Ethyl acetate	1.3701	0.901
(b) Mesitylene	1.4998	0.8642
(c) Nitrobenzene	1.5524	1.21

13–2. The following data were obtained for a series of binary mixtures of benzene and cyclohexane at 20°C.

Wt % Benzene	n_D
10.86	1.38490
36.48	1.41071
55.98	1.43396
70.41	1.45334
82.83	1.47115

(a) Plot the data.

(b) Derive an equation for the line assuming a linear relationship.

(c) Calculate the standard deviation for the slope and about regression.

(d) Calculate the weight percent benzene in a mixture that had a refractive index of 1.43216.

(e) Calculate the standard deviation of the result found in part (d).

(f) The mean of 3 refractive index measurements of a benzene/cyclohexane mixture was 1.46437. Calculate the weight percent benzene.

(g) Calculate the standard deviation of the result found in part (f).

13–3. In binary mixtures of liquids that approach ideal behavior, the refractive index varies linearly with volume percent. Mesitylene and pentane exhibit this behavior. For mesitylene, $n_D^{20} = 1.4998$, the molecular weight is 120.20, and the density is 0.8642 g/mL (all at 20°C). For pentane, $n_D^{20} = 1.3579$, the molecular weight is 72.15, and the density is 0.6262 g/mL. Find the molarity of pentane in a mixture of pentane and mesitylene having a refractive index of (a) 1.3962 and (b) 1.4666.

13–4. Binary liquid mixtures that do not form ideal solutions exhibit refractive indexes which are complicated functions of the composition. One of the simplest functions applies to an acetone/carbon tetrachloride solution whose refractive index is given by $n_D^{20} = ac + b$, where c is the molar concentration of acetone and a and b are constants. For acetone, $n_D^{20} = 1.3588$, the molecular weight is 58.08, and the density is 0.7908 g/mL (all at 20°C). For carbon tetrachloride, $n_D^{20} = 1.4664$, the molecular weight is 153.82, and the density is 1.5942 g/mL. Calculate the refractive index of 3.00 M acetone in CCl_4.

13–5. To a good approximation, the molar refraction of a compound is the sum of refractions contributed by each atom in the compound. Tables of atomic contributions to molar refraction can be found in such reference books as the *Handbook of Chemistry and Physics*. For example, carbon contributes a refraction of 2.591, hydrogen contributes 1.028, oxygen (doubly bonded) contributes 2.122, and chlorine contributes 5.844.
  (a) Calculate the molar refraction of acetone from the data in Problem 13–4 and compare it to the value calculated from atomic contributions.
  (b) Calculate the molar refraction of CCl_4 from atomic contributions.
  (c) Using $n_D^{20} = 1.4664$ and molecular weight = 153.82, calculate the density of CCl_4. Compare your calculated value to the observed value of 1.5942 g/mL.

13–6. What is the difference between unpolarized, plane polarized, circularly polarized, and elliptically polarized light?

13–7. Distinguish between ordinary and extraordinary beams of radiation.

13–8. Define the following terms: (a) optic axis, (b) Nicol prism, (c) quarter-wave plate, (d) double refraction, (e) circular double refraction, (f) circular birefringence, (g) ellipticity, (h) enantiomers, (i) dextrorotatory, and (j) levorotatory.

13–9. What is the difference between optical rotatory dispersion and circular dichroism?

13–10. How can it be determined whether a sample is rotating the plane of polarization of light by +20 deg or −160 deg?

13–11. If the circular birefringence of a solution at 360 nm is 1.0×10^{-5}, what will be the angle of rotation, α_{360}, for a 5-cm path length? If the circular birefringence is 1.0×10^{-5} at 720 nm, what will be α_{720} for a 5-cm path length?

13–12. For sucrose, $C_{12}H_{22}O_{11}$, $[\alpha]_D^{20} = +66.5$ deg.

(a) Calculate the molecular rotation of sucrose.

(b) Calculate the angle of rotation expected for a solution containing sucrose at a concentration of 5.0 g/L in a 10-cm cell.

(c) A solution originally containing 5.0 g of sucrose per liter exhibits $\alpha = +18.6$ deg in a 10-cm cell after a period of heating with acid (reaction on page 399). Calculate the fraction of sucrose which has been hydrolyzed.

14

Nuclear Magnetic Resonance

Spectroscopy

Nuclear magnetic resonance spectroscopy (NMR) is based upon the measurement of absorption of electromagnetic radiation in the radio frequency region of roughly 4 to 600 MHz[1], which corresponds to a wavelength range of about 75 to 0.5 m. In contrast to ultraviolet, visible, and infrared absorption, nuclei of atoms rather than outer electrons are involved in the absorption process. Furthermore, in order to cause nuclei to develop the energy states required for absorption to occur it is necessary to expose the analyte to an intense magnetic field of several thousand gauss.

Nuclear magnetic resonance spectroscopy is one of the most powerful tools available to the chemist and biochemist for elucidating the structure of both organic and inorganic species. It has also proved useful for the quantitative determination of absorbing species.[2]

This chapter is mainly concerned with the theory, instrumentation, and applications of NMR spectroscopy; a somewhat analogous method called *electric spin resonance* is also treated in much less detail.

14A Theory of Nuclear Magnetic Resonance

As early as 1924, Pauli suggested that certain atomic nuclei should have the properties of spin and magnetic moment and that as a consequence, exposure to a magnetic

---

[1] MHz = 10^6 Hz = 10^6 cycles per second.

[2] The following references are recommended for additional study: E. D. Becker, *High Resolution NMR*, 2d ed., Academic Press: New York, 1980; R. Harris, *Nuclear Magnetic Resonance Spectroscopy*, Pitman Publishing: Marshfield, MA, 1983; R. H. Cox and D. E. Leyden, in *Treatise on Analytical Chemistry*, 2d ed., P. J. Elving, M. M. Bursey, and I. M. Kolthoff, Eds., Part I, vol. 10, Wiley: New York, 1983; and D. E. Leyden and R. H. Cox, *Analytical Applications of NMR*, Wiley: New York, 1977.

field would lead to splitting of their energy levels. During the next decade, experimental verification of these postulates was obtained. It was not until 1946, however, that Bloch at Stanford and Purcell at Harvard, working independently, were able to demonstrate that nuclei absorb electromagnetic radiation in a strong magnetic field as a consequence of the energy level splitting induced by the magnetic force. The two physicists shared the 1952 Nobel prize for their work.

In the first five years following the discovery of nuclear magnetic resonance, chemists became aware that molecular environment influences the absorption of radio-frequency radiation by a nucleus in a magnetic field and that this effect can be correlated with molecular structure. Since then, the growth of NMR spectroscopy has been explosive, and the technique has had profound effects on the development of organic, inorganic, and biochemistry. It is doubtful that there has ever been as short a delay between an initial discovery and its widespread application and acceptance.

In common with optical spectroscopy, both classical and quantum mechanics are useful in explaining the nuclear magnetic resonance phenomenon. The two treatments yield identical relationships. Quantum mechanics, however, is more useful for relating absorption frequencies to energy states of nuclei, while classical mechanics is more helpful in providing a physical picture of the absorption process and how it is measured.

14A-1 QUANTUM DESCRIPTION OF NMR

To account for some properties of nuclei, it is necessary to assume that they rotate about an axis and thus have the property of *spin*. Furthermore, it is necessary to postulate that the angular momentum associated with the spin of the particle is an integral or a half-integral multiple of $h/2\pi$, where h is Planck's constant. The maximum spin component for a particular nucleus is its *spin quantum number I*; it is found that a nucleus will then have $(2I + 1)$ discrete states. The component of angular momentum for these states in any chosen direction will have values of $I, I - 1, I - 2, \ldots, - I$. In the absence of an external field, the various states have *identical* energies.

The spin number for the proton is $\frac{1}{2}$; thus two spin states exist corresponding to $I = +\frac{1}{2}$ and $I = -\frac{1}{2}$. Heavier nuclei, being assemblages of various elementary particles, have spin numbers that range from zero (no net spin component) to at least $\frac{11}{2}$. As shown in Table 14-1, the spin number of a nucleus is related to the relative number of protons and neutrons it contains.

Magnetic Properties of Nuclei. Since a nucleus bears a charge, its spin gives rise to a magnetic field that is analogous to the field pro-

---

Table 14-1
Spin Quantum Number for Various Nuclei

Number of Protons	Number of Neutrons	Spin Quantum Number I	Examples
Even	Even	0	$^{12}C, ^{16}O, ^{32}S$
Odd	Even	$\frac{1}{2}$	$^{1}H, ^{19}F, ^{31}P$
		$\frac{3}{2}$	$^{11}B, ^{79}Br$
Even	Odd	$\frac{1}{2}$	^{13}C
		$\frac{3}{2}$	^{127}I
Odd	Odd	1	$^{2}H, ^{14}N$

duced when electricity flows through a coil of wire. The resulting magnetic dipole μ is oriented along the axis of spin and has a value that is characteristic for each type of nucleus.

The interrelation between particle spin and magnetic moment leads to a set of observable magnetic quantum states m given by

$$m = I, I - 1, I - 2, \ldots, -I \qquad (14-1)$$

Energy Levels in a Magnetic Field. When brought into an external magnetic field, a particle possessing a magnetic moment tends to become oriented such that its magnetic dipole—and hence its spin axis—is parallel to the field. The behavior of the particle is somewhat like that of a small bar magnet; when introduced into such a field, the potential energy of each depends upon the orientation of the dipole with respect to the field. The energy of the bar magnet can assume an infinite number of values depending upon its alignment; in contrast, however, the energy of the nucleus is limited to $(2I + 1)$ discrete values (that is, the alignment is limited to $2I + 1$ positions). Whether quantized or not, the potential energy of a magnet in a field is given by the relationship

$$E = -\mu_z H_0 \qquad (14-2)$$

where μ_z is the *component* of magnetic moment in the direction of an external field having a strength of H_0.

The quantum character of nuclei limits the number of possible energy levels to a few. Thus, for a particle with a spin number of I and a magnetic quantum number of m, the energy of a quantum level is given by

$$E = -\frac{m\mu}{I} \beta H_0 \qquad (14-3)$$

where H_0 is the strength of the external field

in gauss (G) and β is a constant called the *nuclear magneton* (5.051×10^{-24} erg G^{-1}); μ is the magnetic moment of the particle expressed in units of nuclear magnetons. The value of μ for the proton is 2.7927 nuclear magnetons.

Turning now to the proton, for which $I = \frac{1}{2}$, we see from Equation 14–1 that this particle has magnetic quantum numbers of $+\frac{1}{2}$ and $-\frac{1}{2}$. The energies of these states in a magnetic field (Equation 14–3) take the following values:

$$m = +\tfrac{1}{2}, \quad E = -\frac{\tfrac{1}{2}(\mu\beta H_0)}{\tfrac{1}{2}} = -\mu\beta H_0$$

$$m = -\tfrac{1}{2}, \quad E = -\frac{-\tfrac{1}{2}(\mu\beta H_0)}{\tfrac{1}{2}} = +\mu\beta H_0$$

These two quantum energies correspond to the two possible orientations of the spin axis with respect to the magnetic field; as shown in Figure 14–1, for the lower energy state ($m = +\frac{1}{2}$) the vector of the magnetic moment is aligned with the field, and for the higher energy state ($m = -\frac{1}{2}$) the alignment is reversed. The energy difference between the two levels is given by

$$\Delta E = 2\mu\beta H_0$$

Also shown in Figure 14–1 are the orientations and energy levels for a nucleus such as ^{14}N, which has a spin number of 1. Here, three energy levels ($m = 1, 0,$ and -1) are found, and the difference in energy between each is $\mu\beta H_0$. In general, the energy differences are given by

$$\Delta E = \mu\beta \frac{H_0}{I} \qquad (14-4)$$

As with other types of quantum states, excitation to a higher nuclear magnetic quantum level can be brought about by absorption of a photon with energy $h\nu$ that just equals ΔE. Thus, Equation 14–4 can be

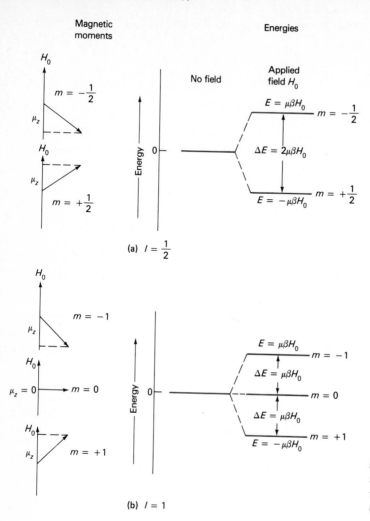

FIGURE 14-1 Magnetic moments and energy levels for a nucleus with a spin quantum number of (**a**) $\frac{1}{2}$ and (**b**) 1.

written as

$$h\nu = \mu\beta \frac{H_0}{I} \qquad (14\text{--}5)$$

EXAMPLE 14-1. Many NMR instruments employ a magnet that provides a field strength of 14,092 G. At what frequency would the proton nucleus absorb in such a field?

Substituting into Equation 14–5 we find

$$\nu = \frac{2.7927 \times 5.051 \times 10^{-24} \times 14,092}{6.6256 \times 10^{-27} \times \frac{1}{2}}$$

$$= 60.0 \times 10^{6} \text{ Hz} = 60.0 \text{ MHz}$$

The foregoing example reveals that radio-frequency radiation of 60.0 MHz will bring about a change in alignment of the magnetic moment of the proton from a direction that parallels the field to a direction that opposes it.

Distribution of Particles Between Magnetic Quantum States. In the absence of a magnetic field, the energies of the magnetic quantum states are identical. Consequently, a large assemblage of protons will contain an identical number of nuclei with $m = +\frac{1}{2}$ and $m = -\frac{1}{2}$. When placed in a field, however, the nuclei tend to orient themselves so that the lower energy state ($m = +\frac{1}{2}$) predominates. It is of interest to calculate the extent of this predominance in a typical NMR experiment. For this purpose, the Boltzmann equation (p. 257) can be written in the form

$$\frac{N_j}{N_0} = \exp\left(\frac{-\Delta E}{kT}\right) \qquad (14\text{–}6)$$

where N_j is the number of protons in the higher energy state ($m = -\frac{1}{2}$), N_0 is the number in the lower state ($m = +\frac{1}{2}$), k is the Boltzmann constant (1.38×10^{-16} erg/deg), T is the absolute temperature, and ΔE is defined by Equation 14–4.

EXAMPLE 14–2. Calculate the relative number of protons in the higher and lower magnetic states when a sample is placed in a 14,092 G field at 20°C.

Substituting Equation 14–4 into 14–6 gives

$$\frac{N_j}{N_0} = \exp\left(\frac{-\mu\beta H_0}{IkT}\right)$$

$$\frac{\mu\beta H_0}{IkT} = \frac{2.79 \times 5.05 \times 10^{-24} \times 14{,}092}{\frac{1}{2} \times 1.38 \times 10^{-16} \times 293}$$

$$= (9.82 \times 10^{-6})$$

$$\frac{N_j}{N_0} = \exp(-9.82 \times 10^{-6}) = 0.999990$$

Thus for exactly 10^6 protons in higher energy states there will be

$$N_0 = 10^6/0.999990 = 1{,}000{,}010$$

in the lower energy state. This figure corresponds to a 10 ppm excess.

Example 14–2 demonstrates that the success of nuclear magnetic resonance measurement depends upon a remarkably small excess (~10 ppm) of lower energy protons. If this excess does not exist, however, no net absorption would be observed because the number of particles excited by the radiation would exactly equal the number producing *induced* emission.

14A-2 CLASSICAL DESCRIPTION OF NMR

To understand the absorption process, and in particular the measurement of absorption, a more classical picture of the behavior of a charged particle in a magnetic field is helpful.

Precession of Particles in a Field. Let us first consider the behavior of a nonrotating magnetic body, such as a compass needle, in an external magnetic field. If momentarily displaced from alignment with the field, the needle will swing in a plane about its pivot as a consequence of the force exerted by the field on its two ends; in the absence of friction, the ends of the needle will fluctuate back and forth indefinitely about the axis of the field. A quite different motion occurs, however, if the magnet is spinning rapidly around its north-south axis. Because of the gyroscopic effect, the force applied by the field to the axis of rotation causes movement not in the plane of the force but perpendicular to this plane; the axis of the rotating particle, therefore, moves in a circular path (or *precesses*) around the magnetic field. This motion, illustrated in Figure 14–2, is similar to the motion of a gyroscope when it is displaced from the vertical by application of a force.

From classical mechanics, it is known that the angular velocity of precession is directly proportional to the applied force and inversely proportional to the angular momentum of the spinning body to which the force is applied. The force on a spinning nucleus in a magnetic field is the product of the field

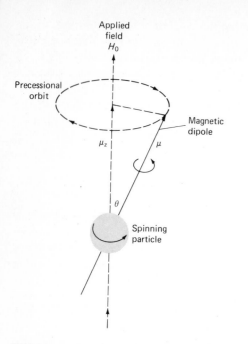

FIGURE 14-2 Precession of a rotating particle in a magnetic field.

strength H_0 and the magnetic moment $\mu\beta$ of the particle, or $\mu\beta H_0$; as noted earlier, the angular momentum is given by $I(h/2\pi)$. Therefore, the precessional velocity ω_0 is

$$\omega_0 = \frac{2\pi\mu\beta}{Ih} \cdot H_0 = \gamma H_0 \qquad (14\text{-}7)$$

where γ is a constant called the *magnetogyric ratio* (or, less appropriately, the gyromagnetic ratio). The magnetogyric ratio expresses the relationship between the magnetic moment and the angular momentum of a rotating particle; that is,

$$\gamma = \frac{\mu\beta}{I(h/2\pi)} \qquad (14\text{-}8)$$

The magnetogyric ratio has a characteristic value for each type of nucleus.

Equation 14–7 can be converted to a frequency of precession ν_0 (the *Larmor frequency*) by division by 2π. Thus,

$$\nu_0 = \frac{\omega_0}{2\pi} = \frac{\gamma H_0}{2\pi} \qquad (14\text{-}9)$$

Equations 14–9 and 14–8 can also be combined to give

$$h\nu_0 = \frac{\mu\beta}{I} H_0 \qquad (14\text{-}10)$$

A comparison of Equations 14–10 and 14–5 suggests that the precessional frequency of the particle derived from classical mechanics is identical to the quantum mechanical frequency of radiant energy required to bring about the transition of a rotating particle from one spin state to another; that is, $\nu_0 = \nu$. Substituting this equality in Equation 14–9 gives a useful relationship between the frequency of absorbed radiation and the strength of the magnetic field:

$$\nu = \frac{\gamma H_0}{2\pi} \qquad (14\text{-}11)$$

Absorption Process. The potential energy E of the precessing particle shown in Figure 14–2 is given by

$$E = -\mu_z H_0 = -\mu H_0 \cos\theta$$

Thus, when radio-frequency energy is absorbed by a nucleus, its angle of precession must change. Hence, we imagine for a nucleus having a spin quantum number of 1/2 that absorption involves a flipping of the magnetic moment that is oriented in the field direction to a state in which the moment is in the opposite direction. The process is pictured in Figure 14–3. In order for the dipole to flip, there must be present a magnetic force at right angles to the fixed field and one with a circular component that can move in phase with the precessing dipole. Circularly polarized radiation (p. 392) of a suitable frequency has these necessary properties; that is, its magnetic vector has a circular component, as represented by the

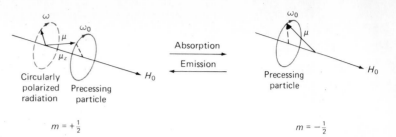

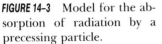

FIGURE 14-3   Model for the absorption of radiation by a precessing particle.

dotted line in Figure 14–3. If the rotational frequency of the magnetic vector of the radiation is the same as the precessing frequency, absorption and flipping can occur. The process is reversible, and the excited particle can thus return to the ground state by reemission of the radiation.

As shown in Figure 13–10 (p. 394), the magnetic vector of plane-polarized radiation can be considered to be composed of two circularly polarized magnetic vectors rotating in opposite directions, in phase, and in a plane perpendicular to the plane of linear polarization. Thus, by irradiating nuclear particles with a beam polarized at 90 deg to the direction of the fixed magnetic field, circularly polarized radiation is introduced in the proper plane for absorption. Only that magnetic component of the beam that rotates in the precessional direction is absorbed; the other half of the beam, being out of phase, passes through the sample unchanged. The process is depicted in Figure 14–4.

14A-3   RELAXATION PROCESSES AND SATURATION

As was shown in Example 14–2, the absorption of radio-frequency radiation by protons in a magnetic field depends upon the presence of a few parts per million excess of particles in the lower magnetic energy state. Upon exposure to a sufficiently intense beam of radiation of appropriate frequency, this tiny excess can, however, be depleted, whereupon the absorption signal rapidly

decreases and approaches zero. This reduction in absorption signal upon exposure to radiation is said to result from *saturation*. Saturation is seldom observed in ultraviolet, visible, and infrared absorption studies, because the ratio of ground state to excited state particles is so enormous that significant depletion of the number of absorbing particles is seldom encountered. Furthermore, numerous efficient, nonradiative relaxation processes exist by which electronically and vibrationally excited particles return to the ground state after a brief period (see pp. 228 to 231).

In NMR spectroscopy, partial or complete saturation is sometimes encountered; the result is a decrease in sensitivity of the mea-

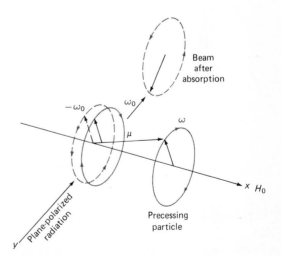

FIGURE 14-4   Absorption of one circular component of a beam that is polarized in the *xy* plane.

surement. One method of minimizing the effects of saturation is by reducing the intensity of the source. In addition, variables that increase the rate of nonradiative relaxation reduce the effects of saturation. Thus, it is of some interest to examine the mechanisms by which a nucleus can return from an excited spin state to its lower energy state.

One obvious relaxation path involves emission of radiation of a frequency corresponding to the energy difference between the states (fluorescence). Radiation theory, however, shows that the probability of re-emission of photons varies as the cube of the frequency and that at radio frequencies, this process does not occur to a significant extent. Thus, in NMR studies, relaxation occurs primarily by nonradiative processes. It is of importance to examine the nature of these processes because an understanding of their mechanisms does permit some control of their rates and thus the quality of the observed absorption signals.

To reduce saturation and produce a readily detectable absorption signal, relaxation should occur as rapidly as possible; that is, the lifetime of the excited state should be small. A second factor—the inverse relationship between the lifetime of an excited state and the width of its absorption line—negates the advantage of very short lifetimes. Thus, when relaxation rates are high, or the lifetimes low, line broadening is observed, which prevents high-resolution measurements. These two opposing factors cause the optimum half-life for an excited species to range from perhaps 0.1 to 1 sec.

Two types of nuclear relaxation are recognized. The first is called *longitudinal* or *spin-lattice* relaxation; the second is termed *transverse* or *spin-spin* relaxation.

Spin-Lattice Relaxation. The absorbing nuclei in an NMR experiment are part of the larger assemblage of atoms that constitutes the sample. The entire assemblage is termed the *lattice,* regardless of whether the sample is a solid, a liquid, or a gas. In the latter two states particularly, the various nuclei comprising the lattice are in violent vibrational and rotational motion, which creates a complex field about each magnetic nucleus. The resulting lattice field thus contains an infinite number of magnetic components, at least some of which must correspond in frequency and phase with the precessional frequency of the magnetic nuclei of interest. These vibrationally and rotationally developed components interact with and convert nuclei from a higher to a lower spin state; the absorbed energy then simply increases the amplitude of the thermal vibrations or rotations. This change corresponds to a minuscule temperature rise for the sample.

Spin-lattice relaxation is a first-order process that can be characterized by a time T_1, which is a measure of the average lifetime of the nuclei in the higher energy state. In addition to depending upon the magnetogyric ratio of the absorbing nuclei, T_1 is strongly affected by the mobility of the lattice. In crystalline solids and viscous liquids, where mobilities are low, T_1 is large. As the mobility increases (at higher temperatures, for example), the vibrational and rotational frequencies increase, thus enhancing the probability for existence of a magnetic fluctuation of the proper magnitude for a relaxation transition; T_1 becomes shorter as a consequence. At very high mobilities, on the other hand, the fluctuation frequencies are further increased and spread over such a broad range that the probability of a suitable frequency for a spin-lattice transition again decreases. The result is a minimum in the relationship between T_1 and lattice mobility.

The spin-lattice relaxation time is greatly shortened in the presence of an element with an unpaired electron which, because of its spin, creates strong fluctuating magnetic fields. A similar effect is caused by nuclei that have spin numbers greater than one-half. These particles are characterized by a nonsymmetrical charge distribution; their rotation also produces a strong fluctuating field that provides yet another pathway for an excited nucleus to give up its energy to the lattice. The marked shortening of T_1 causes line broadening in the presence of such species. An example is found in the NMR spectrum for the proton attached to a nitrogen atom (for ^{14}N, $I = 1$).

Spin-Spin Relaxation and Line Broadening. Several other effects tend to diminish relaxation times and thereby broaden NMR lines. These effects are normally lumped together and described by a spin-spin relaxation time T_2. Values for T_2 are generally so small for crystalline solids or viscous liquids (as low as 10^{-4} sec) as to preclude the use of samples of these kinds for high-resolution spectra unless special techniques are employed. These techniques are described briefly in a later section dealing with NMR studies of solids.

When two neighboring nuclei of the same kind have identical precession rates, but are in different magnetic quantum states, the magnetic fields of each can interact to cause an interchange of states. That is, a nucleus in the lower spin state can be excited while the excited nucleus relaxes to the lower energy state. Clearly, no net change in the relative spin state population (and thus no decrease in saturation) results, but the average lifetime of a particular excited nucleus is shortened. Line broadening is the result.

Two other causes of line broadening should be noted. Both arise if H_0 in Equation 14–11 differs slightly from nucleus to nucleus; under these circumstances, a band of frequencies, rather than a single frequency, is absorbed. One cause for such a variation in the static field is the presence in the sample of other magnetic nuclei whose spins create local fields that may act to enhance or diminish the external field acting on the nucleus of interest. In a mobile lattice, these local fields tend to cancel because the nuclei causing them are in rapid and random motion. In a solid or a viscous liquid, however, the local fields may persist long enough to produce a range of field strengths and thus a range of absorption frequencies. Variations in the static field also result from small inhomogeneities in the field source itself. This effect can be largely offset by rapidly spinning the entire sample in the magnetic field.

14A-4 TYPES OF NMR SPECTRA

Several types of NMR spectra are encountered depending upon the kind of instrument used, the type of nucleus involved, the physical state of the sample, the environment of the analyte nucleus, and the use to which the data is to be put.

Coordinates. The abscissa of an NMR spectrum may be frequency ν, applied field H_0, or the chemical shift parameters δ and τ.

Two methods are available for producing NMR spectra. The first is analogous to the method for obtaining optical spectra; it involves measuring an absorption signal as the electromagnetic frequency is varied; frequency then serves as the abscissa of spectra. No dispersing elements such as prisms or gratings exist for radio frequencies, however. Thus, a variable-frequency oscillator with a linear sweep is required to produce radio frequencies that can be varied continuously over a range of 1 kHz for hydrogen and as great as 10 kHz for such nuclei as ^{13}C and ^{19}F.

An alternative way of collecting NMR data is to employ a constant-frequency radio oscillator and sweep the magnetic field H_0 continuously. Here, it is often more convenient to use H_0 as the abscissa of the spectral plot.

The chemical shift parameters δ and τ, which often serve as abscissa for NMR spectra, are defined in Section 14B–2. The former is directly proportional to frequency while the latter is inversely related to frequency.

Nuclei That Produce Spectra. Well over 200 isotopes have nuclear magnetic properties and thus are capable of yielding spectra. Table 14–2 lists the properties of a few of the common nuclei that have been employed in NMR studies. Note the wide differences in absorption frequency (at a constant field) and sensitivity.

Wide-Line Spectra. Wide-line spectra are those in which the bandwidth of the source of the lines is large enough so that the fine structure due to chemical environment is obscured. Figure 14–5 is a wide-line spectra for a mixture of several isotopes. A single peak is associated with each species. Wide-line spectra are useful for the quantitative determination of isotopes and for studies of the physical environment of the absorbing species. Wide-line spectra are obtained at relatively low magnetic fields.

High-Resolution Spectra. The most widely used NMR spectra are *high resolution* wherein instruments capable of differentiating between very small frequency differences (1

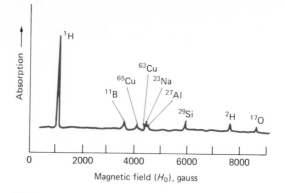

FIGURE 14–5 A low-resolution NMR spectrum of water in a glass container. Frequency = 5 MHz.

ppm or less) are used. Here, for a given isotope, several peaks are commonly encountered as a consequence of chemical environmental effects. Figure 14–6 illustrates two high-resolution spectra for the protons in ethanol. In the upper spectrum, three peaks are observed arising from absorption by the CH_3, CH_2, and OH protons. As shown in

Table 14–2				
Spectral and Magnetic Properties of Some Common Nuclei				
Nucleus	**Spin**	**Absorption Frequency**[a]	**Isotopic Abundance, %**	**Sensitivity**[b]
^1H	$\frac{1}{2}$	60.0	99.98	1.000
^7Li	$\frac{3}{2}$	23.3	92.58	0.293
^{13}C	$\frac{1}{2}$	15.1	1.11	0.0159
^{14}N	1	4.3	99.63	0.0010
^{17}O	$\frac{5}{2}$	8.1	0.037	0.029
^{19}F	$\frac{1}{2}$	56.4	100	0.833
^{23}Na	$\frac{3}{2}$	15.9	100	0.093
^{25}Mg	$\frac{5}{2}$	3.7	10.13	0.0027
^{27}Al	$\frac{5}{2}$	15.6	100	0.206
^{29}Si	$\frac{1}{2}$	11.9	4.70	0.0078
^{31}P	$\frac{1}{2}$	24.3	100	0.066
^{33}S	$\frac{3}{2}$	4.6	0.76	0.0023
^{109}Ag	$\frac{1}{2}$	2.8	48.18	0.0001

[a] MHz in a 14092-G magnetic field.
[b] Sensitivity relative to an equal number of protons at a constant field.

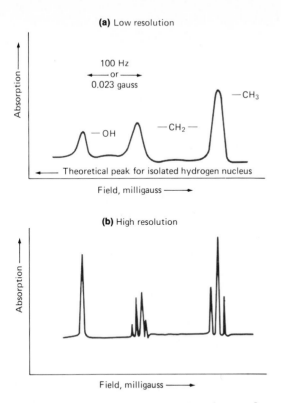

(a) Low resolution

100 Hz
←— or —→
0.023 gauss

—CH₃

—OH

—CH₂—

Theoretical peak for isolated hydrogen nucleus

Field, milligauss ——→

(b) High resolution

Field, milligauss ——→

FIGURE 14-6 NMR spectra of ethanol at a frequency of 60 MHz. **(a)** Resolution $\sim 1/10^6$; **(b)** resolution $\sim 1/10^7$.

Figure 14–6**b**, at higher resolution, two of the three peaks can be resolved into additional peaks.

14B Environmental Effects on NMR Spectra

If the character of NMR spectra was determined solely by *nuclear* properties, only single absorption lines, such as those shown in Figure 14–5, would be observed for each isotope. Spectra of this type would be of some use to chemists for determining isotope concentrations; certainly, however, the amount of information derived from such spectra would not lead to NMR's present role as a major spectroscopic tool.

Fortunately, the frequency of radiation that is absorbed by a given nucleus is strongly influenced by the kind of chemical environment in which it exists. As a consequence, even for simple molecules, a wealth of spectral detail is frequently observed, which can serve as an aid in the elucidation of chemical structures.

Figure 14–6 illustrates the environmental effects on the proton NMR spectra for an organic molecule containing but six protons. At high resolution, these six protons lead to eight characteristic spectral peaks that permit conclusions to be drawn as to the number of protons in the sample and the kind of functional group that contains each proton.

Environmental effects are general and are observed for not only protons but other nuclei such as carbon, fluorine, and phosphorus as well. The discussion that follows centers on proton spectra, the isotope that has been most widely studied. The conclusions, however, apply in most cases to the spectra of other isotopes as well.

14B-1 TYPES OF ENVIRONMENTAL EFFECTS

The spectra for ethyl alcohol, shown in Figure 14–6, illustrate two types of environmental effects. The curve in Figure 14–6**a**, obtained with a lower resolution instrument, shows three proton peaks with areas in the ratio 1 : 2 : 3 (left to right). On the basis of this ratio, it appears logical to attribute the peaks to the hydroxyl, the methylene, and the methyl protons, respectively. Other evidence confirms this conclusion; for example, if the hydrogen atom of the hydroxyl group is replaced by deuterium, the first peak disappears from this part of the spectrum. Thus, small differences occur in the absorption frequency of the proton; such differences depend upon the group to which the hydrogen atom is bonded. This effect is called the *chemical shift.*

The higher resolution spectrum of ethanol, shown in Figure 14–6**b**, reveals that two of the three proton peaks are split into additional peaks. This secondary environmental effect, which is superimposed upon the chemical shift, has a different cause; it is termed *spin-spin splitting*.

Both the chemical shift and spin-spin splitting are important in structural analysis.

Experimentally, the two types of NMR splitting are readily distinguished, for it is found that the peak separations (in units of ν or H_0) resulting from a chemical shift are directly proportional to the field strength or to the oscillator frequency. Thus, if the spectrum in Figure 14–6**a** were to be obtained at 100 MHz rather than at 60 MHz, the horizontal distance between any pair of the peaks would be increased by 100/60 (see Figure 14–7). In contrast, the distance between the fine-structure peaks within a group (lower spectrum) would not be altered by this frequency change.

14B-2 MEASUREMENT OF THE CHEMICAL SHIFT AND SPIN-SPIN SPLITTING

Source of the Chemical Shift. The chemical shift arises from a circulation of the electrons surrounding the nucleus under the influence of the applied magnetic field. This phenomenon is discussed later in greater detail; for the present, it suffices to say that this movement of electrons creates a small magnetic field that ordinarily opposes the applied field. As a consequence, the nucleus is exposed to an effective field that is somewhat smaller (but in some instances, larger) than the external field. The magnitude of the field developed internally is directly proportional to the applied external field, so that we may write

$$H_0 = H_{appl} - \sigma H_{appl} \qquad (14-12)$$
$$= H_{appl}(1 - \sigma)$$

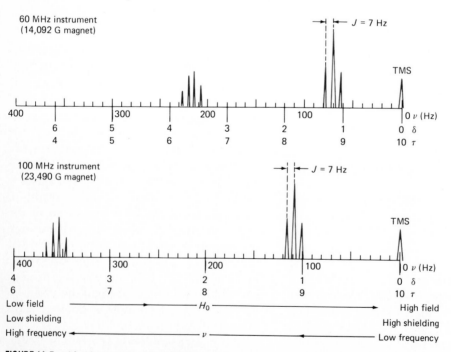

FIGURE 14-7 Abscissa scales for NMR spectra.

where H_{appl} is the applied field and H_0 *is the resultant field which determines the resonance behavior of the nucleus.* The quantity σ is the *shielding parameter,* which is determined by the electron density around the nucleus; electron density depends upon the structure of the compound containing the nucleus.

The shielding parameter for protons in a methyl group is larger than the corresponding parameter for methylene protons; it is even smaller for the proton in an —OH group. For an isolated hydrogen nucleus, the shielding parameter is zero. Thus, in order to bring any of the protons in ethanol into resonance at a given oscillator frequency ν, it is necessary to employ a field H_{appl} that is greater than H_0 (Equation 14–12), the resonance value for the isolated proton. Since σ differs for protons in various functional groups, the required applied field differs from group to group. This effect is shown in the spectrum of Figure 14–6a, where the hydroxyl proton appears at the lowest applied field, the methylene protons next, and finally the methyl protons. Note that all of these peaks occur at an applied field greater than the theoretical one for the isolated hydrogen nucleus, which would lie far to the left in Figure 14–6a. Note also that if the applied field is held constant at a level necessary to excite the methyl proton, an increase in frequency would be needed to bring the methylene protons into resonance.

Source of Spin-Spin Splitting. The splitting of chemical shift peaks can be explained by assuming that the effective field around one nucleus is further enhanced or reduced by local fields generated by *the hydrogen nuclei bonded to an adjacent atom.* Thus, the fine structure of the methylene peak shown in Figure 14–6b can be attributed to the effect of the local fields associated with the adjacent methyl protons. Conversely, the splitting of the methyl peak into three smaller peaks is caused by the adjacent methylene protons. These effects are independent of the applied field and are superimposed on the effects of the chemical shift.

Abscissa Scales for NMR Spectra. The determination of the absolute field strength with the accuracy required for high resolution is difficult or impossible; on the other hand, as will be shown in Section 14C, it is entirely feasible to determine, within a few milligauss, the magnitude of a *change* in field strength. Thus, it is expedient to report the position of resonance absorption peaks relative to the resonance peak for a standard substance that can be measured at essentially the same time. In this way, the effect of fluctuations in the fixed magnetic field is minimized. The use of an internal standard is also advantageous in that chemical shifts can be reported in terms that are independent of the oscillator frequency.

A variety of internal standards have been proposed, but the compound now most generally accepted is tetramethylsilane (TMS), $(CH_3)_4Si$. All of the protons in this compound are identical, and for reasons to be considered later, the shielding parameter for TMS is larger than for most other protons. Thus, the compound provides, at a high applied field, a single sharp peak that is isolated from most of the peaks of interest in a spectrum. In addition, TMS is inert, readily soluble in most organic liquids, and easily removed from samples by distillation (b.p. = 27°C). Unfortunately, TMS is not water soluble; in aqueous media, the sodium salt of 2,2-dimethyl-2-silapentane-5-sulfonic acid (DSS),

$$(CH_3)_3SiCH_2CH_2CH_2SO_3Na$$

may be used in its stead. The methyl protons of this compound produce a peak analogous to that of TMS; the methylene protons give a series of small peaks that are readily identified and can thus be ignored.

The field strength H_{ref} required to produce the TMS resonance line at frequency ν is given by Equation 14–12

$$H_0 = H_{ref}(1 - \sigma_{ref})$$

which can be rewritten as

$$\sigma_{ref} = \frac{H_{ref} - H_0}{H_{ref}} \qquad (14\text{–}13)$$

Similarly, for a given absorption peak of the sample, we may write

$$\sigma_{sple} = \frac{H_{sple} - H_0}{H_{sple}} \qquad (14\text{–}14)$$

where H_{sple} is the field necessary to produce the peak. We then define a *chemical shift parameter* δ as

$$\delta = (\sigma_{ref} - \sigma_{sple}) \times 10^6 \qquad (14\text{–}15)$$

and substitute Equations 14–13 and 14–14 into this expression to obtain

$$\delta = \frac{H_0(H_{ref} - H_{sple})}{H_{sple}H_{ref}} \times 10^6$$

But H_{sple} is very nearly the same as H_0, so that the ratio H_0/H_{sple} is very nearly unity. Thus,

$$\delta = \frac{H_{ref} - H_{sple}}{H_{ref}} \times 10^6 \qquad (14\text{–}16)$$

The quantity δ is dimensionless and expresses the relative shift in parts per million; for a given peak, δ will be the same regardless of whether a 60-, 90-, or 100-MHz instrument is employed. Most proton peaks lie in the δ range of 1 to 12. For other nuclei, the range of chemical shift parameter is greater because of the associated $2p$ electrons. For example, the shift parameter for ^{13}C varies from 0 to 220; for ^{19}F, the range is from -270 to $+65$, while for ^{31}P it is -200 to $+120$.

Another chemical shift parameter τ is defined as

$$\tau = 10 - \delta \qquad (14\text{–}17)$$

Generally, NMR plots have linear scales in δ (and sometimes τ), and the data are plotted with the field increasing from left to right (see Figure 14–7). Thus, if TMS is employed as the reference, its peak will appear on the far right-hand side of the plot, since σ for TMS is large. As shown, the zero value for the δ scale corresponds to the TMS peak and the value of δ increases from right to left. The τ scale, of course, changes in the opposite way. Referring again to Figure 14–7, note that the various peaks appear at the same values of δ and τ in spite of the fact that the two spectra were obtained with instruments having markedly different fixed fields.

For the purpose of reporting spin-spin splitting, it is desirable to utilize scalar units of milligauss or hertz. The latter scale has now come into more general use and is the one shown in Figure 14–7. The position of the reference TMS peak is arbitrarily taken as zero, with frequencies increasing from right to left. The effect of this choice is to make the frequency of a given peak identical with the increase in oscillator frequency that would be needed to bring that proton into resonance if the field were maintained constant at the level required to produce the TMS peak.

It can be seen in Figure 14–7 that the spin-spin splitting in frequency units (J) is the same for the 60-MHz and the 100-MHz instruments. Note, however, that the chemical shift *in frequency units* is enhanced with the higher frequency instrument.

14B-3   THE CHEMICAL SHIFT

As noted earlier, chemical shifts arise from the secondary magnetic fields produced by the circulation of electrons in the molecule. These electronic currents (local *diamagnetic currents*[3]) are induced by the fixed magnetic field and result in secondary fields that may either decrease or enhance the field to which a given proton responds. The effects are complex, and we consider only the major aspects of the phenomenon here. More complete treatments can be found in the several reference works listed under footnote 2, page 407.

Under the influence of the magnetic field, electrons bonding the proton tend to precess around the nucleus in a plane perpendicular to the magnetic field (see Figure 14–8). A consequence of this motion is the development of a secondary field, which opposes the primary field; the behavior here is analogous to the passage of electrons through a wire loop. The nucleus then experiences a resultant field, which is smaller (the nucleus is said to be *shielded* from the full effect of the primary field); as a consequence, the external field must be increased to cause nuclear resonance. The frequency of the precession, and thus the magnitude of the secondary field, is a direct function of the external field.

The shielding experienced by a given nucleus is directly related to the electron density surrounding it. Thus, in the absence of the other influences, shielding would be expected to decrease with increasing electronegativity of adjacent groups. This effect is illustrated by the δ values for the protons in the methyl halides, CH_3X, which lie in the order I (2.16), Br (2.68), Cl (3.05), and F (4.26). Here, iodine (the least electronegative) is the least effective of the halogens in withdrawing electrons from the protons;

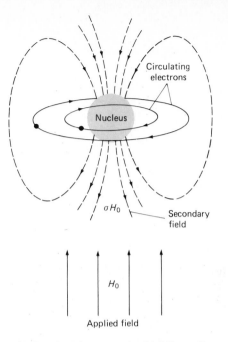

FIGURE 14-8 Diamagnetic shielding of a nucleus.

thus, the electrons of iodine provide the largest shielding effect. Similarly, electron density around the methyl protons of methanol is greater than around the proton associated with oxygen because oxygen is more electronegative than carbon. Thus, the methyl peaks are upfield from the hydroxyl peak. The position of the proton peaks in TMS is also explained by this model, since silicon is relatively electropositive. Finally, acidic protons have very low electron densities, and the peak for the proton in RSO_3H or $RCOOH$ lies far downfield ($\delta > 10$).

Effect of Magnetic Anisotropy. It is apparent from an examination of the spectra of compounds containing double or triple bonds that local diamagnetic effects do not suffice to explain the position of certain proton peaks. Consider, for example, the irregular change in δ values for protons in the following hydrocarbons, arranged in order of increasing acidity (or increased electronegativity of the groups to which the protons are

---

[3] The intensity of magnetization induced in a *diamagnetic* substance is smaller than that produced in a vacuum with the same field. Diamagnetism is the result of motion induced in bonding electrons by the applied field; this motion (a *diamagnetic current*) creates a secondary field that opposes the applied field. *Paramagnetism* (and the resulting *paramagnetic currents*) operates in just the opposite sense.

bonded): $CH_3—CH_3$ (δ = 0.9), $CH_2{=}CH_2$ (δ = 5.8), and $HC{\equiv}CH$ (δ = 2.9). Furthermore, the aldehydic proton RCHO ($\delta \sim 10$) and the protons on benzene ($\delta \sim 7.3$) appear considerably farther downfield than might be expected on the basis of the electronegativity of the groups to which they are attached.

The effects of multiple bonds upon the chemical shift can be explained by taking into account the anisotropic magnetic properties of these compounds. For example, the magnetic susceptibilities[4] of crystalline aromatic compounds have been found to differ appreciably, depending upon the orientation of the ring with respect to the applied field. This anisotropy is readily understood from the model shown in Figure 14–9. Here, the plane of the ring is perpendicular to the magnetic field; in this position, the field can induce a flow of the π electrons around the ring (a ring current). The consequence is similar to that of a current in a wire loop; namely, a secondary field is produced that acts in opposition to the applied

---

[4] The magnetic susceptibility of a substance can be thought of as the extent to which it is susceptible to induced magnetization by an external field.

field. This secondary field, however, exerts a magnetic effect on the protons attached to the ring; as shown in Figure 14–9, this effect is in the direction of the field. Thus, the aromatic protons require a lower external field to bring them into resonance. This effect is either absent or self-canceling in other orientations of the ring.

A somewhat analogous model can be envisioned for the ethylenic or carbonyl double bonds. Here, one can imagine circulation of the π electrons in a plane along the axis of the bond when the molecule is oriented to the field as shown in Figure 14–10a. Again, the secondary field produced acts upon the proton to reinforce the applied field. Thus, deshielding shifts the peak to larger values of δ. With an aldehyde, this effect combines with the deshielding brought about by the electronegative nature of the carbonyl group; a very large value of δ results.

In an acetylenic bond, the symmetrical distribution of π electrons about the bond axis permits electron circulation around the bond (in contrast, such circulation is prohibited by the nodal plane in the electron distri-

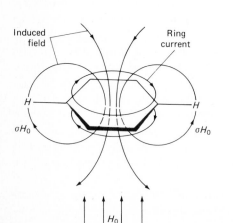

FIGURE 14-9 Deshielding of aromatic protons brought about by ring current.

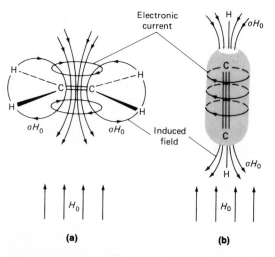

FIGURE 14-10 Deshielding of ethylene and shielding of acetylene brought about by electronic currents.

bution of a double bond). From Figure 14–10**b**, it can be seen that in this orientation the protons are shielded. This effect is apparently large enough to offset the deshielding resulting from the acidity of the protons and from the electronic currents at perpendicular orientations of the bond.

Correlation of Chemical Shift with Structure. The chemical shift is employed for the identification of functional groups and as an aid in determining structural arrangements of groups. These applications are based upon empirical correlations between structure and shift. A number of correlation charts[5] and tables[6]

have been published. Two of these are shown in Figure 14–11 and Table 14–3. It should be noted that the exact values for δ may depend upon the nature of the solvent as well as upon the concentration of the solute. These effects are particularly pronounced for protons involved in hydrogen bonding; an example is the hydrogen atom in the alcoholic functional group.

---

[5] N. F. Chamberlain, *Anal. Chem.*, **1959**, *31*, 56.

[6] R. M. Silverstein, G. C. Bassler, and T. C. Morrill, *Spectrometric Identification of Organic Compounds*, 4th ed., Chapter 4, Wiley: New York, 1981; and L. M. Jackman and S. Sternhall, *Nuclear Magnetic Resonance Spectroscopy*, 2d ed., Pergamon Press: New York, 1969.

Table 14-3
Approximate Chemical Shifts for Certain Methyl, Methylene, and Methine Protons

	δ, ppm		
Structure	**M = CH$_3$**	**M = CH$_2$**	**M = CH**
Aliphatic β substituents			
M—C—Cl	1.5	1.8	2.0
M—C—Br	1.8	1.8	1.9
M—C—NO$_2$	1.6	2.1	2.5
M—C—OH (or OR)	1.2	1.5	1.8
M—C—OC(=O)R	1.3	1.6	1.8
M—C—C(=O)H	1.1	1.7	—
M—C—C(=O)R	1.1	1.6	2.0
M—C—C(=O)OR	1.1	1.7	1.9
M—C—ϕ	1.1	1.6	1.8
Aliphatic α substituents			
M—Cl	3.0	3.5	4.0
M—Br	2.7	3.4	4.1
M—NO$_2$	4.3	4.4	4.6
M—OH (or OR)	3.2	3.4	3.6
M—O—ϕ	3.8	4.0	4.6
M—OC(=O)R	3.6	4.1	5.0
M—C=C	1.6	1.9	—
M—C≡C	1.7	2.2	2.8
M—C(=O)H	2.2	2.4	—
M—C(=O)R	2.1	2.4	2.6
M—C(=O)ϕ	2.4	2.7	3.4
M—C(=O)OR	2.2	2.2	2.5
M—ϕ	2.2	2.6	2.8

Structural type

δ and τ value and range†

| 19 | 18 | 17 | 16 | 15 | 14 | 13 | 12 | 11 | 10 | 9 | 8 | 7 | 6 | 5 | 4 | 3 | 2 | 1 | 0 δ |
| −9 | −8 | −7 | −6 | −5 | −4 | −3 | −2 | −1 | 0 | 1 | 2 | 3 | 4 | 5 | 6 | 7 | 8 | 9 | 10 τ |

1. TMS, 10.0000
2. −CH$_2$−, cyclopropane, 9.78
3. CH$_4$, 9.767
4. ROH, monomer, very dilute solution, *ca* 9.5
5. CH$_3$−C−(saturated), (8.7)9.05–9.15(9.3)
6. R$_2$NH‡, 0.1–0.9 mole fraction in an inert solvent, (7.8)8.4–9.6
7. CH$_3$−C−C−X(X = Cl, Br, I, OH, OR, C=O, N), (8.8)8.90–9.10
8. −CH$_2$− (saturated), 8.65–8.80
9. RSH‡, 8.5–8.9
10. RNH$_2$‡, 0.1–0.9 mole fraction in an inert solvent, (8.2)8.5–8.9
11. −C−H (saturated), 8.35–8.60
12. CH$_3$−C−X (X = F, Cl, Br, I. OH, OR, OAr, N), (8.0)8.1–8.8(9.0)
13. CH$_3$−C=C, 8.1–8.4
14. CH$_3$−C=O, 7.4–7.9(8.1)
15. CH$_3$Ar, 7.5–7.75(7.9)
16. CH$_3$−S−, 7.2–7.9
17. CH$_3$−N, 7.0–7.9
18. H−C≡C−, nonconjugated, 7.35–7.55
19. H−C≡C−, conjugated, 6.9–7.2
20. ArSH‡, 6.0–7.0
21. CH$_3$−O−, (6.0)6.2–6.5(6.7)
22. ArNH$_2$‡, ArNHR‡, and Ar$_2$NH‡, (5.7)6.0–6.6(6.7)
23. ROH‡, 0.1–0.9 mole fraction in an inert solvent, 4.8–7.0
24. CH$_2$=C, nonconjugated, 5.0–5.4
25. HC=C, acyclic, nonconjugated, (4.1)4.3–4.8(4.9)
26. C=C, cyclic, nonconjugated, 4.3–4.8
27. CH$_2$ = C, conjugated, (3.75)4.3–4.7
28. ArOH‡, polymeric association, 2.3–5.5
29. HC=C, conjugated, (2.25)3.3–4.3(4.7)
30. C=C, acyclic, conjugated, (2.9)3.5–4.0(4.5)
31. H−N−C=O, 1.5–4.5
32. ArH, benzenoid, (0.5)2.0–3.4(4.0)
33. ArH, nonbenzenoid, (1.0)1.4–3.8(6.0)
34. RNH$_3^+$, R$_2$NH$_2^+$, and R$_3$NH$^+$ (trifluoroacetic acid solution), 2.3–2.9
35. H−C=O/N, 1.9–2.1
36. H−C=O/O−, 1.8–2.0
37. ArNH$_3^+$, ArRNH$_2^+$, and ArR$_2$NH$^+$ (trifluoroacetic acid solution), 0.5–1.5
38. C=N/OH‡, −0.2–1.2
39. RCHO, aliphatic, α,β-unsaturated, 0.35–0.50
40. RCHO, aliphatic, 0.2–0.3(0.5)
41. ArCHO, (−0.1)0.0–0.3(0.5)
42. ArOH, intramolecularly bonded, (−5.5)−2.5–−0.5
43. −SO$_3$H, −2–−1
44. RCO$_2$H, dimer, in nonpolar solvents, (−3.2)−2.2–−1.0(0.3)
45. Enols, −6–−5

| −9 | −8 | −7 | −6 | −5 | −4 | −3 | −2 | −1 | 0 | 1 | 2 | 3 | 4 | 5 | 6 | 7 | 8 | 9 | 10 |

†Normally, absorptions for the functional groups indicated will be found within the range shown. Occasionally, a functional group will absorb outside this range. Approximate limits for this are indicated by absorption values in parentheses and by shading in the figure.

‡The absorption positions of these groups are concentration-dependent and are shifted to higher τ values in more dilute solutions.

FIGURE 14–11 Absorption positions of protons in various structural environments. (Table taken from J. R. Dyer, *Applications of Absorption Spectroscopy of Organic Compounds*, p. 85, Prentice-Hall, Inc., ©: Englewood Cliffs, NJ, 1965. With permission.)

14B–4 SPIN-SPIN SPLITTING

As may be seen in Figure 14–6, the absorption bands for the methyl and methylene protons in ethanol consist of several narrow peaks that can be separated only with a high-resolution instrument. Careful examination of these peaks shows that the spacing for the three components of the methyl band is identical to that for the four peaks of the methylene band; this spacing in frequency units is called the *coupling constant* for the interaction and is given the symbol J. Moreover, the areas of the peaks in a multiplet approximate an integral ratio to one another. Thus, for the methyl triplet, the ratio of areas is $1:2:1$; for the quartet of methylene peaks, it is $1:3:3:1$.

Origin. It seems plausible to attribute these observations to the effect that the spins of one set of nuclei exert upon the resonance behavior of another. That is to say, a small interaction or coupling exists between the two groups of protons. This explanation presupposes that such coupling takes place via interactions between the nuclei and the bonding electrons rather than through free space. For our purpose, however, the details of this mechanism are not important.

Let us first consider the effect of the methylene protons in ethanol on the resonance of the methyl protons. We must first remember that the ratio of protons in the two possible spin states is very nearly one, even in a strong magnetic field. We can imagine, then, that the two methylene protons in a molecule can have four possible combinations of spin states and that in an entire sample each of these combinations will be approximately equally represented. If we represent the spin orientation of each nucleus with a small arrow, the four states are

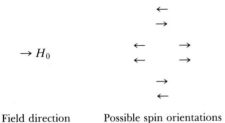

Field direction   Possible spin orientations
of methylene protons

In one combination, the spins of the two methylene protons are paired and aligned against the field, while in a second, the paired spins are reversed; there are also two

combinations in which the spins are opposed to one another. The magnetic effect that is transmitted to the methyl protons on the adjacent carbon atoms is determined by the spin combinations that exist in the methylene group at any instant. If the spins are paired and opposed to the external field, the effective applied field on the methyl protons is slightly lessened; thus, a somewhat higher field is needed to bring them into resonance, and upfield shift results. Spins paired and aligned with the field result in a downfield shift. Neither of the combinations of opposed spin has an effect on the resonance of the methyl protons. Thus, splitting into three peaks results. The area under the middle peak is twice that of either of the other two, since two spin combinations are involved.

Let us now consider the effect of the three methyl protons upon the methylene peak. Possible spin combinations for the methyl protons are

$$
\begin{array}{c}
\leftarrow \quad \leftarrow \\
\leftarrow \quad \rightarrow \\
\rightarrow \quad \rightarrow \\[4pt]
\leftarrow \quad \leftarrow \quad \rightarrow \quad \rightarrow \\
H_0 \rightarrow \quad \leftarrow \quad \rightarrow \quad \leftarrow \quad \rightarrow \\
\leftarrow \quad \leftarrow \quad \rightarrow \quad \rightarrow \\[4pt]
\rightarrow \quad \rightarrow \\
\leftarrow \quad \rightarrow \\
\leftarrow \quad \leftarrow
\end{array}
$$

Here, we have eight possible spin combinations; however, among these are two groups containing three combinations that have equivalent magnetic effects. The methylene peak is thus split into four peaks having areas in the ratio $1:3:3:1$.

The interpretation of spin-spin splitting patterns is relatively simple and straightforward for *first-order* spectra. First-order spectra are those in which the chemical shift between interacting groups of nuclei is large with respect to their coupling constant J. Rigorous first-order behavior requires that

$\Delta\nu/J$ be greater than 20; frequently, however, analysis of spectra by first-order techniques can be accomplished down to value of $\Delta\nu/J$ of somewhat less than 10. The ethanol spectrum shown in Figure 14–7 is an example of a pure first-order spectrum with J for the methyl and methylene peaks being 7 Hz and the separation between the centers of the two multiplets being about 140 Hz.

Interpretation of second-order NMR spectra is difficult and complex and will not be dealt with in this text. It is noteworthy, however, that because $\Delta\nu$ increases with increases in the magnetic field while J does not, spectra obtained with an instrument having a high magnetic field are much more readily interpreted than those produced by a spectrometer with a weaker magnet.

Rules Governing the Interpretation of First-Order Spectra. The following rules govern the appearance of first-order spin-spin spectra.

1. Equivalent nuclei do not interact with one another to give multiple absorption peaks. The three protons in the methyl groups in ethanol give rise to splitting of the adjacent methylene protons only and not to splitting among themselves.

2. Coupling constants decrease with separation of groups, and coupling is seldom observed at distances greater than three bond lengths.

3. The multiplicity of a band is determined by the number n of magnetically equivalent protons on the neighboring atoms and is given by $(n + 1)$. Thus, the multiplicity for the methylene band in ethanol is determined by the number of protons in the adjacent methyl groups and is equal to $(3 + 1)$.

   For nuclei having spins (I) other than 1/2, the multiplicity is given by $(2nI + 1)$.

4. If the protons on atom B are affected by protons on atoms A and C that are nonequivalent, the multiplicity of B is equal to $(n_A + 1)(n_C + 1)$, where n_A and n_C are the number of equivalent protons on A and C, respectively.

5. The approximate relative areas of a multiplet are symmetric around the midpoint of the band and are proportional to the coefficients of the terms in the expansion $(x + 1)^n$. The application of this rule is demonstrated in Table 14–4 and in the examples that follow.

6. The coupling constant is independent of the applied field; thus, multiplets are readily distinguished from closely spaced chemical shift peaks.

Table 14–4
Relative Intensities of First-Order Multiplets ($I = \frac{1}{2}$)

Number of Equivalent Protons, n	Multiplicity, $(n + 1)$	Relative Peak Areas									
0	1					1					
1	2				1		1				
2	3			1		2		1			
3	4		1		3		3		1		
4	5	1		4		6		4		1	
5	6	1	5		10		10		5		1
6	7	1	6	15		20		15	6		1
7	8	1	7	21	35		35	21	7		1

EXAMPLE 14–3. For each of the following compounds, calculate the number of multiplets for each band and their relative areas.

(a) $ClCH_2CH_2CH_2Cl$. The multiplicity of the band associated with the four equivalent protons on the two ends of the molecule would be determined by the number of protons on the central carbon; thus, the multiplicity is $(2 + 1) = 3$ and the areas would be $1:2:1$. The multiplicity of two central methylene protons would be determined by the four equivalent protons at the ends and would thus be $(4 + 1) = 5$. Expansion of $(x + 1)^4$ gives the following coefficients (Table 14–4), which are proportional to the areas of the peaks $1:4:6:4:1$.

(b) $CH_3CHBrCH_3$. The band for the six methyl protons will be made up of $(1 + 1) = 2$ peaks having relative areas of $1:1$; the proton on the central carbon atom has a multiplicity of $(6 + 1) = 7$. These peaks will have areas (Table 14–4) in the ratio of $1:6:15:20:15:6:1$.

(c) $CH_3CH_2OCH_3$. The methyl protons on the right are separated from the other protons by more than three bonds so that only a single peak will be observed for them. The protons of the central methylene group will have a multiplicity of $(3 + 1) = 4$ and a ratio of $1:3:3:1$. The methyl protons on the left have a multiplicity of $(2 + 1) = 3$ and an area ratio of $1:2:1$.

The foregoing examples are relatively simple because all of the protons influencing the multiplicity of any single peak are magnetically equivalent. A more complex splitting pattern results when a set of protons is affected by two or more nonequivalent protons. As an example, consider the spectrum of 1-iodopropane, $CH_3CH_2CH_2I$. If we label the three carbon atoms (a), (b), and (c) from left to right, the chemical shift bands are found at $\delta_{(a)} = 1.02$, $\delta_{(b)} = 1.86$, and $\delta_{(c)} = 3.17$. The band at $\delta_{(a)} = 1.02$ will be split by the two methylene protons on (b) into $(2 + 1) = 3$ peaks having relative areas of $1:2:1$. A similar splitting of the band $\delta_{(c)} = 3.17$ will also be observed. The experimental coupling constants for the two shifts are $J_{(ab)} = 7.3$ and $J_{(bc)} = 6.8$. The band for the methylene protons (b) is affected by two groups of protons which are not magnetically equivalent, as is evident from the difference between $J_{(ab)}$ and $J_{(bc)}$. Thus, invoking rule 4, the number of peaks will be $(3 + 1)(2 + 1) = 12$. In cases such as this, derivation of a splitting pattern, as shown in Figure 14–12, is helpful. Here, the effect of the (a) proton is first shown and leads to four peaks of relative areas $1:3:3:1$ spaced at 7.3 Hz. Each of these is then split into three new peaks spaced at 6.8 Hz, having relative areas of $1:2:1$. (The same final pat-

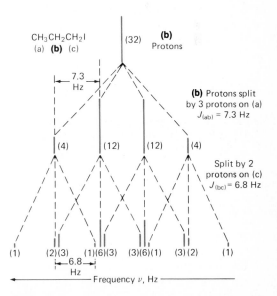

FIGURE 14–12 Splitting pattern for methylene **(b)** protons in $CH_3CH_2CH_2I$. Figures in parentheses are relative areas under peaks.

tern is produced if the original band is first split into a triplet.) At very high resolution, the spectrum for 1-iodopropane exhibits a series of peaks that approximates the series shown at the bottom of Figure 14–12. At lower resolution (so low that the instrument does not detect the difference between $J_{(ab)}$ and $J_{(bc)}$, only six peaks are observed with relative areas of $1 : 5 : 10 : 10 : 5 : 1$.

Second-Order Spectra. Coupling constants are usually smaller than 20, whereas chemical shifts may be as high as 1000 or 60 MHz. Therefore, the splitting behavior described by the rules in the previous section is common. However, when $\Delta\nu/J$ becomes less than perhaps 7, these rules no longer apply. Generally, as $\Delta\nu$ approaches J, the peaks on the inner side of two multiplets tend to be enhanced at the expense of the peaks on the outer side, and the symmetry of each multiplet is thus destroyed, as noted earlier. Analysis of a spectrum under these circumstances is difficult.

Effect of Chemical Exchange on Spectra. Turning again to the spectrum of ethanol (Figure 14–6), it is interesting to consider why the OH proton appears as a singlet rather than a triplet. The methylene protons and the OH proton are separated by only three bonds; coupling should occur to increase the multiplicity of both OH and the methylene peaks. Actually, as shown in Figure 14–13, the expected multiplicity is observed by employing a highly purified sample of the alcohol. Note the triplet OH peaks and the eight methylene peaks in this spectrum. If, now, a trace of acid or base is added to the pure sample, the spectrum reverts to the form shown in Figure 14–6.

The exchange of OH protons among alcohol molecules is known to be catalyzed by both acids and bases, as well as by the impurities that commonly occur in alcohol. It is thus plausible to associate the decoupling observed in the presence of these catalysts to an exchange process. If exchange is rapid, each OH group will have several protons associated with it during any brief period;

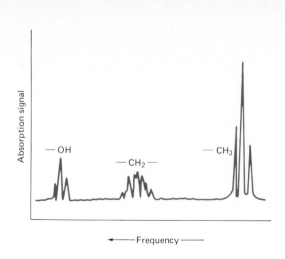

FIGURE 14–13 Spectrum of highly purified ethanol showing additional splitting of OH and CH_2 peaks (compare with Figure 14–6).

within this interval, all of the OH protons will experience the effects of the three spin arrangements of the methylene protons. Thus, the magnetic effects on the alcoholic proton are averaged and a single sharp peak is observed. Spin decoupling always occurs when the exchange frequency is greater than the separation (in frequency units) between the interacting components.

Chemical exchange can affect not only spin-spin spectra but also chemical shift spectra. Purified alcohol/water mixtures have two well-defined and easily separated OH proton peaks. Upon addition of an acid or base as a catalyst, however, the two peaks coalesce to form a single sharp line. Here, the catalyst enhances the rate of proton exchange between the alcohol and the water and thus averages the shielding effect. A single sharp line is obtained when the exchange rate is significantly greater than the separation frequency of the individual lines of alcohol and water. On the other hand, if the exchange frequency is about the same as this frequency difference, shielding is only partially averaged and a broad line results. The correlation of line breadth with exchange rates has provided a direct means for investigating the kinetics of such processes and represents an important application of the NMR experiment.

14B-5 AIDS IN THE ANALYSIS OF COMPLEX SPECTRA

Several methods are available for the simplification of spectra that are too complex for ready analysis. The more common of these are discussed briefly in this section.

Increases in the Magnetic Field. As mentioned earlier, coupling constants are unaffected by increases in the magnetic field, whereas chemical shifts are enhanced. Thus, the use of an instrument with a stronger magnet may convert an uninterpretable second-order spectrum to one that is susceptible to first-order analysis.

Isotopic Substitution. Substitution of deuterium for one or more of the protons in a molecule simplifies the spectrum by removal of the absorption peaks corresponding to the substituted protons. Further simplification may also be observed because the coupling between a deuterium and a proton is significantly less strong than between two protons.

Double Resonance Techniques. Double resonance experiments include a group of techniques in which a nucleus is simultaneously irradiated with two (and sometimes more) radio signals of different frequency. Among these methods are *spin decoupling, spin tickling, nuclear Overhauser effect,* and *internuclear double resonance.* These procedures are used to aid in the interpretation of complex NMR spectra and to enhance the information that can be obtained from spectra.[7] Only the first of these techniques, spin decoupling, will be described here.

Figure 14–14 illustrates the spectral simplification that may accompany spin decoupling. Spectrum *B* shows the absorption as-

sociated with the four protons on the pyridine ring of nicotine. Spectrum *C* was obtained by sweeping the same portion of the spectrum while simultaneously irradiating the sample with a second radio-frequency signal having a frequency corresponding to the absorption peaks of protons (d) and (c) (about 8.6 ppm); the strength of the second signal is sufficient to cause saturation of the signal for these protons. The consequence is a decoupling of the interaction between these two protons and protons (a) and (b). Here, the complex absorption spectra for (a) and (b) collapse to two doublet peaks which arise from coupling between these protons. Similarly, the spectra for (d) and (c) could be simplified by decoupling with a beam having a frequency corresponding to the peaks for protons (a) or (b).

The interaction between dissimilar nuclei can also be decoupled (*heteronuclear* in contrast to *homonuclear* decoupling). An important example is the decoupling of ^{14}N nuclei and protons, which can be accomplished by irradiation of the sample with a beam having a frequency of about 4.3 MHz when the magnetic field is 14,092 G. Absorption of this radiation by the ^{14}N nuclei results in a complete decoupling of their interaction with protons and often produces spectral simplifications. Another important example of heteronuclear decoupling is encountered in ^{13}C NMR where the technique is used to simplify spectra by decoupling protons.

The theory of spin-spin decoupling is complex and beyond the scope of this book.

Chemical Shift Reagent.[8] Important aids in the interpretation of proton NMR spectra are *shift reagents,* which have the effect of dispersing the absorption peaks for certain types of compounds over a much larger frequency range. This dispersion will fre-

---

[7] For a more detailed discussion of double resonance methods, see: E. D. Becker, *High Resolution NMR,* Chapter 9, Academic Press: New York, 1980; R. H. Cox and D. E. Leyden, in *Treatise on Analytical Chemistry,* 2d ed., P. J. Elving, M. M. Bursey, and I. M. Kolthoff, Eds., Part I, vol. 10, pp. 74–87, Wiley: New York, 1983.

[8] For a brief summary of shift reagents, see: *Shift Reagents in NMR, Perkin-Elmer NMR Quarterly,* Number 7, August 1977, The Perkin-Elmer Corporation, Norwalk, CT 06852.

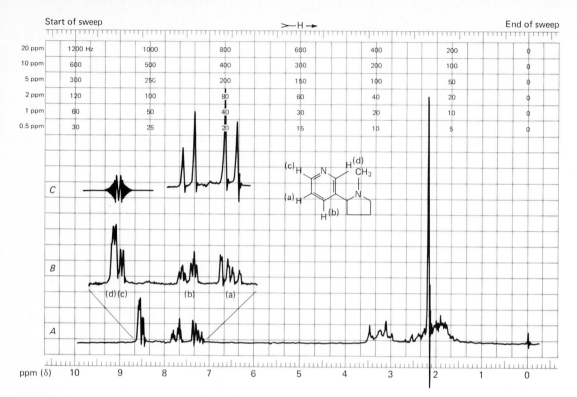

FIGURE 14-14 Effect of spin decoupling on the NMR spectrum of nicotine dissolved in $CDCl_3$. Curve A: the entire spectrum. Curve B: expanded spectrum for the four protons on the pyridine ring. Curve C: spectrum for protons (**a**) and (**b**) when decoupled from (**d**) and (**c**) by irradiation with a second beam that has a frequency corresponding to about 8.6 ppm. (Courtesy Varian Instrument Division, Palo Alto, CA, 94303.)

quently separate otherwise overlapping peaks and permit easier interpretation.

Shift reagents are generally complexes of europium or praseodymium. A typical example is the dipivalomethanato complex of praseodymium(III) [usually abbreviated as Pr(DPM)₃].

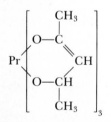

The praseodymium ion in this neutral complex is capable of increasing its coordination by interaction with lone electron pairs.

Therefore, reactions can take place between the complex and molecules containing oxygen, nitrogen, or other atoms that contain free electron pairs.

The DPM complexes of europium and praseodymium are generally employed in nonpolar solvents such as CCl_4, $CDCl_3$, and C_6D_6 to avoid solvent competition with the analyte for electron receptor sites on the metal ion.

Figure 14–15 illustrates the dramatic effect that Pr(DPM)₃ has on the complex spectrum for styrene oxide. Here, the peaks for the protons closest to the oxygen binding site are shifted to higher fields and actually above that for the TMS reference. Note also that the peaks for the ortho hydrogens in the ring are shifted to a greater extent than

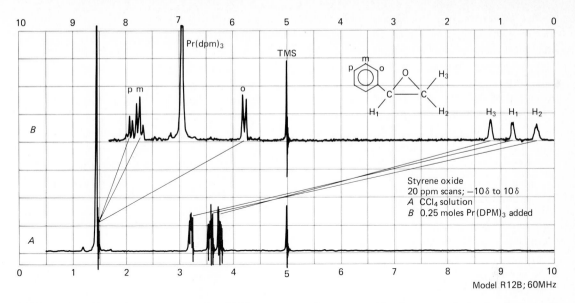

FIGURE 14-15 Effect of Pr(DPM)$_3$ on the NMR spectrum of styrene oxide. Spectrum *A*: in the absence of the reagent. Spectrum *B*: in the presence of the reagent. (Spectra courtesy of Perkin-Elmer, Norwalk, CT.)

the meta or para hydrogen peaks, again as a consequence of their being closer to the metal ion than the other hydrogens. Similar effects are observed with Eu(DPM)$_3$ except that the shifts are to lower fields; here also, the peaks for the protons closest to the metal ion are affected most.

The primary source of the induced chemical shifts is the secondary magnetic field generated by the large magnetic moment of the paramagnetic praseodymium or europium ion. If the geometry of the complex between the analyte and the shift reagent is known, reasonably good estimates can be made as to the extent of shift for various protons.

14C Experimental Methods of NMR Spectroscopy

Nuclear magnetic resonance spectrometers are either *high-resolution* or *wide-line* instruments. Only the former can resolve the fine structures associated with many absorption peaks. High resolution requires the use of magnetic fields greater than 7000 G. Wide-line spectrometers, which employ magnets with field strengths of a few thousand gauss, are much simpler and less expensive than their high-resolution counterparts.

Two types of high-resolution instruments are encountered, *continuous wave* and *Fourier transform* spectrometers. In the former, the absorption signal is monitored continuously as the frequency (or field) is varied. With the Fourier transform instruments, short, regular, repetitive pulses of high-intensity radio-frequency radiation excite all of the nuclei simultaneously within a given frequency range. The output is a time-domain spectrum that can be converted to the frequency domain by a Fourier transformation. Early high-resolution instruments were all continuous-wave spectrometers. By now, Fourier transform instruments are becoming less and less expensive and consequently more widely used.

From an instrumental standpoint, the equipment required for high-resolution

NMR spectroscopy is elaborate, and most chemists are forced to accept only a general understanding of the operating principles of NMR spectrometers. On the other hand, the sample-handling techniques, the interpretation of spectra, and an appreciation of the effects of variables are no more complex than for other types of absorption spectroscopy; these areas are of greatest interest to the chemist.[9]

14C-1 CONTINUOUS WAVE INSTRUMENTS[10]

Figure 14–16 is a simplified schematic diagram showing the important components of a continuous-wave NMR spectrometer. A brief description of each follows.

The Magnet. The sensitivity and resolution of an NMR spectrometer are critically dependent upon the strength and quality of its magnet. Both sensitivity and resolution increase with increases in field strength; in addition, however, the field must be highly

homogeneous and reproducible. These requirements make the magnet by far the most expensive component of an NMR spectrometer.

Spectrometer magnets are of three types: permanent magnets, conventional electromagnets, and superconducting solenoids. Permanent magnets with field strengths of 7046 or 14,092 G are used in several commercial instruments; corresponding oscillator frequencies for proton studies are 30 and 60 MHz. Permanent magnets are highly temperature-sensitive and require extensive thermostating and shielding as a consequence.

Electromagnets are relatively insensitive to temperature fluctuations but require cooling systems to remove the heat generated by the large currents that are required. Elaborate power supplies are also necessary to provide the required stability. Commercial electromagnets generate fields of 14092, 21140 and 23490 G, corresponding to proton absorption frequencies of 60, 90, and 100 MHz.

Superconducting magnets are used in the highest resolution and most expensive instruments. Here, fields as great as 140,000 G are attained, corresponding to a proton frequency of 600 MHz.

---

[9] For a "how to" book on NMR, see: M. L. Martin and G. J. Martin, *Practical NMR Spectroscopy*, Heyden: Philadelphia, 1980.

[10] For a concise description of continuous wave NMR spectrometers, see: D. G. Howery, *J. Chem. Educ.*, **1971**, *48*, A327, A389.

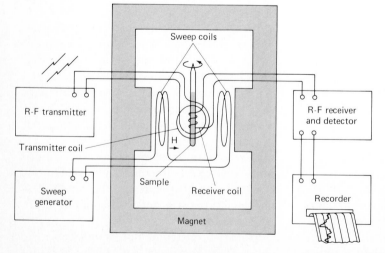

FIGURE 14-16 Schematic diagram of an NMR spectrometer. (Courtesy of Varian Instrument Division, Palo Alto, CA.)

The performance specifications for a spectrometer magnet are stringent. The field produced must be homogeneous to a few parts per billion within the sample area and must be stable to a similar degree for short periods of time. Unfortunately, the inherent stability of most magnets is considerably lower than this figure with variations as large as one part in 10^7 being observed over a period of 1 hr. In order to offset the effect of field fluctuations, a *frequency lock system* is employed in most commercial NMR instruments. Here, a reference nucleus is continuously irradiated and monitored at a frequency corresponding to its resonance maximum at the rated field strength of the magnet. Changes in the intensity of the reference absorption signal, resulting from drifts in the magnetic field, control a feedback circuit, the output of which is fed into coils in the magnetic gap in such a way as to correct for the drift.

Recall that the ratio between the field strengths and resonance frequencies is a constant *regardless of the nucleus involved* (Equation 14–11). Thus, the drift correction for the reference signal is applicable to the signals for all nuclei in the sample area.

Two types of lock systems are encountered. The *external lock* employs a separate sample container and receiver coil for the reference substance (often water); this container must be located as close as possible to the analyte container (usually ~1 cm). In an *internal lock* system, the reference substance is dissolved in the solution containing the sample. Here, the reference compound may be tetramethylsilane (TMS), which also serves as an internal standard (see p. 419). Alternatively, deuterium from the solvent can serve as the reference; in this instance, a second transmitter coil set to the frequency for deuterium is required.

The external lock is simpler to operate, and control is not lost during sample changes. It does not, however, provide as good control over field changes encoun-tered by the sample because of its physical displacement. Thus, the external system provides control on the order of parts in 10^9, while the internal system gives control on the order of parts in 10^{10}.

The Field Sweep Generator. A pair of coils located parallel to the magnet faces (see Figure 14–16) permits alteration of the applied field over a small range. By varying a direct current through these coils, the effective field can be changed by a few hundred milligauss without loss of field homogeneity.

Ordinarily, the field strength is changed automatically and linearly with time, and this change is synchronized with the linear drive of a chart recorder. For a 60-MHz proton instrument, the sweep range is 1000 Hz (235 milligauss) or some integral fraction thereof. For nuclei such as ^{19}F and ^{13}C, sweeps as great as 10 kHz are required.

As was noted earlier, several more recent instruments employ a frequency sweep system in lieu of or in addition to magnetic field sweep.

The Radio-Frequency Source. The signal from a radio-frequency oscillator (transmitter) is fed into a pair of coils mounted at 90 deg to the path of the field. A fixed oscillator of exactly 60, 90, or 100 MHz is ordinarily employed; for high-resolution work, the frequency must be constant to about 1 part in 10^9. The power output of this source is less than 1 W and should be constant to perhaps 1% over a period of several minutes. The radio-frequency output is plane polarized.

Cross-Coil Detectors. Two types of detector systems are encountered, namely cross-coil and single-coil. The two differ significantly in principle, the first being based upon nuclear *induction* and the second upon nuclear *absorption*. The latter more closely resembles the detectors for optical spectroscopy, which we have previously described, in the sense that the attenuation of the source signal by the sample is the quantity measured.

As shown in Figure 14–17**b** and Figure 14–16, the cross-coil configuration employs

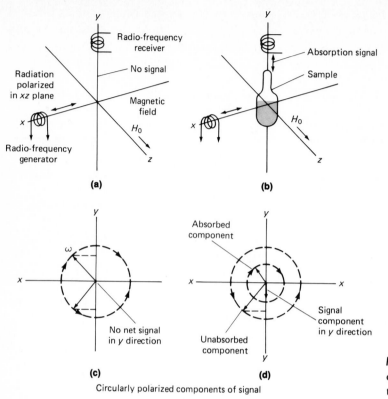

(a)

(b)

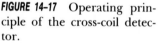

(c)

(d)

Circularly polarized components of signal

FIGURE 14-17 Operating principle of the cross-coil detector.

separate detector and transmitter coils arranged at right angles to one another. In the absence of sample (Figure 14–17**a**), no signal reaches the detector (small leakage currents may be encountered, which must be offset by so-called paddles) because the detector is located at exactly 90 deg to the path of the beam. The presence of an absorbing sample, however, induces radiation that now travels along a path to the detector. The result is a signal that increases in intensity in proportion to the concentration of analyte.

To understand the source of the induced signal it must be realized that the radiation from the antenna of a radio transmitter is plane polarized. With the arrangement shown in Figure 14–17**a**, polarization is in the *xy* direction (the magnetic field is directed along the *z* axis).

As shown in Figure 14–17**c**, the plane-polarized radiation from the source can be resolved into two circularly polarized vectors that rotate in opposite directions to one another in the *xy* plane. Vector addition of these components, regardless of their angular position, indicates that there is no net component *along the y axis*. Thus, no signal is received by a detector located on that axis.

Figures 14–17**b** and 14–17**d** show the effect of a sample positioned at the origin of the three axes. If the source has a frequency that is absorbed by a particular type of nucleus in the sample, the power of one of the two circular components of the beam is diminished (see Figure 14–4). As shown in Figure 14–17**d**, vector addition of the resulting components indicates that the radiation now contains a fluctuating component in the *y* direction, which causes the detector to respond. Thus, the sample serves to couple the generator to the receiver, provided the frequency of radiation corresponds to the precessing frequency of the nuclei in the magnetic field. The extent of coupling, and thus the signal strength, depends upon the number of absorbing nuclei.

Single-Coil Detector. As shown in Figure 14–18, a single coil, surrounding the sample, can serve as both a source and a detector for NMR experiments. Here, with the sample in place, the circuit is tuned with the variable capacitor C. When the external magnetic field brings the sample into resonance, the energy or power absorbed effectively increases the resistance and thus the voltage drop across the parallel capacitor and sample coil. The voltage drop provides a measure of the power absorbed by the sample and thus the absorption of the radiation from the source.

Signal Recorder. The abscissa drive of an NMR recorder is synchronized with the spectrum sweep; often, in fact, the recorder controls the sweep rate. The response of the recorder must be rapid; it is also desirable that the sweep rate be variable.

All modern NMR recorders are equipped with electronic or digital integrators to provide information regarding areas under absorption peaks. Usually, the integral data appear as step functions superimposed on the NMR spectrum (see Figure 14–19). Generally, the area data are reproducible to a few percent relative.

Peak areas are important because they permit estimation of the relative number of absorbing nuclei in each chemical environment (Figure 14–19). This information is vital to the deduction of chemical structure. In addition, peak areas are useful for quantitative analytical work.

Peak heights are not particularly satisfactory measures of concentration because a number of variables that are difficult to control can alter the widths and thus the heights of the peaks. Peak areas, in contrast, are not affected by these variables.

The Sample Holder and Sample Probe. The usual NMR sample cell consists of a 5-mm O.D. glass tube containing about 0.4 mL of liquid.

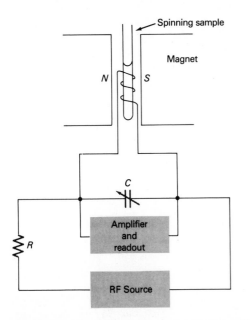

FIGURE 14-18 Single-coil detector and simplified bridge circuit for measurement of absorption.

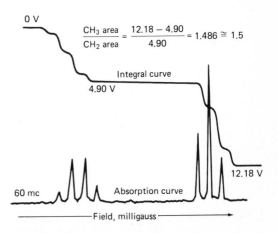

$$\frac{CH_3\ area}{CH_2\ area} = \frac{12.18 - 4.90}{4.90} = 1.486 \cong 1.5$$

FIGURE 14-19 Absorption and integral curve for a dilute ethylbenzene solution (aliphatic region). (Courtesy of Varian Instrument Division, Palo Alto, CA.)

Microtubes for smaller sample volumes are also available.

The sample probe is a device for holding the sample tube in a fixed spot in the field. As will be seen from Figure 14–20, the probe contains not only the sample holder but the sweep, source, and detector coils (for a cross-coil detector) as well, to assure reproducible positioning of the sample with respect to those components; it may also contain the reference cell and liquid used as an external lock. As mentioned earlier, some leakage of power between the source and receiver does occur; the so-called paddles, shown in Figure 14–20, reduce this leakage to a tolerable level.

The sample probe is also provided with an air-driven turbine for rotating the sample tube along its longitudinal axis at a rate of about 4000 rpm. This rotation serves to average out the effects of inhomogeneities in the field; sharper lines and better resolution are obtained as a consequence.

14C-2 *FOURIER TRANSFORM INSTRUMENTS*[11]

Conventional NMR spectroscopy is not very sensitive. Consequently, obtaining good proton spectra for materials available in microgram quantities has been difficult, time-consuming, and sometimes impossible. This lack of sensitivity also seriously inhibited the

applications of NMR spectroscopy to other nuclei, particularly carbon-13. The low natural abundance of this isotope and its relatively small magnetogyric ratio provides an NMR signal that is less by a factor of about 6000 than that for the proton (see Table 14–2). Thus, producing a useful continuous wave spectrum for carbon in unenriched samples required repeated scans and signal averaging for periods of 24 hr or more.

The commercial development of pulsed, Fourier transform NMR spectrometers by several companies since 1970 has resulted in dramatic increases in the sensitivity of NMR measurements; as a result, the routine application of this technique to naturally occurring carbon-13, to protons in microgram quantities of chemical or biological materials, and to such other nuclei as fluorine, phosphorus, and silicon has become widespread. The basis of the increased sensitivity is the same as that discussed in the section on Fourier transform infrared spectroscopy

---

[11] For a more complete discussion of Fourier transform NMR spectroscopy, see: E. D. Becker and T. C. Farrar, *Science*, **1972**, *178*, 361; D. Shaw, *Fourier Transform NMR Spectroscopy*, Elsevier: New York, 1976, R. J. Abraham and P. Loftus, *Proton and Carbon-13 NMR Spectroscopy*, Heyden and Sons: Philadelphia, 1978; and J. W. Akitt, *NMR and Chemistry: An Introduction to the Fourier Transform-Multinuclear Era*, 2d ed., Chapman & Hall: New York, 1983.

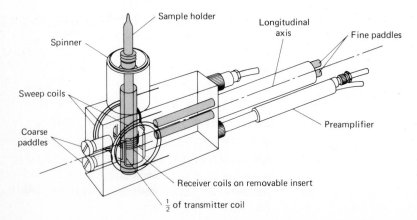

FIGURE 14-20 NMR probe. (Courtesy of Varian Instrument Division, Palo Alto, CA.)

(p. 336). In both, all of the resolution elements of a spectrum are observed in a very brief period by measurements that yield a time-domain rather than a frequency-domain spectrum. A time-domain spectrum can be obtained in a few seconds or less; thus, it becomes practical to replicate spectra hundreds or thousands of times and average the measurements to give a vastly improved signal-to-noise ratio. The frequency-domain spectrum can then be obtained by a Fourier transform employing a digital computer.

Pulsed NMR Spectra.   Three distinctly different types of Fourier transform NMR have been developed, namely *pulsed, stochastic,* and *rapid-scan correlation.* Of the three, the pulse technique is by far the most common and the only one that will be described here.

In pulsed NMR, the sample is irradiated periodically with brief, highly intense pulses of radio-frequency radiation (intense enough to cause saturation for all of the absorbing nuclei), following which the *free induction decay signal*—a characteristic radio-frequency emission signal stimulated by the irradiation—is recorded as a function of time. Typically, pulses of 1 to 10 μsec are employed and the observation time between pulses is approximately 1 sec. Figure 14–21a depicts the time relations for the input signal.

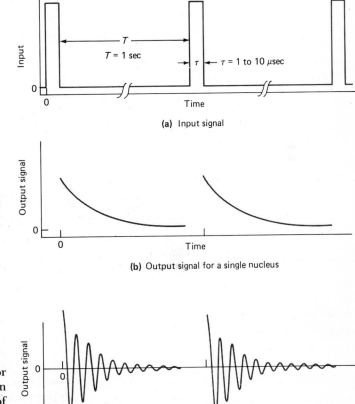

(a) Input signal

(b) Output signal for a single nucleus

(c) Output signal for several types of nuclei

FIGURE 14-21 (a) Input signal for pulsed NMR. **(b)** Free-induction decay signal when a single type of nucleus is present. **(c)** Free-induction decay signal when several types of nuclei are present.

Time-Domain Decay Spectra. The periodic pulses serve the same function as the interferometer in infrared Fourier transform spectroscopy in the respect that they produce a time-domain spectrum during the period between pulses. Two such spectra are shown in Figures 14–21**b** and 14–21**c**. These consist of the free induction decay signals emitted by nuclei as they return to their ground state from the excited state that was induced by the pulse of radiation. The frequencies of the emitted radiation are, of course, identical to the absorption frequencies that would appear in a conventional NMR spectrum for the various types of nuclei present.

Figure 14–21**b** shows the decay signal when the sample contains but a single type of nucleus and the excitation frequency corresponds exactly to the resonance frequency for that nucleus. Figure 14–21**c** depicts the decay signal when more than one type of nucleus is excited. Here, a more complex time-domain response is observed because of interference among the radiations of differing frequencies.

Because the free induction decay curves owe their characteristic appearance to interference among various emitted radiations, it is to be expected that these time-domain spectra will be unique and characteristic for every conceivable combination of frequencies. Thus, just as an infrared time-domain spectrum, obtained with a Michelson interferometer (Section 5I–5), contains all of the information necessary to derive a conventional infrared spectrum, so also does the free induction decay spectrum contain the information needed to yield a normal frequency-domain NMR spectrum. Conversion of the data from the time domain to the frequency domain can be realized with the aid of a digital computer programmed to perform a fast Fourier transform. Figure 14–22 demonstrates the relationship between the two types of spectra. The four peaks shown in the frequency-domain spectrum for carbon-13 arise from the coupling

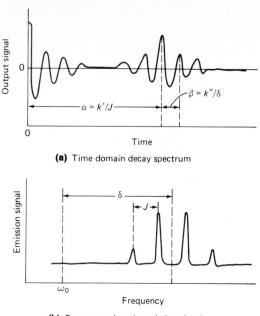

(a) Time domain decay spectrum

(b) Frequency domain emission signal

FIGURE 14-22   Spectra for $H_3^{13}CCl$: (**a**) free-induction decay spectrum; (**b**) conventional spectrum for ^{13}C showing spin-spin coupling between ^{13}C and protons. (Adapted with permission from: T. C. Farrar, *Anal. Chem.,* **1970,** *42*(4), 110A. Copyright 1970 American Chemical Society.)

between the carbon nucleus and the three protons (p. 426). Figure 14–22**a** indicates the parts of the time-domain spectrum that contain information about the chemical shift, δ, and the coupling constant, J.

Frequency Range of a Pulse. Radio-frequency generators, when operated continuously, produce radiation of a single frequency. The NMR experiments we have just described, however, require a range of frequencies sufficiently great to excite nuclei with different resonance frequencies. Fortunately, a sufficiently short pulse of radiation, such as that shown in Figure 14–21**a**, provides a band of frequencies from a monochromatic source. The frequency range of this band is about $1/(4\tau)$ Hz, where τ is the length in seconds of each pulse. Thus, by employing a pulse of 1 μsec with a

15-MHz transmitter, a frequency range of 15 MHz ± 125 kHz would result. This production of *side bands* by pulsing can be understood by reference to Figure 4–4 (p. 94), where it is shown that a rectangular wave form is made up of a series of sine or cosine functions differing from one another by small frequency increments. Thus, a pulse generated by rapidly switching a radio-frequency oscillator from off to on and then to off will consist of an envelope of power having a shape somewhat similar to the positive half of the solid line in Figure 4–4**b**.

Instrument Configuration. The magnets and probes of Fourier transform NMR spectrometers are similar to those described for continuous instruments. In addition, however, a digital computer is required to control the pulses, accumulate the data, and transform the accumulated information for presentation on a cathode-ray tube or chart. Figure 14–23 is a block diagram showing the arrangement of the components in a typical instrument.

14C–3 SAMPLE HANDLING

Liquids. Until recently, high-resolution NMR studies have been restricted to samples that could be converted to a nonviscous liquid state. Most often, solutions of the sample (2 to 15%) are used although liquid samples can also be examined neat if sufficiently nonviscous.

The best solvents for proton NMR spectroscopy contain no protons; from this standpoint, carbon tetrachloride is ideal. The low solubility of many compounds in carbon tetrachloride limits its value, however, and a variety of deuterated solvents are used instead. Deuterated chloroform ($CDCl_3$) and deuterated benzene (C_6D_6) are commonly encountered.

Solids. The application of conventional NMR to crystalline or amorphous solids leads to broad (up to 20 kHz), featureless spectral bands, which provide little useful structural information. Recently, however, several new techniques have been devel-

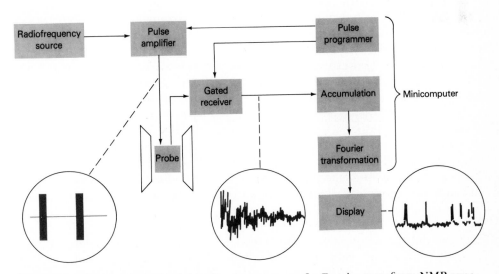

FIGURE 14-23 Block diagram showing the components of a Fourier transform NMR spectrometer. (Reprinted with permission from: G. A. Gray, *Anal. Chem.*, **1975**, *47*, 547A. Copyright 1975 American Chemical Society.)

oped that overcome this band broadening and make possible the acquisition of high-resolution spectra much like the spectra for liquids. The most frequent application of these techniques has been for obtaining ^{13}C spectra of polymers, fossil fuels, and other high-molecular weight substances. A brief discussion of the modifications necessary to produce useful solid NMR spectra will be found in the section devoted to ^{13}C NMR.

14C-4 COST OF NMR INSTRUMENTS

The foregoing brief discussion suggests that NMR instruments are complex. As might be expected, their cost is high, ranging from about $6000 to $300,000 or more. The typical 60-Hz high-resolution instrument employed for routine work in many laboratories is priced from approximately $30,000 to $40,000.

14D Applications of Proton NMR

Unquestionably, the most important applications of proton NMR spectroscopy have been to the identification and structural elucidation of organic, metal-organic, and biochemical molecules. In addition, however, the method often proves useful for quantitative determination of absorbing species.

14D-1 IDENTIFICATION OF COMPOUNDS

An NMR spectrum, like an infrared spectrum, seldom suffices by itself for the identification of an organic compound. However, in conjunction with other observations such as elemental analysis, as well as ultraviolet, infrared, and mass spectra, NMR is a major tool for the characterization of pure compounds. The simple examples that follow give some idea of the kinds of information that can be extracted from NMR spectra.

EXAMPLE 14-4. The NMR spectrum shown in Figure 14–24 is for an organic compound having the empirical formula $C_5H_{10}O_2$. Identify the compound.

An examination of the spectrum suggests the presence of four types of protons. From the integral plot and the empirical formula we deduce that these four types are populated by 3, 2, 2, and 3 protons respectively. The single peak at $\delta = 3.6$ must be due to an isolated, methyl group; upon inspection of Figure 14–11 and Table 14–3, the functional group $CH_3OC(=O)—$ is suggested. The empirical formula and the $2:2:3$ distribution of the remaining protons indicate the presence of an *n*-propyl group as well. The structure $CH_3OC(=O)CH_2CH_2CH_3$ is consistent with all of these observations. In addition, the positions and the splitting patterns of the three remaining peaks are entirely compatible with this hypothesis. The triplet at $\delta = 0.9$ is typical of a methyl group adjacent to a methylene. From Table 14–3, the two protons of the methylene adjacent to the carboxylate peak should yield the observed triplet peak at about $\delta = 2.2$. The other methylene group would be expected to produce a pattern of 12 peaks (3×4) at about $\delta = 1.7$. Only six are observed, presumably because the resolution of the instrument is insufficient.

EXAMPLE 14-5. The spectra shown in Figure 14–25 are for colorless, isomeric, liquids containing only carbon and hydrogen. Identify the two compounds.

The single peak at about $\delta = 7.2$ in the upper figure suggests an aromatic structure; the relative area of this peak corresponds to 5 protons; from this we conclude that we may have a monosubstituted derivative of benzene. The seven peaks for the single proton appearing at $\delta = 2.9$ and the six-proton doublet

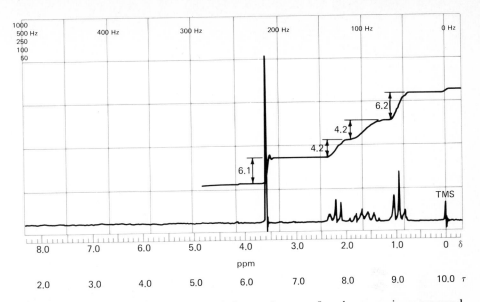

FIGURE 14-24 NMR spectrum and peak integral curve for the organic compound $C_5H_{10}O_2$ in CCl_4. (From R. M. Silverstein, G. C. Bassler, and T. C. Morrill, *Spectrometric Identification of Organic Compounds*, 3d ed., p. 296, Wiley: New York, 1974. Reprinted by permission of John Wiley & Sons, Inc.)

at $\delta = 1.2$ can only be explained by the structure

Thus, we conclude that this compound is cumene.

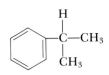

The isomeric compound has an aromatic peak at $\delta = 6.8$; its relative area suggests a trisubstituted benzene, which can only mean that the compound is $C_6H_3(CH_3)_3$. The relative peak areas confirm this diagnosis. We cannot, however, decide which of the three trimethylbenzene derivatives we have from the NMR data.

EXAMPLE 14-6. The spectrum shown in Figure 14–26 is for an organic compound having a molecular weight of 72 and containing carbon, hydrogen, and oxygen only. Identify the compound.

The triplet peak at $\delta = 9.8$ appears (Figure 14–11) to be that of an aliphatic aldehyde, RCHO. If this hypothesis is true, R has a molecular weight of 43, which corresponds to a C_3H_7 fragment. The triplet nature of the peak at $\delta = 9.8$ requires that there be a methylene group adjacent to the carbonyl. Thus, the compound would appear to be

n-butyraldehyde, $CH_3CH_2CH_2CHO$

The triplet peak at $\delta = 0.97$ appears to be

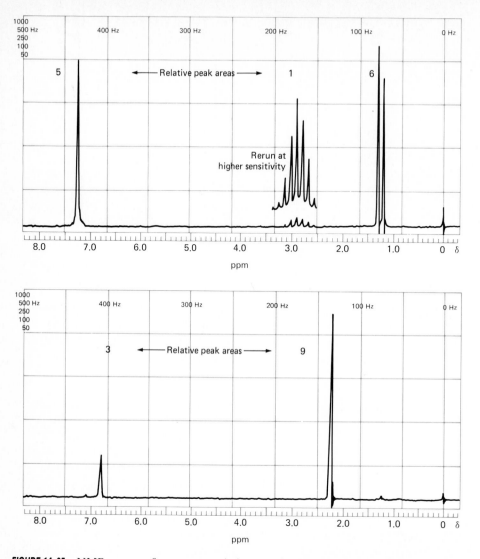

FIGURE 14-25   NMR spectra for two organic isomers in CDCl₃ solution. (Courtesy of Varian Instrument Division, Palo Alto, CA.)

that of the terminal methyl. The protons on the adjacent methylene would be expected to show a complicated splitting pattern of 12 peaks (4 × 3); the grouping of peaks around δ = 1.7 is compatible with this prediction. Finally, the peak for the protons on the methylene group adjacent to the carbonyl should appear as a sextet downfield from the other meth-

ylene proton peaks. The group at δ = 2.4 is consistent with this conclusion.

14D-2   APPLICATION OF PROTON NMR TO QUANTITATIVE ANALYSIS

Quantitative Analysis.   A unique aspect of NMR spectra is the direct proportionality between peak areas and the number of nuclei re-

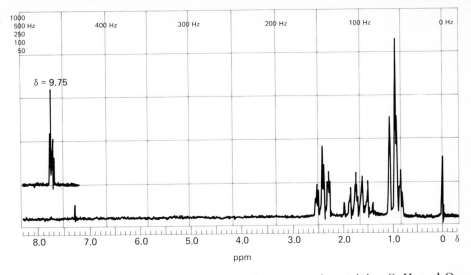

FIGURE 14-26 NMR spectrum of a pure organic compound containing C, H, and O only. (Courtesy of Varian Instrument Division, Palo Alto, CA.)

sponsible for the peak. As a consequence, a quantitative determination of a specific compound does not require pure samples for calibration. Thus, if an identifiable peak for one of the constituents of a sample does not overlap the peaks of the other constituents, the area of this peak can be employed to establish the concentration of the species directly, provided only that the signal area per proton is known. This latter parameter can be obtained conveniently from a known concentration of an internal standard. For example, if the solvent present in a known amount were benzene, cyclohexane, or water, the areas of the single proton peak for these compounds could be used to give the desired information; of course, the peak of the internal standard should not overlap with any of the sample peaks. Organic silicon derivatives are uniquely attractive for calibration purposes, owing to the high upfield location of their proton peaks.

The widespread use of NMR spectroscopy for quantitative work has been inhibited by the cost of the instruments. In addition, the probability that resonance peaks

will overlap becomes greater as the complexity of the sample increases. Often, too, analyses that are possible by the NMR method can be as conveniently accomplished by other techniques.

One of the main problems in quantitative NMR methods is the result of the saturation effect. As we have pointed out, the NMR absorption signal depends upon a very minute excess of nuclei in the lower magnetic energy state and the absorption process tends to depopulate this excess. Whether or not depopulation has a significant effect on the absorption intensity depends upon the relaxation time for the species, the power of the source, and the rate at which the spectrum is scanned. Errors arising from saturation can usually be avoided through control of these variables.

Analysis of Multicomponent Mixtures. Methods for the analysis of many multicomponent mixtures have been reported. For example, Hollis[12] has described a method for the determination of aspirin, phenacetin, and caf-

---

[12] D. P. Hollis, *Anal. Chem.*, **1963**, *35*, 1682.

feine in commercial analgesic preparations. The procedure requires about 20 min, and the relative errors are in the range of 1 to 3%. Chamberlain[13] describes a procedure for the rapid analysis of benzene, heptane, ethylene glycol, and water in mixtures. A wide range of these mixtures was analyzed with a precision of 0.5%.

An important quantitative application of the NMR technique has been to the determination of water in food products, pulp and paper, and agricultural materials. The water in these substances is sufficiently mobile to give a narrow peak suitable for quantitative measurement.[14]

Elemental Analysis. NMR spectroscopy can be employed to determine the total concentration of a given kind of magnetic nucleus in a sample. For example, Jungnickel and Forbes[15] have investigated the integrated NMR intensities of the proton peaks for numerous organic compounds and have concluded that accurate quantitative determinations of total hydrogen in organic mixtures is possible. Paulsen and Cooke[16] have shown that the resonance of fluorine-19 can be used for the quantitative analysis of that element in an organic compound— an analysis that is difficult to carry out by classical methods. For quantitative work, a low-resolution or wide-line spectrometer can be employed.

14E Application of NMR to Isotopes Other Than the Proton

Table 14–2 lists several nuclei which, in addition to the proton, have magnetic moments and can thus be studied by the mag-

netic resonance technique. More than two hundred other isotopes also possess magnetic moments[17]; the spectra for several of these have proven useful for analytical and structural applications.

14E-1 CARBON-13 NMR[18]

Instrumentation. Carbon-13 nuclear magnetic resonance was first studied in 1957, but its widespread use did not begin until the early seventies. The reason for this delay was the time required for the development of instruments sensitive enough to detect the weak NMR signals from the ^{13}C nucleus. This low signal strength is directly related to the low natural isotopic abundance of the isotope (1.1%) and the small magnetogyro ratio, which is about 0.25 that of the proton. These factors combine to make ^{13}C NMR about 6000 times less sensitive than proton NMR.

The most important developments in NMR signal enhancement, which led directly to the explosive growth of ^{13}C magnetic resonance spectroscopy, include higher field strength magnets and Fourier transform instruments. Without these developments, the technique would be restricted to the study of highly soluble low molecular weight solids, neat liquids, and isotopically enriched compounds.

Advantages of Carbon-13 NMR. Carbon-13 NMR has several advantages over proton NMR in

[13] N. F. Chamberlain, in *Treatise on Analytical Chemistry*, I. M. Kolthoff and P. J. Elving, Eds., Part I, vol. 4, p. 1932, Interscience: New York, 1963.

[14] T. M. Shaw and R. H. Elsken, *J. Chem. Phys.*, **1950**, *18*, 1113; *J. Appl. Physics*, **1955**, *26*, 313; and T. M. Shaw, R. H. Elsken, and C. H. Kunsman, *J. Assoc. Offic. Agr. Chemists*, **1953**, *36*, 1070.

[15] J. L. Jungnickel and J. W. Forbes, *Anal. Chem.*, **1963**, *35*, 938.

[16] P. J. Paulsen and W. D. Cooke, *Anal. Chem.*, **1964**, *36*, 1721.

[17] See, for example: E. D. Becker, *High Resolution NMR*, pp. 281–291, Academic Press: New York, 1980.

[18] For a thorough discussion of carbon-13 NMR spectroscopy, see: A. Lombardo and G. C. Levy, in *Treatise on Analytical Chemistry*, 2d ed., P. J. Elving, M. M. Bursey, and I. M. Kolthoff, Eds., Part I, vol. 10, Chapter 2, Wiley: New York, 1983; E. Breitmaier and W. Voelters, *^{13}C NMR Spectroscopy*, 2d ed., Verlag Chemie: New York, 1978; R. J. Abraham and P. Loftus, *Proton and Carbon-13 NMR Spectroscopy*, Heyden: Philadelphia, 1978.

terms of its power to elucidate organic and biochemical structures. First, there is the obvious advantage that ^{13}C NMR provides information about the backbone of molecules rather than about the periphery. In addition, the chemical shifts for ^{13}C in a majority of organic compounds is about 200 ppm, compared with approximately 10 to 15 ppm for the proton; less overlap of peaks is the consequence. Thus, for example, it is often possible to observe individual resonance peaks for each carbon atom in compounds ranging in molecular weight from 200 to 400. Also, homonuclear, spin-spin coupling between carbon atoms is not encountered because in unenriched samples, the probability of two ^{13}C atoms occurring in the same molecule is vanishingly small. Furthermore, heteronuclear spin coupling between ^{13}C and ^{12}C does not occur because the spin quantum number of the latter is zero. Finally, good methods exist for decoupling the interaction between ^{13}C atoms and protons. Thus, generally, the spectrum for a particular type of carbon consists of but a single line.

Spectra for Liquids. Figure 14–27**a** is a proton-decoupled ^{13}C spectrum for a solution of a simple six-carbon organic compound. Here, decoupling was accomplished by irradiation with a beam having a frequency that was centered in the absorption frequency region for the protons present; this beam was broad enough to saturate all of the protons in the sample, thus decoupling them from the carbon atoms. Figure 14–27**b** shows the spectrum obtained without decoupling. Clearly, the presence of protons results in extensive and complex splitting of the ^{13}C peaks.

Figure 14–28 shows some of the chemical shifts which are observed for ^{13}C in various

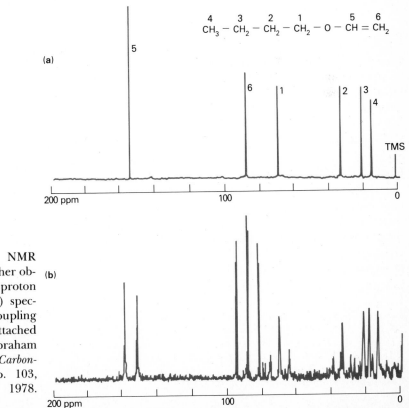

FIGURE 14-27 Carbon-13 NMR spectra for *n*-butylvinylether obtained at 25.2 MHz: (**a**) proton decoupled spectrum; (**b**) spectrum showing effect of coupling between ^{13}C atom and attached protons. (From: R. J. Abraham and P. Loftus, *Proton and Carbon-13 NMR Spectroscopy*, p. 103, Heyden: Philadelphia, 1978. With permission.)

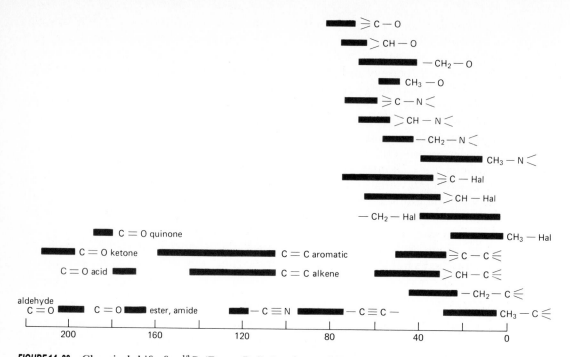

FIGURE 14-28 Chemical shifts for ^{13}C. (From: D. E. Leyden and R. H. Cox, *Analytical Applications of NMR*, p. 196, Wiley: New York, 1977. Reprinted by permission of John Wiley & Sons, Inc.)

chemical environments. As with proton spectra, these shifts are relative to tetramethylsilane and are useful for structural elucidation.

Spectra for Solids.[19] As was noted earlier, NMR spectra for solids have in the past not been very useful for structural studies because of line-broadening, which eliminates or obscures the characteristic sharp individual peaks of NMR. Much of this broadening is attributable to static dipolar interactions, which in liquids average to zero because of the rapid and random motion of molecules. In solids, heteronuclear dipolar interactions between magnetic nuclei, such as ^{13}C and protons, result in characteristic dipolar line

splittings, which depend upon the angle between C—H bonds and the external field. In an amorphous solid, a large number of fixed orientations of these bonds exists and hence a large number of splittings can occur. The broad absorption bands in this instance are made up of the numerous lines arising from these individual dipolar interactions. It is possible to remove dipolar splitting from a ^{13}C spectrum by irradiating the sample at proton frequencies while carrying out the ^{13}C scan. This procedure, called *dipolar decoupling*, is similar to spin decoupling, which was described earlier for liquids, except that a much higher power level is required.

A second type of line broadening for solids is caused by a phenomenon known as *chemical shift anisotropy*. The broadening produced here results from changes in the chemical shift with the orientation of the molecule or part of the molecule with re-

---

[19] See: G. C. Gerstein, *Anal. Chem.*, **1983**, *55*, 781A, 899A; F. P. Miknis, V. J. Bartuska, and G. E. Maciel, *Amer. Lab.*, **1979**, *11* (11), 19; J. Schaefer and E. O. Stejskal, in *Topics in Carbon-13 NMR Spectroscopy*, G. C. Levy, Ed., vol. 3, Wiley: New York, 1979.

spect to an external magnetic field. From both theory and experiment, it has been found possible to remove chemical shift anisotropy by spinning the sample rapidly at an angle of exactly 54.7 deg—a technique called *magic angle spinning*. In effect, the solid acts like a liquid when spun at the magic angle.

One further limitation in ^{13}C Fourier transform NMR is the long spin-lattice relaxation time for excited ^{13}C nuclei in solids. The rate at which the sample can be pulsed is dependent upon the relaxation rate. That is, after each excitation pulse, enough time must elapse for the nuclei to return to the equilibrium ground state. Unfortunately in solids, spin-lattice relaxation times for ^{13}C nuclei are often several minutes; thus, signal averaging sufficient to give a good spectrum would often require several hours or even days.

The problem caused by slow spin-lattice relaxation times is overcome by *cross polarization*, a complicated pulsed technique, which will not be described here.[20]

Instruments are now available commercially which incorporate dipolar decoupling, magic angle spinning, and cross polarization, thus making possible the acquisition of ^{13}C spectrum from solids; other nuclei in solids can also be examined by these techniques.

14E-2 FLUORINE-19 NMR

Fluorine-19 has a spin quantum number of $\frac{1}{2}$ and a magnetic moment of 2.6285 nuclear magnetons. Thus, the resonance frequency of fluorine in similar fields is only slightly lower than that of the proton (56.4 MHz, as compared with 60.0 MHz at 14,092 G). Therefore, with relatively minor changes, a proton NMR spectrometer can be adapted to the study of fluorine resonance.

It is found experimentally that fluorine absorption is also sensitive to the environment; the resulting chemical shifts, however, extend over a range of about 300 ppm compared with a maximum of 20 ppm for the proton. In addition, the solvent plays a much more important role in determining fluorine peak positions than with the proton.

Empirical correlations of the fluorine shift with structure are relatively sparse when compared with information concerning proton behavior. It seems probable, however, that the future will see further developments in this field, particularly for structural investigation of organic fluorine compounds.

14E-3 PHOSPHORUS-31 NMR

Phosphorus-31, with spin number $\frac{1}{2}$, also exhibits sharp NMR peaks with chemical shifts extending over a range of 700 ppm. The resonance frequency of ^{31}P at 14,092 G is 24.3 MHz. Numerous investigations, particularly in the biochemical field, which correlate the chemical shift of the phosphorus nucleus with structure, have been reported.

14E-4 OTHER NUCLEI

A growing body of data has accumulated in the last few years dealing with applications of NMR to a wide variety of isotopes. Among the more widely studied nuclei are ^2D, ^{11}B, ^{23}Na, ^{15}N, ^{29}Si, ^{109}Ag, ^{199}Hg, ^{113}Cd, and ^{207}Pb.

14F Electron Spin Resonance Spectroscopy

Electron spin resonance spectroscopy or ESR (also called electron paramagnetic resonance spectroscopy or EPR) is based upon the absorption of *microwave* radiation by an unpaired electron when it is exposed to a

---

[20] See: A. Pines, M. G. Gibby, and J. S. Waugh, *J. Chem. Phys.*, **1973**, *59*, 569.

strong magnetic field.[21] Species that contain unpaired electrons, and can therefore be detected by ESR spectroscopy, include free radicals, odd-electron molecules, transition-metal complexes, rare-earth ions, and triplet-state molecules.

14F-1 PRINCIPLES OF ESR

The principles of ESR spectroscopy are similar to those of NMR spectroscopy, which were discussed in the previous sections. The electron, like the proton, has a spin quantum number of one-half and thus has two energy levels that differ slightly in energy under the influence of a strong magnetic field. In contrast to the proton, however, the lower energy level corresponds to $m = -\frac{1}{2}$ and the higher to $m = +\frac{1}{2}$; this difference results from the negative charge of the electron.

Applying Equations 14–4 and 14–5 to an unpaired electron yields

$$\Delta E = h\nu = \mu\beta_N \frac{H_0}{I} = g\beta_N H_0 \qquad (14\text{–}18)$$

where g is the *splitting factor* and β_N is the *Bohr magneton*, which has a value of 9.27×10^{-21} erg G^{-1}. The value for g varies with the electron's environment. For a free electron, its value is 2.0023; for an unpaired electron in a molecule or ion, g lies within a few percent of this number.

ESR spectrometers often employ a field of 3400 G. Substituting this value into Equation 14–18 gives

$$\nu = \frac{g\beta_N H_0}{h} = \frac{2.0 \times 9.27 \times 10^{-21} \times 3400}{6.63 \times 10^{-27}}$$

$$= 9.51 \times 10^9 \text{ Hz} \quad \text{or} \quad 9500 \text{ MHz}$$

---

[21] For further details on ESR spectroscopy, see: M. C. R. Symons, *Chemical and Biochemical Aspects of Electron-Spin Resonance Spectroscopy*, Wiley: New York, 1978; I. B. Goldberg and A. J. Bard, in *Treatise on Analytical Chemistry*, 2d ed., P. J. Elving, M. M. Bursey, and I. M. Kolthoff, Eds., Part I, vol. 10, Chapter 3, Wiley: New York, 1983.

Thus, the resonance frequency for an unpaired electron is about 9500 MHz, which lies in the microwave region (see Figure 4–5, p. 95).

14F-2 INSTRUMENTATION

The source of microwave radiation is a klystron tube, which is operated to produce monochromatic radiation having a frequency of about 9500 MHz. The klystron tube is an electronic oscillator in which a beam of electrons is pulsed between a cathode and a reflector. The oscillating output of the klystron is transmitted to a *wave guide* by a loop of wire, which sets up a fluctuating magnetic field (electromagnetic radiation) in the guide. The wave guide, which is a rectangular metal tube, transmits the microwave radiation to the sample, which is generally held in a small quartz tube positioned between the poles of the permanent magnet. Helmholtz coils provide a means for varying the field over the small range in which resonance occurs.

Generally, ESR spectra are recorded in derivative form to enhance sensitivity and resolution. Figure 14–29 contrasts two derivative spectra with their ordinary absorption counterparts.

14F-3 ESR SPECTRA

Most molecules fail to exhibit an ESR spectrum because they contain an even number of electrons; thus, the number in the two spin states is identical (that is, the spins are paired) and the magnetic effects of electron spin are canceled. Such substances are diamagnetic because of the small fields induced from the *orbital precession* of the electrons around the nuclei; these fields act in opposition to the applied field. In contrast, the spin of an unpaired electron induces a field that reinforces the applied field. This paramagnetic effect is much larger than the diamagnetic behavior. Splitting of the energy levels of an electron occurs in a mag-

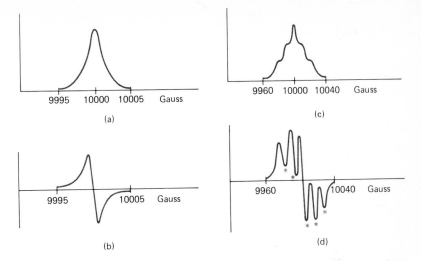

FIGURE 14-29 Comparison of spectral presentations as absorption (**a** and **c**) and the corresponding derivative (**b** and **d**) curves. Note that the shoulders in (**c**) never reach a maximum; consequently the corresponding derivative peaks do not pass through the abscissa. The number of peaks can be obtained by a count of the number of maxima or minima (as shown by the asterisks). (From R. S. Drago, *Physical Methods in Chemistry*, 2d ed., p. 323, W. B. Saunders Co.: Philadelphia, 1977. With permission.)

netic field; this splitting can be observed, as noted earlier, by microwave absorption studies.

The spin of an unpaired electron can couple with the spin of nuclei in the species to give splitting patterns analogous to those observed for nuclear spin-spin coupling. When an electron interacts with n equivalent nuclei, its resonance peak is split into $(2nI + 1)$ peaks, where I is the spin quantum number of the nuclei and n is the number of equivalent nuclei. This process is termed *hyperfine splitting*. Figure 14–30 shows the spectrum that results from hyperfine splitting that occurs in the simplest of all free radicals, the hydrogen atom. Here, the spin of the hydrogen nucleus couples with that of the unpaired electron to produce $(2 \times \frac{1}{2} \times 1 + 1)$ or 2 peaks.

14F-4 APPLICATIONS OF ESR[22]

Figure 14–31 illustrates the application of ESR to the detection of a free radical inter-

mediate in a chemical process. This spectrum was obtained after the addition of base to a solution of hydroquinone in the presence of air. Oxidation of the anion of hydroquinone to give quinone occurs. That is,

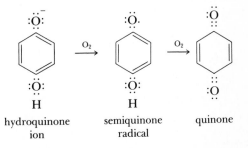

| hydroquinone ion | semiquinone radical | quinone |

Hydroquinone and quinone are not ESR active because neither contains an odd electron. The postulated intermediate does,

[22] For an excellent summary of analytical applications of ESR, see: I. B. Goldberg and A. J. Bard, in *Treatise on Analytical Chemistry*, 2d ed., P. J. Elving, M. M. Bursey, and I. M. Kolthoff, Eds., Part I, vol. 10, pp. 268–285, Wiley: New York, 1983.

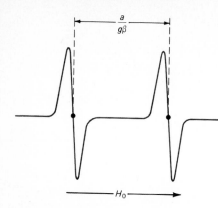

FIGURE 14-30   The ESR spectrum for the hydrogen atom. The hyperfine splitting is $a/g\beta$ gauss. (From R. S. Drago, *Physical Methods in Chemistry*, 2d ed., p. 324, W. B. Saunders Co.: Philadelphia, 1977. With permission.)

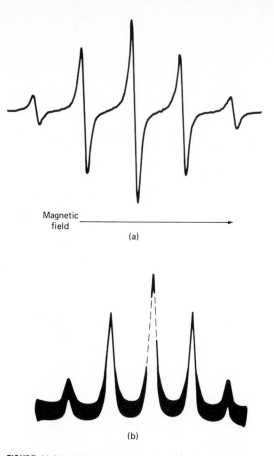

FIGURE 14-31   The ESR spectrum for the semi-quinone radical: (**a**) derivative spectrum; (**b**) absorption spectrum. (Reprinted with permission from: B. Venkataraman and G. K. Fraenkel, *J. Amer. Chem. Soc.*, **1955,** *77,* 2711. Copyright 1955 American Chemical Society.)

however, and the five-peak hyperfine pattern is consistent with a semiquinone radical. Here, the four ring protons are assumed to be equivalent with respect to the interactions of their spins with that of the odd electron; five peaks result. This picture is reasonable since the position of the odd electron is not confined to the oxygen atom as shown. Indeed, several resonance structures can be written in which the odd electron is associated with any of several other atoms in the molecule.

Electron spin has been widely applied to the study of chemical, photochemical, and electrochemical reactions, which proceed via free radical mechanisms. Often, as in the foregoing example, the technique has provided useful structural information about intermediate radicals.

ESR studies are also used to obtain structural information about the transition metals and their complexes. In addition, they have been used for the detection and semiquantitative determination of paramagnetic ions in biological systems in concentrations as low as a few parts per billion.

Another important biological application involves the use of *spin-label reagents*. These compounds are stable molecules that contain an odd electron and, in addition, react selectively with amino acids or other functional groups in a biological system. The ESR spectra of the product then provide information about the environment, such as structural features, polarity, viscosity, phase changes, and chemical reactivity.

PROBLEMS

14–1. Predict the appearance of the high-resolution NMR spectrum of propionic acid.

14–2. Predict the appearance of the high-resolution NMR spectrum of
(a) acetaldehyde.
(b) acetic acid.
(c) ethyl nitrite.

14–3. Predict the appearance of the high-resolution NMR spectrum of
(a) acetone.
(b) methyl ethyl ketone.
(c) methyl *i*-propyl ketone.

14–4. Predict the appearance of the high-resolution NMR spectrum of
(a) cyclohexane.
(b) 1,2-dimethoxyethane, $CH_3OCH_2CH_2OCH_3$.
(c) diethylether.

14–5. Predict the appearance of the high-resolution NMR spectrum of
(a) toluene.
(b) ethyl benzene.
(c) *i*-butane.

14–6. The spectrum in Figure 14–32 is for an organic compound containing a single atom of bromine. Identify the compound.

14–7. The spectrum in Figure 14–33 is for a compound having an empirical formula $C_4H_7BrO_2$. Identify the compound.

14–8. The spectrum in Figure 14–34 is for a compound of empirical formula C_4H_8O. Identify the compound.

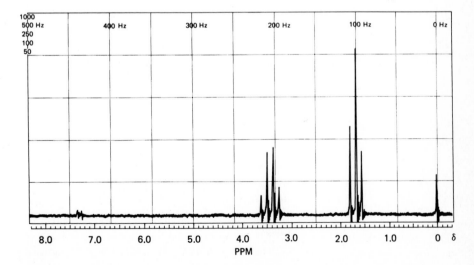

FIGURE 14-32   (Courtesy of Varian Instrument Division, Palo Alto, CA.) See Problem 14–6.

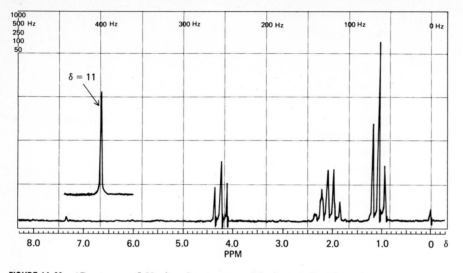

FIGURE 14-33   (Courtesy of Varian Instrument Divsion, Palo Alto, CA.) See Problem 14–7.

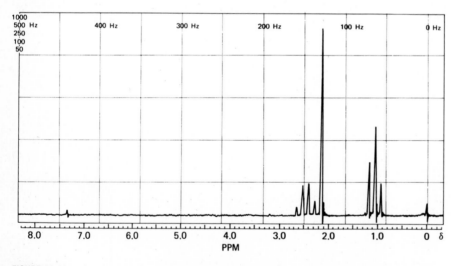

FIGURE 14-34   (Courtesy of Varian Instrument Division, Palo Alto, CA.) See Problem 14–8.

14–9. The spectrum in Figure 14–35 is for a compound having an empirical formula $C_4H_8O_2$. Identify the compound.

14–10. The spectra in Figure 14–36**a** and 14–36**b** are for compounds with empirical formulas C_8H_{10}. Identify the compounds.

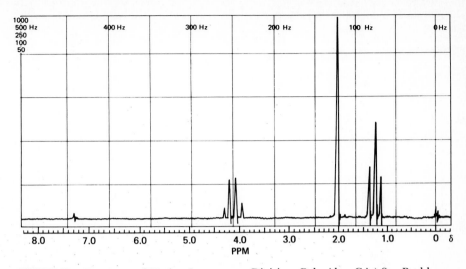

FIGURE 14-35 (Courtesy of Varian Instrument Division, Palo Alto, CA.) See Problem 14–9.

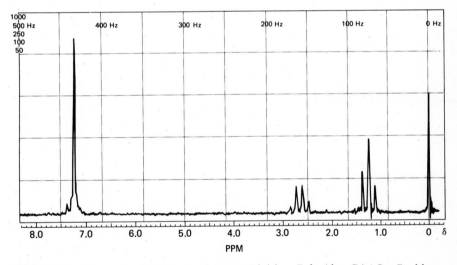

FIGURE 14-36a (Courtesy of Varian Instrument Division, Palo Alto, CA.) See Problem 14–10.

14–11. From the spectrum given in Figure 14–37, deduce the structure of this hydrocarbon.

14–12. From the spectrum given in Figure 14–38, determine the structure of this compound, which is a commonly used pain killer; its empirical formula is $C_{10}H_{13}NO_2$.

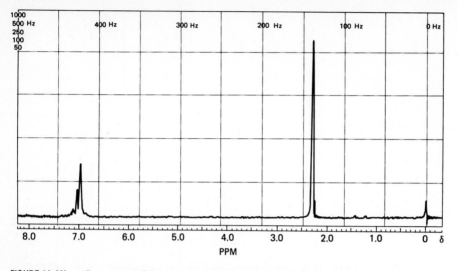

FIGURE 14–36b (Courtesy of Varian Instrument Division, Palo Alto, CA.) See Problem 14–10.

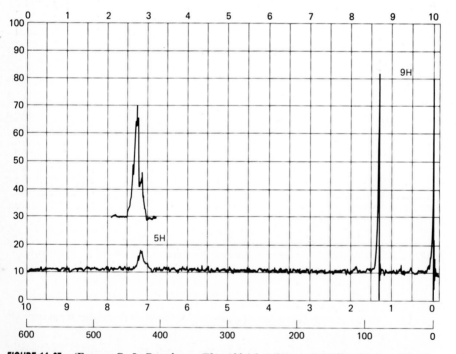

FIGURE 14–37 (From: C. J. Pouchert, *The Aldrich Library of NMR Spectra*, 2d ed., The Aldrich Chemical Company: Milwaukee, WI. With permission.) See Problem 14–11.

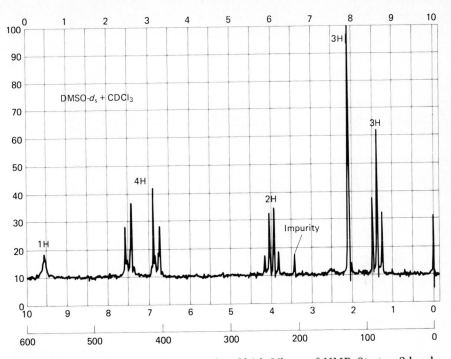

FIGURE 14-38 (From: C. J. Pouchert, *The Aldrich Library of NMR Spectra*, 2d. ed., The Aldrich Chemical Company: Milwaukee, WI. With permission.) See Problem 14–12.

X-Ray Spectroscopy

X-Ray spectroscopy, like optical spectroscopy, is based upon measurement of emission, absorption, scattering, fluorescence, and diffraction of electromagnetic radiation. Such measurements provide much useful information about the composition and structure of matter.[1]

15A Fundamental Principles

X-Rays are defined as short wavelength electromagnetic radiation produced by the deceleration of high-energy electrons or by electronic transitions involving electrons in the inner orbitals of atoms. The wavelength range of X-rays is from perhaps 10^{-5} Å to about 100 Å; conventional X-ray spectroscopy is, however, largely confined to the region of approximately 0.1 Å to 25 Å (1 Å = 0.1 nm = 10^{-10} m).

15A-1 EMISSION OF X-RAYS

For analytical purposes, X-rays are obtained in three ways, namely: (1) by bombardment of a metal target with a beam of high-energy electrons, (2) by exposure of a substance to a primary beam of X-rays in order to generate a secondary beam of fluorescent X-rays, and (3) by employment of a radioactive source whose decay process results in X-ray emission.

X-Ray sources, like ultraviolet and visible emitters, often produce both a continuous and a discontinuous (line) spectrum; both types are of importance in analysis. Continuous radiation is also called *white radiation* or *Bremsstrahlung* (the latter meaning radiation

---

[1] For a more extensive discussion of the theory and analytical applications of X-rays, see: E. P. Bertin, *Introduction to X-Ray Spectrometric Analysis*, Plenum Press: New York, 1978; R. Tertian and F. Claisse, *Principles of Quantitative X-Ray Fluorescence Analysis*, Heyden: London, 1982; H. A. Liebhafsky, H. G. Pfeiffer, E. H. Winslow, and P. D. Zemany, *X-Rays, Electrons and Analytical Chemistry*, Wiley-Interscience: New York, 1972; and R. Jenkins, R. W. Gould, and D. Gedcke, *Quantitative X-Ray Spectrometry*, Marcel Dekker: New York, 1981.

that arises from retardation by particles; such radiation is generally continuous).

Continuous Spectra from Electron Beam Sources. In an X-ray tube, electrons produced at a heated cathode are accelerated toward an anode (the *target*) by a potential as great as 100 kV; upon collision, part of the energy of the electron beam is converted to X-rays. Under some conditions, only a continuous spectrum such as that shown in Figure 15–1 results; under others, a line spectrum is superimposed upon the continuum (see Figure 15–2).

The continuous X-ray spectrum shown in the two figures is characterized by a well-defined, short-wavelength limit (λ_0), which is dependent upon the accelerating voltage V but independent of the target material. Thus, λ_0 for the spectrum produced with a molybdenum target at 35 kV (Figure 15–2) is identical to λ_0 for a tungsten target at the same voltage (Figure 15–1).

The continuous radiation from an electron beam source results from collisions between the electrons of the beam and the atoms of the target material. At each collision,

the electron is decelerated and a photon of X-ray energy is produced. The energy of the photon will be equal to the difference in kinetic energies of the electron before and after the collision. Generally, the electrons in a beam are decelerated in a series of collisions; the resulting loss of kinetic energy differs from collision to collision. Thus, the energies of the emitted X-ray photons vary continuously over a considerable range. The maximum photon energy generated corresponds to the instantaneous deceleration of the electron to zero kinetic energy in a single collision. For such an event, we may write

$$h\nu_0 = \frac{hc}{\lambda_0} = Ve \qquad (15\text{–}1)$$

where Ve, the product of the accelerating voltage and the charge on the electron, is the kinetic energy of all of the electrons in the beam, h is Planck's constant, and c is the velocity of light. The quantity ν_0 is the maximum frequency of radiation that can be produced at voltage V, while λ_0 is the low-wavelength limit for the radiation. This relationship is known as the *Duane-Hunt law*.

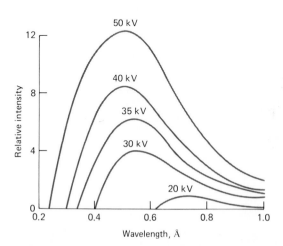

FIGURE 15-1 Distribution of continuous radiation from an X-ray tube with a tungsten target. The numbers above the curves indicate the accelerating voltages.

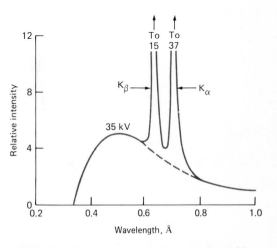

FIGURE 15-2 Line spectrum for a tube with a molybdenum target.

Upon substituting numerical values for the constants and rearranging, Equation 15–1 becomes

$$\lambda_0 = 12{,}398/V \qquad (15\text{–}2)$$

where λ_0 and V have units of ångströms and volts, respectively. It is of interest to note that Equation 15–1 has provided a direct means for the highly accurate determination of Planck's constant.

Characteristic Line Spectra from Electron Beam Sources.   As shown in Figure 15–2, bombardment of a molybdenum target produces intense emission lines at about 0.63 and 0.71 Å; an additional simple series of lines occurs in the longer wavelength range of 4 to 6 Å.

The emission behavior of molybdenum is typical of all elements having atomic numbers larger than 23; that is, the X-ray line spectra are remarkably simple when compared with ultraviolet emission and consist of two series of lines. The shorter wavelength group is called the K series and the other the L series.[2] Elements with atomic numbers smaller than 23 produce only a K series. Table 15–1 presents wavelength data for the emission spectra of a few elements.

A second characteristic of X-ray spectra is that the minimum acceleration voltage required for the excitation of the lines for

each element increases with atomic number. Thus, the line spectrum for molybdenum (atomic number = 42) disappears if the excitation voltage drops below 20 kV. As shown in Figure 15–1, bombardment of tungsten (atomic number = 74) produces no lines in the region of 0.1 to 1.0 Å even at 50 kV. Characteristic K lines appear at 0.18 and 0.21 Å, however, if the voltage is raised to 70 kV.

Figure 15–3 illustrates the linear relationship between the square root of the frequency for a given (K or L) line and the atomic number of the element responsible for the radiation. This property was first discovered by H. G. S. Moseley in 1914.

X-Ray line spectra result from electronic transitions that involve the innermost atomic orbitals. The short-wavelength K series is produced when the high-energy electrons from the cathode remove electrons from those orbitals nearest to the nucleus of

---

[2] For the heavier elements, additional series of lines (M, N, and so forth) are found at longer wavelengths. Their intensities are low, however, and little use is made of them.

The designations K and L arose from the German words kurtz and lang for short and long wavelengths. The additional alphabetical designations were then added for lines occurring at progressively longer wavelengths.

Table 15–1
Wavelengths in Ångström Units of the More Intense Emission Lines For Some Typical Elements

Element	Atomic Number	K Series		L Series	
		α_1	β_1	α_1	β_1
Na	11	11.909	11.617	—	—
K	19	3.742	3.454	—	—
Cr	24	2.290	2.085	21.714	21.323
Rb	37	0.926	0.829	7.318	7.075
Cs	55	0.401	0.355	2.892	2.683
W	74	0.209	0.184	1.476	1.282
U	92	0.126	0.111	0.911	0.720

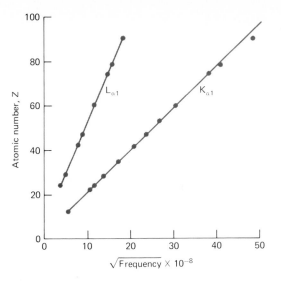

FIGURE 15-3 Relationship between X-ray emission frequency and atomic number ($K_{\alpha 1}$ and $L_{\alpha 1}$ lines).

highest resolution spectrometers (see Figure 15–2).

The energy level diagram in Figure 15–4 would be applicable to any element with sufficient electrons to permit the number of transitions shown. The differences in energies between the levels increase regularly with atomic number because of the increasing charge on the nucleus; therefore, the radiation for the K series appears at shorter wavelengths for the heavier elements (see Table 15–1). The effect of nuclear charge is also reflected in the increase in minimum voltage required to excite the spectra of these elements.

It is important to note that for all but the lightest elements, the wavelengths of characteristic X-ray lines are independent of chemical combination because the transitions responsible for these lines involve electrons that take no part in bonding. Thus, the position of the K_α lines for molybdenum is the same regardless of whether the target is the pure metal, its sulfide, or its oxide.

Fluorescent Line Spectra. Another convenient way of producing a line spectrum is to irradiate the element or one of its compounds with the continuous radiation from an X-ray tube. This process is considered further in a later section.

Radioactive Sources. X-Radiation is often a product of radioactive decay processes. *Gamma rays*, which are indistinguishable from X-rays, owe their production to intranuclear reactions. Many α and β emission processes (see Section 17A–2) leave a nucleus in an excited state, which then releases one or more quanta of gamma rays as it returns to its ground state. *Electron capture* or *K capture* also produces X-radiation. This process involves capture of a K electron (less commonly, an L or an M electron) by the nucleus and formation of an element of the next lower atomic number. As a result of K capture, electronic transitions to the vacated orbital occur, and the X-ray line spectrum of the newly formed element is observed.

the target atom. The collision results in the formation of excited *ions,* which then lose quanta of X-radiation as electrons from outer orbitals undergo transitions to the vacated orbital. As shown in Figure 15–4, the lines in the K series involve electronic transitions between higher energy levels and the K shell. The L series of lines results when an electron is lost from the second principal quantum level, either as a consequence of ejection by an electron from the cathode or from the transition of an L electron to the K level that accompanies the production of a quantum of K radiation. It is important to appreciate that the energy scale in Figure 15–4 is logarithmic. Thus, the energy difference between the L and K levels is significantly larger than that between the M and L levels. The K lines therefore appear at shorter wavelengths. It is also important to note that the energy differences between the transitions labeled α_1 and α_2 as well as those between β_1 and β_2 are so small that only single lines are observed in all but the

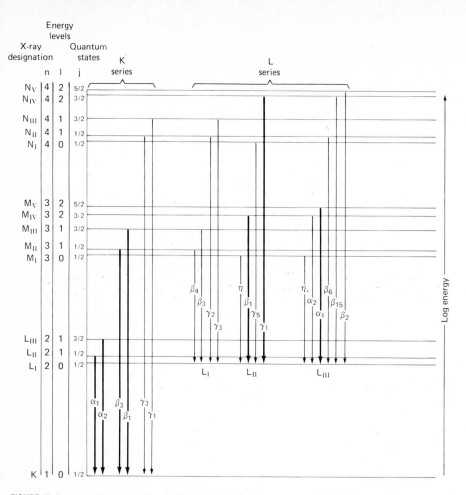

FIGURE 15–4 Partial energy level diagram showing common transitions leading to X-radiation. The most intense lines are indicated by the widest arrows.

The half-lives (p. 505) of K-capture processes range from a few minutes to several thousands of years.

Artificially produced radioactive isotopes provide a very simple source of mono-energetic radiation for certain analytical applications. The best known example is iron-55, which undergoes a K-capture reaction with a half-life of 2.6 years:

$$^{55}\text{Fe} \rightarrow {}^{54}\text{Mn} + h\nu$$

The resulting manganese K_α line at about 2.1 Å has proved to be a useful source for both fluorescence and absorption methods.

Table 15–2 lists some additional common radioisotope sources for X-ray spectroscopy.

15A–2 ABSORPTION OF X-RAYS

When a narrow beam of X-rays is passed through a thin layer of matter, its intensity or power is generally diminished as a consequence of absorption and scattering. The effect of scattering for all but the lightest

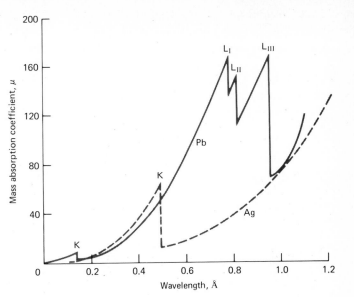

FIGURE 15-5 X-Ray absorption spectra for lead and silver.

elements is ordinarily small and can be neglected in those wavelength regions where appreciable absorption occurs. As shown in Figure 15–5, the absorption spectrum of an element, like its emission spectrum, is simple and consists of a few well-defined absorption peaks. Here again, the wavelengths of the peaks are characteristic of the element and are largely independent of its chemical state.

A peculiarity of X-ray absorption spectra is the appearance of sharp discontinuities, called *absorption edges*, at wavelengths immediately beyond absorption maxima.

Table 15–2
Common Radioisotopic Sources for X-Ray Spectroscopy

Source	Decay Process	Half-Life	Type of Radiation	Energy, keV
$^{3}_{1}$H-Ti[a]	β^-	12.3 years	Continuous	3–10
			Ti-K X-rays	4–5
$^{55}_{26}$Fe	EC[b]	2.7 years	Mn-K X-rays	5.9
$^{57}_{27}$Co	EC	270 days	Fe-K X-rays	6.4
			γ rays	14, 122, 136
$^{109}_{48}$Cd	EC	1.3 years	Ag-K X-rays	22
			γ rays	88
$^{125}_{53}$I	EC	60 days	Te-K X-rays	27
			γ rays	35
$^{147}_{61}$Pm-Al	β^-	2.6 years	Continuous	12–45
$^{210}_{82}$Pb	β^-	22 years	Bi-L X-rays	11
			γ rays	47

[a] Tritium adsorbed on nonradioactive titanium metal.
[b] Electron capture.

The Absorption Process.   Absorption of an X-ray quantum causes ejection of one of the innermost electrons from an atom and the consequent production of an excited ion. In this process, the entire energy $h\nu$ of the radiation is partitioned between the kinetic energy of the electron (the *photoelectron*) and the potential energy of the excited ion. The highest probability for absorption arises when the energy of the quantum is exactly equal to the energy required to remove the electron just to the periphery of the atom (that is, as the kinetic energy of the ejected electron approaches zero).

The absorption spectrum for lead, shown in Figure 15–5, exhibits four peaks, the first occurring at 0.14 Å. The energy of the quantum corresponding to this wavelength exactly matches the energy required to just eject the highest energy K electron of the element; immediately beyond this wavelength, the energy of the radiation is insufficient to bring about removal of a K electron, and an abrupt decrease in absorption occurs. At wavelengths lower than 0.14 Å, the probability of interaction between the electron and the radiation diminishes and results in a smooth decrease in absorption. In this region, the kinetic energy of the ejected photoelectron increases continuously with the decrease in wavelength.

The additional peaks at longer wavelengths correspond to the removal of an electron from the L energy levels of lead. Three sets of L levels, differing slightly in energy, exist (see Figure 15–4); three peaks are, therefore, observed. Another set of peaks, arising from ejections of M electrons, will be located at still longer wavelengths.

Figure 15–5 also shows the K absorption edge for silver, which occurs at 0.485 Å. The longer wavelength for the silver peak reflects the lower atomic number of the element compared with lead.

The Mass Absorption Coefficient.   Beer's law is as applicable to the absorption of X-radiation as to other types of electromagnetic radiation; thus, we may write

$$\ln \frac{P_0}{P} = \mu x$$

where x is the sample thickness in centimeters and P and P_0 are the powers of the transmitted and incident beams. The constant μ is called the *linear absorption coefficient* and is characteristic of the element as well as the number of its atoms in the path of the beam. A more convenient form of Beer's law is

$$\ln \frac{P_0}{P} = \mu_M \rho x \tag{15–3}$$

where ρ is the density of the sample and μ_M is the *mass absorption coefficient*, a quantity that is *independent* of the physical and chemical state of the element. Thus, the mass absorption coefficient for bromine has the same value in gaseous HBr as in solid sodium bromate. Note that the mass absorption coefficient carries units of cm^2/g.

Mass absorption coefficients are additive functions of the weight fractions of elements contained in a sample. Thus,

$$\mu_M = W_A \mu_A + W_B \mu_B + W_C \mu_C + \cdots \tag{15–4}$$

where μ_M is the mass absorption coefficient of a sample containing the weight fractions W_A, W_B, and W_C of elements A, B, and C. The terms μ_A, μ_B, and μ_C are the respective mass absorption coefficients for each of the elements. Tables for mass absorption coefficients for the elements at various wavelengths are found in various handbooks and monographs.[3]

15A–3 X-RAY FLUORESCENCE

The absorption of X-rays produces electronically excited ions that return to their ground state by transitions involving elec-

---

[3] For example, E. P. Bertin, *Principles and Practice of X-Ray Spectrometric Analysis*, 2d ed., pp. 972–976, Plenum Press: New York, 1975.

trons from higher energy levels. Thus, an excited ion with a vacant K shell is produced when lead absorbs radiation of wavelengths shorter than 0.14 Å (Figure 15–5); after a brief period, the ion returns to its ground state via a series of electronic transitions characterized by the emission of X-radiation (fluorescence) of wavelengths identical to those that result from excitation produced by electron bombardment. The wavelengths of the fluorescent lines are always somewhat greater than the wavelength of the corresponding absorption edge, however, because absorption requires a complete removal of the electron (that is, ionization), whereas emission involves transitions of an electron from a higher energy level within the atom. For example, the K absorption edge for silver occurs at 0.485 Å, while the K emission lines for the element have wavelengths at 0.497 and 0.559 Å. When fluorescence is to be excited by radiation from an X-ray tube, the operating voltage must be sufficiently great so that the cutoff wavelength λ_0 (Equation 15–2) is shorter than the absorption edge of the element whose spectrum is to be excited. Thus, to generate the K lines for silver, the tube voltage would need to be

$$V \geq \frac{12398}{0.485} = 25{,}560 \text{ V or } 25.6 \text{ kV}$$

15A-4 DIFFRACTION OF X-RAYS

In common with other types of electromagnetic radiation, interaction between the electric vector of X-radiation and the electrons of the matter through which it passes results in scattering. When X-rays are scattered by the ordered environment in a crystal, interference (both constructive and destructive) takes place among the scattered rays because the distances between the scattering centers are of the same order of magnitude as the wavelength of the radiation. Diffraction is the result.

Bragg's Law.   When an X-ray beam strikes a crystal surface at some angle θ, a portion is scattered by the layer of atoms at the surface. The unscattered portion of the beam penetrates to the second layer of atoms where again a fraction is scattered, and the remainder passes on to the third layer (Figure 15–6). The cumulative effect of this scattering from the regularly spaced centers of the crystal is a diffraction of the beam in much the same way as visible radiation is diffracted by a reflection grating (Section 5D–2). The requirements for X-ray diffraction are: (1) the spacing between layers of atoms must be roughly the same as the wavelength of the radiation and (2) the scattering centers must be spatially distributed in a higher regular way.

In 1912, W. L. Bragg treated the diffrac-

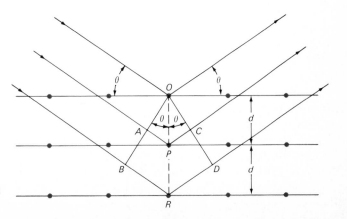

FIGURE 15-6 Diffraction of X-rays by a crystal.

tion of X-rays by crystals as shown in Figure 15–6. Here, a narrow beam strikes the crystal surface at angle θ; scattering occurs as a consequence of interaction of the radiation with atoms located at O, P, and R. If the distance

$$AP + PC = \mathbf{n}\lambda$$

where $\mathbf{n}$ is an integer, the scattered radiation will be in phase at OCD, and the crystal will appear to reflect the X-radiation. But it is readily seen that

$$AP = PC = d \sin \theta \qquad (15\text{–}5)$$

where d is the interplanar distance of the crystal. Thus, we may write that the conditions for constructive interference of the beam at angle θ are

$$\mathbf{n}\lambda = 2d \sin \theta \qquad (15\text{–}6)$$

Equation 15–6 is called the *Bragg equation* and is of fundamental importance. Note that X-rays appear to be reflected from the crystal only if the angle of incidence satisfies the condition that

$$\sin \theta = \frac{\mathbf{n}\lambda}{2d}$$

At all other angles, destructive interference occurs.

15B   Instrument Components

Absorption, emission, fluorescence, and diffraction of X-rays all find applications in analytical chemistry. Instruments for these applications contain components that are analogous in function to the five components of instruments for optical spectroscopic measurement; these components include a source, a device for restricting the wavelength range to be employed, a sample holder, a radiation detector or transducer, and a signal processor and readout. These components differ considerably in detail from their optical counterparts. Their functions, however, are the same, and the ways in which they are combined to form instruments are often similar to those shown in Figure 5–1 (p. 113).

As with optical instruments, both X-ray photometers and spectrophotometers are encountered, the first employing filters and the second monochromators for restricting radiation from the source. In addition, however, a third method is available for obtaining information about isolated portions of an X-ray spectrum. Here, isolation can be achieved electronically with devices that have the power to discriminate between various parts of a spectrum based on the energy rather than the wavelength of the radiation. Thus, X-ray instruments are often described as *wavelength dispersive instruments* or *energy dispersive instruments*, depending upon the method by which they resolve spectra.

15B–1   SOURCES

Three types of sources are encountered in X-ray instruments, namely, tubes, radioisotopes, and secondary fluorescent sources.

The X-Ray Tube.   The most common source of X-rays for analytical work is the X-ray tube (sometimes called a *Coolidge tube*), which can take a variety of shapes and forms; one design is shown schematically in Figure 15–7. Basically an X-ray source is a highly evacuated tube in which is mounted a tungsten filament cathode and a massive anode. The anode generally consists of a heavy, hollow, water-cooled block of copper with a metal target plated on or imbedded in the surface of the copper. Target materials include such metals as tungsten, chromium, copper, molybdenum, rhodium, silver, iron, and cobalt. Separate circuits are used to heat the filament and to accelerate the electrons to the

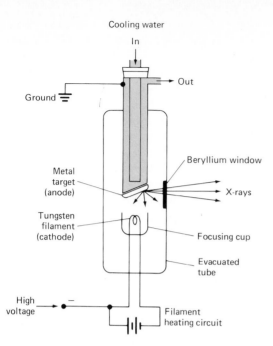

FIGURE 15-7 Schematic diagram of an X-ray tube.

target. The heater circuit provides the means for controlling the intensity of the emitted X-rays while the accelerating potential determines their energy or wavelength. An X-ray tube is normally self-rectifying, and a high-voltage ac source is connected directly to the cathode to provide the accelerating potential.

The production of X-rays by electron bombardment is a highly inefficient process. Less than one percent of the electrical power is converted to radiant power, the remainder being degraded to heat. As a consequence, water cooling of the anodes of X-ray tubes is required.

Radioisotopes. A variety of radioactive substances have been employed as sources in X-ray fluorescence and absorption methods (see Table 15–2). Generally, the radioisotope is encapsulated to prevent contamination of the laboratory and shielded to absorb radiation in all but certain directions.

Many of the best radioactive sources provide simple line spectra; others produce a continuum (see Table 15–2). Because of the shape of X-ray absorption curves, a given radioisotope will be suitable for excitation of fluorescence or for absorption studies for a range of elements. For example, a source producing a line in the region between 0.3 and 0.47 Å would be suitable for fluorescence or absorption studies involving the K absorption edge for silver (see Figure 15–5). Sensitivity would, of course, improve as the wavelength of the source line approaches the absorption edge. Iodine-125 with a line at 0.46 Å would be ideal from this standpoint.

Secondary Fluorescent Sources. In some applications, the fluorescence spectrum of an element that has been excited by radiation from an X-ray tube serves as a source for absorption or fluorescence studies. This arrangement has the advantage of eliminating the continuous component emitted by a primary source. For example, an X-ray tube with a tungsten target (Figure 15–1) could be used to excite the K_α and K_β lines of molybdenum (Figure 15–2). The resulting fluorescence spectrum would then be similar to the spectrum in Figure 15–2 except that the continuum would be absent.

15B-2 FILTERS FOR X-RAY BEAMS

In many applications, it is desirable to employ an X-ray beam that is restricted in its wavelength range. As in the visible region, both filters and monochromators are used for this purpose.

Figure 15–8 illustrates a common technique for producing a relatively monochromatic beam by use of a filter. Here, the K_β line and most of the continuous radiation from the emission of a molybdenum target is removed by a zirconium filter having a thickness of about 0.01 cm. The pure K_α line is then available for analytical purposes. Several other target-filter combinations of this type have been developed, each of

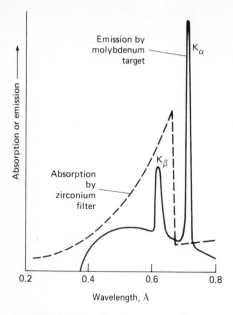

FIGURE 15-8 Use of a filter to produce monochromatic radiation.

which serves to isolate one of the intense lines of a target element. Monochromatic radiation produced in this way is widely used in X-ray diffraction studies. The choice of wavelengths available by this technique is limited by the relatively small number of target-filter combinations that are available.

Filtration of the continuous radiation from an X-ray tube is also feasible with thin strips of metal. As with glass filters for visible radiation, relatively broad bands are transmitted with a significant attenuation of the desired wavelengths.

15B-3 WAVELENGTH DISPERSION WITH MONOCHROMATORS

Figure 15–9 shows the essential components of an X-ray spectrometer. The monochromator consists of a pair of beam collimators, which serve the same purpose as the slits in an optical instrument, and a dispersing element. The latter is a single crystal mounted on a *goniometer* or rotatable table that permits variation and precise determination of the angle θ between the crystal face and the collimated incident beam. From Equation 15–6, it is evident that any given angular setting of the goniometer, only a few wavelengths are diffracted (λ, $\lambda/2$, $\lambda/3$, . . . , λ/n, where $\lambda = 2d \sin \theta$). Thus, an X-ray monochromator does not disperse an entire spectrum simultaneously as does a grating or prism; instead, a particular wavelength is diffracted only when the goniometer is set at an appropriate angle.

In order to derive a spectrum, it is necessary that the exit beam collimator and the detector be mounted on a second table that rotates at twice the rate of the first; that is, as the crystal rotates through an angle θ, the detector must simultaneously move through an angle 2θ. Clearly, the interplanar spacing d for the crystal must be known precisely (Equation 15–6).

The collimators for X-ray monochromators ordinarily consist of a series of closely spaced metal plates or tubes that absorb all but the parallel beams of radiation.

X-Radiation longer than about 2 Å is absorbed by constituents of the atmosphere. Therefore, provision is usually made for a continuous flow of helium through the sample compartment and monochromator when longer wavelengths are required. Alternatively, provisions may be made to evacuate these areas by pumping.

The loss of intensity is high in a monochromator equipped with a flat crystal because as much as 99% of the radiation is sufficiently divergent to be absorbed in the collimators. Increased intensities, by as much as a factor of ten, have been realized by employing a curved crystal surface that acts not only to diffract but also to focus the divergent beam from the source upon the exit collimator.

As illustrated in Table 15–1, most analytically important X-ray lines lie in the region between about 0.1 and 10 Å. A consider-

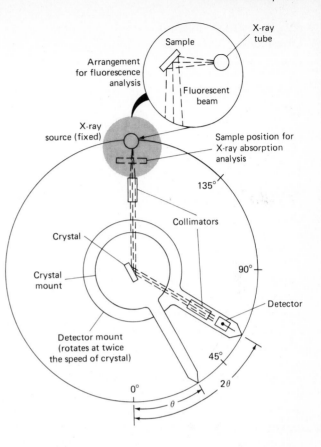

FIGURE 15-9 An X-ray monochromator and detector. Note that the angle of the detector with respect to the beam (2θ) is twice that of the crystal face. For absorption analysis, the source is an X-ray tube and the sample is located in the beam as shown. For emission work, the sample becomes a fluorescent source of X-rays as shown in the insert.

ation of data in Table 15–3, however, leads to the conclusion that no single crystal satisfactorily disperses radiation over this entire range. As a consequence, an X-ray monochromator must be provided with at least two (and preferably more) interchangeable crystals.

The useful wavelength range for a crystal is determined by its lattice spacing d and the problems associated with detection of the radiation when 2θ approaches zero or 180 deg. When a monochromator is set at angles of 2θ that are much less than 10 deg, the amount of polychromatic radiation scattered from the surface becomes prohibitively high. Generally, values of 2θ greater than about 160 deg cannot be measured because the location of the source unit prohibits positioning of the detector at such an angle (see Figure 15–9). The minimum and maximum values for λ_{max} in Table 15–3 were determined from these limitations.

It will be seen from Table 15–3 that a crystal such as ammonium dihydrogen phosphate, with a large lattice spacing, has a much greater wavelength range than a crystal in which this parameter is small. The advantage of large values of d is offset, however, by the consequent lower dispersion. This effect can be seen by differentiation of Equation 15–6, which leads to

$$\frac{d\theta}{d\lambda} = \frac{\mathbf{n}}{2d\cos\theta}$$

Here, $d\theta/d\lambda$, a measure of dispersion, is seen to be inversely proportional to d. Table 15–3 provides dispersion data for the various crystals at their maximum and minimum wavelengths. The low dispersion of ammonium dihydrogen phosphate prohibits its use in the region of low wavelengths; here, a crystal such as topaz or lithium fluoride must be substituted.

15B-4 X-RAY DETECTORS AND SIGNAL PROCESSORS

Early X-ray equipment employed photographic emulsions for detection and measurement of radiation. For reasons of convenience, speed, and accuracy, however, modern instruments are generally equipped with detectors that convert radiant energy into an electrical signal. Three types of transducers are encountered: gas-filled detectors, scintillation counters, and semiconductor detectors. Before considering how each of these devices functions, it is worthwhile to discuss *photon counting*, a signal processing method, which is commonly employed with X-ray detectors as well as detectors of radiation from radioactive sources (Chapter 17). As was mentioned earlier (p. 143), photon counting is also beginning to find use in ultraviolet and visible spectroscopy.

Photon Counting. In contrast to the various photoelectric detectors we have thus far considered, X-ray detectors are usually operated as *photon counters*. In this mode, the individual pulse of electricity produced as a quantum of radiation is absorbed by the transducer and is counted; the power of the beam is then recorded digitally in terms of number of counts per unit of time. This type of operation requires rapid response times for the detector and signal processor with respect to the rate at which quanta are absorbed by the transducer; thus, photon counting is applicable only to beams of relatively low intensity. As the beam intensity increases, the pulse rate becomes greater than the response time of the instrument, and only a steady-state current, which represents an average number of pulses per second, can be measured.

For weak sources of radiation, photon counting generally provides more accurate intensity data than are obtainable by measuring average currents. The improvement can be traced to the fact that signal pulses are generally a good deal larger than the pulses arising from background noise in the source, detector, and associated electronics; separation of the signal from noise can then be achieved with a *pulse height discriminator*,

---

Table 15-3
Properties of Typical Diffracting Crystals

Crystal	Lattice Spacing d, Å	Wavelength Range[a], Å		Dispersion $d\theta/d\lambda$, deg/Å	
		λ_{max}	λ_{min}	at λ_{max}	at λ_{min}
Topaz	1.356	2.67	0.24	2.12	0.37
LiF	2.014	3.97	0.35	1.43	0.25
NaCl	2.820	5.55	0.49	1.02	0.18
EDDT[b]	4.404	8.67	0.77	0.65	0.11
ADP[c]	5.325	10.50	0.93	0.54	0.09

[a] Based on assumption that the measurable range of 2θ is from 160 deg for λ_{max} to 10 deg for λ_{min}.
[b] Ethylenediamine d-tartrate.
[c] Ammonium dihydrogen phosphate.

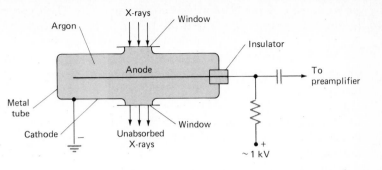

FIGURE 15–10 Cross section of a gas-filled detector.

an electronic device that will be discussed in a later section.

Photon counting is used in X-ray work because the power of available sources is often low. (Photon counting has also been profitably applied to weak sources of ultraviolet and visible radiation such as those encountered in Raman spectrometry.) In addition, photon counting permits spectra to be obtained without the use of a monochromator. This property is considered in the section devoted to energy dispersive systems.

Gas-Filled Detectors. When X-radiation passes through an inert gas such as argon, xenon, or krypton, interactions occur that produce a large number of positive gaseous ions and electrons (ion pairs) for each X-ray quantum. Three types of X-radiation detectors, namely, *ionization chambers*, *proportional counters*, and *Geiger tubes*, are based upon the enhanced conductivity resulting from this phenomenon.

A typical gas-filled detector is shown schematically in Figure 15–10. Radiation enters the chamber through a transparent window of mica, beryllium, aluminum, or Mylar. Each photon of X-radiation may interact with an atom of argon, causing it to lose one of its outer electrons. This *photoelectron* has a large kinetic energy, which is equal to the difference between the X-ray photon energy and the binding energy of the electron in the argon atom. The photoelectron then loses this excess kinetic energy by ionizing several hundred additional atoms of the gas.

Under the influence of an applied potential, the mobile electrons migrate toward the central wire anode while the slower moving cations are attracted toward the cylindrical metal cathode.

Figure 15–11 shows the effect of applied potential upon the number of electrons that reach the anode of a gas-filled detector for

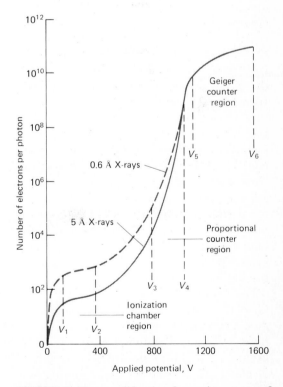

FIGURE 15-11 Gas amplification for various types of gas-filled detectors.

each entering X-ray photon. Three characteristic voltage regions are indicated. At potentials less than V_1, the accelerating force on the ion pairs is low, and the rate at which the positive and negative species separate is insufficient to prevent partial recombination. As a consequence, the number of electrons reaching the anode is smaller than the number produced initially by the incoming radiation.

In the *ionization chamber region* between V_1 and V_2, the number of electrons reaching the anode is reasonably constant and represents the total number formed by a single photon.

In the *proportional counter region* between V_3 and V_4, the number of electrons increases rapidly with applied potential. This increase is the result of secondary ion-pair production caused by collisions between the accelerated electrons and gas molecules; amplification (*gas amplification*) of the ion current results.

In the *Geiger range* V_5 to V_6, amplification of the electrical pulse is enormous but is limited by the positive space charge created as the faster moving electrons migrate away from the slower positive ions. Because of this effect, the number of electrons reaching the anode is independent of the type and energy of incoming radiation and is governed instead by the geometry and gas pressure of the tube.

Figure 5–11 also illustrates that a larger number of electrons is produced by the more energetic, 0.6 Å radiation than the longer wavelength, 5 Å X-rays. Thus, the size of the pulse (the pulse height) is greater for the former than the latter.

The Geiger Tube.   The Geiger tube is a gas-filled detector operated in the voltage region between V_5 and V_6 in Figure 15–11; here, a gas amplification of greater than 10^9 occurs. Each photon produces an avalanche of electrons and cations; the resulting currents are thus large and relatively easy to detect and measure.

The conduction of electricity through a chamber operated in the Geiger region (and in the proportional region as well) is not continuous because the space charge mentioned earlier terminates the flow of electrons to the anode. The net effect is a momentary pulse of current followed by an interval during which the tube does not conduct. Before conduction can again occur, this space charge must be dissipated by migration of the cations to the walls of the chamber. During the *dead time*, when the tube is nonconducting, response to radiation is impossible; the dead time thus represents an upper limit in the response capability of the tube. Typically, the dead time of a Geiger tube is in the range from 50 to 200 μsec.

Geiger tubes are usually filled with argon; a low concentration of an organic substance, often alcohol or methane (a *quenching gas*), is also present to minimize the production of secondary electrons when the cations strike the chamber wall. The lifetime of a tube is limited to some 10^8 to 10^9 counts, by which time the quencher has been depleted.

With a Geiger tube, radiation intensity is determined by counting the pulses of current. The device is applicable to all types of nuclear and X-radiation. However, it lacks the large counting range of other detectors because of its relatively long dead time; its use in X-ray spectrometers is, therefore, limited.

Proportional Counters.   The proportional counter is a gas-filled detector that is operated in the V_3 to V_4 voltage region in Figure 15–11. Here, the pulse produced by a photon is amplified by a factor of 500 to 10,000, but the number of positive ions produced is small enough so that the dead time is only about 1 μsec. In general, the pulses from a proportional counter tube must be amplified before being counted.

The number of electrons per pulse (the *pulse height*) produced in the proportional region depends directly upon the energy of the incoming radiation. A proportional counter can be made sensitive to a restricted range of X-ray frequencies with a *pulse-height analyzer,* which counts a pulse only if its amplitude falls within certain limits. A pulse-height analyzer in effect permits electronic filtration of radiation; its function is analogous to that of a monochromator.

Proportional counters have been widely used as detectors in X-ray spectrometers.

Ionization Chambers. Ionization chambers are operated in the voltage range from V_1 to V_2 in Figure 15–11. Here, the currents are small (10^{-13} to 10^{-16} Å typically) and relatively independent of applied voltage. Ionization chambers are not employed in X-ray spectrometry because of their lack of sensitivity. They do, however, find application in radiochemical measurements, which are considered in the next chapter.

Scintillation Counters. The luminescence produced when radiation strikes a phosphor represents one of the oldest methods of detecting radioactivity and X-rays, and one of the newest as well. In its earliest application, the technique involved the manual counting of flashes that resulted when individual photons or radiochemical particles struck a zinc sulfide screen. The tedium of counting individual flashes by eye led Geiger to the development of gas-filled detectors, which were not only more convenient and reliable but more responsive to radiation as well. The advent of the photomultiplier tube (p. 134) and better phosphors, however, has reversed this trend, and scintillation counting has again become one of the important methods for radiation detection.

The most widely used modern scintillation detector consists of a transparent crystal of sodium iodide that has been activated by the introduction of perhaps 0.2% thallium iodide. Often, the crystal is shaped as a cylinder that is 3 to 4 in. in each dimension; one of the plane surfaces then faces the cathode of a photomultiplier tube. As the incoming radiation traverses the crystal, its energy is first lost to the scintillator; this energy is subsequently released in the form of photons of fluorescent radiation. Several thousand photons with a wavelength of about 400 nm are produced by each primary particle or photon over a period of about 0.25 μsec (the *decay time*). The dead time of a scintillation counter is thus significantly smaller than the dead time of a gas-filled detector.

The flashes of light produced in the scintillator crystal are transmitted to the photocathode of the photomultiplier tube and are in turn converted to electrical pulses that can be amplified and counted. An important characteristic of scintillators is that the number of photons produced in each flash is approximately proportional to the energy of the incoming radiation. Thus, incorporation of a pulse-height analyzer to monitor the output of a scintillation counter forms the basis of energy dispersive photometers, to be discussed later.

In addition to sodium iodide crystals, a number of organic scintillators such as stilbene, anthracene, and terphenyl have been used. In crystalline form, these compounds have decay times of 0.01 and 0.1 μsec. Organic liquid scintillators have also been developed and are used to advantage because they exhibit less self-absorption of radiation than do solids. An example of a liquid scintillator is a solution of *p*-terphenyl in toluene.

Semiconductor Detectors. Semiconductor detectors have assumed major importance as detectors of X-radiation. These devices are sometimes called *lithium drifted silicon* (or *germanium) detectors.*

Figure 15–12 illustrates one form of a lithium drifted detector, which is fashioned from a wafer of crystalline silicon. Three layers exist in the crystal, a *p*-type semiconducting layer that faces the X-ray source, a central *intrinsic* zone, and an *n*-type layer. The outer surface of the *p*-type layer is coated with a thin layer of gold for electrical

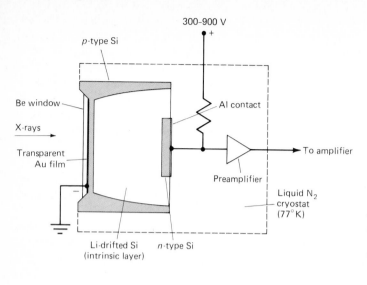

FIGURE 15-12 Vertical cross section of a lithium-drifted silicon detector for X-rays and radioactive radiation.

contact; often, it is also covered with a thin beryllium window which is transparent to X-rays. The signal output is taken from an aluminum layer which coats the n-type silicon; this output is fed into a preamplifier with an amplification factor of about 10. The preamplifier is frequently a field-effect transistor which is made an integral part of the detector.

A lithium drift detector is formed by depositing lithium on the surface of a p-doped silicon crystal. Upon heating to 400 to 500°C, the lithium diffuses into the crystal; because this element easily loses electrons, its presence converts the p region to an n-type. While still at an elevated temperature, a dc potential is applied across the crystal to cause withdrawal of electrons from the lithium layer and holes from the p-type layer. Current passage across the np junction requires migration (or drifting) of lithium *ions* into the p layer and formation of the intrinsic layer where the lithium ions replace the holes lost by conduction. Upon cooling, this central layer has a high resistance relative to the other layers because the lithium ions in this medium are less mobile than the holes they displaced.

The intrinsic layer of a silicon detector

functions in a way that is analogous to argon in the gas-filled detector. Initially, absorption of a photon results in formation of a highly energetic photoelectron, which then loses its kinetic energy by elevating several thousand electrons in the silicon to a conduction band; a marked increase in conductivity results. When a potential is applied across the crystal, a current pulse accompanies the absorption of each photon. In common with a proportional detector, the size of the pulse is directly proportional to the energy of the absorbed photons. In contrast to the proportional detector, however, secondary amplification of the pulse does not occur.

As shown in Figure 15–12, the detector and preamplifier of a lithium drift detector must be thermostated at the temperature of liquid nitrogen (−196°C) to decrease electronic noise to a tolerable level. Furthermore, the performance of the detector degrades badly if allowed to come to room temperature, owing to the tendency of lithium to diffuse rapidly in the silicon. Thus, the liquid nitrogen cryostat is connected to a large (~20 L) Dewar flask, which must be filled every few days.

Distribution of Pulse Heights from X-Radiation Detectors.   To understand the properties of energy dispersive spectrometers, it is important to appreciate that the size of current pulses resulting from absorption of successive X-ray photons of identical energy by the detector will not be exactly the same. Variations arise because the ejection of photoelectrons and their subsequent generation of conduction electrons are random processes governed by the probability law. Thus, a Gaussian distribution of pulse heights around a mean is observed. The breadth of this distribution varies from one type of detector to another, with the drift detector providing a significantly narrower band of pulse heights. It is this property that has made lithium drift detectors so important for energy dispersive X-ray spectroscopy.

15B-5 SIGNAL PROCESSORS

The signal from the preamplifier of an X-ray spectrometer is fed into a linear fast response amplifier whose amplification can be varied by a factor up to 10,000. The result is voltage pulses as large as 10 V.

Pulse-Height Selectors. All modern X-ray spectrometers (wavelength dispersive as well as energy dispersive) are equipped with *discriminators* that reject pulses of about 0.5 V or less (after amplification). In this way, detector and amplifier noise is reduced significantly. In lieu of a discriminator, many instruments are equipped with *pulse-height selectors*, an electronic circuit that rejects not only pulses with heights below some predetermined minimum level but also those above a preset maximum level; that is, it removes all pulses except those that lie within a limited *channel* or *window* of pulse heights. Figure 15–13 provides a schematic diagram of a pulse-height selector and its method of operation. Here, the output pulses from the detector and preamplifier are further amplified and appear as voltage signals (in the 10-V range) as shown in the lower part of the figure. These signals are then fed into

two discriminator circuits each of which can be set to reject any signal below a certain voltage. As shown in the lower part of Figure 15–13, the upper discriminator rejects signal 1, which is smaller than V in voltage, but transmits signals 2 and 3. The lower discriminator, on the other hand, is set to ($V + \Delta V$) and thus rejects all but signal 3. In addition, the lower circuit is so arranged that its output signal is reversed in polarity and thus cancels out signal 3 from the upper circuit in the anticoincidence circuit. As a consequence, only signal 2, with a voltage in the range ΔV, reaches the counter.

Dispersive instruments are often equipped with pulse-height selectors to reject noise and to supplement the monochromator in separating the analyte line from higher order, more energetic radiation that is diffracted at the same crystal setting.

Pulse-Height Analyzers. Pulse-height analyzers consist of one or more pulse-height selectors that are configured in such a way as to provide energy spectra. A single-channel analyzer typically has a voltage range of perhaps 10 V or more with a window of 0.1 to 0.5 V. The window can be manually or automatically adjusted to scan the entire voltage range, thus providing data for an energy dispersion spectrum. Multichannel analyzers contain as few as two or as many as several hundred separate channels, each of which acts as a single channel that is set for a different voltage window. The signal from each channel is then fed to a separate counting circuit thus permitting simultaneous counting and recording of an entire spectrum.

Scalers and Counters. Generally, to obtain convenient counting rates, the output from an X-ray detector is scaled—that is, the number of pulses is reduced by dividing by some multiple of ten (or occasionally two). A brief description of electronic scalers is found in Section 2D–4. Counting of the scaled pulses is now generally carried out with electronic counters such as those described in Section 2D–3.

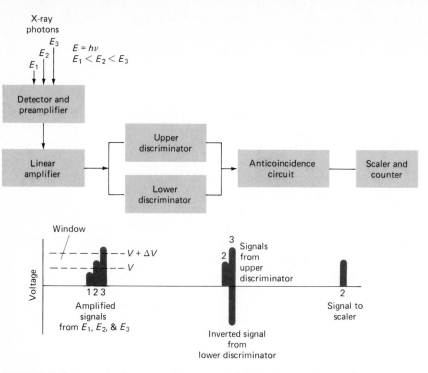

FIGURE 15-13 Schematic diagram of a signal height selector. Upper discriminator rejects voltage below V only while lower discriminator rejects voltage below $(V + \Delta V)$ and inverts remaining signal. Lower plot shows heights of transmitted signals upon exiting from various electronic components.

15C   X-Ray Fluorescence Methods

Although it is feasible to excite an X-ray emission spectrum by incorporating the sample into the target area of an X-ray tube, the inconvenience of this technique discourages its application to many types of materials. Instead, excitation is more commonly brought about by irradiation of the sample with a beam of X-rays from an X-ray tube or a radioactive source. Under these circumstances, the elements in the sample are excited by absorption of the primary beam and emit their own characteristic fluorescent X-rays. This procedure is thus properly called an *X-ray fluorescence* or *emission* method. X-Ray fluorescence is one of the most widely used of all analytical methods

for the qualitative identification of elements having atomic numbers greater than oxygen (>8); in addition, it is often employed for semiquantitative or quantitative elemental analyses as well.

15C-1   INSTRUMENTS

Various combinations of the instrument components discussed in the previous section lead to several recognizable types of X-ray fluorescence instruments. The three basic types are *wavelength dispersive, energy dispersive,* and *nondispersive;* the latter two can be further subdivided depending upon whether an X-ray tube or a radioactive substance serves as a radiation source.

Wavelength Dispersive Instruments.   Wavelength dispersive instruments always employ tubes as a source because of the large energy losses suffered when an X-ray beam is collimated and dispersed into its component wavelengths. Radioactive sources produce X-ray photons at a rate less than 10^{-4} that of an X-ray tube; the added attenuation by a monochromator would then result in a beam that was difficult or impossible to detect and measure accurately.

Wavelength dispersive instruments are of two types, *single-channel* or *sequential,* and *multichannel* or *simultaneous.* The spectrometer shown in Figure 15–9 (p. 467) is a sequential instrument that can be readily employed for X-ray fluorescence analysis; here, the X-ray tube and sample are arranged as shown in the circular insert at the top of the figure. Single-channel instruments may be manual or automatic. The former are entirely satisfactory for the quantitative determination of a few elements. In this application, the crystal and detector are set at the proper angles (θ and 2θ) and counting is continued until sufficient counts have accumulated for precise analyses. Automatic instruments are much more convenient for qualitative analysis, where an entire spectrum must be scanned. Here, the electric drive for the crystal and detector is synchronized with the motor of a recorder while the detector output determines the position of the pen.

Most modern single-channel spectrometers are provided with two X-ray sources; typically, one has a chromium target for longer wavelengths and the other a tungsten target for shorter. For wavelengths longer than 2 Å, it is necessary to remove air between the source and detector by pumping or by displacement with a continuous flow of helium. A means must also be provided for ready interchange of dispersing crystals.

Recording single-channel instruments cost approximately $40,000.

Multichannel dispersive instruments are large, expensive (~$150,000) installations, which permit the simultaneous detection and determination of as many as 24 elements. Here, individual channels consisting of an appropriate crystal and a detector are arranged radially around an X-ray source and sample holder. Ordinarily, the crystals for all or most of the channels are fixed at an appropriate angle for a given analyte line; in some instruments, one or more of the crystals can be moved to permit a spectral scan.

Each detector in a multichannel instrument is provided with its own amplifier, pulse-height selector, scaler, and counter or integrator. These instruments are ordinarily equipped with a computer for instrument control, data processing, and display of analytical results. A determination of 20 or more elements can be completed in a few seconds to a few minutes.

Multichannel instruments are widely used for the determination of several components in materials of industry such as steel, other alloys, cement, ores, and petroleum products.

Both multichannel and single-channel instruments are equipped to handle samples in the form of metals, powdered solids, evaporated films, pure liquids, or solutions. Where necessary, the materials are held in a cell with a Mylar or cellophane window.

Energy Dispersive Instruments.[4] As shown in Figure 15–14, an energy dispersive spectrometer consists of a polychromatic source, which may be an X-ray tube or a radioactive material, a sample holder, a lithium-drifted silicon detector, and the various electronic components required for energy discrimination.

An obvious advantage of energy dispersive systems is the simplicity and lack of moving parts in the excitation and detection components of the spectrometer. Further-

---

[4] See: R. Woldseth, *All You Want To Know About XES,* Kevex Corporation: Foster City, CA, 1973.

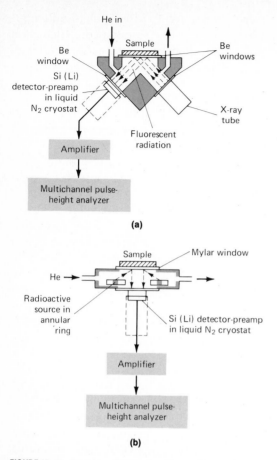

FIGURE 15-14 Energy dispersive X-ray fluorescence spectrometer. Excitation by X-rays from (**a**) an X-ray tube and (**b**) a radioactive substance.

more, the absence of collimators and a crystal diffractor, as well as the closeness of the detector to the sample, result in a 100-fold or more increase in energy reaching the detector. These features permit the use of weaker sources such as radioactive materials or low-power X-ray tubes, which are cheaper and less likely to cause radiation damage to the sample.

In a multichannel, energy dispersive instrument, all of the emitted X-ray lines are measured simultaneously. Increased sensitivity and improved signal-to-noise ratio result from the Fellgett advantage (see p. 150).

The principal disadvantage of energy dispersive systems, when compared with crystal spectrometers, is their lower resolutions at wavelengths longer than about 1 Å (at shorter wavelengths, energy dispersive systems exhibit superior resolution).

Nondispersive Instruments.   Figure 15–15 is a cutaway view of a simple, commercial, nondispersive instrument which has been employed for the routine determination of sulfur and lead in gasoline. For a sulfur analysis, the sample is irradiated by X-rays produced by an iron-55 radioactive source (p. 460); this radiation in turn generates a fluorescent sulfur line at 5.4 Å. The analyte radiation then passes through a pair of adjacent filters and into twin proportional counters. The absorption edge of one of the filters lies just below 5.4 Å, while that of the other is just above it. The difference between the two signals is proportional to the sulfur content of the sample. A sulfur analysis with this instrument requires a counting time of about 1 min. Relative standard deviations of about 1% are obtained for replicate measurements.

15C–2 QUALITATIVE AND SEMIQUANTITATIVE APPLICATIONS OF X-RAY FLUORESCENCE

Figure 15–16 illustrates an interesting qualitative application of the X-ray fluorescence method. Here, the untreated sample, which was excited by radiation from an X-ray tube, was subsequently recovered unchanged. Note that the abscissa for wavelength dispersive instruments is often plotted in terms of the angle 2θ, which can be readily converted to wavelength with knowledge of the crystal spacing of the monochromator (Equation 15–6). Identification of peaks is then accomplished by reference to tables of emission lines of the elements.

Figure 15–17 is a spectrum obtained with an energy dispersive instrument. With such equipment, the abscissa is generally calibrated in channel numbers or energies in keV. Each dot represents the counts col-

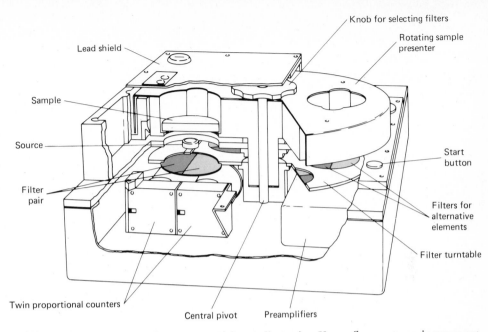

Knob for selecting filters

Rotating sample
presenter

Lead shield

Sample

Source

Filter
pair

Start
button

Filters for
alternative
elements

Filter turntable

Twin proportional counters

Central pivot   Preamplifiers

FIGURE 15-15 Cutaway view of a commercial nondispersive X-ray fluorescence instrument. (Reprinted with permission from: B. J. Price and K. M. Field, *Amer. Lab.*, **1974**, *6* (9), 62. Copyright 1974 by International Scientific Communications, Inc.)

lected by one of the several hundred channels.

Qualitative information, such as that shown in Figures 15–16 and 15–17, can be converted to semiquantitative data by careful measurement of peak heights. To obtain a rough estimate of concentration, the following relationship is used:

$$P_x = P_s W_x \qquad (15\text{--}7)$$

where P_x is the relative line intensity measured in terms of number of counts for a fixed period, and W_x is the weight fraction of the element in the sample. The term P_s is the relative intensity of the line that would be observed under identical counting condi-

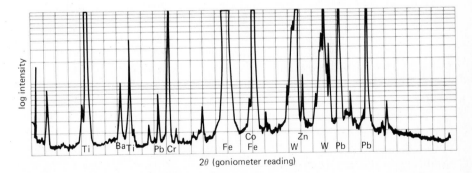

2θ (goniometer reading)

FIGURE 15-16 X-Ray fluorescence spectrum for a genuine bank note recorded with a wavelength dispersive spectrometer. (Taken from: H. A. Liebhafsky, H. G. Pfeiffer, E. H. Winslow, and P. D. Zeman, *X-Ray Absorption and Emission in Analytical Chemistry*, p. 163, Wiley: New York, 1960. Reprinted by permission of John Wiley & Sons, Inc.)

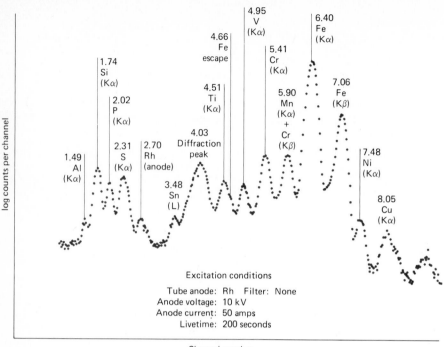

FIGURE 15-17 Spectrum of an iron sample obtained with an energy dispersive instrument with a Rh anode X-ray tube source. The numbers above the peaks are energies in keV. (Reprinted with permission from: J. A. Cooper, *Amer. Lab.*, **1976**, *8* (11), 44. Copyright 1976 by International Scientific Communications, Inc.)

tions if W_x were unity. The value of P_s is determined with a sample of the pure element or a standard sample of known composition.

The use of Equation 15–7, as outlined in the previous paragraph, carries with it the assumption that the emission from the species of interest is unaffected by the presence of other elements in the sample. We shall see that this assumption may not be justified; as a consequence, a concentration estimate may be in error by a factor of two or more. On the other hand, this uncertainty is significantly smaller than that associated with a semiquantitative analysis by optical emission where an order of magnitude error is not uncommon.

15C–3 QUANTITATIVE ANALYSIS

Modern X-ray fluorescence instruments are capable of producing quantitative analyses of complex materials with a precision that equals or exceeds that of the classical wet chemical methods or other instrumental methods. For the accuracy of such analyses to approach this level, however, requires either the availability of calibration standards that closely approach the samples in overall chemical and physical composition or suitable methods for dealing with matrix effects.

Matrix Effects.   It is important to realize that the X-rays produced in the fluorescence process are generated not only from atoms at the surface of a sample but also from atoms well below the surface. Thus, a part of both the incident beam and the resulting fluorescent beam traverse a significant thickness of sample within which absorption and scattering can occur. The extent to which either beam is attenuated depends upon the mass absorption coefficient of the medium, which in turn is determined by the coefficients of *all* of the elements in the sample. Thus, while the net intensity of a line reaching the detector in an X-ray fluorescence measurement depends upon the concentration of the element producing the line, it is also affected by the concentration and mass absorption coefficients of the matrix elements as well.

Absorption effects by the matrix may cause results calculated by Equation 15–7 to be either high or low. If, for example, the matrix contains a significant amount of an element that absorbs either the incident or the emitted beam more strongly than the element being determined, then W_x will be low, since P_s was evaluated with a standard in which absorption was smaller. On the other hand, if the matrix elements of the sample absorb less than those in the standard, high values for W_x result.

A second matrix effect, called the *enhancement effect,* can also yield results that are greater than expected. This behavior is encountered when the sample contains an element whose characteristic emission spectrum is excited by the incident beam, and this spectrum in turn causes a secondary excitation of the analytical line.

Several techniques have been developed to compensate for absorption and enhancement effects in X-ray fluorescence analyses.

Calibration Against Standards. Here, the relationship between the analytical line intensity and the concentration is determined empirically with a set of standards that closely approximate the samples in overall composition. The assumption is then made that absorption and enhancement effects are identical for both samples and standards, and the empirical data are employed to convert emission data to concentrations. Clearly, the degree of compensation achieved in this way depends upon the closeness of the match between samples and standards.

Use of Internal Standards. In this procedure, an element is introduced in known and fixed concentration into both the calibration standards and the samples; the added element must be absent in the original sample. The ratio of the intensities between the element being determined and the internal standard serves as the analytical parameter. The assumption here is that absorption and enhancement effects are the same for the two lines and that use of intensity ratios compensates for these effects.

Dilution of Sample and Standards. Here, both sample and standards are diluted with a substance that absorbs X-rays only weakly (that is, a substance containing elements with low atomic numbers). Examples of such diluents include water; organic solvents containing carbon, hydrogen, oxygen, and nitrogen only; starch; lithium carbonate; alumina; and boric acid or borate glass. By employing an excess of diluent, matrix effects become essentially constant for the diluted standards and samples, and adequate compensation is achieved. This procedure has proved particularly useful for mineral analyses, where both samples and standards are dissolved in molten borax; after cooling, the fused mass is excited in the usual way.

Some Quantitative Applications of X-Ray Fluorescence. With proper correction for matrix effects, X-ray fluorescence spectrometry is perhaps the most powerful tool available to the chemist for the rapid quantitative determination of all but the lightest elements in complex samples. For example, Baird and Henke[5] have demonstrated that nine ele-

---

[5] A. K. Baird and B. L. Henke, *Anal. Chem.,* **1965,** *37,* 727.

ments can be determined in samples of granitic rocks in an elapsed time, including sample preparation, of about 12 min. The precision of the method is better than wet chemical analyses and averages 0.08% relative. It is noteworthy that one of the elements reported is oxygen, which ordinarily can be determined by difference only.

X-ray methods also find widespread application for quality control in the manufacture of metals and alloys. Here, the speed of the analysis permits correction of the composition of the alloy during its manufacture.

X-Ray fluorescence methods are readily adapted to liquid samples. Thus, as mentioned earlier, methods have been devised for the direct quantitative determination of lead and bromine in aviation gasoline samples. Similarly, calcium, barium, and zinc have been determined in lubricating oils by excitation of fluorescence in the liquid hydrocarbon samples. The method is also convenient for the direct determination of the pigments in paint samples.

X-Ray fluorescence methods are being widely applied to the analysis of atmosphere pollutants. For example, one procedure for detecting and determining contaminants involves drawing an air sample through a stack consisting of a micropore filter for particulates and three filter paper disks impregnated with orthotolidine, silver nitrate, and sodium hydroxide, respectively. The latter three retain chlorine, sulfides, and sulfur dioxide in that order. The filters then serve as samples for X-ray fluorescence analysis.

One other indication of the versatility of X-ray fluorescence methods is the choice of this procedure by Russian scientists for the quantitative analysis of rocks on the surface of Venus.[6] The instrument in this case was an energy dispersive spectrometer equipped with an iron-55 and a plutonium-238 source. The plutonium source excited

fluorescence in the lightest elements (Mg, Al, and Si) while the iron source permitted the determination of heavier elements (such as K, Ca, and Ti). The instrument successfully carried out analyses under the severe conditions extant on the surface of Venus—that is, temperatures of 500°C and pressures of 90 atm. Results for eight elements were reported with precisions that varied from about 7% relative for silicon to about 50% for magnesium and manganese.

Summary of the Advantages and Disadvantages of X-Ray Fluorescence Methods. X-Ray fluorescence offers a number of impressive advantages. The spectra are relatively simple; spectral line interference is thus unlikely. Generally, the X-ray method is nondestructive and can be used for the analysis of paintings, archeological specimens, jewelry, coins, and other valuable objects without harm to the sample. Furthermore, analyses can be performed on samples ranging from a barely visible speck to a massive object. Other advantages include the speed and convenience of the procedure, which permits multielement analyses to be completed in a few minutes. Finally, the accuracy and precision of X-ray fluorescence methods often equal or exceed those of other methods.[7]

X-Ray fluorescence methods are generally not so sensitive as the various optical methods that have been discussed earlier in this text. In the most favorable cases, concentrations of a few parts per million can be measured. More commonly, however, the concentration range of the method will be from perhaps 0.01 to 100%. X-Ray fluorescence methods for the lighter elements are inconvenient; difficulties in detection and measurement become progressively worse as atomic numbers become smaller than 23 (vanadium), in part because a competing process, called Auger emission, reduces the

---

[6] See: Y. A. Surko, *et al.*, *Anal. Chem.*, **1982**, *54*, 957A.

[7] For a comparison of X-ray fluorescence and ICP for the analysis of iron oxides, see: R. A. Peterson and D. M. Wheeler, *Amer. Lab.*, **1981**, *13* (10), 138.

fluorescent intensity (p. 491). Present commercial instruments are limited to atomic numbers of 8 (oxygen) or 9 (fluorine). Another disadvantage of the X-ray emission procedure is the high cost of instruments, which ranges from about $5000 for an energy dispersive system with a radioactive source, to more than $100,000 for automated and computerized wavelength dispersive systems.

15D   X-Ray Absorption Methods

In contrast to optical spectroscopy, where absorption methods are of prime importance, X-ray absorption applications are limited when compared with X-ray emission and fluorescence procedures. While absorption measurements can be made relatively free of matrix effects, the required techniques are somewhat cumbersome and time-consuming when compared with fluorescence methods. Thus, most applications are confined to samples in which the effect of the matrix is minimal.

15D-1   DIRECT ABSORPTION METHODS

Direct absorption methods are analogous to optical absorption procedures in which the attenuation of a band or line of X-radiation serves as the analytical parameter. Wavelength restriction is accomplished with a monochromator such as that shown in Figure 15–9 or by a filter technique similar to that illustrated in Figure 15–8. Alternatively, the monochromatic radiation from a radioactive source is employed.

Because of the breadth of X-ray absorption peaks, direct absorption methods are generally useful only when a single element with a high atomic number is to be determined in a matrix consisting of only lighter elements. Examples of applications of this type are the determination of lead in gaso-

line and the determination of sulfur or the halogens in hydrocarbons.

15D-2   THE ABSORPTION EDGE METHOD

The absorption edge method largely avoids the matrix effects that influence the results of both the direct absorption method and the X-ray emission method.

Basis of the Method.[8]   In the absorption edge method the *change* in the mass absorption coefficient of a sample at an absorption edge is employed as a measure of the concentration of the element responsible for the edge. The difference in mass absorption coefficients on either side of the absorption edge is large (see Figure 15–5) when compared with the typical absorption change over the same wavelength region in the absence of an edge. Thus, the difference parameter is not greatly affected by matrix variations except in the very unlikely situation where the matrix contains some other element that also has an absorption edge in the small wavelength region under study.

A Typical Example.   The absorption edge procedure can be performed in a number of ways, one of which is illustrated by Figure 15–18. Here, the analysis of lead is based on its L_{III} edge (λ = 0.95 Å). In order to bracket this edge, the absorption of the sample is measured with radiation of 0.93 and 1.04 Å provided by the fluorescence induced in a sample of sodium bromide. That is, the fluorescent bromine spectrum is excited with the source arrangement employed for emission analysis (see upper part of Figure 15–9). This radiation is then passed through the sample and into a monochromator arranged to measure first the intensity of the K_α line of bromine and then the intensity of the K_β line. In this way the difference in absorption coefficient for the sample can be determined on either side of the lead edge; these data are then readily

---

[8] See: H. W. Dunn, *Anal. Chem.*, **1962**, *34*, 116.

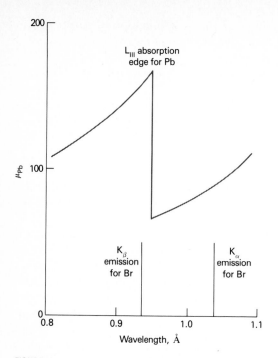

FIGURE 15–18 Data for the absorption edge analysis of lead. Absorption measurements are made with the K_α and K_β lines of bromine.

corrected to give the difference in absorption at the wavelength of the edge. This difference is directly proportional to the concentration of the element and is essentially independent of the matrix. The high intensity and narrow bandwidths associated with bromine fluorescence provide distinct advantages over the direct use of an X-ray tube as the source.

Suitable fluorescent sources for bracketing the absorption edges of some 40 elements have been suggested. Application of the technique to the determination of 10 elements has shown that these analyses can be performed without recourse to standards of any kind; that is, proportionality constants for the concentration-absorption difference relationships can be calculated from literature values of mass absorption coefficients. Matrix variations appear to have little effect on the data and the results are nearly as

accurate as more elaborate procedures employing standards. Typically, relative errors of 2 to 5% are observed.[9]

15E   X-Ray Diffraction Methods

Since its discovery in 1912 by von Laue, X-ray diffraction has provided a wealth of important information to science and industry. For example, much that is known about the arrangement and the spacing of atoms in crystalline materials has been directly deduced from diffraction studies. In addition, such studies have led to a much clearer understanding of the physical properties of metals, polymeric materials, and other solids. X-Ray diffraction is currently of prime importance in elucidating the structures of such complex natural products as steroids, vitamins, and antibiotics. Such applications are beyond the scope of this text.

X-Ray diffraction also provides a convenient and practical means for the qualitative identification of crystalline compounds. This application is based upon the fact that an X-ray diffraction pattern is unique for each crystalline substance. Thus, if an exact match can be found between the pattern of an unknown and an authentic sample, chemical identity can be assumed. In addition, diffraction data sometimes yield quantitative information concerning a crystalline compound in a mixture. The method may provide data that are difficult or impossible to obtain by other means as, for example, the percentage of graphite in a graphite/charcoal mixture.

15E–1   IDENTIFICATION OF CRYSTALLINE COMPOUNDS BY X-RAY DIFFRACTION

Sample Preparation.   For analytical diffraction studies, the crystalline sample is ground to a

[9] E. P. Bertin, R. J. Longobucco, and R. J. Carver, *Anal. Chem.*, **1964**, *36*, 641.

fine homogeneous powder. In such a form, the enormous number of small crystallites are oriented in every possible direction; thus, when an X-ray beam traverses the material, a significant number of the particles can be expected to be oriented in such ways as to fulfill the Bragg condition for reflection from every possible interplanar spacing.

Samples may be held in the beam in thin-walled glass or cellophane capillary tubes. Alternatively, a specimen may be mixed with a suitable noncrystalline binder and molded into an appropriate shape.

Photographic Recording. The classical, and still widely used, method for recording powder diffraction patterns is photographic. Perhaps the most common instrument for this purpose is the *Debye-Scherrer* powder camera, which is shown schematically in Figure 15–19. Here, the beam from an X-ray tube is filtered to produce a nearly monochromatic beam (often the copper or molybdenum K_α line), which is collimated by passage through a narrow tube. The undiffracted

radiation then passes out of the camera via a narrow exit tube. The camera itself is cylindrical and equipped to hold a strip of film around its inside wall. The inside diameter of the cylinder usually is 5.73 or 11.46 cm, so that each lineal millimeter of film is equivalent to 1.0 or 0.5 deg in θ, respectively.

The sample is held in the center of the beam by an adjustable mount.

Figure 15–19**b** depicts the appearance of the exposed and developed film; each set of lines (D_1, D_2, and so forth) represents diffraction from one set of crystal planes. The Bragg angle θ for each line is easily evaluated from the geometry of the camera.

Automatic Diffractometers. In many laboratories, diffraction patterns are obtained with automated instruments similar in design to that shown in Figure 15–9. Here again, the source is an X-ray tube with suitable filters. The powdered sample, however, replaces the single crystal on its mount. In some instances, the sample holder may be rotated in order to increase the randomness of the ori-

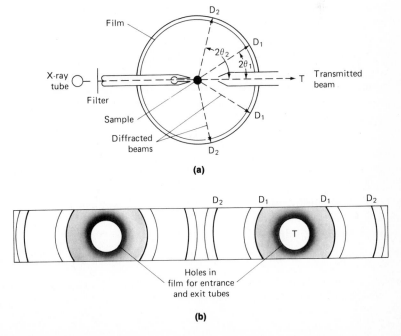

FIGURE 15-19 Schematic diagram of (**a**) a powder camera; (**b**) the film strip after development. D_2, D_1, and T indicate positions of the film in the camera.

entation of the crystals. The diffraction pattern is then obtained by automatic scanning in the same way as for an emission or absorption spectrum. Instruments of this type offer the advantage of considerably higher precision for intensity measurements.

15E-2 INTERPRETATION OF DIFFRACTION PATTERNS

The identification of a species from its powder diffraction pattern is based upon the position of the lines (in terms of θ or 2θ) and their relative intensities. The diffraction angle 2θ is determined by the spacing between a particular set of planes; with the aid of the Bragg equation, this distance d is readily calculated from the known wavelength of the source and the measured angle. Line intensities depend upon the number and kind of atomic reflection centers that exist in each set of planes.

Identification of crystals is empirical. The American Society for Testing Materials (ASTM) markets keysort and IBM cards that provide d spacings and relative line intensities for pure compounds; data for over 25,000 crystalline materials have been compiled.[10] The cards are arranged in order of the d spacing for the most intense line; cards are withdrawn from this file on the basis of a d spacing that lies within a few hundredths of an ångström of the d spacing of the most intense line for the analyte. Further elimination of possible compounds is accomplished by consideration of the spacing for the second most intense line, then the third, and so forth. Ordinarily, three or four spacings serve to identify the compound unambiguously.

If the sample contains two or more crystalline compounds, identification becomes more complex. Here, various combinations of the more intense lines are used until a match can be found.

By measuring the intensity of the diffraction lines and comparing with standards, a quantitative analysis of crystalline mixtures is also possible.

15F The Electron Microprobe

With the electron microprobe method, X-ray emission is stimulated on the surface of the sample by a narrow, focused beam of electrons. The resulting X-ray emission is detected and analyzed with a wavelength or energy dispersive spectrometer.[11]

15F-1 INSTRUMENTS

Figure 15–20 is a schematic diagram of an electron microprobe system. The instrument employs three integrated beams of radiation, namely, electron, light, and X-ray. In addition, a vacuum system is required that provides a pressure of less than 10^{-5} torr and a wavelength- or energy-dispersive X-ray spectrometer (a wavelength dispersive system is shown in Figure 15–20). The electron beam is produced by a heated tungsten cathode and an accelerating anode (not shown). Two electromagnet lenses focus the beam on the specimen; the diameter of the beam lies between 0.1 and 1 μm. An associated optical microscope is used to locate the area to be bombarded. Finally, the fluorescent X-rays produced by the electron beam are collimated, dispersed by a single crystal, and detected by a gas-filled detector. Considerable design effort is required to ar-

---

[10] *Index to the Powder Data File, Amer. Soc. Testing Materials, Spec. Tech. Publ.,* 48L, Philadelphia, 1962.

[11] For a detailed discussion of this method, see: L. S. Birks, *Electron Probe Microanalysis,* 2d ed., Wiley-Interscience: New York, 1971 and K. F. J. Heinrich, *Electron Beam X-Ray Microanalysis,* Van Nostrand: New York, 1981.

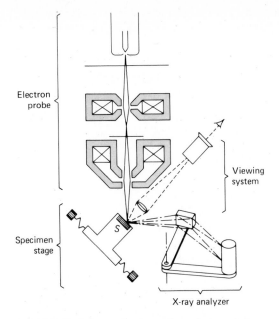

Electron probe

Viewing system

Specimen stage

S

X-ray analyzer

FIGURE 15-20 Schematic view of an electron-micro-probe instrument. (From: D. B. Wittry, in *Treatise on Analytical Chemistry*, I. M. Kolthoff and P. J. Elving, Eds., Vol. 5, Part I, p. 3178, Interscience: New York, 1964. Reprinted by permission of John Wiley & Sons, Inc.)

range the three systems spatially so that they do not interfere with one another.

In addition to the foregoing components, the specimen stage is provided with a mechanism whereby the sample can be moved in two mutually perpendicular directions and rotated as well, thus permitting scanning of the surface.

15F-2 APPLICATIONS

The electron microprobe provides a wealth of information about the physical and chemical nature of surfaces. It has had important applications to phase studies in metallurgy and ceramics, the investigation of grain boundaries in alloys, the measurement of diffusion rates of impurities in semiconductors, the determination of occluded species in crystals, and the study of the active sites of heterogeneous catalysts. In all of these applications, both qualitative and quantitative information about surfaces is obtained.

PROBLEMS

15–1. What is the short-wavelength limit of the continuum produced by an X-ray tube having a silver target and operated at 80 kV?

15–2. What minimum tube voltage would be required to excite the K_β and L_β series of lines for (a) U, (b) K, (c) Rb, (d) W?

15–3. The $K_{\alpha 1}$ lines for Ca, Zn, Zr, and Sn occur at 3.36, 1.44, 0.79, and 0.49 Å, respectively. Calculate an approximate wavelength for the K_α lines of (a) V, (b) Ni, (c) Se, (d) Br, (e) Cd, (f) Sb.

15–4. The L_α lines for Ca, Zn, Zr, and Sn are found at 36.3, 11.9, 6.07, and 3.60 Å, respectively. Estimate the wavelengths for the L_α lines for the elements listed in Problem 15–3.

15–5. The mass absorption coefficient for Ni, measured with the Cu K_α line, is 49.2 cm²/g. Calculate the thickness of a nickel foil that was found to transmit 36.1% of the incident power of a beam of Cu K_α radiation. Assume that the density of Ni is 8.9 g/cm³.

15–6. For Mo K_α radiation (0.711 Å), the mass absorption coefficients for K, I, H, and O are 16.7, 39.2, 0.0, and 1.50 cm²/g, respectively.

(a) Calculate the mass absorption coefficient for a solution prepared by mixing 8.00 g of KI with 92 g of water.

(b) The density of the solution described in (a) is 1.05 g/cm^3. What fraction of the radiation from a Mo K$_\alpha$ source would be transmitted by a 0.50-cm layer of the solution?

15-7. Aluminum is to be employed as windows for a cell for X-ray absorption measurements with the Ag K$_\alpha$ line. The mass absorption coefficient for aluminum at this wavelength is 2.74; its density is 2.70 g/cm^3. What maximum thickness of aluminum foil could be employed to fashion the windows if no more than 2.0% of the radiation is to be absorbed by them?

15-8. A solution of I$_2$ in ethanol had a density of 0.794 g/cm^3. A 1.50-cm layer was found to transmit 27.3% of the radiation from a Mo K$_\alpha$ source. Mass absorption coefficients for I, C, H, and O are 39.2, 0.70, 0.00, and 1.50, respectively.

(a) Calculate the percent I$_2$ present, neglecting absorption by the alcohol.

(b) Correct the results in part (a) for the presence of alcohol.

15-9. Calculate the goniometer setting, in terms of 2θ, required to observe the K$_{\alpha 1}$ lines for Fe (1.76 Å), Se (0.992 Å), and Ag (0.497 Å) when the diffracting crystal is (a) topaz; (b) LiF; (c) NaCl.

15-10. Calculate the goniometer setting, in terms of 2θ, required to observe the L$_{\beta 1}$ lines for Br at 8.126 Å when the diffracting crystal is
(a) ethylenediamine d-tartrate.
(b) ammonium dihydrogen phosphate.

15-11. Calculate the minimum tube voltage required to excite the following lines. The numbers in parentheses are the wavelengths in Å for the corresponding absorption edges.
(a) K lines for Ca (3.064)
(b) L$_\alpha$ lines for As (9.370)
(c) L$_\beta$ lines for U (0.592)
(d) K lines for Mg (0.496)

16

Electron Spectroscopy

Although the basic principles of electron spectroscopy were well understood early in this century, the widespread application of this technique to chemical problems did not occur until relatively recently. An important factor that inhibited studies in the field was the lack of engineering technology necessary for performing high-resolution spectral measurements of electrons having energies varying from a few tenths of an electron volt to several thousand. By the late 1960s, this technology had developed, and commercial electron spectrometers began to appear in the marketplace. With their appearance, an explosive growth in the number of publications devoted to electron spectroscopy occurred.[1]

As shown in Table 16–1, several types of electron spectroscopy have been developed. The first three, X-ray photoelectron, ultraviolet photoelectron, and Auger (pronounced Oh jay') spectroscopy, have received the most attention to date. Several instrument manufacturers offer spectrometers that, with minor adjustments, permit all three types of measurements to be performed. This discussion will be confined to these three techniques.

Electron spectroscopy is a powerful tool for the identification of all of the elements in the periodic table with the exception of hydrogen and helium. More important, the method permits determination of the oxidation state of an element and the type of species to which it is bonded. Finally, the technique provides useful information about the electronic structure of molecules.

Electron spectroscopy has been successfully applied to gases and solids and more recently to solutions and liquids. Because of the poor penetrating power of electrons,

---

[1] References include: *Electron Spectroscopy: Theory, Techniques, and Applications*, 4 vols, C. R. Brundle and A. D. Baker, Eds., Academic Press: New York, 1977–81; T. A. Carlson, *Photoelectron and Auger Spectroscopy*, Plenum Press: New York, 1975; P. K. Ghosh, *Introduction to Photoelectron Spectroscopy*, Wiley: New York, 1983.

however, these methods provide information about solids that is restricted largely to a surface layer that is a few particles thick (20 to 50 Å). Often, the composition of such surface layers is significantly different from the average composition of the entire sample. Indeed, some of the most important and valuable current applications of electron spectroscopy are to the determination of the surface chemistry of solids such as metals, alloys, semiconductors, and heterogeneous catalysts.

16A Principles of Electron Spectroscopy

The principles of the three most important types of electron spectroscopy are illustrated schematically by the first three partial energy diagrams in Figure 16–1. The fourth diagram, showing the basis for X-ray fluorescence, is included because fluorescence is competitive with Auger emission shown in Figure 16–1**c**.

It is important to note the contrast between electron spectroscopy and the other types of spectroscopy we have thus far encountered (including X-ray fluorescence depicted in Figure 16–1**d**). In electron spectroscopy, it is the kinetic energy of emitted electrons that is recorded. The spectrum thus consists of a plot of number of emitted electrons (or the power of the electron beam) as a function of the energy (or the frequency or wavelength) of the emitted electrons (see Figure 16–6). In an electromagnetic spectrum, of course, it is the number of photons (or the power of the electromagnetic beam) that serves as the ordinate with the abscissa also being frequency, wavelength, or energy ($h\nu$).

16A-1 X-RAY PHOTOELECTRON SPECTROSCOPY

X-Ray photoelectron spectroscopy (XPS) is also widely referred to as ESCA (Electron Spectroscopy for Chemical Analysis), a term coined by K. Siegbahn, a Swedish physicist, who was awarded the 1981 Nobel prize for his pioneering work in the development of

Table 16–1
Electron Spectroscopic Methods

Name	Common Abbreviation	Method of Producing Electrons from the Analyte
X-Ray photoelectron spectroscopy	ESCA or XPS	Exposure to monochromatic X-rays
Ultraviolet photoelectric spectroscopy	UPS or PES	Exposure to monochromatic, vacuum-ultraviolet radiation (most commonly the helium(I) line at 58.4 nm)
Auger spectroscopy	AES	Bombardment with electrons or exposure to X-rays (not necessarily monoenergetic)
Ion neutralization spectroscopy	INS	Bombardment with low energy ions such as He^+, Ar^+, or Ne^+
Electron impact spectroscopy	EIP	Bombardment with monoenergetic electrons. Energies of inelastically scattered electrons measured
Penning ionization spectroscopy	PIS	Collision in the gas phase with excited, metastable rare gas atoms

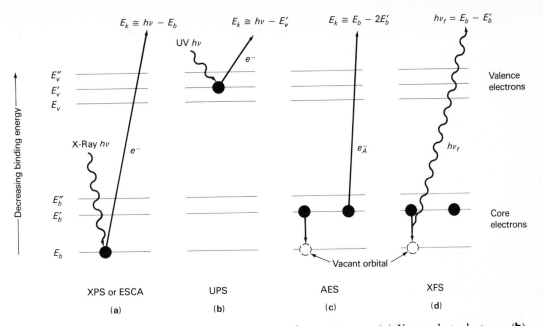

FIGURE 16-1 Schematic comparison of certain types of spectroscopy: (**a**) X-ray photoelectron; (**b**) ultraviolet photoelectron; (**c**) Auger, and (**d**) X-ray fluorescence. Note that the first three are electron spectroscopic procedures requiring the measurement of the kinetic energy of the emitted electron E_k. The fourth, which competes with AES, requires the measurement of the energy of an emitted X-ray photon.

chemical applications of this procedure.[2] Figure 16–1a is a schematic representation of the ESCA process. Here, the three lower lines labeled E_b, E_b', and E_b'' represent energies of the inner shell K or L electrons. The upper three lines represent some of the energy levels of the outer shell or valence electrons. As shown in the illustration, one of the photons of a monochromatic X-ray beam of known energy $h\nu$ displaces an electron e^- from a K orbital E_b. The reaction can be represented by

$$A + h\nu \rightarrow A^{+*} + e^- \qquad (16-1)$$

where A can be an atom, a molecule, or an ion and A^{+*} is an electronically excited ion with a charge one greater than A.

The kinetic energy of the emitted electron E_k is measured in an electron spectrometer. The *binding energy* of the electron E_b can then be calculated by means of the equation

$$E_b = h\nu - E_k - w \qquad (16-2)$$

In this equation, w is the so-called *work function* of the spectrometer, a factor that corrects for the electrostatic environment in which the electron is formed and measured. Various methods are available to determine approximate values for w.

Figure 16–2 shows a low-resolution or

[2] K. Siegbahn, *et al.*, *ESCA: Atomic, Molecular and Solid State Structure by Means of Electron Spectroscopy*, Olmquist and Wiksells: Upsala, 1967 and *ESCA Applied to Free Molecules*, North-Holland Publishing Co.: Amsterdam, 1969. (In English). For a brief description of the history of ESCA, see: K. Siegbahn, *Science*, **1982**, *217*, 111.

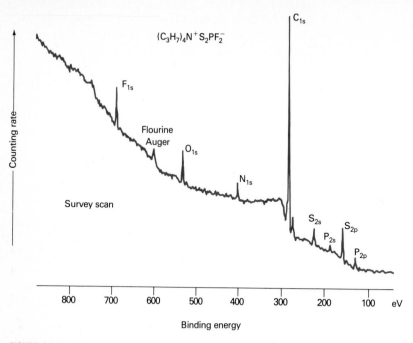

FIGURE 16-2 X-Ray photoelectron spectrum of tetrapropylammoniumdi-fluoridethiophosphate. (Courtesy of DuPont Instrument Systems, Wilmington, DE.)

survey ESCA spectrum consisting of a plot of electron counting rate as a function of binding energy E_b. The analyte consisted of an organic compound made up of six elements. With the exception of hydrogen, well separated peaks for each of the elements can be observed. In addition, a peak for oxygen is present suggesting that some surface oxidation of the compound had occurred. Note that, as expected, the binding energies for 1s electrons increase with atomic number because of the increased positive charge of the nucleus. Note also that more than one peak for a given element can be observed; thus peaks for both 2s and 2p electrons for sulfur and phosphorus can be seen. The large background count arises because associated with each characteristic peak is a tail due to ejected electrons that have lost part of their energy by inelastic collisions within the solid sample. These electrons have less kinetic energy than their nonscattered counterparts and will thus appear at lower kinetic energies or higher binding energies (Equation 16–2). It is evident from Figure 16–2 that ESCA provides a means of qualitative identification of the elements present on the surface of solids.

16A-2 ULTRAVIOLET PHOTOELECTRON SPECTROSCOPY

Equations 16–1 and 16–2 apply not only to XPS but to UPS as well. In the latter case, however, $h\nu$ represents the energy of a vacuum ultraviolet photon. As shown in Figure 16–1**b**, this lower energy radiation generally results in the ejection of outer or valence electrons rather than core electrons. As a consequence, ultraviolet photoelectron spectroscopy provides information about chemical bonds between elements.

For bonding electrons in molecules, the

ionization process depicted by Equation 16–2 can result in the formation of ions in various excited vibrational and rotational states. Here, the binding energy is given by

$$E_b = h\upsilon - E_k - E_\upsilon - E_r - w \qquad (16\text{–}3)$$

where E_υ and E_r, the vibrational and rotational energy levels of the excited ions, are much smaller than E_k. With a high-resolution spectrometer, these small energy differences are detectable in the form of a fine structure for the emission peaks. This fine structure is sometimes helpful in identifying the type of molecular orbital from which the electron is ejected.

16A-3 AUGER SPECTROSCOPY

In contrast to XPS and UPS, Auger spectroscopy is based upon a two-step process in which the first step involves formation of an electronically excited ion A^{+*} by exposing the analyte to a beam of either X-rays or electrons. With X-rays, the reaction shown by Equation 16–1 occurs; for an electron beam, the excitation reaction can be written

$$A + e_i^- \rightarrow A^{+*} + e_i^{'-} + e_A^- \qquad (16\text{–}4)$$

where e_i^- represents an incident electron from the source, $e_i^{'-}$ represents the same electron after it has interacted with A and thus lost some of its energy, and e_A^- represents an electron that is ejected from one of the inner orbitals of A.

As shown in Figures 16–1c and 16–1d, relaxation of the excited ion A^{+*} can occur in either of two ways; that is,

$$A^{+*} \rightarrow A^{++} + e_A^- \qquad (16\text{–}5)$$

or

$$A^{+*} \rightarrow A^+ + h\upsilon_f \qquad (16\text{–}6)$$

Here, e_A^- corresponds to an Auger electron while $h\upsilon_f$ represents a fluorescent photon.

The relaxation process described by Equation 16–6 will be recognized as X-ray fluorescence, which was described in the previous chapter. Note that the energy of the fluorescent radiation $h\upsilon_f$ is independent of the excitation energy. Thus, polychromatic radiation can be used for the excitation step. Auger emission, shown by Equation 16–5 is a radiationless process in which the energy given up in relaxation results in the ejection of an electron (an Auger electron e_A^-) with a kinetic energy E_k. Note that the energy of the Auger electron is *independent* of the energy of the photon or electron which originally created the vacancy in energy level E_b. Thus, as is true in fluorescence spectroscopy, a monoenergetic excitation source is *not* required for excitation.

The kinetic energy of the Auger electron is the difference between the energy released in relaxation of the excited ion ($E_b - E_b'$) and the energy required to remove the second electron from its orbit (E_b'). Thus,

$$E_k = (E_b - E_b') - E_b' = E_b - 2E_b' \qquad (16\text{–}7)$$

Auger emissions are described in terms of the type of orbital transitions involved in the production of the electron. For example, a KLL Auger proton involves an initial removal of a K electron followed by a transition of an L electron to the K orbital with the simultaneous ejection of a second L electron. Other common transitions are LMM and MNN.

Like ESCA spectra, Auger spectra consist of a few characteristic peaks lying in the region of 20 to 1000 eV.

Figure 16–3 shows typical Auger spectra obtained for two samples of a 70% copper/ 30% nickel alloy. Note that the derivative of the counting rate as a function of the kinetic energy of the electron $dN(E)/dE$ serves as the ordinate. Derivative spectra are standard for Auger spectroscopy in order to enhance the small peaks and to repress the effect of the scattered electron background

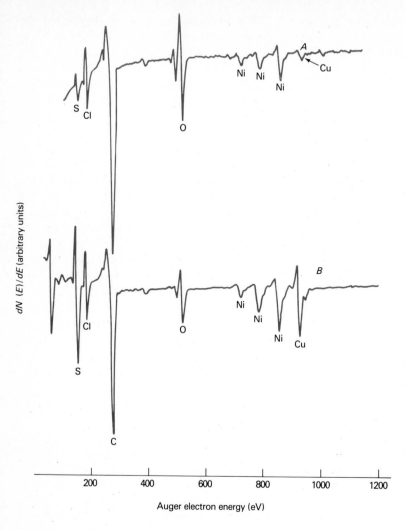

FIGURE 16-3 Auger electron spectra for a 70% Cu/30% Ni alloy: *A*, passivated by anodic oxidation; *B*, not passivated. (Adapted from: G. E. McGuire, *et al.*, *J. Electrochem. Soc.*, 1978, *125*, 1802. Reprinted by permission of the publisher, The Electrochemical Society, Inc.)

radiation. Note that the peaks are well separated making qualitative identification quite straight forward.

Auger electron emission and X-ray fluorescence (Figure 16–1**d**) are competitive processes and their relative rates depend upon the atomic number of the element involved. High atomic numbers favor fluorescence, while Auger emission predominates with atoms of low atomic numbers. As a consequence, X-ray fluorescence is not a very sensitive means for detecting

elements with atomic numbers smaller than about 10.

16B Instrumentation

Instruments for electron spectroscopy are offered by perhaps a dozen instrument manufacturers. These products differ considerably in types of components, configurations, and costs. Some are designed for a single type of application, such as ESCA, while others can be adapted to two or three types of electron spectroscopy by purchase of suitable accessories. All are expensive ($50,000 to $150,000).[3]

Electron spectrometers are made up of components whose functions are completely analogous to those encountered in optical spectroscopic instruments. These components include: (1) a source, (2) a sample holder or container, (3) an analyzer, which has the same function as a monochromator, (4) a detector, and (5) a signal processor and readout. Figure 16–4 shows a typical arrangement of these components.

16B-1 SOURCE AND SAMPLE HOLDER

In most instruments, the sample holder and source form an integral unit from which dis-

crete-energy electrons from the sample are directed through a slit to the electron analyzer. The source or excitation device consists of an X-ray tube, an electron gun, or a gas-discharge tube, depending upon the type of electron spectroscopy to be performed.

X-Ray Sources. The most common X-ray sources for electron spectrometers are X-ray tubes equipped with magnesium or aluminum targets. The K_α lines for these two elements have considerably narrower bandwidths (0.8 to 0.9 eV) than those encountered with higher atomic number targets; narrow bands are desirable since they lead to enhanced resolution. In some instruments, source bandwidths down to 0.3 eV are obtained by means of a crystal monochromater similar to those described in Section 15B–3.

Gas-Discharge Sources. The most important sources for ultraviolet photoelectron spec-

---

[3] For a useful review of electron spectrometers, see: T. A. Carlson, *Photoelectron and Auger Spectroscopy*, Chapter 2, Plenum Press: New York, 1975.

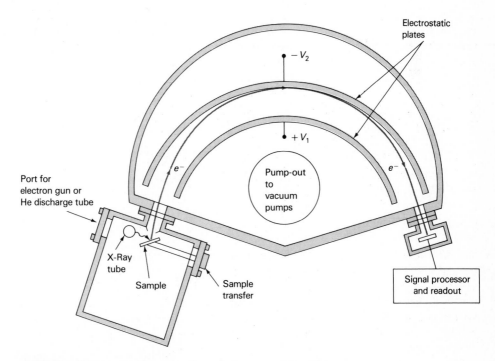

FIGURE 16-4 Schematic diagram of an electron spectrometer.

troscopy are discharge lamps containing helium, neon, or argon at low pressures. The spectra from these lamps are made up of a few sharp lines resulting from relaxation of electrically excited atoms or ions. The most commonly used line is that for helium(I), which occurs at 58.4 nm (21.22 eV).

Electron Guns. Electron guns, which are similar in construction to the electron source in cathode ray tubes (Section 2E–1), produce a beam of electrons with energies of 1 to 10 keV, which can be focused on the surface of a sample for Auger electron studies. One of the special advantages of Auger spectroscopy is its capability for very high spatial-resolution scanning of solid surfaces. Normally electron beams with diameters ranging from 500 to 5 μm are used for this purpose. Guns producing beams of 5 to 10 μm are called *Auger microprobes* and are employed for scanning solid surfaces in order to detect and determine the elemental composition of inhomogeneities.

Sample Compartment. Solid samples are mounted in a fixed position as close to the photon or electron source and the entrance slit of the spectrometer as possible (see Figure 16–4). In order to avoid attenuation of the electron beam, the sample compartment must be evacuated to a pressure of at least 10^{-5} torr. Often, however, much better vacuums (10^{-9} to 10^{-10} torr) are required to avoid contamination of the sample surface by substances such as oxygen or water that react with or are adsorbed on the surface. Furthermore, provisions must be made to clean the sample surface *in situ*. Cleaning may involve bombarding with helium ions from an ion gun, baking the sample at high temperature, mechanical scraping, or sputtering the sample by applying a high voltage under an argon pressure of about 10^{-4} torr. In some instances, the sample may have to be bathed in a reducing atmosphere to free it from oxides.

Gas samples are leaked into the sample area through a slit of such a size as to pro-

vide a pressure of perhaps 10^{-2} torr. Higher pressures lead to excessive attenuation of the electron beam due to inelastic collisions; on the other hand, if the sample pressure is too low, weakened signals are obtained.

16B–2 ANALYZERS

Two basic types of electron analyzers are encountered: (1) retarding-field and (2) dispersion. In the retarding-field instrument shown in Figure 16–5, the electrons from the sample sources pass through two cylindrical grids to an outer collector, which is also cylindrical. The grids are metallic screens, which provide approximately 70% transmission. An increasing potential difference is applied across the grids to retard the electrons flowing from the source to the collector. At a high enough potential difference, electrons of energy e_2 will be retarded and the collector signal will decrease. The collector signal q is amplified, differentiated, and displayed on a recorder as the grid potential is scanned. Retarding-field instruments are relatively simple and efficient but do not have the high resolution of dispersion systems.

Most electron spectrometers are of the dispersion type in which the electron beam

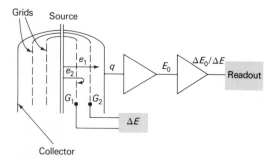

FIGURE 16-5 A retarding-field electron spectrometer. (Reprinted with permission from: D. M. Hercules, *Anal. Chem.*, **1970**, *42*, 27A. Copyright 1970 American Chemical Society.)

is deflected by an electrostatic or magnetic field in such a way that the electrons travel in a curved path (see Figure 16–4). The radius of curvature is dependent upon the kinetic energy of the electron and the magnitude of the field. By varying the field (usually electrostatic), electrons of various kinetic energies can be focused on the detector.

The deflection plates in an electron spectrometer may be cylindrical as in Figure 16–4, spherical, or hemispherical. For cylindrical plates, the relationship between the plate voltage V_1 and V_2 and the energy of the electron E_k is given by

$$V_2 - V_1 = 2E_k R \log(R_1/R_2)$$

where R_1 and R_2 are the radii of the two plates and R is their average.

Typically, pressures in the analyzer of an electron spectrometer are maintained at about 10^{-5} torr.

16B-3 DETECTORS

Most modern electron spectrometers are based upon solid-state, channel electron multipliers, which consist of tubes of glass that have been doped with lead or vanadium. When a potential of several kilovolts is applied across these materials, a cascade or pulse of 10^6 to 10^8 electrons is produced for each incident electron. The pulses are then counted electronically. Several manufacturers are now offering two-dimensional multichannel electron detectors that are analogous in construction and application to the multichannel photon detectors described in Section 5F–3. Here, all of the resolution elements of an electron spectra are monitored simultaneously and the data stored in a computer for subsequent display. The advantages of such a system are similar to those realized with multichannel photon detectors.

16B-4 MAGNETIC SHIELDING

The path of electrons in an analyzer is affected by Earth's magnetic field as well as extraneous magnetic fields likely to be encountered in a laboratory. For high-resolution work these fields must be reduced to about 0.1 mG (the Earth's magnetic field is roughly 500 mG). Two methods are employed for canceling the effects of these fields: ferromagnetic shielding and Helmholtz coils. All commercial spectrometers use the former because it is more simple and compact than the latter. Ferromagnetic shielding is accomplished by surrounding the sample and analyzer areas with two or more layers of a ferromagnetic alloy whose permeability ratio is several orders of magnitude greater than iron.

16C Applications of Electron Spectroscopy

Electron spectroscopy provides qualitative and quantitative information about the elemental composition of matter, particularly of solid surfaces. It also often provides useful structural information.

16C-1 APPLICATIONS OF X-RAY PHOTOELECTRON SPECTROSCOPY

X-Ray photoelectron spectroscopy is the most widely used of the several techniques listed in Table 16–1. Examples of these applications are described in the paragraphs that follow.[4]

Qualitative Elemental Analysis. A low-resolution, wide-scan spectrum (often called a survey

---

[4] In addition to the references listed in footnote 2, see also: H. Windawi and F. F. L. Ho, *Applied Electron Spectroscopy for Chemical Analysis,* Wiley: New York, 1982; D. Briggs, in *Electron Spectroscopy: Theory, Techniques and Applications,* C. R. Brundle and A. D. Baker, Eds., Volume 3, Chapter 6, Academic Press: New York, 1979; T. L. Barr, *Amer. Lab.,* **1978,** *10* (11), 65 and *10* (12), 40.

spectrum) such as that shown in Figure 16–2, serves as the basis for the determination of the elemental composition of samples. With a magnesium or aluminum K_α source all elements except hydrogen and helium emit core electrons having characteristic binding energies. Typically, a survey spectrum encompasses a kinetic energy range of 250 to 1500 eV, which corresponds to binding energies of about 0 to 1250 eV. Every element in the periodic table has one (or more) energy levels that will result in the appearance of peaks in this region. In most instances, the peaks are well resolved and lead to unambiguous identification provided the element is present in concentrations greater than about 0.1%. Occasionally peak overlap is encountered such as O1s/Sb3d or Al2s,2p/Cu3s,3p. Usually, problems due to spectral overlap can be resolved by investigating other spectral regions for additional peaks. Often peaks resulting from Auger electrons are found in ESCA spectra (see, for example, the peak at about 610 eV in Figure 16–2). Such peaks are readily identified by comparing spectra produced by two X-ray sources (usually magnesium and aluminum K_α). Auger peaks remain unchanged on the kinetic energy scale while photoelectric peaks are displaced.

Chemical Shifts and Oxidation States. When one of the peaks of a survey spectrum is examined under conditions of higher energy resolution, the position of the maximum is found to depend to a small degree upon the chemical environment of the atom responsible for the peak. That is, variations in the number of valence electrons, and the type of bonds they form, influence the binding energies of core electrons. The effect of number of valence electrons and thus the oxidation state is demonstrated by the data for several elements shown in Table 16–2. Note that in each case, binding energies increase as the oxidation state becomes more positive. This *chemical shift* can be explained by assuming that the attraction of the nucleus for a core electron is diminished by the presence of outer electrons. When one of these electrons is removed, the effective charge sensed by the core electron is increased; thus an increase in binding energy results.

One of the most important applications of ESCA has been the identification of oxidation states of elements contained in various kinds of inorganic compounds.

Table 16–2
Chemical Shifts as a Function of Oxidation State[a]

Element[b]	Oxidation State									
	−2	−1	0	+1	+2	+3	+4	+5	+6	+7
Nitrogen (1s)	—	*0[c]	—	+4.5[d]	—	+5.1	—	+8.0	—	—
Sulfur (1s)	−2.0	—	*0	—	—	—	+4.5	—	+5.8	—
Chlorine (2p)	—	*0	—	—	—	+3.8	—	+7.1	—	+9.5
Copper (1s)	—	—	*0	+0.7	+4.4	—	—	—	—	—
Iodine (4s)	—	*0	—	—	—	—	—	+5.3	—	+6.5
Europium (3d)	—	—	—	—	*0	+9.6	—	—	—	—

[a] All shifts are in electron volts measured relative to the oxidation states indicated by (*). (Reprinted with permission from D. M. Hercules, *Anal. Chem.*, **1970**, *42*, 28A. Copyright 1970 American Chemical Society.)
[b] Type of electrons given in parentheses.
[c] Arbitrary zero for measurement, end nitrogen in NaN_3.
[d] Middle nitrogen in NaN_3.

Chemical Shifts and Structure. Figure 16–6 illustrates the effect of structure on the position of peaks for an element. Each peak corresponds to the $1s$ electron contained in the carbon atom located directly above it in the structural formula. Here, the shift in binding energies can be rationalized by taking into account the effect of the various functional groups on the effective nuclear charge experienced by the $1s$ core electron. For example, of all of the attached groups, fluorine atoms have the greatest ability to withdraw electron density from the carbon atom. The effective nuclear charge felt by the carbon $1s$ electron is, therefore, a maximum as is the binding energy.

Figure 16–7 indicates the position of peaks for sulfur in its several oxidation states and in various types of organic compounds. The data in the top row clearly demonstrate the effect of oxidation state. Note also in the last four rows of the chart that ESCA discriminates between two sulfur atoms contained in a single ion or molecule. Thus, two peaks are observed for thiosulfate ion ($S_2O_3^{2-}$), suggesting different oxidation states for the two sulfur atoms contained therein.

ESCA spectra provide not only qualitative information about types of atoms present in a compound but also the relative number of each type. Thus, the nitrogen $1s$ spectrum for sodium azide ($Na^+N_3^-$) is made up of two peaks having relative areas in the ratio of $2:1$ corresponding to the two end nitrogens and the center nitrogen respectively.

It is worthwhile pointing out again that the photoelectrons produced in ESCA are incapable of passing through more than perhaps 10 to 50 Å of a solid. Thus, the most important applications of electron spectroscopy, like X-ray microprobe spectroscopy, are for the accumulation of infor-

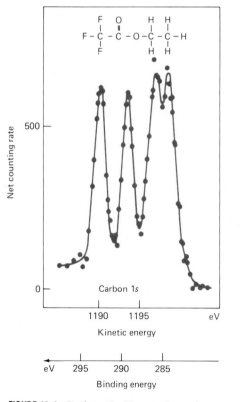

FIGURE 16-6 Carbon $1s$ X-ray photoelectron spectrum for ethyl trifluoroacetate. (From: K. Siegbahn, *et al.*, *ESCA: Atomic, Molecular, and Solid-State Studies by Means of Electron Spectroscopy*, p. 21, Almquist and Wiksells: Upsala, 1967. With permission.)

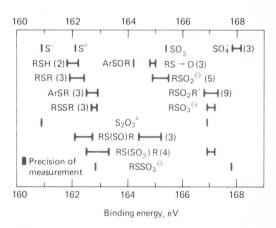

FIGURE 16-7 Correlation chart for sulfur $2s$ electron binding energies. The numbers in parentheses indicate the number of compounds examined. (Reprinted with permission from: D. M. Hercules, *Anal. Chem.*, **1970**, *42* (1), 35A. Copyright 1970 American Chemical Society.)

mation about surfaces. Examples of some of its uses include identification of active sites and poisons on catalytic surfaces, determination of surface contaminants on semiconductors, analysis of the composition of human skin, and study of oxide surface layers on metals and alloys.

It is also evident that the method has a substantial potential in the elucidation of chemical structure (see Figures 16–6 and 16–7); the information obtained appears comparable to that from NMR or infrared spectroscopy. A noteworthy attribute is the ability of ESCA to distinguish among oxidation states of an element.

It is of interest to note that the information obtained by ESCA must also be present in the absorption edge of an X-ray absorption spectrum for a compound. Most X-ray spectrometers, however, do not have sufficient resolution to permit ready extraction of the structural information.

Quantitative Applications. Although several authors have reported using ESCA for quantitative determination of the elemental composition of various inorganic and organic materials, the method has not enjoyed widespread application for this purpose. Both peak intensities and peak areas have been used as the analytical parameter with the relationship between these quantities and concentration being established empirically. Often, internal standards have been recommended. Relative precisions of 3 to 10% have been claimed. For the analysis of solids and liquids, it is necessary to assume that the surface composition of the sample is the same as its bulk composition. For many applications this assumption can lead to significant errors.

16C–2 APPLICATIONS OF ULTRAVIOLET PHOTOELECTRON SPECTROSCOPY

The majority of applications of ultraviolet photoelectron spectroscopy have involved the study of samples in the vapor phase.

Most of these studies have been devoted to the elucidation of electronic effects within molecules, where the technique has provided important information about orbital structure. To date, the strictly analytical applications of UPS have been limited, and it seems doubtful that this type of electron spectroscopy will ever become a major tool for qualitative or quantitative analyses.

16C–3 APPLICATIONS OF AUGER ELECTRON SPECTROSCOPY

Auger and X-ray photoelectron spectroscopy provide similar information about the composition of matter. The methods tend to be complementary rather than competitive, however, with Auger spectroscopy being more reliable and efficient for certain applications and ESCA for others. As mentioned earlier, most instrument manufacturers recognize their complementary nature by making provisions for both kinds of measurements with a single instrument.

The particular strengths of Auger spectroscopy are its sensitivity for atoms of low atomic number, its minimal matrix effects, and above all its high spatial resolution, which permits detailed examination of solid surfaces. To date, Auger spectroscopy has not been extensively used to provide the kind of structural and oxidation state information that was described for ESCA. Quantitative analysis by the procedure is difficult or impossible.

Qualitative Analysis of Solid Surfaces. Typically, an Auger spectrum is obtained by bombarding a small area of a surface (diameter, 5 to 500 μm) with a beam of electrons from a gun. A derivative electron spectrum, such as that shown in Figure 16–3, is then obtained with an analyzer. An advantage of Auger spectroscopy for surface studies is that the low-energy Auger electrons (20 to 1000 eV) are able to penetrate only a few atomic layers (3–20 Å) of solid. Thus, while the electrons from the electron guns penetrate to a considerably greater depth below the sample surface, only those Auger electrons from the first four or five atomic layers escape to reach the analyzer. Consequently, an Auger spectrum is more likely to reflect the true *surface* composition of solids than is an ESCA spectrum.

The two Auger spectra in Figure 16–3 are for samples of a 70% copper/30% nickel alloy, which is often used for structures where saltwater corrosion resistance is required. Corrosion resistance of this alloy is markedly enhanced by preliminary anodic oxidation in a strong solution of chloride. Figure 16–3B is the spectrum of an alloy surface that has been *passivated* in this way. Spectrum A is for another sample of the alloy in which the anodic oxidation potential was not great enough to cause significant passivation. The two spectra clearly reveal the chemical differences between the two samples that account for the greater corrosion resistance of the former. First, the copper to nickel ratio in the surface layer of the nonpassivated sample is approximately that for the bulk, whereas in the passivated material, the nickel peaks completely overshadow the copper peak. Furthermore, the oxygen to nickel ratio in the passivated sample approaches that for pure anodized nickel, which also has a high corrosion resistance. Thus, the resistance towards corrosion of the alloy appears to result from the creation of a surface that is largely nickel oxide. The advantage of the alloy over pure nickel is its significantly lower cost.

Depth Profiling of Surfaces. Depth profiling involves the determination of the elemental composition of a surface as it is being etched away (sputtered) by a beam of argon ions. Either ESCA or Auger spectroscopy can be used for elemental detection although the latter is the more common. Figure 16–8 shows schematically how the process is carried out with a highly focused electron beam called an Auger *microprobe;* the diameter of the microprobe beam is about 5 μm. The

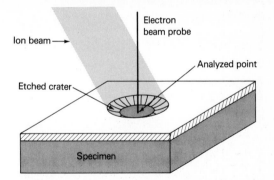

FIGURE 16-8 Schematic representation of the simultaneous use of ion sputter-etching and Auger spectroscopy for determining depth profiles. (Courtesy of Physical Electronic Industries, Inc., Eden Prairie, MN.)

microprobe and etching beams are operated simultaneously with the intensity of one or more (up to six) of the resulting Auger peaks being recorded as a function of time. Since the etching rate is related to time, a depth profile of elemental composition is obtained. Such information is of vital importance in a variety of studies such as corrosion chemistry, catalyst behavior, and properties of semiconductor junctions.

Figure 16–9 gives a depth profile for the copper/nickel alloy described in the previous section (Figure 16–3). Here, the ratio of the peak intensities for copper versus nickel are recorded as a function of sputtering time. Curve A is the profile for the sample that had been passivated by anodic oxidation. With this sample, the copper/nickel ratio is essentially zero for the first ten minutes of sputtering, which corresponds to a depth of about 500 Å. The ratio then rises and approaches that for a sample of alloy that had been chemically etched so that its surface is approximately that of the bulk sample (curve C). The profile for the nonpassivated sample (curve B) resembles that of the chemically etched sample although some evidence is seen for a thin nickel oxide coating.

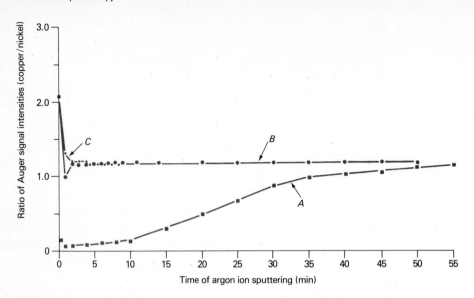

FIGURE 16-9   Auger sputtering profiles for the copper/nickel alloys shown in Figure 16–3; *A*, passivated sample; *B*, nonpassivated sample; *C*, chemically etched sample representing the bulk material. (Adapted from: G. E. McGuire, *et al.*, *J. Electrochem. Soc.*, **1978**, *125*, 1802. Reprinted by permission of the publisher, The Electrochemical Society, Inc.)

Line Scanning.   Line scans are used to characterize the surface composition of solids as a function of distance along a straight line of 100 μm or more. For this purpose, an Auger microprobe is used that produces a beam that can be moved across a surface in a reproducible way. Figure 16–10 shows Auger line scans along the surface of a semiconductor device. In the upper figure, the relative peak amplitude of an oxygen peak is recorded as a function of distance along a line; the lower figure is the same scan when the analyzer was set to a peak for gold.

PROBLEMS

16–1.   Describe the mechanism of the production of an MNN Auger electron.

16–2.   Explain why the information from an ESCA chemical shift must also be contained in an X-ray absorption edge.

16–3.   An ESCA electron was found to have a kinetic energy of 1073.5 eV when a Mg K_α source was employed (λ = 9.8900 Å). The electron spectrometer had a work function of 14.7 eV.
(a)   Calculate the binding energy for the emitted electron.
(b)   If the signal was from a S(2s) electron, was the analyte S^{2-}, S^0, SO_3^{2-}, or SO_4^{2-}?
(c)   What would the kinetic energy have been if an Al K_α source had been used (λ = 8.3393 Å)?

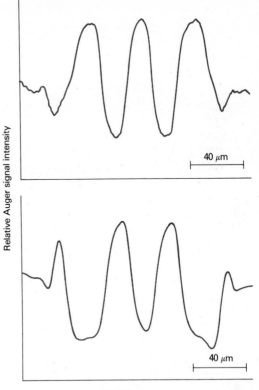

Relative Auger signal intensity

40 μm

40 μm

Linear distance

FIGURE 16-10 Auger line scans for oxygen (top) and gold (bottom) obtained for the surface of a semiconductor device. (Courtesy of Physical Electronics Industries, Inc., Eden Prairie, MN.)

(d) If the ejected electron with the Mg K_α source had been an Auger electron, what would its kinetic energy be with the Al K_α source?

16-4. An ESCA electron was found to have a kinetic energy of 1052.6 eV when ejected with an Al K_α source ($\lambda = 8.3393$ Å) and measured in a spectrometer with a work function of 27.8 eV. The electron is believed to be a N(1s) electron in $NaNO_3$.

(a) What was the binding energy for the electron?

(b) What would be the kinetic energy of the electron if a Mg K_α ($\lambda = 9.8900$ Å) source were used?

(c) How could one be sure that a peak was an ESCA and not an Auger electron peak?

(d) At what binding and kinetic energies would a peak for $NaNO_2$ be expected when the Al K_α source was used with the same spectrometer?

Radiochemical Methods

The ready availability of both natural and artificial radioactive isotopes has made possible the development of analytical methods (radiochemical methods) that are both sensitive and specific.[1] These procedures are often characterized by good accuracy and widespread applicability; in addition, some minimize or eliminate chemical separations that are required in other analytical methods.

Radiochemical methods are of three types. In *activation analysis,* activity is induced in one or more elements of the sample by irradiation with suitable particles (most commonly thermal neutrons from a nuclear reactor); the resulting radioactivity is then measured. In an *isotope dilution* pro-

cedure, a weighed quantity of radioactively tagged analyte having a known activity is added to a measured amount of the sample. After thorough mixing to assure homogeneity, a fraction of the component of interest is isolated and purified; the analysis is then based upon the activity of this isolated fraction. In a *radiometric* method, a radioactive reagent is employed to separate the analyte completely from the bulk of the sample; the activity of the isolated material is then measured. Alternatively, the analyte may be titrated with a radioactive reagent; here, an end point is established by activity measurements.

17A Radioactive Isotopes

With one exception, all atomic nuclei are made up of a collection of protons and neutrons; the exception, of course, is the hydrogen nucleus, which consists of a proton only. The chemical properties of an atom are de-

---

[1] For a detailed treatment of radiochemical methods, see: G. Friedlander, J. W. Kennedy, E. S. Macias, and J. M. Miller, *Nuclear and Radiochemistry,* 3d ed., Wiley: New York, 1981; H. M. Clark, in *Physical Methods of Chemistry,* A. Weissberger and B. W. Rossiter, Eds., Part III D, vol. 1, Chapter 9, Wiley-Interscience: New York, 1972.

termined by its atomic number Z, which is the number of protons contained in its nucleus. The sum of the number of neutrons and protons in a nucleus is the atomic mass. Isotopes of elements are atoms having the same atomic number but a different mass number. That is, the nuclei of isotopes contain the same number of protons but different numbers of neutrons.

Stable isotopes are those that have never been observed to decay spontaneously. *Radioactive isotopes,* in contrast, undergo spontaneous disintegration, which ultimately leads to stable isotopes. The disintegration or *radioactive decay* of isotopes occurs with the emission of electromagnetic radiation in the form of X-rays or γ rays; with the formation of elementary particles such as electrons, positrons, and the helium nucleus; or by *fisson* in which the nuclei break up into smaller nuclei.

17A-1 RADIOACTIVE DECAY PRODUCTS

Table 17–1 lists the more important (from a chemist's viewpoint) types of decay products that make up what is called radioactive radiation. Most can serve as a basis for chemical analysis.

17A-2 DECAY PROCESSES

Several types of radioactive decay processes yield the products listed in Table 17–1.

Alpha Decay. Alpha decay is a common radioactive process encountered with heavier isotopes. Isotopes with mass numbers less than perhaps 150 ($Z \cong 60$) seldom yield alpha particles. The alpha particle is a helium nucleus having a mass of 4 and a charge of $+2$. An example of alpha decay is shown by the equation[2]

$$^{238}_{92}U \rightarrow\, ^{234}_{90}Th + {}^{4}_{2}He$$

Here, uranium-238 is converted to thorium-234, a *daughter* element having an atomic number that is two less than the *parent.*

Alpha particles from a particular decay process are either monoenergetic or are distributed among relatively few discrete energies. They progressively lose their energy as a result of collisions as they pass through matter, and are ultimately converted into helium atoms through capture of two electrons from their surroundings. Their relatively large mass and charge render alpha particles highly effective in producing ion pairs within the matter through which they pass; this property makes their detection and measurement easy. Because of their high mass and charge, alpha particles have a

---

[2] The subscript and superscript for the element refer to its atomic number and mass number, respectively. Thus, the uranium in the equation has an atomic number of 92 and atomic mass of 238.

Table 17–1 Characteristics of Common Radioactive Decay Products			
Product	**Symbol**	**Charge**	**Mass Number**
Alpha particle	α	$+2$	4
Beta particles			
Negatron	β^-	-1	1/1840
Positron	β^+	$+1$	1/1840
Gamma ray	γ	0	0
X-Ray	χ	0	0
Neutron	n	0	1
Neutrino	ν	0	0

low penetrating power in matter. The identity of an isotope that is an alpha emitter can often be established by measuring the length (or range) over which the emitted alpha particles produce ion pairs within a particular medium (often air).

Alpha particles are relatively ineffective for producing artificial isotopes because of their low penetrating power.

Beta Decay.   Any nuclear reaction in which the atomic number changes but the mass number does not is classified as β decay. Three types of β decay are encountered: *negatron formation, positron formation,* and *electron* or *K capture*. Examples of the three processes are:

$$^{14}_{6}C \rightarrow {}^{14}_{7}N + \beta^- + \nu$$

$$^{65}_{30}Zn \rightarrow {}^{65}_{29}Cu + \beta^+ + \nu$$

$$^{48}_{24}Cr \rightarrow {}^{48}_{23}V + \text{X-rays}$$

Here, ν in the first two equations represents a neutrino, a particle of no significance in analytical chemistry. One of the products of the third equation (electron capture) is an X-ray photon, which does have considerable importance in analytical chemistry and has been discussed in Section 15A–1. It should be noted that the X-ray photon is *not a product of the nuclear process*. The capture of an electron by the nucleus, which is the nuclear process, leaves one of the extranuclear orbitals of an atom vacant. X-Ray emission then accompanies the filling of this orbital by an electron from one of the outer orbitals.

Two types of β particles are created by radioactive decay. Negatrons (β^-) are electrons that form when one of the neutrons in the nucleus is converted to a proton. In contrast, the positron (β^+), with the mass of the electron, forms when the number of protons in the nucleus is decreased by one. The positron has a transitory existence, its ultimate fate being annihilation by reaction with an electron to yield two gamma photons.

In contrast to alpha emission, beta decay is characterized by production of particles with a continuous spectrum of energies ranging from nearly zero to some maximum that is characteristic for each decay process. The beta particle is not nearly as effective as the alpha particle in producing ion pairs in matter because of its small mass (about 1/7000 that of an alpha particle); at the same time, its penetrating power is substantially greater. Beta ranges in air are difficult to evaluate because of the high likelihood that scattering will occur. As a result, beta energies are based upon the thickness of an absorber, ordinarily aluminum, required to stop the particle. This thickness is the *range* expressed in mg/cm^2.

Gamma Ray Emission (γ).   Many alpha and beta emission processes leave a nucleus in an excited state, which then returns to the ground state in one or more quantized steps with the release of gamma rays. It is important to note that γ rays, except for their source, are indistinguishable from X-rays. That is, gamma rays are produced by nuclear relaxations while X-rays derive from electronic relaxations. The gamma ray emission spectrum is characteristic for each nucleus and is thus useful for identifying radioisotopes.

Not surprisingly, gamma radiation is highly penetrating. Upon interaction with matter, gamma rays lose energy by three mechanisms; the one that predominates depends upon the energy of gamma photon. With low energy gamma radiation, the *photoelectric effect* predominates. Here, the gamma photon disappears after ejecting an electron from an atomic orbit (usually a K orbit). The photon energy is totally consumed in overcoming the binding energy of the electron and in imparting kinetic energy to the ejected electron. With relatively energetic gamma rays, the *Compton effect* is encountered. In this instance, an electron is also ejected from an atom but acquires only a part of the photon energy. The photon, now with diminished energy, recoils from the electron and can then go on to undergo further Compton or photoelectric interactions. If the gamma photon possesses sufficiently high energy (at last 1.02 MeV), *pair production* can occur. Here, the photon is totally absorbed in creating a positron and an electron in the field surrounding a nucleus.

X-Ray Emission. Two nuclear processes result in loss of inner shell electrons from an atom. X-Rays are then formed from *electronic* transitions in which outer electrons fill the vacancies created by the nuclear process. One of these processes is electron capture, which was discussed in the earlier section on β decay and in Section 15A–1. A second process, which may lead to X-rays, is *internal conversion*, a type of nuclear reaction that competes with or may replace gamma ray emission. In this instance, an electromagnetic interaction between the excited nucleus and an extranuclear electron results in the ejection of an orbital electron with a kinetic energy equal to the difference between the energy of the nuclear transition and the binding energy of the electron (see Section 16A–3). The emission of this so-called *Auger electron* leaves a vacancy in the K or L orbital; X-rays are emitted as the orbital is filled by an electronic transition.

17A–3 RADIOACTIVE DECAY RATES

Radioactive decay is a completely random process. Thus, while no prediction can be made concerning the lifetime of an individual nucleus, the behavior of a large ensemble of like nuclei can be described by the first-order rate expression

$$-\frac{dN}{dt} = \lambda N \tag{17–1}$$

where N represents the number of radioac-

tive nuclei in the sample at time t and λ is the characteristic *decay constant* for a particular radioisotope. Upon rearranging this equation and integrating over the interval between $t = 0$ and $t = t$ (during which the number of radioactive nuclei in the sample decreases from N_0 to N), we obtain

$$\ln \frac{N}{N_0} = -\lambda t \tag{17–2}$$

or

$$N = N_0 e^{-\lambda t} \tag{17–3}$$

The *half-life* of a radioactive isotope is defined as the time required for one-half the number of radioactive atoms in a sample to undergo decay; that is, for N to become equal to $N_0/2$. Substitution of $N_0/2$ for N in Equation 17–2 gives

$$t_{1/2} = \frac{0.693}{\lambda} \tag{17–4}$$

Half-lives of radioactive species range from small fractions of a second to millions of years.

In performing radioactivity measurements, the number of atoms N is generally not evaluated directly nor is the absolute rate of change ($-dN/dt$) usually determined. Instead, the *activity A,* which is proportional to N, is employed where

$$A = cN \quad \text{and} \quad -\frac{dA}{dt} = -c\frac{dN}{dt} \tag{17–5}$$

Here, c is a constant called the *detection coefficient,* which depends upon the nature of the detector, the efficiency of counting disintegrations, and the geometric arrangement of sample and detector. The decay law given by Equation 17–3 can then be written in the form

$$A = A_0 e^{-\lambda t}$$

17A-4 COUNTING STATISTICS[3]

As will be shown in Section 17B, radioactivity is measured by means of a detector that produces a pulse of electricity for each atom undergoing decay. Quantitative information about decay rates is obtained by counting these pulses for a specified period. Table 17–2 shows typical decay data obtained by successive one-minute counts of a radioactive source. Because the decay process is random, considerable variation among the data is observed. Thus, the spread of the data is 55 counts/min and the standard deviation, as defined by Equation 1–6 (Section 1C–3), is 13.7 counts/min.

Although radioactive decay is random, the data, particularly for low counts, are not distributed according to Equation 1–1 (p. 7). That is, the decay process does not follow exactly the Gaussian behavior shown in Figure 1–2. The reason that decay data are not normally distributed lies in the fact that radioactivity consists of a series of discrete events that cannot vary continuously as can the indeterminate errors for which the Gaussian distribution applies. Furthermore, negative counts are not possible. Therefore,

the data cannot be distributed symmetrically about the mean.

In order to describe accurately radioactive behavior, it is necessary to assume a *Poisson distribution*, which is given by the equation

$$y = \frac{\mu^{x_i}}{x_i!} e^{-\mu} \qquad (17\text{–}6)$$

Here, as in Equation 1–1, y is the frequency of occurrence of a given count x_i and μ is the mean for a large set of counting data.[4]

The data plotted in Figure 17–1 were obtained with the aid of Equation 17–6. These curves show the deviation from the true average count $(x_i - \mu)$ that would be expected if 1000 replicate observations were made on the same sample. Curve *A* gives the distribution for a substance for which the true average count μ for a selected period is 5; curves *B* and *C* correspond to samples having the true means of 15 and 35. Note that the *absolute* deviations become greater with in-

---

[3] For a more complete discussion, see: G. Friedlander, J. W. Kennedy, E. S. Macias, and J. M. Miller, *Nuclear and Radiochemistry*, 3d ed., Chapter 9, Wiley: New York, 1981.

[4] In the derivation of this equation it is assumed that the counting period is short with respect to the half-life so that no significant change in the number of radioactive atoms occurs. Further restrictions include a detector that responds to the decay of a single isotope only and an invariant counting geometry so that the detector responds to a constant fraction of the decay events that occur.

Table 17–2
Variations in One-Minute Counts From a Radioactive Source

Minutes	Counts	Minutes	Counts
1	180	7	168
2	187	8	170
3	166	9	173
4	173	10	132
5	170	11	154
6	164	12	167

Total counts = 2004
Average counts/min = $\bar{x}$ = 167

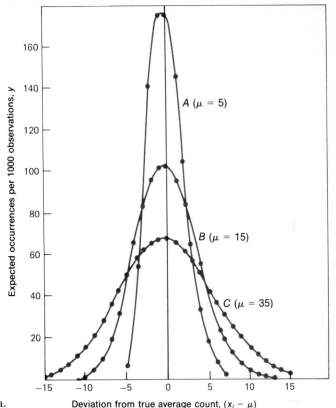

Expected occurrences per 1000 observations, y

A ($\mu = 5$)

B ($\mu = 15$)

C ($\mu = 35$)

Deviation from true average count, ($x_i - \mu$)

FIGURE 17-1 Distribution of counting data.

creases in μ, but the *relative* deviations become smaller. Note also that for the two smaller number of counts the distribution is distinctly not symmetric around the average; this lack of symmetry is a consequence of the fact mentioned earlier that a negative count is impossible, while a finite likelihood always exists that a given count can exceed the average by severalfold.

As the number of counts becomes large (say >100), the distribution of decay data approaches Gaussian behavior, thus permitting the application of many of the techniques described in Section 1C.

Standard Deviation of Counting Data. In Section 1C-3, it was shown that the breadth of a Gaussian curve was determined by the standard deviation σ of the data. In contrast, Equation 17-6 for a Poisson distribution contains no corresponding standard deviation term and indeed, it can be shown that the breadth of curves such as those in Figure 17-1 are dependent only upon the total number of counts for any given period.[5] That is,

$$\sigma_M = \sqrt{M} \qquad (17\text{-}7)$$

where M is the number of counts for any given period and σ_M is the standard deviation for a Poisson distribution.

The relative standard deviation $(\sigma_M)_r$ is given by

$$(\sigma_M)_r = \frac{\sigma_M}{M} = \frac{\sqrt{M}}{M} = \frac{1}{\sqrt{M}} \qquad (17\text{-}8)$$

[5] See footnote 3.

Thus, although the absolute standard deviation increases with the number of counts, the relative standard deviation decreases.

In normal practice, activities of samples are more conveniently expressed in terms of counting rates R (counts per minute or cpm) than in counts for some arbitrary time. Clearly,

$$R = \frac{M}{t} \tag{17-9}$$

where t is the time in minutes required to obtain M counts. The standard deviation in rate units σ_R can be obtained by dividing both sides of Equation 17–7 by t; thus,

$$\sigma_R = \frac{\sigma_M}{t} = \frac{\sqrt{M}}{t}$$

Substitution of Equation 17–9 leads to

$$\sigma_R = \sqrt{\frac{R}{t}} \tag{17-10}$$

In relative terms,

$$(\sigma_R)_r = \frac{\sqrt{R/t}}{R} = \sqrt{\frac{1}{Rt}} \tag{17-11}$$

EXAMPLE 17-1. Calculate the absolute and relative standard deviations in the counting rate for (a) the first entry in Table 17–2 and (b) the mean of all of the data in the table.

(a) Applying Equation 17–10,

$$\sigma_R = \sqrt{\frac{R}{t}} = \sqrt{\frac{180}{1}} = 13.4 \text{ cpm}$$

$$(\sigma_R)_r = \frac{13.4}{180} \times 100 = 7.4\%$$

(b) For the entire set, $R = 2004/12 = 167$ cpm

$$\sigma_R = \sqrt{\frac{167}{12}} = 3.7 \text{ cpm}$$

$$(\sigma_R)_r = \frac{3.7}{167} \times 100 = 2.2\%$$

Confidence Intervals for Counts.　In Section 1C, the confidence interval for a measurement was defined as the limits around a measured quantity within which the true mean can be expected to fall with a stated probability. When the measured standard deviation is believed to be a good approximation of the true standard deviation $(s \rightarrow \sigma)$, the confidence limit *C.L.* is given by Equation 1–7 (p. 11). That is,

$$C.L. \text{ for } \mu = \bar{x} \pm z\sigma$$

For counting rates, this equation takes the form

$$C.L. \text{ for } R = R \pm z\sigma_R \tag{17-12}$$

where z is dependent upon the desired level of confidence. Some values for z are given in Table 1–3 (p. 11).

EXAMPLE 17-2. Calculate the 95% confidence limits for (a) the first entry in Table 17–2 and (b) the mean of all of the data in the table.

(a) In Example 17–1, we found that $\sigma_R = 13.4$ cpm. Table 1–3 reveals that $z = 1.96$ at the 95% confidence level. Thus, for R

$$95\% \ C.L. = 180 + 1.96 \times 13.4$$
$$= 180 \pm 26 \text{ cpm}$$

(b) In this instance, σ_R was found to be 3.7 cpm, and

$$95\% \ C.L. \text{ for } R = 167 \pm 3.7 \times 1.96$$
$$= 167 \pm 7 \text{ cpm}$$

Thus, there are 95 chances in 100 that the true rate for R (for the average of 12 min of counting) lies between 160 and 174 counts/min. For the single count in part (a), 95 out of 100 times the true rate will lie between 154 and 206 counts/min.

Figure 17–2 illustrates the relationship between total counts and tolerable levels of uncertainty as calculated from Equation 17–12. Note that the horizontal axis is logarithmic; it is clear that a tenfold decrease in the relative uncertainty requires an approximately hundredfold increase in the number of counts.

Background Corrections. The count recorded in a radiochemical analysis includes a contribution from sources other than the sample. Background activity can be traced to the existence of minute quantities of radon isotopes in the atmosphere, to the materials used in construction of the laboratory, to accidental contamination within the laboratory, to cosmic radiation, and to the release of radioactive materials into the Earth's atmosphere. In order to obtain a true assay, then, it is necessary to correct the total count for background. The counting period required to establish the background correction frequently differs from that for the sample; as a result, it is more convenient to employ counting rates. Then,

$$R_c = R_x - R_b \qquad (17\text{–}13)$$

where R_c is the corrected counting rate and R_x and R_b are the rates for the sample and the background, respectively.

In order to obtain the standard deviation for R_c in Equation 17–13, we apply Equation 1–15 (p. 15). Thus,

$$\sigma_{R_c}^2 = \left(\frac{\delta R_c}{\delta R_x}\right)^2 \sigma_{R_x}^2 + \left(\frac{\delta R_c}{\delta R_b}\right)^2 \sigma_{R_b}^2 \qquad (17\text{–}14)$$

Taking the partial derivatives of Equation 17–13, holding first R_b and then R_x constant, gives

$$\left(\frac{\delta R_c}{\delta R_x}\right)_{R_b} = 1 \quad \text{and} \quad \left(\frac{\delta R_c}{\delta R_b}\right)_{R_x} = -1$$

Substituting into Equation 17–14 and taking the square root of both sides of the re-

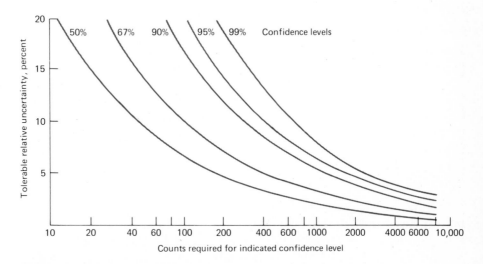

FIGURE 17-2 Relative uncertainty in counting.

sulting expression gives

$$\sigma_{R_c} = \sqrt{\sigma_{R_x}^2 + \sigma_{R_b}^2}$$

Substituting Equation 17–10 yields

$$\sigma_{R_c} = \sqrt{(R_x/t_x) + (R_b/t_b)} \qquad (17\text{–}15)$$

The relative standard deviation $(\sigma_{R_c})_r$ is given by

$$(\sigma_{R_c})_r = \frac{\sqrt{(R_x/t_x) + (R_b/t_b)}}{R_x - R_b} \qquad (17\text{–}16)$$

> EXAMPLE 17–3. A sample yielded 1800 counts in a 10-min period. Background was found to be 80 counts in 4 min. Calculate the absolute uncertainty in the corrected counting rate at the 95% confidence level.
>
> $$R_x = \frac{1800}{10} = 180 \text{ cpm}$$
>
> $$R_b = \frac{80}{4} = 20 \text{ cpm}$$
>
> Substituting into Equation 17–15 yields
>
> $$\sigma_{R_c} = \sqrt{\frac{180}{10} + \frac{80}{4}} = 6.2 \text{ cpm}$$
>
> At the 95% confidence level,
>
> C.L. for R_c = $(180 - 20) \pm 1.96 \times 6.2$
> $= 160 \pm 12 \text{ cpm}$
>
> Here, the chances are 95 in 100 that the true count lies between 148 and 172 counts/min.

Example 17–3 suggests that the contribution of background to the standard deviation is minimized under circumstances where the background count is small with respect to the sample rate. It can be shown that the optimum division of time between

background and sample counting is given by

$$\frac{t_b}{t_x} = \sqrt{\frac{R_b}{R_x}} \qquad (17\text{–}17)$$

Dead Time Correction. In Section 16B–3, it was noted that radiation detectors have recovery times or "dead times" after each count in which they are completely insensitive to incoming photons or particles. The dead time τ for a detector can be determined by calibration with radioactive sources, and the measured counting rate R can then be corrected for the count-loss during dead time. Thus, the corrected counting rate R^* is given by

$$R^* = \frac{R}{1 - R\tau}$$

For example, if a counter with a dead time of 250 μsec yielded a counting rate of 240 counts/sec, the corrected counting rate would be

$$R^* = \frac{240}{1 - 240 \times 250 \times 10^{-6}}$$

$$= 255 \text{ counts/sec}$$

17A–5 UNITS OF RADIOACTIVITY

The *curie* (Ci) is the fundamental unit of radioactivity; it is defined as that quantity of nuclide in which 3.7×10^{10} disintegrations occur per second. Note that the curie is an enumerative quantity only and provides no information concerning the products of the decay process or their energies.

The *millicurie* (mCi) and *microcurie* (μCi) are frequently more convenient units.

17B Instrumentation

Radiation from radioactive sources can be detected and measured in essentially the same way as X-radiation (Sections 15B–4 and 15B–5). Gas-filled detectors, scintillation counters, and semiconductor detectors are all sensitive to alpha, beta, and gamma rays because absorption of these particles produces photoelectrons, which can in turn produce thousands of ion pairs. A detectable electrical pulse is thus produced for each particle reaching the detector.

17B–1 MEASUREMENT OF ALPHA PARTICLES

In order to minimize self-absorption, alpha emitting samples are generally counted as thin deposits prepared by electrodeposition or by distillation and condensation. Often these deposits are then sealed and counted in windowless gas-flow proportional counters or ionization chambers. Alternatively, they are placed immediately adjacent to a solid state detector for counting.

As mentioned earlier, alpha spectra consist of characteristic, discrete energy peaks, which are useful for identification. Pulse-height analyzers (Section 15B–5) permit the derivation of alpha spectra.

17B–2 MEASUREMENT OF BETA PARTICLES

For beta sources having energies greater than about 0.2 MeV, a uniform layer of the sample is ordinarily counted with a thin-windowed Geiger or proportional tube counter. For low-energy beta emitters, such as carbon-14, sulfur-35, and tritium, liquid scintillation counters (p. 471) are preferable. Here, the sample is dissolved in a solution of the scintillating compound. A vial containing the solution is then placed between two photomultiplier tubes housed in a light-tight container. The output from the two tubes is fed into a *coincidence counter,* an electronic device that records a count only when pulses from the two detectors arrive at the same time. The coincidence counter reduces background noise from the detectors and amplifiers because of the low probability of such noise affecting both systems simultaneously.

Because beta spectra are ordinarily continuous, pulse-height analyzers are less useful.

17B–3 MEASUREMENT OF GAMMA RADIATION

Gamma radiation is detected and measured by the methods described in Section 15B for X-radiation. Interference from α- and β-radiation is readily avoided by filtering the beam with a thin window of aluminum or Mylar.

Gamma ray spectrometers are similar to the pulse-height analyzers described in Section 15B–5. Figure 17–3 shows a typical gamma ray spectrum obtained with a 400-channel analyzer. Here, the characteristic peaks for the various elements are superimposed upon a continuum that arises from the Compton effect.

Figure 17–4 is a schematic diagram of a *well-type* scintillation counter that is used for gamma ray counting. Here, the sample is contained in a small vial and placed in a cylindrical hole or well in the scintillating crystal of the counter.

17C Neutron Activation Methods

Activation methods are based upon the measurement of radioactivity that has been induced in samples by irradiation with neutrons or charged particles, such as protons or deuterium or helium-3 ions. Thermal neutrons from a reactor or from a small radioactive decay source are by far the most commonly used particles for inducing radioactivity in a sample. Thus, this discussion will focus on neutron activation methods.[6]

---

[6] Monographs on neutron activation methods include: *Nondestructive Activation Analysis*, S. Amiel, Ed., Elsevier: New York, 1981; D. DeSoete, R. Gybels, and J. Hoste, *Neutron Activation Analysis*, Wiley-Interscience: New York, 1972; P. Kruger, *Principles of Activation Analysis*, Wiley-Interscience: New York, 1971.

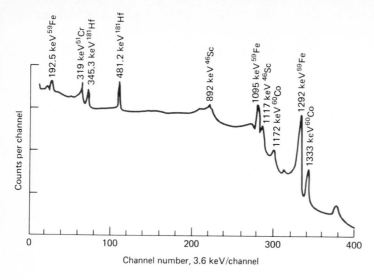

FIGURE 17–3 Gamma ray spectrum of aluminum wire after neutron activation. (Reprinted with permission and taken from: S. G. Prussin, J. A. Harris, and J. M. Hollander, *Anal. Chem.*, **1965,** *37*, 1130. Copyright 1965 American Chemical Society.)

17C–1   NEUTRONS AND NEUTRON SOURCES

The basic characteristics of neutrons are given in Table 17–1. Free neutrons are not stable and decay with a half-life of about

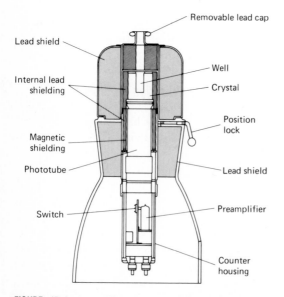

FIGURE 17–4 A well-type scintillation counter. (Courtesy of Texas Nuclear Division, Ramsey Engineering Co., Austin, TX. Formerly Nuclear-Chicago Corporation.)

12.5 min to give protons and electrons. Free neutrons do not, however, generally exist long enough to disintegrate because of their tendency to react with ambient material.

The properties of neutrons are a function of their kinetic energy. Two important categories of neutrons, based upon their tendency to undergo nuclear reactions, are encountered—slow neutrons with kinetic energies of less than about 1 keV and fast neutrons with energies of greater than approximately 0.5 MeV. *Thermal* neutrons, which are neutrons in thermal equilibrium with their environment, are of particular interest; at room temperature, their most probable energy is 0.025 eV.

Reactors.   Nuclear reactors are a source of copious thermal neutrons and are, therefore, widely used for activation analyses. A typical research reactor will have a neutron flux of 10^{11} to 10^{14} $n/(cm^2sec)$. These high densities lead to detection limits that for many elements range from 10^{-3} to 10 μg.

Radioactive Neutron Sources.   Radioactive isotopes are convenient and relatively inexpensive sources of neutrons for activation analyses. Their neutron flux densities range from perhaps 10^5 to 10^8 $n/(cm^2sec)$. As a consequence, detection limits are generally not as good as those in which a reactor serves as a source.

One common radioactive source of neutrons is a transuranium element which undergoes spontaneous fission to yield neutrons. The most common example of this type of source consists of californium-252, which has a half-life of 2.6 years. About 3% of its decay involves spontaneous fission, which yields 3.8 neutrons per fission. Thermal flux densities of about 3×10^7 $n/(cm^2 sec)$ are obtainable from this type of source.

Neutrons can also be produced by preparing an intimate mixture of an alpha emitter such as plutonium, americium, or curium with a light element such as beryllium. A commonly used source of this kind is based upon the reaction

$$^{9}_{4}Be + {}^{4}_{2}He \rightarrow {}^{12}_{6}C + {}^{1}_{0}n + 5.7 \text{ MeV}$$

To produce thermal neutrons, a paraffin container is employed as a moderator, a material that reduces the kinetic energy of the neutrons to about 0.025 keV.

Accelerators. Relatively compact charged particle accelerators are commercially available for the generation of beams of neutrons. A typical generator consists of an ion source that delivers deuterium ions to an area where they are accelerated through a potential of about 150 kV to a target containing tritium absorbed on titanium or zirconium. The reaction is

$$^{2}_{1}H + {}^{3}_{1}H \rightarrow {}^{4}_{2}He + {}^{1}_{0}n$$

The monoenergetic neutrons of about 14 MeV have been found to be particularly useful for the determination of oxygen and nitrogen in a variety of samples. Accelerators can also be made to produce thermal neutrons by introduction of a paraffin modulator in the beam path.

17C-2 INTERACTIONS OF NEUTRONS WITH MATTER

The neutron is a highly effective bombarding particle because its zero charge permits it to approach charged nuclei without interference from coulombic forces. In fact, the most effective neutrons are so-called thermal neutrons, which possess kinetic energies of only about 0.025 eV at room temperature. Thermal neutrons are obtained by passing the radiation from a neutron source such as a reactor through a low-molecular weight moderating solution such as water or deuterium oxide. Collisions of the neutrons with these light nuclei result in a continuous reduction of their energies.

Several types of reactions are encountered when matter is bombarded with neutrons. The two most important are *neutron capture* and *transmutation*.

Neutron Capture. The most common neutron reaction involves capture of a neutron by a nucleus to give an isotope with a mass number greater by one. Prompt γ ray emission accompanies this process. For example,

$$^{23}_{11}Na + {}^{1}_{0}n \rightarrow {}^{24}_{11}Na + \gamma$$

Usually, equations of this type are written in the abbreviated form

$$^{23}_{11}Na(n, \gamma)^{24}_{11}Na$$

Often the products of capture reactions are unstable and undergo β^- decay. For example,

$$^{24}_{11}Na \rightarrow {}^{24}_{12}Mg + \beta^-$$

Transmutation. Another type of nuclear reaction involves absorption of a neutron followed by release of a proton (or occasionally an alpha or deuteron particle). The reaction can be illustrated by

$$^{27}_{13}Al(n, \rho)^{27}_{12}Mg$$

Often the product decays back to the parent element by β^- decay. That is,

$$^{27}_{12}Mg \rightarrow \ ^{27}_{13}Al + \beta^-$$

17C-3 THEORY OF ACTIVATION METHODS

When exposed to a flux of neutrons, the rate of formation of radioactive particles of an element can be shown to be

$$\frac{dN^*}{dt} = N\phi\sigma$$

where dN^*/dt is the formation rate of active particles in nuclei/sec, N is the number of stable target atoms, ϕ is the average flux in neutrons/cm²sec, and σ is the capture cross section in cm²/target atom. The last is a measure of the probability of the nuclei reacting with a neutron at the particle energy employed. Tables of reaction cross section for thermal neutrons list values for σ in *barns* b where $1b = 10^{-24}$ cm²/target atom.

Once formed, the radioactive nuclei decay at a rate $-dN^*/dt$ given by Equation 17–1. That is,

$$\frac{-dN^*}{dt} = \lambda N^*$$

Thus during radiation with a uniform flux of neutrons, the net rate of formation of active particles is

$$\left(\frac{dN^*}{dt}\right)_{net} = N\phi\sigma - \lambda N^*$$

When this equation is integrated from time 0 to t, one obtains

$$N^* = \frac{N\phi\sigma}{\lambda}[1 - \exp(-\lambda t)]$$

Substitution of Equation 17–4 into the exponential term yields

$$N^* = \frac{N\phi\sigma}{\lambda}\left[1 - \exp\left(-\frac{0.693\,t}{t_{1/2}}\right)\right]$$

At the end of the irradiation period, λN^* is equal to $-dN^*/dt$ (Equation 17–1). Thus,

$$\lambda N^* = \frac{-dN^*}{dt}$$

$$= N\phi\sigma\left[1 - \exp\left(-\frac{0.693\,t}{t_{1/2}}\right)\right] = N\phi\sigma S$$

where S is the *saturation factor,* which is equal to one minus the exponential term.

The foregoing equation can be written in terms of experimental activity measurements (p. 505). That is,

$$\lambda A = \frac{-dA}{dt} = N\phi\sigma\left[1 - \exp\left(-\frac{0.693\,t}{t_{1/2}}\right)\right]$$

Figure 17–5 is a plot of activity of an isotope produced by irradiation of a sample at three levels of neutron flux. The abscissa is

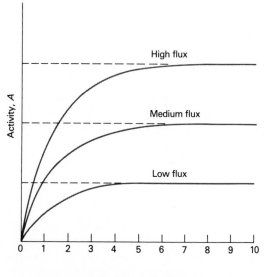

High flux

Medium flux

Low flux

Irradiation time, no. half-lives $(t/t_{1/2})$

FIGURE 17-5 The effect of neutron flux and time upon the activity induced in a sample.

the ratio of the irradiation time to the half-life of the isotope ($t/t_{1/2}$). In each case, the activity approaches a constant value called the *saturation activity*. Here, the rates of formation and disintegration of the isotope approach one another. Clearly, irradiation for periods beyond four or five half-lives for an isotope will result in little improvement in sensitivity.

17C-4 CLASSIFICATION OF ACTIVATION METHODS

Activation methods can be classified in several ways. One is based upon the type of radiation employed for excitation of the sample; slow neutrons, fast neutrons, gamma rays, and various charged particles have been used. As mentioned earlier, most activation methods are based upon thermal neutrons.

A second variable, which is used to characterize activation methods, is the type of emission measured in the final step of the analysis. Here, both beta and gamma radiation have been monitored; the former is often more sensitive but, on the other hand, frequently suffers from being less selective since the radiation is continuous rather than discrete.

Finally, activation methods can be classified as being destructive or nondestructive of the sample. In destructive methods, the irradiated sample is dissolved, and the element of interest is counted after it has been isolated by suitable chemical or physical means; possible interferences from other species made radioactive by the irradiation are thus eliminated. In the nondestructive procedure, the activated sample is counted without preparatory treatment; here, the ability of a gamma ray spectrometer to discriminate among radiation of different energies is called upon to provide the required selectivity. The nondestructive method offers the advantage of great speed. On the other hand, the resolution of a gamma ray spectrometer may be insufficient to eliminate interferences. Also, it does not permit the use of beta emission for completion of the analysis.

17C-5 DESTRUCTIVE METHODS

A destructive method involves decomposition and solution of a known amount of the irradiated sample followed by separation of the analyte from interferences. The isolated material or a fraction thereof is then counted for its beta or gamma activity.

Conventional neutron activation methods involve simultaneous irradiation of the sample and the standard containing a known mass w_s of the analyte in a homogeneous neutron flux. Insofar as the activity that results is proportional to mass, and provided also that the other components of the sample do not produce detectable radioactivity, the weight w_x of the element in the sample is given by

$$w_x = \frac{A_x}{A_s} w_s \qquad (17\text{–}18)$$

where A_x and A_s are the activities of the sample and standard. Generally, however, the neutron flux can be expected to generate activity in elements other than the analyte. Thus, chemical isolation of the species of interest from a solution of the sample often precedes radioassay. When the analyte is present as a trace (as it usually is in a neutron activation analysis), its separation from the major constituents may be difficult and the source of large error. This problem is minimized by introducing a known weight W_x of the analyte to the solution of the irradiated sample as a nonactive *carrier* or *collector*. Separation of the carrier plus the irradiated element ($W_x + w_x$) is then accomplished by precipitation, extraction, ion exchange, or chromatographic means. A weighed quantity w_x' of the isolated material is counted, and the resulting activity a_x is related to the total activity of the original sample A_x by the relationship

$$a_x = A_x \frac{w'_x}{W_x + w_x} \tag{17-19}$$

Ordinarily, the amount of carrier added is several orders of magnitude greater than the weight from the sample; that is, $W_x \gg w_x$. Under this circumstance, Equation 17–19 simplifies to

$$a_x = \frac{A_x w'_x}{W_x} \tag{17-20}$$

The standard sample is treated in an identical way; thus, an analogous expression can be written

$$a_s = \frac{A_s w'_s}{W_s} \tag{17-21}$$

Substituting these expressions into Equation 17–18 yields

$$w_x = w_s \frac{a_x W_x w'_s}{a_s W_s w'_x} \tag{17-22}$$

Where the condition $W_x \gg w_x$ is not satisfied, a more complex equation must be employed.

The Substoichiometric Method.[7] It is experimentally feasible to impose the further conditions that, at the time of assay, $W_x = W_s$ and $w'_x = w'_s$. Equation 17–22 then simplifies to

$$w_x = w_s \frac{a_x}{a_s} \tag{17-23}$$

which forms the basis for the substoichiometric method. Where the mass of collector greatly exceeds that of the radioisotope, W_x

---

[7] For a more detailed treatment, see: J. Ruzicka and J. Stary, *Substoichiometry in Radiochemical Analysis*, Pergamon Press: New York, 1968.

and W_s are essentially identical provided the same weight of collector is added to the sample and the standard. The requirement that the same total amount of species be taken for radioassay (that is, that $w'_x = w'_s$) is analytically unique, in that the same quantity must be taken from solutions of inherently dissimilar concentration. The problem is resolved by introducing a suitable reagent in an amount insufficient *(substoichiometric)* for complete removal of the species of interest from either sample or standard. If the same amount of this reagent is used for both sample and standard, w'_x and w'_s are identical because their amounts are determined by the amount of added reagent.

17C-6 NONDESTRUCTIVE METHOD

In the nondestructive method, a gamma ray spectrometer is used to measure the activities of the sample and the standard after irradiation. The weight of the analyte is then calculated directly from Equation 17–18.

Clearly, success of the nondestructive method requires that the spectrometer be able to isolate the gamma ray signal produced by the analyte from signals arising from the other components. Whether or not an adequate resolution is possible depends upon the complexity of the sample, the presence or absence of elements which produce gamma rays of about the same energy as that of the element of interest, and the resolving power of the spectrometer. Improvements in resolving power, which have been made in the last few years, have greatly broadened the scope of the nondestructive method. At the present time, however, the most selective and sensitive activation methods are still based upon isolation of the analyte. The great advantage of the nondestructive approach is its simplicity in terms of sample handling and speed; to be sure, the required instrumentation is more complex.

17C-7 APPLICATION OF NEUTRON ACTIVATION

Scope. Figure 17–6 illustrates that neutron activation is potentially applicable to the determination of 69 elements. In addition, four of the inert gases form active isotopes with thermal neutrons and thus can also be determined. Finally, three additional elements (oxygen, nitrogen, and yttrium) can be activated with fast neutrons from an accelerator. A list of types of materials to which the method has been applied is impressive and includes metals, alloys, archeological objects, semiconductors, biological specimens, rocks, minerals, and water. Acceptance of evidence developed from activation analysis by courts of law has led to its widespread use in forensic chemistry. Here, the high sensitivity and nondestructive aspect of the method are particularly useful. Most applications have involved the determination of traces of various elements.

Accuracy. The principal errors that arise in activation analyses are due to self-shielding, unequal neutron flux at sample and standard, counting uncertainties, and errors in counting due to scattering, absorption, and differences in geometry between sample and standard. The errors from these causes can usually be reduced to less than 10% relative; uncertainties in the range of 1 to 3% are frequently obtainable.

Sensitivity. The most important characteristic of the neutron activation method is its remarkable sensitivity for many elements. Note in Figure 17–6, for example, that as little as 10^{-5} μg of several elements can be detected. Note also the wide variations in sensitivities among the elements; thus, about 50 μg of iron are required for detection in contrast to 10^{-6} μg for europium.

The efficiency of chemical recovery, if required prior to radioassay, may limit the sensitivity of an activation analysis. Other factors include the sensitivity of the detection equipment for the emitted radiation, the extent to which activity in the sample

decays between irradiation and assay, the time available for counting, and the magnitude of the background count with respect to that for the sample. A high rate of decay is desirable from the standpoint of minimizing the duration of the counting period. Concomitant with high decay rates, however, is the need to establish with accuracy the time lapse between the cessation of irradiation and the commencement of counting. A further potential complication is associated with counting rates that exceed the resolving time of the detecting system; under these circumstances, a correction must be introduced to account for the difference between elapsed (clock) and live (real) counting times.

17D Isotopic Dilution Methods

Isotopic dilution methods, which predate activation procedures, have been and still are extensively applied to problems in all branches of chemistry. These methods are among the most selective available to chemists.

Both stable and radioactive isotopes are employed in the isotopic dilution technique. The latter are the more convenient, however, because of the ease with which the concentration of the isotope can be determined. We shall limit this discussion to methods employing radioactive species.

17D-1 PRINCIPLES OF THE ISOTOPIC DILUTION PROCEDURE

Isotopic dilution methods require the preparation of a quantity of the analyte in a radioactive form. A known weight of this species is then mixed with a weighed quantity of the sample. After treatment to assure homogeneity between the active and nonactive species, a part of the analyte mixture is isolated chemically in the form of a purified compound of known composition. By

Estimated sensitivities of neutron activation methods — Upper numbers correspond to β sensitivities; lower numbers to γ sensitivities (in micrograms).

Element	β	γ
Na	5×10^{-3}	5×10^{-3}
Mg	5×10^{-1}	5×10^{-1}
K	5×10^{-2}	5×10^{-2}
Ca	1.0	5
Sc	1×10^{-2}	5×10^{-2}
Rb	5×10^{-2}	5
Sr	5×10^{-3}	5×10^{-3}
Cs	5×10^{-2}	5×10^{-1}
Ba	5×10^{-1}	1×10^{-1}
La	1×10^{-3}	5×10^{-3}
Ti	5×10^{-1}	5×10^{-2}
V	5×10^{-3}	1×10^{-3}
Cr	—	1
Mn	5×10^{-5}	5×10^{-5}
Fe	50	200
Co	5×10^{-3}	1×10^{-1}
Ni	5×10^{-2}	5×10^{-1}
Cu	1×10^{-3}	1×10^{-3}
Zn	1×10^{-1}	1×10^{-1}
Zr	1	1
Nb	5×10^{-3}	1
Mo	5×10^{-1}	1×10^{-1}
Ru	1×10^{-2}	5×10^{-2}
Rh	1×10^{-3}	1×10^{-4}
Pd	5×10^{-4}	5
Ag	5×10^{-3}	5×10^{-1}
Cd	5×10^{-2}	5×10^{-1}
Hf	—	1
Ta	5×10^{-2}	5×10^{-1}
W	1×10^{-3}	5×10^{-3}
Re	5×10^{-4}	1×10^{-3}
Os	5×10^{-2}	—
Ir	1×10^{-4}	1×10^{-3}
Pt	5×10^{-2}	1×10^{-1}
Au	5×10^{-4}	5×10^{-4}
Hg	—	1×10^{-2}
Ga	5×10^{-3}	5×10^{-3}
In	5×10^{-5}	1×10^{-4}
Al	1×10^{-1}	1×10^{-2}
Si	5×10^{-2}	500
Ge	5×10^{-3}	5×10^{-2}
Sn	5×10^{-1}	5×10^{-1}
Pb	10	—
P	5×10^{-1}	—
As	1×10^{-3}	5×10^{-3}
Sb	5×10^{-3}	1×10^{-4}
Bi	5×10^{-1}	—
S	5	200
Se	—	5
Te	5×10^{-2}	5×10^{-2}
F	—	1
Cl	1×10^{-2}	1×10^{-1}
Br	5×10^{-3}	5×10^{-3}
I	5×10^{-3}	1×10^{-2}
Ce	1×10^{-1}	1×10^{-1}
Pr	5×10^{-4}	5×10^{-2}
Nd	1×10^{-1}	1×10^{-1}
Sm	5×10^{-4}	5×10^{-3}
Eu	5×10^{-6}	5×10^{-4}
Gd	1×10^{-2}	5×10^{-2}
Tb	5×10^{-2}	1×10^{-1}
Dy	1×10^{-6}	5×10^{-6}
Ho	1×10^{-4}	1×10^{-4}
Er	1×10^{-3}	1×10^{-3}
Tm	1×10^{-2}	1×10^{-1}
Yb	1×10^{-3}	1×10^{-3}
Lu	5×10^{-5}	5×10^{-5}
Th	5×10^{-2}	5×10^{-2}
U	5×10^{-3}	5×10^{-3}

FIGURE 17-6 Estimated sensitivities of neutron activation methods. Upper numbers correspond to β sensitivities in micrograms; lower numbers to γ sensitivities in micrograms. In each case samples were irradiated for 1 hr or less in a thermal neutron flux of 1.8×10^{12} neutrons/cm²/sec. (From: V. P. Guinn and H. R. Lukens, Jr., in *Trace Analysis: Physical Methods*, G. H. Morrison, Ed., p. 345, Wiley: New York, 1965. Reprinted by permission of John Wiley & Sons, Inc.)

counting a weighed portion of this product, the extent of dilution of the active material can be calculated and related to the amount of nonactive substance in the original sample. It is important to realize that quantitative recovery of the species is not required. Thus, in contrast to the typical analytical separation, steps can be employed to assure a highly pure product on which to base the analysis. It is this independence from the need for quantitative isolation that leads to the high selectivity of the isotopic dilution method.

Direct Isotope Dilution.   Assume that W_o grams of a radioactive species having an activity of A_o are mixed with a sample containing W_x grams of the inactive substance. After separation and purification, a weight W_r of the species is found to have an activity of A_r. We may then write

$$A_r = \frac{A_o W_r}{W_o + W_x} \qquad (17\text{--}24)$$

which rearranges to

$$W_x = \frac{A_o}{A_r} W_r - W_o \qquad (17\text{--}25)$$

Thus, the weight of the species originally present is obtained from the four measured quantities on the right-hand side of Equation 17–25. Where the activity of the tracer is large, the weight W_o added can be kept small, and Equation 17–25 simplifies to

$$W_x = \frac{A_o}{A_r} W_r \qquad (17\text{--}26)$$

Substoichiometric Isotope Dilution.   A substoichiometric method, analogous to that for activation analysis, can also be used in isotopic dilution experiments. Here, identical amounts W_o of the tracer are added to two solutions that are the same in every respect except that one contains the sample and the other does not. A suitable reagent is then added to

isolate a quantity W_r of the species of interest from each. Care is taken to be sure that the amount of added reagent is, however, less than that required for complete removal of the species; thus, W_r is identical for the two solutions. Equation 17–24 describes the activity of the product from the solution containing the sample; for the solution having no sample, W_x is zero, and Equation 17–24 takes the form

$$A_r' = \frac{A_o W_r'}{W_o} \qquad (17\text{--}27)$$

Recall, however, that conditions have been chosen such that $W_r = W_r'$. As a consequence, division of Equation 17–24 by Equation 17–26 and rearrangement yields

$$W_x = W_o \left(\frac{A_r'}{A_r} - 1 \right) \qquad (17\text{--}28)$$

The substoichiometric procedure is advantageous when the amount recovered, W_r, is so small that its weight is difficult to assess.

17D-2   APPLICATION OF THE ISOTOPIC DILUTION METHOD

The isotopic dilution technique has been employed for the determination of about 30 elements in a variety of matrix materials.[8] Substoichiometric methods have proved useful for determining traces of several me-

---

[8] It is of interest that the dilution principle has also had other applications. One example is its use in the estimation of the size of salmon spawning runs in Alaskan coastal streams. Here, a small fraction of the salmon are trapped, mechanically tagged, and returned to the river. A second trapping then takes place perhaps 10 miles upstream, and the fraction of tagged salmon is determined. The total salmon population is readily calculated from this information and from the number originally tagged. The assumption must, of course, be made that the fish population becomes homogenized during its travel between stations.

tallic elements. For example, fractions of a microgram of cadmium, copper, mercury, or zinc have been determined by a procedure in which the element is isolated for counting by extraction with a substoichiometric amount of dithizone in carbon tetrachloride.

Isotopic dilution procedures have been most widely used for the determination of compounds that are of interest in organic chemistry and biochemistry. Thus, methods have been developed for the determination of such diverse substances as vitamin D, vitamin B_{12}, sucrose, insulin, penicillin, various amino acids, corticosterone, various alcohols, and thyroxine.

Isotopic dilution analysis has had less widespread application since the advent of activation methods. Continued use of the procedure can be expected, however, because of the relative simplicity of the equipment required. In addition, the procedure is often applicable where the activation method fails.

17E Radiometric Methods[9]

One type of radiometric method employs a radioactive reagent of known activity to isolate the analyte from the other components of a sample. After quantitative separation,

the activity of the product is readily related to the amount of the species being determined. Methods for the radiometric determination of more than 30 of the common elements have been described. Examples include the determination of chromium by formation of active silver chromate with radioactive silver ion, precipitation of magnesium or zinc by phosphate containing phosphorus-32, and the determination of fluoride ion by precipitation with radioactive calcium.

Radiometric titrations employ a radioactive compound for preparation of the standard solution. Usually, the reaction between the analyte and the standard involves precipitate formation with the activity of the supernatant liquid being monitored as the titration progresses. An example is the titration of silver ion with a bromide solution that is enriched with radioactive bromine. Until the equivalence point is reached, essentially no radioactivity is found in the supernatant liquid. After equivalence, a linear increase in count as a function of volume is observed. Clearly, precautions are needed to assure prompt coagulation and settling out of the precipitate.

---

[9] See also: T. Brauer and J. Tölgyessy, *Radiometric Titrations*, Pergamon Press: New York, 1967.

PROBLEMS

17–1. Potassium-42 is a β emitter with a half-life of 12.36 hr. Calculate the fraction of this isotope remaining in a sample after (a) 1 hr; (b) 10 hr; (c) 20 hr; (d) 75 hr.

17–2. Calculate the fraction of the following isotopes that remains after 24 hr (half-lifes are given in parentheses):
(a) iron-59 (44.6 days).
(b) titanium-45 (3.09 hr).
(c) calcium-47 (4.54 days).
(d) phosphorus-33 (25.3 days).

17–3. A $PbSO_4$ sample contains 1 microcurie of Pb-200 ($t_{1/2} = 21.5$ hr). What storage period is needed to assure that its activity is less than 0.01 microcurie?

17–4. Estimate the standard deviation and the relative standard deviation associated with counts of (a) 100.0; (b) 750; (c) 7.00×10^3; (d) 2.00×10^4.

17–5. Estimate the standard deviation (in absolute and relative terms) associated with a counting rate of 200 cpm that is observed for (a) 40 sec; (b) 80 sec; (c) 4.0 min; (d) 12.0 min.

17–6. Estimate the absolute and relative uncertainty associated with a measurement involving 800 counts at the
(a) 50% confidence level.
(b) 90% confidence level.
(c) 99% confidence level.

17–7. Estimate the absolute uncertainty at the 90% confidence level for a measurement that involves a total count of (a) 60; (b) 300; (c) 520; (d) 1200.

17–8. Estimate the absolute and relative uncertainty at the 90% confidence level associated with the corrected counting rate obtained from a total counting rate of 300 cpm for 14 min and a background count of
(a) 9 cpm for 2 min.
(b) 9 cpm for 10 min.
(c) 18 cpm for 2 min.
(d) 40 cpm for 2 min.

17–9. The background activity of a laboratory when measured for 3 min was found to be approximately 9 cpm. What total count should be taken in order to keep the relative uncertainty at the 90% confidence level smaller than 5.0%, given a total counting rate of about (a) 90 cpm; (b) 300 cpm; (c) 600 cpm?

17–10. If a total of 25 min is available, calculate the best division of counting time between the background and sample for each of the counting rates in Problem 17–9. What will be the expected relative standard deviation for the analysis if the only significant uncertainty lies in the counting process?

17–11. A 2.00-mL solution containing 0.120 microcurie per milliliter of tritium was injected into the bloodstream of a dog. After allowing time for homogenization, a 1.00-mL sample of the blood was found to have a count corresponding to 15.8 counts per second. Calculate the blood volume of the animal.

17–12. The penicillin in a mixture was determined by adding 0.981 mg of the pure compound having a specific activity of 5.42×10^3 cpm/mg. After equilibration, 0.406 mg of pure crystalline penicillin was isolated. This material had a net activity of 343 cpm. Calculate the mg penicillin in the sample.

17–13. The streptomycin in 500 g of a broth was determined by addition of 1.34 mg of the pure antibiotic containing C-14; the specific activity of this preparation was found to be 223 cpm/mg for a 30-

min count. From the mixture, 0.112 mg of purified streptomycin was isolated, which produced a count of 654 counts in 60.0 min.

(a) Calculate the parts per million streptomycin in the sample.

(b) Calculate the 90% confidence limit for the result assuming the major source of uncertainty is the decay process.

17–14. In order to determine the mercury content of a specimen of animal tissue, a 0.652-g sample of the tissue and a standard solution containing 0.213 μg of Hg as $HgCl_2$ were irradiated for 3 days in a thermal neutron flux of 10^{12} neutrons/cm^2 sec. After irradiation was complete, 25.0 mg of Hg as Hg_2Cl_2 was added to each. Both were digested in a nitric acid/sulfuric acid mixture to oxidize organic material; suitable precautions were taken to avoid loss of mercury by volatilization. Hydrochloric acid was then added, and the $HgCl_2$ formed was distilled from the reaction mixtures. The mercury in each of the distillates was deposited electrolytically on gold foil electrodes, resulting in an increase in weight of 13.5 mg for the sample and 14.6 mg for the standard. The γ activity due to ^{197}Hg was then determined. The sample was found to yield a count of 860 cpm and the standard 1112 cpm; counting time was 10.0 min.

(a) Calculate the ppm Hg in the sample.

(b) Calculate the 95% confidence interval for the result assuming the major source of uncertainty is the decay process.

17–15. Identical electrolytic cells, fitted with silver anodes and platinum cathodes, were arranged in series. Exactly 1.00 mL of a solution containing 4.12×10^{-2} mg of KI labeled with ^{131}I (a β emitter with a half-life of 8.0 days) and 5.00 mL of an HOAc/OAc$^-$ buffer were introduced to one cell. A 5.00-mL aliquot of an iodide-containing sample was added to the acetate buffer contained in the second cell. After passing a substoichiometric quantity of electricity, the anodes were removed and assayed for their β activity.

(a) Calculate the weight of I$^-$ in each milliliter of the sample solution if the activity, determined by a 4.00-min count for the electrode from the cell containing the standard was 4130 cpm, while that from the cell containing the unknown was 3550 cpm.

(b) Calculate the 95% confidence interval for the result assuming the major source of uncertainty is the decay process.

18

Mass Spectrometry

A mass spectrum is obtained by converting components of a sample into rapidly moving gaseous ions and resolving them on the basis of their mass-to-charge ratios.[1] Mass spectrometry is perhaps the most generally applicable of all of the analytical tools available to the scientist in the sense that the technique is capable of providing qualitative and quantitative information about both the atomic and the molecular composition of inorganic and organic materials. Unfortunately, the high cost of purchasing and maintaining mass spectrometers has, to some extent, inhibited their more widespread use.

Mass spectroscopy evolved from studies at the beginning of this century of the behavior of positive ions in magnetic and electrostatic fields. During the following two decades, the method was refined and provided a host of important information concerning the isotopic abundance of various elements.

In about 1940, reliable mass spectrometers first became available from commercial sources. These instruments were designed specifically for the quantitative determination of the components of the complex hydrocarbon mixtures encountered in the petroleum industry. This application was quickly expanded to include other types of organic compounds generated by the chemical industry.

In the middle 1950s, commercial mass spectrometers were also developed for the qualitative and quantitative determination of the elements based upon the mass-to-

---

[1] Reference works on mass spectrometry include: M. E. Rose and R. A. W. Johnstone, *Mass Spectrometry for Chemists and Biochemists*, Cambridge Press: Cambridge, 1982; I. Howe, D. H. Williams, and R. D. Bowen, *Mass Spectrometry: Principles and Applications*, 2d ed., McGraw-Hill: New York, 1981; F. W. McLafferty, *Interpretation of Mass Spectra*, 3d ed., University Science Books: Mill Valley, CA, 1980; and *Practical Mass Spectrometry*, B. S. Middleditch, Ed., Plenum Press: New York, 1979; For brief reviews of recent developments in the field, see: R. G. Cooks, K. L. Busch, and G. L. Glish, *Science*, **1983**, *222*, 273; and W. V. Ligon, Jr., *Science*, **1979**, *205*, 151.

charge ratio of elementary ions formed in an electric spark. The development of spark source spectrometry paralleled the development of two new major industries—electronics and nuclear energy. Both are based upon materials that are sensitive to trace amounts of contaminants. Mass spectrometry provided an important means for detecting and determining the concentration of such contaminants.

Beginning in about 1960, the major thrust in the development of mass spectrometry shifted towards its use for the identification and structural analysis of complex molecules. The basis for this application is the characteristic pattern of ion fragments (each with a differing mass) that are formed when large molecules are ionized. Mass spectra are easier to interpret than IR or NMR spectra in some respects, since they provide information in terms of masses of structural components and usually the molecular weight of the analyte molecule as well. Thus, by the mid-1960s mass spectroscopy had become recognized as a powerful and versatile tool for structural analyses—on a par with infrared and nuclear magnetic resonance.

A recent burgeoning area has been the application of mass spectroscopy to the study of surfaces. Techniques are being developed for the identification and determination of both atomic and molecular species on surfaces and for studying the composi-

tional changes of solids as a function of depth. It seems probable that mass spectroscopy will become a tool of prime importance for studies of this kind.

Finally, it should be noted that the mass spectrometer is finding wider and wider applications as a detector for both gas and liquid chromatography.

18A   The Mass Spectrometer

The principles of mass spectral measurements are simple and easily understood; unfortunately, this simplicity does not extend to the instrumentation. Indeed, the typical high-resolution mass spectrometer is a complex electronic and mechanical device that is expensive ($100,000 to $500,000 or more) in terms of initial purchase as well as operation and maintenance.

18A-1   GENERAL DESCRIPTION OF INSTRUMENT COMPONENTS

Figure 18–1 is a block diagram showing the major components of a mass spectrometer, which are analogous in function to the components of the optical instruments shown in Figure 5–1 (p. 113) and the electron spectrometer shown in Figure 16–4. The source in a mass spectrometer serves to convert the components of a sample into charged particles. Often in this process, extensive frag-

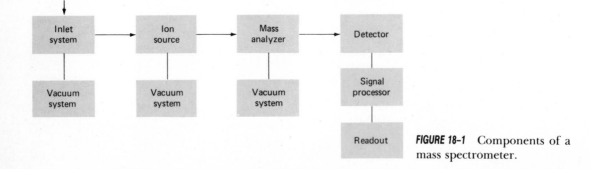

FIGURE 18-1   Components of a mass spectrometer.

mentation of analyte molecules occurs producing a spectrum of charged particles having different mass-to-charge ratios. Both positive and negative particles are produced in the ionization process, and one of these (usually the negative ion) is removed in the source. The mass analyzer is a dispersing device analogous in function to the prism or grating in an optical spectrometer. Here, however, dispersion is based upon the mass of the charged particles.[2] Like its optical counterparts, a mass spectrometer also contains a detector (an ion detector), a signal processor, and a readout. A characteristic feature of mass spectrometry, which is not encountered in most optical methods, is the need to maintain low pressures (10^{-4} to 10^{-8} torr) in all of the components leading up to the detector; thus, like electron spectrometers, elaborate vacuum systems are an important part of mass spectrometers.

Obtaining a mass spectrum with an instrument such as that shown in Figure 18–1 involves the following sequence of steps. (1) A micromole or less of sample is introduced into the source chamber, which is maintained at a pressure of about 10^{-5} torr; most commonly the sample is in the form of a gas, but liquids and solids can also be accommodated. (2) The molecules of the sample are ionized and fragmented by collision with streams of electrons, ions, fast atoms, or photons; alternatively, ionization and fragmentation may be brought about thermally or by a high electrical potential. (3) Positive ions are separated from negative ions by a large negative potential that attracts the former through a slit into the mass analyzer (occasionally, a positive potential is used to reject positive ions and accelerate negative ions into the analyzer). (4) In the analyzer,

the fast moving ions are dispersed and then focused on a detector. As will be shown later, several fundamentally different methods are used for spectral scanning. (5) From the analyzer, ions fall on a collector electrode; the resulting ion currents are amplified and recorded as a function of scan time.

By far, the components most critical to the successful operation of a mass spectrometer are the sample inlet and the ionization systems because these devices must control the efficient formation of ions from the sample without undue degradation of molecular structures.

18A-2 SAMPLE INLET SYSTEMS

The purpose of the inlet system is to permit introduction of a representative sample into the ion source with minimal loss of vacuum. Most modern mass spectrometers are equipped with three types of inlets to accommodate various kinds of samples; these include batch inlets, direct probe inlets, and gas chromatographic inlets. Liquid chromatographic inlets are also beginning to appear on the market.

Batch Inlet Systems. The classical inlet system is the batch type in which the sample is volatilized externally and then allowed to leak into the evacuated ionization region. Figure 18–2 is a schematic diagram of a typical system that is applicable to gaseous and liquid samples having boiling points up to about 500°C. For gaseous samples, a small measured volume of gas is trapped between the two valves enclosing the metering area and is then expanded into the reservoir flask. For liquid samples, a small volume is introduced onto a sintered glass plate by means of a micropipette. The plate is covered with a layer of mercury or liquid gallium to prevent access of air. Sample sizes are adjusted so that the pressure in the reservoir is 10^{-4} to 10^{-5} torr. For samples with boiling points greater than 150°C, the reservoir and tubing must be maintained at an elevated tem-

---

[2] More correctly, the dependence is on the ratio of mass-to-charge (m/e). Generally, however, the ions of interest bear but a single charge; thus, the term mass is frequently used in lieu of the more cumbersome mass-to-charge ratio.

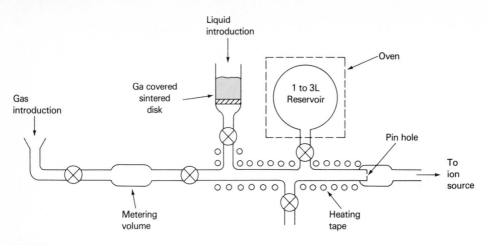

FIGURE 18-2   Schematic diagram of an external sample introduction system.

perature by means of an oven and heating tapes. The maximum temperature of the oven is about 350°C. This maximum limits the system to liquids with boiling points below about 500°C.

The gaseous sample is leaked into the ionization area of the spectrometer via a metal or glass diaphragm containing one or more pinholes having a diameter of 0.005 to 0.02 mm.

The Direct Probe Inlet.   Nonvolatile liquids, thermally unstable compounds, and solids are introduced into the ionization region by means of a sample probe, which is inserted through a vacuum lock. The lock system is such that only a relatively small volume of air has to be evacuated from the system before insertion of the probe into the ionization region. Thus, spectra can be obtained a few minutes after sample introduction. Probes are also used when the quantity of sample is limited because much less sample is wasted than with the batch system. Thus, usable mass spectra can often be obtained with as little as a few nanograms of sample when a probe is used.

Generally, with a probe, the sample is held on the surface of a glass capillary tube, a fine wire, or a small cup. The probe is positioned within a few millimeters of the ionization source and the slit leading to the spectrometer. Usually, provision is made for both cooling and rapid heating of the sample on the probe.

The low pressure of the ionization area and the closeness of the sample to the ionization source and slit often makes it possible to obtain spectra of thermally unstable compounds before major decomposition of the sample has time to occur. The low pressure also leads to greater concentrations of relatively nonvolatile compounds in the ionization area. Thus, the probe permits the study of such nonvolatile materials as carbohydrates, steroids, metal-organic species, and low-molecular-weight polymeric substances.

Chromatographic Inlet Systems.   Sample requirements with regard to volatility and quantity (~1 μmol) are similar for gas chromatography (Chapter 26) and mass spectrometry; thus, the effluent from a chromatographic column can conveniently serve as a sample source. A major problem in interfacing a gas chromatograph with a mass spectrometer arises from the presence of the carrier gas, which dilutes the eluted components enormously and also tends to swamp the pumping system of the spectrometer. Several methods have been developed for overcoming this problem; the principle of one of the most widely used of these is shown in Figure 18–3. Here, the exit gases flow through the nozzle of an all-glass jet separator, which increases the momentum of the heavier analyte molecules so that 50% or more of them travel in a straight path to the skimmer. The light helium atoms, in contrast, are deflected from a straight path by the vacuum and are thus pumped away.

The flow rate from capillary chromographic columns is low enough so that direct coupling of these columns to the ion source is feasible, thus enhancing sensitivity.

Further details on interfacing gas chromatography and mass spectrometry are found in Section 26D–3.

Inlet systems for the effluent from liquid chromatographic columns are much less well developed than those for gas chromatography. A brief description of these devices is found in Section 27C–6.

18A-3 ION SOURCES FOR MOLECULAR STUDIES

As noted earlier, mass spectrometry can be used for studies of the structure and composition of molecular species or for the detection and determination of elements. The nature of the ion sources differs for the two applications. In this section, only those sources that are suited for molecular studies will be considered. Sources for the detection and determination of elements will be described in a later section dealing with elemental analysis.

Historically, ions for mass analysis were produced by bombarding the gaseous sample with a stream of energetic electrons. Despite certain disadvantages, the technique is still of major importance and is the one upon which libraries of mass spectral data are based. During the past two decades, however, several new methods for sample ionization have been developed and are finding more and more use because they offer certain advantages over the classic electron beam source.[3] Table 18–1 lists most of these new methods and the approximate date at which they came into sustained use. Currently, most commercial mass spectrom-

---

[3] For a review of ion sources, see: E. M. Chait, *Anal. Chem.*, **1972**, *44* (3), 77A; G. W. A. Milne and M. J. Lacey, *Crit. Rev. Anal. Chem.*, **1974**, *4*, 45; H. Budzikiewicz, *Angew. Chem. Int. Ed.*, **1981**, *20*, 634; W. F. Lignon, Jr., *Science*, **1979**, *205*, 154–157; R. J. Cotter, *Anal. Chem.*, **1980**, *52*, 1589A; and K. L. Busch and R. G. Cooks, *Science*, **1982**, *218*, 247.

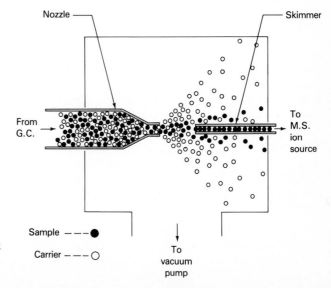

FIGURE 18-3 Schematic diagram of a jet separator. (Courtesy of DuPont Instrument Systems, Wilmington, DE.)

eters are equipped with accessories that permit use of several of these sources interchangeably.

It should be noted that the entries in Table 18–1 fall into two categories. The first are sources in which the sample is first volatilized following which the gaseous components are ionized by collision with electrons or positive ions. The volatilization step may be carried out externally as described for the batch inlet system or internally from a heated probe.

The remaining sources in Table 18–1 are desorption sources. *Desorption ionization* (DI) differs from gaseous ionization in that bulk sample vaporization is dispensed with. Thus, this technique always involves the use of a sample probe. Here, energy in a variety of forms is taken up by the sample causing a direct transfer of ions (and molecules) from a condensed phase into a gaseous ionic phase. A major advantage of desorption ionization is that it permits the examination of nonvolatile and thermally fragile mole-

cules such as those commonly encountered in biochemistry.

18A-4 SOME COMMON ION SOURCES

This section contains details regarding the first five ion sources listed in Table 18–1.

Electron Impact Source (EI). The standard method of producing ions for mass spectrometry involves bombarding the gaseous sample with a beam of energetic electrons. Figure 18–4 is a schematic diagram of a typical electron ionization source or *ion gun*. Here, electrons are produced at heated tungsten or rhenium wires and accelerated to an energy of about 70 eV by a potential impressed across the filament and an anode. The electrons pass through the stream of sample molecules emitted from the molecular leak. With a typical electron current of 100 to 200 μA, perhaps one in every million molecules undergoes ionization; clearly, the process is not very efficient.

The positive ions produced on electron

Table 18–1
Mass Spectroscopic Sources for Molecular Studies

Name	Abbreviation	Type	Ionizing Agent	Date of Sustained Use
Electron ionization	EI	Gas phase	Energetic electrons	
Field ionization	FI	Gas phase	High-potential electrode	
Chemical ionization	CI	Gas phase	Reagent positive ions	1965
Field desorption	FD	Desorption[a]	High-potential electrode	1969
Fast atom bombardment	FAB	Desorption[a]	Energetic atoms	1981
Secondary ion mass spectrometry	SIMS	Desorption[a]	Low fluxes of energetic ions	1977
Plasma desorption	PD	Desorption[a]	High-energy fission fragments from ^{252}Cf	1974
Thermal desorption		Desorption[a]	Heat	1979
Laser desorption	LD	Desorption[a]	Laser beam	1978
Electrohydrodynamic ionization	EHMS	Desorption[a]	High field	1978

[a] Samples as solids, gases, or solutions.

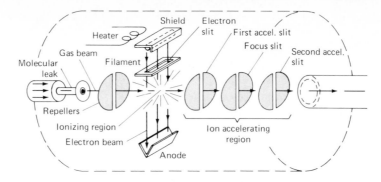

FIGURE 18-4 An ion source. (From: R. M. Silverstein, G. C. Bassler, and T. C. Morrill, *Spectrometric Identification of Organic Compounds,* 4th ed, p. 4, Wiley: New York, 1981. Reprinted by permission of John Wiley & Sons, Inc.)

impact are forced through the slit of the first accelerating plate by a small potential difference between this plate and the repeller. The high potential between the first and second accelerators gives the particles their final velocities; the third slit provides further collimation to the beam. In some spectrometers, the potential impressed between the accelerator slits provides the means whereby particles of a particular mass are focused on the collector.

The electron impact source is sufficiently energetic to leave the *molecular* or *parent ion* (the ion having the same mass as the analyte) in a highly excited vibrational and rotational state. Subsequent relaxation is then often accompanied by extensive fragmentation leading to a large number of positive ions of various masses smaller than that of the original molecule. The complex mass patterns that result are useful for identification. On the other hand, with certain types of molecules, fragmentation results in complete destruction of the molecular ion, which is of prime importance in determining the molecular weight and thus the structure of the analyte. Another limitation of the electron impact source is the need to volatilize the sample, which may result in thermal degradation of some of the compo-

nents before ionization can occur. The effects of thermal decomposition can sometimes be minimized by carrying out the volatilization from a heated probe that is located close to the ion source and the entrance slit of the spectrometer. At the lower pressure of the source area, volatilization occurs at a lower temperature. Furthermore, less time is allowed for thermal decomposition to occur.

Field Ionization (FI) and Field Desorption (FD) Sources. In *field ionization* and *field desorption sources,* ions are formed under the influence of a massive electric field (10^8 V/cm).[4] Such large electrical fields are produced by using high voltages (10 to 20 kV) and specially formed anodes consisting of one or more fine tips or sharp edges having diameters of less than a micrometer. Early field ionization anodes consisted of an array of as many as several hundred microtips fashioned from tungsten, molybdenum, or gold plated copper. Currently, the anode often takes the form of a fine tungsten wire ($\sim$10 μm diameter) on which has been grown carbon dendrites

---

[4] See: M. Anbar and W. H. Aberth, *Anal. Chem.,* **1974,** *46,* 59A; H. D. Beckey, *Principles of Field Ionization and Field Desorption Mass Spectrometry,* Pergamon Press: New York, 1977.

or whiskers by pyrolysis of benzonitrile in a high electric field. The result of this treatment is a growth of many hundreds of carbon microtips surrounding the length of wire.

Field ionization anodes are mounted 0.5 to 2 mm from the cathode, which often also serves as a slit. For volatile samples, a batch inlet system can be used; here, the gaseous analyte is allowed to diffuse into the high-field area around the microtip of the anode. In *field desorption* methods, the anode is held in a probe that can be removed from the sample compartment, coated with a solution of analyte, and remounted.

The electric field in a field ionization source is concentrated at the anode points, and ionization is considered to occur by a quantum mechanical tunneling mechanism in which electrons from the analyte are extracted by the microtips of the anode. In this process (in contrast to electron impact ionization), relatively little vibrational or rotational energy is imparted to the molecule being ionized; thus, less fragmentation occurs, and the molecular ion is often the major product.

Field ionization is a gentle technique in the sense that fragmentation is minimized; the molecular ion and an ion with a mass one greater than the molecular ion [the $(M + 1)^+$ ion] are frequently the major products. In structural investigations, it is often advantageous to obtain spectra with both a field ionization or chemical ionization source (see next section) and an electron impact source; the former provides information about the molecular weight of the substance while the latter leads to characteristic fragmentation patterns, which are useful for identification.

A limitation to field ionization sources is their sensitivity, which is at least an order of magnitude less than electron impact sources; maximum ion currents are of the order of 10^{-11} A.

Chemical Ionization Sources (CI). In chemical ionization, gaseous atoms of the sample (either from a batch inlet or a heated probe) are ionized by collision with positive ions produced by electron bombardment of an excess of a reagent gas.[5] Chemical ionization methods are probably the second most common procedure for producing ions for mass spectrometry.

In order to carry out chemical ionization spectroscopy, it is necessary to modify the electron beam ionization area shown in Figure 18–4 by adding diffusion pump capacity and by reducing the width of the slit to the analyzer section; these measures allow a reagent pressure of about 1 torr to be maintained in the ionization area while still holding a pressure of better than 10^{-5} torr in the analyzer. With these changes, a gaseous reagent is introduced into the ionization region in an amount such that the concentration ratio of sample to reagent is 10^{-3} to 10^{-4}. Because of this large concentration difference, the electron beam reacts nearly exclusively with reagent molecules. Most modern instruments are equipped with a single ionization source that permits either electron bombardment or chemical ionization of samples.

One of the most common reagents is methane, which reacts with high-energy electrons to give several ions such as CH_4^+, CH_3^+, and CH_2^+. The first two predominate and represent about 90% of the reaction products. These ions react rapidly with additional methane molecules as follows:

$$CH_4^+ + CH_4 \rightarrow CH_5^+ + CH_3$$

$$CH_3^+ + CH_4 \rightarrow C_2H_5^+ + H_2$$

Generally, collisions between the sample molecule XH and CH_5^+ or $C_2H_5^+$ are highly reactive and involve proton or hydride

---

[5] For a review of this type of source, see: B. Munson, *Anal. Chem.*, **1977**, *49* (9), 772A; **1971**, *43*, 28A; A. Harrison, *Chemical Ionization Mass Spectrometry*, CRC Press: Boca Raton, Fla., 1983.

transfer. That is,

$$CH_5^+ + XH \rightarrow XH_2^+ + CH_4 \quad \text{proton transfer}$$

$$C_2H_5^+ + XH \rightarrow XH_2^+ + C_2H_4 \quad \text{proton transfer}$$

$$C_2H_5^+ + XH \rightarrow X^+ + C_2H_6 \quad \text{hydride transfer}$$

Note that the first two give the $(M + 1)^+$ ion while the latter produces an ion with a mass one less than the analyte, or the $(M - 1)^+$ ion. With some compounds, an $(M + 29)^+$ peak is also encountered, which results from transfer of a $C_2H_5^+$ ion to the analyte. A variety of other reagents are used for chemical ionization including propane, isobutane, and ammonia. Each produces a somewhat different spectrum with a given analyte.

Usually, chemical ionization produces far less fragmentation than electron ionization but more than field ionization. The sensitivity is significantly greater than field ionization and comparable with or sometimes greater than that for electron impact.

Fast Atom Bombardment Sources (FAB). Fast atom bombardment sources are rapidly assuming a major role in the production of ions for mass spectroscopic studies of high-molecular-weight biologically active species. In fast atom bombardment spectrometry, samples in a condensed state (often in a glycerol matrix) are ionized by bombardment with energetic xenon or argon atoms.[6] Both positive and negative analyte ions are sputtered from the surface of the sample in a desorption process.

A beam of fast atoms is obtained by passing accelerated ions of argon or xenon from an ion gun through a chamber containing argon or xenon atoms at a relatively high pressure ($\sim 10^{-5}$ torr). The speeding ions undergo an electron exchange reaction with the lower energy atoms in which charge neutralization occurs without substantial loss of initial kinetic energy. Thus, a beam

of energetic *atoms* is formed. The lower energy ions from the exchange are readily removed by an electrostatic deflection plate. Fast atom guns are now available from commercial sources, and most older spectrometers can be adapted to their uses. Most newer spectrometers offer accessories permitting this kind of sample excitation.

Fast atom bombardment of organic or biochemical compounds usually produces significant amounts of the parent ion (as well as ion fragments) even for high-molecular-weight and thermally unstable samples. For example, with fast atom bombardment, molecular weights up to 12,500 have been determined and detailed structural information has been obtained for compounds with molecular weights as great as 2800.

18A-5 MASS ANALYZERS

Several methods are available for the resolution of ions with different mass-to-charge ratios. Ideally, the analyzer should distinguish between minute mass differences while still allowing passage of a sufficient number of ions to yield readily measurable ion currents. As with an optical monochromator, to which the analyzer is analogous, these two properties are not entirely compatible, and design compromises must always be made. The principle difference among the various types of spectrometers that are now encountered in the laboratory lies in their systems for separating ions.

Resolution of Mass Spectrometers. The capability of a mass spectrometer to differentiate between masses is usually stated in terms of its resolution $m/\Delta m$, where m and $(m + \Delta m)$ are the masses of two particles that give just separable peaks of equal size. Two peaks are considered to be separated if the height of the valley between them is no more than 10% of their height (sometimes 50% is used as a criterion instead).

The resolution needed in a mass spectrometer depends greatly upon its applica-

---

[6] See: K. L. Rinehart, Jr., *Science*, **1982**, *218*, 254; and M. Barber, *et al.*, *Anal. Chem.*, **1982**, *54*, 645A.

tion. For example, discrimination between particles of the same nominal mass such as $C_2H_4^+$, CH_2N^+, N_2^+, and CO^+ (mass number: 28.0313, 28.0187, 28.0061, and 27.9949, respectively), requires an instrument with a resolution of several thousand. On the other hand, low-molecular-weight particles differing by a unit of mass or more (NH_3^+ and CH_4^+, for example) can be distinguished with an instrument having a resolution smaller than 50. A spectrometer with a resolution of perhaps 250 to 500 is needed for unit mass separation of most organic samples.

Single-Focusing Magnetic Sector Analyzers. Magnetic sector analyzers employ a permanent or electromagnet to cause the ion beam to be deflected into a circular path of 180, 90, or 60 deg. Figure 18–5 shows a 90-deg sector instrument in which ions, formed by electron impact, are accelerated through slit B into the metal analyzer tube, which is maintained at a pressure of about 10^{-7} torr. Particles of different mass can be focused on the exit slit by varying the field strength of the magnet or the accelerating potential between slits A and B. The ions passing through the exit slit fall on a collector electrode, which results in an ion current that is amplified and recorded.

The path described by ions of a given mass in the sector represents a balance between the forces that are acting upon it. The magnetic centripetal force F_M is given by

$$F_M = Hev \qquad (18\text{–}1)$$

where H is the magnetic field strength, v is the particle velocity, and e is the charge on the ion. The balancing centrifugal force F_c can be expressed as

$$F_c = \frac{mv^2}{r} \qquad (18\text{–}2)$$

where m is the particle mass and r is the radius of curvature. Finally, the kinetic energy of the particle E is given by

$$E = eV = \tfrac{1}{2}mv^2 \qquad (18\text{–}3)$$

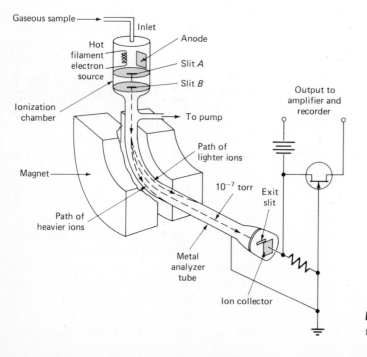

FIGURE 18–5   Schematic diagram of a mass spectrometer.

where V is the accelerating voltage applied in the ionization chamber. Note that all particles of the same charge, regardless of mass, are assumed to acquire the same kinetic energy during acceleration in the electrical field. This assumption is only approximately valid since the ions will possess a small statistical distribution of kinetic energies before acceleration.

A particle must fulfill the condition that F_M and F_c be equal in order to traverse the circular path to the collector; thus, from Equations 18–1 and 18–2

$$Hev = \frac{m\mathrm{v}^2}{r}$$

and

$$\mathrm{v} = \frac{Her}{m} \tag{18-4}$$

Substituting Equation 18–4 into Equation 18–3 and rearranging gives

$$\frac{m}{e} = \frac{H^2 r^2}{2V} \tag{18-5}$$

Equation 18–5 reveals that mass spectra can be obtained by varying one of three variables (H, V, or r) while holding the other two constant. Most modern sector mass spectrometers contain an electromagnet in which ions are sorted by holding V and r constant while varying the current in the magnet and thus H. Older spectrometers, and some smaller ones, on the other hand, are based upon fixed magnets; here, variation in the accelerating voltage V permits spectral scans. In a third type of spectrometer, H and v are held constant while the entire spectrum is recorded simultaneously on a photographic plate (see Figure 18–7); in this instance, r is the variable.

Double-Focusing Analyzers. The resolving power of single-focusing instruments is limited by the small variations in the kinetic energy of

particles of a given species as they leave the ion source. These variations, which arise from the initial statistical distribution of kinetic energies of the neutral molecules, cause a broadening of the ion beam reaching the collector and a loss of resolving power. In the double-focusing instrument, shown schematically in Figure 18–6, the beam is first passed through a radial electrostatic field. This field has the effect of focusing only particles of the same kinetic energy on slit 2, which then serves as the source for the magnetic separator. Resolution of particles differing by small fractions of a mass unit thus becomes possible. The ion currents produced are minute, however, and require large amplification for detection and recording. High-resolution double-focusing mass spectrometers are available commercially; their cost of purchase and maintenance is high.

Compact double-focusing instruments can also be purchased from commercial sources. A typical instrument of this type will have a 6-in. electrostatic sector and a 4-in., 90-deg magnetic mass deflector. Resolutions of about 2500 are common. These instruments, which are relatively compact and less expensive, are often employed as detectors for chromatographic columns.

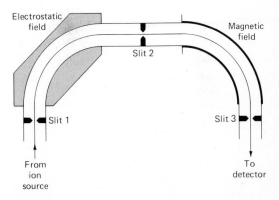

FIGURE 18-6 Design of a double-focusing separator.

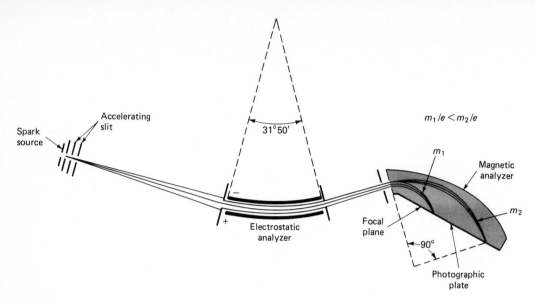

FIGURE 18-7   Double-focusing spark-source mass spectrometer, Mattauch-Herzog type.

Figure 18–7 shows another type of double-focusing instrument, which is based upon Mattauch-Herzog geometry. This geometry results in all ions being focused on a single focal plane regardless of mass-to-charge ratio. Such an arrangement is particularly useful for photographic detection.

Quadrupole Analyzers. The *quadrupole* spectrometer employs four short, parallel metal rods (diameter ~6 mm) arranged symmetrically around the beam (see Figure 18–8). The opposed rods are electrically connected, one pair being attached to the positive side of a variable dc source and the other pair to the negative terminal. In addition, variable radio-frequency ac potentials, which are 180 deg out of phase, are applied to each pair of rods. Neither field acts to

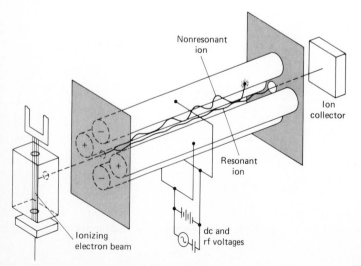

FIGURE 18-8 A quadrupole mass spectrometer. (From: D. Lichtman, *Res. Dev.*, **1964**, *15* (2), 52. With permission.)

accelerate the positive particles ejected from the ion source. The combined fields, however, cause the particles to oscillate about their central axis of travel; only those with a certain mass-to-charge ratio can pass through the array without being removed by collision with one of the rods. Mass scanning is achieved by varying the frequency of the ac supply while holding the potentials constant or by varying the potentials of the two sources while keeping their ratio and also the frequency constant. The latter is more common; the potential changes during a typical scan are illustrated in Figure 18–9. The two diverging straight lines show the variation of the two dc potentials as a function of time during one scan (0 to $+300$ V and 0 to -300 V). During the same time period, the *rf* voltages increase linearly from 0 to ±1500 V. Note that the *rf* signals are 180 deg out of phase.

Quadrupole instruments are compact, less expensive, and more rugged than magnetic focusing instruments.[7] A particular advantage is their low scan times (as low as 0.01 sec per selected scan range). This prop-

---

[7] For a thorough treatment of quadrupole mass spectrometers, see: *Quadrupole Mass Spectrometry and Its Applications,* P. H. Dawson, Ed., Elsevier: New York, 1977.

erty is particularly useful for real-time scanning of chromatographic peaks.

Time-of-Flight Analyzers. In a time-of-flight instrument, positive ions are produced intermittently by bombardment with brief pulses of electrons. These pulses, which are controlled by a grid, typically have a frequency of 10 to 50 kHz and a lifetime of 0.25 μsec. The ions produced in this way are then accelerated by an electrical field pulse that has the same frequency as, but lags behind, the ionization pulse. The accelerated particles pass into a field-free *drift tube* about a meter in length (Figure 18–10). Since all particles entering the tube have the same kinetic energies, their velocities in the drift tube must vary inversely with their masses (Equation 18–4), with the lighter particles arriving at the collector earlier than the heavier ones. Typical flight times are 1 to 30 μsec.

The detector in a time-of-flight mass spectrometer is an electron multiplier tube, similar in principle to a photomultiplier tube (Section 5F–2). The output of this tube is fed across the vertical deflection plates of a cathode ray oscilloscope (Section 2E–1) while the horizontal sweep is synchronized with the accelerator pulses; an essentially instantaneous display of the entire mass spectrum appears on the oscilloscope screen.

From the standpoint of resolution, repro-

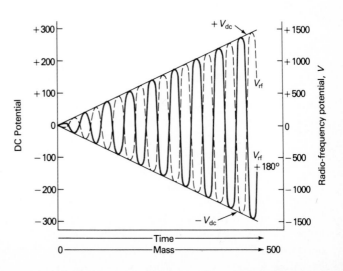

FIGURE 18-9 Voltage relationships during a mass scan with a quadrupole analyzer.

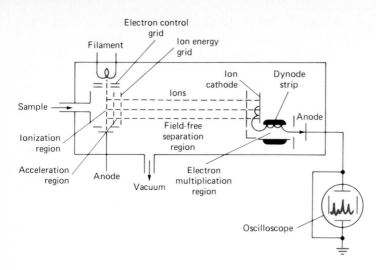

FIGURE 18-10   Schematic diagram of a time-of-flight mass spectrometer.

ducibility, and ease of mass identification, instruments employing time-of-flight mass separators are less satisfactory than those based upon magnetic or quadrupole focusing. On the other hand, several advantages partially offset these limitations. Included among these are ruggedness and ease of accessibility to the ion source, which allows the ready insertion of nonvolatile or heat-sensitive samples. The instantaneous display feature is also useful for the study of short-lived species. In general, time-of-flight instruments are smaller, more mobile, and more convenient to use than their magnetic focusing counterparts. They are much less common than sector or quadrupole analyzers.

18A-6   DETECTORS AND SIGNAL PROCESSORS

Three types of ion detectors are currently in use in mass spectrometers, namely the Faraday cup, solid-state electron multipliers, and ion-sensitive photographic plates.

The Faraday Cup.   The Faraday cup collector consists of a small electrode or plate surrounded by an open-ended reflecting box that prevents the escape of reflected ions and ejected secondary electrons. The plate is connected to ground via a high ohmage resistor (see Figure 18–5). The charge of the positive ions striking the plate is neutralized by a flow of electrons from ground through the resistor, and the resulting potential drop across this element is impressed on a field-effect transistor. The resulting potential is further amplified for display on a strip chart recorder or other readout systems. The main disadvantage of the Faraday detector is the need for a high-impedance amplifier, which limits the speed at which a spectrum can be scanned.

Electron Multipliers.   Virtually all modern mass spectrometers, designed for the analysis of organic compounds, feature electrostatic multipliers for detection. Here, the positive ions from the analyzer are accelerated by a potential difference of 2 to 5 keV and then fall upon the first plate of the multiplier. This plate, which usually consists of a copper/2% beryllium alloy, yields an average of about two electrons for each positive ion. The electrons are then accelerated by a positive potential to a second dynode where electron amplification occurs. After 20 stages, current amplification of about 10^7 is realized.

Ion-Sensitive Photoplates.   Photographic plates coated with silver bromide in gelatin are sensitive to positive ions. The properties and treatment of these plates are similar to those described in Section 10D–5.

Readout Devices. Modern mass spectrometers are often equipped with several different readout devices including analog or digital recorders, oscilloscopes, photographic recording galvanometers, and computerized printers, which produce bar graphs. The potentiometric strip chart recorder is useful for quantitative analysis and for the precise determination of isotope ratios. The response speed of such recorders is slow enough (~0.2 sec), however, to preclude their use in many applications.

One of the characteristics of mass spectra is the great diversity in peak size. Thus, mass spectrometer recorders generally are equipped to respond to several current ranges. A typical set of ranges is shown in Figure 18–11.

18A-7 COMPUTERIZED MASS SPECTROMETERS

Minicomputers and microprocessors are an integral part of most modern mass spec-

trometers.[8] A characteristic of a mass spectrum is the wealth of structural data that it provides. For example, a molecule with a molecular weight of 400 to 500 may be fragmented by an electron beam into as many as 200 different ions, each of which leads to a discrete spectral peak. For a structural determination, the heights and mass numbers of each peak must be determined, stored, and ultimately displayed. Because the amount of information is so large, it is essential that data acquisition and processing be rapid; the computer is ideally suited to these tasks. Moreover, in order for mass spectral data to be useful, several instrumental variables must be closely controlled during data collection. Computers or microprocessors are much more efficient than a human operator in exercising such controls. These considerations have led to the widespread incorporation of mini- and mi-

[8] For a detailed discussion of computerized mass spectrometry, see: J. R. Chapman, *Computers in Mass Spectrometry,* Academic Press: New York, 1978.

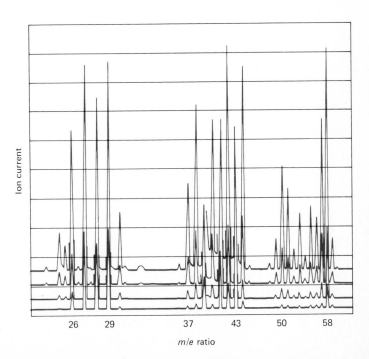

FIGURE 18-11 Mass spectra of *n*-butane recorded simultaneously by four galvanometers of different sensitivities. (From top to bottom, sensitivities are in the ratio 30 : 10 : 3 : 1.)

m/e ratio

crocomputers into modern commercial mass spectrometers.

The interface between a mass spectrometer and a computer usually has provisions for digitizing the amplified ion-current signal plus several other signals that are used for control of instrumental variables. Examples of the latter are source temperature, accelerating voltage, scan rate, and magnetic field strength.

The digitized ion-current signal ordinarily requires considerable processing before it is ready for display. First, the peaks must be normalized, a process whereby the height of each peak relative to some reference peak is calculated. Often the largest peak, called the *base peak,* serves as the reference and is arbitrarily given a peak height of 100 (or sometimes 1000). Each peak must also be assigned a mass number. This assignment is frequently made on the basis of the time of its appearance and the scan rate. Periodic calibration is necessary; for this purpose, perfluorotri-*n*-butylamine (PFTBA) or perfluorokerosene is used. For high-resolution work, the standard may be admitted with the sample. The computer is then programmed to recognize and employ the peaks of the standard as references for mass measurements. For low-resolution instruments, the calibration must generally be obtained separately from the sample because of the likelihood of peak overlaps.

The computer output is often displayed both in digital form and as a bar graph. Figure 18–12 is an example. The odd columns in the digital display list mass number in an increasing order. The even columns contain the corresponding ion currents normalized to the largest peak, which is found at mass 156; the current for this particle is assigned the number 1000, and all other peak heights are relative to this one. Thus, the peak at mass 141 is 82.6% of the base peak.

As with infrared spectroscopy, computer-stored library files of mass spectra are available; some commercial instruments are pro-

grammed to search these files for spectra that match that of the analyte.

18A–8 FOURIER TRANSFORM INSTRUMENTS

The Fourier transform principle has been successfully applied to the design of mass spectrometers. As was the case with infrared and nuclear magnetic resonance (Sections 11C–2 and 14C–2), instruments of this type provide improved signal-to-noise ratios, greater speed, and enhanced sensitivity.[9] A commercial Fourier transform mass spectrometer appeared on the market in the early 1980s.

In order to observe all of the elements of a time-domain mass spectrum simultaneously, it was necessary to abandon the slits and dispersing element of the classical instruments and substitute an *ion cyclotron resonance spectrometer* in their place. This instrument was developed in the mid-1960s to study gas phase reactions based upon the masses of reactants and products.

The Ion Cyclotron Resonance Phenomenon.[10] When a gaseous ion drifts into or is formed in a strong magnetic field, its motion becomes circular in a plane that is perpendicular to the direction of the field. The angular frequency of this motion ω_c can be obtained by rearranging Equation 18–4. That is,

$$\frac{v}{r} = \omega_c = \frac{eH}{m} \qquad (18\text{–}6)$$

where ω_c is the *cyclotron frequency* in radians per unit of time. Note that in a fixed field, the cyclotron frequency depends only upon

[9] See: C. L. Wilkins and M. L. Gross, *Anal. Chem.,* **1981,** *53,* 1661A; C. L. Wilkins, *Anal. Chem.,* **1978,** *50,* 493A; R. T. McIver, Jr., *Amer. Lab.,* **1980,** *12* (11), 18; C. L. Johlman, R. L. White, and C. L. Wilkins, *Mass Spectrom. Revs.,* **1983,** *2,* 389.

[10] For detailed discussions, see: T. A. Lehman and M. M. Bursey; *Ion Cyclotron Resonance Spectrometry,* Wiley: New York, 1976; and M. L. Gross and C. L. Wilkins, *Anal. Chem.,* **1971,** *43* (14), 65A.

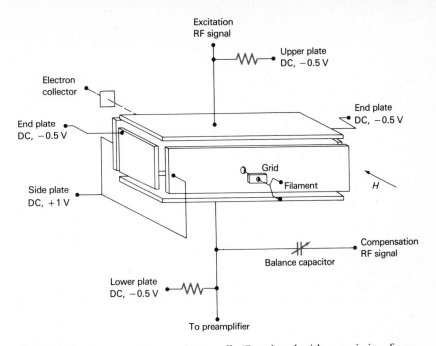

FIGURE 18-14   A trapped ion analyzer cell. (Reprinted with permission from: R. T. McIver, Jr., *Amer. Lab.,* **1980,** *12,* (11), 24. Copyright 1980 by International Scientific Communications, Inc.)

two side plates and a negative voltage on the upper, lower, and two end plates. The ions are accelerated by a radio-frequency signal applied to the upper plate as shown. The lower plate is connected to a preamplifier which amplifies the image current. The dimensions of the trapped ion cell are not crit-ical but are usually a few centimeters on a side.

The ion trap just described is reported to be highly efficient, with storage times up to several minutes having been observed.

The basis of the Fourier transform measurement is illustrated by Figure 18–15.

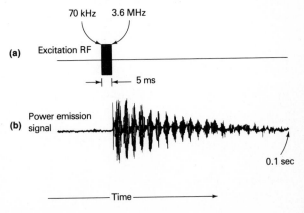

FIGURE 18-15   Schematic diagram showing the timing of (**a**) the radio-frequency signal and (**b**) the transient image signal (lower). (Reprinted with permission from: R. T. McIver, Jr., *Amer. Lab.,* **1980,** *12,* (11), 26. Copyright 1980 by International Scientific Communications, Inc.)

Ions are first generated by a brief electron beam pulse (not shown) and stored in the trapped ion cell. After a brief delay, the trapped ions are subjected to a radio-frequency pulse that increases linearly in frequency during its lifetime. (In Figure 18–15**a**, a pulse of 5 msec is shown during which time the frequency increases linearly from 0.070 to 3.6 MHz.) After the frequency sweep is discontinued, the image current, induced by the various accelerated ion packets, is amplified, digitized, and stored. The time-domain emission signal, shown in Figure 18–15**b**, is then transformed to yield a frequency-domain signal which can be converted to a mass-domain signal via Equation 18–6. Figure 18–16 illustrates the relationship between a time-domain spectrum and its frequency (and mass) domain counterpart.

Fourier transform spectrometers are bulky and expensive instruments. The commercial model employs a superconducting magnet with a nominal field of 1.9 tesla.[11] It has a resolution of 50,000 at mass 1000.

18A–9   COMMERCIAL MASS SPECTROMETERS

Mass spectrometers of various design are available from more than a dozen instrument companies. Their resolutions, mass ranges, and other properties vary widely as do their costs (from a few thousand to several hundred thousand dollars). Table 18–2 lists some of the properties of typical mass spectrometers.

18B   Mass Spectra

Even relatively simple molecules often produce complex mass spectra containing an array of peaks of widely varying intensities (for example, see Figures 18–11 and

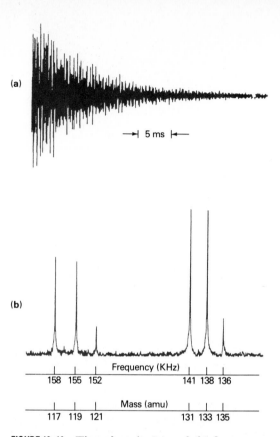

FIGURE 18–16   Time domain (**a**) and (**b**) frequency or mass domain spectrum for 1,1,1,2-tetrachloroethane. (Reprinted with permission from: E. B. Ledford, Jr., *et al., Anal. Chem.,* **1980,** *52,* 466. Copyright 1980 American Chemical Society.)

18–12). The detailed nature of the spectrum for a molecule depends upon its ionization potential, the functional groups it contains, the method of ionization, the sample pressure and temperature, and the instrument design. Complete interpretation of a mass spectrum is seldom possible (or necessary).

18B–1   ELECTRON IMPACT MASS SPECTRA

The most common method for producing ions for mass spectra is by electron impact in which gaseous molecules from the sample

---

[11] 1 tesla = 10,000 Gauss.

are bombarded with energetic (usually 70 eV) electrons. This process produces initially a radical ion called the *molecular* or *parent ion.* For the molecular species M, molecular ion formation is often written as

$$M + e \rightarrow M^{+} + 2e$$

Here, the dot indicates that the ion is a radical formed by loss of an electron. (The simpler formulation M^{+} is used by many mass spectroscopists.)

Generally, molecular ion formation is followed by a series of reactions in which ions having smaller (and sometimes larger) masses than the parent ion are formed. As a consequence, mass spectra are often complex consisting of numerous peaks of varying intensities. Examples of electron impact spectra for two simple organic molecules, ethyl benzene and methylene chloride, are shown in Figure 18–17.

Molecular Ion Peak. The molecular ion or parent ion peak occurs at a mass corresponding to the molecular weight of the original neutral molecule. Thus, for the two spectra in Figure 18–17, molecular ion peaks are observed at 106 for ethyl benzene and at 84 for methylene chloride. The molecular ion peak is, of course, of prime importance in structural determinations because its mass provides the molecular weight of the unknown to the nearest whole number. Unfortunately, it is not always possible to identify the parent ion peak. Indeed, with electron impact ionization, certain less stable molecules yield no molecular ion peak but instead fragmentation peaks (see Figure 18–18b).

The Base Peak. The largest peak in a mass spectrum is called the *base peak.* Often it is assigned an arbitrary height of 100 (or sometimes 1000). The magnitude of the remaining peaks are then all normalized to the base peak. In some instances, individual peak heights are plotted as percentages of the total peak heights, a more informative number, but one that is more laborious to derive.

It should be noted that in each of the spectra in Figure 18–17, the base peak corresponds to a fragment of the molecule, which has a mass significantly less than the molecular weight of the original compound. Thus for ethyl benzene, the base peak occurs at a mass of 91, which corresponds to the ion formed by the loss of a CH_3 group. Similarly, for methylene chloride, the base peak is found at a mass resulting from the loss of one chlorine atom. More often than not, the base peak in electron impact spectra arise from fragments such as these rather than from the original ion.

Table 18–2		
Comparison of Some Typical Commercial Mass Spectrometers		
Type	**Approximate Mass Range**	**Approximate Resolution**
Fourier transform	12–2000	50,000–760,000
Double focusing	8–7200	40,000
	1–4000	20,000
	2–3600	>25,000
Single focusing	1–1400	1,800
	1–900	1,500
Time-of-flight	1–700	150–250
	0–250	130
Quadrupole	1–750	500

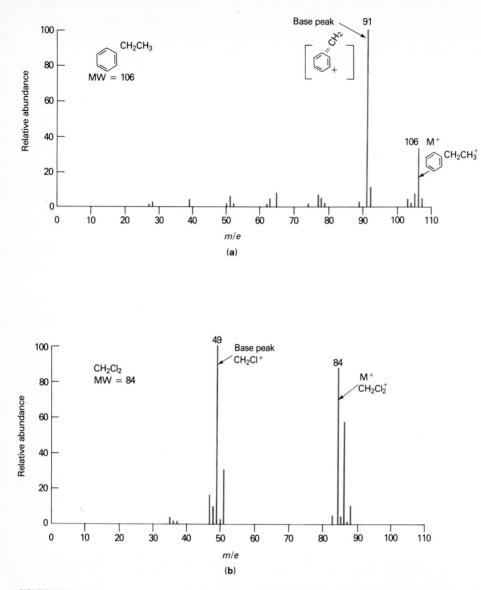

FIGURE 18-17 Electron impact mass spectra of (**a**) ethyl benzene and (**b**) methylene chloride.

Fragmentation and Rearrangement of Excited Radical Ions. As mentioned earlier, the energy of an electron beam far exceeds that required for ionization of organic compounds. Consequently, the radical ions formed initially are in a highly excited state and can undergo fragmentation or rearrangement to give ions of lower (and sometimes higher) masses. Some typical reactions that follow electron impact ionization of a molecule are shown in Table 18–3. Here, A, B, C, and D represent atoms or groups of atoms in the hypothetical molecule ABCD.

Isotope Peaks. It is of interest to note in the

Table 18-3
Some Typical Reactions in an Electron Impact Source

Molecular ion formation $ABCD + e \rightarrow ABCD^{\ddagger} + 2e$

Fragmentation $ABCD^{\ddagger} \rightarrow A^{+} + BCD.$

$$\rightarrow A. + BCD^{+} \longrightarrow BC^{+} + D$$

$$\rightarrow CD. + AB^{+} \begin{array}{l} \rightarrow B + A^{+} \\ \rightarrow A + B^{+} \end{array}$$

$$\rightarrow AB. + CD^{+} \begin{array}{l} \rightarrow D + C^{+} \\ \rightarrow C + D^{+} \end{array}$$

Rearrangement followed
by fragmentation $ABCD^{\ddagger} \rightarrow ADBC^{\ddagger} \begin{array}{l} \rightarrow BC. + AD^{+} \\ \rightarrow AD. + BC^{+} \end{array}$

Collision followed
by fragmentation $ABCD^{\ddagger} + ABCD \rightarrow (ABCD)_2^{\ddagger} \rightarrow BCD. + ABCDA^{+}$

---

spectra shown in Figure 18–17 that peaks occur at masses that are greater than that of the parent ion by one to four mass units. These peaks are attributable to ions having the same chemical but different isotopic compositions. For example, for methylene chloride, the more important isotopic species are $^{12}C^{1}H_2{}^{35}Cl_2$ ($m = 84$), $^{13}C^{1}H_2{}^{35}Cl_2$ ($m = 85$), $^{12}C^{1}H_2{}^{35}Cl^{37}Cl$ ($m = 86$), $^{13}C^{1}H_2{}^{35}Cl^{37}Cl$ ($m = 87$), and $^{12}C^{1}H_2{}^{37}Cl_2$ ($m = 88$). Peaks for each of these species can be seen in Figure 18–17**b**. The size of the various peaks depends upon the relative natural abundance of the isotopes. Table 18–4 lists the more common isotopes for atoms that occur widely in organic compounds. Note that fluorine, phosphorus, and iodine occur only as a single isotope.

Table 18-4
Natural Abundance of Isotopes of Some Common Elements

Element[a]	Most Abundant Isotope	Abundance of Other Isotopes Relative to 100 Parts of the Most Abundant[b]	
Hydrogen	^{1}H	^{2}H	0.015
Carbon	^{12}C	^{13}C	1.08
Nitrogen	^{14}N	^{15}N	0.37
Oxygen	^{16}O	^{17}O	0.04
		^{18}O	0.20
Sulfur	^{32}S	^{33}S	0.80
		^{34}S	4.40
Chlorine	^{35}Cl	^{37}Cl	32.5
Bromine	^{79}Br	^{81}Br	98.0
Silicon	^{28}Si	^{29}Si	5.1
		^{30}Si	3.4

[a] Fluorine (^{19}F), phosphorus (^{31}P), and iodine (^{127}I) have no additional naturally occurring isotopes.
[b] The numerical entries indicate the average number of isotopic atoms present for each 100 atoms of the most abundant isotope; thus, for every 100 ^{12}C atoms there will be an average of 1.08 ^{13}C atoms.

The small peaks for ethyl benzene as mass 107 in Figure 18–17**a** is due to the presence of ^{13}C in some of the molecules. Peaks due to incorporation of two or more ^{13}C atoms are so small as to be undetectable because of the low probability of there being more than one ^{13}C atom in a molecule.

As will be shown in Section 18C–2, isotope peaks sometimes provide a useful means for determining the empirical formula for a compound.

Collision Product Peaks. Ion molecule collisions, such as that shown by the last equation in Table 18–3, can produce peaks at higher mass numbers than that of the molecular ion. At ordinary sample pressures, however, the only important reaction of this type is one in which the collision transfers a hydrogen atom to the ion; an enhanced $(M + 1)^+$ peak results. Proton transfer is a second-order reaction, and the amount of product depends strongly upon the reactant concentration. Consequently, the height of an $(M + 1)^+$ peak of this type increases much more rapidly with increases in sample pressure than do the heights of other peaks; thus detection of this reaction is usually possible.

18B-2   SPECTRA FROM CHEMICAL AND FIELD IONIZATION

Most newer mass spectrometers are designed so that electron ionization and chemical ionization can be carried out interchangeably. In addition, the option of field ionization and field desorption ionization is often provided. In general, these alternate methods yield simpler spectra, which are particularly useful for obtaining molecular weights of compounds that give few or no molecular ions with electron impact.

Chemical Ionization. Figure 18–18 contrasts the electron and chemical ionization spectra for a straight chain aliphatic alcohol possessing ten carbon atoms. The electron impact spectrum (Figure 18–18**b**) shows evidence for rapid and extensive fragmentation of the molecular ion. Thus, no detectable peaks are observed above mass 112, which corresponds to the ion $C_8H_{16}^+$. The base peak is provided by the ion $C_3H_5^+$ at mass 41; other peaks for various C_3 species are grouped around the base peak. A similar series of peaks are found at 14, 28, and 42 mass units greater and correspond to ions with one, two, and three additional CH_2 groups.

Relative to the electron impact spectrum, the chemical ionization spectrum shown in Figure 18–18**a** is simple indeed, consisting of the $(M - 1)^+$ peak, a base peak corresponding to a molecular ion that has lost an OH group, and a series of peaks differing from one another by 14 mass units. As in the electron impact spectrum, these peaks arise from ions formed by cleavage of adjacent carbon-carbon bonds. The reactions shown on page 531 reveal that chemical ionization spectra generally contain well defined $(M + 1)^+$ or $(M - 1)^+$ peaks resulting from the addition or abstraction of a proton in the presence of the reagent ion.

Field Ionization. Figure 18–19 shows three spectra for glutamic acid obtained by (**a**) electron impact ionization, (**b**) field ionization, and (**c**) field desorption ionization. Recall that both field desorption ionization and field ionization are based upon the very high electrical fields generated at a special anode made up of a large number of microtips. For field desorption, the sample is placed directly on the anode, which is then inserted into the beam by means of a sample probe. In contrast, field ionization is carried out by leaking sample vapor over the anode.

With field desorption, the base peak corresponds to a mass that is one greater than the parent ion or the MH^+ species. Only one other small peak at $(M + 2)^+$ is evident. Field ionization also produces a small but distinct MH^+ peak and a base peak that cor-

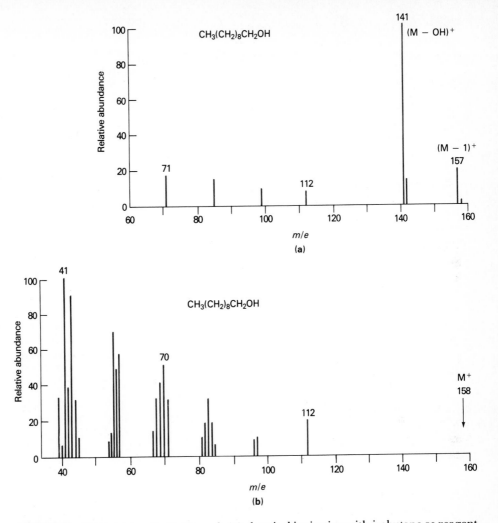

FIGURE 18-18 Mass spectra for 1-decanol: (**a**) chemical ionization with isobutane as reagent gas; (**b**) 70 eV electron impact.

responds to a loss of a water molecule by this ion. A few other fragments are evident, but the spectrum is still simple compared with the electron impact ionization. In the electron impact spectrum (Figure 18–19**a**), the highest observable peak (mass 129) is produced by the ion formed when the molecular ion loses a water molecule. Numerous other fragments are also found at lower masses.

18C Identification of Pure Compounds by Mass Spectrometry

The mass spectrum of a pure compound provides several kinds of data that are useful for its identification. The first is the molecular weight of the compound, and the second is its empirical formula. In addition, study of fragmentation patterns revealed by the mass spectrum often provides informa-

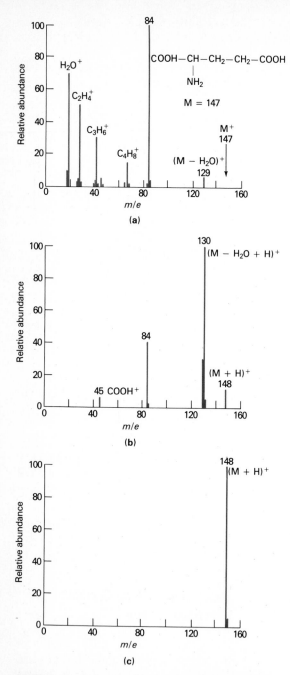

FIGURE 18-19 Mass spectra for glutamic acid: (**a**) electron impact ionization; (**b**) field ionization; and (**c**) field desorption. (From: H. D. Beckey, A. Heindrichs, and H. U. Winkler, *Int. J. Mass Spec. Ion Phys.*, **1970**, *3*, App. 11. With permission.)

tion about the presence or absence of various functional groups. Finally, the actual identity of a compound can often be established by comparison of its mass spectrum with those of known compounds until a close match is realized.

18C-1 MOLECULAR WEIGHT DETERMINATION

For compounds that can be ionized by one of the methods described earlier, the mass spectrometer is an unsurpassed tool for the determination of molecular weight. This determination, of course, requires the identification of the molecular ion peak, or in some cases, the $(M + 1)^+$ or the $(M - 1)^+$ peak. The location of the peak on the abscissa then gives the molecular weight to the nearest whole number—an accuracy that cannot be realized by any other method. It should be noted that the molecular weight found by mass spectroscopy corresponds to a molecule made up of the most abundant isotopes of the various elements it contains. This molecular weight will differ slightly from that calculated from a table of chemical atomic weights or that determined by other types of measurements. This difference is generally small enough to be of no consequence in a qualitative analysis.

A mass spectroscopic molecular weight determination requires the sure knowledge of the position of the molecular ion peak. Caution is therefore always advisable, particularly with electron impact sources, where the parent ion peak may be absent or small relative to impurity peaks. When doubt exists, additional spectra by chemical and field ionization are particularly useful.

Stability of Various Types of Molecular Ions. For a given set of conditions, the intensity of a molecular ion peak depends upon the stability of the ionized particle, a minimum lifetime of about 10^{-5} sec being needed for a particle to reach the collector and be detected. The stability of a radical ion is strongly affected by structure. Thus, the size of molecular ion peaks shows great variability.

In general, the molecular ion is stabilized by the presence of π electron systems that more easily accommodate the loss of an electron. Cyclic structures also give large parent peaks, since rupture of a bond does not necessarily produce two fragments. In general, the stability of molecular ions of various types decreases in the following order: aromatics, conjugated olefins, alicyclics, sulfides, unbranched hydrocarbons, mercaptans, ketones, amines, esters, ethers, carboxylic acids, branched hydrocarbons, and alcohols. These effects are illustrated in Table 18–5, which compares the height of the molecular ion peak for some C_{10} compounds relative to the total peak heights in the spectrum.

Figures 18–18**b** and 18–19**a** are electron impact spectra of compounds with unstable molecular ions. In both instances, the peak with the highest mass occurs well below the mass of the parent ion.

A mass spectrometer will generally exhibit a detectable parent ion peak provided the size of the peak is at least 1% of the total peak heights; for some instruments, this limit is lowered to 0.1%. Thus, perhaps 80 to 90% of all organic compounds yield a detectable molecular ion peak, which is suitable for molecular weight determinations.

Identification of the Molecular Ion Peak. No single method is available for establishing unambiguously that the peak of highest mass number (neglecting small isotope peaks) is indeed produced by the molecular ion. On the other hand, from a series of observations, the experienced mass spectroscopist can ordinarily make this judgment with reasonable assurance of being correct. As mentioned earlier (p. 546), it is frequently possi-

Table 18–5
Variation in Molecular Ion Peak with Structure

Compound	Formula	Relative Peak Height (percent of total peak heights)
Naphthalene		44.3
n-Butylbenzene	C_4H_9	8.3
trans-Decaline		8.2
Diamyl sulfide	$(C_5H_{11})_2S$	3.7
n-Decane	$C_{10}H_{22}$	1.41
Diamylamine	$(C_5H_{11})_2NH$	1.14
Methyl nonanoate	$C_9H_{17}COOCH_3$	1.10
Diamyl ether	$(C_5H_{11})_2O$	0.33
3,3,5-Trimethylheptane	$C_{10}H_{22}$	0.007
n-Decanol	$C_{10}H_{21}OH$	0.002

[a] Taken from K. Biemann, *Mass Spectrometry, Organic Applications*, p. 52, McGraw-Hill Book Company, Inc.: New York, 1962. With permission.

ble to identify an $(M + 1)^+$ peak by observing its behavior as a function of sample size. Determination of whether or not the highest mass peak is indeed the molecular ion is more troublesome. Here, a knowledge of fragmentation patterns for various types of compounds is essential. For example, peaks in the range of $(M - 4)^+$ to $(M - 14)^+$ are unlikely and immediately cast doubt upon the peak at M^+ being caused by the molecular ion. Patterns of this type could only occur as the result of abstraction of several hydrogen atoms or unlikely carbon-containing ions from the molecular ion. Losses by a molecular ion of fragments having masses of 19 to 25 are also seldom encountered unless the compound contains fluorine (here F at mass 19 and HF at mass 20 can be lost). On the other hand, a strong peak at $(M - 18)^+$ or $(M + 1 - 18)^+$ suggests that even a weak peak at M^+ may be the parent ion since, for alcohols and aldehydes, the loss of water is a common occurrence; Figure 18–19**b** provides an example of this effect.

A useful test to distinguish between the parent peak and a prominent $(M - 1)^+$ peak is to observe their relative peak heights as the energy of the electron beam is decreased. This procedure will reduce all peak intensities but will enhance the parent ion peak relative to all other peaks. As mentioned earlier, the widespread availability of chemical and field ionization sources has greatly reduced the problem of positive identification of the molecular ion peak.

18C–2 DETERMINATION OF MOLECULAR FORMULAS

Partial or exact molecular formulas can be determined from the mass spectrum of a compound, provided the molecular ion peak can be identified.

Molecular Formulas from High-Resolution Instruments. A unique formula for a compound can often be derived from the exact mass of the parent ion peak. This application, however, requires a high-resolution instrument capable

of detecting mass differences of a few thousandths of a mass unit. Consider, for example, the molecular weights of the following compounds: purine, $C_5H_4N_4$ (120.044); benzamidine, $C_7H_8N_2$ (120.069); ethyltoluene, C_9H_{12} (120.094); and acetophenone, C_8H_8O (120.157). If the measured mass of the parent ion peak is 120.069, then all but $C_7H_8N_2$ are excluded as possible formulas. Tables that list all reasonable combinations of C, H, N, and O by molecular weight to the third or fourth decimal place have been compiled.[12] A small portion of one of these compilations is shown in Table 18–6. In addition, a table of molecular weights (to the sixth decimal place) of all of the compounds listed in the ninth edition of the *Merck Index* is available.[13]

Formulas from Isotopic Ratios. The data from an instrument that can discriminate between whole mass numbers provides useful information about the formula of a compound, provided only that the molecular ion peak is sufficiently intense that its height and the heights of the $(M + 1)^+$ and $(M + 2)^+$ isotope peaks can be determined accurately. The following example illustrates this type of analysis.

EXAMPLE 18–1. Calculate the ratios of the $(M + 1)^+$ to M^+ peak heights for the following two compounds: dinitrobenzene, $C_6H_4N_2O_4$ (M = 168) and an olefin, $C_{12}H_{24}$ (M = 168).

From Table 18–4, we see that for every 100 ^{12}C atoms there are 1.08 ^{13}C atoms. Since there are six carbon atoms in nitrobenzene, however, we would expect there to be 6.48 (6 × 1.08) molecules of nitrobenzene having one ^{13}C atom for every 100 molecules having none. Thus, from this effect alone the $(M + 1)^+$ peak will be

---

[12] J. H. Beynon and A. E. Williams, *Mass and Abundance Tables for Use in Mass Spectrometry,* Elsevier: Amsterdam, 1963.

[13] *Tables of Molecular Weights,* Merck and Co., Inc.: Rahway, N.J., 1978.

6.48% of the M^+ peak. The isotopes of the other elements also contribute to this peak; we may tabulate their effects as follows:

$$C_6H_4N_2O_4$$

^{13}C	6×1.08	$= 6.48\%$
^2H	4×0.015	$= 0.060\%$
^{15}N	2×0.37	$= 0.74\%$
^{17}O	4×0.04	$= 0.16\%$
	$(M + 1)^+/M^+$	$= 7.44\%$

$$C_{12}H_{24}$$

^{13}C	12×1.08	$= 12.96\%$
^2H	24×0.015	$= 0.36\%$
	$(M + 1)^+/M^+$	$= 13.32\%$

It is seen in this example that a measurement of the ratios of the $(M + 1)^+$ to M^+ peak heights would permit discrimination between two compounds that have identical whole-number mass weights.

Table 18-6
Isotopic Abundance Percentages and Molecular Weights For Various Combinations of Carbon, Hydrogen, Oxygen, and Nitrogen[a]

		Abundance, % M Peak Height		
	Formula	M + 1	M + 2	Molecular Weight
M = 83	C_2HN_3O	3.36	0.24	83.0120
	$C_2H_3N_4$	3.74	0.06	83.0359
	C_3HNO_2	3.72	0.45	83.0007
	$C_3H_3N_2O$	4.09	0.27	83.0246
	$C_3H_5N_3$	4.47	0.08	83.0484
	$C_4H_3O_2$	4.45	0.48	83.0133
	C_4H_5NO	4.82	0.29	83.0371
	$C_4H_7N_2$	5.20	0.11	83.0610
	C_5H_7O	5.55	0.33	83.0497
	C_5H_9N	5.93	0.15	83.0736
	C_6H_{11}	6.66	0.19	83.0861
M = 84	CN_4O	2.65	0.23	84.0073
	$C_2N_2O_2$	3.00	0.43	83.9960
	$C_2H_2N_3O$	3.38	0.24	84.0198
	$C_2H_4N_4$	3.75	0.06	84.0437
	C_3O_3	3.36	0.64	83.9847
	$C_3H_2NO_2$	3.73	0.45	84.0085
	$C_3H_4N_2O$	4.11	0.27	84.0324
	$C_3H_6N_3$	4.48	0.08	84.0563
	$C_4H_4O_2$	4.46	0.48	84.0211
	C_4H_6NO	4.84	0.29	84.0449
	$C_4H_8N_2$	5.21	0.11	84.0688
	C_5H_8O	5.57	0.33	84.0575
	$C_5H_{10}N$	5.94	0.15	84.0814
	C_6H_{12}	6.68	0.19	84.0939
	C_7	7.56	0.25	84.0000

[a] Taken from R. M. Silverstein, G. C. Bassler, and T. C. Morrill, *Spectrometric Identification of Organic Compounds*, 4th ed., p. 49, Wiley: New York, 1981. Reprinted by permission of John Wiley & Sons, Inc.

The use of relative isotope peak heights for the determination of molecular formulas is greatly expedited with the tables developed by Beynon[14]; a portion of a modified form of his tabulations is shown in Table 18–6. Here, a listing for all reasonable combinations of C, H, O, and N is given for mass numbers 83 and 84 (the original tables extend to mass number 500); also tabulated are the heights of the corresponding $(M + 1)^+$ and $(M + 2)^+$ peaks reported as percentages of the M^+ peak. If a reasonably accurate experimental determination of these percentages can be obtained from a spectrum, a likely formula can be readily ascertained. For example, a molecular ion peak at mass 84 and with $(M + 1)^+$ and $(M + 2)^+$ peaks of 5.6 and 0.3% of M^+ suggests a compound having the formula C_5H_8O (Table 18–6).

The isotopic ratio is particularly useful for the detection and estimation of the number of sulfur, chlorine, and bromine atoms in a compound because of the large contribution they make to the $(M + 2)^+$ peak (see Table 18–4). Thus, for example, an $(M + 2)^+$ peak that is about 65% of the M^+ peak would be strong evidence for a molecule containing two chlorine atoms; an $(M + 2)^+$ peak of about 4%, on the other hand, would suggest one atom of sulfur. By examination of the heights of the $(M + 4)^+$ and $(M + 6)^+$ peaks as well, it is sometimes feasible to identify combinations of chlorine and bromine atoms.

The Nitrogen Rule. The *nitrogen rule* also provides information concerning possible formulas of a compound whose molecular weight has been determined. This rule states that all organic compounds with an even molecular weight must contain zero or an even number of nitrogen atoms; all compounds with odd molecular weights must have an odd number of nitrogen atoms. The fragments formed by cleavage of one bond, however, have an odd mass number if they contain zero or an even number of nitrogen atoms; conversely, such fragments have an even mass number if the nitrogens are odd in number. The rule is a direct consequence of the fact that, with the exception of nitrogen, the valency and the mass number of the isotopes of elements that commonly occur in organic compounds are either both even or both odd. The nitrogen rule applies to all covalent compounds containing carbon, hydrogen, oxygen, sulfur, the halogens, phosphorus, and boron.

18C-3 STRUCTURAL INFORMATION FROM FRAGMENTATION PATTERNS

From Figure 18–20, it is evident that fragmentation of even simple molecules produces a large number of ions with different masses. A complex spectrum results, which often permits identification of the parent molecule or at least recognition of likely functional groups in the compound. Systematic studies of fragmentation patterns for pure substances have led to rational fragmentation mechanisms and a series of general rules that are helpful in interpreting spectra.[15] It is seldom possible (or desirable) to account for all of the peaks in a spectrum. Instead, characteristic patterns of fragmentation are sought. For example, the spectrum in Figure 18–20**a** is characterized by clusters of peaks differing in mass by 14. Such a pattern is typical of straight-chain paraffins, in which cleavage of adjacent carbon-carbon bonds results in the observed mass decreases. This same pattern is evident in the left-hand parts of the two lower spec-

---

[14] J. H. Beynon and A. E. Williams, *Mass and Abundance Tables for Use in Mass Spectrometry*, Elsevier: Amsterdam, 1963. See also: R. M. Silverstein, G. C. Bassler, and T. C. Morrill, *Spectrometric Identification of Organic Compounds*, 4th ed., pp. 46–89, Wiley: New York, 1981.

[15] See, for example: R. M. Silverstein, G. C. Bassler, and T. C. Morrill, *Spectrometric Identification of Organic Compounds*, 4th ed., pp. 14–15, Wiley: New York, 1981.

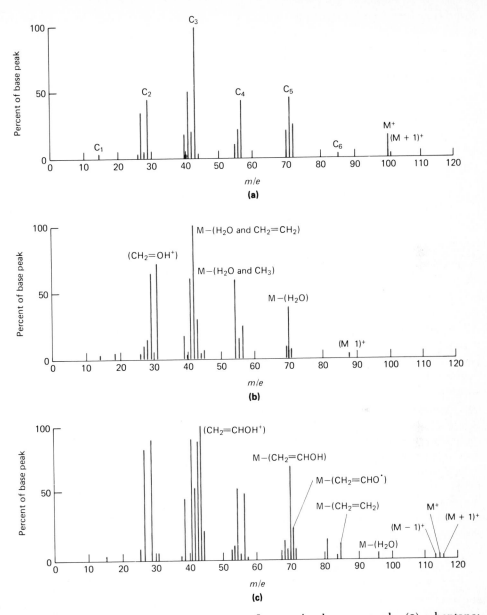

FIGURE 18-20 Electron impact mass spectra of some simple compounds: (**a**) *n*-heptane; (**b**) 1-pentanol; (**c**) *n*-heptanal.

tra as well. Quite generally, the most stable hydrocarbon fragments contain three or four carbon atoms and the corresponding peaks are thus the largest.

Alcohols usually have a very weak or non-existent parent ion peak but lose water to give a strong peak at $(M - 18)^+$ (see Figure 18–20b). Cleavage of the C—C bond next to an oxygen is also common, and primary alcohols always have a strong peak at mass 31 due to the ion $CH_2{=}OH^+$.

The interested reader should refer to one of several references for further generalizations concerning the identification of organic functional groups from mass spectrometric data.[16]

18C-4 IDENTIFICATION OF COMPOUNDS BY COMPARISON OF SPECTRA

Generally, after determining the molecular weight of the analyte and studying its isotopic distribution and fragmentation patterns, the experienced mass spectroscopist is able to narrow the possible structures down to a handful. Final identification is then based upon a comparison of the mass spectra of the unknown with spectra for authentic samples of the suspected compounds. This procedure is based upon the assumptions that: (1) mass fragmentation patterns are unique and (2) experimental conditions can be sufficiently controlled to produce reproducible spectra. The first assumption often is not valid for spectra of stereo- and geometric isomers and occasionally is not valid for certain types of closely related compounds. The probability of different compounds yielding the same spectra becomes markedly smaller as the number of spectral peaks increases. For this reason, electron impact ionization is the method of choice for spectral comparison.

Unfortunately, heights of mass spectral peaks are strongly effected by such variables as the energy of the electron beam, the location of the sample with respect to the beam, the sample pressure and temperature, and the general geometry of the mass spectrometer. As a consequence, significant variations in relative abundance are observed for spectra derived in different laboratories and from different instruments. Nevertheless, it

has proven possible in a remarkably large number of cases to identify unknowns from library spectra obtained with a variety of instruments and operating conditions. Generally, however, it is desirable to confirm the identity of a compound by comparison of its spectra with that of an authentic compound obtained under identical conditions and with the same instrument. Comparison of NMR and IR spectra is also useful for confirmation.

Spectral Libraries.   Several libraries containing printed mass spectral data exist. Two of the largest of these are:

1. *Registry of Mass Spectral Data,* E. Stenhagen, S. Abrahamsson, and F. W. McLafferty. Wiley: New York, 1974, 4 volumes. This collection lists spectral data (*m/e* and relative abundance) for 18,806 different compounds. Spectra are listed in order of molecular weight.

2. *Eight Peak Index of Mass Spectra,* 2d ed., Mass Spectroscopy Data Centre: Aldermaston, Reading, United Kingdom, 1974. Data for 31,101 mass spectra based on the eight major peaks. The data are arranged in three tables: (1) ordered in molecular weight and subordered in formula; (2) ordered in molecular weight and subordered in *m/e* value of the most abundant ion; and (3) arranged according to a particular *m/e* value occurring respectively, as the first, second, and third most abundant ion.

Computerized Library Search Systems.   During the past decade, a great deal of effort has been devoted to the development of libraries in which low-resolution (unit mass), electron impact mass spectra are stored in computer memories.[17] Computerized mass spectrome-

---

[16] R. M. Silverstein, G. C. Bassler, and T. C. Morrill, *Spectrometric Identification of Organic Compounds,* 4th ed., Chapter 2, Wiley: New York, 1981; K. Biemann, *Mass Spectrometry, Organic Chemical Applications,* McGraw-Hill: New York, 1962; H. Budzikiewicz, C. Djerassi, and D. H. Williams, *Interpretation of Mass Spectra of Organic Compounds,* Holden-Day: San Francisco, 1964; and F. W. McLafferty, *Interpretation of Mass Spectra,* 3d ed., University Science Books: Mill Valley, CA, 1980.

[17] For detailed descriptions of some of these systems, see: D. Henneberg, *Adv. in Mass Spectrom.,* **1979,** *8,* 1511; S. R. Heller and G. W. A. Milne, *Amer. Lab.,* **1980,** *12* (3), 33; and S. R. Heller, in *Biochemical Applications of Mass Spectrometry,* First Supplemental Volume, G. R. Waller and O. C. Dermer, Eds., Chapter 6, Wiley: New York, 1980.

ters have also been developed that are able to use these libraries for identification or interpretation of mass spectral data from unknowns.

Two types of libraries exist currently: large comprehensive ones and small ones. The latter contain a few hundred to a few thousand spectra for application to a limited area such as pesticide residues, drugs, or forensics. Small libraries may be a part of the equipment package offered by instrument manufacturers. Often the data are extracted from the large libraries, which are described in the next paragraphs. Alternatively, small libraries may be developed by the instrument user to meet the needs of a given area of interest.

Two large libraries are available. One of these is the Probability Based Matching (PBM) System developed at Cornell University.[18] This system, which contains about 41,000 spectra for approximately 32,000 different compounds, is made available to outside users by the Office of Computer Services, Cornell University through the TYMNET and TELENET computer networking systems.

The second large library, the Mass Spectral Search System (MSSS), is part of the NIH-EPA Chemical Information System (CIS), which contains data bases for not only mass spectra but also carbon-13 NMR, X-ray diffraction data for crystals and powders, mammalian acute toxicity data, and aquatic toxicity data. These libraries can also be accessed by TELENET.[19]

Several methods of using these data files have been developed. One is an interactive option in which the spectroscopist supplies data to the system piece by piece until the number of possible structures has been narrowed to one or a few. If desired, the spectra for the possible compounds can then be printed out for comparison with that of the unknown.

Figure 18–21 shows the printout from an interactive search with MSSS for the identity of a water pollutant that had a base peak at m/e 237 and several other peaks in the m/e ranges of 235 to 237 and 293 to 305. The underlined statements are the responses typed in by the user. Thus the peak search is initiated by the response "peak". The user is then asked to type in the peak location as well as the minimum and maximum possible intensities. The response here is 237 with a range of intensities of 100 to 100, which indicates that the peak is the base peak. After the search, the computer reports that only 23 out of the libraries 32,191 compounds have base peaks at m/e 237. When data for a second peak at m/e of 295 is entered (intensity range 60 to 80), the number of possible compounds is reduced to 2; the third peak at m/e of 293 reduces the possible matches to just 1. The response of "1" to "Next Request" calls for a printout giving the file reference number for the compound; its *Chemical Abstract* registry number; and the molecular weight, formula, and names of the compound.

The remainder of the material in Figure 18–21 is the response to a request that the computer recall from memory and print out a carbon-13 NMR spectrum for lead tetraethyl. The second spectrum can then be used to confirm the identity of the unknown. It should be emphasized that such confirmation is always desirable.

For large numbers of spectra, such as are obtained when a mass spectrometer is coupled with a gas chromatograph for identification of several components of a mixture, batch-type searches are possible. Here, the user's computer system is connected to the library system, and the unknown spectra are

[18] G. M. Pesyna, R. Venkataraghavan, H. E. Dayringer, and F. W. McLafferty, *Anal. Chem.*, **1976**, *48*, 1362.

[19] In January of 1984, it was announced that Wiley's *Registry of Mass Spectral Data* (listed previously in the paragraph labeled **Spectral Libraries**) and MSSS are to be combined and will become available in a variety of industry-standard tape formats. These are to be marketed by Wiley at a basic price of $5000. The tapes will provide direct access to 79,560 mass spectra of 67,128 different compounds.

```
Option? peak

TYPE PEAK,MIN INT,MAX INT
OR TO EXIT, 1 FOR REGN,QI,MW,MF AND NAME

USER:237,100,100
   • REFS     M/Z PEAKS
      23      237

NEXT REQUEST:295,60,80
   • REFS     M/Z PEAKS
       2      237  295

NEXT REQUEST:293,20,50
   • REFS     M/Z PEAKS
       1      237  295  293

NEXT REQUEST:1
        REGN    QI    MW    MF
     78-00-2   652   324   C8H20Pb
```

```
                                            NAME
                                   Plumbane, tetraethyl
                                 —(8CI9CI)
                                 Lead, tetraethyl—
                                 Tetraethyllead
                                 Tetraethylplumbane
                                 TEL
```

```
NEXT REQUEST:
Option? go cnmr
NIH:EPA:NIC CARBON-13 NUCLEAR MAGNETIC RESONANCE
SPECTRAL SEARCH SYSTEM — Version 4.62/4.4 Dec. 1978

Latest news for CNMR . . .
   20 June 79; Version 4.62 Of CNMR Search Software Now Operational

Option? spec
CAS RN: 78002
        C9
        •
        •
        C5
        •
C7•C3•PB•C2•C6
        •
        C4
        •
        C8

    Plumbane, tetraethyl—
WLN: 2 4-PB-
CAS RN = 78-00-2       CMR#      601      MW = 323.35
C8  H20  PB1
EIDGENOSSICHE TECHNISCHE HOCHSCHULE
Solvent: UNKNOWN

SHIFT   MULT  INTENS  ASSIGN
15.6     T      0       2
10.6     Q      0       6
```

FIGURE 18–21   Printout from a computerized library search for the identity of a water contaminant. (Reprinted with permission from: S. R. Heller and G. W. A. Milne, *Amer. Lab.*, **1980**, *12* (3), 38. Copyright 1980 by International Scientific Communications, Inc.)

down loaded into the network for searching. In this case, the library system accepts the complete spectral data for each unknown and then examines the library file sequentially until possible matches are found. These results are reported back to the user; if desired the reference spectra can be printed out for visual comparison by the user.

18C–5   IDENTIFICATION AND DETERMINATION OF COMPONENTS OF MIXTURES

While mass spectrometry is a powerful tool for the identification of pure compounds, its usefulness for direct qualitative (and quantitative) analysis of all but the simplest mixtures is limited because of the immense number of fragments of differing *m/e* values produced in the typical case. Interpretation of the resulting complex spectra is often impossible. For this reason, mass spectrometers are frequently coupled with gas chromatographic (GC/MS) or liquid chromatographic (LC/MS) apparatus for identification of the components as they are emitted from the chromatographic column. Applications of these so-called hyphenated techniques are considered in Sections 26D–3 and 27C–6.

Tandem Mass Spectrometry. Beginning in the late 1970s chemists began to explore the possibility of linking one mass spectrometer to another to give a tandem spectrometric system (MS/MS) that accomplishes much the same task as GC/MS or LC/MS.[20] Here at the outset, all of the components of a mixture are ionized by means of a soft ionization source (chemical or field), which yields largely M^+, $(M + 1)^+$, or $(M - 1)^+$ ions. These ions are then focused successively on an exit slit according to mass. From this slit, the ions pass into a chamber, where fragmentation is induced by collision with gas molecules at a relatively high pressure (10^{-4} torr) or with electrons. The fragments are then separated and detected in the second spectrometer.

For example, consider a hypothetical mixture of the isomers ABCD, BCDA, and other molecules such as IJKL and MNOP. For discriminating between the two isomers, the first spectrometer is set initially to transmit ions having m/e values corresponding to $ABCD^+$ and $BCDA^+$. With fragmentation in the second ionization region, ions such as AB^+, CD^+, BC^+, DA^+, ABC^+, and so forth are formed each of which has a unique value for m/e; thus, identification of the two original isomers is possible in the second spectrometer. The first spectrometer can

also be set to transmit $IJKL^+$ or $MNOP^+$ ions, which in turn yield IJ^+, JKL^+, MNO^+, OP^+, and other characteristic ions that permit identification by the second analyzer.

Instruments for Tandem Mass Spectrometry. Instruments for tandem mass spectrometry are made up of an ion source (usually chemical), a primary ion separator, an ion-molecule collision chamber, and a secondary ion separator and detector. In most early studies, the primary ion separator consisted of a magnetic analyzer while the secondary ion separator was an electrostatic one. That is, the two components of a double-focusing mass spectrometer (Figure 18–6) were used in a reversed configuration. The collision chamber was a region between the two analyzers through which helium or other inert gas was pumped at a pressure of 10^{-2} to 10^{-6} torr; fragmentation and secondary ion formation took place here due to collisions between the fast moving ions and the resident helium atoms. These ionic fragments were then separated in the electrostatic analyzer.

To improve resolution, triple sector and tandem double-focusing instruments are now being used. The former consists of a double-focusing instrument such as that shown in Figure 18–6 coupled through the collision chamber to a magnetic, electrostatic, or quadrupole analyzer. The most elaborate and expensive instruments are based on two double-focusing instruments and a collision chamber. Commercial versions of these instruments are available at a cost of $500,000 to $700,000.

Several instrument manufacturers now offer tandem quadrupole instruments.[21] A schematic diagram of one of these is shown in Figure 18–22. The ion source and first quadrupole analyzer are similar in function and construction to the quadrupole spec-

---

[20] For additional details, see: F. W. McLafferty, *Tandem Mass Spectroscopy,* Wiley: New York, 1983; F. W. McLafferty, *Science,* **1981,** *214,* 280; R. W. Kondrat and R. G. Cooks, *Anal. Chem.,* **1978,** *50,* 81A.

[21] For a description of a tandem quadrupole instrument, see: R. A. Yost and C. G. Enke, *Anal. Chem.,* **1979,** *51,* 1251A.

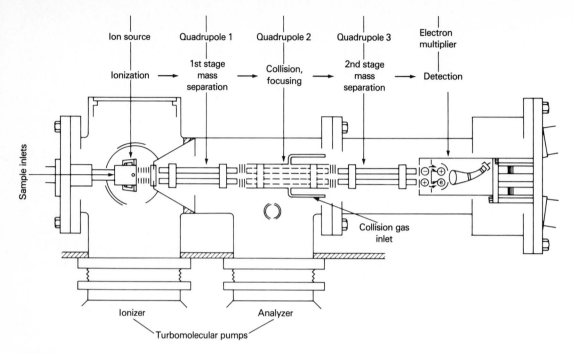

FIGURE 18-22 Schematic of a tandem quadrupole MS/MS instrument. (Courtesy of Finnigan MAT, San Jose, CA.)

trometer shown in Figure 18–8. The collision chamber also contains a quadrupole operated in the radio-frequency mode only (that is, no dc potential is applied across the rods). The radio-frequency only mode provides a highly efficient way of focusing scattered ions. The third quadrupole then serves as an analyzer for separating and detecting the secondary ions formed by collision of the primary ions with helium gas. The instrument provides unit mass resolution over the entire operating range.

Applications of Tandem Mass Spectroscopy. Dramatic progress in the analysis of complex organic and biological mixtures began when the mass spectrometer was first combined with gas chromatography and subsequently with liquid chromatography. The latter is particularly useful for materials of low volatility. Tandem mass spectrometry appears to offer the same advantages of MS/GC and MS/LC and with some additional virtues as well.

Perhaps the most important is speed. Whereas separations on a chromatographic column are achieved in a time scale of a few minutes to hours, equally satisfactory separations in the first of the two mass spectrometers are complete in milliseconds. In addition, the chromatographic techniques require dilution of the sample with large excesses of a mobile phase (and subsequent removal of the mobile phase), which greatly enhances the probability of introduction of interferences. Consequently, tandem mass spectroscopy is potentially more sensitive than either of the hyphenated chromatographic techniques because the chemical noise associated with its use is generally smaller. A current disadvantage of tandem mass spectroscopy with respect to the other two procedures is the greater cost of the equipment required; it is to be hoped that this gap will become smaller as tandem mass spectrometers come into wider use.

To date, tandem mass spectroscopy has been applied for the qualitative and quantitative determinations of the components of a wide variety of complex materials encountered in nature and industry. Some examples include the identification and determination of drug metabolites, insect pheromones, alkaloids in plants, trace contaminants in air, polymer sequences, polychlorodibenzoidioxines, petrochemicals, prostaglandins, diesel exhausts, and odors in air. Tandem mass spectroscopy appears to be a technique that will find wider and wider application among scientists and engineers.

18D Quantitative Applications of Mass Spectroscopy

Quantitative applications of mass spectroscopy fall into three categories based upon the kind of information sought. These categories include (1) the quantitative determination of molecular species or types of molecular species in organic, biological, and occasionally inorganic substances; (2) the determination of the elemental composition of inorganic and less commonly, organic and biological samples; and (3) the quantitative determination of molecular or atomic species on solid surfaces. For category (1), all of the ionization sources listed in Table 18–1 are used. Category (2) methods are largely based upon the spark source although laser, thermal, and secondary ion sources have found occasional use. For category (3) methods, laser, secondary ion, and fast atom bombardment sources have been used.

18D-1 DETERMINATION OF MOLECULAR CONCENTRATIONS

Mass spectrometry has been widely applied to the quantitative determination of one or more components of complex organic (and sometimes inorganic) systems such as those encountered in the petroleum and pharmaceutical industries and in studies of environmental problems.[22] Currently, such analyses are performed by passage of the sample through a chromatographic column and into the spectrometer. With the spectrometer set at a suitable m/e value, the ion current is then recorded as a function of time. This technique is termed *selective ion monitoring*. In some instances, currents at three or four m/e values are monitored in a cyclic manner by rapid switching from one peak to another. The plot of the data, called a *mass chromatogram*, consists of a series of peaks with each peak appearing at a time that is characteristic of one of the several components of the sample that yields ions of the chosen value or values for m/e. Generally, the areas under the peaks are directly proportional to the component concentrations and thus serve as the analytical parameter. In this type of procedure, the mass spectrometer simply serves as a sophisticated detector for quantitative *chromatographic* analyses. Further details on quantitative gas and liquid chromatography are given in Section 25D–2.

In the second type of quantitative mass spectrometry, analyte concentrations are obtained directly from the heights of the mass spectral peaks. For simple mixtures, it is sometimes possible to find peaks at unique m/e values for each component. Under these circumstances, calibration curves of peak heights versus concentration can be prepared and used for analysis of unknowns. More accurate results can ordinarily be realized, however, by incorporating a fixed amount of an internal standard substance in both samples and calibration standards. The ratio of the peak intensity of the analyte species to that of the internal standard is then

---

[22] For a monograph devoted to quantitative mass spectrometry, see: B. J. Millard, *Quantitative Mass Spectrometry,* Heyden: London, 1978.

plotted as a function of analyte concentration. The internal standard tends to reduce uncertainties arising in sample preparation and introduction. With the small samples needed for mass spectroscopy, these uncertainties are often a major source of indeterminate error.

A convenient type of internal standard is a stable, isotope-labeled analog of the analyte. Usually, labeling involves preparation of samples of the analyte in which one or more atoms of deuterium, carbon-13, or nitrogen-15 have been incorporated. It is then assumed that during the analysis, the labeled molecules behave in the same way as the unlabeled ones. The mass spectrometer, of course, easily distinguishes between the two.

Another type of internal standard is a homolog of the analyte that yields a reasonably intense ion peak for a fragment that is chemically similar to the analyte fragment being measured.

With low resolution instruments, it is seldom possible to locate peaks that are unique to each compound of a mixture. In this situation, it is still possible to complete the analysis by collecting intensity data at a number of m/e values that equals or exceeds the number of sample components. For this approach to yield satisfactory results, it is necessary that: (1) each component have a peak that is significantly greater in intensity than peaks for other components at the same m/e value; (2) peak intensities be linearly additive; (3) suitable standards be available; and (4) the sensitivities (ion current per unit partial pressure) be reproducible to perhaps 1% relative.

Calibration.   For volatile compounds, peak heights are found to be directly proportional to partial pressure. Thus, for a mixture containing n compounds, m equations can be written in the form ($m \geq n$)

$$I_1 = A_{11}P_1 + A_{12}P_2 + \ldots A_{1n}P_n$$

$$I_2 = A_{21}P_1 + A_{22}P_2 + \ldots A_{2n}P_n$$

$$\ldots$$

$$I_m = A_{m1}P_1 + A_{m2}P_2 + \ldots A_{mn}P_n$$

where I_m is the relative peak height at mass m in the spectrum of a mixture, P_n is the partial pressure of component n in the mixture, and A_{mn} is the peak height at mass m due to unit pressure of component n. The value of A_{mn} is determined for each component by calibration with a standard at a known partial pressure P_n. Substituting the measured ion currents for a mixture into the equations permits calculation of the partial pressures of each of the components in the sample.

Precision and Accuracy.   The precision of quantitative mass spectral measurements by the procedure just described appears to range between 2 and 5% relative. The analytical accuracy varies considerably depending upon the complexity of the mixture being analyzed and the nature of its components. For gaseous hydrocarbon mixtures containing 5 to 10 components, absolute errors of 0.2 to 0.8 mole percent appear to be typical.

Application.   The early literature dealing with the direct quantitative applications of mass spectrometry is so extensive as to make a summary difficult. The listing of typical applications assembled by Melpolder and Brown[23] demonstrates clearly the versatility of the method. For example, some of the mixtures that can be analyzed without sample heating include natural gas, C_3—C_5 hydrocarbons; C_6—C_8 saturated hydrocarbons; C_1—C_5 alcohols, aldehydes, and ketones; C_1—C_4 chlorides and iodides; fluorocarbons; thiophenes, atmospheric pollu-

---

[23] See: F. W. Melpolder and R. A. Brown, in *Treatise on Analytical Chemistry*, I. M. Kolthoff and P. J. Elving, Eds., Part I, vol. 4, p. 2047, Interscience: New York, 1963. See also: A. L. Burlingame, *et al.*, *Anal. Chem*, **1984,** *56*, 417R; **1982,** *54*, 363R; **1980,** *52*, 214R.

tants; exhaust gases; and many others. By employing higher temperatures, successful analytical methods have been reported for C_{16}—C_{27} alcohols, aromatic acids and esters, steroids, fluorinated polyphenyls, aliphatic amides, halogenated aromatic derivatives, and aromatic nitriles.

Mass spectrometry has also been used for the characterization and analysis of high-molecular-weight polymeric materials. Here, the sample is first pyrolyzed; the volatile products are then admitted into the spectrometer for examination. Alternatively, heating can be performed on the probe of a direct inlet system. Some polymers yield essentially a single fragment; for example, isoprene from natural rubber, styrene from polystyrene, ethylene from polyethylene, and CF_2=$CFCl$ from Kel-F. Other polymers yield two or more products, which depend in amount and kind upon the pyrolysis temperature. Studies of temperature effects can provide information regarding the stabilities of the various bonds, as well as the approximate molecular weight distribution.

Component Type Determination. Because of the complex nature of petroleum products, quantitative data as to *types* of compounds are often more useful than analysis for individual components. Mass spectrometry can provide such information. For example, it has been found that paraffinic hydrocarbons generally give unusually strong peaks at masses 43, 57, 71, 85, and 99. Cycloparaffins and monoolefins, on the other hand, exhibit characteristically intense peaks at masses 41, 55, 69, 83, and 97. Another group of peaks is attributable to cycloolefins, diolefins, and acetylenes (67, 68, 81, 82, 95, and 96). Finally, alkylbenzenes are found to fragment to masses of 77, 78, 79, 91, 92, 105, 106, 119, 120, 133, and 134. A mathematical combination of the peak heights of a set provides an analytical parameter for assessing the concentration of each type of hydrocarbon. Type analyses have been used to characterize the properties and behavior of gasolines, fuel oils, lubricating oils, asphalts, and mixtures of paraffins, olefins, alcohols, and ketones.

18D-2 ELEMENTAL ANALYSIS BY SPARK SOURCE MASS SPECTROMETRY

Spark source mass spectroscopy has become an important method for the determination of the elements—particularly traces of elements—in a variety of matrices including semiconductors, minerals, particulate atmospheric pollutants, lunar rocks, and fossil fuels.[24]

Instruments. In spark source mass spectrometry, the atomic constituents of a sample are converted to gaseous ions for mass analysis by a high potential (~30 kV), radio-frequency spark. The spark is housed in a vacuum chamber located immediately adjacent to the mass analyzer. The chamber is equipped with a separate high-speed pumping system that quickly reduces the internal pressure to about 10^{-8} torr after sample changes. Often the sample serves as one or both electrodes; alternatively, it is mixed with graphite and pressed into an electrode. The gaseous positive ions formed in the spark plasma are accelerated into the analyzer by a dc potential.

A spark source produces ions with a wide range of kinetic energies. Consequently expensive double-focusing mass spectrometers are required. The Mattuch-Herzog type shown in Figure 18–7 is generally employed in order to make photographic detection possible.

Modern spark source instruments are de-

---

[24] For a more complete discussion of applications, see: R. C. Elser, in *Trace Analysis*, J. D. Winefordner, Ed., Chapter 10, Wiley: New York, 1976; and *Trace Analysis by Mass Spectrometry*, A. J. Ahearn, Ed., Academic Press: New York, 1972.

signed to utilize both photographic and electrical detectors. The latter are generally electron multipliers, which were described in Section 18A–6. Often spark source instruments are computer controlled.

Spectra. Spark source mass spectra are much simpler than their atomic emission counterparts, consisting of one major peak for each isotope of an element plus a few weaker lines corresponding to multiply charged ions and dimers or trimers. The presence of these additional ions creates the potential for interference unless the spectrometer is operated at a sufficiently great resolution. For example, in the determination of iron in a silicate matrix, such as with samples of rock or glass, the potential for interference by silicon exists. Here, the analysis is based on the peak for $^{56}Fe^+$, but a detectable peak for $^{28}Si_2^+$ is also observed. Because of the difference in mass defects, however, the absolute mass of $^{56}Fe^+$ is 55.93494 while for $^{28}Si_2^+$ it is 55.95386. The resolving power required to separate these two peaks is

$$m/\Delta m = 56/0.01892 = 3.0 \times 10^3$$

The double-focusing spectrometers required for work with a spark source have resolutions significantly larger than this figure. Thus, it is possible to avoid most interferences of this type.

Qualitative Applications. A spark source mass spectrometer with photographic detection is a powerful tool for qualitative and semi-quantitative analysis. All elements in the periodic table from ^7Li through ^{238}U can be identified in a single excitation. By making multiple exposures, it is possible to determine order of magnitude concentrations for major constituents of a sample as well as constituents in the parts per billion concentration range. Interpretation of spectra on photographic plates does, however, require skill and experience because of the presence

of multiply charged species, polymeric species, and molecular ions.

Quantitative Applications. A radio-frequency spark is not a very reproducible source over short periods. As a consequence, it is necessary to integrate the output signals from a spark for periods ranging from several seconds to hundreds of seconds if good quantitative data are to be obtained. The photoplate is of course an integrating device; with electric detection, provision must be made for current integration. In addition to integration, it is common practice to improve reproducibility by using the ratio of the analyte signal to that of an internal standard as the analytical parameter. Often one of the major elements of the sample matrix is chosen as the standard; alternatively, a fixed amount of a pure compound is added to each sample and each standard used for calibration. In the latter case, the internal standard substance must be absent from the samples. With these precautions, relative standard deviations of a few percent can be realized.

The advantages of spark source mass spectrometry include its high sensitivity, its applicability to a wide range of sample matrices, the large linear range of the output signal (often several orders of magnitude), and the wide range of elements that can be detected and measured quantitatively.

18D-3 *ELEMENTAL ANALYSIS BY ICP/MS*

A recently developed hyphenated technique interfaces an inductively coupled plasma source (Section 10B–3) with a quadrupole mass spectrometer to provide a system for elemental analysis that appears to offer several advantages over ICP/atomic emission methods.[25] A commercial version of this instrument appeared on the market in 1983.[26]

---

[25] R. S. Houk, *et al.*, *Anal. Chem.*, **1980**, *52*, 2283.
[26] See: *Anal. Chem.*, **1983**, *55*, 611A.

In this instrument, positive metal ions, produced in a conventional ICP torch, are sampled through a differentially pumped interface linked to a quadrupole spectrometer. The spectra produced in this way, which are remarkably simple compared with ICP/atomic emission spectra, consist of a simple series of isotope peaks. These spectra are used for quantitative measurements based upon calibration curves, often with an internal standard. Analyses can also be performed by the isotope dilution technique, which was described in Section 17D.

Performance specifications for the commercial instrument include a mass range of 3 to 300 with a resolution of 1, and a dynamic range of 6 orders of magnitude. Over 90% of the elements in the periodic table have been determined with few spectral interferences. Measurement times of 10 sec/element are claimed with detection limits in the 0.1 and 10 ppb range for most elements.

18D-4 ISOTOPE ABUNDANCE MEASUREMENTS

The mass spectrometer was developed initially for the study of isotopic abundance ratios, and the instrument continues to be the most important source for this kind of data. Information regarding the abundance of various isotopes is now employed for a variety of purposes; the determination of formulas or organic compounds cited earlier is one example. Other important applications include analysis by isotope dilution, tracer studies with isotopes, and dating of rocks and minerals by isotopic ratio measurements.[27] The techniques are similar to those described in Chapter 17 for radioactive isotopes. Mass spectrometry, however, permits the extension of isotope detection to nonradioactive species such as ^{13}C, ^{17}O, ^{18}O, ^{15}N, ^{34}S, and numerous others.

In determining isotope concentrations of solids, a thermal ionization source is generally used, although for volatile materials, an electron impact source is also satisfactory. Quadrupole or single sector mass spectrometers suffice since only unit mass resolution is required.

Thermal ionization is based upon the reactions that take place when neutral atoms or molecules are heated to a high temperature on a surface of a metal having a high work function (that is, a high affinity for electrons). At temperatures of about 2000°C, evaporation of the sample is usually accompanied by transfer of electrons from the sample to the metal. The resulting volatile, positively charged ions can then be analyzed with a mass spectrometer. Usually, thermal ionization is carried out by depositing the sample on a fine tungsten wire held in a suitable probe. Vaporization and ionization occurs when the filament is brought rapidly to a high temperature by an electric current.

18D-5 SURFACE ANALYSIS BY MASS SPECTROMETRY

Two instruments for determining the concentration of elements on surfaces have been considered thus far, namely, the electron microprobe (Section 15F) and the Auger microprobe (Section 16C–3). In both, focused beams of electrons (*microprobes*) are employed to cause characteristic emissions by the surface elements. Mass spectroscopy is also capable of providing quantitative surface information. In this instance, ions for mass analysis are formed by bombarding the surface with a focused beam of ions or molecules or by a laser beam. A radio-frequency spark can also be used to examine conducting surfaces.

[27] For a recent description of applications of mass spectral isotope measurements see a series of articles in *Spectra*, **1982**, *8* (4), 3–24. See also: R. M. Caprioli, W. F. Fies, and M. S. Story, *Anal. Chem.*, **1974**, *46*, 453A; and C. Brunnée, *et al.*, *Amer. Lab.*, **1978**, *10* (2), 141.

Atomic Secondary Ion Mass Spectrometry. Secondary ion mass spectrometry (SIMS) is the most highly developed of the mass spectrometric surface methods with several manufacturers offering instruments for this technique. SIMS has proven useful for determining both the atomic and the molecular composition of solid surfaces.[28]

Two types of instruments are encountered: *secondary ion mass analyzers* and *microprobe analyzers*. Both are based upon bombarding the surface of the sample with a beam of 5 to 20 keV ions such as Ar^+, N_2^+, or O_2^+. The ion beam is formed in an *ion gun* in which the gaseous atoms or molecules are ionized by an electron impact source. The positive ions are then accelerated by application of a high dc potential. Impact of these primary ions causes the surface layer of atoms of the sample to be stripped (sputtered) off, largely as neutral atoms. A small fraction, however, form as positive (or negative) secondary ions that are drawn into a spectrometer for mass analysis.

In secondary ion mass analyzers, which serve for general surface analysis and for depth profiling, the primary ion beam diameter ranges from 0.3 to 5 mm. Double-focusing, single-focusing, and quadrupole spectrometers are used for mass sorting. These spectrometers yield qualitative and quantitative information about all of the isotopes (hydrogen through uranium) present on a surface. Sensitivities of 10^{-15} g or better are typical. By monitoring peaks for one or a few isotopes, as a function of time, concentration profiles can be obtained with a depth resolution of 50 to 100 Å.

Ion microprobe analyzers are more sophisticated (and more expensive) instruments that are based upon a focused beam of primary ions that has a diameter of 1 to 2 μm. This beam can be moved across a surface for about 300 μm in both the x and y directions. A microscope is provided to permit visual adjustment of the beam position. Mass analysis is performed with a double-focusing spectrometer. In some instruments, the primary ion beam passes through an additional low-resolution mass spectrometer so that only a single type of primary ion bombards the sample. The ion microprobe version of SIMS permits detailed studies of solid surfaces.

Laser Microprobe Mass Spectrometry. A commercial laser-microprobe mass spectrometer is now available for the study of solid surfaces. Ionization and volatilization is accomplished with a pulsed, neodymium-YAG laser, which produces a 0.5 μm spot of 265 nm radiation. The power density of the radiation within this spot is 10^{10} to 10^{11} W/cm². The power of the beam can be attenuated to 1% by means of a 25-step optical filter. Colinear with the ionization beam is the beam from a second, low-power, He-Ne laser ($\lambda = 613$ nm), which serves as illumination so that the area to be analyzed can be chosen visually. The instrument has an unusually high sensitivity (down to 10^{-20} g), is applicable to both inorganic and organic (including biological) samples, has a spatial resolution of about 1 μm, and produces data at a rapid rate. Some typical applications of this instrument include determination of Na/K concentration ratios in frog nerve fiber, determination of the calcium distribution in retinas, classification of asbestos and coal mine dusts, determination of fluorine distributions in dental hard tissue, analysis of amino acids, and study of polymer surfaces.[29]

[28] See also: C. A. Evans, Jr., *Anal. Chem.*, **1972**, *44*, 67A; W. H. Christie, *Anal. Chem.*, **1981**, *53*, 1240A; and H. A. Storms, *Amer. Lab.*, **1974**, *6* (3), 23.

[29] For further details, see: E. Denoyer, R. Van Grieken, F. Adams, and D. F. S. Natusch, *Anal. Chem.*, **1982**, *54*, 26A, 280A; R. J. Cotter, *Anal. Chem.*, **1984**, *56*, 485A.

18E Molecular Secondary Ion Mass Spectroscopy

In the late 1970s, it was found that by employing lower ion fluxes, the sources developed for atomic secondary ion mass spectroscopy (Section 18D-5) could also be used to obtain mass spectra for molecules.[30] Here, the samples, as solids or liquids or in a solid or liquid matrix, were introduced directly into the beam of the ion gun. The secondary ions from the samples were then separated in a mass analyzer. Molecular SIMS spectra show the same structural specificity that is found in other types of mass spectrometry with the added advantage of being applicable to nonvolatile and thermally unstable samples. For example, a SIMS spectrum for vitamin B-12 exhibits a strong $(M + 1)^+$ peak at mass 1355 and numerous peaks for smaller fragments.

---

[30] See: R. J. Day, S. E. Unger, and R. G. Cooks, *Anal. Chem.*, **1980**, *52*, 557A; S. M. Scheifers, R. C. Holler, K. L. Busch, and R. G. Cooks, *Amer. Lab.*, **1982**, *14* (3), 19; and *Secondary Mass Spectrometry*, A. Benninghoven, *et al.*, Springer-Verlag: New York, 1979.

PROBLEMS

18-1. How do gaseous and desorption sources differ? What are the advantages of each?

18-2. How do the spectra for electron impact, field ionization, and chemical ionization sources differ from one another?

18-3. Describe gaseous field ionization and desorption field ionization sources.

18-4. Why do double-focusing mass spectrometers give narrower peaks and higher resolutions?

18-5. Calculate the resolution required to resolve peaks for
(a) CH_2N^+ (MW = 28.0187) and N_2^+ (MW = 28.0061).
(b) $C_2H_4^+$ (MW = 28.0313) and CO^+ (MW = 27.9949).
(c) $C_3H_7N_3^+$ (MW = 85.0641) and $C_5H_9O^+$ (MW = 85.0653).

18-6. Calculate the resolution required to resolve peaks for
(a) $^{116}Sn^+$ (AtW = 115.90219) and $^{232}Th^{2+}$ (AtW = 232.03800).
(b) $C_2H_6^+$ (MW = 30) and $CH_3NH_2^+$ (MW = 31).
(c) $^{12}C_5H_4NCl^+$ (MW = 113) and $^{13}C_5H_4NCl$ (MW = 114).
(d) Cetyl palmitate$^+$ (MW = 480) and $C_{10}F_{19}^+$ (MW = 481).

18-7. Calculate the ratio of the $(M + 2)^+$ to M^+ peak heights and the $(M + 4)^+$ to M^+ peak heights for
(a) $C_{10}H_6Br_2$.   (b) C_3H_7ClBr.   (c) $C_6H_4Cl_2$.

18-8. When a magnetic sector instrument was operated with an accelerating voltage of 3000 V, a field of 1260 G was required to focus the peak for CH_4^+ on the detector.
(a) What range of field strengths would be required to scan a mass range of 16 to 250 if the accelerating voltage was held constant at 3000 V?
(b) What range of accelerating voltages would be required to scan a mass range of 16 to 250 if the field was held constant at 1260 G?

18–9. If the mass analyzer described in Problem 18–8 were operated at a constant field strength of 1260 G,

(a) what accelerating voltage would be required to focus the parent ion for naphthalene on the detector?

(b) what percent change in accelerating voltage would be required to scan from the M^+ to the $(M + 1)^+$ ion for naphthalene?

18–10. If the mass analyzer in Problem 18–8 were operated at 3000 V,

(a) what field strengths would be required to focus the parent ion for pyridine on the detector?

(b) what percent change in field strength would be required to scan from the M^+ to the $(M + 1)^+$ ion for pyridine?

18–11. Identify the ions responsible for the peaks in the mass spectrum shown in Figure 18–16**b**.

18–12. Identify the ions responsible for the four peaks having greater masses than the M^+ peak in Figure 18–17**b**.

19

An Introduction to Electroanalytical

Chemistry

Electroanalytical chemistry encompasses a group of quantitative analytical methods that are based upon the electrical properties of a solution of the analyte when it is made part of an electrochemical cell. Three types of electroanalytical procedures are encountered. The first involves establishing the relationship between analyte concentration and such electrical quantities as current, potential, resistance (or conductance), capacitance, or charge. In the second, one of these same electrical measurements serves to establish the end point in a titration of the analyte. In the third, an electrical current converts the analyte to a form that can be measured gravimetrically or volumetrically.

Regardless of type, the intelligent application of an electroanalytical method requires an understanding of the basic theory and practical aspects of the operation of electrochemical cells. This chapter is devoted largely to these matters.[1,2]

19A Electrochemical Cells

Electrochemical cells can be conveniently classified as *galvanic* if they are employed to produce electrical energy and *electrolytic* when they consume electricity from an external source. Both find use in electroanalytical chemistry. It is important to appreciate that many cells can be operated in either a galvanic or an electrolytic mode by variation of experimental conditions.

---

[1] Some reference works on electrochemistry and its applications include: A. J. Bard and L. R. Faulkner, *Electrochemical Methods,* Wiley: New York, 1980; J. A. Plambeck, *Electroanalytical Chemistry,* Wiley: New York, 1982; J. J. Lingane, *Electroanalytical Chemistry,* 2d ed., Interscience: New York, 1958; and D. T. Sawyer and J. L. Roberts, Jr., *Experimental Electrochemistry for Chemists,* Wiley: New York, 1974.

[2] For a brief review of recent developments in electrochemical instrumentation, see: J. Osteryoung, *Science,* **1982,** *218,* 261.

567

19A-1 CELL COMPONENTS

An electrochemical cell consists of two conductors called *electrodes,* each immersed in a suitable electrolyte solution. For electricity to flow, it is necessary: (1) that the electrodes be connected externally by means of a metal conductor and (2) that the two electrolyte solutions be in contact to permit movement of ions from one to the other. Figure 19–1 shows an example of a galvanic cell. The fritted glass disk is porous, so that Zn^{2+}, Cu^{2+}, and SO_4^{2-} ions as well as H_2O molecules can move across the junction between the two electrolyte solutions; the disk simply prevents extensive mixing of the contents of the two cell compartments.

Conduction in an Electrochemical Cell.   Electricity is conducted by three distinct processes in various parts of the galvanic cell shown in Figure 19–1. In the copper and zinc electrodes, as well as in the external conductor, electrons serve as carriers, moving from the zinc through the conductor to the copper. Within the two solutions the flow of electricity involves migration of both cations and anions, the former away from the zinc electrode toward the copper and the latter in the reverse direction. *All* ions in the two solutions participate in this process.

A third process occurs at the two electrode surfaces. Here, an oxidation or a reduction reaction provides a mechanism whereby the ionic conduction of the solution is coupled with the electron conduction of the electrode to provide a complete circuit for a flow of electricity. The two electrode processes are described by the equations

$$Zn(s) \rightleftharpoons Zn^{2+} + 2e$$
$$Cu^{2+} + 2e \rightleftharpoons Cu(s)$$

The net cell reaction is the sum of these two *half-cell* reactions:

$$Zn(s) + Cu^{2+} \rightleftharpoons Zn^{2+} + Cu(s)$$

Because this reaction has a strong tendency to proceed to the right, the cell is galvanic and produces a potential of about 1 V under most conditions.

Anode and Cathode.   By definition, the *cathode* of an electrochemical cell is the electrode at which reduction occurs, while the *anode* is the electrode where oxidation takes place. These definitions apply to both galvanic and electrolytic cells.

For the galvanic cell shown in Figure

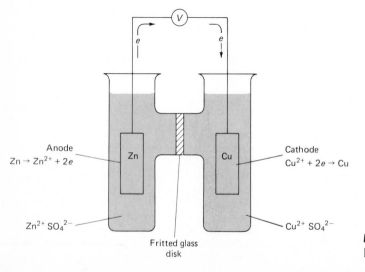

Anode
$Zn \rightarrow Zn^{2+} + 2e$

Cathode
$Cu^{2+} + 2e \rightarrow Cu$

$Zn^{2+}\ SO_4{}^{2-}$

$Cu^{2+}\ SO_4{}^{2-}$

Fritted glass disk

FIGURE 19-1   A galvanic cell with a liquid junction.

19-1, the copper electrode is the cathode and the zinc electrode is the anode. Note that this cell could be caused to behave as an electrolytic cell by imposing a sufficiently large potential from an external source. Under these circumstances, the reactions occurring at the electrodes would be

$$Zn^{2+} + 2e \rightleftharpoons Zn(s)$$

$$Cu(s) \rightleftharpoons Cu^{2+} + 2e$$

Now, the roles of the electrodes are reversed; the copper electrode has become the anode and the zinc electrode the cathode.

Reactions at Cathodes. Some typical cathodic half-reactions are

$$Cu^{2+} + 2e \rightleftharpoons Cu(s)$$

$$Fe^{3+} + e \rightleftharpoons Fe^{2+}$$

$$2H^+ + 2e \rightleftharpoons H_2(g)$$

$$AgCl(s) + e \rightleftharpoons Ag(s) + Cl^-$$

$$IO_4^- + 2H^+ + 2e \rightleftharpoons IO_3^- + H_2O$$

Electrons are supplied for each of these processes from the external circuit via an electrode that does not participate directly in the chemical reaction. In the first process, copper is deposited on the electrode surface; in the second, only a change in oxidation state of a solution component occurs. The third reaction is frequently observed in aqueous solutions that contain no easily reduced species.

The fourth half-reaction is of interest because it can be considered to be the result of a two-step process; that is,

$$AgCl(s) \rightleftharpoons Ag^+ + Cl^-$$

$$Ag^+ + e \rightleftharpoons Ag(s)$$

Solution of the sparingly soluble precipitate occurs in the first step to provide the silver ions that are reduced in the second.

The last half-reaction has been included to demonstrate that a cathodic reaction can involve anions as well as cations.

Reactions at Anodes. Examples of typical anodic half-reactions include

$$Cu(s) \rightleftharpoons Cu^{2+} + 2e$$

$$Fe^{2+} \rightleftharpoons Fe^{3+} + e$$

$$2Cl^- \rightleftharpoons Cl_2(g) + 2e$$

$$H_2(g) \rightleftharpoons 2H^+ + 2e$$

$$2H_2O \rightleftharpoons O_2(g) + 4H^+ + 4e$$

The first half-reaction requires a copper electrode to supply Cu^{2+} ions to the solution. The remaining four half-reactions can take place at any of a variety of inert metal surfaces. To cause the fourth half-reaction to occur, it is necessary to replenish the hydrogen in the solution by bubbling the gas across the surface of the electrode (usually platinum). The reactions can then be formulated as

$$H_2(g) \rightleftharpoons H_2(sat'd)$$

$$H_2(sat'd) \rightleftharpoons 2H^+(aq) + 2e$$

where $H_2(sat'd)$ implies that the solution is kept saturated with hydrogen.

The final reaction, giving oxygen as a product, is a common anodic process in aqueous solutions containing no easily oxidized species.

Liquid Junctions. Cells with a liquid junction, such as that shown at the fritted disk in Figure 19-1, are ordinarily employed to avoid direct reaction between the components of the two half-cells. If the two electrolyte solutions in Figure 19-1 were allowed to mix, a reduction in the cell efficiency would occur as a result of the direct deposition of copper on the zinc. As will be shown later, a small potential called a *junction potential* arises at the interface between two electrolyte solutions that differ in composition.

Occasionally, useful cells can be constructed in which the electrodes share a common electrolyte. An example of a cell without liquid junction is shown in Figure 19–2. Here, the reaction at the silver cathode can be written

$$AgCl(s) + e \rightleftharpoons Ag(s) + Cl^-(aq)$$

Hydrogen is evolved at the platinum anode:

$$H_2(g) \rightleftharpoons 2H^+(aq) + 2e$$

The overall cell reaction is then obtained by multiplying each term in the first equation by 2 and adding. That is,

$$2AgCl(s) + H_2(g) \rightleftharpoons Ag(s) \\ + 2H^+(aq) + 2Cl^-(aq)$$

The direct reaction between hydrogen and solid silver chloride is slow. As a consequence, a common electrolyte can be employed without significant loss of cell efficiency.[3]

---

[3] Occasionally, it is useful to indicate the physical states of one or more of the reactants and products in a reaction. For this purpose, (s) represents the solid state, (l) the liquid, (g) the gaseous, and (aq) the solute in aqueous solution.

The Salt Bridge. For reasons to be discussed later, electrochemical cells are often equipped with a *salt bridge* to separate the electrolytes in the anode and cathode compartments. This device takes a variety of forms. In Figure 19–6 (p. 578), for example, the bridge consists of a U-shaped tube filled with saturated solution of potassium chloride. Such a cell has two liquid junctions; one is between the cathode electrolyte and one end of the bridge while the second is between the anode electrolyte and the other end of the bridge.

Schematic Representation of Cells. To simplify the description of cells, chemists often employ a shorthand notation. For example, the cells shown in Figures 19–1 and 19–2 can be described by

$$Zn|ZnSO_4(xM)|CuSO_4(yM)|Cu$$

$$Pt,H_2(p = 1 \text{ atm})|H^+(0.01\ M), \\ Cl^-(0.01\ M),AgCl(\text{sat'd})|Ag$$

By convention, *the anode and information with respect to the solution with which it is in contact is listed on the left.* Single vertical lines represent phase boundaries at which potentials may develop. Thus, in the first example, a part of the cell potential is associated with the

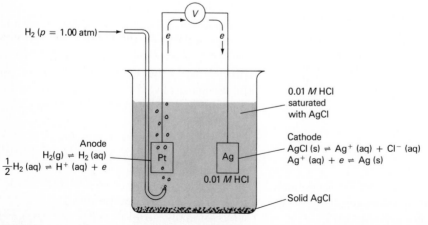

FIGURE 19–2   A galvanic cell without a liquid junction.

phase boundary between the zinc electrode and the zinc sulfate solution. A small potential also develops at liquid junctions; thus, another vertical line is inserted between the zinc and copper sulfate solutions. The cathode is then represented symbolically with another vertical line separating the electrolyte solution from the copper electrode.

In the second cell, only two phase boundaries exist, the electrolyte being common to both electrodes. An equally correct representation of this cell would be

$$Pt|H_2(sat'd),HCl(0.01\ M),$$
$$Ag^+(1.8 \times 10^{-8}\ M)|Ag$$

Here, the molecular hydrogen concentration is that of a saturated solution (in the absence of partial pressure data, 1.00 atm is implied); the indicated molar silver ion concentration was computed from the solubility product constant for silver chloride.

The presence of a salt bridge in a cell is indicated by two vertical lines, implying that a potential difference is associated with each of the two interfaces. Thus, the cell shown in Figure 19–6 (p. 578) would be represented as

$$Pt,H_2(p\ atm)|H^+(xM)\|M^{2+}(yM)|M$$

19A–2 DIRECT CURRENTS IN AN ELECTROCHEMICAL CELL

As noted earlier, electricity is transported within a cell by the migration of ions. In common with metallic conductors, Ohm's law is often obeyed (departures from Ohm's law are discussed in Section 19E–2 devoted to polarization effects). That is,

$$I = \frac{E}{R} \qquad (19–1)$$

where I is the current in amperes, E is the potential difference in volts responsible for movement of the ions, and R is the resistance in ohms of the electrolyte to the current. The resistance depends upon the kinds and concentrations of ions in the solution.

It is found experimentally that under a fixed potential, the rate at which various ions move in a solution differs considerably. For example, the rate of movement (or *mobility*) of the proton is about seven times that for the sodium ion and five times that of the chloride ion. Thus, although all of the ions in a solution participate in conducting electricity, the fraction carried by one ion may differ markedly from that carried by another. This fraction depends upon the relative concentration of the ion as well as its inherent mobility. To illustrate, consider the cell shown in Figure 19–3**a**, which is divided into three imaginary compartments, each containing six hydrogen ions and six chloride ions. Six electrons are then forced into the cathode by a battery, resulting in formation of three molecules of hydrogen; three molecules of chlorine are also produced at the anode (see Figure 19–3**b**). The resulting charge imbalance brought about by the removal of ions from the electrode compartments is offset by migration, with positive ions moving toward the negative electrode and conversely. Because the proton is about five times more mobile than the chloride ion, however, a significant difference in concentration in the outer electrode compartments develops during electrolysis. In effect, five-sixths of the current has resulted from movement of the hydrogen ions and one-sixth from the transport of chloride ions.

It is important to appreciate that the current need not result from transport of the electrode reactants exclusively. Thus, if we were to introduce, say, 100 potassium and nitrate ions into each of the three compartments of the cell under consideration, the charge imbalance resulting from electrolysis could be offset by migration of the added species as well as by the hydrogen and chlo-

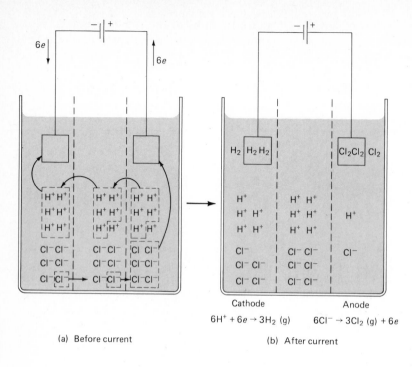

Cathode   Anode
$6H^+ + 6e \rightarrow 3H_2$ (g) $6Cl^- \rightarrow 3Cl_2$ (g) $+ 6e$

(a) Before current

(b) After current

FIGURE 19-3 Changes resulting from a current made up of six electrons.

ride ions. Since the added salt would represent an enormous excess, essentially all of the electricity would be carried within the cell by the potassium and nitrate ions rather than by the reactant ions; only across the electrode surfaces would the current result from the presence of hydrogen and chloride ions.

19A-3 ALTERNATING CURRENTS IN A CELL

When a dc potential is applied to a cell, conduction requires an oxidation reaction at the anode and a reduction at the cathode. The term *faradaic* is sometimes used to denote such currents and processes. Both faradaic and nonfaradaic conduction occurs with an ac potential.

Nonfaradaic Currents.   Nonfaradaic currents involve the formation of an *electrical double layer* at the electrode-solution interface. When a potential is applied to a metallic electrode immersed in an electrolyte, a momentary surge of current creates an excess

(or a deficiency) of negative charge at the surface of the metal. As a consequence of ionic mobility, however, the layer of solution immediately adjacent to the electrode acquires an opposing charge. This effect is illustrated in Figure 19-4a. The charged layer consists of two parts: (1) a compact inner layer, in which the potential decreases linearly with distance from the electrode surface; and (2) a more diffuse layer, in which the decrease is exponential; see Figure 19-4b. This assemblage of charge inhomogeneities is termed the electrical double layer.

The double layer formed by a dc potential involves the development of a momentary current which then drops to zero (that is, the electrode becomes *polarized*) unless some faradaic process occurs. With an alternating current, however, reversal of the charge relationship occurs with each half-cycle as first negative and then positive ions are attracted to the electrode surface. Electrical energy is consumed and converted to frictional heat

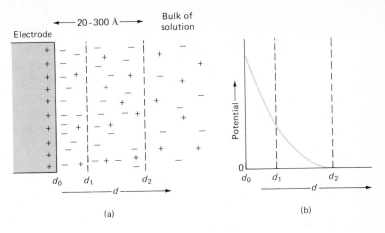

20-300 Å

Bulk of
solution

d_0 d_1 d_2

d

(a)

Potential

0   d_0 d_1 d_2

d

(b)

FIGURE 19-4 Electric double layer formed at electrode surface as a result of an applied potential.

from this ionic movement. Thus, each electrode surface behaves as one plate of a capacitor, the capacitance of which may be large (several hundred to several thousand microfarads per cm²). The capacitance current increases with frequency and with electrode size; by controlling these variables, it is possible to arrange conditions so that essentially all of the alternating electricity flowing through a cell is carried across the electrode interface by this nonfaradaic process.

19A-4 REVERSIBLE AND IRREVERSIBLE CELLS

The galvanic cell shown in Figure 19–2 develops a potential of about 0.46 V. If a battery with a potential somewhat greater than 0.46 V were inserted in the circuit, with its negative terminal connected to the platinum electrode, a reversal in direction of electron flow would occur; the reactions at the two electrodes would thus become

$$2Ag(s) + 2Cl^- \rightleftharpoons 2AgCl(s) + 2e$$

$$2H^+ + 2e \rightleftharpoons H_2(g)$$

Now the silver electrode is the anode and the platinum electrode the cathode. A cell (or an electrode) for which a change in direction of the current causes a reversal of the electrochemical reaction is said to be

chemically reversible. Cells in which a current reversal results in different reactions at one or both electrodes are called *chemically irreversible*. The cell shown in Figure 19–1 is also chemically reversible. If, however, a small amount of dilute acid were introduced into the zinc electrode compartment, the reaction would tend to become irreversible. Here, zinc would not deposit at the cathode upon application of a potential; instead, hydrogen would form by the reaction

$$2H^+ + 2e \rightleftharpoons H_2(g)$$

Thus, the zinc electrode and the cell would be termed chemically irreversible in the presence of acid.

19B Cell Potentials

In order to understand much of the material in the next three sections, the reader should have some understanding of activities and activity coefficients. These topics are reviewed in Appendix 3.

19B-1 THE EFFECT OF CONCENTRATION ON CELL POTENTIALS

This section deals with the effect of concentration (or more exactly activity) of reactants and products on the potential of an electro-chemical cell. We will use as an example the cell illustrated in Figure 19–2 for which the cell reaction is

$$2AgCl(s) + H_2(g) \rightleftharpoons 2Ag(s) + 2Cl^- + 2H^+ \quad (19\text{–}2)$$

The equilibrium constant K for this reaction is given by

$$K = \frac{[H^+]^2[Cl^-]^2}{p_{H_2}} \quad (19\text{–}3)$$

Here, we use the approximate form of the mass law in which the molar concentrations in brackets are assumed to be identical to the activities of the ionic species (see Appendix 3). Note also that an excess of solid silver and silver chloride is assumed and that their concentrations can be considered to be constant and thus do not appear on the right-hand side of the equation.

It is convenient to define a second quantity Q such that

$$Q = \frac{[H^+]_a^2[Cl^-]_a^2}{(p_{H_2})_a} \quad (19\text{–}4)$$

Here, the subscript a indicates that the bracketed terms are instantaneous concentrations and *not equilibrium concentrations*. The quantity Q, therefore, is not a constant, but changes continuously until equilibrium is reached; at that point, Q becomes equal to K and the a subscripts are deleted.

From thermodynamics, it can be shown that the change in free energy ΔG for a cell reaction (that is, the maximum work obtainable at constant temperature and pressure) is given by

$$\Delta G = RT \ln Q - RT \ln K \quad (19\text{–}5)$$

where R is the gas constant (8.316 J mol^{-1} deg^{-1}) and T is the temperature in degrees

K; the term ln refers to the logarithm to the base e. This relationship implies that the magnitude of the free energy for the system is dependent upon how far the system is from the equilibrium state. It can also be shown that the cell potential E_{cell} is related to the free energy of the reaction by the relationship

$$\Delta G = -nFE_{cell} \quad (19\text{–}6)$$

where F is the faraday (96,487 coulombs per chemical equivalent) and n is the number of equivalents of electricity (or moles of electrons) associated with the oxidation-reduction process (in this example, $n = 2$).

Substitution of Equations 19–4 and 19–6 into 19–5 yields, upon rearrangement

$$E_{cell} = -\frac{RT}{nF} \ln \frac{[H^+]_a^2[Cl^-]_a^2}{(p_{H_2})_a} + \frac{RT}{nF} \ln K \quad (19\text{–}7)$$

The last term in this equation is a constant, which is called the *standard potential*, E_{cell}^0 for the cell. That is,

$$E_{cell}^0 = \frac{RT}{nF} \ln K \quad (19\text{–}8)$$

Substitution of Equation 19–8 into Equation 19–7 yields

$$E_{cell} = E_{cell}^0 - \frac{RT}{nF} \ln \frac{[H^+]_a^2[Cl^-]_a^2}{(p_{H_2})_a} \quad (19\text{–}9)$$

Note that the standard potential is equal to the *cell potential when the reactants and products are at unit concentration (more correctly activity) and pressure.*

Equation 19–9 is called the *Nernst equation* in honor of a nineteenth century electro-chemist. It finds wide application in electro-analytical chemistry.

19C   Electrode Potentials

It is useful to think of the cell reaction of an electrochemical cell as being made up of two half-cell reactions, each of which has a characteristic *electrode potential* associated with it. As will be shown later, these electrode potentials measure the driving force for the two half-reactions *when, by convention, they are both written as reductions.* Thus, the two half-cell or electrode reactions for the cell shown in Figure 19–2 are

$$2AgCl(s) + 2e \rightleftharpoons 2Ag(s) + 2Cl^-$$

$$2H^+ + 2e \rightleftharpoons H_2(g)$$

We now assume that electrode potentials E_{AgCl} and E_{H^+} are known for the two half-reactions. To obtain the cell reaction, the second half-reaction is subtracted from the first to give

$$2AgCl(s) + H_2 \rightleftharpoons 2Ag(s) + 2H^+ + 2Cl^-$$

Similarly, the cell potential E_{cell} is obtained by subtracting the electrode potential for the first reaction from the second. That is,

$$E_{cell} = E_{AgCl} - E_{H^+}$$

A more general statement of the last relationship is

$$E_{cell} = E_{cathode} - E_{anode} \qquad (19\text{–}10)$$

where $E_{cathode}$ and E_{anode} are the electrode potentials for the cathodic and anodic half-reactions.

19C–1   NATURE OF ELECTRODE POTENTIALS

At the outset, it should be emphasized that *no* method exists for determining the absolute value of the potential of a single electrode, since all voltage-measuring devices determine only *differences* in potential. One

conductor from such a device is connected to the electrode in question; in order to measure a potential difference, however, the second conductor must be brought in contact with the electrolyte solution of the half-cell in question. This latter contact inevitably involves a solid-solution interface and hence acts as a second half-cell at which a chemical reaction *must also take place* if electricity is to flow. A potential will be associated with this second reaction. Thus, an absolute value for the desired half-cell potential is not realized; instead, what is measured is a combination of the potential of interest and the half-cell potential for the contact between the voltage-measuring device and the solution.

Our inability to measure absolute potentials for half-cell processes is not a serious handicap because *relative half-cell potentials,* measured against a common reference electrode, serve just as well. These relative potentials can be combined to give real cell potentials; in addition, they are useful for calculating equilibrium constants of oxidation-reduction processes.

19C–2   THE STANDARD HYDROGEN ELECTRODE; ELECTRODE POTENTIALS

In order to develop a useful list of relative half-cell or electrode potentials, it is necessary to have a carefully defined reference electrode, which is accepted by the entire chemical community. The *standard hydrogen electrode* (SHE) is such a half-cell. Before defining this half-cell, we shall consider the properties of the hydrogen-gas electrode.

The Hydrogen-Gas Electrode.   The hydrogen-gas electrode was widely used in early electrochemical studies not only as a reference electrode but also as an indicator electrode for the determination of pH. Its composition can be formulated as

$$Pt,H_2(p \text{ atm})|H^+(xM)$$

As suggested by the terms in parentheses, the potential developed at the platinum surface depends upon the hydrogen ion concentration of the solution and upon the partial pressure of the hydrogen employed to saturate the solution with the gas.

Figure 19–5 illustrates the components of a typical hydrogen electrode. The conductor is constructed from platinum foil, which has been *platinized*—that is, coated with a finely divided layer of platinum (called platinum black) by rapid chemical or electrochemical reduction of H_2PtCl_6. The platinum black provides a large surface area to assure that the reaction

$$2H^+ + 2e \rightleftharpoons H_2(g)$$

proceeds rapidly and reversibly at the electrode surface. As was pointed out earlier, the stream of hydrogen serves simply to keep the solution adjacent to the electrode saturated with respect to the gas.

The hydrogen electrode may act as an anode or a cathode, depending upon the half-cell with which it is coupled by means of the salt bridge shown in the figure. Hydrogen is oxidized to hydrogen ions when the electrode is an anode; the reverse reaction takes place as a cathode. Under proper conditions, then, the hydrogen electrode is electrochemically reversible.

The Standard Hydrogen Electrode. The potential of a hydrogen electrode depends upon the temperature, the hydrogen ion concentration (more correctly, the activity) in the solution, and the pressure of the hydrogen at the surface of the electrode. Values for these parameters must be carefully defined in order for the half-cell process to serve as a reference. Specifications for the *standard hydrogen electrode* call for a hydrogen activity of unity and a partial pressure (fugacity) for hydrogen of exactly one atmosphere. *By convention, the potential of this electrode is assigned the value of exactly zero volt at all temperatures.*

Electrode Potentials. Electrode potentials are defined as *cell potentials* for a cell consisting of the electrode in question and the standard hydrogen electrode. It must always be borne in mind that electrode potentials are in fact *relative* potentials, all being referred to a common reference electrode having an assigned potential of zero volt.

Several secondary reference electrodes are more convenient for routine use and are extensively employed in electrode potential measurements; some of these are described in Section 20A of the next chapter. By straightforward calculations, potentials measured with these electrodes are readily converted to electrode potentials in which the standard hydrogen electrode is the reference.

19C–3 MEASUREMENT OF ELECTRODE POTENTIALS

Although the standard hydrogen electrode is the universal standard of reference, it should be understood that the electrode, as described, can never be realized in the labo-

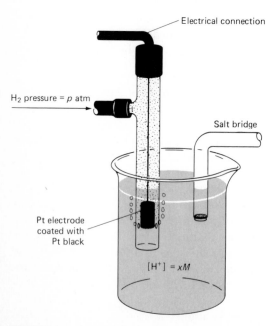

FIGURE 19-5 The hydrogen electrode.

ratory; that is, it is a *hypothetical* electrode to which experimentally determined potentials can be referred only by suitable computation. The reason that the electrode, as defined, cannot be prepared is that chemists lack the knowledge to produce a solution with a hydrogen ion activity of exactly unity; no adequate theory exists to permit evaluation of the activity coefficient of hydrogen ions in a solution in which the ionic strength is as great as unity, as required by the definition (see Appendix 3). Thus, the *concentration* of HCl or other acid required to give a hydrogen ion activity of unity cannot be calculated. Notwithstanding, data for more dilute solutions of acid, where activity coefficients are known, can be used to compute the *hypothetical* potentials at unit activity. Thus, for example, the activity of hydrogen and chloride ions in the cell shown in Figure 19–2 can be derived from the Debye-Hückel relationship; measurements can also be made at lower acid concentrations. These data can then provide, by suitable extrapolation, information about the potential of a hypothetical cell in which the hydrogen and chloride ions have unit activities. For the cell in Figure 19–2, Equation 19–9 then reduces to

$$E_{cell} = E_{cell}^0 - \frac{RT}{nF} \ln \frac{(1.00)^2(1.00)^2}{1.00} = E_{AgCl}^0$$

where E_{AgCl}^0 is the *standard electrode potential* for the half-reaction

$$AgCl(s) + e \rightleftharpoons Ag(s) + Cl^-$$
$$E^0 = 0.222 \text{ V} \qquad (19\text{–}11)$$

We can also imagine a cell with a liquid junction or a salt bridge in which the concentration of hydrogen and chloride ions can be varied independently. For example,

$$\text{Pt,H}_2(1.00 \text{ atm})|\text{H}^+(a_{H^+} = 1.00)\|$$
$$\text{Cl}^-(zM),\text{AgCl(sat'd)}|\text{Ag}$$

where the left half of this cell is the standard hydrogen electrode. The potential for this cell (neglecting junction potential) can be obtained by suitable substitutions into Equation 19–9. That is,

$$E_{cell} = E_{AgCl}$$
$$= E_{AgCl}^0 - \frac{RT}{2F} \ln \frac{[Cl^-]_a^2(1.00)^2}{1.00}$$

or

$$E_{AgCl} = E_{AgCl}^0 - \frac{RT}{F} \ln [Cl^-]_a \qquad (19\text{–}12)$$

This equation shows how the electrode potential E_{AgCl} for the silver/silver chloride electrode varies as a function of chloride ion concentration.

A cell with a salt bridge such as that shown in Figure 19–6 can be used to measure electrode potentials for half-reactions involving the metal and its ion. Here, the reference electrode is shown as a hydrogen electrode in which the acid concentration is sufficiently low to permit calculation of the hydrogen ion activity by means of the Debye-Hückel equation; alternatively, one of the secondary reference electrodes described in Section 20A, whose potential against the standard hydrogen electrode is known, can be substituted. The activity of the metal ion *y* is then varied and the cell potential measured. Suitable extrapolation will then provide a standard electrode potential E_M^0 for the half-reaction

$$M^{2+} + 2e = M(s) \qquad E_M^0 = z \text{ V}$$

Thus, E_M^0 is the potential for the cell when the activities of M^{2+} and H^+ are exactly one. This measurement would be less accurate than the one employing a cell without a liquid junction because of the uncertainty with respect to the junction potentials that exist at the two ends of the salt bridge.

If the metal M in Figure 19–6 is cadmium

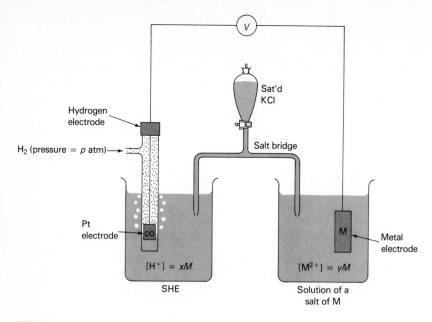

FIGURE 19-6   Schematic diagram of an arrangement for the measurement of electrode potentials against the standard hydrogen electrode.

and the solution is approximately 0.01 M in cadmium ions, the voltage indicated by the measuring device, V, will be about 0.5 V. Moreover, the cadmium will function as the anode; thus, electrons pass from this electrode to the hydrogen electrode via the external circuit. The half-reactions that actually occur in this galvanic cell are

$$Cd(s) \rightleftharpoons Cd^{2+} + 2e \qquad \text{anode}$$

$$2H^+ + 2e \rightleftharpoons H_2(g) \qquad \text{cathode}$$

The overall cell reaction is the sum of these, or

$$Cd(s) + 2H^+ \rightleftharpoons Cd^{2+} + H_2(g)$$

If the cadmium electrode is replaced by a zinc electrode immersed in a solution that is about 0.01 M in zinc ions, a potential of about 0.8 V is observed. The metal electrode is again the anode. The larger voltage reflects the greater tendency of zinc to be

oxidized. The difference between this potential and the one for cadmium is a quantitative measure of the relative strengths of these two metals as reducing agents.

If the half-cell in Figure 19–6 consisted of a copper electrode in a 0.01 M solution of copper(II) ions, a potential of about 0.3 V would develop. However, in distinct contrast to the previous two examples, copper would tend to deposit, and an external electron flow, if allowed, would be from the hydrogen electrode to the copper electrode. The spontaneous cell reaction, then, is the reverse of that in the two cells considered earlier. That is,

$$Cu^{2+} + H_2(g) \rightleftharpoons Cu(s) + 2H^+$$

Thus, metallic copper is a much less effective reducing agent than either zinc, cadmium, or *molecular hydrogen*. As before, the observed potential is a quantitative measure of this strength.

By additional measurements and suitable extrapolations, the data from the experiment just described could be made to yield computed potentials for the system when the activities of M^{2+} and H^+ are unity. These electrode potentials for Cd^{2+}, Zn^{2+}, and Cu^{2+} are then 0.403, 0.763, and 0.337 V, respectively. Note, however, that a need exists to indicate that the copper electrode behaves as a cathode while the zinc and cadmium electrodes function as anodes when coupled to the hydrogen electrode. Positive and negative signs are used to make this distinction, the potential for half-cells such as copper being provided with one sign and the other two electrodes being assigned the opposite. The choice as to which potential will be positive and which will be negative is somewhat arbitrary; however, the sign convention that is chosen must be used consistently.

19C-4 SIGN CONVENTIONS FOR ELECTRODE POTENTIALS

It is perhaps not surprising that the choice of sign for electrode potentials has led to much controversy and confusion in the course of the development of electrochemistry. In 1953, the International Union of Pure and Applied Chemistry (IUPAC), meeting in Stockholm, attempted to resolve this controversy. The sign convention adopted at this meeting is sometimes called the IUPAC or Stockholm convention; there is hope for its general adoption in years to come. We shall always use the IUPAC sign convention.

Any sign convention must be based upon half-cell processes written in a single way— that is, entirely as oxidations or as reductions. According to the IUPAC convention, the term *electrode potential* (or more exactly, *relative electrode potential*) *is reserved exclusively for half-reactions written as reductions.* There is no objection to using the term *oxidation potential* to connote an electrode process written in the opposite sense, but an oxidation potential should never be called an electrode potential. The sign of an oxidation potential will always be opposite to its corresponding electrode or reduction potential.

The sign of the electrode potential is determined by the actual sign of the electrode of interest when it is coupled with a standard hydrogen electrode in a galvanic cell. Thus, a zinc or a cadmium electrode will behave as the anode from which electrons flow through the external circuit to the standard hydrogen electrode. These metal electrodes are thus the negative terminals of such galvanic cells, and their electrode potentials are *assigned* negative values. That is,

$$Zn^{2+} + 2e \rightleftharpoons Zn(s) \qquad E^0 = -0.763 \text{ V}$$
$$Cd^{2+} + 2e \rightleftharpoons Cd(s) \qquad E^0 = -0.403 \text{ V}$$

The potential for the copper electrode, on the other hand, is given a positive sign because the copper behaves as a cathode in a galvanic cell constructed from this electrode and the hydrogen electrode; electrons flow toward the copper electrode through the exterior circuit. It is thus the positive terminal of the galvanic cell and

$$Cu^{2+} + 2e \rightleftharpoons Cu(s) \qquad E^0 = +0.337 \text{ V}$$

It is important to emphasize that electrode potentials and their signs apply to half-reactions *written as reductions.* Both zinc and cadmium are oxidized by hydrogen ion; the spontaneous reactions are thus oxidations. It is evident, then, that the *sign of the electrode potential will indicate whether or not the reduction is spontaneous with respect to the standard hydrogen electrode.* That is, the positive sign for the copper electrode potential means that the reaction

$$Cu^{2+} + H_2(g) \rightleftharpoons 2H^+ + Cu(s)$$

proceeds toward the right under ordinary conditions. The negative electrode potential for zinc, on the other hand, means that the analogous reaction

$$Zn^{2+} + H_2(g) \rightleftharpoons 2H^+ + Zn(s)$$

does not ordinarily occur; indeed, the equilibrium favors the species on the left.

The IUPAC convention was adopted in 1953, but electrode potential data given in some texts and reference works, particularly older ones, are not always in accord with it. For example, in a source of oxidation potential data compiled by Latimer,[4] one finds

$$Zn(s) \rightleftharpoons Zn^{2+} + 2e \qquad E = +0.763 \text{ V}$$

$$Cu(s) \rightleftharpoons Cu^{2+} + 2e \qquad E = -0.337 \text{ V}$$

To convert these oxidation potentials to electrode potentials as defined by the IUPAC convention, one must mentally: (1) express the half-reactions as reductions; and (2) change the signs of the potentials.

The sign convention employed in a table of standard potentials may not be explicitly stated. This information is readily ascertained, however, by referring to a half-reaction with which one is familiar and noting the direction of the reaction and the sign of the potential. Whatever changes, if any, are required to convert to the IUPAC convention are then applied to the remainder of the data in the table. For example, all one needs to remember is that strong oxidizing agents such as oxygen have large positive electrode potentials under the IUPAC convention. That is, the reaction

$$O_2(g) + 4H^+ + 4e \rightleftharpoons 2H_2O \qquad E^0 = +1.23 \text{ V}$$

tends to occur spontaneously with respect to the standard hydrogen electrode. The sign and direction of this reaction in a given table

can then serve as a key to any changes that may be needed to convert all data to the IUPAC convention.

19C-5 EFFECT OF CONCENTRATION ON ELECTRODE POTENTIAL

Equation 19–12 shows how the electrode potential for the silver/silver chloride electrode varies as a function of chloride concentration. Turning to a more general case, consider the half-reaction

$$pP + qQ + \cdots + ne \rightleftharpoons rR + sS$$

where the capital letters represent formulas of reacting species (whether charged or uncharged), e represents the electron, and the lower-case italic letters indicate the number of moles of each species (including electrons) participating in the half-cell reaction. Employing the same arguments that were used in the case of the silver/silver chloride electrode, we obtain

$$E = E^0 - \frac{RT}{nF} \ln \frac{[R]_a^r [S]_a^s \cdots}{[P]_a^p [Q]_a^q \cdots}$$

At room temperature (298°K), the collection of constants in front of the logarithm has units of joules per coulomb or volt. That is,

$$\frac{RT}{nF} = \frac{8.316 \text{ J mol}^{-1} \text{ deg}^{-1} \times 298 \text{ deg}}{n \text{ equiv mol}^{-1} \times 96487 \text{ C equiv}^{-1}}$$

$$= \frac{2.568 \times 10^{-2} \text{ J C}^{-1}}{n} = \frac{2.568 \times 10^{-2}}{n} \text{V}$$

Upon converting from a natural to a base ten logarithm by multiplication by 2.303, the foregoing equation can be written

$$E = E^0 - \frac{0.0591}{n} \log \frac{[R]^r [S]^s \cdots}{[P]^p [Q]^q \cdots} \qquad (19–13)$$

For convenience, we have also deleted the a subscripts, which were inserted earlier as a reminder that the bracketed terms repre-

---

[4] W. M. Latimer, *The Oxidation States of the Elements and Their Potentials in Aqueous Solutions*, 2d ed., Prentice-Hall: Englewood Cliffs, N.J., 1952.

sented nonequilibrium concentrations. Hereafter, *the subscripts will not be used;* the student should, however, be alert to the fact that the quotients that appear in this type of equation are *not equilibrium constants,* despite their similarity in appearance.

To summarize the meaning of the bracketed terms in Equation 19–13: When the substance R is a gas,

[R] = partial pressure in atmospheres

When R is a solute

[R] = concentration in moles per liter

or occasionally

[R] = activity of R, a_R

When R̊ is a pure solid or liquid in excess, or the solvent

[R] = 1.00

Equation 19–13 is a general statement of the *Nernst equation,* which can be applied both to half-cell reactions or cell reactions as was shown on page 577.

19C–6 *THE STANDARD ELECTRODE POTENTIAL, E⁰*

An examination of Equation 19–13 reveals that the constant E^0 is equal to the half-cell potential when the logarithmic term is zero. This condition occurs whenever the activity quotient is equal to unity, one such instance being when the activities of all reactants and products are unity. Thus, the standard potential is often defined as the electrode potential of a half-cell reaction (vs. SHE) when all reactants and products exist at unit activity.

The standard electrode potential is an important physical constant that gives a quantitative description of the relative driving force for a half-cell reaction. Several facts

regarding this constant should be kept in mind. First, the electrode potential is temperature-dependent; if it is to have significance, the temperature at which it is determined must be specified. Second, the standard electrode potential is a relative quantity in the sense that it is really the potential of an electrochemical cell in which the anode is a carefully specified reference electrode—that is, the standard hydrogen electrode—whose potential is *assigned* a value of zero volt. Third, the sign of a standard potential is identical with that of the conductor in contact with the half-cell of interest in a galvanic cell, the other half of which is the standard hydrogen electrode. Finally, the standard potential is a measure of the intensity of the driving force for a half-reaction. As such, it is independent of the notation employed to express the half-cell process. Thus, the potential for the process

$$Ag^+ + e \rightleftharpoons Ag(s) \qquad E^0 = +0.799 \text{ V}$$

although dependent upon the concentration of silver ions, is the same regardless of whether we write the half-reaction as above or as

$$100 \, Ag^+ + 100e \rightleftharpoons 100 \, Ag(s)$$
$$E^0 = +0.799 \text{ V}$$

To be sure, the Nernst equation must be consistent with the half-reaction as it has been written. For the first of these, it will be

$$E = 0.799 - \frac{0.0591}{1} \log \frac{1}{[Ag^+]}$$

and for the second

$$E = 0.799 - \frac{0.0591}{100} \log \frac{1}{[Ag^+]^{100}}$$

Standard electrode potentials are available for numerous half-reactions. Many have been determined directly from voltage measurements of cells in which a hydrogen or other reference electrode constituted the other half of the cell. It is possible, however, to calculate E^0 values from equilibrium studies of oxidation-reduction systems and from thermochemical data relating to such reactions. Many of the values found in the literature were so obtained.[5]

For illustrative purposes, a few standard electrode potentials are given in Table 19–1; a more comprehensive table is found in Appendix 4. The species in the upper left-hand part of the equations in Table 19–1 are most easily reduced, as indicated by the large positive E^0 values; they are therefore the most effective oxidizing agents. Pro-

ceeding down the left-hand side of the table, each succeeding species is a less effective acceptor of electrons than the one above it. The half-cell reactions at the bottom of the table have little tendency to take place as written. On the other hand, they do tend to occur in the opposite sense, as oxidations. The most effective reducing agents, then, are those species that appear in the lower right-hand side of the equations in the table.

A compilation of standard potentials provides the chemist with information regarding the extent and direction of electron-transfer reactions between the tabulated species. On the basis of Table 19–1, for example, we see that zinc is more easily oxidized than cadmium, and we conclude that a piece of zinc immersed in a solution of cadmium ions will cause the deposition of metallic cadmium; conversely, cadmium has little tendency to reduce zinc ions. Table 19–1 also shows that iron(III) is a better oxidizing agent than triiodide ion; therefore, in a solution containing an equilibrium mixture of iron(III), iodide, iron(II), and triiodide

---

[5] Authoritative sources for standard potential data are: W. M. Latimer, *The Oxidation States of the Elements and Their Potentials in Aqueous Solutions*, 2d ed., Prentice-Hall: Englewood Cliffs, N.J., 1952; G. Milazzo and S. Caroli, *Tables of Standard Electrode Potentials*, Wiley-Interscience: New York, 1977; and M. S. Antelman and F. J. Harris, Jr., *Chemical Electrode Potentials*, Plenum Press: New York, 1982.

Table 19–1
Standard Electrode Potentials[a]

Reaction	E^0 at 25°C, V
$Cl_2(g) + 2e \rightleftharpoons 2Cl^-$	+1.359
$O_2(g) + 4H^+ + 4e \rightleftharpoons 2H_2O$	+1.229
$Br_2(aq) + 2e \rightleftharpoons 2Br^-$	+1.087
$Br_2(l) + 2e \rightleftharpoons 2Br^-$	+1.065
$Ag^+ + e \rightleftharpoons Ag(s)$	+0.799
$Fe^{3+} + e \rightleftharpoons Fe^{2+}$	+0.771
$I_3^- + 2e \rightleftharpoons 3I^-$	+0.536
$Hg_2Cl_2(s) + 2e \rightleftharpoons 2Hg(l) + 2Cl^-$	+0.268
$AgCl(s) + e \rightleftharpoons Ag(s) + Cl^-$	+0.222
$Ag(S_2O_3)_2^{3-} + e \rightleftharpoons Ag(s) + 2S_2O_3^{2-}$	+0.010
$2H^+ + 2e \rightleftharpoons H_2(g)$	0.000
$AgI(s) + e \rightleftharpoons Ag(s) + I^-$	−0.151
$PbSO_4(s) + 2e \rightleftharpoons Pb(s) + SO_4^{2-}$	−0.350
$Cd^{2+} + 2e \rightleftharpoons Cd(s)$	−0.403
$Zn^{2+} + 2e \rightleftharpoons Zn(s)$	−0.763

[a] See Appendix 4 for a more extensive list.

ions, we can predict that the latter pair will predominate.

19C-7 CALCULATION OF HALF-CELL POTENTIALS FROM E^0 VALUES

Typical applications of the Nernst equation to the calculation of half-cell potentials are illustrated in the following examples.

EXAMPLE 19-1. What is the potential for a half-cell consisting of a cadmium electrode immersed in a solution that is 0.0150 M in Cd^{2+}?

From Table 19-1, we find

$$Cd^{2+} + 2e \rightleftharpoons Cd(s) \qquad E^0 = -0.403 \text{ V}$$

Thus,

$$E = E^0 - \frac{0.0591}{2} \log \frac{1}{[Cd^{2+}]}$$

Substituting the Cd^{2+} concentration into this equation gives

$$E = -0.403 - \frac{0.0591}{2} \log \frac{1}{0.0150}$$

$$= -0.457 \text{ V}$$

The sign for the potential from the foregoing calculation indicates the direction of the reaction when this half-cell is coupled with the standard hydrogen electrode. The fact that it is negative shows that the reverse reaction

$$Cd(s) + 2H^+ \rightleftharpoons H_2(g) + Cd^{2+}$$

occurs spontaneously. Note that the calculated potential is a larger negative number than the standard electrode potential itself. This follows from mass-law considerations because the half-reaction, *as written,* has less tendency to occur with the lower cadmium ion concentration.

EXAMPLE 19-2 Calculate the potential for a platinum electrode immersed in a solution prepared by saturating a 0.0150 M solution of KBr with Br_2.

Here, the half-reaction is

$$Br_2(l) + 2e \rightleftharpoons 2Br^- \qquad E^0 = 1.065 \text{ V}$$

Note that the term (l) in the equation indicates that the aqueous solution is kept saturated by the presence of an excess of *liquid* Br_2. Thus, the overall process is the sum of the two equilibria

$$Br_2(l) \rightleftharpoons Br_2(sat'd \ aq)$$

$$Br_2(sat'd \ aq) + 2e \rightleftharpoons 2Br^-$$

The Nernst equation for the overall process is

$$E = 1.065 - \frac{0.0591}{2} \log \frac{[Br^-]^2}{1.00}$$

Here, the activity of Br_2 in the pure liquid is constant and equal to 1.00 by definition. Thus,

$$E = 1.065 - \frac{0.0591}{2} \log (0.0150)^2$$

$$= 1.173 \text{ V}$$

EXAMPLE 19-3. Calculate the potential for a platinum electrode immersed in a solution that is 0.0150 M in KBr and $1.00 \times 10^{-3} M$ in Br_2.

Here, the half-reaction used in the preceding example does not apply *because the solution is no longer saturated in Br_2.* Table 19-1, however, contains the half-reaction

$$Br_2(aq) + 2e \rightleftharpoons 2Br^- \qquad E^0 = 1.087 \text{ V}$$

The term (aq) implies that all of the Br_2 present is in solution; that is, 1.087 is the electrode potential for the half-reaction when the Br^- and Br_2 *solution* activities are 1.00 mol/L. It turns out, however, that the solubility of Br_2 in water at 25°C is only about 0.18 mol/L. Therefore, the recorded potential of 1.087 is based on a *hypothetical system that cannot be realized experimentally.* Nevertheless, this potential is useful because it provides the means by which potentials for undersaturated systems can be calculated. Thus,

$$E = 1.087 - \frac{0.0591}{2} \log \frac{[Br^-]^2}{[Br_2]}$$

$$= 1.087 - \frac{0.0591}{2} \log \frac{(1.50 \times 10^{-2})^2}{1.00 \times 10^{-3}}$$

$$= 1.106 \text{ V}$$

Here, the Br_2 activity is 1.00×10^{-3} rather than 1.00, as was the situation when the solution was saturated.

19C-8 ELECTRODE POTENTIALS IN THE PRESENCE OF PRECIPITATION AND COMPLEX-FORMING REAGENTS

As shown by the following example, reagents that react with the participants of an electrode process have a marked effect on the potential for that process.

EXAMPLE 19-4. Calculate the potential of a silver electrode in a solution that is saturated with silver iodide and has an iodide ion activity of exactly 1.00 (K_{sp} for AgI = 8.3×10^{-17}).

$$Ag^+ + e \rightleftharpoons Ag(s) \qquad E^0 = +0.799 \text{ V}$$

$$E = +0.799 - 0.0591 \log \frac{1}{[Ag^+]}$$

We may calculate $[Ag^+]$ from the solubility product constant

$$[Ag^+] = \frac{K_{sp}}{[I^-]}$$

Substituting into the Nernst equation gives

$$E = +0.799 - \frac{0.0591}{1} \log \frac{[I^-]}{K_{sp}}$$

This equation may be rewritten as

$$E = +0.799 + 0.0591 \log K_{sp}$$
$$- 0.0591 \log [I^-] \quad (19-14)$$

If we substitute 1.00 for $[I^-]$ and use 8.3×10^{-17} for K_{sp}, the solubility product for AgI at 25.0°C, we obtain

$$E = -0.151 \text{ V}$$

This example shows that the half-cell potential for the reduction of silver ion becomes smaller in the presence of iodide ions. Qualitatively this is the expected effect because decreases in the concentration of silver ions diminish the tendency for their reduction.

Equation 19–14 relates the potential of a silver electrode to the iodide ion concentration of a solution that is also saturated with silver iodide. *When the iodide ion activity is unity,* the potential is the sum of two constants; it is thus the standard electrode potential for the half-reaction

$$AgI(s) + e \rightleftharpoons Ag(s) + I^- \qquad E^0 = -0.151 \text{ V}$$

where

$$E^0 = +0.799 + 0.0591 \log K_{sp} \qquad (19-15)$$

The Nernst relationship for the silver electrode in a solution saturated with silver iodide can then be written as

$$E = E^0 - 0.0591 \log [I^-]$$
$$= -0.151 - 0.0591 \log [I^-]$$

Thus, when in contact with a solution *saturated with silver iodide,* the potential of a silver electrode can be described *either* in terms of the silver ion concentration (with the standard electrode potential for the simple silver half-reaction) *or* in terms of the iodide ion concentration (with the standard electrode potential for the silver/silver iodide half-reaction). The latter is usually more convenient.

The potential of a silver electrode in a solution containing an ion that forms a soluble complex with silver ion can be treated in a fashion analogous to the foregoing. For example, in a solution containing thiosulfate and silver ions, complex formation occurs:

$$Ag^+ + 2S_2O_3^{2-} \rightleftharpoons Ag(S_2O_3)_2^{3-}$$

$$K_f = \frac{[Ag(S_2O_3)_2^{3-}]}{[Ag^+][S_2O_3^{2-}]^2}$$

where K_f is the *formation constant* for the complex. The half-reaction for a silver electrode in such a solution can be written as

$$Ag(S_2O_3)_2^{3-} + e \rightleftharpoons Ag(s) + 2S_2O_3^{2-}$$

The standard electrode potential for this half-reaction will be the electrode potential when both the complex and the complexing anion are at unit activity. Using the same approach as in the previous example, we find that

$$E^0 = +0.799 + 0.0591 \log \frac{1}{K_f}$$

Data for the potential of the silver electrode in the presence of selected ions are given in the tables of standard electrode potentials in Appendix 4 and Table 19–1. Similar information is also provided for other electrode systems. Such data often simplify the calculation of half-cell potentials.

19C-9 *SOME LIMITATIONS TO THE USE OF STANDARD ELECTRODE POTENTIALS*

Standard electrode potentials are of great importance in understanding electroanalytical processes. There are, however, certain inherent limitations to the use of these data that should be clearly appreciated.

Substitution of Concentrations for Activities. As a matter of convenience, molar concentrations—rather than activities—of reactive species are generally employed in the Nernst equation. Unfortunately, the assumption that these two quantities are identical is valid only in dilute solutions; with increasing electrolyte concentrations, potentials calculated on the basis of molar concentrations can be expected to depart from those obtained by experiment.

To illustrate, the standard electrode potential for the half-reaction

$$Fe^{3+} + e \rightleftharpoons Fe^{2+}$$

is $+0.771$ V. Neglecting activities, we would predict that a platinum electrode immersed in a solution that contained 1 mol/L Fe^{2+} and Fe^{3+} ions in addition to perchloric acid would exhibit a potential numerically equal to this value relative to the standard hydrogen electrode. In fact, however, a potential of $+0.732$ V is observed experimentally when the perchloric acid concentration is 1 M. The reason for the discrepancy is seen if we write the Nernst equation in the form

$$E = E^0 - 0.0591 \log \frac{[Fe^{2+}]f_{Fe^{2+}}}{[Fe^{3+}]f_{Fe^{3+}}}$$

where $f_{Fe^{2+}}$ and $f_{Fe^{3+}}$ are the respective activity coefficients. The activity coefficients of the two species are less than one in this system because of the high ionic strength imparted by the perchloric acid and the iron salts. More important, however, the activity coefficient of the iron(III) ion is smaller than that of the iron(II) ion, inasmuch as the effects of ionic strength on these coefficients increase with the charge on the ion (Appendix 3). As a consequence, the ratio of the activity coefficients as they appear in the Nernst equation would be larger than one and the potential of the half-cell would be smaller than the standard potential.

Activity coefficient data for ions in solutions of the types commonly encountered in oxidation-reduction titrations and electrochemical work are fairly limited; consequently, molar concentrations rather than activities must be used in many calculations. Appreciable errors may result.

Effect of Other Equilibria. The application of standard electrode potentials is further complicated by the occurrence of solvolysis, dissociation, association, and complex-formation reactions involving the species of interest. An example of this problem is encountered in the behavior of the potential of the iron(III)/iron(II) couple. As noted earlier, an equimolar mixture of these two ions in 1 M perchloric acid has an electrode potential of +0.73 V. Substitution of hydrochloric acid of the same concentration alters the observed potential to +0.70 V; a value of +0.6 V is observed in 1 M phosphoric acid. These differences arise because iron(III) forms more stable complexes with chloride and phosphate ions than does iron(II). As a result, the actual concentration of uncomplexed iron(III) in such solutions is less than that of uncomplexed iron(II), and the net effect is a shift in the observed potential.

Phenomena such as these can be taken into account only if the equilibria involved are known and constants for the processes are available. Often, however, such information is lacking; the chemist is then forced to neglect such effects and hope that serious errors do not flaw the calculated results.

Formal Potentials. In order to compensate partially for activity effects and errors resulting from side reactions, Swift[6] has proposed substituting a quantity called the *formal potential E'* in place of the standard electrode potential in oxidation-reduction

calculations. The formal potential of a system is the potential of the half-cell with respect to the standard hydrogen electrode when the *concentrations* of reactants and products are 1 M and the concentrations of any other constituents of the solution are carefully specified. Thus, for example, the formal potential for the reduction of iron(III) is +0.732 V in 1 M perchloric acid and +0.700 V in 1 M hydrochloric acid. Use of these values in place of the standard electrode potential in the Nernst equation will yield better agreement between calculated and experimental potentials, provided the electrolyte concentration of the solution approximates that for which the formal potential was measured. Application of formal potentials to systems differing greatly as to kind and concentration of electrolyte can, however, lead to errors greater than those encountered with the use of standard potentials. The table in Appendix 4 contains selected formal potentials as well as standard potentials; in subsequent chapters, we shall use whichever is the more appropriate.

Reaction Rates. It should be realized that the existence of a half-reaction in a table of electrode potentials does not necessarily imply that a real electrode exists whose potential responds to the half-reaction. Many of the data in such tables have been obtained by calculations based upon equilibrium or thermal measurements rather than from the actual measurement of the potential for an electrode system. For some, no suitable electrode is known; thus, the standard electrode potential for the process,

$$2CO_2 + 2H^+ + 2e \rightleftharpoons H_2C_2O_4$$
$$E^0 = -0.49 \text{ V}$$

has been arrived at indirectly. The electrode reaction is not reversible, and the rate at which carbon dioxide combines to give oxalic acid is negligibly slow. No electrode system is known whose potential varies in the expected way with the ratio of activities of

---

[6] E. H. Swift, *A System of Chemical Analysis*, p. 50, Freeman: San Francisco, 1939.

the reactants and products. Nonetheless, the potential is useful for computational purposes.

19D   Calculation of Cell Potentials from Electrode Potentials

An important use of standard electrode potentials is the calculation of the potential obtainable from a galvanic cell or the potential required to operate an electrolytic cell. These calculated potentials (sometimes called *thermodynamic potentials*) are theoretical in the sense that they refer to cells in which there is essentially no current; additional factors must be taken into account when a flow of electricity is involved.

19D-1   CALCULATION OF THERMODYNAMIC CELL POTENTIALS

As shown earlier (Equation 19–10), the electromotive force of a cell is obtained by combining half-cell potentials as follows:

$$E_{cell} = E_{cathode} - E_{anode}$$

where E_{anode} and $E_{cathode}$ are the *electrode potentials* for the two half-reactions constituting the cell.

Consider the hypothetical cell

$$Zn|ZnSO_4(a_{Zn^{2+}} = 1.00)\|$$
$$CuSO_4(a_{Cu^{2+}} = 1.00)|Cu$$

The overall cell process involves the oxidation of elemental zinc to zinc(II) and the reduction of copper(II) to the metallic state. Because the activities of the two ions are specified as unity, the standard potentials are also the electrode potentials. The cell diagram also specifies that the zinc electrode is the anode. Thus, using E^0 data from Table 19–1,

$$E_{cell} = +0.337 - (-0.763) = +1.100 \text{ V}$$

The positive sign for the cell potential indicates that the reaction

$$Zn(s) + Cu^{2+} \rightarrow Zn^{2+} + Cu(s)$$

occurs spontaneously and that this is a galvanic cell.

The foregoing cell, diagrammed as

$$Cu|Cu^{2+}(a_{Cu^{2+}} = 1.00)\|$$
$$Zn^{2+}(a_{Zn^{2+}} = 1.00)|Zn$$

implies that the copper electrode is now the anode. Thus,

$$E_{cell} = -0.763 - (+0.337) = -1.100 \text{ V}$$

The negative sign indicates the nonspontaneity of the reaction

$$Cu(s) + Zn^{2+} \rightarrow Cu^{2+} + Zn(s)$$

The application of an external potential greater than 1.100 V would be required to cause this reaction to occur.

EXAMPLE 19-5.   Calculate potentials for the following cell employing (a) concentrations and (b) activities:

$$Zn|ZnSO_4(xM),PbSO_4(sat'd)|Pb$$

where $x = 5.00 \times 10^{-4}$, 2.00×10^{-3}, 1.00×10^{-2}, 2.00×10^{-2}, and 5.00×10^{-2}.

(a) In a neutral solution, little HSO_4^- will be formed; thus, we may assume that

$$[SO_4^{2-}] = M_{ZnSO_4} = x = 5.00 \times 10^{-4}$$

The half-reactions and standard potentials are

$$PbSO_4(s) + 2e \rightleftharpoons Pb(s) + SO_4^{2-}$$
$$E^0 = -0.350 \text{ V}$$

$$Zn^{2+} + 2e \rightleftharpoons Zn \qquad E^0 = -0.763 \text{ V}$$

The potential of the lead electrode is given by

$$E_{Pb} = -0.350 - \frac{0.0591}{2} \log 5.00 \times 10^{-4}$$

$$= -0.252 \text{ V}$$

The zinc ion concentration is also $5.00 \times 10^{-4}\ M$ and

$$E_{Zn} = -0.763 - \frac{0.0591}{2} \log \frac{1}{5.00 \times 10^{-4}}$$

$$= -0.860 \text{ V}$$

Since the Pb electrode is specified as the cathode,

$$E_{cell} = -0.252 - (-0.860) = 0.608$$

Cell potentials at the other concentrations can be derived in the same way. Their values are given in column (a) of Table 19–2.

(b) To obtain activity coefficients for Zn^{2+} and SO_4^{2-} we must first calculate the ionic strength with the aid of Equation A3–2 (Appendix). Here, we assume that the concentrations of Pb^{2+}, H^+, and OH^- are negligible with respect to the concentrations of Zn^{2+} and SO_4^{2-}. Thus, the ionic strength is

$$\mu = \tfrac{1}{2}[5.00 \times 10^{-4} \times (2)^2$$
$$+ 5.00 \times 10^{-4} \times (2)^2]$$

$$= 2.00 \times 10^{-3}$$

In Table A3–1, we find for SO_4^{2-}, $\alpha_A = 4.0$ and for Zn^{2+}, $\alpha_A = 6.0$. Substituting these values into Equation A3–3 gives for sulfate ion

$$-\log f_{SO_4} = \frac{0.509 \times 2^2 \times \sqrt{2.00 \times 10^{-3}}}{1 + 0.328 \times 4.0\sqrt{2.00 \times 10^{-3}}}$$

$$= 8.59 \times 10^{-2}$$

$$f_{SO_4} = 0.820$$

$$a_{SO_4} = 0.820 \times 5.00 \times 10^{-4}$$
$$= 4.10 \times 10^{-4}$$

Repeating the calculations employing $\alpha_A = 6.0$ for Zn^{2+} yields

$$f_{Zn} = 0.825$$

$$a_{Zn} = 4.13 \times 10^{-4}$$

The Nernst equation for the Pb electrode now becomes

$$E_{Pb} = -0.350 - \frac{0.0591}{2} \times \log 4.10 \times 10^{-4}$$

$$= -0.250$$

For the zinc electrode

$$E_{Zn} = -0.763 - \frac{0.0591}{2} \times \log \frac{1}{4.13 \times 10^{-4}}$$

$$= -0.863$$

and

Table 19–2
Calculated Potentials for a Cell Based on (a) Concentrations and (b) Activities. (See Example 19–5.)

x	μ	(a) E(calc)	(b) E(calc)	E(exptl)[a]
5.00×10^{-4}	2.00×10^{-3}	0.608	0.613	0.611
2.00×10^{-3}	8.00×10^{-3}	0.572	0.582	0.583
1.00×10^{-2}	4.00×10^{-2}	0.531	0.549	0.553
2.00×10^{-2}	8.00×10^{-2}	0.513	0.537	0.542
5.00×10^{-2}	2.00×10^{-1}	0.490	0.521	0.529

[a] Experimental data from: I. A. Cowperthwaite and V. K. LaMer, *J. Amer. Chem. Soc.*, **1931,** *53*, 4333.

$E_{cell} = -0.250 - (-0.863) = 0.613$ V

Values at other concentrations are listed in column (b) of Table 19–2.

It is of interest to compare the calculated cell potentials shown in the columns labeled (a) and (b) in Table 19–2 with the experimental results shown in the last column. Clearly, the use of activities provides a significant improvement at the higher ionic strengths.

EXAMPLE 19-6. Calculate the potential required to initiate the deposition of copper from a solution that is 0.010 *M* in $CuSO_4$ and contains sufficient sulfuric acid to give a hydrogen ion concentration of 1.0×10^{-4} *M*.

The deposition of copper necessarily occurs at the cathode. Because no easily oxidizable species are present, the anode reaction will involve oxidation of H_2O to give O_2. From the table of standard potentials, we find

$$Cu^{2+} + 2e \rightleftharpoons Cu(s) \qquad E^0 = +0.337 \text{ V}$$

$$O_2(g) + 4H^+ + 4e \rightarrow 2H_2O$$
$$E^0 = +1.229 \text{ V}$$

Thus, for the copper electrode

$$E = +0.337 - \frac{0.0591}{2} \log \frac{1}{0.010}$$

$$= +0.278 \text{ V}$$

Assuming that O_2 is evolved at 1.00 atm, the potential for the oxygen electrode is

$$E = +1.229 - \frac{0.0591}{4}$$

$$\times \log \frac{1}{(1.00)(1.0 \times 10^{-4})^4}$$

$$= +0.993 \text{ V}$$

The cell potential is then

$$E_{cell} = +0.278 - 0.993 = -0.715 \text{ V}$$

Thus, to initiate the reaction

$$2Cu^{2+} + 2H_2O \rightarrow O_2(g) + 4H^+ + 2Cu(s)$$

would require the application of a potential greater than 0.715 V.

19D-2 LIQUID JUNCTION POTENTIAL

When two electrolyte solutions of different composition are brought in contact with one another, a potential develops at the interface. This *junction potential* arises from an unequal distribution of cations and anions across the boundary due to differences in the rates at which these species migrate.

Consider the liquid junction that exists in the system

$$HCl(1 \text{ } M) | HCl(0.01 \text{ } M)$$

Both hydrogen ions and chloride ions tend to diffuse across this boundary from the more concentrated to the more dilute solution, the driving force for this migration being proportional to the concentration difference. The rate at which various ions move under the influence of a fixed force varies considerably (that is, their *mobilities* are different). In the present example, hydrogen ions are several times more mobile than chloride ions. As a consequence, there is a tendency for the hydrogen ions to outstrip the chloride ions as diffusion takes place; a separation of charge is the net result (see Figure 19–7). The more dilute side of the boundary becomes positively charged owing to the more rapid migration of hydrogen ions; the concentrated side, therefore, acquires a negative charge from the

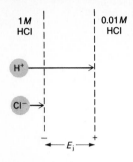

FIGURE 19-7 Schematic representation of a liquid junction showing the source of the junction potential E_j. The length of the arrows corresponds to the relative mobility of the two ions.

excess slower-moving chloride ions. The charge that develops tends to counteract the differences in mobilities of the two ions, and as a consequence, an equilibrium condition soon develops. The junction potential difference resulting from this charge separation may amount to 30 mV or more.

In a simple system such as that shown in Figure 19–7, the magnitude of the junction potential can be calculated from a knowledge of the mobilities of the two ions involved. However, it is seldom that a cell of analytical importance has a sufficiently simple composition to permit such a computation.[7]

It has been found experimentally that the magnitude of the junction potential can be greatly reduced by introduction of a concentrated electrolyte solution (a salt bridge) between the two solutions. The effectiveness of a salt bridge improves not only with concentration of the salt but also as the mobilities of the ions of the salt approach one another in magnitude. A saturated potassium chloride solution is good from both stand-

points, its concentration being somewhat greater than 4 M at room temperature, and the mobility of its ions differing by only 4%. When chloride ion interferes, a concentrated solution of potassium nitrate can be substituted. With such bridges, the junction potential typically amounts to a few millivolts or less, a negligible quantity in many, but not all, analytical measurements.

19E   Effect of Current on Cell Potentials

When electricity is allowed to flow in an electrochemical cell, the measured cell potential normally departs from that derived from thermodynamic calculations such as those shown in Section 19D–1. This departure can be traced to a number of phenomena including ohmic resistance and several *polarization effects* including *charge-transfer overvoltage, reaction overvoltage, diffusion overvoltage,* and *crystallization overvoltage.* Generally, these phenomena have the effect of reducing the potential of a galvanic cell or increasing the potential needed to develop a current in an electrolytic cell.

19E-1   OHMIC POTENTIAL; IR DROP

To develop a current in either a galvanic or an electrolytic cell, a driving force or a potential is required to overcome the resistance of the ions to movement toward the anode and the cathode. Just as in metallic conduction, this force follows Ohm's law and is equal to the product of the current in amperes and the resistance of the cell in ohms. The force is generally referred to as the *ohmic potential,* or the *IR drop.*

The net effect of *IR* drop is to increase the potential required to operate an electrolytic cell and to decrease the measured potential of a galvanic cell. Therefore, the *IR* drop is always *subtracted* from the theoretical

---

[7] For methods for approximating junction potentials, see: A. J. Bard and L. R. Faulkner, *Electrochemical Methods,* pp. 62–72, Wiley: New York, 1980.

cell potential. That is,[8]

$$E_{cell} = E_{cathode} - E_{anode} - IR \qquad (19\text{--}16)$$

EXAMPLE 19-7. The following cell has a resistance of 4.00 Ω. Calculate its potential when it is producing a current of 0.100 A.

$$Cd|Cd^{2+}(0.0100\ M)\|Cu^{2+}(0.0100\ M)|Cu$$

Substitution into the Nernst equation reveals that the electrode potential for the Cu electrode is 0.278 V, while for the Cd electrode it is −0.462 V. Thus, the thermodynamic cell potential is

$$
\begin{aligned}
E &= E_{Cu} - E_{Cd} \\
&= 0.278 - (-0.462) = 0.740\ V
\end{aligned}
$$

$$
\begin{aligned}
E_{cell} &= 0.740 - IR \\
&= 0.740 - (0.100 \times 4.00) = 0.340\ V
\end{aligned}
$$

Note that the emf of this cell drops dramatically in the presence of a current.

EXAMPLE 19-8. Calculate the potential required to generate a current of 0.100 A in the reverse direction in the cell shown in Example 19–7.

$$
\begin{aligned}
E &= E_{Cd} - E_{Cu} \\
&= -0.462 - 0.278 = -0.740\ V
\end{aligned}
$$

$$
\begin{aligned}
E_{cell} &= -0.740 - (0.100 \times 4.00) \\
&= -1.140\ V
\end{aligned}
$$

Here, an external potential greater than 1.140 V would be needed to cause Cd^{2+} to deposit and Cu to dissolve at a rate required for a current of 0.100 A.

19E-2 POLARIZATION

Several important electroanalytical methods are based upon *current-voltage curves,* which

---

[8] Here and in the subsequent discussion we will assume that the junction potential is negligible relative to the other potentials.

are obtained by measuring the variation in current in a cell as a function of its potential. Equation 19–16 predicts that at constant electrode potentials, a linear relationship should exist between the two variables. In fact, departures from linearity are often encountered; under these circumstances, the cell is said to be *polarized.* Polarization may arise at one or both electrodes.

As an introduction to this discussion, it is worthwhile considering current-voltage curves for an ideal polarized and an ideal nonpolarized electrode. Polarization at a single electrode can be studied by coupling it with an electrode that is not readily polarized. Such an electrode is characterized by being large in area and being based on a half-cell reaction that is rapid and reversible. Design details of nonpolarized electrodes will be found in subsequent chapters.

Ideal Polarized and Nonpolarized Electrodes and Cells. The ideal polarized electrode is one in which current remains constant and independent of potential over a considerable range. Figure 19–8**a** is a current-voltage curve for an electrode that behaves ideally in the region between *A* and *B*. Figure 19–8**b** depicts the current-voltage relationship for a depolarized electrode that behaves ideally in the region between *C* and *D*. Here the potential is independent of the current.

Figure 19–9 is a current-voltage curve for a *cell* having electrodes that exhibit ideal nonpolarized behavior between points *A* and *B*. Because of the internal resistance of the cell, the current-voltage curve has a finite slope equal to *R* (Equation 19–16) rather than the infinite slope for the ideal nonpolarized electrode shown in Figure 19–8**b**. Beyond points *A* and *B*, polarization occurs at one or both electrodes resulting in departures from the ideal straight line. The upper half of the curve gives the current-voltage relationship when the cell is operating as an electrolytic cell; the lower half de-

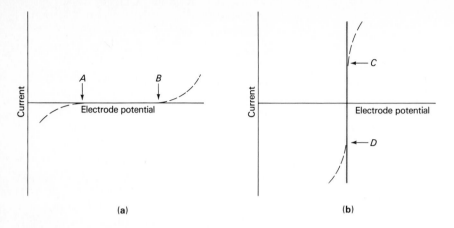

FIGURE 19-8   Current voltage curves for an ideal (**a**) polarized and (**b**) nonpolarized electrode. Dashed lines show departure from ideal behavior by real electrodes.

scribes its behavior as a galvanic cell.[9] Note that when polarization arises in an electrolytic cell, a higher potential is required to achieve a given current. Similarly, polarization of a galvanic cell produces a potential that is lower than expected.

---

[9] Here, the IUPAC current convention is being followed with cathodic currents being given a negative sign. Unfortunately, this convention was not followed in the early development of some techniques. Thus, for historical reasons we will not always follow the IUPAC current sign convention.

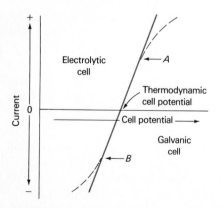

FIGURE 19-9   Current voltage curve for a cell showing ideal nonpolarized behavior between *A* and *B* and polarized behavior (dashed lines).

Sources of Polarization.   Figure 19–10 depicts three regions of a half-cell where polarization can occur. These include the electrode itself, a surface film of solution immediately adjacent to the electrode, and the bulk of the solution. For this half-cell, the overall electrode reaction is

$$Ox + ne \rightleftharpoons Red$$

Any one of the several intermediate steps shown in the figure may, however, limit the rate at which this overall reaction occurs and thus the magnitude of the current. One of these steps in the reaction, called *mass transfer,* involves movement of Ox from the bulk of the solution to the surface film. When this step (or the reverse mass transfer of Red to the bulk) limits the rate of the overall reaction and thus the current, *concentration polarization* is said to exist. Some half-cell reactions proceed by an intermediate chemical reaction in which a species such as Ox' or Red' form; this intermediate is then the actual participant in the electron transfer process. If the rate of formation or decomposition of such an intermediate limits the current, *reaction polarization* is said to be present. In some instances, the rate of a

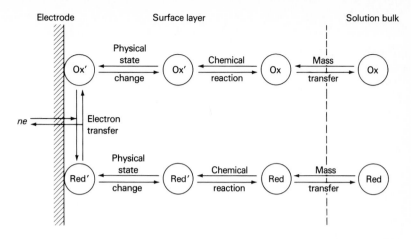

FIGURE 19-10 Steps in the reaction $Ox + ne \rightleftharpoons Red$ at an electrode. Note that the surface layer is only a few molecules thick. (Adapted from: A. J. Bard and L. R. Faulkner, *Electrochemical Methods*, p. 21, Wiley: New York, 1980. Reprinted by permission of John Wiley & Sons, Inc.)

physical process such as adsorption, desorption, or crystallization is current limiting. Here, *adsorption, desorption,* or *crystallization polarization* is occurring. Finally, *charge-transfer polarization* is encountered, where current limitation arises from the slow rate of electron transfer from the electrode to the oxidized species in the surface film or from the reduced species to the electrode. It is not unusual to encounter half-cells in which several types of polarization are occurring simultaneously.

Overvoltage. The degree of polarization of an electrode is measured by the *overvoltage* or *overpotential* η, which is the difference between the actual electrode potential E and the thermodynamic or equilibrium potential E_{eq}. That is,

$$\eta = E - E_{eq} \qquad (19\text{-}17)$$

where $E < E_{eq}$. It is important to realize that polarization always reduces the electrode potential for a system. Thus, as we have indicated, E is always smaller than E_{eq} and η *is always negative.*

Concentration Polarization. Concentration polarization arises when the rate of transport of reactive species to the electrode surface is insufficient to maintain the current demanded by Equation 19–16. With the onset of concentration polarization, a *diffusion overvoltage* develops.

For example, consider a cell made up of an ideal nonpolarized anode and a polarizable cathode consisting of a small cadmium electrode immersed in a solution of cadmium ions. The reduction of cadmium ions is a rapid and reversible process so that when a potential is applied to this electrode, the *surface layer* of the solution comes to equilibrium with the electrode *essentially instantaneously.* That is, a brief current is generated that reduces the surface concentration of cadmium ions to the equilibrium concentration, C_0, given by

$$E = E^0_{Cd} - \frac{0.0591}{2} \log \frac{1}{C_0} \qquad (19\text{-}18)$$

If no mechanism existed for transport of cadmium ions from the bulk of the solution to the surface film, the current would rapidly decrease to zero as the concentration of the film approached C_0. As we shall see, however, several mechanisms do indeed exist that bring cadmium ions from the bulk of the solution into the surface layer at a constant rate. As a consequence, the large initial current decreases rapidly to a constant level that is determined by the rate of ion transport.

It is important to appreciate that for a rapid and reversible electrode reaction, the concentration of the surface layer may always be considered to be the equilibrium concentration, which is determined by the instantaneous electrode potential (Equation 19–18). It is also important to realize that the surface concentration C_0 is often far different from that of the bulk of the solution. That is, while the surface equilibrium is essentially instantaneously achieved, attainment of equilibrium between the electrode and the bulk of the solution often requires minutes or even hours.

For a current of the magnitude required by Equation 19–16 to be maintained, it is necessary that reactant be brought from the bulk of the solution to the surface layer at a rate dC/dt that is given by

$$I = dQ/dt = nF dC/dt$$

where dQ/dt is the rate of flow of electrons in the electrode (or the current I), n is the number of electrons appearing in the half-reaction, and F is the faraday. The rate of concentration change can be written as

$$\frac{dC}{dt} = AJ$$

where A is the surface area of the electrode in square meters (m^2) and J is the concentration flux in mol sec^{-1} m^{-2}. The two equations can then be combined to give

$$I = nFAJ \qquad (19\text{–}19)$$

When this demand for reactant cannot be met by the mass transport process, the IR drop in Equation 19–16 becomes smaller than theoretical, and a diffusion overvoltage appears that just offsets the decrease in IR. That is, with the appearance of concentration polarization, Equation 19–16 becomes

$$E_{cell} = E_{cathode} - E_{anode} - IR + \eta_{cathode}$$

where $\eta_{cathode}$ represents the overpotential associated with the cathode. Note that in this example, it was assumed that the anode was an ideal nonpolarized one. A more general equation is

$$E_{cell} = E_{cathode} - E_{anode} + \eta_{cathode}$$
$$+ \eta_{anode} - IR \qquad (19\text{–}20)$$

where η_{anode} is the anodic overvoltage. Note that the overvoltage associated with each electrode always carries a negative sign and has the effect of reducing the overall potential of the cell.

Mechanisms of Mass Transport. It is important now to investigate the mechanisms by which ions or molecules are transported from the bulk of the solution to a surface layer (or the reverse) because a knowledge of these mechanisms provides insights into how concentration polarization can be prevented or induced as required.

Three forces can be discerned that cause ions or molecules to move toward an electrode surface layer: (1) diffusion, (2) electrostatic attraction, and (3) mechanical or convective forces.

Whenever a concentration gradient develops in a solution, molecules or ions diffuse from the more concentrated to the more dilute region. The rate at which transfer occurs is proportional to the concentration difference. In an electrolysis, a gradient is established as a result of ions being removed from the film of solution adjacent to the cathode. Diffusion then occurs, the rate dC/dt being expressed by the relationship

$$dC/dt = k(C - C_0) \qquad (19-21)$$

where C is the reactant concentration in the bulk of the solution, C_0 is its equilibrium concentration at the electrode surface, and k is a proportionality constant. As shown earlier, *the value of C_0 is fixed by the potential of the electrode and can be calculated from the Nernst equation.* As higher potentials are applied to the electrode, C_0 becomes smaller and smaller and the diffusion rate greater and greater. Ultimately, however, C_0 will become negligible with respect to C; the rate of diffusion then becomes constant. That is, when $C_0 \to 0$

$$dC/dt = kC \qquad (19-22)$$

Under this circumstance, concentration polarization is said to be complete, and the electrode operates as an ideal polarized electrode.

Electrostatic forces also influence the rate at which an ionic reactant migrates to or from an electrode surface. The electrostatic attraction (or repulsion) between a particular ionic species and the electrode becomes smaller as the total electrolyte concentration of the solution is increased. It may approach zero when the reactive species is but a small fraction of the total concentration of ions with a given charge.

Clearly, reactants can be transported to an electrode by mechanical means. Thus, stirring or agitation will aid in decreasing concentration polarization. Convection currents, due to temperature or density differences, also tend to offset concentration polarization.

To summarize, then, concentration polarization is observed when the forces of diffusion, electrostatic attraction, and mechanical mixing are insufficient to transport the reactant to or from an electrode surface at a rate demanded by the theoretical current. Concentration polarization causes the potential of a galvanic cell to be smaller than the value predicted on the basis of the theoretical potential and the *IR* drop. Similarly, in an electrolytic cell, a potential more negative than theoretical is required in order to maintain a given current.

Concentration polarization is important in several electroanalytical methods. In some applications, steps are taken to eliminate it; in others, however, it is essential to the method, and every effort is made to promote its occurrence. The degree of concentration polarization is influenced experimentally by: (1) the reactant concentration; (2) the total electrolyte concentration; (3) mechanical agitation; and (4) the size of the electrodes; as the area toward which a reactant is transported becomes greater, polarization effects become smaller (Equation 19-19).

Other Types of Polarization. Charge transfer polarization and physical polarization, arising from adsorption, desorption, or crystallization, often create problems in electroanalytical measurements and will be discussed here. Reaction polarization will be considered briefly in Chapter 22.

Although exceptions can be cited, some empirical generalizations can be made regarding the magnitude of charge-transfer and physical overvoltages.

1. Overvoltage increases with current density (current density is defined as the amperes per square centimeter of electrode surface).
2. Overvoltage usually decreases with increases in temperature.
3. Overvoltage varies with the chemical composition of the electrode, often being most pronounced with softer metals such as tin, lead, zinc, and particularly mercury.
4. Overvoltage is most marked for electrode processes that yield gaseous products such as hydrogen or oxygen; it is frequently negligible where a metal is being deposited or where an ion is undergoing a change of oxidation state.

5. The magnitude of overvoltage in any given situation cannot be predicted exactly, because it is determined by a number of uncontrollable variables.[10]

Overvoltage associated with the evolution of hydrogen and oxygen is of particular interest to the chemist. Table 19–3 presents data that depict the extent of the phenomenon under specific conditions. The difference between the overvoltage of these gases on smooth and on platinized platinum surfaces is notable. This difference is primarily due to the much larger surface area associated with platinized electrodes, which results in a *real* current density that is significantly smaller than is apparent from the overall dimensions of the electrode. A plati-

nized surface is always employed in construction of hydrogen reference electrodes in order to lower the current density to a point where the overvoltage is negligible.

The high overvoltage associated with the formation of hydrogen permits the electrolytic deposition of several metals that require potentials at which hydrogen would otherwise be expected to interfere. For example, it is readily shown from their standard potentials that rapid formation of hydrogen should occur well below the potential required for the deposition of zinc from a neutral solution. Nevertheless, quantitative deposition of zinc can be attained provided a mercury or copper electrode is used; because of the high overvoltage of hydrogen on these metals, little or no gas is evolved during the electrodeposition.

The magnitude of overvoltage can, at best, be only crudely approximated from empirical information available in the litera-

---

[10] Overvoltage data for various gaseous species at different electrode surfaces are to be found in: *Handbook of Analytical Chemistry*, L. Meites, Ed., p. 5–184, McGraw-Hill: New York, 1963.

Table 19–3
Overvoltage for Hydrogen and Oxygen Formation at Various Electrodes at 25°C[a]

Electrode Composition	Overvoltage (V) (Current Density 0.001 A/cm^2)		Overvoltage (V) (Current Density 0.01 A/cm^2)		Overvoltage (V) (Current Density 1 A/cm^2)	
	H$_2$	O$_2$	H$_2$	O$_2$	H$_2$	O$_2$
Smooth Pt	−0.024	−0.721	−0.068	−0.85	−0.676	−1.49
Platinized Pt	−0.015	−0.348	−0.030	−0.521	−0.048	−0.76
Au	−0.241	−0.673	−0.391	−0.963	−0.798	−1.63
Cu	−0.479	−0.422	−0.584	−0.580	−1.269	−0.793
Ni	−0.563	−0.353	−0.747	−0.519	−1.241	−0.853
Hg	−0.9[b]		−1.1[c]		−1.1[d]	
Zn	−0.716		−0.746		−1.229	
Sn	−0.856		−1.077		−1.231	
Pb	−0.52		−1.090		−1.262	
Bi	−0.78		−1.05		−1.23	

[a] National Academy of Sciences, *International Critical Tables*, vol. 6, pp. 339–340, McGraw-Hill: New York, 1929.
[b] −0.556 V at 0.000077 A/cm^2; −0.929 V at 0.00154 A/cm^2.
[c] −1.063 V at 0.00769 A/cm^2.
[d] −1.126 V at 1.153 A/cm^2.

ture. Calculation of cell potentials in which overvoltage plays a part cannot, therefore, be very accurate. As with diffusion overvoltage, charge-transfer and physical overvolt-ages always carry a negative sign and have the effect of making the observed potential smaller than the thermodynamic electrode potential.

PROBLEMS

19–1. Calculate the electrode potentials of the following half-cells.
  (a) $Ag^+(0.0152\ M)|Ag$
  (b) $Fe^{3+}(2.35 \times 10^{-4}\ M),Fe^{2+}(0.200\ M)|Pt$
  (c) $AgBr(sat'd),Br^-(0.100\ M)|Ag$

19–2. Calculate the electrode potentials of the following half-cells.
  (a) $HCl(3.50\ M)|H_2(0.950\ atm),Pt$
  (b) $IO_3^-(0.235\ M),I_2(5.00 \times 10^{-4}\ M),H^+(2.00 \times 10^{-3}\ M)|Pt$
  (c) $Ag_2CrO_4(sat'd),CrO_4^{2-}(0.0250\ M)|Ag$

19–3. For each of the following half-cells, compare electrode potentials derived from (1) concentration and (2) activity data.
  (a) $HCl(0.0100\ M),NaCl(0.0400\ M)|H_2(1.00\ atm),Pt$
  (b) $Fe(ClO_4)_2(0.0111\ M),Fe(ClO_4)_3(0.0111\ M)|Pt$

19–4. For each of the following half-cells, compare electrode potentials derived from (1) concentration and (2) activity data.
  (a) $Sn(ClO_4)_2(2.00 \times 10^{-5}\ M),Sn(ClO_4)_4(1.00 \times 10^{-5}\ M)|Pt$
  (b) $Sn(ClO_4)_2(2.00 \times 10^{-5}\ M),Sn(ClO_4)_4(1.00 \times 10^{-5}\ M),$
    $NaClO_4(0.0500\ M)|Pt$

19–5. Calculate the potential of a silver electrode in contact with the following:
  (a) a solution that is $0.0200\ M$ in I^- and saturated with AgI.
  (b) a solution that is $0.0060\ M$ in CN^- and $0.0400\ M$ in $Ag(CN)_2^-$.
  (c) the solution that results from mixing 25.0 mL of $0.0400\ M$ KBr with 20.0 mL of $0.200\ M\ Ag^+$.
  (d) the solution that results from mixing 25.0 mL of $0.0400\ M\ Ag^+$ with 20.0 mL of $0.200\ M$ KBr.

19–6. Calculate the electrode potentials for the following systems.
  (a) $Cr_2O_7^{2-}(5.00 \times 10^{-3}\ M),Cr^{3+}(1.00 \times 10^{-2}\ M),H^+(0.200\ M)|Pt$
  (b) $UO_2^+(0.100\ M),U^{4+}(0.200\ M),H^+(0.600\ M)|Pt$

19–7. Calculate the theoretical potential of each of the following cells. Is the cell as written galvanic or electrolytic?
  (a) $Pt|Cr^{3+}(2.00 \times 10^{-4}\ M),Cr^{2+}(1.00 \times 10^{-3}\ M)||Pb^{2+}(6.50 \times 10^{-2}\ M)|Pb$
  (b) $Hg|Hg_2^{2+}(4.00 \times 10^{-2}\ M)||H^+(3.00 \times 10^{-2}\ M),V^{3+}(2.00 \times 10^{-2}\ M),VO^{2+}(6.00 \times 10^{-3}\ M)|Pt$
  (c) $Pt|Fe^{3+}(2.00 \times 10^{-2}\ M),Fe^{2+}(6.00 \times 10^{-5}\ M)||Sn^{2+}(3.50 \times 10^{-2}\ M),Sn^{4+}(1.50 \times 10^{-4}\ M)|Pt$

19–8. Calculate the theoretical potential of each of the following cells. Is the cell galvanic or electrolytic as written?
  (a) $Bi|BiO^+(0.0400\ M),H^+(0.200\ M)||I^-(0.100\ M),AgI(sat'd)|Ag$

(b) $Zn|Zn^{2+}(7.50 \times 10^{-4} M)\|Fe(CN)_6^{4-}(4.50 \times 10^{-2} M),$
$$Fe(CN)_6^{3-}(7.00 \times 10^{-2} M)|Pt$$
(c) $Pt,H_2(0.100 \text{ atm})|HCl(4.50 \times 10^{-4} M),AgCl(\text{sat'd})|Ag$

19–9. Compute E^0 for the process

$$Ni(CN)_4^- + 2e \rightleftharpoons Ni(s) + 4CN^-$$

given that the formation constant for the complex is 1.0×10^{22}.

19–10. The solubility product constant for PbI_2 is 7.1×10^{-9} at 25°C. Calculate E^0 for the process

$$PbI_2(s) + 2e \rightarrow Pb(s) + 2I^-$$

19–11. Calculate the standard potential for the half-reaction

$$BiOCl(s) + 3e \rightleftharpoons Bi(s) + 3OCl^-$$

given that K_{sp} for BiOCl has a value of 8.1×10^{-19}.

19–12. Calculate the standard potential for the half-reaction

$$Al(C_2O_4)_2^- + 3e \rightarrow Al(s) + 2C_2O_4^{2-}$$

if the formation constant for the complex is 1.3×10^{13}.

19–13. From the standard potentials

$$Tl^+ + e \rightleftharpoons Tl(s) \qquad E^0 = -0.336 \text{ V}$$
$$TlCl(s) + e \rightleftharpoons Tl(s) + Cl^- \qquad E^0 = -0.557 \text{ V}$$

calculate the solubility product constant for TlCl.

19–14. From the standard potentials

$$Ag_2SeO_4(s) + 2e \rightleftharpoons 2Ag(s) + SeO_4^{2-} \qquad E^0 = 0.355 \text{ V}$$
$$Ag^+ + e \rightleftharpoons Ag(s) \qquad E^0 = 0.799 \text{ V}$$

calculate the solubility product constant for Ag_2SeO_4.

19–15. A current of 0.0750 A is to be drawn from the cell

$$Pt|V^{3+}(1.0 \times 10^{-5} M),$$
$$V^{2+}(1.65 \times 10^{-1} M)\|Br^-(0.150 M),AgBr(\text{sat'd})|Ag$$

As a consequence of its design, the cell has an internal resistance of 6.74 Ω. Calculate the potential to be expected initially.

19–16. The cell

$$Pt|V(OH)_4^+(1.04 \times 10^{-4} M),VO^{2+}(7.15 \times 10^{-2} M),$$
$$H^+(2.75 \times 10^{-3} M)\|Cu^{2+}(5.00 \times 10^{-2} M)|Cu$$

has an internal resistance of 2.24 Ω. What will be the initial potential if 0.0300 A is drawn from this cell?

19–17. The resistance of the galvanic cell

$$\text{Pt}|\text{Fe(CN)}_6^{4-}(3.60 \times 10^{-2}\ M),$$
$$\text{Fe(CN)}_6^{3-}(2.70 \times 10^{-3}\ M)\|\text{Ag}^+(1.65 \times 10^{-2}\ M)|\text{Ag}$$

is 4.10 Ω. Calculate the initial potential when 0.0106 A is drawn from this cell.

Potentiometric Methods

Since the beginning of this century, potentiometric measurements have been used for the location of end points in titrimetric methods of analysis. Of more recent origin are methods in which ion concentrations are obtained directly from the potential of an ion selective electrode. Such electrodes are relatively free from interference and provide a rapid and convenient means for quantitative estimations of numerous important anions and cations.

The equipment required for potentiometric methods include a *reference electrode, an indicator electrode,* and *a potential measuring device.* The design and properties of each of these components are described in the initial sections of this chapter. Following these discussions, consideration is given to the analytical applications of potentiometric measurements.

20A Reference Electrodes

In many electroanalytical applications, it is desirable that the half-cell potential of one electrode be known, constant, and completely insensitive to the composition of the solution under study. An electrode that fits this description is called a *reference electrode.*[1] Employed in conjunction with the reference electrode will be an *indicator* or *working electrode,* whose response depends upon the analyte concentration.

The ideal reference electrode (1) should be reversible and obey the Nernst equation; (2) should exhibit a potential that is constant with time; (3) should return to its original potential after being subjected to small currents; (4) should exhibit little hysteresis with temperature cycling; and (5) should behave as an ideal nonpolarized electrode (see Figure 19–8b, p. 592). Although no reference electrode completely meets these ideals, several come surprisingly close.

---

[1] For a detailed discussion of reference electrodes, see: D. J. G. Ives and G. J. Janz, *Reference Electrodes,* Academic Press: New York, 1961. For descriptions of typical commercially available reference electrodes, see: R. D. Caton, Jr., *J. Chem. Educ.,* **1973,** *50,* A571; **1974,** *51,* A7; and *The Beckman Handbook of Applied Electrochemistry, 2d ed., Bulletin 7707A,* Beckman Instruments, Inc.: Irvine, CA, 1982.

20A-1 CALOMEL ELECTRODES

Calomel half-cells may be represented as follows:

$$\|Hg_2Cl_2(\text{sat'd}),KCl(xM)|Hg$$

where x represents the molar concentration of potassium chloride in the solution. The electrode reaction is given by the equation

$$Hg_2Cl_2(s) + 2e \rightleftharpoons 2Hg(l) + 2Cl^-$$

The potential of this cell will vary with the chloride concentration x, and this quantity must be specified in describing the electrode.

Table 20–1 lists the composition and the potentials for three commonly encountered calomel electrodes. Note that each solution is saturated with mercury(I) chloride (calomel) and that the cells differ only with respect to the potassium chloride concentration.

The saturated calomel electrode (SCE) is most commonly used by the analytical chemist because of the ease with which it can be prepared. Compared with the other calomel electrodes, however, its temperature coefficient is significantly larger (see Table 20–1). The potential of the saturated calomel electrode at 25°C is 0.244 V.

Several convenient calomel electrodes are available commercially; typical are the two illustrated in Figure 20–1. The body of each electrode consists of an outer glass or plastic tube that is 5 to 15 cm in length and 0.5 to 1.0 cm in diameter. A mercury/mercury(I) chloride paste is contained in an inner tube, which is connected to the saturated potassium chloride solution in the outer tube through a small opening. For electrode (**a**), contact with the second half-cell is made by means of a fritted porcelain plug or a porous asbestos or quartz fiber sealed in the end of the outer tubing. This type of junction has a relatively high resistance (2000 to 3000 Ω) and a limited current-carrying ca-

---

Table 20–1
Potentials of Reference Electrodes in Aqueous Solutions

Temperature, °C	Electrode Potential (V), vs. SHE				
	0.1 Mc Calomel[a]	3.5 Mc Calomel[b]	Saturated[c] Calomel[a]	3.5 $M^{b,c}$ Ag/AgCl	Saturated[b,c] Ag/AgCl
10		0.256		0.215	0.214
12	0.3362		0.2528		
15	0.3362	0.254	0.2511	0.212	0.209
20	0.3359	0.252	0.2479	0.208	0.204
25	0.3356	0.250	0.2444	0.205	0.199
30	0.3351	0.248	0.2411	0.201	0.194
35	0.3344	0.246	0.2376	0.197	0.189
38	0.3338		0.2355		
40		0.244		0.193	0.184

[a] Data from: R. G. Bates, in *Treatise on Analytical Chemistry*, 2d ed., I. M. Kolthoff and P. J. Elving, Eds., Part I, vol 1, p. 793, Wiley: New York, 1978. Reprinted by permission of John Wiley & Sons, Inc.
[b] Data from: D. T. Sawyer and J. L. Roberts, Jr., *Experimental Electrochemistry for Chemists*, p. 42, Wiley: New York, 1974. Reprinted by permission of John Wiley & Sons, Inc.
[c] "M" and "saturated" refer to the concentration of KCl and *not* Hg_2Cl_2.

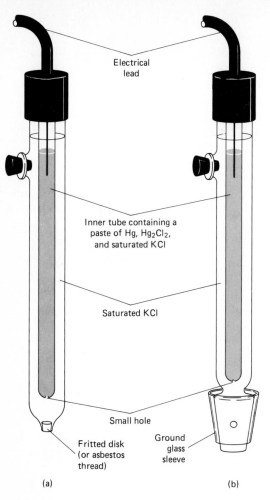

FIGURE 20-1 Typical commercial calomel reference electrodes.

pacity; on the other hand, contamination of the analyte solution due to leakage is minimal. The electrode shown in Figure 20–1**b** has a much lower resistance but tends to leak small amounts of saturated potassium chloride into the sample. Before it is used, the ground glass collar of this electrode is loosened and turned so that a drop or two of the KCl solution flows from the hole and wets the entire inner ground surface. Better electrical contact to the analyte solution is thus established.

20A-2 SILVER/SILVER CHLORIDE ELECTRODES

A reference electrode system analogous to the calomel electrode consists of a silver electrode immersed in a solution of potassium chloride that has been saturated with silver chloride

$$\|AgCl(sat'd),KCl(xM)|Ag$$

The half-reaction is

$$AgCl(s) + e \rightleftharpoons Ag(s) + Cl^-$$

Normally, this electrode is prepared with a saturated or a 3.5 M potassium chloride solution; potentials for these electrodes are given in Table 20–1.

Silver/silver chloride electrodes have the advantage that they can be used at temperatures greater than 60°C, whereas calomel electrodes cannot. On the other hand, mercurous ion reacts with fewer sample components than does silver ion (which can react with proteins for example); such reactions can lead to plugging of the junction between the electrode and the analyte solution.

20A-3 THALLIUM/THALLOUS CHLORIDE ELECTRODE

One manufacturer markets a reference electrode under the name of Thalamid®. This electrode consists of

$$\|TlCl(sat'd),KCl(sat'd)|Tl(40\% \text{ amalgam})$$

It is reported to attain equilibrium potentials much more rapidly after temperature change than calomel or silver/silver chloride electrodes.

20A-4 PRECAUTIONS IN THE USE OF REFERENCE ELECTRODES

In using reference electrodes, such as those shown in Figure 20–1, the level of the internal liquid should always be kept above that of the solution in which it is immersed to prevent contamination of the electrode solution or plugging of the junction due to reaction of analyte ions with the mercury(I) or silver ions. Junction plugging is perhaps the most common source of erratic cell behavior in potentiometric measurements.[2]

With the electrode liquid level above the analyte solutions, some contamination of the sample inevitably occurs. In many instances, the amount of contaminant is so slight as to be of no concern. In determining ions such as chloride, potassium, silver, or mercury, however, precaution must often be taken to avoid this source of error. A common way is to interpose a second salt bridge between the analyte and reference electrode; this bridge should contain a solution having a high concentration of a noninterfering electrolyte, such as potassium nitrate or sodium sulfate. Double junction electrodes based on this design are offered by several instrument makers.

20B Metallic Indicator Electrodes

For purposes of this discussion, it is convenient to divide indicator electrodes into two categories, *metallic* and *membrane*. Metallic electrodes can be further classified as *electrodes of the first kind, electrodes of the second kind, electrodes of the third kind,* and *redox electrodes.*

20B-1 ELECTRODES OF THE FIRST KIND

Electrodes of the first kind are in direct equilibrium with the cation derived from the electrode metal. Here, a single reaction is involved. For example, for a copper elec-

trode, we may write

$$Cu^{2+} + 2e \rightleftharpoons Cu(s)$$

and

$$E = E_{Cu}^0 - \frac{0.0591}{2} \log \frac{1}{[Cu^{2+}]}$$
$$= E_{Cu}^0 - \frac{0.0591}{2} pCu$$

where pCu is the negative logarithm of the copper (II) ion concentration.[3] Thus, the copper electrode provides a direct measure of the pCu of the solution. Other common metals that behave reversibly include silver, mercury, copper, cadmium, zinc, and lead. In contrast, many other metals do not exhibit reversible oxidation-reduction behavior and cannot, as a consequence, serve as satisfactory indicator electrodes for their ions. With such electrodes, the measured potential includes an overpotential that is influenced by a variety of factors including strains, crystal deformations, surface area, and the presence of oxide coatings. Examples of metals behaving irreversibly include

---

[2] For a useful discussion about the care and maintenance of reference electrodes, see: J. E. Fisher, *Amer. Lab.,* **1984,** *16* (6), 54.

[3] The results of potentiometric measurements are usually expressed in terms of a parameter, the *p function,* which is directly proportional to the measured potential. The p function then provides a measure of concentration in terms of a convenient, small, and ordinarily positive, number. Thus, for a solution with a calcium ion concentration of $2.00 \times 10^{-6} M$, we may write

$$pCa = -\log(2.00 \times 10^{-6}) = 5.699$$

Note that as the concentration of calcium increases, its p function decreases. Note also that because the concentration was given to three significant figures, we are entitled to keep this number of figures *to the right of the decimal point* in the computed pCa because these are the only numbers that carry information about the original 2.00. The 5 in the value for pCa provides information about the position of the decimal point in the original number only.

iron, tungsten, nickel, cobalt, and chromium.

20B-2   ELECTRODES OF THE SECOND KIND

A metal electrode can often be made responsive to the concentration of an anion with which its ion forms a precipitate or a stable complex ion. For example, silver can serve as an electrode of the second kind for halide and halide-like anions. Thus, to determine chloride ion, it is only necessary to saturate the surface layer of the analyte solution with silver chloride. The electrode reaction can then be written as

$$AgCl(s) + e \rightleftharpoons Ag(s) + Cl^-   \quad E^0 = 0.222 \text{ V}$$

Application of the Nernst equation gives

$$E = 0.222 - 0.0591 \log [Cl^-]$$
$$= 0.222 + 0.0591 \, pCl$$

A convenient way of preparing a chloride-sensitive electrode is to make a pure silver wire the anode in an electrolytic cell containing potassium chloride. The wire becomes coated with an adherent silver halide deposit, which will rapidly equilibrate with the surface layer of a solution in which it is immersed. Because the solubility of the silver chloride is low, an electrode formed in this way can be used for numerous measurements.

An important electrode of the second kind for measuring the concentration of the EDTA anion Y^{4-} (p. 635) is based upon the response of a mercury electrode in the presence of a small concentration of the stable EDTA complex of Hg(II). The half-reaction for the electrode process can be written as

$$HgY^{2-} + 2e \rightleftharpoons Hg(l) + Y^{4-}   \quad E^0 = 0.21 \text{ V}$$

for which

$$E = 0.21 - \frac{0.0591}{2} \log \frac{[Y^{4-}]}{[HgY^{2-}]}$$

To employ this electrode system, it is necessary to introduce a small concentration of HgY^{2-} into the analyte solution at the outset. The complex is so stable (for HgY^{2-}, $K_f = 6.3 \times 10^{21}$) that its concentration remains essentially constant over a wide range of Y^{4-} concentrations. Therefore, the potential equation can be written in the form

$$E = K - \frac{0.0591}{2} \log [Y^{4-}] = K + \frac{0.0591}{2} \, pY$$

where the constant K is equal to

$$K = 0.21 - \frac{0.0591}{2} \log \frac{1}{[HgY^{2-}]}$$

This electrode is useful for establishing end points for EDTA titrations.

20B-3   ELECTRODES OF THE THIRD KIND

A metal electrode can, under some circumstances, be made to respond to a different cation. It then becomes an electrode of the third kind. As an example, a mercury electrode has been used for the determination of the pCa of calcium-containing solutions. As in the previous example, a small concentration of the EDTA complex of Hg(II) is introduced into the solution. As before, the potential of a mercury electrode in this solution is given by

$$E = K - \frac{0.0591}{2} \log [Y^{4-}]$$

If, in addition, a small volume of a solution containing the EDTA complex of calcium is introduced, a new equilibrium is established, namely

$$CaY^{2-} \rightleftharpoons Ca^{2+} + Y^{4-}   \quad K_f = \frac{[Ca^{2+}][Y^{4-}]}{[CaY^{2-}]}$$

Combining the formation constant expression for CaY^{2-} with the potential expression yields

$$E = K - \frac{0.0591}{2} \log \frac{K_f[CaY^{2-}]}{[Ca^{2+}]}$$

which can be written as

$$E = K - \frac{0.0591}{2} \log K_f[CaY^{2-}]$$
$$- \frac{0.0591}{2} \log \frac{1}{[Ca^{2+}]}$$

If a constant concentration of CaY^{2-} is used in the analyte solution and in the solutions for standardization, we may write

$$E = K' - \frac{0.0591}{2} \text{ pCa}$$

where

$$K' = K - \frac{0.0591}{2} \log K_f[CaY^{2-}]$$

Thus, the mercury electrode has become an electrode of the third kind for calcium ion.

20B-4   METALLIC REDOX INDICATORS

Electrodes fashioned from platinum, gold, palladium, or other inert metals often serve as indicator electrodes for oxidation-reduction systems. For example, the potential of a platinum electrode in a solution containing Ce(III) and Ce(IV) ions is given by

$$E = E^0 - 0.0591 \log \frac{[Ce^{3+}]}{[Ce^{4+}]}$$

Thus, a platinum electrode can serve as the indicator electrode in a titration in which Ce(IV) serves as the standard reagent.

It should be noted, however, that electron transfer processes at inert electrodes are frequently not reversible. As a consequence, inert electrodes do not respond in a predictable way to many of the half-reactions found in a table of electrode potentials. For example, a platinum electrode immersed in a solution of thiosulfate and tetrathionate ions does not develop reproducible potentials because the electron transfer process

$$S_4O_6^{2-} + 2e \rightleftharpoons 2S_2O_3^{2-}$$

is slow and therefore not reversible at the electrode surface.

20C   Membrane Indicator Electrodes[4]

Since the early 1930s, the most convenient way for determining pH has been by measuring the potential difference across a glass membrane separating the analyte solution from a reference solution of fixed acidity. The phenomenon upon which this measurement is based was first recognized by Cremer[5] in 1906 and systematically explored by Haber[6] a few years later. General use of the glass electrode for pH measurements, however, was delayed for two decades until the invention of the vacuum tube permitted the convenient measurement of potentials across glass membranes having resistances of 100 MΩ or more. Systematic studies of the pH sensitivity of glass membranes led ultimately in the late 1960s to the

[4] For further information on this topic, see: *Ion-Selective Methodology*, 2 vol., A. K. Covington, Ed., CRC Press: Boca Raton, Florida, 1979; *Ion Selective Electrodes in Analytical Chemistry*, H. Freiser, Ed., Plenum Press: New York, 1978; T. S. Ma and S. S. M. Hassan, *Organic Analysis Using Ion-Selective Electrodes*, 2 vol., Academic Press: New York, 1982; and *Ion-Selective Electrodes*, R. A. Durst, Ed., National Bureau of Standards Special Publication 314, U.S. Government Printing Office: Washington, D.C., 1969; *Handbook of Electrode Technology*, Orion Research: Cambridge, MA, 1982.

[5] M. Cremer, *Z. Biol.*, **1906**, *47*, 562.

[6] F. Haber and Z. Klemensiewicz, *Z. Phys. Chem.*, **1909**, *67*, 385.

development and marketing of membrane electrodes for two dozen or more ions such as K^+, Na^+, Ca^{2+}, F^-, and NO_3^-.

20C-1 CLASSIFICATION OF MEMBRANE ELECTRODES

As shown in Figure 20–2, membrane electrodes fall into two main categories, those that are responsive to ionic species and those that are applied to the determination of molecular analytes. The former are often called *ion-selective* or *p-ion electrodes*. The term p-ion is derived from the way the data from these electrodes are usually reported—that is, as p-functions such as pH, pCa, or pNO_3 (see footnote 3).

A fundamentally different type of membrane electrode has been developed for the determination of certain molecular species such as carbon dioxide, hydrogen cyanide, glucose, and urea. As shown in Figure 20–2,

two types of electrodes of this class are encountered, gas-sensing probes and enzymatic electrodes. The first sections that follow deal with the several types of ion-selective electrodes listed in Figure 20–2. Following this, gas-sensing and enzyme-substrate electrodes for the determination of molecular species are described.

20C-2 PROPERTIES OF ION-SELECTIVE MEMBRANES

All of the ion-selective membranes shown in Figure 20–2I share common properties, which lead to their sensitivity and selectivity toward certain cations or anions. These properties include:

1. *Minimal solubility.* A necessary property of an ion-selective medium is that its solubility in analyte solutions (usually aqueous) approaches zero. Thus, many membranes are formed from large molecules or molecular aggregates such as silica glasses or polymeric resins. Ionic inorganic compounds of low solubility such as the silver halides can also be converted into membranes.

2. *Electrical conductivity.* A membrane must exhibit some electrical conductivity albeit small. Generally, this conduction takes the form of migration of ions.

3. *Selective reactivity with the analyte.* A membrane or some species contained within the membrane matrix must be capable of selectively binding the analyte ion. Three types of binding are encountered: ion-exchange, crystallization, and complexation. The former two are the most common, and the attention here will be largely focused on these types of bindings.

I. ION-SELECTIVE ELECTRODES
  A. Crystalline Membranes
    1. Single crystal
        Example: LaF_3 for F^-
    2. Polycrystalline or mixed crystal
        Example: Ag_2S for S^{2-} and Ag^+
  B. Noncrystalline Membranes
    1. Glasses
        Examples: silicate glasses for Na^+ and H^+
    2. Liquids
        Examples: liquid ion-exchangers for Ca^{2+} and neutral carriers for K^+
    3. Liquids in B-2 immobilized in a rigid polymer
        Examples: polyvinyl chloride matrix for Ca^{2+} and NO_3^-

II. MOLECULAR-SELECTIVE ELECTRODES
  A. Gas-sensing probes
      Examples: hydrophobic membrane for CO_2 and NH_3
  B. Enzyme substrate electrodes
      Example: urease membrane for blood urea

FIGURE 20–2 Classification of membrane electrodes.

20C-3 ION-SELECTIVE ELECTRODES; PRINCIPLE AND DESIGN

At the outset, the reader should understand that membrane electrodes are *fundamentally different* from metal electrodes both in design and in principle. These differences are pointed out in the paragraphs that follow.

Cell Design. Figure 20–3**a** is a schematic diagram of a membrane cell for the determination of A^{n+}. It consists of a reference electrode (reference electrode 1) and a membrane electrode, both of which are immersed in a test solution in which the activity of A^{n+} is a_1. The membrane electrode consists of an active membrane sealed to one end of a glass or plastic tube. The tube holds a standard solution (the *internal* solution) of A^{n+} with an activity of a_2. Immersed in this

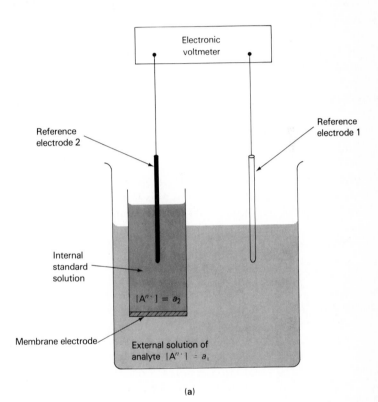

(a)

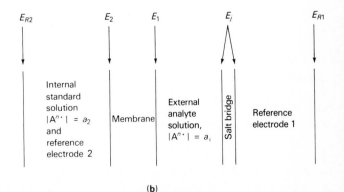

(b)

FIGURE 20-3 (**a**) Schematic diagram of a cell for selective-ion measurement of cation A^{n+}. The membrane electrode consists of the tube holding the membrane, the internal standard solution, and reference electrode 2. (**b**) Diagram of the cell in (**a**) showing sources of potential.

standard solution is reference electrode 2, whose potential is often different from reference electrode 1. The two reference electrodes are connected to an electronic voltmeter having a high internal resistance. Note that the membrane electrode is made up of three components: an ion-selective membrane, a reference electrode, and an internal solution. The last serves a dual function. First, it bathes the internal surface of the membrane with a solution containing a fixed concentration of analyte ions. Second, it serves as part of a reference electrode. For example, the internal solution of a calcium ion electrode is usually a standard solution of calcium chloride that is also saturated with silver chloride. When a silver wire is immersed in the solution, a silver/silver chloride reference electrode is formed. In addition, the calcium ions in the internal solution expose the inner membrane to a constant concentration of the analyte ion.

Electrical Conduction. It is worthwhile noting the different mechanisms of conduction in certain parts of a membrane cell and a cell containing a metal indicator electrode. In both, conduction in the aqueous solutions involves migration of anions and cations. As we have shown in Section 19A–2, the current across the solid-liquid interface of the metal electrode occurs by an oxidation-reduction process. That is, for an electrode fashioned from the metal A,

$$A^{n+} + ne \rightleftharpoons A(s) \qquad (20\text{--}1)$$

In a membrane cell, oxidation or reduction processes are entirely *absent* at the two solid-solution interfaces. Instead, conduction takes place by ion transfer, which is made possible by the ionic nature of the membrane or a species contained in the membrane. For example, one type of membrane is an ion exchanger, which is a substance of limited solubility or miscibility that contains numerous ionic sites capable of interacting with charged species in a solution in contact with the exchanger. A common ion exchanger for ion-selective membranes is a silicate glass, which consists of a three-dimensional infinite network of oxygen atoms held together by silicon to oxygen bonds (see Figure 20–4). Each silicon is bonded to four atoms of oxygen and each oxygen is shared by two SiO_4^{4-} groups. The open regions shown in the structure are occupied by sufficient cations to just neutralize the negative charge of the silicon/oxygen network. Some of these cations are multiply charged (Ca^{2+} or Al^{3+}) and are immobile. Singly charged ions such as Na^+, Li^+, or H^+, on the other hand, exhibit enough mobility in the structure to provide a means by which electricity can be carried through the glass.

Conduction across the two glass-solution interfaces involves transfer of mobile singly

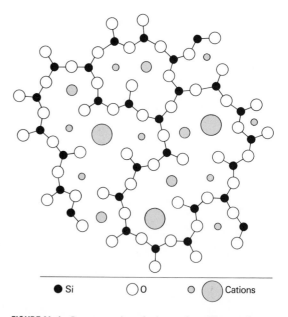

| ● Si | ○ O | ○ ⬤ Cations |

FIGURE 20–4 Cross-sectional view of a silicate glass structure. Note that each silicon is bonded to an additional oxygen atom, either above or below the plane of the paper. (Adapted with permission from: G. A. Perley, *Anal. Chem.*, **1949**, *21*, 395. Copyright 1949 American Chemical Society.)

charged cations, say protons, from the glass to the solution at one interface and from solution to the glass at the other. That is, for a membrane surface in which the singly charged cation sites are occupied largely by protons, the current involves

$$\underset{\substack{\text{membrane} \\ \text{surface 1}}}{H^+Gl^-} \rightleftharpoons \underset{\substack{\text{membrane} \\ \text{surface 1}}}{Gl^-} + \underset{\substack{\text{solution}}}{H^+} \qquad (20\text{–}2)$$

and

$$\underset{\substack{\text{solution}}}{H^+} + \underset{\substack{\text{membrane} \\ \text{surface 2}}}{Gl^-} \rightleftharpoons \underset{\substack{\text{membrane} \\ \text{surface 2}}}{H^+Gl^-} \qquad (20\text{–}3)$$

where Gl^- represents *one* of the many sites on a glass surface that is capable of binding a proton (or other singly charged cations). In the absence of current, the positions of these two equilibria are determined by the relative hydrogen ion concentrations in the internal and external solutions as governed by the mass law. When these positions differ, the surface at which the greater dissociation has occurred will be negative with respect to the other surface. Thus, a potential develops whose magnitude depends upon the *difference* in hydrogen ion concentration on the two sides of the membrane. It is this potential difference, called the *boundary potential*, that serves as the analytical parameter in a potentiometric pH measurement.

20C-4 CELL POTENTIALS

Let us now contrast the sources of potentials in a cell equipped with a metal indicator electrode and one containing a membrane electrode.

Metal Indicator Cell Potential. A typical cell for the determination of the activity a_1 of a species A^{n+} with an indicator electrode fashioned from the metal A often takes the form:

$$A|A^{n+}(a_1)\|\text{reference electrode}$$

where a_1 is the activity of species A^{n+}. The cell potential E_V is given by

$$E_V = E_{ref} + E_j - E_A^0 + \frac{0.0591}{n}\log\frac{1}{a_1}$$

Here, E_{ref} is the potential of the reference electrode and E_j is the sum of two junction potentials one arising at each end of the salt bridge. No IR term is included because potentiometric measurements are always made under conditions in which $I \rightarrow 0$. Often the foregoing equation is simplified to

$$E_V = L + \frac{0.0591}{n}\text{pA} \qquad (20\text{–}4)$$

where L is the sum of E_{ref}, E_j, and $-E_A^0$ and pA is, by definition, the negative logarithm of the activity of the analyte.

Membrane Cell Potential. In Figure 20–3**b**, the membrane cell potential is shown as being made up of five potentials. Two of these, E_{R1} and E_j, are similar to the potentials in the cell just described containing a metal indicator electrode. The potential of the membrane electrode E_M is also composed of three potentials; that is,

$$\begin{aligned} E_M &= E_1 - (E_2 - E_{R2}) \\ &= E_1 - E_2 + E_{R2} = E_b + E_{R2} \qquad (20\text{–}5) \end{aligned}$$

Here, E_1 and E_2 are potentials that develop at the two surfaces on either side of the membrane and E_{R2} is the potential of reference electrode 2. The boundary potential E_b is then defined as

$$E_b = (E_1 - E_2)$$

Note that *neither E_1 nor E_2 arises from a reduction reaction as is the case with a metal-indicator electrode* (Equation 20–1). Instead, these potentials are a measure of the driving forces of ion-exchange reactions such as those shown by Equations 20–2 and 20–3.

The overall potential E_V for the cell

shown in Figure 20–3**a** is simply the difference between the potential of the reference electrode (including the junction potential) and the membrane potential. That is,

$$E_V = E_{R1} + E_j - E_M$$

Substituting Equation 20–5 yields

$$E_V = E_{R1} + E_j - E_b - E_{R2} \qquad (20\text{–}6)$$

or

$$E_V = L' - E_b = L' - (E_1 - E_2) \qquad (20\text{–}7)$$

where L' is a constant made up of E_{R1}, E_{R2}, and E_j.

Membrane Boundary Potential. It can be demonstrated from thermodynamic considerations[7] that E_1 and E_2 in Equation 20–7 are related to analyte activities at each interface by the equations

$$E_1 = j_1 - \frac{0.0591}{n} \log \frac{a_1'}{a_1} \qquad (20\text{–}8)$$

$$E_2 = j_2 - \frac{0.0591}{n} \log \frac{a_2'}{a_2} \qquad (20\text{–}9)$$

where j_1 and j_2 are constants and a_1 and a_2 are activities of A^{n+} in the *solutions* on the external and internal sides of the membrane respectively. Activities a_1' and a_2' are related to the concentrations of A^{n+} at the external and internal *surfaces* of the ion exchanger making up the membrane.

If the two membrane surfaces have the same number of negatively charged sites (as they normally do) from which A^{n+} can dissociate, then the two constants j_1 and j_2 are identical; so also are a_1' and a_2'. With these equalities, substitution of Equations 20–8 and 20–9 into the boundary potential expression gives

---

[7] G. Eisenman, *Biophys. J.*, **1962**, 2 (part 2), 259.

$$E_b = (E_1 - E_2)$$
$$= -\frac{0.0591}{n} \log \frac{1}{a_1} + \frac{0.0591}{n} \log \frac{1}{a_2} \qquad (20\text{–}10)$$

Substitution of this equation into Equation 20–7 gives

$$E_V = L' + \frac{0.0591}{n} \log \frac{1}{a_1} - \frac{0.0591}{n} \log \frac{1}{a_2}$$

Since a_2 is fixed, the term on the right can be combined with L' to yield a single constant L. Rearranging then gives

$$\log \frac{1}{a_1} = \text{pA} = \frac{n(E_V - L)}{0.0591} \qquad (20\text{–}11)$$

It should again be emphasized that the activity-dependent potential of the membrane cell is fundamentally different from that of a cell containing a metal indicator electrode. The latter potential measures the driving force for a *single redox equilibrium* such as that shown in Equation 20–1. The former measures the difference in driving force for *two ion-exchange equilibria*, which in the example under consideration can be written as

$$A^{n+}E_1^{n-} \rightleftharpoons E_1^{n-} + A^{n+} \qquad (20\text{–}12)$$
$$\begin{array}{ccc} \text{membrane} & \text{membrane} & \text{analyte} \\ \text{surface 1} & \text{surface 2} & \text{soln }(a_1) \end{array}$$

$$A^{n+}E_2^{n-} \rightleftharpoons E_2^{n-} + A^{n+} \qquad (20\text{–}13)$$
$$\begin{array}{ccc} \text{membrane} & \text{membrane} & \text{standard} \\ \text{surface 2} & \text{surface 2} & \text{soln }(a_2) \end{array}$$

Here, E_1^{n-} and E_2^{n-} represent n of the multitude of negatively charged sites on each surface of a solid ion-exchange membrane.

Selectivity of Membrane Electrodes. A cation exchanger is capable of undergoing an ion-exchange reaction in which a resident cation is replaced by another kind of cation. The potential of the membrane then reflects the activity of both cations, and selectivity is lost.

For example, when the external membrane containing a univalent cation A^+ is immersed in a solution containing a different singly charged species B^+, an ion-exchange equilibrium may develop which can be written as

$$A^+E_1^- \ + \ B^+ \ \rightleftharpoons \ B^+E_1^- \ + \ A^+$$

membrane	analyte	membrane	analyte
surface 1	soln 1	surface 1	soln 1

The equilibrium constant for this reaction is given by

$$K_{ex} = \frac{a_1 b_1'}{a_1' b_1} \qquad (20\text{-}14)$$

where a_1 and b_1 are activities of A^+ and B^+ in the external solution, and a_1' and b_1' are the activities of the same ions *on the surface of the membrane* facing the analyte solution. The exchange constant K_{ex} depends upon the composition of the membrane and may vary widely from one type of membrane to another.

The effect of B^+ on the emf across the membrane under consideration can be described in quantitative terms by rewriting Equation 20–14 in the form

$$\frac{a_1}{a_1'} = \frac{(a_1 + K_{ex}b_1)}{(a_1' + b_1')} \qquad (20\text{-}15)$$

Substitution of Equation 20–15 into Equation 20–8 and subtraction of Equation 20–9 gives

$$E_b = E_1 - E_2$$
$$= j_1 - j_2 + 0.0591 \log \frac{(a_1 + K_{ex}b_1)a_2'}{(a_1' + b_1')a_2}$$

If the number of sites on each side of the membrane is the same, the activity of A^+ ions on the internal surface is approximately equal to the sum of the activities of the two cations on the external surface; that

is, $a_2' \cong (a_1' + b_1')$. Further, $j_1 \cong j_2$; thus, the foregoing equation reduces to

$$E_b = 0.0591 \log \frac{(a_1 + K_{ex}b_1)}{a_2} \qquad (20\text{-}16)$$

Finally, since a_2 is a constant, we may write

$$E_b = k' + 0.0591 \log(a_1 + K_{ex}b_1) \qquad (20\text{-}17)$$

where $k' = -0.0591 \log a_2$.

It has been shown that when some of the surface sites of a membrane are occupied by a second cation, that a *diffusion potential* also develops that depends upon the relative rates at which the two cations move within the membrane. It has also been shown that a correction can be made for this additional potential by modifying Equation 20–17 as follows[8]:

$$E_b = k' + 0.0591 \log \left[a_1 + K_{ex}\left(\frac{U_B}{U_A}\right)b_1 \right]$$
$$(20\text{-}18)$$

where U_B and U_A are measures of the mobility of B^+ and A^+ in the membrane.

The quantity $K_{ex}U_B/U_A$ in Equation 20–18 is defined as the *selectivity coefficient* $k_{A,B}$ and Equation 20–18 can be written in terms of this coefficient. Thus,

$$E_b = k' + 0.0591 \log(a_1 + k_{A,B}b_1) \qquad (20\text{-}19)$$

The selectivity coefficient is a measure of the interference of ion B in the potentiometric determination of species A. Selectivity coefficients vary from zero (no interference) to numbers significantly greater than 1. For example, if an electrode for ion A responds say 20 times more strongly to ion B, $k_{A,B}$ has a value of 20. If, on the other hand, the response of the electrode to ion C

---

[8] See: G. Eisenman, *Glass Electrodes for Hydrogen and Other Cations*, Chapters 4 and 5, Marcel Dekker: New York, 1967.

is only 0.001 of its response to A (a much more desirable figure), $k_{A,C}$ is 0.001.

Selectivity coefficients, which are widely used to describe the selectivity of electrodes, are dependent upon several variables including total ion concentration and the ratio of the concentration of analyte ion to the interfering ion. Thus, selectivity coefficients are approximations that are accurate to only somewhat better than an order of magnitude.

A more general form of Equation 20–19, which applies to multiply charged species and to solutions containing several cations, can be derived in a similar way. This equation takes the form

$$E_b = k' + \frac{0.0591}{Z_A} \log[a_A + k_{A,B}(a_B)^{Z_A/Z_B}$$
$$+ k_{A,C}(a_C)^{Z_A/Z_C} + \cdots] \qquad (20\text{–}20)$$

Here, a_A, a_B, a_C, . . . refer to activities of cations A, B, and C, . . . in the external solution while Z_A, Z_B, Z_C, . . . are the charges carried by these ions. The constant $k_{A,B}$ is the selectivity coefficient for A in the presence of B, $k_{A,C}$ is the coefficient for A in the presence of ion C, and so forth.

20C–5 THE GLASS ELECTRODE FOR pH MEASUREMENTS

As noted earlier, the glass electrode was the first membrane electrode to be invented, and electrodes for other ions appeared only as an understanding of the response of the pH electrode developed. Figure 20–5 shows a modern cell for the measurement of pH. It consists of a commercial calomel electrode, such as that shown in Figure 20–1**a**, and a glass electrode, both immersed in a solution whose pH is to be measured. The glass electrode is manufactured by sealing a thin, pH-sensitive glass tip to the end of a heavy-walled glass tubing. The resulting bulb is filled with a solution of hydrochloric

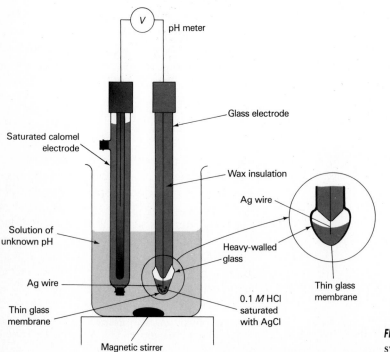

FIGURE 20-5 Typical electrode system for measuring pH.

acid (often 0.1 M) or a buffer solution; this internal solution is saturated with silver chloride. A silver wire, immersed in the solution, forms reference electrode 2 for the system; it is connected via an external lead to one terminal of the potential-measuring device. The calomel electrode, which acts as reference electrode 1, is connected to the other terminal. Note again that the cell contains *two* reference electrodes, each with a potential that is constant and *independent of pH*. One reference electrode is the external calomel electrode; the other is the internal silver/silver chloride electrode, which is a *part* of the glass electrode but is *not* the pH-sensitive component. In fact, it is the *thin membrane at the tip of the electrode* that responds to pH changes.

Composition of Glass Membranes. Much systematic investigation[9] has been devoted to the effects of chemical composition on the sensitivity of glass membranes to protons and other cations, and a variety of compositions are now used commercially. For many years, Corning 015 glass (consisting of approximately 22% Na_2O, 6% CaO, and 72% SiO_2) was widely used. This glass shows an excellent specificity toward hydrogen ions up to a pH of about 9. At higher pH values, however, the membrane becomes somewhat sensitive to sodium and other alkali ions. The Corning 015 glass has been largely supplanted by membranes in which sodium and calcium ions are replaced to varying degrees by lithium and barium. These substitutions enhance the selectivity and lifetime of the electrode.

Hygroscopicity of Glass Membranes. It has been shown that the surfaces of a glass membrane must be hydrated in order to have pH activity. Nonhygroscopic glasses, such as Pyrex and quartz, show no pH function. Even Corning 015 glass shows little pH response after dehydration by storage over a desiccant; its sensitivity is restored, however, after standing for a few hours in water.

It has also been demonstrated experimentally that hydration of a pH-sensitive glass membrane is accompanied by an ion-exchange reaction in which singly charged cations of the glass are exchanged for protons of the solution. The di- and trivalent cations in the silicate structure are much more strongly bonded and are thus not affected. The ion-exchange reaction can be written as

$$H^+ + Na^+Gl^- \rightleftharpoons Na^+ + H^+Gl^- \quad (20\text{--}21)$$
$$\text{soln} \quad \text{membrane} \quad \text{soln} \quad \text{membrane}$$
$$\text{surface} \quad\quad\quad\quad \text{surface}$$

where Gl^- represents one of the numerous cation bonding sites on the glass surface. The equilibrium constant for this process favors incorporation of hydrogen ions into the silicate lattice; as a result, the surface of a well-soaked membrane consists of a layer of silicic acid gel. The thickness of the gel is 10^{-4} to 10^{-5} mm or less.

Asymmetry Potential. If identical solutions and identical reference electrodes are placed on either side of a glass membrane, the boundary potential $E_1 - E_2$ should be zero. However, when this experiment is performed, it is found that a small potential, called the *asymmetry potential*, often develops. Moreover, the asymmetry potential associated with a given glass electrode changes slowly with time.

The causes of the asymmetry potential are obscure; they undoubtedly include such factors as differences in strains established within the two surfaces during manufacture of the membrane, mechanical and chemical attack of the surfaces, and contamination of the outer face during use. The effect of the asymmetry potential on a pH measurement is eliminated by frequent calibration of the

---

[9] For a summary of this work, see: J. O. Isard, "The Dependence of Glass-Electrode Properties on Composition," in *Glass Electrodes for Hydrogen and Other Cations*, G. Eisenman, Ed., Chapter 3, Marcel Dekker: New York, 1967.

electrode against one or more standard buffers of known pH.

The Alkaline Error. In solutions of pH 9 or greater, some glass membranes respond not only to changes in hydrogen ion concentration but also to the concentration of alkali-metal ions. The magnitude of the resulting error for four types of glass membranes is indicated on the right-hand side of Figure 20–6. In each of these curves, the sodium ion concentration was maintained at 1 M, and the pH was varied. Note that the pH error is negative at high pH, which indicates that the electrode is responding to sodium ions as well as to protons. This observation is confirmed by data obtained for solutions with different sodium ion concentrations. Thus, at pH 12, the alkaline error for Corning 015 glass is about −0.7 pH when the sodium ion concentration is 1 M (Figure 20–6) and only about −0.3 pH in solutions that are 0.1 M in this ion.

All singly charged cations cause alkaline errors to a greater or lesser extent. The source of this error can be found in the ear-lier section on selectivity which led to the development of Equation 20–18.

The Acid Error. As shown in Figure 20–6, the typical glass electrode exhibits an error, opposite in sign to the alkaline error, in solutions of pH less than about 0.5; here, readings tend to be too high. The magnitude of the error depends upon a variety of factors and is generally not very reproducible. The causes of the acid error are not well understood.

20C-6 GLASS ELECTRODES FOR THE DETERMINATION OF OTHER CATIONS

The existence of the alkaline error in early glass electrodes led to studies concerning the effect of glass composition on the magnitude of this error. One consequence of this work has been the development of glasses for which the term $K_{ex}(U_B/U_A)b_1$ in Equation 20–18 is small enough so that the alkaline error is negligible below a pH of about 12. Other studies have been directed toward finding glass compositions in which this term is greatly enhanced, in the interests of developing glass electrodes for the determination of cations other than hydrogen. Such applications require that the hydrogen ion activity a_1 in Equation 20–18 be negligible with respect to the second term containing the activity b_1 of the other cation; under this circumstance, the potential of the electrode would be *independent* of pH but would vary with pB in the typical way.

Table 20–2 shows the composition of some glasses that have been developed for the determination of various singly charged cations. Currently, glass electrodes are available commercially for hydrogen, sodium, lithium, ammonium, and total monovalent cations. Because of the strong bonding between multiply charged cations and the silicate ion, it seems unlikely that glasses sensitive to other than singly charged species will ever be developed.

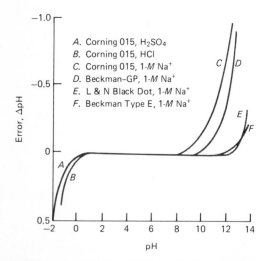

FIGURE 20–6 Acid and alkaline error of selected glass electrodes at 25° C. (From: R. G. Bates, *Determination of pH: Theory and Practice*, p. 316, Wiley: New York, 1964. Reprinted by permission of John Wiley & Sons, Inc.)

20C-7 CRYSTALLINE MEMBRANE ELECTRODES

Two types of crystalline membrane electrodes have been developed, homogeneous and heterogeneous. The latter are prepared by dispersing finely ground crystalline solids in an inert matrix such as silicone rubber, polyvinyl chloride, or paraffin. Generally, the membranes behave selectively toward one of the ions making up the dispersed solid. For example, a dispersion of barium sulfate is responsive to sulfate and barium ions; similarly, a suspension of aluminum phosphate has been used as a sensor for phosphate ions. Compared with homogeneous electrodes, heterogeneous electrodes appear to be less satisfactory in terms of reproducibility, selectivity, life, and linearity

of response. Thus, we will not consider them further.

Homogeneous membranes are manufactured from an ionic compound or a homogeneous mixture of ionic compounds. In some instances the membrane is cut from a single crystal; in others, disks are formed from the finely ground crystalline solid by high pressures or by casting from a melt. The typical membrane has a diameter of about 10 mm and a thickness of 1 or 2 mm. To form an electrode, the membrane is sealed to the end of a tube made from a chemically inert plastic such as Teflon or polyvinyl chloride.

Conductivity of Crystalline Membranes. Most ionic crystals are insulators and do not have sufficient electrical conductivity at room tem-

---

Table 20-2
Properties of Certain Cation-Sensitive Glasses[a]

Principal Cation to be Measured	Glass Composition	Selectivity Coefficient	Remarks
Li^+	$15Li_2O$ $25Al_2O_3$ $60SiO_2$	$K_{Li^+,Na^+} \approx 0.3$, $K_{Li^+,K^+} < 10^{-3}$	Best for Li^+ in presence of H^+ and Na^+
Na^+	$11Na_2O$ $18Al_2O_3$ $71SiO_2$	$K_{Na^+,K^+} \approx 4 \times 10^{-4}$ at pH 11 $K_{Na^+,K^+} \approx 3 \times 10^{-3}$ at pH 7	Nernst type of response to $\sim 10^{-5}$ M Na^+
	$10.4Li_2O$ $22.6Al_2O_3$ $67SiO_2$	$K_{Na^+,K^+} \approx 10^{-5}$	Highly Na^+ selective, but very time-dependent
K^+	$27Na_2O$ $5Al_2O_3$ $68SiO_2$	$K_{K^+,Na^+} \approx 0.05$	Nernst type of response to $< 10^{-4}$ M K^+
Ag^+	$28.8Na_2O$ $19.1Al_2O_3$ $52.1SiO_2$	$K_{Ag^+,H^+} \approx 10^{-5}$	Highly sensitive and selective to Ag^+, but poor stability
	$11Na_2O$ $18Al_2O_3$ $71SiO_2$	$K_{Ag^+,Na^+} < 10^{-3}$	Less selective for Ag^+ but more reliable

[a] Reprinted from G. A. Rechnitz, *Chem. Eng. News*, **1967,** June 12, 149. Copyright 1967 American Chemical Society.

perature to serve as membrane electrodes. Those that are conductive are characterized by having a small ion that is mobile in the solid phase. Examples are fluoride ion in certain rare earth fluorides, silver ion in silver halides and sulfides, and copper(I) ion in cuprous sulfide. In such solids, conduction takes place by a defect mechanism wherein the conducting ion jumps from its normal location in the crystal lattice to a hole present as a defect. The movement of the ion, of course, leaves an oppositely charged hole behind. Since the mobility of an ion in a crystal depends upon its shape, charge, and size, the conduction process is generally limited to a single kind of ion; other ions cannot participate in the process. As a consequence, crystal membranes are generally highly selective.

The Fluoride Electrode. Lanthanum fluoride, LaF_3, is a nearly ideal substance for the preparation of a membrane electrode for the determination of fluoride ion. Fluorides of neodymium and praseodymium have also been used for the same purpose. Although these fluorides are natural conductors, their conductivity can be enhanced by doping with europium fluoride, EuF_2. Membranes are prepared by cutting disks from a single crystal of the doped compound.

The mechanism of the development of a fluoride sensitive potential across a lanthanum fluoride membrane is quite analogous to that described for glass, pH-sensitive membranes. That is, at the two interfaces, ionization creates a charge on the membrane surface. Thus,

$$LaF_3 \rightleftharpoons LaF_2^+ + F^-$$

$$\text{solid} \qquad \text{solid} \qquad \text{solution}$$

The magnitude of the charge is dependent upon the fluoride ion concentration of the solution. Thus, the side of the membrane encountering the lower fluoride ion concen-

tration becomes positive with respect to the other surface; it is this charge difference that provides a measure of the difference in fluoride concentration of the two solutions. The potential of a cell containing a lanthanum fluoride electrode is given by an equation analogous to Equation 20–11. That is,

$$pF = \frac{L - E_V}{0.0591}$$

Note that the sign of the term on the right is reversed since an anion is being determined here.

Commercial lanthanum fluoride electrodes come in various shapes and sizes and are available from several sources. Most are rugged and can be used at temperatures between 0 and 80°C. The response of the fluoride electrode is linear down to $10^{-6} M$ (0.02 ppm), where the solubility of lanthanum fluoride begins to contribute to the concentration of fluoride ion in the analyte solution. The only ion that interferes directly with fluoride measurements is hydroxide ion; this interference becomes serious at a pH greater than eight. At a pH less than five, hydrogen ions also interfere in *total* fluoride determinations; here, undissociated hydrogen fluoride forms to which the electrode is not responsive. In most respects, the fluoride ion electrode approaches the ideal for selective electrodes.

Electrodes Based on Silver Salts. Membranes prepared from single crystals or pressed disks of the various silver halides act selectively toward silver and halide ions. Generally, their behavior is far from ideal, however, owing to low conductivity, low mechanical strength, and a tendency to develop high photoelectric potentials. It has been found, however, that these disadvantages are minimized if the silver salts are mixed with crystalline silver sulfide in an approximately 1 : 1 molar ratio. Homogeneous mixtures are formed from equimolar solutions of sulfide and halide ions by precipitation with silver nitrate. After washing and drying, the product is shaped into disks under a pressure of about 10^5 pounds per square inch. The resulting disk exhibits good electrical conductivity owing to the mobility of the silver ion in the sulfide matrix.

Membranes constructed either from silver sulfide or from a mixture of silver sulfide and another silver salt are useful for the determination of both sulfide and silver ions. Thus, toward silver ions, the electrical response is similar to a metal electrode of the first kind (although the mechanism of activity is totally different). Toward sulfide ions, the electrical response of a silver sulfide membrane is similar to that of an electrode of the second kind (Section 20B−2). Here, upon immersion in the solution to be

analyzed, dissolution of a tiny amount of silver sulfide quickly saturates the film of liquid adjacent to the electrode. The solubility, and thus the silver ion concentration, depends, however, upon the sulfide concentration of the analyte.

Crystalline membranes are also available that consist of a homogeneous mixture of silver sulfide with sulfides of copper(II), lead, or cadmium. Towards these divalent cations, electrodes from these materials have electrical responses similar to electrodes of the third kind (Section 20B−3). It should be noted that these divalent sulfides, by themselves, are not conductors and thus do not exhibit selective-ion activity.

Table 20−3 lists the various solid-state electrodes that are available from commercial sources.

Table 20-3
Commercial Solid-State Electrodes^a

Analyte Ion	Concentration Range, M	Interferences^b
Br^-	10^0 to 5×10^{-6}	mr: 8×10^{-5} CN^-; 2×10^{-4} I^-; 2 NH_3; 400 Cl^-; 3×10^4 OH^-. mba: S^{2-}
Cd^{2+}	10^{-1} to 10^{-7}	Fe^{2+} + Pb^{2+} may interfere. mba: Hg^{2+}, Ag^+, Cu^{2+}
Cl^-	10^0 to 5×10^{-5}	mr: 2×10^{-7} CN^-; 5×10^{-7} I^-; 3×10^{-3} Br^-; 10^{-2} $S_2O_3^{2-}$; 0.12 NH_3; 80 OH^-. mba: S^{2-}
Cu^{2+}	10^{-1} to 10^{-8}	high levels Fe^{2+}, Cd^{2+}, Br^-, Cl^-. mba: Hg^{2+}, Ag^+, Cu^+
CN^-	10^{-2} to 10^{-6}	mr: 10^{-1} I^-; 5×10^3 Br^-; 10^6 Cl^-. mba: S^{2-}
F^-	sat'd to 10^{-6}	0.1 M OH^- gives <10% interference when $[F^-] = 10^{-3}$ M
I^-	10^0 to 5×10^{-8}	mr: 0.4 CN^-; 5×10^3 Br^-; 10^5 S_2O_3; 10^6 Cl^-
Pb^{2+}	10^{-1} to 10^{-6}	mba: Hg^{2+}, Ag^+, Cu^{2+}
Ag^+/S^{2-}	10^0 to 10^{-7} Ag^+ 10^0 to 10^{-7} S^{2-}	Hg^{2+} must be less than 10^{-7} M
SCN^-	10^0 to 5×10^{-6}	mr: 10^{-6} I^-; 3×10^{-3} Br^-; 7×10^{-3} CN^-; 0.13 $S_2O_3^{2-}$; 20 Cl^-; 100 OH^-. mba: S^{2-}

^{a} From: *Handbook of Electrode Technology*, pp. 10–13, Appendix, Orion Research: Cambridge, MA, 1982. With permission.

^{b} mr: maximum ratio $\left(\dfrac{M \text{ interference}}{M \text{ analyte}}\right)$ for no interference.

mba: must be absent.

20C-8 *LIQUID MEMBRANE ELECTRODES*

Liquid membranes are formed from immiscible liquids that selectively bond certain ions. Membranes of this type are particularly important because they permit the direct potentiometric determination of the activities of several polyvalent cations and certain singly charged anions and cations as well.

Early liquid membranes were prepared from immiscible, liquid ion-exchangers, which were retained in a porous inert solid support. As shown schematically in Figure 20–7, a porous, hydrophobic (that is, water repelling) plastic disk (typical dimensions: 3×0.15 mm) served to hold the organic layer between the two aqueous solutions. Wick action caused the pores of the disk to stay filled with the organic liquid from the reservoir in the outer of the two concentric tubes. For divalent cation determinations,

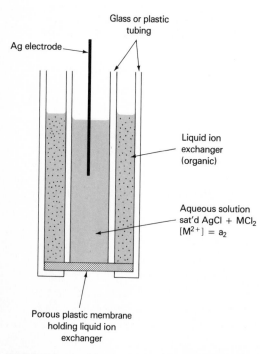

FIGURE 20–7 Liquid membrane electrode sensitive to M^{2+}.

the inner tube contained an aqueous standard solution of MCl_2, where M^{2+} was the cation whose activity was to be determined. This solution was also saturated with AgCl to form a Ag/AgCl reference electrode with the silver lead wire.

As an alternative to the use of a porous disk as a rigid supporting medium, it has recently been found possible to immobilize liquid exchangers in tough polyvinyl chloride membranes. Here, the liquid ion exchanger and polyvinyl chloride are dissolved in a solvent such as tetrahydrofuran. Evaporation of the solvent leaves a flexible membrane, which can be cut and cemented to the end of a glass or plastic tube. Membranes formed in this way behave in much the same way as those in which the ion exchanger is held as an actual liquid in the pores of a disk. Currently, most liquid-membrane electrodes are of this newer type.

The active substances in liquid membranes are of three kinds: (1) cation exchangers, (2) anion exchangers, and (3) neutral macrocyclic compounds, which selectively complex certain cations.

One of the most important liquid membrane electrodes is selective towards calcium ion in approximately neutral media. The active ingredient in the membrane is a cation exchanger consisting of an aliphatic diester of phosphoric acid dissolved in a polar solvent. The diester contains a single acidic proton; thus, two molecules react with the divalent calcium ion to form a dialkyl phosphate with the structure

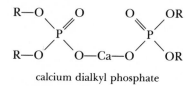

calcium dialkyl phosphate

Here, R is an aliphatic group containing from 8 to 16 carbon atoms. The internal aqueous solution in contact with the exchanger (see Figure 20–7) contains a fixed concentration of calcium chloride and a silver/silver chloride reference electrode. The porous disk (or the polyvinyl chloride membrane) containing the ion-exchange liquid separates the analyte solution from the reference calcium chloride solution. The equilibrium established at each interface can be represented as

$$[(RO)_2POO]_2Ca \rightleftharpoons 2(RO)_2POO^- + Ca^{2+}$$

organic   organic   aqueous

Note the similarity of this equation to Equation 20–2 for the glass electrode. The relationship between potential and pCa is also analogous to that for the glass electrode (Equation 20–11). Thus,

$$pCa = \frac{2(E_V - L)}{0.0591}$$

The calcium membrane electrode has proved to be a valuable tool for physiological studies because this ion plays important roles in nerve conduction, bone formation, muscle contraction, cardiac conduction and contraction, and renal tubular function. At least some of these processes are influenced more by calcium ion *activity* than by calcium ion concentration; activity, of course, is the parameter measured by the electrode.

Another liquid specific-ion electrode of great value for physiological studies is that for potassium, because the transport of nerve signals appears to involve movement of this ion across nerve membranes. Study of the process requires an electrode that can detect small concentrations of potassium ion in the presence of much larger concentrations of sodium. A number of liquid membrane electrodes show promise of meeting these needs; one is based upon the antibiotic valinomycin, an uncharged macrocyclic ether that has a strong affinity for potassium ion. Of great importance is the observation that a liquid membrane consisting of valino-

mycin in diphenyl ether is about 10^4 times as responsive to potassium ion as to sodium ion.[10]

Table 20–4 lists commercially available liquid membrane electrodes. The anion-sensitive electrodes employ a solution of an anion exchanger in an organic solvent. As mentioned earlier, many of the so-called liquid membrane electrodes are in fact solids in which the liquid is held in a polyvinyl chloride matrix. These electrodes are somewhat more convenient to use than the older porous disk electrodes.

20C-9 GAS-SENSING PROBES

During the past two decades, several so-called gas-sensing electrodes have become available from commercial sources. As can be seen in Figure 20–8, these devices are not, however, electrodes but instead are electrochemical cells made up of a specific ion and a reference electrode immersed in an internal solution that is retained by a thin gas-permeable membrane. Thus, *gas-sensing probes* is a more suitable name for these gas sensors.

Gas-sensing probes are remarkably selective and sensitive devices for determining dissolved gases or ions that can be converted to dissolved gases by pH adjustment.

Membrane Probe Design. Figure 20–8 is a schematic diagram showing details of a gas-sensing probe for carbon dioxide. The heart of the probe is a thin, porous membrane, which is easily replaceable. This membrane separates the analyte solution from an internal solution containing sodium bicarbonate and sodium chloride. A pH-sensitive glass electrode having a flat membrane is fixed in position so that a very thin film of the internal solution is sandwiched between it and the gas-permeable membrane. A silver/silver chloride reference electrode is also lo-

---

[10] M. S. Frant and J. W. Ross, Jr., *Science,* **1970,** *167,* 987.

Table 20-4
Liquid Membrane Electrodes[a,b]

Analyte Ion	Concentration Range, M	Interferences[c]
Ca^{2+}	10^0 to 5×10^{-7}	10^{-5} Pb^{2+}; 4×10^{-3} Hg^{2+}, H^+; 6×10^{-3} Sr^{2+}; 2×10^{-2} Fe^{2+}; 4×10^{-2} Cu^{2+}; 5×10^{-2} Ni^{2+}; 0.2 NH_3; 0.2 Na^+; 0.3 $Tris^+$; 0.3 Li^+; 0.4 K^+; 0.7 Ba^{2+}; 1.0 Zn^{2+}; 1.0 Mg^{2+}
BF_4^-	10^0 to 7×10^{-6}	5×10^{-7} ClO_4^-; 5×10^{-6} I^-; 5×10^{-5} ClO_3^-; 5×10^{-4} CN^-; 10^{-3} Br^-; 10^{-3} NO_2^-; 5×10^{-3} NO_3^-; 3×10^{-3} HCO_3^-; 5×10^{-2} Cl^-; 8×10^{-2} $H_2PO_4^-$, HPO_4^{2-}, PO_4^{3-}; 0.2 OAc^-; 0.6 F^-; 1.0 SO_4^{2-}
NO_3^-	10^0 to 7×10^{-6}	10^{-7} ClO_4^-; 5×10^{-6} I^-; 5×10^{-5} ClO_3^-; 10^{-4} CN^-; 7×10^{-4} Br^-; 10^{-3} HS^-; 10^{-2} HCO_3^-; 2×10^{-2} CO_3^{2-}; 3×10^{-2} Cl^-; 5×10^{-2} $H_2PO_4^-$, HPO_4^{2-}, PO_4^{3-}; 0.2 OAc^-; 0.6 F^-; 1.0 SO_4^{2-}
ClO_4^-	10^0 to 7×10^{-6}	2×10^{-3} I^-; 2×10^{-2} ClO_3^-; 4×10^{-2} CN^-, Br^-; 5×10^{-2} NO_2^-, NO_3^-; 2 HCO_3^-, CO_3^{2-}, Cl^-, $H_2PO_4^-$, HPO_4^{2-}, PO_4^{3-}, OAc^-, F^-, SO_4^{2-}
K^+	10^0 to 10^{-6}	3×10^{-4} Cs^+; 6×10^{-3} NH_4^+, Tl^{+1}; 10^{-2} H^+; 1.0 Ag^+, $Tris^+$; 2.0 Li^+, Na^+
Water Hardness ($Ca^{2+} + Mg^{2+}$)	10^{-3} to 6×10^{-6}	3×10^{-5} Cu^{2+}, Zn^{2+}; 10^{-4} Ni^{2+}; 4×10^{-4} Sr^{2+}; 6×10^{-5} Fe^{2+}; 6×10^{-4} Ba^{2+}; 3×10^{-2} Na^+; 0.1 K^+

[a] From: *Handbook of Electrode Technology*, pp. 10–13, Appendix, Orion Research: Cambridge, MA, 1982. With permission.

[b] All of the electrodes except the last are of the newer type in which the liquid ion-exchanger or neutral carrier is supported in a plastic matrix.

[c] The numbers in front of each ion represent the molar concentration of that ion that gives a 10% error when the analyte concentration is 10^{-3} M.

cated in the internal solution. It is the pH of the film of liquid adjacent to the glass electrode that provides a measure of the carbon dioxide content of the analyte solution on the other side of the membrane.

Gas-Permeable Membranes. Two types of membrane material are encountered, microporous and homogeneous. Microporous materials are manufactured from hydrophobic polymers such as polytetrafluoroethylene or polypropylene, which have a porosity (void volume) of about 70% and a pore size of less than 1 μm. Because of the water-repellent properties of the film, water molecules and electrolyte ions are excluded from the pores; gaseous molecules, on the other hand, are free to move in and out of the pores by effusion and thus across this barrier. Typically, microporous membranes are about 0.1 mm in thickness.

Homogeneous films, in contrast, are solid polymeric substances through which the analyte gas passes by dissolving in the membrane, diffusing, and then desolvating into the internal solution. Silicone rubber is the most widely used material for construction. Homogeneous films are generally thinner than microporous ones (0.01 to 0.03 mm) in

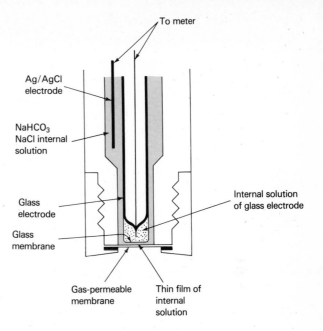

FIGURE 20-8 Schematic diagram of a gas-sensing probe for carbon dioxide.

order to hasten the transfer of gas and thus the rate of response of the system.

Mechanism of Response. When a solution containing dissolved carbon dioxide is brought into contact with the microporous membrane shown in Figure 20–8, the gas *effuses* into the pores, as described by the reaction

$$CO_2(aq) \rightleftharpoons CO_2(g)$$

external	membrane
solution	pores

Because the pores are numerous and small, the equilibrium is rapidly established. The carbon dioxide in the pores, however, is also in contact with the internal solution and a second equilibrium reaction takes place; that is,

$$CO_2(g) \rightleftharpoons CO_2(aq)$$

membrane	internal
pores	solution

As a consequence of the two reactions, the external solution rapidly (in a few seconds to a few minutes) equilibrates with the thin film of internal solution adjacent to the membrane. Here, another equilibrium is established that causes the pH of the internal surface film to change; that is,

$$CO_2(aq) + H_2O \rightleftharpoons HCO_3^- + H^+$$

A glass-reference electrode pair immersed in the film of internal solution then detects the pH change.

The overall reaction for the process just described is obtained by adding the three chemical equations to give

$$CO_2(aq) + H_2O \rightleftharpoons H^+ + HCO_3^-$$

external	internal
solution	solution

The equilibrium constant for the reaction is given by

$$\frac{[H^+][HCO_3^-]}{[CO_2(aq)]_{ext}} = K$$

If the concentration of HCO_3^- in the internal solution is made relatively high so that its concentration is not altered significantly by the carbon dioxide from the sample, then

$$\frac{[H^+]}{[CO_2(aq)]_{ext}} = \frac{K}{[HCO_3^-]} = K_g \qquad (20-22)$$

where K_g is a new constant. This equation can be rearranged to give

$$a_1 \cong [H^+] = K_g[CO_2(aq)]_{ext} \qquad (20-23)$$

where a_1 is the internal hydrogen ion activity.

The potential of the electrode system in the internal solution is dependent upon a_1 as described by Equation 20–11. Substitution of Equation 20–23 into 20–11 yields upon rearrangement

$$E_V = L - 0.0591 \log K_g[CO_2(aq)]_{ext}$$
$$= L'' - 0.0591 \log [CO_2(aq)]_{ext}$$

Thus, the potential of the cell consisting of the internal reference and indicator electrode is determined by the CO_2 concentration of the external solution. Note that *no electrode comes directly in contact* with the analyte. Note also that the only species that will interfere with the measurement are dissolved gases that can pass through the membrane and can additionally affect the pH of the internal solution.

The possibility exists for increasing the selectivity of the gas-sensing electrode by employing an internal electrode that is sensitive to some species other than hydrogen ion; for example, a nitrate-sensing electrode can be used to provide a cell that would be sensitive to nitrogen dioxide. Here, the equilibrium would be

$$2NO_2(aq) + H_2O \rightleftharpoons \underset{\text{internal solution}}{NO_2^- + NO_3^- + 2H^+}$$

external solution · · · · · · · · internal solution

This electrode permits the determination of NO_2 in the presence of gases such as SO_2, CO_2, and NH_3, which would also alter the pH of the internal solution.

Table 20–5 lists gas-sensing probes that are available for purchase. An oxygen-sensitive cell system is also on the market; it, however, is based on a voltammetric measurement and is discussed in Chapter 22.

20C-10 ENZYME ELECTRODES

Since about 1970, considerable effort has been devoted to combining the selectivity of enzyme-catalyzed reactions and selective-ion electrodes to give highly specific procedures for the determination of compounds of biological and biochemical interest.[11] This concept is attractive from several

---

[11] See: D. N. Gray, M. H. Keyes, and P. Watson, *Anal. Chem.*, **1977**, *49*, 1067A; L. D. Bowers and P. W. Carr, *Anal. Chem.*, **1976**, *48*, 544A.

Table 20–5 Commercial Gas-Sensing Probes		
Gas	**Equilibrium in Internal Solution**	**Sensing Electrode**
NH_3	$NH_3 + H_2O \rightleftharpoons NH_4^+ + OH^-$	glass, pH
CO_2	$CO_2 + H_2O \rightleftharpoons HCO_3^- + H^+$	glass, pH
HCN	$HCN \rightleftharpoons H^+ + CN^-$	Ag_2S, pCN
HF	$HF \rightleftharpoons H^+ + F^-$	LaF_3, pF
H_2S	$H_2S \rightleftharpoons 2H^+ + S^{2-}$	Ag_2S, pS
SO_2	$SO_2 + H_2O \rightleftharpoons HSO_3^- + H^+$	glass, pH
NO_2	$2NO_2 + H_2O \rightleftharpoons NO_2^- + NO_3^- + 2H^+$	immobilized ion-exchange, pNO_3

standpoints. First, in principle, complex organic molecules could be determined with the convenience, speed, and ease that characterizes ion-selective measurements of inorganic species. Furthermore, enzymes permit reactions to occur under mild conditions of temperature and pH and at minimal substrate concentrations. Finally, combining the selectivity of the enzymatic reaction and the electrode response should yield procedures that are free from most interferences.

The main limitation to enzymatic procedures is the high cost of enzymes, particularly when used for routine or continuous measurements. This disadvantage has led to the development of immobilized enzyme media in which a small amount of enzyme can be used for the repetitive analysis of hundreds of samples. An extension of this development has been to hold the immobilized enzyme in a membrane layer over the sensor surface of an ion-selective electrode thus forming an enzyme electrode.

Immobilization of enzymes can be accomplished in several ways including physical entrapment in a polymer gel, physical absorption on a porous inorganic support such as alumina, covalent bonding of the enzyme to a solid surface such as glass beads or a polymer, or copolymerization of the enzyme with a suitable monomer.

Apparatus Based on Separate Enzyme and Sensor Functions. Several manufacturers offer instruments in which the immobilized enzyme is separated from the selective electrode or probe. As an example one company offers an instrument for blood urea nitrogen (BUN) determination, an important routine clinical test. In approximately neutral media, urea is hydrolyzed in the presence of the enzyme urease. The reaction is

$$(NH)_2CO + 2H_2O + H^+ \rightarrow 2NH_4^+ + HCO_3^-$$

$$urease$$

$$(20-24)$$

The products can then be measured, after pH adjustment, with a carbon dioxide probe (after acidification) or an ammonia probe (after base addition). The latter is used in this instance. Figure 20–9 is a schematic diagram of the blood-urea instrument. The blood sample (20 μL) is injected into the enzyme cartridge which contains urease adsorbed on porous alumina. Hydrolysis occurs as the sample is carried through the

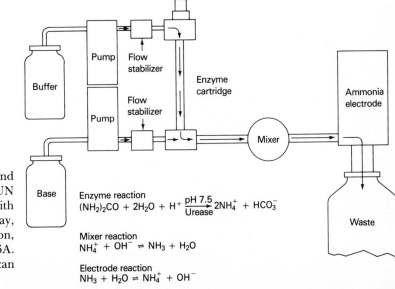

FIGURE 20-9 Flow diagram and reactions of the Kimble BUN analyzer. (Reprinted with permission from: D. N. Gray, M. H. Keyes, and B. Watson, *Anal. Chem.*, **1977**, *49*, 1076A. Copyright 1977 American Chemical Society.)

Enzyme reaction
$$(NH_2)_2CO + 2H_2O + H^+ \xrightarrow[\text{Urease}]{\text{pH 7.5}} 2NH_4^+ + HCO_3^-$$

Mixer reaction
$$NH_4^+ + OH^- \rightleftharpoons NH_3 + H_2O$$

Electrode reaction
$$NH_3 + H_2O \rightleftharpoons NH_4^+ + OH^-$$

cartridge with a pH 7.5 buffer. Upon exiting the cartridge, the ammonia ion is converted quantitatively to ammonia by increasing the pH to 11 or greater by means of a flow of sodium hydroxide. The concentration of ammonia is determined with a gaseous ammonia probe. The output of the probe is monitored electronically and is proportional to the logarithm of the urea concentration. The sensor within the gaseous probe is a pH-sensitive glass electrode.

Other commercial instruments of this type are offered for the routine determination of glucose, lactose, sucrose, galactose, cholesterol, and insecticides (based on their inhibitory effect on cholinesterase). Some of these instruments are based on amperometric (Chapter 22) rather than specific-ion probes.

Enzymatic Electrode. In terms of simplicity and low cost, an attractive concept is that of attaching immobilized enzymes directly to the sensor of a selective-ion electrode or probe (see Figure 20–10).[12] The earliest application of this idea was to the determination of urea based on Equation 20–24. The sensor in Figure 20–10**a** is a glass ammonium ion electrode; that in 20–10**b** is an ammonia gas probe. Unfortunately, both have severe limitations. The glass electrode responds to all

monovalent cations and its selectivity coefficients for NH_4^+ over Na^+ and K^+ are such that interference arises in most media of biological interest (such as blood). The probe shown in Figure 20–10**b** suffers from a different handicap, namely a pH incompatibility between enzyme and sensor. Thus, the enzyme requires a pH of about 7 for maximum catalytic activity, but the sensor's maximum response occurs at a pH that is greater than 8 to 9 (where essentially all of the NH_4^+ has been converted to NH_3).

To date, despite a considerable effort, no commercial enzyme electrodes are available, due at least in part to limitations such as those cited in the previous paragraph.

20C-11 DISPOSABLE MULTILAYER P-ION SYSTEMS

Recently, disposable electrochemical cells, based on p-ion electrodes, have become available. These systems, which have been designed for the routine determination of various ions in clinical samples, are described briefly in Section 29E–3.

20D   Instruments for Measuring Cell Potentials

A prime consideration in the design of an instrument for measuring cell potentials is that its resistance be large with respect to the cell. If it is not, significant errors result as a

---

[12] For reviews of this type of electrode, see: G. Rechnitz, *Anal. Chem.*, **1982**, *54*, 1194A; *Science*, **1981**, *214*, 287.

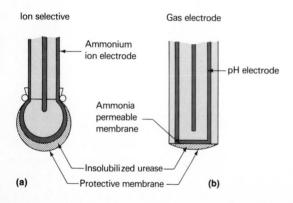

(a) Ion selective — Ammonium ion electrode — Insolubilized urease — Protective membrane

(b) Gas electrode — pH electrode — Ammonia permeable membrane

FIGURE 20–10 Enzyme electrodes for measuring urea. (Reprinted with permission from: D. N. Gray, M. H. Keyes, and B. Watson, *Anal. Chem.*, **1977**, *49*, 1069A. Copyright 1977 American Chemical Society.)

consequence of the *IR* drop in the cell. This effect is demonstrated by the example that follows.

EXAMPLE 20-1. The true potential of a glass/calomel electrode system is 0.800 V; its internal resistance is 20 MΩ. What would be the relative error in the measured potential if the measuring device has a resistance of 100 MΩ?

Here, the circuit can be considered to consist of a potential source E_s and two resistors in series, R_s being that of the source and R_M that of the measuring device. That is,

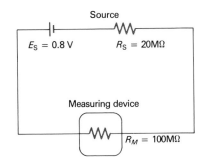

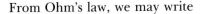

From Ohm's law, we may write

$$E_s = IR_s + IR_M$$

where *I* is the current in this circuit consisting of the cell and the measuring device. The current is then given by

$$I = \frac{0.800}{(20 + 100) \times 10^6} = 6.67 \times 10^{-9} \text{ A}$$

The potential drop across the measuring device (which is the potential indicated by the device, E_i) is IR_M. Thus,

$$E_i = 6.67 \times 10^{-9} \times 100 \times 10^6 = 0.667 \text{ V}$$

and

$$\text{rel error} = \frac{0.667 - 0.800}{0.800} \times 100 = -17\%$$

It is readily shown that to reduce the instrumental error due to *IR* drop to 1% relative, the resistance of the potential measuring device must be about 100 times greater than the cell resistance; for a relative error of 0.1%, the resistance must be 1000 times greater. Since the electrical resistance of cells containing selective-ion electrodes may be 100 MΩ or more, potential measuring devices to be used with these electrodes generally have internal resistance of 10^{12} Ω or more.

It is important to appreciate that an error in potential, such as that shown in Example 20-1 (-0.133 V), would have an enormous effect on the accuracy of a concentration measurement based upon that potential. Thus, as shown on page 629, a 0.001-V uncertainty in potential leads to a relative error of about 4% in the determination of the hydrogen ion concentration of a solution by potential measurement with a glass electrode. An error of the size found in Example 20-1 would result in a concentration uncertainty of two orders of magnitude or more.

Two types of instruments have been employed in potentiometry—the potentiometer and the direct-reading electronic voltmeter. Both instruments are referred to as *pH meters* when their internal resistances are sufficiently high to be used with glass and other membrane electrodes; with the advent of the many new specific ion electrodes, *pIon* or *ion meters* would perhaps be a more descriptive name. Modern ion meters are generally of the direct-reading type; thus, they are the only ones that will be described here.

20D-1 DIRECT-READING INSTRUMENTS

Numerous direct-reading pH meters are available commercially. Generally, these are solid-state devices employing a field-effect transistor or a voltage follower as the first amplifier stage in order to provide the needed high internal resistance. Figure

20–11 gives a schematic diagram of a simple, battery-operated pIon meter that can be built for approximately $50 (exclusive of the electrodes). Here, the output of the ion electrode is connected to a high-resistance field-effect transistor. The output from the operational amplifier is displayed on a meter having a scale extending from −100 to 100 μA. These extremes correspond to a pH range of 1 to 14.

20D–2 COMMERCIAL INSTRUMENTS

A wide variety of ion meters are available from several instrument manufacturers. For example, a 1975 publication[13] listed over one hundred models offered by more than twenty companies. Although the information in this paper is somewhat dated, it does provide a realistic picture of the kinds

[13] J. A. Hauber and C. R. Dayton, *Amer. Lab.*, **1975, 7** (4), 73.

of p-ion meters that are currently available to the scientist. Four categories of meters based on price and readability are described. These include *utility* meters, which are portable, usually battery-operated instruments that currently range in price from $100 to somewhat over $500. Generally, utility meters are readable to 0.1 pH unit or better. *General purpose* meters are line-operated instruments, which are readable to 0.05 pH unit or better. Some offer such features as digital three-digit readout, automatic temperature compensation, scale expansions so that full-scale covers 1.4 units instead of 0 to 14 units, and a millivolt scale. Currently, prices for general purpose meters range from $300 to $900. *Expanded-scale* instruments are generally readable to 0.01 pH unit or better and cost from about $700 to $1500. Most offer full-scale ranges of 0.5 to 2 pH units (as well as 0 to 7 and 0 to 14 ranges), four-digit readout, push-button control, millivolt scale, and automatic tem-

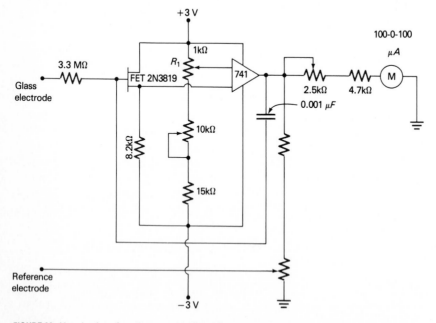

FIGURE 20–11 A simple pIon meter based upon a field-effect transistor and an operational amplifier. (From: D. Sievers, *J. Chem. Educ.*, **1981, 58**, 281. With permission.)

perature compensation. *Research* meters are readable to 0.001 pH unit or better, generally have a five-digit display, and cost in the range of about \$1500 to \$2200. It should be pointed out that the readability of these instruments is usually significantly better than the sensitivity of most ion-selective electrodes.

20E   Direct Potentiometric Measurements

The determination of an ion or molecule by direct potentiometric measurement is rapid and simple, requiring only a comparison of the potential developed by the indicator electrode in the test solution with its potential when immersed in one or more standard solutions of the analyte; insofar as the response of the electrode is specific for the analyte, no preliminary separation steps are required. In addition, direct potentiometric measurements are readily adapted to the continuous and automatic monitoring of ion activities.

Notwithstanding these attractive advantages, the user of direct potentiometric measurements must be alert to limitations that are inherent in the method. An important example is the existence in most potentiometric measurements of a *liquid junction* potential across the salt bridge connecting reference electrode 1 to the analyte solution (E_j in Figure 20–3**b**). For most electroanalytical methods, this junction potential is inconsequential and can be neglected. Unfortunately, however, its existence places a limitation on the accuracy that can be attained from a direct potentiometric measurement.

20E-1   BASIS OF DIRECT POTENTIOMETRY

The direct potentiometric determination of cations is based upon Equation 20–11. For anions, it must be modified by a sign change. That is,

$$\text{pA}_{\text{anion}} = \frac{n(L - E_V)}{0.0591} \qquad (20\text{--}25)$$

It should also be noted that the constant 0.0591 applies at 25°C only. At 5°C it has a value of 0.0552 and at 45°C, 0.0631. Most pH meters have temperature compensating circuits that permit the operator to adjust this constant according to the temperature of the analyte solution. More sophisticated instruments have automatic temperature compensators.

Generally, the constant L in Equations 20–11 and 20–25 *cannot be evaluated from theory* because it contains a junction potential and sometimes an asymmetry potential. Thus, it must be determined experimentally with one or more standard solutions of A. Several methods for performing analyses by direct potentiometry have been developed; all are based, directly or indirectly, upon Equation 20–11 or 20–25.[14]

20E-2   ELECTRODE CALIBRATION METHODS

Equation 20–11 can be rearranged to the form

$$E_V = \frac{0.0591}{n} \text{pA} + L$$
$$= -\frac{0.0591}{n} \log a_1 + L \qquad (20\text{--}26)$$

Thus, at room temperature a plot of the potential of a cell containing a cation electrode as function of pA is in principle a straight line with a slope of 0.0591/n and an intercept of L (note that when A is an anion, the sign of the slope reverses). The behavior of some membrane electrodes follows Equation 20–26 reasonably well; others, however, exhibit slopes at room temperature

---

[14] For additional information, see: *Ion-Selective Electrodes*, R. A. Durst, Ed., National Bureau of Standards Special Publication 314, Chapter 11, U.S. Government Printing Office: Washington, D.C., 1969.

that differ by as much as 0.002 to 0.003 V from $0.0591/n$; furthermore, the slope may change with use.

In the simplest calibration method, the electrode system is immersed in a solution of known pA, and the output of the meter (usually in units of pA) is observed. If the reading differs from the pA of the standard, the assumption is made that this difference is due solely to L in Equation 20–26 (and not to a difference in the slope, $0.0591/n$). Most ion meters are equipped with a potentiometer that provides a counter potential that will bring the reading into agreement with the pA of the standard. For the instrument shown in Figure 20–11, this adjustment is made by means of the potentiometer R_1. Note that the required correction applies at all values of pA providing that Equation 20–26 is followed.

As we have noted earlier, the asymmetry potential for membrane electrodes changes slowly with time. Thus, calibration should be performed at the time that the analyte pA is determined; recalibration is usually desirable every few hours and certainly for day-to-day analyses.

While many of the electrodes described in this chapter exhibit what is known as *Nernstian behavior* (that is, at 25°C, the slopes of their calibration curves are $+0.0591/n$ or $-0.0591/n$), some do not. If any doubt exists about the behavior of a given electrode, at least two standard solutions should be used for calibration, one having a pA greater than that expected for the analyte and the other less. The resulting data give a value for L as well as an empirical value for the slope. Periodic use of at least two calibration standards so that the slope can be determined is often a worthwhile precaution.

The direct electrode calibration method offers the advantages of simplicity, speed, and applicability to the continuous monitoring of pA. Two important disadvantages attend its use, however. One of these is that the results of an analysis are in terms of ac-

tivities rather than concentrations; the other is that the accuracy of a measurement obtained by this procedure is limited by the inherent uncertainty caused by the junction potential (E_j in Figure 20–3b); unfortunately, this uncertainty can never be totally eliminated.

Activity versus Concentration. Electrode response is related to activity rather than to analyte concentration. Ordinarily, however, the scientist is interested in concentration, and the determination of this quantity from a potentiometric measurement requires activity coefficient data. More often than not, activity coefficients will be unavailable because the ionic strength of the solution is either unknown or so high that the Debye-Hückel equation is not applicable. Unfortunately, the assumption that activity and concentration are identical may lead to serious errors, particularly when the analyte is polyvalent.

The difference between activity and concentration is illustrated by Figure 20–12, where the lower curve gives the change in potential of a calcium electrode as a function of calcium chloride concentration (note that the activity or concentration scale is logarithmic). The nonlinearity of the curve is

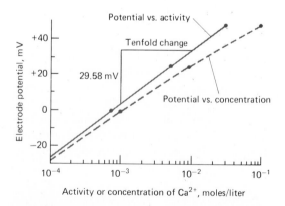

FIGURE 20–12 Response of a calcium ion electrode to variations in the calcium ion concentration and activity of solutions prepared from pure calcium chloride. (Courtesy of Orion Research, Inc., Cambridge, MA.)

due to the increase in ionic strength—and the consequent decrease in the activity coefficient of the calcium ion—as the electrolyte concentration becomes larger. When these concentrations are converted to activities, the upper curve is obtained; note that this straight line has the Nernstian slope of 0.0296 (0.0591/2).

Activity coefficients for singly charged ions are less affected by changes in ionic strength than are coefficients for species with multiple charges. Thus, the effect shown in Figure 20–12 will be less pronounced for electrodes that respond to H^+, Na^+, and other univalent ions.

In potentiometric pH measurements, the pH of the standard buffer employed for calibration is generally based on the activity of hydrogen ions. Thus, the resulting hydrogen ion results are also on an activity scale. If the unknown sample has a high ionic strength, the hydrogen ion *concentration* will differ appreciably from the activity measured.

Inherent Error in the Electrode Calibration Procedure. A serious disadvantage of the electrode calibration method is the existence of an inherent uncertainty that results from the assumption that L in Equation 20–11 remains constant between calibration and analyte determination. This assumption can seldom, if ever, be exactly true because the electrolyte composition of the unknown will almost inevitably differ from that of the solution employed for calibration. The junction potential contained in L (E_j in Equation 20–6) will vary slightly as a consequence, even though a salt bridge is used. This uncertainty will frequently be of the order of 1 mV or more; unfortunately, because of the nature of the potential-activity relationship, such an uncertainty has an amplified effect on the inherent accuracy of the analysis. The magnitude of the uncertainty in analyte concentration can be estimated by differentiating Equation 20–11 with respect to L while holding E_V constant

$$-\log_{10} e \, \frac{da_1}{a_1} = -0.434 \frac{da_1}{a_1} = -\frac{dL}{0.0591/n}$$

$$\frac{da_1}{a_1} = \frac{ndL}{0.0256}$$

Upon replacing da_1 and dL with finite increments and multiplying both sides of the equation by 100, we obtain

$$\frac{\Delta a_1}{a_1} \times 100 = 3.9 \times 10^3 \, n\Delta L = \% \text{ rel error}$$

The quantity $\Delta a_1/a_1$ is the relative error in a_1 associated with an absolute uncertainty ΔL in L. If, for example, ΔL is ±0.001 V, a relative error in activity of about ±$4n$% can be expected. *It is important to appreciate that this uncertainty is characteristic of all measurements involving cells that contain a salt bridge and that this error cannot be eliminated by even the most careful measurements of cell potentials or the most sensitive and precise measuring devices;* nor does it appear possible to devise a method for completely eliminating the uncertainty in L that is the source of this error.

20E-3 CALIBRATION CURVES FOR CONCENTRATION MEASUREMENT

An obvious way of correcting potentiometric measurements to give results in terms of concentration is to make use of an empirical calibration curve such as the lower curve in Figure 20–12. For this approach to be successful, however, it is essential that the ionic composition of the standard closely approximate that of the analyte—a condition that is difficult to realize experimentally for complex samples.

Where electrolyte concentrations are not too great, it is often helpful to swamp both the samples and the calibration standards with a measured excess of an inert electrolyte. Under these circumstances, the added effect of the electrolyte in the sample becomes negligible, and the empirical calibration curve yields results in terms of concentration. This approach has been employed for the potentiometric determination of fluoride in public water supplies with a lanthanum fluoride electrode. Here, both samples and standards are diluted on a 1 : 1 basis with a solution containing sodium chloride, a citrate buffer, and an acetate buffer (this mixture, which fixes both ionic strength and pH, is sold under the name "Total Ionic Strength Adjusting Buffer" or TISAB); the diluent is sufficiently concentrated so that the samples and standards do not differ significantly in ionic strength. The procedure permits a rapid measurement of fluoride ion in the 1-ppm range, with a precision of about 5% relative.

Calibration curves are also useful for electrodes that do not respond linearly to pA.

20E-4   STANDARD ADDITION METHOD

The standard addition method, described in Section 7F–2, is equally applicable to potentiometric determinations. Here, the potential of the electrode system is measured before and after addition of a small volume (or volumes) of a standard to a known volume of the sample. The assumption is made that this addition does not alter the ionic strength and thus the activity coefficient f of the analyte. It is further assumed that the added standard does not significantly alter the junction potential.

The standard addition method has been applied to the determination of chloride and fluoride in samples of commercial phosphors.[15] In this application, solid-state indicator electrodes for chloride and fluoride were used in conjunction with a reference electrode; the added standard contained known quantities of the two anions. The relative standard deviation for the measurement of replicate samples was found to be 0.7% for fluoride and 0.4% for chloride.

When the standard addition method was not used, relative errors for the analyses appeared to range between 1 and 2%.

20E-5   POTENTIOMETRIC pH MEASUREMENTS WITH A GLASS ELECTRODE[16]

The glass electrode is unquestionably the most important indicator electrode for hydrogen ion. It is convenient to use and is subject to few of the interferences that affect other pH-sensing electrodes.

Glass electrodes are available at relatively low cost and in many shapes and sizes. A common variety is illustrated in Figure 20–5; the reference electrode is usually a commercial saturated calomel electrode.

The glass/calomel electrode system is a remarkably versatile tool for the measurement of pH under many conditions. The electrode can be used without interference in solutions containing strong oxidants, reductants, proteins, and gases; the pH of viscous or even semisolid fluids can be determined. Electrodes for special applications are available. Included among these are small electrodes for pH measurements in a drop (or less) of solution or in a cavity of a tooth, microelectrodes which permit the measurement of pH inside a living cell, systems for insertion in a flowing liquid stream to provide a continuous monitoring of pH, a small glass electrode that can be swallowed to indicate the acidity of the stomach contents (the calomel electrode is kept in the mouth), and combination electrodes that contain both indicator and reference electrodes in a single probe.

Summary of Error Affecting pH Measurements with the Glass Electrode.   The ubiquity of the pH meter and the general applicability of the glass electrode tend to lull the chemist into the attitude that any measurement obtained with

---

[15] L. G. Bruton, *Anal. Chem.*, **1971**, *43*, 579.

[16] For a detailed discussion of potentiometric pH measurements, see: R. G. Bates, *Determination of pH: Theory and Practice*, 2d ed., Wiley: New York, 1973.

such an instrument is surely correct. It is well to guard against this false sense of security since there are distinct limitations to the electrode system. These have been discussed in earlier sections and include the following:

1. **The alkaline error.** The ordinary glass electrode becomes somewhat sensitive to alkali-metal ions at pH values greater than 11 to 12.
2. **The acid error.** At a pH less than 0.5, values obtained with a glass electrode tend to be somewhat high.
3. **Dehydration.** Dehydration of the electrode may cause unstable performance and errors.
4. **Errors in unbuffered neutral solutions.** Equilibrium between the electrode surface layer and the solution is achieved only slowly in poorly buffered, approximately neutral solutions. Errors will arise unless time (often several minutes) is allowed for this equilibrium to be established. In determining the pH of poorly buffered solutions, the glass electrode should first be thoroughly rinsed with water. Then, if the unknown is plentiful, the electrodes should be placed in successive portions until a constant pH reading is obtained. Good stirring is also helpful; several minutes should be allowed for the attainment of steady readings.
5. **Variation in junction potential.** It should be reemphasized that this is a fundamental uncertainty in the measurement of pH, for which a correction cannot be applied. Absolute values more reliable than 0.01 pH unit are generally unobtainable. Even reliability to 0.03 pH unit requires considerable care. On the other hand, it is often possible to detect pH *differences* between similar solutions or pH *changes* in a single solution that are as small as 0.001 unit. For this reason, many pH meters are designed to permit readings to less than 0.01 pH unit.
6. **Error in the pH of the standard buffer.** Any inaccuracies in the preparation of the buffer used for calibration, or changes in its composition during storage, will be propagated as errors in pH measurements. A common cause of deterioration is the action of bacteria on organic components of buffers.

20F Potentiometric Titrations

The potential of a suitable indicator electrode is conveniently employed to establish the equivalence point for a titration (a *potentiometric titration*).[17] A potentiometric titration provides different information from a direct potentiometric measurement. For example, the direct measurement of 0.100 M acetic and 0.100 M hydrochloric acid solutions with a pH-sensitive electrode yield widely different pH values because the former is only partially dissociated. On the other hand, potentiometric titrations of equal volumes of the two acids require the same amount of standard base for neutralization.

The potentiometric end point is widely applicable and provides inherently more accurate data than the corresponding method employing indicators. It is particularly useful for titration of colored or turbid solutions and for detecting the presence of unsuspected species in a solution. Unfortunately, it is more time-consuming than a titration that makes use of an indicator unless an automatic titrator is used.

As shown in Figure 20–13, the apparatus for a potentiometric titration can be relatively simple. Ordinarily, the titrant is introduced in large increments initially; additions become smaller as the end point is approached (as indicated by a larger poten-

---

[17] For a recent monograph on this method, see: E. P. Sergeant, *Potentiometry and Potentiometric Titrations*, Wiley: New York, 1984.

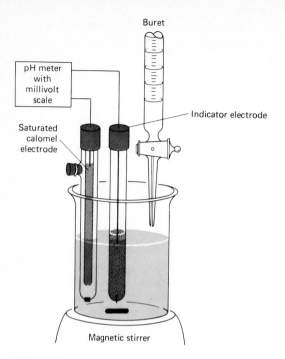

FIGURE 20–13 Apparatus for a potentiometric titration.

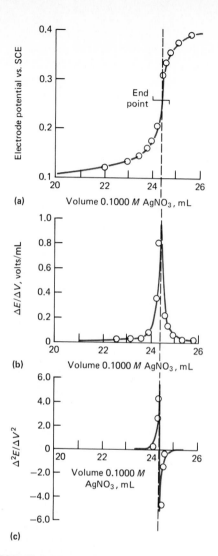

FIGURE 20–14 (**a**) Potentiometric titration curve for 2.433 meq of Cl⁻ with 0.1000 M AgNO₃. (**b**) First-derivative curve. (**c**) Second-derivative curve.

tial change per increment) with the final additions being a drop or less at a time. In many titrations, and particularly near the end point, several seconds must be allowed after each addition to permit attainment of chemical equilibrium. The approach to equilibrium is indicated when a potential drift of less than 1 or 2 mV is observed over a 30-sec period. Good stirring often hastens the approach to equilibrium. Precipitation titrations may require several minutes for equilibration in the end-point region.

The first two columns of Table 20–6 consist of typical potentiometric titration data obtained with the apparatus shown in Figure 20–13. The data near the end point are plotted in Figure 20–14**a**.

20F–1　END-POINT DETERMINATION

Several methods are used to determine the end point for a potentiometric titration. The most straightforward involves a direct plot of potential versus reagent volume, as in Figure 20–14**a**. The midpoint in the steeply rising portion of the curve is then estimated visually and taken as the end point. Various mechanical methods to aid in the establishment of the midpoint have been proposed; it is doubtful, however, that these significantly improve the accuracy.

A second approach is to calculate the

change in potential per unit change in volume of reagent (that is, $\Delta E/\Delta V$), as has been done in column 3 of Table 20–6. A plot of this parameter as a function of average volume leads to a sharp maximum at the end point (see Figure 20–14**b**). Alternatively, the ratio can be evaluated during the titration and recorded directly in lieu of the potential itself. Thus, in column 3 of Table 20–6, it is seen that the maximum is located between 24.3 and 24.4 mL; selection of 24.35 mL would be adequate for most purposes.

Column 4 in Table 20–6 and Figure 20–14**c** show that the second derivative of the data changes sign at the point of inflection in the titration curve. This change is often used as the analytical signal in automatic titrators.

Still another method of locating potentiometric end points involves the use of Gran's plots.[18] Here, an antilog function of the observed potential is plotted against reagent volume. A straight line results that can be extrapolated to zero for the antilog function; the intercept corresponds to the endpoint volume. Only a few points need be taken in the early part of the titration. Semiantilog paper is available from a commercial source[19], which makes possible the acquisition of Gran's plots without computations. The vertical axis of this paper is skewed to eliminate the need to correct for

---

[18] G. Gran, *Analyst,* **1952,** 77, 661.

[19] Orion Research Inc., 380 Putman Ave., Cambridge, MA 02139.

Table 20-6
Potentiometric Titration Data for 2.433 Milliequivalents of Chloride with 0.1000 M Silver Nitrate[a]

Vol AgNO$_3$, mL	E vs. SCE, V	$\Delta E/\Delta V$, V/mL	$\Delta^2 E/\Delta V^2$, V/mL^2
5.0	0.062		
		0.002	
15.0	0.085		
		0.004	
20.0	0.107		
		0.008	
22.0	0.123		
		0.015	
23.0	0.138		
		0.016	
23.50	0.146		
		0.050	
23.80	0.161		
		0.065	
24.00	0.174		
		0.09	
24.10	0.183		
		0.11	
			2.8
24.20	0.194		
		0.39	
			4.4
24.30	0.233		
		0.83	
			−5.9
24.40	0.316		
		0.24	
			−1.3
24.50	0.340		
		0.11	
			−0.4
24.60	0.351		
		0.07	
24.70	0.358		
		0.050	
25.00	0.373		
		0.024	
25.5	0.385		
		0.022	
26.0	0.396		
		0.015	
28.0	0.426		

[a] The electrode system consists of a silver indicator electrode and a saturated calomel electrode. The volume of the analyte solution was 50.00 mL.

volume change, provided the titrant volume is less than 10% of the total at the end point.

20F-2 APPLICATIONS OF POTENTIOMETRIC TITRATIONS

Potentiometric measurements have been employed to establish end points for most precipitation, complex formation, neutralization, and oxidation-reduction titrations.

Precipitation Titrations. The most widely used volumetric precipitation reagent is silver nitrate, which permits determination of the halogens, the halogenoids, mercaptans, sulfides, arsenates, phosphates, and oxalates. Here, a silver indicator electrode or a silver sulfide membrane electrode is used. When dilute solutions are being titrated or where the highest precision is required, reference electrodes such as those shown in Figure 20-1 cannot be used directly because of leakage of chloride ion from the salt bridge. This problem is avoided by immersing the calomel electrode in a concentrated potassium nitrate solution; this solution is then connected to the analyte solution by means of an agar bridge that contains about 3% potassium nitrate. Alternatively, a commercial double junction reference electrode containing a potassium nitrate solution can be used. For reagent and analyte concentrations of 0.1 M or greater, a calomel electrode such as that shown in Figure 20-1a can be immersed directly in the analyte solution without incurring significant errors. Figure 20-14a shows a typical precipitation curve.

An important advantage of the potentiometric method is that it often permits discrimination among components of a mixture that react with a common titrant. Figure 20-15 illustrates this type of application for the determination of iodide, bromide, and chloride ions in an approximately equimolar mixture. During the early stages of the titration, only silver iodide, the least soluble of the three silver halides, precipitates. Indeed, it is readily shown that under

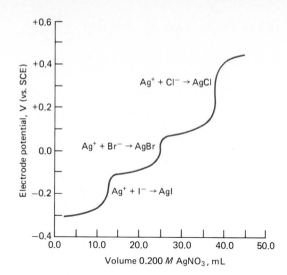

FIGURE 20-15 Potentiometric titration of 2.50 mmol of I$^-$, Br$^-$, and Cl$^-$ with 0.200 M AgNO$_3$.

equilibrium conditions, precipitation of silver bromide will not in theory begin until all but about 0.02% of the iodide originally present has precipitated. Thus, to this point, the titration curve will be identical to that for iodide by itself. Precipitation of silver bromide then occurs; here again, it can be shown that in theory no silver chloride forms until all but about 0.3% of the bromide ion has been removed from solution. Therefore, the titration curve in this region is similar to that for bromide ion alone. Finally, chloride starts to precipitate, and the remainder of the curve is similar to the chloride titration curve in Figure 20-14a.

Complex Formation Titrations. Both metal and membrane electrodes have been applied to the detection of end points for reactions that involve formation of soluble complexes. By far, the most important reagents for complexometric titrations are a group of amino carboxylic acids, of which ethylenediaminetetraacetic acid is an example. This reagent, which is commonly abbreviated as EDTA and given the formula H$_4$Y, has the structure

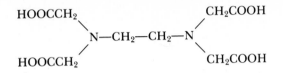

The four carboxylate functional groups, as well as the two amine nitrogens, participate in bond formation with metal ions. Regardless of the charge on the cation, the complex formation reaction can be formulated as

$$M^{n+} + H_4Y \rightleftharpoons MY^{(n-4)} + 4H^+ \qquad (20\text{-}27)$$

An important example of the application of this reagent is the determination of calcium and magnesium ions in hard water. The reaction for calcium is

$$Ca^{2+} + H_4Y \rightarrow CaY^{2-} + 4H^+$$

Potentiometric end points for EDTA ti-

trations can be obtained with a mercury electrode (see Section 20B–2). Reilley and co-workers have made a systematic theoretical and experimental study of the application of this electrode to the potentiometric determination of 29 di-, tri-, and quadrivalent cations with EDTA as the reagent.[20] In these studies, 5 to 500 mg quantities of the cation were titrated with 0.05 or 0.005 M reagent solutions. One drop of a 10^{-3} M solution of HgY^{2-} was employed for each titration.

Figure 20–16 illustrates an application of their procedure to the determination of bismuth, cadmium, and calcium in a mixture. Bismuth(III) is first titrated at a pH of 1.2. At this acidity, neither cadmium nor cal-

---

[20] See: C. N. Reilley and R. W. Schmid, *Anal. Chem.*, **1958**, *30*, 947; and C. N. Reilley, R. W. Schmid, and D. W. Lamson, *Ibid.*, 953.

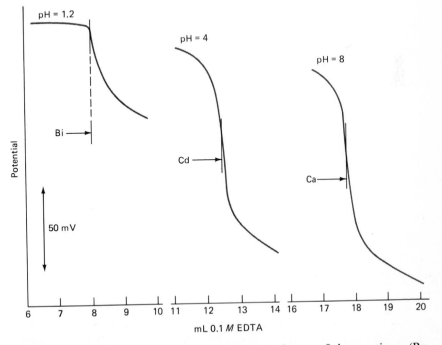

FIGURE 20-16 Potentiometric EDTA titration of a mixture of three cations. (Reprinted with permission from: C. N. Reilley, R. W. Schmid, and D. W. Lamson, *Anal. Chem.*, **1958**, *30*, 957. Copyright 1958 American Chemical Society.)

cium ions react to any significant extent. That is, for these two ions the equilibrium shown by Equation 20–27 lies far to the left. Bismuth, on the other hand, forms a complex that is sufficiently stable to provide a satisfactory end point. After bismuth has been titrated, the solution is brought to a pH of 4 by addition of an acetate/acetic acid buffer, and the titration is continued to an end point for cadmium. Calcium ions do not react appreciably at this pH but can be subsequently titrated in a basic solution obtained by addition of an ammonia/ammonium chloride buffer.

Neutralization Titrations. Potentiometric acid-base titration curves are readily obtained with a glass/calomel electrode system. End points from such curves are particularly useful when colored or turbid solutions must be titrated.

Figure 20–17**a** shows theoretical curves for acids having different dissociation constants when titrated with standard sodium hydroxide. Figure 20–17**b** shows the effect of analyte and reagent concentrations on the titration curve for one of these acids.

Experimental curves, obtained with a glass/calomel electrode system, approximate these theoretical curves closely. Clearly, as the strength of the acid becomes less (and thus the reaction with base less complete) and as the concentration becomes lower, the ease of end-point detection becomes poorer. Furman[21] has indicated that in order to obtain an accuracy of 1% relative in the potentiometric titration of a weak acid with a strong base, it is necessary that

$$M_{\text{NaA}} K_a \geq 10^{-5}$$

where M_{NaA} is the end point concentration of the sodium salt of the acid being titrated and K_a is the dissociation constant for the acid.

The foregoing considerations apply equally to the titration of solutions of weak bases with strong acids.

Potentiometric titrations are advanta-

---

[21] N. H. Furman, in *Treatise on Analytical Chemistry*, I. M. Kolthoff and P. J. Elving, Eds., Part I, vol. 4, p. 2287, Wiley-Interscience: New York, 1967.

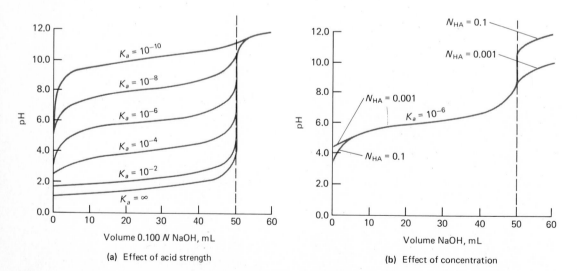

(a) Effect of acid strength

(b) Effect of concentration

FIGURE 20-17 Titration curves for weak acids: (**a**) 50.0 mL of 0.100 N acid in each case; (**b**) 50.0 mL of 0.100 N and 0.001 N HA with 0.100 N and 0.00100 N NaOH.

geously employed for the titration of mixtures of acids (or bases). If the dissociation constants differ sufficiently, it is possible to determine the concentrations of the individual components in such mixtures. Generally, the concentrations of the two components of a mixture of acids (or bases) can be determined if the ratio of their dissociation constants is 10^4 or greater.

Oxidation-Reduction Titrations. A platinum electrode responds rapidly to many important oxidation-reduction couples and develops a potential that depends upon the concentration (strictly, activity) ratio of the reactants and the products of such half-reactions. For example, a platinum electrode system can be used to determine the end point in the titration of iron(II) with a standard solution of cerium(IV). The platinum electrode is responsive to both oxidation-reduction systems that exist in the solution throughout the titration. That is,

$$Ce^{4+} + e \rightleftharpoons Ce^{3+} \qquad E^0 = 1.44 \text{ V}$$

$$Fe^{3+} + e \rightleftharpoons Fe^{2+} \qquad E^0 = 0.771 \text{ V}$$

After each addition of reagent, interaction among the species occurs until reactant and product concentrations are such that the *electrode potentials for the two half-reactions are identical.* That is, at equilibrium,

$$E_{Pt} = 1.44 - \frac{0.0591}{1} \log \frac{[Ce^{3+}]}{[Ce^{4+}]}$$

$$= 0.771 - \frac{0.0591}{1} \log \frac{[Fe^{2+}]}{[Fe^{3+}]}$$

Thus, the measured potential can be thought of as arising from *either* of the two half-cell systems.

The change in the ordinate function in the equivalence-point region of an oxidation-reduction titration becomes larger as the reaction becomes more complete; this effect is identical with that encountered for

other reaction types. Thus, in Figure 20–18, data are plotted for titrations involving a hypothetical analyte that has a standard potential of 0.2 V with several reagents that have standard potentials ranging from 0.4 to 1.2 V; the corresponding equilibrium constants for the reaction range from about 2×10^3 to 9×10^{16}. Clearly, an increase in completeness is accompanied by an increased change in the electrode potential in the end-point region.

The curves in Figure 20–18 were derived for reactions in which the oxidation and reductant each exhibit a one-electron change; where both reactants undergo a two-electron transfer, the change in potential in the region of 24.9 to 25.1 mL is larger by about 0.14 V.

Solutions containing two oxidizing agents or two reducing agents will yield titration curves that contain two inflection points, provided the standard potentials for the two

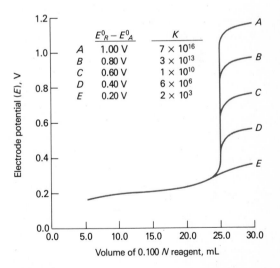

FIGURE 20–18 Titration of 50.0 mL of 0.0500 N A. E_A^0 is assumed to be 0.200 V. From the top, E_R^0 for the reagent is 1.20, 1.00, 0.80, 0.60, and 0.40 V, respectively. Both the reagent and analyte are assumed to undergo a one-electron change.

species are sufficiently different. If this difference is greater than about 0.2 V, the end points are usually distinct enough to permit determination of each component. This situation is quite comparable to the titration, with the same reagent, of two acids with different dissociation constants or of two ions that form precipitates of different solubilities.

20F-3 AUTOMATIC TITRATIONS

In recent years, several automatic titrators based on the potentiometric principle have become available from commercial sources. These instruments are useful where many routine analyses are required.[22] Automatic titrators cannot yield results that are more accurate than those obtained by manual potentiometric techniques; however, they do decrease the operator time needed to perform the titrations and may allow the use of

---

[22] For a comprehensive discussion of automatic titrators, see: G. Svehla, *Automatic Potentiometric Titrations*, Pergamon Press: New York, 1978; and J. K. Foreman and P. B. Stockwell, *Automatic Chemical Analysis*, pp. 44–62, Wiley: New York, 1975.

smaller samples. Thus, automatic titrators often offer significant economic advantages.

Flow Control and Measurement of Reagent Volume. Several methods exist for automatic control of the addition of a reagent and the measurement of its volume. The simplest employs an ordinary buret in which the stopcock is replaced with an electromagnetic pincer device. Here, an elastic plastic tube is inserted between the buret body and the tip. Flow is prevented by pinching the tube between a spring-loaded soft iron piece and a metal wedge. Titrant is introduced by passage of electricity through a solenoid that surrounds the pinching device. In another type of valve, a small piece of iron is sealed into a glass or plastic tube that fits inside the outflow tube of a buret. The two surfaces are ground to form a stopper. Current in a solenoid unseats the stopper and allows a flow of reagent.

The most widely used apparatus for automatic reagent addition consists of a calibrated syringe that is activated by a motor-driven micrometer screw. Such a device is shown in Figure 20–20.

Preset End-Point Titrators. Figure 20–19 is a schematic diagram of the simplest and least ex-

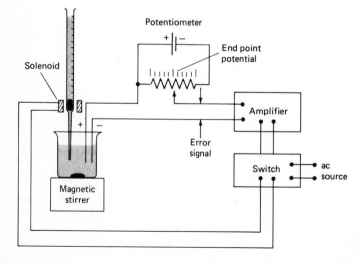

FIGURE 20-19 Automatic preset end-point titrator.

pensive type of automatic titrator. Here, a preset equivalence point potential is applied across the electrodes by means of a calibrated potentiometer. If a difference exists between this potential and that of the electrodes, an "error" signal results. This signal is amplified and closes an electronic switch that permits a flow of electricity through the solenoid-operated valve of the buret. As the error signal approaches zero, current to the solenoid is switched off, and flow of titrant ceases.

Titrators such as the one shown in Figure 20–19 do not respond instantaneously and tend to overshoot the end point. To overcome this problem, some instruments are equipped to superimpose a square-wave signal upon the error signal. The solenoid switch is then set to close only when the net signal exceeds that of the square wave. Titrant is then added in increments that can be controlled by the frequency of the square wave signal.

Second-Derivative Titrators. Second-derivative titrators can also be relatively simple devices; they have the advantage that no preknowledge of the equivalence-point potential is required. The signal processor of these devices contains two electronic derivative circuits (p. 52) in series to convert the amplified signal from the electrode to a voltage proportional to the second derivative of the electrode potential of the indicator electrode. The output is then similar in form to that shown in Figure 20–14**c**, where the sign of the signal changes at the equivalence point. This change in sign then causes a switching device to turn off the flow of titrant.

Recording Titrators. Recording titrators carry a titration beyond the equivalence point while recording a curve for the analyte. In some instruments, the rate of reagent addition is held constant and is synchronized with the chart drive of a millivolt recorder. The pen then records the amplified output potential of the cell as a function of time, which is proportional to the reagent volume.

Often, recording titrators are equipped to perform a titration the way a skilled chemist does—that is, to add titrant rapidly before and after the end point but in small increments as the end point is approached and passed. Figure 20–20 is a simplified block diagram of such an instrument. The switching device in this instrument operates from an amplified error signal that develops when the potential of the electrodes differs by some minimum amount from the potential applied by the potentiometer. The contact of this potentiometer is driven by the motor, which also positions the pen of the recorder. When the error signal is small, the electronic switch activates the motor of the syringe-type buret and the chart drive of the recorder; simultaneously, the potentiometer and pen drive motor is switched off. When sufficient reagent has been added to cause the error signal to exceed some predetermined level, the switching is reversed. Now, the buret and chart drives are off and the potentiometer and pen drives are turned on. Thus, the reagent is added in a series of steps, as shown by the recorded titration curve in Figure 20–20**c**.

Fully Automatic Titrators. A fully automatic titrator is equipped with a turntable that holds a series of samples for titration. After a titration has been completed, the solution is discarded, the vessel and electrodes are rinsed, the buret is refilled, the sample table is rotated, a measured volume of a new sample is introduced into the system, and the titration process is resumed. Such instruments are controlled by microprocessors and usually have computing facilities for calculating and printing out the analytical results. Instruments of this kind are expensive, but their costs can be easily justified for laboratories that must perform large numbers of routine titrations regularly.

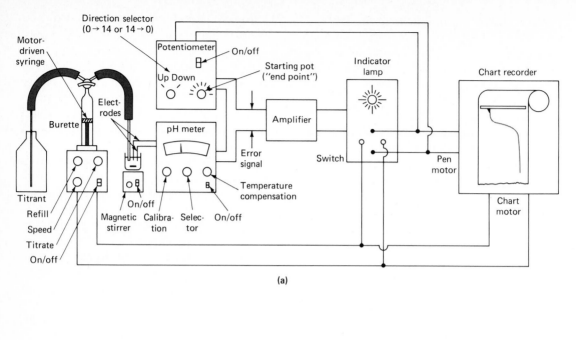

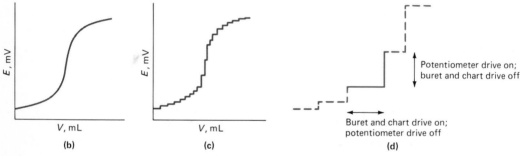

FIGURE 20-20 (**a**) An automated, curve-recording titrator. (**b**) Theoretical titration curve. (**c**) Recorded curve with end-point anticipation. (**d**) Enlarged portion of (**c**) with explanation. (Reprinted with permission from: G. Svehla, *Automatic Potentiometric Titrations*, p. 176, Pergamon Press: New York, 1978. Copyright 1978 Pergamon Press Ltd.)

PROBLEMS   20–1. (a) Calculate the standard potential for the reaction

$$CuSCN(s) + e \rightleftharpoons Cu(s) + SCN^-$$

For CuSCN, $K_{sp} = 4.8 \times 10^{-15}$.

(b) Give a schematic representation of a cell with a copper indicator electrode as an anode and a saturated calomel electrode as a cathode that could be used for the determination of SCN^-.

(c) Derive an equation that relates the measured potential of the cell in (b) to pSCN (assume that the junction potential is zero).

(d) Calculate the pSCN of a thiocyanate-containing solution that is saturated with CuSCN and contained in the cell described in (b) if the resulting potential is 0.411.

20–2. (a) Calculate the standard potential for the reaction

$$Ag_2S(s) + 2e \rightleftharpoons 2Ag(s) + S^{2-}$$

For Ag_2S, $K_{sp} = 6 \times 10^{-50}$.

(b) Give a schematic representation of a cell with a silver indicator electrode as the anode and a saturated calomel electrode as a cathode that could be used for determining S^{2-}.

(c) Derive an equation that relates the measured potential of the cell in (b) to pS (assume that the junction potential is zero).

(d) Calculate the pS of a solution that is saturated with Ag_2S and contained in the cell described in (b) if the resulting potential is 0.594 V.

20–3. The following cell was employed for the determination of $pCrO_4$:

$$Ag|Ag_2CrO_4(sat'd),CrO_4^{2-}(xM)\|SCE$$

Calculate $pCrO_4$ if the cell potential is -0.402 V.

20–4. The following cell was employed to determine the pSO_4 of a solution:

$$Hg|Hg_2SO_4(sat'd),SO_4^{2-}(xM)\|SCE$$

Calculate the pSO_4 is the potential was -0.576 V.

20–5. The formation constant for the mercury(II) acetate complex is

$$Hg^{2+} + 2OAc^- \rightleftharpoons Hg(OAc)_2(aq) \qquad K_f = 2.7 \times 10^8$$

Calculate the standard potential for the half-reaction

$$Hg(OAc)_2(aq) + 2e \rightleftharpoons Hg(l) + 2OAc^-$$

20–6. The standard electrode potential for the reduction of the Cu(II) complex of EDTA is given by

$$CuY^{2-} + 2e \rightleftharpoons Cu(s) + Y^{4-} \qquad E^0 = 0.13 \text{ V}$$

Calculate the formation constant for the reaction

$$Cu^{2+} + Y^{4-} \rightleftharpoons CuY^{2-}$$

20–7. Calculate the potential of the cell (neglecting the junction potential)

$$Hg|HgY^{2-}(2.61 \times 10^{-4}\ M),Y^{4-}(xM)\|SCE$$

where Y^{4-} is the EDTA anion, and the concentration of Y^{4-} is (a) $1.21 \times 10^{-1}\ M$, (b) $1.21 \times 10^{-3}\ M$, and (c) $1.21 \times 10^{-5}\ M$.

$$HgY^{2-} + 2e \rightleftharpoons Hg(l) + Y^{4-} \qquad E^0 = 0.21\ V$$

20–8. The following cell was found to have a potential of 0.124 V:

  membrane electrode for $Cu^{2+}|Cu^{2+}(3.25 \times 10^{-3}\ M)\|SCE$

  When the solution of known copper activity was replaced with an unknown solution, the potential was found to be 0.305 V. What was the pCu of this unknown solution? Neglect the junction potential.

20–9. The following cell was found to have a potential of 1.020 V:

  $Cd|CdX_2(sat'd),X^-(0.0200\ M)\|SCE$

  Calculate the solubility product of CdX_2, neglecting the junction potential.

20–10. The following cell was found to have a potential of 0.515 V:

  $Pt,H_2(1.00\ atm)|HA(0.200\ M),NaA(0.300\ M)\|SCE$

  Calculate the dissociation constant of HA, neglecting the junction potential.

20–11. A 40.00-mL aliquot of 0.0500 M HNO_2 is diluted to 75.0 mL and titrated with 0.0800 M Ce^{4+}. Assume the hydrogen ion concentration is at 1.00 M throughout the titration. (Use 1.44 V for the formal potential of the cerium system.)
   (a) Calculate the potential of the indicator cathode with respect to a saturated calomel reference electrode after the addition of 10.00, 25.00, 40.00, 49.00, 49.9, 50.1, 51.00, 55.00, and 60.00 mL of cerium(IV).
   (b) Draw a titration curve for these data.

20–12. Calculate the potential (vs. SCE) of a lead anode after the addition of 0.00, 10.00, 20.00, 24.00, 24.90, 25.00, 25.10, 26.00, and 30.00 mL of 0.2000 M $NaIO_3$ to 50.00 mL of 0.0500 M $Pb(NO_3)_2$. For $Pb(IO_3)_2$, $K_{sp} = 3.2 \times 10^{-13}$.

20–13. A glass/calomel electrode system was found to develop a potential of -0.0412 V when used with a buffer of pH 6.00; with an unknown solution the potential was observed to be -0.2004 V.
   (a) Calculate the pH and $[H^+]$ of the unknown.
   (b) Assume that L in Equation 20–11 is uncertain by ± 0.001 V as a consequence of a difference in the junction potential between

standardization and measurement. What is the range of $[H^+]$ associated with this uncertainty?

(c) What is the relative error in $[H^+]$ associated with the uncertainty in E_j?

20–14. The following cell was found to have a potential of -0.2714 V:

membrane electrode for $Mg^{2+}|Mg^{2+}(a = 3.32 \times 10^{-3}\ M)\|SCE$

(a) When the solution of known magnesium activity was replaced with an unknown solution, the potential was found to be -0.1901 V. What was the pMg of this unknown solution?

(b) Assuming an uncertainty of ± 0.002 V in the junction potential, what is the range of Mg^{2+} activities within which the true value might be expected?

(c) What is the relative error in $[Mg^{2+}]$ associated with the uncertainty in E_j?

20–15. A 0.400-g sample of toothpaste was boiled with a 50-mL solution containing a citrate buffer and NaCl to extract the fluoride ion. After cooling, the solution was diluted to exactly 100 mL. The potential of a selective ion/calomel system in a 25.0-mL aliquot of the sample was found to be -0.1823 V. Addition of 5.0 mL of a solution containing 0.00107 mg F^-/mL caused the potential to change to -0.1446 V.

Calculate the weight-percent F^- in the sample.

Coulometric Methods

Three electroanalytical methods are based upon electrolytic oxidation or reduction of an analyte for a sufficient period to assure its quantitative conversion to a new oxidation state. These methods are: *constant-potential coulometry, constant-current coulometry* or *coulometric titrations,* and *electrogravimetry.* In electrogravimetric methods, the product of the electrolysis is weighed as a deposit on one of the electrodes. In the two coulometric procedures, on the other hand, the quantity of electricity needed to complete the electrolysis serves as a measure of the amount of analyte present.

The three methods generally have moderate selectivity, sensitivity, and speed; in many instances, they are among the most accurate and precise methods available to the chemist, with uncertainties of a few tenths percent relative being not uncommon. Finally, in contrast to all of the other methods discussed in this text, these three require no calibration against standards; that is, the functional relationship between the quantity measured and the weight of analyte can be derived from theory.

A description of electrogravimetric methods is found in many elementary textbooks.[1] Thus, only the two coulometric methods will be considered in this chapter.

21A Current-Voltage Relationships During an Electrolysis

An electrolysis can be performed in one of three ways: (1) with the applied cell potential held constant, (2) with the electrolysis current held constant, or (3) with the potential of the working electrode held constant. It is useful to consider the consequences of each of these modes of operation. For all three, the behavior of the cell is governed by the relationship

---

[1] For example, see: D. A. Skoog and D. M. West, *Fundamentals of Analytical Chemistry,* 4th ed., Chapter 17, Saunders: Philadelphia, 1982.

$$E_{appl} = E_c - E_a + (\eta_{cc} + \eta_{ck})$$
$$+ (\eta_{ac} + \eta_{ak}) - IR \qquad (21-1)$$

where E_{appl} is the applied potential from an external source and E_c and E_a are the reversible or thermodynamic potentials associated with the cathode and anode, respectively; their values can, of course, be calculated from standard potentials by means of the Nernst equation. The terms η_{cc} and η_{ck} are overvoltages due to concentration polarization and kinetic polarization at the cathode; η_{ac} and η_{ak} are the corresponding anodic overvoltages (see Section 19E–2). It is important to appreciate that the four overvoltages *always carry a negative sign* because they are potentials that must be overcome in order for charge to pass through the cell.[2]

Of the seven potential terms on the right side of Equation 21–1, only E_c and E_a are derivable from theoretical calculations; the others can only be evaluated empirically.

21A–1 OPERATION OF A CELL AT A FIXED APPLIED POTENTIAL

The simplest way of performing an analytical electrolysis is to maintain the applied potential at a more or less fixed level. Ordinarily, however, this procedure has distinct limitations. These limitations can be understood by considering current-voltage relationships in a typical electrolytic cell. For example, consider a cell for the determination of copper(II), which consists of two platinum electrodes, each with a surface area of 150 cm², immersed in 200 mL of a solution that is 0.0220 M with respect to copper(II) ion and 1.00 M with respect to hydrogen ion. The cell resistance is 0.50 Ω. With application of a suitable potential, copper is deposited upon the cathode, and oxygen is

evolved at a partial pressure of 1.00 atm at the anode. The overall cell reaction is

$$Cu^{2+} + H_2O \rightarrow Cu(s) + \tfrac{1}{2}O_2(g) + 2H^+$$

Initial Thermodynamic Potential. Standard potential data for the two half-reactions in the cell under consideration are

$$Cu^{2+} + 2e \rightleftharpoons Cu(s) \qquad E^0 = 0.34 \text{ V}$$
$$\tfrac{1}{2}O_2(g) + 2H^+ + 2e \rightleftharpoons H_2O \qquad E^0 = 1.23 \text{ V}$$

Using the method shown in Example 19–6, the thermodynamic potential for this cell can be shown to be −0.94 V. Thus, no current would be expected at less negative applied potentials; at greater potentials, a linear increase in current should be observed in the absence of kinetic or concentration polarization.

Estimation of Required Potential. For the cell under consideration, kinetic polarization occurs only at the anode where oxygen is evolved, the cathodic reaction being rapid and reversible. That is, η_{ck} in Equation 21–1 is zero. Furthermore, concentration polarization at the anode is negligible at all times because the anodic reactant (water) is in large excess compared with concentration changes brought about by the electrolysis; therefore, the surface layer will never become depleted of reactant, and η_{ac} in Equation 21–1 is also negligible. Finally, it is readily shown by suitable substitutions into the Nernst equation that the increase in hydrogen ion concentration due to the anodic reaction results in a negligible change in electrode potential (<0.01 V); that is, the thermodynamic anode potential will remain constant throughout the electrolysis. Thus, for this example, Equation 21–1 reduces to

$$E_{appl} = E_c - E_a + \eta_{cc} + \eta_{ak} - IR \qquad (21-2)$$

Let us now assume that we wish to operate the cell initially at a current of 1.5 A, which corresponds to a 0.010 A/cm² current den-

---

[2] Throughout this chapter, junction potentials are neglected because they are small enough to be of no pertinence to the discussion.

sity. From Table 19–3, it is seen that η_{ak}, the oxygen overvoltage, will be about -0.85 V. Furthermore, concentration polarization will be negligible at the outset because the concentration of copper ions is initially high. Equation 21–2 then becomes

$$E_{appl} = 0.29 - 1.23 + 0 + (-0.85) \\ - 0.50 \times 1.5 = -2.54 \text{ V}$$

Thus, a rough estimate of the potential required to produce an initial current of 1.5 A is -2.5 V.

Current Changes During an Electrolysis at Constant Applied Potential. It is useful to consider the changes in current in the cell under discussion when the potential is held constant at -2.5 V throughout the electrolysis. Here, the current would be expected to decrease with time owing to depletion of copper ions in the solution as well as to an increase in cathodic concentration polarization. In fact, with the onset of concentration polarization, the current decrease becomes exponential in time. That is,

$$I_t = I_0 e^{-kt}$$

where I_t is the current t min after the onset of polarization and I_0 is the initial current. Lingane[3] has shown that values for the constant k can be computed from the relationship

$$k = \frac{25.8 DA}{V\delta}$$

where D is the diffusion coefficient in cm^2/sec, or the rate at which the reactant diffuses under a unit concentration gradient. The quantity A is the electrode surface area in cm^2, V is the volume of the solution in cm^3, and δ is the thickness of the surface layer in which the concentration gradient exists.

Typical values for D and δ are 10^{-5} cm^2/sec and 2×10^{-3} cm. (The constant 25.8 includes the factor of 60 for converting D to cm^2/min, thus making k compatible with the units of t in the equation for I_t.)

When the initial applied potential is -2.5 V, it is found that concentration polarization, and thus an exponential decrease in current, occurs essentially immediately after application of the potential. Figure 21–1a depicts this behavior; the curve shown was derived for the cell under consideration with the aid of the foregoing two equations. After 30 min, the current has decreased from the initial 1.5 A to 0.08 A; by this time, approximately 96% of the copper has been deposited.

Potential Changes During an Electrolysis at Constant Applied Potential. It is instructive to consider the changes in the various potentials in Equation 21–2 as the electrolysis under consideration proceeds; Figure 21–1b depicts these changes.

As mentioned earlier, the thermodynamic anode potential remains substantially unchanged throughout the electrolysis because of the large excess of the reactant (water) and the small change in the concentration of reaction product (H^+). The reversible cathode potential E_c, on the other hand, becomes smaller (more negative) as the copper concentration decreases. The curve for E_c in Figure 21–1b was derived by substituting the calculated copper concentration after various electrolysis periods into the Nernst equation. Note that the negative shift in potential is approximately linear with time over a considerable period.

The IR drop shown in Figure 21–1b parallels the current changes shown in Figure 21–1a. The negative sign is employed for consistency with Equation 21–2.

As shown by the topmost curve of Figure 21–1b, the oxygen overvoltage η_{ak} also becomes less negative as the current, and thus the current density, falls. The data for this curve were obtained from more extensive

---

[3] See: J. J. Lingane, *Electroanalytical Chemistry,* 2d ed., pp. 223–229, Interscience: New York, 1958.

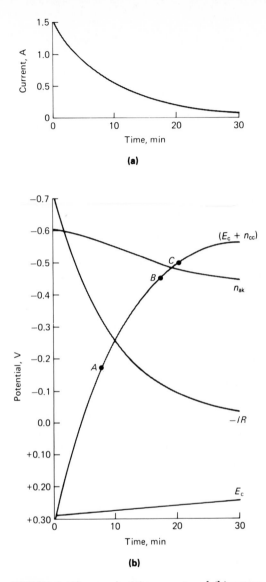

FIGURE 21-1 Changes in (**a**) current and (**b**) potentials during the electrolytic deposition of Cu^{2+}. Points A and B are potentials at which Pb and Cd would begin to codeposit if present. Point C is the potential at which H_2 might begin to form at the cathode. Any of these processes would distort curve A.

compilations that are similar to the data in Table 19–2.

The most significant feature of Figure 21–1**b** is the curve representing the change in *total cathode potential* $(E_c + \eta_{ck})$ as a function of time. It is evident that as IR and η_{ak} become less negative, one or more of the other potentials in Equation 21–2 *must become more negative*. Because of the large excess of reactant and product at the anode, its potential remains substantially constant. Thus, the *only* potentials that can change are those associated with the cathode; as seen from a comparison of the curves labeled E_c and $(E_c + \eta_{cc})$, it is evident that most of this negative drift is a consequence of a rapid increase in concentration polarization (η_{cc} becomes more negative). That is, even with vigorous stirring, copper ions are not brought to the electrode surface at a sufficient rate to prevent polarization. The result is a rapid decrease in IR and a corresponding negative drift of the total cathode potential.

The rapid shift in cathode potential that accompanies concentration polarization often leads to codeposition of other species and loss of selectivity. For example, points A and B on the cathode potential curve indicate the approximate potentials and times when lead and cadmium ions would begin to deposit if present in concentrations about equal to the original copper ion concentrations. Another event that would occur in the absence of lead, cadmium, or other easily reduced species would be the evolution of hydrogen at about point C (this process was not taken into account in deriving the curve in Figure 21–1a).

The interferences just described could be avoided by decreasing the applied potential by several tenths of a volt so that the negative drift of the cathode potential could never reach a level at which the interfering ions react. The consequence, however, is a diminution in current and, ordinarily, an enormous increase in the time required to complete the analysis.

At best, an electrolysis at constant cell potential can only be employed to separate easily reduced cations from those that are more difficult to reduce than hydrogen ion. Evolution of hydrogen would be expected near the end of the electrolysis and would prevent interference by cations that are reduced at more negative potentials.

21A-2 *CONSTANT-CURRENT ELECTROLYSIS*

The analytical electrodeposition under consideration, as well as others, can be carried out by maintaining the current, rather than the applied potential, at a constant level. Here, periodic increases in the applied potential are required as the electrolysis proceeds.

In the preceding section, it was shown that concentration polarization at the cathode causes a decrease in current. Initially, this effect can be partially offset by increasing the applied potential. Electrostatic forces would then postpone the onset of concentration polarization by enhancing the rate at which copper ions are brought to the electrode surface. Soon, however, the solution becomes sufficiently depleted in copper ions so that the forces of diffusion, electrostatic attraction, and stirring cannot keep the electrode surface supplied with sufficient copper ions to maintain the desired current. When this occurs, further increases in E_{appl} cause rapid changes in η_{cc} and thus the cathode potential; codeposition of hydrogen (or other reducible species) then takes place. The cathode potential ultimately becomes stabilized at a level fixed by the standard potential and the overvoltage for the new electrode reaction; further large increases in the cell potential are no longer necessary to maintain a constant current. Copper continues to deposit as copper(II) ions reach the electrode surface; the contribution of this process to the total current, however, becomes smaller and smaller as the deposition becomes more and more nearly complete. The alternative process, such as reduction of hydrogen or nitrate ions, soon predominates. The changes in

cathode potential under conditions of constant current are shown in Figure 21–2.

21A-3 *ELECTROLYSIS AT CONSTANT WORKING ELECTRODE POTENTIALS*

From the Nernst equation, it is seen that a tenfold decrease in the concentration of an ion being deposited requires a negative shift in potential of only $0.0591/n$ V. Electrolytic methods, therefore, are potentially reasonably selective. For example, as the copper concentration of a solution is decreased from $0.10\ M$ to $10^{-6}\ M$, the thermodynamic cathode potential E_c changes from an initial value of $+0.31$ V to $+0.16$ V. In theory, then, it should be feasible to separate copper from any element that does not deposit within this 0.15-V potential range. Species that deposit quantitatively at potentials more positive than $+0.31$ V could be eliminated with a prereduction; ions that require potentials smaller than $+0.16$ V would not interfere with the copper deposition. Thus, if we are willing to accept a reduction in analyte concentration to $10^{-6}\ M$ as a quantitative separation, it follows that divalent ions differing in standard potentials by about

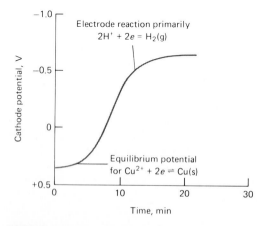

FIGURE 21-2 Changes in cathode potential during the deposition of copper with a constant current of 1.5 A. Here, the cathode potential is equal to $(E_c + \eta_{cc})$.

0.15 V or greater can, theoretically, be separated quantitatively by electrodeposition, provided their initial concentrations are about the same. Correspondingly, about 0.30 and 0.10 V differences are required for univalent and trivalent ions, respectively.

An approach to these theoretical separations, within a reasonable electrolysis period, requires a more sophisticated technique than the ones thus far discussed because concentration polarization at the cathode, if unchecked, will prevent all but the crudest of separations. The change in cathode potential is governed by the decrease in IR drop (Figure 21–1**b**). Thus, where relatively large currents are employed at the outset, the change in cathode potential can ultimately be expected to be large. On the other hand, if the cell is operated at low current levels so that the variation in cathode potential is lessened, the time required for completion of the deposition may become prohibitively long. An obvious answer to this dilemma is to initiate the electrolysis with an applied cell potential that is sufficiently high to ensure a reasonable current; the applied potential is then continuously decreased to keep the cathode potential at the level necessary to accomplish the desired separation. Unfortunately, it is not feasible to predict the required changes in applied potential on a theoretical basis because of uncertainties in variables affecting the deposition, such as overvoltage effects and perhaps conductivity changes. Nor, indeed, does it help to measure the potential across the two electrodes, since such a measurement gives only the overall cell potential, E_{appl}. The alternative is to measure the potential of the working electrode against a third electrode whose potential in the solution is known and constant—that is, a reference electrode. The potential impressed across the working electrode and its counter electrode can then be adjusted to the level that will impart the desired potential to the cathode (or anode) with respect to

the reference electrode. This technique is called *controlled cathode* (or *anode*) *potential electrolysis* or sometimes *potentiostatic electrolysis.*

Experimental details for performing a controlled cathode potential electrolysis are presented in a later section. For the present, it is sufficient to note that the potential difference between the reference electrode and the cathode is measured with a potentiometer or electronic voltmeter. The potential applied between the working electrode and its counter electrode is controlled with a voltage divider so that the cathode potential is maintained at a level suitable for the separation. Figure 21–3 is a schematic diagram of a simple manual apparatus that would permit deposition at a constant cathode potential.

An apparatus of the type shown in Figure 21–3 can be operated at relatively high initial applied potentials to give high currents. As the electrolysis progresses, however, a lowering of the applied potential across AC is required. This decrease, in turn, diminishes the current. Completion of the elec-

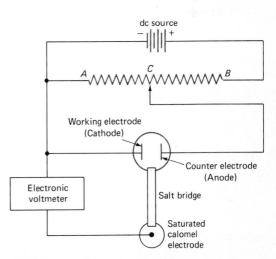

FIGURE 21-3 Apparatus for electrolysis at a controlled cathode potential. Contact C is continuously adjusted to maintain the cathode potential at the desired level.

trolysis will be indicated by the approach of the current to zero. The changes that occur in a typical constant cathode potential electrolysis are depicted in Figure 21–4. In contrast to the electrolytic methods described earlier, this technique demands constant attention during operation. Usually, some provision is made for automatic control; otherwise, the operator time required represents a major disadvantage to the controlled cathode potential method.

21B   An Introduction to Coulometric Methods of Analysis

Coulometry encompasses a group of analytical methods that involve measuring the quantity of electricity (in coulombs) needed to convert the analyte quantitatively to a different oxidation state. In common with gravimetric methods, coulometry offers the advantage that the proportionality constant between the measured quantity (coulombs

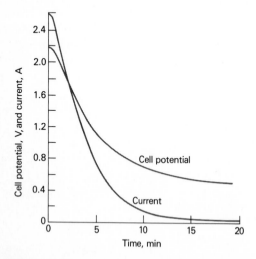

FIGURE 21–4 Changes in applied potential and current during a controlled cathode potential electrolysis. Deposition of copper upon a cathode maintained at −0.36 vs. SCE. (Experimental data from: J. J. Lingane, *Anal. Chem. Acta*, **1948**, *2*, 590. With permission.)

in this case) and the weight of analyte can be derived from known physical constants; thus, calibration or standardization is not ordinarily required. Coulometric methods are often as accurate as gravimetric or volumetric procedures; they are usually faster and more convenient than the former. Finally, coulometric procedures are readily adapted to automation.[4]

21B-1   UNITS FOR QUANTITY OF ELECTRICITY

The quantity of electricity or charge is measured in units of the *coulomb* (C) and the *faraday* (F). *The coulomb is the quantity of charge that is transported in one second by a constant current of one ampere.* Thus, for a constant current of I amperes operating for t seconds, the number of coulombs Q is given by the expression

$$Q = It \qquad (21-3)$$

For a variable current, the number of coulombs is given by the integral

$$Q = \int_0^t I\,dt \qquad (21-4)$$

The faraday is the quantity of electricity that will produce one equivalent of chemical change at an electrode. Since the equivalent in an oxidation-reduction reaction corresponds to the change brought about by one mole of electrons, the faraday is equal to 6.02×10^{23} electrons. One faraday is also equal to 96,487 C.

---

[4] For summaries of coulometric methods, see: E. Bishop, in *Comprehensive Analytical Chemistry*, C. L. Wilson and D. W. Wilson, Eds., vol. IID, Elsevier: New York, 1975; J. J. Lingane, *Electroanalytical Chemistry*, 2d ed., Chapters 19–21, Interscience: New York, 1958; G. W. C. Milner and G. Phillips, *Coulometry in Analytical Chemistry*, Pergamon Press: New York, 1967; and J. T. Stock, *Anal. Chem.*, **1984**, *56*, 1R; **1982**, *54*, 1R; **1980**, *52*, 1R; **1978**, *50*, 1R.

EXAMPLE 21-1. A constant current of 0.800 A was used to deposit copper at the cathode and oxygen at the anode of an electrolytic cell. Calculate the grams of each product that was formed in 15.2 min assuming no other redox reactions.

The equivalent weights are determined from consideration of the two half-reactions

$$Cu^{2+} + 2e \rightarrow Cu(s)$$

$$2H_2O \rightarrow 4e + O_2(g) + 4H^+$$

From Equation 21-3, we find

$$Q = 0.800 \text{ A} \times 15.2 \text{ min} \times 60 \text{ sec/min}$$
$$= 729.6 \text{ A sec} = 729.6 \text{ C}$$

or

$$\frac{729.6 \text{ C}}{96,487 \text{ C/F}} = 7.56 \times 10^{-3} \text{ F}$$
$$\equiv 7.56 \times 10^{-3} \text{ eq}$$

From the definition of the faraday, 7.56×10^{-3} equivalent of copper is deposited on the cathode; a similar quantity of oxygen is evolved at the anode. Therefore,

$$\text{g Cu} = 7.56 \times 10^{-3} \text{ eq Cu} \times \frac{63.5 \text{ g Cu/mol}}{2 \text{ eq Cu/mol}}$$
$$= 0.240$$

and

$$\text{g O}_2 = 7.56 \times 10^{-3} \text{ eq O}_2 \times \frac{32.0 \text{ g O}_2/\text{mol}}{4 \text{ eq O}_2/\text{mol}}$$
$$= 0.0605$$

21B-2 TYPES OF COULOMETRIC METHODS

Two general techniques are used for coulometric analysis, namely *potentiostatic* and *amperostatic*. The first involves maintaining the potential of the *working electrode* (the elec-

trode at which the analytical reaction occurs) at a constant level such that quantitative oxidation or reduction of the analyte occurs without involvement of less reactive species in the sample or solvent. Here, the current is initially high but decreases rapidly and approaches zero as the analyte is removed from the solution (see Figure 21–4). The quantity of electricity required is most commonly measured with an electronic integrator although other charge measuring devices have also been used.

The amperostatic method of coulometry makes use of a constant current, which is continued until an indicator signals completion of the analytical reaction. The quantity of electricity required to attain the end point is then calculated from the magnitude of the current and the time of its passage. The latter method has enjoyed wider application than the former; it is frequently called a *coulometric titration* for reasons that will become apparent later.

A fundamental requirement of all coulometric methods is that the species determined interacts with 100% current efficiency. That is, each faraday of electricity must bring about a chemical change corresponding to one equivalent of the analyte. This requirement does not, however, imply that the analyte must necessarily participate directly in the electron-transfer process at the electrode. Indeed, more often than not, the substance being determined is involved wholly or in part in a reaction that is secondary to the electrode reaction. For example, at the outset of the oxidation of iron(II) at a platinum anode, all of the current results from the reaction

$$Fe^{2+} \rightleftharpoons Fe^{3+} + e$$

As the concentration of iron(III) decreases, however, concentration polarization may cause the anode potential to rise until decomposition of water occurs as a competing process. That is,

$$2H_2O \rightleftharpoons O_2(g) + 4H^+ + 4e$$

The charge required to complete the oxidation of iron(II) would then exceed that demanded by theory. To avoid the consequent error, an unmeasured excess of cerium(III) can be introduced at the start of the electrolysis. This ion is oxidized at a lower anode potential than is water:

$$Ce^{3+} \rightleftharpoons Ce^{4+} + e$$

The cerium(IV) produced diffuses rapidly from the electrode surface, where it then oxidizes an equivalent amount of iron(II):

$$Ce^{4+} + Fe^{2+} \rightarrow Ce^{3+} + Fe^{3+}$$

The net effect is an electrochemical oxidation of iron(II) with 100% current efficiency even though only a fraction of the iron(II) ions are directly oxidized at the electrode surface.

The coulometric determination of chloride provides another example of an indirect process. Here, a silver electrode serves as the anode and silver ions are produced by the current. These cations diffuse into the solution and precipitate the chloride. A current efficiency of 100% with respect to the chloride ion is achieved even though this ion is neither oxidized nor reduced in the cell.

21C Potentiostatic Coulometry

In potentiostatic coulometry, the potential of the working electrode is maintained at a constant level that will cause the analyte to react quantitatively with the current without involvement of other components in the sample. An analysis of this kind possesses all the advantages of an electrogravimetric method and is not subject to the limitation imposed by the need for a weighable product. The technique can therefore be applied to systems that yield deposits with poor physical properties as well as to reactions that yield no solid product at all. For example, arsenic may be determined coulometrically by the electrolytic oxidation of arsenous acid (H_3AsO_3) to arsenic acid (H_3AsO_4) at a platinum anode. Similarly, the analytical conversion of iron(II) to iron(III) can be accomplished with suitable control of the anode potential.

21C-1 INSTRUMENTATION

The instrumentation for potentiostatic coulometry consists of an electrolysis cell, a potentiostat, and an integrating device for determining the number of coulombs by means of Equation 21–4.

Cells. Figure 21–5 illustrates two types of cells that are used for potentiostatic coulometry. The first consists of a platinum gauze working electrode and a platinum wire counter electrode, which is separated from the test solution by a porous tube containing the same supporting electrolyte as the test solution (Figure 21–5a). Separating the counter electrode is sometimes necessary to prevent its reaction products from interfering in the analysis. A saturated calomel reference electrode is joined to the test solution by means of a salt bridge. Often this bridge also contains the same electrolyte as the test solution.

The second type of cell is a mercury pool type. A mercury cathode is particularly useful for separating easily reduced elements as a preliminary step in an analysis. For example, copper, nickel, cobalt, silver, and cadmium are readily separated from ions such as aluminum, titanium, the alkali metals, and phosphates. The precipitated elements dissolve in the mercury; little hydrogen evolution occurs even at high applied potentials because of large overvoltage effects. A coulometric cell such as that shown in Figure 21–5b is also useful for coulometric determination of metal ions and certain types of organic compounds as well.

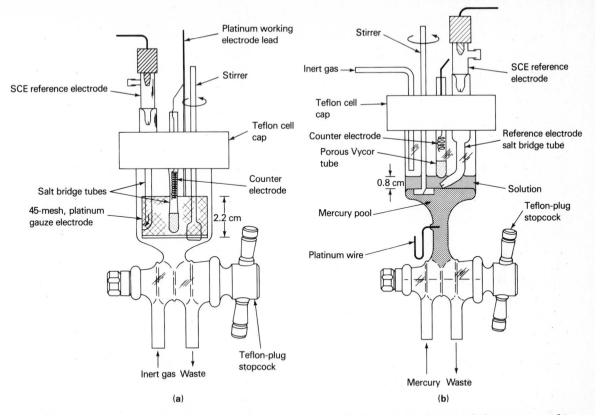

(a)   **(b)**

FIGURE 21-5 Electrolysis cells for potentiostatic coulometry. Working electrode: **(a)** platinum gauze; **(b)** mercury pool. (Reprinted with permission from: J. E. Harrar and C. L. Pomernacki, *Anal. Chem.,* **1973,** *45,* 57. Copyright 1973 American Chemical Society.)

Potentiostats. A potentiostat is an electronic device that maintains the potential of a working electrode at a constant level relative to a reference electrode. Two such devices are shown in Figure 2–20 (p. 49). Figure 21–6 is a schematic diagram of an apparatus for potentiostatic coulometry, which contains a somewhat different type potentiostat. In order to understand the cathode potential control exercised by this system, consider first the operation of the circuit in the absence of the booster amplifier. Note that the input E_1 is to the noninverting terminal of amplifier 1. On page 42, it was shown that in a noninverting circuit of this type, the potentials at the two inputs are always equal; that is, $E_1 = -E_2$. But the poten-

tial applied to the cell consisting of the calomel electrode and the working cathode is $-E_2$, and we may write

$$-E_2 = E_1 = E_c - E_{SCE} - I_2 R_2 \cong E_c - E_{SCE}$$

where E_c and E_{SCE} are the reversible potentials for the two electrodes. Because the internal resistance of the amplifier R_2 is high, current I_2 (and thus $I_2 R_2$) will be negligible, as indicated.

The potential for the cell consisting of the counter electrode and the cathode is given by

$$E_1 = E_c - E_{ce} - I_1 R_1$$

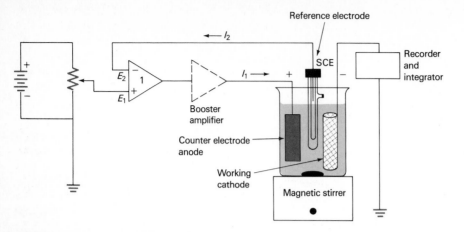

FIGURE 21-6 Schematic diagram of an instrument for potentiostatic coulometry.

where R_1 is the internal resistance of the cell and E_{ce} is the electrode potential of the counter electrode. If the potential of the cathode E_c begins to rise, as a consequence of concentration polarization or an increase in the cell resistance, the amplifier responds by decreasing its output current I_1 until E_1 again equals $-E_2$.

The booster amplifier shown in Figure 21–6 is a noninverting type that provides a much larger current than can be obtained from the single operational amplifier. Its presence has no effect on the cathode control circuit.

Integrators. Most modern apparatus for potentiostatic coulometry employ digital integrators to determine the number of coulombs required to complete an electrolysis.

21C-2 APPLICATION

Controlled potential coulometric methods have been applied to the determination of some 55 elements in inorganic compounds.[5]

Mercury appears to be favored as the cathode, and methods for the deposition of two dozen or more metals at this electrode have been described. The method has found widespread use in the nuclear energy field for the relatively interference-free determination of uranium and plutonium.

The controlled potential coulometric procedure also offers possibilities for the electrolytic determination (and synthesis) of organic compounds. For example, Meites and Meites[6] have demonstrated that trichloroacetic acid and picric acid are quantitatively reduced at a mercury cathode whose potential is suitably controlled:

$$Cl_3CCOO^- + H^+ + 2e \rightarrow Cl_2HCCOO^- + Cl^-$$

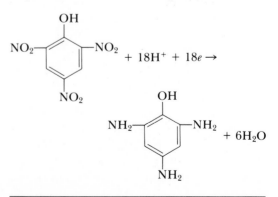

[5] For a summary of the applications, see: J. E. Harrar, *Electroanalytical Chemistry*, A. J. Bard, Ed., vol. 8, Marcel Dekker: New York, 1975; and E. Bishop, in *Comprehensive Analytical Chemistry*, C. L. Wilson and D. W. Wilson, Eds., Vol. IID, Chapter XV, Elsevier: New York, 1975.

[6] T. Meites and L. Meites, *Anal. Chem.*, **1955**, *27*, 1531; **1956**, *28*, 103.

Coulometric measurements permit the analysis of these compounds with a relative error of a few tenths of a percent.

Variable current coulometric methods are frequently used to monitor continuously and automatically the concentration of constituents in gas or liquid streams. An important example is the determination of small concentrations of oxygen.[7] A schematic diagram of the apparatus is shown in Figure 21–7. The porous silver cathode serves to break up the incoming gas into small bubbles; the reduction of oxygen takes place quantitatively within the pores. That is,

$$O_2(g) + 2H_2O + 4e \rightleftharpoons 4OH^-$$

The anode is a heavy cadmium sheet; here the half-cell reaction is

$$Cd(s) + 2OH^- \rightleftharpoons Cd(OH)_2(s) + 2e$$

Note that a galvanic cell is formed so that no external power supply is required. Nor is a potentiostat necessary because the potential

of the working anode can never become great enough to cause oxidation of other species. The electricity produced is passed through a standard resistor and the potential drop is recorded. The oxygen concentration is proportional to the potential, and the chart paper can be made to display the instantaneous oxygen concentration directly. The instrument is reported to provide oxygen concentration data in the range from 1 ppm to 1%.

21D Coulometric Titrations (Amperostatic Coulometry)

A coulometric titration employs a titrant that is electrolytically generated by a constant current. In some analyses, the active electrode process involves only generation of the reagent; an example is the titration of halides by silver ions produced at a silver anode. In other titrations, the analyte may also be directly involved at the generator electrode; an example of the latter is the coulometric oxidation of iron(II)—in part by electrolytically generated cerium(IV) and in part by direct electrode reaction (Section 21B–2). Under any circumstance, the net process must approach 100% current efficiency with respect to a single chemical change in the analyte.

The current in a coulometric titration is carefully maintained at a constant and accurately known level by means of an *amperostat*; the product of this current in amperes and the time in seconds required to reach an end point yields the number of coulombs, which is proportional to the quantity of analyte involved in the electrolysis. The constant current aspect of this operation precludes the quantitative oxidation or reduction of the unknown species entirely at the generator electrode because concentration polarization of the solution is inevitable before the electrolysis can be complete. The electrode potential must then rise if a constant current is to be maintained (Section

---

[7] For further details, see: F. A. Keidel, *Ind. Eng. Chem.*, **1960,** *52,* 490.

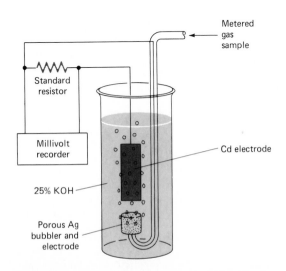

Metered gas sample

Standard resistor

Millivolt recorder

25% KOH

Porous Ag bubbler and electrode

Cd electrode

FIGURE 21-7 An instrument for continuously recording the O_2 content of a gas stream.

21A–2). Unless this potential rise produces a reagent that can react with the analyte, the current efficiency will be less than 100%. In a coulometric titration, then, at least part (and frequently all) of the reaction involving the analyte occurs away from the surface of the working electrode.

A coulometric titration, like a more conventional volumetric procedure, requires some means of detecting the point of chemical equivalence. Most of the end points applicable to volumetric analysis are equally satisfactory here; color changes of indicators, as well as potentiometric, amperometric, and conductance measurements have all been successfully applied.

The analogy between a volumetric and a coulometric titration extends well beyond the common requirement of an observable end point. In both, the amount of analyte is determined through evaluation of its combining capacity—in the one case, for a standard solution and, in the other, for electrons. Similar demands are made of the

reactions; that is, they must be rapid, essentially complete, and free of side reactions.

21D-1 ELECTRICAL APPARATUS

Coulometric titrators are available from several laboratory supply houses. In addition, they can be readily assembled from components available in most laboratories.

Figure 21–8 depicts the principal components of a typical coulometric titrator. Included are a source of constant current and a switch that simultaneously initiates the current and starts an electric timer. Also required is the means for accurately measuring the current; in Figure 21–8, the potential drop across the standard resistor, R_{std} is used for this measurement.

Many electronic or electromechanical amperostats are described in the literature. The ready availability of inexpensive operational amplifiers makes their construction a relatively simple matter (for example, see Figure 2–21, p. 49).

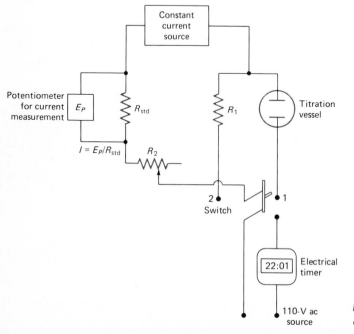

FIGURE 21-8 Schematic diagram of a coulometric titration apparatus.

It is useful to point out the close analogy between the various components of the apparatus shown in Figure 21–8 and the apparatus and solutions employed in a conventional volumetric analysis. The constant current source of known magnitude serves the same function as the standard solution in a volumetric method. The timer and switch correspond closely to the buret, the switch performing the same function as a stopcock. During the early phases of a coulometric titration, the switch is kept closed for extended periods; as the end point is approached, however, small additions of "reagent" are achieved by closing the switch for shorter and shorter intervals. The similarity to the operation of a buret is obvious.

Cells for Coulometric Titrations. A typical coulometric titration cell is shown in Figure 21–9. It consists of a generator electrode at which the reagent is formed and an auxiliary electrode to complete the circuit. The generator electrode, which should have a relatively large surface area, is often a rectangular strip or a wire coil of platinum; a gauze elec-

trode such as the cathode shown in Figure 21–5a can also be employed.

The products formed at the second electrode frequently represent potential sources of interference. For example, the anodic generation of oxidizing agents is often accompanied by the evolution of hydrogen from the cathode; unless this gas is allowed to escape from the solution, reaction with the oxidizing agent becomes a likelihood. To eliminate this type of difficulty, the second electrode is isolated by a sintered disk or some other porous medium.

An alternative to isolation of the auxiliary electrode is a device such as that shown in Figure 21–10 in which the reagent is generated externally. The apparatus is so arranged that flow of the electrolyte continues briefly after the current is discontinued, thus flushing the residual reagent into the titration vessel. Note that the apparatus shown in Figure 21–10 provides either hydrogen or hydroxide ions depending upon which arm is used. The apparatus has also been used for generation of other reagents such as iodine produced by oxidation of iodide at the anode.

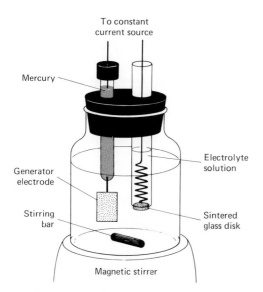

FIGURE 21-9 A typical coulometric titration cell.

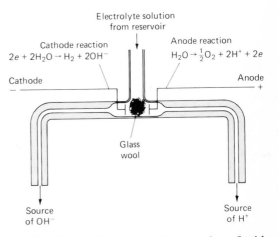

FIGURE 21-10 A cell for external generation of acid and base.

21D-2 APPLICATIONS OF COULOMETRIC TITRATIONS[8]

Coulometric titrations have been developed for all types of volumetric reactions. Selected applications are described in the following paragraphs.

Neutralization Titrations. Both weak and strong acids can be titrated with a high degree of accuracy using hydroxide ions generated at a cathode by the reaction

$$2H_2O + 2e \rightarrow 2OH^- + H_2(g)$$

The cells shown in Figures 21–9 and 21–10 can be employed. A convenient alternative involves substitution of a silver wire as the anode and the addition of chloride or bromide ions to the solution of the analyte. The anode reaction then becomes

$$Ag(s) + Br^- \rightleftharpoons AgBr(s) + e$$

Clearly, the silver bromide will not interfere with the neutralization reaction as would the hydrogen ions that are formed at most anodes.

---

[8] Applications of the coulometric procedure are summarized in E. Bishop, in *Comprehensive Analytical Chemistry*, C. L. Wilson and D. W. Wilson, Eds., Vol. IID, Chapters XVIII–XXIV, Elsevier: New York, 1975.

Both potentiometric and indicator end points can be employed for these titrations. The problems associated with the estimation of the equivalence point are identical with those encountered in a conventional volumetric analysis. A real advantage to the coulometric method, however, is that interference by carbonate ion is far less troublesome; it is only necessary to eliminate carbon dioxide from the solution containing the analyte by aeration with a carbon dioxide free gas before beginning the analysis.

The coulometric titration of strong and weak bases can be performed with hydrogen ions generated at a platinum anode.

$$H_2O \rightleftharpoons \tfrac{1}{2}O_2(g) + 2H^+ + 2e$$

Here, the cathode must be isolated from the solution or external generation must be employed to prevent interference from the hydroxide ions produced at that electrode.

Precipitation and Complex-Formation Titrations. A variety of coulometric titrations involving anodically generated silver ions have been developed (see Table 21–1). A cell, such as that shown in Figure 21–9, can be employed with a generator electrode constructed from a length of heavy silver wire. End points are detected potentiometrically or with chemical indicators. Similar analyses, based upon

Table 21-1		
Summary of Applications of Coulometric Titrations Involving Neutralization, Precipitation, and Complex-Formation Reactions		
Species Determined	**Generator-Electrode Reaction**	**Secondary Analytical Reaction**
Acids	$2H_2O + 2e \rightleftharpoons 2OH^- + H_2$	$OH^- + H^+ \rightleftharpoons H_2O$
Bases	$H_2O \rightleftharpoons 2H^+ + \tfrac{1}{2}O_2 + 2e$	$H^+ + OH^- \rightleftharpoons H_2O$
Cl^-, Br^-, I^-	$Ag \rightleftharpoons Ag^+ + e$	$Ag^+ + Cl^- \rightleftharpoons AgCl(s)$, etc.
Mercaptans	$Ag \rightleftharpoons Ag^+ + e$	$Ag^+ + RSH \rightleftharpoons \underline{AgSR(s)} + H^+$
Cl^-, Br^-, I^-	$2Hg \rightleftharpoons Hg_2^{2+} + 2e$	$Hg_2^{2+} + 2Cl^- \rightleftharpoons \underline{Hg_2Cl_2(s)}$, etc.
Zn^{2+}	$Fe(CN)_6^{3-} + e \rightleftharpoons Fe(CN)_6^{4-}$	$3Zn^{2+} + 2K^+ + 2\overline{Fe(CN)_6^{4-}} \rightleftharpoons$
		$\qquad K_2Zn_3[Fe(CN)_6]_2(s)$
Ca^{2+}, Cu^{2+},	See Equation 21–5	$HY^{3-} + Ca^{2+} \rightleftharpoons CaY^{2-} + H^+$,
$\quad Zn^{2+}$ and Pb^{2+}		etc.

the generation of mercury(I) ion at a mercury anode, have been described.

An interesting coulometric titration makes use of a solution of the amine mercury(II) complex of ethylenediaminetetraacetic acid (H_4Y).[9] The complexing agent is released to the solution as a result of the following reaction at a mercury cathode:

$$HgNH_3Y^{2-} + NH_4^+ + 2e$$
$$\rightleftharpoons Hg + 2NH_3 + HY^{3-} \quad (21-5)$$

Because the mercury chelate is more stable than the corresponding complexes with calcium, zinc, lead, or copper, complexation of these ions will not occur until the electrode process frees the ligand.

Oxidation-Reduction Titrations. Table 21–2 indicates the variety of reagents that can be generated coulometrically and the analyses to which they have been applied. Electrogenerated bromine has proved to be particularly useful among the oxidizing agents and forms the basis for a host of methods.

[9] C. N. Reilley and W. W. Porterfield, *Anal. Chem.,* **1956,** *28,* 443.

Of interest also are some of the unusual reagents not ordinarily encountered in volumetric analysis because of the instability of their solutions; these include dipositive silver ion, tripositive manganese, and the chloride complex of unipositive copper.

Comparison of Coulometric and Volumetric Titrations. Some real advantages can be claimed for a coulometric titration in comparison with the classical volumetric process. Principal among these is the elimination of problems associated with the preparation, standardization, and storage of standard solutions. This advantage is particularly important with labile reagents such as chlorine, bromine, or titanium(III) ion; owing to their instability, these species are inconvenient as volumetric reagents. Their utilization in coulometric analysis is straightforward, however, because they undergo reaction with the analyte immediately after being generated.

Where small quantities of reagent are required, a coulometric titration offers a considerable advantage. By proper choice of current, micro quantities of a substance can be introduced with ease and accuracy; the

Table 21-2
Summary of Applications of Coulometric Titrations Involving Oxidation-Reduction Reactions

Reagent	Generator-Electrode Reaction	Substance Determined
Br_2	$2Br^- \rightleftharpoons Br_2 + 2e$	As(III), Sb(III), U(IV), Tl(I), I^-, SCN^-, NH_3, N_2H_4, NH_2OH, phenol, aniline, mustard gas; 8-hydroxyquinoline
Cl_2	$2Cl^- \rightleftharpoons Cl_2 + 2e$	As(III), I^-
I_2	$2I^- \rightleftharpoons I_2 + 2e$	As(III), Sb(III), $S_2O_3^{2-}$, H_2S
Ce^{4+}	$Ce^{3+} \rightleftharpoons Ce^{4+} + e$	Fe(II), Ti(III), U(IV), As(III), I^-, $Fe(CN)_6^{4-}$
Mn^{3+}	$Mn^{2+} \rightleftharpoons Mn^{3+} + e$	$H_2C_2O_4$, Fe(II), As(III)
Ag^{2+}	$Ag^+ \rightleftharpoons Ag^{2+} + e$	Ce(III), V(IV), $H_2C_2O_4$, As(III)
Fe^{2+}	$Fe^{3+} + e \rightleftharpoons Fe^{2+}$	Cr(VI), Mn(VII), V(V), Ce(IV)
Ti^{3+}	$TiO^{2+} + 2H^+ + e \rightleftharpoons Ti^{3+} + H_2O$	Fe(III), V(V), Ce(IV), U(VI)
$CuCl_3^{2-}$	$Cu^{2+} + 3Cl^- + e \rightleftharpoons CuCl_3^{2-}$	V(V), Cr(VI), IO_3^-
U^{4+}	$UO_2^{2+} + 4H^+ + 2e \rightleftharpoons U^{4+} + 2H_2O$	Cr(VI), Ce(IV)

equivalent volumetric process requires small volumes of very dilute solutions, a recourse that is always difficult.

A single constant-current source can be employed to generate precipitation, complex formation, oxidation-reduction, or neutralization reagents. Furthermore, the coulometric method is readily adapted to automatic titrations, because current control is easily accomplished.

Coulometric titrations are subject to five potential sources of error: (1) variation in the current during electrolysis; (2) departure of the process from 100% current efficiency; (3) error in the measurement of current; (4) error in the measurement of time; and (5) titration error due to the difference between the equivalence point and the end point. The last of these difficulties is common to volumetric methods as well; where the indicator error is the limiting factor, the two methods are likely to have comparable reliability.

With simple instrumentation, currents constant to 0.2% relative are easily achieved; with somewhat more sophisticated apparatus, control to 0.01% is obtainable. In general, then, errors due to current fluctuations are seldom of importance.

Although generalizations concerning the magnitude of uncertainty associated with the electrode process are difficult, current efficiencies of 99.5 to better than 99.9% are often reported in the literature. Currents are readily measured to ±0.1% relative.

To summarize, then, the current-time measurements required for a coulometric titration are inherently as accurate or more accurate than the comparable volume-normality measurements of a classical volumetric analysis, particularly where small quantities of reagent are involved. Often, however, the accuracy of a titration is not limited by these measurements but by the sensitivity of the end point; in this respect, the two procedures are equivalent.

21D-3 AUTOMATIC COULOMETRIC TITRATORS

A number of instrument manufacturers offer automatic coulometric titrators. Most of these employ the potentiometric end point and are similar in construction to the automatic titrators discussed in the previous chapter (Section 20F–3). Here, however, the error signal controls a flow of electricity rather than a flow of liquid reagent. Some of the commercial instruments are multipurpose and can be used for the determination of a variety of species. Others are designed for a single analysis. Examples of the latter include: chloride titrators in which silver ion is generated coulometrically; sulfur dioxide monitors, where anodically generated bromine oxidizes the analyte to sulfate ions; carbon dioxide monitors in which the gas, absorbed in monoethanolamine, is titrated with coulometrically generated base; and water titrators in which Karl Fischer reagent is generated electrolytically.

PROBLEMS

21–1. Lead is to be deposited at a cathode from a solution that is 0.125 M in Pb^{2+} and 0.250 M in $HClO_4$. Oxygen is evolved at a pressure of 0.800 atm at a 30-cm^2 platinum anode. The cell has a resistance of 0.950 Ω.
(a) Calculate the thermodynamic potential of the cell.
(b) Calculate the IR drop if a current of 0.300 A is to be used.
(c) Estimate the O_2 overvoltage.
(d) Estimate the total applied potential required to begin electrodeposition at the specified conditions.

    (e) What potential will be required when the Pb^{2+} concentration is 0.00100 M assuming that all other variables remain unchanged?

21-2. Calculate the minimum difference in standard electrode potentials needed to lower the concentration of the metal M_1 to 1.00×10^{-5} M in a solution that is $1.00 \times 10^{-1} M$ in the less reducible metal M_2, where
    (a) M_1 is univalent and M_2 is divalent.
    (b) M_1 and M_2 are both divalent.
    (c) M_1 is trivalent and M_2 is univalent.
    (d) M_1 is divalent and M_2 is univalent.
    (e) M_1 is divalent and M_2 is trivalent.

21-3. It is desired to separate and determine bismuth, copper, and silver in a solution that is 0.0650 M in BiO^+, 0.175 M in Cu^{2+}, 0.0962 M in Ag^+, and 0.500 M in $HClO_4$.
    (a) Using $1.00 \times 10^{-6} M$ as the criterion for quantitative removal, determine whether or not separation of the three species is feasible by controlled cathode potential electrolysis.
    (b) If any separations are feasible, evaluate the range (vs. SCE) within which the cathode potential should be controlled for the deposition of each.
    (c) Calculate the potential (vs. SCE) needed to deposit the third ion quantitatively.

21-4. Halide ions can be deposited at a silver anode, the reaction being

$$Ag(s) + X^- \rightarrow AgX(s) + e$$

Suppose that a cell was formed by immersing a silver anode in an analyte solution that was 0.0250 M Cl^-, Br^-, and I^- ions and connecting the half-cell to a saturated calomel cathode via a salt bridge.
    (a) Which halide would form first and at what potential? Is the cell galvanic or electrolytic?
    (b) Could I^- and Br^- be separated quantitatively? (Take $1.00 \times 10^{-5} M$ as criteria for quantitative removal of an ion.) If a separation is feasible, what range of cell potential could be used?
    (c) Repeat part (b) for I^- and Cl^-.
    (d) Repeat part (b) for Br^- and Cl^-.

21-5. What cathode potential (vs. SCE) would be required to lower the total $Hg(II)$ concentration of the following solutions to 1.00×10^{-6} M (assume reaction product in each case is elemental Hg):
    (a) an aqueous solution of Hg^{2+}?
    (b) a solution with an equilibrium SCN^- concentration of 0.100 M?

$$Hg^{2+} + 2SCN^- \rightleftharpoons Hg(SCN)_2(aq) \qquad K_f = 1.8 \times 10^7$$

(c) a solution with an equilibrium Br^- concentration of 0.250 M?

$$HgBr_4^- + 2e \rightleftharpoons Hg(l) + 4Br^- \qquad E^0 = 0.223 \text{ V}$$

21-6. Calculate the time needed for a constant current of 0.800 A to deposit 0.100 g of
(a) Tl(III) as the element on a cathode.
(b) Tl(I) as Tl_2O_3 on an anode.
(c) Tl(I) as the element on a cathode.

21-7. At a potential of -1.0 V (vs. SCE), carbon tetrachloride in methanol is reduced to chloroform at a Hg cathode:

$$2CCl_4 + 2H^+ + 2e + 2Hg(l) \rightarrow 2CHCl_3 + Hg_2Cl_2(s)$$

At -1.80 V, the chloroform further reacts to give methane:

$$2CHCl_3 + 6H^+ + 6e + 6Hg(l) \rightarrow 2CH_4 + 3Hg_2Cl_2(s)$$

A 0.801-g sample containing CCl_4, $CHCl_3$, and inert organic species was dissolved in methanol and electrolyzed at -1.0 V until the current approached zero. A coulometer indicated that 9.17 C had been used. The reduction was then continued at -1.80 V; an additional 59.65 C were required to complete the reaction. Calculate the percent CCl_4 and $CHCl_3$ in the mixture.

21-8. A 6.39-g sample of an ant-control preparation was decomposed by wet-ashing with H_2SO_4 and HNO_3. The arsenic in the residue was reduced to the trivalent state with hydrazine. After the excess reducing agent had been removed, the arsenic(III) was oxidized with electrolytically generated I_2 in a faintly alkaline medium:

$$HAsO_3^{2-} + I_2 + 2HCO_3^- \rightarrow HAsO_4^{2-} + 2I^- + 2CO + H_2O$$

The titration was complete after a constant current of 101.1 mA had been passed for 12 min and 36 sec. Express the results of this analysis in terms of the percentage As_2O_3 in the original sample.

21-9. A 0.0145-g sample of a purified organic acid was dissolved in an alcohol/water mixture and titrated with coulometrically generated hydroxide ions. With a current of 0.0324 A, 251 sec were required to reach a phenolphthalein end point. Calculate the equivalent weight of the acid.

21–10. Traces of aniline can be determined by reaction with an excess of electrolytically generated Br_2:

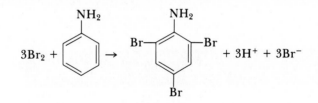

$$3Br_2 + \text{[aniline]} \rightarrow \text{[tribromoaniline]} + 3H^+ + 3Br^-$$

The polarity of the working electrode is then reversed, and the excess bromine is determined by a coulometric titration involving the generation of Cu(I):

$$Br_2 + 2Cu^+ \rightarrow 2Br^- + 2Cu^{2+}$$

Suitable quantities of KBr and copper(II) sulfate were added to a 25.0-mL sample containing aniline. Calculate the micrograms of $C_6H_5NH_2$ in the sample from the accompanying data:

Working Electrode Functioning as	Generation Time (min) with a Constant Current of 1.00 mA
anode	3.46
cathode	0.41

22

Voltammetry and Polarography

Voltammetry comprises a group of electroanalytical methods in which information about the analyte is derived from the measurement of current as a function of applied potential obtained under conditions that encourage polarization of the indicator or working electrode. Generally, the working electrodes in voltammetry are characterized by their small surface area (usually a few square millimeters), which enhances polarization. Such electrodes are generally referred to as *microelectrodes.*

At the outset, it is worthwhile pointing out the basic differences between voltammetry and the two types of electrochemical methods that were discussed in earlier chapters. Voltammetry is based upon the measurement of a current that develops in an electrochemical cell under conditions of complete concentration polarization. In contrast, potentiometric measurements are made at currents that approach zero and where polarization is absent. Voltammetry differs from coulometry in the respect that

with the latter, measures are taken to minimize or compensate for the effects of concentration polarization. Furthermore in voltammetry a minimal consumption of analyte takes place, whereas in coulometry essentially all of the analyte is converted to another state.

Historically, the field of voltammetry developed from the discovery of *polarography* by the Czechoslovakian chemist Jaroslav Heyrovsky[1] in the early 1920s. Polarography, which is still most widely used of all voltammetric methods, differs from the others in the respect that a mercury electrode serves as the microelectrode; usually, this electrode takes the form of a *dropping mercury electrode* (DME). The unique properties of this electrode are discussed in a later section.

By 1950, voltammetry appeared to be a

---

[1] J. Heyrovsky, *Chem. Listy,* **1922,** *16,* 256. Heyrovsky was awarded the 1959 Nobel Prize in chemistry for his discovery and development of polarography.

mature and fully developed technique. The decade from 1955 to 1965, however, was marked by the appearance of several major modifications of the original method, which served to overcome many of its limitations. At about this same time, the advent of low-cost operational amplifiers made possible the development of relatively inexpensive, commercial instruments that incorporated many of these important modifications. The result has been a recent resurgence of interest in the applications of voltammetric methods to the qualitative and quantitative determination of a host of organic and inorganic species.[2]

In the sections that follow, the theory of the classical polarographic method is first considered because this theory is applicable to all polarographic procedures; subsequently, the modifications to the original method, which led to the enhancement of its usefulness, are presented.

22A Theory of Classical Polarography[3]

Classical polarographic measurements are most commonly performed on aqueous solutions containing an analyte in the concentration range of 10^{-2} to 10^{-5} M. Where necessary, as with many organic substances, other solvent systems can be used. A polarographic analysis can easily be performed on 1 to 2 mL of solution; with a little effort, a

---

[2] For a brief summary of several of the new voltammetric techniques, see: J. B. Flato, *Anal. Chem.*, **1972**, *44*, 75A.

[3] The principles and applications of polarography are considered in detail in a number of monographs; see, for example: I. M. Kolthoff and J. J. Lingane, *Polarography*, 2d ed., Interscience: New York, 1952; L. Meites, *Polarographic Techniques*, 2d ed., Interscience: New York, 1965; J. Heyrovsky and J. Kůta, *Principles of Polarography*, Academic Press: New York, 1966; P. Zuman, *Topics in Organic Polarography*, Pergamon Press: New York, 1970; A. M. Bond, *Modern Polarographic Methods in Analytical Chemistry*, Marcel Dekker: New York, 1980.

volume as small as one drop will suffice. As a consequence, classical polarography provides a moderately sensitive means for determining the concentration of many inorganic, organic, and biochemical species. The selectivity of polarographic methods is similar to that of the electrolytic methods discussed in the previous chapter. Relative errors associated with quantitative polarographic methods range from 1 to 3% relative.

22A-1 POLAROGRAPHIC MEASUREMENTS

Polarographic data are obtained by recording the current in a special type of an electrolytic cell as a function of increasing applied potential. The resulting current-voltage curve is called a *polarogram*. Both qualitative and quantitative information are contained in a polarogram.

Cells for Classical Polarography. A cell for classical polarographic measurements can be depicted by the diagram

$$\text{SCE} \| \text{M}^{+n}(xM) | \text{Hg(DME)}$$

Here, the anode is a saturated calomel electrode that is large enough so that it behaves as an ideal nonpolarized electrode (Figure 19–8**b**, Section 19E–2). The cathode of this cell is a dropping mercury electrode which is immersed in the test solution that is xM in the analyte M^{+n}. A dropping mercury electrode consists of a 5 to 20 cm length of fine capillary tubing (i.d. ~0.05 cm) through which mercury is forced under a head of approximately 50 cm. A continuous series of highly reproducible drops are formed having diameters between 0.1 and 1 mm; typical drop times are 2 to 6 sec. Figure 22–11**a** (p. 678) contains a schematic diagram of a dropping mercury electrode.

Current Variations with a Dropping Mercury Electrode. The current in a cell containing a dropping electrode undergoes periodic fluctuations corresponding in frequency to the drop rate.

As a drop breaks, the current falls to zero (see Figure 22–1); it then increases rapidly as the electrode area grows because of the greater surface to which diffusion can occur. The *average current* is the hypothetical *constant* current which in the drop time *t* would produce the same quantity of electricity as the fluctuating current does during this same period. In order to determine the average current, it is necessary to reduce the large fluctuations in the current by a damping device or by sampling the current near the end of each drop, where the change in current with time is relatively small. Damping is ordinarily achieved by employing a well-damped galvanometer or a low-pass filter (Appendix, Section A2–4). As shown in Figure 22–2A, damping limits the oscillations to a reasonable magnitude; the average current (or, alternatively, the maximum current) is then readily determined, provided the drop rate *t* is reproducible. Note the effect of irregular drops in the upper part of curve *A*, probably caused by vibration of the apparatus.

Polarograms. Figure 22–2 shows typical polarograms, one for a solution that is 1.0 *M* in hydrochloric acid and 5×10^{-4} *M* in cadmium ion (curve *A*) while the second is for the acid in the absence of cadmium ion (curve *B*). As is usually the case, the microelectrode used in obtaining the curves in

Figure 22–2 acted as the cathode of an electrolytic cell; that is, it was connected to the negative terminal of the power supply. By convention, the applied potential is given a negative sign under these circumstances. Also by convention, currents are designated as positive when the flow of electrons is from the power supply into the microelectrode.

The step-shaped curve in Figure 22–2A, called a *polarographic wave*, arises from the reaction

$$Cd^{2+} + 2e + Hg \rightleftharpoons Cd(Hg) \qquad (22-1)$$

where Cd(Hg) represents elemental cadmium dissolved in mercury giving an amalgam. The sharp increase in current at about -1 V in both polarograms is caused by the reduction of hydrogen ions to give hydrogen.

For reasons to be considered presently, a polarographic wave suitable for analysis is obtained only in the presence of a large excess of a *supporting electrolyte;* hydrochloric acid serves this function in the present example. Examination of the polarogram for the supporting electrolyte alone reveals that a small current, called the *residual current,* passes through the cell even in the absence of cadmium ions.

A characteristic feature of a polaro-

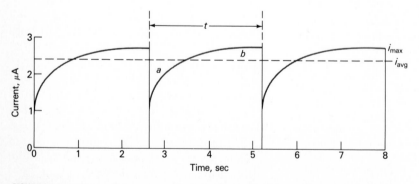

FIGURE 22-1 Effect of drop growth on polarographic current. For i_{avg}, area $a =$ area b and $i_{avg} \cong \frac{6}{7} i_{max}$.

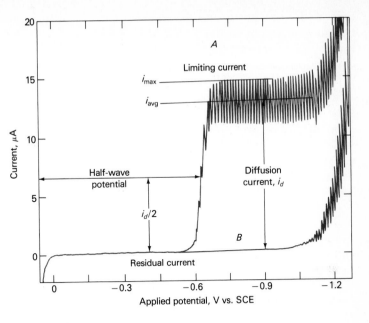

FIGURE 22-2 Polarograms for: *(A)* a 1 *M* solution of HCl that is 5 × 10^{-4} *M* in Cd^{2+} and *(B)* a 1 *M* solution of HCl. From: D. T. Sawyer and J. L. Roberts, Jr., *Experimental Electrochemistry for Chemists*, Wiley: New York, 1974. Reprinted by permission of John Wiley & Sons, Inc.

graphic wave is the region in which the current, after increasing sharply, becomes essentially independent of the applied voltage; this constant current is called a *limiting current*. The limiting current is the result of a restriction in the rate at which the participant in the electrode process can be brought to the surface of the microelectrode; with proper control over experimental conditions, this rate is determined exclusively for all points on the wave by the velocity at which the reactant diffuses. A diffusion-controlled limiting current is given a special name, the *diffusion current*, and is assigned the symbol i_d. Ordinarily, the diffusion current is directly proportional to the concentration of the reactive constituent and is thus of prime importance from the standpoint of analysis. As shown in Figure 22–2, the diffusion current is the difference between the limiting and the residual currents.

One other important quantity, the *half-wave potential*, is the potential at which the current is equal to one-half the diffusion current. The half-wave potential is usually given the symbol $E_{1/2}$; it may permit qualitative identification of the reactant.

22A-2 POLAROGRAPHIC CURRENTS

Limiting currents, such as that shown in Figure 22–2 are observed when the current is limited by the rate at which the reactive species is brought to the electrode surface. As described in Section 19E–2, three forces supply an electrode surface with reactant during an electrolysis: electrostatic forces, mechanical forces (vibration, stirring, convection), and diffusion. In polarography, every effort is made to eliminate the first two and arrange conditions so that the current is totally diffusion controlled. To minimize electrostatic attraction, an inactive supporting electrolyte is added; when its concentration exceeds that of the analyte by 50- to 100-fold, the analyte is surrounded by a highly charged environment and is effectively screened from the charge on the electrode. Mechanical transport of ions to the electrode surface is prevented by avoiding cell vibrations and temperature differentials. With these measures then, the current becomes limited by diffusion rate alone and the current is termed a diffusion current.

The Diffusion Current. Consider the rapid and reversible reduction of an analyte species A at a dropping mercury electrode in a cell such as that shown in Figure 22–11. The product P is, in this case, a soluble ion or molecule. (The arguments that follow apply equally well to a reaction involving amalgam formation as shown in Equation 22–1.) The half-reaction and standard potential are

$$A + ne \rightleftharpoons P \qquad E_A^0 = xV \qquad (22-2)$$

and

$$E_{appl} = E_A^0 - \frac{0.0591}{n} \log \frac{[P]_0}{[A]_0} - E_{SCE} \qquad (22-3)$$

where E_{SCE} is the potential of the saturated calomel electrode. Note that the subscript zero for the two concentration terms denotes that Equation 22–3 applies *only* to the film of solution at the surface of the mercury droplet and generally not to the bulk of the solution. In fact, in the polarographic experiment the bulk concentrations [A] and [P] will usually be quite different from their surface counterparts.

When a potential large enough to cause a significant amount of reaction is applied, a decrease in [A] and an increase in [P] takes place immediately at the mercury-solution interface until Equation 22–3 is satisfied. As a result, concentration gradients are set up and diffusion of A towards the surface film and of P in the opposite direction takes place (here, we will assume that the bulk concentration of P is zero initially and will remain approximately so throughout the electrolysis; that is, so little P is produced at the surface relative to the total volume of solution that the bulk concentration of this species approaches zero at all times). The rates of transport of the two species, R_A and R_P, are proportional to the concentration differences. That is,

$$R_A = k_A'([A] - [A]_0) \qquad (22-4)$$

$$R_P = k_P'([P]_0 - [P]) \cong k_P'[P]_0 \qquad (22-5)$$

where k_A' and k_P' are proportionality constants. When, as specified at the outset, diffusion is the only force causing movement of the two species, the current is proportional to the rate of transport. That is,

$$i = k_A'' R_A = k_A([A] - [A]_0) \qquad (22-6)$$

where $k_A = k_A'' k_A'$. Similarly,

$$i = k_P'' R_P = k_P[P]_0 \qquad (22-7)$$

When E_{appl} becomes sufficiently large so that $[A]_0$ is negligible with respect to the bulk concentration [A] (that is, [A] → 0), Equation 22–6 becomes

$$i = i_d = k_A[A] \qquad (22-8)$$

Here, i_d is the *potential-independent diffusion current, which is directly proportional to analyte concentration in the bulk of the solution* [A]. Quantitative polarography is based upon this equation.

When the current in a cell becomes independent of the applied voltage because of a limitation in the rate at which a reactant can be brought to the surface of an electrode, a state of *complete concentration polarization* is said to exist. With a microelectrode, the current required to achieve this condition is small—typically 3 to 10 μA (microampere) for a $10^{-3} M$ solution. Such current levels do not significantly deplete the solution of reactant. (See Problem 22–7.)

Current and Concentration Relationships for a Planar Microelectrode. Before developing equations for diffusion currents for a dropping mercury electrode, it is instructive to consider a simpler system consisting of a stationary, planar microelectrode operated under conditions in which mass transport occurs by diffusion alone. We will assume that the reduction of the analyte is rapid and reversible, which implies that the instantaneous concentration C_0 of the analyte at the electrode surface is determined by the Nernst equation. Furthermore, we will be concerned with the situation where the applied potential is large enough so that the current corresponds to the limiting current; that is, $C_0 \rightarrow 0$.

Upon initial application of this potential, a relatively large current is generated causing the analyte concentration in the surface layer to rapidly approach zero. A concentration gradient is then set up resulting in diffusion of analyte to the surface layer. Initially this gradient extends only a fraction of a millimeter into the solution (see curve labeled 1 sec, Figure 22–3). With time, however, this gradient, as shown in Figure 22–3, extends farther and farther into the solution. As a consequence, the slope of the curve dC/dx, called the *diffusion gradient*, becomes less and less. The magnitude of the observed current is directly proportional to this gradient. Thus, as shown in Figure 22–4, the current in the planar microelectrode is initially high but decreases rapidly and becomes essentially constant after several minutes. At this time the change in the diffusion gradient or slope has become essentially constant. One of the limitations of a stationary planar electrode is the fact that

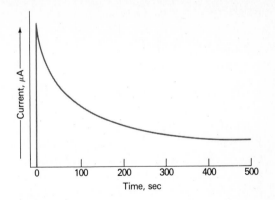

FIGURE 22-4 Current-time relationship in a stationary planar electrode.

constant currents are not achieved until a more or less constant concentration gradient is set up—a process that usually requires several minutes (see Figure 22–4).

The diffusion current at a planar microelectrode is proportional to the number of electrons consumed per mole of reactant (n), the surface area of the electrode A, the diffusion gradient, and the rate at which the reactant migrates under a unit concentration differential. The latter is the diffusion coefficient D, which has units of cm²/sec. Thus,

$$i = nFAD \left(\frac{dC}{dx} \right)_{x=0} \qquad (22–9)$$

where the term in parentheses is the diffusion or concentration gradient at the surface ($x = 0$) at time t, and n and F have their usual meanings. It can be shown that this gradient is given by

$$\left(\frac{dC}{dx} \right)_{x=0} = \frac{C}{\pi^{1/2} D^{1/2} t^{1/2}}$$

where C is the concentration of the analyte in the bulk of the solution. Substitution of this equation into Equation 22–9 gives the *Cottrell equation*, which is applicable to voltammetry and controlled electrode po-

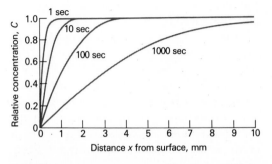

FIGURE 22-3 Concentration profiles for a reactant A diffusing to a planar electrode surface after 1, 10, 100, and 1000 sec of electrolysis.

tential electrolysis. That is,

$$i = \frac{nFACD^{1/2}}{\pi^{1/2}t^{1/2}} = \frac{96,487nACD^{1/2}}{(3.142)^{1/2}t^{1/2}}$$

$$= \frac{54,437nACD^{1/2}}{t^{1/2}} \qquad (22-10)$$

Diffusion Currents at Dropping Electrodes. To extend Equation 22–10 to polarography requires taking into account the spherical shape of the electrode. Thus, the volume V of the dropping electrode *at its maximum* is given by

$$V = (4/3)\pi r^3 = mt/d$$

where r is the radius of the drop in cm, m is the mass flow rate of mercury in mg/sec, d is the density of mercury ($13,600$ mg/cm^3), and t is the drop time in sec. To obtain the surface area A of the drop at its maximum in terms of the readily measurable quantities m and t, we solve this equation for r and substitute into the expression for the area of a sphere, $4\pi r^2$. Thus,

$$A = 4\pi \left(\frac{3}{4\pi} \frac{mt}{13,600} \right)^{2/3}$$

$$= 8.49 \times 10^{-3} m^{2/3} t^{2/3} \text{cm}^2$$

Substitution into Equation 22–10 gives for the maximum current under limiting conditions

$$i_d = 462nD^{1/2}m^{2/3}t^{1/6}C$$

This equation, however, fails to account for the fact that the surface of the electrode is continuously expanding into the solution (in contrast to the fixed electrode for which Equation 22–10 applies). It is possible to show theoretically that a correction for this effect can be made by multiplication of the constant by $\sqrt{7/3}$.[4] Thus,

---

[4] I. M. Kolthoff and J. J. Lingane, *Polarography,* Vol. I, p. 44, Interscience: New York, 1952.

$$(i_d)_{max} = \sqrt{7/3} \times 462nD^{1/2}m^{2/3}t^{1/6}C$$
$$= 706nD^{1/2}m^{2/3}t^{1/6}C$$

This equation is known as the Ilkovic equation in honor of the man who first derived it. It is important to note that the unit for C is *millimoles* per liter and for m is *milligrams* per second. To obtain an expression for the average current rather than the maximum, the constant in the foregoing equation becomes 607 rather than 706. That is,

$$(i_d)_{ave} = 607nD^{1/2}m^{2/3}t^{1/6}C \qquad (22-11)$$

Note that either the average or the maximum current can be used in quantitative polarography.

Effect of Capillary Characteristics on Diffusion Current. The product $m^{2/3}t^{1/6}$ in the Ilkovic equation, called the *capillary constant,* describes the influence of dropping electrode characteristics upon the diffusion current; both m and t are readily evaluated experimentally; comparison of diffusion currents from different capillaries is thus possible.

Two factors, other than the geometry of the capillary itself, play a part in determining the magnitude of the capillary constant. The head that forces the mercury through the capillary influences both m and t such that the diffusion current is directly proportional to the square root of the column height. The drop time t of a given electrode is also affected by the applied potential since the interfacial tension between the mercury and the solution varies with the charge on the drop. Generally, t passes through a maximum at about -0.4 V (vs. SCE) and then falls off fairly rapidly; at -2.0 V, t may be only one-half of its maximum value. Fortunately, the diffusion current varies only as the one-sixth power of the drop time so that, over small potential ranges, the decrease in current due to this variation is negligibly small.

Effect of Temperature on Diffusion Currents. Temperature affects several of the variables that govern the diffusion current for a given species, and its overall influence is thus complex. The most temperature-sensitive factor in the Ilkovic equation is the diffusion coefficient, which ordinarily can be expected to change by about 2.5% per degree. As a consequence, temperature control to a few tenths of a degree is needed for accurate polarographic measurements.

Residual Currents. Figure 22–5 shows a residual current curve (obtained at high sensitivity) for a 0.1 *M* solution of hydrogen chloride. This current has two sources. The first is the reduction of trace impurities that are almost inevitably present in the blank solution; contributors here include small amounts of dissolved oxygen, heavy metal ions from the distilled water, and impurities present in the salt used as the supporting electrolyte.

A second component of the residual current is the so-called *charging* or *condenser current* resulting from a flow of electrons that charge the mercury droplets with respect to the solution; this current may be either negative or positive. At potentials more negative than about −0.4 V, an excess of electrons from the dc source provides the surface of each droplet with a negative charge. These excess electrons are carried down with the drop as it breaks; since each new drop is charged as it forms, a small but continuous current results. At applied potentials smaller than about −0.4 V, the mercury tends to be positive with respect to the solution; thus, as each drop is formed, electrons are repelled from the surface toward the bulk of mercury, and a negative current is the result. At about −0.4 V, the mercury surface is uncharged, and the condenser current is zero. The charging current is a type of *nonfaradaic current* (see Section 19A–3) in the sense that electricity is carried across an electrode-solution interface without an accompanying oxidation-reduction process.

Ultimately, the accuracy and sensitivity of the polarographic method depend upon the magnitude of the nonfaradaic residual current and the accuracy with which a correction for its effect can be determined.

Current Maxima. The shapes of polarograms are frequently distorted by so-called *current maxima* (see Figure 22–6), which are troublesome because they interfere with the accurate evaluation of diffusion currents and half-wave potentials. Although the cause or causes of maxima are not fully understood, there is considerable empirical knowledge of methods for eliminating them. Generally, the addition of traces of such high molecular weight substances as gelatin, Triton X-100 (a commercial surface-active agent), methyl red, or other dyes will cause a maxi-

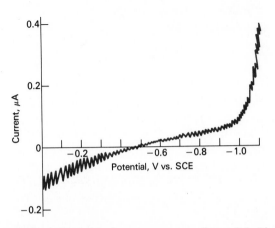

FIGURE 22-5 Residual current curve for a 0.1 *M* solution of HCl.

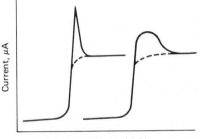

FIGURE 22-6 Typical current maxima.

mum to disappear. Care must be taken to avoid large amounts of these reagents, however, because the excess may reduce the magnitude of the diffusion current. The proper amount of suppressor must be determined by trial and error; the amount required varies widely from analyte to analyte.

Comparison of Currents from Dropping and Stationary Planar Electrodes. Figure 22–4 shows that several minutes may be required to obtain steady currents with a planar electrode after a new voltage setting is made. In contrast, the dropping electrode exhibits constant reproducible currents essentially instantaneously after an applied voltage adjustment. This behavior represents a *major* advantage of the dropping mercury electrode, which accounts for its widespread use.

The rapid achievement of constant currents arises from the highly reproducible nature of the drop formation process and, equally important, the fact that the solution in the electrode area becomes homogenized each time a drop breaks from the capillary. Thus, a concentration gradient is developed only during the brief lifetime of the drop. As we have noted, current changes due to an increase in surface area occur during each lifetime. Changes in the diffusion gradient dC/dx are also occurring during this period. But these changes are entirely reproducible leading to currents that are also highly reproducible.

22A–3 HALF-WAVE POTENTIAL

To obtain a relationship between applied potential and current for the electrode reaction given by Equation 22–2, let us subtract Equation 22–6 from Equation 22–8. Upon rearrangement, we find

$$[A_0] = \frac{i_d - i}{k_A}$$

Substitution of this equation and Equation

22–7 into Equation 22–3 yields upon arrangement

$$E_{appl} = E_A^0 - E_{SCE} - \frac{0.0591}{n} \log \frac{k_A}{k_P}$$
$$- \frac{0.0591}{n} \log \frac{i}{i_d - i} \qquad (22\text{–}12)$$

By definition, $E_{appl} = E_{1/2}$ when

$$i = \frac{i_d}{2}$$

Substituting this relationship in Equation 22–12 reveals that

$$E_{1/2} = E_A^0 - E_{SCE} - \frac{0.0591}{n} \log \frac{k_A}{k_P} \qquad (22\text{–}13)$$

Thus, the half-wave potential is a constant that is related to the standard potential for the half-reaction as well as k_A and k_P. The latter two terms are constants that depend upon the diffusion coefficients for the two species and those characteristics of the dropping electrode that are described by the Ilkovic equation. That is, $k_A = 607\, nD_A^{1/2}m^{2/3}t^{1/6}$; a similar relationship can be written for k_P. Thus,

$$k_A/k_P = \sqrt{D_A/D_P}$$

Substitution of Equation 22–13 into 22–12 yields

$$E_{appl} = E_{1/2} - \frac{0.0591}{n} \log \frac{i}{i_d - i} \qquad (22\text{–}14)$$

As we have seen, i_d can be calculated from the characteristics of the particular electrode and the diffusion coefficient for A. Thus, the current i can be calculated at any desired value of E_{appl} provided $E_{1/2}$ is known. Often the half-wave potential can be estimated from the standard potential for the system.

Examination of Equation 22–13 reveals that the half-wave potential is a reference point on a polarographic wave that is independent of the reactant concentration but directly related to the standard potential for the half-reaction. In practice, the half-wave potential can be a useful quantity for identification of the species responsible for a given polarographic wave.

It is important to note that the half-wave potential may vary considerably with concentration for electrode reactions that are slow and thus irreversible. Under these circumstances, Equation 22–14 is no longer applicable. In fact, behavior according to Equation 22–14 is often used as a criterion for reversibility. Thus, a plot of log $[i/(i_d - i)]$ vs. E_{appl} gives a straight line with a slope of $-0.0591/n$ for reversible reactions. Such a plot permits evaluation of n.

Effect of Complex Formation on Polarographic Waves. We have already seen (Section 19C–8) that the potential for the oxidation or reduction of a metallic ion is greatly affected by the presence of species that form complexes with that ion. It is not surprising, therefore, that similar effects are observed with polarographic half-wave potentials. The data in Table 22–1 show clearly that the half-wave potential for the reduction of a metal complex is generally more negative than that for reduction of the corresponding simple

metal ion. In fact, this negative shift in potential permits the elucidation of the composition of the complex ion and the determination of its formation constant *provided that the electrode reaction is reversible.* Thus, for the reactions

$$M^{n+} + Hg + ne \rightleftharpoons M(Hg)$$

and

$$M^{n+} + xA^- \rightleftharpoons MA_x^{(n-x)+}$$

Lingane[5] derived the following relationship between the molar concentrations of the ligand M_L and the shift in half-wave potential brought about by its presence:

$$(E_{1/2})_c - E_{1/2} = -\frac{0.0591}{n} \log K_f$$
$$-\frac{0.0591x}{n} \log M_L \qquad (22\text{–}15)$$

where $(E_{1/2})_c$ and $E_{1/2}$ are the half-wave potentials for the complexed and uncomplexed cations, respectively, K_f is the formation constant for the complex, and x is the

---

[5] J. J. Lingane, *Chem. Rev.*, **1941**, *29*, 1.

Table 22–1
Effect of Complexing Agents on Polarographic Half-Wave Potentials at the Dropping Mercury Electrode

Ion	Noncomplexing Media	1 *M* KCN	1 *M* KCl	1 *M* NH₃, 1 *M* NH₄Cl
Cd^{2+}	−0.59	−1.18	−0.64	−0.81
Zn^{2+}	−1.00	NR[a]	−1.00	−1.35
Pb^{2+}	−0.40	−0.72	−0.44	−0.67
Ni^{2+}	—	−1.36	−1.20	−1.10
Co^{2+}	—	−1.45	−1.20	−1.29
Cu^{2+}	+0.02	NR[a]	+0.04	−0.24
			−0.22	−0.51

[a] No reduction occurs before involvement of the supporting electrolyte.

molar combining ratio of complexing agent to cation.

Equation 22–15 makes it possible to evaluate the formula for the complex. Thus, a plot of the half-wave potential against log M_L for several ligand concentrations gives a straight line, the slope of which is $0.0591x/n$. If n is known from plots of $\log[i/(i_d - i)]$ versus E_{appl} (p. 673), the combining ratio of ligand to metal ion x is readily calculated. Equation 22–15 can then be employed to calculate K_f.

Polarograms for Irreversible Reactions. Many polarographic electrode processes, particularly those associated with organic systems, are irreversible, which leads to drawn-out and less well-defined waves. The quantitative description of such waves requires an additional term (involving the activation energy of the reaction) in Equation 22–14 to account for the kinetics of the electrode process. Although half-wave potentials for irreversible reactions ordinarily show a dependence upon concentration, diffusion currents remain linearly related to concentration; such processes are, therefore, readily adapted to quantitative analysis.

22A–4 EFFECT OF CELL RESISTANCE ON POLAROGRAMS

Figure 22–7 illustrates the effect of cell resistance on the polarographic wave for a reversible system. At 100 Ω, the IR drop is so small that it does not influence the shape of the wave significantly. With high resistances, on the other hand, the waves become drawn out and ultimately are so poorly defined as to be of little use. The cause of this effect is easily understood by reference to the equation

$$E_{appl} = E_{ind} - E_{SCE} - IR$$

Here, E_{appl} is varied in a regular way while E_{SCE} is constant. If IR is small, the indicator

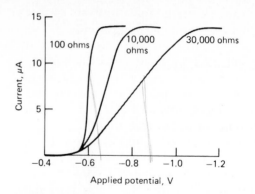

FIGURE 22-7 Effect of cell resistance on a reversible polarographic wave.

electrode then faithfully reflects the variation in E_{appl}. On the other hand, if R is large, a greater and greater fraction of E_{appl} is required to overcome the IR drop with increasing I. Thus, E_{ind} is no longer directly proportional to E_{appl}. A drawn-out wave results.

The use of potentiostatic control has permitted the extension of polarography to solvents with high electrical resistances. Here, three electrodes are employed: a dropping electrode (or other microelectrode), a counter or auxiliary electrode, and a reference electrode. The microelectrode and the counter electrode serve the same purpose as the two electrodes in ordinary polarography. The reference electrode, on the other hand, is employed to measure and control the potential of the microelectrode. The arrangement here is analogous to that described for controlled potential electrolysis (see Figure 21–3 or 21–6). The reference electrode is positioned as close as possible to the dropping electrode, and the potential between the two is then employed to control the applied potential E_{appl} such that the variation of E_{ind} is linear with time. Thus, the abscissa of the polarographic wave becomes E_{ind} rather than E_{appl}. The effect is to produce a wave form similar to that for the lowest resistance curve in Figure 22–7.

22A-5 POLAROGRAMS FOR MIXTURES OF REACTANTS

Ordinarily, the reactants of a mixture will behave independently of one another at a microelectrode; a polarogram for a mixture is thus simply the summation of the waves for the individual components. Figure 22–8 shows the polarogram of a five-component cation mixture. Clearly, a single polarogram may permit the quantitative determination of several elements. Success depends upon the existence of a sufficient difference between succeeding half-wave potentials to permit evaluation of individual diffusion currents. Approximately 0.2 V is required if the more reducible species undergoes a two-electron reduction; a minimum of about 0.3 V is needed if the first reduction is a

one-electron process. The analysis of mixtures is considered in a later section.

22A-6 ANODIC WAVES AND MIXED ANODIC-CATHODIC WAVES

Anodic waves as well as cathodic waves are encountered in polarography. The former are less common because of the relatively small range of anodic potentials that can be covered with the dropping mercury electrode before oxidation of the electrode itself commences. An example of an anodic wave is illustrated in curve *A* of Figure 22–9, where the electrode reaction involves the oxidation of iron(II) to iron(III) in the presence of citrate ion. A diffusion current is obtained at about +0.1 V, which is due to the half-reaction

$$Fe^{2+} \rightleftharpoons Fe^{3+} + e$$

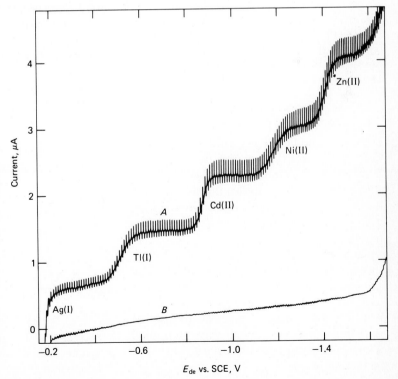

FIGURE 22-8 Polarograms of *A*: approximately 0.1 m*M* each of silver(I), thallium(I), cadmium(II), nickel(II), and zinc(II), listed in the order in which their waves appear, in 1 *M* ammonia/1 *M* ammonium chloride containing 0.002% Triton X-100; *B*: the supporting electrolyte alone. (From: L. Meites, *Polarographic Techniques*, 2d ed, p. 164, Wiley: New York, 1967. Reprinted by permission of John Wiley & Sons, Inc.)

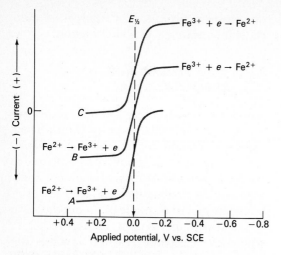

FIGURE 22-9 Polarographic behavior of iron(II) and iron(III) in a citrate medium. Curve *A*: anodic wave for a solution in which $[Fe^{2+}] + 1 \times 10^{-3}$. Curve *B*: anodic-cathodic wave for a solution in which $[Fe^{2+}] = [Fe^{3+}] = 0.5 \times 10^{-3}$. Curve *C*: cathodic wave for a solution in which $[Fe^{3+}] = 1 \times 10^{-3}$.

As the potential is made more negative, a decrease in the anodic current occurs; at about -0.02 V, the current becomes zero because the oxidation of iron(II) ion has ceased.

Curve *C* represents the polarogram for a solution of iron(III) in the same medium. Here, a cathodic wave results from reduction of the iron(III) to the divalent state. The half-wave potential is identical with that for the anodic wave, indicating that the oxidation and reduction of the two iron species are perfectly reversible at the dropping electrode.

Curve *B* is the polarogram of an equimolar mixture of iron(II) and iron(III). The portion of the curve below the zero-current line corresponds to the oxidation of the iron(II); this reaction ceases at an applied potential equal to the half-wave potential. The upper portion of the curve is due to the reduction of iron(III).

22A-7 OXYGEN WAVES

Dissolved oxygen is readily reduced at the dropping mercury electrode; an aqueous solution saturated with air exhibits two distinct waves attributable to this element (see Figure 22–10). The first results from the reduction of oxygen to peroxide:

$$O_2(g) + 2H^+ + 2e \rightleftharpoons H_2O_2$$

The second corresponds to the further reduction of the hydrogen peroxide:

$$H_2O_2 + 2H^+ + 2e \rightleftharpoons 2H_2O$$

As would be expected from stoichiometric considerations, the two waves are of equal height.

These polarographic waves are convenient for determining the concentration of dissolved oxygen; however, the presence of this element often interferes with the accurate determination of other species. Thus, oxygen removal is ordinarily the first step in a polarographic analysis. Deaeration of the solution for several minutes with an inert

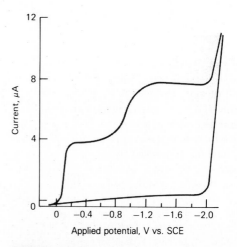

FIGURE 22-10 Polarogram for the reduction of oxygen in an air-saturated 0.1 *M* KCl solution. The lower curve is for oxygen-free 0.1 *M* KCl.

gas (*sparging*) accomplishes this end; a stream of the same gas, usually nitrogen, is passed over the surface during the analysis to prevent reabsorption of oxygen.

22A-8 ADVANTAGES AND DISADVANTAGES OF THE DROPPING MERCURY ELECTRODE

The dropping mercury electrode is the most widely used microelectrode for voltammetry because of its several unique features. The first is the unusually high overvoltage associated with the reduction of hydrogen ions. As a consequence, metal ions such as zinc and cadmium can be deposited from acidic solution even though their thermodynamic potentials suggest that deposition of these metals without hydrogen formation is impossible. A second advantage is that a new metal surface is generated continuously; thus the behavior of the electrode is independent of its past history. In contrast, solid metal electrodes are notorious for their irregular behavior, which is related to adsorbed or deposited impurities. A third unusual feature of the dropping electrode, which has already been described, is that reproducible average currents are *immediately* realized at any given potential regardless of whether this potential is approached from lower or higher settings.

One serious limitation of the dropping electrode is the ease with which mercury is oxidized; this property severely limits the use of the electrode as an anode. At potentials greater than about +0.4 V, formation of mercury(I) occurs giving a wave that masks the curves of other oxidizable species. In the presence of ions that form precipitates or complexes with mercury(I), this behavior occurs at even lower potentials. For example, in Figure 22–2, the beginning of an anodic wave can be seen at 0 V due to the reaction:

$$2Hg + 2Cl^- \rightarrow Hg_2Cl_2(s) + 2e$$

Incidentally, this anodic wave can be used for the determination of chloride ion.

Another important disadvantage of the dropping mercury electrode is the nonfaradaic residual or charging current which limits the sensitivity of the *classical* method to concentrations of about 10^{-5} *M*. At lower concentrations, the residual current is likely to be greater than the diffusion current, a situation that prohibits accurate measurement of the latter. As will be shown later, methods are now available for avoiding this disadvantage.

Finally, the dropping mercury electrode is cumbersome to use and tends to malfunction as a result of clogging.

22B Instrumentation

Instrumentation for classical polarography can be relatively simple and inexpensive, costing a few hundred dollars. On the other hand, a modern instrument capable of performing the various types of modified polarography described in later sections is complicated, often computer controlled, and costs several thousands of dollars.

22B-1 CELLS

A typical general purpose cell for polarographic measurements is shown in Figure 22–11a. The cell itself is a heavy-walled glass container, which is threaded at the top so that it can be screwed into the polypropylene cap that holds the electrodes. The cell walls are tapered so that as much as 30 mL and as little as 2 mL of the analyte solution can be studied. As shown in Figure 22–11b, the cap is fitted with five O-ring adapters to hold electrodes and other accessories. As shown in **a**, one accessory that is always present is a purge tube fitted to a three-way stopcock (not shown) that permits nitrogen to be bubbled through the solution

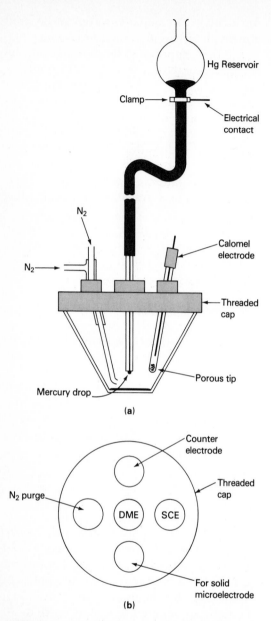

FIGURE 22-11 A dropping mercury electrode and cell: (**a**) cross-section view; (**b**) top view of cap.

for deaeration or over the liquid surface to prevent reabsorption of oxygen.

Modern polarography is generally carried out with three electrodes: a dropping mercury electrode, a reference electrode, and a counter electrode. The last is not shown in Figure 22–11**a** because it is behind the plane of the paper; its position is indicated in Figure 22–11**b**. The counter electrode can be any inert conducting electrode including a mercury pool, a platinum wire, a graphite surface, or a host of other conductors. Generally, the counter electrode does not have to be isolated from the test solution because the quantity of products produced here is small and the time scale of the experiment is such that the likelihood of these products reaching the microelectrode area is negligible.

With the three-electrode arrangement, it is no longer necessary that the reference electrode be large, as in the two-electrode arrangement, because it does not need to carry a significant current.

The simplest cell arrangement of all consists of the dropping electrode and a wire connector to the pool of mercury at the bottom of the cell. This arrangement is capable of producing perfectly satisfactory data for classical polarographic studies.

Dropping Electrodes. A dropping electrode, such as that shown in Figure 22–11, can be purchased from commercial sources. A 10-cm capillary ordinarily has a drop time of 3 to 6 sec under a mercury head of about 50 cm. The tip of the capillary should be as nearly perpendicular to the length of the tube as possible, and care should be taken to assure a vertical mounting of the electrode; otherwise, erratic and nonreproducible drop times and sizes will be observed.

With reasonable care, a capillary can be used for several months or even years. Such performance, however, requires the use of scrupulously clean mercury and the maintenance of a mercury head, no matter how slight, at all times. If solution comes in contact with the inner surface of the tip, malfunction of the electrode is inevitable. For this reason, the head of mercury should always be increased to provide a good flow before the tip is immersed in a solution.

Cleaning of a malfunctioning electrode is usually not very successful in restoring its performance. When not in use, the electrode is immersed in clean mercury after which the head can be reduced.

Mechanically Controlled Dropping Electrodes. As will be seen in later sections, it is often advantageous to be able to dislodge the drop from a mercury electrode at a precisely controllable time. Various types of hammers or drop knockers have been devised for this purpose and are available commercially. An electrode system equipped with one of these devices is shown in Figure 22–12. The mercury is contained in a plastic-lined reservoir about 10 inches above the upper end of the capillary. A compression spring forces the polyurethane tipped plunger against the head of the capillary, thus preventing a flow of mercury. This plunger is lifted upon activation of the solenoid by a signal from the control system. The capillary is much larger in diameter (0.15 mm) than the typical one. As a result the formation of the drop is extremely rapid. After 50, 100, or 200 msec, the valve is closed leaving a full-sized drop in place until it is dislodged by a mechanical knocker that is built into the electrode support block. The three time-intervals produce drops that are described as small, medium, or large. This system has the advantage that the full-sized drop forms quickly and current measurements can be delayed until the surface area is stable and constant. This procedure then largely eliminates the residual or charging current that limits the sensitivity of classical polarography.

22B-2 ELECTRICAL APPARATUS

The electrical apparatus for classical polarographic measurements can be relatively simple, consisting of a voltage divider that permits continuous variation of an applied voltage from 0 to about ± 2.5 V known to about ± 0.01 V. In addition, provision must be made to measure currents over the range between 0.01 and perhaps 100 μA with a precision of ± 0.01 μA. A manual instrument that meets these requirements is easily constructed from equipment available in most laboratories.

Figure 22–13 is a circuit diagram of an electronic instrument for three-electrode polarography. It consists of four components: a *ramp generator,* a potentiostat, a cell containing three electrodes, and a current amplifier and recorder. The purpose of the ramp generator is to provide a continuously varying potential or linear sweep across the indicator/saturated calomel electrode system. Note that the generator circuit is similar to the integration circuit shown in Figure 2–22c (p. 57). The relationship between the constant input potential E_i and the resulting output potential E_o can be obtained from Equation 2–14. That is,

$$E_o = -\frac{E_i}{R_i C_f} \int_0^t dt = -\frac{E_i}{R_i C_f} t$$

Thus, the output potential is determined by E_i and R_i and at any instant is directly proportional to the time that has elapsed after closure of the switch S. A fraction of E_o is applied to the noninverting terminal of the operational amplifier associated with the potentiostat.

The potentiostat shown in Figure 22–13 is similar in design to the one shown in Figure 21–6 and described on page 653. It may be dispensed with for classical polarographic measurements.

The cell consists of a dropping mercury electrode, a saturated calomel electrode (which is part of the control circuit), and a counter electrode. Essentially no current develops in the circuit that includes the reference and indicator electrodes because of the high internal resistance of the operational amplifier. The measured current is thus the output current I from the operational amplifier, which is determined by the potential

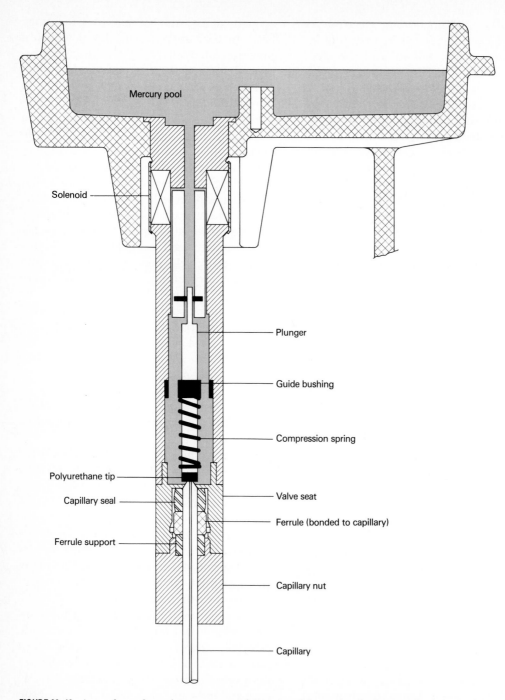

Mercury pool

Solenoid

Plunger

Guide bushing

Compression spring

Polyurethane tip

Valve seat

Capillary seal

Ferrule (bonded to capillary)

Ferrule support

Capillary nut

Capillary

FIGURE 22-12 A modern dropping mercury electrode with mechanical control of drop size and time. (Courtesy of EG&G Princeton Applied Research, Princeton, NJ.)

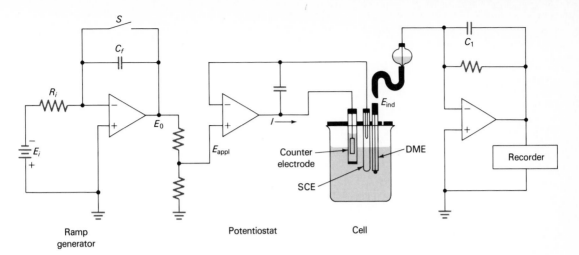

FIGURE 22-13 Schematic diagram showing the components of a modern, three-electrode polarograph and cell.

of the dropping electrode *relative to the saturated calomel electrode.*

The counter electrode in polarographic studies with three electrodes is frequently a pool of mercury in the bottom of the cell. In Figure 22–13, it is depicted as a platinum electrode separated from the analyte solution by a fritted disk.

The capacitor C_1 in the current amplifier circuit is employed to damp the current fluctuations associated with the dropping electrode.

22C Modified Voltammetric Methods

As was mentioned earlier (p. 665), quantitative classical polarography with a dropping mercury electrode is limited to solutions with concentrations greater than about 10^{-5} *M*. This limitation results from the nonfaradaic current associated with the charging of each mercury drop as it forms. Thus, when the ratio of the faradaic current (from the reduction of the analyte) to nonfaradaic current approaches unity, large uncertainties in determining diffusion currents are

inevitable. One of the major goals of the recent modifications of the classical method has been that of increasing the ratio between the faradaic and nonfaradaic currents by suppressing the latter, thus permitting the quantitative determination of species at lower concentrations. Some of these modern adaptations of the classical method are summarized in Figures 22–14 and 22–15 and are described briefly in the paragraphs that follow.[6]

22C-1 *CURRENT-SAMPLED (TAST) POLAROGRAPHY*

A simple modification of the classical polarographic technique, and one that is incorporated into most modern instruments, involves measurement of current only for a period near the end of the lifetime of each drop. Here, a mechanical knocker is generally used to detach the drop after a highly reproducible time interval (usually 0.5 to 5

---

[6] For additional details, see: J. B. Flato, *Anal. Chem.,* **1972,** *44* (11), 75A; and A. M. Bond, *Modern Polarographic Methods in Analytical Chemistry,* Chapters 4–9, Marcel Dekker: New York, 1980.

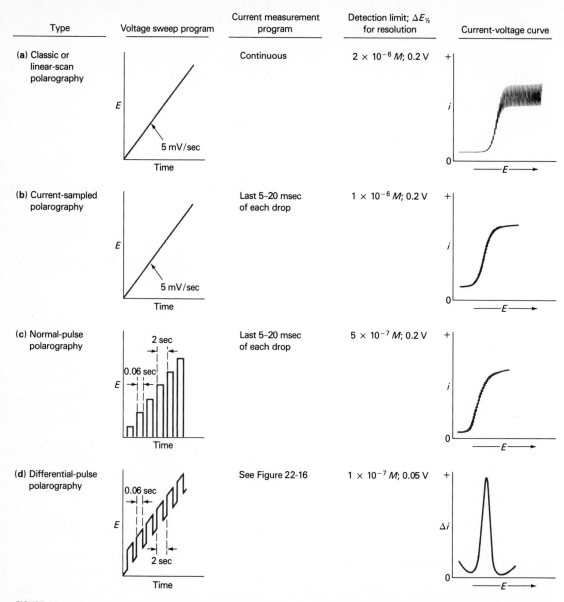

Type	Voltage sweep program	Current measurement program	Detection limit; $\Delta E_{\frac{1}{2}}$ for resolution	Current-voltage curve
(a) Classic or linear-scan polarography	5 mV/sec	Continuous	2×10^{-6} M; 0.2 V	
(b) Current-sampled polarography	5 mV/sec	Last 5-20 msec of each drop	1×10^{-6} M; 0.2 V	
(c) Normal-pulse polarography	2 sec / 0.06 sec	Last 5-20 msec of each drop	5×10^{-7} M; 0.2 V	
(d) Differential-pulse polarography	0.06 sec / 2 sec	See Figure 22-16	1×10^{-7} M; 0.05 V	

FIGURE 22-14 Some types of polarography: (**a**) linear-scan; (**b**) current-sampled; (**c**) normal-pulse; (**d**) differential-pulse.

sec). Because of this feature, the method is sometimes called *tast* polarography (from German *tasten,* to touch). *Current-sampled polarography* is more descriptive than tast polarography and thus to be preferred inas-

much as devices other than the mechanical knocker are now used for drop detachment (see Figure 22–12, for example).

Figure 22–14**b** illustrates the sequence of events in recording a dc current-sampled

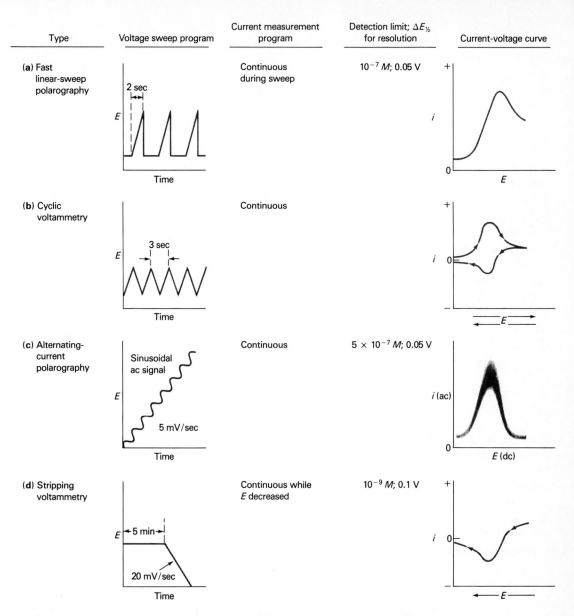

Type	Voltage sweep program	Current measurement program	Detection limit; $\Delta E_{1/2}$ for resolution	Current-voltage curve
(a) Fast linear-sweep polarography	2 sec	Continuous during sweep	10^{-7} M; 0.05 V	
(b) Cyclic voltammetry	3 sec	Continuous		
(c) Alternating-current polarography	Sinusoidal ac signal / 5 mV/sec	Continuous	5×10^{-7} M; 0.05 V	
(d) Stripping voltammetry	5 min / 20 mV/sec	Continuous while E decreased	10^{-9} M; 0.1 V	

FIGURE 22-15 Some types of voltammetry: (**a**) rapid-scan; (**b**) cyclic; (**c**) ac; (**d**) stripping.

polarogram. As in classical polarography, the applied voltage is increased linearly at perhaps 5 mV/sec. Rather than recording the current continuously, however, it is instead sampled for a 5 to 20 msec period just before termination of each drop. Between sampling periods, the recorder is maintained at its last current level by means of a sample and hold circuit.

The advantage of current sampling is that it substantially reduces the large current fluctuations due to the continuous growth and fall of drops that occur with the dropping electrode. Note in Figure 22–1, that the current near the end of the life of a drop is nearly constant, and it is this current only that is recorded in the current-sampled technique. The result is a smoothed curve consisting of a series of steps, which are significantly smaller than the current fluctuations encountered in normal polarography. (Contrast the polarograms in Figures 22–14a and 22–14b, for example.) The improvements in precision and detection limit by current sampling alone are not, unfortunately, as great as might be hoped. For example, Bond and Canterford[7] showed that the detection limits for copper could be lowered from about $3 \times 10^{-6}\ M$ for conventional polarography to $1 \times 10^{-6}\ M$ with the current-sampled method, a marginal change at best.

22C-2 PULSE POLAROGRAPHY

Several types of pulse polarography have been developed beginning in about 1960. The voltage sweep programs for the two most common types of pulse methods, *normal* and *differential*, are shown in Figures 22–14c and **d**. Here, dc voltage pulses are applied for a period of 60 msec (or sometimes less) just before each drop of mercury is detached from the capillary by a mechanical knocker.

Normal Pulse Polarograms. As shown in Figure 22–14c, in normal pulse polarography, the size of successive pulses is increased linearly with time. Like the previous procedure, the current is then sampled during the last 5 to 20 msec of the drop, where increases in current due to the increase in electrode surface area are minimal. Here again, the wide current variations of a classical polarogram are eliminated and a curve similar to that obtained with the normal current-sampled

procedure is recorded. An additional advantage, however, accrues to the pulse techniques, namely enhanced sensitivity. The reasons for this improvement will become apparent shortly.

Differential Pulse Polarograms. In differential pulse polarography, a dc potential, which is increased linearly with time, is applied to the polarographic cell (see Figure 22–14d). As in classical polarography, the rate of increase is perhaps 5 mV/sec. In contrast, however, a dc pulse of an additional 20 to 100 mV is applied for 60 msec just before detachment of the mercury drop from the electrode. Here again, to synchronize the pulse with the drop, the latter is detached at an appropriate time by a mechanical means.

As shown in Figure 22–16, two current measurements are made alternately—one just prior to the dc pulse and one near the end of the pulse. The *difference in current per pulse* (Δi) is recorded as a function of the linearly increasing voltage. A differential curve results consisting of a peak (see Figure 22–17a) the height of which is directly proportional to concentration.

One advantage of the derivative-type polarogram is that individual peak maxima can be observed for substances with half-wave potentials differing by as little as 0.04

[7] A. M. Bond and D. R. Canterford, *Anal. Chem.*, **1972**, *44*, 721.

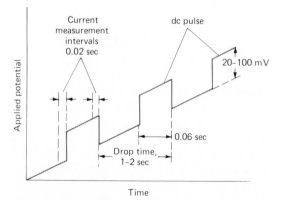

FIGURE 22–16 Voltage program for differential pulsed polarography.

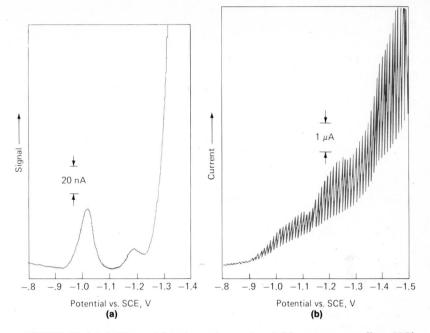

FIGURE 22-17 (**a**) Differential pulse polarogram. 0.36 ppm tetracycline·HCl in 0.1 *M* acetate buffer, pH 4, PAR Model 174 polarographic analyzer, dropping mercury electrode, 50-mV pulse amplitude, 1-sec drop. (**b**) DC polarogram. 180 ppm tetracycline · HCl in 0.1 *M* acetate buffer, pH 4, similar conditions. (Reprinted with permission from: J. B. Flato, *Anal. Chem.*, **1972,** *44* (11), 75A. Copyright 1972 American Chemical Society.)

to 0.05 V; in contrast, classical and normal pulse polarography requires a potential difference of at least 0.2 V for resolution of waves. More important, however, differential pulse polarography increases the sensitivity of the polarographic method significantly. This enhancement is illustrated in Figure 22–17. Note that a classical polarogram for a solution containing 180 ppm of the antibiotic tetracycline gives two barely discernible waves; differential pulse polarography, in contrast, provides well-defined peaks at a concentration level that is 2×10^{-3} that for the classic wave, or 0.36 ppm. Note also that the current scale for Δi is in nA (nanoamperes) or $10^{-3} \mu$A.

The greater sensitivity of pulse polarography, both normal and differential, can be attributed to two sources. The first is an en-

hancement of the faradaic current while the second is a decrease in the nonfaradaic, charging current. To account for the former, let us consider the events that must occur in the surface layer around an electrode as the potential is suddenly increased by 20 to 100 mV. If a reactive species is present in this layer, there will be a surge of current that lowers the reactant concentration to that demanded by the new potential. As the equilibrium concentration for that potential is approached, however, the current decays to a level just sufficient to counteract diffusion; that is, to the diffusion-controlled current. In classical polarography, the initial surge of current is not observed because the time scale of the measurement is long relative to the lifetime of the momentary current. On the other hand, in pulse polarography, the current measurement is made before the surge has completely decayed. Thus, the current measured contains both a diffusion-controlled component and a component that has to do with reducing the surface layer to the concentration demanded by the Nernst expression; the total current is typically several times larger than the diffusion current. It should be noted that when the drop is detached, the solution again becomes homogeneous with respect to the analyte. Thus, at any given voltage, an identical current surge accompanies each voltage pulse.

When the potential pulse is first applied to the electrode, a surge in the nonfaradaic current also occurs as the charge on the drop increases (p. 668). This current, however, decays exponentially with time and approaches zero near the end of the life of a drop when its surface area is changing only slightly (see Figure 22–1). Thus, by measuring currents at this time only, the nonfaradaic residual current is greatly reduced, and the signal-to-noise ratio is larger. Enhanced sensitivity results.

Reliable instruments for pulse polarography are now available commercially at reasonable cost. The method, particularly differential pulse polarography, has thus become an electroanalytical tool of considerable importance.

22C-3 FAST LINEAR-SWEEP POLAROGRAPHY

In fast linear-sweep polarography, the applied potential is swept linearly over a range of perhaps 0.5 V during the lifetime (or part of the lifetime) of a single mercury drop (see Figure 22–15a). Various methods have been developed for synchronizing the sweep time with the drop time so that the voltage sweep starts after a fixed and reproducible number of seconds from the detachment of the previous drop. An oscilloscope is often used to display the current-voltage relationship; for a permanent record, the image on the oscilloscope screen may be photographed. Alternatively, the data can be stored in a computer for subsequent plotting.

The most successful linear-sweep procedure appears to be one in which the voltage sweep occurs over the final 2 to 3 sec of a drop that has a total lifetime of 6 to 7 sec (see Figure 22–15a). As mentioned in the previous section, the residual currents under these circumstances are small when compared with those encountered in ordinary polarography.

As will be seen from Figure 22–15a, a rapid-sweep polarogram also takes the form of a peak or a summit. At the summit, the current is made up to two components. One is the initial current surge required to adjust the concentration of the surface film to an equilibrium state demanded by the Nernst equation. The second is the normal diffusion-controlled current. The first current then decays as equilibrium is approached; ultimately only the diffusion-controlled current remains.

It can be shown that at 25°C the *summit potential* E_s for a reversible process is related to the half-wave potential as follows:

$$E_s = E_{1/2} - 1.1\frac{RT}{nF} = E_{1/2} - \frac{0.065}{n} \quad (22\text{–}16)$$

The current i_s at the summit is given by

$$i_s = 2.69 \times 10^5 n^{3/2} A D^{1/2} C v^{1/2} \quad (22\text{–}17)$$

where A is the area of the electrode in cm^2, v is the rate of scan in volts per second, and the remaining terms are the same as in the Ilkovic equation (Equation 22–11). Generally, a summit current exceeds a diffusion current by a factor of ten or more.

The advantages that can be cited for rapid linear-sweep polarography are enhanced sensitivity, speed, and resolution. Not only are summit currents larger than diffusion currents, but the residual currents are also smaller; therefore, the signal-to-background ratio is considerably more favorable than in normal polarography. Quantitative analysis thus becomes possible for solutions as dilute as 10^{-6} to 10^{-7} M. In addition, the shapes of polarograms are such that accurate current measurements are possible even when the half-wave potentials of two species differ by as little as 0.05 V (see Figure 22–18).

Not surprisingly, rapid linear-sweep equipment is more complex and more expensive than that required for ordinary polarography.

22C-4 CYCLIC VOLTAMMETRY

The apparatus and techniques of cyclic polarography or voltammetry are similar to those for rapid linear-sweep polarography. The difference between the two lies in the voltage sweep program (compare Figures 22–15a and 22–15b). In cyclic voltammetry, the sweep takes a triangular form; that is, the potential is first increased linearly to a peak and then decreased to its starting point

at the same rate. Generally, the cycle is completed in a fraction of a second to a few seconds. With a dropping electrode, the sweep is timed to occur near the end of the lifetime of the drop to minimize the nonfaradaic current.

Figure 22–19 shows cyclic voltammetric curves for three reducible analytes. The solid curve is for a reversible reaction. The cause of cathodic peak A is the same as that for the peaks encountered in rapid linear-sweep polarography. When the potential is first reversed (point B), the current remains positive and is largely due to the diffusion-controlled reduction of the analyte. Ultimately, however, potential C is reached at which the analyte is no longer reduced; the current here is zero. With further positive changes in the potential, oxidation of the previously reduced species begins and proceeds until its concentration reaches zero. The anodic peak D results.

The curve labeled quasireversible is the voltammogram for a system in which the electron transfer process is not instantaneous. Here, the difference in potential between the cathodic and anodic peaks provides a measure of the relative rates of the reduction and oxidation reactions.

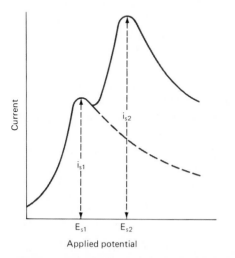

FIGURE 22-18 Rapid linear-sweep polarogram of a two-component mixture. (Taken from: H. Schmidt and M. von Stackelberg, *Modern Polarographic Methods*, p. 27, Academic Press: New York, 1963. With permission.)

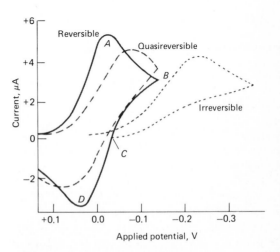

FIGURE 22-19 Cyclic voltammograms.

The third curve in Figure 22–19 is a voltammogram for an irreversible electrode process. Here, only a cathodic peak is observed because the product formed in the initial reduction is not reoxidized at a significant rate.

Cyclic voltammetry, while not used for routine quantitative analyses, has become an important tool for the study of mechanisms and rates of oxidation-reduction processes, particularly in organic and metal-organic systems. Often cyclic voltammograms will reveal the presence of intermediates in oxidation-reduction reactions. Usually, platinum is used for fabrication of microelectrodes used with this technique.

22C-5 SINUSOIDAL ALTERNATING CURRENT POLAROGRAPHY

In the polarographic technique illustrated in Figure 22–15c, a constant, sinusoidal ac potential of a few millivolts is superimposed upon the dc potential normally employed in polarography. As in the conventional experiment, the dc potential is varied over its usual range; here, however, the *sinusoidal alternating current is measured*.

Consider the application of this technique to a solution containing a single, easily reduced species A, which, at a suitable potential, reacts rapidly and reversibly as follows:

$$A + ne \rightleftharpoons P$$

Assume also that an excess of supporting electrolyte is present.

We must first focus our attention on the *faradaic* alternating current in this cell; that is, the current that results from the oxidation-reduction process at the interface. In order for such a current to exist, it is clearly necessary to have both an oxidizable and a reducible species at the microelectrode surface. Significant concentrations of P are present, however, only when the *applied dc potential* corresponds to the steeply rising

portion of the dc polarogram. Thus, no faradaic alternating current will be observed before the decomposition potential (that is, the potential at which a significant amount of P is formed) has been reached. Beyond the decomposition potential, but short of the half-wave potential, the following relationship exists between the *surface* concentration of A and P

$$[A]_0 > [P]_0 > 0$$

and the magnitude of the alternating current will be determined by $[P]_0$. At applied dc potentials slightly greater than the half-wave potential,

$$[P]_0 > [A]_0 > 0$$

and here, the magnitude of the alternating current is controlled by $[A]_0$. At the half-wave potential,

$$[P]_0 = [A]_0$$

and the faradaic current will have reached its maximum value. When a potential corresponding to the diffusion current region is reached, $[A]_0 \rightarrow 0$, and the alternating current will, for lack of a reducible species, again approach zero. Thus, a typical ac polarogram consists of one or more peaks, the maxima occurring at the half-wave potentials for each of the reactive species present (see Figure 22–15c). The relationship between ac and classical dc polarograms is shown in Figure 22–20.

As noted in Section 19A–3, alternating currents also result from a nonfaradaic process that involves charging of the double layer surrounding an electrode. Thus, ac polarographic peaks are superimposed upon a base line of a nonfaradaic current. The double layer structure is, however, affected by the dc component; as a consequence, the base line of an ac polarogram is not horizontal. Fortunately, over a small po-

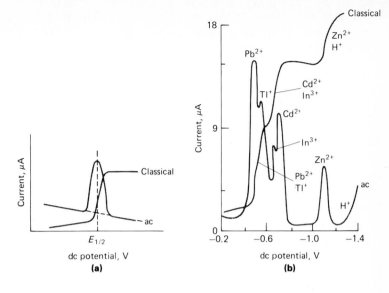

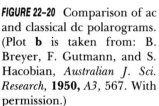

FIGURE 22-20 Comparison of ac and classical dc polarograms. (Plot **b** is taken from: B. Breyer, F. Gutmann, and S. Hacobian, *Australian J. Sci. Research*, **1950**, *A3*, 567. With permission.)

tential range, changes in the nonfaradaic current as a function of dc potential are nearly linear, and a satisfactory correction can be made by extrapolation between the base lines on either side of a peak (see Figure 22–20**a**).

The height of an ac polarographic peak is proportional to bulk concentration of the analyte and, for a reversible reaction, to the square root of frequency. For reversible electrode reactions, ac polarography is more sensitive than the classical procedure and permits quantitative analysis of solutions as dilute as 10^{-6} M.

No ac peak is observed for a totally nonreversible half-reaction (that is, where the rate of the half-reaction is very slow in one direction), since the product generated by the dc current does not provide sufficient reactant for transport of the alternating current during the oxidation half-cycle. For electrode processes where the rate of one or the other half-reaction is less than instantaneous, but still appreciable, the magnitude of the alternating current becomes dependent upon the frequency of the ac source. As the frequency is increased, a condition is ultimately

reached in which the time lapse of a half-cycle is insufficient for completion of the slow half-reaction; the alternating current will then be limited by the reaction rate, and the current maximum will be smaller than for an entirely reversible process. Thus, ac polarography is useful for kinetic studies of electrode reactions.

Alternating current polarography with current having a square wave rather than a sinusoidal form has also been investigated (*square wave polarography*). The advantage of the square wave form arises from the exponential decay of the nonfaradaic current with time and its approach to zero near the end of each half-cycle. By intermittent measurement of the current at the end of each half-cycle only, the base line current becomes vanishingly small.

Alternating current polarography has the advantage of enhanced sensitivity, and in contrast to classical polarography, permits the ready determination of a trace of a reducible species in the presence of a much larger concentration of a substance that gives a wave at lower potentials. The technique is also useful for the study of electrode kinetics. The disadvantage of the procedure is, of course, the complex nature of the equipment required.

22C-6 STRIPPING METHODS

Stripping methods encompass a variety of electrochemical procedures having a common, characteristic initial step.[8] In all of these procedures, the analyte is first collected by a constant electrode potential deposition at a mercury or a solid microelectrode. After an accurately measured period, the electrolysis is discontinued and the deposited analyte is determined by use of one of the voltammetric procedures that have been described earlier in this chapter. During this second step in the analysis, the analyte is redissolved or stripped from the microelectrode; hence the name attached to these methods. Figure 22–15**d** illustrates the voltage program that is followed when linear scanning is used as the second step in an analysis. Often one of the pulse techniques shown in Figures 22–14**c** and **d** is used in place of the simple linear scan.

Stripping methods are of prime importance in trace work because the concentrating aspects of the electrolysis permit the determination of minute amounts of an analyte with reasonable accuracy. Thus, the analysis of solutions in the 10^{-6} to 10^{-9} *M* range becomes feasible by methods that are both simple and rapid.

The stripping method that has had the most widespread application makes use of a micro mercury electrode for the deposition process, followed by anodic voltammetric measurement of the analyte; discussion in

the paragraphs that follow will be largely confined to this particular technique.

Electrodeposition Step.   Ordinarily, only a fraction of the analyte is deposited during the electrodeposition step; hence, quantitative results depend not only upon control of electrode potential but also upon such factors as electrode size, length of deposition, and stirring rate for both the sample and a standard solution employed for calibration.

Microelectrodes for stripping methods have been formed from a variety of materials including gold, silver, platinum, mercury, and glassy carbon. The most popular electrode is the *hanging drop* electrode, which consists of a single drop of mercury in contact with a platinum wire. Figure 22–21 illustrates one method of forming a hanging drop electrode. Here, an ordinary dropping mercury capillary provides a means of transferring a reproducible quantity of mercury (usually 1 to 3 drops) to a Teflon scoop. Note that the dropping mercury capillary does *not* serve as an electrode in this application but merely as a means of dispensing a highly reproducible volume of mercury. The hanging drop electrode is then formed by rotating the scoop and bringing the mercury into contact with a platinum wire sealed in a glass tube. The drop adheres strongly enough so that the solution can be stirred without displacement of the drop; it can, however, be dislodged by tapping the electrode at the completion of the electrolysis. Hanging drop electrodes are also available from several commercial sources. These electrodes consist of a microsyringe with a micrometer for exact control of drop size. The drop is then formed at the tip of a capillary by displacement of the mercury in the syringe controlled delivery system. The system shown in Figure 22–12 is also capable of use as a hanging drop electrode.

To carry out the determination of a metal ion, a fresh hanging drop is formed, stirring is begun, and a potential is applied that is a few tenths of a volt more negative than the

---

[8] For detailed discussions of stripping methods, see: F. Vydra, K. Stulik, and E. Julakova, *Electrochemical Stripping Analysis*, Halsted: New York, 1977; A. M. Bond, *Modern Polarographic Methods in Analytical Chemistry*, Chapter 9, Marcel Dekker: New York, 1980; and W. M. Peterson and R. V. Wong, *Amer. Lab.*, **1981**, *13* (11), 116.

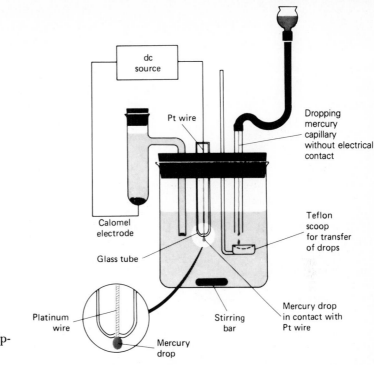

FIGURE 22-21 Apparatus for stripping analysis.

half-wave potential for the ion of interest. Deposition is allowed to occur for a carefully measured period; 5 min usually suffice for solutions that are $10^{-7} M$ or greater, 15 min for $10^{-8} M$ solutions, and 60 min for those that are $10^{-9} M$. It should be emphasized that these times seldom result in complete removal of the ion. The electrolysis period is determined by the sensitivity of the method ultimately employed for completion of the analysis.

Voltammetric Completion of the Analysis. The analyte collected in the hanging drop electrode can be determined by any of several voltammetric procedures. For example, in a linear anodic scan procedure, stirring is discontinued for perhaps 30 sec after termination of the deposition. The voltage is then decreased at a linear fixed rate from its original cathodic value, and the resulting anodic current is recorded as a function of the applied voltage. This linear scan produces a curve of the

type shown in Figure 22–22. In this experiment, cadmium was first deposited from a $1 \times 10^{-8} M$ solution by application of a potential of about -0.9 V (vs. SCE), which is about 0.3 V more negative than the half-wave potential for this ion. After 15 min of electrolysis, stirring was discontinued; 30 sec later, the potential was decreased at a rate of 21 mV/sec. A rapid increase in anodic current occurred at about -0.65 V as a result of the reaction

$$Cd(Hg) \rightarrow Cd^{2+} + Hg + 2e$$

The current decayed after reaching a maximum, owing to depletion of elemental cadmium in the hanging drop. The peak current, after correction for the residual current (curve *B* in Figure 22–22), was directly proportional to the concentration of cadmium ions over a range of 10^{-6} to 10^{-9} M and inversely proportional to deposition

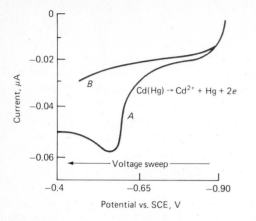

FIGURE 22-22 Curve *A*: Current voltage curve for anodic stripping of cadmium. Curve *B*: Residual current curve for blank. (Reprinted with permission and adapted from: R. D. DeMars and I. Shain, *Anal. Chem.*, **1957**, *29*, 1826. Copyright 1957 American Chemical Society.)

time. This analysis was based on calibration with standard solutions of cadmium ion. With reasonable care, an analytical precision of about 2% relative was obtained.

Most of the other voltammetric procedures shown in Figure 22–14 have also been applied to the stripping step. The most widely used of these appears to be an anodic differential pulse technique. Often, nar-

rower peaks are produced, which is desirable when mixtures are to be analyzed. Another method of obtaining narrower peaks is to use a mercury film electrode. Here, a thin mercury film is electrodeposited on an inert microelectrode such as glassy carbon. Usually, the mercury deposition is carried out simultaneously with the analyte deposition. Because the average diffusion path length from the film to the solution interface is much shorter than that in a drop of mercury, escape of the analyte is hastened; the consequence is narrower and higher voltammetric peaks, which leads to greater sensitivity and better resolution of mixtures. On the other hand, the hanging drop electrode appears to give more reproducible results, especially at higher analyte concentrations. Thus, for most applications the hanging drop electrode is employed. Figure 22–23 is a differential-pulse anodic stripping polarogram for a mixture of cations present at concentrations of 25 ppb showing good resolution and adequate sensitivity for many purposes.

Many other variations of the stripping technique have been developed. For example, a number of cations have been determined by electrodeposition on a platinum cathode. The quantity of electricity required

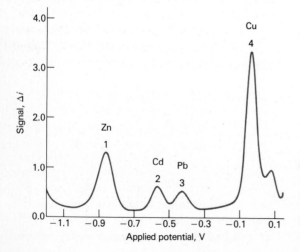

FIGURE 22-23 Differential-pulse anodic stripping voltammogram of 25 ppb zinc, cadmium, lead, and copper. (Reprinted with permission from: W. M. Peterson and R. V. Wong, *Amer. Lab.*, **1981**, *13* (11), 116. Copyright 1981 by International Scientific Communications, Inc.)

to remove the deposit is then measured coulometrically. Here again, the method is particularly advantageous for trace analyses. Cathodic stripping methods for the halides have also been developed. Here, the halide ions are first deposited as mercury(I) salts on a mercury anode. Stripping is then performed by a cathodic current.

22D Application of Polarography

Polarography has been used for the quantitative determination of a wide variety of inorganic and organic species including molecules of biological and biochemical interest. The applicability of polarography to organic species was greatly extended by the development of the potentiostatic methods because this technique made it possible to obtain satisfactory polarograms from nonaqueous solvents having low conductivities.

22D-1 TREATMENT OF DATA

For linear scan polarography, it is common practice to use either the average or the maximum value of the recorder oscillations. Most of the older literature is based upon the former.

Determination of Diffusion Currents. For quantitative work, limiting currents must always be corrected for the residual current. One method involves the use of a blank to derive an experimental residual-current curve. The diffusion current is then evaluated from the difference between the two at some potential in the limiting-current region (see Figure 22–2).

Because the residual current usually increases nearly linearly with applied voltage, it is often possible to correct for the residual current by simple extrapolation as shown in Figure 22–24.

22D-2 QUANTITATIVE POLAROGRAPHY

Calibration Curves. The most straightforward and best method for quantitative polarographic analyses involves preparation of a calibration curve with a series of standard solutions. The composition of these standard solutions should, as nearly as possible, be identical to the analyte solutions. That is,

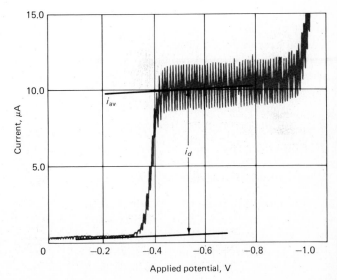

FIGURE 22-24 Extrapolation method for measuring diffusion current. Solution was 0.0010 M in Pb^{2+} and 1 M in $NaClO_4$.

they should contain the same supporting electrolyte, the same amount of maximum suppressor, the same sample matrix elements, and should cover a concentration range that brackets that of the sample. Other variables that must be made constant for sample and standards include temperature, mercury head, and capillary characteristics.

Standard Addition Method. The standard addition method is sometimes employed for samples having complex matrices that are difficult or impossible to duplicate. Here, the diffusion current for an accurately known volume of sample solution is measured. Then, a known amount of the species of interest is introduced (ordinarily as a known volume of a standard solution), and the diffusion current is again evaluated. Provided the relationship between current and concentration is linear, the increase in wave height will permit calculation of the concentration in the original solution.

Accuracy and Precision of Classical Polarography. The accuracy and precision of polarography depends upon the shape of the analyte wave. For the well-defined waves associated with reversible reactions (Figure 22–24, for example), current measurements accurate to 1 to 2% relative can be made. Uncertainties from other sources, such as temperature or drop time variations or instrument noise, lead to overall uncertainties of perhaps 3% relative. For high-precision analysis of serial samples, a precision of 1% relative is often possible.

For nonreversible and ill-defined waves, the error in current measurement may lie in the 5 to 20% relative range.

22D–3 INORGANIC POLAROGRAPHIC ANALYSIS

The polarographic method is widely applicable to the analysis of inorganic substances. Most metallic cations, for example, are reduced at the dropping electrode. Even the alkali and alkaline-earth metals are reducible, provided the supporting electrolyte does not react at the high potentials required; here, the tetraalkyl ammonium halides are useful electrolytes because of their high reduction potentials.

The successful polarographic analysis of cations frequently depends upon the supporting electrolyte that is used. To aid in this selection, tabular compilations of half-wave potential data are available.[9] The judicious choice of anion often enhances the selectivity of the method. For example, with potassium chloride as a supporting electrolyte, the waves for iron(III) and copper(II) interfere with one another; in a fluoride medium, however, the half-wave potential of the former is shifted by about -0.5 V, while that for the latter is altered by only a few hundredths of a volt. The presence of fluoride thus results in the appearance of well separated waves for the two ions.

The polarographic method is also applicable to the analysis of such inorganic anions as bromate, iodate, dichromate, vanadate, selenite, and nitrite. In general, polarograms for these substances are affected by the pH of the solution because the hydrogen ion is a participant in their reduction. As a consequence, strong buffering to some fixed pH is necessary to obtain reproducible data (see p. 695).

For applications of polarography to inorganic analysis, the reader is referred to the monographs by Kolthoff and Lingane and by Meites.[10]

---

[9] See, for example, the references cited in footnote 3, p. 665; another extensive source is *Handbook of Analytical Chemistry*, L. Meites, Ed., McGraw-Hill: New York, 1963.

[10] I. M. Kolthoff and J. J. Lingane, *Polarography*, 2d ed., vol. 2, Interscience: New York, 1952; and L. Meites, *Polarographic Techniques*, 2d ed., Interscience: New York, 1965.

22D-4   ORGANIC POLAROGRAPHIC ANALYSIS

Almost from its inception, the polarographic method has been used for the study and analysis of organic compounds with many papers being devoted to this subject. Several common functional groups are oxidized or reduced at the dropping electrode, thus making possible the determination of a wide variety of organic compounds.[11]

In general, the reactions of organic compounds at a microelectrode are slower and more complex than those for inorganic species. Consequently, theoretical interpretation of the data is more difficult and often impossible; moreover, a much stricter adherence to detail is required for quantitative work. Despite these handicaps, organic polarography has proved fruitful for the determination of structure, for the quantitative analysis of mixtures, and occasionally, for the qualitative identification of compounds.

Effect of pH on Polarograms.   Organic electrode processes ordinarily involve hydrogen ions, the typical reaction being represented as

$$R + nH^+ + ne \rightleftharpoons RH_n$$

where R and RH_n are the oxidized and reduced forms of the organic molecule. Half-wave potentials for organic compounds are therefore markedly pH-dependent. Furthermore, alteration of the pH may result in a change in the reaction product. For example, when benzaldehyde is reduced in a basic solution, a wave is obtained at about -1.4 V, attributable to the formation of benzyl alcohol:

$$C_6H_5CHO + 2H^+ + 2e \rightleftharpoons C_6H_5CH_2OH$$

If the pH is less than 2, however, a wave occurs at about -1.0 V that is just half the size of the foregoing one; here, the reaction involves the production of hydrobenzoin:

$$2C_6H_5CHO + 2H^+ + 2e$$
$$\rightleftharpoons C_6H_5CHOHCHOHC_6H_5$$

At intermediate pH values, two waves are observed, indicating the occurrence of both reactions.

It should be emphasized that an electrode process that consumes or produces hydrogen ions will alter the pH of the solution *at the electrode surface,* often drastically, unless the solution is well-buffered. These changes affect the reduction potential of the reaction and cause drawn-out, poorly defined waves. Moreover, where the electrode process is altered by pH, as in the case of benzaldehyde, nonlinearity in the diffusion current-concentration relationship will also be encountered. Thus, in organic polarography good buffering is generally vital for the generation of reproducible half-wave potentials and diffusion currents.

Solvents for Organic Polarography.   Solubility considerations frequently dictate the use of solvents other than pure water for organic polarography; aqueous mixtures containing varying amounts of such miscible solvents as glycols, dioxane, acetonitrile, alcohols, Cellosolve, or acetic acid have been employed. Anhydrous media such as acetic acid, formamide, diethylamine, and ethylene glycol have also been investigated. Supporting electrolytes are often lithium or tetraalkyl ammonium salts.

Reactive Functional Groups.   Organic compounds containing any of the following functional groups can be expected to produce one or more polarographic waves.

1. *The carbonyl group,* including aldehydes, ketones, and quinones, produce polaro-

---

[11] For a detailed discussion of organic polarographic analysis, see: P. Zuman, *Organic Polarographic Analysis,* Pergamon Press: Oxford, 1964; *Polarography of Molecules of Biological Significance,* W. F. Smyth, Ed., Academic Press: New York, 1979; and *Topics in Organic Polarography,* P. Zuman, Ed., Plenum Press: New York, 1970.

graphic waves. In general, aldehydes are reduced at lower potentials than ketones; conjugation of the carbonyl double bond also results in lower half-wave potentials.

2. *Certain carboxylic acids* are reduced polarographically, although simple aliphatic and aromatic monocarboxylic acids are not. Dicarboxylic acids such as fumaric, maleic, or phthalic acid, in which the carboxyl groups are conjugated with one another, give characteristic polarograms; the same is true of certain keto and aldehydo acids.

3. *Most peroxides and epoxides* yield polarograms.

4. *Nitro, nitroso, amine oxide, and azo groups* are generally reduced at the dropping electrode.

5. *Most organic halogen groups* produce a polarographic wave, which results from replacement of the halogen group with an atom of hydrogen.

6. *The carbon/carbon double bond* is reduced when it is conjugated with another double bond, an aromatic ring, or an unsaturated group.

7. *Hydroquinones and mercaptans* produce anodic waves.

In addition, a number of other organic groups cause catalytic hydrogen waves that can be used for analysis. These include amines, mercaptans, acids, and heterocyclic nitrogen compounds. Numerous applications to biological systems have been reported.[12]

22E Voltammetry at Solid Electrodes

Solid microelectrodes constructed from platinum, gold, glassy carbon, graphite, and other inert conductors have been used for anodic oxidations at potentials where a mercury electrode is oxidized and thus cannot be used. Such electrodes permit the determination of a variety of organic compounds and strong inorganic oxidizing agents.[13] Unfortunately, these solid microelectrodes do not exhibit the reproducible behavior that characterizes the dropping mercury electrode, and their use has been limited. Therefore, the discussion here is confined to a single solid-state electrode, the oxygen electrode, which has found reasonably widespread use. Another solid-state electrode, the rotating platinum electrode, is described in the next section.

An important application of stationary solid microelectrodes is to the determination of oxygen in water, sewage effluents, blood, and other fluids.[14] The electrode consists of a platinum or gold surface on the end of a support rod that is covered with an oxygen-permeable membrane of Teflon or polyurethane. The arrangement is often similar to the gas-sensing electrode shown in Figure 20–8 with the exception that the glass electrode is replaced with a micro platinum electrode. The reference electrode is ordinarily a silver/silver chloride or other silver-based electrode. A potential of about -0.8 V (vs. SCE) is applied across the pair of electrodes, whereupon reduction of oxygen occurs in the film of liquid adjacent to the platinum surface. The electrode process is:

$$O_2 + 2H_2O + 2e \rightleftharpoons H_2O_2 + 2OH^-$$

The current is proportional to the concentration of oxygen in the internal solution, which in turn is proportional to the oxygen

[12] M. Brezina and P. Zuman, *Polarography in Medicine, Biochemistry and Pharmacy*, Interscience: New York, 1958; and *Polarography of Molecules of Biological Significance*, W. F. Smyth, Ed., Academic Press: New York, 1979.

[13] For a review of some of these applications, see: R. N. Adams, in *Treatise on Analytical Chemistry*, I. M. Kolthoff and P. J. Elving, Eds., Part I, vol. 4, Chapter 47, Interscience: New York, 1963.

[14] For a thorough treatment of the oxygen electrode, see: I. Fatt, *Polarographic Oxygen Sensors*, CRC Press: Cleveland, 1976.

content of the external solution adjacent to the electrode. Transport of oxygen occurs by the mechanism described earlier for CO_2 (Section 20C–9).

Oxygen electrodes are available commercially. They could more aptly be termed oxygen sensitive voltammetric cells.

22F Amperometric Titrations

Voltammetric methods can be employed to estimate the equivalence point of titrations, provided at least one of the participants or products of the reaction involved is oxidized or reduced at a microelectrode. Here, the current at some fixed potential is measured as a function of the reagent volume (or of time if the reagent is generated by a constant-current coulometric process). Plots of the data on either side of the equivalence point are straight lines with differing slopes; the end point is established by extrapolation to their intersection.

An amperometric titration is inherently more accurate than voltammetric methods and less dependent upon the characteristics of the microelectrode and the supporting electrolyte. Furthermore, the temperature need not be fixed accurately, although it must be kept constant during the titration. Finally, the substance being determined need not be reactive at the electrode; a reactive reagent or product is equally satisfactory.[15]

22F-1 TYPICAL TITRATION CURVES

Amperometric titration curves typically take one of the forms shown in Figure 22–25. Curve **a** represents a titration in which the analyte reacts at the electrode while the reagent does not. The titration of lead with

<hr/>

[15] For a detailed discussion of amperometric titration, see: J. T. Stock, *Amperometric Titrations*, Interscience: New York, 1965.

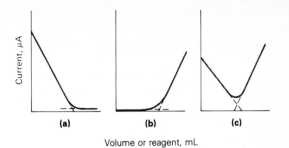

FIGURE 22-25 Typical amperometric titration curves: (**a**) analyte is reduced, reagent is not; (**b**) reagent is reduced, analyte is not; (**c**) both reagent and analyte are reduced.

sulfate or oxalate ions may be cited as an example. The potential, which is applied, is sufficient to give a diffusion current for lead; a linear decrease in current is observed as lead ions are removed from the solution by precipitation. The curvature near the equivalence point reflects the incompleteness of the precipitation reaction in this region. The end point is obtained by extrapolation of the linear portions, as shown.

Figure 22–25**b** is typical of a titration in which the reagent reacts at the microelectrode and the analyte does not; an example is the titration of magnesium with 8-hydroxyquinoline. A diffusion current for the latter is obtained at -1.6 V, whereas magnesium ion is inert at this potential.

Figure 22–25**c** corresponds to the titration of lead ion with a chromate solution at an applied potential greater than -1.0 V. Both lead and chromate ions give diffusion currents, and a minimum in the curve signals the end point. This system would yield a curve resembling that in Figure 22–25**b** at zero applied potential, since only chromate ions are reduced under these conditions.

22F-2 APPARATUS AND TECHNIQUES

Two types of amperometric titration apparatus are encountered. One employs a single polarizable microelectrode coupled to a large nonpolarizable one; the other uses a pair of identical microelectrodes. Methods based on the latter are called biamperometric titrations.

The microelectrode for an amperometric titration may be a dropping mercury electrode, a solid-state electrode of platinum or other inert material, or a rotating platinum electrode.

The Rotating Platinum Electrode. This electrode consists of a 2- to 3-mm length of platinum wire sealed at right angles into a glass tubing. The tube is rotated by a motor at about 600 rpm. It is vital that the speed of the motor be constant in order to obtain reproducible currents.

Voltammetric waves, which are similar to those observed with the dropping electrode, are obtained with the rotating platinum electrode. Here, however, the reactive species is brought to the electrode surface not only by diffusion but also by mechanical mixing. As a consequence, the limiting currents are as much as 20 times larger than those obtained with a microelectrode that is supplied by diffusion only; thus, the rotating electrode has the potential for enhancing sensitivity. Rotating electrodes, like their dropping electrode counterparts, provide steady currents instantaneously. This behavior is in distinct contrast to that of a solid microelectrode in the absence of stirring (see Figure 22–4).

Several factors restrict the widespread application of the rotating platinum electrode for voltammetry. The low hydrogen overvoltage prevents its use as a cathode in acidic solutions. In addition, the high currents obtained cause the electrode to be particularly sensitive to traces of oxygen in the solution. These two factors have largely confined its application to anodic reactions. Limiting currents are often influenced by the previous history of the rotating platinum electrode and are seldom as reproducible as the diffusion currents obtained with a dropping electrode.

Cells. For titration with a single polarized electrode, the cell often consists of a 50- to 100-mL beaker, the analyte solution, the micro indicator electrode, and a large nonpolarizable electrode. The latter may be a large reference electrode, a pool of mercury, or a platinum wire or gauze having a relatively large surface area. Unless a rotating electrode is being used, provision for stirring must also be provided.

For biamperometric titrations, twin platinum or silver electrodes are generally used. These consist of short lengths of wire sealed into glass tubes. It is not necessary that the two electrodes be identical in area.

Electrical Equipment. The electrical equipment for an amperometric titration can be quite simple. It often consists of a voltage divider that produces a dc potential in the 0 to ± 3 V range; the magnitude of this voltage must be known to about 0.1 V. For some titrations, the reference electrode suffices to provide a large enough potential to cause reaction of the analyte at the micro indicator electrode. Under these circumstances, no external power source is needed. Also required is a current measuring device, which operates in the microampere range with a precision of about 1% relative.

Reagent Addition. The standard solution is usually introduced with a micro buret so that the volume of the test solution is not altered appreciably during the titration. Alternatively, the measured currents can be corrected for volume changes. Coulometric generation of reagents offers an alternative to the use of standard solutions. Thus, amperometric end points are frequently encountered in coulometric methods.

22F–3 APPLICATION OF AMPEROMETRIC TITRATIONS WITH ONE POLARIZED ELECTRODE

As shown in Table 22–2, the amperometric end point with one indicator electrode has been largely confined to titrations in which a precipitate or a stable complex is the product. A notable exception is the application of the rotating platinum electrode to titrations with bromate ion in the presence of bromide and hydrogen ions. For example, styrene in an acidic 75% methanol solution can be titrated with a neutral standard solution of KBrO$_3$ that contains an excess of potassium bromide. The reactions are

$$BrO_3^- + 5Br^- + 6H^+ \rightarrow 3Br_2 + 3H_2O$$

$$C_6H_5CH{=}CH_2 + Br_2 \rightarrow C_6H_5CHBrCH_2Br$$

Formation of bromine does not occur in the neutral bromate/bromide standard solution. In the acidic analyte medium, however, quantitative conversion of the bromate to bromine takes place. A titration curve similar to that in Figure 22–25**b** is observed.

22F-4 AMPEROMETRIC TITRATIONS WITH TWO POLARIZED MICROELECTRODES

Although the use of two polarized electrodes for end point detection was first proposed before 1900, almost 30 years passed

before chemists came to appreciate the potentialities of the method.[16] The name *dead-stop end point* was used to describe the technique, and this term is still occasionally encountered. It was not until about 1950 that a clear interpretation of dead-stop titration curves was made.[17]

Titration with Silver Ion. Twin silver microelectrodes permit observation of the end point for various titrations employing silver nitrate as the standard reagent. Consider what happens, for example, when a small potential ($\sim$0.1 V) is applied between two such electrodes during the titration of bromide with silver ions. Short of the equivalence point essentially no current can exist because no easily reduced species is present in the solution. Consequently, electron transfer at the cathode is precluded and that elec-

---

[16] C. W. Foulk and A. T. Bawden, *J. Amer. Chem. Soc.,* **1926,** *48,* 2045.

[17] For an excellent analysis of this type of end point, see: J. J. Lingane, *Electroanalytical Chemistry,* 2d ed., pp. 280–294, Interscience: New York, 1958.

Table 22–2
Applications of Amperometric Titrations with One Polarized Electrode

Reagent	Reaction Product	Type Electrode[a]	Substance Determined
K$_2$CrO$_4$	Precipitate	DME	Pb^{2+}, Ba^{2+}
Pb(NO$_3$)$_2$	Precipitate	DME	SO$_4^{2-}$, MoO$_4^{2-}$, F$^-$, Cl$^-$
8-Hydroxyquinoline	Precipitate	DME	Mg^{2+}, Zn^{2+}, Cu^{2+}, Cd^{2+}, Al^{3+}, Bi^{3+}, Fe^{3+}
Cupferron	Precipitate	DME	Cu^{2+}, Fe^{3+}
Dimethylglyoxime	Precipitate	DME	Ni^{2+}
α-Nitroso-β-naphthol	Precipitate	DME	Co^{2+}, Cu^{2+}, Pd^{2+}
K$_4$Fe(CN)$_6$	Precipitate	DME	Zn^{2+}
AgNO$_3$	Precipitate	RP	Cl$^-$, Br$^-$, I$^-$, CN$^-$, RSH
EDTA	Complex	DME	Bi^{3+}, Cd^{2+}, Cu^{2+}, Ca^{2+}, etc.
KBrO$_3$, KBr	Substitution, addition, or oxidation	RP	Certain phenols, aromatic amines, olefins; N$_2$H$_4$, As(III), Sb(III)

[a] DME: dropping mercury electrode, RP: rotating platinum electrode.

trode is completely polarized. Note that the anode is not polarized because the reaction

$$Ag \rightleftharpoons Ag^+ + e$$

could occur in the presence of a suitable cathodic reactant or depolarizer.

After equivalence has been passed, the cathode becomes depolarized owing to the presence of a significant amount of silver ions, which can react to give silver. That is,

$$Ag^+ + e \rightleftharpoons Ag$$

Current is permitted as a result of this half-reaction and the corresponding oxidation of silver at the anode. The magnitude of the current is, as in other amperometric methods, directly proportional to the concentration of the excess reagent. Thus, the titration curve is similar to that shown in Figure 22–25**b**.

Titration with Bromine and Iodine. An amperometric titration with twin platinum microelectrodes has also been applied to titrations in which iodine or bromine is the titrant. When this technique is employed for the titration of arsenious acid, for example, a curve similar to Figure 22–25**b** is again obtained. No cur-

rent is observed in the early stages of the titration because of cathodic polarization. In contrast to the previous example, however, polarization here is of the kinetic type. That is, even though a cathodic half-reaction involving H_3AsO_4, the titration product, can be written

$$H_3AsO_4 + 2e + 2H^+ \rightleftharpoons H_3AsO_3 + H_2O$$

this process occurs so slowly at the electrode surface that no current can be detected. After equivalence, the cathode becomes depolarized as a consequence of the excess reagent. Thus if iodine is the reagent, the reactions become

cathode: $I_2 + 2e \rightleftharpoons 2I^-$

anode: $2I^- \rightleftharpoons I_2 + 2e$

The current here is proportional to the volume of excess reagent.

The principal advantage of the twin microelectrode procedure is its simplicity. One can dispense with a reference electrode; the only instrumentation needed is a simple voltage divider, powered by a dry cell or a simple power supply, and a galvanometer or microammeter for current detection.

PROBLEMS

22–1. As the terms pertain to a polarographic wave, make a distinction between
    (a) decomposition potential and half-wave potential.
    (b) limiting current and diffusion current.

22–2. How are limiting currents affected in a series of solutions in which
    (a) the concentration of the analyte is increased while the concentration of the supporting electrolyte is held constant?
    (b) the concentration of the supporting electrolyte is decreased while the concentration of the analyte is held constant?

22–3. The following data were obtained from polarograms for a series of Pb^{2+} solutions:

Concn Pb²⁺, mmol/L	Limiting Current, μA	Concn Pb²⁺, mmol/L	Limiting Current, μA
0.00	1.32	3.06	27.91
0.510	5.65	4.08	36.08
1.02	10.70	5.10	45.82
2.04	19.08		

(a) Plot diffusion current versus concentration.
(b) By least squares, derive an equation for the best straight line for the plot in part (a).
(c) Calculate the standard deviations for the slope and about regression.
(d) Calculate the concentration of lead and the absolute and relative standard deviations for each of the following analyses:

	i_d	No. Replicate Measurements
(1)	2.76	2
(2)	7.75	1
(3)	7.75	3
(4)	26.32	3
(5)	40.01	1

22–4. For the data in Problem 22–3, calculate
(a) a mean value for i_d/C.
(b) the absolute and relative standard deviation for the mean.
(c) a theoretical value for i_d/C assuming the data are for average currents, the diffusion coefficient for $Pb^{2+} = 9.8 \times 10^{-6}$ cm²/sec, the reaction product is Pb(Hg), the drop time is 2.88 sec, and the Hg flow rate is 2.63 mg/sec.
(d) the percent error in the theoretical value for i_d/C.

22–5. The following data were obtained for the reduction of Cu(II) from various media:

Supporting Electrolyte	$i_d/(C \, m^{2/3} t^{1/6})$
0.1 M HNO₃	3.24
1.0 M HCl	3.39
0.5 M sodium tartrate	2.24
1 M NH₄Cl/1 M NH₃	3.75

Suggest an explanation for the differences.

22–6. Assume that the data in Problem 22–3 were obtained with an electrode having a drop time of 2.88 sec and a Hg flow of 2.63 mg/sec.

(a) Calculate the diffusion current that would be expected for a solution that was 3.75 mM in Pb^{2+}.

(b) What would be the diffusion current for the same solution as in (a) with a different electrode having a drop time of 6.52 sec and a flow rate of 0.959 mg/sec.

(c) Calculate the lead concentration of a solution giving a diffusion current of 22.12 μA with the electrode described in (b).

(d) Calculate the percent error in the lead determination in (c) if a correction for the change in capillaries had not been made.

22–7. The polarogram for 20.0 mL of solution that was $3.65 \times 10^{-3} M$ in Cd^{2+} gave a wave for that ion with a diffusion current of 31.1 μA. Calculate the percentage change in concentration of the solution if the current in the limiting current region were allowed to continue for (a) 5 min; (b) 10 min; (c) 30 min.

22–8. Calculate the milligrams of cadmium in each milliliter of sample, based upon the following data (corrected for residual current):

		Volumes Used, mL			
Solution	Sample	0.400 M KCl	2.00 $\times$ 10^{-3} M Cd^{2+}	H$_2$O	Current, μA
(a)	15.0	20.0	0.00	15.0	79.7
	15.0	20.0	5.00	10.0	95.9
(b)	10.0	20.0	0.00	20.0	49.9
	10.0	20.0	10.0	10.0	82.3
(c)	20.0	20.0	0.00	10.0	41.4
	20.0	20.0	5.00	5.00	57.6
(d)	15.0	20.0	0.00	15.0	67.9
	15.0	20.0	10.0	5.00	100.3

22–9. The following data were collected for three dropping electrodes. Complete the data for electrodes A and C.

	A	B	C
flow rate, mg/sec	0.982	3.92	6.96
drop time, sec	6.53	2.36	1.37
i_d/C, μA L/mmol^{-1}		4.86	

22–10. Electrode C in Problem 22–9 was employed to study the reduction of an organic compound known to have a diffusion coefficient of 6.2×10^{-6} cm^2/sec. A $4.00 \times 10^{-4} M$ solution of the compound yielded a diffusion current of 9.10 μA. Calculate n for the reaction.

22–11. An organic compound underwent a two-electron reduction at dropping electrode A in Problem 22–9. A diffusion current of 10.1 μA was produced by a 9.6×10^{-4} M solution of the compound. Calculate the diffusion coefficient for the compound.

22–12. The following polarographic data were obtained for the reduction of Pb^{2+} to its amalgam from solutions that were 2.00×10^{-3} M in Pb^{2+}, 0.100 M in KNO_3, and that also had the following concentrations of the anion A^-. From the half-wave potentials, derive the formula of the complex as well as its formation constant.

Concn A^-, M	$E_{1/2}$ vs. SCE, V
0.0000	−0.405
0.0200	−0.473
0.0600	−0.507
0.1007	−0.516
0.300	−0.547
0.500	−0.558

22–13. Shown below is the polarogram for a solution that was 1.0×10^{-4} M in KBr and 0.1 M in KNO_3. Offer an explanation of the wave that occurs at +0.12 V and the rapid change in current that starts at about +0.48 V. Would the wave at 0.12 V have any analytical applications? Explain.

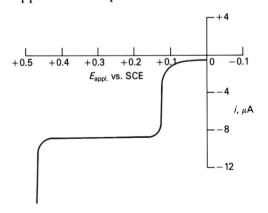

22–14. The following reaction is reversible and has a half-wave potential of −0.349 V when carried out at a dropping mercury electrode from a solution buffered to pH 2.5.

$$Ox + 4H^+ + 4e \rightleftharpoons R$$

Predict the half-wave potential at pH: (a) 1.0; (b) 3.5; (c) 7.0.

Conductometric Methods

Two types of methods are based upon the measurement of the electrical conductance of solutions, namely direct conductance methods and conductometric titrations. The principal advantage of these methods is their simplicity and relatively good sensitivity.[1]

23A Electrical Conductance in Solutions of Electrolytes

23A-1 SOME IMPORTANT RELATIONSHIPS

Conductance. The *conductance G* of a medium is the reciprocal of its electrical resistance R in ohms. Conductance has the units of ohms^{-1} or *siemens*.

Specific Conductance. Conductance is directly proportional to the cross-section area A and inversely proportional to the length l of a uniform conductor. That is,

$$G = \frac{1}{R} = k \frac{A}{l} \tag{23-1}$$

Here, the proportionality constant k is the *specific conductance*. In metric units, k is the conductance of a cube, which is one centimeter on a side. The dimensions of k are ohm^{-1} cm^{-1}.

Equivalent Conductance. The conductance of an electrolyte solution depends upon the number of ions in a solution, their charge, and the rate of their movement under the influence of an electromotive force. The ability of a solute to conduct electricity is expressed in terms of its *equivalent conductance*, Λ, which is defined as the conductance of one gram equivalent of the solute when contained between two electrodes spaced one centimeter apart.[2]

---

[1] For further discussion of conductometric methods, see J. W. Loveland, in *Treatise on Analytical Chemistry*, I. M. Kolthoff and P. J. Elving, Eds., Part I, vol. 4, Chapter 51, Interscience: New York, 1963; and T. Shedlovsky and L. Shedlovsky, in *Physical Methods of Chemistry*, A. Weissberger and B. W. Rossitor, Eds., vol. 1, Part 11A, Chapter 3, Wiley-Interscience: New York, 1971. For an interesting review of the history of conductivity measurements, see: J. T. Stock, *Anal. Chem.*, **1984,** *56,* 561A.

[2] In the context of electrolytic conductance, the equivalent is defined as the gram formula weight of an ionic solute divided by the number of positive or negative charges associated with it. Thus, the equivalent weight of $BaCl_2$ is its formula weight divided by two; for Na_3PO_4, it is the formula weight divided by three.

The equivalent conductance of a solution can be obtained from specific conductance measurements of solutions of known concentrations. Thus, by definition, Λ will be equal to G when one gram equivalent of solute is contained between electrodes that are exactly one centimeter apart. The volume V of the solution (cm^3) that will contain one gram equivalent of solute is given by

$$V = \frac{1000}{C}$$

where C is the concentration in equivalents per cm^3. This volume can also be expressed in terms of the dimensions of the cell

$$V = lA$$

With l fixed by definition at one centimeter,

$$V = A = \frac{1000}{C}$$

Substitution into Equation 23–1 thus gives

$$\Lambda = \frac{1000k}{C} \qquad (23\text{–}2)$$

Equivalent Conductance at Infinite Dilution. The mobility of an ion in solution is governed by four forces. An *electrical force,* equal to the product of the potential of the electrode and the charge of the ion, tends to move the particle toward one of the electrodes. This effect is partially balanced by a *frictional force* that is a characteristic property for each ion. These are the only two effects that play a significant role in determining the conductivity of a solution as it approaches infinite dilution; under these circumstances, the equivalent conductance of a salt is independent of its concentration.

At finite concentrations, two other factors, the *electrophoretic* effect and the *relaxation* effect, become important and cause the equivalent conductance of a substance to decrease as its concentration increases. The behavior of a sodium chloride solution is typical and is shown in Table 23–1.

The electrophoretic effect stems from the motion of the oppositely charged ions surrounding the ion of interest. These ions carry with them molecules of solvent; the motion of the primary particle is thus retarded by the flow of solvent in the opposite direction. The relaxation effect also owes its genesis to movement of the ionic atmosphere surrounding a given particle. Here, however, the ion is slowed by the charge of opposite sign that builds up behind the moving particle.

For strong electrolytes, a linear relationship exists between equivalent conductance and square root of concentration. Extrapolation of this straight-line relationship to zero concentration yields a value for the equivalent conductance at infinite dilution Λ_0. A similar plot for a weak electrolyte is nonlinear, and direct evaluation of Λ_0 is difficult.

At infinite dilution, relaxation and electrophoretic effects become nil; the overall conductance of the solution then consists of the sum of the individual equivalent ionic conductances

$$\Lambda_0 = \lambda_+^0 + \lambda_-^0$$

where λ_+^0 and λ_-^0 are the equivalent ionic conductances of the cation and the anion of the salt at infinite dilution. Individual ionic conductances can be determined from other electrolytic measurements; values for a

Table 23–1
Effect of Concentration on Equivalent Conductance

Concentration of NaCl, eq/L	Λ
0.1	106.7
0.01	118.5
0.001	123.7
infinite dilution	126.4 (Λ_0)

number of common ions are given in Table 23–2. Note that symbols such as $\frac{1}{2}$ Mg^{2+}, $\frac{1}{3}$ Fe^{3+}, and $\frac{1}{2}$ SO_4^{2-} are used to emphasize that the concentration units are *equivalents per liter*.

The differences that exist among the equivalent ionic conductances of various species (Table 23–2) arise primarily from differences in their size and the degree of their hydration.

The equivalent ionic conductance is a measure of the mobility of an ion under the influence of an electric force field and is thus a gauge of its capacity for transporting electricity. For example, the ionic conductance of a potassium ion is nearly the same as that of a chloride ion; therefore electricity passing through a potassium chloride solution is carried nearly equally by the two species. The situation is quite different with hydrochloric acid; because of the greater mobility of the hydrogen ion, a larger fraction of the electricity [350/(350 + 76) = 0.82] is carried by that species in an electrolysis (see Section 19A–2).

Ionic conductance data permit comparison of the relative conductivity of various solutes. Thus, we are justified in saying that 0.01 *M* hydrochloric acid will have a greater conductivity than 0.01 *M* sodium chloride because of the very large ionic conductance of the hydrogen ion. Such conclusions are important in predicting the course of a conductometric titration.

23A–2 ALTERNATING CURRENTS IN A CELL

As was pointed out earlier (Section 19A–3), conduction of dc electricity across a solution/electrode interface is a faradaic process in which oxidation and reduction must occur at the two electrodes. An alternating current, on the other hand, requires no electrochemical reaction at the electrodes; here, electricity flows as a consequence of nonfaradaic processes. Because the changes associated with faradaic conduction can materially alter the electrical characteristics of a cell, conductometric measurements are advantageously based upon nonfaradaic currents.

23B The Measurement of Conductance

A conductance measurement requires a source of electrical power, a cell to contain the solution, and a suitable bridge to measure the resistance of the solution.

23B–1 POWER SOURCES

Use of an alternating current source eliminates the effect of faradaic currents. There are, however, both upper and lower limits to the frequencies that can be employed; audio oscillators that produce signals of about 1000 Hz are the most satisfactory.

For less refined work, an ordinary 60-cycle current, stepped down from 110 to perhaps 10 V, is also used. With such a source, however, faradaic processes often limit the accuracy of the conductance measurements. On the other hand, the convenience and ready availability of 60-cycle power justify its use for many purposes.

Table 23–2
Equivalent Ionic Conductances at 25°C

Cation	λ_+^0	Anion	λ_-^0
H_3O^+	349.8	OH^-	199.0
Li^+	38.7	Cl^-	76.3
Na^+	50.1	Br^-	78.1
K^+	73.5	I^-	76.8
NH_4^+	73.4	NO_3^-	71.4
Ag^+	61.9	ClO_4^-	67.3
$\frac{1}{2}Mg^{2+}$	53.1	$C_2H_3O_2^-$	40.9
$\frac{1}{2}Ca^{2+}$	59.5	$\frac{1}{2}SO_4^{2-}$	80.0
$\frac{1}{2}Ba^{2+}$	63.6	$\frac{1}{2}CO_3^{2-}$	69.3
$\frac{1}{2}Pb^{2+}$	69.5	$\frac{1}{2}C_2O_4^{2-}$	74.2
$\frac{1}{3}Fe^{3+}$	68.0	$\frac{1}{4}Fe(CN)_6^{4-}$	110.5
$\frac{1}{3}La^{3+}$	69.6	—	—

Power sources with frequencies much greater than 1000 Hz create problems in conductance measurements with a bridge. Here, the cell capacitance and stray capacitances in other parts of the circuit cause phase changes in the current for which compensation is inconvenient.

23B-2 RESISTANCE BRIDGES

The Wheatstone bridge arrangement, shown in Figure A1–6 (Appendix 1) is typical of the apparatus used for conductance measurements. The power source S provides an alternating current in the frequency range of 60 to 1000 Hz at a potential of 6 to 10 V. The resistances R_{AC} and R_{BC} can be calculated from the position of contact C. The cell, of unknown resistance R_x, is placed in the upper left arm of the bridge and a precision variable resistance R_s is placed in the right-hand side. A null detector ND is used to indicate the absence of current between D and C. The detector may consist of a pair of ordinary headphones, since the ear is responsive to frequencies in the 1000 Hz range; alternatively, it may be a "magic eye" tube, a cathode-ray tube, or a microammeter.

Capacitance effects within R_x cause an al-ternating-current device such as this to suffer a loss in sensitivity when high resistances are measured; in practice, a variable capacitor across R_s compensates for this effect.[3]

Conductance can also be determined by means of a simple electronic circuit such as that shown in Figure 2–18**a** (p. 47). Here, the conductance is read directly from a dc meter or a recorder.

23B-3 CELLS

Figure 23–1 depicts three common types of cells for the measurement of conductivity. Each contains a pair of electrodes firmly fixed in a constant geometry with respect to one another. The electrodes are ordinarily platinized to increase their effective surface and thus their capacitances; faradaic currents are minimized as a result.

---

[3] The Wheatstone bridge is not very satisfactory for conductance measurements at very high resistances, such as those encountered in nonaqueous solutions and at very low resistances, such as those found in molten salt mixtures. For an analysis of these limitations and a description of a new bipolar pulse technique, which permits accurate measurements over a broad spectrum of conditions, see: D. E. Johnson and C. G. Enke, *Anal. Chem.,* **1970,** *42,* 329.

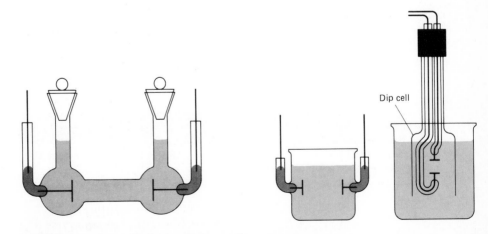

FIGURE 23-1 Typical cells for conductometric measurements.

Determination of the Cell Constant. According to Equation 23–1, the specific conductance k is related to the measured conductance G by the ratio of the distance separating the electrodes to their surface area. This ratio has a fixed and constant value in any given cell and is known as the *cell constant*. Its value is seldom determined directly. Instead, the conductance of a solution whose specific conductance is reliably known is measured; the cell constant can then be calculated. Solutions of potassium chloride are commonly chosen for cell calibration.[4] Typical data are shown in Table 23–3.

Once the value of the cell constant has been determined, conductivity data can be easily converted to terms of specific conductance with the aid of Equation 23–1.

Temperature Control. The temperature coefficient for conductance measurements is about 2%/°C; as a consequence, some temperature control is ordinarily required during a conductometric measurement. Although a constant temperature is necessary, control at some specific level is not required for a successful conductometric titration. For many purposes, it is sufficient to immerse the cell in a reasonably large bath of water or oil, which is at room temperature.

---

[4] G. Jones and B. C. Bradshaw, *J. Amer. Chem. Soc.*, **1933**, *55*, 1780.

23C   Conductometric Titrations

Conductometric measurements provide a convenient means of locating end points in titrations. Sufficient measurements (three to four before and after the equivalence point) are needed to define the titration curve. After being corrected for volume change, the conductance data are plotted as a function of titrant volume. The two linear portions are then extrapolated, the point of intersection being taken as the equivalence point.

The conductometric end point is completely nonspecific. Although it is potentially adaptable to all types of volumetric reactions, the number of its useful applications to oxidation-reduction systems is limited; the substantial excess of hydrogen ions typically needed for such reactions tends to mask conductivity changes associated with the volumetric reaction.

23C-1   ACID-BASE TITRATIONS

Neutralization titrations are particularly well adapted to the conductometric end point because of the large equivalent ionic conductances of hydrogen and hydroxide ions compared with the conductances of the species that replace them.

Titration of Strong Acids or Bases. The solid line in Figure 23–2 represents a curve (corrected for volume change) obtained when hydrochloric acid is titrated with sodium hydroxide. Also plotted are the calculated con-

Table 23–3 Conductance of Solutions for Cell Calibration	
Grams KCl per 1000 g of Solution in Vacuum	**Specific Conductance at 25°C, $ohm^{-1}\ cm^{-1}$**
71.1352	0.111342
7.41913	0.0128560
0.745263	0.00140877

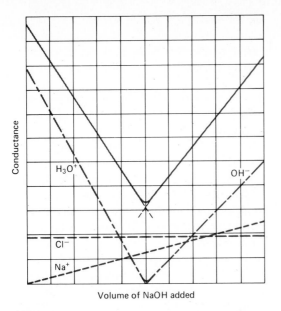

FIGURE 23-2 Conductometric titration of a strong acid with a strong base. The solid line represents the titration curve, corrected for volume change. Broken lines indicate the contribution of the individual species, also corrected for volume change, to the conductance of the solution.

tributions of the individual ions to the conductance of the solution. During neutralization, hydrogen ions are replaced by an equivalent number of less mobile sodium ions; the conductance changes to lower values as a result of this substitution. At the equivalence point, the concentrations of hydrogen and hydroxide ions are at a minimum and the solution exhibits its lowest conductance due largely to sodium and chloride ions. A reversal of slope occurs past the end point as the sodium ion and hydroxide ion concentrations increase. With the exception of the immediate equivalence-point region, an excellent linearity exists between conductance and the volume of base added; as a result, only three or four observations on each side of the equivalence point are needed for an analysis.

The percentage change in conductivity

during the course of the titration of a strong acid or base is the same regardless of the concentration of the solution. Thus, very dilute solutions can be analyzed with an accuracy comparable to more concentrated ones.

Titration of Weak Acids or Bases. Figure 23–3a illustrates application of the conductometric end point to the titration of boric acid ($K_a = 6 \times 10^{-10}$) with strong base. This reaction is so incomplete that a potentiometric or visual indicator end point is unsatisfactory. In the early stages of the titration, a buffer is rap-

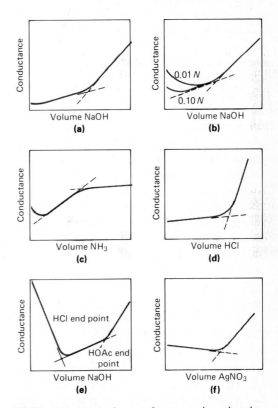

FIGURE 23-3 Typical conductometric titration curves. Titration of: (**a**) a very weak acid ($K_a \approx 10^{-10}$) with sodium hydroxide; (**b**) a weak acid ($K_a \approx 10^{-5}$) with sodium hydroxide (note that for 0.01 N solutions, conductance times 10 is plotted); (**c**) a weak acid ($K_a \approx 10^{-5}$) with aqueous ammonia; (**d**) the salt of a weak acid; (**e**) a mixture of hydrochloric and acetic acids with sodium hydroxide; and (**f**) chloride ion with silver nitrate.

idly established that imparts to the solution a relatively small and nearly constant hydrogen ion concentration. The added hydroxide ions are consumed by this buffer and thus do not directly contribute to the conductivity. A gradual increase in conductance does result, however, owing to the increase in concentration of sodium and of borate ions. With attainment of the equivalence point, no further borate is produced; additions of base then cause a more rapid increase in conductance due to the added, highly mobile hydroxide ions.

Figure 23–3**b** illustrates the titration of a moderately weak acid, such as acetic acid ($K_a \cong 10^{-5}$), with sodium hydroxide. Nonlinearity in the early portions of the titration curve creates problems in establishing the end point; with concentrated solutions, however, the titration is feasible. As before, we can interpret this curve in light of the changes in composition that occur. Here, the solution initially has a moderate concentration of hydrogen ions ($\sim 10^{-3}\ M$). Addition of base results in the establishment of a buffer system and a consequent diminution in the hydrogen ion concentration. Concurrent with this decrease is an increase in the concentration of sodium ions as well as the conjugate base of the acid. These two factors act in opposition to one another. At first, the decrease in hydrogen ion concentration predominates and a decrease in conductance is observed. As the titration progresses, however, the pH becomes stabilized (in the buffer region); the increase in salt content then becomes the more important factor, and a linear increase in conductance results. Beyond the equivalence point, the curve steepens because of the greater ionic conductance of hydroxide ion.

In principle, all titration curves for weak acids or bases contain the general features of Figure 23–3**b**. The ionization of very weak species is so slight, however, that little or no curvature occurs with the establishment of the buffer region (see Figure 23–

3**a**, for example). As the strength of the acid (or base) becomes greater, so also does the extent of the curvature in the early portions of the titration curve. For weak acids or bases with dissociation constants greater than about 10^{-5}, the curvature becomes so pronounced that an end point cannot be distinguished.

Figure 23–3**c** illustrates the titration of the same weak acid as in Figure 23–3**b**, but with aqueous ammonia instead of sodium hydroxide. Here, because the titrant is a weak electrolyte, the curve is essentially horizontal past the equivalence point. Use of ammonia as titrant actually provides a curve that can be extrapolated with less uncertainty than the corresponding curve based upon titration with sodium hydroxide.

Figure 23–3**d** represents the titration curve for a weak base, such as acetate ion, with a standard solution of hydrochloric acid. The addition of strong acid results in formation of sodium chloride and undissociated acetic acid. The net effect is a slight rise in conductance due to the greater mobility of the chloride ion over that of the acetate ion it replaces. After the end point has been passed, a sharp rise in conductance attends the addition of excess hydrogen ions. The conductometric method is convenient for the titration of salts whose acidic or basic character is too weak to give satisfactory end points with indicators.

Figure 23–3**e** is typical of the titration of a mixture of two acids that differ in degree of dissociation. The conductometric titration of such mixtures frequently leads to more accurate results than those obtained with a potentiometric method.

23C–2 PRECIPITATION AND COMPLEX-FORMATION TITRATIONS

Figure 23–3**f** illustrates the conductance changes that occur during the titration of sodium chloride with silver nitrate. The initial additions of reagent in effect cause a substitution of chloride ions by the somewhat less mobile nitrate ions of the reagent; a slight decrease in conductance results. After the reaction is complete a rapid increase in conductance occurs, owing to the addition of excess silver nitrate.

Conductometric methods based upon precipitation or complex-formation reactions are not as useful as those involving neutralization processes. Conductance changes during these titrations are seldom as large as those observed with acid-base reactions because no other ion approaches the conductance of either the hydrogen or the hydroxide ion. Such factors as slowness of reaction and coprecipitation represent further sources of difficulty with precipitation titrations.

23D Applications of Direct Conductance Measurements

Direct conductometric measurements suffer from a lack of selectivity, since any charged species contributes to the total conductance of a solution. On the other hand, the high sensitivity of the procedure makes it an important analytical tool for certain applications. Perhaps the most common application of direct conductometry has been for estimating the purity of distilled or deionized water. The specific conductance of pure water is only about 5×10^{-8} ohm^{-1}cm^{-1}; traces of an ionic impurity will increase the conductance by an order of magnitude or more. Conductance measurements are also widely used to measure the salinity of sea water in oceanographic work.

A more recent and highly important application of direct conductance measurements has been for detection of ions after separation by ion chromatography. This technique is described in Section 27G–3.

Conductance measurements are also employed for determining the concentration of solutions containing a single strong electrolyte, such as solutions of the common alkalis or acids. A nearly linear increase in conductance with concentration is observed for solutions containing as much as 20% by weight of solute. Analyses are based upon calibration curves.

Finally, conductance measurements yield valuable information about association and dissociation equilibria in aqueous solutions—provided, of course, that one or more of the reacting species is ionic.

PROBLEMS   For each of the following problems, derive an approximate titration curve of specific conductance versus volume of reagent, employing the assumption that

(a) 100.0 mL of $1.00 \times 10^{-3} N$ solution is being titrated with $2.00 \times 10^{-2} N$ reagent;*

(b) the volume change occurring as the titration proceeds can be neglected;

(c) the equivalent ionic conductances of the various ions are not significantly different from their equivalent ionic conductances at infinite dilution (Table 23–2).

---

* A 1 N solution contains one equivalent of reactant per liter. Thus, 1 N Ba(OH)$_2$ contains $\frac{1}{2}$ mole of Ba(OH)$_2$ per liter.

Calculate theoretical specific conductances of the mixtures after the following additions of reagent: 0.00, 1.00, 2.00, 3.00, 4.00, 5.00, 6.00, 7.00, and 8.00 mL. Plot the titration curve.

23–1. $Ba(OH)_2$ with HCl

23–2. Phenol with KOH (use $K_a = 1.0 \times 10^{-10}$ and $\lambda^0_- = 30$ for $C_6H_5O^-$). For 0.00 mL reagent, assume that the dissociation of phenol and water are so small that the specific conductance approaches 0.0.

23–3. Sodium acetate with $HClO_4$ (assume K_a for acetic acid $= 1.8 \times 10^{-5}$)

23–4. Propionic acid with NaOH ($\lambda^0_- = 35.8$ for $CH_3CH_2COO^-$ and $K_a = 1.3 \times 10^{-5}$)

23–5. Propionic acid with NH_3 (for NH_3, $K_b = 1.76 \times 10^{-5}$)

23–6. $AgNO_3$ with NaCl

23–7. $AgNO_3$ with LiCl

Thermal Methods

Thermal methods are based upon the measurement of the dynamic relationship between temperature and some property of a system such as mass, heat of reaction, or volume. Wendlandt lists twelve thermal methods; four of the most important of these will be considered in this chapter,[1] namely, *thermogravimetry* (TG), *differential thermal analysis* (DTA), *differential scanning calorimetry* (DSC), and *enthalpimetric methods.*

24A Thermogravimetric Methods

In a thermogravimetric analysis, the mass of sample is recorded continuously as its temperature is increased linearly from ambient to as high as 1200°C. A plot of mass as a function of temperature (a *thermogram*) provides both qualitative and quantitative information.

24A-1 APPARATUS

The apparatus required for a thermogravimetric analysis includes: (1) a sensitive recording analytical balance; (2) a furnace; (3) a furnace temperature controller and programmer; and (4) a recorder that provides a plot of sample mass as a function of temperature. Often, auxiliary equipment is needed to provide an inert atmosphere for the sample.

The Balance. Currently, several thermobalances are available from instrument manufacturers. A schematic diagram of one of these is shown in Figure 24–1. The sample holder in this instrument is housed in a furnace that is thermally isolated from the remainder of the balance. A change in the sample mass causes a deflection of the beam, which interposes a light shutter between a lamp and one of two photodiodes. The re-

---

[1] For a review of thermal methods, see W. W. Wendlandt, *Thermal Methods of Analysis,* 2d ed., Wiley: New York, 1974; T. Meisel and K. Seybold, *Crit. Rev. Anal. Chem.,* **1981,** *12,* 267.

713

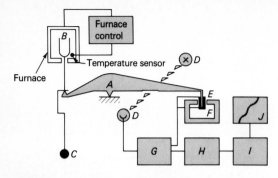

FIGURE 24-1 Components of a thermal balance: *A,* beam; *B,* sample cup and holder; *C,* counterweight; *D,* lamp and photodiodes; *E,* coil; *F,* magnet; *G,* control amplifier; *H,* tare calculator; *I,* amplifier; and *J,* recorder. (Courtesy of Mettler Instrument Corp., Hightstown, NJ.)

sulting imbalance in the photodiode current is amplified and fed into coil *E,* which is situated between the poles of the permanent magnet *F.* The magnetic field generated by the current in the coil restores the beam to its original position. The amplified photodiode current also determines the position of the pen of a recorder. The instrument shown has several weight ranges (1, 10, 100, and 1000 mg) and has a reproducibility of ±10 μg.

Furnaces. The furnace of a thermogravi-

metric apparatus is generally programmed to increase the temperature linearly at predetermined rates (typically, from 0.5 to 25°C/min). The temperature range for most instruments is from ambient to 1200°C. Temperatures are determined by a thermocouple located as close as possible to the sample. Insulation and cooling of the exterior of the furnace is required to avoid heat transfer to the balance.

24A-2 APPLICATIONS

Figure 24-2 is a recorded thermogram obtained by increasing the temperature of pure $CaC_2O_4 \cdot H_2O$ at a rate of 5°C/min. The clearly defined horizontal regions correspond to temperature ranges in which the indicated calcium compounds are stable. This figure illustrates one of the important applications of thermogravimetry, namely, that of defining thermal conditions necessary to produce a pure weighing form for the gravimetric determination of a species.

Figure 24-3**b** illustrates an application of thermogravimetry to the quantitative analysis of a mixture of calcium, strontium, and barium. The three ions are first precipitated as the monohydrated oxalates. The mass in the temperature range between 320 and 400°C is that of the three anhydrous com-

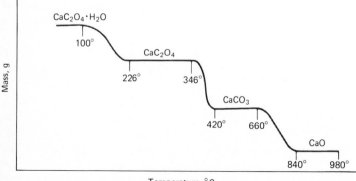

FIGURE 24-2 A thermogram for decomposition of $CaC_2O_4 \cdot H_2O$. (From: S. Peltier and C. Duval, *Anal. Chim. Acta,* **1947,** *1,* 345. With permission.)

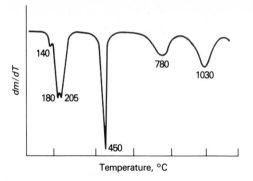

(a) Differential thermogram

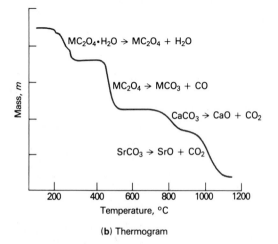

(b) Thermogram

FIGURE 24-3 Decomposition of $CaC_4O_4 \cdot H_2O$, $SrC_2O_4 \cdot H_2O$, and $BaC_2O_4 \cdot H_2O$. From: L. Erdey, G. Liptay, G. Svehla, and F. Paulik, *Talanta*, **1962**, *9*, 490. With permission.)

pounds, CaC_2O_4, SrC_2O_4, and BaC_2O_4, while the mass between about 580 and 620°C corresponds to the weight of the three carbonates. The weight change in the next two steps results from the loss of carbon dioxide, as first CaO and then SrO are formed. Clearly, sufficient data are available ·in the thermogram to calculate the weight of each of the three elements present in the sample.

Figure 24–3**a** is the derivative of the thermogram shown in (**b**). Many modern instruments are equipped with electronic circuitry to provide such a curve as well as the thermogram itself. The derivative curve may reveal information that is not detectable in the ordinary thermogram. For example, the three peaks at 140, 180, and 205°C suggest that the three hydrates lose moisture at different temperatures. However, all appear to lose carbon monoxide simultaneously and thus yield a single sharp peak at 450°C.

Perhaps the most important applications of thermogravimetric methods are found in the study of polymers. Thermograms provide information about decomposition mechanisms for various polymeric preparations. In addition, the decomposition patterns are characteristic for each kind of polymer and can be used for identification purposes. Figure 24–4 shows decomposition patterns for five polymers obtained by thermogravimetry.

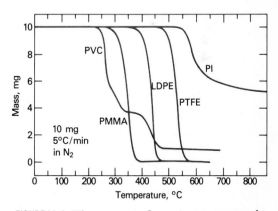

FIGURE 24-4 Thermogram for some common polymeric materials. PVC = polyvinyl chloride; PMMA = polymethyl methacrylate; LDPE = low density polyethylene; PTFE = polytetrafluoroethylene; and PI = aromatic polypyromellitimide. (From: J. Chiu, in *Thermoanalysis of Fiber-Forming Polymers*, R. F. Schwenker, Ed., p. 26, Interscience: New York, 1966. Reprinted by permission of John Wiley & Sons, Inc.)

24B Differential Thermal Analysis and Differential Scanning Calorimetry

In *differential thermal analysis*, the heat absorbed or emitted by a chemical system is observed by measuring the temperature difference between that system and an inert reference compound (often alumina, silicon carbide, or glass beads) as the temperatures of both are increased at a constant rate.[2] In *differential scanning calorimetry*, the sample and a reference substance are also subjected to a continuously increasing temperature; here, however, heat is added to the sample or to the reference as necessary to maintain the two at identical temperatures. The added heat, which is recorded, compensates for that lost or gained as a consequence of endothermic or exothermic reactions occurring in the sample.

---

[2] For a concise review of this method and its applications, see: M. I. Pope and M. D. Judd, *Differential Thermal Analysis,* Heyden: Philadelphia, 1977.

24B-1 GENERAL PRINCIPLES

Figure 24–5 is a differential thermogram obtained by heating calcium oxalate monohydrate in a flowing stream of air. The two minima indicate that the sample becomes cooler than the reference material as a consequence of the heat absorbed by two endothermic processes; equations for these decomposition reactions are shown below the minima. The single maximum indicates that the reaction to give calcium carbonate and carbon dioxide is exothermic. It is noteworthy that when the differential thermogram is obtained in an inert atmosphere, all three reactions are endothermic, and the maximum is replaced by a minimum; here, the reaction product from the decomposition of calcium oxalate is carbon monoxide rather than carbon dioxide (see Figure 24–2).

Sources of Differential Thermogram Peaks. The maxima and minima, such as those appearing in Figure 24–5, are termed peaks. Those appearing above zero are for exothermic changes, which may be the result of physical

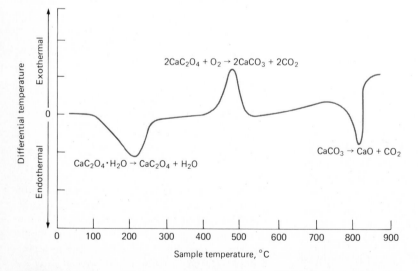

FIGURE 24-5 Differential thermogram of $CaC_2O_4 \cdot H_2O$ in the presence of O_2; the rate of temperature increase was 8°C/min. (From: *Handbook of Analytical Chemistry*, L. Meites, Ed., p. 8–14, McGraw-Hill: New York, 1963. With permission.)

or chemical phenomena. Physical processes that are endothermic include fusion, vaporization, sublimation, absorption, and desorption. Adsorption is generally an exothermic physical change, while crystalline transitions may be either exothermic or endothermic.

Chemical reactions also produce differential peaks; both exothermic and endothermic processes are, of course, possible (for example, see Figure 24–5).

Peak Areas. The peak areas for differential thermograms depend upon the mass of the sample m, the heat or enthalpy ΔH of the chemical or physical process, and certain geometric and heat conductivity factors. These variables are related by the equation[3]

$$A = -\frac{Gm\Delta H}{k} = -k'm\Delta H \qquad (24-1)$$

where A is the peak area ($\Delta T \times$ time), G is a calibration factor that depends upon the sample geometry, and k is a constant related to the thermal conductivity of the sample. The enthalpy ΔH is given a negative sign for an exothermic reaction and a positive one for an endothermic process.

For a given species, k' remains constant provided that a number of variables such as heating rate, particle size of the sample, and placement of the thermocouples are closely controlled. Under these circumstances, Equation 24–1 can be employed to calculate the mass of the analyte from peak areas; here, $k'\Delta H$ can be determined by calibration. Alternatively, Equation 24–1 permits the determination of ΔH for species when k' and m have been measured.

The thermograms obtained by differential scanning calorimetry are similar in appearance to differential thermograms. In this instance, however, the constant k' in

Equation 24–1 becomes independent of the temperature at which the reaction occurs.

24B-2 APPARATUS

The furnace heating programs and recording devices used for differential methods are similar to those employed in thermogravimetry. Indeed, several commercial instruments are designed to permit all three types of thermal analysis.

Figure 24–6 is a schematic diagram showing the components of a typical differential thermal-analyzer system. Weighed quantities of the sample and reference material are held in the small pans labeled S and R. The thermocouple on the right (labeled control TC) controls the rate at which the furnace must be heated in order to provide a linear temperature increase. The sample and reference thermocouples are connected in series. Any current due to a temperature difference between the two is amplified and used to determine the position of a recorder pen. With the switch in position T_S, the sample thermocouple is connected not only to the reference thermocouple but also to a reference junction, which may be at room or ice-bath temperature. The output of this circuit provides a measure of the sample temperature at any instant.

Generally, the sample and reference chamber in a differential thermal apparatus is designed to permit the circulation of inert or reactive gases. Some systems also have the capability for operation at high or low pressures.

For differential scanning calorimetry, individual heaters, located as close as possible to the sample and reference vessels, are provided. When the thermocouples indicate a temperature difference, heat is added to the cooler of the two until temperature equality is restored. The rate of heating required to keep the temperatures equal is recorded as a function of sample temperature. The ordinate of the differential thermogram can

---

[3] See: W. W. Wendlandt, *Thermal Methods of Analysis*, 2d ed., p. 172, Wiley: New York, 1974.

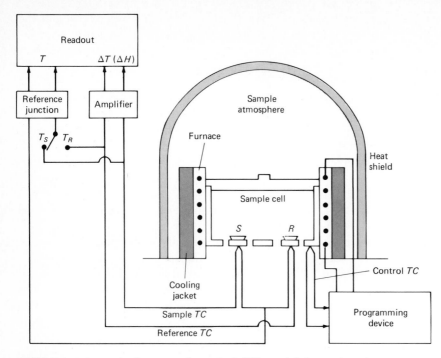

FIGURE 24-6 Schematic diagram of a typical differential thermal analyzer system.

then be expressed in units of calories or millicalories per second. Differential scanning calorimetry thermograms are similar in appearance to Figure 24–5.

Modern thermal instruments are generally of modular design, which permits not only TG, DTA, and DSC but also thermomechanical analysis as well. All are microprocessor controlled and user programmable. Thus, the operator can set operational conditions, input sample information, subtract base-line signals, store or recall thermograms, perform peak temperature and energy analysis, display a derivative curve, compare two curves, and determine peak areas.[4] Computerized libraries of thermal data are also under development.[5]

Figure 24–7 is a printout from a modern thermal analysis instrument. Here, thermal transitions in polyethylene terephthalate are being determined by differential scanning calorimetry. Three major transitions are evident: a glass transition at about 82°C, an exothermic transition at 166°C due to crystallization, and an endothermic melting transition at 250°C. These transitions provide qualitative information on polymer composition and molecular weight; peak integration gives a quantitative measure of heat of fusion and crystallization.

24B-3 APPLICATIONS

Differential thermal methods find widespread use in determining the composition of naturally occurring and manufactured products. The number of applications is impressive and can be appreciated by an examination of a two-volume monograph and re-

---

[4] See for example: R. L. Fyans, *Amer. Lab.*, **1981**, *13* (1), 101; and W. P. Brennan, *et al.*, *Amer. Lab.*, **1983**, *15* (1), 50.

[5] B. Wunderlich, *Amer. Lab.*, **1982**, *14* (6), 28.

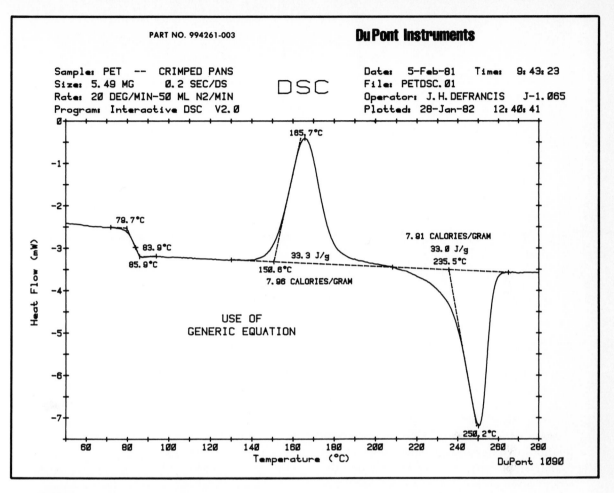

FIGURE 24-7 Differential scanning calorimetry output from a modern thermal instrument showing the thermal transition for polyethylene terephthalate. (Courtesy of DuPont Instrument Systems, Wilmington, DE.)

cent reviews in *Analytical Chemistry*.[6] A few illustrative applications follow.

Inorganic Substances. Differential thermal measurements have been widely used for studies involving the thermal behavior of such inorganic substances as silicates, fer-

rites, clays, oxides, ceramics, and glasses. Information is provided about such processes as fusion, desolvation, dehydration, oxidation, reduction, adsorption, degradation, and solid-state reactions. One of the most important applications is to the generation of phase diagrams and the study of phase transitions. An example is shown in Figure 24–8, which is a differential thermogram for pure sulfur. Here, the peak at 113°C corresponds to a solid-phase change from the rhombic to the monoclinic form, while

---

[6] *Differential Thermal Analysis*, R. C. Mackenzie, Ed., vols. 1 and 2, Academic Press: New York, 1970; W. W. Wendlandt, *Anal. Chem.*, **1984,** *56*, 250R; **1982,** *54*, 97R. For a recent brief listing of applications, see: P. S. Gill, *Amer. Lab.*, **1984,** *16* (1), 39.

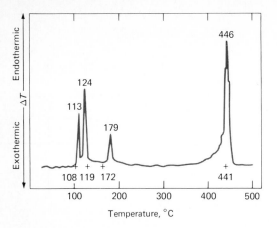

FIGURE 24-8 Differential thermogram for sulfur. (Reprinted with permission from: J. Chiu, *Anal. Chem.*, **1963,** *35*, 933. Copyright 1963 American Chemical Society.)

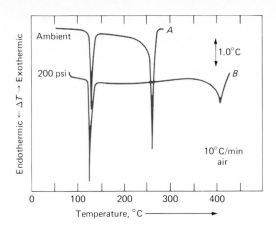

FIGURE 24-9 Differential thermogram for benzoic acid. Curve *A*: at atmospheric pressure; curve *B*: at 200 lbs/in². (From: P. F. Levy, G. Nieuweboer, and L. C. Semanski, *Thermochim. Acta,* **1970,** *1*, 433. With permission.)

the peak at 124°C corresponds to the melting point of the element. Liquid sulfur is known to exist in at least three forms, and the peak at 179°C apparently involves a transition among these. The peak at 446°C corresponds to the boiling point of sulfur.

Organic Compounds. The differential thermal method provides a simple and accurate way of determining the melting, boiling, and decomposition points for organic compounds. Generally, the data appear to be more consistent and reproducible than those obtained with a hot stage or a capillary tube. Figure 24–9 shows thermograms for benzoic acid at (*A*) atmospheric pressure and at (*B*) 200 psi. The first peak corresponds to the melting point and the second to the boiling point of the acid.

Thermal methods have also found widespread use in the pharmaceutical industry for testing the purity of drug samples. An example is shown in Figure 24–10 in which DSC curves are being used to determine the purity of phenacitin preparations. Generally, curves of this type provide purity data with relative uncertainties of ±10%.

Polymers. Differential thermal methods

have been widely applied to the study and characterization of polymeric materials. Figure 24–11 is an idealized thermogram that illustrates the various types of transitions that may be encountered during heating of

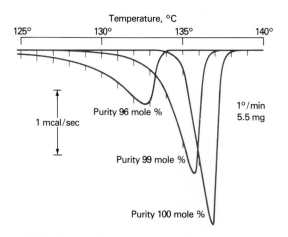

FIGURE 24-10 Differential scanning calorimetry study of samples of the drug phenacetin. (Reprinted with permission from: H.•P. Vaughan and J. P. Elder, *Amer. Lab.,* **1974,** *6* (1), 58. Copyright 1974 by International Scientific Communications, Inc.)

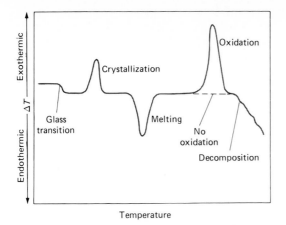

FIGURE 24-11 Schematic differential thermogram showing types of changes encountered with polymeric materials. (From: R. M. Schulken, Jr., R. E. Roy, Jr., and R. H. Cox, *J. Polymer Sci.*, Part C, **1964,** *6,* 18. Reprinted by permission of John Wiley & Sons, Inc.)

24C Enthalpimetric Methods

Enthalpimetric methods are of two main types, namely *thermometric titrations* and *direct injection enthalpy*; the former has also been termed *enthalpimetric titrations.* Thermometric titrations use temperature measurements to establish the end point in a titration; here the analytical parameter is a volume of standard solution. In direct injection enthalpy, an excess of reagent is added as rapidly as possible. The resulting temperature change, which is directly proportional to analyte concentration, is then determined. The equipment for the two procedures can be identical except for the device for introduction of the sample.[7]

The advantages of enthalpimetric methods are several including: their wide applicability to inorganic, organic, and biological

a polymer. Figure 24–12 is a thermogram of a physical mixture of seven commercial polymers. Each peak corresponds to the characteristic melting point of one of the components. Polytetrafluoroethylene (PTFE) has an additional low-temperature peak which arises from a crystalline transition. Clearly, differential thermal methods can be useful for qualitative analysis of polymer mixtures (see also Figure 24–7).

---

[7] For detailed treatments of enthalpimetric methods, see: J. Barthel, *Thermometric Titrations,* Wiley: New York, 1975; G. A. Vaughn, *Thermometric and Enthalpimetric Titrimetry,* Van Nostrand: New York, 1973; J. Jordan, in *Treatise on Analytical Chemistry,* I. M. Kolthoff and P. J. Elving, Eds., Part I, vol. 8, Chapter 91, Interscience: New York, 1968; and H. V. Tyrrell and A. E. Beezer, *Thermometric Titrimetry,* Chapman and Hall: London, 1968. For brief reviews of the subject, see: J. Jordan, *et al., Anal. Chem.,* **1976,** *48,* 427A: A. J. C. L. Hogarth and J. D. Stutts, *Amer. Lab.,* **1981,** *13* (1), 18; P. Sadtler and T. Sadtler, *Amer. Lab.,* **1982,** *14* (4), 86.

FIGURE 24-12 Differential thermogram for a mixture of seven polymers. PTFE = polytetrafluoroethylene; HIPPE = high-pressure (low density) polyethylene; LPPE = low-pressure (high density) polyethylene; PP = polypropylene; POM = polyoxymethylene. (From: J. Chiu, *DuPont Thermogram,* **1965,** *2* (3), 9. With permission.)

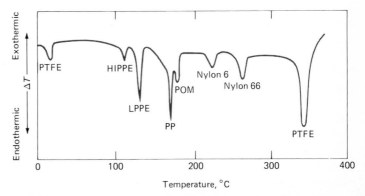

systems; their versatility in the respect that they can be applied to aqueous and non-aqueous solutions, to gases, and to molten salt media; their speed; and their ability to provide fundamental thermodynamic data for reactions (ΔH, ΔG, and ΔS). The precision of thermometric titrations are generally on the order of 1% relative for quantities of analyte in the 1 to 50 mmol range. Direct injection enthalpy measurements are generally somewhat less precise than the titrimetric procedure. On the other hand, they are usually faster, do not require standardized reagents, and offer the possibility of multiple serial determinations.

24C-1 PRINCIPLES OF THERMOMETRIC TITRATIONS

The temperature changes observed during a thermometric titration are the consequence of the heat evolved or absorbed by the reaction between the analyte and the reagent. The heat or enthalpy of a reaction ΔH is given by the familiar thermodynamic expression

$$\Delta H = \Delta G + T\Delta S \qquad (24-2)$$

where T is the temperature, ΔG is the free energy change, and ΔS is the entropy change for the reaction.

Most titrimetric end points, such as the potentiometric ones discussed in Section 20F, require that ΔG in the foregoing equation be a large negative number; only then will the equilibrium between the analyte and reagent lie sufficiently far to the right. This condition is necessary to make the potential changes in the equivalence-point region large enough for accurate location of chemical equivalence.

In contrast, the success of a thermometric titration depends not only upon the magnitude of ΔG but of $T\Delta S$ as well. Thus, if $T\Delta S$ is a large negative number, successful end points may still be realized even though ΔG is zero or even positive. A classic example of

this situation is illustrated by a comparison of the potentiometric and thermometric end points for boric and hydrochloric acids shown in Figure 24–13. For boric acid, ΔG is so small that neutralization is relatively incomplete at chemical equivalence, and no detectable potentiometric end point is observed. On the other hand, ΔH for the neutralization of boric acid is about -10.2 kcal/mol compared with -13.5 for hydrochloric. As a consequence, sharp thermometric end points are realized for both acids (Figure 24–13**b**).

The change in temperature ΔT during a thermometric titration is given by

$$\Delta T = - \frac{n\Delta H}{k} = \frac{Q}{k} \qquad (24-3)$$

where n is the number of moles of reactant, k is the effective heat capacity of the system, and Q is the total amount of heat evolved (or absorbed). Thus, the total change in temperature is directly proportional to the number of moles of analyte.

24C-2 PRINCIPLES OF DIRECT INJECTION ENTHALPIMETRY

The direct injection method is based upon the measurement of ΔT in Equation 24–3 following the addition of a "plug" of reagent to the analyte solution. An excess of reagent is generally used and precise volume measurements are not a prerequisite. Thus, additions are usually performed with a manually operated syringe. The recorder data from this type of analysis take the form shown in Figure 24–14.

24C-3 APPARATUS

Reproducible enthalpimetric data require the use of an automatic instrument such as that shown in Figure 24–15. Several instrument companies offer thermal titration apparatus.

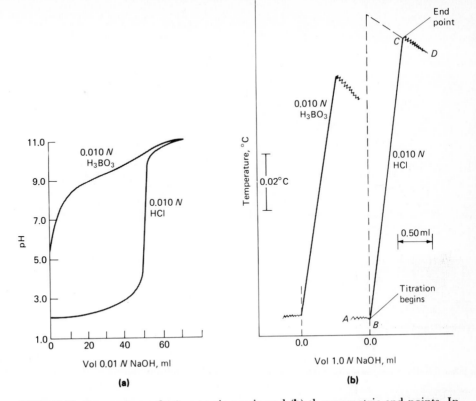

FIGURE 24-13 Comparison of (**a**) potentiometric and (**b**) thermometric end points. In each case, 50.0 mL of acid were titrated. (Thermometric titration curves from: J. Jordan and W. H. Dumbaugh, Jr., *Anal. Chem.*, **1959**, *31*, 212. With permission. Copyright 1959 American Chemical Society.)

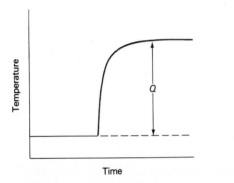

FIGURE 24-14 Temperature changes in a typical direct injection enthalpy experiment.

Reagent Delivery System. For a thermometric titration, reagent is added by means of a screw-driven syringe powered by a motor that is synchronized with the paper drive of the recorder. Generally, the reagent is 50 to 100 times more concentrated than the sample, so that titrant volumes are, at the most, a milliliter or two. Under these circumstances, corrections for dilution or temperature difference between the analyte solution and the reagent are unnecessary.

Reaction Vessel. Enthalpimetric measurements must be carried out under conditions that approach adiabatic, that is, without gain or loss of heat from the surroundings. Thus, the titration vessel is generally a De-

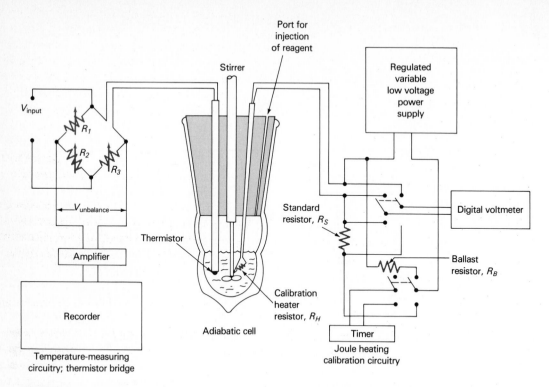

FIGURE 24-15 Schematic diagram of an apparatus for enthalpimetric measurements. (Reprinted with permission from: J. Jordan, *et al.*, *Anal. Chem.*, **1976**, *48*, 430A. Copyright 1976 American Chemical Society.)

war flask or a beaker surrounded by a thick layer of styrofoam insulation. Efficient stirring must be provided. In addition, for a titration, reagent addition must be completed in a brief period (less than 5 min).

Temperature Measurements. Thermistors are generally employed as temperature sensors for enthalpimetric methods because of their large temperature coefficients (about 10 times greater than a thermocouple), their small size, and their rapid response to temperature changes. A thermistor is a sintered metal oxide semiconductor that has, in contrast to other temperature sensors, a negative temperature coefficient of resistance.

As shown in Figure 24–15, changes in resistance appear as a voltage difference across the bridge circuit; the difference is then amplified and recorded. For a thermis-

tor having a resistance of about 2.0 kΩ, the output voltage would correspond to about 0.16 mV per 0.01°C.

24C-4 APPLICATIONS

Figure 24–13**b** reveals that a typical thermometric titration curve is made up of three parts. The region from *A* to *B* gives the temperature of the system before addition of reagent. Theoretically, the slope here should be, but seldom is, zero because of small gains or losses of heat due to stirring and imperfections in the insulation. At point *B*, reagent addition is begun at a controlled rate of several microliters per minute. Point *C* corresponds to the end point in the titration. Ordinarily, conditions are arranged so that point *D* is reached in less than 5 min in order to avoid excess heat leakage to the surroundings. Beyond *C*, a straight line is observed. Its slope may be negative or positive depending upon whether the dilution process is exothermic or endothermic.

Table 24–1 shows some typical applications of thermometric titrations and direct injection enthalpy.[8] An interesting example of usefulness of thermometric titrations for the analysis of mixtures is shown in Figure 24–16. Here, advantage is taken of the fact that the reaction of EDTA (p. 635) with calcium ion is exothermic ($\Delta H = -6.5$ kcal/mol) while with magnesium ion it is endothermic ($\Delta H = +5.5$ kcal/mol). As a consequence, a good end point for each ion is obtained even though the formation constant for the calcium EDTA complex differs from that of magnesium by only 100—a difference that is so small that discrimination between the two ions by potentiometric or colorimetric end points is impossible.

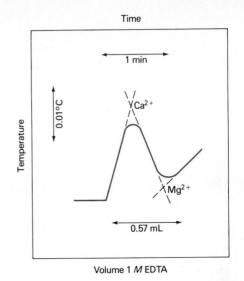

FIGURE 24-16 Thermometric titration of 0.00506 *M* Ca^{2+} and 0.00540 *M* Mg^{2+} with 1.000 *M* EDTA. (From: J. Jordan and T. G. Alleman, *Anal. Chem.,* **1957,** *29,* 12. With permission. Copyright 1957 American Chemical Society.)

Enthalpimetric methods also are beginning to find applications to the determination of species of biochemical and biological interest. Thus, it appears possible to per-

---

[8] For a detailed treatment of applications, see: G. A. Vaughan, *Thermometric and Enthalpimetric Titrimetry,* Chapter 3, Van Nostrand: New York, 1973.

Table 24–1
Some Typical Applications of Enthalpimetric Methods[a]

Analyte	Titrant	Minimum Titratable Concentration, *M*	Precision Relative %	$\Delta H°$ kcal/mol
Acids, $K_a \geq 10^{-10}$	Strong base	0.005	1	-10 to -15
Bases, $K_b \geq 10^{-10}$	Strong acid	0.005	1	—
Divalent cations	EDTA	0.001 to 0.01	0.1 to 1	-13 to $+5$
Ag^+	HCl	0.1	0.3	—
Ca^{2+}	$(NH_4)_2C_2O_4$	0.01	1	-6.1
Fe^{2+}	Ce^{4+}	0.001	1	-24
	$Cr_2O_7^{2-}$	0.006	0.5	-27
	MnO_4^-	0.003	1	-28
$Fe(CN)_6^{4-}$	Ce^{4+}	0.001	1	-10
Ti^{3+}	Ce^{4+}	0.002	2	-30

[a] Data taken from *Handbook of Analytical Chemistry,* L. Meites, Ed., pp. **8**–5 to **8**–7, McGraw-Hill: New York, 1963. With permission.

form analyses by enzyme-catalyzed reaction on serum and, in some cases, whole blood, without deproteinization.[9] Enthalpimetry has also been used to determine enzyme activity[10] and for the study of proteins and lipids.[11]

[9] See: C. D. McGlothlin and J. Jordan, *Anal. Chem.,* **1975,** *47,* 786, 1479; *Clin. Chem.,* **1975,** *21,* 741.

[10] J. K. Grime, in *Comprehensive Analytical Chemistry,* Volume XII, Part B, Chapter 8, N. D. Jespersen, Ed., Elsevier: New York, 1982.

[11] D. J. Eatough, *et al.,* in *Comprehensive Analytical Chemistry,* Volume XII, Part B, Chapter 7, N. D. Jespersen, Ed., Elsevier: New York, 1982.

PROBLEMS

24–1. Describe what quantity is measured and how the measurement is performed for each of the following techniques: (a) thermogravimetric analysis; (b) differential thermal analysis; (c) differential scanning calorimetry; and (d) thermometric titration.

24–2. A 0.6025-g sample was dissolved, and the Ca^{2+} and Ba^{2+} ions present were precipitated as $BaC_2O_4 \cdot H_2O$ and $CaC_2O_4 \cdot H_2O$. The oxalates were then heated in a thermogravimetric apparatus leaving a residue that weighed 0.5713 g in the 320 to 400°C range and 0.4673 g in the 580 to 620°C range. Calculate the % Ca and % Ba in the sample.

24–3. Why are the two low-temperature endotherms in Figure 24–9 coincident whereas the high temperature peaks are displaced from each other?

24–4. The reaction of Al^{3+} with ethylenediaminetetraacetic acid (EDTA) is exothermic by 10.9 kcal/mol. The reaction of Ni^{2+} with EDTA is endothermic by 7.2 kcal/mol. The stability constants for the EDTA complexes of Al^{3+} and Ni^{2+} are $10^{16.1}$ and $10^{18.6}$, respectively. Draw a schematic diagram showing the expected thermometric titration curve for the reaction of an equimolar mixture of Al^{3+} and Ni^{2+} with EDTA. Show how you would locate the two end points.

24–5. The iron-transport protein, transferrin, found in human blood serum, has a molecular weight of 81,000 and is capable of binding 0, 1, or 2 iron(III) ions. When subjected to differential scanning calorimetry, transferrin (and many other proteins) exhibits endotherms corresponding to thermal denaturation (unfolding) of part or all of the protein. In the absence of iron, a solution of transferrin exhibited two endotherms at 62° and 72°C. When one mole of iron per mole of transferrin was added in the form of ferric nitrilotriacetate at pH 7.5, the thermogram of this solution exhibited endotherms at 76° and 88°C, with only a minor endotherm at 62°C. When a second mole of iron was added, the sample produced just a single endotherm at 89°C. Suggest some possible interpretations of these results in terms of structural equivalence of the sites and the sequence of iron binding to the sites.

25

An Introduction to

Chromatographic Separations

Generally, methods for chemical analysis are, at best, selective; few, if any, are truly specific. Consequently, the separation of the analyte from potential interferences is an important step in many analytical procedures. Without question, the most widely used means of performing analytical separations is *chromatography*, a procedure that finds application to all branches of science. Column chromatography was invented and named by the Russian botanist Mikhail Tswett shortly after the turn of the century. He employed the technique to separate various plant pigments such as chlorophylls and xanthophylls by passing solutions of these compounds through a glass column packed with finely divided calcium carbonate. The separated species appeared as colored bands on the column, which accounts for the name he chose for the method (Greek *chroma* meaning color and *graphein* meaning to write).

The applications of chromatography have grown explosively in the last four decades, owing not only to the development of several new types of chromatographic techniques but also to the growing need by scientists for better methods for characterizing complex mixtures. The tremendous impact of these methods on science is attested by the 1952 Nobel prize that was awarded to A. J. P. Martin and R. L. M. Synge for their discoveries in the field. Perhaps more impressive is a list of twelve Nobel prize awards between 1937 and 1972 that were based upon work in which chromatography played a vital role.[1]

25A A General Description of Chromatography

Chromatography encompasses a diverse and important group of separation methods that permit the scientist to separate, isolate, and identify related components of complex

---

[1] See: L. S. Ettre, in *High-Performance Liquid Chromatography*, C. Horváth, Ed., Vol. 1, p. 4, Academic Press: New York, 1980.

mixtures; many of these separations are impossible by other means.[2]

The term "chromatography" is difficult to define rigorously, owing to the variety of systems and techniques to which it has been applied. All of these methods, however, make use of a *stationary phase* and a *mobile phase*. Components of a mixture are carried through the stationary phase by the flow of the mobile one; separations are based on differences in migration rates among the sample components.

25A-1 CLASSIFICATION OF CHROMATOGRAPHIC METHODS

Chromatographic methods can be categorized in two ways. The first is based upon the physical means by which the stationary and mobile phases are brought into contact. In *column chromatography*, the stationary phase is held in a narrow tube through which the mobile phase is forced under pressure or by gravity. In *planar* chromatography, the stationary phase is supported on a flat plate or in the interstices of a paper; here, the mobile phase moves through the stationary phase by capillary action or under the influence of gravity. The discussion in this and the next two chapters focuses on column chromatography. Chapter 28 is devoted to planar methods. It is important to point out here, however, that the equilibria upon which the two types of chromatography are based are identical and that the theory developed for column chromatography is readily adapted to planar as well.

A more fundamental classification of chromatographic methods is one based

upon the types of mobile and stationary phases and the kinds of equilibria involved in the transfer of solutes between phases. Table 25–1 lists two general categories of chromatography: *liquid chromatography* and *gas chromatography*. In the former, the mobile phase is a liquid; in the latter it is a gas. As shown in column 2 of the table, several specific chromatographic methods fall into each of the two general categories. It should be noted here that several procedures are employed to fix a stationary liquid phase in place. For example, a thin film of the liquid may be adsorbed onto the surface of a finely divided, inert solid, or retained in the pores or interstices of such a solid. Alternatively, the liquid may be retained by adsorption on the inner walls of a capillary tubing. Ideally, the solid plays no direct part in the separation, serving only as a support for the liquid. In most cases, however, the nature of the solid does have an effect on the separation.

It is also noteworthy that liquid chromatography can be performed in columns and on plane surfaces; gas chromatography, on the other hand, is restricted to column procedures.

25A-2 LINEAR CHROMATOGRAPHY

All chromatographic separations are based upon differences in the extent to which solutes are partitioned between the mobile and the stationary phase. The equilibrium involved can be described quantitatively by means of a *partition coefficient K*, which for chromatography, is defined as

$$K = \frac{C_S}{C_M} \qquad (25\text{–}1)$$

Here, C_S is the molar analytical concentration of a solute in the stationary phase and C_M is its concentration in the mobile phase. In the ideal case, the partition ratio is constant over a wide range of solute concentrations; that is, C_S is directly proportional to

[2] General references on chromatography include: *Chromatography*, 3d ed., E. Heftmann, Ed., Van Nostrand-Reinhold: New York, 1975; B. L. Karger, L. R. Snyder, and C. Horváth, *An Introduction to Separation Science*, Wiley: New York, 1973; R. Stock and C. B. F. Rice, *Chromatographic Methods*, 3d ed., Chapman & Hall: London, 1974; and *Chromatographic and Allied Methods*, O. Mikeš, Ed., Wiley: New York, 1979.

C_M. Often, however, nonlinear relationships are encountered. Some typical distribution curves are shown in Figure 25–1.

The ideal relationship, shown by curve C in this figure, is often approximated by distribution equilibria between two immiscible liquids, provided association or dissociation reactions do not occur in one of the solvents. Where such reactions do exist, a relationship similar to B or D is more likely. For example, if the stationary phase is water and the mobile phase is benzene, a curve of type B is observed for a solute consisting of a weak organic acid. Only the undissociated acid is soluble in benzene. In the aqueous solution, however, appreciable amounts of both the undissociated acid and its conjugate base are present; moreover, the ratio between these species is concentration-dependent. Thus, at low concentrations, a

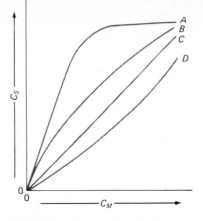

FIGURE 25-1 Typical distribution curves. C_S is the concentration of solute in the stationary phase and C_M its concentration in the mobile phase.

Table 25-1
Classification of Column Chromatographic Methods

General Classification	Specific Method	Stationary Phase	Type Equilibrium
Liquid chromatography (LC) (Mobile phase: liquid)	Liquid-liquid (LLC)	Liquid adsorbed on a solid	Partition between immiscible liquids
	Liquid-solid (LSC)	Solid	Adsorption
	Liquid-bonded phase (LBC)	Organic species bonded to a solid surface	Partition/adsorption
	Ion-exchange (IEC)	Ion-exchange resin	Ion-exchange
	Gel-permeation (GPC)	Liquid in interstices of a polymeric solid	Partition/sieving
Gas chromatography (GC) (Mobile phase: gas)	Gas-liquid (GLC)	Liquid adsorbed on a solid	Partition between gas and liquid
	Gas-solid (GSC)	Solid	Adsorption
	Gas-bonded phase	Organic species bonded to a solid surface	Partition/adsorption

smaller fraction of the total acid is available for distribution between the two solvents, and the partition ratio is greater. If, on the other hand, water represented the mobile phase, a curve such as *D* would be expected. Curves of type *B* are also commonly encountered in gas-liquid equilibria, where a gas is the mobile phase.

Curve *A* is a typical *adsorption isotherm*, which relates the amount of solute adsorbed on the surface of a solid to the solute concentration in the solution that contacts the solid. At high solute concentrations, all adsorption sites on the solid surface are occupied; the extent of adsorption then becomes independent of the solute concentration. Curves of this type can be described by the relationship

$$C_S = KC_M^n$$

where *n* is a constant. Note that this equation reverts to Equation 25–1 when *n* is unity.

The plots in Figure 25–1 suggest that over limited concentration ranges, and particularly at low concentrations, it is possible to assume a linear relationship between C_S and C_M without introducing serious error. Fortunately, much of chromatography is performed under conditions of this kind, thus greatly simplifying the derivation of mathematical expressions for the separation process. Chromatography carried out under conditions in which *K* is more or less constant is termed *linear chromatography*. The discussions that follow will deal exclusively with separations of this type.

25A-3   COLUMN CHROMATOGRAPHY

Figure 25–2 is a schematic representation of how two components A and B are resolved by column chromatography. The method shown is by far the most common one and is called *elution chromatography*, where *elution* is defined as a process whereby a solute is

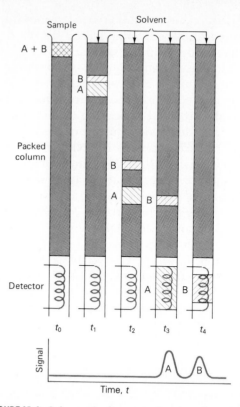

FIGURE 25-2 Schematic diagram showing the separation of components A and B of a mixture by column elution chromatography. The lower figure shows the output of the signal detector at the various stages of elution shown in the upper part of the figure.

washed through a column by additions of fresh solvent. As shown in the figure, a single portion of the sample, dissolved in the mobile phase, is introduced at the head of a column (time t_0 in Figure 25–2), whereupon the components of the sample distribute themselves between the two phases. Introduction of additional mobile phase (the *eluent*) forces the solvent containing a part of the sample down the column, where further partition between the mobile phase and fresh portions of the stationary phase occurs (time t_1). Simultaneously, partitioning between the fresh solvent and the stationary phase takes place at the site of the original sample. Continued additions of solvent carry solute molecules down the column in a continuous series of transfers between the mobile and the stationary phases. Because solute movement can only occur in the mobile phase, however, the average *rate* at which a solute migrates *depends upon the fraction of time it spends in that phase.* This fraction is small for solutes with partition coefficients that favor retention in the stationary phase (compound B in Figure 25–2, for example) and is large where retention in the mobile phase is more likely (component A). Ideally, the resulting differences in rates cause the components in a mixture to separate into bands located along the length of the column (see time t_2 in Figures 25–2 and 25–3). Isolation of the separated species is then accomplished by passing a sufficient quantity of mobile phase through the column to cause the individual bands to pass out the end, where they can be collected (times t_3 and t_4 in Figure 25–2).

If a detector that responds to solute concentration is placed at the end of the column and its signal is plotted as a function of time (or of volume of the added mobile phase), a series of symmetric peaks is obtained, as shown in the lower part of Figure 25–2. Such a plot, called a *chromatogram,* is useful for both qualitative and quantitative analysis. The positions of peaks on the time axis may serve to identify the components of the sample; the areas under the peaks provide a quantitative measure of the amount of each component.

25A-4 THEORIES OF ELUTION CHROMATOGRAPHY

Figure 25–3 shows concentration profiles for solutes A and B at an early and late stage of elution from the chromatographic column shown in Figure 25–2.[3] The partition coefficient for B is the larger of the two; thus, B lags during the migration process. It is apparent that movement down the column increases the distance between the two bands. At the same time, however, broadening of both bands takes place, which lowers the efficiency of the column as a separating device. Band broadening is unavoidable; fortunately, however, under ideal conditions, it occurs more slowly than band separation. Thus, as shown in Figure 25–3, a clean resolution of species is often possible provided the column is sufficiently long.

It has been found that a number of chemical and physical phenomena influence the

---

[3] Note that the relative positions of bands for A and B in the concentration profile shown in Figure 25–3 appear to be the reverse of the peak in the chromatogram shown in the lower part of Figure 25–2. The difference, of course, is that the abscissa in the former is distance along the column while in the latter it is time. Thus in the chromatogram in Figure 25–2, the *front* of a peak lies to the left and the *tail* to the right; for the concentration profile or band the reverse obtains.

FIGURE 25-3 Concentration profiles for solutes A and B at different times in their migration down a column. The times t_1 and t_2 are indicated in Figure 25–2.

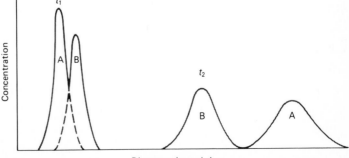

rates of band separation and of band broadening. As a consequence, improved separations can often be realized by control of variables that increase the former and/or decrease the latter. Figure 25–4 illustrates how each of these methods can lead to a more efficient separation of a two-component mixture.

From the foregoing discussion, it is apparent that a useful theory of chromatography must be able to account not only for relative rates at which solutes migrate but also for rates at which bands broaden during migration. The original theory of chromatography, called the *plate theory*, was able to account for the variables that influence migration rates in quantitative terms. Unfortunately, however, its utility was limited because it failed to describe the effects of the numerous variables responsible for zone broadening. As a consequence, the plate theory has now been supplanted by the *kinetic* or *rate theory*, which is capable of accounting for the latter variables as well.

It is useful, however, to consider the plate theory briefly in order to indicate the genesis of two terms employed in the kinetic theory. The plate theory, which was originally adapted by Martin and Synge from the the-

ory of distillation columns,[4] envisages a chromatographic column as being composed of a series of discrete but contiguous, narrow, horizontal layers called *theoretical plates*. At each plate, equilibration of the solute between the mobile and the stationary phase is assumed to take place. Movement of the solute and solvent is then viewed as a series of stepwise transfers from one plate to the next.

The efficiency of a chromatographic column as a separation device improves as the number of equilibrations increases; that is, as the number of theoretical plates increases. Thus, the *number of theoretical plates*, *N*, is used as measure of column efficiency. A second efficiency term, the *height equivalent of a theoretical plate*, *H* (or sometimes HETP) is the thickness of one plate. Clearly, the product of these two terms is the column length *L*. That is,

$$L = NH$$

which rearranges to

$$N = L/H \tag{25–2}$$

[4] A. J. P. Martin and R. L. M. Synge, *Biochem. J.*, **1941**, *35*, 1358.

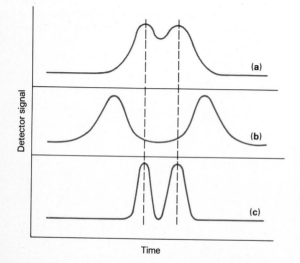

Detector signal

(a)

(b)

(c)

Time

FIGURE 25–4 Two-component chromatograms illustrating two methods of improving separations: (**a**) original chromatogram with overlapping peaks; improvements brought about by (**b**) an increase in band separation; and (**c**) a decrease in band spread.

Note that H decreases as the efficiency of a column becomes greater. That is, as H becomes smaller, the number of equilibrations that occur in a given length of column becomes larger. It is important to note that H and N are retained as efficiency parameters in the rate theory and that Equation 25–2 continues to apply. It should be appreciated, however, that *a plate as a physical entity does not exist* in a column. Thus, the plate and the plate height should be viewed as criteria for column efficiency only.

25B The Rate Theory of Chromatography

The rate theory of chromatography successfully describes the variables that influence the time at which an elution band appears at the end of a chromatographic column as well as the variables that affect the width of the eluted peak.[5]

25B-1 MEASURES OF COLUMN EFFICIENCY

Examination of the peaks in a typical chromatogram (Figure 25–2) or the bands on columns (Figure 25–3) reveals a similarity to normal error or Gaussian curves (Figure 1–2, p. 7), which are obtained when replicate values of a measurement are plotted as a function of the frequency of their occurrence.[6] As was shown in Section 1C–3, normal error curves can be rationalized by assuming that the uncertainty associated with any single measurement is the summation of a much larger number of small, individually undetectable and random uncertainties, each of which has an equal probability of being positive or negative. The most common occurrence is for these uncertainties to cancel one another, thus leading to the mean value. With less likelihood, the summation may cause results that are greater or smaller than the mean. The consequence is a symmetric distribution of data around the mean value. In a similar way, the typical Gaussian shape of a chromatographic band can be attributed to the additive combination of the random motions of the myriad solute particles in the chromatographic band or zone.

It is instructive to first consider the behavior of an individual solute particle, which, during migration, undergoes many thousands of transfers between the stationary and the mobile phase. The time it spends in either phase after a transfer is highly irregular and depends upon it accidentally gaining sufficient thermal energy from its environment to accomplish a reverse transfer. Thus, in some instances, the residence time in a given phase may be transitory; in others, the period may be relatively long. Recall that the particle is eluted *only during residence in the mobile phase;* as a result, its migration down the column is also highly irregular. Because of variability in the residence time, the average rate at which individual particles move relative to the mobile phase varies considerably. Certain individual particles travel rapidly by virtue of their accidental inclusion in the mobile phase for

---

[5] For a detailed presentation of the rate theory, see: J. C. Giddings, *Dynamics of Chromatography*, Part I, Marcel Dekker: New York, 1965; J. C. Giddings, in *Chromatography*, 3d ed., E. Heftmann, Ed., Chapter 3, Van Nostrand-Reinhold: New York, 1975; and R. P. W. Scott, *Contemporary Liquid Chromatography*, (Vol. XI of *Techniques of Chemistry*, A. Weissberger, Ed.), Chapter 2, John Wiley & Sons: New York, 1976. For a shorter presentation, see: J. C. Giddings, *J. Chem. Educ.*, **1958**, *35*, 588; **1967**, *44*, 704.

[6] Often, chromatographic peaks are nonideal and exhibit *tailing* or *fronting*. In the former case the tail of the peak, appearing to the right on the chromatogram, is drawn out while the front is steepened. With fronting, the reverse is the case. One cause of tailing

and fronting is a nonlinear distribution relationship. Thus tailing is associated with a convex distribution curve such as curves A and B in Figure 25–1; fronting results from the distribution relationship shown in curve D of the same figure. Distortions of this kind are undesirable because they lead to poorer separations and less reproducible elution times.

a majority of the time. Others, in contrast, may lag because they happen to have been incorporated in the stationary phase for a greater-than-average time. The consequence of these random individual processes is a symmetric spread of velocities around the mean value, which represents the behavior of the average and most common particle.

The breadth of a band increases as it moves down the column because more time is allowed for spreading to occur. Thus, the zone breadth is directly related to residence time in the column and inversely related to the velocity at which the mobile phase flows.

Variance and Standard Deviation as Measures of Column Efficiency. As shown in Section 1C–3, the breadth of a Gaussian curve is directly related to the variance σ^2 or the standard deviation σ of a measurement. Because chromatographic bands are also in principle Gaussian, it is convenient to define the efficiency of a column in terms of variance per unit length of column. By definition, then,

$$H = \sigma^2/L \qquad (25\text{--}3)$$

where H, the plate height in centimeters, is the measure of efficiency and L is the column length, also in centimeters. Note that σ^2 then carries units of cm^2. Substituting 25–2 and rearranging provide an alternate description of efficiency of a column. That is,

$$N = L^2/\sigma^2 \qquad (25\text{--}4)$$

where N is the number of plates contained in a column of length L. Clearly, column efficiency increases as the number of plates it contains becomes larger or as the plate height becomes smaller.

Experimental Evaluation of N and H. Figure 25–5 shows a typical chromatogram with time as the abscissa. Here, t_R, the *retention time,* is defined as the time required after sample injection for the solute peak to appear at the end of the column. As will be shown momentarily, the variance of this peak can be obtained by a simple graphical procedure. This variance will, however, be in units of sec^2 and is usually designated as τ^2 to distinguish it from σ^2, which has units of cm^2.

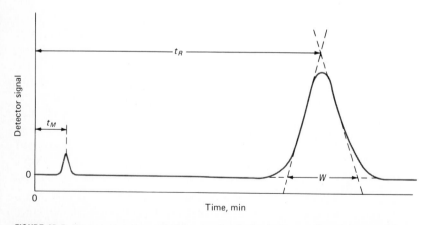

FIGURE 25-5 Determination of the standard deviation τ from a chromatographic peak. Here, $W = 4\tau$; t_R is the retention for a solute that is retained by the column packing, and t_M is the time for one that is not. Thus, t_M is equal approximately to the time required for a molecule of the mobile phase to pass through the column.

The two standard deviations τ and σ are related by

$$\tau = \frac{\sigma}{L/t_R} \tag{25–5}$$

where L/t_R is the average linear velocity of the solute in centimeters per second.

Figure 25–5 illustrates a simple means for approximating τ from an experimental chromatogram. Tangents at the inflection points on the two sides of the Gaussian peak are extended to form a triangle with the abscissa. The area of this triangle can be shown to be approximately 96% of the total area under the peak. In Section 1C–3, it was shown that about 96% of the area under a Gaussian peak is included within plus or minus two standard deviations ($\pm 2\sigma$) of its maximum. Thus, the intercepts shown in the figure occur at approximately $\pm 2\tau$ from the maximum, and

$$W = 4\tau$$

where W is the magnitude of the base of the triangle shown in Figure 25–5. Substituting this relationship into Equation 25–5 yields with rearranging

$$\sigma = \frac{LW}{4t_R}$$

Substitution into Equation 25–3 gives

$$H = \frac{LW^2}{16t_R^2}$$

To obtain N, we substitute into Equation 25–2 and rearrange, giving

$$N = 16 \left(\frac{t_R}{W} \right)^2 \tag{25–6}$$

Thus, N can be calculated from two time measurements t_R and W; to obtain H, the length of the column packing L must also be known.

Another method for approximating N, which some believe to be usually more reliable, is to determine $W_{1/2}$, the width of a peak at half its maximum height. The number of theoretical plates is then given by

$$N = 5.54 \left(\frac{t_R}{W_{1/2}} \right)^2 \tag{25–7}$$

The two parameters N and H are widely used in the literature and by instrument manufacturers as measures of column performance. It is important to appreciate that for these parameters to be meaningful in comparing two columns, it is essential that they be determined with the *same compound*.

25B-2 BAND BROADENING

Band broadening has been shown to be the consequence of the finite rate at which several mass transfer processes occur during migration of a solute down a column. Some of these rates are controllable by adjustment of experimental variables, thus permitting improvement in separations. The most important variables affecting separation efficiency are particle size of the packing, thickness of the immobilized film (when the stationary phase is an adsorbed liquid), viscosity of the mobile phase, temperature, and linear velocity of the mobile phase. Decreases in the first three of these variables generally lead to improved separations or decreased plate heights, while decreases in temperature have the opposite effect. Usually, plots of plate height versus mobile phase velocity exhibit a minima where maximum separation efficiency is achieved.

Over the last thirty years, an enormous amount of theoretical and experimental effort has been devoted to developing quantitative relationships describing the effects of the aforementioned variables on plate heights for various types of columns. Perhaps a dozen or more mathematical expressions for calculating plate height have been put forward and applied with varying degrees of success. It is apparent that none of these is entirely adequate to explain the complex physical interactions and effects that lead to zone broadening. Some of the equations, though imperfect, have been of considerable use, however, in pointing the way toward improved column performance. One of these is presented here.

The efficiency of most chromatographic columns can be approximated by the expression

$$H = B/u + C_S u + C_M u \qquad (25\text{–}8)$$

where H is the plate height in centimeters and u is the linear velocity of the mobile phase in centimeters per second. The quan-

tities B, C_S, and C_M are mass transfer coefficients, which are related to column and solute properties by the equations shown in Table 25–2.[7]

Longitudinal Diffusion (B/u). Longitudinal diffusion results from the tendency of molecules

---

[7] Theoretical studies of zone broadening in the 1950s by Dutch chemical engineers led to the *van Deemter equation*, which can be written in the form

$$H = A + B/u + Cu$$

Here, the constants, A, B, and C are coefficients of eddy diffusion, longitudinal diffusion, and mass transfer, respectively. The van Deemter equation is of considerable historic interest; its modernized version takes the form of Equation 25–8 (see: S. J. Hawkes, *J. Chem. Educ.*, **1983**, *60*, 393).

Table 25–2
Kinetic Processes That Contribute to Peak Broadening

Name of Process	Term in Equation 25–8	Relationship to Column[a] and Solute Properties
Longitudinal diffusion	B/u	$\dfrac{B}{u} = \dfrac{2\psi D_M}{u}$
Mass transfer to and from stationary phase[b]	$C_S u$	$C_S u = \dfrac{q k' d_f^2 u}{(1 + k')^2 D_S}$
Mass transfer in the mobile phase	$C_M u$	$C_M u = \dfrac{f(d_p^2, u)}{D_M} u$

[a] D_M and D_S: diffusion coefficients for the solute in the mobile and stationary phases respectively (cm²/sec).
u: linear velocity of the mobile phase (cm/sec).
k': capacitance factor (see Equation 25–15 for definition).
d_p: particle diameter of packing (cm).
d_f: film thickness of adsorbed liquid of the stationary phase (cm).
f: function of.
ψ, q, w: constants.
[b] The equation shown applies to liquid stationary phases only. For solid stationary phases at which adsorption takes place

$$C_S = \frac{2 t_d k' u}{(1 + k')^2}$$

where t_d is the average desorption time of the analyte from the surface; t_d is equal to $1/k_d$, where k_d is the first-order rate constant for desorption.

to migrate from the concentrated center of a band toward more dilute regions on either side; that is, in and opposed to the direction of flow. Although this type of diffusion occurs both in the mobile and stationary phases, it is only in the former that significant band broadening occurs, and then only when the mobile phase is a gas. That is, except for gas chromatography, B in Equation 25–8 is for all practical purposes zero.

As shown in Table 25–2, the contribution of longitudinal diffusion to plate height is directly proportional to the rate of diffusion of the analyte in the mobile phase and inversely proportional to the linear velocity of the solvent. The latter effect is to be expected since lower velocities provide a greater residence time during which diffusion can occur. The constant ψ is the obstruction factor, which recognizes that longitudinal diffusion is hindered by the packing. With packed columns, a typical value for ψ is about 0.6; for coated capillary columns, which contain no packing, its value is unity.

Mass Transfer To and From the Stationary Phase ($C_S u$). The rate at which analyte molecules transfer into and out of the stationary phase often influences the efficiency of a chromatographic column. For a liquid stationary phase, this rate is controlled by the rate of diffusion of solute to the surface of the immobilized liquid where desorption can occur. Thus, as seen in Table 25–2, the contribution of stationary phase mass transfer to plate height is inversely proportional to the diffusion coefficient of the solute in the stationary phase. For a solid stationary phase, the modified form of the equation for C_S shown in footnote b obtains. Here, the rate-controlling step is the rate of desorption as given by the average desorption time t_d or the reciprocal of the first order rate constant for this process.

For both types of stationary phase, the mass transfer term increases linearly with increases in velocity of the mobile phase be-cause at higher rates, less time is allowed for mass transfer to occur thus giving lower efficiencies. Note also that for liquid stationary phases, thin films are desirable (d_f^2 small) since the distance the diffusing solute must travel to get back to the interface for desorption is minimized under this circumstance.

The constant q is a shape factor that depends upon the way the immobilized liquid is dispersed. For liquids uniformly distributed on solid particles, $q = 2/3$. For ion-exchange particles, which may be regarded as uniform liquid spheres supported by a gel matrix, $q = 2/15$. For a liquid held in the rod-shaped fibers of a paper, $q = 1/2$.

Mass Transfer in the Mobile Phase ($C_M u$). The mass transfer processes that occur in the mobile phase are sufficiently complex to defy complete analysis, at least to date. On the other hand, a good qualitative understanding of the variables affecting zone broadening from this cause exists, and this understanding has led to vast improvements in all types of chromatographic columns. In Table 25–2, it is seen that the mass transfer coefficient is a function of the square of the particle diameter of the packing, and of the solvent velocity. Note that the contributions of mobile-phase mass transfer to plate height is the product of the coefficient of mass transfer C_M (which is a function of solvent velocity) and the velocity of the solvent. Thus, the net contribution of $C_M u$ to plate height is not linear (see Figure 25–7) but bears a complex dependency on solvent velocity.

Zone broadening in the mobile phase arises in part from the multitude of pathways by which a molecule (or ion) can find its way through a packed column. As shown in Figure 25–6, the length of these pathways may differ significantly; thus, the residence time in the column for molecules of the same species is also variable. Solute molecules then reach the end of the column over a time interval, which leads to a broadened band. This effect, which is sometimes

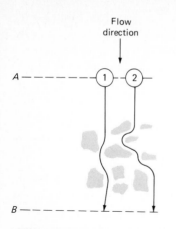

Flow direction

FIGURE 25-6 Typical pathways of two solute molecules during elution. Note that distance traveled by molecule 2 is greater than that traveled by molecule 1. Thus, molecule 2 would arrive at *B* later than molecule 1.

called *eddy diffusion*, would be independent of solvent velocity if it were not partially offset by ordinary diffusion, which results in molecules being transferred from a stream following one pathway to that following another. If the velocity of flow is very slow, a large number of these transfers will occur so that each molecule in its movement down the column samples numerous flow paths, spending a brief time in each. As a consequence, the rate at which each molecule moves down the column tends to approach that of the average. Thus, at low mobile-phase velocities, the molecules are not significantly dispersed by the multiple-path nature of the packing. At moderate or fast velocities, however, sufficient time is not available for diffusion averaging to occur, and band broadening due to the different path lengths is observed. At sufficiently high velocities, the effect of eddy diffusion becomes independent of flow velocity.

Superimposed upon the eddy diffusion effect is one that arises from stagnant pools of the mobile phase, which are retained in the stationary phase. Thus, when a solid serves as the stationary phase, its pores are filled with static volumes of mobile phase. Solute molecules must then diffuse through these stagnant pools before transfer can occur between the *moving* mobile phase and the solid surface. This situation applies not only to solid stationary phases but also to liquid stationary phases immobilized on porous solids because the immobilized liquid does not usually fully fill the pores.

The presence of stagnant pools of mobile phase slows the exchange process and results in a contribution to the plate height that is directly proportional to the mobile phase velocity and inversely proportional to the diffusion coefficient for the solute in the mobile phase. Increases in diameter of the particle d_p also have a large effect because of the increase of internal volume that accompanies increases in particle size.

As shown in Table 25–1, the various processes just described result in the mobile phase coefficient C_M being dependent upon some complex function of the square of the diameter of the particles in a column packing. The coefficient C_M is also inversely proportional to the diffusion coefficient of the solute in the mobile phase.

Effect of Mobile Phase Velocity on Plate Height. Figure 25–7 shows the variation of the three terms in Equation 25–8 as a function of mobile phase velocity. The top curve is the summation of these various effects. Note that an optimum velocity exists at which the plate height is a minimum and the separation efficiency is at a maximum.

Effects of Other Variables on Plate Height. From the relationship in Table 25–2, it is apparent that an important controllable variable that affects the efficiency of columns is the diameter of particles making up the packing. This effect is demonstrated by the data shown in Figures 26–1 and 27–2. Variables that increase the rate of diffusion of the analyte in the mobile phase (that is, D_M in the expression for C_M) also lead to reduced plate heights and less zone broadening. Thus, low viscosity mobile phases and elevated tem-

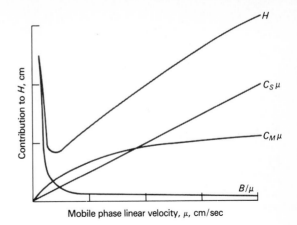

FIGURE 25-7 Contribution of various mass transfer coefficients to plate height H of a column. $C_S u$ arises from rate of mass transfer to and from the stationary phase; $C_M u$ comes from a limitation in the rate of mass transfer in the mobile phase; and B/u is associated with longitudinal diffusion.

peratures generally give improved column efficiencies. Finally, when liquid stationary phases are employed, the thickness of the liquid film should be minimized since C_S in Equation 25–8 is proportional to the square of this variable (see Table 25–2).

25B-3   THERMODYNAMIC VARIABLES AFFECTING BAND SEPARATION

The previous section dealt with kinetic variables that influence band broadening and hence separation efficiency. This section treats the second set of variables that determines whether or not a clean separation of two solutes is possible, namely equilibrium or thermodynamic variables.

Retention Times. For the chromatogram shown in Figure 25–5, zero on the time axis corresponds to the instant the sample is injected onto the column packing and elution is started. The peak at t_M is for a species that is *not* retained by the column packing; its rate of motion will be the same as the average rate for the molecules of the mobile phase. The *retention time, t_R,* for the solute responsible for the second peak is the time required for that peak to reach the detector at the end of a column.

The average linear rate of migration $\bar{v}$ of the solute is given by

$$\bar{v} = \frac{L}{t_R} \tag{25–9}$$

Similarly, the average linear rate of movement u of molecules of the mobile phase will be

$$u = \frac{L}{t_M} \tag{25–10}$$

Relationship Between Retention Time and Partition Coefficient. It is of interest to relate the retention time of a solute to its partition coefficient between the stationary and mobile phases (Equation 25–1). As we have mentioned earlier, a solute migrates only when it is in the mobile phase. Thus, the rate of migration can be expressed as a fraction of the velocity of the mobile phase. That is,

$$\bar{v} = u \left(\begin{array}{c} \text{fraction of time the solute} \\ \text{spends in the mobile phase} \end{array} \right)$$

This fraction, however, equals the average number of moles of the solute in the mobile phase at any instant compared with the total number of moles in the column. That is,

$$\bar{v} = u \times \frac{\text{no. moles solute in mobile phase}}{\text{total number moles solute}}$$

or

$$\bar{v} = u \frac{C_M V_M}{C_M V_M + C_S V_S} = u \left(\frac{1}{1 + C_S V_S / C_M V_M} \right)$$

where C_M and C_S are the molar concentrations of the solute in the mobile and stationary phases, respectively; similarly, V_M and V_S are the total volumes of the two phases in the column.

Substitution of Equation 25–1 into this expression gives

$$\bar{v} = u \left(\frac{1}{1 + K V_S / V_M} \right) \qquad (25\text{--}11)$$

The Capacity Factor.   The *capacity factor* k' is an important constant that is related to the migration rate of the solute. It is defined as

$$k' = \frac{K V_S}{V_M} \qquad (25\text{--}12)$$

Substitution of Equation 25–12 into 25–11 yields

$$\bar{v} = u \left(\frac{1}{1 + k'} \right) \qquad (25\text{--}13)$$

In order to show how k' can be derived from a chromatogram, we substitute Equations 25–9 and 25–10 into expression 25–13, which gives

$$\frac{L}{t_R} = \frac{L}{t_M} \left(\frac{1}{1 + k'} \right) \qquad (25\text{--}14)$$

This equation rearranges to

$$k' = \frac{t_R - t_M}{t_M} = \frac{t'_R}{t_M} \qquad (25\text{--}15)$$

where t'_R, the *adjusted retention time*, is equal to $(t_R - t_M)$. As shown in Figure 25–5, t_R and t_M are readily obtained from a chromatogram.

The capacity factor is widely used to characterize column performance.

The Selectivity Factor.   The ability of a column to resolve two solutes is of prime interest in chromatography. Clearly, this property is related to the relative magnitude of the partition coefficients for the two species. The *selectivity factor*, α, of a column for two solutes is defined as

$$\alpha = \frac{K_B}{K_A} \qquad (25\text{--}16)$$

where K_B is the partition coefficient for the more strongly retained solute B and K_A is the constant for the less strongly held or more rapidly moving species A. By this definition, α must always be greater than unity.

For a given column, substitution of Equation 25–12 into 25–16 and rearrangement provides a relationship between the selectivity factor and the capacity factor. That is,

$$\alpha = \frac{k'_B}{k'_A} \qquad (25\text{--}17)$$

where k'_B and k'_A are the capacity factors for B and A, respectively. Substitution of Equation 25–15 gives an expression that permits the determination of α from an experimental chromatogram. That is,

$$\alpha = \frac{(t_R)_B - t_M}{(t_R)_A - t_M} \qquad (25\text{--}18)$$

Column Resolution.   Figure 25–8 shows chromatograms for species A and B on three columns having different resolving powers. The *resolution*, R_S, for each of these columns is defined as

$$R_S = \frac{\Delta Z}{W_A/2 + W_B/2} = \frac{2\Delta Z}{W_A + W_B}$$

$$= \frac{2[(t_R)_B - (t_R)_A]}{W_A + W_B} \qquad (25\text{--}19)$$

All of the terms in this equation are defined in Figure 25–8.

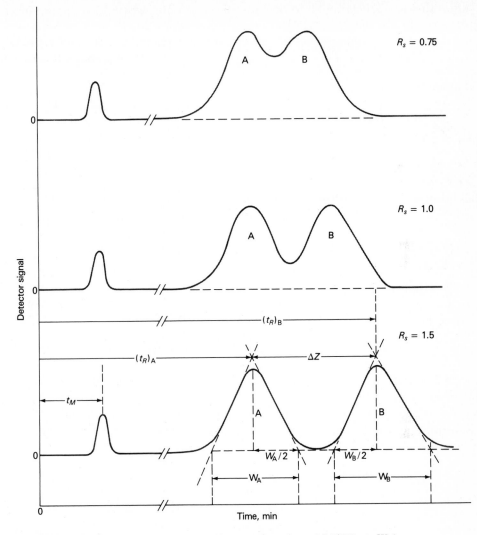

FIGURE 25-8   Separations at three resolutions. Here, $R_S = 2\Delta Z/(W_A + W_B)$.

It is evident from Figure 25–8 that a resolution of 1.5 gives an essentially complete separation of A and B, whereas a resolution of 0.75 does not. At a resolution of 1.0, zone B contains about 4% A and conversely; at a resolution of 1.5, the overlap is about 0.3%. The resolution for a given stationary phase can be improved by lengthening the column, thus increasing the number of plates. An adverse consequence of the added

plates, however, is an increase in the time required for the resolution.

Relationship Between Resolution and Column Properties.   It is useful to develop a mathematic relationship between the resolution of a column and the capacity factors k'_A and k'_B for two solutes, the selectivity factor α, and the number of plates N making up the column. To this end, we will assume that we are dealing with two solutes A and B having retention times that are close enough to one another that we can assume

$$W_A \cong W_B \cong W$$

Equation 25–19 then takes the form

$$R_S = \frac{(t_R)_B - (t_R)_A}{W}$$

Equation 25–6 permits the expression of W in terms of $(t_R)_B$ and N, which can then be substituted into the foregoing equation to give

$$R_S = \frac{(t_R)_B - (t_R)_A}{(t_R)_B} \times \frac{\sqrt{N}}{4}$$

Substitution of Equation 25–15 and rearrangement leads to an expression for R_S in terms of the capacity factors for A and B. That is,

$$R_S = \frac{k'_B - k'_A}{1 + k'_B} \times \frac{\sqrt{N}}{4}$$

Let us eliminate k'_A from this expression by substituting Equation 25–17 and rearranging. Thus,

$$R_S = \frac{\sqrt{N}}{4}\left(\frac{\alpha - 1}{\alpha}\right)\left(\frac{k'_B}{1 + k'_B}\right) \qquad (25–20)$$

Often it is desirable to calculate the number of theoretical plates required to achieve a desired resolution. An expression for this quantity is obtained by rearranging Equation 25–20 to give

$$N = 16R_S^2\left(\frac{\alpha}{\alpha - 1}\right)^2\left(\frac{1 + k'_B}{k'_B}\right)^2 \qquad (25–21)$$

Simplified forms of Equations 25–20 and 25–21 are sometimes encountered where these equations are applied to a pair of solutes whose partition coefficients are similar enough to make their separation difficult.

Thus, when $K_A \cong K_B$, it follows from Equation 25–12 that $k'_A \cong k'_B = k'$ and from Equation 25–16, $\alpha \to 1$. With these approximations, Equations 25–20 and 25–21 reduce to

$$R_S = \frac{\sqrt{N}}{4}(\alpha - 1)\left(\frac{k'}{1 + k'}\right) \qquad (25–22)$$

$$N = 16R_S^2\left(\frac{1}{\alpha - 1}\right)^2\left(\frac{1 + k'}{k'}\right)^2 \qquad (25–23)$$

where k' is the average of k'_A and k'_B.

Relationship Between Resolution and Elution Time. Before considering in detail the significance of the four equations just derived, it is worthwhile developing an equation for a related performance characteristic for a column, namely the time required to complete the separation of solutes A and B. Clearly, what is desired in chromatography is the highest possible resolution in the shortest possible elapsed time. Unfortunately, these two properties cannot both be maximized under the same conditions, and a compromise must always be struck.

The time for completion of a separation is determined by the velocity $\bar{v}_B$ of the slower moving solute, as given in Equation 25–9. That is,

$$\bar{v}_B = \frac{L}{(t_R)_B}$$

Combining this expression with 25–13 and 25–2 yields after rearranging

$$(t_R)_B = \frac{NH(1 + k'_B)}{u}$$

where $(t_R)_B$ is the time required to bring the peak for B to the end of the column when the velocity of the mobile phase is u. When this equation is combined with Equation 25–21 and rearranged, we find that

$$(t_R)_B = \frac{16R_S^2 H}{u}\left(\frac{\alpha}{\alpha - 1}\right)^2\frac{(1 + k'_B)^3}{(k'_B)^2} \qquad (25–24)$$

25B–4 OPTIMIZATION OF COLUMN PERFORMANCE

Equations 25–20 and 25–24 are significant because they serve as guides to the choice of conditions that are likely to allow the user of chromatography to achieve the sometimes elusive goal of a clean separation in a minimum of time. An examination of these equations reveals that each is made up of three parts. The first, which is related to the kinetic effects that lead to band broadening, consists of $\sqrt{N}$ or H/u. The second and third terms are related to the thermodynamics of the constituents being separated—that is, to the relative magnitude of their distribution coefficients and the volumes of the mobile and stationary phases. The second term in Equations 25–20 and 25–24, which is the quotient containing α, is a selectivity term that depends solely upon the properties of the two solutes. The third term, which is the quotient containing k_B', depends upon the properties of both the solute and the column.

In seeking optimum conditions for achieving a desired separation, it must be kept in mind that the fundamental parameters, α, k', and N (or H) can be adjusted more or less independently. Thus, α and k' can be varied most easily by varying temperature or the composition of the mobile phase. Less conveniently, a different type of column packing can be employed. As we have seen, it is possible to change N by changing the length of the column, and H by altering the flow rate of the mobile phase, the particle size of the packing, the viscosity of the mobile phase (and thus D_M or D_S), and the thickness of the film of adsorbed liquid constituting the stationary phase (see Table 25–2).

Variation in N. An obvious way to improve resolution is to increase the number of plates in the column (Equation 25–20). As shown by Example 25–1, which follows, this expedient is usually expensive in terms of time required to complete the separation

unless the increase in N is realized by a reduction in H rather than by lengthening the column.

EXAMPLE 25–1. Substances A and B were found to have retention times of 16.40 and 17.63 min, respectively, on a 30.0-cm column. An unretained species passed through the column in 1.30 min. The peak widths (at base) for A and B were 1.11 and 1.21 min, respectively. Calculate: (a) the column resolution; (b) the average number of plates in the column; (c) the plate height; (d) the length of column required to achieve a resolution of 1.5; (e) the time required to elute substance B on the longer column; and (f) the plate height required for a resolution of 1.5 on the original 30-cm column and in the original time. Substituting into Equation 25–19 gives

(a) $R_S = 2(17.63 - 16.40)/(1.11 + 1.21)$
$= 1.06$

(b) Equation 25–6 permits computation of N. Thus,

$$N = 16\left(\frac{16.40}{1.11}\right)^2 = 3493$$

and

$$N = 16\left(\frac{17.63}{1.21}\right)^2 = 3397$$

$$N_{av} = (3493 + 3397)/2 = 3445$$
$$= 3.4 \times 10^3$$

(c) $H = L/N = 30.0/3445$
$= 8.7 \times 10^{-3}$ cm

(d) k' and α do not change with increasing N and L. Thus, substituting N_1 and N_2 into Equation 25–21 and dividing one of the resulting equations by the other yields

$$\frac{(R_S)_1}{(R_S)_2} = \frac{\sqrt{N_1}}{\sqrt{N_2}}$$

where subscripts 1 and 2 refer to the original and the longer columns, respectively. Substituting the appropriate values for N_1, $(R_S)_1$, and $(R_S)_2$ gives

$$\frac{1.06}{1.5} = \frac{\sqrt{3445}}{\sqrt{N_2}}$$

$$N_2 = 3445\left(\frac{1.5}{1.06}\right)^2 = 6.9 \times 10^3$$

Substitution into Equation 25–2 yields

$$L = N \times H$$
$$= 6.9 \times 10^3 \times 8.7 \times 10^{-3} = 60 \text{ cm}$$

(e) Substituting $(R_S)_1$ and $(R_S)_2$ into Equation 25–24 and dividing yields

$$\frac{(t_R)_1}{(t_R)_2} = \frac{(R_S)_1^2}{(R_S)_2^2} = \frac{17.63}{(t_R)_2} = \frac{(1.06)^2}{(1.5)^2}$$

and

$$(t_R)_2 = 35 \text{ min}$$

Thus, to obtain the improved resolution of 1.5 requires that the time of separation be doubled.

(f) Substituting H_1 and H_2 into Equation 25–24 and dividing one of the resulting equations by the second gives

$$\frac{(t_R)_B}{(t_R)_B} = \frac{(R_S)_1^2}{(R_S)_2^2} \times \frac{H_1}{H_2}$$

where the subscripts 1 and 2 refer to the original and new plate heights, respectively. Rearranging gives

$$H_2 = H_1 \frac{(R_S)_1^2}{(R_S)_2^2} = 8.7 \times 10^{-3} \frac{(1.06)^2}{(1.5)^2}$$
$$= 4.3 \times 10^{-3} \text{ cm}$$

Thus to achieve a resolution of 1.5 in 17.63 min on a 30-cm column, the plate height would need to be halved.

Variation in H. In Example 25–1e, it is shown that a significant improvement in resolution can be achieved at no cost in time if the plate height can be reduced. Table 25–2 reveals the variables that are available for accomplishing this end. Note that decreases in particle size of the packing lead to marked improvements in H. For liquid mobile phases, where B/u is ordinarily negligible, reduced plate heights can also be achieved by reducing the solvent viscosity thus increasing the diffusion coefficient in the mobile phase. One way of accomplishing this reduction is by increasing the temperature.

Variation in k_B'. Often, a separation can be improved significantly by manipulation of the capacity factor k_B'. Increases in k_B' generally enhance resolution (but at the expense of elution time). To determine the optimum range of values for k_B', it is convenient to write Equation 25–20 in the form

$$R_S = Q \frac{k_B'}{1 + k_B'}$$

and Equation 25–24 as

$$(t_R)_B = Q' \frac{(1 + k_B')^3}{(k_B')^2}$$

where Q and Q' contain the rest of the terms in the two equations. Figure 25–9 is a plot of R_S/Q and $(t_R)_B/Q'$ as a function of k_B', assuming that Q and Q' remain approximately constant. It is clear that values of k_B' greater than about 10 are to be avoided because they provide little increase in resolution but markedly increase the time required for separations. The minimum in the elution time curve occurs at a value of k_B' of about 2. Often, then, the optimal value of k_B', taking into account both resolution and expended time, lies in the range of 1 to 5.

Usually, the easiest way to improve resolution is by optimizing k'. For gaseous mobile phases, k' can often be improved by temperature increase. For liquid mobile phases, change in the solvent composition

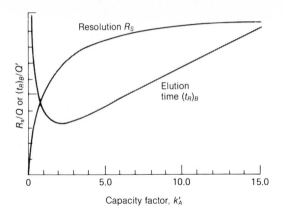

FIGURE 25–9 Effect of capacity factor k_A' on resolution R_S and elution time $(t_R)_B$. It is assumed that Q and Q' remain constant with variation in k_B'.

often permits manipulation of k' in such a way as to yield better separations. An example of the dramatic effect that relatively simple solvent changes can bring about is demonstrated in Figure 25–10. Here, modest variations in the methanol/water ratios convert unsatisfactory chromatograms (**a** and **b**) to ones with well-separated peaks for each

component (**c** and **d**). For most purposes, the chromatogram shown in (**c**) would be the best since it shows adequate resolution in a minimum time.

Variation in α. When α approaches unity, optimizing k' and increasing N are not sufficient to give a satisfactory separation of two solutes in a reasonable time. Under this circumstance, a means must be sought to increase α while still maintaining k' in the optimum range of 1 to 10. Several options are available; in decreasing order of their desirability as determined by promise and convenience, the options include: (1) changing the composition of the mobile phase, (2) changing the pH of the mobile phase, (3) changing the column temperature, (4) changing the composition of the stationary phase, and (5) using special chemical effects.

An example of the use of option (1) has been reported for the separation of anisole ($C_6H_5OCH_3$) and benzene.[8] With a mobile

[8] L. R. Snyder and J. J. Kirkland, *Introduction to Modern Liquid Chromatography,* 2d ed., p. 75, Wiley: New York, 1979.

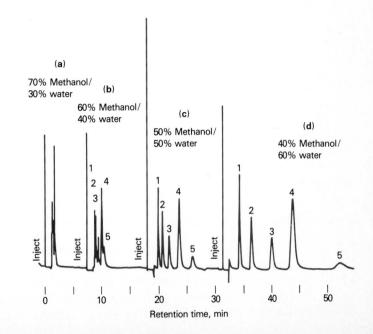

FIGURE 25–10 Effect of solvent variation on chromatograms. Analytes: (1) 9,10-anthraquinone; (2) 2-methyl-9,10-anthraquinone; (3) 2-ethyl-9,10-anthraquinone; (4) 1,4-dimethyl-9,10-anthraquinone; (5) 2-*t*-butyl-9,10-anthraquinone. (Courtesy of DuPont Instrument Systems, Wilmington, DE.)

phase that was a 50% mixture of water and methanol, k' for the two solutes was 4.5 and 4.7, respectively, while α was only 1.04. Substitution of an aqueous mobile phase containing 37% tetrahydrofuran gave k' values of 3.9 and 4.7 and an α value of 1.20. Peak overlap was significant with the first solvent system and negligible with the second.

For separations involving ionizable acids or bases, alteration in pH of the mobile phase often allows manipulation of α values without major changes in k'; enhanced separation efficiencies result.

A less convenient, but often highly effective method, of improving α while still maintaining values for k' in their optimal range, is to alter the chemical composition of the stationary phase. To take advantage of this option, most laboratories which carry out chromatographic separations frequently, will maintain several columns, which can be interchanged with a minimum of effort.

Increases in temperature usually cause increases in k' but have little effect on α values in liquid-liquid and liquid-solid chromatography. In contrast, with ion-exchange chromatography, the effect of temperature can have large enough effects to make exploration of this option worthwhile before resorting to a change in column packing.

A final method for enhancing resolution is to incorporate into the stationary phase a species that complexes or otherwise interacts with one or more components of the sample. A well-known example of the use of this option arises where an adsorbent impregnated with a silver salt improves the separation of olefins as a consequence of the formation of complexes between silver ions and organic unsaturated compounds.

25B-5 *THE GENERAL ELUTION PROBLEM*

In Figure 25–11 are hypothetical chromatograms for a six-component mixture made up of three pairs of components having widely differing distribution coefficients and thus capacity factors. In chromatogram

(**a**) conditions have been adjusted so that capacity factors for components 1 and 2 (k'_1 and k'_2) are in the optimal range of 2 to 5. The corresponding factors for the other components are, however, far larger than the optimum. Thus, the peaks for 5 and 6 appear only after an inordinate time; furthermore, these peaks are so broadened that they may be difficult to identify unambiguously.

As shown in chromatogram (**b**), changing conditions to optimize the separation of components 5 and 6, bunches the peaks for the first four components to the point where their resolution is unsatisfactory. Here, however, the total elution time is ideal.

A yet third set of conditions, in which k' values for components 3 and 4 are optimal, results in chromatogram (**c**). Again, separation of the other two pairs is not entirely satisfactory.

The phenomenon illustrated in Figure 25–11 is encountered often enough to be given a name—the *general elution problem*. A common solution to this problem is to change conditions that determine the values of k' as the separation proceeds. These changes may be performed in a stepwise manner or continuously. Thus, for the mixture shown in Figure 25–11, conditions at the outset could be those producing chromatogram (**a**). Immediately after elution of components 1 and 2, however, conditions could be changed to those that were optimal for separating 3 and 4 (as in chromatogram **c**). With the appearance of peaks for these components, the elution could be completed under conditions used for producing chromatogram (**b**). Often such a procedure leads to satisfactory peaks for all of the components in a mixture in minimal time.

For liquid chromatography, variations in k' are brought about by variations in the composition of the mobile phase during elution (*gradient elution* or *solvent programming*). For gas chromatography, temperature increases (*temperature programming*) serve to achieve optimal conditions for separations.

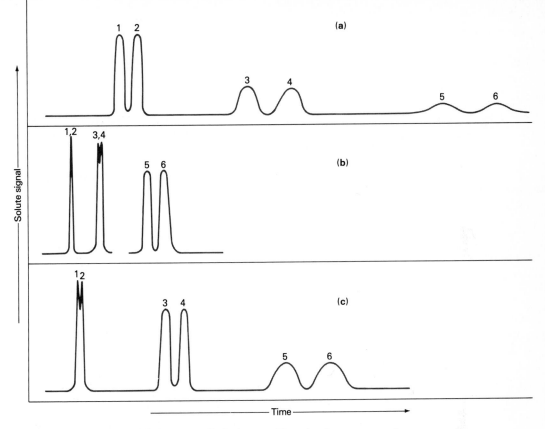

FIGURE 25-11   Illustration of the general elution problem in chromatography.

25C   Summary of Important Relationships for Chromatography

The number of quantities, terms, and relationships employed in chromatography is large and often confusing. Table 25–3 serves to summarize the most important definitions and equations that will be used in the next three chapters.

25D   Qualitative and Quantitative Analysis by Chromatography

Chromatography has grown to be the premiere method for separating closely related chemical species. In addition, it can be em-

ployed for qualitative identification and quantitative determination of separated species. This section considers some of the general characteristics of chromatography as a tool for completion of an analysis.

25D-1   QUALITATIVE ANALYSIS

A chromatogram provides only a single piece of qualitative information about each species in a sample, namely, its retention time or its position on the stationary phase after a certain elution period. Additional data can, of course, be derived from chromatograms involving different mobile and stationary phases and various elution temperatures. Still, the number of data points for a species obtainable by chromatography is small compared with the number provided by a single IR, NMR, or mass spectrum. Furthermore, spectral abscissa data can be determined with a much higher precision than can their chromatographic counterpart (t_R).

The foregoing should not be interpreted to mean that chromatography lacks important qualitative applications. Indeed, it is a

Table 25-3
Summary of the Most Important Chromatographic Quantities and Relationships

Experimental Quantities

Name	Symbol	Determined from
Migration time, nonretained species	t_M	Chromatogram (see Figure 25–8)
Retention times, species A and B	$(t_R)_A, (t_R)_B$	Chromatogram (see Figure 25–8)
Adjusted retention time, species A	$(t'_R)_A$	$(t'_R)_A = (t_R)_A - t_M$
Peak widths, species A and B	W_A, W_B	Chromatogram (see Figure 25–8)
Length of column packing	L	Direct measurement
Flow rate	F	Direct measurement
Volume of stationary phase	V_S	Packing preparation data
Concentration of solute in mobile and stationary phases	C_M, C_S	Analysis and preparation data

Derived Quantities

Name	Calculation of Derived Quantities	Relationship to Other Quantities
Linear mobile phase velocity	$u = L/t_M$	
Volume of mobile phase	$V_M = t_M F$	
Capacity factor	$k' = (t_R - t_M)/t_M$	$k' = \dfrac{KV_S}{V_M}$
Partition coefficient	$K = \dfrac{k' V_M}{V_S}$	$K = \dfrac{C_S}{C_M}$
Selectivity factor	$\alpha = \dfrac{(t_R)_B - t_M}{(t_R)_A - t_M}$	$\alpha = \dfrac{k'_B}{k'_A} = \dfrac{K_B}{K_A}$
Resolution	$R_S = \dfrac{2[(t_R)_B - (t_R)_A]}{W_A + W_B}$	$R_S = \dfrac{\sqrt{N}}{4}\left(\dfrac{\alpha - 1}{\alpha}\right)\left(\dfrac{k'_B}{1 + k'_B}\right)$
Number of plates	$N = 16\left(\dfrac{t_R}{W}\right)^2$	$N = 16 R_S^2 \left(\dfrac{\alpha}{\alpha - 1}\right)^2 \left(\dfrac{1 + k'_B}{k'_B}\right)^2$
Plate height	$H = L/N$	
Retention time	$(t_R)_B = \dfrac{16 R_S^2 H}{u}\left(\dfrac{\alpha}{\alpha - 1}\right)^2 \dfrac{(1 + k'_B)^3}{(k'_B)^2}$	

widely used tool for recognizing the presence or absence of components of mixtures containing a limited number of possible species whose identities are known. For example, 30 or more amino acids in a protein hydrolysate can be detected with a relatively high degree of certainty by means of a chromatogram. Even here, however, confirmation of identity requires spectral or chemical investigation of the isolated components. Note, however, that positive spectroscopic identification would ordinarily be impossible on as complex a sample as the foregoing without a preliminary chromatographic separation. Thus, chromatography is often a vital precursor to qualitative spectroscopic analyses.

It is important to note that while chromatograms may not lead to positive identification of species present in a sample, they often provide sure evidence of the *absence* of certain compounds. Thus, if the sample does not produce a peak at the same retention time as a standard run under identical conditions, it can be assumed that the compound in question is absent (or is present at a concentration level below the detection limit of the procedure).

25D-2 QUANTITATIVE ANALYSIS

Chromatography owes its precipitous growth during the past two decades in part to its speed, simplicity, relatively low cost, and wide applicability as a separating tool. It is doubtful, however, if its use would have become as widespread had it not been for the fact that it can also provide useful quantitative information about the separated species. It is important, therefore, to discuss some of the quantitative aspects that apply to all types of chromatography.

Quantitative column chromatography is based upon a comparison of either the height or the area of the analyte peak with that of one or more standards. For planar or flat-bed chromatography, the area covered by the separated species serves as the analytical parameter. If conditions are properly controlled, these parameters vary linearly with concentration.

Analyses Based on Peak Height. The height of a chromatographic peak is obtained by connecting the base lines on either side of the peak by a straight line and measuring the perpendicular distance from this line to the peak. This measurement can ordinarily be made with reasonably high precision and yields accurate results, provided variations in column conditions do not alter the peak widths during the period required to obtain chromatograms for sample and standards. The variables that must be controlled closely are column temperature, eluent flow rate, and rate of sample injection. In addition, care must be taken to avoid overloading the column. The effect of sample injection rate is particularly critical for the early peaks of a chromatogram. Relative errors of 5 to 10% due to this cause are not unusual with syringe injection.

Analyses Based on Peak Areas. Peak areas are independent of broadening effects due to the variables mentioned in the previous paragraph. From this standpoint, therefore, areas are a more satisfactory analytical parameter than peak heights. On the other hand, peak heights are more easily measured and, for narrow peaks, more accurately determined.

Most modern chromatographic instruments are equipped with digital electronic integrators which permit precise estimation of peak areas. Older instruments were often equipped with a mechanical integrator called a ball and disc integrator. If such equipment is not available, a manual estimate must be made. A simple method, which works well for symmetric peaks of reasonable widths, is to multiply the height of the peak by its width at one-half the peak height. Other methods involve the use of a planimeter or cutting out the peak and determining its weight relative to the weight of a known area of recorder paper. McNair and Bonelli measured the precision of these various techniques on chromatograms for ten replicate samples.[9] They reported the following relative standard deviations: electronic integration, 0.44%; ball and disc integration, 1.3%; weight of paper, 1.7%; height times width at one-half height, 2.6%; and planimeter, 4.1%.

Calibration with Standards. The most straightforward method for quantitative chromatographic analyses involves the preparation of a series of standard solutions that approximate the composition of the unknown. Chromatograms for the standards are then obtained and peak heights or areas are plotted as a function of concentration. A plot of the data should yield a straight line passing through the origin; analyses are based upon this plot. Frequent restandardization is necessary for highest accuracy.

The most important source of error in analyses by the method just described is usually the uncertainty in the volume of sample; occasionally the rate of injection is also a factor. Ordinarily, samples are small (~1

μL), and the uncertainties associated with injection of a reproducible volume of this size with a microsyringe may amount to several percent relative. The situation is exacerbated in gas-liquid chromatography, where the sample must be injected into a heated sample port; here, evaporation from the needle tip may lead to large variations in the volume injected.

Errors in sample volume can be reduced to perhaps 1 to 2% relative by means of a rotary sample valve such as that shown in Figure 25–12. Here, the sample loop ACB in (a) is filled with sample; rotation of the valve by 45 deg then introduces a reproducible volume of sample (the volume originally contained in ACB) into the mobile-phase stream.

The Internal Standard Method. The highest precision for quantitative chromatography is obtained by use of internal standards because the uncertainties introduced by sample injection are avoided. In this procedure, a carefully measured quantity of an internal standard substance is introduced into each standard and sample, and the ratio of analyte to internal standard peak areas (or heights) serves as the analytical parameter. For this method to be successful, it is necessary that the internal standard peak be well

[9] H. M. McNair and E. J. Bonelli, *Basic Gas Chromatography*, p. 158, Varian Aerograph: Walnut Creek, CA, 1968.

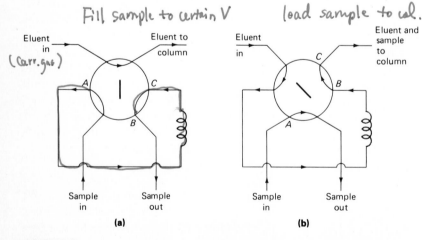

FIGURE 25–12   A rotary sample value: (**a**) valve position for filling sample loop *ACB* and (**b**) for introduction of sample into column.

separated from the peaks of all other components of the sample ($R_S > 1.25$); the standard peak should, on the other hand, appear close to the analyte peak. With a suitable internal standard, precisions of 0.5 to 1% relative are reported.

The Area Normalization Method. Another approach that avoids the uncertainties associated with sample injection is the area normalization method. Complete elution of all components of the sample is required. In the normalization method, the areas of all eluted peaks are computed; after correcting these areas for differences in the detector response to different compound types, the concentration of the analyte is found from the ratio of its area to the total area of all peaks. The following example illustrates the procedure.

EXAMPLE 25-2. The following area data were obtained from a chromatogram of a mixture of butyl alcohols (the detector sensitivity corrections were obtained in separate experiments with known amounts of pure alcohols).

Alcohol	Peak Area, cm^2	Detector Response Factor	Corrected Areas, cm^2
n-butyl	2.74	0.603	1.652
i-butyl	7.61	0.530	4.033
s-butyl	3.19	0.667	2.128
t-butyl	1.66	0.681	1.130
			8.943

Each entry in column 4 is the product of the data in columns 2 and 3. To normalize,

% n-butyl = 1.652 × 100/8.943 = 18.5

% i-butyl = 4.033 × 100/8.943 = 45.1

% s-butyl = 2.128 × 100/8.943 = 23.8

% t-butyl = 1.130 × 100/8.943 = 12.6

100.0

25E Computerized Chromatography

Most modern instruments for chromatography are equipped with microprocessors for controlling such operating parameters as column and detector temperature, eluent flow rate, sample injection time, and sample temperature. As a consequence, chromatograms can be obtained with little or no human control. The most sophisticated instruments are completely automatic. Here, several dozen samples are contained in a sample turntable. After each analysis, the column is returned automatically to its initial condition, the sample turntable is rotated, a new sample is withdrawn and injected into the column, and the chromatogram is obtained and stored in memory. The data are presented as a chromatogram and in tabular form with retention times and relative peak areas printed out. Some instruments also print out the names of possible compounds producing each peak. Figure 25–13 shows a typical printout from an automatic gas-chromatographic instrument. At the top is the chromatogram with the retention times printed to the right of each peak. Below this are found the actual and expected retention times and the areas for each peak. The calculated amounts of each component are based upon calibration data relative to an internal standard. The peak at 5.94 min corresponds to the standard. At the bottom of the printout are the experimental conditions for the run.

Several instrument manufacturers offer small, modular computers that are specifically designed for processing the output signals from chromatographic instruments. Figure 25–14 is a schematic diagram showing the components of one such unit, which is capable of handling data from four chromatographic columns. The analog signal, after digitization, passes to an operating program stored in ROM (read only memory), which constantly monitors and oper-

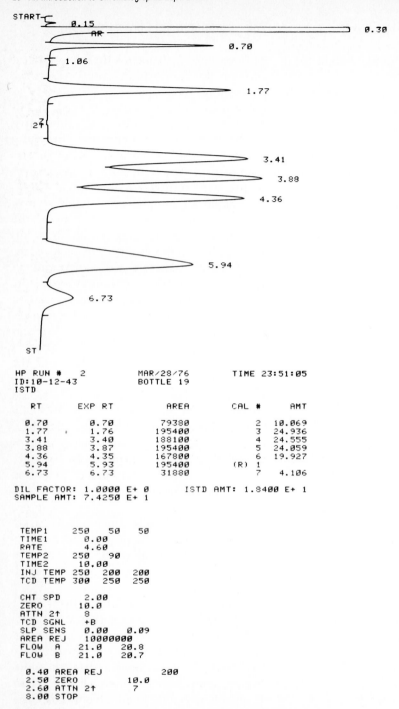

START
0.15
AR
0.30
0.70
1.06
1.77
2↑
3.41
3.88
4.36
5.94
6.73
ST

HP RUN #   2   MAR/28/76   TIME 23:51:05
ID:10-12-43   BOTTLE 19
ISTD

   RT   EXP RT   AREA   CAL #   AMT

  0.70   0.70   79380   2   10.069
  1.77   1.76   195400   3   24.936
  3.41   3.40   188100   4   24.555
  3.88   3.87   195400   5   24.059
  4.36   4.35   167800   6   19.927
  5.94   5.93   195400   (R) 1
  6.73   6.73   31880   7   4.106

DIL FACTOR: 1.0000 E+ 0   ISTD AMT: 1.8400 E+ 1
SAMPLE AMT: 7.4250 E+ 1

TEMP1   250   50   50
TIME1   0.00
RATE   4.60
TEMP2   250   90
TIME2   10.00
INJ TEMP 250   200   200
TCD TEMP 300   250   250

CHT SPD   2.00
ZERO   10.0
ATTN 2↑   8
TCD SGNL   +B
SLP SENS   0.00   0.09
AREA REJ 10000000
FLOW A   21.0   20.8
FLOW B   21.0   20.7

  0.40 AREA REJ   200
  2.50 ZERO   10.0
  2.60 ATTN 2↑   7
  8.00 STOP

FIGURE 25-13 Data output from a modern automated chromatographic instrument. (Courtesy of Hewlett-Packard Company, Palo Alto, CA.)

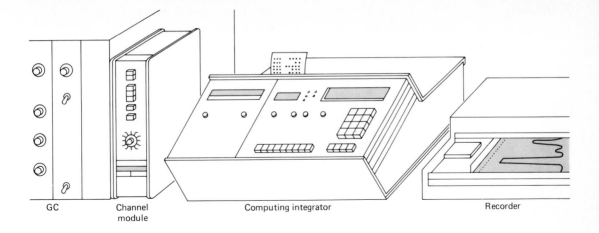

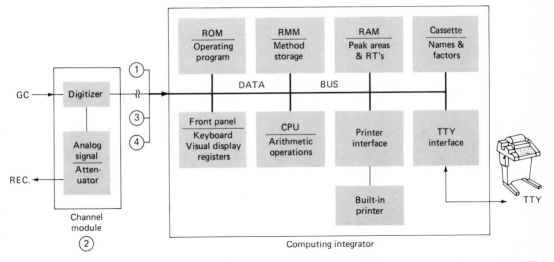

FIGURE 25-14 A computing integrator for chromatography. (Reprinted with permission from: J. M. Gill, *Chromatography*, Series 1, vol. 1, p. 188, International Scientific Communications, Inc.: Fairfield, CT. Copyright 1974 International Scientific Communications, Inc.)

ates on the data. Here, peaks are detected and their areas and retention times determined. The arithmetic operations are performed by the CPU. These processed data are then transferred to the RAM (random access memory) module for storage until the end of the analysis. The RMM (read mostly memory) module contains programs for optimizing the data treatment. Instructions as to the choice of programs are entered by means of the front panel keyboard. Each of the three memories consists of a single chip. The contents of the ROM are fixed and cannot be changed. The contents of the RMM are also protected but can be changed under certain circumstances. Data can be readily entered or removed from the random access memory.

PROBLEMS

25–1. The following data apply to a column for liquid chromatography:

length of packing	24.7 cm
flow rate	0.313 mL/min
V_M	1.37 mL
V_S	0.164 mL

A chromatogram of a mixture of species A, B, C, and D provided the following data:

	Retention Time, min	Width of Peak Base (W), min
nonretained	3.1	—
A	5.4	0.41
B	13.3	1.07
C	14.1	1.16
D	21.6	1.72

Calculate
(a) the number of plates from each peak.
(b) the mean and the standard deviation for N.
(c) the plate height for the column.

25–2. From the data in Problem 25–1, calculate for A, B, C, and D
(a) the capacity factor.
(b) the partition coefficient.

25–3. From the data in Problem 25–1, for species B and C, calculate
(a) the resolution.
(b) the selectivity factor, α.
(c) the length of column necessary to give a resolution of 1.5.
(d) the time required to separate B and C with a resolution of 1.5.

25–4. From the data in Problem 25–1 for species C and D, calculate
(a) the resolution.
(b) the length of column required to give a resolution of 1.5.

25–5. The following data were obtained by gas-liquid chromatography on a 40-cm packed column:

Compound	t_R, min	$W_{1/2}$, min
air	1.9	—
methylcyclohexane	10.0	0.76
methylcyclohexene	10.9	0.82
toluene	13.4	1.06

Calculate
(a) an average number of plates from the data.
(b) the standard deviation for the average in (a).
(c) an average plate height for the column.

25–6. Referring to Problem 25–5, calculate the resolution for
(a) methylcyclohexene and methylcyclohexane.
(b) methylcyclohexene and toluene.
(c) methylcyclohexane and toluene.

25–7. If a resolution of 1.5 was desired in resolving methylcyclohexane and methylcyclohexene in Problem 25–5,
(a) how many plates would be required?
(b) how long would the column have to be if the same packing were employed?
(c) what would be the retention times for methylcyclohexene on the column in Problem 25–5b?

25–8. If V_S and V_M for the column in Problem 25–5 were 19.6 and 62.6 mL, respectively, and a nonretained air peak appeared after 1.9 min, calculate the
(a) capacity factor for each of the three compounds.
(b) partition coefficient for each of the three compounds.
(c) selectivity factor for methylcyclohexane and methylcyclohexene.
(d) selectivity for methylcyclohexene and toluene.

25–9. List variables that lead to (a) band broadening and (b) band separation.

25–10. What would be the effect on a chromatographic peak of introducing the sample at too slow a rate?

25–11. From distribution studies species M and N are known to have partition coefficients between water and hexane of 6.01 and 6.20 ($K = [M]_{H_2O}/[M]_{hex}$). The two species are to be separated by elution with hexane in a column packed with silica gel containing adsorbed water. The ratio V_S/V_M for the packing is known to be 0.422.
(a) Calculate the capacity factor for each of the solutes.
(b) Calculate the selectivity factor.
(c) How many plates will be needed to provide a resolution of 1.5?
(d) How long a column is needed if the plate height of the packing is 2.2×10^{-3} cm?
(e) If a flow rate of 7.10 cm/min is employed, what time will be required to elute the two species?

25–12. Repeat the calculations in Problem 25–11 assuming $K_M = 5.81$ and $K_N = 6.20$.

25–13. The relative peak areas obtained from a gas chromatogram of a mixture of methyl acetate, methyl propionate, and methyl n-butyrate were 17.6, 44.7, and 31.1, respectively. Calculate the percentage of each compound if the respective relative detection responses were 0.65, 0.83, and 0.92.

25–14. The relative areas for the five gas chromatographic peaks shown in Figure 26–11e are given below. Also shown are the relative responses of the detector to the five compounds. Calculate the percentage of each component in the mixture.

Compound	Peak Area, Relative	Detection Response, Relative
A	27.6	0.70
B	32.4	0.72
C	47.1	0.75
D	40.6	0.73
E	27.3	0.78

26

Gas Chromatography

In gas chromatography (GC), the sample is vaporized and injected onto the head of a chromatographic column. Elution is brought about by the flow of an inert gaseous mobile phase. In contrast to most other types of chromatography, the mobile phase does not interact with molecules of the analyte, its only function being to transport those molecules through the packing.

Gas-solid chromatography is based upon a solid stationary phase in which retention of analytes is the consequence of physical adsorption. Gas-solid chromatography has limited application owing to semipermanent retention of active or polar molecules and severe tailing of elution peaks (a consequence of the nonlinear character of adsorption isotherms; see Figure 25–1A). Thus, this technique has not found wide application except for the separation of certain low molecular weight gaseous species; it will, therefore, be discussed only briefly at the end of this chapter.

The concept of *gas-liquid chromatography* (GLC) was first enunciated in 1941 by Martin and Synge, who were also responsible for the development of liquid-liquid partition chromatography. More than a decade was to elapse, however, before the value of gas-liquid chromatography was demonstrated experimentally.[1] Three years later, in 1955, the first commercial apparatus for gas chromatography appeared on the market. Since that time, the growth in applications of this technique has been phenomenal. For example, by 1965 over 2000 publications on gas chromatography were appearing annually; by 1976, this number had grown to 3000.[2]

[1] A. J. Jones and A. J. P. Martin, *Analyst,* **1952,** 77, 915.

[2] For monographs on GLC, see: J. A. Perry, *Introduction to Analytical Gas Chromatography*, Marcel Dekker: New York, 1981; *Modern Practice of Gas Chromatography*, R. L. Grob, Ed., Wiley-Interscience: New York, 1977; J. Q. Walker, M. T. Jackson, Jr., and J. B. Maynard, *Chromatographic Systems*, 2d ed., Part II, Academic Press: New York, 1977.

26A   Principles of Gas-Liquid Chromatography

The general principles of chromatography, which were developed in Section 25B, and the mathematical relationships summarized in Section 25C are applicable to gas chromatography with only a minor modification that arises from the compressibility of gaseous mobile phases.

26A-1   RETENTION VOLUMES

To take into account the effects of pressure and temperature in gas chromatography, it is sometimes useful to use retention volumes rather than the retention times that were employed in Section 25B. The relationship between the two is

$$V_R = t_R F \qquad (26\text{--}1)$$

and

$$V_M = t_M F \qquad (26\text{--}2)$$

where F is the average volumetric flow rate within the column, V and t are retention volumes and times respectively, and the subscripts R and M refer to species that are retained and not retained on the column. The average flow rate is not directly measurable, however; instead, only the rate of gas flow as it exits the column is conveniently determined experimentally. Normally, this rate is measured by means of a soap-bubble meter, which is described on page 760. The average flow rate F is then

$$F = F_m \times \frac{T_C}{T} \times \frac{(P - P_{H_2O})}{P} \qquad (26\text{--}3)$$

where T_C is the column temperature in degrees K, T is the temperature at the meter, F_m is the measured flow rate, and P is the gas pressure at the end of the column. Usually P and T are the ambient pressure and temperature. In the soap-bubble meter, the gas becomes saturated with water. Thus, the pressure must be corrected for the vapor pressure of water, P_{H_2O}.

Both V_R and V_M depend upon the average pressure *within the column*—a quantity that lies intermediate between the inlet pressure P_i and the outlet pressure P (atmospheric pressure). *Corrected retention volumes V_R^0 and V_M^0*, which correspond to volumes at the average column pressure are obtained from the relationships

$$V_R^0 = jt_R F \text{ and } V_M^0 = jt_M F \qquad (26\text{--}4)$$

The quantity j is readily calculated from the relationship

$$j = \frac{3[(P_i/P)^2 - 1]}{2[(P_i/P)^3 - 1]} \qquad (26\text{--}5)$$

The specific retention volume is then defined as

$$\begin{aligned} V_g &= \frac{V_R^0 - V_M^0}{W} \times \frac{273}{T_C} \\ &= \frac{jF(t_R - t_M)}{W} \times \frac{273}{T_C} \end{aligned} \qquad (26\text{--}6)$$

where W is the weight of the mobile phase, a quantity determined at the time of column preparation, and T_C is the column temperature in °K.

26A-2   RELATIONSHIP BETWEEN V_g AND K

It is of interest to relate V_g to the partition coefficient K. To do so, we substitute the expression relating t_R and t_M to k' (Equation 25–15, p. 740) into 26–6, which gives

$$V_g = \frac{jF t_M k'}{W} \times \frac{273}{T_C}$$

Combining this expression with Equation 26–4 yields

$$V_g = \frac{V_M^0 k'}{W} \times \frac{273}{T_C}$$

Substituting Equation 25–12 (p. 740) for k' gives (here, V_M^0 and V_M are identical)

$$V_g = \frac{K V_S}{W} \times \frac{273}{T_C}$$

The density of ρ_S of the liquid on the stationary phase is given by

$$\rho_S = \frac{W}{V_S}$$

Thus,

$$V_g = \frac{K}{\rho_S} \times \frac{273}{T_C} \tag{26–7}$$

Note that V_g at a given temperature depends only upon the partition coefficient of the solute and the density of the liquid making up the stationary phase. As such, it should in principle be a useful parameter for identifying species.

The literature contains large numbers of specific retention volumes; unfortunately these data are widely scattered and often unreliable.

26A-3 EFFECT OF MOBILE-PHASE FLOW RATE

Equation 25–8 and the relationships shown in Table 25–2 (p. 736) are fully applicable to gas chromatography. The longitudinal diffusion term (B/u) is more important in gas-liquid chromatography, however, than in other chromatographic processes because of the much larger diffusion rates in gases (10^4 times greater than liquids). As a consequence, the minimum shown for curve H in Figure 25–7 is often considerably broadened in gas chromatography (see Figure 26–1).

26B Instruments for Gas-Liquid Chromatography

Numerous instruments for gas-liquid chromatography are on the market, which vary in sophistication and price range (several hundreds to several thousands of dollars). The basic components of these instruments

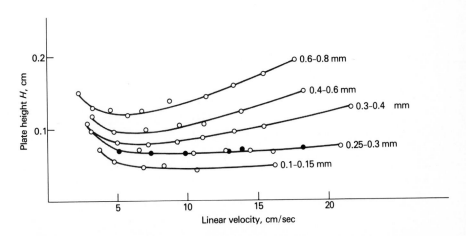

FIGURE 26-1 Effect of particle size on plate height. The numbers to the right are particle diameters in millimeters. (From: J. Boheman and J. H. Purnell, in *Gas Chromatography 1958*, D. H. Desty, Ed., Academic Press: New York, 1958. With permission of Butterworth and Co., Woburn, MA.)

are shown in Figure 26–2. A description of each component follows.

26B-1 CARRIER GAS SUPPLY

Carrier gases, which must be chemically inert, include helium, argon, nitrogen, and hydrogen; helium is the most widely used. As will be shown later, the choice of gases is usually dictated by the type of detector used. Associated with the gas supply are pressure regulators, gauges, and flowmeters. In addition, the carrier gas system often contains a molecular sieve to remove water or other impurities.

Flow rates are controlled by a pressure regulator. Inlet pressures usually range from 10 to 50 psi (above room pressure), which lead to flow rates of 25 to 150 mL/min. Generally, it is assumed that flow rates will be constant if the inlet pressure remains constant. Flow rates can be established by a

rotameter at the column head; this device, however, is not as accurate as a simple soap-bubble meter, which, as shown in Figure 26–2, is located at the end of the column. A soap film is formed in the path of the gas when a rubber bulb containing an aqueous solution of soap or detergent is squeezed; the time required for this film to move between two graduations on the buret is measured and converted to volumetric flow rate.

26B-2 SAMPLE INJECTION SYSTEM

Column efficiency requires that the sample be of suitable size and be introduced as a "plug" of vapor; slow injection of oversized samples causes band spreading and poor resolution. A microsyringe is used to inject liquid samples through a rubber or silicone diaphragm or septum into a heated sample port located at the head of the column (the sample port is ordinarily about 50°C above

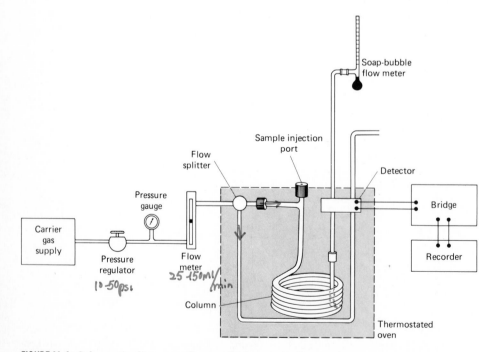

FIGURE 26-2 Schematic diagram of a gas chromatograph.

the boiling point of the least volatile component of the sample). For ordinary analytical columns, sample sizes vary from a few tenths of a microliter to 20 μL. Capillary columns require much smaller samples ($\sim 10^{-3}$ μL); here, a sample splitter system is employed to deliver only a small fraction of the injected sample to the column head, with the remainder going to waste.

Gas samples are best introduced by means of a sample valve such as that shown in Figure 25–12 (p. 750). Solid samples are introduced as solutions or, alternatively, are sealed into thin-walled vials that can be inserted at the head of the column and punctured or crushed from the outside.

26B–3 PACKED COLUMNS

Two types of columns are encountered in gas-liquid chromatography, *packed* and *open tubular*. The former, which are discussed in this section, can accommodate larger samples and are generally more convenient to use. The latter, which are described in Section 26B–4, are equally important, however, because of their unparalleled resolution.

Column Dimensions. Present-day packed columns are fabricated from glass or metal (stainless steel, copper, or aluminum) tubes that typically have lengths of 2 to 3 m and inside diameters of 2 to 4 mm. The tubes are ordinarily formed as coils having diameters of roughly 15 cm; this configuration allows them to be fitted into ovens for thermostating.

Types of Solid Supports. The solid support serves to hold the liquid stationary phase in place so that as large a surface area as possible is exposed to the mobile phase. The ideal support would consist of small, uniform, spherical particles with good mechanical strength and a specific surface area of at least 1 m^2/g. In addition, the material should be inert at elevated temperatures

and be uniformly wetted by the liquid phase. No substance that meets all of these criteria perfectly is yet available.

The first, and still the most widely used, supports for gas chromatography were prepared from naturally occurring diatomaceous earth, which is made up of the skeletons of thousands of species of single-celled plants that inhabited ancient lakes and seas. Such plants received their nutrients and disposed of their wastes via molecular diffusion through their pores. As a consequence, their remains are well-suited as support materials because gas chromatography is also based upon the same kind of molecular diffusion.

Two types of supports are derived from diatomaceous earth. The first, which is generally known by the trade name of Chromosorb P, is prepared by crushing, blending, and briquetting the diatomaceous earth as it comes from the ground followed by heating at over 900°C. The resulting bricks, which are also used for high-temperature insulation, are then ground and separated by screening into particles having uniform diameters. The second type of support, called Chromosorb W or G, is prepared by mixing the diatomaceous earth with a sodium carbonate flux before heating at about 900°C. This product is more rugged than the first and, in addition, shows less tendency to adsorb solutes. Unfortunately, its specific surface area is only about 1 m^2/g compared with 4 m^2/g for the Chromosorb P.

Particle Size of Supports. As in liquid chromatography, the efficiency of a column increases rapidly with decreasing particle diameter of the packing (Figure 26–1). The pressure difference required to maintain a given flow-rate of carrier gas, however, varies inversely as the square of the particle diameter; the latter relationship has placed lower limits on the size of particles employed in gas chromatography because it is not convenient to use pressure differences that are greater than about 30 psi. As a result, the usual support particles are 60 to 80 mesh (250 to 170 μm) or 80 to 100 mesh (170 to 149 μm).

Column efficiency increases as the range of particle size decreases. As a consequence, some supports are now being sold with 10- rather than 20-mesh ranges.

Adsorption on Solid Supports. A problem that has plagued gas chromatography from its inception has been the physical adsorption on support surfaces of polar or polarizable analyte species such as alcohols or aromatic hydrocarbons. Adsorption results in distorted peaks, which are broadened and often exhibit a tail. It has been established that adsorption is the consequence of silanol groups that form on the surface of silicates by reaction with moisture. Thus, a fully hydrolyzed silicate surface has the structure

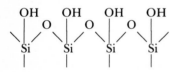

The SiOH groups on the support surface have a strong affinity for polar organic molecules and tend to retain them by adsorption.

Support materials can be deactivated by silanization with dimethylchlorosilane (DMCS). The reaction is

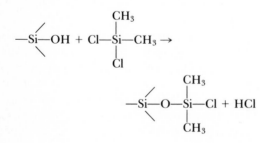

Upon washing with alcohol, the second chloride is replaced by a methoxy group. That is,

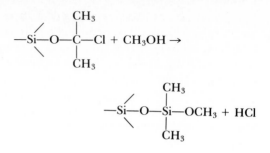

Silanized supports may still show a residual adsorption, which apparently arises from mineral impurities in the diatomaceous earth. Acid washing prior to silanization removes these impurities. Packings that have been acid washed and silanized are sold with the designation —AW—DMCS.

Another silanization reagent is hexamethyldisilazane, $(CH_3)_3$ $SiNHSi(CH_3)_3$. Packings treated in this way are designated by —HMDS.

Several other types of supports have been developed and are sold commercially, including 40 to 60 mesh Teflon particles, 60 to 80 mesh etched glass beads, and porous polymer beads. None of these has had the general acceptance of the classical diatomaceous earth based packings.

Liquid Phase. Desirable properties for the immobilized liquid phase in a gas-liquid chromatographic column include: (1) *low volatility* (ideally, the boiling point of the liquid should be at least 100° higher than the maximum operating temperature for the column); (2) *thermal stability;* (3) *chemical inertness;* and (4) *solvent characteristics* such that k' and α values for the solutes to be resolved fall within a suitable range.

A myriad of solvents have been proposed as stationary phases in the course of the development of gas-liquid chromatography. By now, only a handful—perhaps a dozen or less—suffice for most applications. The proper choice among these solvents is often critical to the success of a separation. Qualitative guidelines exist for making this choice, but in the end, the best stationary phase can only be determined in the laboratory.

The properties of the various common stationary phases are discussed in Section 26C.

Column Preparation. Several techniques are employed for coating support particles with the stationary liquid phase. In one method, a known weight of support is mixed with 3 to 4 times its volume of a volatile solvent containing a weight of the stationary phase that is 1 to 10% of the weight of the support particles. The solvent is then evaporated, often in a rotary stripper, to give a dry, freeflowing solid, which is slowly poured into the column with gentle tapping.

A properly prepared column will have 250 to 1000 plates per foot and may be used for several hundred analyses. Numerous types of prepacked columns are available from commercial sources. Typical costs range from $50 to $150.

26B-4 OPEN TUBULAR COLUMNS

Open tubular or capillary columns were first used as early as 1957, when it became apparent from theoretical considerations that such columns should provide separations that were unprecedented in terms of speed and number of theoretical plates. At that time, it was demonstrated in several laboratories that columns having 300,000 plates or more were practical. Despite such spectacular applications, capillary columns did not find widespread use until more than two decades after their invention. The reasons for the delay appear to be several including: small sample capacity, the fragile nature of the column, the mechanical problems associated with sample introduction and connection of the column to the detector, the difficulty in coating the column reproducibly, the short lifetime of poorly prepared columns, the tendency of columns to clog, and patents, which limited commercial development to a single manufacturer (the

original patent expired in 1977). By the late 1970s these problems had become manageable and several instrument companies were offering capillary columns at a reasonable cost. As a consequence, a major growth in the use of open tubular columns has occurred in the last few years.[3]

Types of Columns. Open tubular columns are of three types, namely, fused silica (FSOT), wall-coated (WCOT), and porous-layer (PLOT) or support-coated (SCOT) open-tubular columns. Table 26-1 compares some important properties of each with those for packed columns.

Wall-coated columns are simply capillary tubes coated with a thin layer of the stationary phase. Early columns were constructed of stainless steel, aluminum, copper, or plastic. Subsequently glass was used. Often the glass is etched with gaseous hydrochloric acid, strong aqueous hydrochloric acid, or potassium hydrogen fluoride to give a rough surface, which bonds the stationary phase more tightly. Often, too, the inner surfaces of glass columns are treated with silanizing agents (p. 762) or other deactivating reagents.

In porous-layer, open tubular columns, the inner surface of the capillary is lined with a thin film ($\sim$30 μm) of finely divided diatomaceous earth. The presence of this layer permits the column to carry several times as much stationary phase; correspondingly larger samples can then be accommodated.

The newest capillary columns, which first appeared in 1979, are the fused silica columns. Their manufacture is based upon techniques developed for the production of optical fibers. Silica capillaries have much thinner walls than their glass counterparts; the outside diameters are about 0.3 to 0.4 mm and inside 0.1 to 0.3 mm. The tubes are

---

[3] For a recent monograph on open tubular columns, see: M. L. Lee, J. Yang, and K. D. Bartle, *Open Tubular Column Gas Chromatography*, Wiley: New York, 1984.

given added strength by an outside protective polyimide coating, which is applied as the capillary tubing is being drawn. The resulting columns are quite flexible and can be bent into coils having diameters of a few inches. Silica open tubular columns are available commercially (as are metal and glass wall-coated columns) and appear to have several important advantages such as physical strength, lower reactivity toward solids, and flexibility. For most applications, they are replacing the older types of open tubular columns.

Column Coating. The walls of open tubular columns are coated with a film of stationary liquid phase that varies in thickness from 0.1 to 1 μm. The liquids are the same as those used for packed columns, which are discussed in Section 26C–1. Several methods have been developed for coating the walls of capillary columns. In the dynamic method, a dilute solution of the stationary phase in a volatile solvent is gently driven through the column under a light gas pressure. In the static method, the column is filled with the solution of the stationary phase. With one end of the column capped, the solvent is then slowly removed under vacuum.

Initially, problems were encountered in coating silica columns with the more polar stationary phases. It appears now, however, that these problems have been or will soon be solved, thus providing a full spectrum of fused-silica, wall-coated open tubular columns.

26B–5 COLUMN THERMOSTATING

Column temperature is an important variable that must be controlled to a few tenths of a degree for precise work. Thus, the column is ordinarily housed in a thermostated oven. The optimum column temperature depends upon the boiling point of the sample and the degree of separation required. Roughly, a temperature equal to or slightly above the average boiling point of a sample results in an elution period of reasonable time (2 to 30 min). For samples with a broad boiling range, it is often desirable to employ temperature programming

Table 26-1
Properties and Characteristics of Typical Gas Chromatographic Columns

	Type Column[a]			
	FSOT	WCOT	PLOT	Packed
Length	10–100 m	10–100 m	10–100 m	1–6 m
Inside diameter	0.1–0.3 mm	0.25–0.75 mm	0.5 mm	2–4 mm
Efficiency	2000–4000 plates/m	1000–4000 plates/m	600–1200 plates/m	500–1000 plates/m
Sample size	10–75 ng	10–1000 ng	10–1000 ng	$10–10^6$ ng
Relative pressure	Low	Low	Low	High
Relative speed	Fast	Fast	Fast	Slow
Chemical inertness	Best ⟶			Poorest
Flexible ?	Yes	No	No	No

[a] FSOT: Fused-silica, open tubular column.
  WCOT: Wall-coated, open tubular column.
   PLOT: Porous-layer, open tubular column (also called support-coated, open tubular SCOT).

whereby the column temperature is increased either continuously or in steps as the separation proceeds. Figure 26–3c shows the improvement in a chromatogram brought about by temperature programming.

In general, optimum resolution is associated with minimal temperature; the cost of lowered temperature, however, is an increase in elution time and therefore the time required to complete an analysis. Figures 26–3a and b illustrate this principle.

26B-6 DETECTORS

Dozens of detectors have been investigated and used during the development of gas chromatography. Only four, however, have found widespread use: *thermal conductivity, flame ionization, thermionic,* and *electron capture.* After listing the desirable characteristics of detectors for gas chromatography, each of these four detectors will be discussed briefly. In addition, detection by means of mass spectrometers and infrared

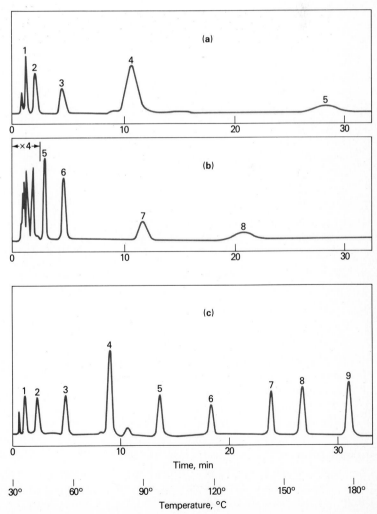

FIGURE 26-3 Effect of temperature on gas chromatograms. (**a**) Isothermal at 45°C; (**b**) isothermal at 145°C; (**c**) programmed at 30 to 180°C. (From: W. E. Harris and H. W. Habgood, *Programmed Temperature Gas Chromatography*, p. 10, Wiley: New York, 1966. Reprinted by permission of John Wiley & Sons, Inc.)

spectrophotometers will be considered in Section 26D–3.

Characteristics of the Ideal Detector.   The ideal detector for gas chromatography would have the following characteristics:

1. Adequate sensitivity. Just what constitutes adequate sensitivity cannot be described in quantitative terms. For example, the sensitivities of the four most popular detectors, alluded to earlier, differ by a factor of 10^7. Yet all are clearly adequate for certain tasks; the least sensitive are not, however, satisfactory for other applications. In general, the sensitivities of present-day detectors lie in the range of 10^{-8} to 10^{-15} g solute/sec.
2. Good stability and reproducibility.
3. A linear response to solutes that extends over several orders of magnitude.
4. A temperature range from room temperature to perhaps 400°C.
5. A short response time that is independent of flow rate.
6. High reliability and ease of use. The detector should, to the extent possible, be foolproof in the hands of inexperienced operators.
7. Similarity in response toward all solutes or alternatively a highly predictable and selective response toward one or more classes of solutes.
8. Nondestructive of sample.

Needless to say, no detector meets all of these desiderata, and it seems unlikely that such a detector will ever be designed.

Thermal Conductivity Detectors.   A very early detector for gas chromatography, and one that still finds wide application, is based upon changes in the thermal conductivity of the gas stream brought about by the presence of analyte molecules. This device is sometimes called a *katharometer*. The sensing element of a katharometer is an electrically heated source whose temperature at constant electrical power depends upon the thermal conductivity of the surrounding gas. The

heated element may be a fine platinum, gold, or tungsten wire or, alternatively, a semiconducting thermistor. The resistance of the wire or thermistor gives a measure of the thermal conductivity of the gas; in contrast to the wire detector, the thermistor has a negative temperature coefficient. Figure 26–4 is a cross-section view of a commercial thermoconductivity detector.

In chromatographic applications, a double detector system is usually employed, with one element being placed in the gas stream *ahead* of the sample injection chamber and the other immediately beyond the column. Alternatively, the gas stream may be split as shown in Figure 26–2. In either case, the thermal conductivity of the carrier gas is canceled, and the effects of variation in flow rate, pressure, and electrical power are minimized. The resistances of the twin detectors are usually compared by incorporating them into two arms of a simple Wheatstone bridge circuit such as that shown in Figure 26–5.

A modulated single-filament thermal conductivity detector was introduced in 1979; this device offers higher sensitivity, freedom from base-line drift, and reduced equilibration time. Here, the analytical and reference gases are passed alternately over a tiny filament held in a ceramic detector cell, which has a volume of only 5 μL. The gas-switch-

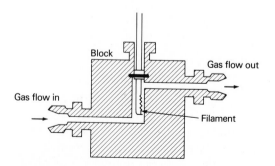

FIGURE 26–4 A typical thermoconductivity detector. (Courtesy of Varian Instrument Division, Palo Alto, CA.)

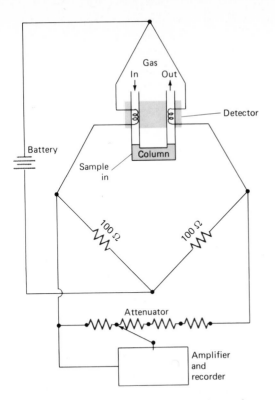

FIGURE 26–5 Schematic diagram of a thermal conductivity detector-recorder system.

ing device operates at a frequency of 10 Hz. The output from the filament is thus a 10-Hz electrical signal whose amplitude is proportional to the difference in thermal conductivity of the analytical and reference gases. Because the amplifier circuit responds only to a 10-Hz signal, thermal noise in the system is largely eliminated.

The thermal conductivities of hydrogen and helium are roughly six to ten times greater than those of most organic compounds. Thus, in the presence of even small amounts of organic materials, a relatively large decrease in the thermal conductivity of the column effluent takes place; consequently, the detector undergoes a marked rise in temperature. The conductivities of other carrier gases more closely resemble those of organic constituents; therefore, a thermal conductivity detector dictates the use of hydrogen or helium.

The advantage of the thermal conductivity detector is its simplicity, its large linear dynamic range ($\sim 10^5$), its general response to both organic and inorganic species, and its nondestructive character, which permits collection of solutes after detection. A limitation of the katharometer is its relatively low sensitivity ($\sim 10^{-8}$ g solute/mL carrier gas). Other detectors exceed this sensitivity by factors as large as 10^4 to 10^7. It should be noted that the low sensitivity of thermal detectors preclude their use in conjunction with capillary columns because of the very small samples that can be accommodated by such columns.

Flame Ionization Detectors. Most organic compounds, when pyrolyzed at the temperature of a hydrogen/air flame, produce ionic intermediates that provide a mechanism by which electricity can be carried through the plasma. With a burner such as that shown in Figure 26–6, the charged species are attracted to and captured by a collector; an ion current results, which can be amplified and recorded. The electrical resistance of a flame plasma is high (perhaps 10^{12} Ω), and the resulting currents are, therefore, minuscule; an electrometer must be employed for their measurement.

The ionization of carbon compounds in a flame is a poorly understood process, although it is observed that the number of ions produced is roughly proportional to the number of *reduced* carbon atoms in the plasma. Functional groups, such as carbonyl, alcohol, halogen, and amine, yield fewer ions or none at all. In addition, the detector is insensitive towards noncombustible gases such as H_2O, CO_2, SO_2, and NO_x. These properties make the flame ionization detector a most useful general detector for the analysis of most organic samples including those that are contaminated with water and the oxides of nitrogen and sulfur.

The flame ionization detector exhibits a

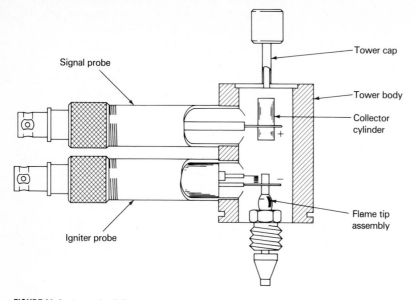

Signal probe

Tower cap

Tower body

Collector
cylinder

+

−

Flame tip
assembly

Igniter probe

FIGURE 26-6 A typical flame ionization detector. (Courtesy of Varian Instrument Division, Palo Alto, CA.)

high sensitivity ($\sim 10^{-13}$ g/mL), large linear response range ($\sim 10^{7}$), and low noise. It is generally rugged and easy to use. For these reasons, it is one of the most popular detectors. It is normally the detector used with capillary columns. A disadvantage of the flame ionization detector is that it is destructive of sample.

Thermionic Detectors.   The thermionic detector is selective towards organic compounds containing phosphorus and nitrogen. Its response to a phosphorus atom is approximately 10 times greater than to a nitrogen atom and 10^{4} to 10^{6} larger than a carbon atom. Compared with the flame ionization detector, the thermionic detector is approximately 500 times more sensitive for phosphorus-containing compounds and 50 times more sensitive for nitrogen-bearing species. These properties make thermionic detection particularly useful for detecting and determining the many phosphorus-containing pesticides.

A thermionic detector is similar in struc-

ture to the flame detector shown in Figure 26–6. The hydrogen from the column passes through the flame tip assembly, is mixed with excess air, and then flows around an electrically heated rubidium silicate bead, which is maintained at about 180 V with respect to the collector. The heated bead forms a plasma having a temperature of 600 to 800°C. Exactly what occurs in the plasma to produce unusually large numbers of ions from phosphorus or nitrogen containing molecules is not understood; but large ion currents result, which are useful for determining compounds containing these two elements.

Electron-Capture Detectors.   Electron-capture detectors operate in much the same way as a proportional counter for measurement of X-radiation (Section 15B–4). Here, the effluent from the column is passed over a β-emitter, such as nickel-63 or tritium (adsorbed on platinum or titanium foil). An electron from the emitter causes ionization of the carrier gas (often nitrogen) and the production of a burst of electrons. In the absence of organic species, a constant standing current between a pair of electrodes results from this ionization process. The current decreases, however, in the presence of those organic molecules that tend to capture electrons. The response is nonlinear unless the potential across the detector is pulsed.

The electron-capture detector is selective in its response, being highly sensitive toward molecules containing electronegative functional groups such as halogens, peroxides, quinones, and nitro groups. It is insensitive toward functional groups such as amines, alcohols, and hydrocarbons. An important application of the electron-capture detector has been for the detection and determination of chlorinated insecticides.

Electron-capture detectors are highly sensitive and possess the advantage of not altering the sample significantly (in contrast to the flame detector). On the other hand, their linear response range is usually limited to about two orders of magnitude.

26C The Stationary Phase

The proper choice of stationary phase is vital to the success of a gas-chromatographic separation. Generally, this choice is based upon polarity parameters of the stationary phase relative to those of the sample constituents; unfortunately these parameters are imperfect guides. As a consequence, optimal separation conditions can only be realized finally by trial-and-error experiments.

26C-1 POLARITY PARAMETERS IN GAS CHROMATOGRAPHY

Three related measures of polarity have found widespread use in gas-liquid chromatography, namely *retention indexes, Rohrschneider constants,* and *McReynolds constants.* The latter two, which are closely related, are derived from retention indexes.[4]

The Retention Index. The retention index I was first defined by Kovats in 1958 as a parameter for identifying solutes from chromatograms.[5] The retention index for any given solute can be derived from a chromatogram of a mixture of that solute with at least two normal paraffins having retention times that bracket that of the solute. That is, normal paraffins are the standards upon which the retention index scale is based. By definition, the retention index for a normal paraffin is equal to 100 times the number of carbons in the compound regardless of the column packing, the temperature, or other chromatographic conditions. The retention index for all compounds other than normal paraffins vary, often several hundredfold, with column variables.

It has long been known that within a homologous series, a plot of the logarithm of adjusted retention time t'_R ($t'_R = t_R - t_M$) versus the number of carbon atoms is linear, provided the lowest member of the series is excluded. Such a plot for C_4 to C_9 normal paraffin standards is shown in Figure 26–7. Also indicated on the ordinate are log retention times for three compounds on the same column and at the same temperature. Their retention indexes are then obtained by multiplying the corresponding abscissa values by 100. Thus, the retention index for toluene is 749 while for benzene, it is 644.

Normally a graphical procedure is not required in determining retention indexes. Instead adjusted retention data are derived by interpolation from a chromatogram of a mixture of the solute of interest and two normal paraffin standards, one having a retention time less than that of the solute and one with a greater retention time.

---

[4] For an excellent review of these parameters and their uses, see: L. S. Ettre, *Chromatographia*, **1973**, *6*, 489 and **1974**, *7*, *39*, 261.

[5] E. Kovats, *Helv. Chim. Acta*, **1958**, *41*, 1915. A computerized retention index library for four types of fused silica columns is now available. See: J. F. Sprouse and A. Varano, *Amer. Lab.*, **1984**, *16* (9), 54.

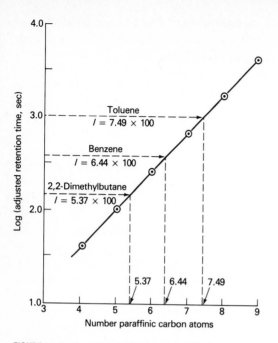

FIGURE 26–7 Graphical illustration of the method for determining retention indexes for three compounds. Stationary phase: squalane. Temperature: 60°C.

The retention index system has the advantage of being based upon readily available reference materials that cover a wide boiling range. In addition, the temperature dependence of retention indexes is relatively small.

It is important to reiterate that the retention index *for a normal paraffin* is independent of temperature and column packing.

Thus, I for heptane, by definition, is always 700. In contrast, the retention index of all other solutes may, and often does, vary widely from one column to another. For example, the retention index for dioxane on a squalane stationary phase is 651. With diisodecylphthalate as the stationary phase, it is 779 while with diethyleneglycolsuccinate, it is 1363.

Retention Index Differences. The data just cited suggests a way of comparing polarities of stationary phases. That is, the least polar of the three, squalane, could be chosen as a reference material and retention index differences ΔI calculated. That is,

Stationary Phase	Relative Polarity, ΔI
Squalane	$661 - 661 = 0$
Diisodecylphthalate	$779 - 661 = 118$
Diethyleneglycolsuccinate	$1363 - 661 = 702$

Here, ΔI provides a measure of increasing polarity for the three stationary phases.

Rohrschneider/McReynolds Constants. The data in Table 26–2 demonstrate that relative polarities, as measured by retention index differences, are solute dependent. Thus, OV–17 is next to the lowest in polarity when 1-butanol serves as the solute but is next to the highest with benzene. This variation in polarity toward two compounds as different as benzene and 1-butanol is not surprising since interaction of solutes with solvents

Table 26–2
Retention Index Differences for Two Solutes on Five Column Packings

Stationary Phase	Benzene, ΔI	1-Butanol, ΔI
Squalane	0	0
Di(2-ethylhexyl)sebacate	75	168
Ucon Oil LB-550-X	116	271
OV-17 (phenylmethylsilicone oil)	132	158
DEGS (diethyleneglycolsuccinate)	495	705

takes several forms. In 1966, L. Rohrschneider[6] suggested that five different reference compounds would suffice to fully characterize stationary phases for the effects of these various interactions. The first column in Table 26–3 shows the five standards proposed by Rohrschneider. A second set of reference compounds, suggested by McReynolds in 1970[7], are shown in the third column. Note that the two sets differ only in the chain-length of the alcohol, ketone, and nitro standards. These changes were made for convenience; the normal paraffins that bracket 1-butanol, methyl-*n*-propyl ketone, and nitropropane are always liquids, whereas those that bracket ethanol, methylethyl ketone, and nitromethane are sometimes gases, which are inconvenient to mix with liquid solutes. A further difference between the two systems is that Rohrschneider divided values of ΔI by 100; McReynolds did not. McReynolds constants are more often quoted than Rohrschneider.

Examples of McReynolds constants for

some common stationary phases are found in Table 26–4. It should be noted that squalane, which heads the list, is somewhat limited in its practical applications because its use is restricted to temperatures lower than about 150°C. Nevertheless, squalane is important because it is the standard nonpolar stationary phase against which all others are compared. Where a practical separation demands a low polarity stationary phase, a silicone solvent such as SE-30 is used.

Significance of McReynolds Constants. As shown in Table 26–3, the five McReynolds constants are symbolized by X′, Y′, Z′, U′, and S′. Each is related to one type of interaction between a solute and the stationary phase. The magnitude of each of these constants provides a measure of the strength of this type of interaction for a group of compounds represented by the standard. Thus, as shown in the fourth column of Table 26–3, aromatics and olefins have the same polarity characteristics as benzene. As a consequence, compounds containing these functionals will be most strongly retained by a stationary phase having a large value of X′. Similarly, a stationary phase exhibiting a

---

[6] L. Rohrschneider, *J. Chromatogr.*, **1966**, *22*, 6.

[7] W. O. McReynolds, *J. Chromatogr. Sci.*, **1970**, *8*, 685.

Table 26–3
Test Substances for Rohrschneider and McReynolds Constants and Groups Represented

Rohrschneider Constants		McReynolds Constants		Groups Represented by Test Substance
Test Substances	ΔI Symbol	Test Substances	ΔI Symbol	
Benzene	X_p	Benzene	X′	Aromatics, olefins
Ethanol	Y_p	1-Butanol	Y′	Alcohols, phenols, acids
Methylethyl ketone	Z_p	Methyl-*n*-propyl ketone	Z′	Ketones, ethers, aldehydes, esters, epoxides, dimethylamino derivatives
Nitromethane	U_p	Nitropropane	U′	Nitro and nitrile derivatives
Pyridine	S_p	Pyridine	S′	Bases, aromatic N-heterocyclics, dioxane

Table 26–4
Some Common Stationary Phases[a]

Name	Maximum Temp., °C	Chemical Nature	McReynolds Number,[b] ΔI				
			X'	Y'	Z'	U'	S'
Squalane	150	Cycloparaffin	0	0	0	0	0
SE-30	350	Dimethylsilicone	15	53	44	64	41
OV-7	350	Phenylmethylsilicone (20% phenyl)	69	113	111	171	128
DC-710	300	Phenylmethylsilicone (50% phenyl)	107	149	153	228	190
QF-1	250	Trifluoropropylmethylsilicone	144	233	355	463	305
Carbowax-20M	250	Polyethyleneglycol	322	536	368	572	510
DEG-Succinate	225	Diethyleneglycolsuccinate	492	733	581	833	791
TCEP	175	1,2,3-Tris(2-cyanoethoxyl)propane	593	857	752	1025	915

[a] Data from: J. A. Perry, *Introduction to Analytical Gas Chromatography*, p. 64, Marcel Dekker: New York, 1981. By courtesy of Marcel Dekker, Inc.
[b] Test Solutes: X' = benzene; Y' = 1-butanol; Z' = methyl-*n*-propyl ketone; U' = nitromethane; and S' = pyridine.

large Y' and a small Z' will retard alcohols more strongly than ketones. Conversely, when Y' is smaller than Z', ketones will be preferentially retained.

26C-2   USES OF McREYNOLDS CONSTANTS

McReynolds constants are widely used to characterize the multitude of stationary phases that are now available. Thus, most commercial producers of these materials provide McReynolds numbers in their descriptive catalogs. These data are used in two ways.

Comparison of Stationary Phases. One of the important uses of McReynolds numbers is for recognizing similar liquid phases in order to avoid duplication of effort in testing the efficiency of columns for a given separation. With hundreds of packings now available, the ability to recognize columns that have similar performance characteristics is of major importance. It is noteworthy that stationary phases with widely different chemical composition may exhibit remarkably similar chromatographic performance characteristics. For example consider the three listed in Table 26–5. It would be pointless to substitute one of these columns for another in the hopes of improving a separation. For all practical purposes, they would yield identical chromatograms. On the other hand, if

Table 26–5
McReynolds Constants for Three Stationary Phases

	X'	Y'	Z'	U'	S'
Polyoxyethelene monostearate	191	382	244	380	333
Poly(propylene glycol)	202	394	253	392	341
Oleyl poly(ethyleneoxy)ethanol	202	395	251	395	344

one of these stationary phases had a higher boiling point or greater thermal stability, that one could be chosen with full confidence that its performance would mimic the others under similar conditions.

Selection of Stationary Phases. It is important to understand the circumstances under which McReynolds constants are useful guides in selecting a stationary phase for a given separation, and equally important when these constants are of no help at all. In general, McReynolds numbers provide useful information for selecting columns for separating compounds that differ in functionality, such as alcohols from ethers, nitro compounds from nitriles, or olefins from paraffins. They are of no value as a guide to separating homologous compounds or isomers. Some examples of the use of McReynolds constants follow.

EXAMPLE 26–1. Which of the following three columns should provide the best separation of saturated from unsaturated fatty acid esters: DC-710, DEG-succinate, or diglycerol? McReynolds constants for the first two are shown in Table 26–4. For the last, they are 371, 826, 560, 676, and 854 respectively.

In Table 26–3, we see that for esters, a packing with a high Z' constant is desirable. On this basis the DC-710 can be eliminated. For separating saturated from unsaturated functional groups, however, a high value of X' is also desirable. On this score the DEG-succinate is clearly superior to diglycerol.

EXAMPLE 26–2. The two insecticides, dieldrin and aldrin have structures

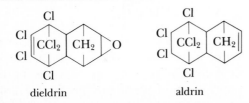

dieldrin   aldrin

An SE-30 and a QF-1 column are available. Which should be chosen?

The dieldrin has an epoxide group, which should lead to its preferential retention on a column with a high Z' value. The QF-1 is clearly better on this score.

26C–3 CHOICE OF STATIONARY PHASE

Separation of Homologs. In Section 26C–1, it was noted that for a homologous series, a linear relationship exists between the number of carbon atoms (excluding the C_1 member) and the logarithm of the adjusted retention time. A similar plot is obtained if boiling point is substituted for number of carbon atoms. To separate homologs, then, a stationary phase is chosen that will readily dissolve the sample. Thus for paraffinic hydrocarbons, squalane or SE-30 could be used. For a series of aliphatic ketones, a packing with a high Z' would be suitable; for example, Carbowax 20M or a DEG-succinate. In either case, elution would be expected in the order of boiling point or molecular weight.

Separation of Compounds Containing Different Functional Groups. As shown by the foregoing examples, the selection of a stationary phase for mixtures of molecules having similar boiling points but different functional groups is relatively straightforward if McReynolds numbers are available. Even without such data, the general principle of "like dissolves like" is frequently applicable. For example, for separating toluene from cyclohexane, an aromatic stationary phase such as benzyldiphenyl would retard toluene preferentially. The elution order could be reversed by employing a hydrocarbon phase such as squalane or nonpolar methylsilicone such as SE-30.

26C–4 CHEMICALLY BONDED STATIONARY PHASE

The idea of chemically binding a stationary phase to an inert solid surface is attractive from several standpoints, and a good deal of effort was expended in the 1970s in trying to develop chemically bonded stationary phases. In 1980, patents describing the successful preparation of bonded-phase supports for gas chromatography were issued to the Dow Chemical Company who in turn licensed their manufacture under the name of Permabond Supports. Several of these packings having different temperature ranges and polarities are now being marketed.[8]

Chemical bonding is carried out on hydrolyzed silica particles whose surfaces are covered with silanol groups. The reaction of one of these groups with silicon tetrachloride produces a chlorosilinated silica by the reaction

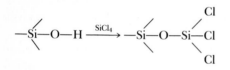

The product of this reaction is then treated with a polyol having a molecular weight of 3000 or more. Thus,

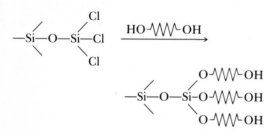

The alcohol groups on the surface of this product can be further modified by treatment with various reagents to produce surfaces having various polarities.

Chemically bonded supports provide a uniform layer of organic phase that has highly reproducible properties. The mono-molecular nature of this layer should lead to rapid equilibration of solutes between the two phases, which, as shown in Section 25B–2, leads to improved efficiencies (recall that the C_{su} term in Equation 25–8 is directly proportional to the square of the thickness of the layer of the stationary phase medium). Another advantage of the chemically bonded supports are their thermal stability, which for some extends to 250°C and for others to 350°C. A disadvantage of the present generation of chemically bonded packings is their sensitivity to degradation by oxygen and water.

26D Applications of Gas-Liquid Chromatography

In evaluating the importance of GLC, it is necessary to distinguish between the two roles the method plays. The first is as a tool for performing separations; in this capacity, it is unsurpassed when applied to complex organic, metal-organic, and biochemical systems. The second, and distinctly different function, is that of providing the means for completion of an analysis. Here, retention times or volumes are employed for qualitative identification, while peak heights or areas provide quantitative information. For qualitative purposes, GLC is much more limited than most of the spectroscopic methods considered in earlier chapters. As a consequence, an important trend in the field has been in the direction of combining the remarkable fractionation qualities of GLC with the superior identification properties of such instruments as mass, infrared, and NMR spectrometers.

26D-1 QUALITATIVE ANALYSIS

Gas chromatograms are widely used as criteria of purity for organic compounds. Contaminants, if present, are revealed by the appearance of additional peaks; the areas

[8] See: J. N. Driscoll and I. S. Krull, *Amer. Lab.,* **1983,** *15* (5), 42.

under these peaks provide rough estimates of the extent of contamination. The technique is also useful for evaluating the effectiveness of purification procedures.

In theory, retention times should be useful for the identification of components in mixtures. In fact, however, the applicability of such data is limited by the number of variables that must be controlled in order to obtain reproducible results. Nevertheless, gas chromatography provides an excellent means of confirming the presence or absence of a suspected compound in a mixture, provided an authentic sample of the substance is available. No new peaks in the chromatogram of the mixture should appear upon addition of the known compound, and enhancement of an existing peak should be observed. The evidence is particularly convincing if the effect can be duplicated on different columns and at different temperatures.

We have seen (Section 25B–3) that the selectivity factor α for compounds A and B is given by the relationship

$$\alpha = \frac{K_B}{K_A} = \frac{(t_R)_B - t_M}{(t_R)_A - t_M} = \frac{(t_R')_B}{(t_R')_A}$$

If a standard substance is chosen as compound B, then α can provide an index for identification of compound A, which is largely independent of column variables other than temperature; that is, numerical tabulations of selectivity factors for pure compounds relative to a common standard can be prepared and then used for the characterization of solutes. The amount of such data available in the literature is presently limited.

26D–2 QUANTITATIVE ANALYSIS

The detector signal from a gas-liquid chromatographic column has had wide use for quantitative and semiquantitative analyses. An accuracy of 1 to 3% relative is attainable under carefully controlled conditions. As with most analytical tools, reliability is directly related to the control of variables; the nature of the sample also plays a part in determining the potential accuracy.

The general discussion of quantitative chromatographic analysis given in Section 25D–2 applies to gas chromatography as well as to other types; therefore, no further consideration of this topic is given here.

26D–3 GAS-LIQUID CHROMATOGRAPHY WITH SELECTIVE DETECTORS

Gas chromatography is often coupled with the selective techniques of spectroscopy and electrochemistry. The resulting so-called hyphenated methods provide the chemist with powerful tools for identifying the components of complex mixtures.[9]

In early hyphenated methods, the eluates from the chromatographic column were collected as separate fractions in a cold trap, a nondestructive and nonselective detector being employed to indicate their appearance. The composition of each fraction was then investigated by nuclear magnetic resonance, infrared, or mass spectroscopy, or by electroanalytical measurements. A serious limitation to this approach was the very small (usually micromolar) quantities of solute contained in a fraction; nonetheless, the general procedure proved useful for the qualitative analysis of many multicomponent mixtures.

A second general method, which now finds widespread use, involves the application of a selective detector to monitor the column effluent continuously. Generally, these procedures require computer control of instruments and computer memory for storage of spectral data for subsequent display as spectra and chromatograms.

---

[9] For a review on hyphenated methods, see: T. Hirschfeld, *Anal. Chem.*, **1980,** *52,* 297A; C. L. Wilkins, *Science,* **1983,** *222,* 291.

Gas Chromatography/Mass Spectrometry. Several instrument manufacturers offer gas-chromatographic equipment that can be directly interfaced with rapid-scan mass spectrometers of various types.[10] As was pointed out in Section 18A–2, the flow rate from capillary columns is generally low enough that the column output can be fed directly into the ionization chamber of the mass spectrometer. For packed columns, however, a separator such as that shown in Figure 18–3 must be employed to remove most of the carrier gas from the analyte.

Most quadrupole and magnetic sector mass spectrometers are offered with accessories that permit interfacing with gas-chromatographic equipment. In addition, the Fourier transform mass spectrometer described in Section 18A–8 has been coupled to gas-liquid columns.[11] Its speed and high sensitivity are particularly advantageous for this application.

In the last decade, several mass spectrometers designed specifically as gas chromatographic detectors have appeared on the market. Generally, these are compact quadrupole instruments, which are less expensive ($25,000 to 50,000) and easier to use and maintain than the multipurpose mass spectrometers described in Chapter 18.[12]

Recently, a related instrument, called an *ion trap detector* (ITD), has been announced.[13] In this instrument, ions are created from the eluted sample by electron impact and stored in a radio-frequency field (see Figure 26–8). The trapped ions are then ejected from the storage area to an electron multiplier detector. The ejection is controlled so that scanning on the basis of mass-to-charge ratio is possible. The physical principle upon which this scanning is based has not been published at the date of this writing. The ion trap detector is remarkably compact and less expensive than quadrupole instruments.

Mass spectrometric detectors ordinarily have several display modes, which fall into two categories: real time and computer reconstructed. Within each of these categories are a choice of total ion current chromatograms (a plot of the sum of all ion currents as a function of time), selected ion current chromatograms (a plot of ion currents for one or a few ions as a function of time), and mass spectra of various peaks. Real time mass spectra appear on an oscilloscope screen equipped with mass markers; the mass chromatogram may appear on the oscilloscope screen or as a real-time recorder plot. After a separation is complete, computer-reconstructed chromatograms can be displayed on the screen or can be printed out. Reconstructed mass spectra for each peak can also be displayed or printed. Some instruments are further equipped with spectral libraries for compound identification.

Gas chromatography/mass spectrometry instruments have been used for the identification of hundreds of components that are present in natural and biological systems. For example, these procedures have permitted characterization of the odor and flavor components of foods, identification of water pollutants, medical diagnosis based on breath components, and studies of drug metabolites.

An example of one application of GC/MS is shown in Figure 26–9. The upper figure is a computer-reconstructed mass chromatogram of a sample trapped from an en-

---

[10] For additional information, see: G. M. Message, *Practical Aspects of Gas Chromatography/Mass Spectrometry,* Wiley: New York, 1984; W. McFadden, *Techniques of Combined Gas Chromatography/Mass Spectrometry,* Wiley-Interscience: New York, 1973; J. F. Holland, *et al., Anal. Chem.,* **1983,** *55,* 997A; and C. Fenselau, *Anal. Chem.,* **1977,** *49,* 563A.

[11] E. B. Ledford, *et al., Anal. Chem.,* **1980,** *52,* 2450.

[12] See: N. Gochman, L. J. Bowie, and D. N. Bailey, *Anal. Chem.,* **1979,** *51,* 525A.

[13] G. C. Stafford, Jr., P. E. Kelley, and D. C. Bradford, *Amer. Lab.,* **1983,** *15* (6), 51; S. A. Borman, *Anal. Chem.,* **1983,** *55,* 726A.

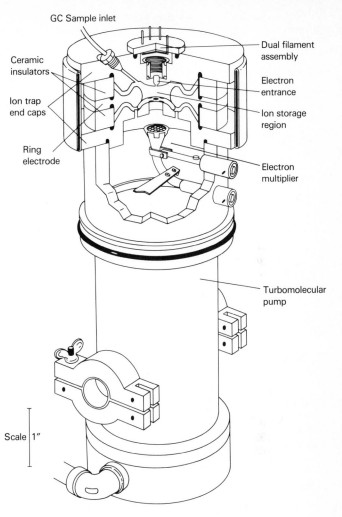

Labels on figure:
GC Sample inlet
Ceramic insulators
Ion trap end caps
Ring electrode
Dual filament assembly
Electron entrance
Ion storage region
Electron multiplier
Turbomolecular pump
Scale 1″

FIGURE 26-8 Schematic diagram of the ion trap detector. (Reprinted with permission from: G. C. Stafford, Jr., P. E. Kelley, and D. C. Bradford, *Amer. Lab.,* **1983,** *15* (6), 51. Copyright 1983 by International Scientific Communications, Inc.)

vironmental chamber during the combustion of cloth treated with a fire-retarding chemical. The ordinate here is the total ion current while the abscissa is retention time. The lower figure is the computer-reconstructed mass spectrum of peak 12 on the chromatogram. Here, relative ion currents are plotted as a function of mass number.

Gas Chromatography/Infrared Spectroscopy. Coupling capillary column gas chromatographs with Fourier transform infrared spectrometers provides a potent means for separating and identifying the components of difficult mix-

tures. Several instruments of this type are now offered commercially.[14]

As with GC/MS, the interface between the column and the detector is critical. In this instance, a narrow light pipe having a length of a few millimeters to several centimeters and an inside diameter of 1 to 3 mm is connected to the column by means of narrow

[14] See: C. L. Wilkins, *Science,* **1983,** *222,* 291; P. R. Griffiths, J. A. de Haseth, and L. V. Azarraga, *Anal. Chem.,* **1983,** *55,* 1361A; S. A. Borman, *Anal. Chem.,* **1982,** *54,* 901A.

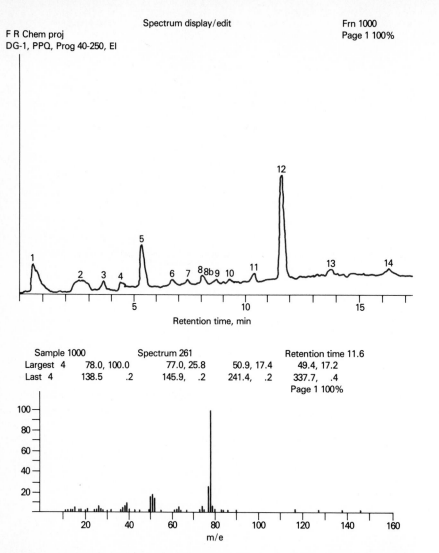

FIGURE 26-9 Typical output from a GC/MS instrument. The upper curve is a computer-reconstructed chromatogram. The peaks correspond to: (1) air, (2) water, (3) hydrogen cyanide, (4) unknown, (5) acetaldehyde, (6) ethanol, (7) acetonitrile, (8) acetone, (8-b) unknown, (9) carbon disulfide, (10) unknown, (11) unknown, (12) benzene, (13) toluene, (14) xylene. The lower plot is the computer-reconstructed mass spectrum for peak 12 (benzene). (Reprinted with permission from: T. W. Sickels and D. T. Stafford, *Amer. Lab.*, **1977**, *12* (5), 17. Copyright 1977 by International Scientific Communications, Inc.)

tubing. The light pipe, a version of which is shown schematically in Figure 26–10, consists of a Pyrex tube that is internally coated with gold. Transmission of radiation occurs by multiple reflections off the wall. Often

the light pipe is heated in order to avoid condensation of the sample components. Light pipes of this type are designed to maximize the path length for enhanced sensitivity while minimizing the dead volume to

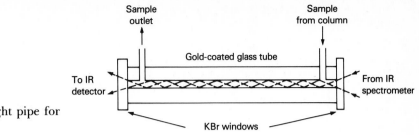

FIGURE 26-10 A typical light pipe for GC/IR instruments.

lessen band broadening. The radiation detectors are generally highly sensitive, liquid-nitrogen-cooled mercury-cadmium telluride devices. Scanning is triggered by the output from a nondestructive chromatographic peak detector and begins after a brief delay to allow the component to travel from the detector region to the infrared cell. The spectral data are digitized and stored in a computer from which printed spectra are ultimately derived.

Difficulty is sometimes encountered when attempts are made to compare spectra for the gaseous effluents from a column with library spectra that have been obtained with liquid or solid samples. Gaseous spectra contain rotational fine structure, which is absent in liquid or solid spectra; significant differences in appearance are the result.

As with GC/MS, digitized spectral libraries and search systems are being developed to handle the enormous amount of data that are produced by MS/FTIR instruments, even with relatively simple samples.[15]

26E Gas-Solid Chromatography

Gas-solid chromatography is based upon adsorption of gaseous substances on solid surfaces. Distribution coefficients are generally much larger than those for gas-liquid chromatography. Thus, gas-solid chromatography is useful for the separation of species that are not retained by gas-liquid columns, such as the components of air, hydrogen sulfide, carbon disulfide, nitrogen oxides, carbon monoxide, carbon dioxide, and the rare gases.

Two types of adsorbents are encountered, molecular sieves and porous polymers.

26E-1 MOLECULAR SIEVES

Molecular sieves are aluminum silicate ion exchangers, whose pore size depends upon the kind of cation present. Commercial preparation of these materials are available in particle sizes of 40/60 mesh to 100/120 mesh. The sieves are classified according to the maximum diameter of molecules that can enter the pores. Commercial molecular sieves come in pore sizes of 4, 5, 10, and 13 Å. Molecules smaller than these dimensions penetrate into the interior of the particles where adsorption takes place. For such molecules, the surface area is enormous when compared with the area available to larger molecules. Thus molecular sieves can be used to separate small molecules from large. For example, a 6 ft, 5 Å packing at room temperature will easily separate a mixture of helium, oxygen, nitrogen, methane, and carbon monoxide in the order given. Figure 26–11**a** shows a typical molecular sieve chromatogram.

26E-2 POROUS POLYMERS

Porous polymer beads of uniform size are manufactured from styrene cross-linked

---

[15] For a description of one of these systems, see: S. R. Lowry and D. A. Huppler, *Anal. Chem.*, **1981**, *53*, 889.

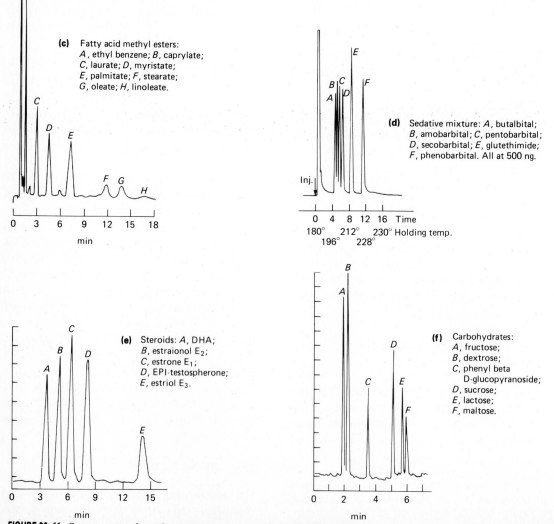

(a) Exhaust mixture: A, 35% H_2; B, 25% CO_2;
C, 1% O_2; D, 1% N_2; E, 1% C_2; F, 30% CH_4;
G, 3% CO; H, 1% C_3; I, 1% C_4; J, 1% $i\text{-}C_5$;
K, 1% $n\text{-}C_5$.

(b) Pesticides: A, Lindane; B, Heptachlor;
C, Aldrin; D, Dieldrin; E, DDT.
A–D, 0.3 ng; E, 3.0 ng.

(c) Fatty acid methyl esters:
A, ethyl benzene; B, caprylate;
C, laurate; D, myristate;
E, palmitate; F, stearate;
G, oleate; H, linoleate.

(d) Sedative mixture: A, butalbital;
B, amobarbital; C, pentobarbital;
D, secobarbital; E, glutethimide;
F, phenobarbital. All at 500 ng.

(e) Steroids: A, DHA;
B, estraionol E_2;
C, estrone E_1;
D, EPI-testospherone;
E, estriol E_3.

(f) Carbohydrates:
A, fructose;
B, dextrose;
C, phenyl beta
D-glucopyranoside;
D, sucrose;
E, lactose;
F, maltose.

FIGURE 26-11 Some examples of gas-chromatographic separations.

with divinylbenzene (Section 27H–2). These materials are sold under the trade name of Porapak.[16] The pore size of these beads is uniform and is controlled by the amount of cross-linking. The Porapaks have found considerable use in the separation of gaseous polar species such as hydrogen sulfide, oxides of nitrogen, water, carbon dioxide, methanol, and vinyl chloride.

26F   Examples of Applications of Gas Chromatography

Figure 26–11 illustrates some typical applications of gas chromatography to analytical

---

[16] Waters Associates, Inc., Milford, Massachusetts.

problems. All are based upon gas-liquid equilibria except the one shown in Figure 26–11**a**. Here, a gas-liquid column is followed by a gas-solid molecular sieve. The former retains only the carbon dioxide and passes the remaining gases at rates corresponding to the carrier rate. When the carbon dioxide is eluted from the first column, a switch directs the flow around the second column briefly to avoid permanent adsorption on the molecular sieve. After the carbon dioxide signal has returned to zero, the flow is switched back through the second column, thereby permitting elution of the remainder of the sample components.

Pertinent data on column conditions for the various chromatograms in Figure 26–11 are found in Table 26–6.

Table 26-6
Conditions for Chromatograms Shown in Figure 26-11

Chromatogram	Column[a]	Packing	Detector[b]	Temperature, °C	Carrier	Flow, mL/min
(a)	$10' \times \frac{1}{8}''$ S	Chromosorb 102	TCD	65–200	He	30
	$5' \times \frac{1}{8}''$ S	Molecular sieve 5A				
(b)	$6' \times \frac{1}{4}''$ G	1.5% OV-17 on Chromosorb G	ECD	220		
(c)	$6' \times \frac{1}{8}''$ S	Chromosorb W	TCD	190	He	30
(d)	$6' \times \frac{1}{4}''$ G	1.5% OV-17 on HP Chromosorb G	FID	180–230		
(e)	$6' \times 3.4$ mm G	5% OV-210 + 2.5% OV-17 on Supelcoport	ECD	260	A	
(f)	$6' \times 3.4$ mm	2% OV-17 on Chromosorb W	FID	105–325	N_2	40

[a] S = stainless steel; G = glass.
[b] TCD = thermal conductivity; ECD = electron capture; FID = flame ionization.

PROBLEMS

26–1. A GLC column was operated under the following conditions:

Column: 1.10 m × 2.0 mm packed with Chromosorb P; weight of stationary liquid added, 1.40 g; density of liquid, 1.02 g/mL

Pressures: inlet, 25.1 psi above room; room, 748 torr

Measured outlet flow rate: 24.3 mL/min

Temperature: room, 21.2°C; column, 102.0°C

Retention times: air, 18.0 sec; methyl acetate, 1.98 min; methyl propionate, 4.16 min; methyl n-butyrate, 7.93 min

Peak widths at base: 0.19, 0.39, and 0.79 min, respectively

Calculate

(a) the average flow rate in the column.

(b) the corrected retention volumes for air and the three esters.

(c) the specific retention volumes for the three components.

(d) partition coefficients for each of the esters.

(e) a corrected retention volume and a retention time for methyl n-hexanoate.

26–2. From the data in Problem 26–1, calculate

(a) k' for each compound.

(b) α values for each adjacent pair of compounds.

(c) the average number of theoretical plates and plate height for the column.

(d) the resolution for each adjacent pair of compounds.

26–3. For the chromatogram described in Problem 26–1, the areas of the three peaks were found to be 16.4, 45.2, and 30.2 in the order of increasing retention time. Calculate the percent of each compound if the relative detector response for the three species was 0.60, 0.78, and 0.88, respectively.

26–4. The stationary liquid in the column described in Problem 26–1 was didecylphthalate, a solvent of intermediate polarity. If a nonpolar solvent such as a silicone oil had been used instead, would the retention times for the three compounds be larger or smaller? Why?

26–5. Corrected retention times for ethyl, n-propyl, and n-butyl alcohols on a column employing a packing coated with silicone oil are 0.69, 1.51, and 3.57. Predict retention times for the next two members of the homologous series.

26–6. A chromatographic column was operated under the following conditions:

Column: 1.25 m × 2.0 mm packed with Chromosorb P; weight of paraffin stationary liquid, 2.43 g; density, 0.796 g/mL

Pressures: inlet, 18.3 psi above room; room, 768 torr

Outlet flow rate: 31.3 mL/min

Temperatures: room, 19.6°C; column, 90.3°C

Retention time: air, 0.395 min

Retention times and peak widths: *i*-propylamine, 4.59 and 0.365 min; *n*-propylamine, 4.91 and 0.382 min

Calculate

(a) the average flow rate in the column.

(b) the corrected retention volumes for air and the two amines.

(c) the specific retention volumes for the two amines.

(d) partition coefficients for the two amines.

26–7. From the data in Problem 26–6, calculate

(a) k' for each amine.

(b) an α value.

(c) an average number of theoretical plates in the column.

(d) the plate height for the column.

(e) the resolution for the two amines.

(f) the length of column required to achieve a resolution of 1.5.

(g) the time required to achieve a resolution of 1.5, assuming the original linear flow rate.

26–8. What would be the effect of the following on the plate height of a column? Explain.

(a) Increasing the weight of the stationary phase relative to the packing weight

(b) Decreasing the rate of sample injection

(c) Increasing the injection port temperature

(d) Increasing the flow rate

(e) Reducing the particle size of the packing

(f) Decreasing the column temperature

26–9. Calculate the retention index for each of the following compounds:

Compound	$(t_R - t_M)$
(a) Propane	1.29
(b) *n*-Butane	2.21
(c) *n*-Pentane	4.10
(d) *n*-Hexane	7.61
(e) *n*-Heptane	14.08
(f) *n*-Octane	25.11
(g) Toluene	16.32
(h) Butane-2	2.67
(i) *n*-Propanol	7.60
(j) Methylethyl ketone	8.40
(k) Cyclohexane	6.94
(l) *n*-Butanol	9.83

27

High-Performance Liquid

Chromatography

This chapter deals with the four basic types of column chromatography in which the mobile phase is a liquid. The four include: *partition chromatography*, which is conveniently subdivided into *bonded-phase* and *liquid-liquid; adsorption* or *liquid-solid chromatography; ion-exchange chromatography;* and *exclusion* or *gel chromatography*. It is noteworthy that these same types of chromatography can also be performed on a flatbed or plane. Planar chromatography is treated in Chapter 28.

Early liquid chromatography, including Tswett's original work, was carried out in glass columns with diameters of 1 to 5 cm and lengths of 50 to 500 cm. To assure reasonable flow rates, the diameter of the particles of the solid stationary phase was usually in the 150 to 200 μm range. Even then, flow rates were low, amounting to a few tenths of a milliliter per minute. Thus separation times were long—often several hours. Attempts to speed up the classic procedure by

application of vacuum or by pumping were not effective, however, because increases in flow rates acted to increase plate heights beyond the minimum in the typical plate height versus flow rate curve (see Figure 25–7); decreased efficiencies were the result.

Early in the development of liquid chromatography, it was realized that major increases in column efficiency could be expected to accompany decreases in the particle size of packings. It was not until the late 1960s, however, that the technology for producing and using packings with particle diameters as small as 10 μm was developed. This technology required sophisticated instruments, which contrasted markedly with the simple glass columns of classic liquid chromatography. The name *high-performance liquid chromatography* (HPLC) is often employed to distinguish these newer procedures from the classic methods, which still find considerable use for preparative purposes. This chapter deals exclusively with HPLC.[1]

27A Scope of HPLC

High-performance liquid chromatography is unquestionably the fastest growing of all of the analytical separation techniques with annual sales of HPLC equipment approaching the billion dollar mark. The reasons for this explosive growth is attributable to the

---

[1] A large number of books on liquid chromatography are available. Among these are: L. R. Snyder and J. J. Kirkland, *Introduction to Modern Liquid Chromatography,* 2d ed., Wiley: New York, 1979; R. J. Hamilton and P. A. Sewell, *Introduction to High Performance Liquid Chromatography,* 2d ed., Chapman and Hall: New York, 1982; and *High-Performance Liquid Chromatography,* C. Horváth, Ed., vols. 1 and 2, Academic Press: New York, 1980.

sensitivity of the method, its ready adaptability to accurate quantitative determinations, its suitability for separating nonvolatile species or thermally fragile ones, and above all, its widespread applicability to substances that are of prime interest to industry, to many fields of science, and to the public. Examples of such materials include: amino acids, proteins, nucleic acids, hydrocarbons, carbohydrates, drugs, terpenoids, pesticides, antibiotics, steroids, metal-organic species, and a variety of inorganic substances.

Figure 27–1 reveals that the various liquid chromatographic procedures tend to be complementary insofar as their areas of applications are concerned. Thus, for solutes having molecular weights greater than 10,000, exclusion chromatography is often used although it is now becoming possible to handle such compounds by reversed-phase

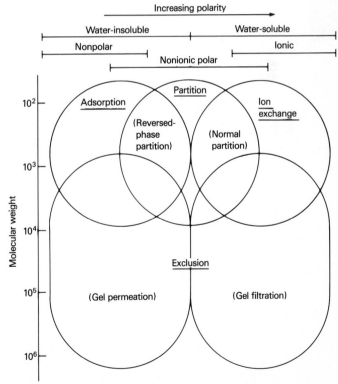

FIGURE 27-1 Applications of liquid chromatography. (From: D. L. Saunders, in *Chromatography,* 3d ed, E. Heftmann, Ed., p. 81, Van Nostrand Reinhold: New York, 1975. With permission.)

partition chromatography as well. For lower molecular weight ionic species, ion-exchange chromatography is widely used. Small polar but nonionic species are best handled by partition methods. In addition, this procedure is frequently useful for separating members of a homologous series. Adsorption chromatography is often chosen for separating nonpolar species, structural isomers, and compound classes such as aliphatic hydrocarbons from aliphatic alcohols.

27B   Column Efficiency in Liquid Chromatography

The discussion on band broadening in Section 25B–2 is generally applicable to liquid chromatography. The present section illustrates the important effect of stationary phase particle size and describes two additional sources of zone spreading that are sometimes of considerable importance in liquid chromatography.

27B–1   EFFECT OF PARTICLE SIZE OF PACKINGS

An examination of the mobile-phase mass transfer coefficient in Table 25–2 reveals that C_M in Equation 25–8 is directly related to the square of the diameter of the particles making up a packing. As a consequence, the efficiency of an HPLC column should improve dramatically as the particle size is decreased. Figure 27–2 is an experimental demonstration of this effect, where it is seen that a reduction of particle size from 45 to 6 μm results in a 10-fold or more decrease in plate height. – good

It is noteworthy that none of the plots in this figure exhibits the minimum that is pre-

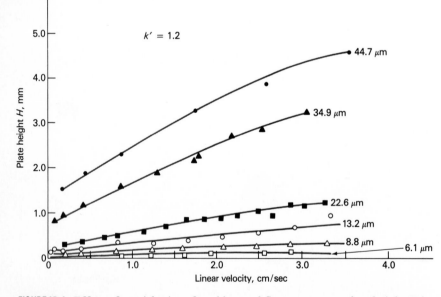

FIGURE 27–2   Effect of particle size of packing and flow rate upon plate height H in liquid chromatography. Column dimensions: 30 cm × 2.4 mm. Solute: N,N-diethyl-*n*-aminoazobenzene. Mobile phase: mixture of hexane, methylene chloride, isopropyl alcohol. (From: R. E. Majors, *J. Chromatogr. Sci.*, **1973**, *11*, 92. With permission.)

dicted by Equation 25–8. Such minima are, in fact, observable in liquid chromatography (see Figure 27–27) but usually at flow rates that are too low for most practical applications.

27B-2 EXTRA-COLUMN BAND BROADENING IN LIQUID CHROMATOGRAPHY

In liquid chromatography, significant band broadening sometimes occurs outside the column packing itself. This so-called *extra-column band broadening* occurs as the solute is carried through open tubes such as those found in the injection system, the detector region, and the piping connecting the various components of the system. Here, broadening arises from differences in flow rates between layers of liquid adjacent to the wall and in the center of the tube. As a consequence, the center part of a solute band moves more rapidly than the peripheral part. In gas chromatography, extra-column spreading is largely offset by diffusion. Diffusion in liquids, however, is significantly slower, and band broadening of this type often becomes noticeable.

It has been shown that the contribution of extra-column effects H_{ex} to the total plate height is given by[2]

$$H_{ex} = \frac{\pi r^2 u}{24 \, D_M} \qquad (27\text{--}1)$$

where u is the linear flow velocity in mL/min, r is the radius of the tube in cm, and D_M is the diffusion coefficient of the solute in the mobile phase in cm^2/sec.

Extra-column broadening can become quite serious when micro columns are used. Here, every effort must be made to reduce the radius of the extra-column components.

27B-3 EFFECT OF SAMPLE SIZE ON COLUMN EFFICIENCY

Figure 27–3 shows the effect of sample weight (μg sample/g packing) on column efficiency for various types of liquid chromatography. Note the superior performance of reversed-phase, bonded packings (see Section 27E–2) compared with liquid-liquid and adsorption packings.

---

[2] R. P. W. Scott and P. Kucera, *J. Chromatogr. Sci.,* **1971,** *9,* 641.

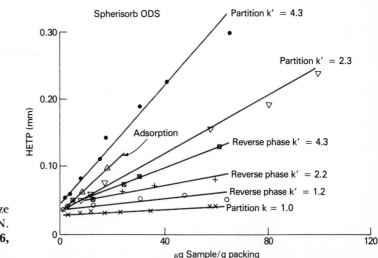

FIGURE 27-3 Effect of sample size on plate height. (From: J. N. Done, *J. Chromatogr. Sci.,* **1976,** *125,* 54. With permission.)

27C   Instruments for Liquid Chromatography

In order to realize reasonable eluent flow rates with packings in the 3 to 10 μm particle sizes, which are common in modern liquid chromatography, pumping pressures of several hundred atmospheres are required. As a consequence of these high pressures, the equipment required for HPLC tends to be more elaborate and expensive than that encountered in other types of chromatography.

Figure 27–4 is a schematic diagram showing the important components of a typical high-performance liquid chromatograph; each is discussed in the paragraphs that follow.[3]

27C-1   MOBILE PHASE RESERVOIRS AND SOLVENT TREATMENT SYSTEMS

A modern HPLC apparatus is equipped with one or more glass or stainless steel reservoirs, each of which contains 500 mL or more of a solvent. The reservoirs are often equipped with a means of removing dissolved gases—usually oxygen and nitrogen—that interfere by forming bubbles in the column and the detector systems. These

---

[3] For a detailed discussion of HPLC systems, see: N. A. Parris, *Instrumental Liquid Chromatography*, 2d ed., Elsevier: New York, 1984; H. Engelhardt, *High Performance Liquid Chromatography*, Springer-Verlag: New York, 1978; and D. J. Runser, *Maintaining and Troubleshooting HPLC Systems*, Wiley: New York, 1981.

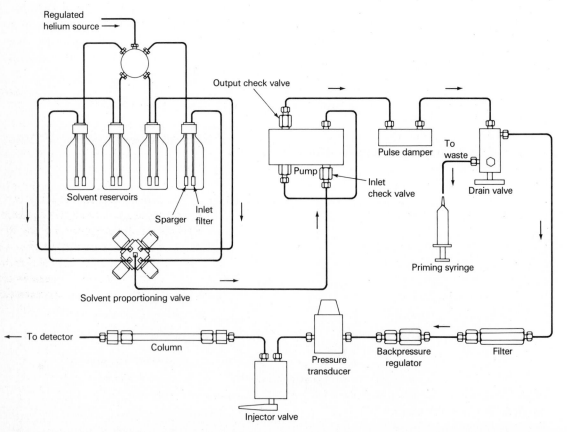

FIGURE 27–4   Schematic of an apparatus for HPLC. (Courtesy of Perkin-Elmer Corporation, Norwalk, CT.)

bubbles cause band spreading; in addition, they often interfere with the performance of the detector. Degassers may consist of a vacuum pumping system, a distillation system, devices for heating and stirring the solvents or, as shown in Figure 27–4, systems for *sparging* in which the dissolved gases are swept out of the solution by fine bubbles of an inert gas of low solubility. Often the systems also contain a means of filtering dust and particulate matter from the solvents. It is not necessary that the degassers and filters be integral parts of the HPLC system as shown in Figure 27–4. For example, a convenient way of treating solvents before introduction into the reservoir is to filter them through a millipore filter under vacuum. This treatment removes gases as well as suspended matter.

A separation that employs a single solvent of constant composition is termed an *isocratic elution*. Frequently, separation efficiency is greatly enhanced by *gradient elution*. Here two (and sometimes more) solvent systems that differ significantly in polarity are employed. After elution is begun, the ratio of the two solvents is varied in a programmed way, sometimes continuously and sometimes in a series of steps. Modern HPLC equipment is often equipped with devices that introduce solvents from two or more reservoirs into a mixing chamber at rates that vary continuously; the volume ratio of the solvents may then be altered linearly or exponentially with time.

Figure 27–5 illustrates the advantage of a gradient eluent in the separation of a mixture of chlorobenzenes. Isocratic elution with a 50 : 50 (v/v) methanol/water solution yielded curve (**b**). Curve (**a**) is for gradient elution, which was initiated with a 40 : 60 mixture of the two solvents; the methanol concentration was then increased at the rate of 8%/min. Note that gradient elution shortened the time of separation significantly without sacrifice in resolution of the early peaks. Note that gradient elution pro-

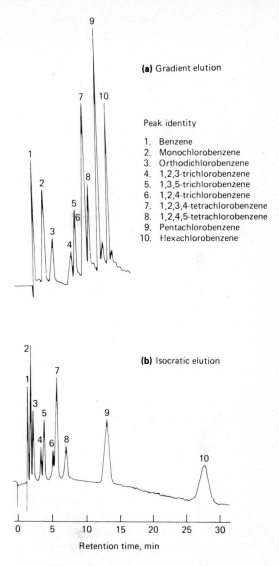

(a) Gradient elution

Peak identity

1. Benzene
2. Monochlorobenzene
3. Orthodichlorobenzene
4. 1,2,3-trichlorobenzene
5. 1,3,5-trichlorobenzene
6. 1,2,4-trichlorobenzene
7. 1,2,3,4-tetrachlorobenzene
8. 1,2,4,5-tetrachlorobenzene
9. Pentachlorobenzene
10. Hexachlorobenzene

(b) Isocratic elution

Retention time, min

FIGURE 27-5 Improvement in separation efficiency by gradient elution. Column: 1 m × 2.1 mm id, precision-bore stainless; packing: 1% Permaphase® ODS. Sample: 5 μL of chlorinated benzenes in isopropanol. Detector: UV photometer (254 nm). Conditions: temperature, 60°C, pressure, 1200 psi. (From: J. J. Kirkland, *Modern Practice of Liquid Chromatography*, p. 88, Interscience: New York, 1971. Reprinted by permission of John Wiley & Sons, Inc.)

duces effects that are similar to those produced by temperature programming in gas chromatography.

27C-2   PUMPING SYSTEMS

A principal problem that had to be overcome in the development of HPLC was the design of suitable pumps. The requirements were severe and included: (1) the generation of pressures of up to 6000 psi (lbs/in^2); (2) pulse-free output; (3) flow rates ranging from 0.1 to 10 mL/min; (4) flow control and flow reproducibility of 0.5% relative or better; and (5) corrosion-resistant components (seals of stainless steel or Teflon).

Three types of pumps, each with its own set of advantages and disadvantages, are encountered, namely reciprocating pumps, syringe or displacement type pumps, and pneumatic or constant pressure pumps.

Reciprocating Pumps.   Reciprocating pumps, which are currently most widely used, usually consist of a small chamber in which the solvent is pumped by the back and forth motion of a motor-driven piston; the solvent is in direct contact with the piston. As an alternative, pressure may be transmitted to the solvent via a flexible diaphragm which in turn is hydraulically pumped by a reciprocating piston. Reciprocating pumps have the disadvantage of producing a pulsed flow, which must be damped because its presence is manifested as base-line noise on the chromatogram. The advantages of reciprocating pumps include their small internal volume (35 to 400 μL), their high output pressures (up to 10,000 psi), their ready adaptability to gradient elution, and their constant flow rates, which are largely independent of column back-pressure and solvent viscosity.

Displacement Pumps.   Displacement pumps usually consist of large, syringe-like chambers equipped with a plunger that is activated by a screw-driven mechanism powered by a stepping motor. Displacement

pumps also produce a flow that tends to be independent of viscosity and back pressure. In addition the output is pulse free. Disadvantages include limited solvent capacity ($\sim$250 mL) and considerable inconvenience when solvents must be changed.

Pneumatic Pumps.   In the simplest pneumatic pumps, the mobile phase is contained in a collapsible container housed in a vessel that can be pressurized by a compressed gas. Pumps of this type are inexpensive and pulse free; they suffer from limited capacity and pressure output as well as a dependence of flow rate on solvent viscosity and column back-pressure. In addition, they are not amenable to gradient elution and are limited to pressures less than about 2000 psi.

It should be noted that the high pressures generated by the pumping devices do not constitute an explosion hazard because liquids are not very compressible. Thus, rupture of a component of the system results only in solvent leakage. Of course, such leakage may constitute a fire hazard.

Flow Control and Programming Systems.   As part of their pumping systems, many commercial instruments are equipped with computer controlled devices for measuring the flow rate by determining the pressure drop across a restrictor located at the pump outlet. Any difference in signal from a preset value is then used to increase or decrease the speed of the pump motor. Most instruments also have a means for varying the composition of the solvent either continuously or in a stepwise fashion. For example, the instrument shown in Figure 27–4 contains a proportioning valve that permits mixing of up to four solvents in a preprogrammed and continuously variable way.

27C-3   SAMPLE INJECTION SYSTEMS

Often, the limiting factor in the precision of liquid chromatographic measurement lies in the reproducibility with which samples can be introduced onto the column packing.

The problem is exacerbated by band broadening, which accompanies overloading columns. Thus, the volumes used must be minuscule—a few tenths of a microliter to perhaps 500 μL. Furthermore, it is convenient to be able to introduce the sample without depressurizing the system.

Syringe Injections. The earliest and simplest means of sample introduction was syringe injection through a self-sealing, elastomeric septum. For this purpose, micro syringes designed to withstand pressures up to 1500 psi are used. Unfortunately, the reproducibility of syringe injection is rarely better than 2 to 3% and is often considerably worse.

Stop-Flow Injection. *Stop-flow* is a second type of syringe injection in which no septum is used. Here the solvent flow is stopped momentarily, a fitting at the column head is removed, and the sample is injected directly onto the head of the column packing. After replacing the fitting, the system is again pressurized. The advantage of this technique is its simplicity.

Sampling Valves. Although syringe injection finds general application because of its convenience, the most widely used sampling devices are now sampling valves or loops, such as the one shown in Figure 25–12, page 750.[4] These devices are often an integral part of liquid-chromatographic equipment and have interchangeable loops providing a choice of sample sizes from 5 to 500 μL. Sampling loops of this type permit the introduction of samples at pressures up to 7000 psi with precisions of a few tenths percent relative. Micro sample injection valves, with sampling loops having volumes of 0.5 to 5 μL are also available.

27C-4 LIQUID-CHROMATOGRAPHIC COLUMNS

Liquid-chromatographic columns are ordinarily constructed from smooth-bore stainless steel tubing although heavy-walled glass tubing is occasionally encountered. The latter is restricted to pressures that are lower than about 600 psi. Hundreds of packed columns differing in size and packing are available from several manufacturers. Their costs range from $200 to $800 generally.

Analytical Columns. The majority of liquid-chromatographic columns range in length from 10 to 30 cm. Normally, the columns are straight with added length, where needed, being gained by coupling two or more columns together. Occasionally coiled columns are encountered although some loss in efficiency results from this configuration. The inside diameter of liquid columns is often 4 to 10 mm; the most common particle size of packings is 5 or 10 μm. Perhaps the most common column currently in use is one that is 25 cm in length, 4.6 mm in inside diameter, and packed with 5 μm particles. Columns of this type contain 40,000 to 60,000 plates/meter.

Recently, manufacturers have been producing high-speed, high-performance micro columns, which have smaller dimensions than those just described.[5] Such columns may have inside diameters that range from 1 to 4.6 mm and be packed with 3 or 5 μm particles. Often, their lengths are as short as 3 to 7.5 cm. Such columns contain as many as 100,000 plates/meter and have the advantage of speed and minimal solvent consumption. The latter property is of considerable importance because the high-purity solvents required for liquid chromatography are expensive. Figure 27–6 shows chromatograms obtained with (**a**) a conventional column and (**b**) a 7.5-cm column packed

[handwritten margin notes: most common; 4-10 id; 25 cm; 5 μm; microcolu; 1-4.6 id; 3-7.5 cm; 3-5 μm; adv; speed; minimal solv]

---

[4] For descriptions of some commercial systems, see: E. Walcek and D. Ball, *Amer. Lab.*, **1982**, *14* (9), 118; M. C. Harvey and S. D. Stearns, *Amer. Lab.*, **1982**, *14* (5), 68.

[5] See: *Microcolumn High-Performance Liquid Chromatography*, P. Kucera, Ed., Elsevier: New York, 1984.

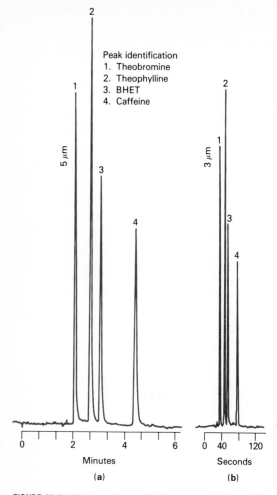

FIGURE 27-6 Comparison of chromatograms with (a) a conventional 25.0-cm column packed with 5.0 μm particles and (b) a 7.5-cm column with 3.0 μm particles. (Courtesy of Beckman Instruments Inc., Fullerton, CA.)

with 3.0 μm particles. Note that satisfactory resolution is realized in both instances but that the separation with the shorter column required only one-third the time with a concomitant saving in solvent.

Guard Columns. Often, a short guard column is introduced before the analytical column to increase the life of the analytical column by removing particulate matter and contaminants from the solvents. In addition, in liq-

uid-liquid chromatography, the guard column serves to saturate the mobile phase with the stationary phase so that losses of this solvent from the analytical column are minimized. The composition of the guard-column packing should be similar to that of the analytical column; the particle size is usually larger, however, to minimize pressure drop.

Column Thermostats. For many applications, close control of column temperature is not necessary and columns are operated at ambient temperature. Often, however, better chromatograms are obtained by maintaining column temperatures constant to a few tenths degree centigrade. Most modern commercial instruments are now equipped with forced air ovens that permit the use of controlled temperature from a few degrees above ambient to 100 to 150°C. Columns may also be fitted with water jackets fed from a constant temperature bath to give precise temperature control.

27C-5 TYPES OF COLUMN PACKINGS

Two basic types of packings have been used in liquid chromatography, *pellicular* and porous particle. The former consist of spherical, nonporous, glass or polymer beads with typical diameters of 30 to 40 μm. A thin, porous layer of silica, alumina, or an ion-exchange resin is deposited on the surface of these beads. For some applications, an additional coating is applied, which consists of a liquid stationary phase that is held in place by adsorption. Alternatively, the beads may be treated chemically to give an organic surface layer.

For most applications, pellicular particles have been displaced by porous micro particles having diameters ranging from 3 to 10 μm. The particles are composed of silica, alumina, or an ion-exchange resin with silica being by far the most common. Silica particles are synthesized by agglomerating submicron silica particles under conditions that lead to larger particles having highly uniform diameters. The resulting particles are often coated with thin organic films, which are chemically or physically bonded to the surface.

27C-6 DETECTORS

Unlike gas chromatography, liquid chromatography has no detectors that are as versatile, as universal in applicability, and as reliable as the flame ionization and thermal conductivity detectors described in Section 26B–6. A major challenge in the development of liquid chromatography has been in detector improvement.

Characteristics of the Ideal Detector. The ideal detector for liquid chromatography should have all of the properties listed on page 766 for gas chromatography with the exception that the liquid chromatography detector need not be responsive over as great a temperature range. In addition, an HPLC detector should have minimal internal volume in order to reduce zone broadening (Section 7B–2).

Types of Detectors. Liquid chromatographic detectors are of three types. General detectors respond to a mobile-phase bulk property which is modulated by the presence of solutes. In contrast, specific detectors respond to some property of the solutes that is not possessed by the mobile phase. Finally, some detectors show a general response to solutes after removal of the mobile phase by volatilization.

Table 27–1 lists the most common detectors for HPLC and some of their more important properties. A recent survey[6] of 365 published papers in which liquid chromatography played an important role revealed that 71% were based upon detection by UV absorption, 15% by fluorescence, 5.4% by refractive index, 4.3% by electrochemical measurements, and another 4.3% by other measurements. Of the UV absorption detectors, 39% were based upon one of the emission lines of mercury, 13% upon filtered radiation from a deuterium source, and 48% upon radiation emitted from a grating monochromator.

Cell Configuration for Absorbance Measurement. Figure 27–7 is a schematic diagram of a typical, Z-shaped, flow-through cell for absorbance

---

[6] See: *Anal. Chem.*, **1982**, *54*, 327A.

Table 27–1
Characteristics of Liquid Chromatography Detectors[a]

Detector Basis	Type[b]	Maximum Sensitivity[c]	Flow Rate Sensitive?	Temperature Sensitivity	Useful with Gradient?	Available Commercially?
UV absorption	S	2×10^{-10}	No	Low	Yes	Yes
IR absorption	S	10^{-6}	No	Low	Yes	Yes
Fluorometry	S	10^{-11}	No	Low	Yes	Yes
Refractive index	G	1×10^{-7}	No	$\pm 10^{-4}$ °C	No	Yes
Conductometric	S	10^{-8}	Yes	2%/°C	No	Yes
Mass spectrometry	G	10^{-10}	No	None	Yes	Yes
Electrochemical	S	10^{-12}	Yes	1.5%/°C	No	Yes
Radioactivity	S	—	No	None	Yes	No

[a] Most of these data were taken from L. R. Snyder and J. J. Kirkland, *Introduction to Modern Liquid Chromatography*, 2d ed., p. 162, Wiley-Interscience: New York, 1979. Reprinted by permission of John Wiley & Sons, Inc.
[b] G = general; S = selective.
[c] Sensitivity for a favorable sample in g/mL.

From column

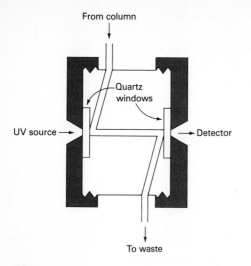

FIGURE 27-7 Ultraviolet detector cell for HPLC.

measurements on eluents from a chromatographic column. In order to minimize extracolumn band broadening, the volume of such a cell is kept as small as possible. Thus, typically, volumes are limited to 1 to 10 μL and cell lengths to 2 to 10 mm. Most cells of this kind are restricted to pressures no greater than about 600 psi. Consequently, a pressure reduction device is often required.

Many absorbance detectors are double-beam devices in which one beam passes through the eluent cell and the other through a filter to reduce its intensity. Matched photoelectric detectors are then used to compare the intensities of the two beams. Alternatively, a chopped beam system similar to that shown in Figure 9–17**b** (p. 269) is used in conjunction with a single phototube. In either case, the chromatogram consists of a plot of the log of the ratio of the two transduced signals as a function of time. Single-beam instruments are also encountered. Here, intensity measurements of the solvent system are stored in a computer memory and ultimately recalled for the calculation of absorbance.

Ultraviolet Absorbance Detectors Based Upon Filters. The simplest UV absorption detectors are filter

photometers (Section 7D–1) with a mercury lamp as the source. Most commonly the intense line at 254 nm is isolated by filters; with some instruments, lines at 250, 313, 334, and 365 can also be employed by substitution of filters. Obviously this type of detector is restricted to solutes that absorb at one of these wavelengths. As shown in Section 7B, several organic functional groups and a number of inorganic species exhibit broad absorption bands that encompass one or more of these wavelengths.

Deuterium or tungsten filament sources with interference filters also provide a simple means of detection of absorbing species as they are eluted from a column. Some modern instruments are equipped with filter wheels containing several filters that can be rapidly switched to detect various species as they are eluted. Such devices are particularly useful for repetitive, quantitative analyses where the qualitative composition of the sample is known so that a sequence of appropriate filters can be chosen. Often, the filter changes are computer controlled.

Absorbance Measurements with Monochromators. Most HPLC manufacturers offer detectors that consist of a scanning spectrophotometer with grating optics. Some are limited to ultraviolet radiation; others encompass both ultraviolet and visible radiation. Several operational modes can be chosen. For example, the entire chromatogram can be obtained at a single wavelength; alternatively, when eluent peaks are sufficiently separated in time, different wavelengths can be chosen for each peak. Here again, computer control is often used to select the best wavelength for each eluent. Where entire spectra are desired for identification purpose, the flow of eluent can be stopped for a sufficient period to permit scanning the wavelength region of interest.

The most powerful ultraviolet spectrophotometric detectors are diode-array instruments such as the one described in Section 7D–3 and Figure 7–15.[7] Several manufacturers offer such instruments, which permit collection of data for an entire spectrum in approximately one second. Thus, spectral data for each chromatographic peak can be collected and stored as it appears at the end of the column. One form of presentation of the spectral data, which is helpful in identification of species and for choosing conditions for quantitative determination, is a three-dimensional plot such as that shown in Figure 27–8. Here, spectra were obtained at successive five-second intervals. The appearance and disappearance of each of the three steroids in the eluent is clearly evident.

Infrared Detectors. Two types of infrared detectors are offered commercially. The first is similar in design to the instrument shown in Figure 11–13 (p. 339) with wavelength scanning being provided by three semicircular filter wedges. The range of this instrument

is from 2.5 to 14.5 μm or 4000 to 690 cm^{-1}.

The second, and much more sophisticated, type of infrared detector is based upon Fourier transform instruments similar to those discussed in Section 11C–2. Several of the manufacturers of Fourier transform instruments offer accessories that permit their use as HPLC detectors.

Infrared detector cells are similar in construction to those used with ultraviolet radiation except that windows are constructed of sodium chloride or calcium fluoride. Cell lengths range from 0.2 to 1.0 mm and volumes from 1.5 to 10 μL.

The simpler infrared instruments can be operated at one or more single wavelength settings; alternatively, the spectra for peaks can be scanned by stopping the flow at the time of elution. The Fourier transform instruments are used in an analogous way to the diode-array instrument for ultraviolet absorbance measurement described in the previous section.

A major limitation to the use of infrared detectors lies in the low transparency of

---

[7] See: J. C. Miller, S. O. George, and B. G. Willis, *Science*, **1982,** *218,* 241; S. A. Borman, *Anal. Chem.,* **1983,** *55,* 836A.

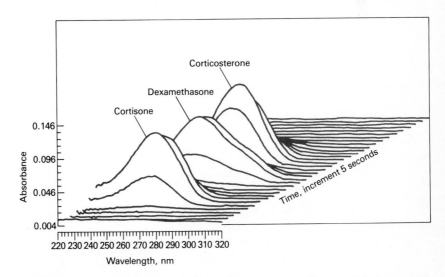

FIGURE 27–8 Absorption spectra of the eluent from a mixture of three steroids taken at 5-sec intervals. (Courtesy of Hewlett-Packard Company, Palo Alto, CA.)

many useful solvents. For example, the broad infrared absorption bands for water and the alcohols largely preclude the use of this detector for many applications.

Fluorescence Detectors.   Fluorescence detectors for HPLC are similar in design to the fluorometers and spectrofluorometers described in Section 8B. In most, fluorescence is observed by a photoelectric detector located at 90 deg to the excitation beam. The simplest detectors employ a mercury excitation source and one or more filters to isolate a band of emitted radiation. More sophisticated instruments are based upon a Xenon source and employ a grating monochromator to characterize the fluorescent radiation. Future developments in fluorescence detectors will probably be based upon tunable laser sources, which should lead to enhanced sensitivity and selectivity.[8]

As was pointed out in Chapter 8, an inherent advantage of fluorescence methods is their high sensitivity, which is typically greater by an order of magnitude or two than most absorbance procedures. This advantage has been exploited in liquid chromatography for the separation and determination of the components of samples that fluoresce. As was pointed out in Section 8C–2, fluorescent compounds are fre-

quently encountered in the analysis of such materials as pharmaceuticals, natural products, clinical samples, and petroleum products. Often, the number of fluorescing species can be enlarged by preliminary treatment of samples with reagents that form fluorescent derivatives. For example, dansylchloride (5-dimethylaminonapthalene-1-sulphonyl chloride), which reacts with primary and secondary amines, amino acids, and phenols to give fluorescent compounds, has been widely used for the detection of amino acids in protein hydrolyzates.

Refractive-Index Detectors.   Figure 27–9 is a schematic diagram of a differential refractive-index detector in which the solvent passes through one-half of the cell on its way to the column; the eluate then flows through the other chamber. The two compartments are separated by a glass plate mounted at an angle such that bending of the incident beam occurs if the two solutions differ in refractive index. The resulting displacement of the beam with respect to the photosensitive surface of a detector causes variation in the output signal, which, when amplified and recorded, provides the chromatogram.

Refractive-index detectors have the significant advantage of responding to nearly all solutes. That is, they are general detectors analogous to flame detectors in gas chromatography. In addition, they are reliable and unaffected by flow rate.

[8] See: R. B. Green, *Anal. Chem.*, **1983**, *55*, 20A; E. S. Yeung and N. J. Sepaniak, *Anal. Chem.*, **1980**, *52*, 1465A.

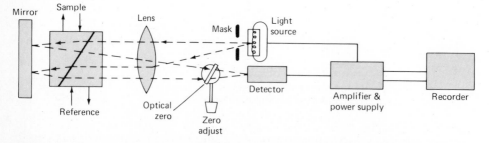

FIGURE 27–9 Schematic diagram of a differential refractive-index detector. (Courtesy of Waters Associates, Inc., Milford, MA 01757.)

Electrochemical Detectors.[9] Electrochemical detectors of several types are currently available from instrument manufacturers. These devices are based upon four of the methods described in Chapters 21, 22, and 23—that is, amperometry, polarography, coulometry, and conductometry.

Although electroanalytical procedures have as yet not been exploited to the extent of optical detectors, they appear to offer advantages, in many instances, of high sensitivity, simplicity, convenience, and widespread applicability. This last property is illustrated in Figure 27–10, which depicts the potential ranges at which oxidation or reduction of 16 organic functional groups occur. In principle, then, species containing any of these groups could be detected by amperometric, polarographic, or coulometric procedures. Thus, electrochemical detection would appear to have the potential for fulfilling a long-time need of HPLC,

namely a sensitive general or universal detector.

A variety of HPLC/electrochemical detector cells have been described in the literature and several are available from commercial sources. Figure 27–11 is an example of a simple thin-layer type of flow-through cell for amperometric detection. Here, the electrode surface is part of a channel wall formed by sandwiching a 50 μm Teflon gasket between two machined blocks of Kel-F plastic. The indicator electrode is platinum, gold, glassy carbon, or a carbon paste. A reference electrode, and often a counter electrode, are located downstream from the indicator electrode block. The cell volume is 1 to 5 μL. A useful modification of this cell, which is available commercially, includes two working electrodes, which can be operated in series or in parallel.[10] The former configuration, in which the eluent flows first over one electrode and then the second, requires that the analyte undergo a reversible oxidation (or reduction) at the upstream

[9] For short reviews of electrochemical detectors, see: P. T. Kissinger, *J. Chem. Educ.*, **1983,** *60,* 308; I. S. Krull, *et al., Amer. Lab.,* **1983,** *15* (2), 57; K. Bratin, *et al., Amer. Lab.,* **1984,** *16* (5), 33.

[10] D. A. Roston, R. E. Shoup, and P. T. Kissinger, *Anal. Chem.,* **1982,** *54,* 1417A.

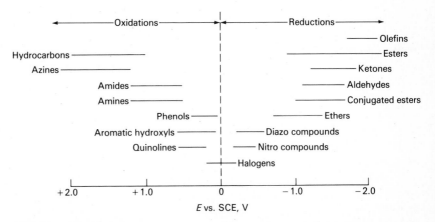

FIGURE 27–10 Potentially detectable organic functional groups by electroanalytical measurements. The horizontal lines show the range of oxidation or reduction potentials wherein compounds containing the indicated functional groups are electroactive.

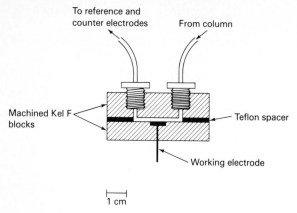

To reference and
counter electrodes   From column

Machined Kel F
blocks   Teflon spacer

Working electrode

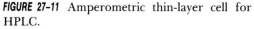
1 cm

FIGURE 27-11 Amperometric thin-layer cell for HPLC.

electrode. The second electrode then operates as a cathode, (or an anode) to determine the oxidation (or reduction) product. This arrangement enhances the selectivity of the detection system. An interesting application of this system is for the detection and determination of the components in mixtures containing both thiols and disulfides. Here, the upstream mercury electrode reduces the disulfides at about -1.0 V. That is,

$$RSSR + 2H^+ + 2e \rightarrow 2RSH$$

A downstream mercury electrode is then oxidized in the presence of the thiols from the original sample as well as those formed at the upstream electrode. That is,

$$2RSH + Hg(l) \rightarrow Hg(SR)_2(s) + 2H^+ + 2e$$

In the parallel configuration, the two electrodes are rotated so that the axis between them is at 90 deg to the stream flow. The two can then be operated at different potentials (relative to a downstream reference electrode), which often gives an indication of peak purity. Alternatively, one electrode can be operated as a cathode and the other as an anode thus making possible detection of both oxidants and reductants.

Polarographic detectors have also been

described in the literature. For example, an accessory is available for the mechanically controlled dropping electrode shown in Figure 22–12 (p. 680); this device directs the eluent flow around the mercury droplet.[11] The potential of the electrode is then maintained at a suitable level during elution. Plots of current versus time provide elution patterns for species that are reduced at the chosen potential.

Conductometric and coulometric[12] detectors are also available from instrument manufacturers. The former are discussed in Section 27G–3.

Mass Spectrometric Detectors. A fundamental problem in coupling liquid chromatography with mass spectrometry is the enormous mismatch between the relatively large solvent volumes from the former and the vacuum requirements of the latter. Several interfaces have been developed for solving this problem.[13] In one, which is available commercially, the eluent from the column is

---

[11] For details, see: S. K. Vohra, *Amer. Lab.,* **1981,** *13* (5), 66.

[12] For a description of series and parallel coulometric detectors, see: R. W. Andrews, *et al., Amer. Lab.,* **1982,** *14* (10), 140.

[13] See: P. J. Arpino and G. Guiochon, *Anal. Chem.,* **1979,** *51,* 682A. See also: *Spectra,* **1983,** *9* (1). The entire issue of this Finnigan MAT publication is devoted to LC/MS interfaces.

split, with only a tiny fraction being introduced directly into the mass spectrometer. Direct liquid introduction systems appear to hold considerable promise when used in conjunction with the new microbore columns, which typically have flow rates of 10 to 50 μL/min. In a second type of interface, which is also sold commercially, the effluent is deposited on a continuous, moving-belt or -wire that transports the solvent and analyte to a heated chamber for removal of the former by volatilization. Following solvent evaporation, the analyte residues on the belt or wire pass into the ion source area, where desorption-ionization occurs.

A new and promising interface, which is now available commercially, is called a *thermospray*.[14] A thermospray interface permits direct introduction of the total effluent from a column at flow rates as high as 2 mL/min. With this interface, the liquid is vaporized as it passes through a stainless steel heated capillary tube to form an aerosol jet of solvent and analyte molecules. In the spray, the analyte is ionized through a charge exchange mechanism with a salt, such as ammonium acetate, which is incorporated in the eluent. Thus, the thermospray is not only an interface but also an ionization source. This new interface is applicable only to polar analyte molecules and polar mobile phases that will dissolve a salt such as ammonium acetate. With these limitations, the thermospray interface provides spectra for a wide range of nonvolatile and thermally stable compounds such as peptides and nucleotides. Detection limits down to 1 to 10 picograms have been reported.

Computer control and data storage is generally used with mass spectrometric detectors. Both real-time and computer-reconstructed chromatograms and spectra of the eluted peaks can be obtained.

27D Mobile Phases

In contrast to gas chromatography, many different types of mobile phases are available for liquid chromatography. Furthermore, the proper choice among these liquids is often critical to the success of a separation. Unfortunately, theories of mobile phase (and packing) interaction with respect to a given set of solutes are imperfect, and at best a scientist can only narrow the choice to one or two classes of solvents. Thus, in the end, it is frequently necessary to perform a series of trial-and-error experiments in which chromatograms are obtained with various solvents and solvent combinations until a satisfactory chromatogram is obtained.

27D-1 DESIRABLE MOBILE-PHASE QUALITIES

In addition to being a good solvent for the samples being studied, a mobile phase for liquid chromatography must possess certain other qualities that include:

1. High purity to avoid introduction of peaks that may overlap analyte peaks.
2. Ready availability at reasonable cost.
3. A boiling point that is 20 to 50°C above the column temperature.
4. Low viscosity. Generally the viscosity should be no greater than 0.5 cP.[15]
5. Low reactivity to avoid chemical interaction with the solutes or polymerization in the presence of the column packing. This requirement generally precludes the use of aldehydes, olefins, many sulfur-containing compounds, ketones, and certain nitrocompounds.
6. For liquid-liquid chromatography, immiscibility with the stationary phase.
7. Compatibility with the detector. Thus, for refractive-index detection, the sol-

---

[14] M. L. Vestal, *Anal. Chem.*, **1983**, *55*, 750, 1741.

[15] The poise, P, is a common unit of viscosity expressed in terms of dyne-sec/cm^2.

vent refractive index should be significantly different from that of the solutes.
For absorbance detection, the solvent obviously should not absorb at the wavelengths to be used.

8. Limited flammability and toxicity.

The foregoing general requirements
greatly reduce the number of candidates for
mobile phases in liquid chromatography.
The list is further reduced by the need for a
solvent in which the capacity factors k' for
the sample components lie in a suitable
range (see Section 25B–4). For a two- or
three-component sample, it is realistic to
search for a solvent in which k' lies in the
ideal range of 2 to 5. For a multicomponent
mixture this range often must be extended
to 0.5 to 20 to provide a time scale large
enough to include separated peaks for all of
the constituents. Usually, several solvents or
solvent combinations can be found that will
provide suitable capacity factors. The choice
among these then depends upon which provides the best selectivity factors (α) for the
sample components.

27D-2 TYPES OF MOBILE PHASE

Mobile phases for partition, adsorption, and
ion-exchange chromatography are interactive in the sense that they interact electro-
statically with sample components. As a
result, retention times may be strongly influenced by solvent type. In contrast, mobile
phases for exclusion (and gas) chromatography are non-interactive. Consequently, retention times with these techniques are
largely independent of the composition of
the mobile phase.

Mobile- and Stationary-Phase Polarities. Successful
chromatography with interactive solvents
requires a proper balance of intermolecular
forces among the three active participants in
the separation process—the solute, the mobile phase, and the stationary phase. These

intermolecular forces are described qualitatively in terms of the relative polarity of
each of the three reactants. The polarities of
various organic functional groups in increasing order are: hydrocarbons < ethers
< esters < ketones < aldehydes < amides <
amines < alcohols. Water is more polar than
compounds containing any of the preceding
functional groups.

As a rule, most good chromatographic
separations are achieved by matching the
polarity of the analytes to that of stationary
phase; a mobile phase of considerably different polarity is then used. This procedure
is generally more successful than one in
which the polarities of the solute and mobile
phase are matched but different from that
of the stationary phase. Here, the stationary
phase often cannot compete successfully for
the sample components; retention times
then become too short for practical application. At the other extreme, of course, is the
situation where the polarity of the solute
and stationary phases are too much alike
and totally different from that of the mobile
phase. Here, retention times become inordi-
nately long.

In summary, then, polarities for solute,
mobile phase, and stationary phase must be
carefully blended if good separations are to
be realized in a reasonable time.

Normal- and Reversed-Phase Chromatography. Two
types of chromatography are distinguishable based upon the relative polarities of the
mobile and stationary phases. Early work in
liquid chromatography was based upon
highly polar stationary phases such as silica
or alumina surfaces or polar liquids, such as
triethyleneglycol supported by silica particles; a relatively nonpolar solvent such as
hexane or i-propylether then served as the
mobile phase. For historic reasons, this type
of chromatography is now referred to as
normal-phase chromatography. In *reversed-phase
chromatography*, the stationary phase is
nonpolar, often a hydrocarbon, and the mo-

bile phase is relatively polar (such as water, methanol, or acetonitrile).[16] In normal-phase chromatography, the *least* polar component is eluted first; *increasing* the polarity of the mobile phase then *decreases* the elution time. In contrast, in the reversed-phase method, the *most* polar component appears first and *increasing* the mobile phase polarity *increases* the elution time. These relationships are illustrated in Figure 27–12.

Table 27–2 lists important physical properties for some of the most widely used chromatographic mobile phases. The *polarity index* (P') and *eluent strength* (ε^0) listed in the last two columns are polarity parameters that aid in the choice of mobile phase for partition and liquid-solid chromatography, respectively. Their use will be described in the sections that follow.

---

[16] For a monograph on reversed-phase chromatography, see: A. M. Krstulovic and P. R. Brown, *Reversed-Phase High Performance Liquid Chromatography*, Wiley: New York, 1982.

27E Partition Chromatography

Partition chromatography has become the most widely used of all of the liquid chromatographic procedures. In the past, most of the applications have been to non-ionic, polar compounds of low to moderate molecular weight (usually < 3000). Recently, however, methods have been developed (derivatization and ion-pairing) that have extended partition separations to ionic compounds; it is possible that in the future partition chromatography may replace ion-exchange as the method of choice for separating many charged species.

Partition chromatography can be subdivided into *liquid-liquid* and *bonded-phase* chromatography. The difference in these techniques lies in the method by which the stationary phase is held on the support particles of the packing. With liquid-liquid, retention is by physical adsorption while with bonded-phase, covalent bonds are involved. Early partition chromatography was exclusively of the liquid-liquid type; by now, how-

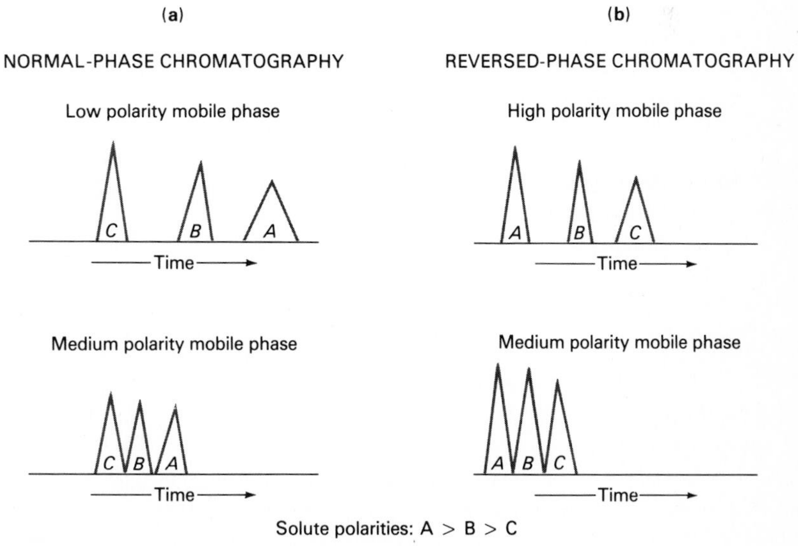

(a)

NORMAL-PHASE CHROMATOGRAPHY

Low polarity mobile phase

Medium polarity mobile phase

(b)

REVERSED-PHASE CHROMATOGRAPHY

High polarity mobile phase

Medium polarity mobile phase

Solute polarities: A > B > C

FIGURE 27–12 The relationship between polarity and elution times for normal-phase and reversed-phase chromatography.

ever, the bonded-phase method has become predominate, with liquid-liquid separations being relegated to certain special applications.

27E-1 MOBILE PHASE SELECTION FOR PARTITION CHROMATOGRAPHY

In Section 25B–4, three methods were described for improving the resolution of a chromatographic column; each is based upon varying one of the three parameters (N, k', and α) contained in Equation 25–20. In liquid chromatography, the capacity factor k' is experimentally the most easily manipulated of the three because of the strong dependence of this constant upon the com-

position of the mobile phase. As noted earlier, for optimal performance, k' should be in the ideal range between 2 and 5; for complex mixtures, however, this range must be expanded to perhaps 0.5 to 20. The larger range unfortunately, leads to longer times for complete separations.

Sometimes, adjustment of k' alone does not suffice to produce individual peaks with no overlap; variation in selectivity factors α must then be resorted to. Here again, the simplest way of bringing about changes in α is by altering the mobile-phase composition, taking care, however, to keep k' within a reasonable range. Alternatively, α can be changed by choosing a different column packing.

Table 27–2
Properties of Common Chromatographic Mobile Phases

Solvent	Refractive Index[a]	Viscosity, cP[b]	Boiling Point, °C	Polarity Index, P'	Eluent Strength,[c] ε^0
Fluoroalkanes[d]	1.27–1.29	0.4–2.6	50–174	<-2	−0.25
Cyclohexane	1.423	0.90	81	0.04	−0.2
n-Hexane	1.372	0.30	69	0.1	0.01
1-Chlorobutane	1.400	0.42	78	1.0	0.26
Carbon tetrachloride	1.457	0.90	77	1.6	0.18
i-Propyl ether	1.365	0.38	68	2.4	0.28
Toluene	1.494	0.55	110	2.4	0.29
Diethyl ether	1.350	0.24	35	2.8	0.38
Tetrahydrofuran	1.405	0.46	66	4.0	0.57
Chloroform	1.443	0.53	61	4.1	0.40
Ethanol	1.359	1.08	78	4.3	0.88
Ethyl acetate	1.370	0.43	77	4.4	0.58
Dioxane	1.420	1.2	101	4.8	0.56
Methanol	1.326	0.54	65	5.1	0.95
Acetonitrile	1.341	0.34	82	5.8	0.65
Nitromethane	1.380	0.61	101	6.0	0.64
Ethylene glycol	1.431	16.5	182	6.9	1.11
Water	1.333	0.89	100	10.2	Large

[a] At 25°C.
[b] Centipoise at 25°C. See footnote 15.
[c] On Al_2O_3. Multiplication by 0.8 gives ε^0 on SiO_2.
[d] Properties depend upon molecular weight. Range of data given.

Solvent-Solute Interactions in the Mobile Phase. To understand how k' for a solute can be made smaller (or sometimes larger) by solvent variation, it is desirable to understand the nature of the interactions between solvent and solute that determine k'. These interactions arise from *dispersion* or *London forces, dipole forces, dielectric forces,* and *molecular complexing.*[17]

Dispersion Interactions. Dispersion interactions between solute and solvent molecules occur because the motions of electrons in a molecule are random. As a result, momentary asymmetric configurations constantly occur in which one part of the molecule becomes negatively charged with respect to the other. This momentary *polarization* of a solute molecule will induce polarization in an adjacent solvent molecule. As a consequence, a momentary electrostatic attraction develops. Dispersion forces are always present in any solute-mobile phase or solute-stationary phase system. They are the only attractive force between nonpolar species. Dispersion interactions are strongest between molecules that are readily polarized or deformed. To a first approximation, refractive index provides a measure of polarizability. Generally, then, strong dispersion interaction can be expected between solvent and solute molecules having high refractive indexes. This interaction is manifested in the large solubilities of polarizable solutes in solvents of high refractive index.

Dipole Interactions. When both solvent and solute molecules have permanent dipoles (that is, permanent charge asymmetry), as shown by high dipole moments, dipolar interactions occur, which lead to enhanced solubility and lower values for k'. These interactions are electrostatic and involve orientation of molecules with respect to one another.

Dielectric Interactions. Dielectric interactions arise from the electrostatic attraction between an ionized solute and a solvent with a high permanent dipole. These interactions are manifested in the high solubilities of ionic and ionizable solutes in polar solvents such as water or methanol.

Molecular Complex Formations. The most common example of molecular complexing is hydrogen bond formation between a solute proton donor and a mobile-phase proton acceptor or the reverse. Another example is found in the formation of silver ion complexes of olefins.

Solvent Strength and the Capacity Factor. Solvents that interact strongly with solutes as a result of four mechanisms just described are often termed "strong" solvents or polar solvents. Several indexes have been developed for quantitatively describing the polarity of solvents. The most useful of these for partition chromatography appears to be the polarity index P', which was developed by Snyder.[18] This parameter is based upon solubility measurements for the substance in question in dioxane (a low dipole proton acceptor), nitromethane (a high dipole proton acceptor), and ethyl alcohol (a high dipole proton donor). From these measurements, an adjusted partition coefficient K'_g for each solvent is obtained. The *polarity index* is then defined as

$$P' = \log (K'_g)_e + \log (K'_g)_d + \log (K'_g)_n \quad (27-2)$$

where the subscripts e, d, and n refer to ethanol, dioxane, and nitromethane, respectively.

An inspection of Table 27–2 reveals that values for the polarity index vary from 10.2 for the highly polar water molecule to -2 for the nonpolar fluoroalkanes. Any desired

---

[17] For a detailed discussion of these forces and their effects on chromatographic selectivity, see: S. H. Langer and R. J. Sheehan, *Advances in Anal. Chem.,* **1968,** *6,* 291.

[18] L. R. Snyder, *J. Chromatogr. Sci.,* **1978,** *16,* 223.

polarity index between these limits can be achieved by mixing two appropriate solvents. Thus, the polarity index P'_{AB} of a mixture of solvents A and B is given by

$$P'_{AB} = \phi_A P'_A + \phi_B P'_B \qquad (27-3)$$

where P'_A and P'_B are polarity indexes of the two solvents and ϕ_A and ϕ_B are the volume fractions of each.

In Section 25B–4, it was pointed out that one of the easiest ways of improving the chromatographic resolution of two species is by manipulation of the capacity factor k'. For liquid chromatography, variations in k' are brought about by varying the polarity index of the mobile phase. Here, adjustment of P' is easily accomplished by the use of mobile phases that consist of a mixture of two solvents. Typically, a 2-unit change in P' results (very roughly) in a 10-fold change in k'. That is, for a normal-phase separation

$$\frac{k'_2}{k'_1} = 10^{(P'_1 - P'_2)/2} \qquad (27-4)$$

where k'_1 and k'_2 are initial and final values of k' for a solute and P'_1 and P'_2 are the corresponding values for P'. For a reversed-phase column,

$$\frac{k'_2}{k'_1} = 10^{(P'_2 - P'_1)/2} \qquad (27-5)$$

It should be emphasized that these equations apply only approximately. Nonetheless they can be useful (see Example 27–1).

EXAMPLE 27-1. In a reversed-phase column, a solute was found to have a retention time of 31.3 min while an unretained species required 0.48 min for elution when the mobile phase was 30% (by volume) methanol and 70% water. Calculate (a) k' and (b) a water/methanol composition that should bring k' to a value of about 5.

(a) Application of Equation 25–15 yields

$$k' = (31.3 - 0.48)/0.48 = 64$$

(b) To obtain P' for the mobile phase we substitute polarity indexes for methanol and water from Table 27–2 into Equation 27–3 to give

$$P' = 0.30 \times 5.1 + 0.70 \times 10.2 = 8.7$$

Substitution of this result into Equation 27–5 gives

$$\frac{5}{64} = 10^{(P'_2 - 8.7)/2}$$

Taking the log of both sides of this equation gives

$$-1.11 = (P'_2 - 8.7)/2 = 0.5\,P'_2 - 4.35$$
$$P'_2 = 6.5$$

Letting x be the volume fraction methanol in the new solvent mixture and substituting again into Equation 27–3, we find

$$6.5 = x \times 5.1 + (1 - x)10.2$$
$$x = 0.73 \text{ or } 73\%$$

Thus, a 73% methanol/27% water mixture should provide the desired value of k'.

Mobile-Phase Selectivity. In many cases, adjusting k' to a suitable level is all that is needed to give a satisfactory separation. When two bands still overlap, however, the selectivity factor α for the two species must be made larger. Such a change can be brought about most conveniently by changing the chemical nature of the mobile phase while holding the predetermined value of k' more or less the same. For example, if the solvent mixture described in Example 27–1 did not provide a satisfactory separation of two species, one of the two solvents could be replaced with a chemically different one. For example, the methanol might be replaced by nitromethane. Here, k' for a nitromethane/water mixture could be determined and the ratio of the two solvents required to give a suitable k' calculated. This mixture would then be tested for improved resolution.

Classification of Solvents. In order to systematize the foregoing trial-and-error procedure, Snyder[19] has developed a classification of solvents based upon their properties as proton donors or acceptors and their permanent dipole strengths. Here, the corrected partition coefficients K'_g mentioned earlier are used to define three selectivity parameters x_e, x_d, and x_n, where

$$x_e = \log(K'_g)_e/P'$$
$$x_d = \log(K'_g)_d/P'$$
$$x_n = \log(K'_g)_n/P'$$

The subscripts refer to the three reference solutes, ethanol, dioxane, and nitromethane respectively, and P' is the polarity index for the mobile phase under consideration.

When the three selectivity parameters for a variety of common solvents are plotted on triangular graph, as in Figure 27–13, eight classes of mobile phases are distinguishable which differ in their dipoles and their properties as proton donors and proton acceptors. Table 27–3 lists the types of mobile phases that appear in each class. Note that solvents with low P' values, such as aliphatic hydrocarbons, carbon tetrachloride, and fluorinated hydrocarbon compounds, are not classified in this system.

The mobile-phase selectivity triangle is

---

[19] L. R. Snyder, *J. Chromatogr. Sci.*, **1978**, *16*, 223.

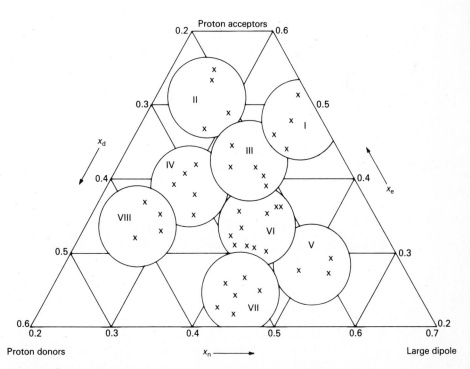

FIGURE 27-13 Mobile-phase classification triangle. Each **x** represents the position of one mobile phase. (From: L. R. Snyder, *J. Chromatogr. Sci.*, **1978**, *16*, 227. With permission.)

useful for systematically varying the composition of solvent systems to find a mobile phase giving satisfactory α values. For example, as a first step in a normal-phase separation, a solvent system such as chloroform/ n-hexane might be chosen on the basis of sample solubility. The composition would then be varied, as in Example 27–1, to find a mixture producing a k' value in the 2 to 5 range. If this mixture did not give good peak separations, the more polar component (chloroform) would then be changed. Normally, the more powerful interactions between solute and a polar solvent will overwhelm similar interactions with nonpolar ones. Thus, it is usually more profitable to change the polar rather than the nonpolar component. In choosing a substitute for chloroform, which belongs in class VIII, it would be wise to choose a solvent from a class that is most distant from VIII in the selectivity triangle in Figure 27–13—that is, group I or V. Turning to Table 27–3, we see that a possible choice might be an aliphatic ether from group I. Various mixtures of the ether and hexane would then be tested until a suitable value of k' was realized. If this composition still did not provide a satisfactory chromatogram, the process would be repeated with a group V solvent such as methylene chloride. It should be noted that moving to a new class of solvents is usually much more effective in changing α than choosing a second solvent within the same group.

For reversed-phase chromatography, a four-solvent optimization procedure has been developed that is also based on the selectivity triangle.[20] Here, three compatible solvents are used for adjustment of α values. These include methanol, acetonitrile, and tetrahydrofuran from groups VI, III, and II, respectively. Water is then used to adjust the strength of the mixture in such a way as to yield a suitable value for k'. For normal-phase operations a similar four-solvent system is used in which the selectivity solvents are ethyl ether, methylene chloride, and chloroform; the solvent strength adjustment is then made with n-hexane. With these four-solvent systems, optimization is said to be possible with a minimal number of experiments.

Figure 27–14 illustrates the systematic, four-solvent approach to the development of a separation of six steroids by reversed-phase chromatography. Figures 27–14a and b show the results from initial experiments to determine the minimum value of k' required; these experiments were performed by varying the acetonitrile to water ratio. At $k' = 5$, elution is over so rapidly that peak separation cannot, under any cir-

---

[20] See: R. Lehrer, *Amer. Lab.*, **1981,** *13* (10), 113; and J. L. Glajch, *et al.*, *J. Chromatogr. Sci.*, **1980,** *199*, 57.

Table 27–3 Classes of Mobile Phases	
I.	Aliphatic ethers and alkyl amines.
II.	Aliphatic alcohols.
III.	Tetrahydrofuran, pyridine derivatives, dimethylsulfoxide, amides (except formamide).
IV.	Formamide, acetic acid, benzyl alcohol, glycols.
V.	Methylene chloride, 1, 2-dichloroethane.
VI.	Halogenated alkanes, esters, ketones, dioxanes, nitriles, aniline.
VII.	Benzene and benzene derivatives, aliphatic nitro compounds.
VIII.	Chloroform, m-cresol, water, fluoroalkanols.

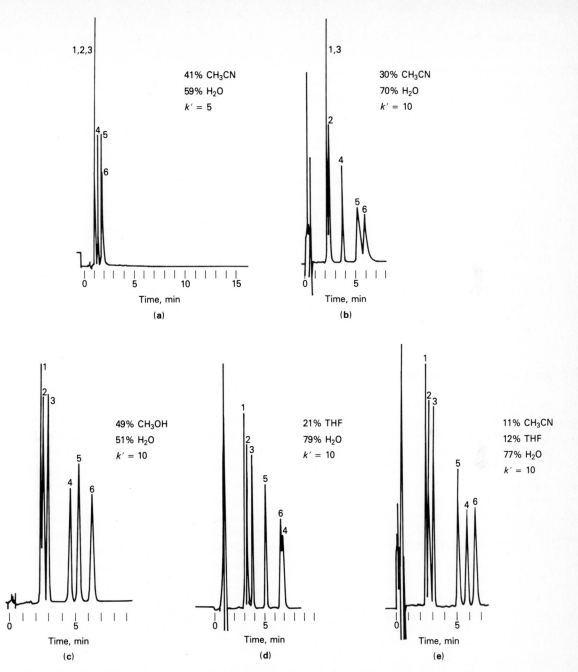

FIGURE 27-14 Systematic approach to the separation of six steroids. The use of water to adjust k' is shown in (**a**) and (**b**). The effect of varying α at constant k' are shown in (**b**), (**c**), (**d**), and (**e**). Column: 0.4×150 mm packed with 5 μm C_8 bonded, reversed-phase particles. Temperature: 50°C. Flow rate: 3.0 cm³/min. Detector: UV-254 nm. THF = tetrahydrofuran. CH_3CN = acetonitrile. Compounds: 1) prednisone; 2) cortisone; 3) hydrocortisone; 4) dexamethasone; 5) corticosterone; 6) cortoexolone. (Courtesy of DuPont Instrument Systems, Wilmington, DE.)

cumstance, be expected. With k' increased to 10, in contrast, room exists on the time scale for discrete peaks; here, however, the α value for components 1 and 3 (and to a lesser extent 5 and 6) is not great enough for satisfactory resolution. Further experiments were then performed with the goal of finding better α values; in each case, the water concentration was adjusted to a level that yielded $k' = 10$. The results of experiments with methanol/water and tetrahydrofuran/water mixtures are shown in Figures 27–14c and d. Several additional experiments involving the systematic variation of pairs of the organic solvents were performed (in each case adjusting k' to 10 with water). Finally, the mixture shown in Figure 27–14e was chosen as the best mobile phase for the separation of the particular group of compounds.

27E-2 PACKINGS FOR PARTITION CHROMATOGRAPHY

As was mentioned earlier, column packing for partition chromatography is of two types with bonded-phase now being by far the more popular. In fact, it has been estimated that over 75% of all liquid chromatographic separations (including ion-exchange and exclusion) are performed on bonded columns.

Comparison of Bonded-Phase and Liquid-Liquid Type Packings. The most important advantages of bonded-phase packings are attributable to the enhanced stability that results from chemically bonding the stationary phase to the support. In contrast to physical bonding, losses of the stationary phase owing to solubility in the mobile phase are not encountered. To minimize losses in liquid-liquid chromatography, it is necessary to saturate the mobile phase with the stationary phase, usually by passage through a guard column. Even with this precaution, periodic recoating of liquid-liquid packings is necessary.

A second advantage of bonded-phase packings is that they permit gradient elution, again because of their stability. Gradient separations are not practical with physically bonded phases because of solubility losses.

Finally, bonded-phase packings having a variety of polarities have now been developed, thus permitting both normal-phase and reversed-phase separations.

Liquid-liquid chromatography does offer certain advantages that accounts for its continued, if limited, use. Among these are: high capacity, which permits separations of large samples without serious loss of efficiency; broad selectivity resulting from the wide choice of stationary phases; high reproducibility because the support plays little role in the separation process; and convenience of column renewal because the packing can be readily stripped and recoated thus leading to long column life. In contrast, bonded-phase packings have a limited lifetime after which the packings must be replaced. Most replacement columns cost from $100 to $300.

Siloxane Packings. Almost all bonded-phase packings are prepared from rigid silica, or silica-based, compositions. These solids are formed as uniform, porous, mechanically sturdy particles commonly having diameters of 3, 5, or 10 μm. The surface of fully hydrolyzed silica (hydrolyzed by heating with 0.1 M HCl for a day or two) is made up of chemically reactive silonal groups. That is,

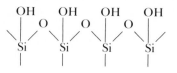

Typical silica surfaces contain about 8 μmol/m² of OH groups.

The most useful bonded-phase particles are siloxanes formed by reaction of the hydrolyzed surface with an organochlorosilane. For example,

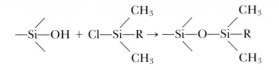

where R is generally a straight chain alkyl group. In many commercial preparations the alkyl is either an *n*-octyl or an *n*-decaoctyl group.

Surface coverage by silanization is limited to 4 μmol/m^2 or less because of steric effects. The unreacted SiOH groups, unfortunately, impart an undesirable polarity to the surface, which may lead to tailing of chromatographic peaks, particularly for basic solutes. To lessen this effect, siloxane packings are frequently *capped* by further reaction with chlorotrimethylsilane which, because of its smaller size, can bond many of the unreacted silanol groups.

Silanization is also carried out with the bifunctional alkylmethyldichlorosilane and the trifunctional alkyltrichlorosilane, which are more reactive than the monochloro reagent. Between 1 and 2 mol of SiOH groups are bonded by 1 mol of these reagents. The products, therefore, contain unreacted chloro groups, which hydrolyze upon further treatment to produce an increased population of silanol groups on the surface. Their presence also leads to tailing of peaks and lowered capacities.

Figure 27–15 illustrates the effect of chain length of the alkyl group upon performance. As expected, longer chains produce packings that are more retentive. In addition, longer chain lengths permit the use of larger samples. For example, the maximum sample size for a C$_{18}$ packing is roughly double that for a C$_4$ preparation under similar conditions.

It has been estimated that more than three-quarters of all HPLC separations are currently being carried out on octyl- or octyldecylsiloxane packings. With such preparations, the long chain hydrocarbon groups are aligned parallel to one another and perpendicular to the particle surface giving a brush- or bristle-like structure. The mechanism by which these surfaces retain solute molecules is at present not entirely clear. Some scientists believe that from the standpoint of the solute molecules, the brush behaves as a liquid hydrocarbon medium. Others propose that the bonded-phase imbibes components of the mobile phase to form what amounts to a conventional immobilized liquid stationary phase. Still others prefer to view the brush surface as a modified surface at which adsorption occurs. The molecules of the mobile phase then compete with the analyte molecules for position on the organic surface. Regardless of the detailed mechanism of retention, a bonded coating can be treated as if it were a conventional, physically retained liquid.

In most applications of reversed-phase chromatography, elution is carried out with a highly polar mobile phase such as an aqueous solution containing various concentrations of such solvents as methanol, acetonitrile, or tetrahydrofuran. In this mode, care must be taken to avoid pH values greater than about 7.5 because hydrolysis of the siloxane takes place, which leads to degradation or destruction of the packing.

Detailed discussions of preparation and properties of siloxane bonded surfaces are available in several sources.[21]

Other Types of Packings.   Two other types of reactions that are used for preparing bonded-phase surfaces involve ester formation and formation of silicon/carbon or silicon/nitrogen bonds. Typical reactions are

---

[21] For example, see: W. R. Melander and C. Horváth, in *High Performance Liquid Chromatography*, C. Horváth, Ed., vol. 2, pp. 123–162, Academic Press: New York, 1980. For brief discussions, see: I. Halasz, *Anal. Chem.*, **1980,** 52, 1393A; B. L. Karger and R. W. Giese, *Anal. Chem.*, **1978,** 50, 1048A.

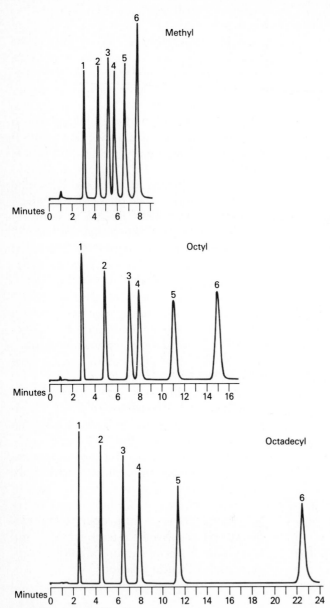

Operating conditions
Instrument: IBM LC/9533
Columns: IBM's Octadecyl, Octyl,
  and Methyl (all 4.5 × 250mm)
Mobile phase: 50/50 Methanol/Water
Flow rate: 1.0 mL/min
Detection: IBM LC/9522 (254 nm)

Peak identification
 1. Uracil
 2. Phenol
 3. Acetophenone
 4. Nitrobenzene
 5. Methyl benzoate
 6. Toluene

FIGURE 27-15 Effect of chain-length on performance of reversed-phase siloxane columns packed with 5 μm particles. (Courtesy of IBM Instruments Inc., Danbury, CT.)

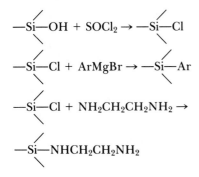

and

where Ar represents an aromatic group.

Numerous polar packings for normal-phase chromatography have been developed based upon these reactions (and the earlier reaction as well). Several of these products are available commercially. To obtain polar bonded-phases, organic functional groups such as esters, ethers, diols, nitriles, or amines are incorporated into the molecules bonded to the surface. Polarities differ considerably among the various functional groups, which provides a range of α values for various solute pairs.

Effect of Sample Volume and Weight in Partition Chromatography. As mentioned earlier, resolution is dependent upon sample volume. This effect is shown in Table 27–4 which relates maximum permissible sample volume V_s to peak volume V_p (flow rate times peak width) for several levels of resolution loss. The example that follows illustrates the use of this table.

EXAMPLE 27-2. Two adjacent chromatographic peaks were found to have average peak widths of 18 sec with a solvent flow rate of 1.9 mL/min. What maximum sample size can be used if the loss in resolution is to be kept below (a) 5% and (b) 10%?

Here, $V_p = \dfrac{18 \text{ sec} \times 1.9 \text{ (mL/min)}}{60 \text{ sec/min}}$

$= 0.570 \text{ mL or } 570 \ \mu\text{L}$

(a) For 5% resolution, we see from Table 27–4,

$$V_s = 0.25 \times 570 = 140 \ \mu\text{L}$$

(b) For 10% resolution,

$$V_s = 0.44 \times 570 = 250 \ \mu\text{L}$$

The effect of sample weight on resolution is shown in Figure 27–3, where the superiority of reversed-phase, bonded packings are evident. Nevertheless, liquid-liquid packings, loaded with a maximum amount of a nonviscous stationary phase, are more tolerant toward heavy loading and usually permit the use of larger samples before *severe* efficiency losses are encountered. For this reason, liquid-liquid packings are often employed for preparative chromatographic separations.

Effect of pH. Often, in reversed-phase chromatography with aqueous solutions, it is desirable to add buffering reagents to suppress solute ionization, which leads to tailing. Furthermore, gradient elution involving changes in pH is sometimes helpful in improving separation of acidic analytes.

Table 27–4
Maximum Sample Volume for Partition Chromatography[a]

$V_s =$	Permissible Resolution Loss, %
0.25 V_p	5
0.33 V_p	8
0.44 V_p	10
0.57 V_p	20

[a] From: L. R. Snyder and J. J. Kirkland, *Introduction to Modern Chromatography*, 2d ed., p. 290, Wiley: New York, 1979. Reprinted by permission of John Wiley & Sons, Inc.

It should be stressed, however, that bonded-phase silica packings cannot be used with solutions having pH values much greater than 7 because of stripping of the organic functional group.

Temperature. Because aqueous solvents are relatively viscous, reversed-phase separations are sometimes performed at elevated temperatures (50 to 80°C). These elevated temperatures not only increase sample solubility but also decrease k' (roughly a twofold decrease in k' accompanies a 30° increase in temperature).

Applications of Partition Chromatography. Reversed-phase bonded packings, when used in conjunction with highly polar solvents (often aqueous), approach the ideal, universal system for liquid chromatography. Because of their wide range of applicability, their convenience, and the ease with which k' and α can be altered by manipulation of aqueous mobile phases, these packings are frequently applied before all others for exploratory separations with new types of samples.

Both bonded-phase and liquid-liquid chromatography separate on the basis of types and numbers of functional groups contained in sample molecules. Partition chromatography is particularly useful for homologs and benzologs, which differ in molecular weight.

Table 27–5 lists a few typical examples of the multitude of uses of partition chromatography in various fields. A more complete picture of the ubiquity of this technique can be obtained by consulting recent reviews in *Analytical Chemistry*.[22]

Figure 27–16 illustrates two of many thousands of applications of bonded-phase, partition chromatography to the analysis of consumer and industrial materials.

Derivative Formation. In some instances, it is useful to convert the components of a sample to a derivative before, or sometimes after, chromatographic separation is undertaken. Such treatment may be desirable (1) to reduce the polarity of the species so that partition rather than adsorption or ion-exchange columns can be used, (2) to increase the detector response for all of the sample components, and (3) to selectively enhance the detector response to certain components of the sample.

Figure 27–17 illustrates the use of derivatives to reduce polarity and enhance sensitivity. The sample was made up of 30 amino acids of physiological importance. Heretofore such a separation would be performed on an ion-exchange column with photometric detection based upon post-column reaction of the amino acids with a colorimetric reagent such as ninhydrin. The chromato-

---

[22] R. E. Majors, H. G. Barth, and C. H. Lochmüller, *Anal. Chem.*, **1984**, *56*, 300R; **1982**, *54*, 323R; H. F. Walton, *Anal. Chem.*, **1980**, *52*, 15R.

Table 27–5
Typical Applications of Partition Chromatography

Field	Typical Mixtures
Pharmaceuticals	Antibiotics, Sedatives, Steroids, Analgesics
Biochemical	Amino acids, Proteins, Carbohydrates, Lipids
Food products	Artificial sweetners, Antioxidants, Aflatoxins, Additives
Industrial chemicals	Condensed aromatics, Surfactants, Propellants, Dyes
Pollutants	Pesticides, Herbicides, Phenols, PCBs
Forensic chemistry	Drugs, Poisons, Blood alcohol, Narcotics
Clinical medicine	Bile acids, Drug metabolites, Urine extracts, Estrogens

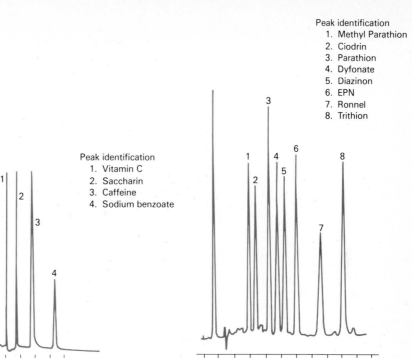

Peak identification
1. Methyl Parathion
2. Ciodrin
3. Parathion
4. Dyfonate
5. Diazinon
6. EPN
7. Ronnel
8. Trithion

Peak identification
1. Vitamin C
2. Saccharin
3. Caffeine
4. Sodium benzoate

Time (min)
(a)

Minutes
(b)

FIGURE 27–16 Typical applications of bonded-phase chromatography. (**a**) Soft-drink additives. Column: 4.6 to 250 mm packed with polar (nitrile) bonded-phase packings. Isocratic solvent: 6% HOAC/94% H_2O. Flow rate: 1.0 cm^3/min. (Courtesy of DuPont Instrument Systems, Wilmington, DE.) (**b**) Organophosphate insecticides. Column 4.5 × 250 mm packed with 5 μm, C_8, bonded-phase particles. Gradient: 67% CH_3OH/33% H_2O to 80% CH_3/20% H_2O. Flow rate: 2 mL/min. (Courtesy of IBM Instruments Inc., Danbury, CT.) Both used 254 nm UV detectors.

gram shown in Figure 27–17 was obtained by automatic, *pre*-column derivative formation with orthophthalaldehyde. The substituted isoindoles[23] formed by the reaction exhibit intense fluorescence at 425 nm, which permits detection down to a few picomoles (10^{-12} mol). Furthermore, the polarity of the derivatives are such that separa-

tion on a C_{18} reversed-phase packing becomes feasible. The advantages of this newer procedure are speed and smaller sample size.

Ion-Pair Chromatography. Ion-pair chromatography has extended the applications of bonded-phase chromatography to highly polar, multiply charged, and strongly basic samples. Both the normal-phase and the reversed-phase mode have been used, although the latter is more popular. For many applications, ion-pair chromatography ap-

[23] For the structure of these compounds, see: S. S. Simons, Jr. and D. F. Johnson, *J. Amer. Chem. Soc.*, **1976**, *98*, 7098.

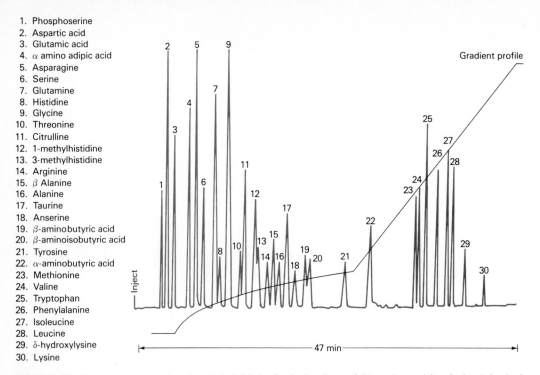

1. Phosphoserine
2. Aspartic acid
3. Glutamic acid
4. α amino adipic acid
5. Asparagine
6. Serine
7. Glutamine
8. Histidine
9. Glycine
10. Threonine
11. Citrulline
12. 1-methylhistidine
13. 3-methylhistidine
14. Arginine
15. β Alanine
16. Alanine
17. Taurine
18. Anserine
19. β-aminobutyric acid
20. β-aminoisobutyric acid
21. Tyrosine
22. α-aminobutyric acid
23. Methionine
24. Valine
25. Tryptophan
26. Phenylalanine
27. Isoleucine
28. Leucine
29. δ-hydroxylysine
30. Lysine

FIGURE 27-17 Chromatogram of orthophthalaldehyde derivatives of 30 amino acids of physiological importance. Column: 5 μm C_{18}, reversed-phase. Solvent A: 0.05 M Na_2HPO_4, pH 7.4, 96:2:2 $CH_3OH/THF/H_2O$. Solvent B: 65:35 CH_3OH/H_2O. Fluorescence detector: excitation 334 nm; emission 425 nm. (Reprinted with permission from: R. Pfiefer, *et al., Amer. Lab.,* **1983,** *15* (3), 86. Copyright 1983 by International Scientific Communications, Inc.)

pears to be replacing ion-exchange separations because of the higher efficiency, speed, and reproducibility of the former.

When reversed-phase packings are used, the mobile phase in ion-pair chromatography consists of an aqueous buffer containing an organic solvent such as methanol or acetonitrile and a *counter-ion* of opposite charge to the analyte. A counter-ion is an ion that combines with the analyte ion to form an *ion-pair,* which is a neutral species that is soluble in nonaqueous solvents. Some common counter-ions that have been used for reversed-phase separations are shown in Table 27–6.

As an example, consider the reversed-phase elution of carboxylic acids from a dimethylsiloxane bonded packing with a pH 7.4 buffered aqueous solution of tetrabu-

tylammonium ion. As a first approximation, the equilibrium between the aqueous and the bonded organic phases can be written in the form

$$A_{aq}^- + B_{aq}^+ \rightleftharpoons A^-B_{org}^+$$

where A^- represents the carboxylate anion and B^+ is the tetrabutylammonium ion $(C_4H_9)_4N^+$ and A^-B^+ is the ion-pair.

A concentration equilibrium constant E for this reaction takes the form

$$E = \frac{[A^-B^+]_{org}}{[A^-]_{aq}[B^+]_{aq}}$$

where E will depend upon pH, ionic strength, and concentration of any organic solvent present. Note that $[A^-B^+]_{org}/[A^-]_{aq}$

corresponds to C_S/C_M in Equation 25–1. Thus,

$$E = K/[\text{B}^+]_{\text{aq}} \qquad (27\text{–}6)$$

where K is the partition coefficient for A⁻ between the stationary and mobile phases. Substitution of this relationship into Equation 25–12 gives

$$k' = E[\text{B}^+]_{\text{aq}} V_S/V_M \qquad (27\text{–}7)$$

Thus, the capacity factor k' is readily varied by adjusting the concentration of the counter-ion.

Figure 27–18 shows a typical ion-pair chromatogram obtained on a reversed-phase octadecylsiloxane packing. The sample was a sustained-release antihistamine containing four amines. The counter-ion was the heptane sulfonate ion $\text{C}_7\text{H}_{15}\text{SO}_3^-$.

27F Adsorption Chromatography

Adsorption or liquid-solid chromatography is the classic form of liquid chromatography first introduced by Tswett at the beginning of this century. In more recent times, it has been adapted to and become an important member of HPLC methods.

The only stationary phases that are used for liquid-solid HPLC are silica and alumina with the former being preferred for most but not all applications because of its higher sample capacity and its wider range of useful forms. With a few exceptions, the adsorption characteristics of the two substances parallel one another. With both, the order of retention times is: olefins < aromatic hydrocarbons < halides, sulfides < ethers < nitro-compounds < esters ≈ aldehydes ≈ ketones < alcohols ≈ amines < sulfones < sulfoxides < amides < carboxylic acids. Note that the groups having the smallest values of k' head this list.

27F-1 SOLVENT SELECTION FOR ADSORPTION CHROMATOGRAPHY

In liquid-solid chromatography, the only variable available to optimize k' and α is the composition of the mobile phase (in contrast to partition chromatography, where the column packing has a pronounced effect on α). Fortunately, in adsorption chromatogra-

---

Table 27–6
Systems for Reversed-Phase Ion-Pair Chromatograms

Sample	Mobile Phase	Counter-Ion	Type Stationary Phase
Amines	0.1 M HClO₄/H₂O/acetonitrile	ClO_4^-	BP[a]
	H₂O/CH₃OH/H₂SO₄	$\text{C}_{12}\text{H}_{25}\text{SO}_3^-$	BP
Carboxylic acids	pH 7.4	$(\text{C}_4\text{H}_9)_4\text{N}^+$	BP
	pH 7.4	$(\text{C}_4\text{H}_9)_4\text{N}^+$	L[b]
Sulfonic acids	H₂O/C₃H₇OH	$(\text{C}_{16}\text{H}_{33})(\text{CH}_3)_3\text{N}^+$	BP
	pH 7.4	$(\text{C}_4\text{H}_9)_4\text{N}^+$	L[b]
	pH 3.8	Bis-(2-ethylhexyl)phosphate	L[c]
Dyes	pH 2–4; H₂O/CH₃OH	$(\text{C}_4\text{H}_9)_4\text{N}^+$	BP

[a] Bonded-phase.
[b] Adsorbed 1-pentanol.
[c] Adsorbed bis-(2-ethylhexyl)phosphoric acid/CHCl₃.

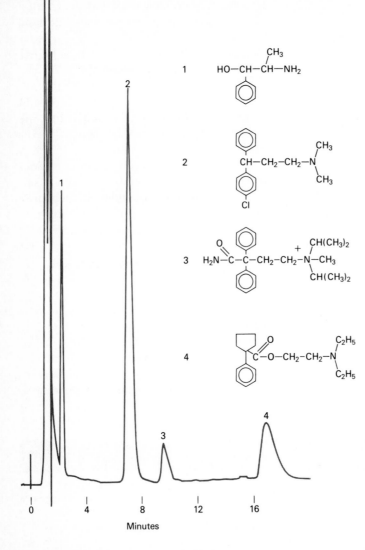

FIGURE 27–18 Chromatogram of a sustained-release, antihistamine capsule (ornade spanule). Column: 3.9 × 300 mm packed with 10 μm octadecylsiloxane coated particle. Mobile phase: 0.005 M $C_7H_{15}SO_3H$, 1% HOAc, 30% acetonitrile in H_2O. Detector: UV at 235 nm. Flow rate: 2.0 mL/min. (Reprinted with permission from: S. R. Bakalyar, *Amer. Lab.,* **1978,** *10* (6), 53. Copyright 1978 by International Scientific Communications, Inc.)

phy, enormous variations in resolution and retention time accompany variations in the solvent system, and only rarely can a suitable mobile phase not be found.

Solvent Strength. It has been found that the polarity index P', which was described in Section 27E–1, can also serve as a rough guide to the strengths of solvents for adsorption chromatography. A much better index, however, is the *eluent strength* ε^0, which is the adsorption energy per unit area of solvent.[24] This parameter depends upon the adsorbent, with ε^0 values for silica being about 0.8 of those on alumina. The values for ε^0 in the last column of Table 27–2 are

---

[24] See: L. R. Snyder, *Principles of Adsorption Chromatography*, Chapter 8, Marcel Dekker: New York, 1968.

for alumina. Note that solvent-to-solvent differences in ε^0 roughly parallel those for P'.

Choice of Solvent Systems. The techniques for choosing a solvent system for adsorption chromatography are similar to those used in separations based upon partition. That is, two compatible solvents are chosen, one of which is too strong (ε^0 too large) and the other of which is too weak. A suitable value for k' is then obtained by varying the volume ratio of the two. It has been found that an increase in ε^0 value by 0.05 unit usually decreases all k' values by a factor of 3 to 4. Thus, enormous variations in k' are possible, and some binary system involving the solvents in Table 27-2 can be found that will give adequate retention times for nearly any sample. Unfortunately ε^0 does not vary linearly with volume ratios, as was the case with P' in partition chromatography. Thus, calculating an optimal mixture is more difficult. Graphs have been developed, however, that relate ε^0 to composition for a number of common binary solvent mixtures.[25]

When overlapping peaks are encountered, exchanging one strong solvent for another while holding k' more or less constant, will change α values and often provide the desired resolution. This trial-and-error approach is tedious and sometimes unsuccessful. Methods for systematizing and shortening the search have been developed.[26] One is to carry out preliminary scouting by means of thin-layer chromatography, the open-bed version of adsorption chromatography. Under similar conditions the thin-layer retardation factor R_f (Section 28B-2) is related to k' by the relationship

$$k' = \frac{(1 - R_f)}{R_f}$$

Determining R_f is often much easier and faster than obtaining k' from a column chromatogram.

27F-2   APPLICATIONS OF ADSORPTION CHROMATOGRAPHY

Figure 27-1 shows that adsorption chromatography is best suited for nonpolar compounds having molecular weights less than perhaps 5000. Although some overlap exists between adsorption and partition chromatography, the methods tend to be complementary.

Generally, liquid-solid chromatography is best suited to samples that are soluble in nonpolar solvents and correspondingly have limited solubility in aqueous solvents such as those used in the reversed-phase partition procedure. As with partition chromatography, compounds with differing kinds or numbers of functional groups are usually separable. A particular strength of adsorption chromatography, which is not shared by other methods, is its ability to differentiate among the components of isomeric mixtures. Table 27-7 compares selectivities of partition and adsorption chromatography for several types of analytes. Note that resolution of homologs and benzologs is generally better with reversed-phase partition chromatography. Separation of isomers, however, is usually better with the adsorption procedure.

Figure 27-19 illustrates a typical application of adsorption chromatography.

27G   Ion-Exchange Chromatography

Ion-exchange chromatography was the first liquid chromatographic procedure to be adapted to automation in the early 1960s. The applications at that time were to the

---

[25] For this and other practical aspects of adsorption chromatography, see: D. L. Saunders, *J. Chromatogr. Sci.*, **1977**, *15*, 372.

[26] See: L. R. Snyder and J. J. Kirkland, *Introduction to Modern Liquid Chromatography*, 2d ed., pp. 365–389, Wiley: New York, 1979.

Table 27-7
Comparison of Selectivities of Adsorption and Reversed-Phase Chromatography[a]

Separation of	Compound	Adsorption α	Reversed-Phase α
Homologs			
	R = C_1	4.8	3.3
	C_2		6.5
	C_4	4.1	17
	C_{10}	3.6	
Benzologs		1.2	1.4
		1.1	1.8
Isomers		12.5	1.06
		1.8	
		3.4	
	1,2,3,4 dibenzanthrocene, $C_{22}H_{14}$/Picene $C_{22}H_{14}$	20	

[a] Data from: L. R. Snyder and J. J. Kirkland, *Introduction to Modern Liquid Chromatography*, 2d ed., pp. 357–358, Wiley: New York, 1979. Reprinted by permission of John Wiley & Sons, Inc.

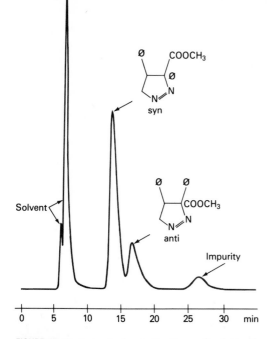

FIGURE 27-19 A typical application of adsorption chromatography: separations of *syn-* and *anti*-pyrazoline. Column: 100×0.3 cm pellicular silica. Mobile phase: 50% methylene chloride/isooctane. Temperature: ambient. Flow rate: 0.25 mL/min. Detector: UV, 254 nm. (Courtesy of Hewlett-Packard Company, Palo Alto, CA.)

separation and determination of amino acids in various physiological fluids and proteins. These separations were carried out at low pressure and typically required 10 to 70 hr to complete. By now of course, ion-exchange separations are performed under HPLC conditions. As mentioned earlier, ion-pair chromatography is now taking over many of the applications that were formerly reserved for ion-exchange. Nevertheless, for certain applications, the older procedure is still the method of choice.

27G–1 ION-EXCHANGE EQUILIBRIA

Ion-exchange processes are based upon exchange equilibria between ions in solution and ions of like sign on the surface of an essentially insoluble, high-molecular-weight solid. Natural ion-exchangers, such as clays and zeolites, have been recognized and used for several decades. Synthetic ion-exchange resins were first produced in the mid-1930s for water softening, water deionization, and solution purification; more recently, these synthetic resins have found general use for analytical separation of ions. The most common active sites for cation exchange resins are the sulfonic acid group $-SO_3^-H^+$, a strong acid, and the carboxylic acid group $-COOH$, a weak acid. Anionic exchangers contain tertiary amine groups $-N(CH_3)_3^+OH^-$ or primary amine groups $-NH_3OH$; the former is a strong base and the latter a weak one.

When a sulfonic acid ion-exchanger is brought in contact with an aqueous solvent containing a cation M^{x+}, an exchange equilibria is set up that can be described by

$$x\underset{\text{solid}}{RSO_3^-H^+} + \underset{\text{solution}}{M^{x+}} \rightleftharpoons$$

$$\underset{\text{solid}}{(RSO_3^-)_xM^{x+}} + \underset{\text{solution}}{xH^+}$$

where $RSO_3^-H^+$ represents one of many sulfonic acid groups attached to a large polymer molecule. Similarly a strong base exchanger interacts with the anion A^{x-} as shown by the reaction

$$x\underset{\text{solid}}{RN(CH_3)_3^+OH^-} + \underset{\text{solution}}{A^{x-}} \rightleftharpoons$$

$$\underset{\text{solid}}{[RN(CH_3)_3^+]_xA^{x-}} + \underset{\text{solution}}{xOH^-}$$

As an example of the application of the mass-action law to ion-exchange equilibria, we will consider the reaction between a singly charged ion B^+ with a sulfonic acid resin held in a chromatographic column. Initial retention of B^+ ions at the head of the column occurs because of the reaction

$$RSO_3^-H^+(s) + B^+(aq) \rightleftharpoons$$
$$RSO_3^-B^+(s) + H^+(aq) \quad (27\text{--}8)$$

Here, the (s) and (aq) emphasize that the system contains a solid and a liquid phase. Elution with a dilute solution of hydrochloric acid shifts the equilibrium in Equation 27–8 to the left causing part of the B^+ ions in the stationary phase to be transferred into the mobile phase. These ions then move down the column in a series of transfers between the stationary and mobile phases.

The equilibrium constant K_{ex} for the exchange reaction shown in Equation 27–8 takes the form

$$\frac{[RSO_3^- B^+]_s [H^+]_{aq}}{[RSO_3^- H^+]_s [B^+]_{aq}} = K_{ex} \qquad (27\text{–}9)$$

Here, $[RSO_3^- B^+]_s$ and $[RSO_3^- H^+]_s$ are concentrations (strictly activities) of B^+ and H^+ *on the solid surface.* Rearranging yields

$$\frac{[RSO_3^- B^+]_s}{[B^+]_{aq}} = K_{ex} \frac{[RSO_3^- H^+]_s}{[H^+]_{aq}} \qquad (27\text{–}10)$$

During the elution, the aqueous concentration of hydrogen ions is much larger than the concentration of the singly charged B^+ ions in the mobile phase. Furthermore, the exchanger has an enormous number of exchange sites relative to the number of B^+ ions being retained. Thus, the overall concentrations $[H^+]_{aq}$ and $[RSO_3^- H^+]_s$ are not affected significantly by shifts in the equilibrium 27–8. Therefore, when $[RSO_3^- H^+]_s \gg [RSO_3^- B^+]_s$ and $[H^+]_{aq} \gg [B^+]_{aq}$ the right-hand side of Equation 27–10 is substantially constant, and we can write

$$\frac{[RSO_3^- B^+]_s}{[B^+]_{aq}} = K = \frac{C_S}{C_M} \qquad (27\text{–}11)$$

where K is a constant that corresponds to the distribution coefficient as defined by Equation 25–1. All of the equations in Table 25–3 (Section 25C) can then be applied to ion-exchange chromatography in the same way as to the other types, which have already been considered.

Note that K_{ex} in Equation 27–9 represents the affinity of the resin for the ion B^+ *relative* to another ion (here, H^+). Where K_{ex} is large, a strong tendency exists for the solid phase to retain B^+; where K_{ex} is small, the reverse obtains. By selecting a common reference ion such as H^+, distribution ratios for different ions on a given type of resin can be experimentally compared. Such experiments reveal that polyvalent ions are much more strongly held than singly charged species. Within a given charge group, however, differences appear that are related to the size of the hydrated ion as well as to other properties. Thus, for a typical sulfonated cation exchange resin, values for K_{ex} decrease in the order $Tl^+ > Ag^+ > Cs^+ > Rb^+ > K^+ > NH_4^+ > Na^+ > H^+ > Li^+$. For divalent cations, the order is $Ba^{2+} > Pb^{2+} > Sr^{2+} > Ca^{2+} > Ni^{2+} > Cd^{2+} > Cu^{2+} > Co^{2+} > Zn^{2+} > Mg^{2+} > UO_2^{2+}$.

For anions, K_{ex} decreases in the order $SO_4^{2-} > C_2O_4^{2-} > I^- > NO_3^- > Br^- > Cl^- > HCO_2^- > CH_3CO_2^- > OH^- > F^-$. This sequence is somewhat dependent upon type of resin and reaction conditions and should thus be considered only approximate.

27G-2 ION-EXCHANGE PACKINGS

Historically, ion-exchange chromatography was performed on small, porous beads formed during emulsion copolymerization of styrene and divinylbenzene. The presence of divinylbenzene (usually ~8%) results in cross-linking, which imparts mechanical stability to the beads. In order to make the polymer active towards ions, acidic or basic functional groups are then bonded chemically to the structure. The most common groups are sulfonic acid and quaternary amines.

Figure 27–20 shows the structure of a strong acid resin. Note the cross-linking that holds the linear polystyrene molecules together. The other types of resins have similar structures except for the active functional group.

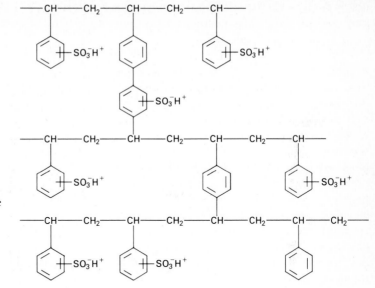

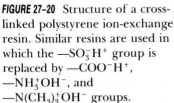

FIGURE 27-20 Structure of a cross-linked polystyrene ion-exchange resin. Similar resins are used in which the $-SO_3^-H^+$ group is replaced by $-COO^-H^+$, $-NH_3^+OH^-$, and $-N(CH_3)_3^+OH^-$ groups.

Porous polymeric particles are not entirely satisfactory for chromatographic packings because of the slow rate of diffusion of analyte molecules through the micropores of the polymer matrix and because of the compressibility of the matrix. To overcome this problem, two newer types of packings have been developed and are in more general use than the porous polymer type. One is a pellicular bead packing in which the surface of a relatively large, (30 to 40 μm) nonporous, spherical, glass or polymer bead is coated with a synthetic ion-exchange resin. A second type of packing is prepared by coating porous microparticles of silica, such as those used in adsorption chromatography, with a thin film of the exchanger. With either type, faster diffusion in the polymer film leads to enhanced efficiency. On the other hand, the sample capacity of these particles is less, particularly for the pellicular type.

27G–3 APPLICATIONS OF ION-EXCHANGE CHROMATOGRAPHY

The mobile phase in ion-exchange chromatography must have the same general properties that are required for other types of chromatography. That is, it must dissolve the sample, have a solvent strength that leads to reasonable retention times (correct k' values), and interact with solutes in such a way as to lead to selectivity (suitable α values). The mobile phases in ion-exchange chromatography are aqueous solutions, which may contain moderate amounts of methanol or other water miscible organic solvents; these solvents also contain ionic species, often in the form of a buffer. Solvent strength and selectivity are determined by the kind and concentration of these added ingredients. In general, the ions of the mobile phase compete with analyte ions for the active sites on the ion-exchange packing.

Inorganic Applications; Ion Chromatography.[27]   Ion-exchange chromatography was first exploited for the analysis of inorganic species in the 1940s and 1950s with several notable suc-

---

[27] For a brief review of inorganic chromatography, see: H. Small, *Anal. Chem.,* **1983,** *55,* 235A. For a detailed description of ion chromatography, see: F. C. Smith, Jr. and R. C. Chang, *The Practice of Ion Chromatography,* Wiley: New York, 1983.

cesses such as the separation of rare earth and isotope ions—separations that previously had been difficult or impossible.

While ion-exchange chromatography was acknowledged as a powerful separatory tool, its widespread application for the determination of inorganic species was inhibited by the lack of a good, general detector, which would permit quantitative determination of ions on the basis of chromatographic peak areas. Conductivity detectors are an obvious choice for this task. They can be highly sensitive, they are universal for charged species, and, as a general rule, they respond in a predictable way to concentration changes. Furthermore, such detectors are simple, inexpensive to construct and maintain, easy to miniaturize, and ordinarily give prolonged, trouble-free service. The only limitation to conductivity detectors proved to be a serious one, which delayed their general use until the 1970s. This limitation arises from the high electrolyte concentration required to elute most analyte ions in a reasonable time. As a consequence, the conductivity from the mobile-phase components tends to swamp that from analyte ions, thus greatly reducing the detector sensitivity.

In 1975, the problem of high eluent conductance was solved[28] by the introduction of a so-called eluent suppressor column immediately following the ion-exchange column. The suppressor column is packed with a second ion-exchange resin which effectively converts the ions of the solvent to a molecular species of limited ionization without affecting the analyte ions. For example, when cations are being separated and determined, hydrochloric acid is chosen as the eluting reagent, and the suppressor column is an anion exchange resin in the hydroxide form. The product of the reaction in the suppressor is water. That is,

$$H^+(aq) + Cl^-(aq) + Resin^+OH^-(s) \rightarrow$$
$$Resin^+Cl^-(s) + H_2O$$

The analyte cations are, of course, not retained by this second column.

For anion separations, the suppressor packing is the acid form of a cation exchange resin. Here, sodium bicarbonate or carbonate may serve as the eluting agent. The reaction in the suppressor is then

$$Na^+(aq) + HCO_3^-(aq) + Resin^-H^+(s) \rightarrow$$
$$Resin^-Na^+(s) + H_2CO_3(aq)$$

Here, the largely undissociated carbonic acid does not contribute to the conductivity.

An inconvenience associated with the original suppressor columns was the need to regenerate them periodically (typically, every 8 to 10 hr) in order to convert their packings back to the original acid or base form. Recently, however, fiber membrane suppressors have become available that operate continuously. For example, where sodium carbonate or bicarbonate are to be removed, the eluent is passed through a hollow fiber constructed from sulfonated polyethylene. Acid for regeneration flows continuously around the outside of the fiber, which acts as a kind of ion-exchange membrane. That is, the sodium ions from the eluent exchange with hydrogen ions on the inner surface of the sulfonated resin, where they can migrate to the other surface for exchange with hydrogen ions of the regenerating reagent. Hydrogen ions migrate in the reverse direction thus preserving electrical neutrality.

Figure 27–21 shows two applications of ion chromatography based upon a suppressor column and conductometric detection. In each case, the ions were present in the parts per million range, and the sample sizes were 50 μL in one case and 100 μL in the other. The method is particularly important for anion analysis because no other rapid

[28] H. Small, T. S. Stevens, and W. C. Bauman, *Anal. Chem.*, **1975**, *47*, 1801.

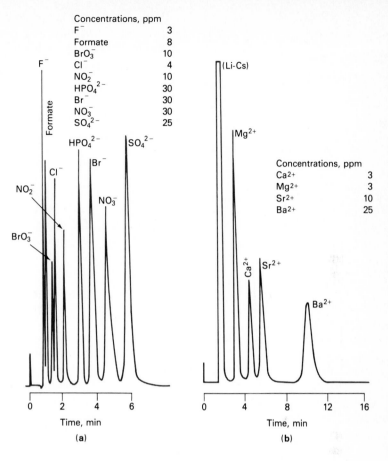

FIGURE 27-21 Typical applications of ion chromatography. (**a**) Separation of anions on an anion exchange column. Eluent: 0.0028 M NaHCO$_3$/ 0.0023 M Na$_2$CO$_3$. Sample size: 50 μL. (**b**) Separation of alkaline earth ions on a cation exchange column. Eluent: 0.025 M phenylenediamine dihydrochloride/0.0025 M HCl. Sample size: 100 μL. (Courtesy of Dionex, Inc., Sunnyvale, CA.)

and convenient method for handling mixtures of this type now exists.

Recently, equipment has become available for ion chromatography in which no suppressor column is used. This approach depends upon the small differences in conductivity between the eluted sample ions and the prevailing eluent ions. To amplify these differences, low-capacity exchangers are used and specially selected eluents, which have low equivalent conductances, are chosen.[29]

An indirect photometric method that permits the separation and detection of nonabsorbing anions and cations has recently been described.[30] Here also, no suppressor column is used, but instead, anions or cations that do absorb are used to displace the analyte ions from the column. When the analyte ions are displaced from the exchanger, their place is taken by an equal number of eluent ions. Thus, the absorbance of the eluate decreases as analyte ions exit from the column. Figure 27-22 shows a chromatogram obtained by this procedure. Here, the eluent was a dilute solution of di-

[29] See: J. S. Fritz, D. T. Gjerde, and C. Phlandt, *Ion Chromatography*, Huthig: Heidelberg, 1982; J. S. Fritz, D. J. Gjerde, and R. M. Becker, *Anal. Chem.,* **1980,** *52,* 1519.

[30] H. Small and T. E. Miller, *Anal. Chem.,* **1982,** *54,* 462.

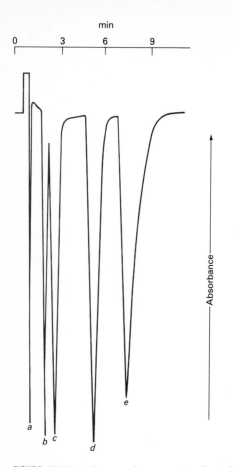

min

0   3   6   9

— Absorbance —

a

b *c*

d

e

FIGURE 27–22 Indirect photometric detection of several anions by elution. Eluent: 10^{-3} M disodium phthalate, 10^{-3} M boric acid, pH 10. Flow rate: 5 mL/min. Sample volume: 0.02 mL. UV detector. Sample ions: (**a**) 18 μg carbonate; (**b**) 1.4 μg chloride; (**c**) 3.8 μg phosphate; (**d**) 5 μg azide; (**e**) 10 μg nitrate. (Reprinted with permission from: H. Small, *Anal. Chem.*, **1983**, *55*, 240A. Copyright 1983 American Chemical Society.)

sodium phthalate, phthalate ion being the ultraviolet-absorbing displacing ion.

Organic and Biochemical Applications. Ion-exchange chromatography has been applied to a variety of organic and biochemical systems including drugs and their metabolites, serums, food preservatives, vitamin mixtures, sugars, and pharmaceutical preparations.

An example of one of these applications is shown in Figure 27–23 in which 1×10^{-8} mol each of 17 amino acids was separated on a cation-exchange column.

27H   Size-Exclusion Chromatography

Size-exclusion chromatography is the newest of the liquid chromatographic procedures.[31] It has also been called gel, gel filtration, and gel permeation chromatography. Size-exclusion chromatography is a powerful technique that is particularly applicable to high-molecular weight species (see Figure 27–1).

27H–1   PRINCIPLES OF SIZE-EXCLUSION CHROMATOGRAPHY

Packings for size-exclusion chromatography consist of small ($\sim$10 μm) silica or polymer particles containing a network of uniform pores into which solute and solvent molecules can diffuse. While in the pores, molecules are effectively trapped and removed from the flow of the mobile phase. The average residence time in the pores depends upon the effective size of the analyte molecules. Molecules that are larger than the average pore size of the packing are excluded and thus suffer essentially no retention; such species are the first to be eluted. Molecules having diameters that are significantly smaller than the pores can penetrate throughout the pore maze and are thus entrapped for the greatest time; these are last to be eluted. Between these two extremes are intermediate-size molecules whose average penetration into the pores of the packing depends upon their diameters. Within this group, fractionation occurs, which is directly related to molecular size and to some

---

[31] For a monograph on this technique, see: W. Yau, J. Kirkland, and D. Bly, *Modern Size-Exclusion Liquid Chromatography*, Wiley: New York, 1979.

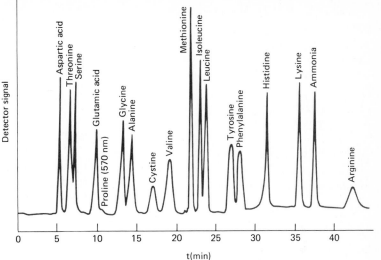

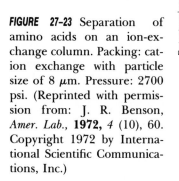

FIGURE 27-23 Separation of amino acids on an ion-exchange column. Packing: cation exchange with particle size of 8 μm. Pressure: 2700 psi. (Reprinted with permission from: J. R. Benson, *Amer. Lab.,* **1972,** *4* (10), 60. Copyright 1972 by International Scientific Communications, Inc.)

extent molecular shape. Note that size-exclusion separations differ from the other procedures we have been considering in the respect that no chemical or physical interaction between analytes and the stationary phase are involved. Indeed, every effort is made to avoid such interactions because they lead to impaired column efficiencies.

27H-2 COLUMN PACKING

Two types of packing for size-exclusion chromatography are encountered, organic gels and silica-based particles. The latter have the advantages of greater rigidity, which leads to easier packing; greater stability, which permits the use of a wider range of solvents including water; more rapid equilibration with new solvents; and stability at higher temperatures. The disadvantages of silica-based particles include their tendency to retain solutes by adsorption and their potential for catalyzing the degradation of solute molecules.

Most early size-exclusion chromatography was carried out on cross-linked styrene-

divinylbenzene copolymers similar in structure (except that the sulfonic acid groups are absent) to that shown in Figure 27–20. The pore size of these polymers is controlled by the extent of cross-linking and hence the relative amount of divinylbenzene present during manufacture. As a consequence, polymeric packings having several different average pore sizes are marketed. Originally, styrene-divinylbenzene gels were hydrophobic and thus could only be used with nonaqueous mobile phases. Now, however, hydrophilic gels are available making possible the use of aqueous solvents for the separation of large, water-soluble molecules such as sugars. These hydrophilic gels are sulfonated divinylbenzenes or polyacrylamides.

Porous glasses and silica particles, having average pore sizes ranging from 40 Å to 2500 Å, are now available commercially. In order to reduce adsorption, the surface of these particles are often modified by reaction with organic substituents. For example, the surface of one hydrophilic packing has the structure

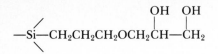

Table 27–8 lists the properties of some typical commercial size-exclusion packings.

27H–3 THEORY OF SIZE-EXCLUSION CHROMATOGRAPHY

The Distribution Coefficient. The total volume V_t of a column packed with a porous polymer or silica gel is given by

$$V_t = V_g + V_i + V_o \qquad (27-12)$$

where V_g is the volume occupied by the solid matrix of the gel, V_i is the volume of solvent held in its pores, and V_o is the free volume outside the gel particles. Assuming no mixing or diffusion, V_o also represents the theoretical volume of solvent required to transport through the column those components too large to enter the pores of the gel. In fact, however, some mixing and diffusion will occur, and as a consequence the nonretained components will appear in a Gaussian-shaped band with a concentration maximum at V_o. For components small enough to enter freely into the pores of the gel, band maxima will appear at the end of the column at an eluent volume corresponding to $(V_i + V_o)$. Generally, V_i, V_o, and V_g are of the same order of magnitude; thus, a gel column permits separation of the large molecules of a sample from the small with a minimal volume of eluate.

Molecules of intermediate size are able to transfer into some fraction K of the solvent held in the pores; the elution volume V_e for these retained molecules is

$$V_e = V_o + KV_i \qquad (27-13)$$

Equation 27–13 applies to all of the solutes on the column. For molecules too large to enter the gel pores, $K = 0$ and $V_e = V_o$; for molecules that can enter the pores unhindered, $K = 1$ and $V_e = (V_o + V_i)$. In deriving Equation 27–13, the assumption was made that no interaction, such as adsorption, occurs between the solute molecules and the gel surfaces. With adsorption, the amount of interstitially held solute will in-

	Table 27–8		
	Properties of Typical Commercial Packings for Size-Exclusion Chromatography		
Type	**Particle Size, μm**	**Average Pore Size, Å**	**Molecular Weight Exclusion Limit[a]**
Polystyrene-divinylbenzene	10	10^2	700
		10^3	(0.1 to 20) $\times$ 10^4
		10^4	(1 to 20) $\times$ 10^4
		10^5	(1 to 20) $\times$ 10^5
		10^6	(5 to > 10) $\times$ 10^6
Silica	10	125	(0.2 to 5) $\times$ 10^4
		300	(0.03 to 1) $\times$ 10^5
		500	(0.05 to 5) $\times$ 10^5
		1000	(5 to 20) $\times$ 10^5

[a] Molecular weight above which no retention occurs.

crease; with small molecules; K will then be greater than unity.

Equation 27–13 rearranges to

$$K = (V_e - V_o)/V_i = C_S/C_M \qquad (27\text{–}14)$$

where K is the distribution coefficient for the solute (see Equation 25–1). Values of K range from zero for totally excluded large molecules to unity for small molecules. The distribution coefficient is a valuable parameter for comparing data from different packings. In addition, it makes possible the application of all of the equations in Table 25–3 to exclusion chromatography.

Calibration Curves. The useful molecular weight range for a size-exclusion packing is conveniently shown by means of a calibration curve such as that shown in the upper part of Figure 27–24. Here, molecular weight, which is directly related to the size of solute molecules, is plotted against retention volume V_R, where V_R is the product of the retention time and the volumetric flow rate. Note that the ordinate scale is logarithmic. The *exclusion limit* defines the molecular weight of a species beyond which no retention occurs. All species having greater molecular weight than the exclusion limit are so large that they are not retained and elute

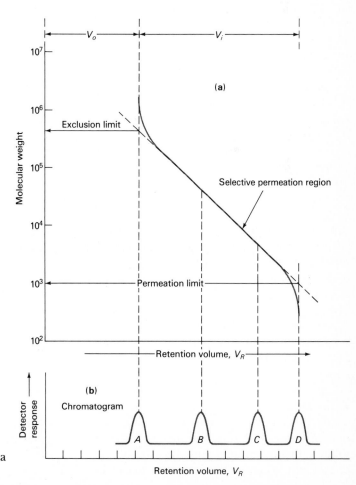

FIGURE 27-24 Calibration curve for a size-exclusion column.

together to give peak *A* in the chromatogram shown in Figure 27–24**b**. The *permeation limit* is the molecular weight below which the solute molecules can penetrate into the pores completely. All molecules below this molecular weight are so small that they elute as the single band labeled *D*. As molecular weights decrease from the exclusion limit, solute molecules spend more and more time, on the average, in the particle pores and thus move progressively more slowly. It is in the selective permeation region that fractionation occurs yielding individual solute peaks such as *B* and *C* in the chromatogram.

Experimental calibration curves, similar in appearance to the hypothetical one in Figure 27–24 are readily obtained by means of standards. Often, such curves are supplied by manufacturers of packing materials.

27H–4 APPLICATION OF SIZE-EXCLUSION CHROMATOGRAPHY

Size-exclusion methods are subdivided into *gel filtration* and *gel permeation* chromatography. The former use aqueous solvents and hydrophilic packings. The latter are based upon nonpolar organic solvents and hydrophobic packings. The methods are complementary in the sense that the one is applied to water-soluble samples and the other to substances soluble in less polar organic solvents.

Comparison of Size-Exclusion to Other Types of Liquid Chromatography. The most important advantages of exclusion procedures include: (1) short and well-defined separation times [all solutes leave the column between V_o and ($V_o + V_i$) in Equation 27–12 and Figure 27–24]; (2) narrow bands, which lead to good sensitivity; (3) freedom from sample loss because solutes do not interact with the stationary phase; and (4) absence of column deactivation brought about by interaction of solute with the packing.

The disadvantages are: (1) only a limited number of bands can be accommodated because the time scale of the chromatogram is short; and (2) inapplicability to samples of similar size, such as isomers. Generally, a 10% difference in molecular weight is required for reasonable resolution.

Examples of Typical Uses. One useful application of the size-exclusion procedure is to the separation of high-molecular weight, natural-product molecules from low-molecular weight species and from salts. For example, a gel with an exclusion limit of several thousand can clearly separate proteins from amino acids and low-molecular weight peptides.

Another useful application of gel permeation chromatography is to the separation of homologs and oligomers. These applications are illustrated by the two examples shown in Figure 27–25. The first shows the separation of a series of fatty acids ranging in molecular weight from 116 to 344 on a polystyrene-based packing with an exclusion limit of 1000. The second is a chromatogram of a commercial epoxy resin, again on a polystyrene packing. Here, *n* refers to the number of monomeric units (mw = 284) in the molecules.

Figure 27–26 illustrates an application of size-exclusion chromatography to the determination of glucose, sucrose, and fructose in four types of fruit juices. The packing, which had an exclusion limit of 1000, was a cross-linked polystyrene polymer made hydrophilic by sulfonation. A 25-cm column with this packing contained 7600 plates at 80°C, the temperature used.

Another major application of size-exclusion chromatography is to the rapid determination of molecular weight or molecular weight distribution of larger polymers or natural products. Here, the elution volumes of the sample are compared with elution

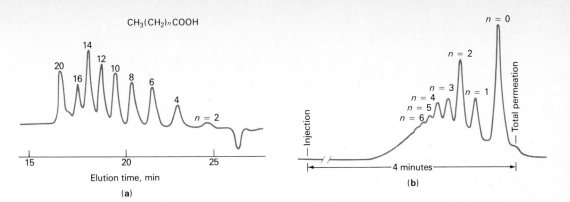

FIGURE 27-25 Application of size-exclusion chromatography. (**a**) Separation of fatty acids. Column: polystyrene based, 7.5 × 600 mm, with exclusion limit of 1×10^4. Mobile phase: tetrahydrofuran. Flow rate: 1.2 mL/min. Detector: refractive index. (**b**) Analysis of a commercial epoxy resin ($n = $ number monomeric units in the polymer). Column: porous silica 6.2 × 250 mm. Mobile phase: tetrahydrofuran. Flow rate: 1.3 mL/min. Detector: UV absorption. (Courtesy of DuPont Instrument Systems, Wilmington, DE.)

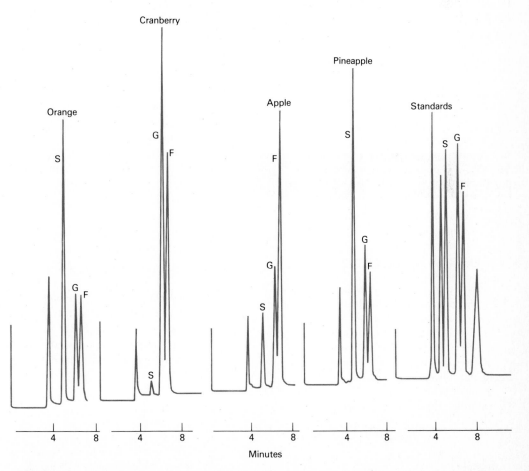

FIGURE 27-26 Determination of glucose (G), fructose (F), and sucrose (S) in canned juices. (Courtesy of Perkin-Elmer Corporation, Norwalk, CT.)

volumes for a series of standard compounds that possess the same chemical characteristics.

27I   Comparison of High-Performance Liquid Chromatography with Gas-Liquid Chromatography

Table 27–9 provides a comparison between high-performance liquid chromatography (HPLC) and gas-liquid chromatography. When either is applicable, gas-liquid chromatography offers the advantage of speed and simplicity of equipment. On the other hand, HPLC is applicable to nonvolatile substances (including inorganic ions) and thermally unstable materials whereas gas-liquid chromatography is not. Often, the two methods are complementary.

Table 27–9
Comparison of High-Performance Liquid and Gas-Liquid Chromatography

Characteristics possessed by both methods:
    Efficient, highly selective, and widely
      applicable
    Only small sample required
    May be nondestructive of sample
    Readily adapted to quantitative analysis
Particular advantages of high-performance
  liquid chromatography:
    Can accommodate nonvolatile and
      thermally unstable samples
    Generally applicable to inorganic ions
Particular advantages of gas-liquid
  chromatography:
    Simple and inexpensive equipment
    Rapid
    Unparalled resolution (with capillary
      columns)
    Easily interfaced with MS

27J   Supercritical-Fluid Chromatography

Supercritical-fluid chromatography is a kind of hybrid of gas and liquid chromatography that combines some of the best features of each. This technique is still in its infancy, and it is not clear at this time how important a branch of chromatography it will become. Its potential appears to be considerable, however.[32]

27J–1   IMPORTANT PROPERTIES OF SUPERCRITICAL FLUIDS

At the critical temperature of a substance, its vapor and its liquid have the same density, and the substance is referred to as a *supercritical fluid*. Above the critical temperature, a gas cannot be liquified no matter how high the pressure. Table 27–10 compares certain properties of supercritical fluids to those of typical gases and liquids. The properties chosen are those that are of importance in gas, liquid, and supercritical-fluid chromatography.

Critical temperatures for fluids that have been used for chromatography vary widely from about 30°C to over 200°C. Lower critical temperatures are advantageous from several standpoints. Thus, much of the work to date has focused on such supercritical fluids as carbon dioxide, ethane, and nitrous oxide, which have critical temperatures of 31, 32, and 37°C. Note that these temperatures, and the pressures at these temperatures, are well within the operating conditions of ordinary HPLC.

An important property of supercritical fluids, which is related to their high densities (0.2 to 0.5 g/cm^3), is their ability to dis-

---

[32] For reviews of this technique, see: L. G. Randall, *Sep. Sci. Technol.*, **1982**, *17*, 1; U. van Wasen, I. Swaid, and G. M. Schneider, *Angew. Chem. Int. Ed. Engl.*, **1980**, *19*, 575; D. R. Gere, *Science*, **1983**, *222*, 253; J. C. Fjeldsted and M. L. Lee, *Anal. Chem.*, **1984**, *56*, 619A.

solve large, nonvolatile molecules. For example, supercritical carbon dioxide readily dissolves *n*-alkanes containing from 5 to 22 carbon atoms, di-*n*-alkylphthalates in which the alkyl groups contain 4 to 16 carbon atoms, and various polycyclic aromatic hydrocarbons made up of several rings. It is perhaps noteworthy that certain important industrial processes are based upon the high solubility of organic species in supercritical carbon dioxide. For example, this medium has been employed for extracting caffeine from coffee beans to give decaffeinated coffee and for extracting nicotine from cigarette tobacco.

27J-2 OPERATING VARIABLES AND INSTRUMENTATION

As mentioned earlier, the pressures and temperatures required for creating supercritical fluids derived from several common gases and liquids lie well within the operating limits of ordinary HPLC equipment. In order to adapt such equipment to supercritical applications, it is thus only necessary to provide an independent means for controlling the internal pressure of the system. One instrument manufacturer now offers an adapter for this purpose.[33]

Effect of Pressure. Pressure changes in supercritical chromatography have a pronounced effect on k'. For example, increasing the carbon dioxide pressure in a packed column from about 70 atm to 90 atm was found to reduce the elution time for hexadecane from about 25 min to 5 min. This effect is general and, in fact, a kind of gradient elution is practiced in which the column pressure is increased linearly as the elution proceeds. The results here are analogous to those obtained with temperature-gradient elution in gas chromatography and solvent-gradient elution in liquid chromatography.

Stationary Phases. To date, most supercritical-fluid chromatography has been carried out with column packings that are commonly used in liquid chromatography. In addition, capillary supercritical-fluid chromatography has been performed with stationary phases that consist of organic films bonded to capillary tubing.[34] Because of the low viscosity of supercritical fluids, long columns (50 m or more) can be used. The advantages here are similar to the advantages encountered in capillary gas chromatography—that is, very high resolutions in a reasonable elapsed time.

Mobile Phases. The most widely used mobile phase for supercritical-fluid chromatography is carbon dioxide. It is an excellent solvent for a variety of organic molecules. In

---

[33] Hewlett-Packard Company, Avondale, PA 19311.

[34] See: M. Novotny, *et al.*, *Anal. Chem.*, **1981**, *53*, 407A.

Table 27-10
Comparison of Properties of Supercritical Fluids with Liquids and Gases (All of the data are order-of-magnitude only)

	Gas (STP)	Supercritical Fluid	Liquid
Density (g/cm^3)	$(0.6-2) \times 10^{-3}$	$0.2-0.5$	$0.6-1.6$
Diffusion coefficient (cm^2/sec)	$(1-4) \times 10^{-1}$	$10^{-3}-10^{-4}$	$(0.2-2) \times 10^{-5}$
Viscosity (g cm^{-1} sec^{-1})	$(1-3) \times 10^{-4}$	$(1-3) \times 10^{-4}$	$(0.2-3) \times 10^{-2}$

addition, it transmits in the ultraviolet, is odorless, nontoxic, readily available, and remarkably inexpensive when compared with other chromatographic solvents. Carbon dioxide's critical temperature of 31°C, and its pressure of 72.9 atm at the critical temperature, permits a wide selection of temperatures and pressures without exceeding the operating limits of modern HPLC equipment. In some applications, polar organic modifiers such as methanol are introduced in small concentrations ($\sim$1%) to modify α values for analytes.

A number of other substances have served as mobile phases for supercritical chromatography including ethane, pentane, dichlorodifluoromethane, diethyl ether, and tetrahydrofuran.

Detectors. Several liquid chromatographic detectors have been applied to supercritical fluids as well. The most widely used is ultraviolet absorption. Refractive index, flame ionization, and mass spectrometry have also been used.

27J–3 SUPERCRITICAL-FLUID CHROMATOGRAPHY VERSUS OTHER COLUMN METHODS

The information in Table 27–10, and other data as well, reveal that several physical properties of supercritical fluids are intermediate between gases and liquids. As a consequence, this new type of chromatography combines some of the characteristics of both gas and liquid chromatography. Thus, like gas chromatography, supercritical-fluid chromatography is inherently faster than liquid chromatography because of the lower viscosity and higher diffusion rates in supercritical fluids. High diffusivity, however, leads to longitudinal band spreading (p. 736), which is a significant factor with gas but not liquid chromatography. Thus, the intermediate diffusivities and viscosities of supercritical fluids should result in faster

separations than are achieved with liquid chromatography accompanied by lower zone spreading than is encountered in gas chromatography.

Figures 27–27 and 27–28 compare the performance characteristics of a packed column when elution is performed with supercritical carbon dioxide and a conventional mobile phase. In Figure 27–27 it is seen that at a linear mobile-phase velocity of 0.6 cm/sec, the supercritical column yields a plate height of 0.013 mm while the plate height with a conventional eluent is three times as large or 0.039 mm. Thus, a reduction in peak width by a factor of three should be realized. Alternatively, there is a gain of four in linear velocity at a plate height corresponding to the minima in the HPLC curve; this gain would result in

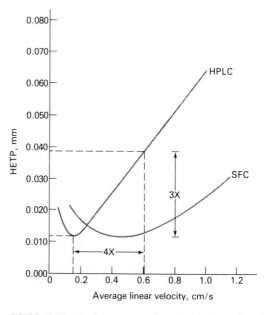

FIGURE 27-27 Performance characteristics of a 5 μm ODS column when elution is carried out with a conventional mobile phase (HPLC) and supercritical carbon dioxide SFC. (From D. R. Gere, *Application Note 800–3,* Hewlett-Packard Company, 1983. With permission.)

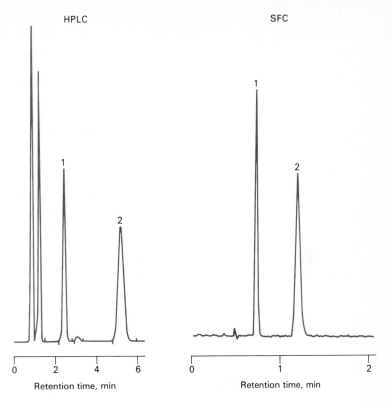

FIGURE 27-28 Comparison of chromatograms obtained by conventional partition chromatography (HPLC) and supercritical fluid chromatography (SFC). Column: 20 cm × 4.6 mm packed with 10 μm reversed-phase bonded packing. Analytes: 1) biphenyl; 2) terphenyl. For HPLC, mobile phase 65/35% CH_3OH/H_2O; flow rate 4 mL/min; linear velocity 0.55 cm/sec; sample size 10 μL. For SFC, mobile phase CO_2; flow rate 5.4 mL/min; linear velocity 0.76 cm/sec; sample size 3 μL. (From D. R. Gere, T. J. Stark, and T. N. Tweeten, *Application Note 800-4*, Hewlett-Packard Company, 1983. With permission.)

a reduction of analysis time by a factor of 4. These advantages are reflected in the two chromatograms shown in Figure 27–28.

It is worthwhile comparing the role of the mobile phase in gas, liquid, and supercritical-fluid chromatography. Ordinarily in gas chromatography, this phase serves but one purpose—zone movement. As we have seen in earlier parts of this chapter, in liquid chromatography, the mobile phase provides not only transport of solute molecules but also interactions with solutes that influence selectivity factors (α). When a molecule dissolves in a supercritical medium, the process resembles volatilization but at a much lower temperature than under normal circumstances. Thus, at a given temperature, the vapor pressure for a large molecule in a su-

percritical fluid may be 10^{10} times greater than in the absence of that fluid. As a consequence, high-molecular weight compounds, thermally unstable species, polymers, and biological molecules can be brought into a much more fluid state than a normal liquid solution of these same molecules. Interactions between solute molecules and the molecules of a supercritical fluid must occur to account for their solubility in these media. The solvent power is thus a function of the chemical composition of the fluid. Therefore, in contrast to gas chromatography, the possibility exists for varying α by changing the mobile phase.

The solvent power of a supercritical fluid is also related in a complex way to the gas density and thus the gas pressure. One manifestation of this relationship is the general existence of a threshold density below which no solution of solute occurs. Beyond this threshold, the solubility increases rapidly and then ultimately levels off. As mentioned earlier, this density effect makes possible gradient elution by pressure variation (gradient elution by temperature variation is also possible).

27J-4 APPLICATIONS

Supercritical-fluid chromatography appears to have a potential niche in the spectrum of column-chromatographic methods because it can handle larger molecules than gas chromatography with higher efficiencies than liquid chromatography. It is also considerably easier to interface with mass spectrometers than is its liquid counterpart. Its potential applications based upon the high resolution of coated capillary columns may be particularly significant. Accurate assessment of its ultimate utility cannot be made at this time.

A summary of applications of supercritical fluids to separations can be found in a review article by Randall (see footnote 32). Its use as a routine analytical technique is described by Rawdon and Norris.[35]

---

[35] M. G. Rawdon and T. A. Norris, *Amer. Lab.*, **1984**, *16* (5), 17.

PROBLEMS

27–1. List the kinds of substances to which each of the following kinds of chromatography is the most applicable:
  (a) gas-liquid
  (b) liquid-partition
  (c) reversed-phase partition
  (d) ion-exchange
  (e) gel-permeation
  (f) gel-filtration
  (g) gas-solid
  (h) liquid-adsorption

27–2. Describe three general methods for improving resolution in partition chromatography.

27–3. Describe a way of manipulating the capacity factor of a solute in partition chromatography.

27–4. How can the selectivity factor be manipulated in (a) gas chromatography and (b) liquid chromatography?

27–5. In preparing a benzene/acetone gradient for an alumina HPLC column, is it desirable to increase or decrease the proportion of benzene as the column is eluted?

27–6. Define the following terms:
  (a) sparging
  (b) isocratic elution
  (c) gradient elution
  (d) stop-flow injection
  (e) pellicular packing
  (f) light pipe
  (g) reversed-phase packing

27–7. What is meant by the linear response range of a detector?

27–8. List the differences in properties and roles of the mobile phases in gas, liquid, and supercritical-fluid chromatography. How do these differences influence the characteristics of the three methods?

27-9. In a normal-phase partition column, a solute was found to have a retention time of 29.1 min, while an unretained sample had a retention time of 1.05 min when the mobile phase was 50% by volume chloroform and 50% n-hexane. Calculate (a) k' for the solute and (b) a solvent composition that would bring k' down to a value of about 10.

27-10. The mixture of solvents in Problem 27-9 did not provide a satisfactory separation of two solutes. Suggest how the mobile phase might be altered to improve the resolution.

27-11. Suggest a type of liquid chromatography that would be suitable for the separation of

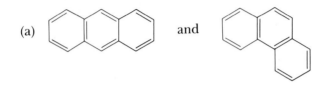

(a)   and

(b) CH_3CH_2OH and $CH_3CH_2CH_2OH$.
(c) Ba^{2+} and Sr^{2+}.
(d) C_4H_9COOH and $C_5H_{11}COOH$.
(e) high-molecular-weight glucosides.

27-12. On a silica gel column, a compound was found to have a retention time of 28 min when the mobile phase was toluene. Which solvent, carbon tetrachloride or chloroform, would be more likely to shorten the retention time? Explain.

27-13. For a normal-phase separation, predict the order of elution of
(a) n-hexane, n-hexanol, benzene.
(b) ethyl acetate, diethyl ether, nitrobutane.

27-14. For a reversed-phase separation, predict the order of elution of the solutes in Problem 27-13.

27-15. Estimate the partition coefficient for compounds B and C in Figure 27-24 if the retention volume for compound A was 5.1 mL and for compound D was 14.2 mL.

27-16. What is a mass chromatogram and how is it obtained?

27-17. Two chromatographic peaks were found to have average peak widths of 9 sec with a flow rate of 2.5 mL/min. What is the maximum volume of sample that can be taken if the loss of resolution is to be less than (a) 5% and (b) 20%?

27-18. For the three chromatograms shown in Figure 27-15, estimate the resolution between compounds (a) 5 and 6; (b) 3 and 4.

27–19. For the three chromatograms shown in Figure 27–15 assume that the elution time for an unretained species is 1 min.
   (a) Estimate k' for compound 6.
   (b) Estimate α for compounds 5 and 6.
   (c) Estimate α for compounds 3 and 4.
   (d) Estimate the number of plates in each column from peak 6.
   (e) Calculate the plate height for each column.

28

Planar Chromatography

Planar chromatography takes three forms, namely thin-layer (TLC), paper (PC), and electrochromatography. Planar chromatography differs from its column counterparts in the respect that the stationary phase is a flat, relatively thin layer of material, which may be self-supporting or may be supported on a glass, plastic, or metal plate. The mobile phase moves through the stationary phase by capillary action, sometimes assisted by gravity or an electrical potential. Planar chromatography is sometimes called two-dimensional, although this description is not entirely correct inasmuch as the stationary phase does have a finite thickness.

Historically, paper chromatography, in which a sheet or strip of heavy filter paper serves as the stationary phase, was first used for separations in the middle of the nineteenth century. It was not until the late 1940s, however, that the usefulness of this medium for separations was fully appreciated and exploited.

Thin-layer chromatography, in which the stationary phase consists of a narrow layer of finely divided coated or uncoated particles supported on a glass plate, was first introduced in 1928 as "spot chromatography". Widespread use of this technique did not, however, begin until 1956 with the development of convenient spreaders for the preparation of thin-layer plates.

Currently, most planar chromatography is based upon the thin-layer technique, which is faster, has better resolution, and is more sensitive than its paper counterpart. Consequently, this chapter is devoted largely to thin-layer methods.

28A Scope of Thin-Layer Chromatography

In terms of theory, kinds of stationary and mobile phases, and applications, thin-layer and liquid chromatography are remarkably similar. In fact, as mentioned in Chapter 27, thin-layer plates can be profitably applied for developing optimal conditions for sepa-

837

ration by column liquid chromatography. Here, the speed and low cost of the thin-layer technique is a significant advantage. Many chromatographers believe that thin-layer chromatography should always precede column experiments.

By now, equipment is available for thin-layer work from several commercial suppliers and at relatively low cost. Furthermore, instruments for quantitative measurement of separated species have been sufficiently improved so that they rival liquid chromatographic data in precision and accuracy. As a consequence, it is believed that at least as many analyses are performed by TLC as by HPLC.[1] Nevertheless, the former has not had the publicity devoted to it as has the latter, perhaps because TLC is so mechanically simple.

Thin-layer chromatography has become the workhorse of the drug industry for the all-important determination of product purity. It has also found widespread use in clinical laboratories and is the backbone of many biochemical and biological studies. Finally, it finds widespread use in the laboratories of chemical industry.[2]

28B Principles of Thin-Layer Chromatography

Typical thin-layer separations are performed on a glass plate that is coated with a thin and adherent layer of finely divided particles, which constitute the stationary phase. These particles are similar in type to most of the packings described in Chapter 27 for liquid chromatography.

28B-1 DEVELOPMENT OF CHROMATOGRAMS

Generally in thin-layer chromatography, samples are introduced onto the coated plates as dilute solutions (0.1 to 1 $\mu g/\mu L$) in a volatile solvent. A capillary tube or a micropipet is employed to deliver a drop of this solution to a spot that is 1 to 2 cm from the end of the plate (see dotted circles in Figure 28–1a). The position of the samples is indicated by marking an *origin line* on the plate with a pencil. After evaporation of the solvent (often with a hair dryer), the chromatogram is *developed* by the flow of mobile phase over the surface. Details of the development step are given in Section 28C–4. The development process is analogous to the elution step in column chromatography. After development is judged to be complete, the flow of mobile phase is discontinued and the position of the *solvent front* is marked. After evaporation of the solvent, the position of the separated species is located by a variety of methods, which will be described later. One common method involves spraying the plate with a colorimetric reagent, which, as shown in Figure 28–1a, permits location of the analytes visually.

Figure 28–1b illustrates one of the methods for performing quantitative TLC. Here, the absorbance of the developed chromatogram is recorded as a function of distance from the origin line. The areas or heights of the two solute peaks then serve as the basis for quantitative analysis.

28B-2 PERFORMANCE CHARACTERISTICS OF THIN-LAYER PLATES

Most of the terms and the relationships developed for column chromatography in Section 25B can, with slight modification, be applied to thin-layer chromatography as

---

[1] T. H. Mauch II, *Science,* **1982,** *216,* 161.

[2] Two monographs devoted to the principles and applications of TLC are: B. Fried and J. Sherma, *Thin-Layer Chromatography,* Marcel Dekker: New York, 1982; J. C. Touchstone, *Practice of Thin-Layer Chromatography,* 2d ed., Wiley: New York, 1983. For briefer reviews, see: D. C. Fenimore and C. M. Davis, *Anal. Chem.,* **1981,** *53,* 253A; H. J. Issaq and E. W. Barr, *Anal. Chem.,* **1977,** *49,* 83A; D. Rogers, *Amer. Lab.,* **1984,** *16* (6), 65.

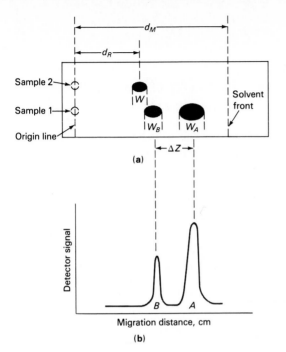

FIGURE 28-1 Thin-layer chromatograms. **(a)** Thin-layer chromatograms for two samples; **(b)** densitometric scan of sample 1, which contained two components, *A* and *B*.

well. One new term, the *retardation factor* or the R_F factor, is required.

The Retardation Factor. The thin-layer chromatogram for a single solute is shown as chromatogram 2 in Figure 28-1a. The retardation factor for this solute is given by

$$R_F = \frac{d_R}{d_M} \qquad (28-1)$$

where d_R and d_M are linear distances measured from the origin line. Values for R_F can vary from one for solutes that are not retarded to a value that approaches zero. It should be noted that if the spots are not symmetric, as they are in Figure 28-1, the measurement of d_R is based on the position of maximum intensity.

Plate Height and Number of Plates. The efficiency of a thin-layer plate is given by its plate height

H or its number of plates N. As in column chromatography (Equation 25-3)

$$H = \sigma^2/L$$

where σ^2 is the variance in *units of distance*. Turning again to chromatogram 1 in Figure 28-1, let us consider the plate height at distance d_R from the origin. For this length of plate,

$$H = \sigma^2/d_R \qquad (28-2)$$

The number of plates N contained in the distance d_R is d_R/H or

$$H = d_R/N$$

Substitution of this relationship into Equation 28-2 gives upon rearrangement

$$N = \frac{d_R^2}{\sigma^2} \qquad (28-3)$$

As in a column chromatogram, the distribution of a thin-layer spot tends to be Gaussian, and the width W of the spot is 4σ. That is,

$$W = 4\sigma$$

Substitution of this relationship into Equation 28-3 gives upon rearrangement

$$N = 16 \left(\frac{d_R}{W}\right)^2 \qquad (28-4)$$

Thus, the number of theoretical plates for a thin-layer bed is readily obtained from a chromatogram.

The Capacity Factor. All of the equations in Table 25-3 (p. 748) are readily adapted to thin-layer chromatography. In order to apply these equations it is only necessary to relate d_R and d_M as defined in Figure 28-1 to t_R and t_M, which are defined in Figure 25-5 (p. 734). To arrive at these relationships, consider the single solute that appears in chromatogram 2 in Figure 28–1. Here, t_M and t_R correspond to times required for the mobile phase and the solute to travel a fixed distance, in this case d_R. For the mobile phase, this time is equal to the distance divided by its linear velocity u, or

$$t_M = d_R/u \qquad (28\text{–}5)$$

The solute does not reach this same point, however, until the mobile phase has traveled the distance d_M. Therefore,

$$t_R = d_M/u \qquad (28\text{–}6)$$

Substitution of Equations 28–5 and 28–6 into Equation 25–15 yields

$$k' = \frac{d_M - d_R}{d_R} \qquad (28\text{–}7)$$

The capacity factor k' can also be expressed in terms of the retardation factor by rewriting Equation 28–7 in the form

$$k' = \frac{1 - d_R/d_M}{d_R/d_M} = \frac{1 - R_F}{R_F} \qquad (28\text{–}8)$$

Resolution. Resolution R_S in thin-layer chromatography is defined by an equation that is analogous to Equation 25–19 for liquid chromatography. Thus, for the two solutes A and B in sample 1 of Figure 28–1**a**,

$$R_S = \frac{2\Delta Z}{W_A + W_B} = \frac{2[(d_R)_A - (d_R)_B]}{W_A + W_B} \qquad (28\text{–}9)$$

Equation 25–20 then applies. Thus,

$$R_S = \frac{\sqrt{N}}{4}\left(\frac{\alpha - 1}{\alpha}\right)\left(\frac{k'}{1 + k'}\right) \qquad (28\text{–}10)$$

Here, the selectivity factor is given by (Equations 25–17 and 28–7)

$$\alpha = \frac{k'_B}{k'_A} = \frac{d_M - (d_R)_B}{d_M - (d_R)_A} \times \frac{(d_R)_A}{(d_R)_B} \qquad (28\text{–}11)$$

28C The Practice of Thin-Layer Chromatography

The steps involved in a thin-layer chromatographic separation include preparation or acquisition of thin-layer plates, application of samples onto the plates, development of chromatograms, and determination of the location of the separated species. Quantitative chromatography requires the additional step of determining the amounts of the separated analytes. The equipment required for these steps can be remarkably simple and inexpensive.

28C-1 THIN-LAYER PLATES

The efficiency of a thin-layer separation depends on having small particles of uniform size distributed in a thin, dense layer of uniform thickness. When thin-layer chromatography was first introduced, the first step in a separation involved plate preparation, and various devices were developed to aid in forming films of constant thickness. While these devices are still available from equipment supply houses, most scientists now find it more economic to buy precoated plates.

Commercial Precoated Plates. Since about 1967, several manufacturers have offered a variety of precoated plates, which are usually glass, although plastic and aluminum backings are also available. The common plate sizes in centimeters are 5 × 20, 10 × 20, 20 × 20, and 20 × 40. Layer thicknesses vary from 100 to 2000 μm with 100, 200, and 250 μm being the most widely used for analytical separations. The thicker films are used for preparative chromatography. Depending upon the type of coating, a 10 × 20 precoated plate costs from $1 to $3. It should be noted, however, that such a plate can accommodate 30 or more samples, which makes the cost per sample nominal.

Commercial thin-layer plates come in two categories, conventional and high-performance. The former have thicker layers (200 to 250 μm) of particles having nominal diameters of 20 μm or greater. High-performance plates usually have film thicknesses of 100 μm and particle diameters of 5 μm or less; the distribution of particle sizes on a given plate is also smaller. Recommended sample volumes with the conventional analytical plates are 0.5 to 5 μL, while for high-performance plates they are 0.05 to 0.5 μL.

High-performance plates, as their name implies, provide sharper separations in shorter times. Thus, an ordinary plate typically will exhibit 2000 theoretical plates in 12 cm with a development time of 25 min. The corresponding figures for a high-performance plate are 4000 theoretical plates in 3 cm requiring 10 min for development. A disadvantage of high-performance plates is the small sample size that can be accommodated.

28C-2 STATIONARY AND MOBILE PHASES

Types of Stationary Phases.   Many of the stationary phases used in thin-layer chromatography are similar to those described earlier in the discussions on adsorption, normal- and reversed-phase partition, ion-exchange, and size-exclusion column chromatography. In addition, however, cellulose-based stationary phases find considerable use in the separation of hydrophilic molecules such as carbohydrates, amino acids, nucleic acid derivatives, and inorganic species. Cellulose plates behave very much like the paper strips used in paper chromatography. Usually, separations on cellulose plates and paper involve partition equilibria in which the stationary phase is water held in the interstices of the cellulose. Some treated cellu-

loses are available, however, that are hydrophobic and can be used for reversed-phase separations. An ion-exchange type cellulose is also marketed.

28C-3 SAMPLE APPLICATION

Sample application is perhaps the most critical aspect of thin-layer chromatography, particularly for quantitative measurements. Usually the sample, as a 0.01 to 1% solution, is applied as a spot 1 to 2 cm from the edge of the plate. For best separation efficiency, the spot should have a minimal diameter—about 5 mm for qualitative work and smaller for quantitative analyses. For dilute solutions, three or four repetitive applications are used with drying in between.

Manual application of samples is performed by touching a capillary tube containing the sample to the plate or by use of a hypodermic syringe. A number of mechanical dispensers, which increase the precision and accuracy of sample application, are now offered commercially.

28C-4 CHROMATOGRAM DEVELOPMENT

Typical developing chambers are illustrated in Figure 28–2. Development takes place in a closed chamber that is saturated with solvent vapor. In some instances it is desirable to equilibrate the plate with this vapor before development begins. The divider in the bottom of the vessel in (**a**) permits presaturation of the plate with the left chamber empty and the right containing solvent. When it is desired to begin developing, the vessel can be tipped so that the solvent is transferred to the chamber holding the plate. Note that the sample is always applied a sufficient distance from the edge of the plate so that it is never immersed in the developer.

Figure 28–3 illustrates the separation of amino acids in a mixture by development in

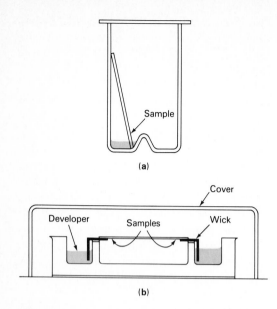

FIGURE 28-2 Two types of developing chambers: (**a**) ascending flow; (**b**) horizontal flow. Here, samples are placed on either end of the plate and developed toward the middle thus doubling the number of samples that can be accommodated.

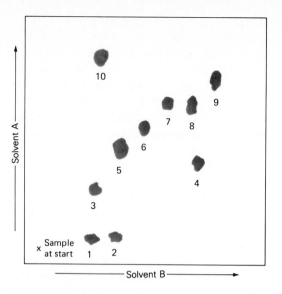

FIGURE 28-3 Two-dimensional thin-layer chromatogram (silica gel) of some amino acids. Solvent A: toluene/2-chloroethanol/pyridine. Solvent B: chloroform/benzyl alcohol/acetic acid. Amino acids: (1) aspartic acid; (2) glutamic acid; (3) serine; (4) β-alanine; (5) glycine; (6) alanine; (7) methionine; (8) valine; (9) isoleucine; and (10) cysteine.

two directions *(two-dimensional planar chromatography)*. The sample was placed in one corner of a square plate and development was performed in the ascending direction with solvent A. This solvent was then removed by evaporation, and the plate was rotated 90 deg, following which ascending development with solvent B was performed. After solvent removal, the positions of the amino acids were determined by spraying with ninhydrin, a reagent that forms a pink to purple product with amino acids. The spots were identified by comparison of their positions with those of standards.

28C-5   METHODS FOR LOCATING ANALYTES AFTER SEPARATION

Several methods are available for location (often called *visualization*) of noncolored species after separation on a thin-layer plate.

Physical Methods.   Two common physical methods for visualization are based on absorption and fluorescence with ultraviolet and visible radiation. Here, a short (254 nm) and a long (365 nm) wavelength mercury lamp are required as well as a tungsten filament bulb. Because the concentrations are low, viewing is carried out in a darkroom or a darkened viewing cabinet.

A convenient way of viewing ultraviolet absorbing species is to incorporate an inorganic phosphor, such as cadmium or zinc sulfide, in the original coating (many commercial plates are so prepared and labeled as F or UV). When such plates are irradiated, ultraviolet absorbing species appear as dark spots on a yellow-green background. This type of visualization is sometimes called fluorescence quenching.

A number of reagents are available that

convert certain types of analytes to fluorescing (or absorbing) derivatives, which can be visualized by irradiation with ultraviolet lamps.

Chemical Methods. Touchstone and Dobbins[3] list 207 reagents that react with various types of solutes on a thin-layer plate to yield colored or fluorescent species for visualization. Several of these are general in the sense that they produce detectable products with most organic species. Perhaps the most common general reagent is iodine which produces a light brown color with most organic compounds. The color is developed by placing the plate in a closed vessel containing a few crystals of iodine. After exposure, the spots are usually outlined with a pencil because the color fades with exposure to air. Iodine can also be sprayed onto plates as a 1% solution.

Concentrated sulfuric acid, sometimes mixed with nitric acid, sodium dichromate, or acetic anhydride, is another common general reagent. It reacts with organic species to form a black char that is readily detected. Usually the acid is sprayed onto the surface of the plate. In some instances the plate must be heated.

Several pH indicators have been used to visualize acidic or basic constituents such as carboxylic acids.

Examples of selective reagents include cerium(IV) for alcohols, 2,4-dinitrophenylhydrazine for aldehydes and ketones, diethylamine for barbituates, and *p*-toluenesulfonic acid for flavonoids, indoles, and steroids. Application of these reagents is carried out by spraying or by dipping the plate in a vessel containing a solution of the reagent. Sometimes, subsequent heat treatment is necessary. Both colored and fluorescent species are formed.

---

[3] J. C. Touchstone and M. F. Dobbins, *Practice of Thin-Layer Chromatography*, pp. 170–214, Wiley: New York, 1978.

28C-6 QUALITATIVE THIN-LAYER CHROMATOGRAPHY

The data from a single chromatogram usually do not provide sufficient information to permit identification of the various species present in a mixture because of the variability of R_F values with sample size, the thin-layer plate, and the conditions extant during development. In addition, the possibility always exists that two quite different solutes may exhibit identical or nearly identical R_F values under a given set of conditions.

Variables That Influence R_F. At best, R_F values can be reproduced to but two significant figures; among several plates, one significant figure may be a more valid statement of precision. The most important factors that determine the magnitude of R_F include thickness of the stationary phase, moisture content of the mobile and stationary phases, temperature, degree of saturation of the developing chamber with mobile-phase vapor, and sample size. Complete control of these variables is generally not practical. Partial amelioration of their effects can often be realized, however, by substituting a relative retention factor R_X for R_F, where

$$R_X = \frac{\text{travel distance of analyte}}{\text{travel distance of a standard substance}}$$

Use of Authentic Substances. A method which often provides tentative identification of the components of a sample is to apply to the plate the unknown and solutions of purified samples of species likely to be present in that unknown. A match in R_F values between a spot for the unknown and that for a standard provides strong evidence as to the identity of one of the components of the sample (see Figure 28–7). Confirmation is always necessary, however; one convenient confirmatory test is to repeat the experiment with different stationary and mobile phases as well as with different visualization reagents.

Elution Methods. The identity of separated analyte species can also be confirmed or determined by a scraping and dissolution technique. Here, the area containing the analyte is scraped from the plate with a razor or a spatula and the contents collected on a piece of glazed paper. After transfer to a test tube or other container, the analyte is dissolved with a suitable solvent and separated from the stationary phase by centrifugation or filtration. Identification is then carried out by such techniques as mass spectrometry, nuclear magnetic resonance, or infrared spectroscopy.

28C-7 QUANTITATIVE THIN-LAYER CHROMATOGRAPHY

Quantitative as well as qualitative data can be obtained from thin-layer chromatographic separations. With proper attention to detail, the precision and accuracy of such measurements can be made to approach those for column methods. Generally, to obtain the best results, it is necessary to apply several standards to the plate containing the samples so that development and quantitative measurement of samples and standards are performed under identical conditions.

Quantitative thin-layer chromatography is performed either by elution or *in situ*. With the former, scraping and dissolving, as described in the previous section, serve to separate the analyte from the stationary phase. Quantitative analysis is then carried out by spectroscopic or electroanalytical procedures. Dissolution is often incomplete; concurrent calibrations, however, at least partially compensate for the resulting loss of analyte.

In situ procedures involve visual or spectroscopic measurement of the spots without removing them from the thin-layer plate. This technique is becoming the more common method for quantitative analysis because of its speed and convenience. Quantitative *in situ* procedures involve applying identical volumes of sample and standard side-by-side to the thin-layer plate followed by simultaneous development.

In Situ Visual Methods. The simplest *in situ* technique involves visual matching of the size and/or color of the spots for the unknown with several standards. The results are generally semiquantitative with relative uncertainties of perhaps 20 to 50%.

Somewhat better accuracy can be achieved by measurement of spot areas with a planimeter, by outlining on tracing paper followed by cutting and weighing of the paper, or by tracing onto graph paper and counting squares. These techniques may reduce the uncertainty by a factor of two or three.

In Situ Spectroscopic Methods. Several manufacturers offer instruments called *densitometers* that permit spectroscopic measurements to be made on developed thin-layer plates. Generally, these instruments are of a scanning type in which the plate is moved at a constant speed through a beam of radiation. The plate drive and recorder motors are then synchronized so that a plot is obtained of the optical signal as a function of distance along the length of the plate. Figure 28–4 shows a chromatogram derived from densitometric absorbance measurements based upon reflected radiation. The peak heights or peak areas from chromatograms of this type are used for quantitative determination of the concentration of analytes.

Densitometers. As was true of the optical instruments described in earlier parts of this book, the components making up commercial densitometers vary widely in design, sophistication, and cost. The simplest instruments employ a mercury lamp and filters as a source of ultraviolet radiation (254 and 366 nm). Visible radiation is provided by tungsten filament bulbs and glass or interference filters. The more sophisticated instruments use deuterium or xenon lamps in conjunction with a grating monochromator. Both photocells and photomultiplier tubes are used for detectors.

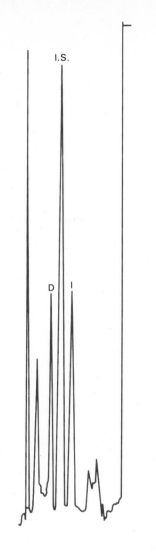

I.S.

D I

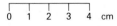

0 1 2 3 4 cm

FIGURE 28-4 A typical thin-layer chromatogram obtained by absorption/reflectance scanning. D = desipramine; I = imapramine; IS = butaperazine internal standard. D and I were extracted from a 1.0 mL sample of blood plasma at concentrations of 100 mg/mL. (Reprinted with permission from: D. C. Fenimore and C. M. Davis, *Anal. Chem.,* **1981,** *53,* 262A. Copyright by 1981 American Chemical Society.)

Figure 28–5 illustrates two optical designs of typical densitometers. The instrument shown in (**a**) is a double-beam device for measuring absorbance either by reflected radiation or transmitted radiation. Note that the motion of the thin-layer plate is perpendicular to the plane of the page. One of the beams is aligned with the chromatogram while the second scans an unused portion of the plate. The absorbance is then the negative logarithm of the ratio of the outputs of the two photomultiplier signals.

Absorbance measurements based upon transmitted radiation are usually more satisfactory than those from reflected radiation because the former mode provides a more nearly linear relationship between peak area or peak height and concentration; in addition, sensitivities tend to be greater. The transmission mode is, however, restricted to radiation having wavelengths greater than about 400 nm as a consequence of absorption by the glass thin-layer plate.

Figure 28–5**b** illustrates the design of an instrument for determining fluorescing species. Most instruments of this kind employ a mercury or xenon lamp and filters as a source of excitation energy. The fluorescent radiation is then passed through a cut-off filter to black out scattered radiation from the source.

The instrument in Figure 28–5**b** can also be applied to measure fluorescence quenching. Here, a fluorescing substance, such as cadmium sulfide, is incorporated into the stationary phase. In the presence of spots of analyte that absorb the excitation wavelength, a decrease in fluorescent intensity is observed; the reduction in intensity then serves as a measure of the amount of analyte present. In this case, the chromatograms consist of a series of minima rather than maxima.

The precision of *in situ* measurements with a densitometer generally ranges from 1 to 5% relative. Analyses based upon peak areas are usually more precise. On the other

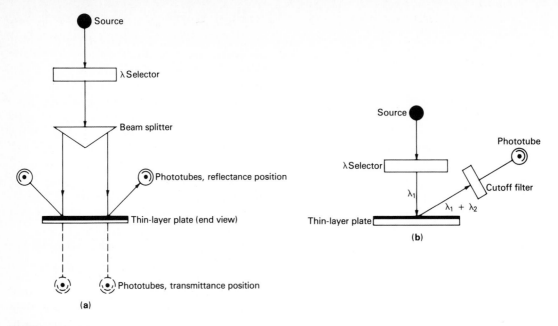

FIGURE 28-5 Schematics of thin-layer instruments: (**a**) double-beam densitometer for absorbance measurements either by reflectance or transmittance; (**b**) single-beam instrument for the measurement of fluorescence or fluorescence quenching.

hand, peak heights are less subject to interferences from the presence of other species.

For high-performance thin-layer chromatography, it is often possible to measure quantities of fluorescing species in the 10 to 100 pg range. With absorbance measurements, the measurable concentration is 10 to 1000 times larger.

28D   Applications of Thin-Layer Chromatography

As was mentioned earlier in this chapter, the applications of thin-layer chromatography parallel those of partition, adsorption, ion-exchange, and exclusion chromatography as performed in a column under high-performance conditions. Consequently, the discussions on applications of these techniques in Sections 27E through 27H apply here as well. It is, however, of some importance to point out the circumstances under

which HPTLC is the method of choice and conditions where HPLC is favored.

Thin-layer chromatography is generally the better choice for:

1. Exploratory experiments, where suitable stationary and mobile phase compositions are being sought. Here, the speed and low cost of HPTLC are particularly valuable.
2. Situations requiring a high sample throughput. Here, the advantage of HPTLC arises from the fact that 20 or 30 separations can be performed simultaneously.
3. Situations where cost is an important factor. As noted earlier, HPTLC capital equipment, maintenance, and operating supply costs are significantly lower than for HPLC.
4. Samples for which detection is a problem. Here, the static nature of HPTLC provides a wider choice of methods for

detecting and measuring the separated species.

High-performance liquid chromatography is favored for:

1. Separations requiring the highest number of theoretical plates. Here, the high pressures of HPLC permit the use of a much larger stationary-phase bed and thus a higher number of total plates.
2. High-speed separations for a limited number of samples. As noted in Chapter 27, separations on a modern HPLC column often can be completed in one or two minutes or less (for example, see Figure 27–6**b**) whereas HPTLC separations typically require ten minutes or more. It is important to mention again that for a large number of samples, the time per sample for HPTLC separations may be significantly smaller than for HPLC.

With recent improvements in *in situ* quantitative measurements, the accuracy and precision of HPTLC analyses appear to approach that obtainable with HPLC. Indeed, the inherent accuracy of thin-layer methods would seem to be greater than that of column methods because standards and samples can be processed and measured in parallel.

28E Paper Chromatography

Separations by paper chromatography are performed in the same way as those on thin-layer plates. Usually, special papers, which are highly purified and reproducible as to porosity and thickness, are employed. Such papers contain sufficient adsorbed water that normal paper chromatography can be classified as a liquid-liquid type. Other liquids can be made to displace the water, however, thus providing a different type of stationary phase. For example, paper treated with silicone or paraffin oil permits re-

versed-phase paper chromatography, in which the mobile phase is a polar solvent. Also available commercially are special papers that contain an adsorbent or an ion-exchange resin, thus permitting adsorption and ion-exchange paper chromatography.

Generally, thin-layer chromatography is superior to paper chromatography in resolution, compactness of spots, and speed. An obvious virtue of paper as a separation medium is its low cost. Thus the cost of a sheet of filter paper is approximately one-tenth that for a thin-layer plate of equivalent size. Another advantage of paper is the ease with which analyte dissolutions can be carried out. Here it is only necessary to cut out the spots, a process that is simpler and less troublesome than scraping the stationary phase from a thin-layer plate.[4]

A perusal of current literature dealing with planar chromatography reveals that thin-layer plates are now considerably more widely used than paper.[5]

28F Electrophoresis and Electrochromatography

Electrophoresis is defined as the migration of particles through a solution under the influence of an electrical field. Historically, the particles referred to were colloidal and owed their charge to adsorbed ions. This definition has now become too restrictive, and the term electrophoresis is presently applied both to the migration of individual ions and to colloidal aggregates. Electrophoretic methods provide a powerful means of fractionating the components of a mixture, be they aggregated or monodispersed.

Electrophoretic methods can be subdi-

---

[4] For a comparison of paper and thin-layer chromatography, see: G. Zweig and J. Sherma, *J. Chromatogr. Sci.*, **1973**, *11*, 279.

[5] See: J. Sherma and B. Fried, *Anal. Chem.*, **1984**, *56*, 48R; **1982**, *54*, 45R; **1980**, *52*, 276R.

vided into two categories, depending upon whether the separation is carried out in the absence or presence of a supporting or stabilizing medium. In the *free-solution method,* the sample solution is introduced as a band at the bottom of a U-tube filled with a buffered liquid. A field is applied by means of electrodes located near the tube ends; the differential movement of the charged particles toward one or the other electrode is observed. Separations occur as a result of differences in migration rates; these rates in turn are related to the charge-to-mass ratios and the inherent mobilities of the species in the medium. The free-solution method was perfected and applied to the separation of proteins by Tiselius; for this work, he received the 1948 Nobel prize.

Many of the experimental difficulties associated with free-solution electrophoresis are avoided if the separations are carried out in a stabilizing medium such as a paper, a layer of finely divided solid, or a column packed with a suitable solid. Here, the experimental techniques closely resemble the various chromatographic methods discussed earlier, with the additional parameter of the superimposed electrical field. Depending upon the properties of the medium, the separations may result primarily from the electrophoretic effect or from a combination of electrophoresis and adsorption, ion-exchange, or other distribution equilibria. Methods based upon electrophoresis in a stabilizing medium bear a variety of names, including *electrochromatography, zone electrophoresis electromigration,* and *ionophoresis.* Our discussion is limited to electrochromatography.[6]

---

[6] For a detailed discussion, see: R. J. Wieme and H. Michlin, *Chromatography,* 3d ed., E. Heftmann, Ed., Chapters 10 and 11, Reinhold: New York, 1974; and J. R. Sargent and S. G. George, *Methods in Zone Electrophoresis,* 3d ed., BDH Chemicals, Ltd.: Poole, England, 1975.

28F-1 EXPERIMENTAL METHODS OF ELECTROCHROMATOGRAPHY

The solid media employed in electrochromatography are as numerous and varied as those encountered in the other chromatographic methods. Examples include paper, cellulose acetate membranes, cellulose powders, starch gels, ion-exchange resins, glass powders, and agar gels. Depending upon the physical nature of the solid, separations are performed on strips of paper or membranes, in columns, in trays, or in thin layers supported by glass or plastic. Despite certain disadvantages, filter paper and cellulose acetate are the most widely used stabilizing materials; we will focus our attention on paper electrochromatography.

Figure 28–6 is a schematic diagram illustrating two of the many ways for performing a paper electrophoresis. In Figure 28–6**a**, a strip of paper is held horizontally between two containers filled with a buffer mixture; the paper is well soaked with buffer and evaporation is prevented by housing the apparatus in an air-tight container. The sample is introduced as a band at the center of the strip, and a dc potential of 100 to 1000 V is applied across the two electrodes. The latter are sufficiently isolated from the paper to prevent the electrode reactions from altering the composition of the buffer on the paper. Currents in the milliampere range are ordinarily observed. After a suitable electrolysis period, the paper is removed, dried, and the component bands are detected by applying suitable colorimetric reagents.

Many other modifications of the apparatus shown in Figure 28–6**a** have been described, and several are available commercially. In some, the paper is supported (either horizontally or at some suitable angle) between two glass or plastic plates; the plates may be cooled to dissipate the heat generated by the current. In all of these de-

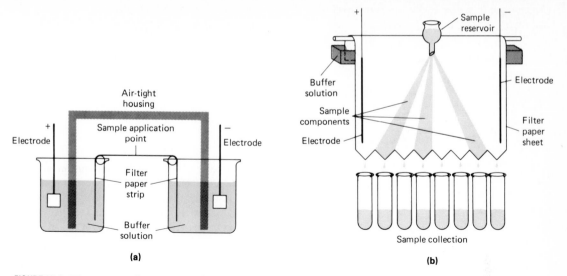

FIGURE 28–6 Two types of apparatus for paper electrochromatography.

vices, the rate at which components migrate is controlled primarily by their charge-to-mass ratios and their mobilities, with adsorption or other equilibria having only a secondary influence.

Figure 28–6**b** illustrates an apparatus for a type of two-dimensional electrochromatography. Here, the paper is held vertically and the sample components are carried down the sheet by the flow of a buffered solvent. Separation then occurs in the vertical direction as a consequence of differences in distribution ratios between the mobile and the fixed phases. In addition, however, a field is applied at right angles to the solution flow, and differential electromigration along the horizontal axis occurs as a consequence. Thus, the various species in the sample describe a radial path from the point at which the sample is introduced.

The technique illustrated in Figure 28–6**b** has been employed for both analytical and preparatory purposes. In the former, an amount of buffer just equal to that retained by the paper is allowed to pass before the

experiment is discontinued; the paper is then removed, dried, and the detection reagents are applied. In preparatory applications, the buffer flow is continued and the various fractions are collected as shown in Figure 28–6**b**.

28F–2 APPLICATIONS OF ELECTROCHROMATOGRAPHY

Electrochromatographic methods are indispensible to the clinical chemist and the biochemist, who use them for fractionating an amazing variety of biological materials. The widest application has perhaps been in clinical diagnosis; here, electrochromatography makes possible the separation of proteins and other large molecules contained in serum, urine, spinal fluid, gastric juices, and other body fluids.

Electrochromatography has also been applied widely by biochemists to fractionate smaller molecules such as alkaloids, antibiotics, nucleic acids, vitamins, natural pigments, steroids, amino acids, carbohydrates, and organic acids.

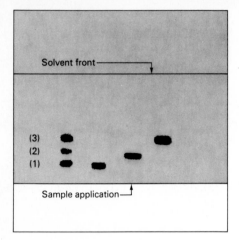

FIGURE 28-7 A thin-layer plate used for separation of three carbamate insecticides: (1) carbofuran, (2) carbaryl, and (3) metalkamate. Conditions: samples, 5 μL of solutions containing 1 mg/mL of each component in $CHCl_3$. Mobile phase: toluene/isopropyl ether (70/30% v/v).

PROBLEMS

28–1. The thin-layer plate in Figure 28–7 has dimensions of 20 × 20 cm. The row of spots on the left is a chromatogram for a mixture of three carbonates. The remaining three chromatograms are for standard solutions containing 1 mg/mL of each of the individual compounds. Careful measurement of the spots on the right yielded the following data in centimeters:

$$d_M = 9.8$$
$$(d_R)_1 = 1.7 \qquad W_1 = 0.57$$
$$(d_R)_2 = 2.5 \qquad W_2 = 0.54$$
$$(d_R)_3 = 3.9 \qquad W_3 = 0.75$$

Calculate
(a) the retardation factor for each solute.
(b) the efficiency of the medium in terms of the number of plates for each compound.
(c) the plate height from each spot.

28–2. For the chromatogram in Problem 28–1, calculate
(a) the capacity factor for each solute.
(b) the selectivity factor for each solute pair.
(c) the resolution for each solute pair.

28–3. Several authors have suggested that the most efficient separation region with thin-layer plates is in the R_F range of 0.3 to 0.7. To what range of capacity factor in liquid chromatography does this range correspond?

29

Automated Methods of Analysis

One of the major developments in analytical chemistry during the last three decades has been the appearance of commercial automatic analytical systems, which provide analytical data with a minimum of operator intervention. Initially, these systems were designed to fulfill the needs of clinical laboratories, where perhaps thirty or more species are routinely determined for diagnostic and screening purposes. The number of such analyses demanded by modern medicine is enormous[1]; the need to keep their cost at a reasonable level is obvious. These two considerations motivated the development of early automatic analytical systems. By now, such instruments find application in such diverse fields as the control of industrial processes and the routine determination of a wide spectrum of species in air,

water, soils, and pharmaceutical and agricultural products.[2]

29A An Overview of Automatic Instruments and Automation

At the outset, it should be noted that the International Union of Pure and Applied Chemists recommends that a distinction be made between *automatic* and *automated* systems. By its terminology, automatic devices perform certain desired functions without human intervention. For example, an automatic acid-base titrator adds reagent to a solution and simultaneously records pH as a function of volume of reagent. In contrast, an automated instrument contains one or more feedback systems that control the

---

[1] For example, in 1976, about 100 million samples were analyzed in domestic clinical laboratories. See: Snyder, *et al., Anal. Chem.,* **1976,** *48,* 1942A.

[2] For a thorough treatment of automatic analysis, see: J. K. Foreman and P. B. Stockwell, *Automatic Chemical Analysis,* Wiley: New York, 1975.

course of the analysis. Thus, some automated titrators compare the potential of a glass electrode to its theoretical potential at the equivalence point and use the difference to control the rate of addition of acid or base. While this distinction between automatic and automated may be useful, it is not one that is followed by the majority of workers in the field. Nor is it followed in this presentation.

29A-1 DISCRETE AND CONTINUOUS INSTRUMENTS

Automatic analytical instruments are of two types, *discrete* and *continuous*; occasionally, a combination of the two is encountered. In a discrete instrument, individual samples are maintained as separate entities and kept in separate vessels throughout such unit operations as sampling, dilution, reagent addition, mixing, centrifugation, and transportation to the measuring device. In continuous systems, in contrast, the sample becomes a segment of a flowing stream where the several unit operations just mentioned take place as the sample is carried from the injection point to a flow-through measuring unit, and thence to waste. Both discrete and continuous instruments are generally computer controlled.

Several examples of partially automatic or fully automatic systems of the discrete type have already been described in early chapters. Among these are the automatic infrared instrument described in Section 11C–5 and the automatic titrators illustrated in Figures 20–19 and 20–20. The only example of a continuous instrument thus far encountered is the instrument for coulometric determination of oxygen shown in Figure 21–7 (p. 655); consequently, somewhat more space in this chapter will be devoted to continuous instruments. This extra emphasis should not, however, be construed as a measure of the relative importance of the two types of systems. Both are widely used and are of equal importance.

29A-2 A COMPARISON OF DISCRETE AND CONTINUOUS ANALYTICAL SYSTEMS

Discrete automatic instruments ordinarily are capable of higher sample throughputs than their continuous counterparts. For example, commercial automatic photometers of the discrete type may provide as many as 300 determinations per hour, whereas the output of the average continuous photometric systems is often no more than one-third to one-tenth that number.

Because discrete instruments are based upon the use of individual containers, cross-contamination among samples is totally eliminated. On the other hand, interactions among samples is always a concern in continuous systems, particularly as the rate of sample throughput is increased. Here, special precautions are required to minimize sample contamination.

Continuous analyzers are generally mechanically simpler than their discrete counterparts. Indeed, in many continuous systems, the only moving parts are peristaltic pumps, which are inexpensive and reliable. In contrast, discrete systems often have a number of moving parts such as syringes, valves, and mechanical devices for transporting samples from one part of the system to another. As a consequence, discrete devices tend to be more expensive to purchase and to maintain.

Continuous analyzers are generally favored when pretreatment of the samples by filtration, distillation, dialysis, or solvent extraction is required. The design problems in incorporating these processes in a flowing stream are generally less formidable than those encountered where samples are treated individually. As a consequence, continuous systems are often the choice when an analysis requires separation stages between sampling and measurement.

Continuous analyzers are often based upon transport of solutions of the sample and reagents through flexible tubing. Thus, such instruments are limited to samples and solvents that do not react with or corrode the tubing. For discrete instruments, a wide choice of construction material exists, and corrosion problems can normally be circumvented.

29A-3 ADVANTAGES AND DISADVANTAGES OF AUTOMATIC ANALYSES

In the proper context, automated instruments offer a major economic advantage because of their savings in labor costs. For this advantage to be realized, however, it is necessary that the volume of work for the instrument be large enough to offset the original capital investment, which is often large, and the extensive effort that is usually required to put the automatic system in full operation. For laboratories in which large numbers of routine analyses are performed daily, the savings realized by automation can be enormous. With respect to savings in labor costs, it is worthwhile noting that most automated instruments require less skilled, and thus less expensive, operating personnel; on the other hand, more skilled supervisors may be necessary.

A second major advantage of automated instruments is their speed, which is frequently significantly greater than that of manual devices. Indeed, this speed often makes possible the continuous monitoring of the composition of products a they are being manufactured. This information in turn permits alteration of conditions to improve quality or yield. Continuous monitoring is also useful in medicine where analytical results can be used to determine a patient's current condition and his response to therapy.

A third advantage of automation is that a well-designed analyzer can usually produce more reproducible results over a long period of time than can an operator employing a manual instrument. Two reasons can be cited for the higher precision of an automated device. First, machines do not suffer from fatigue, which has been demonstrated to adversely effect the results obtained manually, particularly near the end of a working day. A second important contributing factor to precision is the high reproducibility of the timing sequences of automated instruments—a reproducibility that can seldom be matched in human manual operations. For example, automatic analyzers permit the use of colorimetric reactions that are incomplete or that produce products whose stabilities are inadequate for manual measurement. Similarly, separation techniques, such as solvent extraction or dialysis, where analyte recoveries are incomplete, are still applicable when automated systems are used. In both instances, the high reproducibility of the timing of the operational sequences assures that samples and standards are processed in exactly the same way and for exactly the same length of time.

29B Automation of Sampling and Preliminary Sample Treatment

Most quantitative analytical methods can be broken down into several steps or unit operations including:

1. sample preparation, which may involve such processes as grinding, homogenizing by mixing or stirring, and drying.
2. sample definition, which involves taking a measured weight or volume of the sample.
3. dissolution of the sample, which may involve heating with a solvent, decomposition with a flux, ignition in air, or treatment with acid, base, oxidizing, or reducing reagents.
4. separation of interferences by such procedures as filtration, extraction, dialysis, distillation, or chromatography.
5. completion of the analysis by measuring some physical property that varies as a function of analyte concentration.

6. calculation and reporting the concentrations of the analytes.

For some analyses, one or more of these steps is not required.

In a totally automatic method, all of the unit operations just listed are performed without human intervention. That is, with a totally automatic instrument, an unmeasured quantity of the untreated sample is introduced into the device and ultimately an analytical result is produced as a print-out or graph. Such instruments are available for certain limited types of samples, but their number is few. Certainly, no totally automatic instrument exists that can accommodate a wide variety of sample matrices and compositions.

This section is devoted to instruments that have been developed and are available for performing the four steps that are preliminary to the final measurement step.[3] At the outset, it should be noted that instruments for automation of steps (1) and (3) in the foregoing list are not nearly as highly developed and as readily available as are those for steps (4), (5), and (6). Sophisticated automated sample definition devices (step 2) are widely available for liquids and gases; the situation is far less satisfactory for solids. As a consequence, the bottleneck in the goal of full automation often lies in the problems associated with grinding and homogenizing solid samples and weighing them out. In many instances, for both solid and liquid samples, the decomposition and solution steps limit the degree of automation that is currently possible.

29B-1 SAMPLING AND SAMPLE DEFINITION OF LIQUIDS AND GASES

Most automatic instruments are designed to handle liquid or gaseous samples. Solids must be weighed out and dissolved manu-ally to give solutions that can then be treated automatically. Several dozen devices for sampling liquids and gases are currently available from instrument manufacturers. For discrete instruments, three methods for introducing and diluting a known volume of sample are encountered. The first, which is widely used in automatic chromatography, consists of an injection valve equipped with sample loops of known volumes; Figure 25–12 (p. 750) illustrates this type of device.

Reversible Pump Samplers. Figure 29–1**a** illustrates the principle of reversible pump samplers. This device consists of a movable probe, which is a syringe needle or a piece of fine plastic tubing supported by an arm that periodically lifts the tip of the needle or tube from the sample container and positions it over a second container in which the analysis is performed. This motion is synchronized with the action of a reversible peristaltic pump. As shown in Figure 29–1**a**, with the probe in the sample container, the pump moves the liquid from left to right for a brief period. The probe is then lifted and positioned over the container on the right, and the direction of pumping is reversed. Pumping is continued in this direction until the sample and the desired volume of diluent have been delivered. The probe then returns to its original position to sample the next container. Needless to say, the sample volume is always kept small enough so that none of the sample ever reaches the pump or diluent container.

As shown in Figure 29–5, the reversible pump sampler is often used in conjunction with a rotating sample table. Tables of this kind generally accommodate 40 or more samples in plastic or glass cups or tubes. The rotation of the table is synchronized with the moving arm of the sampler so that samples are withdrawn sequentially.

Syringe-Based Samplers. Figure 29–1**b** illustrates a typical syringe-based sampler and diluter. Here again a movable sample probe is used. With the probe in the sample cup,

---

[3] For a review of automated sample preparation, see: D. A. Burns, *Anal. Chem.*, **1981**, *53*, 1403A.

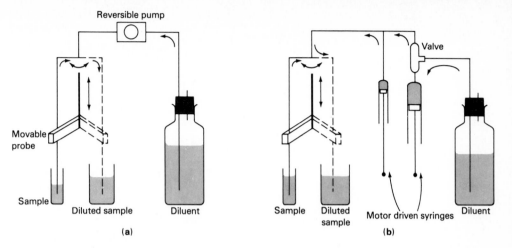

FIGURE 29-1 Automatic samplers: (**a**) reversible pump type; (**b**) syringe type.

the screw-driven syringe on the left withdraws a fixed volume of sample. Simultaneously, the syringe on the right withdraws a fixed volume of diluent (usually larger). The valve shown in the figure permits these two processes to go on independently. When the probe moves over the container on the right, both syringes empty, dispensing the two liquids into the analytical vessel.

Generally, syringe-type injectors are driven by computer-controlled stepping motors, which force the liquid out of the syringe body in series of identically sized pulses. Thus, for example, a 1-mL syringe powered by a motor that requires 1000 steps to empty the syringe is controllable to 1×10^{-3} mL or 1 μL. For a 5000-step motor, the precision would be 0.02 μL.

29B-2 AUTOMATIC HANDLING OF SOLIDS

With solids, sample preparation, definition, and dissolution involves such unit operations as grinding, homogenizing, drying, weighing, igniting, fusing, and treating with solvents. Each of these individual procedures has been automated. Only recently, however, are instruments beginning to appear that can be programmed to perform

several of these unit operations sequentially and without operator intervention. Generally, such instruments require the use of small laboratory robots, which are just beginning to appear on the market.[4] Figure 29-2 is a schematic diagram of an automated laboratory system based on one of these robots. Central to the system is a horizontal arm, which is mounted on two vertical pillars and has four degrees of freedom in its movement. Its 360-deg rotational motion covers a circumference of about 150 inches; its maximum reach is then 24 inches. The device is equipped with a tonged hand, which has a 360-deg wrist-like motion that permits manipulating vials or tubes, pouring of liquids or solids, and shaking and swirling liquids in tubes. Consequently, the arm and hand are capable of carrying out many of the manual operations that are performed by the laboratory chemist. One important feature of the device is its ability to

---

[4] For a description of laboratory robots, which are now available commercially, and their performance characteristics, see: R. Dessy, *Anal. Chem.,* **1983,** *55,* 1100A, 1232A; G. L. Hawk, J. N. Little, and F. H. Zenie, *Amer. Lab.,* **1982,** *14* (6), 98. See also: B. J. McGratten and D. J. Macero, *Amer. Lab.,* **1984,** *16* (9), 16.

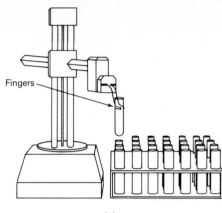

(a)

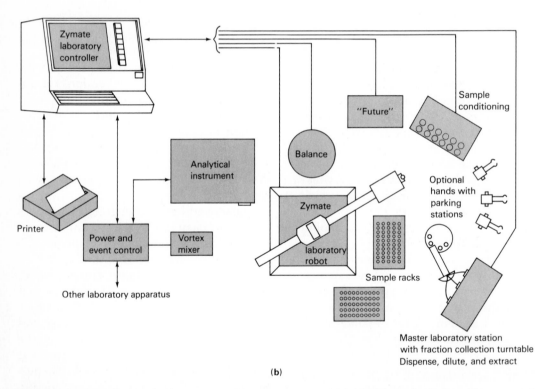

(b)

FIGURE 29-2 A robotic laboratory system: (**a**) robot arm and hand; (**b**) total system. (Courtesy of Zymark Corporation, Hopkinton, MA.)

change hands. Thus, the robot can leave its tonged hand on the table and attach a syringe in its place for pipetting liquids.

The robotic system is controlled by a mi-croprocessor that can be user-programmed. Thus, the instrument can be instructed to bring samples to the master laboratory station where they can be diluted, filtered, partitioned, and treated with reagents. The device can also be instructed to heat and shake samples, dispense liquids, and collect fractions from a column. In addition, the robot can be interfaced with an automatic electric balance for weighing samples. Future developments that are planned include capability for centrifugation, homogenizing, and sample injection into a gas-chromatographic column.

One application of this device has been to the routine screening of newly synthesized compounds for electrochemical activity.[5] Here, the robot cleans, fills, and deaerates the electrochemical cell, introduces samples, performs standard additions, starts and stops the polarographic analyzer, and records data. Analyses were performed with a relative standard deviation of 2%.

Another robotic system has also been described that totally automates the acquisition of pH titration curves for solid samples.[6] This system performs automatically and sequentially weighing out of sample, dissolution, dilution, pH meter calibration, acquisition of titration data, selection of end point, and finally, report generation. Undoubtedly, more and more instruments of this kind can be expected in the near future.

29B-3 AUTOMATIC SAMPLE DECOMPOSITION

The analysis of many materials require their preliminary decomposition under rigorous conditions. Thus, the determination of the elemental composition of organic compounds usually requires that the sample be combusted in oxygen; oxidized with a strong reagent, such as concentrated sulfuric acid; or fused with a solid flux, such as sodium peroxide. Similarly, in order to de-

termine the composition of rocks and minerals, fusion at high temperatures with sodium carbonate or boric oxide is often necessary to rupture the strong silicate bonds and produce products that are soluble in aqueous solvents.

Only a few automatic devices for sample decomposition are available from instrument manufacturers. The most widely used are for decomposing organic compounds for the analysis of such elements as carbon, hydrogen, oxygen, nitrogen, and sulfur. Generally, such instruments are fully automated and require only the introduction of a weighed sample. One instrument of this type is described in Section 29D-3.

Automatic instruments are also available for evolving and determining oxygen, nitrogen, hydrogen, and sulfur in various solid samples such as ferrous metals, soils, shales, and limestones. Again these instruments are fully automated except for sample weighing.

29B-4 AUTOMATIC SEPARATIONS

The separation procedures employed in analytical laboratories have been automated with varying degrees of success. Automated apparatus for filtration or centrifugation has not been developed. In contrast, devices for dialysis, extraction, and distillation have proven more amenable to automation and are encountered as part of certain widely used instruments for continuous methods of analysis. Without question, the various forms of chromatography discussed in earlier chapters are much more readily adapted to automation than the classical separation procedures.

Automated Chromatography. Fully automated and partially automated systems for both gas and high-performance liquid chromatography are now available from several instrument manufacturers. In a typical, fully automated system, liquid samples and standards

---

[5] M. L. Dittenhafer and J. D. McClean, *Anal. Chem.*, **1983**, *55*, 1242A.

[6] G. D. Owens and R. J. Eckstein, *Anal. Chem.*, **1982**, *54*, 2347.

are loaded into small, screw-cap vials (~1 to 2 mL), which are held in a mechanical device that transports them in succession to the sampler. Generally, the caps are equipped with plastic septa, which can then be pierced with the sampling needle of an automatic sampler. In some instruments, vials are labeled with a bar-coded label, which identifies the sample and the operating procedure to be used. A built-in bar-code reader then automatically identifies each vial as it is sampled and transmits the information to a computer, which controls the operating conditions and stores the information for correlation with the output data. Both syringe and loop-type samplers are encountered. Generally, automatic chromatographic instruments are fully controlled by a user-programmed computer, which allows wide variation in the operating conditions. In most instances, sample injection is held up until the instrument controls indicate that the entire system has come to equilibrium with respect to the input variables.

Completely automated thin-layer chromatography has not been widely practiced. On the other hand, several instrument manufacturers offer automatic devices for sample application and automatic densitometers for quantitative measurements after development of the chromatogram.

Nonchromatographic Separations. Equipment for separations based upon dialysis, extraction, distillation, and filtration is available for use with the continuous flow apparatus; some of these devices are described briefly in the next section.

29C   Continuous Flow Methods

Two types of automated continuous flow methods are encountered: *segmented flow methods,* in which the analytical stream is divided into discrete segments by periodic injection of bubbles of air and nonsegmented

flow procedures in which the analytical stream is unbroken. The latter is generally termed a *flow injection analysis* (FIA).

29C-1   SEGMENTED FLOW METHODS

The first continuous automated analyzer was described by L. T. Skeggs in 1957.[7] A commercial version of this instrument, the Technicon AutoAnalyzer®, became available that same year. It very quickly gained widespread use in clinical laboratories throughout the world and must be considered one of the major developments in analytical chemistry in the last few decades. By now, the number of literature references to the AutoAnalyzer runs into several thousands.[8]

Figure 29–3 is a schematic diagram of a single-channel AutoAnalyzer used for the analysis of one of the constituents of blood. Ordinarily several of these channels are arranged in parallel, each being dedicated to a single type of analysis. With multichannel instruments, a sampler module is included, which dilutes the sample and partitions it into aliquots for introduction in each of the several channels.

Sample and Reagent Transport System. The heart of a continuous flow instrument, be it segmented or nonsegmented, is the peristaltic proportioning pump system. A peristaltic pump is a device in which a fluid (liquid or gas) is squeezed through plastic tubing by metal rollers mounted on a pair of parallel continuous chains. Figure 29–4 illustrates the operating principle of the peristaltic pump. Here, the spring-loaded platen

---

[7] L. T. Skeggs, *Am. J. Clin. Pathol.,* **1957,** *28,* 311.

[8] For a comprehensive review of the AutoAnalyzer literature, see: W. B. Furman, *Continuous Flow Analysis,* Marcel Dekker: New York, 1976. For practical details of AutoAnalyzer instrumentation, see: W. A. Coakley, *Handbook of Automated Analysis,* Marcel Dekker: New York, 1981.

1. Sample photocell
2. Reference photocell
3. Flowcell
4. Light source

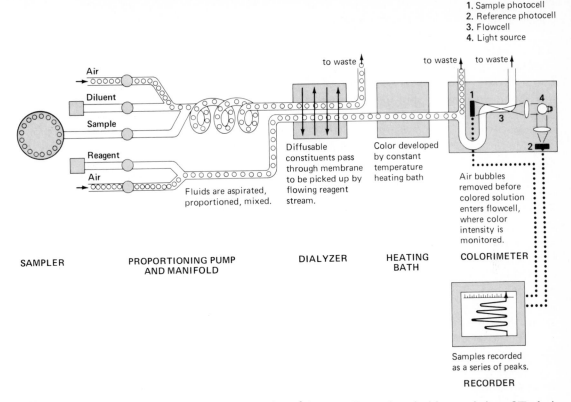

to waste

to waste   to waste

Air

Diluent

Sample

Reagent

Air

Fluids are aspirated,
proportioned, mixed.

Diffusable
constituents pass
through membrane
to be picked up by
flowing reagent
stream.

Color developed
by constant
temperature
heating bath

Air bubbles
removed before
colored solution
enters flowcell,
where color
intensity is
monitored.

SAMPLER

PROPORTIONING PUMP
AND MANIFOLD

DIALYZER

HEATING
BATH

COLORIMETER

Samples recorded
as a series of peaks.

RECORDER

FIGURE 29–3 A single channel Technicon AutoAnalyzer® System. (Reproduced with permission of Technicon Instruments Corporation. TECHNICON, SMA, and AUTOANALYZER are registered trademarks of Technicon Instruments Corporation.)

pinches the tubing against one of the rollers at all times thus forcing a continuous flow of fluid through the tubing. Generally, peristaltic pumps are driven by a constant speed

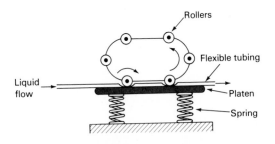

Rollers

Flexible tubing

Liquid
flow

Platen

Spring

FIGURE 29–4 Schematic of a peristaltic proportionating pump. ·

motor and are capable of delivering remarkably reproducible volumes of liquid. The volume is controlled by the inside diameter of the tubing and the length of pumping time. A wide variety of tube sizes are available commercially that permit aspiration rates as little as 0.015 mL/min and as great as 3.9 mL/min.

It should be noted that the volume delivered by a peristaltic pump changes gradually with tubing age. For heavily used systems, replacing tubing as often as once a week is recommended. It is also noteworthy that standard tubing is attacked by strong acids and several organic solvents. Special tubing has been developed to accommodate the former but common solvents such as chloroform, acetone, and pyridine cannot be used even with these resistant materials.

Sample Definition.   Figure 29–5 illustrates how sampling is performed in continuous flow instruments. In this application, a *multichannel proportionating pump* is used. This device consists of a single peristaltic pump with long enough rollers to move fluids simultaneously through three side-by-side tubes. (Modern peristaltic pumps can accommodate as many as 28 tubes.) The volumetric ratio of the fluids is determined by the tube size. Thus, typically, the ratio of sample to diluent flow might be 1 : 5. In order to separate samples in a continuous flow instrument, it is necessary to introduce a fixed volume of wash liquid between samples. As shown in the figure, a movable probe, similar to that shown in Figure 29–1**a**, is employed, which alternates between the sample container held on a rotating table and a vessel that is continuously replenished with fresh wash liquid. It should be noted that the typical sampling table can accommodate 40 or more sample vials.

Segmentation.   An important feature of the apparatus shown in Figure 29–5 is that provision is made to pump regularly spaced air bubbles into each of the flowing streams. The spacing of the bubbles is such that each sample is carried through the system in several successive liquid segments. Before final measurement, the stream passes through a debubbler wherein the segments are recombined (see Figure 29–3). Segmenting the sample tends to maintain a sharp concentration profile at the leading and following edges of each sample. In the absence of bubbles, tailing along the tube walls occurs thereby enhancing the probability of contaminating the following sample. A second purpose of the bubbles is to promote mixing of sample and reagent. As shown in Figure 29–3, each segment is inverted as it rises and falls through the turns of a mixing coil; maximum mixing efficiency occurs when the length of each segment is less than half the coil diameter.

Separations in Continuous Flow Analyzers.   The instrument shown in Figure 29–3 contains a dialysis module in which the small analyte ions and molecules diffuse through a membrane

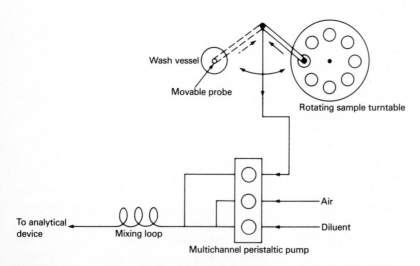

FIGURE 29–5 Sampler handling and pumping system for an AutoAnalyzer.

into a segmented stream of a colorimetric reagent. Larger molecules remain in the original stream and are carried to waste. The membrane is supported between two Lucite or Kel-F plates in which congruent channels have been cut to accommodate the two stream flows. The transfer of smaller species through this membrane is quite incomplete (less than 50%). Thus, successful quantitative analysis requires close control of temperature and flow rates for both samples and standards.

Another common separation technique used in continuous flow methods is extraction. Here, the two immiscible liquids are mixed in a mixing coil and then passed into a separator in the form of a horizontal glass tube containing an inner tube that partially removes either the lighter or heavier part of the stream and carries it to the detector module. Two of several types of phase separators for the AutoAnalyzer are shown in Figure 29–6.

It is important to reiterate that none of the separation procedures in continuous flow methods is ever complete. The lack of completeness is of no consequence, however, because unknowns and standards are treated in an identical way. As was pointed out earlier, the timing sequences in automatic instruments are sufficiently reproducible so that loss of precision does not accompany incomplete reactions as is the case with manual operations.

Detectors. A wide variety of detectors have been used with the AutoAnalyzer including atomic absorption and emission instruments, fluorometers, electrochemical systems, refractometers, spectrophotometers, and colorimeters. The last is by far the most common and the one that is shown in Figure 29–3. In this instance, the photometer is of a twin-beam design in which a part of the radiation passes through a tubular flow cell to a photoelectric detector; the second beam falls on a reference photocell after collimation.

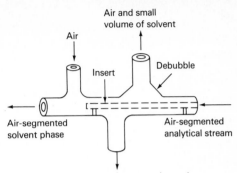

(a)

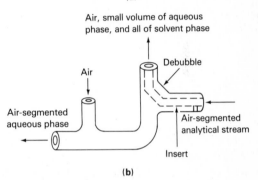

(b)

FIGURE 29-6 Two types of separators for continuous flow extractions: (**a**) for use with a solvent less dense than aqueous phase when the organic phase is to be retained; (**b**) for use with a less dense solvent when the aqueous phase is to be retained. (Reprinted from: W. A. Coakley, *Handbook of Automated Analysis*, p. 117, by courtesy of Marcel Dekker, Inc.)

Figure 29–7 shows the output from a colorimetric detector coupled to a single-channel analyzer. Here, each peak corresponds to one sample while the minima are for the segments of wash liquid, which were incorporated between samples.

Figure 29–8 shows the computer-generated readout from a 12-channel AutoAnalyzer employed for the routine analysis of blood. The shaded areas show the range of concentrations that are considered normal for the population. A modern AutoAnalyzer for routine blood screening is capable

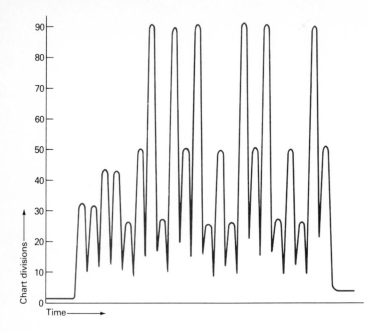

FIGURE 29-7 Output from a single-channel continuous flow analyzer. Here, NO_3^- plus NO_2^- in waste water are being determined by a colorimetric method. Each peak represents a 39-sec sampling time while each minimum corresponds to a 6-sec injection of wash water. (Reprinted with permission from: J. Salpeter and F. LaPerch, *Amer. Lab.*, **1981**, *13* (9), 80. Copyright 1981 by International Scientific Communications, Inc.)

of analyzing 200 serum samples[9] per day for 20 different constituents. That is, the instrument performs 4000 separate analyses each day. Interspersed with the 200 samples are typically 100 standards, which serve for calibration.

Applications of Segmented Flow Analyzers.   As noted earlier, a major area of application of segmented flow analyzers has been in clinical laboratories, where these instruments are the workhorses for routine clinical tests. In addition, however, there now exists a vast literature having to do with the applications of segmented flow instruments in environmental, pharmaceutical, food, agricultural, and metallurgy areas, both for laboratory and process-monitoring uses. Thus, it has been estimated that continuous flow automated methods are being used in industry for the determination of over 300 different species in more than 1000 types of sample

matrices.[10] Two comprehensive bibliographies of these applications are available.[11]

29C-2   FLOW INJECTION METHODS

Flow injection analysis (FIA) is a continuous flow method in which highly precise sample volumes are introduced into a stream that is unsegmented by air bubbles. Perhaps the first use of an unsegmented stream as a medium in which to perform analyses appeared in 1970 when Nagy, Feher, and Pungor[12] described a system in which samples were injected into a flowing stream and were carried through a magnetically stirred chamber and thence to a flow-through

---

[9] Blood serum is the clear fluid remaining after clotting has occurred.

[10] A. Conetta, W. T. Dorsheimer, and M. J. F. DuCros, *Amer. Lab.*, **1981**, *13* (9), 116.

[11] *Technicon AutoAnalyzer Bibliography 1957/1967*, Technicon Corporation, Tarrytown, N.Y., 1968; and *Technicon AutoAnalyzer Pharmaceutical Bibliography*, Technicon Industrial Systems, Tarrytown, N.Y., 1971.

[12] G. Nagy, Z. Feher, and E. Pungor, *Anal. Chim. Acta*, **1970**, *52*, 47.

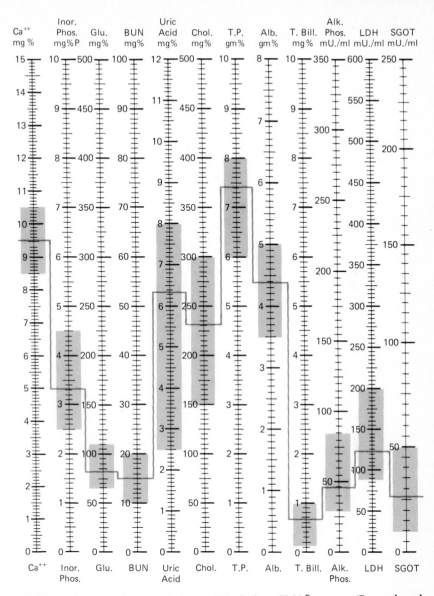

FIGURE 29–8 Readout from a 12-channel Technicon SMA® system. (Reproduced with permission of Technicon Instruments Corporation. TECHNICON, SMA, and AUTOANALYZER are registered trademarks of Technicon Instruments Corporation.)

voltammetric cell. The detector output took the form of transient peaks, which served as a basis for quantitative analysis. Five years later Stewart, Beecher, and Hare in the United States and Ruzicka and Hansen in Denmark[13] simultaneously modified this technique by dispensing with the mixing chamber and using flow-induced sample dispersion to provide contact between analyte and reagent. This method avoided excess sample dilution, which accompanied the mechanical stirring in the earlier procedure. By now, it has become apparent that the flow injection method is an important new way of carrying out automatic wet chemical analyses in a rapid and efficient way.[14] Equipment for flow injection measurements, which is relatively simple, has currently become available from several instrument manufacturers.

Instrumentation. Instruments for flow injection analysis are a hybrid of segmented flow analyzers and high-performance liquid chromatographs. Indeed, the early work in Denmark was performed with the former while the initial studies in the United States were carried out with modified liquid-chromatographic apparatus.

Figure 29–9 is a schematic diagram showing details of a typical flow injection instrument. In this device, as with segmented flow analyzers, sample and reagents are transported through flexible plastic tubing by the action of a peristaltic pump. In other designs, depulsed positive displacement pumps are used. Typically, narrower tubing (often 0.5 mm i.d.) than is found in seg-

mented analyzers is employed. All of the various flow-type detectors of segmented flow analyzers are equally applicable to flow injection measurements. In addition, separations by extraction or dialysis are possible using the same types of devices that were described in Section 29C–1.

Advantages of Flow Injection Measurement. Flow injection equipment differs markedly from its segmented flow counterpart in the respect that no air bubble segmentation is used. Before the advent of flow injection methods, it was believed that air bubbles were vital to the success of continuous flow techniques, they being necessary to prevent excess sample dispersion, to promote turbulent mixing, and to scrub the walls of the conduit thus preventing cross-contamination between samples. In fact, however, excess dispersion or dilution and cross-contamination are nearly completely avoided in a properly designed system without air bubbles. In addition, mixing of reagents and sample occurs rapidly, albeit by a different mechanism.

The absence of air bubbles imparts several important advantages to flow injection measurements including: (1) higher analysis rates (typically 100 to 300 samples/hr); (2) enhanced response times (often less than 1 min between sample injection and recorder response); (3) much more rapid start-up and shut-down times (less than 5 min for each); and (4) except for the injection system, simpler and more flexible equipment. The last two advantages are of particular importance because they make it feasible and economic to apply automated measurements to a relatively few samples of a non-routine kind. That is, no longer are continuous flow methods restricted to situations where the number of samples is large and the analytical method highly routine.

Injectors. Sample sizes for flow injection procedures range from 5 to 200 μL with 10 to 30 μL being typical for most applications. For a successful analysis, it is vital that the

---

[13] K. K. Stewart, G. R. Beecher, and P. E. Hare, *Anal. Biochem.*, **1976**, *70*, 167; J. Ruzicka and E. H. Hansen, *Anal. Chim. Acta*, **1975**, *78*, 145.

[14] For a monograph dealing with flow injection analysis, see: J. Ruzicka and E. H. Hansen, *Flow Injection Analysis*, Wiley: New York, 1981. For brief reviews of the method, see: J. Ruzicka, *Anal. Chem.*, **1983**, *55*, 1041A; K. K. Stewart, *Anal. Chem.*, **1983**, *55*, 931A; C. B. Ranger, *Anal. Chem.*, **1981**, *53*, 20A; and D. Betteridge, *Anal. Chem.*, **1978**, *50*, 832A.

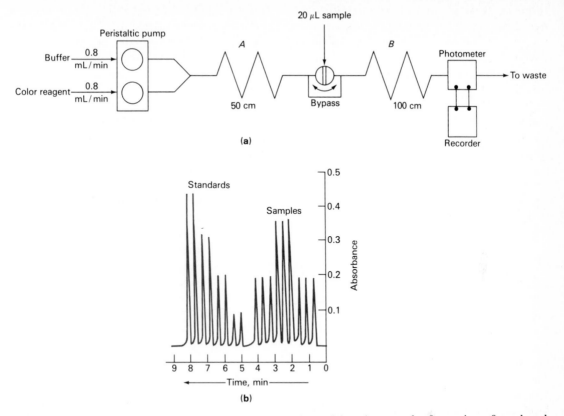

FIGURE 29-9 (**a**) Flow injection apparatus for determining calcium in water by formation of a colored complex with *o*-cresolphalein complexone at pH 10. All tubing had an inside diameter of 0.5 mm. *A* and *B* are reaction coils having the indicated lengths. (**b**) Recorder output. Three sets of curves at right are for triplicate injections of three samples. Four sets of peaks on the left are for duplicate injections of standards containing 5, 10, 15, and 20 ppm calcium. (From: E. H. Hansen, J. Ruzicka, and A. K. Ghose, *Anal. Chim. Acta,* **1978,** *100,* 151. With permission.)

sample solution be injected rapidly as a pulse or plug of liquid; in addition, the injections must not disturb the flow of the carrier stream. The requirements here are significantly more rigorous than is the case with air-segmented flow because dispersion of the sample is more likely to occur in the absence of bubbles. To date, the most satisfactory injector systems are based upon sampling valves similar to those encountered in chromatography. A typical design is shown in Figure 29–9. With the sampling loop in the position shown, the flow of reagents continues through the bypass. When

sample has been injected into the loop and the valve turned 90 deg, the sample enters the flow as a single, well-defined zone. For all practical purposes, flow through the bypass ceases with the valve in this position because the diameter of the sample loop is significantly greater than that of the bypass tubing.

The need for a relatively sophisticated and expensive sampling valve, which is subject to wear and malfunction, has represented a major disadvantage of the flow injection method with respect to the air segmented procedure. Recently, however, a remarkably simple method of introducing a fixed volume of sample, based on two synchronized peristaltic pumps, has been described. As shown in Figure 29–10, the flow from pump 1 is stopped while sample is injected by the action of pump 2. The volume of the injected sample is $\pi r^2 L$, where L is the length of a small section of tubing of diameter r. After injection is completed, pump 2 is shut off and simultaneously pump 1 is turned on. With this device, the relative standard deviation for repetitive chloride determinations was only slightly poorer than with standard rotary valves (0.5% versus 0.3%).

Principles of Flow Injection Analysis. Immediately after injection with a sampling valve, the sample zone in a flow injection apparatus has the rectangular concentration profile shown in Figure 29–11a. As it moves through the tubing, band broadening or *dispersion* takes place. The shape of the resulting zone is determined by two phenomena. The first is convection arising from laminar flow in which the center of the fluid moves more rapidly than the liquid adjacent to the walls thus creating the parabolic-shaped front and the skewed zone profile shown in

Figure 29–11b. Broadening also occurs as a consequence of diffusion. Two types of diffusion can, in principle, occur—radial or perpendicular to the flow direction and longitudinal or parallel to the flow. It has been shown that the latter is of no significance in narrow tubing, whereas radial is always important under this circumstance. In fact, at low flow rates it may be the major source of dispersion. When such conditions exist, the Gaussian-shaped distribution shown in Figure 29–11d is approached. In fact, flow injection analyses are usually performed under conditions in which dispersion by both convection and radial diffusion occurs; peaks like that in Figure 29–11c are then obtained. Here, the radial dispersion from the walls toward the center serves the important function of essentially freeing the walls of analyte and thus eliminating cross-contamination between samples.

Dispersion. Dispersion D is defined by the equation

$$D = C_0/C \qquad (29\text{--}1)$$

where C_0 is the analyte concentration of the injected sample and C is the peak concentra-

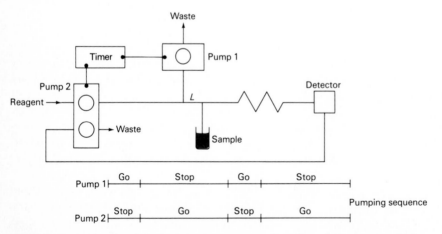

FIGURE 29–10 A simple sample injection system for flow injection analysis. (Adapted from J. Ruzicka and E. H. Hansen, *Anal. Chim. Acta*, **1983**, *145*, 13. With permission.)

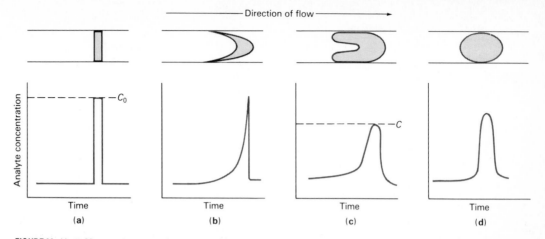

FIGURE 29-11 Effects of convection and diffusion on concentration profiles of analytes at the detector: (**a**) no dispersion; (**b**) dispersion by convection; (**c**) dispersion by convection and radial diffusion; (**d**) dispersion by diffusion. (Reprinted with permission from: D. Betteridge, *Anal. Chem.,* **1978,** *50,* 836A. Copyright 1978 American Chemical Society.)

tion at the detector (see Figure 29–11**c**). Dispersion is readily measured by injecting a dye solution of known concentration C_0 and then measuring the absorbance in the flow-through cell. After calibration, C is calculated from Beer's law.

Dispersion is influenced by three interrelated and controllable variables, namely sample volume, tube length, and pumping rate. The effect of sample volume on dispersion is shown in Figure 29–12**a**; here, the other two variables were held constant. Note that at large sample volumes, the dispersion becomes unity. Under these circumstances, no mixing of sample and carrier takes place, and thus no sample dilution has occurred. Most flow injection analyses, however, involve interaction of the sample with the carrier or an injected reagent. Here, dispersion greater than unity is necessary. For example, a dispersion of 2 would be required if sample and carrier are to be mixed in a 1:1 ratio.

The dramatic effect of sample volume on peak height shown in Figure 29–12**a** emphasizes the need for highly reproducible injection volumes when dispersions of 2 and

greater are used. Other conditions also must be closely controlled if good precision is to be obtained.

Figure 29–12**b** demonstrates the effect of tube length on dispersion when sample size and pumping rate are constant. Here, the number above each peak gives the length of sample travel in centimeters.

In the flow injection literature, the terms limited, medium, and large dispersion are frequently encountered where these adjectives refer to dispersions of 1 to 3, 3 to 10, and greater than 10, respectively. As shown in the next sections, methods based on all three types of dispersion have been developed.

Limited-Dispersion Application. Limited-dispersion flow injection techniques have found considerable application for high-speed feeding of such detector systems as flame atomic absorption and emission as well as inductively coupled plasma. As these sources are normally used, sample aliquots are aspirated directly into the flame or plasma, and a steady state signal is measured. With the flow injection procedure, in contrast, a blank reagent is pumped

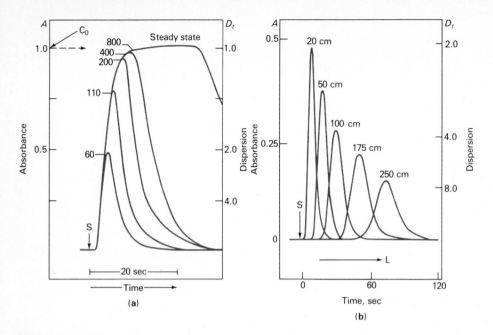

FIGURE 29-12 Effect of sample volume and length of tubing on dispersion. (**a**) Tube length: 20 cm; flow rate: 1.5 mL/min; indicated volumes are in μL. (**b**) Sample volume: 60 μL; flow rate: 1.5 mL/min. (From J. Ruzicka and E. H. Hansen, *Anal. Chim. Acta,* **1980,** *114,* 21. With permission.)

through the system to the source continuously to give a base-line output; samples are then injected periodically, and the resulting transient analyte signals are recorded. Sampling rates of up to 300 samples an hour have been reported.

Limited-dispersion injection has also been used with electrochemical detectors such as specific-ion electrodes and voltammetric microelectrodes. The justification for using flow injection methods for obtaining such data as pH, pCa, or pNO$_3$ is the small sample size required ($\sim$25 μL) and the short measurement time ($\sim$10 sec). That is, measurements are made well before steady-state equilibria are established, which for many p-ion electrodes may require a minute or more. With flow injection measurements, transient signals for sample and standards provide equally accurate analytical data. For example, it has been reported that pH measurements on blood serum can be accom-

plished at a rate of 240/hr with a precision of $\pm$ 0.002 pH. Results are displayed within 5 sec after sample injection.

In general, limited-dispersion conditions are realized by reducing as much as possible the distance between injector and detector, slowing the pumping speed, and increasing the sample volume. Thus, for the pH measurements just described, the length of 0.5 mm tubing was only 10 cm, and the sample size was 30 μL.

Medium-Dispersion Method. Most flow injection procedures are operated under conditions of medium dispersion because time must be allowed for mixing of sample with carrier and reagent solutions to produce products that absorb, fluoresce, or otherwise provide a detector response. Figure 29–13a is a flow diagram of the simplest of all medium-dispersion flow injection systems. Here, a colorimetric reagent for chloride ion is pumped directly into the sampling valve and thence through a 50-cm reactor coil where the reagent diffuses into the sample plug and produces a colored product by the sequence of reactions

$$Hg(SCN)_2(aq) + 2Cl^- \rightleftharpoons HgCl_2(aq) + 2SCN^-$$

$$Fe^{3+} + SCN^- \rightleftharpoons Fe(SCN)^{2+}$$

red

The recorder output for a series of standards containing from 5 to 75 ppm of chloride are shown on the left of Figure 29–13**b**. Note that four injections of each standard were made to demonstrate the reproducibility of the system. This process took 23 min, which corresponds to a sampling rate of 130 samples/hr. The two curves on the right of the figure are high-speed recorder scans of samples containing 30 (R_{30}) and 75 (R_{75}) ppm chloride. These curves demonstrate that less than 1% of an analyte is present in the flow cell after 28 sec, the time of the next injection (S_2). This system has been successfully used for the routine determination of chloride ion in brackish and waste waters as well as in serum samples.

Figure 29–9 (p. 865) illustrates a somewhat more complicated system for the colorimetric determination of calcium in serum, milk, and drinking water. Here, a borax buffer and a color reagent are combined in a 50-cm mixing coil A prior to sample injection. The recorder output for three samples in triplicate and four standards in duplicate are shown in (**b**) of the figure.

Figure 29–14 illustrates a yet more complicated apparatus designed for the spectrophotometric determination of caffeine in drug preparations after extraction of the caffeine into chloroform. The chloroform solvent, after cooling in an ice bath to minimize evaporation, is mixed with the alkaline sample stream in a modified AutoAnalyzer T-tube. After passing through the 2-m extraction coil, the mixture enters a T-tube separator, which is differentially pumped so that about 35% of the organic phase passes

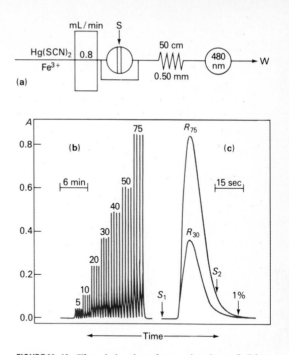

FIGURE 29–13 Flow injection determination of chloride; (**a**) flow diagram; (**b**) recorder readout for quadruplicate runs on standards containing 5 to 75 ppm of chloride ion; (**c**) fast scan of two of the standards to demonstrate the low analyte carryover (less than 1%) from run to run. Note that the point marked 1% corresponds to where the response would just begin for a sample injected at time S_2. (From: J. Ruzicka and E. H. Hansen, *Flow Injection Methods*, p. 8. Wiley: New York, 1981. Reprinted by permission of John Wiley & Sons, Inc.)

into the flow cell, the other 65% accompanying the aqueous solution to waste. In order to avoid contaminating the flow cell with water, Teflon fibers, which are not wetted by water, were twisted into a thread and inserted in the inlet to the T-tube in such a way as to form a smooth downward bend. The chloroform flow then follows this bend to the photometer cell.

Stopped-Flow Methods. It was noted earlier that dispersion decreases with flow rate. In fact, it has been found that dispersion ceases almost entirely when flow is stopped. This fact

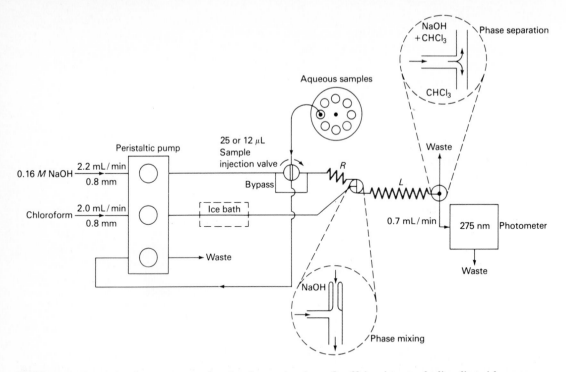

FIGURE 29–14 Flow injection apparatus for the determination of caffeine in acetylsalicyclic acid preparations. With the valve rotated at 90 deg, the flow in the bypass is essentially zero because of its small diameter. R and L are Teflon coils with 0.8 mm inside diameters. L has a length of 2 m while the distance from the injection point through R to the mixing point is 0.15 m. (Adapted from: B. Karlberg and S. Thelander, *Anal. Chim. Acta,* **1978,** *98,* 2. With permission.)

has been exploited to increase the sensitivity of measurements by allowing time for reactions to go further toward completion without dilution of the sample zone by dispersion. In this type of application, a timing device is required to turn the pump off at precisely timed and regular intervals.

A second application of the stop-flow technique is for kinetic measurements. In this application, the flow is stopped with the reaction mixture in the flow cell where the changes in the concentration of reactants or products can be followed as a function of time. This technique was first applied to the enzymatic determination of glucose based upon the use of the enzyme glucose dehy-

drogenase.[15] The reaction is carried out in the presence of the coenzyme nicotinamide-adenine dinucleotide, which serves as the chromophoric agent ($\lambda_{max} = 340$ nm). As many as 120 samples/hr can be analyzed in this way. The procedure has the considerable virtue of consuming less than one unit of the expensive enzyme per sample.

Flow Injection Titrations. Titrations can also be performed continuously in a flow injection apparatus. Here, the injected sample is combined with a carrier in a mixing chamber that promotes large dispersion. The mix-

---

[15] J. Ruzicka and E. H. Hansen, *Anal. Chim. Acta,* **1979,** *106,* 207.

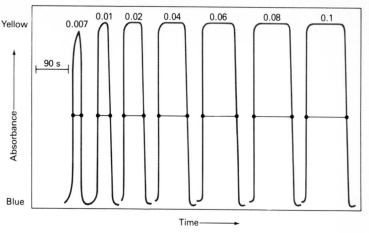

FIGURE 29–15 Flow injection titration of HCl with 0.001 *M* NaOH. The molarities of the HCl solutions are shown at the top of the figure. The indicator was bromothymol blue. The time interval between the points is a measure of the acid concentration. (From: J. Ruzicka, E. H. Hansen, and H. Mosbaek, *Anal. Chim. Acta*, **1980,** *114,* 29. With permission.)

ture is then transported to a confluence fitting, where it is mixed with the reagent, which contains an indicator. If the detector is set to respond to the color of the indicator in the presence of excess analyte, peaks such as those shown in Figure 29–15 are obtained. Here, an acid is being titrated with a standard solution of sodium hydroxide, which contains bromothymol blue indicator. With injection of samples, the solution changes from blue to yellow and remains yellow until the acid is consumed and the solution again becomes blue. As shown in the figure, the concentration of analyte is determined from the widths of the peaks at half height. Titrations of this kind can be performed at a rate of 60 samples/hr.

29D Discrete Methods

A wide variety of discrete analyzers are offered by numerous instrument manufacturers. Some of these instruments are fully automated; others are partially so. Some have been designed for a specific analysis only—for example, the determination of nitrogen in organic compounds. Others can perform a variety of analyses of a given general type.

For example, several automatic titrators are available that can perform neutralization, precipitation, complex formation, and oxidation-reduction titrations as directed by a user-programmed computer.

Several discrete automated instruments have been described in early chapters. A few additional examples are provided here.

29D-1 THE CENTRIFUGAL FAST SCAN ANALYZER

A type of batch analyzer that is capable of analyzing as many as 16 samples simultaneously for a single constituent is based upon the use of a centrifuge to mix the samples with a reagent and to transfer the mixtures to cells for photometric or spectrophotometric measurement.[16] The system is such that conversion from one type of reagent to another is usually easy.

The principle of the instrument is seen in Figure 29–16, which is a cross-sectional view of the circular, plastic rotor of a centrifuge. The rotor has 17 dual compartments ar-

---

[16] For a brief description of a typical system, see: R. L. Coleman, W. D. Shults, M. T. Kelley, and J. A. Dean, *Amer. Lab.,* **1971,** *3* (7), 26; and C. D. Scott and C. A. Burtis, *Anal. Chem.,* **1973,** *45* (3), 327A.

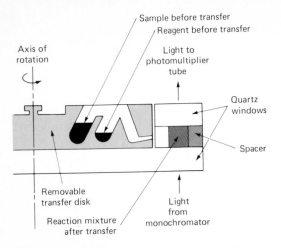

Axis of rotation

Sample before transfer

Reagent before transfer

Light to photomultiplier tube

Quartz windows

Spacer

Removable transfer disk

Light from monochromator

Reaction mixture after transfer

FIGURE 29-16 Rotor for a centrifugal fast analyzer. (Reprinted with permission from: R. L. Coleman, W. D. Shults, M. T. Kelley, and J. A. Dean, *Amer. Lab.*, **1971**, *3* (7), 26. Copyright 1971 by International Scientific Communications, Inc.)

ranged radially around the axis of rotation. Samples and reagents are pipetted automatically into 16 of the compartments as shown; solvent and reagent are measured into the seventeenth. When the rotor reaches a rotation rate of about 350 rpm, reagent and liquids in the 17 compartments are mixed simultaneously and carried into individual cells located at the outer edge of the rotor; these cells are equipped with horizontal quartz windows. Mixing is hastened by drawing air through the mixtures. Radiation from an interference-filter photometer or a spectrophotometer passes through the cells and falls upon a photomultiplier tube. For each rotation, a series of electrical pulses are produced, 16 for the samples and 1 for the blank. Between each of these pulses is a signal corresponding to the dark current.

The successive signals from the instrument are collected in the memory of a dedicated microcomputer for subsequent manipulation. Signal averaging can be employed to optimize the signal-to-noise ratio.

One of the most important applications of the centrifugal fast analyzer is for the determination of enzymes. Ordinarily, enzyme analyses are based upon the catalytic effect of the analyte upon a reaction that involves formation or consumption of an absorbing species. Here, a calibration curve relating the rate of appearance or disappearance of the absorbing species as a function of enzyme concentration serves as the basis for the analysis. The centrifugal analyzer permits the simultaneous determination of the rates of 16 reactions under exactly the same conditions; thus, 16 simultaneous enzyme analyses are feasible.

29D-2 AN AUTOMATIC MULTIPURPOSE ANALYZER

Automatic instruments that can be user-programmed to perform a variety of analyses are now beginning to appear in the market place. Figure 29–17 shows the sample-handling module of a remarkably versatile instrument of this kind, which is capable of carrying out not only most titrimetric procedures but colorimetric and specific-ion measurements as well. The fluid-handling system actually has two additional syringes thus making it possible to introduce five different liquids into the reaction vessel. These syringes are powered by stepping motors, which deliver the full capacity of the syringes (1, 5, 10, or 20 mL) in 2000 or 8000 identical pulses. Thus, with the 1-mL syringe, each step corresponds to 0.5 or 0.13 μL. The reactor vessel shown in the figure is Teflon lined and can be rotated for mixing at a rate of 0 to 2400 rpm. The vessel accommodates as many as three detectors, which may be specific-ion or amperometric electrodes or a colorimetric detector.

A colorimetric module for the instrument is also available. It consists of 1- or 10-cm

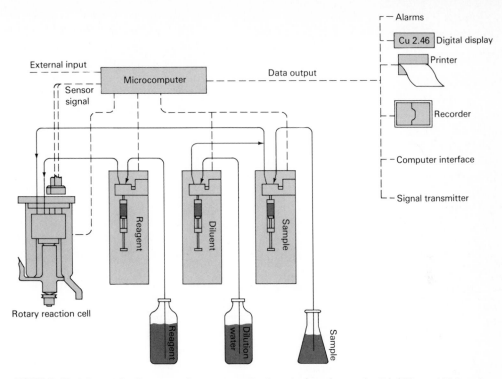

FIGURE 29-17 Schematic diagram of an automatic chemical analyzer, the DigiChem 4000 series programmable chemical analyzer. (Courtesy of Ionics Inc., Watertown, MA.)

cells, a series of 7 interference filters covering a wavelength range of 400 to 700 nm, and a photoelectric detector.

The instrument is controlled by a computer with 4K of RAM and 9K of EPROM memory. These permit programmable dispensing, diluting, mixing, and measuring of liquids from the five syringes. In addition, up to 99 user-programmed analytical routines can be stored in memory for later use. Each of these, when called, will then supervise the automatic analysis of up to 200 samples, provided the proper reagents and detectors are in place. The computer can also be programmed to perform electrode standardizations, standard addition procedures, and to print out titration curves, first and second derivatives curves, and Gran's plots.

29D-3 AN AUTOMATIC ELEMENTAL ANALYZER

Several manufacturers produce automatic instruments for analyzing organic compounds for one or more of the common elements including carbon, hydrogen, oxygen, sulfur, and nitrogen. All of these instruments are based upon high-temperature oxidation of the organic compounds, which converts the elements of interest to gaseous molecules. In some instruments, the gases are separated on a chromatographic column while in others separations are based upon specific absorbents. In either case, thermal conductivity detection serves to complete the determinations. Often these instruments are equipped with devices that automatically load the weighed samples into the combustion area.

Figure 29–18 is a schematic diagram of an automatic instrument for the determination of carbon, hydrogen, and nitrogen. Here, samples are oxidized at 900°C under static conditions in a pure oxygen environment that produces a gaseous mixture of carbon dioxide, carbon monoxide, water, elemental nitrogen, and oxides of nitrogen. After 2 to 6 min in the oxygen environment, the products are swept with a stream of helium through a 750°C tube furnace where hot copper reduces the oxides of nitrogen to the element and also removes the oxygen as copper oxide. Additional copper oxide is also present to convert carbon monoxide to the dioxide. Halogens are removed by a silver wool packing.

The products from the reaction furnace pass into a mixing chamber where they are brought to a constant temperature of 75°C, a volume of 300 mL, and a pressure of 2 atm. The resulting homogenous mixture is then analyzed by passing it through a series of three precision thermal conductivity detectors, each detector consisting of a pair of sensing cells.

Between the first pair of cells is a magnesium perchlorate absorption trap that removes water. The differential signal then serves as a measure of the hydrogen in the sample. Carbon dioxide is removed in a second absorption trap. Again, the differential signal between the second pair of cells is a measure of carbon in the sample. The remaining gas, consisting of helium and nitrogen, passes through the third detector cell. The output of this cell is compared to that of a reference cell through which pure helium flows. The potential across this pair of cells is related to the amount of nitrogen in the sample.

For oxygen analysis, the reaction tube is replaced by a quartz tube filled with platinized carbon. When the sample is pyrolyzed in helium and swept through this tube, all of the oxygen is converted to carbon monoxide, which is then converted to carbon dioxide by passage over hot copper oxide. The remainder of the procedure is the same as just described, with the oxygen concentration being related to the differential signal before and after absorption of the carbon dioxide.

For a sulfur analysis, the sample is combusted in an oxygen atmosphere in a tube packed with tungstic oxide or copper oxide. Water is removed by a dehydrating reagent located in the cool zone of the same tube. The dry sulfur dioxide is then separated and determined by the differential signal at what is normally the hydrogen detection bridge. In this instance, however, the sulfur

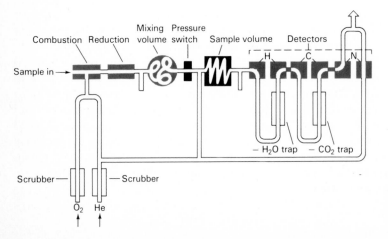

FIGURE 29-18 An automatic C, H, and N analyzer. (Courtesy of Perkin-Elmer, Norwalk, CT.)

dioxide is absorbed by a silver oxide reagent.

The instrument shown in Figure 29–18 can be fully automated, whereby up to 60 weighed samples contained in small capsules are loaded into a carousel sample tray for automatic sampling.

29E　Automatic Analysis Based on Multilayer Films

During the past decade, a technology has been developed for performing the various steps in a quantitative analysis automatically in discrete films arranged in multilayers supported on transparent, disposable plates having the size of a postage stamp. Here, a small drop of the sample (10 to 50 μL) is placed on the top layer of the multilayer film element where it spreads rapidly and uniformly. Water and low-molecular-weight components diffuse from the spreading layer through one or more reagent layers where the analyte interacts to produce a colored (or occasionally a fluorescent) product, which is then determined by reflectance photometry. To date, this technology has been applied to the routine determination of blood metabolites, such as glucose; serum enzymes, such as lactate dehydrogenase; and therapeutic drugs in samples of blood. In addition, a modification of the technique has been developed that permits the potentiometric determination of the electrolyte components, such as potassium, in blood serums. As noted earlier, hundreds of millions of such determinations are performed each year in clinical laboratories throughout the world, generally by means of automatic continuous flow or discrete instruments. The cost of such instruments is large, which has required their location in centralized laboratories, where their expense can be amortized over a large volume of samples. The advent of multilayer film elements means that it is now economic to perform

some of these routine clinical analyses automatically in decentralized locations—even in the individual physician's office and perhaps ultimately in the home.[17] In fact, glucose monitoring kits are now available for use by insulin-dependent diabetic patients.

29E-1　GENERAL PRINCIPLES

The development of multilayer film elements for chemical analysis is an outgrowth of the technology developed by the color photographic industry for producing multilayer films wherein complex chemistries take place. For example, a typical instant color film is made up of as many as 15 separate layers having thicknesses of 1.5 to 5 μm. One of these layers contains a developing fluid, which, when released, diffuses through the remaining layers wherein a series of complex chemical reactions occur that ultimately lead to the deposition of blue, green, and red dyes in those areas that have been photosensitized by exposure in a camera.

This same technology makes it feasible to perform automatic chemical analyses based upon a sequence of physical or chemical reactions. For example, the product of a chemical reaction produced in the first layer may be separated from interferences by selective diffusion through a second layer to a third layer where further reaction can occur. Each layer of a multilayer film thus offers a separate domain in which a chemical reaction or a physical separation can be carried out that is similar to ones employed in standard analytical procedures. For each type of analysis, a complete set of reagents is miniaturized in a disposable dry form. Reconstitution of reagents and many other manual manipulations are replaced by the

---

[17] For review articles describing this technology, see: B. Walter, *Anal. Chem.*, **1983**, *55*, 499A; H. G. Curme, *et al.*, *Clin. Chem.*, **1978**, *24*, 1335; R. W. Spayd, *et al.*, *Clin. Chem.*, **1978**, *24*, 1343.

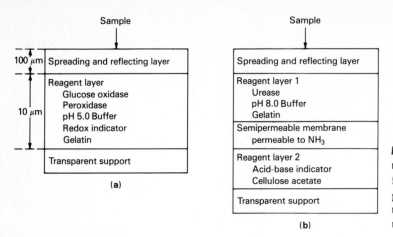

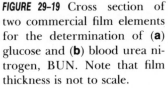

FIGURE 29-19 Cross section of two commercial film elements for the determination of (**a**) glucose and (**b**) blood urea nitrogen, BUN. Note that film thickness is not to scale.

single step of applying the sample to the film element. The advantages of such systems are obvious.

29E-2 FILM STRUCTURES

Most multilayer film elements consist of a transparent support, one or more reaction layers, a reflective layer, and a spreading or metering layer. Figure 29–19**a** is a cross-sectional diagram of a film element, which is commercially available, for the determination of glucose in serum. The chemistry that occurs in the reagent layer is described by the reactions

$$\text{glucose} + O_2 + 2H_2O$$
$$\xrightarrow[\text{oxidase}]{\text{glucose}} \text{gluconic acid} + 2H_2O_2$$

$$2H_2O_2 + \text{indicator}_{(\text{reduced})}$$
$$\xrightarrow{\text{peroxidase}} \text{indicator}_{(\text{oxid})} + 4H_2O$$

The reaction product is an oxidized dye, which absorbs strongly at 495 nm.

The reagent layer contains, in addition to the oxidation-reduction indicator, the enzymes glucose oxidase and peroxidase, and a pH 5.0 buffer. All are immobilized in a gelatin binder, which is approximately 10

μm thick. This film is supported on a rigid, transparent plastic film.

The spreading or metering layer upon which a drop of sample is placed, is typically 100 μm thick and is made up of cellulose acetate in which titanium dioxide is dispersed as a reflectant. The spreading layer serves three purposes. The first is to reflect the radiation from a source back through the reagent and support layers to the detector of a photometer. Its second function is to cause the sample to spread into a uniform layer. During the spreading, which takes 5 to 10 sec, most of the fluid movement is lateral until all of the liquid is contained in the pore structure. As a result of this rapid lateral spreading with little penetration, the amount of fluid and analyte per unit area is relatively independent of drop volume. Thus, a 10% variation in sample size produces only a 1% change in concentration of analyte per unit of area. Furthermore, the concentration of analyte is found to be uniform throughout the width of the spot. A third function of the spreading layer is to retain cells, crystals, and particulate matter as well as large molecules such as proteins.

The determination of blood glucose with the plate just described is remarkably simple and can be fully automated. A 10 μL sample is introduced onto the plate, which is then held in an incubator at 37 ($\pm$0.05)°C for 7 min. The reflectance at 495 nm is measured with the instrument described in the next section.

Figure 29–19**b** depicts a somewhat more complicated film element for the determination of serum urea nitrogen. Here, two reagent layers are separated by a semipermeable layer of cellulose acetate butyrate that passes ammonia but excludes carbon dioxide and hydroxide ions from the acid-base indicator layer. The reactions in this instance are

$$(NH_2)_2CO + H_2O \xrightarrow[\text{pH 8.0}]{\text{urease}} 2NH_3 + CO_2$$

$$NH_3 + \underset{\text{colorless}}{HIn} \rightarrow NH_4^+ + \underset{\text{colored}}{In^-}$$

29E-3 INSTRUMENTATION

Quantitative measurements of the products of multilayer film separations and reactions have been based on reflectance photometry, specific-ion potentiometry, and fluorescence.

Reflective Photometer. Figure 29–20 is a schematic diagram of a manual instrument for measuring the diffuse reflectance from a multilayer film. Here, the sample while still in an incubator is illuminated by filtered radiation of a wavelength absorbed by the analyte. The source beam, which is at 45 deg to the film element, is viewed normal to that element. This geometry minimizes front surface reflection. The diameter of the illuminated spot is about 2.5 mm. Detection is by means of a photomultiplier tube.

In reflectance spectroscopy, two types of reflection are encountered: specular or mirror-like, in which the angles of incidence and reflection are identical, and diffuse, which is reflection from a matte structure. The latter is the one that serves as the basis for reflectance spectroscopy. Diffuse reflection is not a surface phenomenon but results from scattering, transmission, and absorption interactions of the radiation in the volume of the illuminated film. Absorption within the volume reduces the reflected intensity.

Reflectance data are usually expressed in terms of the *percent reflectance* (% R), which is analogous to percent transmittance in absorption spectroscopy. Thus,

$$\% R = \frac{I_s}{I_r} \times 100$$

where I_s is the intensity of the beam reflected from the sample and I_r is the intensity from a reference standard, usually barium sulfate. As with transmittance, reflectance decreases nonlinearly with increases in the concentration of absorbing species. Several algorithms have been developed for linearlizing this relationship. The specific algorithm used depends upon the reflection characteristic of the particular multilayer film, the nature of the illumination, and the geometry of the instrument.

Both manual and fully automated reflectance instruments are now available. With the latter, the operator needs only provide the sample and specify the tests to be performed. The instrument then chooses the appropriate film element from a cassette, se-

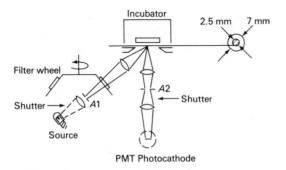

FIGURE 29–20 A manual reflectance photometer for use with thin-film elements. (From: H. G. Curme, *et al.*, *Clin. Chem.*, **1978**, *24*, 1336. With permission.)

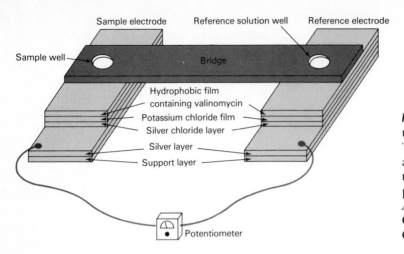

FIGURE 29-21 Multilayer selective-ion cell for potassium. The dimensions of the device are 2.8 × 2.4 cm by 150 μm thickness. (Reprinted with permission from: B. Walter, *Anal. Chem.*, **1983**, *55*, 508A. Copyright 1983 American Chemical Society.)

lects the appropriate radiation filter, calibrates the instrument, applies the sample, and prints out the result. This instrument is capable of performing more than 500 determinations per hour involving any of 16 available tests.

Potentiometry. Multilayer film technology has been extended to the manufacture of single-test, disposable, ion-selective membrane systems for the determination of potassium, sodium, and chloride ions as well as the determination of carbon as carbonate. A schematic diagram of the cell used for potassium ion determination in serum samples

Table 29-1
Performance Data For Some Thin-Film Elements[a]

Analyte	Dynamic Range	Precision, % Relative Standard Deviation
Albumin	72.5–869 μmol/L	4.9
Ammonia	0.01–12 mmol/L	5
Bilirubin	3.4–445 μmol/L	2.5
Calcium	0.25–4.0 mmol/L	1.5
Carbon dioxide	5–55 mmol/L	5–7
Chloride	50–175 mmol/L	1.5
Cholesterol	0.39–14.3 mmol/L	5.2
Creatinine	4.4–1459 μmol/L	4
Glucose	1.1–34.7 mmol/L	2.1
Potassium	1–14 mmol/L	2.0
Sodium	75–250 mmol/L	1.3
Triglycerides	0–6.5 mmol/L	2.7
Urea nitrogen	0.7–42.8 mmol/L	3.3
Uric acid	29.7–1010 μmol/L	2.3

[a] From: T. L. Shirey, *Clin. Biochem.*, **1983**, *16*, 147. With permission.

is shown in Figure 29–21. The dimensions of the device are 2.8 by 2.4 cm. Here, two identical film elements are coupled by a paper salt bridge. Approximately 10 μL of sample is placed in one well and 10 μL of standard potassium solution in the other. These solutions rapidly diffuse laterally to activate the salt bridge. The top layer of the film element, which is the ion-selective membrane, is made up of valinomycin in a hydrophobic plastic. As was described earlier (Section 20C–8), valinomycin selectively binds potassium ions, which causes a potential to develop across the film interface. The remaining three layers in each element make up a silver/silver chloride reference electrode.

Unlike conventional specific-ion measurements, the potential of the sample cell is referred directly to that of a cell containing the standard. The potassium concentration in the unknown is then obtained from the potential difference between these two cells. The performance of this and the other ion-selective cells appears to be comparable with that of conventional specific-ion electrode systems.

29E-4 PERFORMANCE AND APPLICATIONS

Thin-film elements and detectors are now offered by several manufacturers for a growing number of clinical tests. Currently, the number of different types of assays is about three dozen. Table 29–1 lists some representative determinations that can be performed with these devices. Generally, the volume of sample is small (10 to 50 μL), and the time required to complete an analysis is short—usually 1 to 10 min. The dynamic ranges shown in the table are sufficiently large to accommodate approximately 98% of the samples encountered in a clinical laboratory. Samples outside this range can usually be handled by suitable dilutions.

Extensive performance testing of these new devices generally reveals a good correlation between the data they produce and the results by standard procedures. Precision of 1 to 10% relative are reported depending upon the type of test, which again is comparable with the data from automated standard methods.

Appendix 1

Simple Electrical Measurements

This appendix describes simple devices for measuring electrical currents, voltages, and resistances. A familiarity with Ohm's and Kirchhoff's laws, which can be found in an elementary physics text, is required.

A1-1 GALVANOMETERS

The classical instrument for measuring direct currents and voltages is the galvanometer, a device that was invented nearly a century ago and still finds widespread use in the laboratory. The galvanometer is based upon the current-induced motion of a coil suspended in a fixed magnetic field. This arrangement is called a *D'Arsonval movement* or *coil.*

Figure A1–1 shows construction details of a typical D'Arsonval meter. The current to be measured passes through a rectangular coil wrapped about a cylindrical soft iron core that pivots between a pair of jeweled bearings; the coil is mounted in the air space between the pole pieces of a permanent magnet. The current-dependent magnetic field that develops in the coil interacts with the permanent magnetic field in such a way as to cause rotation of the coil and deflection of the attached pointer; the deflection is directly proportional to the current present in the coil.

Meters with D'Arsonval movements are generally called *ammeters, milliammeters,* or *microammeters,* depending upon the current range they have been designed to measure. Their accuracies vary from 0.5 to 3% of full scale, depending upon the quality of the meter.

More sensitive galvanometers are obtained by suspending the coil between the poles of a magnet by a fine vertical filament that has very low resistance to rotation. A small mirror is attached to the filament; a light beam, reflected off the mirror to a scale, detects and measures the extent of rotation during current passage. Galvanometers can be constructed that will detect currents as small as 10^{-10} A.

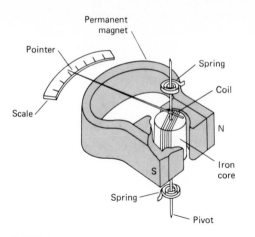

FIGURE A1-1 Current meter. (Taken from R. J. Smith, *Circuits, Devices, and Systems.* New York: Wiley, 1976. With permission.)

Current Measurement. Figure A1–2 shows how a D'Arsonval meter is used to measure an unknown current I_L. Here, the meter is represented by the *equivalent circuit* shown in (**a**) in which the electrical resistance of the coil is represented by R_M. As shown in Figure A1–2**b**, the meter is placed in series with the

remainder of the circuit for a current measurement. Unless the resistance of the meter is much smaller than that of the source R_L, the presence of the meter will inevitably reduce the current thus introducing an error. The relative magnitude of the error is given by

$$\text{rel error} = (I_M - I_L)/I_L$$

where I_L is the current when the switch is in position 1 and I_M is the measured current when the switch is moved to position 2. From Ohm's law, we may write

$$I_L = V_L/R_L$$
$$I_M = V_L/(R_L + R_M)$$

Substituting these relationships into the first equation and rearranging gives

$$\text{rel error} = -R_M/(R_L + R_M) \qquad (A1\text{–}1)$$

Thus, if R_L is 100 Ω and the meter resistance is 50 Ω

$$\text{rel error} = -50/(100 + 50)$$
$$= -0.33 \quad \text{or} \quad -33\%$$

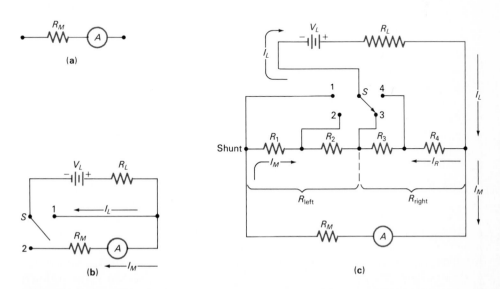

FIGURE A1-2 Measurement of current with a D'Arsonval meter. (**a**) Equivalent current of meter. (**b**) Circuit for measurement of I_L. (**c**) Meter with an Aryton shunt.

If, on the other hand, R_L is 100 kΩ

$$
\begin{aligned}
\text{rel error} &= -50/(100 \times 10^3 + 50) \\
&= -5.0 \times 10^{-4} \quad \text{or} \quad -0.05\%
\end{aligned}
$$

The Aryton Shunt. As shown in Figure A1–2c, the *Aryton shunt* is a network of resistors, which varies the range of a current meter. The example that follows demonstrates how the resistors can be chosen.

EXAMPLE A1–1. Assume that the meter in Figure A1–2c registers full scale with a current of 50 μA; its internal resistance R_M is 3000 Ω. Derive values for the four resistors such that the meter ranges will be 0 to 0.1, 1, 10, and 100 mA.

For each setting of the contact S, the circuit is equivalent to two resistances in parallel. One resistance is made up of the meter and the resistors to the left of the contact in Figure A2–1c; the other consists of the remaining resistors to the right of the contact. The potential across parallel resistances is, of course, the same. Thus,

$$
I_M(R_M + R_{\text{left}}) = I_R R_{\text{right}}
$$

where R_M is the resistance of the meter, R_{left} is the resistance of the resistors to the left of the contact, and R_{right} is the resistance to the right. We may also write, from Kirchhoff's current law

$$
I_L = I_M + I_R
$$

which can be combined with the first equation to eliminate I_R. Thus, after rearrangement,

$$
\frac{I_L}{I_M} = \frac{R_M + R_{\text{left}} + R_{\text{right}}}{R_{\text{right}}} = \frac{R_M + R_T}{R_{\text{right}}}
$$

where R_T is the total resistance of the shunt ($R_{\text{left}} + R_{\text{right}}$).

With the contact in position 1, $R_{\text{left}} = 0$ and $R_{\text{right}} = R_T$; here, the greatest fraction of the current will pass through the meter. Therefore, setting 1 corresponds to the lowest current range, where the full meter reading ($I_M = 50$ μA) should correspond to 0.1 mA ($I_L = 0.1$ mA). Substituting these values into the derived equation yields

$$
\frac{0.1 \times 10^{-3}\ \text{A}}{50 \times 10^{-6}\ \text{A}} = \frac{3000 + 0 + R_{\text{right}}}{R_{\text{right}}}
$$

and

$$
R_{\text{right}} = R_T = 3000\ \Omega
$$

Substituting this value for R_T in the working equation gives

$$
\frac{I_L}{I_M} = \frac{R_M + R_T}{R_{\text{right}}} = \frac{3000 + 3000}{R_{\text{right}}}
$$

or

$$
R_{\text{right}} = 6000 \times I_M/I_L
$$

$$
= 6000 \times 50 \times 10^{-6}/I_L = \frac{0.300}{I_L}
$$

Employing this relationship, we obtain

Setting	I_L	R_{right}
1	0.1×10^{-3}	$3000 = (R_1 + R_2 + R_3 + R_4)$
2	1×10^{-3}	$300 = (R_2 + R_3 + R_4)$
3	10×10^{-3}	$30 = (R_3 + R_4)$
4	100×10^{-3}	$3 = R_4$

Thus, $R_4 = 3\ \Omega$, $R_3 = (30 - 3) = 27\ \Omega$, $R_2 = (300 - 30) = 270\ \Omega$, and $R_1 = (3000 - 300) = 2700\ \Omega$.

Voltage Measurement. A current meter can be used to measure an unknown voltage by placing the meter across (in parallel with) the unknown voltage as shown in Figure

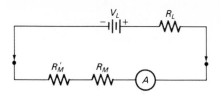

FIGURE A1-3 Voltage measurement with a D'Arsonval meter.

A1–3. Here, the voltage is represented as a potential source V_L with an internal or load resistance of R_L. The current in the meter is then

$$I = V_L/(R_L + R_M + R'_M) \qquad (A1-2)$$

where R'_M is an added resistor chosen so that the full scale of the meter corresponds to a convenient maximum voltage. If $(R_M + R'_M)$ is made much larger than R_L, this equation reduces to

$$I \cong V_L/(R_M + R'_M)$$

and the scale of the meter can be calibrated directly in volts by adjusting R'_M to a suitable value.

EXAMPLE A1-2. What should be the size of R'_M in Figure A1–3 if the full scale reading of the 50 μA meter is to correspond to exactly 2.00 V? The meter resistance is 3000 Ω.
Here,

$$50 \times 10^{-6} \text{ A} = 2.00 \text{ V}/(R_M + R'_M) \ \Omega$$

$$R_M + R'_M = 2.00/50 \times 10^{-6}$$
$$= 40 \times 10^4 \ \Omega \quad \text{or} \quad 40 \text{ k}\Omega$$

$$R'_M = 40 \text{ k}\Omega - 3000 \times 10^{-3} \text{ k}\Omega = 37 \text{ k}\Omega$$

If the load resistance in Equation A1–2 is not small with respect to the meter resistance (plus any resistance placed in series with the meter) a *loading error* develops. The magnitude of this error is given by

rel error $= (I_M - I_L)/I_L$

where I_L is the current in the meter when $(R_M + R'_M) \gg R_L$ and I_M is the actual current in the presence of a significant load resistance. From Ohm's law, we may write

rel error

$$= \left(\frac{V_L}{R_M + R'_M + R_L} - \frac{V_L}{R_M + R'_M}\right) \Big/ \left(\frac{V_L}{R_M + R'_M}\right)$$

or

$$\text{rel error} = \frac{R_M + R'_M}{R_M + R'_M + R_L} - 1 \qquad (A1-3)$$

EXAMPLE A1-3. Calculate the relative loading error for the circuit described in Example A1–2 if the resistance of the load is (a) 1 kΩ and (b) 50 kΩ.
Here, the resistance of the meter circuit is $(R_M + R'_M)$ or 40 kΩ.

(a) Applying Equation A1–3

$$\text{rel error} = \frac{40}{40 + 1} - 1$$
$$= -0.024 \quad \text{or} \quad -2.4\%$$

(b) $\quad \text{rel error} = \dfrac{40}{40 + 50} - 1$
$$= -0.56 \quad \text{or} \quad -56\%$$

To summarize, for accurate current measurements, the resistance of a meter must be small with respect to the resistance of the source. In contrast, for accurate potential readings, the resistance must be significantly larger than the resistance of the source.

Resistance Measurement with a Current Meter. Figure A1–4 is a diagram of a circuit for measuring an unknown resistance R_L. With the switch in position 1, the variable resistor R_V is adjusted until A reads full scale. Here,

$$V_S = I_1(R_V + R_M)$$

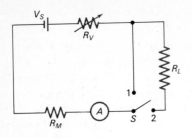

FIGURE A1-4 Resistance measurement with a D'Arsonval meter.

In position 2,

$$V_S = I_2(R_V + R_M + R_L)$$

Dividing one equation by the other yields upon rearrangement

$$R_L = \left(\frac{I_1}{I_2} - 1\right)\left(R_V + R_M\right) \qquad \text{(A1–4)}$$

If, for example, a 1.00 mA meter is employed and $(R_V + R_M)$ is 2000 Ω,

$$R_L = \left(\frac{1.00 \times 10^{-3}}{I_2} - 1\right) 2000 \ \Omega$$

Clearly, the relationship between current and resistance is nonlinear.

A1-2 NULL INSTRUMENTS

Instruments based upon comparison of the system of interest with a standard are widely employed for chemical measurements; such devices are called *null* instruments. Null instruments are generally more accurate and precise than direct-reading instruments, in part because a null measurement generally causes less disturbance to the system being studied. Moreover, the readout devices employed can often be considerably more rugged and less subject to environmental effects than a direct-reading device.

Potentiometers. The potentiometer is a null instrument that permits accurate measurement of potentials while drawing a mini-

mum current from the source under study. A typical laboratory potentiometer employs a linear voltage divider to attenuate a reference voltage until a null point is reached. In its simplest form, the divider consists of a uniform resistance wire mounted on a meter stick. A sliding contact or wiper permits variation of the output voltage. More conveniently, the divider is a precision, wire-wound resistor formed in a helical coil. A continuously variable wiper that can be moved from one end of the helix to the other provides a variable voltage. Ordinarily, the divider is powered by a line operated dc power supply or by mercury batteries, which provide a potential somewhat larger than that which is to be measured.

Figure A1–5 is a sketch of a laboratory potentiometer employed for the measurement of an unknown potential V_L by comparing it with a standard potential V_S. Here, two null-point measurements are required. Switch P is first closed to provide a current in the slide wire and a potential drop across

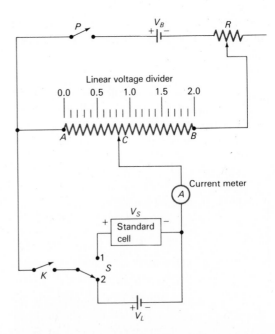

FIGURE A1-5 Circuit diagram for a laboratory potentiometer for measuring V_L.

AC. The standard cell is next placed in the circuit by moving switch *S* to position 1. When the tapping key *K* is closed momentarily, a current will be indicated by the galvanometer *G*, unless the output from the standard cell is identical with the potential drop V_{AC} between *A* and *C*. The position of contact *C* is then varied until a null condition is achieved, as indicated by the galvanometer when *K* is closed.

The foregoing process is repeated with *S* in position 2 so that the unknown cell is in the circuit. For both adjustments, the magnitude of the potential V_{AC} between points *A* and *C* is given by

$$V_{AC} = V_B \frac{R_{AC}}{R_{AB}} = \frac{\not{k}AC}{\not{k}AB} = \frac{AC}{AB}$$

Here, *AC* and *AB* are the linear distances from *A* to points *B* and *C*, respectively, while *k* is the proportionality constant relating length of the conductor to its resistance. When *S* is at position 1, we may write

$$V_S = V_B \frac{AC_S}{AB}$$

where AC_S represents the linear distance between *A* and *C* as shown on the scale. At position 2,

$$V_L = V_B \frac{AC_L}{AB}$$

Dividing these equations gives, upon rearrangement,

$$V_L = V_S \frac{AC_L}{AC_S} \qquad (A1\text{--}5)$$

Thus, V_L is obtained from the potential of the standard cell and the two slide-wire settings.

The slide wire shown in Figure A1–5 is readily calibrated to read directly in volts, thus avoiding the necessity of calculating V_L

each time. Calibration is accomplished by setting the slide-wire reading to the voltage of the standard cell. Then, with that cell in the circuit, the potential across *AB* is adjusted by means of the variable resistor *R* until a null is achieved. The slide-wire setting will then give V_L directly when *C* is adjusted to the null position with the unknown voltage.

Current Measurements with a Potentiometer. The null method is readily applied to the determination of current. A small precision resistor is placed in series in the circuit and the potential drop across the resistor is measured with a potentiometer. The current can then be calculated by means of Ohm's law.

The Wheatstone Bridge. The Wheatstone bridge shown in Figure A1–6 provides another example of a null device, applied here to the measurement of resistance. The power source *S* provides an alternating current at a potential of 6 to 10 V. The resistances R_{AC} and R_{CB} are determined from the total resistances of the linear voltage divider, *AB*, and the position of *C*. The upper right arm of the bridge contains a series of precision resistors that provide a choice of several resistance ranges. The unknown resistance R_x is placed in the upper-left arm of the bridge.

A null detector *ND* is employed to indicate an absence of current between *D* and *C*.

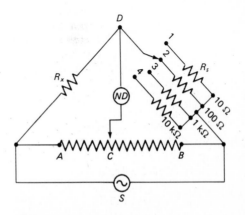

FIGURE A1-6 A Wheatstone bridge for resistance measurements.

The detector often consists of a pair of ordinary earphones with an ac signal of about 1000 Hz being employed. The human ear is sensitive in this frequency range. Alternatively, the detector may be a cathode-ray tube or an ac microammeter.

To measure R_x, the position of C is adjusted to a minimum as indicated by the null detector. Then

$$V_{AC} = V_{AB} \frac{AC}{AB}$$

For the circuit ADB, we may write

$$V_{AD} = V_{AB} \frac{R_x}{R_x + R_s}$$

But at null, $V_{AC} = V_{AD}$. Thus, the two equations can be combined to give, upon rearrangement,

$$R_x = \frac{R_s AC}{AB - AC} = R_s \frac{AC}{BC} \qquad (A1-6)$$

Appendix 2

Alternating Currents and

Electrical Reactance

Whenever the current in an electrical circuit is increased or decreased, energy is required to charge the electrical and magnetic fields associated with the flow of charge. As a consequence, there develops a counterforce or *reactance* that tends to counteract the current change. Two types of reactance can be recognized, namely *capacitance* and *inductance*.

Capacitance and inductance are present in all circuit elements including switches, junctions, conductors, and resistors. Ordinarily, reactance in these components is undesirable, and every effort is made to minimize it. The effects of reactance are particularly pronounced when the rate of current change is large (such as with high-frequency ac signals). With direct or low-frequency alternating current, reactance can frequently be made negligible with a minimal effort.

As will be shown later, capacitance and inductance are often introduced deliberately into a circuit to accomplish useful functions. It is the purpose of this appendix to describe some of these applications. At the outset, however, a brief review of the properties of alternating currents will be presented.

A2–1 ALTERNATING CURRENT

The electrical output from transducers of analytical signals often fluctuates periodically. As shown in Figure A2–1, these fluctuations can be represented by a plot of the instantaneous current or potential as a function of time. The *period* t_p for the signal is the time required for the completion of one cycle.

The reciprocal of the period is the *frequency f* of the signal. That is,

$$f = 1/t_p$$

The unit of frequency is the *hertz*, Hz, which is defined as one cycle per second.

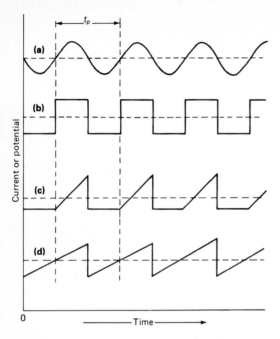

Current or potential

FIGURE A2-1 Examples of periodic signals: (a) sinusoidal, (b) square wave, (c) ramp, and (d) sawtooth.

Sine-Wave Currents. The sinusoidal wave (Figure A2–1a) is the most frequently encountered type of periodic electrical signal. A common example is the alternating current produced by rotation of a coil in a magnetic field (as in an electrical generator). Here, if the instantaneous current or voltage produced by a generator is plotted as a function of time, a sine wave results.

A pure sine wave is conveniently represented as a vector of length I_p (or V_p), which is rotating counterclockwise at a constant angular velocity ω. The relationship between the vector representation and the sine-wave plot is shown in Figure A2–2a. The vector rotates at a rate of 2π radians in the period t_p; thus, the angular velocity is given by

$$\omega = \frac{2\pi}{t_p} = 2\pi f \qquad \text{(A2–1)}$$

If the vector quantity is current or voltage, the instantaneous current i or instantaneous voltage v at time t is given by[1] (see Figure A2–2b)

$$i = I_p \sin \omega t = I_p \sin 2\pi ft \qquad \text{(A2–2)}$$

or alternatively

$$v = V_p \sin \omega t = V_p \sin 2\pi ft \qquad \text{(A2–3)}$$

where I_p and V_p, the maximum or peak current and voltage, are called the *amplitude A* of the sine wave.

Figure A2–3 shows two sine waves having different amplitudes. The two waves are also out of *phase* by 90 deg or $\pi/2$ radians. The phase difference is called the *phase angle,* which arises from one vector leading or lagging a second vector by this amount. A more generalized equation for a sine wave, then, is

$$\begin{aligned} i &= I_p \sin (\omega t + \phi) \\ &= I_p \sin (2\pi ft + \phi) \end{aligned} \qquad \text{(A2–4)}$$

where ϕ is the phase angle in radians from some *reference* sine wave. An analogous equation in terms of voltage is

$$v = V_p \sin (2\pi ft + \phi) \qquad \text{(A2–5)}$$

The current or voltage associated with a sinusoidal current may be expressed in several ways. The simplest is the peak amplitude I_p (or V_p), which is the maximum instantaneous current or voltage during a cycle; the peak-to-peak value, which is $2I_p$ or $2V_p$ is also encountered occasionally. The *root mean square* or *rms current* in an ac circuit

---

[1] In treating currents that change with time, it is useful to symbolize the instantaneous current, voltage, and charge with the lower case letters i, v, and q. On the other hand, capital letters are used for steady current, voltage, or charge, or for a specifically defined variable quantity such as peak voltage and current; that is, V_p and I_p.

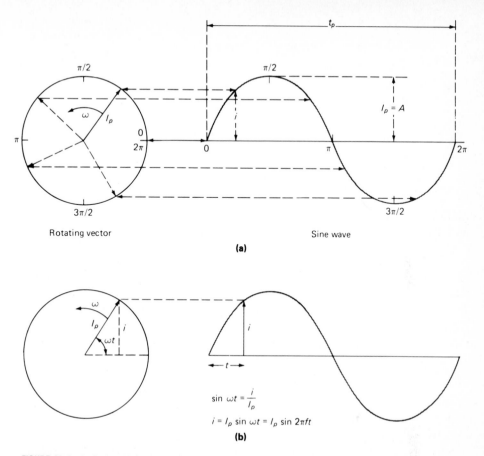

(a)

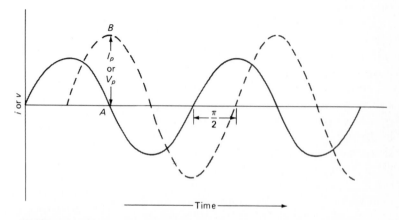

$$\sin \omega t = \frac{i}{I_p}$$

$$i = I_p \sin \omega t = I_p \sin 2\pi f t$$

(b)

FIGURE A2-2 Relationship between a sine wave of period t_p and amplitude I_p and the corresponding vector of length I_p rotating at an angular velocity of ω radians/second or a frequency of f Hz.

FIGURE A2-3 Sine waves with different amplitudes (I_p or V_p) and out of phase by 90 deg or $\pi/2$ radians.

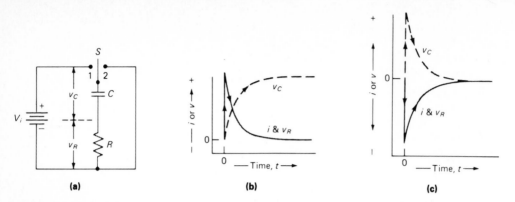

FIGURE A2–4 (**a**) A series *RC* circuit. Time response of circuit when switch *S* is in positions (**b**) 1 and (**c**) 2.

will produce the same heating in a resistor (that is, have the same power consumption[2]) as a direct current of the same magnitude. Thus, the rms current is important in power calculations. The rms current is given by

$$I_{rms} = \sqrt{\frac{I_p^2}{2}} = 0.707\, I_p \quad \text{and} \quad V_{rms} = 0.707\, V_p$$

A2–2 CAPACITORS

Capacitors and inductors are introduced into electronic circuits to provide a variety of useful functions. Insofar as possible, however, modern electronic design avoids the use of inductors because of their bulk and the fact that they are not amenable to printed circuit technology. Consequently, this discussion will be limited largely to the properties of capacitors.

Construction Features. A typical capacitor consists of a pair of conductors separated by a thin layer of a *dielectric* substance—that is, by an electrical insulator, which contains essentially no mobile, charged species that can

conduct electricity. The simplest capacitor is made up of two sheets of metal foil separated by a thin film of a dielectric such as air, oil, plastic, mica, paper, ceramic, or metal oxide. Except for air and mica capacitors, the two layers of foil plus the insulator are usually folded or rolled into a compact package and sealed to prevent atmospheric deterioration.

Capacitance. In order to describe the properties of a capacitor, it is useful to consider the dc circuit shown in Figure A2–4**a**, which consists of a battery V_i, a resistor R, and a capacitor C in series. The capacitor is symbolized by a pair of parallel lines. A circuit of this kind is generally called a *series RC circuit.*

When the switch S is closed to position 1, electrons flow from the negative terminal of the battery into the lower conductor or *plate* of the capacitor. Simultaneously, electrons are repelled from the upper plate and flow toward the positive terminal of the battery. This movement constitutes a momentary current, which decays to zero, however, owing to the potential difference that builds up across the plates and prevents the continued flow of electrons. When the current ceases, the capacitor is *charged*.

If the switch now is moved from 1 to 2, electrons will flow from the negatively

---

[2] Electrical power P is the rate of electrical work in joules per second or *watts* W. The power dissipated by a direct current is given by $P = IV$. From Ohm's law

$$P = I^2R = V^2/R$$

charged lower plate of the capacitor through the resistor R to the positive upper plate. Again, this movement constitutes a current which decays to zero as the potential between the two plates disappears; here, the capacitor is said to be *discharged*.

A useful property of a capacitor is its ability to store an electrical charge for a period and then to give it up upon demand. For example, if S in Figure A2–4a is first held at 1 until C is charged and is then moved to a position *between* 1 and 2, the capacitor will remain in a charged condition for an extended period. Upon moving S to 2, discharge occurs as before.

The quantity of electricity, Q, required to charge a capacitor fully depends upon the area of the plates, their shape, the spacing between them, and the dielectric constant for the material that separates them. In addition, the charge, Q, is directly proportional to the applied voltage. That is,

$$Q = CV \qquad (A2–6)$$

where V is the applied potential in volts and Q is the quantity of charge in coulombs; the proportionality constant C is the *capacitance* of a capacitor in *farads*, F. One farad then corresponds to one coulomb of charge per applied volt. Most of the capacitors used in electronic circuitry have capacitances in the microfarad (10^{-6} F) to picofarad (10^{-12} F) ranges.

Capacitance is important in ac circuits, particularly because a voltage that varies with time gives rise to a time-varying charge—that is, a *current*. This behavior is seen by differentiating Equation A2–6 to give

$$\frac{dq}{dt} = C\frac{dv_C}{dt} \qquad (A2–7)$$

where v_C is the instantaneous voltage across the capacitor. By definition, the current i is the rate of movement of charge; that is, $dq/dt = i$. Thus,

$$i = C\frac{dv_C}{dt} \qquad (A2–8)$$

It is important to note that the current in a capacitor is zero when the applied voltage is time independent—that is, for a direct current. Because a direct current refers to a steady state, the initial transient current that charges the capacitor is of no significance in considering the overall effect of a capacitor on a current that has a dc component.

Rate of Current and Potential Changes in an RC Circuit.   The rate at which a capacitor is charged or discharged is finite. Consider, for example, the circuit shown in Figure A2–4a. At any instant after the switch is moved to position 1, the sum of the voltage across C and R (v_C and v_R) must equal the input voltage V_i. Thus,

$$V_i = v_C + v_R \qquad (A2–9)$$

Because V_i is constant, the increase in v_C that accompanies the charging of the capacitor must be exactly offset by a decrease in v_R.

Substitution of Ohm's law and Equation A2–6 into this equation gives, upon rearrangement,

$$V_i = \frac{q}{C} + iR \qquad (A2–10)$$

Differentiating with respect to time t yields (here V_i is a constant dc potential)

$$0 = \frac{dq/dt}{C} + R\frac{di}{dt} \qquad (A2–11)$$

Here again, we have used lower case letters to represent instantaneous charge and current.

As noted earlier, $dq/dt = i$. Substituting this expression into Equation A2–11 yields upon rearrangement

$$\frac{di}{i} = -\frac{dt}{RC}$$

Integration between the limits of the initial current I_{init} and i gives

$$i = I_{init}e^{-t/RC} \tag{A2-12}$$

In order to obtain a relationship between the instantaneous voltage across the resistor, Ohm's law is employed to replace i and I_{init} in Equation A2–12. Thus,

$$v_R = V_i e^{-t/RC} \tag{A2-13}$$

Substitution of this expression into Equation A2–9 yields upon rearrangement

$$v_C = V_i (1 - e^{-t/RC}) \tag{A2-14}$$

Note that the product RC that appears in the last three equations has the units of time; since $R = v_R/i$ and $C = q/v_C$,

$$RC \equiv \frac{volt}{coulomb/second} \times \frac{coulomb}{volt} = second$$

The term RC is called the *time constant* for the circuit.

The following example illustrates the use of the three equations that were just derived.

EXAMPLE A2-1. Values for the components in Figure A2–4a are $V_i = 10$ V, $R = 1000$ Ω, $C = 1.00$ μF or 1.00×10^{-6} F. Calculate: (a) the time constant for the circuit; and (b) i, v_C, and v_R after two time constants ($t = 2RC$) have elapsed.

(a) Time constant = RC
   $= 1000 \times 1.00 \times 10^{-6}$
   $= 1.00 \times 10^{-3}$ sec   or   1.00 msec
(b) Substituting Ohm's law and $t = 2.00$ msec into Equation A2–12 reveals

---

$$i = \frac{V}{R}e^{-t/RC} = \frac{10.0}{1000}e^{-2.00/1.00}$$

$$= 1.35 \times 10^{-3} \quad \text{or} \quad 1.35 \text{ mA}$$

We find from Equation A2–13 that

$$v_R = 10.0\, e^{-2.00/1.00} = 1.35 \text{ V}$$

and by substituting into Equation A2–14

$$v_C = 10.0\, (1 - e^{-2.00/1.00}) = 8.65 \text{ V}$$

Figure A2–4b shows the changes in i, v_R, and v_C that occur during the charging cycle of an RC circuit. These plots were based upon the data given in the example just considered. Note that v_R and i assume their maximum values the instant the switch is moved to 1. At the same instant, on the other hand, the voltage across the capacitor increases rapidly from zero and ultimately approaches a constant value. For practical purposes, a capacitor is considered to be fully charged after $5RCs$ have elapsed. At this point, the current will have decayed to less than 1% of its initial value ($e^{-5RC/RC} = e^{-5} = 0.0067 \cong 0.01$).

When the switch in Figure A2–4a is moved to position 2, the battery is removed from the circuit and the capacitor becomes a source of current. The flow of charge, however, will be in the opposite direction from what it was previously. Thus,

$$dq/dt = -i$$

The initial potential will be that of the battery. That is,

$$V_C = V_i$$

Employing these equations and proceeding as in the earlier derivation, we find that for the discharge cycle

$$i = -\frac{V_C}{R}e^{-t/RC} \tag{A2-15}$$

$$v_R = -V_C\, e^{-t/RC} \qquad\qquad (A2\text{–}16)$$

and

$$v_C = V_C e^{-t/RC} \qquad\qquad (A2\text{–}17)$$

Figure A2–4**c** shows how these variables change with time.

It is important to note that in each cycle, the change in voltage across the capacitor is *out of phase with and lags behind* that of the current and the potential across the resistor.

A2-3 RESPONSE OF SERIES RC CIRCUITS TO SINUSOIDAL INPUTS

In this section, the response of series RC circuits to a sinusoidal ac voltage signal will be considered. The input signal v_s is described by Equation A2–3.

Phase Changes Brought About by a Capacitive Element. If the switch and battery in the RC circuit shown in Figure A2–4**a** is replaced with a sinusoidal ac source, the capacitor continuously stores and releases charge thus permitting a continuous flow of ac electricity. The presence of the capacitor would alter the current in two ways, however. First, a phase difference ϕ between the current and voltage would be introduced as a consequence of the finite time required to charge and discharge the capacitor (see Figures A2–4**b** and A2–4**c**). Second, the flow of charge would be impeded to some degree, thus leading to a smaller current.

The magnitude of the phase shift in a *pure* capacitance circuit (an imaginary circuit having no resistance) is readily derived by combining Equations A2–2 and A2–8 to give after rearrangement

$$C\,\frac{dv_C}{dt} = I_p \sin 2\pi ft$$

At time $t = 0$, $v_C = 0$. Thus, upon rearranging this equation and integrating between times 0 and t, we obtain

$$v_C = \frac{I_p}{C} \int_0^t \sin 2\pi ft\ dt = \frac{I_p}{2\pi fC}(-\cos 2\pi ft)$$

But from trigonometry, $-\cos x = \sin(x - 90)$. Therefore, we may write

$$v_C = \frac{I_p}{2\pi fC} \sin(2\pi ft - 90) \qquad (A2\text{–}18)$$

By comparison with Equation A2–5, it is evident that $I_p/(2\pi fC) = V_p$ and Equation A2–18 can be written in the form

$$v_C = V_p \sin(2\pi ft - 90) \qquad (A2\text{–}19)$$

The instantaneous current, however, is given by Equation A2–3. That is,

$$i = I_p \sin 2\pi ft$$

It is evident from comparing the last two equations that the voltage across a pure capacitor resulting from a sinusoidal input current is sinusoidal but lags behind the current by 90 deg (see Figure A2–5). As will be shown later, this lag is less in a real circuit, which also contains resistance.

Reactance of a Capacitor. Like a resistive element, a capacitance in a circuit impedes the flow of electricity and thus causes a reduction in the magnitude of the current. This effect results from the energy consumed in charging the capacitor; in contrast to a resistance, however, charging does not involve a permanent loss of energy as heat. Here, the energy consumed in the charging process returns to the system during discharge.

Ohm's law can be applied to capacitive impedance and takes the form

$$\frac{V_p}{I_p} = X_C \qquad\qquad (A2\text{–}20)$$

where X_C is the *capacitive reactance* of the capacitor, a property that is analogous to the resistance of a resistor. A comparison of

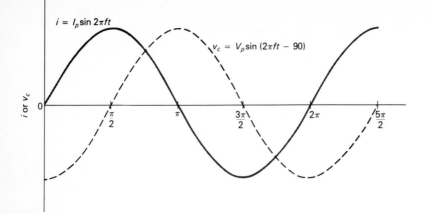

FIGURE A2-5 Sinusoidal current (i) and voltage (V_C) signals in a capacitor.

Equations A2–18 and A2–19 shows, however, that

$$V_p = \frac{I_p}{2\pi fC}$$

Thus, the capacitive reactance is given by

$$X_C = \frac{V_p}{I_p} = \frac{I_p}{I_p \, 2\pi fC} = \frac{1}{2\pi fC} \qquad \text{(A2–21)}$$

It is readily shown that X_C has the dimensions of ohms.

It should also be noted that in contrast to R, capacitive reactance *is frequency dependent* and becomes less at higher frequency; at zero frequency X_C becomes infinite so that a capacitor acts as an insulator toward a direct current (neglecting the momentary initial charging current).

EXAMPLE A2-2. Calculate the reactance of a 0.020 μF (2.0 × 10⁻⁸ F) capacitor at a frequency of 3.0 MHz and 3.0 kHz.

Substituting 3.0 MHz or 3 × 10⁶ Hz into Equation A2–21 yields

$$X_C = \frac{1}{2 \times 3.14 \times 3.0 \times 10^6 \times 2.0 \times 10^{-8}}$$

$$= 2.7 \ \Omega$$

At 3.0 kHz or 3 × 10³ Hz,

$$X_C = \frac{1}{2 \times 3.14 \times 3.0 \times 10^3 \times 2.0 \times 10^{-8}}$$

$$= 2{,}700 \ \Omega \quad \text{or} \quad 2.7 \ \text{k}\Omega$$

Impedance in a Series RC Circuit. The *impedance Z* of an *RC* circuit is made up of two components, namely the resistance of the resistor and the reactance of the capacitor. Because of the phase shift with the latter, however, the two cannot be combined directly but must be added vectorially as shown in Figure A2–6a. Here the phase angle for R is chosen as zero. As we have shown, the phase angle for a pure capacitive element is −90 deg. Thus the X_C vector is drawn at right angles to and extends down from the R vector. It is evident from the figure that the

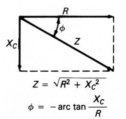

$$Z = \sqrt{R^2 + X_C^2}$$

$$\phi = -\text{arc tan} \frac{X_C}{R}$$

FIGURE A2-6 Vector diagram for series *RC* circuit.

quantity Z, called the *impedance*, is given by

$$Z = \sqrt{R^2 + X_C^2} \qquad (A2\text{-}22)$$

The phase angle is

$$\phi = \arctan \frac{X_C}{R} \qquad (A2\text{-}23)$$

To show the frequency dependence of the impedance and of the phase angle, we can substitute Equation A2–21 into A2–22 and A2–23 giving

$$Z = \sqrt{R^2 + (1/2\pi fC)^2} \qquad (A2\text{-}24)$$

and

$$\phi = \arctan \frac{1}{2\pi fRC} \qquad (A2\text{-}25)$$

Note that the extent to which the voltage lags the current in an RC circuit (ϕ) is dependent upon both the frequency and the circuit resistance.

Ohm's law for a series RC circuit can be written as

$$I_p = \frac{V_p}{Z} = \frac{V_p}{\sqrt{R^2 + (1/2\pi fC)^2}} \qquad (A2\text{-}26)$$

EXAMPLE A2-3. A sinusoidal ac source having a peak voltage of 20 V was placed in series with a 1.5×10^4 Ω resistor and a 0.00800 μF condenser. Calculate the peak current, the phase angle, and the voltage drop across each of the components if the frequency of the source was (a) 750 Hz and (b) 75.0 kHz.

(a) At 750 Hz, we find by substituting into Equation A2–21

$$X_C = \frac{1}{2\pi fC}$$

$$= \frac{1}{2\pi \times 750 \times 8.00 \times 10^{-9}}$$

$$= 2.65 \times 10^4 \ \Omega$$

From Equation A2–22, we find

$$Z = \sqrt{(1.50 \times 10^4)^2 + (2.65 \times 10^4)^2}$$

$$= 3.05 \times 10^4 \ \Omega$$

Substituting into Equation A2–26, yields

$$I_p = 20.0/3.05 \times 10^4 = 6.56 \times 10^{-4} \text{ A}$$

To obtain ϕ, we employ Equation A2–23. Thus,

$$\phi = \arctan \frac{X_C}{R}$$

$$= \arctan \frac{2.65 \times 10^4}{1.50 \times 10^4}$$

$$= 60.5 \text{ deg}$$

Application of the equation for a voltage divider gives

$$(V_p)_R = 20.0 \times \frac{R}{Z}$$

$$= \frac{20.0 \times 1.50 \times 10^4}{3.05 \times 10^4}$$

$$= 9.84 \text{ V}$$

$$(V_p)_C = 20.0 \times \frac{X_C}{Z}$$

$$= \frac{20.0 \times 2.65 \times 10^4}{3.05 \times 10^4}$$

$$= 17.4 \text{ V}$$

where $(V_p)_R$ and $(V_p)_C$ are the peak voltage drops across the resistor and capacitor respectively.

(b) Proceeding in a similar way, the following data are obtained for a 75.0 kHz current.

$$X_C = 2.65 \times 10^2 \ \Omega \qquad \phi = 1.01 \text{ deg}$$
$$Z = 1.50 \times 10^4 \ \Omega \qquad (V_p)_R = 20.0 \text{ V}$$
$$I_p = 1.33 \times 10^{-3} \text{ A} \qquad (V_p)_C = 0.353 \text{ V}$$

Several noteworthy properties of a series RC circuit are illustrated by the results obtained in Example A2–3. First, the sum of the peak voltages for the resistor and the capacitor are not equal to the peak voltage of the source. At the lower frequency, for example, the sum is 27.2 V compared with 20.0 V for the source. This apparent anomaly is understandable when it is realized that the peak voltage occurs in the resistor at an earlier time than in the capacitor because of the voltage lag in the latter. At any time, however, the sum of the *instantaneous* voltages across the two elements would equal that of the source.

A second important point shown by the data in Example A2–3 is that the reactance of the capacitor is two orders of magnitude greater at the lower frequency. As a consequence, the impedance at the higher frequency is largely associated with the resistor and the current is significantly greater. Associated with the lowered reactance is the much smaller voltage drop across the capacitor (0.35 V compared with 17.4 V).

Finally, the magnitude of the voltage lag in the capacitor is of interest. At the lower frequency this lag amounted to approximately 60 deg while at the higher frequency it was only about 1 deg.

A2–4 FILTERS

Series RC circuits are often used as filters to attenuate high-frequency signals while passing low-frequency components (a *low-pass filter*), or, alternatively to reduce low-frequency components while passing the high (a *high-pass filter*). Figure A2–7 shows how a series RC circuit can be arranged to give a high- and a low-pass filter. In each case, the input and output are indicated as the voltages $(V_p)_i$ and $(V_p)_o$.

High-Pass Filters. In order to employ an RC circuit as a high-pass filter, the output voltage is taken across the resistor R (see Figure A2–7a). The peak current in this circuit can be found by substituting into Equation A2–26. Thus,

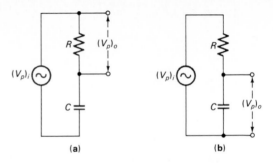

FIGURE A2-7 Filter circuits: (**a**) a high-pass RC filter; (**b**) a low-pass RC filter.

$$I_p = \frac{(V_p)_i}{Z} = \frac{(V_p)_i}{\sqrt{R^2 + (\frac{1}{2\pi fC})^2}} \qquad \text{(A2–27)}$$

Since the voltage drop across the resistor is in phase with the current,

$$I_p = \frac{(V_p)_o}{R}$$

The ratio of the peak output to the peak input voltage is obtained by dividing the second equation by the first and rearranging. Thus,

$$\frac{(V_p)_o}{(V_p)_i} = \frac{R}{\sqrt{R^2 + (\frac{1}{2\pi fC})^2}}$$
$$= \frac{1}{\sqrt{1 + (\frac{1}{2\pi fRC})^2}} \qquad \text{(A2–28)}$$

A plot of this ratio as a function of frequency for a typical high-pass filter is shown as curve A in Figure A2–8a. Note that frequencies below 20 Hz have been largely removed from the input signal.

Low-Pass Filters. For the low-pass filter shown in Figure A2–7b, we may write

$$(V_p)_o = I_p X_C$$

Substituting Equation A2–21 gives upon rearranging

$$I_p = 2\pi fC(V_p)_o$$

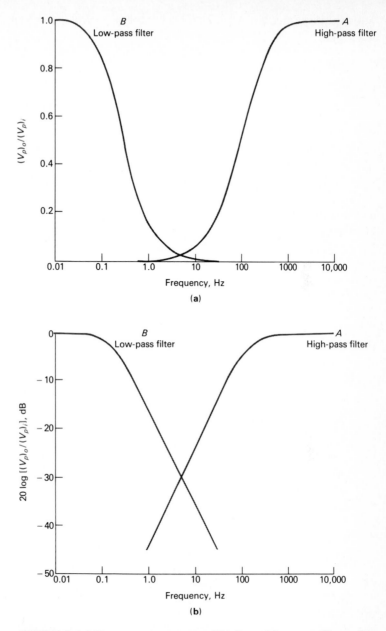

FIGURE A2-8 (**a**) Frequency response of high- and low-pass filters. (**b**) Bode diagram for a high- and low-pass filter. For high-pass filter, $R = 10\ k\Omega$ and $C = 0.1\ \mu F$. For low-pass filter, $R = 1\ M\Omega$ and $C = 1\ \mu F$.

Substituting Equation A2–27 and rearranging yields

$$\frac{(V_p)_o}{(V_p)_i} = \frac{1}{2\pi fC\sqrt{R^2 + (\frac{1}{2}\pi fC)^2}}$$

$$= \frac{1}{\sqrt{(2\pi fRC)^2 + 1}} \quad \text{(A2–29)}$$

Curve *B* in Figure A2–8a shows the frequency response of a typical low-pass filter; the data for the plot were obtained with the aid of Equation A2–29. In this case, direct and low-frequency currents are effectively removed.

Figure A2–8b shows *Bode diagrams* or plots for the two filters just described. Plots of this kind are widely encountered in the electronics literature to show the frequency dependence of input/output ratios for various circuits (amplifiers and filters for example). The quantity $20 \log [(V_p)_o/(V_p)_i]$ gives the gain of an amplifier or a filter in *decibels*, dB.

Low- and high-pass filters are of great importance in the design of electronic circuits.

A2-5 BEHAVIOR OF RC CIRCUITS WITH PULSED INPUTS

When a pulsed input is applied to an *RC* circuit, the voltage outputs across the capacitor and resistor take various forms, depending upon the relationship between the width of the pulse and the time constant for the circuit. These effects are illustrated in Figure A2–9 where the input is a square wave having a pulse width of T_p seconds. The second column shows the variation in capacitor potential as a function of time, while the third column shows the change in resistor potential at the same times. In the top set of plots (Figure A2–9a), the time constant of the circuit is much greater than

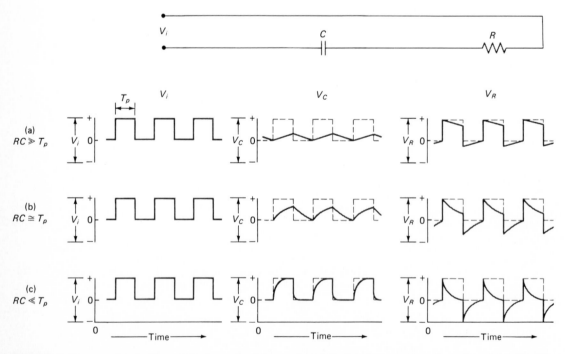

FIGURE A2-9 Output signals V_R and V_C for pulsed input signal V_i; (**a**) time constant $\gg$ pulse width T_p; (**b**) time constant $\cong$ pulse width; (**c**) time constant $\ll$ pulse width.

the input pulse width. Under these circumstances, the capacitor can become only partially charged during each pulse. It then discharges as the input pulse potential returns to zero; a sawtooth output results. The output of the resistor under these circumstances rises instantaneously to a maximum value and then decreases essentially linearly during the pulse lifetime.

The bottom set of graphs (Figure A2–9c) illustrates the two outputs when the time constant of the circuit is much shorter than the pulse width. Here, the charge on the capacitor rises rapidly and approaches full charge near the end of the pulse. As a consequence, the potential across the resistor rapidly decreases to zero after its initial rise. When V_i goes to zero, the capacitor discharges immediately; the output across the resistor peaks in a negative direction and then quickly approaches zero.

These various output wave forms find applications in electronic circuitry. The sharply peaked voltage output shown in Figure A2–9c is particularly important in timing and trigger circuits.

Appendix 3

Activity Coefficients

The relationship between the *activity* a_M of a species and its molar concentration [M] is given by the expression

$$a_M = f_M[M] \qquad (A3-1)$$

where f_M is a dimensionless quantity called the *activity coefficient*. The activity coefficient, and thus the activity of M varies with the *ionic strength* of a solution such that the employment of a_M instead of [M] in an electrode potential calculation, or in the other equilibrium calculations, renders the numerical value obtained independent of the ionic strength. Here, the ionic strength μ is defined by the equation

$$\mu = \tfrac{1}{2}(m_1 Z_1^2 + m_2 Z_2^2 + m_3 Z_3^2 + \cdots) \qquad (A3-2)$$

where m_1, m_2, m_3, $\cdots$ represent the molar concentration of the various ions in the solution and Z_1, Z_2, Z_3, $\cdots$ are their respective charges. Note that an ionic strength calcula-

tion requires taking account of *all* ionic species in a solution, not just the reactive ones.

Example A3–1. Calculate the ionic strength of a solution that is 0.0100 M in NaNO$_3$ and 0.0200 M in Mg(NO$_3$)$_2$.

Here, we will neglect the contribution of H$^+$ and OH$^-$ to the ionic strength since their concentrations are so low compared with that of the two salts. The molarities of Na$^+$, NO$_3^-$, and Mg^{2+} are 0.0100, 0.0500, and 0.0200 respectively. Then

$$
\begin{aligned}
m_{\text{Na}^+} \times (1)^2 &= 0.0100 \times 1 &= 0.0100 \\
m_{\text{NO}_3^-} \times (1)^2 &= 0.0500 \times 1 &= 0.0500 \\
m_{\text{Mg}^{2+}} \times (2)^2 &= 0.0200 \times 2^2 &= \underline{0.0800} \\
&& \text{Sum} = 0.1400
\end{aligned}
$$

$$\mu = \tfrac{1}{2} \times 0.1400 = 0.0700$$

Properties of Activity Coefficients. Activity coefficients have the following properties:

1. The activity coefficient of a species can be thought of as a measure of the effec-

tiveness with which that species influences an equilibrium in which it is a participant. In very dilute solutions, where the ionic strength is minimal, ions are sufficiently far apart that they do not influence one another's behavior. Here, the effectiveness of a common ion on the position of equilibrium becomes dependent only upon its molar concentration and independent of other ions. Under these circumstances, the activity coefficient becomes equal to unity and [M] and a in Equation A3–1 are identical. As the ionic strength becomes larger, the behavior of an individual ion is influenced by its nearby neighbors. The result is a decrease in effectiveness of the ion in altering the position of chemical equilibria. Its activity coefficient then becomes less than unity. We may summarize this behavior in terms of Equation A3–1. At moderate ionic strengths, $f_M < 1$; as the solution approaches infinite dilution $(\mu \rightarrow 0)$, $f_M \rightarrow 1$ and thus $a_M \rightarrow$ [M].

At high ionic strengths, the activity coefficients for some species increase and may even become greater than one. The behavior of such solutions is difficult to interpret; we shall confine our discussion to regions of low to moderate ionic strengths (that is, where $\mu < 0.1$).

The variation of typical activity coefficients as a function of ionic strength is shown in Figure A3–1.

2. In dilute solutions, the activity coefficient for a given species is independent of the specific nature of the electrolyte, and depends only upon the ionic strength.

3. For a given ionic strength, the activity coefficient of an ion departs further from unity as the charge carried by the species increases. This effect is shown in Figure A3–1. The activity coefficient of an uncharged molecule is approximately one, regardless of ionic strength.

4. Activity coefficients for ions of the same charge are approximately the same at

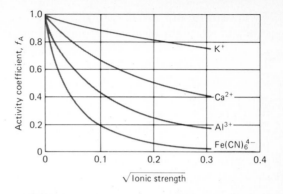

FIGURE A3–1 Effect of ionic strength on activity coefficients.

any given ionic strength. The small variations that do exist can be correlated with the effective diameter of the hydrated ions.

5. The product of the activity coefficient and molar concentration of a given ion describes its effective behavior in all equilibria in which it participates.

Experimental Evaluation of Activity Coefficients. Although activity coefficients for individual ions can be calculated from theoretical considerations, their experimental measurement is, unfortunately, impossible. Instead, only a mean activity coefficient for the positively and negatively charged species in a solution can be derived.

For the electrolyte A_mB_n, the mean activity coefficient $f_\pm$ is defined by the equation

$$f_\pm = (f_A^m \times f_B^n)^{1/(m+n)}$$

The mean activity coefficient can be measured in any of several ways, but it is impossible experimentally to resolve this term into the individual activity coefficients f_A and f_B. For example, if A_mB_n is a precipitate, we can write

$$K_{sp} = [A]^m[B]^n \times f_A^m \times f_B^n$$
$$= [A]^m[B]^n \times f_\pm^{(m+n)}$$

By measuring the solubility of A_mB_n in a solution in which the electrolyte concentration approaches zero (that is, where f_A and $f_B \rightarrow 1$), we could obtain K_{sp}. A second solubility measurement at some ionic strength, μ_1 would give values for [A] and [B]. These data would then permit the calculation of $f_A^m \times f_B^n = f_{\pm}^{(m+n)}$ for ionic strength μ_1. It is important to understand that there are insufficient experimental data to permit the calculation of the *individual* quantities f_A and f_B, however, and that there appears to be no additional experimental information that would permit evaluation of these quantities. This situation is general; the *experimental* determination of individual activity coefficients appears to be impossible.

The Debye-Hückel Equation. In 1923, P. Debye and E. Hückel derived the following theoretical expression, which permits the calculation of activity coefficients of ions:[1]

$$-\log f_A = \frac{0.509\, Z_A^2 \sqrt{\mu}}{1 + 0.328\, \alpha_A \sqrt{\mu}} \qquad \text{(A3–3)}$$

where

f_A = activity coefficient of the species A

Z_A = charge on the species A

μ = ionic strength of the solution

α_A = the effective diameter of the hydrated ion in ångström units

[1] P. Debye and E. Hückel, *Physik. Z.*, **1923**, *24*, 185.

Table A3–1
Activity Coefficients for Ions at 25°C[a]

Ion	α_A Effective Diameter, Å	Activity Coefficients at Indicated Ionic Strengths				
		0.001	0.005	0.01	0.05	0.1
H_3O^+	9	0.967	0.933	0.914	0.86	0.83
Li^+, $C_6H_5COO^-$	6	0.965	0.929	0.907	0.84	0.80
Na^+, IO_3^-, HSO_3^-, HCO_3^-, $H_2PO_4^-$, $H_2AsO_4^-$, OAc^-	4–4.5	0.964	0.928	0.902	0.82	0.78
OH^-, F^-, SCN^-, HS^-, ClO_3^-, ClO_4^-, BrO_3^-, IO_4^-, MnO_4^-	3.5	0.964	0.926	0.900	0.81	0.76
K^+, Cl^-, Br^-, I^-, CN^-, NO_2^-, NO_3^-, $HCOO^-$	3	0.964	0.925	0.899	0.80	0.76
Rb^+, Cs^+, Tl^+, Ag^+, NH_4^+	2.5	0.964	0.924	0.898	0.80	0.75
Mg^{2+}, Be^{2+}	8	0.872	0.755	0.69	0.52	0.45
Ca^{2+}, Cu^{2+}, Zn^{2+}, Sn^{2+}, Mn^{2+}, Fe^{2+}, Ni^{2+}, Co^{2+}, Phthalate^{2-}	6	0.870	0.749	0.675	0.48	0.40
Sr^{2+}, Ba^{2+}, Cd^{2+}, Hg^{2+}, S^{2-}	5	0.868	0.744	0.67	0.46	0.38
Pb^{2+}, CO_3^{2-}, SO_3^{2-}, $C_2O_4^{2-}$	4.5	0.868	0.742	0.665	0.46	0.37
Hg_2^{2+}, SO_4^{2-}, $S_2O_3^{2-}$, CrO_4^{2-}, HPO_4^{2-}	4.0	0.867	0.740	0.660	0.44	0.36
Al^{3+}, Fe^{3+}, Cr^{3+}, La^{3+}, Ce^{3+}	9	0.738	0.54	0.44	0.24	0.18
PO_4^{3-}, $Fe(CN)_6^{3-}$	4	0.725	0.50	0.40	0.16	0.095
Th^{4+}, Zr^{4+}, Ce^{4+}, Sn^{4+}	11	0.588	0.35	0.255	0.10	0.065
$Fe(CN)_6^{4-}$	5	0.57	0.31	0.20	0.048	0.021

[a] From J. Kielland, *J. Amer. Chem. Soc.*, **1937**, *59*, 1675.

The constants 0.509 and 0.328 are applicable to solutions at 25°C; other values must be employed at different temperatures.

Unfortunately, considerable uncertainty exists regarding the magnitude of α_A in Equation A3–3. Its value appears to be approximately 3 Å for most singly charged ions so that for these species, the denominator of the Debye-Hückel equation reduces to approximately $(1 + \sqrt{\mu})$. For ions with higher charge, α_A may be larger than 10 Å. It should be noted that the second term of the denominator becomes small with respect to the first when the ionic strength is less than 0.01; under these circumstances, uncertainties in α_A are of little significance in calculating activity coefficients.

Kielland[2] has calculated values of α_A for numerous ions from a variety of experimental data. His "best values" for effective diameters are given in Table A3–1. Also presented are activity coefficients calculated from Equation A3–3 using these values for the size parameter.

For ionic strengths up to about 0.01, activity coefficients from the Debye-Hückel equation lead to results from equilibrium calculations that agree closely with experiment; even at ionic strengths of 0.1, major discrepancies are generally not encountered. At higher ionic strengths, however, the equation fails, and experimentally determined mean activity coefficients must be employed. Unfortunately, many electrochemical calculations involve solutions of high ionic strength for which no experimental activity coefficients are available. Concentrations must thus be employed instead of activities; uncertainties that vary from a few percent relative to an order of magnitude may be expected.

---

[2] J. Kielland, *J. Amer. Chem. Soc.*, **1937,** *59,* 1675.

Appendix 4

Some Standard and Formal Electrode Potentials

Half-reaction[a]	E^0, V	Formal potential, V
$Ag^+ + e \rightleftharpoons Ag(s)$	+0.799	0.228, 1 M HCl; 0.792, 1 M HClO$_4$; 0.77, 1 M H$_2$SO$_4$
$AgBr(s) + e \rightleftharpoons Ag(s) + Br^-$	+0.073	
$AgCl(s) + e \rightleftharpoons Ag(s) + Cl^-$	+0.222	0.228, 1 M KCl
$Ag(CN)_2^- + e \rightleftharpoons Ag(s) + 2CN^-$	−0.31	
$Ag_2CrO_4(s) + 2e \rightleftharpoons 2Ag(s) + CrO_4^{2-}$	+0.446	
$AgI(s) + e \rightleftharpoons Ag(s) + I^-$	−0.151	
$Ag(S_2O_3)_2^{3-} + e \rightleftharpoons Ag(s) + 2S_2O_3^{2-}$	+0.017	
$Al^{3+} + 3e \rightleftharpoons Al(s)$	−1.662	
$H_3AsO_4 + 2H^+ + 2e \rightleftharpoons H_3AsO_3 + H_2O$	+0.559	0.577, 1 M HCl, HClO$_4$
$Ba^{2+} + 2e \rightleftharpoons Ba(s)$	−2.906	
$BiO^+ + 2H^+ + 3e \rightleftharpoons Bi(s) + H_2O$	+0.320	
$BiCl_4^- + 3e \rightleftharpoons Bi(s) + 4Cl^-$	+0.16	
$Br_2(l) + 2e \rightleftharpoons 2Br^-$	+1.065	1.05, 4 M HCl
$Br_2(aq) + 2e \rightleftharpoons 2Br^-$	+1.087[b]	
$BrO_3^- + 6H^+ + 5e \rightleftharpoons \frac{1}{2}Br_2(l) + 3H_2O$	+1.52	
$BrO_3^- + 6H^+ + 6e \rightleftharpoons Br^- + 3H_2O$	+1.44	
$Ca^{2+} + 2e \rightleftharpoons Ca(s)$	−2.866	
$C_6H_4O_2$ (quinone) $+ 2H^+ + 2e \rightleftharpoons C_6H_4(OH)_2$	+0.699	0.696, 1 M HCl, HClO$_4$, H$_2$SO$_4$
$2CO_2(g) + 2H^+ + 2e \rightleftharpoons H_2C_2O_4$	−0.49	
$Cd^{2+} + 2e \rightleftharpoons Cd(s)$	−0.403	
$Ce^{4+} + e \rightleftharpoons Ce^{3+}$		1.70, 1 M HClO$_4$; 1.61, 1 M HNO$_3$; 1.44, 1 M H$_2$SO$_4$; 1.28, 1 M HCl

Half-reaction[a]	E^0, V	Formal potential, V
$Cl_2(g) + 2e \rightleftharpoons 2Cl^-$	+1.359	
$HClO + H^+ + e \rightleftharpoons \frac{1}{2}Cl_2(g) + H_2O$	+1.63	
$ClO_3^- + 6H^+ + 5e \rightleftharpoons \frac{1}{2}Cl_2(g) + 3H_2O$	+1.47	
$Co^{2+} + 2e \rightleftharpoons Co(s)$	−0.277	
$Co^{3+} + e \rightleftharpoons Co^{2+}$	+1.808	
$Cr^{3+} + e \rightleftharpoons Cr^{2+}$	−0.408	
$Cr^{3+} + 3e \rightleftharpoons Cr(s)$	−0.744	
$Cr_2O_7^{2-} + 14H^+ + 6e \rightleftharpoons 2Cr^{3+} + 7H_2O$	+1.33	
$Cu^{2+} + 2e \rightleftharpoons Cu(s)$	+0.337	
$Cu^{2+} + e \rightleftharpoons Cu^+$	+0.153	
$Cu^+ + e \rightleftharpoons Cu(s)$	+0.521	
$Cu^{2+} + I^- + e \rightleftharpoons CuI(s)$	+0.86	
$CuI(s) + e \rightleftharpoons Cu(s) + I^-$	−0.185	
$F_2(g) + 2H^+ + 2e \rightleftharpoons 2HF(aq)$	+3.06	
$Fe^{2+} + 2e \rightleftharpoons Fe(s)$	−0.440	
$Fe^{3+} + e \rightleftharpoons Fe^{2+}$	+0.771	0.700, 1 M HCl; 0.732, 1 M HClO$_4$; 0.68, 1 M H$_2$SO$_4$
$Fe(CN)_6^{3-} + e \rightleftharpoons Fe(CN)_6^{4-}$	+0.36	0.71, 1 M HCl; 0.72, 1 M HClO$_4$, H$_2$SO$_4$
$2H^+ + 2e \rightleftharpoons H_2(g)$	0.000	−0.005, 1 M HCl, HClO$_4$
$Hg_2^{2+} + 2e \rightleftharpoons 2Hg(l)$	+0.788	0.274, 1 M HCl; 0.776, 1 M HClO$_4$; 0.674, 1 M H$_2$SO$_4$
$2Hg^{2+} + 2e \rightleftharpoons Hg_2^{2+}$	+0.920	0.907, 1 M HClO$_4$
$Hg^{2+} + 2e \rightleftharpoons Hg(l)$	+0.854	
$Hg_2Cl_2(s) + 2e \rightleftharpoons 2Hg(l) + 2Cl^-$	+0.268	0.244, sat'd KCl; 0.282, 1 M KCl; 0.334, 0.1 M KCl
$Hg_2SO_4(s) + 2e \rightleftharpoons 2Hg(l) + SO_4^{2-}$	+0.615	
$HO_2^- + H_2O + 2e \rightleftharpoons 3OH^-$	+0.88	
$I_2(s) + 2e \rightleftharpoons 2I^-$	+0.5355	
$I_2(aq) + 2e \rightleftharpoons 2I^-$	+0.615[b]	
$I_3^- + 2e \rightleftharpoons 3I^-$	+0.536	
$ICl_2^- + e \rightleftharpoons \frac{1}{2}I_2(s) + 2Cl^-$	+1.056	
$IO_3^- + 6H^+ + 5e \rightleftharpoons \frac{1}{2}I_2(s) + 3H_2O$	+1.196	
$IO_3^- + 6H^+ + 5e \rightleftharpoons \frac{1}{2}I_2(aq) + 3H_2O$	+1.178[b]	
$IO_3^- + 2Cl^- + 6H^+ + 4e \rightleftharpoons ICl_2^- + 3H_2O$	+1.24	
$H_5IO_6 + H^+ + 2e \rightleftharpoons IO_3^- + 3H_2O$	+1.601	
$K^+ + e \rightleftharpoons K(s)$	−2.925	
$Li^+ + e \rightleftharpoons Li(s)$	−3.045	
$Mg^{2+} + 2e \rightleftharpoons Mg(s)$	−2.363	
$Mn^{2+} + 2e \rightleftharpoons Mn(s)$	−1.180	
$Mn^{3+} + e \rightleftharpoons Mn^{2+}$		1.51, 7.5 M H$_2$SO$_4$
$MnO_2(s) + 4H^+ + 2e \rightleftharpoons Mn^{2+} + 2H_2O$	+1.23	1.24, 1 M HClO$_4$
$MnO_4^- + 8H^+ + 5e \rightleftharpoons Mn^{2+} + 4H_2O$	+1.51	
$MnO_4^- + 4H^+ + 3e \rightleftharpoons MnO_2(s) + 2H_2O$	+1.695	
$MnO_4^- + e \rightleftharpoons MnO_4^{2-}$	+0.564	
$N_2(g) + 5H^+ + 4e \rightleftharpoons N_2H_5^+$	−0.23	
$HNO_2 + H^+ + e \rightleftharpoons NO(g) + H_2O$	+1.00	
$NO_3^- + 3H^+ + 2e \rightleftharpoons HNO_2 + H_2O$	+0.94	0.92, 1 M HNO$_3$
$Na^+ + e \rightleftharpoons Na(s)$	−2.714	

Half-reaction[a]	E^0, V	Formal potential, V
$Ni^{2+} + 2e \rightleftharpoons Ni(s)$	-0.250	
$H_2O_2 + 2H^+ + 2e \rightleftharpoons 2H_2O$	$+1.776$	
$O_2(g) + 4H^+ + 4e \rightleftharpoons 2H_2O$	$+1.229$	
$O_2(g) + 2H^+ + 2e \rightleftharpoons H_2O_2$	$+0.682$	
$O_3(g) + 2H^+ + 2e \rightleftharpoons O_2(g) + H_2O$	$+2.07$	
$Pb^{2+} + 2e \rightleftharpoons Pb(s)$	-0.126	-0.14, 1 M $HClO_4$; -0.29, 1 M H_2SO_4
$PbO_2(s) + 4H^+ + 2e \rightleftharpoons Pb^{2+} + 2H_2O$	$+1.455$	
$PbSO_4(s) + 2e \rightleftharpoons Pb(s) + SO_4^{2-}$	-0.350	
$PtCl_4^{2-} + 2e \rightleftharpoons Pt(s) + 4Cl^-$	$+0.73$	
$PtCl_6^{2-} + 2e \rightleftharpoons PtCl_4^{2-} + 2Cl^-$	$+0.68$	
$Pd^{2+} + 2e \rightleftharpoons Pd(s)$	$+0.987$	
$S(s) + 2H^+ + 2e \rightleftharpoons H_2S(g)$	$+0.141$	
$H_2SO_3 + 4H^+ + 4e \rightleftharpoons S(s) + 3H_2O$	$+0.450$	
$S_4O_6^{2-} + 2e \rightleftharpoons 2S_2O_3^{2-}$	$+0.08$	
$SO_4^{2-} + 4H^+ + 2e \rightleftharpoons H_2SO_3 + H_2O$	$+0.172$	
$S_2O_8^{2-} + 2e \rightleftharpoons 2SO_4^{2-}$	$+2.01$	
$Sb_2O_5(s) + 6H^+ + 4e \rightleftharpoons 2SbO^+ + 3H_2O$	$+0.581$	
$H_2SeO_3 + 4H^+ + 4e \rightleftharpoons Se(s) + 3H_2O$	$+0.740$	
$SeO_4^{2-} + 4H^+ + 2e \rightleftharpoons H_2SeO_3 + H_2O$	$+1.15$	
$Sn^{2+} + 2e \rightleftharpoons Sn(s)$	-0.136	-0.16, 1 M $HClO_4$
$Sn^{4+} + 2e \rightleftharpoons Sn^{2+}$	$+0.154$	0.14, 1 M HCl
$Ti^{3+} + e \rightleftharpoons Ti^{2+}$	-0.369	
$TiO^{2+} + 2H^+ + e \rightleftharpoons Ti^{3+} + H_2O$	$+0.099$	0.04, 1 M H_2SO_4
$Tl^+ + e \rightleftharpoons Tl(s)$	-0.336	-0.551, 1 M HCl; -0.33, 1 M $HClO_4$, H_2SO_4
$Tl^{3+} + 2e \rightleftharpoons Tl^+$	$+1.25$	0.77, 1 M HCl
$UO_2^{2+} + 4H^+ + 2e \rightleftharpoons U^{4+} + 2H_2O$	$+0.334$	
$V^{3+} + e \rightleftharpoons V^{2+}$	-0.256	-0.21, 1 M $HClO_4$
$VO^{2+} + 2H^+ + e \rightleftharpoons V^{3+} + H_2O$	$+0.359$	
$V(OH)_4^+ + 2H^+ + e \rightleftharpoons VO^{2+} + 3H_2O$	$+1.00$	1.02, 1 M HCl, $HClO_4$
$Zn^{2+} + 2e \rightleftharpoons Zn(s)$	-0.763	

[a] Sources for E^0 values: G. Milazzo, S. Caroli, and V. K. Sharma, *Tables of Standard Electrode Potentials*. Wiley: New York, 1978. Source of formal potentials: E. H. Swift and E. A. Butler, *Quantitative Measurements and Chemical Equilibria*. W. H. Freeman and Company: San Francisco. Copyright © 1972.

[b] These potentials are hypothetical because they correspond to solutions that are 1.00 M in Br_2 or I_2. The solubilities of these two compounds at 25°C are 0.18 M and 0.0020 M, respectively. In saturated solutions containing an excess of $Br_2(l)$ or $I_2(s)$, the standard potentials for the half-reactions $Br_2(l) + 2e \rightleftharpoons 2Br^-$ or $I_2(s) + 2e \rightleftharpoons 2I^-$ should be used. On the other hand, at Br_2 and I_2 concentrations less than saturation, these hypothetical electrode potentials should be employed.

Appendix 5

Compounds Recommended for the Preparation of Standard Solutions of Some Common Elements[a]

Element	Compound	FW	Solvent[b]	Notes
Aluminum	Al metal	26.98	Hot dil HCl	a
Antimony	$KSbOC_4H_4O_6 \cdot \frac{1}{2}H_2O$	333.93	H_2O	c
Arsenic	As_2O_3	197.84	dil HCl	i,b,d
Barium	$BaCO_3$	197.35	dil HCl	
Bismuth	Bi_2O_3	465.96	HNO_3	
Boron	H_3BO_3	61.83	H_2O	d,e
Bromine	KBr	119.01	H_2O	a
Cadmium	CdO	128.40	HNO_3	
Calcium	$CaCO_3$	100.09	dil HCl	i
Cerium	$(NH_4)_2Ce(NO_3)_6$	548.23	H_2SO_4	
Chromium	$K_2Cr_2O_7$	294.19	H_2O	i,d
Cobalt	Co metal	58.93	HNO_3	a
Copper	Cu metal	63.55	dil HNO_3	a
Fluorine	NaF	41.99	H_2O	b
Iodine	KIO_3	214.00	H_2O	i
Iron	Fe metal	55.85	HCl, hot	a
Lanthanum	La_2O_3	325.82	HCl, hot	f
Lead	$Pb(NO_3)_2$	331.20	H_2O	a
Lithium	Li_2CO_3	73.89	HCl	a
Magnesium	MgO	40.31	HCl	
Manganese	$MnSO_4 \cdot H_2O$	169.01	H_2O	g
Mercury	$HgCl_2$	271.50	H_2O	b
Molybdenum	MoO_3	143.94	l M NaOH	

Element	Compound	FW	Solvent[b]	Notes
Nickel	Ni metal	58.70	HNO_3, hot	a
Phosphorus	KH_2PO_4	136.09	H_2O	
Potassium	KCl	74.56	H_2O	a
	$KHC_8H_4O_4$	204.23	H_2O	i,d
	$K_2Cr_2O_7$	294.19	H_2O	i,d
Silicon	Si metal	28.09	NaOH, concd	
	SiO_2	60.08	HF	
Silver	$AgNO_3$	169.87	H_2O	a
Sodium	NaCl	58.44	H_2O	i
	$Na_2C_2O_4$	134.00	H_2O	i,d
Strontium	$SrCO_3$	147.63	HCl	a
Sulfur	K_2SO_4	174.27	H_2O	
Tin	Sn metal	118.69	HCl	
Titanium	Ti metal	47.90	H_2SO_4, 1:1	a
Tungsten	$Na_2WO_4 \cdot 2H_2O$	329.86	H_2O	h
Uranium	U_3O_8	842.09	HNO_3	d
Vanadium	V_2O_5	181.88	HCl, hot	
Zinc	ZnO	81.37	HCl	a

[a] The data in this table were taken from a more complete list assembled by B. W. Smith and M. L. Parsons, *J. Chem. Educ.*, **1973**, *50*, 679. Unless otherwise specified, compounds should be dried to constant weight at 110°C.

[b] Unless otherwise specified, acids are concentrated analytical grade.

a Approaches primary standard quality.

b Highly toxic.

c Loses $\frac{1}{2}$ H_2O at 110°C. After drying, fw = 324.92. The dried compound should be weighed quickly after removal from the desiccator.

d Available as a primary standard from the National Bureau of Standards.

e H_3BO_3 should be weighed directly from the bottle. It loses 1 H_2O at 100°C and is difficult to dry to constant weight.

f Absorbs CO_2 and H_2O. Should be ignited just before use.

g May be dried at 110°C without loss of water.

h Loses both waters at 110°C, fw = 293.82. Keep in desiccator after drying.

i Primary standard.

Answers to Problems

Chapter 1

1-1. (a) $\bar{x}$ = 61.43, 3.25, 12.10, 2.65, 9.985. (b) s = 0.11, 0.020, 0.057, 0.21, 0.019; s_r = 1.7, 6.1, 4.7, 79, 1.9 ppt. (c) N = 2, 5, 1, 3, 4. (d) CV = 0.17, 0.61, 0.47, 7.9, 0.19%.

1-2. (a) −0.28, −0.03, −0.13, −0.10, −1.3. (b) −0.45, −0.91, −1.1, −3.6, −11%.

1-3. (a) −2%. (b) −0.5%. (c) −0.2%. (d) −0.1%.

1-4. (a) 10 g. (b) 2.1 g. (c) 1.3 g. (d) 0.87 g.

1-5. −25 ppt.

1-6. (a) 0.10, 0.099, 0.11, 0.13, 0.10 mg Ca/100 mL. (b) 0.11 mg Ca/100 mL.

1-7. s = 0.15, 0.14, 0.17, 0.18 μg Pb/m^3; pooled s = 0.16 μg Pb/m^3.

1-8. (a) ±0.064 ppb. (b) ±0.037 ppb. (c) ±0.029 ppb.

1-9. (a) ±0.029 ppm. (b) 12 measurements.

1-10. (a) ±0.76 ppm. (b) 7 measurements.

1-11. ±0.82 ppm.

1-12. 15.6 ± 0.5 mg K/100 mL.

1-13. (a) ±0.00014, ±0.00041%, 33.8083 ± 0.0001. (b) ±0.00014, ±0.29%, 0.0489 ± 0.0001. (c) ±0.00022, 0.017%, 1.3401 ± 0.0002. (d) 0.019 × 10^{-5}, 0.96%, 2.00(±0.02) × 10^{-5}.

1-14. (a) 0.46, 1.9%, 24.5 ± 0.5. (b) 0.012, 0.13%, 9.31 ± 0.01. (c) 4.7, 0.86%, 543 ± 5. (d) 0.0068 × 10^{-7}, 0.43%, 1.571(±0.007) × 10^{-7}.

1-15. (a) 0.0052, 0.27%, 1.891 ± 0.005. (b) 0.016, 1.4%, 1.20 ± 0.02. (c) 0.00019, 0.40%, 0.0461 ± 0.0002. (d) 0.071, 0.30%, 23.66 ± 0.07. (e) 0.057 × 10^5, 1.0%, 5.70(±0.06) × 10^5.

1-16. (a) 0.0020, 2.943 ± 0.002. (b) 0.00035, −0.3048 ± 0.0004. (c) 1.0 × 10^2, 4.4(±0.1) × 10^3. (d) 0.030 × 10^{-8}, 6.44(±0.03) × 10^{-8}.

1-17. (a) Plotting meter reading, R, versus mg SO$_4^{2-}$/L, C_x, yields a straight line. (b) R = 0.232C_x + 0.162. (d) s_β = 0.017, s_r = 0.28. (e) 15.1 mg SO$_4^{2-}$/L, s_c = 1.4 mg SO$_4^{2-}$/L, CV = 9.3%; (f) 15.1 mg SO$_4^{2-}$/L, s_c = 0.81 mg SO$_4^{2-}$/L, CV = 5.4%. (g) 3.5 mg SO$_4^{2-}$/L. (h) calib. sens. = 0.232, analyt. sens. = 2.3.

1–18. (a) Plotting E vs. pCa yields a straight line. (b) $E = -29.7$ pCa $+ 92.9$. (c) $s_\beta = 0.68$. (d) $s_r = 2.1$. (e) pCa $= 2.44$, $s_c = 0.080$, $(s_c)_r = 3.3\%$. (f) For $m = 2$, $s_c = 0.061$ pCa, $(s_c)_r = 2.5\%$. For $m = 8$, $s_c = 0.043$ pCa, $(s_c)_r = 1.8\%$. (g) $[Ca^{2+}] = 3.6 \times 10^{-3}$ M. (h) For $m = 2$, $s_c = 0.51 \times 10^{-3}$ M, $(s_c)_r = 14\%$. For $m = 8$, $s_c = 0.36 \times 10^{-3}$ M, $(s_c)_r = 9.9\%$.

1–19. (a) $\mu A = 5.57$ mmol/L $+ 0.902$. (c) $s_r = 0.40$, $s_\beta = 0.096$. (d) 0.969 and 4.78 mmol/L. (e) For Sample 1 and $m = 1$, $s_c = 0.086$ mmol/L, $(s_c)_r = 8.8\%$; for $m = 4$, 0.058 mmol/L and 6.0%; for Sample 2 and $m = 1$, $s_c = 0.084$ mmol/L, $(s_c)_r = 1.8\%$; for $m = 4$, 0.056 mmol/L and 1.2%. (f) For Sample 1, calib. sens. $= 5.57$ and analyt. sens. $= 25$. For Sample 2, calib. sens. $= 5.57$ and analyt. sens. $= 12$. (g) $C_m = 0.19$ mmol MVK/L.

Chapter 2

2–1. (a) -27.3 mV. (b) 0.910 μA. (c) 0.910 μA.

2–2. -0.04%.

2–3. See figure.

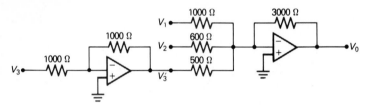

2–4. See figure.

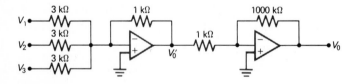

2–5. See figure.

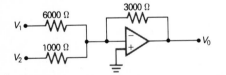

2–6. See figure, where $R_f = 1.00$ kΩ and $R_1 = 250$ Ω.

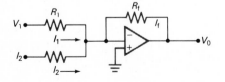

2–7. (a) $V_o = \dfrac{V_1 R_{f1} R_{f2}}{R_1 R_4} + \dfrac{V_2 R_{f1} R_{f2}}{R_2 R_4} - \dfrac{V_3 R_{f2}}{R_3}$. (b) $V_o = V_1 + 4V_2 - 40V_3$.

2–8. $V_o = 10V_1 + 6V_2 - 3V_3 - 2V_4$.

2-9. $V_o = -(5V_1 + 20V_2)\int_0^t dt$.

2-10. $V_o = (R_{k1}/R_1)(V_2 - V_1) = V_2 - V_1$ (Equation 2-7 applies).

2-11. $V_o = -V_i R_{CB}/R_{AB}$.

2-12. 60.0 cm.

2-13. See figure.

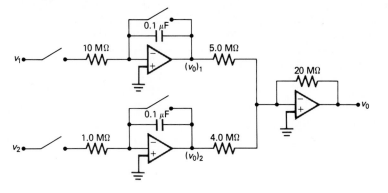

2-14. See figure.

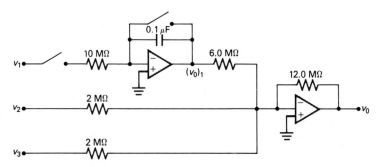

2-15. V_o after 1, 3, 5, and 7 sec would be 8, 24, 40, and 56 mV, respectively.

2-16. (a) $S/N = 870$. (b) 12 measurements.

2-17. (a) $S/N = 5.3$. (b) 29 measurements.

2-18. (a) 30.6 Ω. (b) See figure.

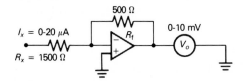

2-19. (a) 10.5 Ω. (b) Similar circuit to Figure **g** with $R_f = 200$ Ω and $\beta = 18.0$.

Chapter 4

4-1. (a) 1.1×10^{18} Hz, 3.7×10^7 cm^{-1}. (b) 1.421×10^{15} Hz, 4.739×10^4 cm^{-1}. (c) 4.318×10^{14} Hz, 1.440×10^4 cm^{-1}. (d) 2.83×10^{13} Hz, 943 cm^{-1}. (e) 2.04×10^{13} Hz, 680 cm^{-1}. (f) 1.61×10^{10} Hz, 0.538 cm^{-1}.

4-2. (a) 251.3 cm. (b) 9.157×10^4 cm. (c) 297 cm. (d) 3.156 cm.

4–3. (a) 7.4×10^{-9}, 9.412×10^{-12}, 2.860×10^{-12}, 1.87×10^{-13}, 1.35×10^{-13}, 1.07×10^{-16} erg/photon. (b) 1.1×10^5, 135.3, 41.18, 2.70, 1.95, 1.54×10^{-3} kcal/mol. (c) 4.6×10^3, 5.877, 1.786, 0.117, 8.44×10^{-2}, 6.67×10^{-5} eV.

4–4. (a) 7.903×10^{-19}, 2.169×10^{-21}, 6.69×10^{-19}, 6.29×10^{-17} erg/photon. (b) 1.138×10^{-5}, 3.122×10^{-8}, 9.63×10^{-6}, 9.06×10^{-4} kcal/mol. (c) 4.934×10^{-7}, 1.354×10^{-9}, 4.18×10^{-7}, 3.93×10^{-5} eV.

4–5.

	(a)	(b)	(c)	(d)
n_D	1.0014	1.4459	1.890	1.360
v_i, cm/sec	2.994×10^{10}	2.073×10^{10}	1.586×10^{10}	2.204×10^{10}
λ_i, Å	5882	4073	3116	4331
ν, Hz	5.090×10^{14}	5.090×10^{14}	5.090×10^{14}	5.090×10^{14}

4–6. (a) 10.1%. (b) 6.4%.

Chapter 5

5–1. The effective bandwidth of a monochromator is given by $\Delta\lambda_{eff} = wD^{-1}$. For a prism, D^{-1} increases as λ becomes longer. Thus, the slit width w must be varied if $\delta\lambda_{eff}$ is to be constant. For a grating, D^{-1} is essentially independent of λ. Therefore, variation in w is unnecessary.

5–2. Qualitative analysis requires narrow slits for high resolution. Quantitative analysis is usually based upon wider slits, which lead to higher S/N ratios.

5–3. (a) 725 nm. (b) 1.45 μm. (c) 2.90 μm.

5–4. (a) 1.46×10^7 W/m^2. (b) 9.10×10^5. (c) 5.69×10^4.

5–5. (a) 1010 nm and 967nm. (b) 3.86×10^6 and 4.61×10^6 W/m^2.

5–6. Spontaneous emission occurs when an excited species loses its energy by fluorescence. Incoherent radiation is produced. Stimulated emission is produced when externally produced photons strike excited species of matching energy. Coherent radiation is the result.

5–7. A four-level system requires less pumping to produce a population inversion.

5–8. Width of the emitted band in wavelength units when measured at one-half the peak intensity.

5–9. (a) 1.69 μm. (b) 4.54×2 μm, 4.54×3 μm, and so forth.

5–10. For first-order interference, the thickness of the dielectric layer should decrease continuously along the 10.0 cm length from 0.265 μm to 0.152 μm.

5–11. The higher angular dispersion of glass (see Table 5–1) would lead to greater linear dispersion and better resolving power for prisms of the same size.

5–12. 446 lines/mm.

5–13. $\lambda/\Delta\lambda = 720$, 2.8 cm^{-1}.

5–14. (a) 5.89 and 2.94 μm. (b) 10.6 and 5.3 μm. (c) 15.4 and 7.7 μm.

5–15.

		Source	Wavelength Selector	Sample Holder	Detector
(a)		W lamp	Grating or glass prism	Silicate glass	Photomultiplier
(b)		Globar	Grating, 50 lines/m	KBr windows	Golay cell
(c)		W lamp	Glass filter (green)	Pyrex test tube	Photocell
(d)		Nichrome wire	Interference filter	TlBr/TlI windows	Thermocouple
(e)		Gas/O$_2$ flame	Grating or quartz prism	Flame	Photomultiplier
(f)		Ar or Xe lamp	Grating, 3000 lines/m	KBr windows	Photomultiplier
(g)		W lamp	Grating or glass prism	Glass windows	Photoconductor

5–16. $f = 1.9$.

5–17. $f = 3.1$, light-gathering power of lens in 5–16 is 2.6 times greater.

5–18. (a) 2.50×10^4. (b) 0.50 and 0.25 nm/mm.

5–19. (a) 0.77 nm/mm. (b) 6.0×10^4. (c) 0.0093 nm.

5–20. A silicon diode detector is a reverse-biased pn junction. Photons striking the depletion layer create holes and electrons, giving a current that is proportional to the number of photons.

5–21. (a) A spectroscope consists of a monochromator with an eyepiece that moves along the focal plane permitting visual identification of spectral lines. (b) A spectrograph is a monochromator with a photographic plate or film mounted on the focal plane for photographic recording of spectra. (c) A spectrophotometer is a monochromator with a photodetector mounted behind a slit on the focal plane.

5–22. (a) 8.33×10^4 Hz. (b) 3.57×10^4 Hz. (c) 3.33×10^3 Hz. (d) 1.25×10^3 Hz.

5–23. (a) 2.1 cm. (b) 0.313 cm.

Chapter 6

6–1. (a) 61.4%. (b) 9.44%. (c) 96.4%. (d) 28.2%. (e) 13.6%. (f) 1.85%.

6–2. (a) 0.487. (b) 0.664. (c) 0.046. (d) 1.559. (e) 0.210. (f) 0.523.

6–3. (a) 78.3%. (b) 30.7%. (c) 98.2%. (d) 53.1%. (e) 36.9%. (f) 13.6%.

6–4. (a) 0.788. (b) 0.965. (c) 0.347. (d) 1.86. (e) 0.511. (f) 0.824.

6–5. (a) 57.1%. (b) 46.6%. (c) 94.9%. (d) 16.6%. (e) 78.5%. (f) 54.8%.

6–6.

	A	$\%T$	ε	b, cm	c, M	c	a
(a)		44.7	3.70×10^2			147 mg/L	1.32×10^{-3}
(b)	0.684		3.96×10^1			1.93×10^3 ppm	1.41×10^{-4}
(c)		6.14	7.00×10^1		8.93×10^{-4}	4.87 g/L	
(d)	0.597			0.367		8.40 mg/100 mL	0.194
(e)	1.452	3.53			1.19×10^{-5}		9.82×10^{-2}
(f)	0.876	13.3				22.0 mg/L	2.73×10^{-2}
(g)	0.254		1.06×10^5		2.40×10^{-5}		3.78×10^{-1}
(h)		10.6			1.75×10^{-4}	4.91×10^{-2} g/L	1.99×10^1
(i)	0.074		1.77×10^5		3.36×10^{-7}	9.40×10^{-2} mg/L	
(j)	1.112			0.124		2.73×10^3 mg/L	3.28×10^{-3}

6–7. 6.70×10^{-3} M.

6–8. 3.05×10^{-4} M.

6–9. (a) 1.35. (b) 4.52×10^{-3}. (c) 1.65.

6–10. (a) 0.205. (b) 62.4%. (c) 1.08×10^{-5} to 1.61×10^{-4} M.

6–11. 1.66×10^{-5} to 3.64×10^{-4} M.

6–12. 3.79×10^{-6} to 5.69×10^{-5} M.

6–13.

$M_{Cr_2O_7^{2-}}$	$[CrO_4^{2-}]$	$[Cr_2O_7^{2-}]$	A_{345}	A_{390}	A_{400}
4.00×10^{-4}	3.055×10^{-4}	2.473×10^{-4}	0.827	1.649	0.621
3.00×10^{-4}	2.551×10^{-4}	1.724×10^{-4}	0.654	1.353	0.512
2.00×10^{-4}	1.961×10^{-4}	1.019×10^{-4}	0.470	1.018	0.388
1.00×10^{-4}	1.216×10^{-4}	3.919×10^{-5}	0.266	0.614	0.236

6–14.	Concn, M	A (a)	A (b)	A (c)	A (d)
	4.00×10^{-4}	1.200	1.181	1.089	0.935
	3.00×10^{-4}	0.900	0.891	0.844	0.756
	2.00×10^{-4}	0.600	0.596	0.576	0.533
	1.00×10^{-4}	0.300	0.299	0.292	0.276

6–15. 6.48×10^3.

6–16. 2.51 ppm.

6–17. (a) 1.84×10^3. (b) 1.16×10^{-2}. (c) 2.53×10^{-4} M. (d) 0.201. (e) 0.396.

6–18. (a) 0.320. (b) 0.478. (c) 3.00 cm. (d) 1.56×10^{-5} M. (e) 4.50×10^{-6} M.

6–19. (a) 33.8%. (b) 0.471. (c) 0.697. (d) 0.114.

6–20. 21.2 ppm.

6–21. (a) 11%. (b) 4.9%. (c) 2.7%. (d) 1.4%. (e) 1.7%. (f) 2.9%. (g) 11%.

Chapter 7

7–1. 5.43×10^{-5} to 1.19×10^{-3} M.

7–2. 1.23×10^{-5} to 2.47×10^{-4}.

7–3. (a) Plotting the data yields a straight line. (b) $A = -1.01 \times 10^{-3} + 3.95 \times 10^{-2}$ ppm Fe. (c) $s_r = 3.3 \times 10^{-3}$. (d) $s_\beta = 1.1 \times 10^{-4}$.

7–4. (a) $3.65 \pm (0.10)$ and $3.65(\pm 0.08)$ ppm Fe. (b) $17.12(\pm 0.09)$ and $17.12(\pm 0.06)$ ppm Fe. (c) $1.75(\pm 0.11)$ and $1.75(\pm 0.08)$ ppm Fe. (d) $25.57(\pm 0.09)$ and $25.57(\pm 0.06)$ ppm Fe. (e) $38.31(\pm 0.10)$ and $38.31(\pm 0.08)$ ppm Fe. (f) $13.85(\pm 0.09)$ and $13.85(\pm 0.06)$ ppm Fe.

7–5. $\Sigma V_s = 50.00$, $\Sigma A = 3.117$, $\Sigma V_s^2 = 750.00$, $\Sigma A^2 = 2.312591$, $\Sigma V_s A = 40.70$; $S_{xx} = 250.00$, $S_{yy} = 0.3694532$, $S_{xy} = 9.530$; $\beta = 0.03812$, $\alpha = 0.2422$; ppm Fe $= (\alpha \times 11.1)/(\beta \times 10.0) = 7.05$; $s_r = 0.045$, $s_\beta = 0.0029$, $s_c = 0.2$ ppm Fe.

7–6. 20.1 ppm.

7–7. 8.26×10^{-4}%.

7–8.	(a)	(b)	(c)	(d)	(e)
$M_{Ni^{2+}}$	7.6×10^{-5}	1.02×10^{-4}	1.83×10^{-4}	7.0×10^{-5}	6.8×10^{-5}
$M_{Co^{2+}}$	1.01×10^{-4}	9.1×10^{-5}	5.9×10^{-5}	1.87×10^{-4}	1.43×10^{-4}

7–9. The respective molar concentrations of A and B are: (a) 7.54×10^{-5} and 2.03×10^{-5}. (b) 2.93×10^{-5} and 5.30×10^{-5}. (c) 5.55×10^{-5} and 3.84×10^{-5}. (d) 4.44×10^{-5} and 4.34×10^{-5}.

7–10. (a) At 485 nm, $\varepsilon_{In} = 104$ and $\varepsilon_{HIn} = 908$; at 625 nm, $\varepsilon_{In} = 1646$ and $\varepsilon_{HIn} = 352$. (b) 2.07×10^{-6}. (c) 3.735. (d) 5.633 and 2.33×10^{-6}. (e) $A_{475} = 0.091$ and $A_{625} = 0.306$.

7–11. (a) ± 1.5%. (b) ± 1.5%. (c) ± 9.1%. (d) ± 2.9%. (e) ± 77%. (f) ± 53%.

7–12. (a) Until the equivalence point, the absorbance would increase linearly with volume of reagent. Beyond equivalence, the absorbance would be constant. (b) Same as part a. (c) Until the equivalence point, the absorbance would decrease linearly with volume of reagent. Beyond equivalence, the absorbance would be constant and approximately zero. (d) Same as part a.

7–13. With a small amount of Cu^{2+} present, the analyte solution would show virtually no absorbance until after equivalence when the absorbance would increase rapidly.

7–14. 1.8×10^8.

7–15. 1.53×10^6.

Chapter 8

8–1. (a) *Fluorescence* is the process in which a molecule, excited by the absorption of radiation, emits a photon while descending from an excited electronic state to a lower state of the same spin multiplicity (that is, a singlet → singlet transition). (b) *Phosphorescence* is the process in which a molecule, excited by the absorption of radiation, emits a photon while descending from an excited electronic state to a lower state of different spin multiplicity (that is, a triplet → singlet transition). (c) *Resonance fluorescence* is observed when excited atoms in a vapor emit radiation of the same frequency as that used to excite the atoms. (d) A *singlet state* is one in which the spins of the electrons of an atom or molecule are all paired, so there is no net spin angular momentum. (e) A *triplet state* is one in which the electrons of an atom or molecule are unpaired, so that their angular moments add to give a net non-zero moment. (f) *Vibrational relaxation* is the radiation-less process by which a molecule descends to a lower electronic state without emitting radiation. (g) *Internal conversion* is the intermolecular process in which a molecule crosses to a lower electronic state without emitting radiation. (h) *External conversion* is a radiationless process in which a molecule falls to a lower electronic state while transferring energy to the solvent or other solutes. (i) *Intersystem crossing* refers to the process in which a molecule in one spin state changes to a different spin state (for example, singlet → triplet) with nearly the same total energy. (See the horizontal arrow at the upper right side of Figure 8–1. (j) *Predissociation* occurs when a molecule changes from a higher electronic state to an upper vibrational level of a lower electronic state in which the vibration energy is great enough to produce bond breaking. (k) *Dissociation* occurs when radiation promotes a molecule directly up to a state with sufficiently high vibrational energy for a bond to break. (l) *Quantum yield* is the fraction of excited molecules undergoing the process of interest. For example, the quantum yield for phosphorescence is the fraction of all molecules which have absorbed radiation that goes on to phosphoresce. (m) *Chemiluminescence* is a process by which radiation is produced as a consequence of a chemical reaction.

8–2. In a fluorescence spectrum, the exciting wavelength is held constant and the emission is measured as a function of wavelength. In an excitation spectrum, the emission is measured at one wavelength while the exciting wavelength is varied. The excitation spectrum most closely resembles an absorption spectrum since the emission intensity is usually proportional to the absorbance of the molecule.

8–3. In spectrofluorometry, the sensitivity can be enhanced by increasing the power of the excitation source or by amplifying the analytical signal. These measures are not applicable to spectrophotometry because it is based upon measurement of a difference ($A = \log P_0 - \log P$). Increases in source intensity or amplification increase both P_0 and P but not the difference.

8–4. Fluorescein is more rigid than phenolphthalein, which should lead to stronger fluorescence.

8–5. The fluorescence should be greatest in 1-chloropropane. The bromine or iodine atom of the other solvents promotes intersystem crossing (the "heavy atom effect") and decreases the population of the S_1 state.

8–6. Fluorescence is greater at pH 3 because at pH 10 a significant fraction of phenol is present as phenolate ion, which fluoresces less strongly than phenol itself (Table 8–1).

8–7. (a) A plot of the data reveals a linear relationship between intensity I and concentration C. (b) $I = 3.6 \times 10^{-4} + 22.35\,C$. (c) $s_\beta = 0.27$ and $s_r = 0.17$. (d) 0.544 μM. (e) $s_c = 0.0084$ μM. (f) $s_c = 0.0054$ μM.

8–8. (a) A plot of the data shows a linear relationship between the fluorometer reading F and volume V_s of standard. (b) $F = 6.188 + 1.202\,V_s$. (c) $s_\beta = 0.017$ and $s_r = 0.15$. (d) 1.13 ppm. (e) 0.032 ppm.

8–9. (a) The fluoride ion reacts with and decomposes the aluminum complex to form AlF_6^{3-}. Neither the fluoride complex nor the uncomplexed organic reagent fluoresce. Thus, a decrease in fluorescence with increasing fluoride concentration is observed. (b) A plot of the data reveals a linear relationship between the meter reading M and the volume of standard V_s. (c) $M = 68.2 - 13.22\ V_s$. (d) $s_B = 0.19$ and $s_r = 0.43$. (e) 10.3 ppb F^-.

8–10. $3.07 \times 10^{-5}\ M$.

8–11. Assume that the luminescent intensity L is proportional to the partial pressure of S_2^*. Then we may write: $L = k[S_2^*]$ and $K = [S_2^*][H_2O]^4/([SO_2]^2[H_2]^4)$ where the bracketed terms are all partial pressures and k and K are constants. The two equations can be combined to give upon rearrangement $[SO_2] = ([H_2O]^2/[H_2]^2)(L/kK)^{1/2}$. In a hydrogen-rich flame, the partial pressure of H_2O and H_2 should be more or less constant. Thus, $[SO_2] = k'L^{1/2}$.

Chapter 9

9–1. In atomic emission spectroscopy, light emitted from the hot sample is used as a measure of the concentration of the desired element. In atomic fluorescence spectroscopy, a light source is used to promote the atoms in the vaporized sample to excited states. Fluorescence from the excited atoms is then measured.

9–2. (a) *Atomization* is the process in which a sample is vaporized and decomposed to its atoms, usually by heat. (b) *Pressure broadening* refers to the broadening of atomic line widths at higher concentrations of atoms in a flame. (c) *Doppler broadening* arises because atoms moving toward or away from the monochromator give rise to absorption or emission lines at slightly different frequencies. (d) In a *turbulent flow nebulizer* (Figure 9–10), the entire sample is aspirated directly into the flame. (e) In a *laminar flow nebulizer* (Figure 9–11), the sample is nebulized by passing the oxidant over the tip of a capillary, which is immersed in the sample. After mixing with a fuel, the resulting aerosol passes a series of baffles, which remove the larger drops before the flame is reached. (f) A *hollow cathode lamp* (Figure 9–16) has a tungsten anode and a cylindric-shaped cathode containing the element of interest. The element is sputtered from the cathode into the gas phase, where the element emits its characteristic radiation. (g) *Sputtering* is the process in which gaseous cations bombard a cathode and eject atoms from the cathode into the gas phase. (h) *Self-absorption* refers to the absorption of radiation by unexcited atoms in the gas phase of a hollow cathode lamp or other source. (i) *Spectral interference* is encountered when the absorption or emission of a non-analyte species overlaps a peak being used for the determination of the analyte. (j) *Chemical interference* is the result of any process which decreases or increases the absorption or emission characteristics of the analyte. (k) A *radiation buffer* is a substance added in excess to both sample and standards, which swamps out the effect of the sample matrix on the analyte emission or absorption. (l) A *releasing agent* is a cation which preferentially reacts with a species that would otherwise react with the analyte to cause a chemical interference. (m) *Protective agents* prevent interference by forming stable and volatile products with the analyte. (n) An *ionization suppressor* provides a high concentration of electrons in the flame. These electrons suppress ionization of the analyte.

9–3. The CaOH spectra arise from a multitude of molecular states involving molecular vibration and rotation.

9–4. The population of excited molecules, from which emission arises, is very sensitive to the flame temperature. The population of ground state molecules, from which absorption and fluorescence originate, is not very sensitive to temperature.

9–5. The absorbance of Cr decreases with increasing flame height because chromium oxides are formed to a greater and greater extent as the Cr rises through the flame. The Ag absorbance increases as the silver becomes more atomized as it rises through the flame. Silver oxides are not readily formed. Magnesium exhibits a maximum as a result of both effects mentioned above opposing each other.

9–6. The nonflame atomizer requires less sample and keeps the sample in the optical beam for a much longer time than does a flame.

9–7. Source modulation is employed to distinguish between the component of light arising from the source and the component of light arising from the flame background.

9–8. The continuous radiation from the deuterium lamp is passed through the flame alternately with the hollow cathode lamp beam. By comparing the power of the beams, it is possible to continuously correct the signal power for the variable attenuation encountered during sample aspiration in the flame.

9–9. The alcohol reduces the surface tension of the solution thus leading to smaller droplets, a greater number of which then reach the flame in a given unit of time. Thus, a greater number of Ni atoms are present at any instant.

9–10. At higher currents, more unexcited atoms are formed in the sputtering process. These atoms generally have less kinetic energy than the excited ones. The Doppler broadening of their absorption lines is therefore less than the broadening of the emission lines of the faster moving excited atoms. Thus, only the center of the emission line is attenuated by self absorption.

9–11. (a) $(N_j)_{2cm}/(N_j)_{2cm}$ = 1.0. (b) $(N_j)_{3cm}/(N_j)_{2cm}$ = 2.1. (c) $(N_j)_{4cm}/(N_j)_{2cm}$ = 1.7. (d) $(N_j)_{5cm}/(N_j)_{2cm}$ = 1.1.

9–12. (a) Sulfate complexes of Fe(II) and Fe(III) are less readily atomized. Thus, the concentration of Fe atoms in the flame is smaller. (b) (1) use higher temperatures, (2) add a releasing agent such as La(III), (3) add a protective reagent such as EDTA.

9–13. 0.053 Å.

9–14. For Na, $\lambda \simeq 5893$ and $E_j = 3.37 \times 10^{-12}$ erg; for Mg^+, $\lambda \simeq 2800$ and $E_j = 7.09 \times 10^{-12}$ erg. N_j/N_0 for Na and Mg^+ are respectively: (a) 2.3×10^{-5} and 5.1×10^{-11}. (b) 6.1×10^{-4} and 5.1×10^{-8}. (c) 2.2×10^{-3} and 7.2×10^{-7}. (d) 5.0×10^{-2} and 5.5×10^{-4}.

9–15. N_{4s}/N_{3s} = (a) 1.7×10^{-5}. (b) 1.1×10^{-2}.

9–16. 0.101 ± 0.005 μg Pb/mL.

9–17. The first peak is probably due to volatile absorbing species in the sample. The second peak is likely due to scattering by smoke formed during ashing.

9–18. The nonlinearity is caused by ionization of the U.

9–19. An internal standard improves precision by compensating for uncontrollable variables.

9–20. 0.504 ppm Pb.

9–21. (a) A plot of the instrument reading R versus concentration of Na C_x is linear. (b) R = 3.18 + 0.920 C_x. (c) s_β = 0.015 and s_r = 0.96. (d) For A, B, and C respectively, % Na_2O = 0.257, 0.389, 0.736; s_c = 0.013, 0.012, 0.015 % Na_2O; $(s_c)_r$ = 49, 32, 20 ppt.

9–22. (a) A plot of the data reveals a linear relationship between volume of standard V_s and A. (b) A = 0.202 + 8.81×10^{-3} mL std. (c) $s_\beta = 4.1 \times 10^{-5}$ and $s_r = 1.3 \times 10^{-3}$. (d) 28.0 ppm Cr. (e) 0.22 ppm Cr.

Chapter 10

10–1. An internal standard is a substance that responds to uncontrollable variables in a similar way as the analyte. It is introduced into, or is present in, both standards and samples in a fixed amount. The ratio of the analyte signal to the internal standard then serves as the analytical signal.

10–2. Flame atomic absorption methods require a separate lamp for each element, which is not convenient when several elements are to be determined.

10–3. The temperature of a spark plasma is so great ($\sim$40,000°K) that most atoms present are ionized. In a lower temperature arc ($\sim$4000°K) only the lighter elements are ionized to any significant extent. In a plasma, the high concentration of electrons prevents extensive ionization of the analyte atoms.

10–4. To improve the *S/N* ratio by removing the background radiation from the plasma.

10–5. 4.0 and 1.2 Å/mm.

10–6. To reduce the intensity of cyanogen bonds.

Chapter 11

11–1. (a) 1.9×10^6 dynes/cm. (b) 2.1×10^3 cm^{-1}.

11–2. (a) 4.81×10^5 dynes/cm. (b) 2.1×10^3 cm^{-1}.

11–3. (a) Inactive. (b) Active. (c) Active. (d) Active. (e) Inactive. (f) Active. (g) Inactive.

11–4. For C—H, calc $\sigma = 3.02 \times 10^3$ cm^{-1} and experimental $\sigma = 2.8$ to 3.0×10^3 cm^{-1}. For C—D, 2.22×10^3 cm^{-1}.

11–5. Vinyl alcohol, $CH_2{=}CH{-}CH_2OH$.

11–6.

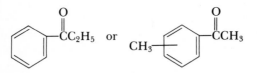

The spectrum is, in fact, that of o-methyl acetophenone.

11–7. Acrolein, $CH_2{=}CH{-}CHO$ with H_2O contaminant.

11–8. Propanenitrile, CH_3CH_2CN.

11–9. Diphenylamine, $C_6H_5NHC_6H_5$.

11–10. Probably an aliphatic alcohol. The spectrum is, in fact, that of 2-hexanol.

11–11. Benzaldehyde, C_6H_5CHO.

11–12. An aliphatic ester. The spectrum is, in fact, that of t-butylacetate.

11–13. 7.1×10^{-3} cm.

11–14. 0.21 cm.

11–15. (a) 25 cm. (b) 0.25 cm.

Chapter 12

12–1. (a) Stokes lines at 641.7, 645.6, 651.7, 664.9, and 666.1 nm. Anti-Stokes lines at 624.2, 620.5, 614.9, 603.7, and 602.7 nm. (b) Stokes lines at 493.2, 495.6, 499.2, 506.8, and 507.6 nm. Anti-Stokes lines at 482.9, 480.6, 477.3, 470.5, and 469.9 nm.

12–2. $I_{Ar}/I_{He/Ne} = 2.82$.

12–3. For studying compounds that are photochemically unstable.

12–4. (a) At 20°C, $I_{AS}/I_S = 0.117$ and at 40°C, $I_{AS}/I_S = 0.135$. (b) $I_{AS}/I_S = 0.0110$ and 0.0147. (c) $I_{AS}/I_S = 4.24 \times 10^{-4}$ and 6.96×10^{-4}.

12–5. (a) $p = 0.77$. (b) $p = 0.012$, polarized. (c) $p = 0.076$, polarized. (d) $p = 0.76$.

Chapter 13

13–1. (a) 0.2511 and 22.12. (b) 0.3402 and 40.89. (c) 0.264 and 32.5.

13–2. (a) A plot of the data reveals a linear relationship between wt % C_6H_6 and refractive index. (b) $n_D = 1.36942 + 1.1963 \times 10^{-3}$ wt % C_6H_6. (c) $s_\beta = 5.0 \times 10^{-5}$ and $s_r = 2.9 \times 10^{-3}$. (d) 52.4 wt %. (e) 2.6 wt %. (f) 79.4 wt %. (g) 2.1 wt %.

13–3. (a) 6.34 M. (b) 2.03 M.

13–4. 1.4427.

13–5. (a) 16.16 and 16.06. (b) 25.97. (c) 1.642 g/mL.

13–6. In *unpolarized light,* the electric vectors of the radiation oscillate in all directions. In *plane polarized light,* all the electric vectors oscillate in a single plane. In *circularly polarized light,* the electric vector describes a circular helix as the light travels. Viewed end on, the vector describes a circle (see Figure 13–8). This condition arises when the two perpendicular components of plane polarized light are one-quarter wavelength out of phase. If the two components are between zero and one-quarter or between one-quarter and one-half wavelength out of phase, *elliptical polarization* (Figure 13–9) results.

13–7. The *ordinary beam* is that polarized beam which travels at the same velocity along the optic axis and perpendicular to the optic axis of an anisotropic crystal. The *extraordinary beam* travels at a different rate in each of these directions.

13–8. (a) The *optic axis* of a crystal is that axis about which there is a symmetric distribution of particles. Polarized light travels along this axis at a uniform speed, regardless of the direction of polarization. (b) A *Nicol prism* (Figure 13–5) uses two pieces of calcite to produce plane polarized radiation by transmitting the extraordinary beam and deflecting the ordinary beam. (c) A *quarter wave plate* retards one of the two components of plane polarized light by one-quarter wavelength, thus producing circularly polarized light. (d) *Double refraction* refers to the property of an anisotropic crystal, in which the two characteristic refractive indexes can separate two components of incident light. (e) *Circular double refraction* is the property of optically active substances whose refractive indexes for right and left circularly polarized light are different. (f) *Circular birefringence* is the difference between the refractive indexes for right and left circularly polarized light $(n_l - n_d)$. When plane polarized light passes through a medium for which $n_l \neq n_d$, the light becomes elliptically polarized. (g) The *ellipticity* is defined as the angle θ in Figure 13–15. (h) *Enantiomers* are mirror image isomers of the same compound, which cannot be superimposed on each other. Each enantiomer rotates the plane of polarized light in the opposite direction. (i) The *dextrorotatory* component of plane polarized light is that circularly polarized component which exhibits clockwise rotation as the beam approaches the observer. (j) The *levorotatory* component exhibits counterclockwise rotation.

13–9. In *optical rotatory dispersion,* the wavelength dependence of the molecular rotation of a compound is measured. In *circular dicroism,* the wavelength dependence of $(\varepsilon_l - \varepsilon_d)$ is measured.

13–10. The dependence of the angle on the sample concentration will distinguish the two possibilities. For example, if the concentration is increased by 25%, the rotation will be either +25 deg or −200 deg. These two rotations are no longer indistinguishable.

13–11. $\alpha_{360} = 250$ deg; $\alpha_{720} = 125$ deg.

13–12. (a) 228 deg. (b) 33.2 deg. (c) 33.5%.

Chapter 14

14–1. Singlet at $\delta = 11$ to 12, quartet at $\delta = 2.2$, and triplet at $\delta = 1.1$.

14–2. (a) Doublet at $\delta = 2.2$ and quartet at $\delta = 9.7$ to 9.8. (b) Singlet at $\delta = 2.2$ and singlet at $\delta = 11$ to 12. (c) Triplet at $\delta = 1.6$ and quartet at $\delta = 4.4$.

14–3. (a) A single peak at $\delta = 2.1$. (b) A singlet at $\delta = 2.1$, a quartet at $\delta = 2.4$, and a triplet at $\delta = 1.1$. (c) A singlet at $\delta = 2.1$, 7 peaks centered at $\delta = 2.6$, and a doublet at $\delta = 1.1$.

14–4. (a) A single peak at $\delta = 1.2$ to 1.4. (b) Singlet peaks at $\delta = 3.2$ and 3.4 with peak areas in the ratio of 3:2 respectively. (c) A triplet at $\delta = 1.2$ and a quartet at $\delta = 3.4$.

14–5. (a) Singlets at $\delta = 2.2$ and at 6.5 to 8 with peak areas in the ratio of 3:5. (b) A triplet at $\delta = 1.1$, a quartet at $\delta = 2.6$, and a singlet at $\delta = 6.5$ to 8. (c) A doublet at $\delta = 0.9$ to 1 and ten peaks at 1.5.

14–6. CH_3CH_2Br.

14–7. $CH_3CH_2\overset{\displaystyle Br}{\underset{\displaystyle |}{C}}HCOOH$.   14–8. $CH_3CH_2\overset{\displaystyle O}{\overset{\displaystyle \|}{C}}CH_3$.   14–9. $CH_3\overset{\displaystyle O}{\overset{\displaystyle \|}{C}}OC_2H_5$.

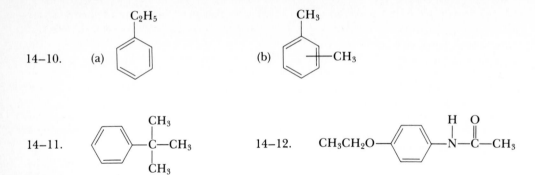

14–10.   (a)   [C₂H₅ benzene structure]   (b)   [CH₃ benzene with CH₃ structure]

14–11.   [benzene with C(CH₃)₃ structure]   14–12. CH_3CH_2O—[benzene]—N–C–CH_3 (with H and O)

Chapter 15

15–1. 0.155 Å.

15–2. For K_β and L_β series respectively, V = (a) 112 and 17.2 kV. (b) 3.59 kV and no L_β lines. (c) 15.0 and 1.75 kV. (d) 67.4 and 9.67 kV.

15–3. (a) 2.5 Å. (b) 1.6 Å. (c) 1.1 Å. (d) 1.0 Å. (e) 0.54 Å. (f) 0.47 Å.

15–4. (a) 24 Å. (b) 14 Å. (c) 8.9 Å. (d) 8.3 Å. (e) 4.0 Å. (f) 3.5 Å.

15–5. 2.3×10^{-3} cm.

15–6. (a) 3.94 cm²/g. (b) 0.126.

15–7. 2.73×10^{-3} cm.

15–8. (a) 1.39%. (b) 0.261%.

15–9. For Fe, Se, and Ag respectively, 2θ = (a) 80.9, 42.9, and 21.1 deg. (b) 51.8, 28.5, and 14.2 deg. (c) 36.4, 20.3, and 10.1 deg.

15–10. (a) 135 deg. (b) 99.5 deg.

15–11. (a) 4.05 kV. (b) 1.32 kV. (c) 20.9 kV. (d) 25.0 kV.

Chapter 16

16–1. An M electron is ejected by X-radiation or an electron beam. An N electron then descends to the M orbital while ejecting a second electron as an Auger electron.

16–2. The ESCA binding energy is the minimum energy required to remove an inner electron from its orbital to a region where it no longer feels the nuclear charge. The absorption edge results from this same transition. Thus, in principle it is possible to observe chemical shifts by either type of measurement.

16–3. (a) 165.4 eV. (b) SO_3^{2-}. (c) 1306.6 eV. (d) 1073.5 eV.

16–4. (a) 406.3 eV. (b) 819.5 eV. (c) By observing the peak with sources of differing energy (such as Al and Mg X-ray tubes). Auger peaks would not have different kinetic energies with the two sources. (d) 403.4 eV.

Chapter 17

17–1. (a) 0.945. (b) 0.571. (c) 0.326. (d) 0.0149.

17–2. (a) 0.985. (b) 0.00459. (c) 0.858. (d) 0.973.

17–3. 143 hr or 5.95 days.

17–4. σ_M and $(\sigma_M)_r$ respectively = (a) 10 counts and 10%. (b) 27 counts and 3.7%. (c) 84 counts and 1.2%. (d) 141 counts and 0.71%.

17–5. σ_R and $(\sigma_R)_r$ respectively = (a) 17 cpm or 8.7%. (b) 12 cpm or 6.1%. (c) 7.1 cpm or 3.5%. (d) 4.1 cpm or 2.0%.

17–6. (a) ±19 counts and ±2.4%. (b) ±46 counts and ±5.8%. (c) ±73 counts or ±9.1%.

17–7. (a) ±13 counts. (b) ±28 counts. (c) ±37 counts. (d) ±57 counts.

17–8. (a) ±8.4 cpm and ±2.9%. (b) ±7.8 cpm and ±2.7%. (c) ±9.0 cpm and ±3.2%. (d) ±10.6 cpm and ±4.1%.

17–9. (a) 2.61×10^3 counts. (b) 1.19×10^3 counts. (c) 1.12×10^3 counts.

17–10. (a) 3.1%. (b) 1.4%. (c) 0.93%.

17–11. 562 mL.

17–12. 5.32 mg.

17–13. (a) 3.46 ppm. (b) 3.5 ± 0.2 ppm.

17–14. (a) 0.273 ppm. (b) 0.273 ± 0.008 ppm.

17–15. (a) 5.41×10^{-3} mg I$^-$/mL. (b) $5.4\ (\pm\ 0.12) \times 10^{-3}$ mg I$^-$/mL.

Chapter 18

18–1. In a gaseous ionization source, the sample is first volatilized (by heating if necessary) and then transmitted to the ionization area for ion formation. In a desorption source, a probe is used and ionization takes place directly from the condensed form of the sample. The advantages of desorption ionization are that it can be applied to high molecular weight and thermally unstable samples. The advantages of gaseous sources are their simplicity and speed (no need to use probe and wait for probe area to be pumped out).

18–2. The most fragmentation and thus the most complex spectra are encountered with electron impact ionization. Field ionization produces the simplest spectra. Chemical and electron ionization produce greater sensitivities than does field ionization.

18–3. Both field ionization and field desorption ionization are performed at anodes containing numerous sharp tips so that very high electrical fields are realized. In field ionization, the sample is volatilized before ionization, whereas field desorption takes place at an anode that has been coated with the sample. The latter requires the use of a sample probe.

18–4. The resolution of a single focusing analyzer is often limited by the inherent spread in kinetic energies of the sample molecules. As a consequence, the accelerated ions have slightly different final velocities, which result in peak broadening. In double-focusing instruments, the first analyzer focuses ions of the same kinetic energy on the slit of the second analyzer. Thus, the band broadening due to different initial kinetic energies is eliminated.

18–5. (a) 2.22×10^3. (b) 769. (c) 7.09×10^4.

18–6. (a) 993. (b) 30. (c) 113. (d) 480.

18–7. (a) 196%. (b) 130.5%. (c) 65%.

18–8. (a) 1260 to 4981 G. (b) 192 to 3000 V.

18–9. (a) 375 V. (b) −0.77%.

18–10. (a) 2800 G. (b) 0.61%.

18–11. M = 131 due to $^{35}Cl_3CCH_2^+$, M = 133 due to $^{37}Cl^{35}Cl_2CCH_2^+$, M = 135 due to $^{37}Cl_2{}^{35}ClCCH_2^+$, M = 117 due to $^{35}Cl_3C^+$, M = 119 due to $^{37}Cl^{35}Cl_2C^+$, M = 121 due to $^{37}Cl_2{}^{35}ClC^+$.

18–12. M = 84 due to $^{35}Cl_2{}^{12}CH_2^+$, M = 85 due to $^{35}Cl_2{}^{13}CH_2^+$, M = 86 due to $^{37}Cl^{35}Cl^{12}CH_2^+$, M = 87 due to $^{37}Cl^{35}Cl^{13}CH_2^+$, M = 88 due to $^{37}Cl_2{}^{12}CH_2^+$.

Chapter 19

19–1. (a) 0.692V. (b) 0.598 V. (c) 0.132 V.
19–2. (a) 0.033 V. (b) 0.999 V. (c) 0.493 V.
19–3. (a1) −0.118 V. (a2) −0.122 V. (b1) 0.771 V. (b2) 0.751 V.
19–4. (a1) 0.145 V. (a2) 0.143 V. (b1) 0.145 V. (b2) 0.125 V.
19–5. (a) −0.051 V. (b) −0.13 V. (c) 0.729 V. (d) 0.143 V.
19–6. (a) 1.25 V. (b) 0.299 V.
19–7. (a) 0.288 V; galvanic. (b) −0.599 V; electrolytic. (c) −0.836 V; electrolytic.
19–8. (a) −0.357 V; electrolytic. (b) 1.227 V; galvanic. (d) 0.588 V; galvanic.
19–9. −0.900 V.
19–10. −0.367 V.
19–11. −0.036 V.
19–12. −1.920 V.
19–13. 1.82×10^{-4}.
19–14. 9.4×10^{-16}.
19–15. 0.121 V.
19–16. −0.298 V.

Chapter 20

20–1. (a) −0.325 V. (b) $Cu|SCN^-(xM),CuSCN(sat'd)\|SCE$.
　　(c) $-\log SCN = pSCN = (E_{SCE} - E^0_{CuSCN} - E_{cell})/0.0591 = 9.63 - (E_{cell}/0.0591)$. (d) 2.67.
20–2. (a) −0.66 V. (b) $Ag|Ag_2S(sat'd),S^{2-}(xM)\|SCE$.
　　(c) $pS = 2(-E_{cell} - E^0_{Ag_2S} + E_{SCE})/0.0591 = 30.6 - (2E_{cell}/0.0591)$. (d) 10.5.
20–3. 6.77.
20–4. 6.94.
20–5. 0.605 V.
20–6. 1.0×10^7.
20–7. (a) 0.113 V. (b) 0.054 V. (c) −0.005 V.
20–8. 8.63.
20–9. 9.5×10^{-17}.
20–10. 3.90×10^{-5}.

20–11. (a)

Vol, mL	E, V	Vol, mL	E, V	Vol, mL	E, V
10.00	−0.68	49.00	−0.75	51.00	−1.10
25.00	−0.70	49.90	−0.78	55.00	−1.14
40.00	−0.71	50.10	−1.04	60.00	−1.16

20–12. (a)

Vol, mL	E, V	Vol, mL	E, V	Vol, mL	E, V
0.00	0.408	24.00	0.455	25.10	0.528
10.00	0.417	24.90	0.484	26.00	0.587
20.00	0.433	25.00	0.499	30.00	0.627

20–13. (a) 3.31 and 4.90×10^{-4} M. (b) 4.68×10^{-4} to 5.13×10^{-4}. (c) 4.7% and −4.5%.
20–14. (a) 5.232 and $[Mg^{2+}] = 6.85 \times 10^{-6}$. (b) 5.02×10^{-6} to 6.85×10^{-6} M. (c) +17% and −14%.
20–15. 0.00127%.

Chapter 21

21–1. (a) -1.345 V. (b) 0.285 V. (c) -0.85 V. (d) -2.48 V. (e) -2.54 V.
21–2. (a) 0.266 V. (b) 0.118 V. (c) 0.040 V. (d) 0.089 V. (e) 0.128 V.
21–3. (a) Not feasible because Bi deposits before quantitative deposition of Cu. (b) Ag can be separated from Cu and Bi by keeping the cathode potential between 0.200 and 0.071 V vs. SCE. (c) -0.059 V vs. SCE.
21–4. (a) Iodide forms at 0.300 V vs. SCE; galvanic. (b) Separation possible; 0.100 to 0.076 V vs. SCE. (c) Separation possible; 0.100 to -0.073 V vs. SCE. (d) Not feasible.
21–5. (a) 0.433 V. (b) 0.277 V. (c) -0.127 V.
21–6. (a) 2.95 min. (b) 1.97 min. (c) 0.984 min.
21–7. 1.82% CCl_4 and 1.66% $CHCl_3$.
21–8. 0.613%.
21–9. 172 g/equiv.
21–10. 29.4 μg.

Chapter 22

22–1. (a) The decomposition potential is the potential on a current-voltage curve at which the current first becomes greater than the residual current that is associated with a blank. The half-wave potential is the potential on a current-voltage curve that corresponds to a current that is exactly one-half the diffusion current. (b) A limiting current is a current that is essentially constant and independent of applied potential. It arises as a consequence of a limitation in the rate at which reactants can be brought to an electrode surface. A diffusion current is a current whose magnitude is limited by the rate at which reactants *diffuse* to the electrode surface. In order to observe a diffusion current it is necessary to minimize the transport of reactant to the electrode by mechanical mixing and electrostatic attraction.
22–2. (a) Generally, a limiting current increases linearly with analyte concentration. (b) If the analyte bears a charge that is opposite to that of the electrode, decreases in electrolyte concentration result in increases in the limiting current. If the charges are the same, however, the effect is reversed. For an uncharged analyte, limiting currents are not affected by the concentration of the electrolyte.
22–3. (a) A plot of the data reveals a linear relationship between i_d and mmol/L Pb^{2+}. (b) $i_d = 0.188 + 8.61\ C_{Pb}$. (c) $s_B = 0.10$; $s_r = 0.42$.

(d)	mmol Pb^{2+}/mL	s_c	$(s_c)_r$		mmol Pb^{2+}/mL	s_c	$(s_c)_r$
(1)	0.299	0.049	16%	(4)	3.03	0.035	1.1%
(2)	0.878	0.057	6.5%	(5)	4.62	0.058	1.3%
(3)	0.878	0.041	4.6%				

22–4. (a) $\bar{x} = 8.72$. (b) $s = 0.25$; $s_r = 2.9\%$. (c) 8.64. (d) -0.92%.
22–5. In the HNO_3, Cu^{2+} ions are present as aquo complexes. In the other media they are present in the form of chloride, tartrate, and ammine complexes. The relative diffusion coefficients for the complexes vary in the order ammine > chloro > aquo > tartrate.
22–6. (a) 32.7 μA. (b) 19.1 μA. (c) 4.34 mM. (d) -41%.
22–7. (a) -0.05%. (b) -0.14%. (c) -0.41%.
22–8. (a) 0.369. (b) 0.346. (c) 0.144. (d) 0.314 mg/mL.
22–9. For A, $i_d/C = 2.29$; for C, $i_d/C = 6.50$.
22–10. 4.

22–11. 4.1×10^{-5} cm²/sec.

22–12. 7.2×10^{5}.

22–13. Anodic wave at 0.12 V results from $2Hg + 2Br^- \rightleftharpoons Hg_2Br_2(s) + 2e$. Wave at > 0.48 V results from oxidation of the electrode: $2Hg \rightarrow Hg_2^{2+} + 2e$.

22–14. (a) -0.260 V. (b) -0.408 V. (c) -0.615 V.

Chapter 23

	Specific Conductance $k \times 10^4$						
Vol, mL	23–1	23–2	23–3	23–4	23–5	23–6	23–7
0.00	2.63	—	0.910	0.415	0.415	1.33	1.33
1.00	2.38	0.244	0.979	0.330	0.218	1.31	1.29
2.00	2.14	0.504	1.06	0.413	0.437	1.29	1.24
3.00	1.89	0.790	1.16	0.548	0.655	1.26	1.19
4.00	1.64	1.12	1.33	0.699	0.874	1.24	1.15
5.00	1.40	1.49	1.66	0.861	1.09	1.22	1.10
6.00	2.25	1.92	2.26	1.36	1.10	1.47	1.33
7.00	3.10	2.38	3.00	1.86	1.11	1.72	1.56
8.00	3.96	2.88	3.78	2.35	1.11	1.97	1.79

Chapter 24

24–1. (a) In *thermogravimetric analysis,* a sample is heated and its mass is measured as a function of temperature. (b) In *differential thermal analysis,* the heat absorbed or emitted by the sample is observed by measuring the temperature difference between the sample and an inert reference as the temperature of each is increased at a constant rate. (c) In *differential scanning calorimetry,* the sample and reference are heated such that each is kept at the same (increasing) temperature. The difference in heat needed to keep the sample and reference at the same temperature is recorded and represents the heat of exothermic or endothermic processes which occur during the increase in temperature. (d) In a *thermometric titration,* the temperature of a solution being titrated is measured as a function of the volume of reagent added.

24–2. 18.1% Ca and 22.4% Ba.

24–3. The melting point of benzoic acid is not very pressure dependent. The boiling point, however, increases as the pressure increases.

24–4.

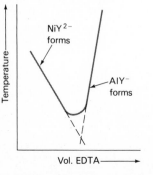

24–5. One interpretation is that the two domains of the protein are not equivalent in structure and iron binding. In the absence of iron, one domain unfolds at 62°C and the other unfolds at 72°C. When one equivalent of iron is added, it binds mainly to the 62°C site and raises the denaturation temperature from 62 to 88°C. The other domain is slightly affected, with its denaturation temperature raised to 76°C. When both sites are occupied by iron the two domains unfold at the same temperature (89°C).

Chapter 25

25–1. (a) $N_A = 2775$; $N_B = 2472$; $N_C = 2364$; $N_D = 2523$. (b) $\overline{N} = 2.5 \times 10^3$; $s = 0.2 \times 10^3$. (c) 9.8×10^{-3} cm.

25–2. (a) $k'_A = 0.74$; $k'_B = 3.3$; $k'_C = 3.5$; $k'_D = 6.0$. (b) $K_A = 6.2$; $K_B = 27$; $K_C = 30$; $K_D = 50$.

25–3. (a) 0.72. (b) 1.1. (c) 105 cm. (d) 60 min.

25–4. (a) 5.2. (b) 2.9 cm.

25–5. (a) 2.7×10^3 plates. (b) 0.14×10^3 plates. (c) 0.015 cm.

25–6. (a) 1.1. (b) 2.7. (c) 3.7.

25–7. (a) 5.4×10^3 plates. (b) 79 cm. (c) 21.6 min.

25–8. (a) $k'_1 = 4.3$; $k'_2 = 4.7$; $k'_3 = 6.1$. (b) $K_1 = 14$; $K_2 = 15$; $K_3 = 19$. (c) $\alpha_{2,1} = 1.11$. (d) $\alpha_{3,2} = 1.28$.

25–9. (a) Variables that lead to band broadening: (1) very low or very high flow rates; (2) high viscosity; (3) low temperature; (4) large particle size for packing; (5) for liquid stationary phases, thick films; (6) long columns; (7) slow introduction of sample; (8) large samples. (b) Variables that may lead to band separation: (1) packings that produce partition coefficients that differ significantly; (2) increasing the length of the packing; (3) variation in solvent composition; (4) choosing an optimum temperature; (5) changing pH of the mobile phase; (6) incorporating a species in the stationary phase that selectively complex certain analytes.

25–10. Slow sample introduction leads to band broadening.

25–11. (a) $k'_M = 2.54$; $k'_N = 2.62$. (b) $\alpha_{N,M} = 6.20/6.01 = 1.03$. (c) 8.1×10^4 plates. (d) 1.8×10^2 cm. (e) 91 min.

25–12. (a) $k'_M = 2.45$; $k'_N = 2.62$. (b) $\alpha_{N,M} = 1.07$. (c) 1.6×10^4 plates. (d) 35 cm. (e) 18 min.

25–13. 14.8% methyl acetate; 48.1% methyl propionate; 37.1% n-methyl butyrate.

25–14. 15.0% A; 18.1% B; 27.4% C; 23.0% D; 16.5% E.

Chapter 26

26–1. (a) 30.2 mL/min. (b) $V_M^0 = 8.4$ mL; $(V_R^0)_1 = 55.3$ mL; $(V_R^0)_2 = 116.2$ mL; $(V_R^0)_3 = 221.5$ mL. (c) $(V_g)_1 = 24.4$ mL; $(V_g)_2 = 56.1$ mL; $(V_g)_3 = 111$ mL. (d) $K_1 = 34.2$; $K_2 = 78.6$; $K_3 = 156$. (e) $t_r = 32.4$ min and $V_R^0 = 905$ mL.

26–2. (a) $k'_1 = 5.6$; $k'_2 = 13$; $k'_3 = 25$. (b) $\alpha_{2,1} = 2.3$; $\alpha_{3,2} = 2.0$. (c) $= 1.7 \times 10^3$ plates; 0.064 cm/plate. (d) $(R_S)_{2,1} = 7.5$; $(R_S)_{3,2} = 6.4$. (e) 13.7% methyl acetate; 49.2% methyl propionate; 37.1% methyl n-butyrate.

26–4. The retention times would be smaller since the three compounds are relatively polar and would thus be less compatible with a nonpolar solvent.

26–5. t_r for C_5 alcohol = 8.03 min; for C_6 alcohol = 18.3 min.

26–6. (a) 38.0 mL/min. (b) $V_M^0 = 14.1$ mL; $(V_R^0)_1 = 164$ mL; $(V_R^0)_2 = 176$ mL; (c) $(V_g)_1 = 46.4$ mL; $(V_g)_2 = 50.1$ mL. (d) $K_1 = 49.2$; $K_2 = 53.1$.

26–7. (a) $k'_1 = 10.6$; $k'_2 = 11.4$. (b) $\alpha = 1.08$. (c) 2.59×10^3 plates. (d) 0.0483 cm. (e) 0.86. (f) 4.1 m. (g) 16 min.

26–8. (a) Increasing V_S/V_m leads to an increase in the film thickness d_f. This increase causes a marked increase in $C_S\mu$ (see Table 25–2) in Equation 25–8 and thus H. (b) Reducing the rate of sample injection will lead to band broadening because all of the molecules do not start down the column at the same instant; reduced efficiency and an increase in H results. (c) Increasing the injection port temperature will tend to decrease H because the sample evaporation rate will increase; the effect is the same as in part (b). (d) Increasing the flow rate may cause either increases or decreases in H depending upon the flow rate (see Equation 25–8). (e) Reducing particle size increases the surface area and thus decreases film thickness d_f in the $C_S\mu$ term in Equation 25–8 (see Table 25–2). A decrease in particle size also makes the $C_M\mu$ term smaller. Both effects then lead to a smaller plate height. (f) The diffusion rates D_M and D_S increase with temperature. In Table 25–2, it is seen that the $C_S\mu$ and $C_M\mu$ terms become smaller as a consequence while the B/μ term becomes larger. In most cases, a reduction in plate height accompanies temperature increases.

26–9. (a) 300. (b) 400. (c) 500. (d) 600. (e) 700. (f) 800. (g) 726. (h) 431. (i) 629. (j) 616. (k) 585. (l) 642.

Chapter 27

27–1. (a) Gas-liquid chromatography is applicable to volatile, thermally stable, non-ionic species. (b) Liquid partition chromatography is applicable to non-ionic but polar organic compounds with molecular weights up to perhaps 10,000. It is useful for separations based on type and number of functional groups in a molecule as well as homologs and benzologs. (c) Reversed-phase partition chromatography, used in conjunction with strongly polar solvents, is applicable to a wide variety of species having molecular weights of 10,000 or less. It is useful for separations based on number and type of functional groups contained in a molecule as well as homologs and benzologs. (d) Ion-exchange chromatography is used for separation of inorganic, organic, and biological species that carry a charge. (e) Cell permeation chromatography is used for the separation of high molecular weight ($>10^3$ to 10^4) nonpolar organic species that are soluble in organic solvents. It is also useful for separation of homologs and oligomers as well as for the determination of molecular weight distributions of high molecular weight polymers. (f) Gel filtration chromatography is applicable to moderate to high molecular weight, water soluble polar species. (g) Gas-solid chromatography has found widespread use for the separation of small gaseous molecules such as helium, oxygen, and nitrogen from one another and from higher molecular weight species. With polymer bead packings, separation of polar gaseous molecules such as CO_2 and oxides of nitrogen is also possible. (h) Liquid adsorption chromatography is useful for the separation of nonpolar species of low or moderate molecular weight (<5000 to 10,000). It is particularly useful for separating mixtures of isomers.

27–2. Three methods for improving resolution include: (1) adjustment of k'_A and k'_B by employing a multicomponent mobile phase and varying the ratio of the components to find an optimal mixture; (2) variation in the chemical composition of the solvent system in such a way as to make α larger; (3) employing a different packing in which α is greater.

27–3. In partition chromatography, k' is conveniently varied by using a two (or more) component system and varying the ratio of the components.

27–4. In gas-liquid chromatography, α is generally varied by varying the column packing. For liquid chromatography, both column packing and chemical composition of the mobile phase can be varied to yield better α values.

27–5. In absorption chromatography on an alumina packing, the mobile phase is generally increased in polarity as the elution proceeds. Thus, the ratio of benzene to acetone should be decreased.

27–6. (a) Sparging involves removal of dissolved gases from a liquid by deaerating it with an inert, nonsoluble gas such as helium or nitrogen. (b) Isocratic elution is a technique used in HPLC in which elution is performed with a solvent of constant composition. (c) In gradient elution, the mobile phase consists of a mixture of two or more solvents whose volume ratios are varied as the elution proceeds. (d) Stop-flow injection is a method for introducing a sample onto an HPLC column in which the flow is stopped, a fitting at the head of the column is removed, and the sample is injected directly onto the head of the column. After replacing the fitting, the flow is resumed. (e) A pellicular packing is made up of small, inert spherical beads of uniform size, which have been coated with a uniform layer of the stationary phase component. (f) A light pipe is a flow-through cell for measuring the infrared absorption of the eluent from an HPLC column. It consists of a narrow (1 to 3 mm) Pyrex tubing that is internally coated with gold. Typically the tube is several centimeters in length and is equipped with KBr windows. Transmission of the *IR* radiation occurs by multiple reflections off the walls. (g) A reversed-phase packing is a bonded packing in which the stationary phase is nonpolar—often a hydrocarbon.

27–7. The linear response range of a detector is the analyte concentration range over which the detector signal varies linearly with concentration.

27–8. In gas chromatography, the mobile phase serves simply to carry the sample through the stationary phase; it does not interact appreciably with the components of the sample. In both liquid and supercritical fluid chromatography the mobile phase serves not only to transport the sample but also to undergo dispersion interactions with sample components. Thus, in the latter two methods, the efficiency of separation can often be enhanced by suitable changes in the composition of the mobile phase. With gas chromatography, on the other hand, only temperature programming of the mobile phase can be used to improve separations.

27–9. (a) 26.7. (b) 71% $CHCl_3$ and 29% *n*-hexane.

27–10. Here the chloroform solvent, which is a type VIII, might be replaced by a type I or type III solvent (see Figure 27–13 and Table 27–3). Thus, an aliphatic ether or tetrahydrofuran might be used in conjunction with the *n*-hexane.

27–11. (a) Adsorption. (b) Partition. (c) Ion-exchange. (d) Normal phase partition. (e) Exclusion.

27–12. $CHCl_3$ which has a greater eluent strength than CCl_4.

27–13. (a) *n*-Hexane, benzene, hexanol. (b) Diethyl ether, ethyl acetate, nitrobutane.

27–14. (a) *n*-Hexanol, benzene, *n*-hexane. (b) Nitrobutane, ethyl acetate, diethyl ether.

27–15. $K_B = 0.38$ and $K_C = 0.73$.

27–16. A mass chromatogram is a plot of total ion current or ion current for one or more ions in a mass spectrogram as a function of chromatographic retention time. It is obtained by passing the eluent from a chromatographic column into a mass spectrometer and measuring the total ion current or the ion current when the spectrometer is set to one or more *m/e* ratios.

27–17. (a) 0.94 mL. (b) 2.1 mL.

27–18. For methyl, octyl, and octyldecyl, (a) $(R_S)_{6,5} = 1.5, 3.9, 12$. (b) $(R_S)_{4,3} = 1.1, 1.7, 3.2$.

27–19. For methyl, octyl, and octyldecyl, (a) $k' = 6.7, 13.9, 21.4$. (b) $\alpha_{6,5} = 1.2, 1.4, 2.1$. (c) $\alpha_{4,3} = 1.1, 1.2, 1.3$. (d) $N = 1.5 \times 10^3, 2.5 \times 10^3, 6.6 \times 10^3$ plates. (e) $H = 0.017, 0.010, 0.005$ cm.

Chapter 28

28–1. For compounds 1, 2, and 3 respectively, (a) $R_F = 0.17, 0.26, 0.40$. (b) $N = 1.4 \times 10^2, 3.4 \times 10^2, 4.3 \times 10^2$ plates. (c) 0.012, 0.0073, 0.0090 cm.

28–2. (a) $k_1' = 4.8, k_2' = 2.9, k_3' = 1.5$. (b) $\alpha_{2,1} = 1.6, \alpha_{3,1} = 3.2, \alpha_{3,2} = 1.9$. (c) $(R_S)_{2,1} = 1.4, (R_S)_{3,1} = 3.3, (R_S)_{3,2} = 2.2$.

28–3. $k' = 2.3$ to 4.3.

Index

Page numbers in *italics* refer to illustrations; page numbers followed by *t* refer to tables.

Abbé refractometer(s), 384–385, *385*

Abscissa scale(s), for NMR spectra, *418*, 419–420

Absolute refractive index n_{vac}, 382

Absorbance, *A*, measurements of, 161, 161*t*, 171*t*
  with HPLC detectors, 794–795, *794–795*

Absorbing species, quantitative analysis of, 209

Absorption, effect of temperature on, in atomic spectroscopy, 258–259
  in laser action, *117*, 118
  X-ray methods of, 460–462, *461*, 481–482, *482*

Absorption analysis, cleaning of cells for, 210
  filters for, 198
  hollow cathode lamps for, 267–268, *267*
  procedures in, 209–213
  standard addition method in, 210–212, *212*
  wavelength selection for, 210

Absorption band, 108

Absorption characteristic(s), of aromatic compounds, 188–189, *189*, 189*t*

Absorption edge(s), 461

Absorption edge method, in X-rays, 481–482, *482*

Absorption filter(s), 123

Absorption measurement(s), effect of slit width on, 173–174, *173*
  instruments for, in ultraviolet, visible, and infrared regions, 193–207, 325–340
  quantitative aspects of, 162–175

Absorption of electromagnetic radiation, 105–109, *106–107*
  by atoms, 106 (*see also* Atomic absorption)
  by molecules, 106–108, *107* (*see also* Molecular absorption)
  relaxation processes in, 109

Absorption of ultraviolet and visible radiation, by inorganic species, 183–193

Absorption spectra, 106 (*see also* specific types)
  atomic, 254–255 (*see also* specific types)
  in ultraviolet-visible regions, 208
  methods of plotting, 189–190, *189–190*
  of bovine albumin, *215*

Absorption spectroscopy, 113, *113*, 160–181

Absorption spectroscopy (*continued*).
  molecular ultraviolet-visible, 182–224
  absorbing species in, 183–193
  molar absorptivity in, 182–183
  qualitative analysis by, 207–208, *208*, 208*t*
  quantitative analysis by, 208–215, *212–215*

Absorptivity, *a*, in absorption measurement, 161–162, 161*t*

Accumulator(s), definition of, 82

Acetaldehyde, effect of solvent on absorption spectra of, *208*

Acetone, as solvent for ultraviolet-visible regions, 208*t*

Acetylene flame(s), temperature of, 257*t*

Acid-base titration(s), 708–710, *709*

Acid error, glass electrode, 614, *614*, 630–631

Actinide ion(s), absorption by, 189–190

Activation analysis, 502, 511 (*see also* Neutron activation analysis)

Activity coefficient(s), f_M, 900–903, *901*, 902*t*

Adsorption chromatography, 815–817, *816*
  applications of, 817, 818*t*, *819*

I.1

Adsorption isotherm(s), 730

Alkaline error, glass electrode, 614, *614*, 630–631

Alkene, absorption characteristics of, 187*t*

Alkyne, absorption characteristics of, 187*t*

Alloy junction transistor(s), *34*

Alpha decay, 503–504

Alpha particle(s), measurement of, 511

Alphanumeric display(s), 64–65, *65*

Alternating current(s), 887–890, *888–889*
  in electrochemical cells, 572–573, *573*

Alternating current polarography, *683*, 688–690, *689*

Amici prism, 384, *384*

Amido group(s), absorption characteristics of, 187*t*

Amperometric titration(s), 697–700
  applications of, 698–700, 699*t*
  apparatus and techniques for, 697–698

Amperostat(s), 49, *49*, 655

Amperostatic coulometry, 651, 655–660, *656–657*, 658*t*, 659*t* (see also Coulometric titrations)

Amplification, of light, in lasers, 118, *119*
  of transducer signals, 43–48, *44–45*, *47*
  with bipolar transistors, 35–36, *35*

Amplifier(s), chopper, 70–72, *70–71*
  lock-in, 72–73, *72*
  operational, 39–54 (see also Operational amplifiers)

Amplitude *A*, of electromagnetic radiation, *91*, 92

Analog computer(s), 81

Analog filtering, for noise reduction, 69–70, *69*

Analog signal(s), 54–55

Analog-to-digital converter(s), 61, *62*

Analytical instrument(s), 3–5, *3*, *4t*

Analytical method(s), types of, 2–3, 2*t*

Analytical signal(s), 1, 2*t*

Analyzer(s), automatic elemental, 873–875, *874*
  automatic multipurpose, 872–873, *873*
  centrifugal fast scan, 871–872, *872*
  for electron spectroscopy, 494–495, *494*

Analyzer(s) (*continued*).
  for mass spectrometers, 531–536, *532–536*
  microprobe, 564
  secondary ion mass, 564

Anharmonic oscillator(s), 322–323

Anisotropic crystal(s), 386–390, *388–390*, 389*t*
  effect of, on polarized radiation, 392–393
  refractive index for, 389, 389*t*

Anisotropy, magnetic effect of, 421–423, *422*

Anode, of electrochemical cells, 568–569

Anodic wave(s), in polarography, 675–676, *675*

Anodic-cathodic wave(s), in polarography, 675–676, *676*

Anomalous dispersion, 100 (see also Dispersion)

Anthracene, excitation spectrum of, 237, *237*

Antilogarithm(s), of input signals, *51*, 52

Anti-Stokes shift(s), 364

Arc source(s), for emission spectroscopy, 292–293, 306–314
  instruments for, 308–310, *309–310*
  quantitative applications of, 312–314
  sample handling in, 306–307, *306*

Arc spectra, for emission spectroscopy, 307

Area normalization method, for chromatographic analysis, 751

Argon ion laser(s), 120

Argon plasma methods, of atomization, 251*t*

Argon plasma source(s), for emission spectroscopy, 294

Aromatic compound(s), absorption by, in the ultraviolet, 188–189, *189*, 189*t*

Aryton shunt(s), for electrical measurements, *881*, 882

Asymmetry potential, glass electrodes and, 613–614

Atomic absorption, 106

Atomic absorption method(s), radiation sources for, 266–268, *267*

Atomic absorption spectra, 254–255

Atomic absorption spectroscopy, 266–279
  analytical techniques for, 277–279
  application of, 279
  detection limits for, 286*t*, 304*t*

Atomic absorption spectroscopy (*continued*).
  instruments for, 268–270, *269*
  interferences affecting, 270–273, *271*, *273*
  role of organic solvents in, 278
  sequential spectrometer for, 298, *298*
  vs. atomic emission spectroscopy, 285–286

Atomic emission spectra, 254, *255*

Atomic emission spectroscopy, 279–286
  analytical techniques for, 284–285
  detection limits for, 286*t*, 304*t*
  instruments for, 280–281, *280*
  interferences in, 282–284, *283–285*
  multielement analysis in, instruments for, 281–282, *282*
  vs. atomic absorption spectroscopy, 285–286

Atomic fluorescence spectra, 255

Atomic fluorescence spectroscopy, 286–288
  application of, 288
  detection limits for, 286*t*, 304*t*
  instruments for, 286–287
  interferences in, 287–288
  plasma atomization in, 304–306, *305*

Atomic secondary ion mass spectrometry (SIMS), in surface analysis, 564

Atomic spectra, sources of, 251–256, *252–253*, *255*

Atomic spectroscopy, 250–291
  atomizers for, 261–266, *261–262*, *264–265*
  flame atomization in, 257–261
  methods for, 251*t*
  theory of, 251–257, *252–253*, *255–256*

Atomization, definition of, 250
  in ICP source, 295–296

Atomizer(s), for atomic spectroscopy, 261–266, *261–262*, *264–265*

Attenuation, in infrared absorbance spectroscopy, 327–328

Auger electron spectroscopy, 488*t*, 491–492, *492*
  applications of, 498–500, *499*
  depth profiling by, 499–500, *499–500*
  line scanning in, 500, *501*
  qualitative analysis by, 498–499

Auger microprobe(s), 494

Automatic titrator(s), 638–640, *638*

Automation, continuous flow analyzers in, 858–871

Automation (*continued*).
  of analyses, 851–879
    advantages and disadvantages of, 853
    based on multilayer films, 875–879, *876–878*, 878*t*
    (*see also* Multilayer films)
    instruments for, 852–853
  of handling solids, 855–857, *856*
  of instruments, in infrared absorption spectroscopy, 332
  of sampling, 853–858, *855–856*
Azo group(s), absorption characteristics of, 187*t*

Background correction method(s), in atomic absorption spectroscopy, 272–273, *273*
  in radiochemical analysis, 509–510
Band broadening, extra-column, in HPLC, 787
  in elution chromatography, 735–739, 736*t*, *738–739*
Band separation, thermodynamic variables affecting, 739–742, *741*
Band spectra, in atomic emission spectroscopy, 256–257
  correction for, 283, *284*
  interference from, 283, *284*
Bandwidth, of filters, 120–121, *121*
  of monochromators, 121, 132–133
Barrier-layer cell(s), 137–138, *137*
Base-line method, in quantitative infrared absorption analysis, 352, *352*
Base peak, of mass spectrum, 538, 543, *544*
Batch inlet system(s), for mass spectrometers, 525–526, *526*
Beam attenuation, in infrared spectrophotometers, 327–328
Beckman Acculab infrared spectrophotometer, 331
Beckman DU spectrophotometer, 200–201, *201*
Beer's law, 162–164, *163*
  application to mixtures, 164
  deviations from, 165–166, *166*, 166*t*
    in infrared absorption analysis, 351
  effect of stray radiation on, 167–168, *168*
  limitations to, 164–168, *166–168*
Bending vibration(s), 318
Benzene, as solvent for ultraviolet-visible regions, 208*t*
  effect of substitution on fluorescence of, 233*t*

Beta decay, in radiochemical analysis, 504
Beta particle(s), measurement of, 511
Binary counter(s), 58–60, *58–60*
Binary number system, 55
Bioluminescence, 244
Bipolar transistor(s), 33–36, *34–35*
Bit(s), definition of, 81
Black-body radiation, 109, *110*
Bode diagram(s), for filters, *897*, *898*
Bohr magneton B_n, 448
Bolometer(s), 328
Bonded-phase chromatography, 801–802, 808
  applications of, 812, *813*
Bovine albumin, absorption spectra of, *215*
Boxcar averaging, 74–75, *75*
Bragg equation, 464
Bragg's law, for X-ray diffraction, 463–464
Bremsstrahlung, 456–457
Bromine, amperometric titrations with, 700
Bus(es), computer, 84
Byte, definition of, 81

Calcium ion, liquid membrane electrode for, 618–619, 620*t*
Calibration, of quantitative mass spectral measurements, 560
Calibration curve(s), for direct potentiometric measurements, 629–630
  for quantitative polarography, 693–694
  in atomic absorption analysis, 279
  in size-exclusion chromatography, 827–828, *827*
  uncertainties in, 19–22, *19*, 21*t*
  with ICP source, 303–304, *303*
Calibration standard(s), for atomic absorption analysis, 278
  for chromatographic analysis, 750, *750*
  for quantitative X-ray analysis, 479
Calomel electrode(s), 601–602, 601*t*, *602*
Capacitance, in electronic circuits, 890–891, *890*
Capacitive reactance, X_C, 893–894
Capacitor(s), in electronic circuits, 890–893, *890*
Capacity factor, k', chromatographic, 740, 748*t*
Capacity factor(s), in thin-layer chromatography, 839–840

Capillary constants, for DME, 670
Capture cross section, for photons, 163
Carbon-13 NMR, 436, 444–447, *447–446*
Carbon dioxide, gas-sensing electrode for, 621–622, *621*, 622*t*
Carbonyl group(s), absorption characteristics of, 187*t*
Carboxyl group(s), absorption characteristics of, 187*t*
Carrier gas(es), for gas-liquid chromatography, 760, *760*
Cathode, of electrochemical cells, 568–569
Cathode-ray tube(s), 62–63, *63*
Cell(s), alternating currents in, and conductometric measurements, 706
  amperometric thin-layer, for HPLC, 797–798, *798*
  electrochemical, 567–573 (*see also* Electrochemical cells)
  for classical polarography, 665
  for conductance measurements, 707–708, 707, 708*t*
  for coulometric titrations, 657, *657*
  for fluorescence analysis, 238
  for polarographic analysis, 677–679, *678*
  for potentiostatic coulometry, 652, *653*
  ideal polarized and nonpolarized, 591–592, *592*
  irreversible, 573
  reversible, 573
  trapped ion analyzer, 540–541, *541*
Cell configuration, for absorbance measurement in HPLC, 793–794, *794*
Cell constant, determination of, for conductance measurements, 708, 708*t*
Cell-in/cell-out method, for quantitative infrared absorption analysis, 352
Cell potential(s), 573–574
  calculation of, 587–589, 588*t*
  effect of concentration on, 573–574
  effect of current on, 590–597, *592–593*, 596*t*
  instruments for measuring, 624–627, *626*
  with metallic and membrane indicator electrodes, 609–612
Central processing unit (CPU), 83–84

Centrifugal fast scan analyzer(s), 871–872, *872*
Charge-transfer absorption, 193
Charging current(s), 671
Chemical analysis, definition of, 1
Chemical instrumentation, micro-computers and microproces-sors in, 80–89
Chemical interference(s), in atomic absorption spectros-copy, 270, 273–277, *276t*, *277*
in atomic emission spectros-copy, 283
Chemical ionization mass spectra, 546, *547*
Chemical ionization source(s), for mass spectrometers, 528t, 530–531
Chemical shift(s), for carbon-13, 445–446, *446*
in X-ray photoelectron spec-troscopy, 496–498, *496t*
proton, in NMR spectroscopy, 417, 420–423, *421–422*, *423t*, *424*
source of, 418–419
Chemical shift anisotropy, 446
Chemical shift reagent(s), in NMR spectroscopy, 429–431
Chemiluminescence, analytical applications of, 245–246
measurement of, 245
Chemiluminescence spectroscopy, 225–226, 244–246
Chemistry, electroanalytical, 567–599 (see also Electroanalytical chemistry)
Chopper amplifier(s), for noise reduction, 70–72, *70–71*
Chopping, signal, 70–72, *70–71*
Chromatogram(s), 731
for gas-liquid and gas-solid equilibria, 781, *781t*
molecular sieve, 779, *780*, *781t*
Chromatographic analysis, appli-cations of, 747–751, *750*
Chromatographic inlet system(s), for mass spectrometers, 526–527, *527*
Chromatographic separation(s), 727–756
gas, *780*, *781t*
Chromatography, 727–756 (see also specific types)
adsorption, 815–817, *816*
automated, 857–858
band broadening in, 735–739, *736t*, *738–739*
computerized, 751–753, *752–753*
gas-liquid, 757–783 (see also Gas-liquid chromatography)

Chromatography (*continued*).
general elution problem in, 746, *747*
ion-exchange, 817–825 (see also Ion-exchange chromatogra-phy)
optimization of column per-formance in, 743–746, *745*
paper, 847
partition, 801–815 (see also Partition chromatography)
planar, 837–850 (see also Thin-layer chromatography)
rate theory of, 733–747
types of, 728–733, *729–732*, 729t, 748t (see also specific types, e.g., Linear chroma-tography, Column chroma-tography)
Chromophore(s), conjugation of, effect of, 186–188, *188t*
in molecular ultraviolet-visible absorption spectroscopy, 183–184
organic, 186, *187t*, *188t*
Circuit(s), flip-flop, 58–60
operational amplifier, 41–43, *42* (see also specific types, e.g., Feedback, Voltage-follower)
RC, 890–899 (see also *RC* cir-cuits)
switching, 52–54, *53*
Circular birefringence, *(n_l − n_d)*, 395
wavelength dependence of, 399
Circular dichroism, 399–404
applications of, 403–404
curves for, 400–401, *401*
instrumentation for, 403, *403*
Circular double refraction, 394–396, 395
Circularly polarized light, *392*, *393*
Circularly polarized radiation, 392–393, *392*
Coefficient of variation, *CV*, 8
Coherence, of electromagnetic radiation, 99
Coherent anti-Stokes Raman spectroscopy (CARS), applica-tions of, 377
Coincidence counter, 511
Collector, in radiochemical analy-sis, 515–516
Collision product peak(s), 545t, 546
Colorimeter, definition of, 145
Column(s), efficiency of, for HPLC, 786–787, *786–787*
for HPLC, 791–792, *792*
gas-liquid chromatographic, 761–764, *764t*

Column chromatography, 728, 730–731, *730*
Column efficiency, chromato-graphic, 733–735, *734*
Column packing, in size-exclusion chromatogaphy, 825–826, 826t
types of, for HPLC, 792–793
Column performance, chromato-graphic, optimization of, 743–746, *745*
Column resolution, *R_S*, chromato-graphic, 740–742, *741*, 748t
Common terminal, of bipolar transistor, 34
Complex formation, effect of, on polarographic waves, 673–674, *673t*
Complex-formation titration(s), conductometric, 710–711
coulometric, 658–659, *659t*
potentiometric, 634–636, *635*
Complex spectra, aids in analysis of, 429–431
Compound(s), aromatic, absorp-tion by, in the ultraviolet region, 188–189, *189t,189t*
for standard solutions of com-mon elements, 907–908
identification of, in NMR spec-troscopy, 440–442, *441–443*
Compton effect, 504–505
Computer(s), analog vs. digital, 81
applications of, 88–89, *88*
components of, 83–87, *84–86*
hardware and software for, 82
input/output systems of, 87
operational modes of, 82–83, *82*
Computer memory, 84–87, *85–86*
Computer programming, 87–88, *88*
Computer search system(s), for infrared spectra, 350–351
for mass spectra, 554–556, *556*
Computer terminology, 81–82
Concave grating(s), 126
Concentration, effect of, on fluo-rescent intensity, 235
Concentration polarization, 593–595
Condenser current(s), 671
Conductance, *G*, 704
applications of direct measure-ments of, 711
measurements of, 706–708, *707*, 708t
Conduction, in electrochemical cells, 568
Conductometric method(s), 704–712

Conductometric titration(s), 708–711, *709*

Confidence interval(s), 9–13, 11*t*, 13*t*
for counts, in radiochemical analysis, 508–509, *509*

Constant-current electrolysis, 648

Constant-current source(s), 49, *49*

Constant-potential coulometry, 644

Constant-potential source(s), 48–49, *49*

Constant-voltage source(s), 48–49

Constructive interference, 95

Continuous flow analyzer(s), automated, 858–871

Continuous source(s), in optical spectroscopy, 115

Continuous-source correction method, in atomic absorption spectroscopy, 271–272, *271*

Continuous spectra, 109

Continuous wave nuclear magnetic resonance (NMR) spectrometer(s), 431, 432–436, *432, 434–436*

Controlled cathode potential electrolysis, 649–650, *649–650*

Converter(s) (*see* specific types)

Coolidge tube(s), 464–465, *465*

Cornu prism, 127–128, *127*

Correlation chart(s), 344, *345*
limitations to, 346–350

Correlation method(s), for processing data, 76

Correlation table(s), for chemical shifts, in NMR spectroscopy, 423, *423t*

Cotton effect, 399, *400*

Cottrell equation, 669–670

Coulomb, *C*, 650

Coulometric analysis, 644–663
(*see also* Potentiostatic, Amperostatic coulometry)
methods in, 651–652

Coulometric titration(s), 644, 651, 655–660, *656–657*, 658*t*, 659*t*
applications of, 658–660, 658*t*, 659*t*
automatic, 660
external generation of reagents in, 657, *657*
vs. volumetric titrations, 659–660

Coulometric titrator(s), 656–657, *656–657*

Coulometry, potentiostatic, 652–655, *653–655* (*see also* Potentiostatic coulometry)

Count(s), confidence intervals for, in radiochemical analysis, 508–509, *509*

Counter(s), 57–60, *57–60* (*see also* specific types, e.g., Binary, Decimal)
for X-ray spectrometers, 473
well-type scintillation, 511, *512*

Counting statistic(s), in radiochemical analysis, 506–510, 506*t*, *507, 509*

Coupling constant, 425*t*

Critical angle θ_c, 383–384, *384*

Critical angle refractometer(s), 383–384, *384*

Cross-coil detectors, in NMR spectrometers, 432, 433–435, *434*

Crystal(s), diffraction of X-rays by, 468*t* (*see also* Bragg equation)

Crystalline compound(s), identification of, by X-ray diffraction, 482–484, *483*

Crystalline membrane electrode(s), 615–617, 617*t*

Crystal-field theory, 190

Curie, 510

Current(s), alternating, 887–890, *888–889*
in electrochemical cells, 572–573, *573*
changes in, during electrolysis, 646, *647*
diffusion, 667, *667* (*see also* Diffusion currents)
direct, in electrochemical cells, 571–572, *572*
effect of, on cell potentials, 590–597, *592–593*, 596*t*
faradaic, 572
in polarographic waves, 667–672, *669, 671* (*see also* Polarographic waves)
limiting, 667, *667* (*see also* Limiting currents)
measurement of, 43–45, *44*
nonfaradaic, 572–573, *573*
residual, 671, *671* (*see also* Residual currents)
sine-wave, 888–890, *888–889*

Current maxima, in polarograms, 671–672, *671*

Current rectifier(s), semiconductor, 32

Current-sampled polarography, 681–684, *682–683*

Current-voltage curve(s), in electroanalytical chemistry, 591

Current-voltage relationship(s), during electrolysis, 644–650, *647–650*

Cut-off filter(s), 123

Cyclic voltammetry, *683*, 687–688, *687*

Cyclohexane, as solvent for ultraviolet-visible regions, 208*t*

D'Arsonval meter(s), for electrical measurements, 881–882, *881–884*

D'Arsonval movement, 880

dc plasma (DCP) source, for emission spectroscopy, 294, 296–297, *297*

dc potential(s), in electrochemical cells, 572–573

Deactivation, of excited molecular states, 228–231

Dead time correction method(s), in radiochemical analysis, 510

Debye-Hückel equation, 902–903

Debye-Scherrer powder camera, 483, *483*

Decay, radioactive (*see* Radioactivity)

Decay constant, 505

Decimal counter(s), 60, *60*

Densitometer(s), for thin-layer chromatography, 844

Depolarization ratio, 370

Derivative curve(s), vs. transmittance curves, *214*

Derivative spectrophotometry, 213–215, *214–215*

Desorption ionization, 528, 528*t*

Desorption source(s), field, for mass spectrometers, 528*t*, 529–530

Destructive interference, 95

Detection coefficient, 505

Detection limit, of an instrument, 23–24

Detector(s), for continuous flow analysis, 861–862, *862–863*
for electron spectroscopy, 495
for fluorometers, 238
for gas-liquid chromatography, 765–769, *766–768*
for HPLC, 793–799, 793*t*, *794–798*
for mass spectrometers, 536–537, *537*
for NMR spectrometers, 433–435, *434–435*
for supercritical-fluid chromatography, 832
for X-rays, 468–473, *469, 472*
heat, 143
radiation, for optical spectroscopy, 135–143 (*see also* specific types)
selective, for gas-liquid chromatography, 775–779, *777–779*
semiconductor, 471–472, *472*
thermal for infrared radiation, 328–329, *329*

Detection limit(s), for atomic absorption spectroscopy, 279, 280*t*

Deuterium lamp(s), 194

Diamagnetic shielding, 421, *421*

Dielectric, 890

Diethyl ether, as solvent for ultra-violet-visible regions, 208*t*

Difference amplifier(s), for reduction of transducer noise, 69

Differential scanning calorimetry, 716–721, *719–720*

Differential thermal analysis, 716–721, *716*, *718*, *720–721*

Differential thermogram(s), 716–717, *716*, *718*, *720–721*

Diffraction, of electromagnetic radiation, 96–99, *97–98*
of X-rays, 463–464, *463*
patterns of, interpretation of, 484
X-ray, methods of, 482–484, *483*

Diffusion, longitudinal, 736–737, 736*t*, *739*

Diffusion current(s), determination of, 693, *693*

Diffusion currents, effects of capillary characteristics and temperature on, 670–671
equation for, 668
in polarographic waves, 667, *667*

Diffusion gradient, 669

Digital computer(s), 81

Digital electronic(s), 54–62

Digital filtering, 75–76

Digital signal(s), 54–55

Digital-to-analog converter(s), 61

Dilution method(s), for quantitative X-ray analysis, 479

Diode(s), 32

1,4-Dioxane, as solvent for ultra-violet-visible regions, 208*t*

Dipolar decoupling, 446

Dipole change(s), during molecular rotation and vibrations, 317–318, 318*t*

Direct absorption method(s), in X-ray analyses, 481

Direct current(s), in electrochemical cells, 571–572, *572*

Direct injection enthalpimetry, 721, 722, *723*

Direct potentiometric measurement(s), 627–631, *628*

Direct probe inlet(s), for mass spectrometers, 526

Dispersion, in FIA, 866–869, *867–870*
of electromagnetic radiation, 100–101, *100*

Dispersion (*continued*).
of monochromator(s), grating, 129
prism, 129–130, *130–131*

Dissociation, of excited molecules, 230

Dissociation reaction(s), in flames, 274–275

Dissolved oxygen, effect of, on fluorescence, 235

Distribution coefficient, in size-exclusion chromatography, 826–827

DME (*see* Dropping mercury electrode)

Doping, of semiconductor crystals, 31

Doppler broadening, 256

Double-beam instrument(s), for absorption measurements, 175–177, *176*
for Fourier transform spectroscopy, 335, *335*
for ultraviolet-visible regions, 202–203, *202–203*
vs. single-beam instruments, 177–178

Double-beam photometer, for absorption spectroscopy, 196, *197*

Double-beam spectrophotometer(s), for atomic absorption spectroscopy, 269–270, *269*
for infrared absorption spectroscopy, 330, *330*
multichannel, 205–207, *206*

Double bond region, in infrared spectrum, 346

Double dispersing spectrophotometer(s), 203, *204*

Double-focusing mass spectrometer(s), 533–534, *533–534*, 543*t*

Double resonance technique(s), in NMR spectroscopy, 429

Doublet state, definition of, 227

Dropping mercury electrode (DME), 664
advantages and disadvantages of, 677
current variations with, 665–666, *666*
diffusion currents at, 670
in polarography, 678–679, *678*, *680*

Dual-channel instrument(s), for absorption measurements, 176–177, *176*

Dual wavelength spectrophotometry, 214–215, *214–215*

Duane-Hunt law, 457

Dye laser(s), 120

Echelle grating, 127

Echelle grating monochromator(s), in atomic emission spectroscopy, 282
in emission spectroscopy, 300–302, *300–302*, 301*t*

Echellette grating, 125–126, *125*

Eddy diffusion, zone broadening by, 738

Electric arc method(s), of atomization, 251*t*

Electric spark method(s), of atomization, 251*t*

Electrical conductance, in membrane cell(s), 608–609, *608*
in solutions of electrolytes, 704–706, 705*t*, 706*t*

Electrical double layer, in non-faradaic currents, 572, *573*

Electrical force(s), effect of, on ionic mobility, 705

Electrical measurement(s), instruments for, 880–886

Electricity, units for quantity of, in coulometric analysis, 650–651

Electroanalytical chemistry, 567–599

Electrochemical cell(s), 567–573
components of, 568–571, *568*, *570*
conduction in, 568
direct currents in, 571–572, *572*
irreversibility of, 573
potential of, 573–574
reversibility of, 573
schematic representation of, 570–571

Electrochemical detector(s), for HPLC, 797–798, *797–798*

Electrochromatography, 848–849, *849*

Electrode(s), calomel, 601–602, 601*t*, *602* (*see also* Calomel electrodes)
crystalline membrane, 615–617, 617*t*
enzyme, 622–624, *623–624*
gas-sensing membrane, 619–622, *621*, 622*t*
glass, 612–614, *612*, *614* (*see also* Glass electrodes)
hydrogen-gas, 575–576, *576*
indicator, 600
in electrochemical cells, 568–571, *568*, *570*
liquid membrane, 618–619, *618*, 620*t*
membrane indicator, 605–624 (*see also* Membrane indicator electrodes)
metallic indicator, 603–605 (*see*

Electrode(s) (*continued*).
   also Metallic indicator electrodes)
   polarization in, 591–597, *592–593*, *596t*
   reference, 600–603, *601t*, *602* (*see also* Reference electrodes)
   rotating platinum, and voltammetric waves, 698
   silver/silver chloride, 602
   standard hydrogen, 575–576, *576*
   thallium/thallous chloride, 602
   working, 651
Electrode calibration method(s), for direct potentiometric measurements, 627–629, *628*
Electrodeless discharge lamp(s), for atomic fluorescence analysis, 268
Electrode potential(s), calculation of, 587–590, *588t*
   effect of competing equilibria on, 586
   effect of concentration on, 580–581
   effect of precipitation and amplexation reagents on, 584–585
   IUPAC sign convention for, 579–580
   limitations to use of, 585–587
   measurement of, 576–579, *578*
   of electrochemical cells, 575–587
   reaction rates and, 586–587
   sign convention for, 579–580
   standard, 581–583, *582t*
   standard and formal, 904–906
Electrogravimetric analysis, 644
Electrolysis, at constant working electrode potentials, 648–650, *648–650*
   controlled cathode potential, 649–650, *649–650*
   constant-current, 648
   current-voltage relationships during, 644–650, *647–650*
Electrolytic cell(s), 567 (*see also* Electrochemical cells)
Electromagnetic radiation, absorption of, 105–109, *106–107* (*see also* Absorption of electromagnetic radiation)
   coherent, 99
   diffraction of, 96–99, *97–98*
   dispersion of, 100–101, *100*
   emission of, 109–111, *110*
   interaction with matter, 90–111
   monochromatic, 91, *91*
      diffraction of, *98*
   polarization of, 103–104, *103*

Electromagnetic radiation (*continued*).
   photoelectric effect of, 105
   quantum-mechanical properties of, 104–111, *106–107, 110*
   reflection of, 101–102
   refraction of, 101, *101*
   scattering of, 102
   transmission and refraction of, 99–101, *100–101*
   wave properties of, 91–95, *91–94*
Electromagnetic spectrum, 95–96, *95*, *96t*
Electromagnetic wave(s), superposition of, 93–95, *93–94*
Electrometer, definition of, 45
Electron(s), in absorbing species, 183–193
Electron beam source(s), for X-rays, 457–459
Electron capture, in X-rays, 459
Electron-capture detector(s), for gas-liquid chromatography, 765, 768–769
Electron gun(s), 494
Electron impact mass spectra, 542–546, *543t*, *544*, *545t*
Electron impact source(s), for mass spectrometers, 528–529, *528t*, *529*
   reactions in, *545t*
Electron microprobe, in X-ray spectroscopy, 484–485, *485*
Electron multiplier(s), for mass spectrometers, 536
Electron spectrometer(s), 492–495, *493–494*
   sample holders in, 494
   sources in, 493–494
Electron spectroscopy, 487–501
   applications of, 495–500
   instrumentation in, 492–495, *493–494*
   methods of, *488t*
   principles of, 488–492, *489–490, 492*
   quantitative analysis in, 498
Electron spin, 226–227
Electron spin resonance (ESR), and absorption, 108–109
Electron spin resonance (ESR) spectroscopy, 447–450, *449–450*
Electronic(s), digital, 54–62
Electronic(s), elementary, 30–79 (*see also* specific devices, e.g., Semiconductors, Transistors)
Electronic counter(s), 55–60, *55t*, *57–60*
Electronic transition(s), induced by absorption of radiation, 185–193

Electrophoresis, 847–848
Electrophoretic effect, on ionic mobility, 705
Electrostatic force(s), mass transport by, 595
Electrothermal atomization, 250–291 (*see also* Atomic spectroscopy)
Electrothermal atomizer(s), 263–266, *264–265*
Electrothermal method(s), of atomization, *251t*
Elemental analysis, by ICP/MS, 562–563
   by spark source mass spectrometry, 561–562
Elliptically polarized radiation, 393, *393*
Eluent(s), in column chromatography, 730
Eluent strength, ε^0, 816
Elution, 730, *730*
   in thin-layer chromatography, 844
Elution chromatography, 730, *730*
   theories of, 731–733, *731–732*
Emission, effect of temperature on, in atomic spectroscopy, 257–258
Emission method, 474
Emission of electromagnetic radiation, 109–111, *110* (*see also* Atomic emission spectroscopy, Emission spectroscopy, Fluorescence spectroscopy)
Emission spectra, atomic, 254, *255* (*see also* specific types)
   in fluorescence spectroscopy, 235–236, *236*
   wavelengths for, *458t*
Emission spectroscopy, 113, *113*, 292–314 (*see also* Atomic emission spectroscopy)
   arc and spark sources for, 306–314 (*see also* Arc sources, Spark sources)
   instruments for, 297–303, *298–302*, *301t*
   plasma sources for, 294–304
   sources for, qualitative applications of, 310–311, *312*
   spectra, sources of, 293
End point(s), potentiometric, 632–634, *632, 633t*
Energy dispersive instrument(s), for X-ray fluorescence, 464, 475–476, *476, 478*
Energy level diagram, for photoluminescent molecules, 227–228, *228*
   in atomic spectroscopy, 251–254, *252–253, 255*

Enhancement effect, 479
Ensemble averaging, 73–74, *73–74*
Enthalpimetric titration(s) (*see* Thermometric titrations)
Environmental noise, 68, *68*
Enzyme electrode(s), 622–624, *623–624*
Equivalent conductance, 704–706, 705*t*, 706*t*
  effect of concentration on, 705–706, 705*t*
ESCA spectroscopy (*see* X-Ray photoelectron spectroscopy)
Ethanol, as solvent for ultraviolet-visible regions, 208*t*
  NMR spectra of, 417
Excitation spectra, in fluorescence spectroscopy, 226–228, *228*, 235–236, *236*
  in phosphorescence spectroscopy, 226–228, *228*
Excited molecular state(s), 105, 226–228, *228*
  deactivation of, 228–231
Exclusion limit, 827
External conversion(s), deactivation by, 230
External lock system, for NMR spectrometers, 433
Extraordinary beam, 388

Faradaic current(s), 572
  in ac polarography, 688
Faraday, *F*, 650
Faraday cup, in mass spectrometers, 536
Far-infrared region, 355
Fast atom bombardment source(s), for mass spectrometers, 528*t*, 531
Fast linear-sweep polarography, *683*, 686–687, *687*
Feedback circuit(s), in operational amplifiers, 41–43, *42–43*
FIA (*see* Flow injection analysis)
Fiber optics, for optical spectroscopy, 144–145, *145*
Field desorption source(s), for mass spectrometers, 528*t*, 529–530
Field effect transistor(s) (FET), 36–37, *36*
Field ionization source(s), for mass spectrometers, 528*t*, 529–530
Field ionization spectra, 546–547, *548*
Field sweep generator, NMR spectrometers, *432*, 434

Filter(s), for absorption analysis, 120–123, *121–123*, 198
  for current fluctuations, 38, *38*
  for electrical signals, 896–898, *896–897*
  for X-ray beams, 465–466, *466*
  in fluorometers, 238
  in infrared absorption spectroscopy, 326–327
  interference, 121–122, *121–122*
Filter photometer(s), infrared, 337–338, *338*
Filter wedge(s), 327
Fingerprint region, in infrared spectrum, 346
First-order indicator electrode(s), 603–604
First-order nuclear magnetic resonance (NMR) spectra, rules governing, 426–428, 426*t*, *427*
Flame(s), dissociation reactions in, 274–275
  ionization in, 275–276, 276*t*, 277
  structures of, 259–261, *259–260*
  temperature of, in atomic spectroscopy, 257–259, 257*t*
Flame absorption analysis, accuracy of, 279
Flame absorption spectra, 256–257, *256*
Flame atomization, in atomic spectroscopy, 257–261
Flame atomizer(s), 261–263, *261–262*
Flame emission spectra, 254, *255*, 256–257
Flame emission spectroscopy, 279–286 (*see also* Atomic emission spectroscopy)
Flame ionization detector(s), for gas-liquid chromatography, 765, 767–768, *768*
Flame method(s), of atomization, 251*t*
Flame profile(s), absorbance, 260, *260*
  emission, 260–261, *260*
  temperature, 259–260, *259*
Flame spectroscopy, 250–291 (*see also* Atomic spectroscopy)
Flicker (1/*f*) noise, 67
  effect of, on measurement of transmittance, 172
Flip-flop circuit(s), 58–60
Flow analyzer(s), automated continuous, 858–871
  automated segmented, 858–862, *859–862*

Flow injection analysis (FIA), 858, 862–871
  dispersion in, 866–869, *867–870*
  instrumentation for, 864–866, *865–866*
  principles of, 866
Flow injection titration(s), 870–871, *871*
Fluorescence, deactivation processes in, 228–231
  effect of structural rigidity in, 233–234
  effect of temperature on, in atomic spectroscopy, 258–259
  factors affecting, 233–235
  in electromagnetic radiation, 110–111
  structures of compounds in, 232–233, 233*t*
  theory of, 226–236
  transition types in, 231–232
  X-ray, 462–463
    methods of, 474–481 (*see also* X-Ray fluorescence)
Fluorescence detector(s), for HPLC, 796
Fluorescence spectra, atomic, 255 (*see also* specific types)
Fluorescence spectroscopy, 113–114, *113*, 225–244 (*see also* Fluorescence)
  application to inorganic species by, 242, 243*t*
  application to organic species by, 242–243
  instruments for measuring, 237–241
  methods of, application of, 241–244, 243*t*
  plasma atomization in, 304–306, *305*
Fluorescent emission, energy diagram for, *368*
Fluorescent intensity, effect of concentration on, 235
Fluorescent source(s), secondary, in X-ray spectroscopy, 465
Fluorescent spectra, in X-rays, 459
Fluoride electrode(s), 616
Fluorine-19, NMR of, 451
Fluorometer(s), 236–237
  components of, *236*
  definition of, 146
  designs of, 238–239, *239*
Formal potential *E'*, 586
Fourier transform optical spectrometer(s), 157
  resolution of, 156

Fourier transform spectrometer(s), advantages of, 336
for far-infrared region, 355
for infrared absorption spectroscopy, 332–336, *332–335*
in mass spectrometry, 538–542, *540–542*, 543t
NMR, 431–432, 436–439, *437–439*
performance characteristics of, 336
Fourier transform spectroscopy, instruments for, 149–150
Fourier transformation(s), 95
of interferograms, 154–156, *155*
Forward biasing, 33
Fragmentation pattern(s), in mass spectrometry, 544, 545t
structural information from, 552–554, *553*
Free induction decay signal, in Fourier transform NMR, 437–438, *438*
Frequency, ν, of electromagnetic radiation, 91
Frequency lock system, for NMR spectrometers, 437
Fresnel rhomb, 402–403, *403*
Frictional force(s), effect of, on ionic mobility, 705
Fuel regulator(s), for flame spectroscopy, 262–263
Functional group analysis, in ultraviolet-visible regions, 208
Furnace(s), for thermogravimetric analysis, 714

Galvanic cell(s), 567, *568* (see also Electrochemical cells)
Galvanometer(s), for electrical measurements, 880–884, *881, 883–884*
Gamma ray(s), emission of, in radiochemical analysis, 504
measurement of, 511
Gamma ray spectrometer(s), 511
Gamma ray spectrum, *512*
Gas(es), chemiluminescent analysis of, 245–246
emission of electromagnetic radiation by, 109–110
Gas chromatography, 728, 729t
Gas-filled detector(s), for X-rays, 469–470, *469*
Gas-liquid chromatography, 757–783
applications of, 774–779, *777–779*, 781, 781t
comparison with HPLC, 830, 830t
instruments for, 759–769

Gas-liquid chromatography (*continued*).
interfacing of, with infrared spectroscopy, 777–779, *779*
polarity parameters in, 769–772, *770, 770t, 771t, 772t*
selective detectors for, 775–779, *777–779*
stationary phase of, 769–774
with mass spectrometry, 776–777, *777–778*
Gas-sensing membrane electrode(s), 619–622, *621, 622t*
Gas-solid chromatography, 757, 779, 781, *780, 781t*
application of, 781, 781t
Geiger tube(s), for X-ray detection, 469–470
General elution problem, in chromatography, 746, *747*
Germanium detector(s), for X-rays, 471–472, *472*
Germanium semiconductor(s), properties of, 31–32
Glass electrode(s), composition of, 613
for univalent cations, 614, 615t
hygroscopicity of, 613
pH measurements with, 612–614, *612, 614*
potentiometric pH measurements with, 630–631
Globar(s), 326
Golay detector(s), 329
Goniometer(s), 466
Gradient elution, 789, *789*
Grating(s), reflection, for infrared absorption spectroscopy, 327
Grating monochromator(s), 124–127, *124–125* (see also specific types, e.g., Echellette grating, Concave grating, Holographic grating)
in spectrofluorometers, 238
Ground connection(s), in amplifier circuits, 41
Ground state, 105
Grounding, for noise reduction, 69
Group frequencies, in infrared absorption spectroscopy, 343

Hadamard transform spectrometer(s), 332
Half-cell potential(s), 575–587 (see also Electrode potentials)
calculation of, from E^0 values, 583–584
Half-shadow devices, for polarimeters, 398

Half-wave potential, in polarographic analysis, 667, *667*, 672–674, 673t
Harmonic oscillator(s), 318–323, *319*
Heat detector(s), for infrared radiation, 143
Helium/neon laser(s), 120
High-performance liquid chromatography (HPLC), 784–836 (see also specific types, e.g., Partition chromatography, Size-exclusion chromatography)
applications of, *785*
column efficiency for, 786–787, *786–787*
columns for, 791–792, *792*
comparison with gas-liquid chromatography, 830, 830t
detectors for, 793–799, 793t, *794–798*
extra-column band broadening in, 787
instruments for, 788–799
mobile phases of, 799–801, *801, 802t*
High-resolution nuclear magnetic resonance spectrometer(s), 431–432
High-resolution spectra, 416–417, *417*
Hole(s), in semiconductors, 31
Hollow cathode lamp(s), as sources for atomic absorption analysis, 267–268, *267*
Holographic grating(s), 127
HPLC (see High-performance liquid chromatography)
Hydride generation technique(s), in atomic absorption analysis, 278–279
Hydrogen flame(s), temperature of, 257t
Hydrogen-gas electrode(s), 575–576, *576*
Hydrogen stretching region, in infrared spectrum, 344
Hyperfine splitting, 449, *450*

Ilkovic equation, 670
Image displacement refractometer(s), 385
Impedance, Z, in RC circuits, 894–896, *894*
Incandescent wire source, 326
Incoherence, of electromagnetic radiation, 99
Indeterminate uncertainties, distribution of, 7–9, *7*
Indicator electrode(s), 600, 603–624

Inductively coupled plasma (ICP) source, for emission spectroscopy, 294–296, *294–296*
in emission spectroscopy, calibration curves with, 303–304, *303*
detection limits for, 304*t*
sequential spectrometer for, 298, *298*
Infrared absorption spectroscopy, 315–361
continuous radiation sources for, 325–326
dispersive instruments for, 330–332, *330–331*
instruments for, 325–340
nondispersive instruments for, 337–339, *338*
photoacoustic instruments for, 336–337
qualitative applications of, 343–351
quantitative analysis by, 339–340, *339*
quantitative applications of, 351–355, *352–353*, 354*t*
sample handling in, 340–343
theory of, 316–325
Infrared detector(s), 795–796
Infrared emission spectroscopy, 355
Infrared radiation, continuous sources of, 325–326
detectors for, 328–329, *329*
Infrared region, instruments for absorption measurement in, 193–207
Infrared spectra, collections of, 350
comparison with Raman spectra, 362–363, 367–370, *369*
computer search systems for, 350–351
Infrared spectral region(s), 344–350, *348–349*
Infrared spectroscopy, with gas-liquid chromatography, 777–779, *779*
Infrared spectrum, regions of, 316*t*
Inlet system(s), for mass spectrometers, 525–527, *526–527*
Inorganic analysis, polarographic, 694
Inorganic anion(s), absorption in the ultraviolet region by, 189
Inorganic species, chemiluminescent analysis of, 246
fluorometric determination of, 242, 243*t*
Raman spectra of, 373–374
Input/output system(s), for computers, 87

Input transducer(s), 4, 4*t*
Instrument(s), sensitivity and detection limit for, 22–24 (*see also* specific types, e.g., Single-beam instruments, Double-beam instruments)
Instrumental measurement(s), precision and accuracy in, 5–6
types of errors in, 6
uncertainties in, 5–22
Instrumental methods, 2–3, 2*t*
Instrumental noise, 66–68, *68*
effect on spectrophotometric analyses, 168–173, 170*t*–171*t*, *171–172*
Integrator(s), for potentiostatic coulometry, 654
Intensity, *I*, of electromagnetic radiation, 92
Interconversion, between systems, 55–56, 55*t*
Interference filter(s), 121–122, *121–122*
for infrared analysis, 326–327
Interference wedge(s), 122–123
Interferogram(s), 153
Fourier transformation of, 154–156, *155*
Interferometer(s), for infrared absorption spectroscopy, 332–336, *332–334*
for measuring refractive index, 383
Michelson, 152–154, *153*
Internal conversion, deactivation by, 229–230
Internal lock system, for NMR spectrometers, 433
Internal-reflectance spectroscopy, 342–343, *343*
Internal standard(s), for quantitative X-ray analysis, 479
in emission spectroscopy, 312–314
Internal standard method, for chromatographic analysis, 750–751
Intersystem crossing, deactivation by, 230
Inverting and noninverting terminals, in amplifiers, 41
Iodine, amperometric titrations with, 700
Ion-exchange chromatography, 817–825
applications of, 821–824, *823–825*
equilibria in, 819–820
packings in, 820–821, *820*
Ion cyclotron resonance spectrometer(s), 538–540, *540*
Ion gun(s), for mass spectrometers, 528

Ionization, desorption, 528, 528*t*
in flames, 275–276, 276*t*, 277
in ICP source, 295–296
Ionization chamber(s), for X-ray detection, 469, 471
Ionization source(s), chemical, for mass spectrometers, 528*t*, 530–531
field, for mass spectrometers, 528*t*, 529–530
Ionization suppressor(s), in atomic absorption spectroscopy, 276
Ion-pair chromatography, 813–815, 815*t*
Ion-selective membrane electrode(s), principle and design of, 606–609, *607–608*
properties of, 606, *606*
Ion-sensitive photoplate(s), for mass spectrometers, 536–537
Ion source(s), for molecular studies, 527–531, 528*t*, *529*
Ion trap detector, 776, *777*
IR drop, of electrochemical cells, 590–591
Irreversible cell(s), 573
Irreversible reaction(s), polarographic waves for, 674
Isocratic elution, 789
Isotope(s), radioactive, 502–510, 503*t*, 506*t*, *507*, *509* (*see also* Radioactive isotopes)
Naturally occurring, 545*t*
Isotope abundance(s), measurements of, 563
with mass spectrometry, 551*t*
Isotope dilution, 502
methods of, 517–520
Isotope peak(s), of mass spectra, 544–546, *544*, 545*t*
Isotope ratio(s), determination of molecular formulas from, in mass spectrometry, 550–552, 551*t*
Isotope substitution, in NMR spectroscopy, 429
IUPAC sign convention, for electrode potentials, 579–580

Jet separator(s), 527, *527*
Johnson noise, 67
Junction potential(s), 569–570

K capture, in X-rays, 459–460
K series, of X-rays, 458
Katharometer, 766
Kimble BUN analyzer, 623–624, *623*
Kinetic process(es), in chromatographic peak broadening, 736*t*

Kinetic theory, of elution chromatography, 732, 735–739, 736t, 738–739 (see also Rate theory)

L section filter(s), 38
L series, of X-rays, 458
Laboratory recorder(s), 63–64, 64
Laminar flow burner(s), for atomic spectroscopy, 261–262, 262
Lanthanide ion(s), absorption by, 189
Larmor frequency, 412
Laser(s) as sources in optical spectroscopy, 115–120, 116–117, 119
  carbon dioxide, as source, in infrared absorption spectroscopy, 326
  excitation, as source for photoluminescence, 238
Laser excitation, in emission spectroscopy, 308
Laser-fringe reference system, in Fourier transform spectrometers, 334, 334
Laser microprobe(s), 308
Laser microprobe mass spectrometry, in surface analysis, 564
Laser system(s), three- and four-level, 118–119, 119
Lattice, 414
Ligand(s), effect on absorption maxima, for transition-metal ions, 192t
Ligand field strength, 192, 192
Light amplification in lasers, 118, 119
Light pipe(s), for gas-liquid chromatographic instruments, 777–779, 779
Limiting current(s), in polarographic waves, 667, 667
Line broadening, in NMR spectra, 415
Line scanning, in Auger electron spectroscopy, 500, 501
Line source(s), in optical spectroscopy, 115
Line width(s), in atomic spectra, 255–256
Linear chromatography, 728–730, 729, 748t
Lippich prism, 398
Liquid(s), for infrared absorption analysis, 341–342, 341
Liquid chromatography, 728, 729t
  high-performance, 784–836 (see also High-performance liquid chromatography HPLC)
Liquid junction(s), 569–570, 570

Liquid junction potential(s), 589–590, 590
Liquid-liquid chromatography, 801–802, 808
Lithium-drifted silicon detector(s), for X-rays, 471, 472, 472
Littrow prism, 127, 128
Lock-in amplifier(s), 72–73, 72
Logarithm(s), of input signals, 51, 52
Longitudinal relaxation, 414
Luminescence procedure(s), 225

Magic angle spinning, 447
Magnet(s), in continuous wave NMR spectrometers, 432–433, 432
Magnetic anisotropy, effect of, 421–423, 422
Magnetic field strength, in NMR spectroscopy, 429
Magnetic quantum level(s), transition among, 409–410, 410
Magnetic shielding for electron spectroscopy, 495
Magnetogyric ratio, 412
Majority carrier(s), of current in semiconductors, 31
Mass absorption coefficient, 462
Mass analyzer(s), 531–536, 532–536
Mass chromatogram, 559
Mass spectra, 542–547, 543t, 544, 545t, 547–548
  chemical and field ionization, 546–547, 547–548
  comparison of, in identification of compounds, 554–556, 556
  library sources of, 554–556, 556
Mass spectrometer(s), 524–542, 524
  commercial, 542, 543t
  computerized, 537–538, 539
  detectors and signal processors for, 536–537, 537
  for determining molecular formulas, 550–552, 551t
  for determining molecular weight, 548–550, 549t
  inlet systems in, 525–527, 526–527
  instrument components of, 524–525, 524
  ion sources for, 527–531, 528t, 529
  resolution of, 531–532
Mass spectrometric detector(s), for HPLC, 798–799
Mass spectrometry, 523–566
  analyzers for, 531–536, 532–536

Mass spectrometry (continued).
  for surface analysis, 563–564
  Fourier transform spectrometers in, 538–542, 540–542, 543t
  identification of pure compounds by, 547–559
  molecular secondary ion, 565
  quantitative applications of, 559–565
  with gas-liquid chromatography, 776–777, 777–778
Mass transfer, in polarization, 592
  zone broadening by, 736t, 737–738, 738–739
Mass transport, mechanisms of, in polarization, 594–595
Matrix effect(s), in quantitative X-ray analysis, 478–479
Mattauch-Herzog geometry, 534
McReynolds constant(s), 770–773, 771t, 772t, 773t
Mechanical oscillator(s), vibrational frequency of, 320–321
Membrane indicator electrode(s), 605–624
  cell potential with, 609–612
  classification of, 606, 606
Memory, computer, 84–87, 85–86
Mercury arc, in infrared absorption spectroscopy, 326
Metal vapor lamp(s), as line sources in optical spectroscopy, 115
Metallic indicator electrode(s), 603–605
  cell potential with, 609
Metallic redox indicator(s), 605
Michelson interferometer(s), modulation of optical radiation with, 152–154, 153
Microcomputer(s), in chemical instrumentation, 80–89
Microcurie, 512
Microelectrode(s), in voltammetry, 664
  planar, current and concentration relationships for, 668–670, 669
Microprobe analyzer(s), 564
Microprocessor(s), in chemical instrumentation, 80–89
Millicurie, 510
Minority carrier(s), of current in semiconductors, 31
Mixtures of absorbing substance(s), analysis, of, 212–213, 213
Mobile phase(s), flow rate of, in gas-liquid chromatography, 748t, 759

Mobile phase(s) (*continued*).
for supercritical-fluid chromatography, 831–832
of HPLC, 799–801, *801*, 802*t*
selection of, for partition chromatography, 802–808, *805*, 806*t*, *807*
Mobile phase reservoir(s), for HPLC, 788–790, *789*
Modulation, for noise reduction, 70, *70*
to reduce interferences from radiation sources, 268
Molar absorptivity, ε, in absorption measurement, 161–162, 161*t*
in ultraviolet-visible absorption spectroscopy, 182–183
Molar refraction R, 383
Molecular absorption, 106–108, *107*
Molecular concentration(s), determination of, in mass spectrometry, 559–561
Molecular fluorescence spectroscopy, 225–244 (*see also* Fluorescence spectroscopy)
Molecular formula(s), determination by mass spectrometry, 550–552, 551*t*
Molecular ion(s), of mass spectrum, 529, 543, 545*t*
stability of types of, 548–549, 549*t*
Molecular ion peak(s), identification of, 549–550
of mass spectrum, 543
Molecular orbital(s) and absorption of radiation, 184–185, *184–185*
Molecular rotation, [*M*], 397
Molecular secondary ion mass spectrometry (SIMS), 565
Molecular sieve(s), for gas-solid chromatography, 779, *780*, 781*t*
Molecular spectra, in atomization, 256–257, *256*
Molecular vibration(s), frequency of, 321
normal modes of, 324
quantum treatment of, 321–323
types of, 318, *318*
Molecular weight, determination by mass spectrometry, 548–550, 549*t*
Monochromatic radiation, filters in, 465–466, *466*
Monochromator(s), components of, 124, *124*
for absorption analysis, 120, 123–135

Monochromator(s) (*continued*).
grating vs. prism, 134–135, *135*
in infrared absorption spectroscopy, 326–327
light-gathering power of, 131–132
performance characteristics of, 128–132, *130–131*
resolving power of, 130–131
types of, 124, *124* (*see also* specific types, e.g., Echelle grating monochromator, Grating monochromator, Bunsen prism monochromator)
wavelength dispersion with, 466–468, *467*, 468*t*
Monochromator slit(s), 132–134, *132–133*
MOSFET transistor(s), 36–37, *36*
Mull(s), 342
Multichannel absorption instrument(s), 205–207, *206*
Multichannel photon detector(s), for optical spectroscopy, 140–143, *141–142*
Multichannel proportionating pump(s), 860, *860*
Multichannel spectrophotometer(s), in emission spectroscopy, 298–300, *299*
Multielement analysis(es), by atomic emission spectroscopy, instruments for, 281–282, *282*
Multilayer film(s), applications of, 878*t*, 879
automatic analysis based on, 875–879, *876–878*, 878*t*
instrumentation for, 877–879, *877–878*, 878*t*
potentiometry in, 878–879, *878*
Multiplex instruments, 332–336, *332–335* (*see also* Fourier transform spectrometers)
for optical spectroscopy, 147*t*, 148–157

Narrow-band electronic filter(s), for noise reduction, 70
Nebulizer(s), for atomic spectroscopy, 260
Negative feedback circuit(s), in operational amplifiers, 42–43, *42–43*
Nephelometric analysis, 379–381, 381*t*
applications of, 381, 381*t*
instruments for, 380
scattering in, 379–381
Nernst equation, 574, 581
applications of, to half-cell potentials, 583–584
Nernst glower(s), 326

Nerstian behavior, 628
Neutralization titration(s), coulometric, 658, 658*t*
potentiometric, 636–637, *636*
Neutron(s), 503*t*, 512–513
accelerators for, 513
interactions with matter, 513–514
Neutron activation analysis, 511–517, *512*, *514*
application of, 517
destructive methods in, 515–516
nondestructive methods in, 516
sensitivity of, 517, *518*
theory of, 514–515, *514*
Neutron capture, 513
n-Hexane, as solvent for ultraviolet-visible regions, 208*t*
Nicol prism(s), 389–390, *389–390*
polarimeters and, 397–398
Nitrate group(s), absorption characteristics of, 187*t*
Nitrogen rule, in mass spectrometry, 552
Nitro group(s), absorption characteristics of, 187*t*
Nitroso group(s), absorption characteristics of, 187*t*
NMR (*see* Nuclear magnetic resonance)
Noise, effect on precision of current measurement, 65–68, *65*, *66*, *68*
effect on precision of spectrophotometric analyses, 168–173, 170*t*–171*t*, *171–172*
Noise reduction, hardware devices for, 69–73, *69–72*
software for, 73–76, *73–75*
Nonabsorbing species, photometric determination of, 209
Nondispersive instrument(s), for X-ray fluorescence, 476, *477*
Nonfaradaic current(s), 572–573, *573*
as basis for conductometric measurements, 706
in polarography, 671
Nonresonance fluorescence, 110–111
Normal dispersion, 100, *100* (*see also* Dispersion)
Normal mode(s), of molecular vibrations, 324
Normal-phase chromatography, 800–801, *801*
npn transistor(s), 33–34, *34*
Nuclear magnetic resonance (NMR), and absorption, 108
classical description of, 411–413, *412–413*

Nuclear magnetic resonance (NMR) (*continued*).
  relaxation processes and saturation in, 413–415
  theory of, 407–417
Nuclear magnetic resonance (NMR) spectra, abscissa scales for, *418*, 419–420
  environmental effects on, 417–431
  line broadening in, 415
  types of, 415–417, *416–417*, 416*t*
Nuclear magnetic resonance (NMR) spectrometer(s), carbon-13 and, 436
Nuclear magnetic resonance (NMR) spectroscopy, 407–455
  applications of, 440–444
  application to isotopes other than protons, 444–447
  carbon-13, 444–447, *445–446*
  continuous wave spectrometers in, 431, 432–436
  Fourier transform spectrometers in, 431–432
  instruments for, 431–440
  quantum description of, 408–411, 408*t*, *410*
  sample handling in, 439–440
Nuclear magneton, 409
Nuclear relaxation, in a magnetic field, 413–415
Nuclei, magnetic properties of, 416, 416*t*
Null instrument(s), for electrical measurements, 884–886, *884–885*

Ohmic potential, of electrochemical cells, 590–591
Ohm's law, in electrochemical cells, 571
Operational amplifier(s), 39–54
  application to mathematical operations, 50–52, *51*
  application to voltage and current control, 48–49, *49*
  general characteristics of, 39–41
  symbols for, 40–41, *41*
Operational amplifier circuit(s), 41–43, *42*
Optic axis, of anisotropic crystal, 387, *388*
Optical activity, in polarimetric analysis, 396
Optical density, in emission spectroscopy, 309, *309*
Optical design(s), of instruments, 145–157, 147*t*
Optical multichannel analyzer(s), 143

Optical rotation, mechanism of, 397
  variables affecting, 396–397
Optical rotatory dispersion, 387–390, 399–404
  applications of, 403–404
  curves for, 400, *401*
  instrumentation for, 401–402
Optical rotatory power, 387
Optical spectroscopic method(s), 112–115, *113–114*
Optical spectroscopy, containers for, 135
  instruments for, 112–159
    components, of, 113–115, *113–114*
    radiation detectors for, 135–143
    radiation sources for, 115–120, *116–117, 119*
Optrode(s), 145
Orbital energy(ies), of *d*-electrons, 190–192, *191–192*
Order of interference, 98
Ordinary beam, 388
Organic analysis, polarographic, 695–696
Organic functional group(s), and polarographic waves, 695–696
Organic solvent(s), role of, in atomic absorption spectroscopy, 278
Organic species, chemiluminescent analysis of, 246
  fluorometric determination of, 242–243
  Raman spectra of, 374–376, *374–375*
Oscillator(s), anharmonic, 322–323
  harmonic, 318–323, *319*
Output signal, in electrothermal atomizers, 265–266, *265*
Overpotential *v*, in polarization, 593
Overvoltage, in polarizaion, 593, 595–596, 596*t*
Oxidant regulator(s), for flame spectroscopy, 262–263
Oxidation-reduction titration(s), coulometric, 659, 659*t*
  potentiometric, 637–638, *637*
Oxidation state(s), chemical shifts and, in ESCA spectroscopy, 496–498, 496*t*
Oxygen wave(s), in polarographic analysis, 676–677, *676*

Pair production, 505
Paper chromatography, 847
Parent ion(s), of mass spectrum, 529, 543

Particle size, effect of, on efficiency of HPLC column, 786–787, *786*
Partition chromatography, 801–815
  applications of, 812, 812*t*
  mobile phase selection for, 802–808, *805*, 806*t*, *807*
  packings for, 808–815
Partition coefficient, K, 728
  chromatographic, 739–740, 748*t*
  in gas-liquid chromatography, 758–759
Peak(s), analytical uses of, in chromatographic analysis, 748*t*, 749–750
  collision product, 545*t*, 546
  molecular ion and base, in mass spectrum, 543, *544*
Period, p, of electromagnetic radiation, 91
Peristaltic pump(s), in continuous flow analyzers, 858, *859–860*
pH, effect of, on fluorescence, 235–236
  potentiometric measurement of, with glass electrodes, 612–614, *612, 614*
Phenanthrene, spectra for, *236*
Phosphorescence, deactivation by, 230–231
  deactivation processes in, 228–231
  electromagnetic radiation from, 110–111
  theory of, 226–236
Phosphorescence spectroscopy, 113–114, *113*, 225–244 (*see also* Phosphorescence)
  instruments for measuring, 237–241
  methods of, application of, 241–244, 243*t*
Phosphorimeter(s), designs of, 240–241, *241*
Phosphorimetric method(s), 243–244
Phosphorus-31, NMR of, 447
Photoacoustic infrared instruments, 336–337
Photoacoustic spectroscopy, 217–220
Photochemical reaction, definition of, 183
Photoconductivity detector(s), 140
Photoelectric detector(s), in emission spectroscopy, 310
Photoelectric effect, 105
Photoemissive surface(s), for phototubes, 138–139, *138*
Photographic detection, of emission spectra, 310–311

Photoluminescence, 225 (*see also* Fluorescence spectroscopy, Phosphorescence spectroscopy)

Photoluminescence measurement(s), in liquid chromatography, 244

Photoluminescence method(s), applications of, 241–244, *243t*

Photometer, definition of, 45, 145–146
    for absorption analysis, 175–178, *176*, 196–198, *197–198*
    for atomic emission spectroscopy, 280–281
    infrared filter, 337–338, *338*
    nondispersive, 338–339, *338*
    reflective, for multilayer films, 877–878, *877*

Photometric analysis, procedures in, 209–213 (*see also* Absorption analysis)

Photometric titration(s), 215–217

Photomultiplier tube(s), 139–140, *139*

Photon(s), of electromagnetic radiation, 90, 104

Photon counter(s), for X-ray spectrosopy, 468

Photon counting, 143–144, 468–469

Photon detector(s), 137–140, *137–139*
    multichannel, 140–143, *141–142*

Photoplate(s), ion-sensitive, for mass spectrometers, 536–537

Phototube(s), vacuum, 138–139, *138*

Photovoltaic cell(s), 137

Planar chromatography, 728, 837–850 (*see also* Thin-layer chromatography)

Planar microelectrode(s), current and concentration relationships for, 668–670, *669*

Plasma atomization, in atomic fluorescence spectroscopy, 305–306, *305*
    instrument design and performance characteristics in, 305–306, *305*

Plasma source(s), for emission spectroscopy, 292–293, 294–304
    quantitative applications of, 303–304, *303t, 304, 304t*

Plate(s), height of, effect of particle size on, *759*
    effect of sample size on, in HPLC, *787*
    effects of variables on, 738–739, *739, 748t*

Plate(s) (*continued*).
    in thin-layer chromatography, 839
    theoretical, 732
    thin-layer, 840–841

Plate calibration curves, in emission spectroscopy, 309, *309*

Plate theory, of elution chromatography, 732

pn junction(s), 32

pnp transistor(s), 33–34, *34–35*, 35

Pockels cell, 403

Poisson distribution, 506

Polarimeter(s), 397–398, *398*

Polarimetry, 386–399
    applications of, 398–399
    interference in, 390–394, *391–394*
    optical activity in, 396
    sample handling in, 398

Polarity, parameter, in gas-liquid chromatography, 769–772, *770, 771t, 772t*

Polarization, concentration, 593–595
    in electroanalytical chemistry, 591–597, *592–593, 596t*
    sources of, 592–593, *593*
    mass transport in, 594–595
    of electromagnetic radiation, 103–104, *103*
    overvoltage from, 593, 595–596, *596t*

Polarized radiation, interference effects with, 390–394, *391–394*

Polarogram(s), 665, 666–667, *667* (*see also* Polarographic waves)
    effect of pH on, 695

Polarographic analysis, 665–667, *666*
    cells for, 677–679, *678*
    effect of temperature on, 670–671
    inorganic, 694
    of mixtures, 675, *675*
    organic, 695–696
    oxygen waves in, 676–677, *676*

Polarographic current(s), 667–672, *669, 671*
    drop growth and, 666, *666*

Polarographic wave(s), 666–667, *667*
    current maxima in, 671–672, *671*
    effect of cell resistance on, 674–675, *674*
    effect of complex formation on, 673–674, *673t*
    for irreversible reactions, 674
    organic functional groups and, 695–696

Polarography, 664–703 (*see also* Polarographic analysis, Polarographic currents, Polarographic waves)
    applications of, 693–696, *693*
    classical, accuracy and precision of, 694
    electrical apparatus for, 679–681, *681*
    theory of, 665–677
    current-sampled, 681–684, *682–683*
    DME, in 678–679, *678, 680* (*see also* Dropping mercury electrodes)
    fast linear-sweep, *683,* 686–687, *687*
    instrumentation for, 677–681, *678, 680–681*
    pulse, *682,* 684–686, *684–685*
    quantitative, 693–694
    sinusoidal alternating current (ac), *683,* 688–690, *689*
    stripping methods in, *683,* 690–693, *691–692*

Polychromatic radiation, effect on Beer's law, 166–167, *167, 168*

Polymer(s), differential thermal analysis for, 720–721, *721*
    porous, for gas-solid chromatography, 779–781

Population inversions in lasers, 118, *119*

Potassium ion, liquid membrane electrode for, 619, *620t*

Potential(s), cell, 573–574 (*see also* Cell potentials)
    changes in, during electrolysis, 646–648, *647*
    electrode, of electrochemical cells, 575–587 (*see also* Electrode potentials)
    formal, 586
    half-cell, 575–587 (*see also* Electrode potentials)
    half-wave, in polarographic analysis, 667, *667,* 672–674, *673t*
    liquid junction, 589–590, *590*
    ohmic, of electrochemical cells, 590–591

Potential energy, of harmonic oscillators, 319–320, *319*

Potential measurement(s), 45

Potentiometer(s), 624–631
    and laboratory recording, 63–64, *64*
    for electrical measurements, 884–885, *884*

Potentiometric analysis, 600–643
    direct measurements in, 627–631, *628* (*see also* Direct

Potentiometric analysis (*continued*).
  potentiometric measurements)
  instruments for, 624–627, *626*
  pH measurements with a glass electrode in, 630–631
Potentiometric titration(s), 631–640
  applications of, 634–638, *635–637*
  automatic, 638–640, *638*
  end-point determinations for, 632–634, *632*, 633*t*
  of mixtures, 634–636, *635*
Potentiometry, with multilayer films, 878–879, *878*
Potentiostat(s), definition of, 48
  for constant electrode potential coulometric analysis, 653–654, *654*
Potentiostatic coulometry, 652–655 *653–655*
  application of, 654–655, *655*
  instrumentation for, 652–654, *653–654*
Potentiostatic electrolysis, 649–650, *649–650* (see also Controlled cathode potential electrolysis)
Potentiostatic method, in coulometric analysis, 651
Power source(s), for conductance measurements, 607–707
Power supplies, for laboratory instruments, 37–39, *37–39*
Power transformer(s), 37–38, *37*
Precession, of particles in a magnetic field, 411–413, *412–413*
Precipitation, effect of, on electrode potentials, 584–585
Precipitation titration(s), conductometric, 710–711
  coulometric, 658–659, 658*t*
  potentiometric, 634, *634*
Precision, of quantitative mass spectral measurements, 560
Predissociation, deactivation by, 230
Preset end-point titrator(s), potentiometric, 638–639, *638*
Pressure, effect of, on refractive index, 383
  on supercritical-fluid chromatography, 831
Pressure broadening, 256
Principle of superposition, 93–95, *93–94*
Prism(s), Amici, 384, *384* (see also Amici prisms)
  Nicol, 389–390, *389–390* (see also Nicol prisms)

Prism monochromator(s), 127–128, *127*, 128*t*
Proportional counter(s), for X-ray detection, 469, 470–471
Proton nuclear magnetic resonance spectroscopy (see Nuclear magnetic resonance (NMR) spectroscopy)
Pulsed input(s), effect of, on RC circuits, 898–899, *898*
Pulsed nuclear magnetic resonance spectra, 437, *437*
Pulse-height analyzer(s), for X-ray spectrometers, 473
Pulse-height analyzer, in X-ray detection, 471
Pulse-height discriminator, 468
Pulse-height distribution, from X-ray detectors, 472–473
Pulse-height selector(s), for X-ray spectrometers, 473
Pulse polarography, 682, 684–686, *684–685*
Pumping, in laser action, *117*, 118
Pumping system(s), for HPLC, 790
Pyroelectric detector(s), 328–329

Quadrupole mass spectrometer(s), 534–535, *534–535*, 543*t*
Qualitative analysis (see specific methods)
Quanta, of electromagnetic radiation, 104
Quantitative analysis (see specific methods)
Quantum efficiency, for fluorescence processes, 232
Quantum-mechanical properties, of electromagnetic radiation, 104–111, *106–107*, *110*
Quantum yield, for fluorescent processes, 231
Quarter-wave plate(s), 393

Radiant power *P*, of electromagnetic radiation, 92
Radiation, electromagnetic (see Electromagnetic radiation)
Radiation buffer(s), 271
Radiation detector(s), for optical spectroscopy, 135–143
Radiation source(s), for atomic absorption spectroscopy, 266–268, *267*
  in infrared absorption spectroscopy, 325–326
  in optical spectroscopy, 115–120, *116–117*, *119*
Radioactive decay processes, as source of X-rays, 459–460
Radioactive isotope(s), 502–510, 503*t*, 506*t*, *507*, *509*

Radioactive neutron(s), sources of, 512–513
Radioactivity, decay processes in, 503–505
  decay products in, 503*t*
  decay rates in, 505
  units of, 510
Radiochemical analysis, 502–522
  counting in, 506–510, 506*t*, *507*, 511
  instrumentation for, 510–511
  isotopic dilution in, 517–520
  neutron activation in, 511–517, *512*, *514*
  radiometric methods in, 520
  substoichiometric method in, 516
Radio-frequency source, in NMR spectrometers, *432*, 433
Radioiotope(s), as source in X-ray spectroscopy, 465
Radioisotopic substance(s), for X-ray spectroscopy, 461*t*
Radiometric method(s), in radiochemical analysis, 520
Radiometric titration(s), 520
RAM (random access memory), 86
Raman scattering, 102–103
  normal, mechanism of, 364–367, *366*
Raman spectra, comparison with infrared spectra, 362–363, 367–370, *369*
  depolarization ratios in, 370–371, *370*
  excitation of, 363–364, *363*
  intensity of peaks in, 370
Raman spectroscopy, 362–378
  applications of, 373–377, *374–375*
  instrumentation for, 371–373, *372*
  theory of, 363–371
Rate theory, of elution chromatography, 732, 733–747
Rayleigh scattering, 102, 362
  mechanism of, 364, *366*
RC circuit(s), changes of current and potential in, *890*, 891–893
  effect of pulsed inputs on, 898–899, *898*
  impedance in, 894–896, *894*
  response of, to sinusoidal inputs, 893–896, *894*
  time constants for, 892
Reaction vessel(s), for thermometric titrations, 723–724
Readout device(s), 62–65, *63–65*
  for mass spectrometers, 537, *537*
  for optical spectroscopy, 143–144

Reagent(s), complex-forming, effect of, on electrode potentials, 584–585

Reagent delivery system(s), for thermometric titrations, 723

Recording titrator(s), potentiometric, 639, *640*

Rectifier(s), 38, *38*

Reference electrode(s), for potentiometric analysis, 600–603, 601*t*, *602*
precautions in use of, 602–603

Reference junction, for thermocouples, 328

Reflection, of electromagnetic radiation, 101–102

Reflection grating(s), for infrared absorption spectroscopy, 327

Reflective photometer(s), for multilayer films, 877–878, *877*

Refraction, of electromagnetic radiation, 99–101, *100–101*
specific and molar, 382–383

Refractive index, *n*, definition of, 100
detectors, for HPLC, 796, *796*
for anisotropic crystals, 389, 389*t*
instruments for measuring, 383–385, *384–385*
measurements of, 382
variables affecting, 383

Refractometer(s), 383–385, *384–385*

Refractometry, 382–385, *384–385*
applications of, 385–386

Register(s), definition of, 82

Relaxation effect, on ionic mobility, 705

Relaxation processes, in absorption of electromagnetic radiation, 109
in NMR spectroscopy, 413–414

Residual current(s), 671, *671*
in polarographic waves, 666, *667*

Resistance, electrical measurement of, with D'Arsonval meters, 883–884, *884*

Resistance bridge(s), for conductance measurements, 707

Resistance or conductance measurement(s), 45–46

Resolution, in thin-layer chromatography, 740–742, 840

Resonance fluorescence, 110, 226

Resonance line(s), absorption of, 266, *267*

Resonance radiation, 226

Resonance Raman scattering, 367, *368*

Resonance Raman spectroscopy, applications of, 376–377

Retardation factor, R_F, in thin-layer chromatography, 839

Retention index, *I*, in gas-liquid chromatography, 769–770, *770*, 770*t*

Retention time, t_R, chromatographic, 734–735, *734*, 739–740, 748*t*

Retention volume(s), V_R in gas-liquid chromatography, 758

Reversed-phase chromatography, 800–801, *801*, *810*, 818*t*
ion-pair, systems for, 814–815, 815*t*

Reversible cell(s), 573

Reversible pump sampler(s), automatic, 854, *855*

Rohrschneider/McReynolds constant(s), 770–772, 771*t*, 772*t*

ROM (read only memory), 86–87

Rotating platinum electrode(s), and voltammetric waves, 698

Rotational transition(s), molecular, in infrared absorption spectroscopy, 317

Rowland circle, 298

Sadtler collection, of infrared data, 350–351

Salt bridge(s), for electrochemical cells, 570

Sample(s), decomposition of, automatic, 857

Sample application, in thin-layer chromatography, 841

Sample handling, in NMR spectroscopy, 439–440

Sample holder(s), for NMR spectrometers, 435–436, *436*

Sample illumination system(s), for Raman spectroscopy, 371–373, *372*

Sample injection, for gas-liquid chromatography, 760–761
of ICP source, for emission spectroscopy, 295, *295*

Sample injection system(s), for HPLC, 790–791

Sample preparation, for atomic absorption analysis, 277–278

Sample probe, for NMR spectrometers, 435–436, *436*

Sample size, effect of, on efficiency of HPLC, 787, *787*

Sampling, automation of, 853–858, *855–856*
in continuous flow analyzers, 860, *860*

Saturated calomel electrode (SCE), 601, 601*t*

Saturation, in NMR spectroscopy, 413–415

Saturation factor, *S*, in radiochemical analysis, 514

Scaler(s), and electromechanical counters, 60–61
for X-ray spectrometers, 473

Scattered radiation, effect of, at wavelength extremes of a spectrophotometer, 174–175, *174*

Scattering, analytical methods based on, 379–381 (*see also* Nephelometric analysis, Turbidimetric analysis)
effect of wavelength on, 380
of electromagnetic radiation, 102
methods of, applications of, 381, 381*t*

Scattering spectroscopy, 113–114, *113*

Scintillation counter(s), in X-ray detection, 471

Secondary ion mass analyzer(s), 564

Secondary ion mass spectrometry (SIMS), atomic, in surface analysis, 564

Second-derivative titrator(s), potentiometric, 639

Second-order indicator electrode(s), 604

Second-order nuclear magnetic resonance (NMR) spectra, 428

Segmentation, in continuous flow analyzers, 860, *860*

Segmented flow analyzer(s), applications of, 862
automated, 858–862, *859–862*

Selection rule(s), for vibrational transitions, 320

Selective ion monitoring, 559

Selectivity factor, α, chromatographic, 740, 748*t*

Self-absorption, in atomic emission spectroscopy, 283–284, *285*

Semiconductor(s), 31–37, *32–33*

Semiconductor detector(s), for X-rays, 471–472, *472*

Semiconductor diode(s), 32–33, *32–33*

Semiquantitative method(s), in emission spectroscopy, 314

Sensitivity, of an instrument, 22–23

Separation(s), automatic, 857–858
in continuous flow analyzers, 860–861, *861*

Separation method(s), 1–2, 2*t*, 3

Sequential spectrometer(s), 298, *298*
Series *RC* circuit(s), 890, *890*
Servosystem, definition of, 63
SHE (*see* Standard hydrogen electrode)
Shielding, for noise reduction, 69
Shift reagent(s), 429–431, *431*
Shot noise, 67
Sigma (σ), methods for good approximations of, 10–11
Signal(s), 54–55 (*see also* specific types)
Signal averaging, 74, *74*
Signal chopping, 70–72, *70–71*
Signal generator(s), 4 4*t*
Signal processor(s), 4, 4*t*
  for mass spectrometers, 536–537, *537*
  for optical spectroscopy, 143–144
  for X-ray spectrometers, 473
Signal recorder(s), in NMR spectrometers, 435, *435*
Signal shaper(s), 57–58, *57*
Signal-to-noise, (*S/N*), enhancement, 68–76
Signal-to-noise, (*S/N*), ratio(s), 66, *66*
Silicon diode(s), as photodetectors, 140, 142–143, *142*
Silicon semiconductor(s), properties of, 31–32
Silver ion, amperometric titrations with, 699–700
Silver/silver chloride electrode(s), 602
Simple harmonic motion, 318
SIMS (*see* Secondary ion mass spectrometry)
Sine-wave current(s), 888–890, *888–889*
Single-beam instrument(s), for absorption measurements, 175, *176*
  for ultraviolet-visible regions, 200–201, *201*
  vs. double-beam instruments, 177–178
Single-beam photometer, for absorption spectroscopy, 196–197, *197*
Single-beam spectrophotometer(s), computerized, for ultraviolet-visible regions, 201–202
  for atomic absorption spectroscopy, 269, *269*
Single-coil detectors, in NMR spectrometers, 435, *435*
Single-focusing mass spectrometer(s), 532–533, *532*, 543*t*
Singlet state, definition of, 227

Sinusoidal alternating current (ac) polarography, *683*, 688–690, *689*
Size-exclusion chromatography, 824–830
  applications of, 828–830, *829*
  column packing in, 825–826, *826t*
Slit(s), for monochromators, 132–134, *132–133*
Slit width(s), effect on absorption measurements, 173–174, *173*
Solid(s), automatic handling of, 855–857, *856*
  for infrared absorption analysis, 342
Solid-state electrode(s), 616–617, 617*t*
Solvent(s), classification of, for mobile phase selection, in partition chromatography, 805–808, *805*, 806*t*, *807*
  effect of, on fluorescence, 234
  for adsorption chromatography, 815–817, *816*
  for infrared absorption analysis, 340–341, *340*
  for organic polarography, 695
  for ultraviolet-visible regions, 207–208, *208*, 208*t*
Solvent treatment system(s), for HPLC, 788–790, *789*
Source flicker noise, effect on measurement of transmittance, 172 (*see also* Flicker noise)
Spark source(s), for emission spectroscopy, 292–293, 306–314, *308*
  instruments for, 308–310, *309–310*
  qualitative applications of, 310–311, *312*
  quantitative applications of, 312–314
  sample handling in, 306–307, *306*
Spark source mass spectrometry, elemental analysis by, 561–562
Spatial design(s), for optical instruments, 147–148, 147*t*
Specific conductance, *k*, 704
Specific refraction, 382–383
Specific rotation, $[\alpha]_\lambda^T$, 396
Spectra (*see also* specific types, e.g., NMR spectra, Absorption spectra, Emission spectra)
  ESR, 448–449, *450*
  mass, 542–547 (*see also* Mass spectra)
  purity of, in monochromators, 128–129

Spectra (*continued*).
  sources of, in emission spectroscopy, 293
  spark source mass, 562
  time-domain decay, in Fourier transform NMR, 438, *438*
Spectral interference(s), in atomic absorption spectroscopy, 270–273, *271*, *273*
  in atomic emission spectroscopy, 282–283, *282*
Spectrofluorometer(s), 236–237
  components of, *236*, 237–238
  definition of, 146
  designs of, 239–240, *240*
  spectra information from, 240
Spectrograph(s), in emission spectroscopy, 310, *310–311*
Spectrometer(s), for Raman spectroscopy, 372
  Fourier transform, 332–336, *332–335* (*see also* Fourier transform spectrometers)
  NMR, 431–440
Spectrometry, mass, 523–566 (*see also* Mass spectrometry)
Spectrophotometer(s), dispersive infrared, 330–332, *330–331*
  double dispersing, 203, *204*
  effect of scattered radiation on, 174–175, *174*
  for absorption measurements, 175–178, *176*
  for atomic absorption spectroscopy, 269–270, *269*
  for atomic emission spectroscopy, 280, *284*
  for ultraviolet-visible regions, 199–204
Spectrophotometric analysis, effect of, instrumental noise on precision of, 168–173, 170*t*, 171*t*, *171–172*
  procedures in, 209–213 (*see also* Absorption analysis)
Spectrophotometry, ultraviolet-visible, qualitative analysis by, 207–208, *208*, 208*t*
Spectroscope(s), definitions, 145
Spectroscopic method(s), 96*t*
Spectroscopy, absorption, 160–181 (*see also* Absorption spectroscopy)
  molecular ultraviolet-visible, 182–224
  atomic, 250–291 (*see also* Atomic spectroscopy)
  atomic emission, 279–286 (*see also* Atomic emission spectroscopy)
  definition of, 96

Spectroscopy (*continued*).
  electron, 487–501 (*see also* Electron spectroscopy)
  emission, 279–286 (*see also* Emission spectroscopy)
  far-infrared, 355
  infrared absorption, 315–361 (*see also* Infrared absorption spectroscopy)
  infrared emission, 355
  methods of, 112–115, *113–114* (*see also* specific types, e.g., Emission spectroscopy, Absorption spectroscopy)
  NMR, 407–455 (see also Nuclear magnetic resonance (NMR) spectroscopy)
  optical, instruments, for, 112–159 (*see also* Optical spectroscopy)
  Raman, 362–378 (*see also* Raman spectroscopy)
  X-ray, 456–486 (*see also* X-ray spectroscopy)
Spin decoupling, in NMR spectra, 429, *430*
Spin-label reagent(s), 450
Spin-lattice relaxation, 414–415
Spin quantum number, I, 408, *408t*
Spin-spin relaxation, 414–415
Spin-spin relaxation, time, T_2, 415
Spin-spin splitting, in NMR spectra, 418–420, 424–429, *426t*, *427–428*
  source of, 419
Splitting factor, 448
Spontaneous emission, in laser action, *117*, 118
Sputtering, in hollow cathode lamps, 267
Square planar configuration, splitting of d-orbitals in, 191
Square wave polarography, 689
Stable isotope(s), 503
Standard addition method, for direct potentiometric measurement, 630
  for quantitative polarography, 694
  in absorption analysis, 210–212, *212*
  in atomic absorption analysis, 279
Standard deviation, σ, as measure of column efficiency, 734, *734*
  definition of, 7–8
  for small sets of data, 8–9
  of counting data, in radiochemical analysis, 507–508, *507*

Standard electrode potential E^0, 581–582, *582t* (*see also* Electrode potentials)
Standard error, 8
Standard hydrogen electrode (SHE), 575–576, *576*
Standard potential, 574
Stationary phase(s), chemically bonded, 773–774
  for supercritical-fluid chromatography, 831
  in thin-layer chromatography, 841
  in gas-liquid chromatography, 769–774
  selection of, 773
    uses of McReynolds constants in, 772–773, *772t*
Stimulated emission, in laser action, *117*, 118
Stokes shift, 111, 226
  in Raman spectroscopy, 364–366
Stopped-flow method(s), 869–870
Stray radiation, effect on Beer's law, 167–168, *168*
Stretching vibration(s), mechanical model of, 318–321, *319*
  molecular, 318
Stripping analysis, in polarography, *683*, 690–693, *691–692*
Substoichiometric method, in radiochemical analysis, 516
  isotope dilution, 519
Sulfur, differential thermogram for, 719–720, *720*
Summing point, 42
Supercritical-fluid chromatography, 830–834
  applications of, 834
  vs. other column methods, 832–834, *832–833*
Superposition, principle of, 93–95, *93–94*
Surface analysis, by mass spectrometry, 563–564
Syringe-based sampler(s), automatic, 854–855, *855*

Tandem mass spectrometry, 557–559, *558*
Temperature, changes in, during thermometric titrations, 722, 724
  effect of, on atomic spectra, 259–261, *259–260*
    on conductance measurements, 708
    on diffusion currents, 670–671
    on fluorescence, 234
    on gas-liquid chromatography, 764–765, *765*

Temperature (*continued*).
    on partition chromatography, 812
    on refractive index, 383
    on supercritical-fluid chromatography, 830–831
  in ICP source, 295–296, *296*
  of columns in HPLC, 792
Temperature profile(s), of flames, 259–260, *259*
Temporal instrument(s), for optical spectroscopy, 146–147, *147t*
Tetrahedral configuration, splitting of d-orbitals in, 191
Thallium/thallous chloride electrode(s), 602
Thermal conductivity detector(s), for gas-liquid chromatography, 766–767, *766–767*
Thermal detector(s), for infrared radiation, 328–329, *329*
Thermal ionization, in determination of isotope concentrations, 563
Thermal method(s), 713–726 (*see also* Thermometric titrations)
Thermal neutron(s), 512
Thermal radiation, emission of, 109
Thermionic detector(s), for gas-liquid chromatography, 765, 768
Thermobalance(s), 713–714, *714*
Thermocouple(s), 328, *329*
  function(s), 328
Thermodynamic cell potential(s), calculation of, 587–589, *588t* (*see also* Cell potentials)
Thermogram(s), 713
  differential, 716–717, *716*, *718*, *720–721*
Thermogravimetric analysis, 713–715, *714–715*
  applications of, 714–715, *714–715*
Thermometric analysis, enthalpimetric methods in, 721–726
Thermometric titration(s), 721–726
  apparatus for, 722–724, *723*
  applications of, 724–726, *725*, *725t*
Thermospray source, 799
Thin-layer chromatography, applications of, 846–847
  chromatogram development in, 841–842, *842*
  performance characteristics of, 838–840, *839*
  qualitative, 843–844
  quantitative, 844–846, *845–846*

Third-order indicator electrode(s), 604–605
Time constant(s), in *RC* circuits, 892
Time-domain decay spectra, in Fourier transform NMR, 438, *438*
Time-domain mass spectrum, Fourier transform spectrometers in, 538–542, *542*
Time domain spectra, acquisition of, 151–152, *152*
Time domain spectroscopy, 150–151, *151*
Time-of-flight mass spectrometer(s), 535–536, *536*, 543*t*
Titration(s), automatic, 638–640, *638*
  conductometric, 708–711, *709*
  coulometric, 655–660 (*see also* Coulometric titrations)
  coulometric vs. volumetric, 659–660
  flow injection, 870–871, *871*
  photometric, 215–217, *216–217*
  potentiometric, 631–640 (*see also* Potentiometric titrations)
  radiometric, 520
  thermometric, 721–726 (*see also* Thermometric titrations)
Titration curve(s), photometric, 215–217, *216–217*
Transducer(s), definition of, 4
  examples of, 4 (*see also* specific types)
  radiation, types of, 136–137, *136*
Transducer output(s), comparison of, 46–48, *47*
Transducer signal(s), amplification and measurement of, 43–48, *44–45*, *47*
Transformer(s), power, 37–38, *37*
Transistor(s), 33–37, *34–36*
Transition(s), electronic, in molecular ultraviolet and visible radiation, 185–193
Transition-metal ion(s), absorption of radiation by, 190–192, *190–192*, 192*t*
Transmission, of electromagnetic radiation, 99–101, *100–101*
Transmittance, *T*, in absorption measurement, 160–162, *161*, 161*t*, 171*t*
  effect of instrumental noise on, 169–171, 171*t*
Transmittance curve(s), vs. derivative curves, *214*
Transmutation, 513–514
Transverse relaxation, 414

Trapped ion analyzer cell, 540–541, *541*
Triple bond region, in infrared spectrum, 346
Triplet state, 111, 227
Tungsten filament lamp(s), 194–195
  in infrared absorption spectroscopy, 326
Turbidimetric analysis, 379–381, 381*t*
  applications of, 381, 381*t*
  instruments for, 380–381
  scattering in, 379–381
Turbidity coefficient, τ, and scattering, 380
Turbulent flow burner(s), for atomic spectroscopy, 261, *261*
Turner Model 110 fluorometer, *239*
Two-line correction method, in atomic absorption spectroscopy, 271

Ultraviolet photoelectron spectroscopy, 488*t*, 490–491
  applications of, 498
Ultraviolet photometer(s), for absorption spectroscopy, 198
Ultraviolet-visible region(s), absorption spectroscopy in, 182–224
  spectrophotometers for, 199–204

Vacuum phototube(s), 138–139, *138*
Variance, as measure of column efficiency, 734
Vibration(s), bending, 318
  molecular, frequency of, 321
  types of, 318, *318*
  quantum treatment of, 321–323
  stretching, 318
Vibrational coupling, 324–325
Vibrational frequency, of mechanical oscillator, 320–321
Vibrational mode(s), and Raman effect, 367–370, *369*
  in infrared absorption spectroscopy, 323–324
Vibrational quantum number, 321
Vibrational relaxation, deactivation by, 229
Vibrational-rotational transition(s), molecular, infrared absorption spectroscopy, 317–318, *318*
Visible photometer(s), 196–197
Visible-ultraviolet region(s) (*see* Ultraviolet-visible regions)

Voltage, measurement of, with D'Arsonval meters, 882–883, *883*
Voltage-follower circuit, 43, *43*
Voltage measurement(s), 45, *45*
Voltage regulator(s), 38–39, *39*
Voltammetry, 664–703 (*see also* Polarography, Polarographic analysis, Polarographic waves)
  amperometric titrations in, 697–700
  cyclic, *683*, 687–688, *687*
  modified methods of, 681–693
  stripping analysis in, *683*, 690–693, *691–692*
  types of, *683*, 686–693, *687*, *689*, *691–692*
  with solid electrodes, 696–697
Volumetric titration(s), vs. coulometric titrations, 659–660

Water, as solvent for ultraviolet-visible regions, 208*t*
Wave(s), mathematical description of, *91*, 92–93
  polarographic (*see* specific types, e.g., Anodic waves, Polarographic waves)
Wavelength(s), for intense X-ray emission lines, 458*t*
Wavelength dispersive instrument(s), for X-ray fluorescence, 474–475, *477*
  X-ray, 464
Wavelength dispersion, with monochromators, 466–468, *467*, 468*t*
Wavelength λ, effect of, on refractive index, 383
  on scattering, 380
  of electromagnetic radiation, 91–92, *91*
Wavelength modulation, devices for, 214, *214*
Wavelength selection, for absorption analysis, 120, 135, 210
Wavenumber σ of electromagnetic radiation, 92
Wave properties, of electromagnetic radiation, 91–95, *91–94*
Wave train(s), of electromagnetic radiation, 99
Wedge(s) filter, 327
  interference, 122–123
Well-type scintillation counter, 511, *512*
Wheatstone bridge(s), 885–886, *885*
White-light interferometer system, in Fourier transform spectrometers, 334, *334*
White radiation, 456

Wide-line nuclear magnetic resonance spectrometer(s), 431
Wide-line spectra, 416, *416*
Work function, *w*, 489
 of a metal surface, 105
Working electrode(s), 651

Xenon arc lamp(s), 195
 as source in fluorescence spectroscopy, 238
X-ray(s), absorption of, 460–462, *461*
 diffraction of, 463–464, *463*
 emission of, 456–460, *457*, *458t*, *459–460*, 461*t*
  in radiochemical analysis, 505
X-ray absorption, methods of, 481–482, *482*

X-ray beam(s), filters for, 465–466, *466*
X-ray detector(s), 468–473, *469*, 472
 pulse-height distribution from, 472–473
X-ray diffraction, methods of, 482–484, *483*
X-ray fluorescence, 462–463, 474
 (*see also* X-ray spectroscopy)
 instrumentation for, 474–476, *476*
 methods of, 474–481
  advantages and disadvantages of, 480–481
  qualitative applications of, 476–478, *477–478*
  quantitative analysis by, 478–481
X-ray monochromator(s), 466–468, *467*

X-ray photoelectron spectroscopy (ESCA), 489–490, *489–490*
 application of, 495–498
X-ray radiation, emission of, 110
X-ray spectroscopy, 456–486
 instrumentation for, 464–473
 principles of, 456–464
 radioisotopic sources for, 461*t*
X-ray tube(s), 464–465, *465*

Zeeman effect, in atomic absorption spectroscopy, 272–273, *273*
Zener breakdown voltage, for semiconductor diodes, 33
Zener diode(s), in voltage regulators, 39
Zone broadening, in elution chromatography, 735–739, *736t*, *738–739*

Conversion Factors for Electromagnetic Radiation

(To convert data in units of x shown in the first column to units indicated in the remaining columns, multiply or divide as shown.)

Units of x	Frequency, Hz	Wavenumber, cm^{-1}	Energy, kcal/mol	Energy, erg/particle	Energy, eV	Wavelength, cm
Hz	$1.000x$	$3.336 \times 10^{-11}x$	$9.537 \times 10^{-14}x$	$6.626 \times 10^{-27}x$	$4.136 \times 10^{-15}x$	$\dfrac{2.998 \times 10^{10}}{x}$
cm^{-1}	$2.998 \times 10^{10}x$	$1.000x$	$2.859 \times 10^{-3}x$	$1.986 \times 10^{-16}x$	$1.240 \times 10^{-4}x$	$\dfrac{1.000}{x}$
kcal/mol	$1.049 \times 10^{13}x$	$3.498 \times 10^{2}x$	$1.000x$	$6.947 \times 10^{-14}x$	$4.338 \times 10^{-2}x$	$\dfrac{2.859 \times 10^{-3}}{x}$
erg/particle	$1.509 \times 10^{26}x$	$5.035 \times 10^{15}x$	$1.439 \times 10^{13}x$	$1.000x$	$6.241 \times 10^{11}x$	$\dfrac{1.986 \times 10^{-16}}{x}$
eV	$2.418 \times 10^{14}x$	$8.067 \times 10^{3}x$	$2.305 \times 10^{1}x$	$1.602 \times 10^{-12}x$	$1.000x$	$\dfrac{1.240 \times 10^{-4}}{x}$
cm	$\dfrac{2.998 \times 10^{10}}{x}$	$\dfrac{1.000}{x}$	$\dfrac{2.859 \times 10^{-3}}{x}$	$\dfrac{1.986 \times 10^{-16}}{x}$	$\dfrac{1.240 \times 10^{-4}}{x}$	$1.000x$
nm	$\dfrac{2.998 \times 10^{17}}{x}$	$\dfrac{1.000 \times 10^{7}}{x}$	$\dfrac{2.859 \times 10^{4}}{x}$	$\dfrac{1.986 \times 10^{-9}}{x}$	$\dfrac{1.240 \times 10^{3}}{x}$	$1.000 \times 10^{-7}x$

Symbols for Common Physical and Chemical Quantities

A	absorbance, area	S/N	signal to noise
a	absorptivity, activity	T	transmittance, temperature
C	capacitance, concentration	t	time
D	diffusion coefficient	V	dc voltage
d	diameter, spacing	V	volume
deg	angular degree	v	ac voltage
E	electrical potential, energy	v	velocity
e	electron	X	reactance
f	frequency, activity coefficient	$\bar{x}$	mean
G	conductance, free energy	Z	impedance
H	magnetic field, enthalpy		
I	dc current		
i	ac current		
K	equilibrium constant	ε	(epsilon) molar absorptivity
L	inductance	λ	(lambda) wavelength
n	refractive index, number of equivalents	μ	(mu) mean
n	spectral order	ν	(nu) frequency
P	radiant or electrical power	ρ	(rho) density
Q	quantity of dc electricity	σ	(sigma) standard deviation, wavenumber
q	quantity of ac electricity	τ	(tau) period
R	electrical resistance, gas constant	ϕ	(phi) phase angle
s	standard deviation	ω	(omega) angular velocity